International Poisonous Plants Checklist

An Evidence-Based Reference

International Poisonous Plants Checklist

An Evidence-Based Reference

D. Jesse Wagstaff

CRC Press
Taylor & Francis Group
Boca Raton London New York

CRC Press is an imprint of the
Taylor & Francis Group, an **informa** business

CRC Press
Taylor & Francis Group
6000 Broken Sound Parkway NW, Suite 300
Boca Raton, FL 33487-2742

Printed in the United States of America on acid-free paper
10 9 8 7 6 5 4 3 2 1

International Standard Book Number-13: 978-1-4200-6252-6 (Hardcover)

Library of Congress Cataloging-in-Publication Data

Wagstaff, D. Jesse, 1935-
International poisonous plants checklist : an evidence-based reference / D. Jesse Wagstaff.
p. cm.
Includes bibliographical references and index.
ISBN 978-1-4200-6252-6 (hardback : alk. paper) 1. Poisonous plants--Toxicology. 2. Poisonous plants--Identification. I. Title.

QK100.A1W34 2008
615.9'52--dc22 2008008708

Visit the Taylor & Francis Web site at
http://www.taylorandfrancis.com

and the CRC Press Web site at
http://www.crcpress.com

WARNING

Failure to find a particular plant in this checklist should not be interpreted as evidence that the plant is not toxic. Only a small fraction of plants have been examined for toxicity. On the other hand, many food and drug plants listed here are generally recognized to be safe but can be toxic in uncommon situations. Dosages and poisoning circumstances such as species susceptibilities should be determined from the original literature. Questions regarding plant intoxications should be directed to poison control centers and health care providers.

CONTENTS

ACKNOWLEDGMENTS

Contributions of several individuals are acknowledged including Wayne Binns (deceased), Claude Culvenor, John Kingsbury, Kenneth Lampe (deceased), as well as numerous botanists, toxicologists, librarians, and technical assistants who helped sharpen the focus, provided moral support, supplied helpful materials, and provided access to information sources. Particular gratitude is expressed to John Wiersema for including numerous poisonous plants in GRIN and for providing the model for this book by his publication, together with his coauthor Blanca León, of *World Economic Plants: A Standard Reference*. Manuscript reviewers are Marion R. Cooper, Elizabeth A. Dauncey, Terry D. Jacobson, Lynn F. James, Anthony W. Johnson, Anthony P. Knight, Ross A. McKenzie, Kip E. Painter, James A. Pfister, and Richard J. Schmidt. They are thanked for their efforts. Gratitude is also expressed for those anonymous reviewers selected by the editor.

INTRODUCTION

Increasing international trade has focused attention on the inadequacy of regional information sources. Poisonous plant literature is geographically fragmented and is of variable reliability. As Claus et al. (1970) said, "Probably no field of scientific endeavor exists in which it is more difficult to separate fact from fiction than in the field of poisonous plants." Confusion can be reduced through use of reliable information and universally-recognized plant names. Objectives of this checklist are:

1. To define the known set of toxic vascular plants of the world using accepted botanic names.
2. To support toxicity information for each plant from references to primary literature.
3. To cross-reference selected synonyms and common names.

This book is intended as an international reference for toxicologists, food safety regulators, researchers, laboratory diagnosticians, poison control center personnel, economic botanists, food safety scientists, trade officials, herbalists, physicians, veterinarians, and others who need information about plant toxicology. The subject is addressed as a subset of toxicology. Botany, chemistry, and other fields contribute but they are not the focus.

A poisonous plant is here defined as a plant which, by means of toxic substances it contains, produces adverse effects in humans or economic animals such as livestock, laboratory animals, pets, wild animals, birds, fish, and bees. Most poisoning cases result from ingestion or other type of exposure to a plant for nutritional, therapeutic, or recreational intent. Other cases are due to experimental exposure, and a few are due to malicious intent. Vascular plants include ferns, fern allies, cycads, and flowering plants. Nonvascular plants are excluded, as are surface contamination, concentrated plant products, pharmaceuticals, and most recreational drugs.

A statement that a plant is or is not toxic is only hearsay unless supported by reliable evidence, i.e, primary (original) sources. A primary source is a record created near the time of an event by an eyewitness qualified by training and experience. In contrast, a secondary source is a record made by a non-witness, an unqualified witness, or made long after an event, when memory may be fallible. Or even worse, a document is copied or extracted from prior documents in a chain stretching backwards through several steps of one author passing on material from prior authors until often the original report can not be located.

Knowledge of plant toxicity has always been important but has not always been reliable. Prehistoric peoples learned which plants to eat and which to avoid. Certain toxic plant products were used to poison fish or to coat arrow tips. This knowledge, as part of their oral tradition, was not recorded. In the Biblical account, the fruit of one tree produced life, while another caused death (Genesis, Holy Bible). Dioscorides (Gunther 1959) wrote about medicinal and toxic properties of specific plants from sources made available to Greek scholars from lands of the expansive Hellenic Empire. His work was widely quoted and incorporated into books about poisonous plants down to the early part of the twentieth century. But even in his day, he complained of mistakes and falsehoods in the literary sources available to him. Nearly a millennium and a half passed before there was much further progress. In fact there was retrogression as ancient records were destroyed, as typified by the sacking and burning of libraries at Alexandria, Baghdad, and Rome. Mysticism tended to replace direct observation and experimentation as the bases for knowledge about plants.

Finally in the Renaissance new ideas were put forth. The writings in the early 1500s of the Swiss physician Paracelsus were pivotal. He reasoned that effects of a plant on the body were due to particular chemical components of the plant and that the type and magnitude of the effects were proportional to the amount of the plant substance administered (Pachter 1961). Richard Mead, in *A Mechanical Account of Poisons* (1702), in several essays expressed regret that knowledge was extremely defective. Over time, authors began to reject from their lists of poisonous plants some that previous authors had included. The 19th century saw the application of the scientific method to the toxicology of plants when the chemist Orfila (1814–1815) used experimental animals, mainly dogs, to demonstrate the toxicity of 45 species of plants.

Outside of Europe, science in this field developed slowly. It was not until the 1890s that scientists such as Victor King Chesnut were employed to study poisonous plants (Wagstaff 2004). Kingsbury (1964) reviewed the history of poisonous plant publications in detail. International communication expanded with the advent of the International Symposia on Poisonous Plants, which were started in 1977 as a joint effort of the United States and Australia and have been held since then at about 4-year intervals and later expanded to cover other areas of the world (Acamovic et al. 2004; Colegate & Dorling 1994; Garland & Barr 1998; James et al. 1992; Keeler et al. 1978; Painter et al. 2007; Seawright et al. 1985).

Botanic names are sometimes called scientific or Latin names. There are two purposes for using botanic names. The first is to satisfy the need for an international standard of identity to support worldwide communication and commerce. Second, the system of plant classification, being based upon common ancestry and genetic connections, provides a rational basis for studying and using plant relationships. The accepted botanic names in this checklist were taken from Wiersema and Leon's (1999) *World Economic Plants: A Standard Reference* and updates to this work in the Germplasm Information Network online database (GRIN) maintained by the U.S. Department of Agriculture. Plant names not found in GRIN were taken from Missouri Botanical Garden's Vascular Topics online database (VAST) or from the International Plant Names Index (INI), another online database. A final comprehensive check of these sources for the plants named in this publication was performed in August 2007.

A botanic name is comprised of two parts, genus and species, to which is appended the name or abbreviation of the author(s) who first validly published the name and authors who taxonomically reclassified the plant. For example, Abrus precatorius L. is the name published by Linnaeus for the precatory bean. Application of an accepted name to a plant specimen is based upon a description written and published in accord with rules of the International Botanical Nomenclature Committee (Greuter et al. 1994) and upon comparison to a dried, or otherwise preserved, specimen of the plant, called the type, which is filed in an internationally recognized herbarium. When a plant has been given different scientific names, rules of international taxonomic procedures are applied to determine the name with precedence, and the other names become synonyms.

Four checklists of poisonous plants of the world have been published. The first, *Poisonous Plants of All Countries*, was published in 1905 by Bernhard-Smith, an English physician interested in plants. The second was published in 1911 as a section of the book *A Manual of Poisonous Plants* by the American botanist Pammel. The third was a 1923 revision of Bernhard-Smith's earlier work. The fourth, *Poisonous Plants of the World*, was distributed in mimeographed form in 1949 (third revision in 1951) by Moldenke, a taxonomist at the New York Botanical Garden. However, none of these checklists are comprehensive, nor are any of them based upon primary evidence of toxicity.

The checklist is organized alphabetically like a dictionary. There are two types of entries. First, main entries are the toxic plants themselves. And second, selected synonyms and common names are cross-referenced to the main entries. The header of each main entry is the accepted botanic name in bolded italicized print, followed by plant author(s) in non-bolded print, and then the plant family in square brackets. For example, the main entry for poison ivy is "***TOXICODENDRON RADICANS*** (L.) Kuntze [Anacardiaceae]." Beneath each header are selected botanic synonyms and common names. Following the nomenclatural information, the main entry is completed by primary citations supporting toxicity of the plant. In synonym cross-references the synonym is referred to the accepted botanic name with an equal (=) sign. An example is "Rhus radicans L. = Toxicodendron (L.) Kuntze." In common name cross-references, the common name is referred to the accepted botanic name by the character string "–see– ." An example is "poison ivy –see– ***Toxicodendron radicans*** (L.) Kuntze." An occasional note is placed at the bottom of a main entry.

Literature citation lists for some plants are extensive but are not exhaustive. Likewise, the number of synonyms is limited to those common in poisonous plant literature. Attempts have been made in the past to standardize common names of plants, but these have not been universally accepted. The approach here is to list the most prevalent common names used in poisonous plant literature rather than evaluate acceptability of any particular name. Most of these names are from the English language together with some from other languages that use the Roman alphabet. A few local and tribal plant names are included. The array of common names in this checklist is not exhaustive. Such a list would be virtually endless and continually changing.

Bibliographic files obtained from John Kingsbury of Cornell University and Kenneth Lampe of the American Medical Association were the initial sources of references for a relational bibliographic database. To these were added poisonous plant references from TOXLINE, MEDLINE, and other bibliographic sources. Because of the international and multidisciplinary nature of the material, major reference works from various geographic areas of the world such as those listed below in the general references were consulted. Confirmation of toxicity of each plant was obtained from original publications. Each document—book, monograph, or periodical article—was indexed by keyterms. Document-type keyterms were primary for original articles, review when the information was referenced, and secondary when it was not an original report and no references were given. Document subjects indexing terms were case report, experimental exposure, epidemiology, chemical analysis, botany, pharmacology, mechanism, metabolism, animal behavior, and weed control. Citations in this checklist were indexed as primary document type and case report, experimental exposure, or epidemiology document subject. The first major output from the bibliographic database was a poisonous plant subfile of TOXLINE produced in 1987 for the U.S. National Library of Medicine (Wagstaff et al 1989).

Inclusion in this checklist of virtually all common animal feed plants and human food plants as well as many plants that are sources of herbal products widely used as dietary supplements, tonics, and therapeutic agents needs to be explained. The rate of adverse effects among users of these products, which are generally considered to be safe, is low. But because of the vast number of exposures, even this low rate of adverse effects results in some poisoning cases. Natural variation in populations results in some plants of high potency and some humans and animals of high susceptibility. Many other factors such as food processing inadequacies may lead to intoxications.

Plants are toxic only under particular circumstances. Failure to find a plant name in this checklist should not be taken as proof that the plant is not toxic. The vast majority of plants have not been studied for toxicity. And some reports of toxicity may not have come to the attention of the author of this publication or he may not have properly evaluated them. Readers are encouraged to report deficiencies in this publication and to assist in rectification of identified problems.

Care should be exercised in interpreting this checklist. A plant may have poisoning potential because of toxic substances it contains but it is included only if reports of actual poisoning cases have come the attention of this compiler. An exposure to a plant may not have been observed but was assumed from circumstantial evidence. Another issue is that adverse effects following an exposure may have been caused by factors other than the plant. Occasionally, a reported plant was not included in this checklist because the name given was so ambiguous that it could not be identified even after using reasonable assumptions. The standard is a voucher specimen of the plant collected from the lot to which the humans or animals were exposed that is deposited in an internationally recognized herbarium. Reference to such a voucher is uncommon in poisonous plant reports. This is not to say that plants named in reports are necessarily erroneous; many authors consult a trained botanist but fail to identify that person and the methods of plant identification. Much progress has been made in plant toxicology, but reporting deficiencies still occur (Wagstaff 2007).

GENERAL REFERENCES

Aslani MR (2004) *Poisonous plants of Iran and their effects on animals*. Ferdowsi University Press. Mashhad, Iran.

Baldwin RE (1979) *Hawaii's poisonous plants*. Petroglyph Press. Hilo, Hawaii.

Bernhard-Smith A (1905) *Poisonous plants of all countries*. Bristol, J Wright and Co. London.

Bernhard-Smith A (1923) *Poisonous plants of all countries*, 2nd ed. Bailliere Tindall Cox. London.

Bible, King James Version. Cambridge University Press. Cambridge.

Burrows GE, Tyrl RJ (2001) *Toxic plants of North America*. Iowa State University Press. Ames, Iowa.

Blohm H (1962) *Poisonous plants of Venezuela*. Harvard University Press. Cambridge, Massachusetts.

Bruneton J (1999) *Toxic plants. Dangerous to humans and animals*. Intercept. Andover, UK.

Caius JF (1986) *Medicinal and poisonous plants of India*. Scientific Publ. Jodhpur, India.

Cañizo Gomez JD, Martinez Vazquez M (1964) Plantas toxicas para el ganado en praderas y pastizales. *Min Agric Manuales Tecnicos* Ser M #26. Madrid.

Claus EP, Tyler VE Jr, Brady LR (1970) *Pharmacognosy*, 6th ed. Lea & Febiger. Philadelphia.

Colegate SM, Dorling PR (eds.) (1994) *Plant-associated toxins. Agricultural, phytochemical and ecological aspects*. CABI. Wallingford, UK.

Connor HE (1977) *The poisonous plants in New Zealand*. EC Keating. Wellington.

Cooper MR, Johnson AW, Dauncey E (2003). *Poisonous plants and fungi: An illustrated guide*, 2nd ed. The Stationery Office. London.

Cornevin C (1887) *Des plantes vénéneuses et des empoisonnements qu'elles déterminent*. Firmin-Didot. Paris.

Escobar N (1972) *Flora tóxica de Panamá*. Editorial Universitaria. Panama.

Everist SL (1981) *Poisonous plants of Australia*, rev ed. Angus and Robertson. Sydney.

Frohne D, Pfander HJ (2005) *Poisonous plants: A handbook for doctors, pharmacists, toxicologists, biologists, and veterinarians*, 2nd ed. Timber Press. Portland, Oregon.

Gillis W (1971) The systematics and ecology of poison-ivy and the poison-oaks (Toxicodendron, Anacardiaceae). *Rhodora* 73:72–159.

Garland T, Barr AC (1998) *Toxic plants and other natural toxicants*. CABI. New York.

Greuter W, Barrie FR, Burdet HM, et al. (1994) *International code of botanical nomenclature (Tokyo Code)*, adopted by the Fifteenth International Botanical Congress. Koeltz Scientific Books. Königstein.

Gunther RT (1959) *The Greek herbal of Dioscorides*. Hafner. New York.

James LF, Keeler RF, Bailey JM Jr, et al. (1992) *Poisonous plants: Proceedings of the Third International Symposium*. Iowa State University Press. Ames, Iowa.

Keeler RF, Van Kampen KR, James LF (1978) *Effects of poisonous plants on livestock*. Academic Press. New York.

Kellerman TS, Botha CJ, Coetzer JA, et al. (2005) *Plant poisonings and mycotoxicoses of livestock in southern Africa*, 2nd ed. Oxford University Press. Oxford.

Kingsbury JM (1964) *Poisonous plants of the United States and Canada*. Prentice-Hall. Englewood Cliffs, New Jersey.

Lampe KF, McCann MA (1985). *AMA handbook of poisonous and injurious plants*. Chicago Review Press. Chicago.

Mead R (1702) *A mechanical account of poisons*. R South. London

Mitchell JC, Rook AJ (1979) *Plant dermatology*. Greengrass. Vancouver.

Moldenke HN (1951) *Poisonous plants of the world*, 3rd ed. of a mimeographed report. New York Botanical Garden.

Mulligan GA, Munro DB (1990) *Poisonous plants of Canada*. Agriculture Canada. Ottawa.

Nelson LS, Shih RD, Balick MJ (2007) *Handbook of poisonous and injurious plants*, 2nd ed. Springer Science and Business Media, New York.

Núñez Meléndez E (1989) *Plantas venenosas de Puerto Rico y las que producen dermatitis*. University of Puerto Rico. Rio Piedras.

Oakes AJ, Butcher JO (1962) *Poisonous and injurious plants of the U.S. Virgin Islands*. USDA Misc Publ #882.

Orfila MJ (1814-1815) *Traité des poisons tiré des rèsgnes minéral, végétal et animal ou toxicologie général consideree sous les rapports de la pathologie et des la medecine legale*. Crochard. Paris.

Pachter HM (1961) *Paracelsus: Magic into science*. Collier. New York.

Pammel LH (1911) *A manual of poisonous plants*. Torch Press. Cedar Rapids, Iowa.

Painter KE, Wierenga TL, Pfister JA (2007) *Poisonous plants: Global research and solutions*. CABI, Wallingford, UK.

Roth L. Daunderer M, Kormann K (1994) *Giftpflanzen-Pflanzengifte*. Nikol Verlagsgesellschaft. Hamburg.

Sandberg F (1980) Medicinal and toxic plants from Equatorial Africa: A pharmacologic approach. *J Ethnopharm* 2(2):105–108.

Seawright AA, Hegarty MP, James LF, et al. (1985) *Plant toxicology. Queensland Poisonous Plant Committee*. Yeerongpilly, Queensland.

Stoilis E, Brikas P, Calaboukas C (1967) Poisonous and noxious plants of Greece. *Hellenike Kteniatrike* 10:149–167.

Tokarnia CH, Döbereiner J, Peixoto PV (2000) *Plantas tóxicas do Brasil*. Editoria Helianthus, Rio de Janeiro.

Verdcourt B, Trump EC (1969) *Common poisonous plants of East Africa*. Collins. London.
Wagstaff DJ (2004) Victor King Chesnut (1867–1938), a poisonous plant pioneer. In: Acamovic T, Stewart CS, Pennycott TW (eds.) *Poisonous plants and related poisons*. CABI. Wallingford, UK, pp. 120–122.
Wagstaff DJ (2007) Evaluation of plant poisoning reports. In: Painter KE, Wierenga TL
Wagstaff DJ, Wagstaff AT, Goshorn JC (1989) A poisonous plant file in TOXLINE. *Toxicon* 27(2):259–263.
Wiersema JH, León B (1999) *World economic plants: A standard reference*. CRC Press. Boca Raton, Florida.
Zolla C (1982) *Plantas tóxicas de México*. Instituto Mexicano del Seguro Social. Mexico City Instituto.

AUTHOR

D. Jesse Wagstaff has worked many years in food safety and development of poisonous plant bibliographic databases. His interest started when he worked with Wayne Binns and Lynn James at the U.S. Department of Agriculture's Poisonous Plant Laboratory before veterinary training at Cornell University, where John Kingsbury taught poisonous plants. After earning a Ph.D., Wagstaff taught toxicology at the University of Missouri. He researched safety of animal feeds and drugs and evaluated epidemiologic and safety aspects of human foods at the U.S. Food and Drug Administration (FDA). Although he is in emeritus status, the search for additional documents goes on. His collection of poisonous plant documents is the world's largest.

Plant toxicology data has been obtained from interagency projects of the FDA, the U.S. National Library of Medicine (NLM), and Oak Ridge National Laboratory. Further development has been funded by FDA, NLM, and the U.S. National Institutes of Health. Databases were developed with Kenneth Lampe of the American Medical Association. Additional material has come from personal searches at the Library of Congress, the U.S. National Agricultural Library, the Smithsonian Institution Herbarium, and libraries at Arizona State University, Brigham Young University, Colorado State University, Cornell University, Montana State University, University of California at Davis, University of Nevada at Reno, University of Utah, University of Washington, and Utah State University. Extensive use has been made of interlibrary loans, online dabases such as Pubmed and Agricola, and library catalogs.

A

Aalhorn –see– *Sambucus nigra* L.

aam –see– *Mangifera indica* L.

Aaron –see– *Arum maculatum* L.

aasvoëbossie –see– *Pteronia pallens* L. f.

abachi –see– *Triplochiton scleroxylon* K. Schum.

ABACOPTERIS MULTILINEATA (Wall. ex Hook.) Ching [Thelypteridaceae]

Citations:

Devi S, Yasmeen, Singh J, et al. (1989) Patch testing animals to allergenic fern spores. J Toxicol Cutaneous Ocul Toxicol 8(2):167-172.

abang –see– *Milicia excelsa* (Welw.) C. C. Berg

Abassian boxwood –see– *Buxus sempervirens* L.

ABELMOSCHUS ESCULENTUS (L.) Moench [Malvaceae]

Synonyms:

Hibiscus esculentus L.

Common Names:

bhindi; gumbo; lady's-finger; okra

Citations:

Matsushita T, Aoyama K, Manda F, et al. (1989) Occupational dermatoses in farmers growing okra (Hibiscus esculentus L.). Contact Dermatitis 21(15):321-325.

Singh R, Siddiqui MA, Baruah MC (1978) Plant dermatitis in Delhi. Indian J Med Res 68(Oct):650-655.

Sinha SM, Pasricha JS, Sharma RC, et al. (1977) Vegetables responsible for contact dermatitis of the hands. Arch Dermatol 113(6):776-779.

abendlandischer Lebensbaum –see– *Platycladus orientalis* (L.) Franco

ABIES BALSAMEA (L.) Mill. [Pinaceae]

Common Names:

balsam fir

Citations:

Kappes LO (1948) Balsam as a cause of contact dermatitis in a florist. Ann Allergy 6:(Jan-Feb)21-22.

abnormal leaf flower –see– *Phyllanthus abnormis* Baill.

abnormal leg flower –see– *Phyllanthus abnormis* Baill.

abrepuño –see– *Centaurea solstitialis* L.

abrin –see– *Abrus precatorius* L.

abrojos –see– *Tribulus terrestris* L.

ABRUS PRECATORIUS L. [Fabaceae]

Common Names:

abrin; bejuco-de-peonia; black-eye Susan; Buddhist rosary bead; coral bead plant; crab's-eye; gidee gidee; habet-el-arus; Indian bead; Indian licorice; jequirity bean; jequirity pea; jumbee bead; love bean; love pea; lucky bean; minnie minnies; Paternostererbse; prayer bead; prayer bean; precatory bean; precatory pea; ratti; red bead vine; rosary bean; rosary pea; Santa Juane; weather plant; wild licorice

Citations:

Anonymous (1955) Abrin, lethal jewelry. AMA Arch Ind Health 12:468-469.

Anonymous (1969) Rosary bean kills child, 2. Miami Herald. Nov 7:2C.

Barri ME, El Dirdiri NI, Abu Damir H, et al. (1990) Toxicity of Abrus precatorius in Nubian goats. Vet Hum Toxicol 32(6):541-545.

Bukhari AQ, Ahmad S, Qureshi S (1976) Oral toxicology of Abrus precatorius Linn. J Pak Med Assoc 26(1):4-6.

Burda A, Wahl M, Fischbein C, et al. (2004) Atypical poisonings with botanicals raise suspicion of malicious activity. Vet Hum Toxicol 46(6):341.

Carvalho EC (1987) Sensibilidade de aves domésticas (Gallus gallus domesticus L.) à abrina, toxalbumina de Abrus precatorius L. - Aspectos anatomopatológicos na toxicose experimental. Arq Fluminenses Med Vet 2(3):89.

Davis JH (1978) Abrus precatorius (rosary pea). The most common lethal plant poison. J Fla Med Assoc 65(3):188-191.

Desai VB, Sirsi M (1964) The effect of Abrus precatorius on pregnancy of mice. Curr Sci 33(19):585-587.

Fernando C (2001) Poisoning due to Abrus precatorius (jequirity bean). Anaesthesia 56(12):1178-1180.

Frohne D, Schmoldt A, Pfänder HJ (1984) Die Paternostererbse - Keineswegs harmlos. Dtsch Apoth Ztg 124(43):2109-2113.

Genest K, Lavalle A, Nera E (1971) Comparative acute toxicity of Abrus precatorius and Ormosia seeds in animals. Arzneimittelforschung 21(6):888-889.

Guggisberg M (1968) A propos d'une curieuse intoxication par des grains de chapelet (Abrus precatorius). Rev Med Suisse Romande 88(3):206-208.

Hart M (1963) Hazards to health. Jequirity-bean poisoning. N Engl J Med 268(16):885-886.

Kinamore PA, Jaeger RW, de Castro FJ, et al. (1980) Abrus and ricinus ingestion: Management of three cases. Clin Toxicol 17(3):401-405.

Moraillon R, Mutel M, Henry N, et al. (1978) A propos d'un cas d'intoxication mortelle d'un chien par la graine d'Abrus precatorius. Rec Med Vet Ec Alfort 154(3):223-225.

Niyogi SK (1977) Elevation of enzyme levels in serum due to Abrus precatorius (jequirity bean) poisoning. Toxicon 15(6):577-580.

Omer SA, Ibrahim FH, Khalid SA, et al. (1992) Toxicological interactions of Abrus precatorius and Cassia senna in the diet of Lohmann broiler chicks. Vet Hum Toxicol 34(4):310-313.
Simpson KS, Banerjee PC (1932) Cases of poisoning in the horse with ratti seeds (Abrus precatorius) by oral administration. Indian J Vet Sci Anim Husb 2:59-65.
Swanson-Biearman B, Dean BS, Krenzelok EP (1992) Failure of whole bowel irrigation to decontaminate the GI tract following massive jequirity bean ingestion. Vet Hum Toxicol 34(4):352.
Tokarnia CH, Döbereiner J, Monteiro MC (1970) Intoxicação experimental em bovinos pela semente de Abrus precatorius. Pesq Agric Bras 5(3):441-452.

absinthe –see– *Artemisia absinthium* L.

absinthium –see– *Artemisia absinthium* L.

Abutilon avicennae Gaertn. = Abutilon theophrasti Medik.

abutilon hemp –see– *Abutilon theophrasti* Medik.

ABUTILON THEOPHRASTI Medik. [Malvaceae]

Synonyms:

Abutilon avicennae Gaertn.

Common Names:

abutilon hemp; American jute; butter print; button print; button weed; Chinese jute; chingma; cottonweed; Indian mallow; velvet weed; velvetleaf

Citations:

Dugan GM, Gumbmann MR (1990) Toxicological and nutritional evaluation of velvetleaf seed: Subchronic 90-day feeding study and protein efficiency ratio assay. Food Chem Toxicol 28(2):95-99.
Keshavarz K, Park K (1992) The use of velvetweed in laying ration causes formation of eggs with pasty yolks. Poult Sci 71(Suppl 1):54.

Abyssinian kale –see– *Crambe abyssinica* Hochst. ex R. E. Fr.

Abyssinian tea –see– *Catha edulis* (Vahl) Forssk. ex Endl.

ACACIA BERLANDIERI Benth. [Fabaceae]

Common Names:

guajillo

Citations:

Boughton IB, Hardy WT (1940) Feeding trials of suspected plants. Texas Agric Exp Sta Annu Rep 53:236-237.
Boughton IB, Hardy WT (1941) Feeding trials of suspected plants. Texas Agric Exp Sta Annu Rep 54:159.

ACACIA BINERVIA (J. C. Wendl.) J. F. Macbr. [Fabaceae]

Synonyms:

Acacia glaucescens Willd.

Common Names:

coast myall; river myall; Sally wattle

Citations:

Seddon HR, White HC (1928) River myall or sally wattle (Acacia glaucescens.) Proved poisonous to stock. Agric Gaz New South Wales 39:668-670.
Seddon HR, White HC (1929) Acacia glaucescens ("river myall" or "sally wattle") proved poisonous to stock. New South Wales Dep Agric Vet Res Rep 1927-1928(5):96-99.

ACACIA CATECHU (L. f.) Willd. [Fabaceae]

Common Names:

black catechu; catechu; cutch tree; khair

Citations:

Kulakkattolickal AT (1987) Piscicidal plants of Nepal: Preliminary toxicity screening using grass carp (Ctenopharyngodon idella) fingerlings. J Ethnopharmacol 21(1):1-9.

Acacia cyanophylla Lindl. = Acacia saligna (Labill.) H. L. Wendl.

acacia falsa –see– *Robinia pseudoacacia* L.

ACACIA GEORGINAE F. M. Bailey [Fabaceae]

Common Names:

Georgina gidyea; gidyea tree

Citations:

Barnes JE (1958) Georgina poisoning of cattle in the Northern Territory. Aust Vet J 34(Sep):281-290.
Bell AT, Newton LG, Everist SL, et al. (1955) Acacia georginae poisoning of cattle and sheep. Aust Vet J 31(Oct):249-257.
Whittem JH, Murray LR (1963) The chemistry and pathology of Georgina River poisoning. Aust Vet J 39(Apr):168-173.

ACACIA ×GIRAFFAE Willd. [Fabaceae]

Common Names:

giraffe thorn; kameelboom; kameeldoring; mimosa; Transvaal camelthorn; Transvaal kameeldoring

Citations:

Steyn DG (1936) Recent investigations into the toxicity of known and unknown poisonous plants in the Union of South Africa. Onderstepoort J Vet Sci Anim Indus 7(1):169-178.

ACACIA GLAUCA (L.) Moench [Fabaceae]

Synonyms:

Acacia villosa (Sw.) Willd.

Citations:

Dunham LJ, Sheets RH, Morton JF (1974) Proliferative lesions in cheek pouch and esophagus of hamsters treated with plants from Curacao, Netherland Antilles. J Natl Cancer Instit 53(5):1259-1269.

Acacia glaucescens Willd. = Acacia binervia (J. C. Wendl.) J. F. Macbr.

ACACIA JACQUEMONTII Benth. [Fabaceae]

Common Names:

kikar

Citations:

Singh R, Siddiqui MA, Baruah MC (1978) Plant dermatitis in Delhi. Indian J Med Res 68(Oct):650-655.

ACACIA LASIOPETALA Oliv. [Fabaceae]

Common Names:

Natal camel thorn

Citations:

Steyn DG (1934) The camel-thorn tree and other Acacia species as stockfeed. Risk of stock being poisoned. Farming South Africa 9(Mar):89-90, 106.

ACACIA LEUCOPHLOEA (Roxb.) Willd. [Fabaceae]

Common Names:

babul; kikar; rajma; raunja; reunja; safed kikar

Citations:

Gupta I, Nauriyal MM (1966) Acacia leucophloea Willd. (raunja) poisoning in livestock. Indian Vet J 43(6):538-543.

Prasad J, Singh AP, Rekib A (1977) Hydrocyanic acid poisoning in grazing sheep and goat on Acacia leucophloea (reunja). Indian Vet J 54(Sep):748-751.

Vihan VS, Panwar HS (1987) A note on toxicity of Acacia leucophloea (rajma) in sheep. Indian J Anim Res 21(1):53-55.

ACACIA MELANOXYLON R. Br. [Fabaceae]

Common Names:

Australian blackwood; blackwood; Tasmanian blackwood

Citations:

Burry JN (1969) The value of patch testing: A review of 363 cases of allergic contact dermatitis. Med J Aust 1(Jun 14):1124-1231.

Correia O, Barros MA, Mesquita-Guimarães J (1992) Airborne contact dermatitis from the woods Acacia melanoxylon and Entandrophragma cylindricum. Contact Dermatitis 27(5):343-344.

Tilsley DA (1990) Australian blackwood dermatitis. Contact Dermatitis 23(6):40-41.

ACACIA NILOTICA (L.) Delile subsp. kraussiana (Benth.) Brenan [Fabaceae]

Citations:

Terblanche M, Pienaar JG, Bigalke R, et al. (1967) Acacia nilotica (L.) Del. subsp. kraussiana (Benth.) Brenan as a poisonous plant in South Africa. J S Afr Vet Assoc 38(1):57-63.

ACACIA PENDULA A. Cunn. ex G. Don [Fabaceae]

Common names:

myall

Citations:

Hindmarsh WL (1937) Mortality in sheep due to the ingestion of the foliage of myall trees (Acacia pendula) infested with the boree moth (Teara contraria). Aust Vet J 13:124-125.

ACACIA PENNATA (L.) Willd. [Fabaceae]

Common Names:

aila; arfoo

Citations:

Kulakkattolickal AT (1987) Piscicidal plants of Nepal: Preliminary toxicity screening using grass carp (Ctenopharyngodon idella) fingerlings. J Ethnopharmacol 21(1):1-9.

ACACIA SALICINA Lindl. [Fabaceae]

Common Names:

black Sally wattle; black wattle; cooby; doolan

Citations:

McCosker PJ, Hunt SE (1966) Suspected poisoning of cattle with Acacia salicina. Aust Vet J 42(Sep):355.

ACACIA SALIGNA (Labill.) H. L. Wendl. [Fabaceae]

Synonyms:

Acacia cyanophylla Lindl.

Citations:

Allen JG, Creeper JH, Forshaw D, et al. (2004) Plant-associated diseases, either new or new to the state, encountered over the last decade (1991-2001) in Western Australia. In: Acamovic T, Stewart CS, Pennycott TW (eds.) Poisonous plants and related toxins. CABI. Wallingford, UK. pp. 540-547.

ACACIA ×SENEGAL (L.) Willd. [Fabaceae]

Synonyms:

Acacia verek Guill. & Perr.

Common Names:

gum acacia; gum arabic

Citations:

Fötisch K, Fäh J, Wüthrich B, et al. (1998) IgE antibodies specific for carbohydrates in a patient allergic to gum arabic (Acacia senegal). Allergy 53(11):1043-1051.

Ilchyshyn A, Smith AG (1985) Gum arabic sensitivity associated with epidemic hysteria dermatologica. Contact Dermatitis 13(4):282-283.

Sprague PH (1942) Bronchial asthma due to sensitivity to gum acacia. Can Med Assoc J 47:253.

Acacia verek Guill. & Perr. = Acacia senegal (L.) Willd.

Acacia villosa (Sw.) Willd. = Acacia glauca (L.) Moench

acajou –see– *Anacardium occidentale* L.; *Semecarpus ater* (G. Forst.) Vieill.

acajou-jaune-du-Congo –see– *Sarcocephalus diderrichii* De Wild.

acalypha –see– *Acalypha indica* L.

ALYPHA INDICA L. [Euphorbiaceae]

Common Names:

acalypha; hierba-de-cancer; Indian acalypha; machelikoane; three-seed mercury

Citations:

Fernando R (2002) The national poisons information centre in Sri Lanka: The first ten years. J Toxicol Clin Toxicol 40(5):551-555.

Senanayake N, Sanmuganthan PS (1996) Acute intravascular haemolysis in glucose-6-phosphate dehydrogenase deficient patients following ingestion of herbal broth containing Acalypha indica. Trop Doct 26(1):32.

Steyn DG (1937) Recent investigations into the toxicity of known and unknown poisonous plants in the Union of South Africa. VIII. Onderstepoort J Vet Sci Anim Indus 9:573-582.

Steyn DG (1965) An investigation into cases of suspected poisoning in Africans in northern Rhodesia. S Afr Med J 39(16):344-350.

ACANTHOSPERMUM HISPIDUM DC. [Asteraceae]

Common Names:

dashin yawo; hurab ekhowsa; star bur

Citations:

Ali B, Adam SE (1978) Effects of Acanthospermum hispidum on goats. J Comp Pathol 88(4):533-544.

Ali B, Adam SE (1978) Toxicity of Acanthospermum hispidum to mice. J Comp Pathol 88(3):443-448.

acebo –see– *Ilex aquifolium* L.

acedera –see– *Rumex crispus* L.

acedera común –see– *Rumex acetosa* L.

acederilla –see– *Oxalis acetosella* L.

ACER MACROPHYLLUM Pursh [Aceraceae]

Common Names:

big-leaf maple; Oregon maple

Citations:

Meeham J (1889) Poisonous plants. Garden Forest 2(57):154.

ACER RUBRUM L. [Aceraceae]

Common Names:

Carolina maple; meadow maple; red maple; scarlet maple; soft maple; swamp maple; water maple; yellow maple

Citations:

De Witt SF, Bedenice D, Mazan MR (2004) Hemolysis and Heinz body formation associated with ingestion of red maple leaves in two alpacas. J Am Vet Med Assoc 225(4):578-583.

Divers TJ, George LW, George JW (1982) Hemolytic anemia in horses after the ingestion of red maple leaves. J Am Vet Med Assoc 180(3):300-302.

George LW, Divers TJ, Mahaffey EA, et al. (1982) Heinz body anemia and methemoglobinemia in ponies given red maple (Acer rubrum L.) leaves. Vet Pathol 19(5):521-533.

Long PH, Payne JW (1984) Red maple-associated pulmonary thrombosis in a horse. J Am Vet Med Assoc 184(8):977-978.

McConnico RS, Brownie CF (1992) The use of ascorbic acid in the treatment of 2 cases of red maple (Acer rubrum)-poisoned horses. Cornell Vet 82(3):293-300.

Plumlee KH (1991) Red maple toxicity in a horse. Vet Hum Toxicol 33(1):66-67.

Stair EL, Edwards WC, Burrows GE, et al. (1993) Suspected red maple (Acer rubrum) toxicosis with abortion in two Percheron mares. Vet Hum Toxicol 35(3):229-230.

Tennant B, Dill SG, Glickman LT, et al. (1980) Red maple poisoning in horses associated with acute hemolytic anemia and methemoglobinemia. Proc Am Assoc Equine Pract 26:243-250.

Tennant B, Dill SG, Glickman LT, et al. (1981) Acute hemolytic anemia, methemoglobinemia, and Heinz body formation associated with ingestion of red maple leaves by horses. J Am Vet Med Assoc 179(2):143-150.

ache-des-marrais –see– *Apium graveolens* L.

ache douce –see– *Apium graveolens* L.

ache olorante –see– *Apium graveolens* L.

Achillea borealis Bong. = Achillea millefolium L.

Achillea lanulosa Nutt. = Achillea millefolium L.

ACHILLEA MILLEFOLIUM L. [Asteraceae]

Synonyms:

Achillea borealis Bong.; ***Achillea lanulosa*** Nutt.

Common Names:

milfoil; nosebleed; plumajillo; Schafgarbe; thousand-leaf; yarrow

Citations:

Davies MG, Kersey PJ (1986) Contact allergy to yarrow and dandelion. Contact Dermatitis 14(4):256-257.

Gans O (1929) Ueber die Dermatitis durch Achillea millefolium. Ein Beitrag zur Aetiologie der beim Freibaden entstehenden Hautausschläge. Dtsch Med Wochenschr 55(29):1213-1215.

Hausen BM (1979) The sensitizing capacity of Compositae plants. III. Test results and cross-reactions in Compositae-sensitive patients. Dermatologica 159(1):1-11.

Philadelphy A (1928) Eine bullöse Hauterkrankung, hervorgerufen durch Achillea millefolium. Wien Klin Wochenschr 41:88-89.

ACHNATHERUM ROBUSTUM (Vasey) Barkworth [Poaceae]

Synonyms:

Stipa robusta (Vasey) Scribn.; ***Stipa vaseyi*** Scribn.; ***Stipa viridula*** Trin. var. robusta Vasey

Common Names:

Mexican sleepy grass; robust needlegrass; sleepy grass; Vasey's-stipa grass

Citations:

Bailey V (1903) Sleepy grass and its effect on horses. Science 17(427):392-393.

Marsh CD, Clawson AB (1929) Sleepy grass (Stipa vaseyi) as a stock-poisoning plant. U S Dep Agric Tech Bull #114:19 pp.

Smalley HE, Crookshank HR (1976) Toxicity studies on sleepy grass, Stipa robusta (Vasey) Scribn. Southwestern Vet 29(1):35-39.

ACHYRANTHES ASPERA L. [Amaranthaceae]

Common Names:

apamarg; colic weed; hug melose; man-better-man; rabo-de-gato

Citations:

Han ST, Lin CC (2003) Cardiac toxicity caused by Achyranthes aspera. Vet Hum Toxicol 45(4):212-213.

Sanchez Perere LL, Alfonso HA, Noa M, et al. (1995) Intoxication due to Achyranthes aspera L. Vet Hum Toxicol 37(6):582.

ackee –see– *Blighia sapida* K. D. Koenig

Ackerkrone –see– *Agrostemma githago* L.

Ackerrettich –see– *Raphanus sativus* L.

Ackerrittersporn –see– *Consolida regalis* Gray

Ackersenf –see– *Sinapis arvensis* L.

Ackerwicke –see– *Vicia sativa* L.

Ackerwinde –see– *Convolvulus arvensis* L.

ACMENA SMITHII (Poir.) Merr. & L. M. Perry [Myrtaceae]

Synonyms:

Eugenia smithii Poir.

Common Names:

lily pilly

Citations:

Rosner SA, Nurse DS (1984) Contact dermatitis to lily pilly (Eugenia smithii): Report of a case. Aust J Dermatol 25(1):25-26.

Acokanthera friesiorum Markgr. = Acokanthera schimperi (A. DC.) Oliv.

Acokanthera longiflora Stapf = Acokanthera oppositifolia (Lam.) Codd

ACOKANTHERA OPPOSITIFOLIA (Lam.) Codd [Apocynaceae]

Synonyms:

Acokanthera longiflora Stapf; ***Acokanthera venenata*** auct.

Common Names:

arrow poison bush; bushman's-poison; gifboom; Hottentot poison bush; ouabain; wabaio

Citations:

Mugera GM (1970) Toxic and medicinal plants of East Africa. I. Bull Epizootic Dis Afr 18(4):377-387.

ACOKANTHERA SCHIMPERI (A. DC.) Oliv. [Apocynaceae]

Synonyms:

Acokanthera friesiorum Markgr.

Common Names:

Hottentot poison bush; ouabain; wabaio

Citations:

Cassels BK (1985)Analysis of a Maasai arrow poison. J Ethnopharmacol 14(2-3):273-281.

Mugera GM (1970) Toxic and medicinal plants of East Africa. I. Bull Epizootic Dis Afr 18(4):377-387.

Acokanthera venenata auct. = Acokanthera oppositifolia (Lam.) Codd

aconite –see– *Aconitum carmichaelii* Debeaux; *Aconitum ferox* Wall. ex Ser.; *Aconitum kusnezoffii* Rchb.; *Aconitum napellus* L.

Aconitum angelicum Stapf = Aconitum napellus L.

ACONITUM BRACHYPODUM Diels [Ranunculaceae]

Citations:

Tai YT, But PP, Young K, et al. (1992) Cardiotoxicity after accidental herb-induced aconite poisoning. Lancet 340(8830):1254-1256.

Zhang YG, Huang GZ (1988) Poisoning by toxic plants in China. Report of 19 autopsy cases. Am J Forensic Med Pathol 9(4):313-319.

ACONITUM ×CAMMARUM L. [Ranunculaceae]

Citations:

Cheung K; Hinds JA; Duffy P (1989) Detection of poisoning by plant-origin cardiac glycoside with the Abbott TDx analyzer. Clin Chem 35(2):295-297.

Pohjalainen T, Elomaa E, Hoppu K (2003) Cardiovascular disorders induced by Aconitum species. J Toxicol Clin Toxicol 41(4):515-516.

ACONITUM CARMICHAELII Debeaux [Ranunculaceae]

Common Names:

aconite; bushi; chuanwu; fo tzu; fu tse; wutou; zhi-chuan-wu

Citations:

But PP Tai YT, Young K (1994) Three fatal cases of herbal aconite poisoning. Vet Hum Toxicol 39(3):212-215.

Chan TY, Chan JC, Tomlinson B, et al. (1994) Poisoning by Chinese herbal medicines in Hong Kong: A hospital-based study. Vet Hum Toxicol 36(6):546-547.

Chan TY, Tomlinson B, Chan WW, et al. (1993) A case of acute aconitine poisoning caused by chuanwu and caowu. J Trop Med Hyg 96(1):62-63.

Chan TY, Tomlinson B, Critchley JA, et al. (1994) Herb-induced aconitine poisoning presenting as tetraplegia. Vet Hum Toxicol 36(2):133-134.

Chan WY, Ng TB, Lu JL (1995) Effects of decoctions prepared from Aconitum carmichaelii, Aconitum kusnezoffii and Tripterygium wilfordii on serum lactate dehydrogenase activity and histology of liver, kidney, heart, and gonad. Hum Exp Toxicol 14(6):489-493.

Chen SS, Hung DZ, Deng JF, et al. (1992) Aconite poisoning in a herb preparation. Toxicon 30:497-498.

Dickens P, Tai YT, But PP, et al. (1994) Fatal accidental aconitine poisoning following ingestion of Chinese herbal medicine: A report of two cases. Forensic Sci Int 67(1):55-58.

Fitzpatrick AJ, Crawford M, Allan RM, et al. (1994) Aconite poisoning managed with a ventricular assist device. Anaesth Intensive Care 22(6):714-717.

Liu Y, Wolf LR, Zhu W (1997) Epidemiology of adult poisoning at China Medical University. J Toxicol Clin Toxicol 35(2):175-180.

Tai YT, But PP, Young K, et al. (1992) Cardiotoxicity after accidental herb-induced aconite poisoning. Lancet 340(8830):1254-1256.

Tai YT, Lau CP, But PP, et al. (1992) Bidirectional tachycardia induced by herbal aconite poisoning. Pacing Clin Electrophysiol 15(5):831-839.

Zhang ZZ (1983) [Experimental study on "18 against compatibilities" of Chinese traditional medicine: Acute toxicity of Aconitum carmichaelii and Pinellia ternata.] Bull Chin Materia Medica 8(4):33-34.

ACONITUM CARMICHAELII Debeaux var. truppelianum (Ulbr.) W. T. Wang & P. K. Hsiao [Ranunculaceae]

Synonyms:

Aconitum chinense Paxton

Common Names:

Chinese aconite root

Citations:

Zhang YG, Huang GZ (1988) Poisoning by toxic plants in China. Report of 19 autopsy cases. Am J Forensic Med Pathol 9(4):313-319.

Aconitum chinese Paxton = Aconitum carmichaelii Debeaux var. truppelianum (Ulbr.) W. T. Wang & P. K. Hsiao

Aconitum excelsum Rchb. = Aconitum lycoctonum L. subsp. lycoctonum

ACONITUM FEROX Wall. ex Ser. [Ranunculaceae]

Common Names:

aconite; bish poison; Indian aconite; Nepaul aconite

Citations:

Mahajani SS, Joshi RS, Gangar VU, et al. (1990) Some observations on the toxicity and antipyretic activity of crude and processed aconite roots. Planta Med 56:665.

ACONITUM KUSNEZOFFII Rchb. [Ranunculaceae]

Common Names:

aconite; caowu; zhi-cao-wu

Citations:

But PP Tai YT, Young K (1994) Three fatal cases of herbal Aconite poisoning. Vet Hum Toxicol 39(3):212-215.

Chan TY, Chan JC, Tomlinson B, et al. (1994) Poisoning by Chinese herbal medicines in Hong Kong: A hospital-based study. Vet Hum Toxicol 36(6):546-547.

Chan TY, Tomlinson B, Chan WW, et al. (1993) A case of acute aconitine poisoning caused by chuanwu and caowu. J Trop Med Hyg 96(1):62-63.

Chan TY, Tomlinson B, Critchley JA, et al. (1994) Herb-induced Aconitine poisoning presenting as tetraplegia. Vet Hum Toxicol 36(2):133-134.

Chan WY, Ng TB, Lu JL (1995) Effects of decoctions prepared from Aconitum carmichaelii, Aconitum kusnezoffii and Tripterygium wilfordii on serum lactate dehydrogenase activity and histology of liver, kidney, heart, and gonad. Hum Exp Toxicol 14(6):489-493.

Chen SS, Hung DZ, Deng JF, et al. (1992) Aconite poisoning in a herb preparation. Toxicon 30:497-498.

Dickens P, Tai YT, But PP, et al. (1994) Fatal accidental aconitine poisoning following ingestion of Chinese herbal medicine: A report of two cases. Forensic Sci Int 67(1):55-58.

Fitzpatrick AJ, Crawford M, Allan RM, et al. (1994) Aconite poisoning managed with a ventricular assist device. Anaesth Intensive Care 22(6):714-717.

Tai YT, But PP, Young K, et al. (1992) Cardiotoxicity after accidental herb-induced aconite poisoning. Lancet 340(8830):1254-1256.

Tai YT, Lau CP, But PP, et al. (1992) Bidirectional tachycardia induced by herbal aconite poisoning. Pacing Clin Electrophysiol 15(5):831-839.

ACONITUM LYCOCTONUM L. subsp. lycoctonum [Ranunculaceae]

Synonyms:

Aconitum excelsum Rchb.; ***Aconitum septentrionale*** Koelle

Citations:

Leine N (1988) Forgiftning med tyrihjelm (lusehatt) Aconitum septentrionale. Norsk Veterinaertidsskrift 100(10):736-737.

ACONITUM LYCOCTONUM L. subsp. vulparia (Rchb.) Nyman [Ranunculaceae]

Synonyms:

Aconitum vulparia Rchb.

Common Names:

farkasölö sisakvirág; Gelbereisenhut; Gelber Sturmhut; monkshood; Wolfs-Eisenhut; wolfsbane

Citations:

Puschner B, Booth MC, Tor ER, et al. (2002) Diterpenoid alkaloid toxicosis in cattle in the Swiss Alps. Vet Hum Toxicol 44(1):8-10.

Puschner B, Booth MC, Tor ER, et al. (2004) Delphinium alkaloid toxicosis in cattle from Switzerland. In: Acamovic T, Stewart CS, Pennycott TW (eds.) Poisonous plants and related toxins. CABI. Wallingford, UK. pp. 38-43.

ACONITUM NAPELLUS L. [Ranunculaceae]

Synonyms:

Aconitum angelicum Stapf

Common Names:

aconite; bachnag; bear's-foot; Blauer Eisenhut; Blauer Sturmhut; blue rocket; Echter Eisenhut; Echter Sturmhut; European aconite; Fischerkappe; friar's-cap; helmet flower; monk's-cowl; monkshood; Reiterkappe; soldier's-cap; Taubenkutsche; Turk's-cap; Venus'-chariot; wolfsbane

Citations:

Birtanov Y, Toibaeva G, Birtanov A (2000) Acute aconite poisoning: A review of 219 cases. J Toxicol Clin Toxicol 38(5):579.

Crandon K, Thompson JP (2003) A fatal case of monkshood poisoning. J Toxicol Clin Toxicol 41(4):519-520.

Edwards EP (1925) Aconite poisoning. Vet Rec 5(Jul 4):559.

Feldkamp A, Köster B, Weber HP (1991) Tödliche Vergiftung durch Blauen Eisenhut (Aconitum napellus). Monatsschr Kinderheilkd 139(6):366-367.

Felgenhauer N, Zilker T, Dorfmann N (1999) Severe intoxication with Aconitum. J Toxicol Clin Toxicol 37(3):416.

Gohel DR, Patel BI, Agarwal SB (1989) Aconite intoxication and cardiac arrythmias. J Assoc Physicians India 37(3):245.

Imazio M, Belli R, Pomari F, et al. (2000) Malignant ventricular arrythmias due to Aconitum napellus seeds. Circulation 102(23):2907-2908.

Moritz F, Compagnon P, Kaliszczak IG, et al. (2005) Severe acute poisoning with homemade Aconitum napellus capsules: Toxicokinetic and clinical data. J Toxicol Clin Toxicol 43(7):873-876.

Northall FS, Dauncey EA, Butler JM (2003) An overview of plant and fungal poisonings in the UK, and some interesting cases. J Toxicol Clin Toxicol 41(4):518-519.

Puschner B, Booth MC, Tor ER, et al. (2002) Diterpenoid alkaloid toxicosis in cattle in the Swiss Alps. Vet Hum Toxicol 44(1):8-10.

Puschner B, Booth MC, Tor ER, et al. (2004) Delphinium alkaloid toxicosis in cattle from Switzerland. In: Acamovic T, Stewart CS, Pennycott TW (eds.) Poisonous plants and related toxins. CABI. Wallingford, UK. pp. 38-43.

Singh S, Fadnis PP, Sharma BK (1986) Aconite poisoning. J Assoc Physicians India 34(11):825-826.

Solway RI (1949) Aconite intoxication and myocardial infarction. Connecticut State Med J 13:727-728.

Taylor H (1925) Poisoning by the aconite plant. Vet Rec 5(Jun 27):533.

Toibaeva GM, Birtanov YA, Birtanov AB (2001) Acute aconite poisoning: A review of 219 cases. J Toxicol Clin Toxicol 39(3):302.

Travis AD, Gummin DD, McCann P, et al. (2002) Monkshood-induced dysrhythmia treated with magnesium. J Toxicol Clin Toxicol 40(5):646.

Aconitum septentrionale Koelle = Aconitum lycoctonum L. subsp. lycoctonum

Aconitum vulparia Rchb. = Aconitum lycoctonum L. subsp. vulparia (Rchb.) Nyman

acoro bastardo –see– *Iris pseudoacorus* L.

acorus –see– *Acorus calamus* L.

ACORUS CALAMUS L. [Acoraceae]

Common Names:

acorus; calamus; Kalmus; sweet cane; sweet cinnamon; sweet flag; sweet root; vegetable calomel

Citations:

Gross MA, Jones WI, Cook EL, et al. (1967) Carcinogenicity of oil of calamus. Proc Am Assoc Cancer Res 8:24.

Thomas M, Anandan S, Kuruvilla PJ, et al. (2000) Profile of hospital admissions following acute poisoning – Experiences from a major teaching hospital in south India. Adverse Drug React Toxicol Rev 19(4):313-317.

Vargas CP, Wolf LR, Gamm SR, et al. (1998) Getting to the root (Acorus calamus) of the problem. J Toxicol Clin Toxicol 36(3):259-260.

acrid buttercup –see– *Ranunculus acris* L.

Acroptilon repens (L.) DC. = Rhaponticum repens (L.) Hidalgo

ACROSTICHUM AUREUM L. [Pteridaceae]

Citations:

Devi S, Yasmeem, Singh J, et al. (1989) Patch testing animals to allergenic fern spores. J Toxicol Cutaneous Ocul Toxicol 8(2):167-172.

ACTAEA RACEMOSA L. [Ranunculaceae]

Synonyms:

Cimicifuga racemosa (L.) Nutt.

Common Names:

black cohosh; black snakeroot; bugbane; cimicifuga; macrotys; rattleroot; squawroot

Citations:

Litovitz TL, Felberg L, Soloway RA, et al. (1995) 1994 Annual Report of the American Association of Poison Control Centers Toxic Exposure Surveillance System. Am J Emerg Med 12(5):551-597.

Thomsen M, Vitetta L, Sali A, et al. (2004) Acute liver failure associated with the use of herbal preparations containing black cohosh. Med J Aust 180(11):598-600.

ACTAEA RUBRA (Aiton) Willd. [Ranunculaceae]

Synonyms:

Actaea spicata L. var. rubra Aiton

Common Names:

actée rouge; baneberry; coralberry; doll's-eyes; red baneberry; snakeberry; western baneberry; yerba-del-Peco

Citations:

Bacon AE (1903) An experiment with the fruit of red baneberry. Rhodora 5(51):77-79.

Actaea spicata L. var. rubra Aiton = Actaea rubra (Aiton) Willd.

actée rouge –see– *Actaea rubra* (Aiton) Willd.

Actinea odorata (DC.) Kuntze = Hymenoxys odorata DC.

Actinea richardsonii (Hook.) Kuntze = Hymenoxys richardsonii (Hook.) Cockerell

ACTINIDIA CHINENSIS Planch. [Actinidiaceae]

Common Names:

kiwi

Citations:

Zina AM, Bundino S (1983) Contact urticaria to Actinidia chinensis. Contact Dermatitis 9(1):85.

ACTINIDIA DELICIOSA (A. Chev.) C. F. Liang & A. R. Ferguson [Actinidiaceae]

Common Names:

kiwi

Citations:

Gall H, Kalveram KJ, Forck G, et al. (1994) Kiwi fruit allergy: A new birch pollen-associated food allergy. J Allergy Clin Immunol 94(1):70-76.

Gastaminza G, Bernaola G, Camino ME (1998) Acute pancreatitis caused by allergy to kiwi fruit. Allergy 53(11):1104-1105.

Mancuso G, Berdondini RM (2001) Oral allergy syndrome from kiwi fruit after a lover's kiss. Contact Dermatitis 45(1):41.

Voitenko V, Poulsen LK, Nielsen L, et al. (1997) Allergenic properties of kiwi-fruit extract: Cross-reactivity between kiwi-fruit and birch-pollen allergens. Allergy 52(2):136-143.

Adam-and-Eve –see– *Arum maculatum* L.

addad –see– *Chamaeleon gummifera* (L.) Cass.

addar –see– *Sorghum halepense* (L.) Pers.

adder's-meat –see– *Arum maculatum* L.

adder's-root –see– *Arum maculatum* L.

adelfa –see– *Nerium oleander* L.

ADENIA GUMMIFERA (Harv.) Harms [Passifloraceae]

Citations:

Neame PB, Pillay VK (1964) Spontaneous hypoglycaemia, hepatic and renal necrosis following the intake of herbal medicines. S Afr Med J 38(part 2):729-732.

ADENIA HONDALA (Gaertn.) W. J. de Wilde [Passifloraceae]

Synonyms:

Adenia palmata (Lam.) Engl.; ***Modecca palmata*** Lam.

Common Names:

hondala

Citations:

Fernando R (2002) The national poisons information centre in Sri Lanka: The first ten years. J Toxicol Clin Toxicol 40(5):551-555.

Fernando R, Fernando DN (1990) Poisoning with plants and mushrooms in Sri Lanka: A retrospective hospital based study. Vet Hum Toxicol 32(6):579-581.

Lucas GN (1997) Plant poisoning: A hospital-based study in Sri Lanka. Indian J Pediatr 64(4):495-502.

Munasinghe DR, Karunanayake A, Nimalasuriya A, et al. (1971) Poisoning with Adenia palmata. Ceylon Med J 16(3):182-186.

Adenia palmata (Lam.) Engl. = Adenia hondala (Gaertn.) W. J. de Wilde

ADENIA VOLKENSII Harms [Passifloraceae]

Citations:

Verdcourt B, Trump EC (1969) Common poisonous plants of East Africa. Collins. London.

ADENOSTYLES ALLIARIAE A. Kern. [Asteraceae]

Common Names:

Alpendost

Citations:

Sperl W, Stuppner H, Gassner I, et al. (1995) Reversible hepatic veno-occlusive disease in an infant after consumption of pyrrolizidine-containing herbal tea. Eur J Pediatr 154(2):112-116.

adhered –see– *Agrostemma githago* L.

Adina cordifolia (Roxb.) Hook. f. ex Brandis = Haldina cordifolia (Roxb.) Ridsdale

Adlerblume –see– *Consolida regalis* Gray

Adlerfarn –see– *Pteridium aquilinum* (L.) Kuhn

ADONIS AESTIVALIS L. [Ranunculaceae]

Common Names:

Adonisröschen; nyári hérics; Sommeradonisröschen; summer adonis; summer pheasant's-eye

Citations:

Kummer H (1952) Vergiftungen bei Pferden durch Adonis in Luzerneheu. Tierarztl Umsch 7:430-431.

Woods LW, Filigenzi MS, Booth MC, et al. (2004) Summer pheasant's eye (Adonis aestivalis) poisoning in three horses. Vet Pathol 41(3):215-220.

ADONIS ANNUA L. [Ranunculaceae]

Synonyms:

Adonis phoenicea Bercht. & J. Presl

Common Names:

autumn adonis; bird's-eye; pheasant's-eye; Teufelsauge

Citations:

Degen A (1932) Adonis-Vergiftung. Fortschr Landwirtschaft 7:556-558.

ADONIS MICROCARPA DC. [Ranunculaceae]

Common Names:

pheasant's-eye

Citations:

Davies RL, Whyte PB (1989) Adonis microcarpa (pheasant's eye) toxicity in pigs fed field pea screenings. Aust Vet J 66(5):141-143.

adonis morado –see– *Duranta erecta* L.

Adonis phoenicea Bercht. & J. Presl = Adonis annua L.

Adonisröschen –see– *Adonis aestivalis* L.

AEGLE MARMELOS (L.) Corrêa [Rutaceae]

Citations:

Arseculeratne SN, Gunatilaka AA, Panabokke RG (1985) Studies on medicinal plants of Sri Lanka. Part 14. Toxicity of some traditional medicinal herbs. J Ethnopharmacol 13:323-335.

Kulakkattolickal AT (1987) Piscicidal plants of Nepal: Preliminary toxicity screening using grass carp (Ctenopharyngodon idella) fingerlings. J Ethnopharmacol 21(1):1-9.

AESCHYNANTHUS PULCHELLA Hort. ex Hanst. [Gesneriaceae]

Citations:

van Ketel WG (1979) Occupational contact dermatitis due to Codiaeum variegatum and possibly to Aeschynanthus pulcher. Dermatosen 27(5):141-142.

AESCHYNOMENE INDICA L. [Fabaceae]

Common Names:

Buddha pea

Citations:

Haraguchi M, Zambronio F, Górniak SL, et al. (2003) Neurotoxicity to pigs and rodents from different fractions of Aeschynomene indica seeds. Vet Hum Toxicol 45(4):177-179.

Oliveira FN, Barros RR, Rissi DR, et al. (2004) Focal symmetrical encephalomalacia in swine from ingestion of Aeschynomene indica seeds. Vet Hum Toxicol 46(6):309-311.

Oliveira FN, Rech RR, Rissi DR, et al. (2005) Intoxicação em suínos pela ingestão de sementas de Aeschynomene indica (Leg. Papilionoideae). Pesq Vet Bras 25(3):135-142.

Riet-Correa F, Timm CD, Barros SS, et al. (2003) Symmetric focal degeneration in the cerebellar and vestibular nuclei in swine caused by ingestion of Aeschynomene indica seeds. Vet Pathol 40(3):311-316.

AESCULUS CALIFORNICA (Spach) Nutt. [Sapindaceae]

Common Names:

California buckeye; horse chestnut

Citations:

Vansell GB (1926) Buckeye poisoning of the honey bee. California Agric Exp Sta Circ #301:12 pp.

AESCULUS FLAVA Sol. [Sapindaceae]

Synonyms:

Aesculus octandra Marshall

Common Names:

buckeye; sweet buckeye; yellow buckeye

Citations:

Magnusson RA, Whittier WD, Veit HP, et al. (1983) Yellow buckeye (Aesculus octandra Marsh) toxicity in calves. Bovine Pract 18(Nov):195-199.

AESCULUS GLABRA Willd. [Sapindaceae]

Synonyms:

Pavia glabra Spach

Common Names:

buckeye; fetid buckeye; horse chestnut; Kastanie; Ohio buckeye; spring buckeye; Texas buckeye; western buckeye; white buckeye

Citations:

Casteel SW, Johnson GC, Wagstaff DJ (1992) Aesculus glabra intoxication in cattle. Vet Hum Toxicol 34(1):55.

Creek RD, Vasaitis V (1963) The toxicity of the Ohio Buckeye to the chick. Poult Sci 42(6):1473-1474.

Hansen AA (1928) Toxic trees. North Am Vet 9(10):49-53.

Kornheiser KM (1983) Buckeye poisoning in cattle. Vet Med Small Anim Clin 78(5):769-770.

Mayer H, Wacker R, Dalchow W (1986) Phytotoxikosen durch Kastanien, Oleander, Eicheln und Herbstzeitlose bei verschiedenen Zoo- und Wildtieren. Tierarztl Umsch 41(3):169-178.

AESCULUS HIPPOCASTANUM L. [Sapindaceae]

Common Names:

buckeye; European horse chestnut; horse chestnut; Ignatius bean; marronier d'Inde; marton d'Inde; Pferdekastanie; Roßkastanie; venoplant; wilde Kastanie

Citations:

Jaspersen-Schib R, Theus L, Quirguis-Oeschger M, et al. (1996) Wichtige Pflanzenvergiftungen in der Schweiz 1966-1994. Schweiz Med Wochenschr 126(25):1085-1098.

Matyunas N, Krenzelok E, Jacobsen T, et al. (1997) Horse chestnut (Aesculus spp.) ingestion in the United States: 1985-1994. J Toxicol Clin Toxicol 35:527-528.

Popp W, Horak F, Jager S, et al. (1992) Horse chestnut (Aesculus hippocastanum) pollen: A frequent cause of allergic sensitization in urban children. Allergy 47(4 Part 2):380-383.

Ramón MF, Ballesteros S, Martínez-Arrieta R, et al. (2003) Plant and herb exposures in Spain. J Toxicol Clin Toxicol 41(4):514.

Schweitzer H (1952) Tödliche Saponinvergiftung durch Genuß von Roßkastanien. Med Klin 47(20):683-685.

Spoerke DG, Temple AR (1978) One year's experience with potential plant poisonings reported to the Intermountain Regional Poison Control Center. Vet Hum Toxicol 20(2):85-89.

Wälli F, Grob PJ, Müller-Schoop SJ (1981) Pseudo-(venocuran-)lupus: Eine kleine Episode Medizingeschichte. Schweiz Med Wochenschr 111(38):1398-1405.

Williams MC (1982) Poisonous plants. II. Poisonous plants of home and garden. Weeds Today 13(4):4-5.

AESCULUS INDICA (Wall. ex Cambess.) Hook. [Sapindaceae]

Common Names:

Indian horse chestnut

Citations:

Sood AK, Sharma M, Katoch BS (1990) Improvement in the nutritive value of Indian horse-chestnut kernel and cassia seeds by chemical treatments for feeding of chicks. Indian J Anim Sci 60(11):1364-1369.

Aesculus octandra Marshall = Aesculus flava Sol.

AESCULUS PAVIA L. [Sapindaceae]

Common Names:

buckeye; firecracker plant; horse chestnut; red buckeye; scarlet buckeye; small buckeye

Citations:

Cary CA (1922) Poisonous action of red buckeye on horses, mules, cattle, hogs and fish. Alabama Agric Exp Sta Bull #218:20 pp.

AETHUSA CYNAPIUM L. [Apiaceae]

Common Names:

ass's-parsley; dog parsley; dog poison; false parsley; falso perejil; fool's-cicely; fool's-parsley; Gartenschierling; Glanzpeterlein; Glanzpetersilie; Hundsdill; Hundspetersilie; Krötenpeterlein; lesser hemlock; Petersilienschierling; petite ciguë; small hemlock

Citations:

Barr AG, Davies CS (1963) An unusual case of poisoning in a sow and litter. Vet Rec 75(17):457.

Bubien MZ, Magat A, Delatour P (1970) Fréquence des intoxications des animaux en Pologne. Bull Soc Vet Med Comp Lyon 72(2):167-173.

Seltmann L (1931) Über die toxische Wirkung der Hundspetersilie (Aethusa cynapium). Med Klin 27(8):281.

Swart FW (1975) Vergiftiging van geiten door hondspeterselie. Tijdschr Diergeneeskd 100(18):989-990.

Teuscher E, Greger H, Adrian V (1990) Untersuchungen zur Toxizität von Aethusa cynapium L., der Hundspetersilie. Pharmazie 45(7):537-538.

African Bermuda grass –see– *Cynodon nlemfuensis* Vanderyst

African bird pepper –see– *Capsicum annuum* L.

African chili –see– *Capsicum frutescens* L.

African coffee tree –see– *Ricinus communis* L.

African cucumber –see– *Momordica charantia* L.

African eggplant –see– *Solanum dasyphyllum* Schumach. & Thonn.

African hemp –see– *Sparrmannia africana* L. f.

African lilac tree –see– *Melia azedarach* L.

African mahogany –see– *Khaya anthotheca* (Welw.) C. DC.; *Khaya ivorensis* A. Chev.

African maple –see– *Triplochiton scleroxylon* K. Schum.

African marigold –see– *Tagetes erecta* L.

African milk bush –see– *Synadenium grantii* Hook. f.

African oxeye daisy –see– *Callilepis laureola* DC.

African pepper –see– *Capsicum frutescens* L.

African poison ivy –see– *Smodingium argutum* E. Mey. ex Sond.

African rose –see– *Papaver rhoeas* L.

African rue –see– *Peganum harmala* L.

African sorghum –see– *Sorghum bicolor* (L.) Moench

African stargrass –see– *Cynodon dactylon* (L.) Pers.; *Cynodon nlemfuensis* Vanderyst

African teak –see– *Milicia excelsa* (Welw.) C. C. Berg

African violet –see– *Saintpaulia ionantha* H. Wendl.

African whitewood –see– *Triplochiton scleroxylon* K. Schum.

African yohimbine –see– *Pausinystalia johimbe* (K. Schum.) Pierre ex Beille

Afrikanische Schwarznuß –see– *Mansonia altissima* (A. Chev.) A. Chev.

Afrikanischer Birnbaum –see– *Tieghemella heckelii* Pierre ex A. Chev.

Afrikanischer Buchsbaum –see– *Sarcocephalus diderrichii* De Wild.

Afrikanisches Mahagoni –see– *Khaya ivorensis* A. Chev.; *Khaya senegalensis* (Desr.) A. Juss.

afrormosia –see– *Pericopsis elata* (Harms) Meeuwen

Afrormosia elata Harms = Pericopsis elata (Harms) Meeuwen

afufa oyibo –see– *Solanum incanum* L.

Afzelia –see– *Afzelia africana* Sm. ex Pers.

AFZELIA AFRICANA Sm. ex Pers. [Fabaceae]

Common Names:

Afzelia; oven

Citations:

Dantin-Gallego J, Armayor AF, Riesco J (1952) Some new toxic woods: Some new manifestations of toxicity. Ind Med Surg 21(2):41-46.

Oleffe JA, Sporcq J, Hublet P (1975) Epidemiological study of the wood industry in Belgium. Contact Dermatitis 1(5):315-316.

agathi –see– *Sesbania grandiflora* (L.) Pers.

AGAVE AMERICANA L. [Agavaceae]

Common Names:

American agave; American aloe; century plant; maguey

Citations:

Kerner J, Mitchell J, Maibach HI (1973) Irritant contact dermatitis from Agave americana L. Incorrect use of sap as "hair restorer." Arch Dermatol 108(1):102-103.

Ricks MR, Vogel PS, Elston DM, et al. (1999) Purpuric agave dermatitis. J Am Acad Dermatol 40(2 Part 2):356-358.

Villarreal R, Martínez O, Berumen U Jr (1985) Phytobezoar from the stem ("quiote") of the cactus Agave americana: Report of case. Am J Gastroenterol 80(11):838-840.

Agave lecheguilla Torr. = Agave lechuguilla Torr.

AGAVE LECHUGUILLA Torr. [Agavaceae]

Synonyms:

Agave lecheguilla Torr.

Common Names:

century plant; lecheguilla; lechuguilla; little lettuce

Citations:

Burrows GE, Stair EL (1990) Apparent Agave lecheguilla intoxication in Angora goats. Vet Hum Toxicol 32(3):259-260.

Camp BJ, Bridges CH, Hill DW, et al. (1988) Isolation of a steroidal sapogenin from the bile of a sheep fed Agave lecheguilla. Vet Hum Toxicol 30(6):533-535.

Jungherr E (1931) Lechuguilla fever of sheep and goats; a form of swellhead in West Texas. Cornell Vet 21(3):227-242.

Mathews FP (1937) Lechuguilla (Agave lecheguilla) poisoning in sheep, goats, and laboratory animals. Texas Agric Exp Sta Bull #554:36 pp.

Mathews FP (1938) An experimental investigation of lechuguilla poisoning. Arch Pathol 25:661-683.

Mathews FP (1938) Lechuguilla (Agave lecheguilla) poisoning in sheep and goats. J Am Vet Med Assoc 93(Sep):168-175.

Schmidt H, Jungherr E (1930) Swellhead of sheep and goats. Texas Agric Exp Sta Annu Rep 43:9-11.

AGAVE SISALANA Perrine [Agavaceae]

Common Names:

sisal

Citations:

Figueiredo LJ (1975) Estudo experimental da toxidade resíduo do sisal Agave sisalana perrine para bovinos. Arq Esc Vet Univ Fed Minas Gerais 27:391-392.

AGAVE TEQUILANA F. A. C. Weber [Agavaceae]

Citations:

Salinas ML, Ogura T, Soffchi L (2001) Irritant contact dermatitis caused by needle-like calcium oxalate crystals, raphides, in Agave tequilana among workers in tequila distilleries and agave plantations. Contact Dermatitis 44(2):94-96.

agba –see– *Gossweilerodendron balsamiferum* (Vermoesen) Harms

Agelaea nitida Sol. ex Planch. = Agelaea pentagyna (Lam.) Baill.

Agelaea obliqua (P. Beauv.) Baill. = Agelaea pentagyna (Lam.) Baill.

AGELAEA PENTAGYNA (Lam.) Baill. [Connaraceae]

Synonyms:

Agelaea nitida Sol. ex Planch.; ***Agelaea obliqua*** (P. Beauv.) Baill.; ***Agelaea trifolia*** (Lam.) Baill.

Common Names:

esura

Citations:

Vickery B, Vickery ML (1974) The toxicity of some members of the Connaraceae family. Br Vet J 130(2):41-43.

Agelaea trifolia (Lam.) Baill. = Agelaea pentagyna (Lam.) Baill.

AGERATINA ADENOPHORA (Spreng.) R. M. King & H. Rob. [Asteraceae]

Synonyms:

Eupatorium adenophorum Spreng.; ***Eupatorium glandulosum*** Michx.

Common Names:

catweed; Crofton weed; croton weed; hemp agrimony; Mexican devil; sticky agrimony; sticky eupatorium; whitetop

Citations:

Anonymous (1953) Numimbah horse disease caused by Crofton weed. Agric Gaz New South Wales 63(Feb 1):110.

Katoch R, Sharma OP, Dawra RK, et al. (2000) Hepatotoxicity of Eupatorium adenophorum to rats. Toxicon 38(2):309-314.

Kaushal V, Dawra RK, Sharma OP, et al. (2001) Biochemical alterations in the blood plasma of rats associated with hepatotoxicity induced by Eupatorium adenophorum. Vet Res Commun 25(7):601-608.

O'Sullivan BM (1979) Crofton weed (Eupatorium adenophorum) toxicity in horses. Aust Vet J 55(1):19-21.

O'Sullivan BM (1985) Investigations into Crofton weed (Eupatorium adenophorum) toxicity in horses. Aust Vet J 62(1):30-32.

Sani Y, Harper PA, Cook RL, et al. (1992) The toxicity of Eupatorium adenophorum for the liver of the mouse. In: James LF, Keeler RF, Bailey JM Jr, et al. (eds.) Poisonous plants. Iowa State Univ Press. Ames, Iowa. pp. 626-629.

AGERATINA ALTISSIMA (L.) R. M. King & H. Rob. [Asteraceae]

Synonyms:

Eupatorium ageratoides L. f.; ***Eupatorium rugosum*** Houtt.; *Eupatorium urticaefolium* Reichard

Common Names:

boneset; deerwort; fall poison; Indian sanicle; poolroot; poolwort; richweed; squawweed; thoroughwort; white sanicle; white snakeroot; white snakeweed; whitetop

Citations:

Bulger HA, Smith FM, Steinmeyer A (1928) Milk sickness and the metabolic disturbances in white snakeroot poisoning. JAMA 91(25):1964-1966.

Couch JF (1927) The toxic constituent of richweed or white snakeroot (Eupatorium urticaefolium). J Agric Res 35(6):547-576.

Curtis RS, Wolf FA (1917) Eupatorium ageratoides, the cause of trembles. J Agric Res 9(11):397-404.

Doyle JT (1947) Milk sickness. N C Med J 8(Jul):404-410.

Doyle LP, Walkey FL (1923) White snakeroot poisoning in livestock. Indiana Agric Exp Sta Bull #270:15 pp.

Finno CJ, Valberg SJ, Wünschmann A, et al. (2006) Seasonal pasture myopathy in horses in the midwestern United States: 14 cases (1998-2005). J Am Vet Med Assoc 229(7):1134-1141.

Greer M (1925) Milk sickness. Ill Med J 48:75-78.

Hansen AA (1924) A unique experiment with white snakeroot. J Am Vet Med Assoc 65:224-226.

Hansen AA (1924) Poison in the woods. The strange story of white snakeroot, which science has finally found guilty of causing many mysterious deaths. Am Forests Forest Life 30:163-191.

Hansen AA (1927) Relation of white snakeroot to human milksickness (field studies). J Indiana State Med Assoc 20(May):182-187.

Hardin RH (1934) Milk sickness. Va Med Monthly 61:528-531.

Hartmann AF Sr, Hartmann AF Jr, Purkerson ML, et al. (1963) Tremetol poisoning - Not yet extinct. JAMA 185(9):706-709.

Knight HW (1935) Milk sickness due to white snake root (Eupatorium urticaefolium). Southern Med Surg 97:142.

Marsh CD, Clawson AB (1917) Eupatorium urticaefolium as a poisonous plant. J Agric Res 11(13):699-715.

Moseley EL (1906) The cause of trembles in cattle, sheep and horses and of milk-sickness in people. Ohio Naturalist 6(4):463-470.

Moseley EL (1906) The cause of trembles in cattle, sheep and horses, and of milk-sickness in people. Ohio Naturalist 6(5):477-483.

Nielsen EL, Eveleth DF (1941) White snakeroot poisoning in Arkansas. Vet Med 36(3):136-137.

Olson CT, Keller WC, Gerken DF, et al. (1984) Suspected tremetol poisoning in horses. J Am Vet Med Assoc 185(9):1001-1003.

Pammel LH (1923) Milksickness or trembles caused by white snakeroot. Vet Med 18:437-440.

Richardson WP (1931) The milk sickness or trembles. Southern Med Surg 93:731.

Sackett WG (1919) The connection of milksickness with the poisonous qualities of white snakeroot (Eupatorium urticaefolium). J Infect Dis 24:231-259.

Sanders M (1983) White snakeroot poisoning in a foal: A case report. Equine Vet Sci 3(4):128-131.

Smetzer DL, Coppock RW, Ely RW, et al. (1983) Cardiac effects of white snakeroot intoxication in horses. Equine Pract 5(2):26-32.

Stotts R (1984) White snakeroot toxicity in dairy cattle. Vet Med Small Anim Pract 79(1):118-120.

Thompson LJ (1989) Depression and choke in a horse: Probable white snakeroot toxicosis. Vet Hum Toxicol 31(4):321-322.

Walsh WE (1926) "Milk Sick," or white snakeroot poisoning. JAMA 87(8):555-556.

White JL, Shivaprasad HL, Thompson LJ, et al. (1985) White snakeroot (Eupatorium rugosum) poisoning. Clinical effects associated with cardiac and skeletal muscle lesions in experimental equine toxicoses. In: Seawright AA, Hegarty MP, James LF, et al. (eds.) Plant toxicology. Queensland Poisonous Plant Committee. Yeerongpilly, Queensland. pp. 411-422.

Wilkinson GR (1925) Milk sickness. South Med J 18(11):797-780.

Wolf FA, Curtis RS, Kaupp BF (1917-1918) Studies on trembles or milksickness and white snake root. J Am Vet Med Assoc 52:820-827.

Wolf FA, Curtis RS, Kaupp BF (1918) A monograph on trembles or milksickness and white snakeroot. North Carolina Agric Exp Sta Tech Bull #15:85 pp.

AGERATINA RIPARIA (Regel) R. M. King & H. Rob. [Asteraceae]

Synonyms:

Eupatorium riparium Regel

Common Names:

creeping croftonweed; mist flower

Citations:

Gibson JA, O'Sullivan BM (1984) Lung lesions in horses fed mist flower (Eupatorium riparium). Aust Vet J 61(8):271.

AGERATINA WRIGHTII (A. Gray) R. M. King & H. Rob. [Asteraceae]

Synonyms:

Eupatorium wrightii A. Gray

Common Names:

Wright's-eupatorium

Citations:

Hershey AL (1949) Another poisonous species of Eupatorium. Colorado Wyoming Acad Sci J 4(4):52.

ageratum –see– *Ageratum houstonianum* Mill.

AGERATUM CONYZOIDES L. [Asteraceae]

Common Names:

bakarai bish; billygoat weed; floss flower; goat weed; nilphulia; whiteweed

Citations:

Purohit K (1962) Nilphulia (Ageratum conyzoides) poisoning in sheep. Indian Vet J 39:553-556.

AGERATUM HOUSTONIANUM Mill. [Asteraceae]

Synonyms:

Ageratum mexicanum Sims

Common Names:

ageratum; blue billygoat weed; celestina azul; floss flower

Citations:

Alfonso HA, Rivera M, Aparicio JM, et al. (1989) Intoxicación natural y experimental en bovinos con Ageratum houstonianum Mill (celestina azul). Rev Cubana Cienc Vet 20(1):113-119.

Noa M, Sánchez LM, Durand R (2004) Ageratum houstonianum toxicosis in Zebu cattle. Vet Hum Toxicol 46(4):193-194.

Ageratum mexicanum Sims = Ageratum houstonianum Mill.

agnus castus –see– *Vitex agnus-castus* L.

AGRIMONIA EUPATORIA L. [Rosaceae]

Common Names:

agrimony

Citations:

O'Donovan WJ (1942) Dermatitis bullosa striata pratensis. Agrimony dermatitis. Br J Dermatol Syph 54:39-46.

agrimony –see– *Agrimonia eupatoria* L.

AGROSTEMMA GITHAGO L. [Caryophyllaceae]

Synonyms:

Lychnis githago (L.) Scop.

Common Names:

Ackerkrone; Adhered; bastard nigelle; cockle; corn campion; corn cockle; cow cockle; crown-of-the-field; Kornrade; la-nielle-des-blés; licheta; mullein pink; neguillon; nielle; old-maid's-pink; purple cockle; rose campion; wild savages

Citations:

Brandl J (1908) Über Sapotoxin von Agrostemma githago. III. Fütterungsversuche an Schweinen, Kanichen, Rindern und Pferden. Naunyn Schmiedebergs Arch Exp Pathol Pharmakol 59:299-310.

Bubien MZ, Magat A, Delatour P (1970) Fréquence des intoxications des animaux en Pologne. Bull Soc Vet Med Comp Lyon 72(2):167-173.

Debré R, Névot A (1938) Toxicité expérimentale des graines de nielle (Lychnis githago) chez le singe (Macacus rhesus). C R Soc Biol 127:1094-1096.

Heuser GF, Schumacher AE (1942) The feeding of corn cockle to chickens. Poult Sci 21:86-93.

Kotz J (1964) Badania nad morfologia i patogeneza zatruć kąkolem u drobiu. I. Historia zatruć kąkolem. Med Weter 20(4):200-204.

Kotz J (1965) Badania nad morfologia i patogeneza zatruć kąkolem u drobiu. II. Badania własne. Med Weter 21(3):143-150.

Kotz J (1965) Badania nad morfologia i patogeneza zatruć kąkolem u drobiu. III. Badania własne. Med Weter 21(9):520-524.

Kotz J (1965) Badania nad morfologia i patogeneza zatruć kąkolem u drobiu. IV. Badania własne. Med Weter 21(12):730-734.

Quigley GD, Waite RH (1931) Miscellaneous feeding trials with poultry. I. The effects of corn cockle on poultry. Maryland Agric Exp Sta Bull #325:343-354.

Roberg M (1950) Konjunktivitis durch Kornradesamen. Klin Monatsbl Augenheilkd 116:425-426.

Smith RA, Miller RE, Lang DG (1997) Presumptive intoxication of cattle by corn cockle, Agrostemma githago (L) Scop. Vet Hum Toxicol 39(4):250.

aguacate –see– *Persea americana* Mill. var. americana

ague tree –see– *Sassafras albidum* (Nutt.) Nees

Ahlkirsche –see– *Prunus padus* L.

ahouai –see– *Thevetia peruviana* (Pers.) K. Schum.

aila –see– *Acacia pennata* (L.) Willd.

ailanthus –see– *Ailanthus altissima* (Mill.) Swingle

AILANTHUS ALTISSIMA (Mill.) Swingle [Simaroubaceae]

Synonyms:

Ailanthus glandulosa Desf.

Common Names:

ailanthus; Chinese sumach; copel tree; heavenwood; paradise tree; tree-of-heaven

Citations:

Blumstein GI (1943) Sensitivity to Ailanthus pollen with report of two clinically sensitive patients. J Allergy 14(4):329-334.

Ailanthus glandulosa Desf. = Ailanthus altissima (Mill.) Swingle

air plant –see– *Kalanchoe daigremontiana* Raym.-Hamet & H. Perrier; *Kalanchoe pinnata* (Lam.) Pers.

ajenjo –see– *Artemisia absinthium* L.

aji caballero –see– *Capsicum frutescens* L.

aji picante –see– *Capsicum frutescens* L.

ajo –see– *Allium sativum* L.

AJUGA DECUMBENS Thunb. [Lamiaceae]

Citations:

Chen TZ, Den JF, Tsai WJ (1990) Digitalis-like intoxication resulted from Ajuga decumbens Thunb consuming: A case report. Am Assoc Poison Control Cent 1990:111.

ak –see– *Calotropis procera* (Aiton) W. T. Aiton

akác –see– *Robinia pseudoacacia* L.

akada –see– *Calotropis procera* (Aiton) W. T. Aiton

akanda –see– *Calotropis gigantea* (L.) W. T. Aiton

Akanta –see– *Rauvolfia vomitoria* Afzel.

Akazie –see– *Robinia pseudoacacia* L.

akee –see– *Blighia sapida* K. D. Koenig

akee apple –see– *Blighia sapida* K. D. Koenig

akpu –see– *Manihot esculenta* Crantz

akwuosa –see– *Dichapetalum barteri* Engl.

Alant –see– *Inula helenium* L.

Alaska pea –see– *Pisum sativum* L.

albariã –see– *Drimia maritima* (L.) Stearn

albescens –see– *Pycnanthemum albescens* Torr. & A. Gray

ALBIZIA ADINOCEPHALA (Donn. Sm.) Britton & Rose ex Record [Fabaceae]

Citations:

Grant G, More LJ, McKenzie NH, et al. (1995) Nutritional and haemagglutination properties of several tropical seeds. J Agric Sci 124:437-445.

ALBIZIA JULIBRISSIN Durazz. [Fabaceae]

Common Names:

mimosa; pink siris; silk tree

Citations:

Robinson GH, Burrows GE, Holt EM, et al. (1998) Evaluation of the toxic effects of the legumes of mimosa (Albizia julibrissin) and identification of the toxicant. In: Garland T, Barr AC (eds.) Toxic plants and other natural toxicants. CABI. New York. pp. 453-458.

ALBIZIA LEBBECK (L.) Benth. [Fabaceae]

Synonyms:

Mimosa sirissa Roxb.

Common Names:

lebbeck tree; siris; sirisha

Citations:

Dwatmadgi, Teleni E, Bird AR, et al. (1992) Nutritive value of Albizia lebbeck supplements for growing sheep. Aust J Exp Agric 32(3):273-278.

Grant G, More LJ, McKenzie NH, et al. (1995) Nutritional and haemagglutination properties of several tropical seeds. J Agric Sci 124:437-445.

ALBIZIA TANGANYICENSIS Baker f. [Fabaceae]

Common Names:

red paperbark tree

Citations:

Basson PA, Adelaar TF, Naudé TW, et al. (1970) Albizia poisoning: Report of the first outbreak and some experimental work in South Africa. J S Afr Vet Assoc 41(2):117-130.

Jackson JJ, Needham AJ, Lawrence JA (1968) Some recent investigations into Rhodesian toxic plants. Proc Trans Rhodesia Sci Assoc 53:9-12.

ALBIZIA VERSICOLOR Welw. ex Oliv. [Fabaceae]

Common Names:

poison pod albizia

Citations:

Gummow B, Bastianello SS, Labuschagne L, et al. (1992) Experimental Albizia versicolor poisoning in sheep and its successful treatment with pyridoxine hydrochloride. Onderstepoort J Vet Res 59(2):111-118.

Gummow B, Erasmus GL (1990) Pyridoxine (a vitamin B6) and its derivative pyridoxal as treatment for Albizia versicolor poisoning in guinea-pigs. Onderstepoort J Vet Res 57(2):109-114.

Jackson JJ, Needham AJ, Lawrence JA (1968) Some recent investigations into Rhodesian toxic plants. Proc Trans Rhodesia Sci Assoc 53:9-12.

Needham AJ, Lawrence JA (1966) The toxicity of Albizia versicolor. Rhodesia Agric J 63(6):137-140.

Soldan AW, van Inzen C, Edelsten RM (1996) Albizia versicolor poisoning of sheep and goats in Malawi. J S Afr Vet Assoc 67(4):217-221.

Van Rensburg HJ (1967) Albizia versicolor Welw.: A poisonous plant. Rhodesia Zambia Malawi J Agric Res 5:109-110.

alcanfor –see– *Cinnamomum camphora* (L.) J. Presl

ALCEA ROSEA L. [Malvaceae]

Synonyms:

Althaea rosea (L.) Cav.

Common Names:

hollyhock; Stockrose

Citations:

Hofstee AW (1989) Acute vergiftigingen bij dieren in Nederland over de periode 1985-1988. Tijdschr Diergeneeskd 114(22):1154-1158.

alder buckthorn –see– *Frangula alnus* Mill.

alder-leaf mountain mahogany –see– *Cercocarpus montanus* Raf.

alecrim –see– *Holocalyx balansae* Micheli

alecrim-de-campinas –see– *Holocalyx balansae* Micheli

ALECTRYON OLEIFOLIUS (Desf.) S. T. Reynolds [Sapindaceae]

Synonyms:

Heterodendrum oleifolium Desf.

Common Names:

boonaree; boonery; bullock bush; red heart-of-bullock bush; rosewood; western rosewood

Citations:

Ramsay AA, Henry M (1929) Rosewood (Heterodendrum oleaefolium) and native fuchsia (Eremophila maculata): Two poisonous plants. Agric Gaz New South Wales 40:834-837.

alefoot –see– *Glechoma hederacea* L.

alehoof –see– *Glechoma hederacea* L.

alerce –see– *Fitzroya cupressoides* (Molina) I. M. Johnst.

aleurit-des-molluques –see– *Aleurites moluccanus* (L.) Willd.

Aleurites fordii Hemsl. = Vernicia fordii (Hemsl.) Airy Shaw

ALEURITES MOLUCCANUS (L.) Willd. [Euphorbiaceae]

Synonyms:

Aleurites trilobus J. R. Forst. & G. Forst.

Common Names:

aleurit-des-molluques; arbol-de-India; Bengal walnut; Camiri nut; candleberry; candlenut; country walnut; Indian walnut; Jamaican walnut; kukui nut; lumbán; lumbang; nogal-de-la-India; nuez-de-India; otaheite walnut; Singapore nut; varnish tree

Citations:

Emmel MW (1947) The toxic principle of the species Aleurites. J Am Vet Med Assoc 111(Nov):386-387.

Emmel MW (1947) The toxic principle of the tung tree. Florida Agric Exp Sta Bull #431:35 pp.

Aleurites montanus (Lour.) E. H. Wilson = Vernicia montana Lour.

Aleurites trilobus J. R. Forst. & G. Forst. = Aleurites moluccanus (L.) Willd.

Aleurites trisperma Blanco = Reutealis trisperma (Blanco) Airy Shaw

Alexandrian clover –see– *Trifolium alexandrinum* L.

Alexandrian senna –see– *Senna alexandrina* Mill.

alfalfa –see– *Medicago sativa* L.

alfalfa dodder –see– *Cuscuta epithymum* (L.) L.

alfilaria –see– *Erodium cicutarium* (L.) L'Her.

alfilerillo –see– *Erodium cicutarium* (L.) L'Her.

alfombrilla –see– *Drymaria arenarioides* Humb. & Bonpl. ex Schult.

Algarroba –see– *Prosopis glandulosa* Torr.; *Prosopis juliflora* (Sw.) DC.; *Vicia monantha* Retz.

Algerian ivy –see– *Hedera helix* L. subsp. canariensis (Willd.) Cout.

algerita –see– *Berberis aquifolium* Pursh

algodão-de-seda –see– *Calotropis procera* (Aiton) W. T. Aiton

algodón –see– *Gossypium hirsutum* L.

algodón-de-seda –see– *Calotropis gigantea* (L.) W. T. Aiton

aligustre –see– *Ligustrum vulgare* L.

alkali grass –see– *Anticlea elegans* (Pursh) Rydb.; *Toxicoscordion nuttallii* (A. Gray) Rydb.; *Toxicoscordion paniculatum* (Nutt.) Rydb.

alkali hymenoxys –see– *Hymenoxys lemmonii* (Greene) Cockerell

alkali mallow –see– *Malva parviflora* L.

alkali weed –see– *Isocoma plurifolia* (Torr. & A. Gray) Greene

all heal –see– *Panax ginseng* C. A. Mey.; *Valeriana officinalis* L.

allamanda –see– *Allamanda cathartica* L.

ALLAMANDA CATHARTICA L. [Apocynaceae]

Synonyms:

Allamanda grandiflora (Aubl.) Lam.; ***Orelia grandiflora*** Aubl.

Common Names:

allamanda; bejuco-de-San Jose; buttercup; canario; cantiva; flor-de-barbero; golden trumpet; jalapa; jasmin amarillo; San Jose; San Jose amarillo; trompeta-de-oro; yellow allamanda

Citations:

Tokarnia CH, Armién AG, Peixoto PV, et al. (1996) Estudo experimental sobre a toxidez de algumas plantas ornamentais em bovinos. Pesq Vet Bras 16(1):5-20.

Allamanda grandiflora (Aubl.) Lam. = Allamanda cathartica L.

alleluia –see– *Oxalis acetosella* L.

alligator pear –see– *Persea americana* Mill. var. americana

alligator weed –see– *Alternanthera philoxeroides* (Mart.) Griseb.

Allium ascalonicum auct. = Allium cepa L. var. aggregatum G. Don

ALLIUM CANADENSE L. [Alliaceae]

Common Names:

Canadian garlic; meadow garlic; meadow leek; wild garlic; wild onion

Citations:

Pierce KR, Joyce JR, England RB, et al. (1972) Acute hemolytic anemia caused by wild onion poisoning in horses. J Am Vet Med Assoc 160(3):323-327.

Pipal FJ (1918) A suspected case of stock poisoning by wild onion (Allium canadense). Proc Indiana Acad Sci 1917:139-143.

ALLIUM CEPA L. [Alliaceae]

Common Names:

cebola; cipolle; løg; onion; ui; uie; uien; wild onion; Zwiebel

Citations:

Burgess JF (1952) Occupational dermatitis due to onion and garlic. Can Med Assoc J 66(3):275.

Burks JW Jr (1954) Classic aspects of onion and garlic dermatitis in housewives. Ann Allergy 12(5):592-596.

Farkas MC, Farkas JN (1974) Hemolytic anemia due to ingestion of onions in a dog. J Am Anim Hosp Assoc 10(1):65-66.

Fighera RA, Souza TM, Langohr I, et al. (2002) Intoxicação experimental por cebola, Allium cepa (Liliaceae), em gatos. Pesq Vet Bras 22(2):79-84.

Franken P, van Beukelen P, Blok G (1980) Uien: Geen paardevoer. Tijdschr Diergeneeskd 105(13):529-534.

Gill PA, Sergeant ES (1981) Onion poisoning in a bull. Aust Vet J 57(10):484.

Goldsmith WW (1909) Onion poisoning in cattle. J Comp Pathol Ther 22:151.

Gruhzit OM (1931) Anemia of dogs produced by feeding of the whole onions and of onion fractions. Am J Med Sci 181:812-815.

Harvey JW, Rackear D (1985) Experimental onion-induced hemolytic anemia in dogs. Vet Pathol 22(4):387-392.

Hjorth N, Roed-Petersen J (1976) Occupational protein contact dermatitis in food handlers. Contact Dermatitis 2(1):28-42.

Hothi DS, Arneja JS, Chawla JS (1980) Onion (Allium cepa) poisoning in bullocks. Indian Vet J 57(Aug):690-692.

Hutchison TW (1977) Onions as a cause of Heinz body anaemia and death in cattle. Can Vet J 18(12):358-360.

Imada O (1980) [Experimental studies on the onion poisoning in dogs - Relation between the quantity of supplied onion and onset of the disease.] Azabu Univ Vet Med Bull 1(2):271-287.

Kaplan AJ (1995) Onion powder in baby food may induce anemia in cats. J Am Vet Med Assoc 207(11):1405.

Kay JM (1983) Onion toxicity in a dog. Mod Vet Pract 64(6):477-478.

Kirk JH, Bulgin MS (1979) Effects of feeding cull domestic onions (Allium cepa) to sheep. Am J Vet Res 40(3):397-399.

Knight AP, Lassen D, McBride T, et al. (2000) Adaptation of pregnant ewes to an exclusive onion diet. Vet Hum Toxicol 42(1):1-4.

Kobayashi K (1981) Onion poisoning in the cat. Feline Pract 11(1):22-27.

Kodama H, Yamaashi K (1976) [Acute haemolytic anaemia in dogs fed onions.] J Jpn Vet Med Assoc 29(9):482-484.

Koger LM (1956) Onion poisoning in cattle. J Am Vet Med Assoc 129(Jul 15):75.

Kongshammer GB (1989) Onion poisoning resulting in Heinz body anaemia in a dog. Dansk Veterinaertidsskrift 72(16):948.

Lazarus AE, Rajamani S (1968) Poisoning due to onion spoilage in cattle. Indian Vet J 45(10):877-880.

Lincoln SD, Howell ME, Combs JJ, et al. (1992) Hematologic effects and feeding performance in cattle fed cull domestic onions (Allium cepa). J Am Vet Med Assoc 200(8):1090-1094.

Maede Y (1977) High concentration of blood glutathione in dogs with acute hemolytic anemia. Jpn J Vet Sci 39(2):187-189.

Mandelli G, Persiani G (1964) Osservazione su di un episodio di avvelenamento da cipolla (Allium cepa) in bovini di allevamento. Clin Vet (Milano) 87(1):1-8.

Neves RG (1964) Dermatito de contacto por bulbo de Liliáceas. An Bras Derm Sifilogr 39(4):23-31.

Nordio CB (1952) Anemia mielotossica da alimentazione con cipolle (Allium cepa). Boll Soc Ital Biol Sper 28(1):53-56.

Ogawa E, Shinoki T, Akahori F, et al. (1986) Effect of onion ingestion on anti-oxidizing agents in dog erythrocytes. Nippon Juigaku Zasshi 48(4):685-691.

Pekelder JJ, Akkermans JP (1985) Uienvergiftiging. Tijdschr Diergeneeskd 110(1):31.

Pérez-Calderón R, Gonzalo-Garijo MA, Fernández de Soria R (2002) Exercise-induced anaphylaxis to onion. Allergy 57(8):752-753.

Petersson U, Petersson L, Björnehammar U (1984) Lökförgiftning på racing greyhound. Sven Vet 36(14):697.

Rae HA (1999) Onion toxicosis in a herd of beef cows. Can Vet J 40(1):55-57.

Robertson JE, Christopher MM, Rogers QR (1998) Heinz body formation in cats fed baby food containing onion powder. J Am Vet Med Assoc 212(8):1260-1266.

Schlotthauser CF, Berryman GH (1943) Production of anemia in dogs due to feeding onions. J Am Vet Med Assoc 102(791):109-111.

Sebrell WH (1930) An anemia of dogs produced by feeding onions. Public Health Rep 45(21):1175-1191.

Segalini E (1966) Osservazioni cliniche sulla intossicazione da Allium cepa nei bovini. Veterinaria (Milano) 15:45-48.

Solter P, Scott R (1987) Onion ingestion and subsequent Heinz body anemia in a dog: A case report. J Am Anim Hosp Assoc 23(5):544-546.

Spice RN (1976) Hemolytic anemia associated with ingestion of onions in a dog. Can Vet J 17(7):181-183.

Stallbaumer M (1981) Onion poisoning in a dog. Vet Rec 108(24):523-524.

Thorp F Jr, Harshfield GS (1939) Onion poisoning in horses. J Am Vet Med Assoc 94:52-53.

van der Kolk JH (2000) Onion poisoning in a herd of dairy cattle. Vet Rec 147(18):517-518.

Van Hecke E (1977) Contact allergy to onion. Contact Dermatitis 3(3):167-168.

van Ketel WG, de Haan P (1978) Occupational eczema from garlic and onion. Contact Dermatitis 4(1):53-54.

van Schouwenburg S (1982) Hemolitiese anemie in 'n miniatuur dachschund veroorsaak deur inname van groot hoeveelhede uie (Allium cepa). J S Afr Vet Assoc 53(3):212.

Verhoeff J, Hajer R, van den Ingh TS (1985) Onion poisoning of young cattle. Vet Rec 117:497-498.

Yamoto O, Maede Y (1992) Susceptibility to onion-induced hemolysis in dogs with hereditary high erythrocyte reduced glutathione and potassium concentrations. Am J Vet Res 53(1):134-137.

Note:

Onion is named *Allium cepa* L. var. *cepa* in some publications

ALLIUM CEPA L. var. aggregatum G. Don [Alliaceae]

Synonyms:

Allium ascolonicum auct.

Common Names:

shallot

Citations:

Crespo R, Chin RP (2004) Effect of feeding green onions (Allium ascalonicum) to white Chinese geese (Threskiornis spinicollis). J Vet Diagn Invest 16(4):321-325.

ALLIUM PORRUM L. [Alliaceae]

Common Names:

Lauchpflanze; leek

Citations:

Armentia A, Lombardero M, Fernàndez S, et al. (2005) Occupational rhinitis to leek (Allium porrum). Allergy 60(1):132-133.

Cadot P, Tits G, Bussels L, et al. (2001) Asthma and hand dermatitis to leek. Allergy 56(2):192-193.

Hjorth N, Roed-Petersen J (1976) Occupational protein contact dermatitis in food handlers. Contact Dermatitis 2(1):28-42.

Veien NK, Hattel T, Justesen O, et al. (1983) Causes of eczema in the food industry. Derm Beruf Umwelt 31(3):84-86.

ALLIUM SATIVUM L. [Alliaceae]

Common Names:

ajo; diente-de-ajo; garlic; Graserich; Knoblauch; Knobleig; Knoflack; look; rustic treacle; stinking rose; wild garlic

Citations:

Al Bekairi AM, Shah AH, Qureshi S (1990) Effect of Allium sativum on epididymal spermatozoa, estradiol-treated mice and general toxicity. J Ethnopharmacol 29(2):117-125.

Alnaqeeb MA, Thomson M, Bordia T, et al. (1996) Histopathological effects of garlic on liver and lung of rats. Toxicol Lett 85(3):157-164.

Bleumink E, Douglas HM, Klokke AH, et al. (1972) Allergic contact dermatitis to garlic. Br J Dermatol 87(1):6-9.

Bleumink E, Nater JP (1972) Garlic as a possible cause of contact dermatitis. Contact Dermatol Newsl 12(Aug):310-311.

Bleumink E, Nater JP (1973) Contact dermatitis to garlic; Crossreactivity between garlic, onion and tulip. Arch Dermatol Forsch 247(2):117-124.

Bojs G, Svensson Å (1988) Contact allergy to garlic used for wound healing. Contact Dermatitis 18(3):179-181.

Burgess JF (1952) Occupational dermatitis due to onion and garlic. Can Med Assoc J 66(3):275.

Burks JW Jr (1954) Classic aspects of onion and garlic dermatitis in housewives. Ann Allergy 12(5):592-596.

Campolmi P, Lombardi P, Lotti T, et al. (1982) Immediate and delayed sensitization to garlic. Contact Dermatitis 8(5):352-353.

Couturier P, Bousquet J (1982) Occupational allergy secondary to inhalation of garlic dust. J Allergy Clin Immunol 70(2):145.

Dixit VP, Joshi S (1982) Effects of chronic administration of garlic (Allium sativum Linn) on testicular function. Indian J Exp Biol 20(7):534-536.

Edelstein AJ (1950) Dermatitis caused by garlic. Arch Derm Syphilol 61(1):111.

Ekeowa-Anderson AL, Shergill B, Goldsmith P (2007) Allergic contact dermatitis to garlic. Contact Dermatitis 56(3):174-175.

Falleroni AE, Zeiss CR, Levitz D (1981) Occupational asthma secondary to inhalation of garlic dust. J Allergy Clin Immunol 68(2):156-160.

Fehri B, Aiache JM, Korbi S, et al. (1991) Effets toxiques engendrés par une administration réitérée d'Allium sativum L. J Pharm Belg 46(6):363-374.

Fernández de Corres L, Corrales JL, Muñoz D, et al. (1984) Dermatitis alérgicas de contacto por plantas. Allergol Immunopathol (Madr) 12(4):313-319.

Henson GE (1940) Garlic: An occupational factor in the etiology of bronchial asthma. Case report. J Fla Med Assoc 27(2):86.

Hubbard VG, Goldsmith P (2005) Garlic-fingered chefs. Contact Dermatitis 52(3):165-166.

Kaplan B, Schewach Millet M, Yorav S (1990) Factitial dermatitis induced by application of garlic. Int J Dermatol 29(1):75-76.

Lee TY, Lam TH (1991) Contact dermatitis due to topical treatment with garlic in Hong Kong. Contact Dermatitis 24(3):193-196.

Lembo G, Balato N, Patruno C, et al. (1991) Allergic contact dermatitis due to garlic (Allium sativum). Contact Dermatitis 25(5):330-331.

Lybarger JA, Gallagher JS, Pulver DW, et al. (1982) Occupational asthma induced by inhalation and ingestion of garlic. J Allergy Clin Immunol 69(5):448-454.

Mitchell JC (1980) Contact sensitivity to garlic (Allium). Contact Dermatitis 6(5):356-357.

Neves RG (1964) Dermatito de contacto por bulbo de Liliáceas. An Bras Derm Sifilogr 39(4):23-31.

Pires G, Pargana E, Loureiro V, et al. (2002) Allergy to garlic. Allergy 57(10):957-958.

Pirogova EP, Katiukhina ZD (1970) [Artificial dermatitis caused by garlic.] Vestn Dermatol Venerol 44(12):53-54.

Rose KD, Croissant PD, Parliament CF, et al. (1990) Spontaneous spinal epidural hematoma with associated platelet dysfunction from excessive garlic ingestion: A case report. Neurosurgery 26:880-882.

van Ketel WG, de Haan P (1978) Occupational eczema from garlic and onion. Contact Dermatitis 4(1):53-54.

Note:

Garlic is named *Allium sativum* L. *sativum* in some publications.

ALLIUM SCHOENOPRASUM L. [Alliaceae]

Common Names:

chives; ezonegi

Citations:

Hjorth N, Roed-Petersen J (1976) Occupational protein contact dermatitis in food handlers. Contact Dermatitis 2(1):28-42.

Kobayashi T (1950) Studies on the histo-pathologic changes of experimental cases of the "Ezonegi-poisoning" in horses. Jpn J Vet Sci 12(5):209-210.

ALLIUM URSINUM L. [Alliaceae]

Common Names:

Bärenlauch; ransoms; Waldknoblauch; wild garlic; Zigeunerlauch

Citations:

Stevens H (1984) Suspected wild garlic poisoning in sheep. Vet Rec 115(14):363.

ALLIUM VALIDUM S. Watson [Alliaceae]

Common Names:

Pacific onion; swamp onion; wild onion

Citations:

James LF, Binns W (1966) Effects of feeding wild onions (Allium validum) to bred ewes. J Am Vet Med Assoc 149(5):512-514.

Van Kampen KR, James LF, Johnson AE (1970) Hemolytic anemia in sheep fed wild onion (Allium validum). J Am Vet Med Assoc 156(3):328-332.

allspice –see– *Pimenta dioica* (L.) Merr.

almendro macho –see– *Andira inermis* (W. Wright) Kunth ex DC.

almendro montes –see– *Andira inermis* (W. Wright) Kunth ex DC.

almendro seal –see– *Andira inermis* (W. Wright) Kunth ex DC.

almendron –see– *Metopium toxiferum* (L.) Krug & Urb.

almond –see– *Prunus dulcis* (Mill.) D. A. Webb

almortas –see– *Lathyrus sativus* L.

alocasia –see– *Colocasia esculenta* (L.) Schott

ALOCASIA CUCULLATA (Lour.) G. Don [Araceae]

Common Names:

nai habarala; panu habarala

Citations:

Deng JF, Lin TJ, Kao WF, et al. (1997) The difficulty in handling poisonings associated with Chinese traditional medicine: A poison control center experience for 1991-1993. Vet Hum Toxicol 39(2):106-114.

Goonasekera CD, Vasanthatkilake VW, Ratnatunga N, et al. (1993) Is Nai Habarala (Alocasia cucullata) a poisonous plant? Toxicon 31(6):813-816.

Alocasia denudata Engl. = Alocasia longiloba Miq.

Alocasia indica (Lour.) Spach = Alocasia macrorrhizos (L.) G. Don

ALOCASIA LONGILOBA Miq. [Araceae]

Synonyms:

Alocasia denudata Engl.; ***Alocasia lowii*** Hook. f.

Common Names:

keladi; keladi chandek

Citations:

Teck LK (1967) Keladi poisoning in Singapore children. J Singapore Paediat Soc 9(1):54-57.

Alocasia lowii Hook. f. = Alocasia longiloba Miq.

ALOCASIA MACRORRHIZOS (L.) G. Don [Araceae]

Synonyms:

Alocasia indica (Lour.) Spach; ***Colocasia indica*** (Lour.) Kunth; ***Colocasia macrorrhizos*** (L.) Schott & Endl.

Common Names:

ape; Australian cabbage; biga; creek lily; cungevoi; cunjeboi; cunjeboy; cunjevoi; elephant's-ear; giant elephant's-ear; giant taro; green arum; hai yu; kuan-yin-lien; lang du; lao-hu-yu; Queensland maple; spoon lily; tu-chiao-lien

Citations:

Chan TY, Chan LY, Tam LS, et al. (1995) Neurotoxicity following the ingestion of a Chinese medicinal plant, Alocasia macrorrhiza. Hum Exp Toxicol 14(9):727-728.

Cleland JB (1931) Plants, including fungi, poisonous or otherwise injurious to man in Australia. Series III. Med J Aust 2(Dec 19):775-778.

Lin TJ, Hung DZ, Hu WH, et al. (1998) Calcium oxalate is the main toxic component in clinical presentations of Alocasia macrorrhiza (L) Schott and Endl poisonings. Vet Hum Toxicol 40(2):93-95.

Tang EW, Law RW, Lai JS (2006) Corneal injury by wild taro. Clin Exp Ophthalmol 34(9):895-896.

ALOE ARBORESCENS Mill. [Asphodelaceae]

Synonyms:

Aloe arborescens Mill. var. natalensis (J. M. Wood & M. S. Evans) A. Berger

Common Names:

Barbados aloe; candelabra aloe; medicinal aloe; octopus; torch plant

Citations:

Nakamura T, Kotajima S (1984) Contact dermatitis from Aloe arborescens. Contact Dermatitis 11(1):51.

Shoji A (1982) Contact dermatitis to Aloe arborescens. Contact Dermatitis 8(3):164-167.

Aloe arborescens Mill. var. natalensis (J. M. Wood & M. S. Evans) A. Berger = Aloe arborescens Mill.

Aloe barbadensis Mill. = Aloe vera (L.) Burm. f.

aloe capensis –see– *Aloe vera* (L.) Burm. f.

aloe vera –see– *Aloe vera* (L.) Burm. f.

ALOE VERA (L.) Burm. f. [Asphodelaceae]

Synonyms:

Aloe barbadensis Mill.; ***Aloe vulgaris*** Lam.

Common Names:

aloe capensis; aloe vera; Barbados aloe; curaco aloe; Indian aloe; medicinal aloe; Mediterranean aloe; sempervivum; sinkle Bible; star cactus; zabila; zavila

Citations:

Bacis G, Moretti M, Chinaglia D, et al. (2004) Fatal lipoid pneumonitis after inhalation of Aloe vera extract. J Toxicol Clin Toxicol 42(4):559.

Aloe vulgaris Lam. = Aloe vera (L.) Burm. f.

ALOYSIA CITRODORA Palau [Verbenaceae]

Synonyms:

Aloysia triphylla (L'Her.) Britton; ***Lippia citriodora*** Kunth

Common Names:

lemon verbena

Citations:

Ciganda C, Laborde A (2003) Herbal infusions used for induced abortion. J Toxicol Clin Toxicol 41(3):235-239.

Aloysia triphylla (L'Her.) Britton = Aloysia citrodora Palau

Alpendost –see– *Adenostyles alliariae* A. Kern.

Alpenkreuzkraut –see– *Senecio alpinus* (L.) Scop.

Alpenveilchen –see– *Cyclamen persicum* Mill.

alpine golden-chain –see– *Laburnum alpinum* (Mill.) J. Presl

alpine kalmia –see– *Kalmia microphylla* (Hook.) A. Heller

alpine laurel –see– *Kalmia microphylla* (Hook.) A. Heller

alpine violet –see– *Cyclamen persicum* Mill.

ALPINIA GALANGA (L.) Sw. [Zingiberaceae]

Synonyms:

Maranta galanga L.

Citations:

Hong SJ, Chang CH (2006) Erythema multiforme-like generalized allergic contact dermatitis caused by Alpinia galanga. Contact Dermatitis 54(2):118-120.

Alraunwurzel –see– *Mandragora officinarum* L.

Alsatian clover –see– *Trifolium hybridum* L.

alsike clover –see– *Trifolium hybridum* L.

alstonia –see– *Alstonia constricta* F. Muell.

ALSTONIA CONSTRICTA F. Muell. [Apocynaceae]

Common Names:

alstonia; bitter bark; Peruvian bark; quinine bush

Citations:

Copeland HM, Seddon HR (1931) Poisoning of sheep by quinine bush or bitter bark (Alstonia constricta). Agric Gaz New South Wales 42:925-926.

ALSTROEMERIA LIGTU L. [Alstroemeriaceae]

Common Names:

Inca lily; Inkalilie

Citations:

Bjorkner, BE (1982) Contact allergy and depigmentation from Alstroemeria. Contact Dermatitis 8:178-184.

Hausen BM, Prater E, Schubert H (1983) The sensitizing capacity of Alstroemeria cultivars in man and guinea pig. Remarks on the occurrence, quantity and irritant and sensitizing potency of their constituents tuliposide A and tulipalin A and (α-methylene-γ-butyrolactone). Contact Dermatitis 9(1):46-54.

Illuminati R, Russo R, Guerra L, et al. (1988) Occupational airborne contact dermatitis in a florist. Contact Dermatitis 18(4):246.

Marks JG Jr (1988). Allergic contact dermatitis to Alstroemeria. Arch Dermatol 124: 914-916.

Mascarenhas R, Robalo-Cordeiro M, Fernandes B, et al. (2001) Allergic and irritant occupational contact dermatitis from alstroemeria. Contact Dermatitis 44(3):196-197.

Rycroft RJ, Calnan CD (1981). Alstroemeria dermatitis. Contact Dermatitis 7(5): 284.

alta fescue –see– *Festuca arundinacea* Schreb.

altamisa –see– *Artemisia vulgaris* L.

ALTERNANTHERA PHILOXEROIDES (Mart.) Griseb. [Amaranthaceae]

Common Names:

alligator weed

Citations:

Bourke CA, Rayward D (2003) Photosensitisation in dairy cattle grazing alligator weed (Alternanthera philoxeroides) infested pastures. Aust Vet J 81(6):361-362.

Althaea rosea (L.) Cav. = Alcea rosea L.

alui –see– *Anacardium occidentale* L.

amandier –see– *Prunus dulcis* (Mill.) D. A. Webb

amapalo amarillo –see– *Epipremnum pinnatum* (L.) Engl.

amaranth –see– *Amaranthus retroflexus* L.

AMARANTHUS BLITOIDES S. Watson [Amaranthaceae]

Common Names:

prostrate amaranth; prostrate pigweed; spreading pigweed

Citations:

Ramos JJ, Ferrer LM, García L, et al. (2005) Polioencephalomalacia in adult sheep grazing pastures with prostrate pigweed. Can Vet J 46(1):59-61.

AMARANTHUS BLITUM L. [Amaranthaceae]

Citations:

Cheeke PR, Carlsson R, Kohler GO (1981) Nutritive value of leaf protein concentrates prepared from amaranthus species. Can J Anim Sci 61:199-204.

AMARANTHUS CAUDATUS L. [Amaranthaceae]

Synonyms:

Amaranthus edulis Speg.

Citations:

Cheeke PR, Carlsson R, Kohler GO (1981) Nutritive value of leaf protein concentrates prepared from amaranthus species. Can J Anim Sci 61:199-204.

Takken A, Connor JK (1985) Some toxicological aspects of grain amaranth for pigs. In: Seawright AA, Hegarty MP, James LF, et al. (eds.) Plant toxicology. Queensland Poisonous Plant Committee. Yeerongpilly, Queensland. pp.170-177.

AMARANTHUS CRUENTUS L. [Amaranthaceae]

Synonyms:

Amaranthus paniculatus L.

Common Names:

panicle amaranth; redshank; wild amaranth

Citations:

Cheeke PR, Carlsson R, Kohler GO (1981) Nutritive value of leaf protein concentrates prepared from amaranthus species. Can J Anim Sci 61:199-204.

Amaranthus edulis Speg = Amaranthus caudatus L.

AMARANTHUS FLAVUS L. [Amaranthaceae]

Citations:

Cheeke PR, Carlsson R, Kohler GO (1981) Nutritive value of leaf protein concentrates prepared from amaranthus species. Can J Anim Sci 61:199-204.

Amaranthus gangeticus L. = Amaranthus tricolor L.

Amaranthus gracilis Desf., nom. nud. = Amaranthus viridis L.

AMARANTHUS HYBRIDUS L. [Amaranthaceae]

Common Names:

carelessweed; green pigweed; pigweed; quelite morado; redroot; smooth pigweed

Citations:

Duckworth RH (1975) Poisoning of cattle by Amaranthus. N Z Vet J 23(7):154-155.

Jeppesen QE (1966) Bovine perirenal disease associated with pigweed. J Am Vet Med Assoc 149(1):22.

Amaranthus hybridus L. subsp. quitensis (Kunth) Costea & Carretero = Amaranthus quitensis Kunth

AMARANTHUS HYPOCHONDRIACUS L. [Amaranthaceae]

Citations:

Alfaro MA, Ramírez R, Martínez A, et al. (1987) Evaluacion de diferentes niveles de harina de amaranto (partes vegetativas), en sustitucion de harina de alfalfa para conejos en crecimiento. Arch Latinoam Nutr 37(1):174-185.

Cheeke PR, Carlsson R (1978) Evaluation of several crops as sources of leaf meal: Composition, effect of drying procedure, and rat growth response. Nutr Rep Int 18(4):465-473.

Cheeke PR, Carlsson R, Kohler GO (1981) Nutritive value of leaf protein concentrates prepared from amaranthus species. Can J Anim Sci 61:199-204.

Amaranthus mangostanus L. = Amaranthus tricolor L.

AMARANTHUS PALMERI S. Watson [Amaranthaceae]

Common Names:

bledo; careless weed; Palmer's-amaranth; Palmer's-pigweed; pigweed; redroot

Citations:

Castle RB (1944) Successful therapy for cyanogen poisoning in cattle. Vet Med 39:37-38.

Thornber JJ (1919) Work on poison plants. Arizona Agric Exp Sta Annu Rep 30:428-431.

Amaranthus paniculatus L. = Amaranthus cruentus L.

AMARANTHUS QUITENSIS Kunth [Amaranthaceae]

Synonyms:

Amaranthus hybridus L. subsp. quitensis (Kunth) Costea & Carretero

Common Names:

pigweed; yuyo colorado

Citations:

Cursack HA, Romano LA (1967) Posible intoxicacion subaguda con nitratos y nitritos. El "yuyo colorado" (Amaranthus hybridus, var. quitensis) como especie sospechosa de ser tóxica para el ganado vacuno. Gaceta Vet 29:69-74.

Duffy SJ, León EA, Gavier MD, et al. (1985) Intoxicacion por Amaranthus quitensis (Yuyo colorado) en bovinos. Vet Argent 2(20):942, 944-949.

Rodríguez Armesto R, Grande HA, de Baroni AC, et al. (1989) Intoxicacion por Amaranthus quitensis en vaquillonas Holando Argentino. Vet Argent 6(60):692, 694-700.

Salles MS, Barros CS, Lemos RA, et al. (1991) Perirenal edema associated with Amaranthus spp poisoning in Brazilian swine. Vet Hum Toxicol 33(6):616-617.

AMARANTHUS RETROFLEXUS L. [Amaranthaceae]

Common Names:

amaranth; careless weed; green amaranthus; green pigweed; pigweed; Prince-of-Wales feather; quelite; redroot; redroot pigweed; reflexed amaranthus; rough pigweed; szörös disznóparéj

Citations:

Brakenridge DT (1956) Nitrate poisoning caused by turnips and red root. N Z Vet J 4:165-166.

Buck WB, Preston KS, Abel M, et al. (1965-1966) Common weeds associated with perirenal edema in swine. Proc Am Coll Vet Toxicol 1965:24-28.

Buck WB, Preston KS, Abel M, et al. (1966) Perirenal edema in swine: A disease caused by common weeds. J Am Vet Med Assoc 148(12):1525-1531.

Buck WM [sic], Preston KS, Abel M (1965) Common weeds as a cause of perirenal edema in swine. Iowa State Univ Vet 27(3):105-108.

Casteel SW, Johnson GC, Miller MA, et al. (1994) Amaranthus retroflexus (redroot pigweed) poisoning in cattle. J Am Vet Med Assoc 204(7):1068-1070.

Cho SW, Lee CS (1984) [Histopathological observations of the natural case and experimental occurrence of perirenal edema in pigs.] Korean J Vet Res 24(2):173-182.
Ivana F, Georgescu C (1970) Observații asupra unor cazuri de intoxicație cu Amaranthus retroflexus la bovine. Rev Zootehnie Med Vet 20(11):64-66.
Jeppesen QE (1966) Bovine perirenal disease associated with pigweed. J Am Vet Med Assoc 149(1):22.
Kerr LA, Kelch WJ (1998) Pigweed (Amaranthus retroflexus) toxicosis in cattle. Vet Hum Toxicol 40(4):216-218.
Osweiler GD, Buck WB, Bicknell EJ (1969) Production of perirenal edema in swine with Amaranthus retroflexus. Am J Vet Res 30(4):557-566.
Rae CA, Binnington BD (1995) Amaranthus retroflexus (redroot pigweed) poisoning in lambs. Can Vet J 36(7):446.
Sanko RE (1975) Perirenal edema in swine - Caused by ingestion of Amaranthus retroflexus (pigweed). Vet Med Small Anim Clin 70(1):42-43.
Šenk L, Pogačnik M, Princes I, et al. (1985) Pojava perirenalnog edema (nefropatije) kod svinja u sr Sloveniji. Veterinarski Glasnik 39(12):1277-1282.
Spearman G, Johnson K (1989) Redroot pigweed toxicosis in cattle. Can Vet J 30(3):255-256.
Stuart BP, Nicholson SS, Smith JB (1975) Perirenal edema and toxic nephrosis in cattle, associated with ingestion of pigweed. J Am Vet Med Assoc 167(10):949-950.
Torres MB, Kommers GD, Dantas AF, et al. (1997) Redroot pigweed (Amaranthus retroflexus) poisoning of cattle in southern Brazil. Vet Hum Toxicol 39(2):94-96.
Zhokovski N (1966) [Poisoning and abortion caused by Amaranthus retroflexus in ewes.] Veterinarna Sbirka 63(11):26.

AMARANTHUS SPINOSUS L. [Amaranthaceae]

Common Names:

hogweed; needle bur; quelite espinoso; spiny amaranth; spiny pigweed; stickerweed; thorny amaranthus, thorny pigweed

Citations:

Peixoto PV, Brust LA, Brito MF, et al. (2003) Intoxicação natural por Amaranthus spinsosus (Amaranthaceae) em ovinos no Sudeste do Brasil. Pesq Vet Bras 23(4):179-184.

AMARANTHUS TRICOLOR L. [Amaranthaceae]

Synonyms:

Amaranthus gangeticus L.; ***Amaranthus mangostanus*** L.

Citations:

Cheeke PR, Carlsson R, Kohler GO (1981) Nutritive value of leaf protein concentrates prepared from amaranthus species. Can J Anim Sci 61:199-204.

AMARANTHUS VIRIDIS L. [Amaranthaceae]

Synonyms:

Amaranthus gracilis Desf., nom. nud.

Common Names:

bledo blanco; chaulai; green amaranth; white amaranth

Citations:

Rivera M, Aguilo R, Lorenzo M, et al. (1984) Reporte de intoxicación en bovinos jóvenes por el Amaranthus viridis (bledo blanco). Rev Cubana Cienc Vet 15(3-4):335-338.
Salles MS, Barros CS, Lemos RA, et al. (1991) Perirenal edema associated with Amaranthus spp poisoning in Brazilian swine. Vet Hum Toxicol 33(6):616-617.
Singh R, Siddiqui MA, Baruah MC (1978) Plant dermatitis in Delhi. Indian J Med Res 68(Oct):650-655.

amargo –see– *Lagenaria siceraria* (Molina) Standl.

amber –see– *Hypericum perforatum* L.

ambroisie –see– *Dysphania ambrosioides* (L.) Mosyakin & Clemants

AMBROSIA ARTEMISIIFOLIA L. [Asteraceae]

Common Names:

annual ragweed; bitterweed; hogweed; low ragweed; short ragweed; western ragweed

Citations:

Brunsting LA, Anderson CR (1934) Ragweed dermatitis. A report based on eighteen cases. JAMA 103(17):1285-1290.
Brunsting LA, Williams DH (1936) Ragweed (contact) dermatitis. JAMA 106(18):1533-1535.
Epstein S (1960) Role of dermal sensitivity in ragweed contact dermatitis. Arch Dermatol 82(Jul):48-55.
Jordon JW, Campbell PC, Osborne ED (1942) Ragweed dermatitis among workers in the flour and grain industries. Arch Derm Syphilol 46:721-724.
Möller H, Spirén A, Svensson Å, et al. (2002) Contact allergy to the Asteraceae plant Ambrosia artemisiifolia L. (ragweed) in sesquiterpenelactone-sensitive patients in southern Sweden. Contact Dermatitis 47(3):157-160.

AMBROSIA DELTOIDEA (Torr.) W. W. Payne [Asteraceae]

Common Names:

triangle-leaf bursage

Citations:

Schumacher MJ, Silvis NG (2003) Airborne contact dermatitis from Ambrosia deltoidea (triangle-leaf bursage). Contact Dermatitis 48(4):212-216.

AMBROSIA ELATIOR L. [Asteraceae]

Common Names:

short ragweed

Citations:

Sutton RL (1919) Ragweed dermatitis. JAMA 73(19):1433-1437.

AMBROSIA PSILOSTACHYA DC. [Asteraceae]

Common Names:

western ragweed

Citations:

Brunsting LA, Williams DH (1936) Ragweed (contact) dermatitis. JAMA 106(18):1533-1535.

Ford RM (1963) Ragweed pollinosis. Med J Aust 1(May 11):712.

AMBROSIA TRIFIDA L. [Asteraceae]

Common Names:

giant ragweed; great ragweed; ironweed; tall ragweed

Citations:

Brunsting LA, Williams DH (1936) Ragweed (contact) dermatitis. JAMA 106(18):1533-1535.

Sutton RL (1919) Ragweed dermatitis. JAMA 73(19):1433-1437.

amche –see– *Toxicodendron striatum* (Ruiz & Pav.) Kuntze

AMELANCHIER ALNIFOLIA (Nutt.) Nutt. ex M. Roem. [Rosaceae]

Common Names:

Saskatoon serviceberry; western serviceberry

Citations:

Majak W, Udenberg T, Clark LJ, et al. (1980) Toxicity of Saskatoon serviceberry to cattle. Can Vet J 21(3):74-76.

Quinton DA (1985) Saskatoon serviceberry toxic to deer. J Wildl Manage 49(2):362-364.

American agave –see– *Agave americana* L.

American aloe –see– *Agave americana* L.

American arbor vitae –see– *Thuja occidentalis* L.

American arum –see– *Dieffenbachia seguine* (Jacq.) Schott

American beech –see– *Fagus grandifolia* Ehrh.

American cancer –see– *Phytolacca americana* L.

American cocklebur –see– *Xanthium strumarium* L. var. canadense (Mill.) Torr. & A. Gray

American coffeebean –see– *Gymnocladus dioicus* (L.) K. Koch

American coffeeberry –see– *Gymnocladus dioicus* (L.) K. Koch

American custard apple –see– *Asimina triloba* (L.) Dunal

American dogweed –see– *Verbesina encelioides* (Cav.) Benth. & Hook. f. ex A. Gray

American ephedra –see– *Ephedra viridis* Coville

American false hellebore –see– *Veratrum viride* Aiton

American fly poison –see– *Amianthium muscitoxicum* (Walter) A. Gray

American ginseng –see– *Panax ginseng* C. A. Mey.

American hellebore –see– *Veratrum viride* Aiton

American hemp –see– *Apocynum cannabinum* L.

American holly –see– *Ilex opaca* Aiton

American ivy –see– *Parthenocissus quinquefolia* (L.) Planch.

American jute –see– *Abutilon theophrasti* Medik.

American laurel –see– *Kalmia latifolia* L.

American mandrake –see– *Podophyllum peltatum* L.

American manna grass –see– *Glyceria grandis* S. Watson

American May apple –see– *Podophyllum peltatum* L.

American mistletoe –see– *Phoradendron leucarpum* (Raf.) Reveal & M. C. Johnst.

American nightshade –see– *Phytolacca americana* L.

American papaw –see– *Asimina triloba* (L.) Dunal

American pennyroyal –see– *Hedeoma pulegioides* (L.) Pers.

American purging nut –see– *Jatropha curcas* L.

American rubber plant –see– *Peperomia obtusifolia* (L.) A. Dietr.

American sanfoin –see– *Galega officinalis* L.

American sassafras –see– *Sassafras albidum* (Nutt.) Nees

American sneezeweed –see– *Helenium amarum* (Raf.) H. Rock

American stinging nettle –see– *Urtica dioica* L.

American water hemlock –see– *Cicuta maculata* L.

American white hellebore –see– *Veratrum viride* Aiton

American wormseed –see– *Dysphania ambrosioides* (L.) Mosyakin & Clemants

American wormwood –see– *Dysphania ambrosioides* (L.) Mosyakin & Clemants

Amerikanische Grenadill –see– *Brya ebenus* (L.) DC.

Amerikanischer Efeu –see– *Parthenocissus quinquefolia* (L.) Planch.

Amerikanischer Lebensbaum –see– *Thuja occidentalis* L.

Amerikanisches Mahagoni –see– *Swietenia mahagoni* (L.) Jacq.

amerikansk olieplante –see– *Ricinus communis* L.

AMIANTHIUM MUSCITOXICUM (Walter) A. Gray [Melanthiaceae]

Synonyms:

Chrosperma muscitoxicum (Walter) Kuntze; ***Zigadenus muscitoxicum*** (Walter) Regel

Common Names:

American fly poison; black snakeroot; crow poison; death camas; fall poison; fly poison; gray fly poison; St. Elmos'-feather; stagger grass; stagger plant

Citations:

Anonymous (1912) Poisonous-plant studies. U S Dep Agric Annu Rep 1911:61-62, 92, 277-280.

Marsh CD, Clawson AB, Marsh H (1918) Stagger grass (Chrosperma muscaetoxicum) as a poisonous plant. U S Dep Agric Bull #710:4 pp.

amila –see– *Phyllanthus emblica* L.

AMMI MAJUS L. [Apiaceae]

Common Names:

atarillal; bishop's-weed; falsa visnaga; greater ammi; Knorpelmöhre; lace flower; meadow sweet; Queen Anne's-lace

Citations:

Cheymol J (1958) Ammi majus et brunissement épidérmique. Therapie 13(6):974-978.
Dollahite JW, Younger RL, Hoffman GO (1978) Photosensitization in cattle and sheep caused by feeding Ammi majus (greater Ammi; Bishop's-Weed). Am J Vet Res 39(1):193-197.
Egyed MN, Malkinson M, Shlosberg A (1974) Observations on the experimental poisoning of young geese with Ammi majus. Avian Pathol 3(2):79-87.
Egyed MN, Shlosberg A, Eilat A (1977) Photosensitizing effects of furocoumarin containing weeds on domestic birds. Vet Hum Toxicol 19(1):11-13.
Egyed MN, Shlosberg A, Eilat A, et al. (1974) Photosensitization in dairy cattle associated with the ingestion of Ammi majus. Refu Vet 31(3):128-131.
Egyed MN, Shlosberg A, Eilat A, et al. (1975) Acute and chronic manifestations of Ammi majus induced photosensitisation in ducks. Vet Rec 97(11):198-199.
Egyed MN, Shlosberg A, Eilat A, et al. (1975) Chronic lesions in geese photosensitized by Ammi majus. Avian Dis 19(4):822-826.
Egyed MN, Singer L, Eilat A, et al. (1975) Eye lesions in ducklings fed Ammi majus seeds. Zentralbl Veterinarmed A 22(9):764-768.
Eilat A, Malkinson M, Shlosberg A, et al. (1974) A field outbreak of photosensitization in goslings caused by the ingestion of Ammi majus. Refu Vet 31(2):83-86.
Kiistala R, Mäkinen-Kiljunen S, Heikkinen K, et al. (1999) Occupational allergic rhinitis and contact urticaria caused by bishop's weed (Ammi majus). Allergy 54(6):635-639.
Méndez MC, Riet-Correa F, Schild AL, et al. (1991) Fotossensibilização em bovinos causada por Ammi majus (Umbelliferae) no Rio Grande do Sul. Pesq Vet Bras 11(1-2):17-19.
Odriozola E (1984) Fotosensibilizacion y queratoconjunctivitis en rumiantes por consumo de semillas de falsa viznaga (Ammi majus). Vet Argent 1(7):684-688.
Schild AC, Motta AC, Riet-Correa F, et al. (2004) Photosensitization in cattle in southern Brazil. In: Acamovic T, Stewart CS, Pennycott TW (eds.) Poisonous plants and related toxins. CABI. Wallingford, UK. pp. 162-166.
Shlosberg A, Egyed MN (1978) Ammi majus induced photosensitization in chickens and turkey poults. Refu Vet 35(4):159-161.
Shlosberg A, Egyed MN, Eilat A (1974) The comparative photosensitizing properties of Ammi majus and Ammi visnaga in goslings. Avian Dis 18(4):544-550.
Sidi E, Bourgeois-Gavardin J (1953) Mise au point du traitment du vitiligo par L'Ammi majus. Presse Med 61(21):436-441.
Witzel DA, Dollahite JW, Jones LP (1978) Photosensitization in sheep fed Ammi majus (Bishop's weed) seed. Am J Vet Res 39(2):319-320.
Yeruham I, Lemberg D, Natan A, et al. (1988) Photosensitization in sheep due to ingestion of Vicia hay contaminated by Ammi majus. Isr J Vet Med 44(2):147.

AMMI VISNAGA (L.) Lam. [Apiaceae]

Common Names:

bishop's-weed; bisnaga; biznaga; khella; khilla; khillala; toothpick ammi

Citations:

Egyed MN, Shlosberg A, Eilat A (1975) The susceptibility of young chickens, ducks, and turkeys to the photosensitizing effect of Ammi visnaga seeds. Avian Dis 19(4):830-833.
Ibrahim IA, El Badwi SM, Gadir WS, et al. (2004) Susceptibility of Bovans chicks to low dietary levels of Ammi visnaga and Artemisia herba-alba. Vet Hum Toxicol 46(2):67-69.
Shlosberg A, Egyed MN, Eilat A (1974) The comparative photosensitizing properties of Ammi majus and Ammi visnaga in goslings. Avian Dis 18(4):544-550.
Trenchi H (1960) Ingestion of Ammi visnaga seeds and photosensitization - The cause of vesicular dermatitis in fowls. Avian Dis 4(3):275-280.

Amomum cardamomum L. = Elettaria cardamomum (L.) Maton

AMORPHOPHALLUS KONJAC K. Koch [Araceae]

Synonyms:

Amorphophallus rivieri Durieu ex Carrière; ***Conophallus konjak*** Schott

Common Names:

devil's-tongue; konjac; leopard palm; snake palm; umbrella arum

Citations:

Yoshizawa K (1971) [Ocular injury caused by bulbs of Amorphophallus konjac K. Koch.] Acta Soc Ophthalmol Jpn 75(1):50-58;134-142.

Amorphophallus rivieri Durieu ex Carrière = Amorphophallus konjac K. Koch

Ampelopsis quinquefolia Michx. = Parthenocissus quinquefolia (L.) Planch.

Amsinckia arizonica Suksd. = Amsinckia intermedia Fisch. & C. A. Mey.

Amsinckia echinata A. Gray = Amsinckia intermedia Fisch. & C. A. Mey.

Amsinckia intactilis J. F. Macbr. = Amsinckia intermedia Fisch. & C. A. Mey.

AMSINCKIA INTERMEDIA Fisch. & C. A. Mey. [Boraginaceae]

Synonyms:

Amsinckia arizonica Suksd.; ***Amsinckia echinata*** A. Gray; ***Amsinckia intactilis*** J. F. Macbr.; ***Amsinkia microphylla*** Suksd.; ***Amsinkia nana*** Suksd.; ***Amsinkia rigida*** Suksd.

Common Names:

buckthorn; coast buckthorn; coast fiddleneck; fiddleneck; finger weed; fireweed; ironweed; rancher's-fiddleneck; tarweed; yellow burweed; yellow forget-me-not; yellow tarweed

Citations:

Fowler ME (1968) Pyrrolizidine alkaloid poisoning in calves. J Am Vet Med Assoc 152(8):1131-1137.

Kennedy PC (1957) Tarweed poisoning in swine. J Am Vet Med Assoc 130(Apr 1):305-306.

McCulloch EC (1940) Hepatic cirrhosis of horses, swine and cattle due to the ingestion of seeds of the tarweed, Amsinkia intermedia. J Am Vet Med Assoc 96(754):5-18.

McCulloch EC (1942) The use of grain containing tarweed (Amsinckia intermedia) seed as poultry feed. J Am Vet Med Assoc 101(Dec):481-483.

Muth OH (1941) An attempt to determine the toxicity of Amsinckia intermedia (tarweed) for fattening lambs. J Am Vet Med Assoc 99(773):145-146.

Woolsey JH Jr, Jasper DE, Cordy DR, et al. (1952) Two outbreaks of hepatic cirrhosis in swine in California, with evidence incriminating the tarweed, Amsinckia intermedia. Vet Med 47:55-58.

Amsinkia microphylla Suksd. = Amsinckia intermedia Fisch. & C. A. Mey.

Amsinkia nana Suksd. = Amsinckia intermedia Fisch. & C. A. Mey.

Amsinkia rigida Suksd. = Amsinckia intermedia Fisch. & C. A. Mey.

amte –see– *Toxicodendron striatum* (Ruiz & Pav.) Kuntze

Amur River privet –see– *Ligustrum obtusifolium* Siebold & Zucc. subsp. suave (Kitag.) Kitag.

amyroot –see– *Apocynum cannabinum* L.

anabasia –see– *Anabasis aphylla* L.

ANABASIS APHYLLA L. [Chenopodiaceae]

Common Names:

anabasia

Citations:

Shatrov AP (1962) [Poisoning of sheep with Anabasis aphylla and A. solsa [sic].] Veterinariia Moscow 39(8):49-54.

ANABASIS SALSA C. A. (Mey.) Benth. ex Volkens [Chenopodiaceae]

Citations:

Shatrov AP (1962) [Poisoning of sheep with Anabasis aphylla and A. solsa [sic].] Veterinariia Moscow 39(8):49-54.

ANACARDIUM OCCIDENTALE L. [Anacardiaceae]

Synonyms:

Cassuvium pomiferum Lam.

Common Names:

acajou; alui; balubad; cacahuil; cajú; cajuil; cashew; cashew apple; jocote marañón; marañón; noix d'acajou; palu-de-cashupete; pommier cajou; ului

Citations:

Aber R, Marks J, DeMelfi T, et al. (1983) Dermatitis associated with cashew nut consumption - Pennsylvania. MMWR Morb Mortal Wkly Rep 32(9):129-130.

Bedello PG, Goitre M, Cane D, et al. (1985) Allergic contact dermatitis to cashew nut. Contact Dermatitis 12(4):235.

Bedi BM (1971) Cashewnut dermatitis (report of two cases). Indian J Dermatol 16(3):63-64.

Diogenes MJ, Morais SM, Cavalho FF (1996) Contact dermatitis among cashew nut workers. Contact Dermatitis 35(2):114-115.

Downing JG, Gurney SW (1940) Dermatitis from cashew nut shell oil. J Ind Hyg Toxicol 22(5):169-174.

Hamilton TK, Zug KA (1998) Systemic contact dermatitis to raw cashew nuts in a pesto sauce. Am J Contact Dermat 9(1):51-54.

Kilburn SA, Warner JO (2001) Clinical features of cashew allergy. Allergy 56(3):252-253.

Kligman AM (1958) Cashew nut shell oil for hyposensitization against Rhus dermatitis. Arch Dermatol 78(3):359-363.

Lockey SD (1944) Cashew nut oil dermatitis. Ann Allergy 2(1):22-25.

Marks JG Jr, DeMelfi T, McCarthy MA, et al. (1984) Dermatitis from cashew nuts. J Am Acad Dermatol 10(4):627-631.

McNairy DJ (1959) Contact dermatitis from cashew nut shell novelties. Ariz Med 16(5):361-362.

Orris L (1958) Cashew nut dermatitis. N Y State J Med 58(17 part 1):2799-2800.

Quercia O, Rafanelli S, Marsigli L, et al. (1999) Unexpected anaphylaxis to cashew nut. Allergy 54(8):895-897.

Ratner JH, Spencer SK, Grainge JM (1974) Cashew nut dermatitis. An example of internal-external contact-type hypersensitivity. Arch Dermatol 110(6):921-923.

Reginella RF, Fairfield JC, Marks JG Jr (1989) Hyposensitization to poison ivy after working in a cashew nut shell oil processing factory. Contact Dermatitis 20(12):274-279.

Rosen T, Fordice DB (1994) Cashew nut dermatitis. South Med J 87(4):543-546.

Schwartz L, Birmingham DJ, Campbell PC Jr, et al. (1945) Skin hazards - in the manufacture and use of cashew nut shell liquid - formaldehyde resins. Indus Med 14:500-506.

Srinivas CR, Pasricha JS (1990) Dermatological problems of workers employed in cashew nut factories. Contact Dermatitis 22(5):192.

Anacardium orientale auct. ex Steud. = Semecarpus anacardium L. f.

ANADENANTHERA COLUBRINA (Vell.) Brenan var. cebil (Griseb.) Altschul [Fabaceae]

Synonyms:

Anadenanthera macrocarpa (Benth.) Brenan; ***Piptadenia macrocarpa*** Benth.

Common Names:

angico; cebil colorado

Citations:

Brito MF, Franca TN, Oliveira KD, et al. (2000) Estudos experimentais em coelhos com plantas cianogênicas. Pesq Vet Bras 20(2):65-70.

Tokarnia CH, Peixoto PV, Brito MF, et al. (1999) Estudos experimentais com plantas cianogênicas em bovinos. Pesq Vet Bras 19(2):84-90.

Anadenanthera macrocarpa (Benth.) Brenan = Anadenanthera colubrina (Vell.) Brenan var. cebil (Griseb.) Altschul

ANAGALLIS ARVENSIS L. [Myrsinaceae]

Synonyms:

Anagallis caerulea Lam.; ***Anagallis foemina*** Mill.; ***Anagallis orientalis*** hort. ex Fisch. Mey. & Avé-Lall.; ***Anagallis pulchella*** Salisb.

Common Names:

bird's-eye; blue pimpernel; dharati dhak; morgeline; mouron-des-champs; murajes; pimpernel; poison chickweed; poison weed; poor-man's-weatherglass; red chickweed; red pimpernel; roode muur; rooimuur; scarlet pimpernel; shepherd's-calendar; shepherd's-clock; shepherd's-delight; shepherd's-weatherglass

Citations:

Pullar EM (1939) Studies on five suspected poisonous plants. Aust Vet J 15:19-23.

Rivero R, Zabala A, Gianneechini R, et al. (2001) Anagallis arvensis poisoning in cattle and sheep in Uruguay. Vet Hum Toxicol 43(1):27-30.

Rothwell JT, Marshall DJ (1986) Suspected poisoning of sheep by Anagallis arvensis (scarlet pimpernel). Aust Vet J 63(9):316.

Schneider DJ (1978) Fatal ovine nephrosis caused by Anagallis arvensis L. J S Afr Vet Assoc 49(4):321-324.

Anagallis caerulea Lam. = Anagallis arvensis L.

Anagallis foemina Mill. = Anagallis arvensis L.

Anagallis orientalis hort. ex Fisch. Mey. & Avé-Lall. = Anagallis arvensis L.

Anagallis pulchella Salisb. = Anagallis arvensis L.

anamú –see– *Petiveria alliacea* L.

ANANAS COMOSUS (L.) Merr. [Bromeliaceae]

Synonyms:

Ananas sativus Schult. & Schult. f.; ***Bromelia ananas*** L.; ***Bromelia comosus*** L.

Common Names:

pineapple

Citations:

Kabir I, Speelman P, Islam A (1993) Systemic allergic reaction and diarrhoea after pineapple ingestion. Trop Geogr Med 45(2):77-79.

Polunin I (1951) Pineapple dermatosis. Br J Dermatol 63(12):441-455.

Schilling R, Der Marderosian A, Speaker J (1980) Incidence of plant poisonings in Philadelphia noted as poison information calls. Vet Hum Toxicol 22(3):148-150.

Ananas sativus Schult. & Schult. f. = Ananas comosus (L.) Merr.

ananbeam –see– *Euonymus europaeus* L.

anantamul –see– *Hemidesmus indicus* (L.) W. T. Aiton

anda –see– *Joannesia princeps* Vell.

anda assu –see– *Joannesia princeps* Vell.

Anderson's-larkspur –see– *Delphinium andersonii* A. Gray

ANDIRA INERMIS (W. Wright) Kunth ex DC. [Fabaceae]

Synonyms:

Andira jamaicensis Urb.; ***Geoffroea inermis*** W. Wright; ***Geoffroea jamaicensis*** W. Wright

Common Names:

almendro macho; almendro montes; almendro seal; bastard cabbage; bastard mahogany; black blossom cherry; cabbagebark; carne asada; chaperno; cocu; cornwood; cuilinbaca; maca colorada; macallo; partridge; Rebhuhnholz

Citations:

Figueroa V, Sutherland TM (1972) "Muerte súbita" (sudden death) in cattle. 5. The role of toxic plants. Rev Cubana Cienc Agric Eng Ed 6(1):53-59.

Glander KE (1977) Poison in a monkey's Garden of Eden. Nat Hist 86(Mar):35-41.

Andira jamaicensis Urb. = Andira inermis (W. Wright) Kunth ex DC.

andrachne –see– *Andrachne decaisnei* Benth.

ANDRACHNE DECAISNEI Benth. [Phyllanthaceae]

Synonyms:

Leptopus decaisnei (Benth.) Pojark.

Common Names:

andrachne

Citations:

Churchward RE, Gurney EH (1938) A plant poisonous to sheep. Queensland Agric J 50:180-184.

Andromeda floribunda Pursh = Pieris floribunda (Pursh) Benth. & Hook. f.

Andropogon arundinaceus Scop. = Sorghum halepense (L.) Pers.

Andropogon citratus DC. = Cymbopogon citratus (DC.) Stapf

Andropogon halepensis (L.) Brot. = Sorghum halepense (L.) Pers.

Andropogon nardus L. = Cymbopogon nardus (L.) Rendle

Andropogon sorghum (L.) Brot. = Sorghum bicolor (L.) Moench

ANEMONE PATENS L. [Ranunculaceae]

Synonyms:

Pulsatilla hirsutissima Britton; ***Pulsatilla patens*** (L.) Mill.

Common Names:

pasqueflower; prairie crocus; pulsatilla

Citations:

Aaron TH, Muttitt EL (1964) Vesicant dermatitis due to prairie crocus (Anemone patens L.). Arch Dermatol 90(Aug):168-171.

Vance JC (1982) Toxic plants of Minnesota: Skin toxicity of the prairie crocus (Anemone patens L.). Minn Med 65(3):149-151, 188.

ANEMONE SYLVESTRIS L. [Ranunculaceae]

Common Names:

snowdrop anemone

Citations:

Eikhler IN (1965) [Intoxication by "inebriating" poisonous honey.] Vopr Pitan 24(5):65-67.

anès estrella –see– *Illicium verum* Hook. f.

ANETHUM GRAVEOLENS L. [Apiaceae]

Synonyms:

Anethum sowa Roxb. ex Fleming; ***Peucedanum graveolens*** (L.) Benth. & Hook. f.

Common Names:

dill

Citations:

Knudson EA, Kroon S (1988) In vitro and in vivo phototoxicity of furocoumarin-containing plants. Clin Exp Dermatol 13(2):92-96.

Monteseirín J, Pérez-Formoso JL, Hérnandez M, et al. (2003) Contact urticaria from dill. Contact Dermatitis 48(5):275.

Monteseirín J, Pérez-Formoso JL, Sánchez-Hernández MC, et al. (2002) Occupational contact dermatitis to dill. Allergy 57(9):866-867.

Anethum sowa Roxb. ex Fleming = Anethum graveolens L.

angela –see– *Moringa oleifera* Lam.

angelica –see– *Angelica archangelica* L.

ANGELICA ARCHANGELICA L. [Apiaceae]

Synonyms:

Angelica officinalis Moench

Common Names:

angelica; Engelwurz

Citations:

Knudson EA, Kroon S (1988) In vitro and in vivo phototoxicity of furocoumarin-containing plants. Clin Exp Dermatol 13(2):92-96.

Kuske H (1939) Experimentelle untersuchungen zur Photosensibilisierung der Haut durch pflanzliche Wirkstoffe. 1. Mitteilung. Lichtsensibilisierung durch Furocumarine als Ursache verschiedener phytogener Dermatosen. Arch Derm Syphilol 178:112-123.

Kuske H (1940) Perkutane Photosensibilisierung durch pflanzliche Wirkstoffe. Dermatologica 82:273-338.

Angelica officinalis Moench = Angelica archangelica L.

angel's-tears –see– *Brugmansia arborea* (L.) Lagerh.

angel's-trumpet –see– *Brugmansia arborea* (L.) Lagerh.; *Brugmansia* ×*candida* Pers.; *Brugmansia suaveolens* (Humb. & Bonpl. ex Willd.) Bercht. & J. Presl; *Datura stramonium* L.

angel's-trumpet lily –see– *Brugmansia arborea* (L.) Lagerh.

angel's-wings –see– *Caladium bicolor* (Aiton) Vent.

angico –see– *Anadenanthera colubrina* (Vell.) Brenan var. cebil (Griseb.) Altschul

änglatrumpet –see– *Brugmansia suaveolens* (Humb. & Bonpl. ex Willd.) Bercht. & J. Presl

anguna –see– *Dregea volubilis* (L. f.) Benth. ex Hook. f.

anhalonium –see– *Lophophora williamsii* (Lem. ex Salm-Dyck) J. M. Coult.

anil indigo –see– *Indigofera suffruticosa* Mill.

anileira –see– *Indigofera suffruticosa* Mill.

aninga para –see– *Dieffenbachia seguine* (Jacq.) Schott

anis –see– *Illicium anisatum* L.; *Pimpinella anisum* L.

anis hinono –see– *Foeniculum vulgare* Mill.

anise –see– *Pimpinella anisum* L.

ankra –see– *Calotropis procera* (Aiton) W. T. Aiton

Anna Maria –see– *Hyacinthus orientalis* L.

ANNONA CHERIMOLA Mill. [Annonaceae]

Common Names:

cherimoya; sugar apple; sweet sop

Citations:

Sánchez-Guerrero IM, Escudero AI, Tortosa JA, et al. (2000) Anaphylaxis to cherimoya. Allergy 55(10):976-977.

ANNONA MURICATA L. [Annonaceae]

Common Names:

anonas; guanábana; laguana; sorsaka; soursop

Citations:

Dunham LJ, Sheets RH, Morton JF (1974) Proliferative lesions in cheek pouch and esophagus of hamsters treated with plants from Curacao, Netherland Antilles. J Natl Cancer Instit 53(5):1259-1269.

ANNONA RETICULATA L. [Annonaceae]

Common Names:

bullock's-heart; custard apple; Jamaica apple

Citations:

Gonzalo MA, Moneo I, Ventas P, et al. (1997) IgE-mediated hypersensitivity to custard-apple. Allergy 52(5):597.

Sánchez-Morillas L, Moneo I, Sedano E, et al. (2003) Oral allergic syndrome after ingestion of custard apple. Allergy 58(3):260-261.

ANNONA SQUAMOSA L. [Annonaceae]

Common Names:

atis; custard apple; sitaphal; sugar apple; sweet sop

Citations:

Visweswariah K, Jayaram M, Krishnaprasad NK, et al. (1971) Toxicological studies of the seeds of Annona squamosa. Indian J Exp Biol 9(4):519-521.

Annona triloba L. = Asimina triloba (L.) Dunal

annual bitterweed –see– *Hymenoxys odorata* DC.

annual mercury –see– *Mercurialis annua* L.

annual ragweed –see– *Ambrosia artemisiifolia* L.

annual rye –see– *Secale cereale* L.

annual ryegrass –see– *Lolium multiflorum* Lam.; *Lolium rigidum* Gaudin

annual sage –see– *Salvia reflexa* Hornem.

annual sunflower –see– *Helianthus annuus* L.

annual sweet pea –see– *Lathyrus odoratus* L.

ANOECTOCHILUS FORMOSANUS Hayata [Orchidaceae]

Citations:

Deng JF, Lin TJ, Kao WF, et al. (1997) The difficulty in handling poisonings associated with Chinese traditional medicine: A poison control center experience for 1991-1993. Vet Hum Toxicol 39(2):106-114.

anonas –see– *Annona muricata* L.

antbush –see– *Senna occidentalis* (L.) Link

ANTHEMIS ARVENSIS L. [Asteraceae]

Common Names:

corn chamomile; scentless chamomile

Citations:

Möslein P (1963) Pflanzen als Kontakt-Allergene. Derm Beruf Umwelt 11:24-32.

ANTHEMIS COTULA L. [Asteraceae]

Synonyms:

Maruta cotula (L.) DC.

Common Names:

camomile; dog camomile; dog chamomile; dog daisy; dog fennel; Dorfpflanze; Hundskamille; stinking chamomile; stinking Mayweed

Citations:

Benner MH, Lee HJ (1973) Anaphylactic reaction to chamomile tea. J Allergy Clin Immunol 52(5):307-308.

Krantz W (1938) Ueber eine bei Erntearbeitern beobachtete bullöse Dermatitis. Munch Med Wochenschr 85(28):1057-1060.

Rowe AH (1934) Camomile (Anthemis cotula) as a skin irritant. J Allergy 5:383-388.

Sequeira JH (1921) A case of bullous eruption caused by May-weed. Lancet 2:560.

Sternberg L (1937) Contact dermatitis. Cases caused by oil of cloves and by oil of camomile tea (Anthemis cotula). J Allergy 8:185-186.

Anthemis nobilis L. = Chamaemelum nobile (L.) All.

ANTHOXANTHUM ODORATUM L. [Poaceae]

Common Names:

sweet vernal grass

Citations:

Bartol JM, Thompson LJ, Minnier SM, et al. (2000) Hemorrhagic diathesis, mesenteric hematoma, and colic associated with ingestion of sweet vernal grass in a cow. J Am Vet Med Assoc 216(10):1569-1570, 1605-1608.

Dwyer CJ, Downing GM, Gabor LJ (2003) Dicoumarol toxicity in neonatal calves associated with the feeding of sweet vernal (Anthoxanthum odoratum) hay. Aust Vet J 81(6):332-335.

Pritchard DG, Markson LM, Brush PJ, et al. (1983) Haemorrhagic syndrome of cattle associated with the feeding of sweet vernal (Anthoxanthum odoratum) hay containing dicoumarol. Vet Rec 113(4):78-84.

Runciman DJ, Lee AM, Reed KF, et al. (2002) Dicoumarol toxicity in cattle associated with ingestion of silage containing sweet vernal grass (Anthoxanthum odoratum). Aust Vet J 80(1-2):28-32.

ANTHRISCUS SYLVESTRIS (L.) Hoffm. [Apiaceae]

Common Names:

cow parsley; Wiesenkerbel; wild chervil; wild parsley

Citations:

Bellringer HE (1949) Phyto-photo-dermatitis. Br Med J 1(Jun 4):984-986.

ANTIARIS TOXICARIA Lesch. [Moraceae]

Common Names:

antsjar; ches tennent; dart poison; hypo; ipoh tree; Javanischer Giftbaum; Malayan arrow poison; pohon upas; pohun ipoh; pokok ipoh; upas tree

Citations:

Fasal P (1945) Cutaneous diseases in the tropics. A clinical study based on observations in Malaya. Arch Derm Syphilol 51(3):163-171.

Ho LM, Cheong I, Jalil HA (1996) Rhabdomyolysis and acute renal failure following blowpipe dart poisoning. Nephron 72(4):676-678.

ANTICLEA ELEGANS (Pursh) Rydb. [Melanthiaceae]

Synonyms:

Zigadenus coloradensis Rydb.; ***Zigadenus elegans*** Pursh; ***Zigadenus glaucus*** (Nutt.) Nutt.

Common Names:

alkali grass; crow poison; death camas; elegant death camas; glaucous zigadenus; meadow death camas; mountain death camas; swamp camas; white camas; white death camas; wild onion

Citations:

Marsh CD, Clawson AB (1922) The death camas species, Zygadenus paniculatus and Z. elegans, as poisonous plants. U S Dep Agric Bull #1012:25 pp.

Marsh CD, Clawson AB, Marsh H (1915) Zygadenus, or death camas. U S Dep Agric Bull #125:46 pp.

antigopher plant –see– *Euphorbia lathyris* L.

antler tree –see– *Rhus typhina* L.

antsjar –see– *Antiaris toxicaria* Lesch.

apamarg –see– *Achyranthes aspera* L.

ape –see– *Alocasia macrorrhizos* (L.) G. Don

ape nut –see– *Gluta renghas* L.

Apfel –see– *Malus domestica* Borkh.

Apfelsine –see– *Citrus sinensis* (L.) Osbeck

apio –see– *Apium graveolens* L.

APIUM GRAVEOLENS L. [Apiaceae]

Synonyms:

Celeri graveolens (L.) Britton

Common Names:

ache douce; ache olorante; ache-des-marrais; apio; céleri; celeriac; celery; sedano; selino; Sellerie; wild celery; Zeller

Citations:

Agathos M (1980) Occupational dermatitis from celeriac. Contact Dermatitis 6(3):225.

Anonymous (1985) Phytophotodermatitis among grocery workers - Ohio. MMWR Morb Mortal Wkly Rep 34(1):11-13.

Austad J, Kavli G (1983) Phototoxic dermatitis caused by celery infected by Sclerotinia sclerotiorum. Contact Dermatitis 9(6):448-451.

Berkley SF, Hightower AW, Beier RC, et al. (1986) Dermatitis in grocery workers associated with high natural concentrations of furanocoumarins in celery. Ann Intern Med 105(3):351-355.

Birmingham DJ, Key MM, Tubich GE, et al. (1961) Phototoxic bullae among celery harvesters. Arch Dermatol 83(Jan):127-141.

Borghijs A, Roelandts R (1984) Phototoxic dermatitis from Sclerotinia sclerotiorum infected celery. Contact Dermatitis 11(1):59.

Déchamp C, Michel J, Deviller P, et al. (1984) Choc anaphylactique au céleri et sensibilisation à l'ambroisie et à l'armoise. Allergie croisée ou allergie concomitante? Presse Med 13(14):871-874.

Forsbeck M, Ros AM (1979) Anaphylactoid reaction to celery. Contact Dermatitis 5(3):191.

Gelfand HH (1936) Hypersensitiveness to celery. Report of a case of celery dermatitis. J Allergy 7:590-593.

Groot BJ, Belinfante-Van Gelder ME, Jans HW (1992) Een epidemie van dermatitis door bleekselderij. Ned Tijdschr Geneeskd 136(25):1210-1213.

Henry SA (1933) Celery itch: Dermatitis due to celery in vegetable canning. Br J Dermatol Syph 45:301-309.

Henry SA (1938) Further observations on dermatitis due to celery in vegetable canning. Br J Dermatol Syph 50:342-351.

Knudson EA, Kroon S (1988) In vitro and in vivo phototoxicity of furocoumarin-containing plants. Clin Exp Dermatol 13(2):92-96.

Kremser M, Lindemayr W (1983) Sellerieallergie (Selleriekontakturikariasyndrom) und Zusammenhänge mit Allergien gegen andere Pflanzenantigene. Wien Klin Wochenschr 95(23):838-843.

Ljunggren B (1990) Severe phototoxic burn following celery ingestion. Arch Dermatol 126(10):1334-1336.

Montgomery JF, Oliver RE, Poole WS (1987) A vesiculobullous disease in pigs resembling foot and mouth disease. 1. Field cases. N Z Vet J 35:21-26.

Montgomery JF, Oliver RE, Poole WS, et al. (1987) A vesiculo-bullous disease in pigs resembling foot and mouth disease. 2. Experimental reproduction of the lesion. N Z Vet J 35:27-30.

Pauli G, Bessot JC, Braun PA, et al. (1988) Celery allergy: Clinical and biological study of 20 cases. Ann Allergy 60(3):243-246.

Pauli G, Bessot JC, Dietemann-Molard A, et al. (1985) Celery sensitivity: Clinical and immunological correlations with pollen allergy. Clin Allergy 15(3):273-279.

Rose MH, Altman LC (1985) Anaphylaxis after ingestion of raw celery. Ann Allergy 54(2):166.

Seligman PJ, Mathias CG, O'Malley MA, et al. (1987) Phytophotodermatitis from celery among grocery store workers. Arch Dermatol 123(11):1478-1482.

Silverstein SR, Frommer DA, Dobozin B, et al. (1986) Celery-dependent exercise-induced anaphylaxis. J Emerg Med 4(3):195-199.

Wiswell JG, Irwin JW, Guba EF, et al. (1948) Contact dermatitis of celery farmers. J Allergy 19:396-402.

Wüthrich B, Dietschi R (1985) Das «Sellerie-Karotten-Beifuss-Gewürz-Syndrom»: Hauttest- und RAST-Ergebnisse. Schweiz Med Wochenschr 115(11):358-364.

Note:

Celery is named *Apium graveolens* L. var. *dulce* (Mill.) Pers. in some publications

Apium petroselinum L. = Petroselinum crispum (Mill.) Nyman ex A. W. Hill

Aplopappus heterophyllus (A. Gray) S. F. Blake = Isocoma plurifolia (Torr. & A. Gray) Greene

APOCYNUM CANNABINUM L. [Apocynaceae]

Common Names:

American hemp; amyroot; bitter root; black Indian hemp; bowman's-root; Canadian hemp; chanvre-de-Canada; dogbane; glabrous hemp; hemp dogbane; Indian hemp; milkweed; rheumatism root; rheumatism weed; wild cotton

Citations:

Bania T, Hoffman RS, Howland MA, et al. (1993) Accidental Indian hemp (Apocyneacea [sic] cannabinum) cardiac glycoside toxicity. Vet Hum Toxicol 35(4):328.

apple –see– *Malus domestica* Borkh.

apple-of-Peru –see– *Datura inoxia* Mill.; *Datura stramonium* L.; *Nicandra physalodes* (L.) Gaertn.

apple-of-Sodom –see– *Solanum anguivi* Lam.; *Solanum carolinense* L.; *Solanum pseudocapsicum* L.

apple-of-the-tropics –see– *Mangifera indica* L.

apple-of-Tolguacha –see– *Datura stramonium* L.

apricot –see– *Prunus armeniaca* L.

apricot vine –see– *Passiflora incarnata* L.

Aprikose –see– *Prunus armeniaca* L.

Arabian jasmine –see– *Jasminum sambac* (L.) Aiton

Arabian myrrh –see– *Commiphora myrrha* (Nees) Engl.

ARACHIS HYPOGAEA L. [Fabaceae]

Common Names:

earthnut; Erdnuß; groundnut; monkey nut; peanut

Citations:

Bernard H, Paty E, Mondoulet L, et al. (2003) Serological characteristics of peanut allergy in children. Allergy 58(12):1285-1292.

Levy Y, Broides A, Segal N, et al. (2003) Peanut and tree nut allergy in children: Role of peanut snacks in Israel? Allergy 58(11):1206-1207.

McCarrison R (1934) The goitrogenic action of soya-bean and ground-nut. Indian J Med Res 21:179-181.

Arachnoides adiantiformis (Forst) Tindale = Rumohra adiantiformis (G. Forst.) Ching

Aragallus lambertii (Pursh) Greene = Oxytropis lambertii Pursh

Aragallus sericeus (Nutt. ex Torr. & A. Gray) Greene = Oxytropis sericea Nutt.

Aragallus spicatus (Hook.) Rydb. = Oxytropis campestris (L.) DC. var. spicata Hook.

aralia –see– *Fatsia japonica* (Thunb.) Decne. & Planch.; *Tetrapanax papyrifer* (Hook.) K. Koch

Aralie –see– *Fatsia japonica* (Thunb.) Decne. & Planch.

aralu –see– *Terminalia chebula* Retz.

aran –see– *Hypericum perforatum* L.

arandi –see– *Ricinus communis* L.

arandon –see– *Wikstroemia ovata* C. A. Mey.

Araujia hortorum E. Fourn. = Araujia sericifera Brot.

ARAUJIA SERICIFERA Brot. [Apocynaceae]

Synonyms:

Araujia hortorum E. Fourn.

Common Names:

bladder flower; cruel vine; moth plant; Peruvian creeper; rubber vine; white moth plant

Citations:

Gaig P, Gázquez V, Lombardero M, et al. (2005) Moth plant (Araujia sericifera) allergy. Allergy 60(8):1092-1093.

Hart L (1940) Poisonous to fowls. Seeds of the plant Araujia sericifera. Agric Gaz New South Wales 51:472-474.

arbol-de-India –see– *Aleurites moluccanus* (L.) Willd.

arbol-de-la-Muerte –see– *Hippomane mancinella* L.

arbol-de-las-perlas –see– *Moringa oleifera* Lam.

arbol-de-quitasol –see– *Melia azedarach* L.

árbol-del-paraíso –see– *Elaeagnus angustifolia* L.

arbol santo –see– *Jatropha curcas* L.

arbor vitae –see– *Thuja occidentalis* L.; *Thuja plicata Donn* ex D. Don

arbre-à-suif –see– *Triadica sebifera* (L.) Small

arbre corail –see– *Jatropha multifida* L.

ARCHIDENDRON JIRINGA (Jack) I. C. Nielsen [Fabaceae]

Synonyms:

Pithecellobium jiringa (Jack) Prain; ***Pithecellobium lobatum*** Benth.

Common Names:

djaring; djengkol; djenkol; djenkol bean; Djenkolbohne; djering; genkol bean; jengkol bean; jenkol; jenkol bean; jering; kra niang; krakos; ma niang; niang

Citations:

Eiam-Ong S, Sitprija V, Saetang P, et al. (1987) Djenkol bean nephrotoxicity in Southern Thailand. In: Gopalakrishnakone P, Tan CK (eds.) Progress in venom and toxin research. National Univ. Singapore. pp. 628-632.

H'ng PK, Nayar SK, Lau WM, et al. (1991) Acute renal failure following jering ingestion. Singapore Med J 32(2):148-149.

Reimann HA, Sukaton RU (1956) Djenkol bean poisoning (djenkolism): A cause of hematuria and anuria. Am J Med Sci 232(2):172-174.

Segasothy M, Swaminathan M, Kong NC, et al. (1995) Djenkol bean poisoning (djenkolism): An unusual cause of acute renal failure. Am J Kidney Dis 25(1):63-66.

Vachvanichsanong P, Lebel L (1997) Djenkol beans as a cause of hematuria in children. Nephron 76(1):39-42.

Wirya IW, Muhidin TT, Alatas H, et al. (1987) β_2-microglobulin in renal function of patients with jengkol intoxication. Paediatr Indones 27(7-8):155-162.

Yong M, Cheong I (1995) Jering-induced acute renal failure with blue urine. Trop Doct 25(1):31.

Arctic poppy –see– *Papaver nudicaule* L.

Arctic sweet clover –see– *Melilotus albus* Medik.

ARCTIUM LAPPA L. [Asteraceae]

Synonyms:

Lappa major Gaertn.; ***Lappa vulgaris*** Hill

Common Names:

bardana; beggar's-buttons; burdock; great burdock; große Klette; harebur; Klettenhaare; lappa

Citations:

Bruhn AM (1938) Klinische und experimentelle Untersuchungen über Augenschädigungen durch Klettenhaare. Klin Monatsbl Augenheil Augenarztl Fortb 101:730-741.

ARCTIUM MINUS (Hill) Bernh. [Asteraceae]

Synonyms:

Arctium pubens Bab.

Common Names:

burdock; lesser burdock

Citations:

Havener WH, Falls HF, McReynolds WU (1955) Burdock bur ophthalmia. Arch Ophthalmol 53(2):260-263.

Arctium pubens Bab. = Arctium minus (Hill) Bernh.

ARCTOTHECA CALENDULA (L.) Levyns [Asteraceae]

Synonyms:

Cryptostemma calendula (L.) Druce

Common Names:

Capeweed

Citations:

Hand EA (1944) Contact dermatitis due to Capeweed. Arch Derm Syphilol 49:331-332.

Harris DJ, Rhodes HA (1969) Nitrate and nitrite poisoning in cattle in Victoria. Aust Vet J 45(12):590-591.

Van der Walt SJ, Steyn DG (1940) Recent investigations into the toxicity of known and unknown poisonous plants in the Union of South Africa, X. Onderstepoort J Vet Res 15(1-2):261-277.

ARECA CATECHU L. [Arecaceae]

Common Names:

areca nut; betel nut; Betelkerne; cutch; noix d'arec; pinang wang; sorte-de-noix-d'arec

Citations:

Babu S, Bhat RV, Kumar PU, et al. (1996) A comparative clinico-pathological study of oral submucous fibrosis in habitual chewers of pan masala and betel quid. J Toxicol Clin Toxicol 34(3):317-322.

Chiang WT, Yang CC, Deng JF, et al. (1998) Cardiac arrhythmia and betel nut chewing - Is there a causal effect? Vet Hum Toxicol 40(5):287-289.

Deng JF, Ger J, Tsai WJ, et al. (2001) Acute toxicities of betel nut: Rare but probably overlooked events. J Toxicol Clin Toxicol 39(4):355-360.

Hung DZ, Deng JF (1998) Acute myocardial infarction temporally related to betel nut chewing. Vet Hum Toxicol 40(1):25-28.

Löhrer F, Kaiser R (1999) Biogene Suchtmittel. Neue Konsumgewohnheiten bei jungen Abhängigen? Nervenarzt 70(11):1029-1033.

Merchant A, Husain SS, Hosain M, et al. (2000) Paan without tobacco: An independent risk factor for oral cancer. Int J Cancer 86:128-131.

Sarma AB, Chakrabarti J, Chakrabarti A, et al. (1992) Evaluation of pan masala for toxic effects on liver and other organs. Food Chem Toxicol 30(2):161-163.

areca nut –see– *Areca catechu* L.

arfoo –see– *Acacia pennata* (L.) Willd.

argemone –see– *Argemone mexicana* L.

ARGEMONE MEXICANA L. [Papaveraceae]

Common Names:

argemone; briuma dandie; cardo santo; chicalote; devil's-fig; kateli; Mexican poppy; Mexican prickly-poppy; pila dhatura; poppy weed; prickly poppy; prickly yellow poppy; satayanashi; satyanasi; thistle root; yellow prickly poppy; yellow thistle

Citations:

Brink AJ, Lewis CM, Weber HW (1965) Myocardiopathy in Argemone mexicana poisoning. S Afr Med J 39(1):108-114.

Chaudhuri MN, Chaudhuri RN (1965) Epidemic dropsy in West Dinajpur. Bull Calcutta Sch Trop Med 13(2):37-38.

Dogra J, Sharma KC (1986) Epidemic dropsy: Atypical presentation. Am J Med 81(6):1115.

Gomber S, Daral TS, Sharma PP, et al. (1994) Epidemic dropsy in Trans Yamuma areas of Delhi and U.P. Indian Pediatr 31(6):671-674.

Hakim SA, Jehangir RP (1977) Argemone oil poisoning. J Trop Med Hyg 80(7):149-151.

Hart L (1941) The toxicity of seeds of Argemone mexicana for fowls. Aust Vet J 17(Apr):69-71.

Krishnamachari KA, Satyanarayana K (1972) Epidemic dropsy in Andhra Pradesh. Indian J Med Res 60(5):741-746.

Kumar A, Husain F, Das M, et al. (1992) An out-break of epidemic dropsy in the Barabanki District of Uttar Pradesh, India: A limited trial for the scope of antioxidants in the management of symptoms. Biomed Environ Sci 5:251-256.

Kumar L, Chugh KS, Singhal PC, et al. (1975) Argemone mexicana poisoning in north India. J Assoc Physicians India 23(3):205-208.

Misra NP, Varma P, Jain SC, et al. (1984) Clinico-pathological and experimental study of epidemic dropsy. J Assoc Physicians India 32(11):943-947.

Mohan M, Sood NN, Dayal Y, et al. (1984) Ocular and clinico-epidemiological study of epidemic dropsy. Indian J Med Res 80(Oct):449-456.

Norton JH, O'Rourke PK (1980) Oedema disease in chickens caused by Mexican poppy (Argemone mexicana) seed. Aust Vet J 56(4):187-189.

Pahwa R, Chatterjee VC (1989) The toxicity of Mexican poppy (Argemone mexicana L.) seeds to rats. Vet Hum Toxicol 31(6):555-558.

Ramasastri BV, Babu S (1975) A study on toxicity of argemone oil in experimental animals. Indian J Med Res 63(9):1353-1356.
Sainani GS, Rajkondawar VL, Wechalekar MD, et al. (1972) Epidemic dropsy in Chandrapur, Maharashtra: Epidemiological and clinical study. J Assoc Physicians India 20(4):301-306.
Sandhu HS, Tiwari SC, Naik G, et al. (1982) An epidemiological study of an outbreak of epidemic dropsy in Itarsi (MP). Indian J Public Health 26(1):10-15.
Sanghvi LM, Misra SN, Bose TK (1960) Cardiovascular manifestations in Argemone mexicana poisoning (epidemic dropsy). Circulation 21(Jun):1096-1106.
Sharma KC, Panwogra J, Banerjee S, et al. (1986) Epidemic dropsy in Rajasthan (India) - A clinical study. Indian J Nutr Diet 23(2):41-44.
Shenolikar IS, Rukmini C, Krishnamachari KA, et al. (1974) Sanguinarine in the blood and urine of cases of epidemic dropsy. Food Cosmet Toxicol 12(5-6):699-702.
Singh R, Faridi MM, Singh K, et al. (1999) Epidemic dropsy in the eastern region of Nepal. J Trop Pediatr 45(1):8-13.
Sood NN, Sachdev MS, Mohan M, et al. (1985) Epidemic dropsy following transcutaneous absorption of Argemone mexicana oil. Trans R Soc Trop Med Hyg 79(4):510-512.
Tandon RK, Singh DS, Arora RR, et al. (1975) Epidemic dropsy in New Delhi. Am J Clin Nutr 28(8):883-887.
Tandon RK, Tandon HD, Nayak NC, et al. (1976) Liver in epidemic dropsy. Indian J Med Res 64(7):1064-1069.
Upreti KK, Das M, Kumar A, et al. (1989) Biochemical toxicology of argemone oil. IV. Short-term oral feeding response in rats. Toxicology 58:285-298.

ARGYRANTHEMUM FRUTESCENS (L.) Sch. Bip. [Asteraceae]

Common Names:

marguerite daisy

Citations:

Paulsen E (1998) Occupational dermatitis in Danish gardeners and greenhouse workers (II). Etiological factors. Contact Dermatitis 38(1):14-19.

ARGYREIA NERVOSA (Burm. f.) Bojer [Convolvulaceae]

Common Names:

Hawaiian baby woodrose; woolly morning-glory

Citations:

Furbee RB, Curry SC, Kunkle DB (1991) Ingestion of Argyreia nervosa (Hawaiian baby woodrose) seeds. Vet Hum Toxicol 33(4):370.
Hentschel H, Bergmann I, Lampe J, et al. (2000) Two cases of ingestion of Hawaiian baby wood rose (Argyreia nervosa). J Toxicol Clin Toxicol 38(2):234.
Hruby K (2002) Poisoning from natural drugs rediscovered for recreational abuse. J Toxicol Clin Toxicol 40(3):276-277.

ARIOCARPUS RETUSUS Scheidw. [Cactaceae]

Common Names:

false peyote

Citations:

Furst PT (1971) Ariocarpus retusus, the "false peyote" of Huichol tradition. Econ Bot 25(2):182-187.

Arisaema acuminatum Small = Arisaema triphyllum (L.) Schott

Arisaema atrorubens (Aiton) Blume = Arisaema triphyllum (L.) Schott

Arisaema pusillum (Peck) Nash = Arisaema triphyllum (L.) Schott

Arisaema quinatum (Nutt.) Schott = Arisaema triphyllum (L.) Schott

Arisaema stewardsonii Britton = Arisaema triphyllum (L.) Schott

ARISAEMA TRIPHYLLUM (L.) Schott [Araceae]

Synonyms:

Arisaema acuminatum Small; ***Arisaema atrorubens*** (Aiton) Blume; ***Arisaema pusillum*** (Peck) Nash; ***Arisaema quinatum*** Nutt.; ***Arisaema stewardsonii*** Britton; ***Arum atrorubens*** Aiton; ***Arum triphyllum*** L.

Common Names:

bog onion; brown dragon; cuckoo plant; devil's-ear; dragon turnip; dragonroot; Indian jack-in-the-pulpit; Indian turnip; jack-in-the-pulpit; marsh turnip; meadow turnip; memory root; parson-in-the-pulpit; pepper turnip; priest's-pentle; small jack-in-the-pulpit; starchwort; wake robin; wild pepper

Citations:

Mrvos R, Krenzelok EP, Jacobsen TD (2001) Toxidromes associated with the most common plant ingestions. Vet Hum Toxicol 43(6):366-369.
Schilling R, Der Marderosian A, Speaker J (1980) Incidence of plant poisonings in Philadelphia noted as poison information calls. Vet Hum Toxicol 22(3):148-150.

aristolochia –see– *Aristolochia debilis* Siebold & Zucc.

Aristolochia bracteata Retz. = Aristolochia bracteolata Lam.

ARISTOLOCHIA BRACTEOLATA Lam. [Aristolochiaceae]

Synonyms:

Aristolochia bracteata Retz.

Common Names:

worm killer

Citations:

Barakat SE, Wasfi IA, Adam SE (1983) The toxicity of Aristolochia bracteata in goats. Vet Pathol 20(5):611-616.
El Dirdiri NI, Barakat SE, Adam SE (1987) The combined toxicity of Aristolochia bracteata and Cadaba rotundifolia to goats. Vet Hum Toxicol 29(2):133-137.

ARISTOLOCHIA DEBILIS Siebold & Zucc. [Aristolochiaceae]

Common Names:

aristolochia

Citations:

Levi M, Guchelaar HJ, Woerdenbag HJ, et al. (1998) Acute hepatitis in a patient using a Chinese herbal tea - A case report. Pharm World Sci 20(1):43-44.

ARISTOLOCHIA FANGCHI Y. C. Wu ex L. D. Chow & S. M. Hwang [Aristolochiaceae]

Common Names:

guang-fang-ji

Citations:

Schmeiser HH, Bieler CA, Wiessler M, et al. (1996) Detection of DNA adducts formed by aristolochic acid in renal tissue from patients with Chinese herbs nephropathy. Cancer Res 56(9):2025-2028.

Stengel B, Jones E (1998) Insuffisance rénale terminal associée à la consommation d'herbes chinoises en France. Nephrologie 19(1):15-20.

Vanherweghem LJ (1998) Misuse of herbal remedies: The case of an outbreak of terminal renal failure in Belgium. J Altern Complement Med 4(1):9-13.

Vanherweghem JL, Tielemans C, Simon J, et al. (1995) Chinese herbs nephropathy and renal pelvic carcinoma. Nephrol Dial Transplant 10(2):270-273.

ARISTOLOCHIA MOLLISSIMA Hance [Aristolochiaceae]

Citations:

Lo SH, Wong KS, Arlt VM, et al. (2005) Detection of herba aristolochia mollissemae in a patient with unexplained nephropathy. Am J Kidney Dis 45(2):407-410.

ARISTOLOCHIA PISTOLOCHIA L. [Aristolochiaceae]

Citations:

Pena JM, Borras M, Ramos J, et al. (1996) Rapidly progressive interstitial renal fibrosis due to a chronic intake of a herb (Aristolochia pistolochia) infusion. Nephrol Dial Transplant 11(7):1359-1360.

arka –see– *Calotropis gigantea* (L.) W. T. Aiton; *Calotropis procera* (Aiton) W. T. Aiton

armoire –see– *Artemisia herba-alba* Asso

armoire heche blanche –see– *Artemisia herba-alba* Asso

armoise –see– *Artemisia absinthium* L.

armoise commune –see– *Artemisia vulgaris* L.

Armoracia lapathifolia Gilib. ex Usteri = Armoracia rusticana P. Gaertn. et al.

ARMORACIA RUSTICANA P. Gaertn. et al. [Brassicaceae]

Synonyms:

Armoracia lapathifolia Gilib. ex Usteri; ***Cochlearia armoracea*** L.; ***Nasturtium armoracia*** (L.) Fr.; ***Radicula armoracia*** (L.) B. L. Rob.; ***Rorippa armoracia*** (L.) Hitchc.

Common Names:

Fleischkraut; horseradish; Kren; Mährrettig; Märek; Meerrettich; Pfefferwurzel; rábano magistro; rábano ragisco; rábano rusticano; rábano silvestre; raiz forte; red cole

Citations:

Hackett W (1917) Poisoning of cattle with horse-radish. J Comp Pathol Ther 30:138.

Rubin HR, Wu AW (1988) The bitter herbs of Seder: More on horseradish horrors. JAMA 259(13):1943.

Spitzer DE (1988) Horseradish horrors: Sushi syncope. JAMA 259(2):218-219.

arn tree –see– *Sambucus nigra* L.

arnica –see– *Arnica montana* L.

ARNICA LONGIFOLIA D. C. Eaton [Asteraceae]

Citations:

Hausen BM, Herrmann HD, Willuhn G (1978) The sensitizing capacity of compositae plants. I. Occupational contact dermatitis from Arnica longifolia Eaton. Contact Dermatitis 4(1):3-10.

ARNICA MONTANA L. [Asteraceae]

Common Names:

arnica; Arnika; Bergwohlverleih; Fallkraut; leopardbane; medicinal leopardbane; mountain alkanet; mountain tobacco; Wohlverleih; wolfbane

Citations:

Hausen BM (1978) Identification of the allergens of Arnica montana L. Contact Dermatitis 4(5):308.

Hausen BM (1979) The sensitizing capacity of compositae plants. III. Test results and cross-reactions in compositae-sensitive patients. Dermatologica 159(1):1-11.

Hausen BM (1980) Arnicaallergie. Hautarzt 31(1):10-17.

Hausen BM, Herrmann HD, Willuhn G (1978) The sensitizing capacity of compositae plants. I. Occupational contact dermatitis from Arnica longifolia Eaton. Contact Dermatitis 4(1):3-10.

Pirker C, Moslinger T, Koller DY, et al. (1992) Cross-reactivity with Tagetes in Arnica contact eczema. Contact Dermatitis 26(4):217-219.

Reider N, Komericki P, Hausen BM, et al. (2001) The seamy side of natural medicines: Contact sensitization to arnica (Arnica montana L.) and marigold (Calendula officinalis L.). Contact Dermatitis 45(5):269-272.

Rudzki E, Grzywa Z (1977) Dermatitis from Arnica montana. Contact Dermatitis 3(5):281-282.

Topliff A, Grande G (2000) Significant toxicity after the ingestion of arnica. J Toxicol Clin Toxicol 38(5):518.

ARNICA SACHALINENSIS (Regel) A. Gray [Asteraceae]

Citations:

Paßreiter CM, Florack M, Willuhn G, et al. (1988) Allergische Kontaktdermatitis auf Asteraceae. Identifizierung eines 8,9-epoxythymoldiesters als Kontakallergen von Arnica sachalinensis. Dermatosen 36(3):79-82.

Arnika –see– *Arnica montana* L.

aroeira –see– *Lithraea brasiliensis* Marchand; *Schinus terebinthifolius* Raddi

ARONIA spp. [Rosaceae]

Citations:

Lamminpaa A, Kinos M (1996) Plant poisonings in children. Hum Exp Toxicol 15(3):245-249.

Aronsbeer –see– *Arum maculatum* L.

Aronstab –see– *Arum maculatum* L.

Aronstaub –see– *Arum maculatum* L.

ARRABIDAEA BILABIATA (Sprague) Sandwith [Bignoniaceae]

Common Names:

chibata; gibata; xibata

Citations:

Döbereiner J, Peixoto PV, Tokarnia CH (1984) Intoxicação experimental por Arrabidaea bilabiata (Bignoniaceae) em coelhos. Pesq Vet Bras 4(3):89-96.

Döbereiner J, Tokarnia CH, Silva MF (1983) Intoxicação por Arrabidaea bilabiata (Bignoniaceae) em bovinos na Região Amazônica do Brasil. Pesq Vet Bras 3(1):17-24.

Jabour FF, Seixas JN, Tokarnia CH, et al. (2006) Variação da toxidez de Arrabidaea bilabiata (Bignoniaceae) em coelhos. Pesq Vet Bras 26(3):171-176.

ARRABIDAEA ELEGANS (Vell.) A. H. Gentry [Bignoniaceae]

Synonyms:

Pseudocalymma elegans (Vell.) Kuhlm.

Citations:

Tavares MI, Rezende AM, Döbereiner J (1974) Intoxicação experimental por Pseudocalymma elegans em coelhos e cobaios. Pesq Agric Bras Vet 9(7):91-94.

Tokarnia CH, Döbereiner J, Canella CF, et al. (1969) Intoxicação experimental por Pseudocalymma elegans (Vell.) Kuhlm. em bovinos. Pesq Agric Bras 4:195-204.

ARRABIDAEA JAPURENSIS (A. DC.) Bureau & K. Schum. [Bignoniaceae]

Citations:

Döbereiner J, Tokarnia CH (1983) Intoxicação experimental por Arrabidaea japurensis (Bignoniaceae) em coelhos. Pesq Vet Bras 3(3):95-97.

Tokarnia CH, Döbereiner J (1981) Intoxicação por Arrabidaea japurensis (Bignoniaceae) em bovinos em Roraima. Pesq Vet Bras 1(1):7-17.

arrebenta boi –see– *Solanum aculeatissimum* Jacq.

arrow crotalaria –see– *Crotalaria sagittalis* L.

arrow poison bush –see– *Acokanthera oppositifolia* (Lam.) Codd

arrowgrass –see– *Triglochin maritima* L.

arrowhead vine –see– *Syngonium podophyllum* Schott

arrowroot –see– *Arum maculatum* L.

arrowroot-de-Floride –see– *Zamia integrifolia* L. f.

arruda –see– *Ruta graveolens* L.

arsenic bush –see– *Senna ×floribunda* (Cav.) H. S. Irwin & Barneby

ARTEMISIA ABSINTHIUM L. [Asteraceae]

Common Names:

absinthe; absinthium; ajenjo; armoise; madderwort; Magenkraut; mingwort; mugwort; old woman; warmot; Wermut; wermuth; wormwood

Citations:

Hamouda C, Amamou M, Thabet H, et al. (2000) Plant poisonings from herbal medication admitted to a Tunisian toxicologic intensive care unit, 1983-1998. Vet Hum Toxicol 42(3):137-141.

Lundh K, Hindsén M, Gruvberger B, et al. (2006) Contact allergy to herbal teas derived from Asteraceae plants. Contact Dermatitis 54(4):196-201.

Palmer M, Haller C, McKinney P, et al. (1999) Botanicals and other dietary supplements: Adverse events by age. J Toxicol Clin Toxicol 37(5):609.

Weisbord SD, Soule JB, Kimmel PL (1997) Poison on line - Acute renal failure caused by oil of wormwood purchased through the Internet. N Engl J Med 337(12):825-827.

Artemisia arbuscula Nutt. subsp. nova (A. Nelson) G. H. Ward = Artemisia nova A. Nelson

ARTEMISIA FILIFOLIA Torr. [Asteraceae]

Common Names:

old-man sagebrush; sand sagebrush; sandsage

Citations:

Mathews FP (1941) Poisonous plants in the Davis Mountains. Texas Agric Exp Sta Annu Rep 54:93.

ARTEMISIA HERBA-ALBA Asso [Asteraceae]

Common Names:

armoire; armoire heche blanche; sheeh

Citations:

Al-Khazraji SM, Al-Shamaony LA, Twaij HA (1993) Hypoglycaemic effect of Artemisia herba alba. I. Effect of different parts and influence of the solvent on hypoglycaemic activity. J Ethnopharmacol 40:163-166.

Hamouda C, Amamou M, Thabet H, et al. (2000) Plant poisonings from herbal medication admitted to a Tunisian toxicologic intensive care unit, 1983-1998. Vet Hum Toxicol 42(3):137-141.

Ibrahim IA, El Badwi SM, Gadir WS, et al. (2004) Susceptibility of Bovans chicks to low dietary levels of Ammi visnaga and Artemisia herba-alba. Vet Hum Toxicol 46(2):67-69.

Artemisia incompta Nutt. = Artemisia ludoviciana Nutt.

ARTEMISIA LUDOVICIANA Nutt. [Asteraceae]

Synonyms:

Artemisia incompta Nutt.; ***Artemisia ludoviciana*** Nutt. var. mexicana (Willd. ex Spreng.) D. D. Keck; ***Artemisia mexicana*** Willd. ex Spreng.

Common Names:

estafiate; Louisiana wormwood; prairie sage; romerillo; sagewort; western mugwort; white sage

Citations:

Mitchell JC, Geissman TA, Dupuis C, et al. (1971) Allergic contact dermatitis caused by Artemisia and Chrysanthemum species. The role of sesquiterpene lactones. J Invest Dermatol 56(2):98-101.

Artemisia ludoviciana Nutt. subsp. mexicana (Willd. ex Spreng.) D. D. Keck = Artemisia ludoviciana Nutt.

Artemisia mexicana Willd. ex Spreng. = Artemisia ludoviciana Nutt.

ARTEMISIA NOVA A. Nelson [Asteraceae]

Synonyms:

Artemisia arbuscula Nutt. subsp. nova (A. Nelson) G. H. Ward; ***Artemisia tridentata*** Nutt. subsp. nova (A. Nelson) H. M. Hall & Clem.

Common Names:

black sagebrush

Citations:

Johnson AE (1978) Tetradymia toxicity - A new look at an old problem. In: Keeler RF, Van Kampen KR, James LF, et al. (eds.) Effects of poisonous plants on livestock. Academic Press. New York. pp. 209-216.

ARTEMISIA PONTICA L. [Asteraceae]

Common Names:

Roman wormwood

Citations:

Berlin R, Smilkstein M (1996) Wormwood oil@toxic.ing. J Toxicol Clin Toxicol 34:583.

ARTEMISIA TAURICA Willd. [Asteraceae]

Citations:

Gadzhiev KS (1989) [Poisoning of lambs with Artemisia taurica in Azerbaidzhan.] Veterinariia Moscow 6:13-14.

ARTEMISIA TRIDENTATA Nutt. [Asteraceae]

Synonyms:

Seriphidium tridentatum (Nutt.) W. A. Weber

Common Names:

big sagebrush; chamiso hediondo; sagebrush

Citations:

Johnson AE, James LF, Spillett J (1976) The abortifacient and toxic effects of big sagebrush (Artemisia tridentata) and juniper (Juniperus osteosperma) on domestic sheep. J Range Manag 29(4):278-280.

Artemisia tridentata Nutt. subsp. nova (A. Nelson) H. M. Hall & Clem. = Artemisia nova A. Nelson

ARTEMISIA VULGARIS L. [Asteraceae]

Common Names:

altamisa; armoise commune; Beifuß; California mugwort; Fliegenkraut; Flöhkraut; mugwort; tithwan; wormwood

Citations:

Hausen BM (1979) The sensitizing capacity of compositae plants. III. Test results and cross-reactions in compositae-sensitive patients. Dermatologica 159(1):1-11.
Hausen BM, Schulz KH (1978) Polyvalente Kontaktallergie bei einer Floristin. Derm Beruf Umwelt 26(5):175-176.
Jimeno L, Duffort O, Serrano C, et al. (2004) Monoclonal antibody-based ELISA to quantify the major allergen of Artemisia vulgaris pollen, Art v 1. Allergy 59(9):995-1001.
Kurz G, Rapaport MJ (1979) External/internal allergy to plants (Artemisia). Contact Dermatitis 5(6):407-408.

artichauts –see– *Cynara cardunculus* L.

artichoke –see– *Cynara cardunculus* L.

Artischocke –see– *Cynara cardunculus* L.

arum –see– *Dieffenbachia seguine* (Jacq.) Schott

Arum atrorubens Aiton = Arisaema triphyllum (L.) Schott

arum lily –see– *Zantedeschia aethiopica* (L.) Spreng.

ARUM MACULATUM L. [Araceae]

Common Names:

Aaron; Adam-and-Eve; adder's-meat; adder's-root; Aronsbeer; Aronstab; Aronstaub; arrowroot; bloody-man's-finger; bobbing Joan; bobbins; bulls-and-cows; calf's-foot; cuckoopint; friar's-cowl; Gefleckter Aronstab; kings-and-queens; lamb-in-a-pulpit; lilygrass; lords-and-ladies; mandrake; nightingales; pied-de-veau; Portland sago; Portland starchroot; spotted arum; starchwort; wake robin; wild arum; Zehrwurz

Citations:

Dabija G, Domilescu C, Nemteanu S (1968) Über die Giftigkeit des Aronstabs (Arum maculatum) für Tiere. Arch Vet 4(1-2):157-168.
Kanngiesser F (1916) Ueber die Giftigkeit der Aronsbeeren (Arum maculatum). Zeit Med 29:595-597.
Kyle RA (1983) Poisoning of sheep by lords and ladies. Vet Rec 113(1):23.
O'Moore LB (1955) Arum maculatum poisoning in cattle. Ir Vet J 9:146-147.

Stahl E, Kaltenbach U (1965) Die basischen Inhaltsstoffe des Aronstabes (Arum maculatum L.). Arch Pharm Ber Dtsch Pharm Ges 298(9):599-604.

Wolfle J, Kowalewski S (1995) Epidemiology of ingestions in a regional poison control center over twenty years. Vet Hum Toxicol 37(4):367-368.

arum-of-the-Antilles –see– *Dieffenbachia seguine* (Jacq.) Schott

Arum triphyllum L. = Arisaema triphyllum (L.) Schott

ARUNDO DONAX L. [Poaceae]

Common Names:

cane reed

Citations:

Friedman SJ, Connolly SM (1986) Clarinettist's cheilitis. Cutis 38(3):183-184.

Inoue A, Shoji A, Yashiro K (1998) Saxophonist's cane reed cheilitis. Contact Dermatitis 39(1):37.

Ruiz-Hornillos FJ, Alonso E, Zapatero L, et al. (2007) Clarinetist's cheilitis caused by immediate-type allergy to cane reed. Contact Dermatitis 56(11):243-245.

arvi –see– *Colocasia esculenta* (L.) Schott

Asaemia axillaris (Thunb.) Harv. ex O. Hoffm. = Asaemia minuta (L. f.) K. Bremer

ASAEMIA MINUTA (L. f.) K. Bremer [Asteraceae]

Synonyms:

Asaemia axillaris (Thunb.) Harv. ex O. Hoffm.

Common Names:

vuursicktebos

Citations:

Coetzer JA, Bergh T (1983) Photosensitivity in South Africa. IV. Pathological changes in the liver in ovine photosensitivity caused by the plant Asaemia axillaris (Thunb.) Harv. ex Jackson. Onderstepoort J Vet Res 50(1):55-58.

Kellerman TS, Basson PA, Naudé TW, et al. (1973) Photosensitivity in South Africa. 1. A comparative study of Asaemia axillaris (Thunb.) Harv. ex Jackson and Lasiospermum bipinnatum (Thunb.) Druce poisoning in sheep. Onderstepoort J Vet Res 40(3):115-126.

Van der Walt SJ, Steyn DG (1939) Recent investigations into the toxicity of known and unknown poisonous plants in the Union of South Africa. IX. Onderstepoort J Vet Sci Anim Indus 12:335-366.

asafetida –see– *Ferula assa-foetida* L.

asafoetida –see– *Ferula assa-foetida* L.

asarabacca –see– *Asarum europaeum* L.

ASARUM EUROPAEUM L. [Aristolochiaceae]

Common Names:

asarabacca; Brechwurz; cabaret; Europäische Haselwurz; Hafelkraut; Hafelmünch; Hafelührlein; Haselwurz; Hasenöhrlein; hazelwort; hazelwurz; Lederkraut; oreille d'homme; Rebelwurz; Schwarzkraut; Schwarzwurz; wild nard

Citations:

Brändle W, Gurtner B, Wegmann T (1969) Hemiparese bei Abortversuch mit Haselwurtzteeabsud (Asarum europaeum). Praxis 58(27):868-869.

Jaspersen-Schib R, Theus L, Quirguis-Oeschger M, et al. (1996) Wichtige Pflanzenvergiftungen in der Schweiz 1966-1994. Schweiz Med Wochenschr 126(25):1085-1098.

asbos –see– *Psilocaulon absimile* N. E. Br.

Asclepias asperula (Decne.) Woodson subsp. capricornu (Woodson) Woodson = Asclepias asperula (Decne.) Woodson var. decumbens (Nutt.) Shinners

ASCLEPIAS ASPERULA (Decne.) Woodson var. decumbens (Nutt.) Shinners [Asclepiadaceae]

Synonyms:

Asclepias asperula (Decne.) Woodson subsp. capricornu (Woodson) Woodson; ***Asclepiodora decumbens*** (Nutt.) A. Gray

Common Names:

spider antelopehorn

Citations:

Anonymous (1935) Investigation of stock-poisoning plants. U S Dep Agric Bur Anim Indus Chief Rep 1935:39-40.

ASCLEPIAS BRACHYSTEPHANA Engelm. [Asclepiadaceae]

Common Names:

short-crown milkweed

Citations:

Rowe LD, Ohlenbusch PD, Dollahite JW (1970) Toxicity of Asclepias brachystephana (shortcrown milkweed) for sheep. Southwestern Vet 23(Spring):219-221, 224.

Asclepias cornuti Decne. = Asclepias syriaca L.

ASCLEPIAS CURASSAVICA L. [Asclepiadaceae]

Common Names:

blood flower; Kittie McWanie; milkweed; oficial-de-sala; red cotton; redhead cottonbush; redtop; scarlet milkweed; silkweed

Citations:

Chakraborty S, Siegenthaler J, Buchi ER (1995) Corneal edema due to Asclepias curassavica. Arch Ophthalmol 113(8):974-975.

Tokarnia CH, Brito MF, Cunha BR (2001) Intoxicação experimental por Asclepias curassavica (Asclepiadaceae) em bovinos. Dados complementares. Pesq Vet Bras 21(1):1-4.

Tokarnia CH, Döbereiner J, Canella CF (1972) Intoxicação experimental em bovinos por Asclepias curassavica. Pesq Agric Bras Vet 7:31-39.

ASCLEPIAS ERIOCARPA Benth. [Asclepiadaceae]

Common Names:

broad-leaf milkweed; California milkweed; Indian milkweed; woolly-pod milkweed

Citations:

Baxter CM (1944) Broadleaf milkweed poisoning. Cornell Vet 34(3):256-259.

Benson JM, Seiber JN, Bagley CV, et al. (1979) Effects on sheep of the milkweeds Asclepias eriocarpa and A. labriformis and of cardiac glycoside-containing derivative material. Toxicon 17(2):155-165.

Benson JM, Seiber JN, Keeler RF, et al. (1978) Studies on the toxic principle of Asclepias eriocarpa and Asclepias labriformis. In: Keeler RF, Van Kampen KR, James LF, et al. (eds.) Effects of poisonous plants on livestock. Academic Press. New York. pp. 273-284.

Marsh CD, Clawson AB (1923) Woolly-pod milkweed: A dangerous stock-poisoning plant. U S Dep Agric Circ #272:3 pp.

Marsh CD, Clawson AB (1924) The woolly-pod milkweed (Asclepias eriocarpa) as a poisonous plant. U S Dep Agric Bull #1212:13 pp.

Vail EL (1942) Woolly-pod or broad-leafed milkweed (Asclepias eriocarpa) poisoning of rabbits. North Am Vet 23(Aug):539-542.

Asclepias florida N. E. Br. = Gomphocarpus fruticosa (L.) W. T. Aiton

Asclepias fruticosa L. = Gomphocarpus fruticosa (L.) W. T. Aiton

Asclepias galioides auct. Amer. = Asclepias subverticillata (A. Gray) Vail

ASCLEPIAS HIRTELLA (Pennell) Woodson [Asclepiadaceae]

Citations:

Burrows GE, Ogden L, Tyrl RJ (1990) Intoxication due to Asclepias sp. (milkweeds). Toxicon 28(6):603-604.

Ogden L, Burrows GE, Tyrl RJ, et al. (1992) Experimental intoxication in sheep by Asclepias. In: James LF, Keeler RF, Bailey JM Jr, et al. (eds.) Poisonous plants. Iowa State Univ Press. Ames, Iowa. pp. 495-499.

ASCLEPIAS INCARNATA L. [Asclepiadaceae]

Common Names:

rose silkweed; swamp milkweed; water nerveroot; white Indian hemp

Citations:

Burrows GE, Ogden L, Tyrl RJ (1990) Intoxication due to Asclepias sp. (milkweeds). Toxicon 28(6):603-604.

Ogden L, Burrows GE, Tyrl RJ, et al. (1992) Experimental intoxication in sheep by Asclepias. In: James LF, Keeler RF, Bailey JM Jr, et al. (eds.) Poisonous plants. Iowa State Univ Press. Ames, Iowa. pp. 495-499.

ASCLEPIAS LABRIFORMIS M. E. Jones [Asclepiadaceae]

Common Names:

labriform milkweed

Citations:

Anonymous (1936) Investigations of stock-poisoning plants. U S Dep Agric Bur Anim Indus Chief Rep. 1936:44-45.

Anonymous (1939) Investigations of stock-poisoning plants. U S Dep Agric Bur Anim Indus Chief Rep. 1939:59-60.

Benson JM, Seiber JN, Bagley CV, et al. (1979) Effects on sheep of the milkweeds Asclepias eriocarpa and A. labriformis and of cardiac glycoside-containing derivative material. Toxicon 17(2):155-165.

Benson JM, Seiber JN, Keeler RF, et al. (1978) Studies on the toxic principle of Asclepias eriocarpa and Asclepias labriformis. In: Keeler RF, Van Kampen KR, James LF, et al. (eds.) Effects of poisonous plants on livestock. Academic Press. New York. pp. 273-284.

Burrows GE, Ogden L, Tyrl RJ (1990) Intoxication due to Asclepias sp. (milkweeds). Toxicon 28(6):603-604.

Ogden L, Burrows GE, Tyrl RJ, et al. (1992) Comparison of Asclepias species based on their toxic effects on chickens. In: James LF, Keeler RF, Bailey JM Jr, et al. (eds.) Poisonous plants. Iowa State Univ Press. Ames, Iowa. pp. 500-505.

Ogden L, Burrows GE, Tyrl RJ, et al. (1992) Experimental intoxication in sheep by Asclepias. In: James LF, Keeler RF, Bailey JM Jr, et al. (eds.) Poisonous plants. Iowa State Univ Press. Ames, Iowa. pp. 495-499.

ASCLEPIAS LATIFOLIA (Torr.) Raf. [Asclepiadaceae]

Common Names:

broad-leaf milkweed; le chones; milkweed

Citations:

Burrows GE, Ogden L, Tyrl RJ (1990) Intoxication due to Asclepias sp. (milkweeds). Toxicon 28(6):603-604.

Mathews FP (1932) Miscellaneous poisonous plants. Texas Agric Exp Sta Annu Rep 45:11-12.

Ogden L, Burrows GE, Tyrl RJ, et al. (1992) Experimental intoxication in sheep by Asclepias. In: James LF, Keeler RF, Bailey JM Jr, et al. (eds.) Poisonous plants. Iowa State Univ Press. Ames, Iowa. pp. 495-499.

Tunnicliff EA, Cory VL (1930) Broad-leafed milkweed (Asclepias latifolia) poisonous for sheep and goats. J Am Vet Med Assoc 77:165-168.

ASCLEPIAS MEXICANA Cav. [Asclepiadaceae]

Common Names:

Mexican whorled milkweed; narrow-leaf whorled milkweed

Citations:

Campbell HW (1931) Poisoning in chickens with whorled milkweed. J Am Vet Med Assoc 79:102-104.

Campbell HW (1931) The whorled milkweed as a poisonous plant for poultry. California Dep Agric Bull 20:577-582.

Fleming CE (1920) Poisonous range plants. Nevada Agric Exp Sta Annu Rep 1919:39-43.

Fleming CE, Peterson NF, Miller MR, et al. (1920) The narrow-leaved milkweed (Asclepias mexicana) and the broad-leaved or showy milkweed (Asclepias speciosa), Plants poisonous to livestock in Nevada. Nevada Agric Exp Sta Bull #99:32 pp.

Marsh CD, Clawson AB (1921) The Mexican whorled milkweed (Asclepias mexicana) as a poisonous plant. U S Dep Agric Bull #969:16 pp.

Asclepias phillipsiae N. E. Br. = Gomphocarpus fruticosa (L.) W. T. Aiton

ASCLEPIAS PUMILA (A. Gray) Vail [Asclepiadaceae]

Common Names:

dwarf poison milkweed; Great Plains whorled milkweed; low whorled milkweed; plains whorled milkweed; whorled milkweed

Citations:

Marsh CD, Clawson AB (1921) Poisonous properties of the whorled milkweeds Asclepias pumila and A. verticillata var. geyeri. U S Dep Agric Bull #942:14 pp.

ASCLEPIAS SPECIOSA Torr. [Asclepiadaceae]

Common Names:

broad-leaf milkweed; showy milkweed

Citations:

Burrows GE, Ogden L, Tyrl RJ (1990) Intoxication due to Asclepias sp. (milkweeds). Toxicon 28(6):603-604.

Fleming CE, Peterson NF, Miller MR, et al. (1920) The narrow-leaved milkweed (Asclepias mexicana) and the broad-leaved or showy milkweed (Asclepias speciosa), Plants poisonous to livestock in Nevada. Nevada Agric Exp Sta Bull #99:32 pp.

Ogden L, Burrows GE, Tyrl RJ, et al. (1992) Comparison of Asclepias species based on their toxic effects on chickens. In: James LF, Keeler RF, Bailey JM Jr, et al. (eds.) Poisonous plants. Iowa State Univ Press. Ames, Iowa. pp. 500-505.

Ogden L, Burrows GE, Tyrl RJ, et al. (1992) Experimental intoxication in sheep by Asclepias. In: James LF, Keeler RF, Bailey JM Jr, et al. (eds.) Poisonous plants. Iowa State Univ Press. Ames, Iowa. pp. 495-499.

ASCLEPIAS SUBVERTICILLATA (A. Gray) Vail [Asclepiadaceae]

Synonyms:

Asclepias galioides auct. Amer.

Common Names:

bedstraw milkweed; beeweed; horsetail milkweed; narrow-leaf milkweed; western whorled milkweed; whorled milkweed

Citations:

Burrows GE, Ogden L, Tyrl RJ (1990) Intoxication due to Asclepias sp. (milkweeds). Toxicon 28(6):603-604.

Clark JG (1979) Whorled milkweed poisoning. Vet Hum Toxicol 21(6):431.

Marsh CD, Clawson AB, Couch JF, et al. (1920) The whorled milkweed (Asclepias galioides) as a poisonous plant. U S Dep Agric Bull #800:40 pp.

May WL (1920) Whorled milkweed. The worst stock-poisoning plant in Colorado. Colorado Agric Exp Sta Bull #255:39 pp.

Ogden L, Burrows GE, Tyrl RJ, et al. (1992) Comparison of Asclepias species based on their toxic effects on chickens. In: James LF, Keeler RF, Bailey JM Jr, et al. (eds.) Poisonous plants. Iowa State Univ Press. Ames, Iowa. pp. 500-505.

Ogden L, Burrows GE, Tyrl RJ, et al. (1992) Experimental intoxication in sheep by Asclepias. In: James LF, Keeler RF, Bailey JM Jr, et al. (eds.) Poisonous plants. Iowa State Univ Press. Ames, Iowa. pp. 495-499.

Robinson GH, Burrows GE, Holt EM, et al. (1998) Investigation of the neurotoxic compounds in Asclepias subverticillata, western-whorled milkweed. In: Garland T, Barr AC (eds.) Toxic plants and other natural toxicants. CABI. New York. pp. 435-439.

Stiles GW (1942) Poisoning of turkey poults from whorled milkweed (Asclepias galioides). Poult Sci 21(May):263-270.

ASCLEPIAS SYRIACA L. [Asclepiadaceae]

Synonyms:

Asclepias cornuti Decne.

Common Names:

selyemkóró; silkweed; summer milkweed

Citations:

Sályi G, Petri Á (1987) A szarvasmarhák selyemkóró - (Asclepias syriaca) mérgesése. Magyar Allator Lapja 42(1):56-58.

ASCLEPIAS TUBEROSA L. [Asclepiadaceae]

Common Names:

butterfly milkweed; butterfly weed; Canadian root; Indian nosy; orange apocynum; orange milkweed; orange swallowwort; orangeroot; pleurisy root; rubberroot; silkweed; tuberroot; whiteroot; windroot; yellow milkweed

Citations:

Burrows GE, Ogden L, Tyrl RJ (1990) Intoxication due to Asclepias sp. (milkweeds). Toxicon 28(6):603-604.

Ogden L, Burrows GE, Tyrl RJ, et al. (1992) Comparison of Asclepias species based on their toxic effects on chickens. In: James LF, Keeler RF, Bailey JM Jr, et al. (eds.) Poisonous plants. Iowa State Univ Press. Ames, Iowa. pp. 500-505.

Ogden L, Burrows GE, Tyrl RJ, et al. (1992) Experimental intoxication in sheep by Asclepias. In: James LF, Keeler RF, Bailey JM Jr, et al. (eds.) Poisonous plants. Iowa State Univ Press. Ames, Iowa. pp. 495-499.

ASCLEPIAS VERTICILLATA L. [Asclepiadaceae]

Common Names:

dwarf milkweed; eastern whorled milkweed; horsetail milkweed; spider milkweed; whorled milkweed

Citations:

Burrows GE, Ogden L, Tyrl RJ (1990) Intoxication due to Asclepias sp. (milkweeds). Toxicon 28(6):603-604.

Glover GH (1917) The whorled milkweed. Am J Vet Med 12:303-304, 308.

Glover GH, Newsom IE, Robbins WW (1918) A new poisonous plant, the whorled milkweed (Asclepias verticillata). Colorado Agric Exp Sta Bull #246:16 pp.

Marsh CD, Clawson AB (1921) Poisonous properties of the whorled milkweeds Asclepias pumila and A. verticillata var. geyeri. U S Dep Agric Bull #942:14 pp.

Ogden L, Burrows GE, Tyrl RJ, et al. (1992) Comparison of Asclepias species based on their toxic effects on chickens. In: James LF, Keeler RF, Bailey JM Jr, et al. (eds.) Poisonous plants. Iowa State Univ Press. Ames, Iowa. pp. 500-505.

Ogden L, Burrows GE, Tyrl RJ, et al. (1992) Experimental intoxication in sheep by Asclepias. In: James LF, Keeler RF, Bailey JM Jr, et al. (eds.) Poisonous plants. Iowa State Univ Press. Ames, Iowa. pp. 495-499.

Pammel LH (1919) The whorled milkweed. Am J Vet Med 14:135-136.

Sprowls RW (1982) Horsetail milkweed intoxications in horses and cattle. Southwestern Vet 35(1):15.

ASCLEPIAS VIRIDIS Walter [Asclepiadaceae]

Common Names:

green milkweed; Ozark milkweed; spider milkweed

Citations:

Burrows GE, Ogden L, Tyrl RJ (1990) Intoxication due to Asclepias sp. (milkweeds). Toxicon 28(6):603-604.

Ogden L, Burrows GE, Tyrl RJ, et al. (1992) Comparison of Asclepias species based on their toxic effects on chickens. In: James LF, Keeler RF, Bailey JM Jr, et al. (eds.) Poisonous plants. Iowa State Univ Press. Ames, Iowa. pp. 500-505.

Ogden L, Burrows GE, Tyrl RJ, et al. (1992) Experimental intoxication in sheep by Asclepias. In: James LF, Keeler RF, Bailey JM Jr, et al. (eds.) Poisonous plants. Iowa State Univ Press. Ames, Iowa. pp. 495-499.

Smith RA, Scharko P, Bolin D, et al. (2000) Intoxication of sheep exposed to Ozark milkweed (Asclepias viridis Walter). Vet Hum Toxicol 42(6):349-350.

Asclepiodora decumbens (Nutt.) A. Gray = Asclepias asperula (Decne.) Woodson var. decumbens (Nutt.) Shinners

ash tree –see– *Fraxinus excelsior* L.

Asian ginseng –see– *Panax ginseng* C. A. Mey.

Asiatische Alpenrose –see– *Rhododendron ponticum* L.

ASIMINA TRILOBA (L.) Dunal [Annonaceae]

Synonyms:

Annona triloba L.

Common Names:

American custard apple; American papaw; custard apple; dog apple; dunal; false banana; fetid shrub; monim; papaw; pawpaw; poor-man's-banana

Citations:

Barber MA (1905) Poisoning due to the papaw (Asimina triloba). JAMA 45(Dec 30):2013-2014.

Douglass BW (1918) A case of papaw poisoning and an idiosyncrasy. Indianapolis Med J 21:120.

Williams BG (1909) Case of dermatitis herpetiformis following papaw poisoning. JAMA 53(Dec 4):1916.

asin asin –see– *Sauropus androgynus* (L.) Merr.

asparagus –see– *Asparagus officinalis* L.

asparagus fern –see– *Asparagus officinalis* L.

ASPARAGUS OFFICINALIS L. [Asparagaceae]

Common Names:

asparagus; asparagus fern; Spargel

Citations:

Brenning (1920) Ein Fall von Dermatitis durch Spargelsaft. Dermatol Wochenschr 71:851.

Escribano MM, Muñoz-Bellido FJ, Serrano P, et al. (1998) Acute urticaria after ingestion of asparagus. Allergy 53(6):622-623.

Hajos B, Mohrmann BH (1929) Spargelidiosyncrasie. Klin Wochenschr 8(22):1024-1027.

aspidium –see– *Dryopteris filix-mas* (L.) Schott

ASPIDOSPERMA GOMEZIANUM A. DC. [Apocynaceae]

Common Names:

peroba-da-campos

Citations:

de Jong JC, Lenstra JB, Vermeer DJ (1951) Eczema due to the wood of Peroba da Campos: Isolation of the allergen. Acta Derm Venereol 31:108-110.

assacu –see– *Hura crepitans* L.

ass's-parsley –see– *Aethusa cynapium* L.

Aster simplex Willd. = Symphyotrichum lanceolatum (Willd.) G. L. Nesom

Aster umbellatus Mill. = Doellingeria umbellata (Mill.) Nees

asthma herb –see– *Chamaesyce hirta* (L.) Millsp.

Asthmador –see– *Atropa belladonna* L.; *Datura stramonium* L.

ASTRAGALUS ADSURGENS Pall. [Fabaceae]

Common Names:

standing milk vetch

Citations:

Zuyu L, Xueqin F, Zihua Z, et al. (1987) [The effects of feeding diets containing Astragalus adsurgens Pall and Coronilla varia L. on broiler chickens.] Acta Vet Zootech Sin 18(3):157-162.

Astragalus amphidoxus Blank. = Astragalus miser Douglas ex Hook.

Astragalus argillosus M. E. Jones = Astragalus flavus Nutt. ex Torr. & A. Gray

ASTRAGALUS ATROPUBESCENS J. M. Coult. & Fisch. [Fabaceae]

Common Names:

Kelsey's-milk vetch

Citations:

Cronin EH, Williams MC, Olsen JD (1981) Toxicity and control of Kelsey milkvetch. J Range Manag 34(3):181-183.

ASTRAGALUS BERGII Hieron. [Fabaceae]

Common Names:

garbancillo; yerba loca

Citations:

Giusti L (1934) Experiencias sobre la acción fisiológica de la Wedelia glauca y del Astragalus bergii. Rev Argent Agron 1:223-228.

Astragalus bigelovii A. Gray = Astragalus mollissimus Torr. var. thompsoniae (S. Watson) Barneby

ASTRAGALUS BISULCATUS (Hook.) A. Gray [Fabaceae]

Synonyms:

Astragalus bisulcatus (Hook.) A. Gray var. bisulcatus; ***Astragalus bisulcatus*** (Hook.) A. Gray var. haydenianus (A. Gray) Barneby; ***Astragalus diholcos*** Tidestr.; ***Astragalus haydenianus*** A. Gray; ***Diholcos bisulcatus*** (Hook.) Rydb.; ***Diholcos micranthus*** Rydb.

Common Names:

Hayden's-poison vetch; poison vetch; two-groove locoweed; two-groove milk vetch; two-groove poison vetch

Citations:

Baker DC, James LF, Hartley WJ, et al. (1989) Toxicosis in pigs fed selenium-accumulating Astragalus plant species or sodium selenate. Am J Vet Res 50(8):1396-1399.

Beath OA, Draize JH, Eppson HF (1932) Three poisonous vetches. Wyoming Agric Exp Sta Bull #189:23 pp.

Beath OA, Lehnert EH (1917) The poisonous properties of the two-grooved milk vetch (Astragalus bisulcatus). Wyoming Agric Exp Sta Bull #112:59-65.

Hartley WJ, James LF, Broquist H, et al. (1985) Pathology of experimental locoweed and selenium poisoning in pigs. In: Seawright AA, Hegarty MP, James LF, et al. (eds.) Plant toxicology. Queensland Poisonous Plant Committee. Yeerongpilly, Queensland. pp. 141-149.

James LF, Van Kampen KR, Hartley WJ (1983) Astragalus bisulcatus - A cause of selenium or locoweed poisoning? Vet Hum Toxicol 25(2):86-89.

Panter KE, Hartley WJ, James LF, et al. (1996) Comparative toxicity of selenium from selano-DL-ethionine, sodium selenate, and Astragalus bisulcatus in pigs. Fundam Appl Toxicol 32:217-223.

Van Kampen KR, James LF (1978) Manifestation of intoxication by selenium-accumulating plants. In: Keeler RF, Van Kampen KR, James LF, et al. (eds.) Effects of poisonous plants on livestock. Academic Press. New York. pp. 135-138.

Astragalus bisulcatus (Hook.) A. Gray var. bisulcatus = Astragalus bisulcatus (Hook.) A. Gray

Astragalus bisulcatus (Hook.) A. Gray var. haydenianus (A. Gray) Barneby = Astragalus bisulcatus (Hook.) A. Gray

Astragalus campestris (Torr. & A. Gray) A. Gray = Astragalus convallarius Greene

Astragalus campestris L. var. diversifolius (A. Gray) J. F. Macbr. = Astragalus diversifolius A. Gray

ASTRAGALUS CANADENSIS L. [Fabaceae]

Synonyms:

Astragalus carolinianus L.; ***Phaca canadensis*** (L.) MacMill.

Common Names:

Canadian milk vetch

Citations:

James LF, Hartley WJ, Williams MC, et al. (1980) Field and experimental studies in cattle and sheep poisoned by nitro-bearing Astragalus or their toxins. Am J Vet Res 41(3):377-382.

Williams MC, James LF (1975) Toxicity of nitro-containing Astragalus to sheep and chicks. J Range Manag 28(4):260-263.

ASTRAGALUS CANADENSIS L. var. brevidens (Gand.) Barneby [Fabaceae]

Citations:

Williams MC, James LF (1974) Toxicity of several nitrite-bearing Astragalus species. Proc West Soc Weed Sci 27:4.

Astragalus carolinianus L. = Astragalus canadensis L.

ASTRAGALUS CIBARIUS E. Sheld. [Fabaceae]

Common Names:

browse milk vetch

Citations:

Williams MC, James LF (1974) Toxicity of several nitrite-bearing Astragalus species. Proc West Soc Weed Sci 27:4.

Williams MC, James LF (1975) Toxicity of nitro-containing Astragalus to sheep and chicks. J Range Manag 28(4):260-263.

ASTRAGALUS CICER L. [Fabaceae]

Common Names:

cicer milk vetch

Citations:

Marten GC, Ehle FR, Ristau EA (1987) Performance and photosensitization of cattle related to forage quality of four legumes. Crop Sci 27(Jan-Feb):138-145.

Marten GC, Jordan RM, Ristau EA (1990) Performance and adverse response of sheep during grazing of four legumes. Crop Sci 30(Jul-Aug):860-866.

Walter-Hansen K, Ruth G, Schwartz S, et al. (1986) Photosensitivity in calves and sheep fed cicer milkvetch (Astragalus cicer). Proc Am Assoc Vet Lab Diagn 29:209-220.

Astragalus confertiflorus A. Gray = Astragalus flavus Nutt. ex Torr. & A. Gray

ASTRAGALUS CONVALLARIUS Greene [Fabaceae]

Synonyms:

Astragalus campestris (Torr. & A. Gray) A. Gray; ***Homalobus campestris*** Torr. & A. Gray

Common Names:

lesser rushy milk vetch; timber milk vetch; timber poison vetch

Citations:

Beath OA, Draize JH, Eppson HF (1932) Three poisonous vetches. Wyoming Agric Exp Sta Bull #189:23 pp.

Williams MC, James LF (1974) Toxicity of several nitrite-bearing Astragalus species. Proc West Soc Weed Sci 27:4.

Williams MC, James LF (1975) Toxicity of nitro-containing Astragalus to sheep and chicks. J Range Manag 28(4):260-263.

Astragalus decumbens (Nutt. ex Torr. & A. Gray) A. Gray var. serotinus (J. G. Cooper) M. E. Jones = Astragalus miser Douglas ex Hook. var. serotinus (A. Gray ex J. G. Cooper) Barneby

Astragalus diholcos Tidestr. = Astragalus bisulcatus (Hook.) A. Gray

Astragalus diphysus A. Gray = Astragalus lentiginosus Douglas ex Hook. var. diphysus (A. Gray) M. E. Jones

ASTRAGALUS DIVERSIFOLIUS A. Gray [Fabaceae]

Synonyms:

Astragalus campestris L. var. diversifolius (A. Gray) J. F. Macbr.

Common Names:

mesic milk vetch

Citations:

Williams MC, James LF (1974) Toxicity of several nitrite-bearing Astragalus species. Proc West Soc Weed Sci 27:4.

Williams MC, James LF (1975) Toxicity of nitro-containing Astragalus to sheep and chicks. J Range Manag 28(4):260-263.

Astragalus earlei Greene ex Rydb. = Astragalus mollissimus Torr. var. earlei (Greene ex Rydb.) Tidestr.

ASTRAGALUS EMORYANUS (Rydb.) Cory [Fabaceae]

Synonyms:

Astragalus emoryanus (Rydb.) Cory var. emoryanus; ***Astragalus emoryanus*** (Rydb.) Cory var. terlinguensis (Cory) Barneby

Common Names:

Emory's-locoweed; Emory's-milk vetch; peavine; red-stem peavine

Citations:

James LF, Hartley WJ, Williams MC, et al. (1980) Field and experimental studies in cattle and sheep poisoned by nitro-bearing Astragalus or their toxins. Am J Vet Res 41(3):377-382.

Mathews FP (1938) Poisonous plants in the Davis Mountains area. Texas Agric Exp Sta Annu Rep 51:13-14.

Mathews FP (1940) The toxicity of red-stemmed peavine (=Astragalus emoryanus) for cattle, sheep and goats. J Am Vet Med Assoc 97:125-134.

Mathews FP (1941) Poisonous plants in the Davis Mountains. Texas Agric Exp Sta Annu Rep 54:93.

Williams MC, James LF (1975) Toxicity of nitro-containing Astragalus to sheep and chicks. J Range Manag 28(4):260-263.

Williams MC, James LF (1976) Poisoning in sheep from emory milkvetch and nitro compounds. J Range Manag 29(2):165-167.

Williams MC, James LF, Bond BO (1979) Emory milkvetch (Astragalus emoryanus var emoryanus) poisoning in chicks, sheep, and cattle. Am J Vet Res 40(3):403-406.

Astragalus emoryanus (Rydb.) Cory var. emoryanus = Astragalus emoryanus (Rydb.) Cory

Astragalus emoryanus (Rydb.) Cory var. terlinguensis (Cory) Barneby = Astragalus emoryanus (Rydb.) Cory

ASTRAGALUS FALCATUS Lam. [Fabaceae]

Common Names:

Russian-sickle milk vetch; sicklepod milk vetch

Citations:

James LF, Hartley WJ, Williams MC, et al. (1980) Field and experimental studies in cattle and sheep poisoned by nitro-bearing Astragalus or their toxins. Am J Vet Res 41(3):377-382.

Williams MC, James LF (1975) Toxicity of nitro-containing Astragalus to sheep and chicks. J Range Manag 28(4):260-263.

Williams MC, James LF, Bleak AT (1976) Toxicity of introduced nitro-containing Astragalus to sheep, cattle, and chicks. J Range Manag 29(1):30-33.

ASTRAGALUS FLAVUS Nutt. ex Torr. & A. Gray [Fabaceae]

Synonyms:

Astragalus argillosus M. E. Jones; ***Astragalus confertiflorus*** A. Gray

Common Names:

yellow milk vetch

Citations:

Anonymous (1935) Investigation of stock-poisoning plants. U S Dep Agric Bur Anim Indus Chief Rep 1935:39-40.

Astragalus glaber Michx. = Astragalus michauxii (Kuntze) F. J. Herm.

Astragalus haydenianus A. Gray = Astragalus bisulcatus (Hook.) A. Gray

ASTRAGALUS HORNII A. Gray [Fabaceae]

Common Names:

Horn's-milk vetch

Citations:

Vasey G (1874) Plants poisonous to cattle in California. U S Dep Agric Rep 1874:159-160.

Astragalus hylophilus (Rydb.) A. Nelson = Astragalus miser Douglas ex Hook. var. hylophilus (Rydb.) Barneby

ASTRAGALUS LENTIGINOSUS Douglas ex Hook. [Fabaceae]

Common Names:

blue locoweed; freckled milk vetch; speckled locoweed; spotted locoweed

Citations:

Balls LD, James LF (1973) Effect of locoweed (Astragalus spp.) on reproductive performance of ewes. J Am Vet Med Assoc 162(4):291-292.

Daniel PF, Warren CD, James LF, et al. (1985) Characterization of oligosaccharides from locoweed-poisoned sheep and comparison with bovine mannosidosis. In: Seawright AA, Hegarty MP, James LF, et al. (eds.) Plant toxicology. Queensland Poisonous Plant Committee. Yeerongpilly, Queensland. pp. 290-300.

Ellis LC, James LF, McMullen RW, et al. (1985) Reduced progesterone and altered cotyledonary prostaglandin values induced by locoweed (Astragalus lentiginosus) in sheep. Am J Vet Res 46(9):1903-1907.

Hartley WJ, James LF (1973) Microscopic lesions in fetuses of ewes ingesting locoweed (Astragalus lentiginosus). Am J Vet Res 34(2):209-211.

Hartley WJ, James LF (1975) Fetal and maternal lesions in pregnant ewes ingesting locoweed (Astragalus lentiginosus). Am J Vet Res 36(6):825-826.

Hartley WJ, James LF (1978) Summary of experimental Astragalus lentiginosus intoxication in the pregnant ewe. In: Keeler RF, Van Kampen KR, James LF, et al. (eds.) Effects of poisonous plants on livestock. Academic Press. New York. pp. 368-369.

Hartley WJ, James LF, Broquist H, et al. (1985) Pathology of experimental locoweed and selenium poisoning in pigs. In: Seawright AA, Hegarty MP, James LF, et al. (eds.) Plant toxicology. Queensland Poisonous Plant Committee. Yeerongpilly, Queensland. pp. 141-149.

James LF (1971) Lesions in neonatal lambs resulting from maternal ingestion of locoweed. Cornell Vet 61(4):667-670.

James LF (1972) Effect of locoweed on fetal development: Preliminary study in sheep. Am J Vet Res 33(4):835-840.

James LF (1976) Effect of locoweed (Astragalus lentiginosus) feeding on fetal lamb development. Can J Comp Med Vet Sci 40(4):380-384.

James LF, Foote W, Nye W, et al. (1978) Effects of feeding Oxytropis and Astragalus pollen to mice and Astragalus seeds to rats. Am J Vet Res 39(4):711-712.

James LF, Hartley WJ (1977) Effects of milk from animals fed locoweed on kittens, calves, and lambs. Am J Vet Res 38(8):1263-1265.

James LF, Olsen JD, Sharma RP (1977) Locoweed poisoning in sheep: Electroencephalographic and brain amine changes. Clin Toxicol 11(1):53-60.

James LF, Shupe JL, Binns W, et al. (1967) Abortive and teratogenic effects of locoweed on sheep and cattle. Am J Vet Res 28(126):1379-1388.

James LF, Van Kampen KR (1971) Effects of locoweed intoxication on the genital tract of the ram. Am J Vet Res 32(8):1253-1256.

James LF, Van Kampen KR (1976) Effects of locoweed toxin on rats. Am J Vet Res 37(7):845-846.

James LF, Van Kampen KR, Hartley WJ (1970) Comparative pathology of Astragalus (locoweed) and Swainsona poisoning in sheep. Pathol Vet 7(2):116-125.

James LF, Van Kampen KR, Johnson AE (1970) Physiopathologic changes in locoweed poisoning of livestock. Am J Vet Res 31(4):663-672.

James LF, Van Kampen KR, Staker GR (1969) Locoweed (Astragalus lentiginosus) poisoning in cattle and horses. J Am Vet Med Assoc 155(3):525-530.

Keeler RF, James LF (1971) Experimental teratogenic lathyrism in sheep and further comparative aspects with teratogenic locoism. Can J Comp Med Vet Sci 35(4):332-337.

Keeler RF, James LF (1971) Failure of dietary supplementation to prevent the abortions and congenital malformations of lathyrism and locoism in sheep. Can J Comp Med Vet Sci 35(4):342-345.

Keeler RF, James LF, Binns W, et al. (1967) An apparent relationship between locoism and lathyrism. Can J Comp Med Vet Sci 31(12):334-341.

Nelson BK, James LF, Sharma RP, et al. (1977) Subtle postnatal effects of locoweed in rats. Toxicol Appl Pharmacol 41(1):139-140.

Nelson BK, James LF, Sharma RP, et al. (1980) Locoweed embryotoxicity in rats. Clin Toxicol 16(2):149-166.

Panter KE, James LF, Hartley WJ (1989) Transient testicular degeneration in rams fed locoweed (Astragalus lentiginosus). Vet Hum Toxicol 31(1):42-46.

Panter KE, James LF, Nielson D, et al. (1988) The relationship of Oxytropis sericea (green and dry) and Astragalus lentiginosus with high mountain disease in cattle. Vet Hum Toxicol 30(4):318-323.

Pfister JA, Stegelmeier BL, Gardner DR, et al. (2003) Grazing of spotted locoweed (Astragalus lentiginosus) by cattle and horses in Arizona. J Anim Sci 81(9):2285-2293.

Rhees RW, James LF, Van Kampen KR (1978) Ultrastructural observations following locoweed (Astragalus lentiginosus) poisoning in sheep. Fed Proc 37(3):501.

Sharma RP, James LF, Molyneux RJ (1984) Effect of repeated locoweed feeding on peripheral lymphocytic function and plasma proteins in sheep. Am J Vet Res 45(10):2090-2093.

Stegelmeier BL, James LF, Panter KE, et al. (1995) Serum swainsonine concentration and α-mannosidase activity in cattle and sheep ingesting Oxytropis sericea and Astragalus lentiginosus (locoweeds). Am J Vet Res 56(2):149-154.

Stegelmeier BL, Ralphs MH, Gardner DR, et al. (1994) Locoweed intoxication in range cattle and sheep: Serum α-mannosidase activity and clinicopathologic alterations. In: Colegate SM, Dorling PR (eds.) Plant-associated toxins. Agricultural, phytochemical and ecological aspects. CABI. New York. pp. 501-506.

Stegelmeier BL, Snyder PW, James LF, et al. (1998) The immunologic and toxic effects of chronic locoweed (Astragalus lentiginosus) intoxication in cattle. In: Garland T, Barr AC (eds.) Toxic plants and other natural toxicants. CABI. New York. pp. 285-290.

Tulsiani DR, Broquist HP, James LF, et al. (1984) The similar effects of swainsonine and locoweed on tissue glycosidases and oligosaccharides of the pig indicate that the alkaloid is the principal toxin responsible for the induction of locoism. Arch Biochem Biophys 224(2):594-600.

Tulsiani DR, Broquist HP, James LF, et al. (1988) Production of hybrid glycoproteins and accumulation of oligosaccharides in the brain of sheep and pigs administered swainsonine or locoweed. Arch Biochem Biophys 264(2):607-617.

Van Kampen KR, James LF (1969) Pathology of locoweed poisoning in sheep. Pathol Vet 6(5):413-423.

Van Kampen KR, James LF (1970) Pathology of locoweed (Astragalus lentiginosus) poisoning in sheep. Sequential development of cytoplasmic vacuolation in tissues. Pathol Vet 7(6):503-508.

Van Kampen KR, James LF (1971) Ovarian and placental lesions in sheep from ingesting locoweed (Astragalus lentiginosus). Vet Pathol 8(3):193-199.

Van Kampen KR, James LF (1972) Sequential development of the lesions in locoweed poisoning. Clin Toxicol 5(4):575-580.

Vansell GH (1934) Adult bees found dying on spotted loco. J Econ Entomol 27(3):635-637.

Vasey G (1874) Plants poisonous to cattle in California. U S Dep Agric Rep 1874:159-160.

Walkley SU, James LF (1984) Locoweed-induced neuronal storage disease characterized by meganeurite formation. Brain Res 324(1):145-150.

Warren CD, Bugge B, Daniel PF, et al. (1989) Locoweed toxicosis in sheep: Oligosaccharides accumulated in fetal and maternal tissues. In: James LF, Elbein AD, Molyneux RJ, et al. (eds.) Swainsonine and related glycosidase inhibitors. Iowa State Univ Press. Ames, Iowa. pp. 344-359.

Warren CD, Sadeh S, Daniel PF, et al. (1983) Induced mannosidosis-excretion of oligosaccharides by locoweed-intoxicated sheep. FEBS Lett 163(1):99-103.

ASTRAGALUS LENTIGINOSUS Douglas ex Hook. var. diphysus (A. Gray) M. E. Jones [Fabaceae]

Synonyms:

Astragalus diphysus A. Gray; ***Astragalus macdougali*** E. Sheld.; ***Cystium diphysum*** (A. Gray) Rydb.

Common Names:

blue locoweed; freckled milk vetch; rattleweed; speckled milk vetch; spotted locoweed; tall locoweed

Citations:

Marsh CD (1919) The locoweed disease. U S Dep Agric Farmers Bull #1054:24 pp.

Pfister JA; Stegelmeier BL; Gardner DR; et al. (2003) Grazing of spotted locoweed (Astragalus lentiginosus) by cattle and horses in Arizona. J Anim Sci 81:2285-2293.

ASTRAGALUS LENTIGINOSUS Douglas ex Hook. var. wahweapensis S. L. Welsh [Fabaceae]

Common Names:

spotted locoweed; Wahweap milk vetch

Citations:

Ralphs MH, James LF, Nielsen DB, et al. (1988) Cattle grazing Wahweap milkvetch in southeastern Utah. J Anim Sci 66(12):3124-3130.

Astragalus lusitanicus Lam. = Erophaca baetica (L.) Boiss.

Astragalus macdougali E. Sheld. = Astragalus lentiginosus Douglas ex Hook. var. diphysus (A. Gray) M. E. Jones

ASTRAGALUS MICHAUXII (Kuntze) F. J. Herm. [Fabaceae]

Synonyms:

Astragalus glaber Michx.; ***Tium michauxii*** (Kuntze) Rydb.

Common Names:

Michaux's-milk vetch; poison vetch

Citations:

Duncan WH, Piercy PL, Starling RJ (1955) Toxicological studies of southeastern plants. I. Leguminosae. Econ Bot 9(3):243-255.

Williams MC, Yost GS, Stermitz FR (1977) Miserotoxin, a toxic compound in Astragalus michauxii. Phytochemistry 16(9):1438-1439.

ASTRAGALUS MISER Douglas ex Hook. [Fabaceae]

Synonyms:

Astragalus amphidoxus Blank.

Common Names:

timber milk vetch; weedy milk vetch

Citations:

James LF, Hartley WJ, Williams MC, et al. (1980) Field and experimental studies in cattle and sheep poisoned by nitro-bearing Astragalus or their toxins. Am J Vet Res 41(3):377-382.

Williams MC, Van Kampen KR, Norris FA (1969) Timber milkvetch poisoning in chickens, rabbits, and cattle. Am J Vet Res 30(12):2185-2190.

ASTRAGALUS MISER Douglas ex Hook. var. hylophilus (Rydb.) Barneby [Fabaceae]

Synonyms:

Astragalus hylophilus (Rydb.) A. Nelson

Common Names:

forest milk vetch; green timber milk vetch; timber milk vetch; Yellowstone milk vetch

Citations:

Anonymous (1933) Investigation of stock-poisoning plants. U S Dep Agric Bur Anim Indus Chief Rep 1933:34-35.

Newsom IE, Cross F, McCrory BR, et al. (1936) Timber milk vetch as a poisonous plant. Colorado Agric Exp Sta Bull #425:42 pp.

ASTRAGALUS MISER Douglas ex Hook. var. oblongifolius (Rydb.) Cronquist [Fabaceae]

Common Names:

poison vetch; Rydberg's-weedy milk vetch; timber milk vetch; Wasatch milk vetch

Citations:

Williams MC (1989) Toxicological investigations on Toano, Wasatch, and stinking milkvetches. J Range Manag 42(5):366-368.

Williams MC, Binns W (1967) Toxicity of Astragalus miser Dougl., var. oblongifolius (Rydb) Cronquist. Weeds 15(4):359-362.

ASTRAGALUS MISER Douglas ex Hook. var. serotinus (A. Gray ex J. G. Cooper) Barneby [Fabaceae]

Synonyms:

Astragalus decumbens (Nutt. ex Torr. & A. Gray) A. Gray var. serotinus (A. Gray ex J. G. Cooper) M. E. Jones; ***Astragalus palliseri*** A. Gray; ***Astragalus serotinus*** A. Gray ex J. G. Cooper

Common Names:

Columbia milk vetch; Cooper's-milk vetch; late locoweed; late milk vetch; Palliser poison vetch; timber milk vetch

Citations:

Majak W, Neufeld R, Corner J (1980) Toxicity of Astragalus miser var. serotinus to the honeybee. J Agric Res 19(3):196-199.

Mosher GA, Krishnamurti CR, Kitts WD (1971) Physiological effects of timber milk vetch, Astragalus miser var. serotinus, on sheep. Can J Anim Sci 51(Aug):465-474.

Nicholson HH (1963) The treatment of timber milk-vetch poisoning among cattle and sheep. Can J Anim Sci 43(2):237-240.

ASTRAGALUS MOLLISSIMUS Torr. var. earlei (Greene ex Rydb.) Tidestr. [Fabaceae]

Synonyms:

Astragalus earlei Greene ex Rydb.

Common Names:

Big Bend locoweed; Earle's-locoweed; garbancillo

Citations:

Anonymous (1931) Investigation of stock-poisoning plants. U S Dep Agric Bur Anim Indus Chief Rep 1931:56-57.

Mathews FP (1932) Locoism in domestic animals. Texas Agric Exp Sta Bull #456:28 pp.

ASTRAGALUS MOLLISSIMUS Torr. var. mollissimus [Fabaceae]

Synonyms:

Astragalus mollissimus Torr.; ***Astragalus simulans*** Cockerell

Common Names:

crazyweed; locoweed; purple locoweed; Texas locoweed; woolly locoweed

Citations:

Crawford AC (1908) Laboratory work on loco-weed investigations. U S Dep Agric Bur Plant Indus Bull #121(Part 3):39-40.

James LF, Van Kampen KR (1971) Acute and residual lesions of locoweed poisoning in cattle and horses. J Am Vet Med Assoc 158(5):614-618.

Kirkpatrick JG, Burrows GE (1990) Locoism in horses. Vet Hum Toxicol 32(2):168-169.

Klench JP (1888) Rattleweed or loco-disease. Am Vet Rev 12:395-402.

Marsh CD (1908) Results of loco-weed investigations in the field. U S Dep Agric Bur Plant Indus Bull #121(Part 3):37-38.

Marsh CD (1909) The locoweed disease of the plains. U S Dep Agric Bur Anim Indus Bull #112:130 pp.

Marsh CD (1919) The locoweed disease. U S Dep Agric Farmers Bull #1054:24 pp.

Marshall HT (1904) Loco-weed disease of sheep. Bull Johns Hopkins Hosp 15(158):181-182.

McIlwraith CW, James LF (1982) Limb deformities in foals associated with ingestion of locoweed by mares. J Am Vet Med Assoc 181(3):255-258.

Oehme FW, Bailie WE, Hulbert LC (1968) Astragalus mollissimus (locoweed) toxicosis of horses in western Kansas. J Am Vet Med Assoc 152(3):271-278.

Peters AT, Sturdevant LB (1908) Loco weed poisoning in horses. Nebraska Agric Exp Sta Annu Rep 21:74-107.

Stalker M (1886) The "Loco" plant and its effect on animals. U S Dep Agric Bur Anim Indus Annu Rep 3:271-276.

Stegelmeier BL, Molyneux RJ, Elbein AD, et al. (1995) The lesions of locoweed (Astragalus mollissimus), swainsonine, and castanospermine in rats. Vet Pathol 32(3):289-298.

Stegelmeier BL, Molyneux RJ, James LF (1994) The pathology of swainsonine and locoweed (Astragalus mollissimus) in rodents. Vet Pathol 31:620.

Stegelmeier BL, Ralphs MH, Gardner DR, et al. (1994) Locoweed intoxication in range cattle and sheep: Serum α-mannosidase activity and clinicopathologic alterations. In: Colegate SM, Dorling PR (eds.) Plant-associated toxins. Agricultural, phytochemical and ecological aspects. CABI. New York. pp. 501-506.

Stegelmeier BL, Ralphs MH, Gardner DR, et al. (1994) Serum α-mannosidase activity and the clinicopathologic alterations of locoweed (Astragalus mollissimus) intoxication in range cattle. J Vet Diagn Invest 6(4):473-479.

ASTRAGALUS MOLLISSIMUS Torr. var. thompsoniae (S. Watson) Barneby [Fabaceae]

Synonyms:

Astragalus bigelovii A. Gray; ***Astragalus thompsoniae*** S. Watson

Common Names:

Thompson's-woolly locoweed

Citations:

Buck WB, James LF, Binns W (1961) Changes in serum transaminase activities associated with plant and mineral toxicity in sheep and cattle. Proc Am Coll Vet Toxicol 1961:13-24.

Buck WB, James LF, Binns W (1961) Changes in serum transaminase activities associated with plant and mineral toxicity in sheep and cattle. Cornell Vet 51(Oct):568-585.

Astragalus nigrescens (Hook.) A. Gray = Astragalus tenellus Pursh

Astragalus palliseri A. Gray = Astragalus miser Douglas ex Hook. var. serotinus (A. Gray ex J. G. Cooper) Barneby

ASTRAGALUS PRAELONGUS E. Sheld. [Fabaceae]

Synonyms:

Astragalus recedens (Rydb.) Ced. Porter; ***Jonesiella recedens*** Rydb.

Common Names:

stinking milk vetch; straight-stem locoweed

Citations:

Baker DC, James LF, Hartley WJ, et al. (1989) Toxicosis in pigs fed selenium-accumulating Astragalus plant species or sodium selenate. Am J Vet Res 50(8):1396-1399.

Hartley WJ, James LF, Broquist H, et al. (1985) Pathology of experimental locoweed and selenium poisoning in pigs. In: Seawright AA, Hegarty MP, James LF, et al. (eds.) Plant toxicology. Queensland Poisonous Plant Committee. Yeerongpilly. Queensland. pp. 141-149.

James LF, Smart RA, Shupe JL, et al. (1982) Suspected phytogenic selenium poisoning in sheep. J Am Vet Med Assoc 180(12):1478-1481.

Williams MC (1989) Toxicological investigations on Toano, Wasatch, and stinking milkvetches. J Range Manag 42(5):366-368.

ASTRAGALUS PTEROCARPUS S. Watson [Fabaceae]

Common Names:

wing milk vetch

Citations:

James LF, Hartley WJ, Williams MC, et al. (1980) Field and experimental studies in cattle and sheep poisoned by nitro-bearing Astragalus or their toxins. Am J Vet Res 41(3):377-382.

Williams MC, James LF (1974) Toxicity of several nitrite-bearing Astragalus species. Proc West Soc Weed Sci 27:4.

Williams MC, James LF (1975) Toxicity of nitro-containing Astragalus to sheep and chicks. J Range Manag 28(4):260-263.

ASTRAGALUS PUBENTISSIMUS Torr. & A. Gray [Fabaceae]

Common Names:

Green River milk vetch; locoweed

Citations:

Balls LD, James LF (1973) Effect of locoweed (Astragalus spp.) on reproductive performance of ewes. J Am Vet Med Assoc 162(4):291-292.

Buck WB, James LF, Binns W (1961) Changes in serum transaminase activities associated with plant and mineral toxicity in sheep and cattle. Proc Am Coll Vet Toxicol 1961:13-24.

Buck WB, James LF, Binns W (1961) Changes in serum transaminase activities associated with plant and mineral toxicity in sheep and cattle. Cornell Vet 51(Oct):568-585.

James LF, Bennett KL, Parker KG, et al. (1968) Loco plant poisoning in sheep. J Range Manag 21(6):360-365.

James LF, Binns W (1967) Blood changes associated with locoweed poisoning. Am J Vet Res 28(125):1107-1110.

James LF, Keeler RF, Binns W (1969) Sequence in the abortive and teratogenic effects of locoweed fed to sheep. Am J Vet Res 30(3):377-380.

James LF, Shupe JL, Binns W, et al. (1967) Abortive and teratogenic effects of locoweed on sheep and cattle. Am J Vet Res 28(126):1379-1388.

James LF, Van Kampen KR, Hartley WJ (1970) Comparative pathology of Astragalus (locoweed) and Swainsona poisoning in sheep. Pathol Vet 7(2):116-125.

James LF, Van Kampen KR, Johnson AE (1970) Physiopathologic changes in locoweed poisoning of livestock. Am J Vet Res 31(4):663-672.

Van Kampen KR, James LF (1969) Pathology of locoweed poisoning in sheep. Pathol Vet 6(5):413-423.

Astragalus recedens (Rydb.) Ced. Porter = Astragalus praelongus E. Sheld.

Astragalus serotinus A. Gray ex J. G. Cooper = Astragalus miser Douglas ex Hook. var. serotinus (A. Gray ex J. G. Cooper) Barneby

ASTRAGALUS SILIQUOSUS Boiss. [Fabaceae]

Citations:

Williams MC, James LF, Bleak AT (1976) Toxicity of introduced nitro-containing Astragalus to sheep, cattle, and chicks. J Range Manag 29(1):30-33.

Astragalus simulans Cockerell = Astragalus mollissimus Torr. var. mollissimus

ASTRAGALUS TENELLUS Pursh [Fabaceae]

Synonyms:

Astragalus nigrescens (Hook.) A. Gray; ***Homalobus clementis*** Rydb.; ***Homalobus multiflorus*** (Pursh) Torr. & A. Gray; ***Homalobus stipitatus*** Rydb.; ***Homalobus strigulosus*** Rydb.

Common Names:

pulse milk vetch

Citations:

Anonymous (1934) Investigation of stock-poisoning plants. U S Dep Agric Bur Anim Indus Chief Rep 1934:38.

ASTRAGALUS TETRAPTERUS A. Gray [Fabaceae]

Common Names:

four-wing milk vetch; poison vetch

Citations:

Marsh CD, Clawson AB (1920) Astragalus tetrapterus, a new poisonous plant of Utah and Nevada. U S Dep Agric Circ #81:6 pp.

Williams MC, James LF (1975) Toxicity of nitro-containing Astragalus to sheep and chicks. J Range Manag 28(4):260-263.

Astragalus thompsoniae S. Watson = Astragalus mollissimus Torr. var. thompsoniae (S. Watson) Barneby

ASTRAGALUS TOANUS M. E. Jones [Fabaceae]

Common Names:

Toano milk vetch

Citations:

Williams MC (1989) Toxicological investigations on Toano, Wasatch, and stinking milkvetches. J Range Manag 42(5):366-368.

Williams MC, James LF (1975) Toxicity of nitro-containing Astragalus to sheep and chicks. J Range Manag 28(4):260-263.

ASTRAGALUS WOOTONII E. Sheld. [Fabaceae]

Common Names:

bladderpod locoweed; garbancillo; rattleweed; western locoweed; Wooton's-milk vetch

Citations:

Anonymous (1931) Investigation of stock-poisoning plants. U S Dep Agric Bur Anim Indus Chief Rep 1931:56-57.

James LF, Van Kampen KR (1974) Effect of protein and mineral supplementation on potential locoweed (Astragalus spp.) poisoning in sheep. J Am Vet Med Assoc 164(10):1042-1043.

Mathews FP (1932) Locoism in domestic animals. Texas Agric Exp Sta Bull #456:5-28.

Nelson BK, James LF, Sharma RP, et al. (1977) Subtle postnatal effects of locoweed in rats. Toxicol Appl Pharmacol 41(1):139-140.

Nelson BK, James LF, Sharma RP, et al. (1980) Locoweed embryotoxicity in rats. Clin Toxicol 16(2):149-166.

Staley EE (1978) An approach to treatment of locoism in horses. Vet Med Small Anim Clin 73(9):1205-1206.

ASTROLEPIS COCHISENSIS (Goodd.) D. M. Benham & Windham [Pteridaceae]

Synonyms:

Notholaena sinuata (Lag. ex Sw.) Kaulf. var. cochisensis (Goodd.) Weath

Common Names:

cloak fern; false cloak fern; jimmy fern

Citations:

Mathews FP (1941) Poisonous plants in the Davis Mountains. Texas Agric Exp Sta Annu Rep 54:93.

Mathews FP (1942) Fern (Notholaena sinuata, var. crenata) poisoning in sheep, goats, and cattle - The so-called "jimmies" of the Trans-Pecos. Texas Agric Exp Sta Bull #611:5-15.

Mathews FP (1945) A comparison of the toxicity of Notholaena sinuata and N. sinuata var. cochisensis. Rhodora 47(564):393-395.

ASTRONIUM URUNDEUVA (Allemão) Engl. [Anacardiaceae]

Citations:

Vilar MD, Diógenes MJ, Vilar JL, et al. (2004) Contact dermatitis associated with Astronium urundeuva (Allemão) Engl., a traditional medicinal plant from Brazil. Contact Dermatitis 51(5-6):311.

ATALAYA HEMIGLAUCA (F. Muell.) F. Muell. ex Benth. [Sapindaceae]

Common Names:

whitewood

Citations:

McConnell JD, Barnes JE (1956) The toxicity of the fruits of Atalaya hemiglauca ("Whitewood") for horses. Aust Vet J 32(Apr):74-76.

Murnane D (1927) "Walkabout," or Kimberley horse disease. J CSIRO Aust 1:168-173.

Murnane D (1929) Kimberley horse disease. J CSIRO Aust 2:110-111.

Murnane D (1953) The toxicity of Atalaya hemiglauca (whitewood) for horses. Aust Vet J 29(Jul):188-190.

Murnane D, Ewart AJ (1928) Kimberley horse disease (walkabout disease). Aust CSIRO Bull #36:61 pp.

Rose AL, Gardner CA, McConnell JD, et al. (1957) Field and experimental investigation of "walk-about" disease of horses (Kimberly horse disease) in Northern Australia. Crotalaria poisoning in horses. Part II. Aust Vet J 33(Mar):49-62.

Whittem JH (1968) Experimental whitewood, Atalaya hemiglauca, poisoning in the horse. Aust Vet J 44(Sep):426.

atarillal –see– *Ammi majus* L.

ATELEIA GLAZIOVEANA Baill. [Fabaceae]

Citations:

García y Santos MC, Schild AL, Barros SS, et al. (2004) Lesões perinatais em bovinos na intoxicação experimental por Ateleia glazioviana (Leg. Papilionoideae). Pesq Vet Bras 24(4):178-184.

Gava A, Barros CS (2001) Field observations of Atelia glazioviana poisoning in cattle in southern Brazil. Vet Hum Toxicol 43(1):37-41.

Gava A, Barros CS, Pilati C, et al. (2001) Intoxicação por Ateleia glazioviana (Leg. Papilionoideae) em bovinos. Pesq Vet Bras 21(2):49-59.

Raffi MB, Barros RR, Braganca JF, et al. (2004) The pathogenesis of reproductive failure induced in sheep by the ingestion of Ateleia Glazioviana. Vet Hum Toxicol 46(5):233-238.

Stigger AL, Barros CS, Langohr IM, et al. (2001) Intoxicação experimental por Ateleia glazioviana (Leg. Papilionoideae) em ovinos. Pesq Vet Bras 21(3):98-108.

ATHANASIA TRIFURCATA (L.) L. [Asteraceae]

Common Names:

klaaslouwbossie; kouterbossie

Citations:

Kellerman TS, Coetzer JA, Schneider DJ, et al. (1983) Photosensitivity in South Africa. III. Ovine hepatogenous photosensitivity caused by the plant Athanasia trifurcata L. (Asteraceae). Onderstepoort J Vet Res 50(1):45-53.

atis –see– *Annona squamosa* L.

Atlasholz –see– *Chloroxylon swietenia* DC.

Atractylis gummifera L. Chamaeleon gummifera (L.) Cass.

ATRIPLEX CANESCENS (Pursh) Nutt. [Chenopodiaceae]

Common Names:

chamiza; four-wing saltbush; wingscale

Citations:

Fleming CE (1920) Poisonous range plants. Nevada Agric Exp Sta Annu Rep 1919:39-43.

James LF, Smart RA, Shupe JL, et al. (1982) Suspected phytogenic selenium poisoning in sheep. J Am Vet Med Assoc 180(12):1478-1481.

ATRIPLEX CONFERTIFOLIA (Torr. & Frem.) S. Watson [Chenopodiaceae]

Common Names:

shadscale

Citations:

Fleming CE, Miller MR, Vawter LR, et al. (1931) Poisonous range plants. Nevada Agric Exp Sta Annu Rep #1928:21-22.

ATRIPLEX HORTENSIS L. [Chenopodiaceae]

Common Names:

garden orache; Gartenmelde; mountain orach; orache; orashe; Wilder Spinat; Zuckermelde

Citations:

Cheeke PR, Carlsson R (1978) Evaluation of several crops as sources of leaf meal: Composition, effect of drying procedure, and rat growth response. Nutr Rep Int 18(4):465-473.

Tyszlukiewicz D, Żelazowski K (1964) Zmiany skórne wywołane spożyciem łobody ogrodowej. Pol Tyg Lek (Wars) 19(30):1166-1167.

ATRIPLEX PATULA L. [Chenopodiaceae]

Common Names:

orache

Citations:

Scheuer-Karpin R (1948) Poisoning by food plants. Lancet 254(Apr 10):574-575.

ATRIPLEX ROSEA L. [Chenopodiaceae]

Common Names:

red orache; redscale; Rosenmelde

Citations:

Fleming CE (1920) Poisonous range plants. Nevada Agric Exp Sta Annu Rep 1919:39-43.

ATROPA BELLADONNA L. [Solanaceae]

Common Names:

Asthmador; banewort; belladone; belladonna; black cherry; daftberry; deadly nightshade; death's-herb; dwale; dwayberry; galnebaer; great morel; morelle furieuse; naughty-man's-cherry; nettleberry; poison blackcherry; sleeping nightshade; Tollkirsche; Waldnachtschatten; Wolfskirsche

Citations:

Anonymous (1846) Deaths from nightshade berries. Lancet 2:251-252.

Anonymous (1846) Poisoning with the deadly nightshade. Lancet 2:280.

Anonymous (1948) Deadly nightshade poisoning. Lancet 2(Sep 25):513-514.

Anonymous (1948) Nightshade poisoning. Lancet 2(Sep 11):438.

Çaksen T, Odabaş D, Akbayram S, et al. (2003) Deadly nightshade (Atropa belladonna) intoxication: An analysis of 49 children. Hum Exp Toxicol 22(12):665-668.

Cummins BM, Obetz SW, Wilson MR Jr (1968) Belladonna poisoning as a facet of psychodelia. JAMA 204(11):1011.

Firth D, Bentley JR (1921) Belladonna poisoning from eating rabbit. Lancet 2(Oct 29):901.

Gabel MC (1968) Purposeful ingestion of belladonna for hallucinatory effects. J Pediatr 72(6):864-866.

Galizia EJ (1983) Clinical curio: Hallucinations in elderly tea drinkers. Br Med J 287(Oct 1):979.

Golwalla A (1965) Multiple extrasystoles: An unusual manifestation of belladonna poisoning. Dis Chest 48:83-84.

Hartmeier SH, Steurer J (1996) Mydriasis, Tachykardie. Schweiz Rundsch Med Prax 85(15):495-498.

Joll ME (1916) Three cases of belladonna poisoning. Lancet 2(Oct 7):647.

Joshi P, Wicks AC, Munshi SK (2003) Recurrent autumnal psychosis. Postgrad Med J 79(930):239-240.

Koff M (1966) Poisoning from ingestion of asthma "powders." JAMA 198(9):1034.
Lange A, Toft P (1990) Forgiftning med galnebær, Atropa belladonna. Ugeskr Laeger 152(15):1096.
Löhrer F, Kaiser R (1999) Biogene Suchtmittel. Neue Konsumgewohnheiten bei jungen Abhängigen? Nervenarzt 70(11):1029-1033.
Minors EH (1948) Five cases of belladonna poisoning. Br Med J 2(Sep 11):518-519.
Muller DJ (1967) Unpublicized hallucinogens. The dangerous belladonna alkaloids. JAMA 202(7):650-651.
Northall FS, Dauncey EA, Butler JM (2003) An overview of plant and fungal poisonings in the UK, and some interesting cases. J Toxicol Clin Toxicol 41(4):518-519.
Pestalozzi BC, Caduff F (1986) Gruppenvergiftung mit Tollkirschentee. Schweiz Med Wochenschr 116(27-28):924-926.
Plackova S, Caganova B (1998) Acute intoxications by mushrooms and plants in Slovakia. J Toxicol Clin Toxicol 36(5):452-453.
Schneider R, Lutum R, Kintz P, et al. (1996) Plasma and urine concentrations of atropine after the ingestion of cooked deadly nightshade berries. J Toxicol Clin Toxicol 34(1):113-117.
Schvartsman S, Marcondes E (1965) Intoxicações acidentais agudas na infancias. Revisão de 208 casos. Rev Paul Med 66(1):24-39.
Smith HC, Taussig RA, Peterson PC (1956) Deadly nightshade poisoning in swine. Case reports of two herds. J Am Vet Med Assoc 129(Aug 1):116-117.
Southgate HJ, Egerton M, Dauncey EA (2000) Lessons to be learned: A case study approach. Unseasonal severe poisoning of two adults by deadly nightshade (Atropa belladonna). J R Soc Health 120(2):127-130.
Testasecca D, Caputi C, Pavoni PA (1978) Su di un caso di avvelenamento da bacche di belladonna. Clin Ter 86(3):277-280.
Tiossa GG, Pavliuk VT (1994) [Emergency care and intensive therapy in acute poisonings by belladonna berries.] Voen Med Zh 2(2):38-39.
Tita B, Bolle P, Martinoli L, et al. (1988) A comparative study of Atropa belladonna and atropine on an animal model of urinary retention. Pharmacol Res Commun 20(Suppl 5):55-58.
Trabattoni G, Visintini D, Terzano GM, et al. (1984) Accidental poisoning with deadly nightshade berries: A case report. Hum Toxicol 3(6):513-516.
Wilcox WP Jr (1967) More cases of atropinism. N Engl J Med 277(22):1209.

ATTALEA SPECIOSA Mart. ex Spreng. [Arecaceae]

Synonyms:

Orbignya phalerata Mart.

Common Names:

babassu

Citations:

Gaitan E, Cooksey RC, Legan J, et al. (1994) Antithyroid effects in vivo and in vitro of babassu and mandioca: A staple food in goiter areas of Brazil. Eur J Endocrinol 131(21):138-144.

aubergine –see– *Solanum melongena* L.

aucuba –see– *Aucuba japonica* Thunb.

AUCUBA JAPONICA Thunb. [Garryaceae]

Common Names:

aucuba; blotch-leaf laurel; gold-dust plant; gold-leaf plant; Japanese aucuba; Japanese laurel; spotted laurel; variegated laurel

Citations:

Leveau AM, Durand M, Paris RR (1979) Sur la toxicite des fruits de l'Aucuba japonica (Cornacees). Plantes Med Phytotherap 13(3):199-204.

aulne noir –see– *Frangula alnus* Mill.

auricula –see– *Primula auricula* L.

Ausdauerndes Bingelkraut –see– *Mercurialis perennis* L.

Australian asthma herb –see– *Chamaesyce hirta* (L.) Millsp.

Australian flame tree –see– *Sterculia foetida* L.

Australian bean tree –see– *Castanospermum australe* A. Cunn. & C. Fraser ex Hook.

Australian black bean tree –see– *Castanospermum australe* A. Cunn. & C. Fraser ex Hook.

Australian blackwood –see– *Acacia melanoxylon* R. Br.

Australian bush nut –see– *Macadamia integrifolia* Maiden & Betche; *Macadamia tetraphylla* L. A. S. Johnson

Australian cabbage –see– *Alocasia macrorrhizos* (L.) G. Don

Australian finger cherry –see– *Rhodomyrtus macrocarpa* Benth.

Australian indigo –see– *Indigofera australis* Willd.

Australian ivy palm –see– *Schefflera actinophylla* (Endl.) Harms

Australian nettle –see– *Dendrocnide moroides* (Wedd.) Chew

Australian nut palm –see– *Cycas revoluta* Thunb.

Australian phalaris –see– *Phalaris aquatica* L.

Australian pine –see– *Casuarina cristata* Miq.; *Casuarina equisetifolia* L.

Australian poison bush –see– *Gastrolobium grandiflorum* F. Muell.

Australian silk oak –see– *Grevillea robusta* A. Cunn. ex R. Br.

Australian snakeweed –see– *Chamaesyce hirta* (L.) Millsp.

Australian tea tree –see– *Melaleuca alternifolia* (Maiden & Betche) Cheel

Australian umbrella tree –see– *Schefflera actinophylla* (Endl.) Harms

Australische Brennessel –see– *Dendrocnide moroides* (Wedd.) Chew

Austrian pea –see– *Pisum sativum* L.

autumn adonis –see– *Adonis annua* L.

autumn crocus –see– *Colchicum autumnale* L.

autumn sneezeweed –see– *Helenium autumnale* L.

avasa –see– *Tephrosia purpurea* (L.) Pers.

ave grace –see– *Ruta graveolens* L.

aveia –see– *Avena sativa* L.

aveia louca –see– *Phalaris angusta* Nees ex Trin.

aveia-de-sangue –see– *Phalaris angusta* Nees ex Trin.

avellana-purgante-de-Santo Domingo –see– *Jatropha multifida* L.

AVENA SATIVA L. [Poaceae]

Common Names:

aveia; Hafer; oats

Citations:

Allen AL, Townsend HG, Doige CE (1998) Ingestion of nitrate-containing plants as a possible risk factor for congenital hypothyroidism in foals. In: Garland T, Barr AC (eds.) Toxic plants and other natural toxicants. CABI. New York. pp. 334-338.

Bradley WB, Beath OA, Eppson HF (1939) Oat hay poisoning. Science 89(Apr 21):365.

de Paz Arranz S, Pérez Montero A, Zapatero Remón L, et al. (2002) Allergic contact urticaria to oatmeal. Allergy 57(12):1215.

Dempster JG (1981) Contact dermatitis from bran and oats. Contact Dermatitis 7(2):122.

Newsom IE, Stout EN, Thorp F Jr, et al. (1937) Oat hay poisoning. J Am Vet Med Assoc 90(43):66-75.

Sippel WL, Burnside JE (1954) Oat dermatitis. Georgia Vet 6(2):3-4.

Solomons B (1971) Sensitization to oats and barley. Contact Dermatol Newsl 10(Jul):231.

Thorp F Jr (1938) Further observations on oat hay poisoning. J Am Vet Med Assoc 92:159-170.

averil –see– *Narcissus pseudonarcissus* L.

AVERRHOA CARAMBOLA L. [Oxalidaceae]

Common Names:

star fruit

Citations:

Chang CH, Yeh JH (2004) Non-convulsive status epilepticus and consciousness disturbance after star fruit (Averrhoa carambola) ingestion in a dialysis patient. Nephrology (Carlton) 9(6):362-365.

Chang CT, Chen YC, Fang JT, et al. (2002) Star fruit (Averrhoa carambola) intoxication: An important cause of consciousness disturbance in patients with renal failure. Ren Fail 24(3):379-382.

Neto MM (1998) Intoxication by star fruit (Averrhoa carambola) in six dialysis patients? Nephrol Dial Transplant 13(3):570-572.

Neto MM, da Costa JA, Garcia-Cairasco N, et al. (2003) Intoxication by star fruit (Averrhoa carambola) in 32 uraemic patients: Treatment and outcome. Nephrol Dial Transplant 18(1):120-125.

Tse KC, Yip PS, Lam MF, et al. (2003) Star fruit intoxication in uraemic patients: Case series and review of the literature. Intern Med J 33(7):314-316.

avocado –see– *Persea americana* Mill. var. americana

avocado pear –see– *Persea americana* Mill. var. americana

avordiré –see– *Turraeanthus africanus* (Welw. ex C. DC.) Pellegr.

awa –see– *Piper methysticum* G. Forst.

axwort –see– *Securigera varia* (L.) Lassen

ayahuasca –see– *Banisteriopsis caapi* (Spruce ex Griseb.) C. V. Morton

ayahuasca roja –see– *Banisteriopsis caapi* (Spruce ex Griseb.) C. V. Morton

ayan –see– *Distemonanthus benthamianus* Baill.

AZADIRACHTA INDICA A. Juss. [Meliaceae]

Common Names:

Chinese tallow tree; margosa tree; neem; neempathi; nim tree; nimb

Citations:

Ali BH (1987) The toxicity of Azadirachta indica leaves in goats and guinea pigs. Vet Hum Toxicol 29(1):16-19.

Ali BH, Salih AM (1982) Suspected Azadirachta indica toxicity in a sheep. Vet Rec 111(21):494.

Bhandari DS, Joshi MS (1974) The effect of feeding deoiled neem cake on health of sheep. Indian Vet J 51(Sep-Oct):659-660.

Christopher KJ, Ahmed MN, Sastry GA (1976) Effects of feeding deoiled neem seed cake to chicken. Indian J Vet Pathol 1(1):27-30.

Ibrahim IA, Omer SA, Ibrahim FH, et al. (1992) Experimental Azadirachta indica toxicosis in chicks. Vet Hum Toxicol 34(3):221-224.

Joshi AR, Ahamed RN, Pathan KM, et al. (1996) Effect of Azadirachta indica leaves on testis and its recovery in albino rats. Indian J Exp Biol 34(11):1091-1094.

Lai SM, Lim KW, Cheng HK (1990) Margosa oil poisoning as a cause of toxic encephalopathy. Singapore Med J 31:463-465.

Rahman MF, Siddiqui MK, Jamil K (2001) Effects of Vepacide (Azadirachta indica) on aspartate and alanine aminotransferase profiles in a subchronic study with rats. Hum Exp Toxicol 20:243-249.

Sadagopan VR, Johri TS, Reddy V, et al. (1982) Feeding value of neem seed meal for starter chicks. Indian Vet J 59(Jun):462-465.

Sinniah D, Baskaran G (1981) Margosa oil poisoning as a cause of Reye's syndrome. Lancet 1(8218):487-489.

Sinniah D, Baskaran G, Looi AH, et al. (1982) Reye-like syndrome due to margosa oil poisoning: Report of a case with postmortem findings. Am J Gastroenterol 77(3):158-161.

azafran –see– *Crocus sativus* L.

azalea –see– *Rhododendron indicum* (L.) Sweet; *Rhododendron simsii* Planch.

azobe tree –see– *Lophira alata* Banks ex C. F. Gaertn.

azufaifa –see– *Ziziphus jujuba* Mill.

B

ba dau –see– *Croton tiglium* L.

babassu –see– *Attalea speciosa* Mart. ex Spreng.

babul –see– *Acacia leucophloea* (Roxb.) Willd.

baby rubber plant –see– *Peperomia obtusifolia* (L.) A. Dietr.

baby's-breath –see– *Gypsophila paniculata* L.

BACCHARIDASTRUM TRIPLINERVIUM (Less.) Cabrera [Asteraceae]

Citations:

Langohr IM, Gava A, Barros CS (2005) Intoxicação por Baccharidastrum triplinervium (Asteraceae) em bovinos. Pesq Vet Bras 25(4):235-238.

baccharis –see– *Baccharis glomeruliflora* Pers.

BACCHARIS CORIDIFOLIA DC. [Asteraceae]

Common Names:

mio mio; oromerillo; romerillo

Citations:

Döbereiner J, Rezende AM, Tokarnia CH (1976) Intoxicação experimental por Baccharis coridifolia em coelhos. Pesq Agric Bras Vet 11(9):27-35.

Rissi DR, Rech RR, Fighera RA, et al. (2005) Intoxicação espontânea por Baccharis coridifolia em bovinos. Pesq Vet Bras 25(2):111-114.

Rozza DB, Raymundo DL, Corrêa AM, et al. (2006) Intoxicação espontânea por Baccharis coridifolia (Compositae) em ovinos. Pesq Vet Bras 26(1):21-25.

Ruiz LF (1930) Tres plantas toxicas de la flora Argentina Bol Minist Agric Nac Rep Argentina 29(Mar):45-55.

Tokarnia CH, Döbereiner J (1975) Intoxicação experimental em bovinos por "Mio-Mio," Baccharis coridifolia. Pesq Agric Bras Vet 10(8):79-97.

Tokarnia CH, Döbereiner J (1976) Intoxicação experimental em ovinos por "Mio-Mio," Baccharis coridifolia. Pesq Agric Bras Vet 11(9):19-26.

Varaschin MS, Alessi AC (2003) Poisoning of mice by Baccharis coridifolia: An experimental model. Vet Hum Toxicol 45(1):42-44.

Varaschin MS, Barros CS, Jarvis BB (1998) Intoxicação experimental por Baccharis coridifolia (Compositae) em bovinos. Pesq Vet Bras 18(2):69-74.

BACCHARIS ERIGEROIDES DC. [Asteraceae]

Citations:

Andrade SO, Camargo WV, Fernandes N (1963) II. Investigações sôbre plantas tóxicas no Estado de São Paulo. Arq Inst Biol (Sao Paulo) 30(Oct):189-203.

BACCHARIS GLOMERULIFLORA Pers. [Asteraceae]

Common Names:

baccharis; groundsel bush; silverling; southern baccharis

Citations:

Duncan WH, Piercy PL, Feurt SD, et al. (1957) Toxicological studies of southeastern plants. II. Compositae. Econ Bot 11:75-85.

BACCHARIS HALIMIFOLIA L. [Asteraceae]

Common Names:

consumption weed; eastern baccharis; groundsel bush; groundsel tree; salt groundsel; sea myrtle; silverling

Citations:

Duncan WH, Piercy PL, Feurt SD, et al. (1957) Toxicological studies of southeastern plants. II. Compositae. Econ Bot 11:75-85.

Manley GD, Edds GT, Sundlof SF (1982) Cattle deaths from poisonous plant. Florida Vet J 11(Dec):20.

Baccharis megapotamica Spreng. var. weirii (Baker) G. M. Barroso = Baccharis megapotamica Spreng.

BACCHARIS MEGAPOTAMICA Spreng. [Asteraceae]

Synonyms:

Baccharis megapotamica Spreng. var. weirii (Baker) G. M. Barroso

Citations:

Driemeier D, Cruz C, Loretti A (2000) Baccharis megapotamica var weirii poisoning in Brazilian cattle. Vet Hum Toxicol 42(4):220-221.

Tokarnia CH, Peixoto PV, Gava A, et al. (1992) Intoxicação experimental por Baccharis megapotamica var. megapotamica e var. weirii (Compositae) em bovinos. Pesq Vet Bras 12(1-2):19-31.

BACCHARIS PTERONIOIDES DC. [Asteraceae]

Common Names:

hierba-de-pasmo; yerba-de-pasmo; yerba manza

Citations:

Marsh CD, Clawson AB, Eggleston WW (1920) Baccharis pteronioides as a poisonous plant of the Southwest. J Am Vet Med Assoc 57:430-434.

BACCHARIS STENOCEPHALA Baker [Asteraceae]

Citations:

Andrade SO, Camargo WV, Fernandes N (1963) II. Investigações sôbre plantas tóxicas no Estado de São Paulo. Arq Inst Biol (Sao Paulo) 30(Oct):189-203.

bachnag –see– *Aconitum napellus* L.

Badekraut –see– *Levisticum officinale* W. D. J. Koch

badoh –see– *Ipomoea violacea* L.; *Turbina corymbosa* (L.) Raf.

badoh negro –see– *Ipomoea tricolor* Cav.

bagin –see– *Derris trifoliata* Lour.

bagpod –see– *Sesbania vesicaria* (Jacq.) Elliott

bagpod sesbane –see– *Sesbania vesicaria* (Jacq.) Elliott

bagpod sesbania –see– *Sesbania vesicaria* (Jacq.) Elliott

bagseed –see– *Sesbania vesicaria* (Jacq.) Elliott

bahera –see– *Terminalia bellirica* (Gaertn.) Roxb.

Bahia –see– *Picradeniopsis oppositifolia* (Nutt.) Rydb.

Bahia grass –see– *Paspalum dilatatum* Poir.; *Paspalum notatum* Flüggé

Bahia oppositifolia (Nutt.) DC. = Picradeniopsis oppositifolia (Nutt.) Rydb.

Bahia rosewood –see– *Dalbergia nigra* (Vell.) Allemão ex Benth.

baijiaolian –see– *Podophyllum pleianthum* Hance

baileya –see– *Baileya multiradiata* Harv. & A. Gray ex Torr.

BAILEYA MULTIRADIATA Harv. & A. Gray ex Torr. [Asteraceae]

Common Names:

baileya; cloth-of-gold; desert baileya; desert marigold; many-ray Baileya; paper flower

Citations:

Dollahite JW (1960) Desert baileya poisoning in sheep, goats, and rabbits. Texas Agric Exp Sta Prog Rep #2149:4 pp.

Mathews FP (1932) Miscellaneous poisonous plants. Texas Agric Exp Sta Annu Rep 45:11-12.

Mathews FP (1933) The toxicity of Baileya multiradiata for sheep and goats. J Am Vet Med Assoc 83(Nov):673-679.

Mathews FP, Schmidt H (1934) Miscellaneous poisonous plants. Texas Agric Exp Sta Annu Rep 47:12-13.

bajiaolian –see– *Podophyllum peltatum* L.

bajra –see– *Pennisetum glaucum* (L.) R. Br.

bakarai bish –see– *Ageratum conyzoides* L.

bakario –see– *Indigofera linnaei* Ali

bakkain –see– *Melia azedarach* L.

balewort –see– *Papaver somniferum* L.

BALFOURODENDRON RIEDELIANUM (Engl.) Engl. [Rutaceae]

Common Names:

Piquiá marfim

Citations:

Freise FW (1936) Vergiftungen durch brasilianische Werkhölzer. II. Jacareúba-Holz und Seidenholz. Sammlung Vergiftungsfallen 7(C33):61-72.

bali bali –see– *Euphorbia tirucalli* L.

ball nettle –see– *Solanum carolinense* L.

ball nightshade –see– *Solanum carolinense* L.

balloon cotton –see– *Gomphocarpus fruticosa* (L.) W. T. Aiton

balsam –see– *Euphorbia drummondii* Boiss.

balsam fir –see– *Abies balsamea* (L.) Mill.

balsam-of-pine –see– *Pinus sylvestris* L.

balsam-of-spruce –see– *Picea abies* (L.) H. Karst.

balsam-of-tolu –see– *Myroxylon balsamum* (L.) Harms

balsam pear –see– *Momordica charantia* L.

balsamino –see– *Momordica charantia* L.

balubad –see– *Anacardium occidentale* L.

balugo –see– *Entada phaseoloides* (L.) Merr.

bambatsi grass –see– *Panicum coloratum* L. var. makariense Gooss.

bamboo –see– *Bambusa vulgaris* Schrad. ex J. C. Wendl.

bambu –see– *Bambusa vulgaris* Schrad. ex J. C. Wendl.

BAMBUSA VULGARIS Schrad. ex J. C. Wendl. [Poaceae]

Common Names:

bamboo; bambu; caña brava

Citations:

Barbosa JD, Oliveira CM, Duarte MD, et al. (2006) Poisoning of horses by bamboo, Bambusa vulgaris. J Equine Vet Sci 26 (9):393-398.

Schiff BL (1951) Contact dermatitis caused by bamboo. Arch Derm Syphilol 64(1):66-67.

ban –see– *Quercus leucotrichophora* A. Camus

banana –see– *Musa ×paradisiaca* L.

Banane –see– *Musa ×paradisiaca* L.

bandera –see– *Euphorbia pulcherrima* Willd. ex Klotzsch

bandjiebos –see– *Tylecodon wallichii* (Harv.) Toelken

baneberry –see– *Actaea rubra* (Aiton) Willd.

banewort –see– *Atropa belladonna* L.

Bangar nut –see– *Sterculia foetida* L.

Bangkok teak –see– *Tectona grandis* L. f.

bangor nut –see– *Sterculia foetida* L.

BANISTERIOPSIS CAAPI (Spruce ex Griseb.) C. V. Morton [Malpighiaceae]

Synonyms:

Banisteriopsis inebrians C. V. Morton

Common Names:

ayahuasca; ayahuasca roja; caapi; yage

Citations:

Flores FA, Lewis WH (1978) Drinking the South American hallucinogenic ayahuasca. Econ Bot 32(Apr-Jun):154-156.

Löhrer F, Kaiser R (1999) Biogene Suchtmittel. Neue Konsumgewohnheiten bei jungen Abhängigen? Nervenarzt 70(11):1029-1033.

Banisteriopsis inebrians C. V. Morton = Banisteriopsis caapi (Spruce ex Griseb.) C. V. Morton

BANISTERIOPSIS LONGIALATA (Nied.) B. Gates [Malpighiaceae]

Synonyms:

Banisteriopsis rusbyana (Nied.) C. V. Morton

Citations:

Der Marderosian AH, Pinkley HV, Dobbins MF 4th (1968) Native use and occurrence of N,N-dimethyltryptamine in the leaves of Banisteriopsis rusbyana. Am J Pharm 140(5):137-147.

Flores FA, Lewis WH (1978) Drinking the South American hallucinogenic ayahuasca. Econ Bot 32(Apr-Jun):154-156.

Banisteriopsis rusbyana (Nied.) C. V. Morton = Banisteriopsis longialata (Nied.) B. Gates

Bank's-grevillea –see– *Grevillea banksii* R. Br.

banok –see– *Sarcolobus globosus* Wall.

banucalad nut –see– *Reutealis trisperma* (Blanco) Airy Shaw

BAPTISIA ALBA (L.) Vent. [Fabaceae]

Common Names:

white wild indigo; wild indigo

Citations:

Duncan WH, Piercy PL, Starling RJ (1955) Toxicological studies of southeastern plants. I. Leguminosae. Econ Bot 9(3):243-255.

BAPTISIA AUSTRALIS (L.) R. Br. [Fabaceae]

Common Names:

blue false indigo; blue wild indigo; false indigo

Citations:

Burrows GE, Schwab RP, Stein LE, et al. (1998) Comparison of the reproductive effects of Baptisia australis, Iva annua and Sophora nuttalliana in rats. In: Garland T, Barr AC (eds.) Toxic plants and other natural toxicants. CABI. New York. pp. 297-302.

Hansen AA (1930) Indiana plants injurious to livestock. Purdue Agric Exp Sta Circ #175:38 pp.

bar foot –see– *Helleborus foetidus* L.

barba ursului –see– *Equisetum arvense* L.

Barbados aloe –see– *Aloe arborescens* Mill.; *Aloe vera* (L.) Burm. f.

Barbados purging nut –see– *Jatropha curcas* L.

BARBAREA VULGARIS R. Br. [Brassicaceae]

Common Names:

bitter water cress; winter cress; yellow rocket

Citations:

Hansen AA (1930) Indiana plants injurious to livestock. Purdue Agric Exp Sta Circ #175:38 pp.

barbasco - see - *Deguelia ulilis* (A.C.Sm.) A.M.G. Azevedo; *Verbascum thapsus* L.

barbatimão –see– *Stryphnodendron adstringens* (Mart.) anon.; *Stryphnodendron coriaceum* Benth.; *Stryphnodendron obovatum* Benth.

Barbey's-larkspur –see– *Delphinium barbeyi* (Huth) Huth

bardana –see– *Arctium lappa* L.

Bärenfuß –see– *Helleborus foetidus* L.; *Helleborus viridis* L.

Bärenklau –see– *Heracleum mantegazzianum* Sommier & Levier

Bärenlauch –see– *Allium ursinum* L.

barilla –see– *Halogeton glomeratus* (M. Bieb.) C. A. Mey.

barley –see– *Hordeum vulgare* L.

baron's-mercury –see– *Mercurialis annua* L.

barr tree –see– *Sambucus nigra* L.

barra –see– *Terminalia bellirica* (Gaertn.) Roxb.

BARRINGTONIA ACUTANGULA (L.) Gaertn. [Lecythidaceae]

Common Names:

freshwater mangrove; Indian oak

Citations:

Chakraborty DP, Nandy AC, Philipose MT (1972) Barringtonia acutangula (L.) Gaertn. as a fish poison. Indian J Exp Biol 10(1):78-80.

barseem –see– *Trifolium alexandrinum* L.

baru –see– *Sorghum halepense* (L.) Pers.

Basin goldenrod –see– *Solidago spectabilis* (D. C. Eaton) A. Gray

basket plant –see– *Callisia fragrans* (Lindl.) Woodson

basora cora –see– *Melochia tomentosa* L.

Bassia echinopsila (F. Muell.) F. Muell = Sclerolaena anisacanthoides (F. Muell.) Domin

BASSIA HYSSOPIFOLIA (Pall.) Kuntze

[Chenopodiaceae]

Common Names:

five-hook bassia; smother weed

Citations:

James LF, Williams MC, Bleak AT (1976) Toxicity of Bassia hyssopifolia to sheep. J Range Manag 29(4):284-285.

BASSIA SCOPARIA (L.) A. J. Scott

[Chenopodiaceae]

Synonyms:

Kochia scoparia (L.) Schrad.

Common Names:

belvedere; burningbush; fireball; fireweed; kochia; Mexican fireweed; morenita; poor-man's-alfalfa; summer cypress

Citations:

Coxworth EC, Salmon RE (1972) Kochia seed as a component of the diet of turkey poults: Effects of different methods of saponin removal or inactivation. Can J Anim Sci 52(Dec):721-729.

Dickie CW, Berryman JR (1979) Polioencephalomalacia and photosensitization associated with Kochia scoparia consumption in range cattle. J Am Vet Med Assoc 175(5):463-465.

Dickie CW, James LF (1983) Kochia scoparia poisoning in cattle. J Am Vet Med Assoc 183(7):765-768.

Galitzer SJ, Oehme FW (1979) Studies of the comparative toxicity of Kochia scoparia (L.) Schrad (fireweed). Toxicol Lett 3:43-49.

Kirkpatrick JG, Helman RG, Burrows GE, et al. (1998) Transient hepatoxicity in sheep grazing Kochia scoparia. In: Garland T, Barr AC (eds.) Toxic plants and other natural toxicants. CABI. New York. pp. 504-508.

Rankins DL Jr, Smith GS (1991) Nutritional and toxicological evaluations of Kochia hay (Kochia scoparia) fed to lambs. J Anim Sci 69(7):2925-2931.

Rankins DL Jr, Smith GS, Hallford DM (1991) Altered metabolic hormones, impaired nitrogen retention, and hepatotoxicosis in lambs fed Kochia scoparia hay. J Anim Sci 69(7):2932-2940.

Rankins DL Jr, Smith GS, Hallford DM (1991) Effects of metoclopramide on steers fed Kochia scoparia hay. J Anim Sci 69(9):3699-3705.

Rankins DL Jr, Smith GS, Hallford DM (1991) Serum constituents and metabolic hormones in sheep and cattle fed Kochia scoparia hay. J Anim Sci 69(7):2941-2946.

Sprowls RW (1981) Problems observed in horses, cattle, and sheep grazing kochia. Proc Am Assoc Vet Lab Diagn 24:397-405.

Thilsted J, Hibbs C, Kiesling H, et al. (1989) Kochia (Kochia scoparia) toxicosis in cattle: Results of four experimental grazing trials. Vet Hum Toxicol 31(1):34-41.

bastard anise –see– *Illicium anisatum* L.

bastard cabbage –see– *Andira inermis* (W. Wright) Kunth ex DC.

bastard clover –see– *Trifolium hybridum* L.

bastard feverfew –see– *Parthenium hysterophorus* L.

bastard hellebore –see– *Helleborus viridis* L.

bastard lentil –see– *Vicia ervilia* (L.) Willd.

bastard mahogany –see– *Andira inermis* (W. Wright) Kunth ex DC.; *Dalbergia retusa* Hemsl.

bastard nigelle –see– *Agrostemma githago* L.

Bastardklee –see– *Trifolium hybridum* L.

Bathurst bur –see– *Xanthium spinosum* L.; *Xanthium strumarium* L.

bat's-wing fern –see– *Histiopteris incisa* (Thunb.) J. Sm.

batu –see– *Croton tiglium* L.

baura –see– *Trianthema portulacastrum* L.

bay –see– *Laurus nobilis* L.

bay laurel –see– *Laurus nobilis* L.

be-still tree –see– *Thevetia peruviana* (Pers.) K. Schum.

beach apple –see– *Hippomane mancinella* L.

beach bean –see– *Canavalia rosea* (Sw.) DC.

beach laurel –see– *Leucothoe davisiae* Torr. ex A. Gray

bead tree –see– *Melia azedarach* L.

beaked nightshade –see– *Solanum rostratum* Dunal

bean –see– *Phaseolus vulgaris* L.

bean caper –see– *Zygophyllum fabago* L.

bean tree –see– *Castanospermum australe* A. Cunn. & C. Fraser ex Hook.; *Laburnum anagyroides* Medik.

bean trefoil –see– *Laburnum anagyroides* Medik.

bear corn –see– *Veratrum viride* Aiton

beargrass –see– *Nolina microcarpa* S. Watson; *Nolina texana* S. Watson

bearded darnel –see– *Lolium temulentum* L.

bear's-ears –see– *Primula auricula* L.

bear's-foot –see– *Aconitum napellus* L.; *Helleborus foetidus* L.; *Helleborus niger* L.; *Helleborus viridis* L.

bearwood –see– *Metopium toxiferum* (L.) Krug & Urb.

beaver poison –see– *Cicuta douglasii* (DC.) J. M. Coult. & Rose; *Cicuta maculata* L.; *Cicuta virosa* L.

bedstraw milkweed –see– *Asclepias subverticillata* (A. Gray) Vail

beech –see– *Fagus sylvatica* L.

beech drops –see– *Orobanche minor* Sm.

beechnut –see– *Fagus grandifolia* Ehrh.; *Fagus sylvatica* L.

beef wood tree –see– *Casuarina equisetifolia* L.

beefsteak plant –see– *Perilla frutescens* (L.) Britton

beesbossie –see– *Chrysocoma ciliata* L.

beeskaroo –see– *Chrysocoma ciliata* L.

beet –see– *Beta vulgaris* L.

beeweed –see– *Asclepias subverticillata* (A. Gray) Vail

beggar's-buttons –see– *Arctium lappa* L.

beggar's-lice –see– *Cynoglossum officinale* L.

BEGONIA ARGENTEA Linden [Begoniaceae]

Citations:

Mrvos R, Krenzelok EP, Jacobsen TD (2001) Toxidromes associated with the most common plant ingestions. Vet Hum Toxicol 43(6):366-369.

behen –see– *Podophyllum peltatum* L.

Beifuß –see– *Artemisia vulgaris* L.

Beinholz –see– *Ligustrum vulgare* L.

Beinved –see– *Ilex aquifolium* L.

Beinwell –see– *Symphytum officinale* L.

Beißbeere –see– *Daphne mezereum* L.

bejuco-de-estrella –see– *Pedilanthus tithymaloides* (L.) Poit.

bejuco-de-pan –see– *Dalechampia scandens* L.

bejuco-de-peonia –see– *Abrus precatorius* L.

bejuco-de-San Jose –see– *Allamanda cathartica* L.

bejuco marrullero –see– *Pentalinon luteum* (L.) B. F. Hansen & Wunderlin

belder root –see– *Oenanthe crocata* L.

beleño negro –see– *Hyoscyamus niger* L.

beleric myrobalan –see– *Terminalia bellirica* (Gaertn.) Roxb.

bell pepper –see– *Capsicum annuum* L.

bell rose –see– *Narcissus pseudonarcissus* L.

bella gutta tree –see– *Semecarpus anacardium* L. f.

bella sombra –see– *Phytolacca dioica* L.

belladone –see– *Atropa belladonna* L.

belladonna –see– *Atropa belladonna* L.

BELLIS PERENNIS L. [Asteraceae]

Common Names:

English daisy

Citations:

Hausen BM (1979) The sensitizing capacity of compositae plants. III. Test results and cross-reactions in compositae-sensitive patients. Dermatologica 159(1):1-11.

bellyache plant –see– *Jatropha gossypiifolia* L.

belvedere –see– *Bassia scoparia* (L.) A. J. Scott

ben –see– *Moringa oleifera* Lam.

ben dock –see– *Oenanthe crocata* L.

Bengal gam –see– *Cicer arietinum* L.

Bengal gram –see– *Cajanus cajan* (L.) Millsp.; *Cicer arietinum* L.

Bengal walnut –see– *Aleurites moluccanus* (L.) Willd.

Benjamin tree –see– *Ficus benjamina* L.

benweed –see– *Jacobaea vulgaris* Gaertn.

bequilla –see– *Sesbania exaltata* (Raf.) Rydb. ex A. W. Hill

BERBERIS AQUIFOLIUM Pursh [Berberidaceae]

Synonyms:

Mahonia aquifolium (Pursh) Nutt.

Common Names:

algerita; feuilles-de-houx; mahonia; Mahonie; mountain grape; mountain graperoot; Oregon grape; Oregon holly grape; racine-de-mahonia; trailing mahonia

Citations:

Lahde G (1973) Berberinvergiftung beim Banteng (Bos javanicus javanicus). In: Ippen R et al. (eds.) Erkrank der Zoot XV. Int Symp, Kolmarden Akad Verlag. pp. 131-133.

Spoerke DG, Temple AR (1978) One year's experience with potential plant poisonings reported to the Intermountain Regional Poison Control Center. Vet Hum Toxicol 20(2):85-89.

BERBERIS DARWINII Hook. [Berberidaceae]

Citations:

Donovan D (1975) Barberri darwinii - A poisonous plant? J Ir Med Assoc 68(16):414.

BERBERIS WILSONIAE Hemsl. [Berberidaceae]

Citations:

Hopkinss TR (1926) Barberry poisoning. Florists Exchange & Horticultural Trade World 61:1485.

berceaux-de-la-vierge –see– *Clematis vitalba* L.

berenghenas –see– *Solanum melongena* L.

berg slangkop –see– *Drimia depressa* (Baker) Jessop

bergamot –see– *Citrus bergamia* Risso & Poit.

bergamot mint –see– *Mentha ×piperita* L. nothosubsp. citrata (Ehrh.) Briq.

Bergamotte –see– *Citrus bergamia* Risso & Poit.

Berglorbeer –see– *Kalmia angustifolia* L.

Bergpfeffer –see– *Daphne mezereum* L.

Bergwohlverleih –see– *Arnica montana* L.

BERKHEYOPSIS BECHUANENSIS S. Moore [Asteraceae]

Citations:

Van der Walt SJ, Steyn DG (1943) Recent investigations into the toxicity of plants, etc. XIII. Onderstepoort J Vet Sci Anim Indus 18(1-2):207-224.

BERKHEYOPSIS ECHINUS (Less.) O. Hoffm. [Asteraceae]

Citations:

Van der Walt SJ, Steyn DG (1943) Recent investigations into the toxicity of plants, etc. XIII. Onderstepoort J Vet Sci Anim Indus 18(1-2):207-224.

berloque –see– *Citrus bergamia* Risso & Poit.

Bermuda buttercup –see– *Oxalis pes-caprae* L.

Bermuda oxalis –see– *Oxalis pes-caprae* L.

Bermudagrass –see– *Cynodon dactylon* (L.) Pers.

berry rue –see– *Cneoridium dumosum* (Nutt. ex Torr. & A. Gray) Hook. f. ex Baill.

BERSAMA ABYSSINICA Fresen. [Melianthaceae]

Citations:

Gourlay RN, Harker KW (1960) Bersama abyssinica poisoning: The clinical and pathological picture. J Comp Pathol 70(Oct):464-474.
Mugera GM (1970) Toxic and medicinal plants of East Africa. II. Bull Epizootic Dis Afr 18(4):389-403.
Verdcourt B, Trump EC (1969) Common poisonous plants of East Africa. Collins. London.

BERSAMA SWYNNERTONI Baker f. [Melianthaceae]

Citations:

Shone DK (1966) Poisonous plants of Rhodesia. Part I. Rhodesia Zambia Malawi J Agric Res 4:81-94.

berseem clover –see– *Trifolium alexandrinum* L.

BERTEROA INCANA (L.) DC. [Brassicaceae]

Common Names:

hoary alyssum; hoary false alyssum

Citations:

Ellison SP (1992) Possible toxicity caused by hoary alyssum (Berteroa incana). Vet Med 87(May):473-475.
Geor RJ, Becker RL, Kanara EW, et al. (1992) Toxicosis in horses after ingestion of hoary alyssum. J Am Vet Med Assoc 201(1):63-67.
Hovda LR, Rose ML (1993) Hoary alyssum (Berteroa incana) toxicity in a herd of broodmare horses. Vet Hum Toxicol 35(1):39-40.
Pammel LH (1921) Hoary alyssum. Vet Med 16:45.

BERTHOLLETTIA EXCELSA Bonpl. [Lecythidaceae]

Common Names:

Brazil nut; Para nut

Citations:

Asero R (2002) Birch and ragweed pollinosis north of Milan: A model to investigate the effects of exposure to "new" airborne allergens. Allergy 57(11):1063-1066.
Borja JM, Bartolome B, Gomez E, et al. (1999) Anaphylaxis from Brazil nut. Allergy 54(9):1007-1008.
Hjorth N, Roed-Petersen J (1976) Occupational protein contact dermatitis in food handlers. Contact Dermatitis 2(1):28-42.
Markson LS (1942) Dermatitis from seed and oil of Bertholletia excelsa (Brazil nut). Arch Derm Syphilol 46:831-832.

BERULA ERECTA (Huds.) Coville [Apiaceae]

Synonyms:

Berula thunbergii (DC.) H. Wolff; ***Sium thunbergii*** DC.

Common Names:

cow cress; creeping cut-leaf water parsnip; creeping water parsnip; narrow-leaf water parsnip; water parsnip

Citations:

Fernández de Corres L, Corrales JL, Muñoz D, et al. (1984) Dermatitis alérgicas de contacto por plantas. Allergol Immunopathol (Madr) 12(4):313-319.
Mogg AO (1927) Vlei poisoning. S Afr J Sci 24:269-277.

Berula thunbergii (DC.) H. Wolff = Berula erecta (Huds.) Coville

besedin palm –see– *Caryota mitis* Lour.

Besenginster –see– *Cytisus scoparius* (L.) Link

Besentrauch –see– *Cytisus scoparius* (L.) Link

besom –see– *Cytisus scoparius* (L.) Link

bessiboom –see– *Melia azedarach* L.

BETA VULGARIS L. [Chenopodiaceae]

Common Names:

beet; beterraba; betteraves; chard; fodder beet; Futterrübe; mangel; mangel wurzel; mangold; Mangold Rüben; remolacha; rote Bete; Rübe; Runkelrübe; silver beet; sugar beet; Swiss chard; Zuckerrübe

Citations:

Alexeyev NP (1976) [Parathyroid glands in metabolism of mineral substances in bulls fed pulp.] Selskok Biol 11(1):96-102.
Alibaşoğlu M, Ertürk E, Meriç I (1973) Kesif pancar posasiyla beslenen danalarda görülen encephalo-ocular syndrome. Vet Fakul Dergisi Ankara Univ 20:239-255.
Apostolov S, Toteshev I (1961) [Mass poisoning in cattle with beet.] Veterinarna Sbirka 58(6):14.
Ertürk E, Tekeli O, Gürocak B, et al. (1978) Semirtme amaciyla yüksek oranda kuru pancar posasi kapsayan rasyonlara beslenen Holstein-Friesian danalarinda saptanan sinirsel bozukluklar. Vet Fakul Dergisi Ankara Univ 25:99-104.
Fritzsch R (1966) Beitrag zur Futterintoxikation beim Rind durch Giftstoffe obligater Futterpflanzen. Monatsh Veterinarmed 21(9):327-331.
Fritzsch W (1960) Verdorbene Rübenblattsilage - Ursache einer akuten Erkrankung in einem Rinderbestand. Monatsh Veterinarmed 15:802-804.
González-Mancebo E, Alfaya T, Pulido Z, et al. (2000) Swiss chard-induced asthma. Allergy 55(5):511-512.
Gorb TV, Maksakov VY (1962) [Action of sugarbeet tops on cattle.] Veterinariia Moscow 39(2):66-68.
Gorišek J (1960) O procesu z grušavanja krviio kalciju u krvi i mokraći kod bolesti uzrokovane lišćem šećerne repe. Vet Arch 30(11-12):300-306.
Gorišek J (1963) Stoffwechselstörungen bei Milchkühen im Zusammenhang mit der Verfütterung von Zuckerrübenblatt. Proc World Vet Cong 2:1343-1344.
Gratzl E (1960) Eine durch Mangelfütterung (insbesondere Thiaminmangel) bedingte Encephalomyelopathie bei Pferden in Österreich. Wien Tierarztl Monatsschr 47(1):25-51.

Kachur MI (1972) [A case of cattle poisoning by sugar beets.] Veterinariia Moscow 49(10):106.
Krasnov VA (1965) [Mass poisoning of cows with sugar beet.] Veterinariia Moscow 42(9):69-70.
Lochkarev VA (1974) [Poisoning of cattle by sugar beet and maize cobs.] Veterinariia Moscow 51(9):97-99.
McIntosh IG, Nielson RL, Robinson WD (1943) Mangel poisoning in pigs. N Z J Agr 66:341-343.
Nielsen K, Kroght P, Møller T (1975) Forgiftning hos kvæg, forårsaget af parti sukkerroeaffald. Nord Vet Med 27(9):401-410.
Nový J (1960) K alimentárním intoxikacím vyvolaným řepnou siláží u skotu. Vet Cas 9(5):439-446.
O'Connor JG (1951) Sugar beet poisoning? Ir Vet J 5:259-260.
Pammel LH (1919) Frozen beet tops. Am J Vet Med 14:244.
Parkinson JG (1985) Fodder beet problems. Vet Rec 116(13):354.
Penny RH (1954) Suspected poisoning by fodder-beet in the bovine. Vet Rec 66(9):134.
Pinkiewicz E, Madej E, Samorek M (1963) Frische Futterrüben als Ursache von Ketose des Rindes. Proc World Vet Cong 2:1377-1378.
Price EL (1954) Fodder-beet poisoning in cattle. Vet Rec 66(12):182.
Rossow N, Czarnetzki G (1965) Geschlegeltes Zuckerrübenblatt als mögliche Krankheitsursache beim Rind. Monatsh Veterinarmed 20(23):952-957.
Savage A (1949) Nitrate poisoning from sugar beet tops. Can J Comp Med Vet Sci 13(1):9-10.
Scarisbrick R (1954) Acid indigestion in a sheep fed on mangolds. Vet Rec 66(9):131-132.
Shevtsova IN (1965) [Sugar beet poisoning in cows.] Veterinariia Moscow 42(3):72.
Simesen MG, Konggaard SP (1970) Eksperimentelle undersøgelser vedrørende roeforgiftning hos kvæg. Nord Vet Med 22(3):174-185.
Snoz GV, Ho VN, Trubnikov PM (1973) [Changes in the liver of cattle fattened on beet pulp.] Veterinariia Moscow 50(11):85-87.
Van der Walt SJ, Steyn DG (1946) Recent investigations into the toxicity of plants, etc. XV. Onderstepoort J Vet Sci Anim Indus 21(1):45-55.
Williams VJ, Coup MR (1959) Preliminary studies on the toxicity of fodder beet to sheep. N Z Vet J 7:8-14.
Worden AN, Bunyan J, Pickup J (1954) A fatal hypocalcemia-like syndrome in dairy cows following the excess consumption of fodder-beet. Vet Rec 66(9):133-134.
Zdelar F, Mitin V, Bišćan J, et al. (1967) Istraživanja o gušavosti goveda u hrvatskoj. IV. Utjeca svježeg lišća šećerne repe na funkciju štitnjače u krava muzara. Veterinarski Arhiv 37:208-218.
Zindler GA, Colovos GC (1950) Anthocyaninuria and beet allergy. Ann Allergy 8(5):603-617, 694.

Note:

Beet and other plants of this taxon are named *Beta vulgaris* L. subsp. *vulgaris* in some publications.

bété –see– *Mansonia altissima* (A. Chev.) A. Chev.

betek –see– *Carica papaya* L.

betel nut –see– *Areca catechu* L.

Betelkerne –see– *Areca catechu* L.

beterraba –see– *Beta vulgaris* L.

beteta –see– *Euphorbia pulcherrima* Willd. ex Klotzsch

Bethabara –see– *Tabebuia serratifolia* (Vahl) G. Nicholson

betoine aquatique –see– *Scrophularia auriculata* L.

betteraves –see– *Beta vulgaris* L.

BETULA LENTA L. [Betulaceae]

Common Names:

black birch; sweet birch

Citations:

Shelley WB (1964) Birch pollen and aspirin psoriasis. JAMA 189(13):985-988.

bhang –see– *Cannabis sativa* L.

bhindi –see– *Abelmoschus esculentus* (L.) Moench

bibba nut –see– *Semecarpus anacardium* L. f.

bibha –see– *Semecarpus anacardium* L. f.

bichi nut –see– *Semecarpus anacardium* L. f.

BIDENS PILOSA L. [Asteraceae]

Common Names:

cobbler's-pegs

Citations:

Mirvish SS, Salmasi S, Lawson TA, et al. (1985) Test of catechol, tannic acid, Bidens pilosa, croton oil, and phorbol for cocarcinogenesis of esophageal tumors induced in rats by methyl-n-amylnitrosamine. J Natl Cancer Instit 74(6):1283-1290.

Bicuculla canadensis (Goldie) Millsp. = Dicentra canadensis (Goldie) Walp.

Bicuculla cucullaria (L.) Millsp. = Dicentra cucullaria (L.) Bernh.

bidh-el-ghoul –see– *Mandragora officinarum* L.

Big Bend locoweed –see– *Astragalus mollissimus* Torr. var. earlei (Greene ex Rydb.) Tidestr.

Big Bend lupine –see– *Lupinus sericeus* Pursh

big galleta –see– *Pleuraphis rigida* Thurb.

big ivy –see– *Kalmia latifolia* L.

big-leaf maple –see– *Acer macrophyllum* Pursh

big-purge nut –see– *Jatropha curcas* L.

big sagebrush –see– *Artemisia tridentata* Nutt.

biga –see– *Alocasia macrorrhizos* (L.) G. Don

Bigelowia hartwegi A. Gray = Isocoma plurifolia (Torr. & A. Gray) Greene

bigfield fern –see– *Bowenia serrulata* (W. Bull) Chamb.

Bikukulla canadensis (Walp.) Druce = Dicentra canadensis (Goldie) Walp.

Bikukulla cucullaria Millsp. = Dicentra cucullaria (L.) Bernh.

billygoat weed –see– *Ageratum conyzoides* L.

Bilsen –see– *Hyoscyamus niger* L.

Bilsenbohne –see– *Hyoscyamus niger* L.

Bilsenkraut –see– *Hyoscyamus niger* L.

Bilsensee –see– *Hyoscyamus niger* L.

bindii –see– *Soliva sessilis* Ruiz & Pav.

bindweed –see– *Convolvulus arvensis* L.

bindwith –see– *Clematis vitalba* L.

bindy-eye –see– *Soliva sessilis* Ruiz & Pav.

Bingelkraut –see– *Mercurialis annua* L.

bini-da-zugu –see– *Jatropha curcas* L.; *Jatropha gossypiifolia* L.

binjai –see– *Mangifera caesia* Jack

bird cherry –see– *Prunus padus* L.

bird flower –see– *Crotalaria laburnifolia* L.

bird-of-paradise –see– *Caesalpinia gilliesii* (Hook.) D. Dietr.

bird pepper –see– *Capsicum frutescens* L.

birdlime thistle –see– *Chamaeleon gummifera* (L.) Cass.

bird's-eye –see– *Adonis annua* L.; *Anagallis arvensis* L.

bird's-foot indigo –see– *Indigofera linnaei* Ali

bird's-foot trefoil –see– *Lotus corniculatus* L.

birdseedgrass –see– *Phalaris canariensis* L.

Birdsville indigo –see– *Indigofera linnaei* Ali

birdwoodgrass –see– *Cenchrus setiger* Vahl

Birne –see– *Pyrus communis* L.

bish poison –see– *Aconitum ferox* Wall. ex Ser.

bishop's-weed –see– *Ammi majus* L.; *Ammi visnaga* (L.) Lam.

bisnaga –see– *Ammi visnaga* (L.) Lam.

bistort –see– *Bistorta officinalis* Delarbre

BISTORTA OFFICINALIS Delarbre [Polygonaceae]

Synonyms:

Polygonum bistorta L.

Common Names:

bistort

Citations:

Salgues R (1961) Le genre Polygonum L. (Polygonacées). Etudes chimiques et toxicologiques. Les faits nématologiques de l'empoisonnement expérimental. Qualitas Plant Mat Veg 8:367-395.

bitamo –see– *Pedilanthus tithymaloides* (L.) Poit.

biting clematis –see– *Clematis vitalba* L.

biting persicaria –see– *Persicaria hydropiper* (L.) Spach

Bittekonvall –see– *Maianthemum bifolium* (L.) F. W. Schmidt

bitter actinea –see– *Hymenoxys odorata* DC.

bitter almond –see– *Prunus dulcis* (Mill.) D. A. Webb

bitter apple –see– *Citrullus colocynthis* (L.) Schrad.; *Cucumis myriocarpus* Naudin; *Ecballium elaterium* (L.) A. Rich.; *Solanum incanum* L.; *Solanum panduriforme* E. Mey.

bitter bark –see– *Alstonia constricta* F. Muell.

bitter bush –see– *Chrysocoma ciliata* L.

bitter cassava –see– *Manihot esculenta* Crantz

bitter cucumber –see– *Momordica charantia* L.

bitter damsoe –see– *Simarouba amara* Aubl.

bitter gourd –see– *Citrullus colocynthis* (L.) Schrad.; *Momordica charantia* L.; *Momordica dioica* Roxb. ex Willd.

bitter lupin –see– *Lupinus angustifolius* L.

Bitter Mandel –see– *Prunus dulcis* (Mill.) D. A. Webb

bitter melon –see– *Citrullus lanatus* (Thunb.) Matsum. & Nakai; *Momordica charantia* L.

bitter nightshade –see– *Solanum dulcamara* L.

bitter orange –see– *Citrus aurantium* L.

bitter rubberweed –see– *Hymenoxys odorata* DC.; *Helenium amarum* (Raf.) H. Rock; *Helenium autumnale* L.

bitter vetch –see– *Lathyrus sativus* L.; *Vicia ervilia* (L.) Willd.

bitter water cress –see– *Barbarea vulgaris* R. Br.

bitterappel –see– *Solanum kwebense* N. E. Br.

bitterbos –see– *Chrysocoma ciliata* L.

bitterbossie –see– *Chrysocoma ciliata* L.

bitterkaroo –see– *Chrysocoma ciliata* L.

Bitterklee –see– *Menyanthes trifoliata* L.

bitterroot –see– *Apocynum cannabinum* L.; *Menyanthes trifoliata* L.

Bittersüß –see– *Solanum dulcamara* L.

Bittersüßer Nachtschatten –see– *Solanum dulcamara* L.

bittersweet –see– *Hymenoxys odorata* DC.; *Solanum dulcamara* L.

bittersweet nightshade –see– *Solanum dulcamara* L.

bitterweed –see– *Ambrosia artemisiifolia* L.; *Helenium amarum* (Raf.) H. Rock; *Helenium autumnale* L.; *Hymenoxys odorata* DC.; *Senecio glabellus* Poir.

bitterweed actinea –see– *Hymenoxys odorata* DC.

bitterwort –see– *Gentiana lutea* L.

biznaga –see– *Ammi visnaga* (L.) Lam.

blaargif –see– *Dichapetalum cymosum* (Hook.) Engl.

blaasoppies –see– *Gomphocarpus fruticosa* (L.) W. T. Aiton

black acacia –see– *Robinia pseudoacacia* L.

black alder –see– *Frangula alnus* Mill.

black bean –see– *Castanospermum australe* A. Cunn. & C. Fraser ex Hook.; *Erythrophleum chlorostachys* (F. Muell.) Baill.; *Phaseolus vulgaris* L.

black bindweed –see– *Fallopia convolvulus* (L.) Á. Löve

black birch –see– *Betula lenta* L.

black blossom cherry –see– *Andira inermis* (W. Wright) Kunth ex DC.

black bryony –see– *Dioscorea communis* (L.) Caddick & Wilkin

black catechu –see– *Acacia catechu* (L. f.) Willd.

black chebulic –see– *Terminalia chebula* Retz.

black cherry –see– *Atropa belladonna* L.; *Prunus laurocerasus* L.; *Prunus serotina* Ehrh.; *Prunus virginiana* L.

black cohosh –see– *Actaea racemosa* L.

black cumin –see– *Nigella sativa* L.

black datura –see– *Datura metel* L.

black dogwood –see– *Frangula alnus* Mill.

black elderberry –see– *Sambucus nigra* L.

black-eye root –see– *Dioscorea communis* (L.) Caddick & Wilkin

black-eye pea –see– *Vigna unguiculata* (L.) Walp.

black-eye Susan –see– *Abrus precatorius* L.

black greasewood –see– *Sarcobatus vermiculatus* (Hook.) Torr.

black heart –see– *Prunus myrtifolia* (L.) Urb.

black hellebore –see– *Helleborus niger* L.

black henbane –see– *Hyoscyamus niger* L.

black Indian hemp –see– *Apocynum cannabinum* L.

black jack pine –see– *Pinus ponderosa* C. Lawson

black kidney bean –see– *Phaseolus vulgaris* L.

black laurel –see– *Leucothoe davisiae* Torr. ex A. Gray

black locust –see– *Robinia pseudoacacia* L.

black mercury –see– *Toxicodendron radicans* (L.) Kuntze

black mulberry –see– *Morus nigra* L.

black mustard –see– *Brassica nigra* (L.) W. D. J. Koch

black nightshade –see– *Solanum nigrum* L.; *Solanum ptycanthum* Dunal

black nut –see– *Semecarpus anacardium* L. f.

black pea –see– *Pisum sativum* L.; *Swainsona luteola* F. Muell.

black pepper –see– *Piper nigrum* L.

black persimmon –see– *Diospyros texana* Scheele

black pigweed –see– *Trianthema portulacastrum* L.

black poisonwood –see– *Metopium brownei* (Jacq.) Urb.; *Metopium toxiferum* (L.) Krug & Urb.

black sagebrush –see– *Artemisia nova* A. Nelson

black Sally wattle –see– *Acacia salicina* Lindl.

black scours –see– *Goodia lotifolia* Salisb.; *Goodia medicaginea* F. Muell.

black snakeroot –see– *Actaea racemosa* L.; *Amianthium muscitoxicum* (Walter) A. Gray

black-spine nightshade –see– *Solanum anguivi* Lam.

black varnish tree –see– *Gluta renghas* L.

black vomit nut –see– *Jatropha curcas* L.

black walnut –see– *Juglans nigra* L.

black wattle –see– *Acacia salicina* Lindl.

black wild cherry –see– *Prunus serotina* Ehrh.

blackberry nightshade –see– *Solanum nigrum* L.

blackbrush –see– *Flourensia cernua* DC.

blackthorn –see– *Prunus spinosa* L.

blackwood –see– *Acacia melanoxylon* R. Br.

blackwort –see– *Symphytum officinale* L.

bladder cottonbush –see– *Gomphocarpus fruticosa* (L.) W. T. Aiton

bladder flower –see– *Araujia sericifera* Brot.

bladder soapwort –see– *Vaccaria hispanica* (Mill.) Rauschert

bladderpod –see– *Sesbania vesicaria* (Jacq.) Elliott

bladderpod locoweed –see– *Astragalus wootonii* E. Sheld.

Blaue Ilge –see– *Iris pseudoacorus* L.

Blaue Lupine –see– *Lupinus angustifolius* L.

Blauer Eisenhut –see– *Aconitum napellus* L.

Blauer Sturmhut –see– *Aconitum napellus* L.

Blauregen –see– *Wisteria sinensis* (Sims) DC.

bledo –see– *Amaranthus palmeri* S. Watson

bledo blanco –see– *Amaranthus viridis* L.

bleeding heart –see– *Dicentra cucullaria* (L.) Bernh.; *Lamprocapnos spectabilis* (L.) Fukuhara

blessed milk thistle –see– *Silybum marianum* (L.) Gaertn.

BLIGHIA SAPIDA K. D. Koenig [Sapindaceae]

Common Names:

ackee; akee; akee apple; faux acajou; fisa; gwanja kusa; heuvo vegetal; Ishin; isin; okpu; seso vegetal

Citations:

Anonymous (1992) Toxic hypoglycemic syndrome: Jamaica, 1989-1991. MMWR Morb Mortal Wkly Rep 41(4):53-55.

Arnold LE (1944) Vomiting sickness (or ackee poisoning). Govt Printer, Kingston.

Arnold LE (1947) Vomiting sickness - or ackee poisoning. Jamaica Med Rev 1:26-55.

Barennes H, Valea I, Boudat AM, et al. (2004) Early glucose and methylene blue are effective against unripe ackee apple (Blighia sapida) poisoning in mice. Food Chem Toxicol 42(5):809-815.

Belson M, Joskow R, Kaiser R, et al. (2001) An investigation of ackee fruit poisoning in Haiti. J Toxicol Clin Toxicol 39(5):547.

Chambers HD (1953) The syndrome called "vomiting sickness." West Indian Med J 2(1):37-42.

Connal A, Ralston W (1918) Some experiments with the fruit of Blighia sapida in Nigeria. J Trop Med Hyg 21(8):81-84.

Doughty DD, Larson E (1960) Tissue changes in experimental ackee poisoning. Pathological changes produced in rabbits. Trop Geogr Med 12(Sep):243-250.

Escoffery CT, Shirley SE (2004) Fatal poisoning in Jamaica: A coroner's autopsy study from the University Hospital of the West Indies. Med Sci Law 44(2):116-120.

Evans KL, Arnold LE (1938) Experimental studies of poisoning with ackee (Blighia sapida). Trans R Soc Trop Med Hyg 32(3):355-362.

Fitzmaurice LW (1953) The vomiting sickness of Jamaica. West Indian Med J 2:93-124.

Foungbe S, Naho Y, Declume C (1986) Etude expérimentale de la toxicité des arilles de Blighia sapida (Sapindacées) en rapport avec l'intoxication des enfants de Katiola (Côte-d'Ivoire). Ann Pharm Fr 44(6):509-515.

Fox HC, Miller DS (1960) Ackee toxin: A riboflavin antimetabolite? Nature 186(4724):561-562.

Golden KD, Kean EA, Terry SI (1984) Jamaican vomiting sickness: A study of two adult cases. Clin Chim Acta 142(3):293-298.

Hill KR, Bras G, Clearkin KP (1955) Acute toxic hypoglycaemia occurring in the vomiting sickness of Jamaica: Morbid anatomical aspects. West Indian Med J 4(2):91-104.

Jelliffe DB, Stuart KL (1954) Acute toxic hypoglycaemia in the vomiting sickness of Jamaica. Br Med J 1(4853):75-77.

Jordan EO, Burrows W (1937) The vomiting sickness of Jamaica, B. W. I. and its relation to akee poisoning. Am J Hyg 25:520-545.

Kean BH (1943) Death due to akee poisoning in Panama. Am J Trop Med 23:339-341.

Larson E, Wynn MF, Lynch SJ, et al. (1953) Some further studies on the akee. Q J Florida Acad Sci 16(3):151-156.

Larson J, Vender R, Camuto P (1994) Cholestatic jaundice due to ackee fruit poisoning. Am J Gastroenterol 89(9):1577-1578.

McIntosh RM, Andrews J (1971) Blighia sapida. Toxic effects on renal morphology and function in rat. N Y State J Med 71(13):1639-1643.

Meda HA, Diallo B, Buchet JP, et al. (1999) Epidemic of fatal encephalopathy in preschool children in Burkina Faso and consumption of unripe ackee (Blighia sapida) fruit. Lancet 353(9152):536-540.

Moya J (2001) Ackee (Blighia sapida) poisoning in the Northern Province, Haiti, 2001. Epidemiol Bull 22(2):8-9.

Quere M, Ogouassangni A, Bokossa A, et al. (1999) Methylene blue and fatal encephalopathy from ackee fruit poisoning. Lancet 353(9164):1623.

Resiere D, Megarbane B, Gueye PN, et al. (2001) Acute poisoning with ackee fruit in Jamaica. J Toxicol Clin Toxicol 39(3):311-312.

Scott HH (1916) On the "vomiting sickness" of Jamaica. Ann Trop Med Parasitol 10:1-79.

Stuart KL, Jelliffe DB, Hill KR (1955) Acute toxic hypoglycaemia occurring in the vomiting sickness of Jamaica (Clinical aspects). J Trop Pediatr 1(2):69-87.

Tanaka K (1973) Isovaleric acidaemia and its induction in experimental animals by hypoglycin A. In: Hommes FA, Van den Berg CJ (eds.) Inborn errors of metabolism. Academic Press. New York. pp. 269-289.

Tanaka K, Kean EA, Johnson B (1976) Jamaican vomiting sickness. Biochemical investigation of two cases. N Engl J Med 295(9):461-467.

blind-eyes –see– *Papaver rhoeas* L.

blind-your-eyes –see– *Excoecaria parvifolia* Müll. Arg.

blindgrass –see– –see– *Stypandra imbricata* R. Br.

blinkblaar –see– *Dichapetalum cymosum* (Hook.) Engl.

blister buttercup –see– *Ranunculus sceleratus* L.

blister plant –see– *Ranunculus acris* L.

blister weed –see– *Thamnosma texana* (A. Gray) Torr.

blood flower –see– *Asclepias curassavica* L.

blood lily –see– *Scadoxus multiflorus* (Martyn) Raf.

bloodroot –see– *Sanguinaria canadensis* L.

bloodwort –see– *Sanguinaria canadensis* L.

bloody finger –see– *Digitalis purpurea* L.

bloody-man's-finger –see– *Arum maculatum* L.

blotch-leaf laurel –see– *Aucuba japonica* Thunb.

blou oliebloom –see– *Datura stramonium* L.

blou stinkblaar –see– *Datura stramonium* L.

blou tulp –see– *Moraea setacea* Ker-Gawl.

bloubos –see– *Diospyros lycioides* Desf. subsp. lycioides

blousaad –see– *Panicum schinzii* Hack.

blousaadgras –see– *Megathyrsus maximus* (Jacq.) B. K. Simon & S. W. L. Jacobs

blue bean –see– *Lupinus argenteus* Pursh

blue bells-of-England –see– *Hyacinthoides non-scripta* (L.) Chouard ex Rothm.

blue billygoat weed –see– *Ageratum houstonianum* Mill.

blue bindweed –see– *Solanum dulcamara* L.

blue bottle –see– *Hyacinthoides non-scripta* (L.) Chouard ex Rothm.

blue canarygrass –see– *Phalaris coerulescens* Desf.

blue chicory –see– *Cichorium intybus* L.

blue cohosh –see– *Caulophyllum thalictroides* (L.) Michx.

blue devil –see– *Echium vulgare* L.

blue elderberry –see– *Sambucus cerulea* Raf.

blue euphorbia –see– *Euphorbia caerulescens* Haw.

blue false indigo –see– *Baptisia australis* (L.) R. Br.

blue gum eucalyptus –see– *Eucalyptus globulus* Labill.

blue heliotrope –see– *Heliotropium amplexicaule* Vahl

blue jack oak –see– *Quercus incana* W. Bartram

blue larkspur –see– *Delphinium nuttallianum* Pritz.

blue locoweed –see– *Astragalus lentiginosus* Douglas ex Hook. *Astragalus lentiginosus* Douglas ex Hook. var. diphysus (A. Gray) M. E. Jones

blue lupin –see– *Lupinus angustifolius* L.

blue morning-glory –see– *Ipomoea tricolor* Cav.

blue nightshade –see– *Solanum dulcamara* L.

blue oak –see– *Quercus douglasii* Hook. & Arn.

blue pea –see– *Lupinus leucophyllus* Douglas ex Lindl.

blue periwinkle –see– *Vinca major* L.

blue pimpernel –see– *Anagallis arvensis* L.

blue rocket –see– *Aconitum napellus* L.

blue sage –see– *Salvia reflexa* Hornem.

blue sailors –see– *Cichorium intybus* L.

blue-star morning-glory –see– *Ipomoea violacea* L.

blue thistle –see– *Echium vulgare* L.

blue tulip –see– *Moraea setacea* Ker Gawl.

blue tulp –see– *Moraea polystachya* (Thunb.) Ker Gawl.

blue vervain –see– *Verbena officinalis* L.

blue water –see– *Ipomoea tricolor* Cav.

blue wild indigo –see– *Baptisia australis* (L.) R. Br.

bluebell –see– *Hyacinthoides non-scripta* (L.) Chouard ex Rothm.

blueberry elderberry –see– *Sambucus mexicana* C. Presl ex DC.

blueweed –see– *Drimia sanguinea* (Schinz) Jessop; *Echium vulgare* L.; *Perilla frutescens* (L.) Britton

Blumenkohl –see– *Brassica oleracea* L. var. botrytis L.

Blutkraut –see– *Chelidonium majus* L.

Blutschierling –see– *Conium maculatum* L.

boar's-foot –see– *Helleborus viridis* L.

bobbing Joan –see– *Arum maculatum* L.

bobbins –see– *Arum maculatum* L.

BOBGUNNIA MADAGASCARIENSIS (Desv.) J. H. Kirkbr. & Wiersema [Fabaceae]

Synonyms:

Swartzia madagascariensis Desv.

Common Names:

snake bean

Citations:

Kurki J, Gibson JD (1985) As a cause of red urine in cattle. Zimbabwe Vet J 16(1-2):23.

Perchman GE (1978) Toxicity of Swartzia madagascariensis Desv. J S Afr Vet Assoc 49(4):362.

Bocksdorn –see– *Lycium barbarum* L.

bodark –see– *Maclura pomifera* (Raf.) C. K. Schneid.

bodhangero –see– *Lagerstroemia parviflora* Roxb.

bog arum –see– *Calla palustris* L.

bog asphodel –see– *Narthecium ossifragum* (L.) Huds.

bog bean –see– *Menyanthes trifoliata* L.

bog laurel –see– *Kalmia microphylla* (Hook.) A. Heller

bog onion –see– *Arisaema triphyllum* (L.) Schott

boggabri –see– *Chenopodium carinatum* R. Br.

boh-gol-zhee –see– *Cullen corylifolium* (L.) Medik.

Bohne –see– *Phaseolus coccineus* L.; *Phaseolus vulgaris* L.

Bohnenbaum –see– *Laburnum anagyroides* Medik.

Bohnenstrauch –see– *Laburnum anagyroides* Medik.

boi –see– *Sinomenium acutum* (Thunb.) Rehder & E. H. Wilson

bois d'arc –see– *Maclura pomifera* (Raf.) C. K. Schneid.

bois-de-pavana –see– *Croton tiglium* L.

bois-des-molluques –see– *Croton tiglium* L.

bois eniorant –see– *Deguelia utilis* (A. C. Sm.) A. M. G. Azevedo

bois gentil –see– *Daphne mezereum* L.

bois-gris-jaunàtre –see– *Sarcocephalus diderrichii* De Wild.

bois joli –see– *Daphne mezereum* L.

bois mulatre –see– *Metopium brownei* (Jacq.) Urb.

bois purgatif –see– *Croton tiglium* L.

Bokhara Klee –see– *Melilotus albus* Medik.

bokuintjie –see– *Moraea setacea* Ker Gawl.

boldo –see– *Peumus boldus* Molina

boldus –see– *Peumus boldus* Molina

Bolmört –see– *Hyoscyamus niger* L.

bolsa-de-pastor –see– *Capsella bursa-pastoris* (L.) Medik.

BOMBAX BREVICUSPE Spreng. [Bombacaceae]

Citations:

Oleffe JA, Sporcq J, Hublet P (1975) Epidemiological study of the wood industry in Belgium. Contact Dermatitis 1(5):315-316.

BOMBAX CHEVALIERI Pellegr. [Bombacaceae]

Citations:

Oleffe JA, Sporcq J, Hublet P (1975) Epidemiological study of the wood industry in Belgium. Contact Dermatitis 1(5):315-316.

bomubomu –see– *Calotropis procera* (Aiton) W. T. Aiton

bonduc –see– *Gymnocladus dioicus* (L.) K. Koch

boneset –see– *Ageratina altissima* (L.) R. M. King & H. Rob.; *Symphytum officinale* L.

Bongkrek –see– *Cocos nucifera* L.

bonnet-de-prêtre –see– *Euonymus europaeus* L.

boobialla –see– *Myoporum acuminatum* R. Br.; *Myoporum tetrandrum* (Labill.) Domin

boobyalla –see– *Myoporum insulare* R. Br.

boon tree –see– *Sambucus nigra* L.

boonaree –see– *Alectryon oleifolius* (Desf.) S. T. Reynolds

boonery –see– *Alectryon oleifolius* (Desf.) S. T. Reynolds

BOOPHONE DISTICHA (L. f.) Herb. [Amaryllidaceae]

Synonyms:

Buphane disticha (L. f.) Herb.; ***Buphane toxicaria*** Herb.

Common Names:

candelabra flower; Cape poison bulb; gifbloom; gifbol; kafir onion; lesoma; oxkiller; seeroogblom; sore-eye flower

Citations:

du Plooy WJ, Swart L, van Huysteen GW (2001) Poisoning with Boophane disticha: A forensic case. Hum Exp Toxicol 20:277-278.

Gordon I (1947) A case of fatal buphanine poisoning. Clin Proc 6:90-93.

Laing RO (1979) Three cases of poisoning by Boophane disticha. Cent Afr J Med 25(12):265-266.

Nyazema NZ (1986) Herbal toxicity in Zimbabwe. Trans R Soc Trop Med Hyg 80(3):448-450.

Steyn DG (1934) Recent investigations into the toxicity of known and unknown poisonous plants in the Union of South Africa. Onderstepoort J Vet Sci Anim Indus 3(1):125-130.

boor tree –see– *Sambucus nigra* L.

booroo molie –see– *Goodia lotifolia* Salisb.

bootlace plant –see– *Pimelea trichostachya* Lindl.

booty tree –see– *Sambucus nigra* L.

bopple nut –see– *Macadamia integrifolia* Maiden & Betche; *Macadamia tetraphylla* L. A. S. Johnson

BORASSUS FLABELLIFER L. [Arecaceae]

Common Names:

palmyra palm

Citations:

Arseculeratne SN, Grieg JB, Sirisinha S (1984) The immunosuppressive effect of palmyrah (Borassus flabellifer) flour is not associated with its neurotoxic fraction. Asian Pac J Allergy Immunol 2(1):13-16.

Arseculeratne SN, Panabokke RG, Tennekoon GE, et al. (1971) Toxic effects of Borassus flabellifer (palmyrah palm) in rats. Br J Exp Pathol 52(5):524-537.

Panabokke RG, Arseculeratne SN (1976) Veno-occlusive lesions in the liver of rats after prolonged feeding with palmyrah (Borassus flabellifer) flour. Br J Exp Pathol 57(2):189-199.

Sumudunie KA, Jansz ER, Jayasekera S, et al. (2004) The neurotoxic effect of palmyrah (Borassus flabellifer) flour re-visited. Int J Food Sci Nutr 55(8):607-614.

Borgia's-bouquet –see– *Pimelea trichostachya* Lindl.

borico –see– *Plumbago scandens* L.

borrachero –see– *Brugmansia* ×*candida* Pers.

borsteuphorbia –see– *Euphorbia leuconeura* Boiss.

BOSCIA FOETIDA Schinz [Capparaceae]

Common Names:

noeniebossie; oumiedbos; stinkbossie

Citations:

Van der Walt SJ, Steyn DG (1943) Recent investigations into the toxicity of plants, etc. XIII. Onderstepoort J Vet Sci Anim Indus 18(1-2):207-224.

BOSWELLIA SERRATA Roxb. ex Colebr. [Burseraceae]

Citations:

Acebo E, Ratón JA, Sautúa S, et al. (2004) Allergic contact dermatitis from Boswellia serrata extract in a naturopathic cream. Contact Dermatitis 51(2):91-92.

bottle gourd –see– *Lagenaria siceraria* (Molina) Standl.

bottlebrush –see– *Crotalaria burkeana* Benth.; *Equisetum arvense* L.

botuje –see– *Jatropha curcas* L.; *Jatropha multifida* L.

botuje papa –see– *Jatropha gossypiifolia* L.

bouquet corail –see– *Jatropha multifida* L.

bourdaine –see– *Frangula alnus* Mill.

bourse-à-pasteur –see– *Capsella bursa-pastoris* (L.) Medik.

bourtree –see– *Sambucus nigra* L.

bouton d'or –see– *Ranunculus acris* L.

bowwood –see– *Maclura pomifera* (Raf.) C. K. Schneid.

BOWDICHIA NITIDA Spruce ex Benth. [Fabaceae]

Common Names:

sebipira; sicopira; sucupira

Citations:

Gonçalo, S (1992) Allergic contact dermatitis from Bowdichia nitida (sucupira) wood. Contact Dermatitis 26(3):205.

Hausen BM, Simatupang MH, Kingreen JC (1972) Untersuchungen zur Überempfindlichkeit gegen Sucupira - und Palisanderholz. Derm Beruf Umwelt 20(1):1-7.

BOWENIA SERRULATA (W. Bull) Chamb. [Zamiaceae]

Common Names:

bigfield fern; Byfield fern

Citations:

Anderson JL, Hall WT (1964) Neurotoxic effects from cycad leaves. Fed Proc 23:1349.

Hall WT, McGavin MD (1968) Clinical and neuropathological changes in cattle eating the leaves of Macrozamia lucida or Bowenia serrulata (Family Zamiaceae). Pathol Vet 5(1):26-34.

Seawright AA, Oelrichs PB, Ng JC, et al. (1998) The toxicity of the Australian cycad Bowenia serrulata to cattle. In: Garland T, Barr AC (eds.) Toxic plants and other natural toxicants. CABI. New York. pp. 447-452.

bowman's-root –see– *Apocynum cannabinum* L.

box –see– *Buxus sempervirens* L.

boxberry –see– *Gaultheria procumbens* L.

boxchechem –see– *Metopium brownei* (Jacq.) Urb.

boxthorn –see– *Lycium barbarum* L.

boxwood –see– *Buxus sempervirens* L.

boys-and-girls –see– *Mercurialis annua* L.

Brachiaria brizantha (Hochst. ex A. Rich) Stapf = Urochloa brizantha (Hochst. ex A. Rich.) R. D Webster

Brachiaria decumbens Stapf = Urochloa decumbens (Stapf) R. D. Webster

Brachiaria purpurescens (Raddi) Henry = Urochloa mutica (Forssk.) T. Q. Nguyen

Brachiaria radicans Napper = Urochloa arrecta (Hack. ex T. Durand & Schinz) Morrone & Zuloaga

Brachiaria ruziziensis R. Germ. & C. M. Evrard = Urochloa ruziziensis (R. Germ. & C. M. Evrard) Crins

BRACHYACHNE CONVERGENS (F. Muell.) Stapf [Poaceae]

Common Names:

gulf stargrass; Kimberley couch; native couchgrass; spidergrass

Citations:

Anonymous (1940) A native couch grass dangerous to stock. Queensland Agric J 54:267.

BRACHYCHITON POPULNEUS (Schott & Endl.) R. Br. [Malvaceae]

Synonyms:

Sterculia diversifolia G. Don

Common Names:

kurrajong

Citations:

Setchell BP, McInnes P, Christie DG (1964) Poisoning of sheep with anthelmintic doses of carbon tetrachloride. IV. Effect of stinkwort (Inula graveolens). Aust Vet J 40(Jan):30-31.

BRACHYGLOTTIS REPANDA J. R. Forst. & G. Forst. [Asteraceae]

Common Names:

rangiora

Citations:

Mortimer PH, White EP (1967) Hepatotoxic substance in Brachyglottis repanda. Nature 214(5094):1255-1256.

bracken fern –see– *Pteridium aquilinum* (L.) Kuhn

brahmi –see– *Centella asiatica* (L.) Urb.

brake fern –see– *Pteridium aquilinum* (L.) Kuhn

braken fern –see– *Pteridium aquilinum* (L.) Kuhn

bran –see– *Triticum aestivum* L.

branched larkspur –see– *Consolida regalis* Gray

branched onion weed –see– *Trachyandra divaricata* (Jacq.) Kunth

bransbossie –see– *Chrysocoma ciliata* L.

Brassaia actinophylla Endl. = Schefflera actinophylla (Endl.) Harms

Brassica alba (L.) Rabenh. = Sinapis alba L.

Brassica arvensis (L.) Rabenh. = Sinapis arvensis L.

Brassica campestris L. = Brassica rapa L. subsp. campestris (L.) A. R. Clapham

BRASSICA JUNCEA (L.) Czern. [Brassicaceae]

Common Names:

brown mustard; Indian mustard; leaf mustard; oriental mustard; sarson

Citations:

Katamoto H, Nishiguchi S, Harada K, et al. (2001) Suspected oriental mustard (Brassica juncea) intoxication in cattle. Vet Rec 149(7):215-216.

Kernaleguen A, Smith RA, Yong CW (1989) Acute mustard seed toxicosis in beef cattle. Can Vet J 30(6):524.

Poulsen E (1958) Forgiftning med myrosinasefri sennepsskrå hos kvæg. Nord Vet Med 10:487-497.

Semalulu SS, Rousseaux CG (1989) Suspected oriental mustard seed (Brassica juncea) poisoning in cattle. Can Vet J 30(7):595-596.

Brassica kaber (DC.) L. C. Wheeler = Sinapis arvensis L.

Brassica napus L. subsp. oleifera (Delile) Sinskaya = Brassica napus L. var. napus

BRASSICA NAPUS L. var. napobrassica (L.) Rchb. [Brassicaceae]

Synonyms:

Brassica rutabaga DC. ex H. Lév.

Common Names:

rutabaga; Swede; Swede turnip

Citations:

Bobek S (1968) Właściwości wolotwórcze i przeciwwolotwórcze żółtej brukwi (Brassica rutabaga). Endokrynol Pol 19(1):47-60.

Debackere M, Hoorens J, Haustraete KH (1966) Vergiftiging door koolzaad - En raapzaadschroot. Vlaams Diergeneesk Tijdschr 35(9-10):393-399.

Kennedy TH, Purves HD (1941) Studies on experimental goitre. I. The effect of Brassica seed diet on rats. Br J Exp Pathol 22(5):241-244.

Simesen MG, Konggaard SP (1970) Eksperimentelle undersøgelser vedrørende roeforgiftning hos kvæg. Nord Vet Med 22(3):174-185.

BRASSICA NAPUS L. var. napus [Brassicaceae]

Synonyms:

Brassica napus L. subsp. oleifera (Delile) Sinskaya

Common Names:

canola; cole; colza; green canola; rape; Raps; rapsfromel; winter rape

Citations:

Allen CE, Dow DS (1952) The biological assessment of the value of rapeseed oil meal as a dietary component. Sci Agric 32:403-410.

Allen JG, Creeper JH, Forshaw D, et al. (2004) Plant-associated diseases, either new or new to the state, encountered over the last decade (1991-2001) in Western Australia. In: Acamovic T, Stewart CS, Pennycott TW (eds.) Poisonous plants and related toxins. CABI. Wallingford, UK. pp. 540-547.
Anderson GH, Harris L, Rao AV, et al. (1976) Trace mineral deficiencies in rats caused by feeding rapeseed flours during growth, gestation and lactation. J Nutr 106:1166-1174.
Bachmann M, Theus R, Lüthy J, et al. (1985) Vorkommen von goitrogenen Stoffen in Milch. 1. Übergang von Goitrin in die Milch von Kühen bei Verfütterung von Rapsextraktionsschrot. Z Lebensm Unters Forsch 181(5):375-378.
Bäckgren AW, Jönsson G (1969) Blood and bone marrow studies in cattle feeding on Brassica species. Acta Vet Scand 10(4):309-318.
Baglioni T, Locatelli A, Longo L, et al. (1966) Ulteriori osservazioni sulla azione della alimentazione con colza nel bovino. Arch Vet Ital 17(5):343-356.
Baglioni T, Locatelli A, Longo L, et al. (1968) Rilievi sull' alimentazione esclusiva con colza. Nota I. Osservazioni sul coniglio. Clin Vet (Milano) 91(2):16-22.
Barnouin J, Chalus T, Lescourret F (1992) Increased perinatal French dairy calf mortality associated with fresh rape in the prepartum diet. Prev Vet Med 12:111-120.
Barnouin J, Paccard P (1988) Facteurs de risque nutritionnels de la pathologie hépatique dans les troupeaux bovins laitiers en France. Can Vet J 29(11):915-920.
Bell JM, Williams K (1953) Growth depressing factors in rapeseed oilmeal. Can J Agric Sci 33(May-Jun):201-209.
Bhatnagar MK, Singh A, Yamashiro S, et al. (1977) Ultrastructural alterations in the heart of rats fed rapeseed oil diets. Anat Histol Embryol 6:366.
Bhatnagar MK, Yamashiro S, David LL (1979) Ultrastructural study of liver of turkeys fed rapeseed products. Anat Histol Embryol 8:182.
Blakeley RM, Anderson RW (1948) Studies with rape-seed oilcake meal. I. The effect of various levels of rape-seed oilcake meal in the diet on the weight of the thyroid glands of turkey poults. Sci Agric 28(Sep):393-397.
Blakeley RM, Anderson RW (1948) Studies with rape-seed oilcake meal. II. The effect of the inclusion of protamone in the diet, on the thyroid enlargement induced by the feeding of rapeseed oilcake meal to turkey poults. Sci Agric 28(Sep):398-402.
Bromidge ES, Wells JW, Wight PA (1985) Elevated bile acids in the plasma of laying hens fed rapeseed meal. Res Vet Sci 39(3):378-382.
Clandinin DR, Bayly L (1960) Rapeseed oil meal studies. 2. Effects of feeding rapeseed oil meal on the structure of the thyroid glands and chickens. Poult Sci 39:1239-1240.
Crawshaw HA (1953) Rape blindness. Vet Rec 65(16):254.
Dalton PJ (1953) Rape blindness. Vet Rec 65(19):298.
de Wildt DJ, Speijers GJ (1984) Influence of dietary rapeseed oil and erucic acid upon myocardial performance and hemodynamics in rats. Toxicol Appl Pharmacol 74(1):99-108.
Dixon PM, McGorum B (1990) Oilseed rape and equine respiratory disease. Vet Rec 126(23):585.
Eklund A, Agren G, Langler T, et al. (1971) Rapeseed protein fractions. II. Chemical composition and biological quality of a lipid-protein concentrate from rapeseed (Brassica napus L.). J Sci Food Agric 22:653-657.
Evans ET (1951) Kale and rape poisoning in cattle. Vet Rec 63(19):348-349.
Friend DW, Gilka F, Corner AH (1975) Growth, carcass quality and cardiopathology of boars and gilts fed diets containing rapeseed and soybean oils. Can J Anim Sci 55(4):571-578.
Friend DW, Kramer JK, Corner AH (1976) Growth and cardiopathology in boars fed rapeseed oil. Can J Anim Sci 56(Jun):361-364.
González N, Haardt E, Potoćnjak J (1964) El afrecho de raps en la alimentacion de aves (aves de 18 semanas a 18 meses). Bol Prod Anim (Chile) 2(2):101-120.
Grandhi RR, Slinger SJ, Summers JD (1977) Productive performance and liver lesions in two strains of laying hens receiving two rapeseed meals. Poult Sci 56(6):1904-1908.
Hall AE (1954) Fatal hemorrhages in chickens, possibly due to rape pasture. Michigan State Coll Vet 14(Spring):148-149.
Hawrysh ZJ, Clandinin DR, Robblee AR, et al. (1975) Influence of rapeseed meal on the odor and flavor of eggs from different breeds of chickens. Can Inst Food Sci Technol J 8(1):51-54.
Huey IB (1958) Rape poisoning in cattle. Ir Vet J 12:83.
Hulan HW, Hunsaker WG, Kramer JK, et al. (1976) The development of dermal lesions and alopecia in male rats fed rapeseed oil. Can J Physiol Pharmacol 54(1):1-6.
Hulan HW, Kramer JK, Mahadevan S, et al. (1976) Effect of cold stress on rapeseed oil fed rats. Lipids 11(1):6-8.
Israels ED, Papas A, Campbell LD, et al. (1979) Prevention by menadione of the hepatotoxic effects in chickens fed rapeseed meal. Observations on coagulation factors and cytochrome P-450. Gastroenterology 76(3):584-589.
Iwarsson K (1973) Rapeseed meal as a protein supplement for dairy cows. I. The influence on certain blood and milk parameters. Acta Vet Scand 14(4):570-594.
Iwarsson K, Ekman L, Everitt BR, et al. (1973) The effect of feeding rapeseed meal on thyroid function and morphology in growing bulls. Acta Vet Scand 14(4):610-629.
Iwarsson K, Nilsson, PO (1973) Rapeseed meal as a protein supplement for dairy cows. II. Investigations in rats on the goitrogenic properties of milk from cows fed rapeseed meal. Acta Vet Scand 14(4):595-609.
Jackson N (1969) Toxicity of rapeseed meal and its use as a protein supplement in the diet of two hybrid strains of caged laying hens. J Sci Food Agric 20(12):734-740.
Jackson N (1970) Algerian and French rapeseed meals as a protein source for caged laying hens, with observations on their toxic effects. J Sci Food Agric 21(10):511-516.
Josefsson E, Uppström B (1976) Influence of glucosinolates and native enzymes on the nutritional value of low-glucosinolate rapeseed meal. J Sci Food Agric 27(5):433-437.
Kennedy TH, Purves HD (1941) Studies on experimental goitre. I. The effect of Brassica seed diet on rats. Br J Exp Pathol 22(5):241-244.
Kratzer FH, Davis PN, Williams DE, et al. (1954) Factors influencing the growth of chicks and poults fed rations containing rapeseed oil meal. J Nutr 53:407-418.
Lo MT, Hill DC (1972) Effect of dietary rapeseed meal on the serum protein of rats. Can J Physiol Pharmacol 49(12):1100-1105.
Lodhi GN, Renner R, Clandinin DR (1969) Studies on the metabolizable energy of rapeseed meal for growing chickens and laying hens. Poult Sci 48(3):964-970.

Loew FM, Doige CE, Manns JG, et al. (1976) Evaluation of dietary rapeseed protein concentrate flours in rats and dogs. Toxicol Appl Pharmacol 35(2):257-267.
Marangos AG, Hill R (1977) The influence of rapeseed and mustardseed meals on reproductive efficiency in gilts. Br Vet J 133(1):46-55.
March BE, Biely J, Soong R (1972) Rapeseed meal in the chicken breeder diet. Effects on production, mortality, hatchability, and progeny. Poult Sci 51(5):1589-1596.
March BE, Biely J, Soong R (1975) The effects of rapeseed meal fed during the growing and/or laying periods on mortality and egg production in chickens. Poult Sci 54(6):1875-1882.
March BE, Bragg DB, Soong R (1978) Low erucic acid, low glucosinolate rapeseed meal, with and without added gums in the layer diet. Poult Sci 57:1599-1604.
March BE, Soong R (1976) Mortality and production characteristics of laying chickens fed high- and low-erucic acid rapeseed oils. Poult Sci 55(4):1557-1560.
Martland MF, Butler EJ, Fenwick GR (1984) Rapeseed induced liver haemorrhage, reticulolysis and biochemical changes in laying hens: The effects of feeding high and low glucosinolate meals. Res Vet Sci 36(3):298-309.
McCutcheon JS, Umermura T, Bhatnagar MK, et al. (1976) Cardiopathogenicity of rapeseed oils and oil blends differing in erucic, linoleic, and linolenic acid content. Lipids 11(7):545-552.
McKinnon PJ, Bowland JP (1979) Effects of feeding low and high glucosinolate rapeseed meals and soybean meal on thyroid function of young pigs. Can J Anim Sci 59(3):589-596.
Meding B (1985) Immediate hypersensitivity to mustard and rape. Contact Dermatitis 13(2):121-122.
Monkiewicz J, Kinal S, Łuczak W (1977) Poziom witaminy A w wątrobie kurczqt rzeźnych a zawartość substancji toksycznych w śrutach rzepakowych. Med Weter 33(2):104-106.
Nordfeldt S, Gellerstedt N, Falkmer S (1954) Studies of rapeseed meal and its goitrogenic effects on pigs; A nutritional and histopathological study. Acta Pathol Microbiol Scand 35(2):217-236.
O'Driscoll J (1958) Rape poisoning in cattle. Ir Vet J 12:82.
Onderscheka K, Tataruch F, Steineck T, et al. (1987) Gehäufte Rehwildverluste nach Aufnahme von 00-Raps. Z Jagdwiss 33(3):191-205.
Onderscheka K, Tataruch F, Steineck T, et al. (1987) Untersuchungen über die durch Rapsaufnahme bedingten Rehwildverluste. Wien Tierarztl Monatsschr 74(11):369-379.
O'Neil JB (1957) Rapeseed oil meal as a vegetable protein supplement in the diet of laying and breeding hens. Poult Sci 36:1146.
Papas A, Campbell LD, Cansfield PE (1979) A study of the association of glucosinolates to rapeseed meal-induced haemorrhagic liver in poultry and the influence of supplemental vitamin K. Can J Anim Sci 59(Mar):133-144.
Papas A, Ingalls JR, Campbell LD (1979) Studies on the effects of rapeseed meal on thyroid status of cattle, glucosinolate and iodine content of milk and other parameters. J Nutr 109(7):1129-1139.
Pearson AW, Butler EJ, Curtis RF, et al. (1978) Effects of rapeseed meal on laying hens (Gallus domesticus) in relation to fatty liver-haemorrhagic syndrome and egg taint. Res Vet Sci 25(3):307-313.
Pearson AW, Butler EJ, Curtis RF, et al. (1979) Effect of rapeseed meal on hepatic trimethylamine oxidase activity in the domestic fowl in relation to egg taint. J Sci Food Agric 30(3):291-298.
Pearson AW, Butler EJ, Curtis RF, et al. (1979) Effect of rapeseed meal on trimethylamine metabolism in the domestic fowl in relation to egg taint. J Sci Food Agric 30(8):799-804.
Pearson AW, Butler EJ, Curtis RF, et al. (1979) Rapeseed meal and egg taint: Demonstration of the metabolic defect in male and female chicks. Vet Rec 104:318-319.
Pearson AW, Butler EJ, Fenwick GR (1979) Rapeseed meal and liver damage: Effect on plasma enzyme activities in chicks. Vet Rec 105(9):200-201.
Pearson AW, Butler EJ, Fenwick GR (1979) Rapeseed meal goitrogens and egg taint. Vet Rec 104(Feb 24):168.
Pearson AW, Greenwood NM, Butler EJ, et al. (1980) Low glucosinolate rapeseed meals and egg taint. Vet Rec 106:560.
Pearson AW, Greenwood NM, Butler EJ, et al. (1983) Biochemical changes in layer and broiler chickens when fed on a high-glucosinolate rapeseed meal. Br Poult Sci 24(3):417-427.
Perrett DR (1947) Suspected rape poisoning in cattle. Vet Rec 59(49):674.
Pettit JH, Slinger SJ, Evans EV, et al. (1944) The utilization of sunflower seed oil meal, wheat distillers' dried grains and rapeseed oil meal in poultry rations. Can J Agric Sci 24(5):201-213.
Purves HD (1943) Studies on experimental goitre. IV. The effect of di-iodotyrosine and thyroxine on the goitrogenic action of Brassica seeds. Br J Exp Pathol 24(5):171-173.
Rotkiewicz T, Koska J, Kozłowski M, et al. (1981) Über die Eignung nach verschiedenen Verfahren hergestellter Rapsextraktionsschrote für die Broilermast. 2. Zur strumigenen Wirkung von Rapsextraktionsschrot. Nahrung 25(4):371-378.
Russel AJ (1967) A note on goitre in lambs grazing rape (Brassica napus). Anim Prod 9(1):131-133.
Schellner HP (1987) Raps als mögliche Ursache für Hasen- und Rehsterben. Tierarztl Umsch 42(11):902-904.
Schmid A, Schmid H (1992) Rapsvergiftung wildlebender Pflanzenfresser. Tierarztl Prax 20(3):321-325.
Schofield FW (1947) The constant occurrence of macrocytic anemia in cattle feeding on rape. Ontario Vet Coll Rep 29:122-125.
Schoon HA, Brunckhorst D, Fehlberg U (1989) 00-Raps-Vergiftung beim Rehwild. Prakt Tierarzt 70(12):50-52.
Seth PC, Clandinin, DR (1973) Effect of including rapeseed meal in the ration of broiler-type chickens on the incidence of perosis and the ineffectiveness of supplemental manganese. Poult Sci 52(3):1158-1160.
Sharpe GL, Larsson KS, Leidén SÅ (1975) Toxicological and teratological studies of a rapeseed protein diet in rats and mice. Nutr Metab 18(5-6):245-257.
Smith TK, Campbell LD (1976) Rapeseed meal glucosinolates: Metabolism and effect on performance in laying hens. Poult Sci 55(3):861-867.
Takamori O (1963) [Studies on experimental goiter. IV. Effect of rapeseed feeding on the thyroid and adrenal glands of chickens and rabbits.] Jpn J Vet Sci 25:227-232.
Thomas D, Robblee AR, Clandinin DR (1978) Effects of low and high glucosinolate rapeseed meals on productive performance, egg quality, composition of liver and incidence of haemorrhagic liver in laying birds. Br Poult Sci 19(4):449-454.

Timms LM (1983) Forms of leg abnormality observed in male broilers fed on a diet containing 12.5 per cent rapeseed meal. Res Vet Sci 35(2):182-189.
Turner CW (1946) Effect of rapeseed on the thyroid of the chick. Poult Sci 25:186-187.
Turner CW (1948) Effect of rapeseed oil meal on the thyroid of the chick. Poult Sci 27:118-120.
Umemura T, Slinger SJ, Bhatnagar MK, et al. (1978) Histopathology of the heart from rats fed rapeseed oils. Res Vet Sci 25(3):318-322.
Umemura T, Yamashiro S, Bhatnagar MK, et al. (1977) Intravascular fat globules in the heart of pigs fed rapeseed meals. Res Vet Sci 23(1):59-61.
Umemura T, Yamashiro S, Bhatnagar MK, et al. (1977) Liver fibrosis of the turkey on rapeseed products. Res Vet Sci 23(2):139-145.
van der Veen HE, Hart PC (1952) Onderzoekingen betreffende de anti-thyreoide werking van koolzaad en raapzaad. Landbouwkwundig Tijdschr 7:461-472.
Vermunt JJ, West DM, Cooke MM (1993) Rape poisoning in sheep. N Z Vet J 41:151-152.
Wheeler JL, Park RJ, Spurway RA, et al. (1974) Variation in the effects of forage rape on meat flavour in sheep. J Agric Sci 83:569-571.
Wiesen B, Kincaid RL, Hillers JK, et al. (1990) The use of rapeseed screenings in diets for lactating cows and subsequent effects on milk yield and composition. J Dairy Sci 73(12):3555-3562.
Wight PA, Shannon DW, McCorquodale CC, et al. (1987) Experimental systems which modify and simulate rapeseed-induced liver haemorrhages in in-lay hens. Res Vet Sci 43(3):351-366.
Wight PA, Wells JW, Shannon DW (1986) Liver haemorrhages induced by rapeseed meal: Incidence in adult male and female fowls. Br Poult Sci 27(2):247-252.
Witz WM, Carpenter MM, Hayward JW (1950) Nutritional studies with rapeseed meal. Poult Sci 29:786.
Yamashiro S, Bast T (1978) Ultrastructure of livers of broiler chickens fed diets containing rapeseed meal. Res Vet Sci 25(1):21-24.
Yamashiro S, Bhatnagar MK, Scott JR, et al. (1975) Fatty haemorrhagic liver syndrome in laying hens on diets supplemented with rapeseed products. Res Vet Sci 19(3):312-321.
Yamashiro S, Geissinger HD, Bast T, et al. (1979) Effects of rapeseed products on turkey heart. Anat Histol Embryol 8:187.
Yamashiro S, Umemura T, Bhatnagar MK, et al. (1977) Haemorrhagic liver syndrome of broiler chickens fed diets containing rapeseed products. Res Vet Sci 23(2):179-184.
Yule WJ, McBride RL (1978) Rapeseed meals in broiler diets: Effect on performance and sensory evaluation of carcasses. Br Poult Sci 19(4):543-548.

Note:

Rape is named *Brassica napus* L. var. *napus* or *Brassica rapa* L. subsp. *campestris* (L.) A. R. Clapham in some publications.

BRASSICA NAPUS L. var. pabularia (DC.) Rchb. [Brassicaceae]

Common Names:

rape kale

Citations:

Paxman PJ, Hill R (1974) The goitrogenicity of kale and its relation to thiocyanate content. J Sci Food Agric 25(3):329-337.

BRASSICA NIGRA (L.) W. D. J. Koch [Brassicaceae]

Synonyms:

Sinapis nigra L.

Common Names:

black mustard; brown mustard; moutarde noire; mustard; Schwarzer Senf; Senf

Citations:

Benazzi P, Soldati G (1969) Avvelenamento da «Brassica nigra Koch» (Sinapis nigra Linneo) nei bovini. Atti Soc Ital Buiatria 1:145-152.
Dannaker CJ, White IR (1987) Cutaneous allergy to mustard in a salad maker. Contact Dermatitis 16(6):212-214.
Gulbransen G, Esernio Jenssen D (1998) Aspiration of black mustard. J Toxicol Clin Toxicol 36(6):591-593.
Hercus CE, Purves HD (1936) Studies on endemic and experimental goitre. J Hyg (Lond) 36:182-203.
Jahn S, Seffner W (1962) Auswirkung der Verfütterung von Senfkuchen an Schafe. Arch Tierernahr 12:11-16.
Malet A, Valero A, Lluch M, et al. (1993) Hypersensitivity to mustard seed. Allergy 48(1):62-63.
Meding B (1985) Immediate hypersensitivity to mustard and rape. Contact Dermatitis 13(2):121-122.
Monreal P, Botey J, Pena M, et al. (1992) Mustard allergy. Two anaphylactic reactions to ingestion of mustard sauce. Ann Allergy 69(4):317-320.
Niinimäki A, Hannuksela M (1981) Immediate skin test reactions to spices. Allergy 36(7):487-493.
Roub JF (1902) Sinapis nigra poisoning. Am Vet Rev 26:437-439.

BRASSICA OLERACEA L. var. botrytis L. [Brassicaceae]

Common Names:

Blumenkohl; cauliflower; cavolfiore

Citations:

Quirce S, Madero MF, Fernández-Nieto M, et al. (2005) Occupational asthma due to the inhalation of cauliflower and cabbage vapors. Allergy 60(7):969-970.
Sinha SM, Pasricha JS, Sharma RC, et al. (1977) Vegetables responsible for contact dermatitis of the hands. Arch Dermatol 113(6):776-779.
Van Ketel WG (1975) A cauliflower allergy. Contact Dermatitis 1(5):324-325.

BRASSICA OLERACEA L. var. capitata L. [Brassicaceae]

Common Names:

cabbage; cavolo cappaccio; drumhead cabbage; Kohl; repolho; Weißkraut

Citations:

Blum F (1950) Studien zum Kropfproblem. IV. Mitteilung. Ueber pflanzliche Kropferreger und Versuche zu ihrer Umzüchtung. Schweiz Med Wochenschr 80(6):142-144.

Burrows GE, Morton DJ, Basudde CD (1990) The effects of feeding cabbage (Brassica oleracea) and Rhodesian ragwort (Senecio latifolius) on the disposition of antipyrine and acetaminophen in sheep. Toxicon 28(6):603.

Calnan CD (1981) Contact urticaria from cabbage (Brassica). Contact Dermatitis 7(5):279.

Clegg FG, Evans RK (1962) Haemoglobinaemia of cattle associated with the feeding of Brassicae species. Vet Rec 74(44):1169-1176.

Langer P (1964) Study of chemical representatives of the goitrogenic activity of raw cabbage. Physiol Bohemoslov 13(6):542-549.

Moraga BL, Scheidegger A (1987) Intoxicacion de bovinos con repello (Brassica oleracea var. capitata). Agric Tec Chile 47(4):427-430.

Parodi A, Guenet JL (1965) Goitre enzootique chez des chevreaux. Rec Med Vet Ec Alfort 141:57-61.

Quirce S, Madero MF, Fernández-Nieto M, et al. (2005) Occupational asthma due to the inhalation of cauliflower and cabbage vapors. Allergy 60(7):969-970.

Taljaard TL (1993) Cabbage poisoning in ruminants. J S Afr Vet Assoc 64(2):96-100.

Yalçin E, Özlem MB, Turgut K, et al. (1987) Maltiz keçisi ve oğlaklarinda guvatr. Vet Fakul Dergisi Ankara Univ 34(2):241-250.

BRASSICA OLERACEA L. var. gemmifera Zenker [Brassicaceae]

Common Names:

Brussels sprouts; choux fourrager; couve-de-Bruxelas; Rosenkohl

Citations:

Clegg FG, Evans RK (1962) Haemoglobinaemia of cattle associated with the feeding of Brassicae species. Vet Rec 74(44):1169-1176.

de Groot AP, Willems MI, de Vos RH (1991) Effects of high levels of Brussels sprouts in the diet of rats. Food Chem Toxicol 29(12):829-837.

BRASSICA OLERACEA L. var. italica Plenck [Brassicaceae]

Common Names:

broccoli

Citations:

Sánchez-Guerrero IM, Escudero AI (1998) Occupational contact dermatitis to broccoli. Allergy 53(6):621-622.

BRASSICA OLERACEA L. var. medullosa Thell. [Brassicaceae]

Common Names:

Cansons kale; chou moellier; chou moullier; Maris Kestrel kale; Markstammkohl; marrow stem kale

Citations:

Clegg FG, Evans RK (1962) Haemoglobinaemia of cattle associated with the feeding of Brassicae species. Vet Rec 74(44):1169-1176.

Fix HP, Rockstroh A (1964) Über den Einfluß von Markstammkohl auf Milchleistung und Blutbild (Fütterungsversuch). Monatsh Veterinarmed 19:888-891.

Grant CA, Holtenius P, Jönsson G, et al. (1968) Kale anaemia in ruminants. I. Survey of the literature and experimental induction of kale anaemia in lactating cows. Acta Vet Scand 9(2):126-140.

Grant CA, Holtenius P, Jönsson G, et al. (1968) Kale anaemia in ruminants. II. Observations on kale-fed sheep. Acta Vet Scand 9(2):141-150.

Greenhalgh JF, Sharman GA, Aitken JN (1970) Kale anaemia. II. Further factors concerned in the occurrence of the disease under experimental conditions. Res Vet Sci 11(3):232-238.

Greenhalgh JF, Sharman GA, Aitken JN, et al. (1967) Haemolytic anaemia in cattle fed on kale. Anim Prod 9:286.

Hill FI, Ebbett PC (1997) Polioencephalomalacia in cattle in New Zealand fed chou moellier (Brassica oleracea). N Z Vet J 45:37-39.

Kennedy TH, Purves HD (1941) Studies on experimental goitre. I. The effect of Brassica seed diet on rats. Br J Exp Pathol 22(5):241-244.

Mason RW, Lucas P (1983) Acute poisoning in cattle after eating old non-viable seed of chou moellier (Brassica oleracea convar. acephala). Aust Vet J 60(9):272-273.

Paxman PJ, Hill R (1974) The goitrogenicity of kale and its relation to thiocyanate content. J Sci Food Agric 25(3):329-337.

Piatkowski B, Steger H (1965) Der Einfluß verschiedener Futterzusätze auf die Verdaulichkeit des Markstammkohles und die Auswirkungen auf das rote Blutbild bei Rindern. Arch Tierernahr 15(6):447-453.

Pickard DW, Crighton DB (1967) An investigation into the possible oestrogenic effect of kale. Br Vet J 123(2):64-69.

Steger H, Piatkowski B, Busch B (1965) Der Einfluß von Heißlufttrocknung und Silierung auf den anaemischen Faktor des Markstammkohls. Arch Tierernahr 15(6):455-459.

Steger H, Piatkowski B, Busch B, et al. (1968) Weitere Untersuchungen über die Wirkung von frischem und siliertem Markstammkohl auf die Verdauung und die Blutzusammensetzung. Arch Tierernahr 18(4):331-338.

Tucker EM (1969) The onset of anaemia and the production of haemoglobin C in sheep fed on kale. Br Vet J 125(9):472-479.

Williams HL, Hill R, Alderman G (1965) The effects of feeding kale to breeding ewes. Br Vet J 121(1):2-17.

BRASSICA OLERACEA L. var. ramosa DC. [Brassicaceae]

Common Names:

Canson kale; thousand-head kale

Citations:

Greenhalgh JF, Sharman GA, Aitken JN, et al. (1967) Haemolytic anaemia in cattle fed on kale. Anim Prod 9:286.

Paxman PJ, Hill R (1974) The goitrogenicity of kale and its relation to thiocyanate content. J Sci Food Agric 25(3):329-337.

Wright E, Sinclair DP (1958) The goitrogenic effect of thousand-headed kale on adult sheep and rabbits. N Z J Agric Res 1:477-485.

BRASSICA OLERACEA L. var. viridis L. [Brassicaceae]

Common names:

kale

Citations:

Boyd H, Reed HC (1961) Investigation into the incidence and causes of infertility in dairy cattle - Influence of kale feeding, milk production and management factors associated with 'farming intensity'. Br Vet J 117:192-200.

Czajkowski Z, Brzozowski A, Małecki J (1968) Wpływ dużych dawek kapusty pastewnej na wydajnośći mleczną i na niektóre właściwośc obwodowej krwi krów rasy N.C.B. Med Weter 24(8):496-500.

Dunbar GM, Chambers TA (1963) Suspected kale poisoning in dairy cows. Vet Rec 75(22):566-567.

Evans ET (1951) Kale and rape poisoning in cattle. Vet Rec 63(19):348-349.

Greenhalgh JF, Aitken JN, Gunn JB (1972) Kale anaemia. 3. A survey of kale feeding practices and anaemia in cattle on dairy farms in England and Scotland. Res Vet Sci 13(1):15-21.

Greenhalgh JF, Sharman GA, Aitken JN (1969) Further studies on the occurrence of kale anaemia. Anim Prod 11:272.

Greenhalgh JF, Sharman GA, Aitken JN (1969) Kale anaemia. I. The toxicity to various species of animal of three types of kale. Res Vet Sci 10(1):64-72.

Melrose DR, Brown BB (1962) Some observations on the possible effect of kale feeding on fertility in dairy cattle. J Reprod Fertil 4:232.

Paxman PJ, Hill R (1974) The goitrogenicity of kale and its relation to thiocyanate content. J Sci Food Agric 25(3):329-337.

Penny RH, David JE, Wright AI (1961) Heinz-Ehrlich bodies associated with kale feeding. Vet Rec 73(30):747-748.

Penny RH, David JS, Wright AI (1964) Observations on the blood picture of cattle, sheep and rabbits fed on kale. Vet Rec 76(38):1053-1059.

Penny RH, Wright AI (1963) Suspected kale poisoning in dairy cows. Vet Rec 75(24):642-643.

Shand A (1953) The goitrogenic factor in kale. Proc Conf Metabolic Disorders 23:58-65.

Sinclair DP, Andrews ED (1954) Goitre in new-born lambs. N Z Vet J 2:72-79.

Sinclair DP, Andrews ED (1958) Prevention of goitre in new-born lambs from kale-fed ewes. N Z Vet J 6:87-95.

Smith RH, Earl CR, Matheson NA (1974) The probable role of S-methylcysteine sulphoxide in kale poisoning in ruminants. Biochem Soc Trans 2:101-104.

BRASSICA RAPA L. subsp. campestris (L.) A. R. Clapham [Brassicaceae]

Synonyms:

Brassica campestris L.

Common Names:

canola; Chinese cabbage; nabo silvestre; novet; turnip rape

Citations:

Bachmann M, Theus R, Lüthy J, et al. (1985) Vorkommen von goitrogenen Stoffen in Milch. 1. Übergang von Goitrin in die Milch von Kühen bei Verfütterung von Rapsextraktionsschrot. Z Lebensm Unters Forsch 181(5):375-378.

Bäckgren AW, Jönsson G (1969) Blood and bone marrow studies in cattle feeding on Brassica species. Acta Vet Scand 10(4):309-318.

Bell JM, Williams K (1953) Growth depressing factors in rapeseed oilmeal. Can J Agric Sci 33(May-Jun):201-209.

Friend DW, Corner AH, Kramer JK, et al. (1975) Growth, cardiopathology and cardiac fatty acids of swine fed diets containing soybean oil or low erucic acid rapeseed oil. Can J Anim Sci 55(Mar):49-59.

Friend DW, Gilka F, Corner AH (1975) Growth, carcass quality and cardiopathology of boars and gilts fed diets containing rapeseed and soybean oils. Can J Anim Sci 55(4):571-578.

Friend DW, Kramer JK, Corner AH (1976) Growth and cardiopathology in boars fed rapeseed oil. Can J Anim Sci 56(Jun):361-364.

González N, Haardt E, Potoćnjak J (1964) El afrecho de raps en la alimentacion de aves (aves de 18 semanas a 18 meses). Bol Prod Anim (Chile) 2(2):101-120.

Grandhi RR, Slinger SJ, Summers JD (1977) Productive performance and liver lesions in two strains of laying hens receiving two rapeseed meals. Poult Sci 56(6):1904-1908.

Hawrysh ZJ, Clandinin DR, Robblee AR, et al. (1975) Influence of rapeseed meal on the odor and flavor of eggs from different breeds of chickens. Can Inst Food Sci Technol J 8(1):51-54.

Hulan HW, Hunsaker WG, Kramer JK, et al. (1976) The development of dermal lesions and alopecia in male rats fed rapeseed oil. Can J Physiol Pharmacol 54(1):1-6.

Hulan HW, Kramer JK, Mahadevan S, et al. (1976) Effect of cold stress on rapeseed oil fed rats. Lipids 11(1):6-8.

Hulan HW, Kramer JK, Mahadevan S, et al. (1976) Relationship between erucic acid and myocardial changes in male rats. Lipids 11(1):9-15.

Israels ED, Papas A, Campbell LD, et al. (1979) Prevention by menadione of the hepatotoxic effects in chickens fed rapeseed meal. Observations on coagulation factors and cytochrome P-450. Gastroenterology 76(3):584-589.

Iwarsson K, Ekman L, Everitt BR, et al. (1973) The effect of feeding rapeseed meal on thyroid function and morphology in growing bulls. Acta Vet Scand 14(4):610-629.

Iwarsson K, Nilsson, PO (1973) Rapeseed meal as a protein supplement for dairy cows. II. Investigations in rats on the goitrogenic properties of milk from cows fed rapeseed meal. Acta Vet Scand 14(4):595-609.

Jackson N (1969) Toxicity of rapeseed meal and its use as a protein supplement in the diet of two hybrid strains of caged laying hens. J Sci Food Agric 20(12):734-740.

Jackson N (1970) Algerian and French rapeseed meals as a protein source for caged laying hens, with observations on their toxic effects. J Sci Food Agric 21(10):511-516.

Josefsson E, Uppström B (1976) Influence of glucosinolates and native enzymes on the nutritional value of low-glucosinolate rapeseed meal. J Sci Food Agric 27(5):433-437.

Kennedy TH, Purves HD (1941) Studies on experimental goitre. I. The effect of Brassica seed diet on rats. Br J Exp Pathol 22(5):241-244.

Kramer JK, Hulan HW (1977) Changes in the acyl and alkenyl group composition of cardiac phospholipids in boars fed corn oil or rapeseed oil. Lipids 12(2):159-164.

Kramer JK, Hulan HW, Mahadevan S, et al. (1975) Brassica campestris var. span: II. Cardiopathogenicity of fractions isolated from span rapeseed oil when fed to male rats. Lipids 10(9):511-516.

Kratzer FH, Davis PN, Williams DE, et al. (1954) Factors influencing the growth of chicks and poults fed rations containing rapeseed oil meal. J Nutr 53:407-418.
Leung P, March BE (1976) The thyroidal response to chronic goitrogenic stimulation and the persistence of effects of early goitrogenic stimulation. Can J Physiol Pharmacol 54(4):583-589.
Lo MT, Hill DC (1972) Cyano compounds and goitrin in rapeseed meal. Can J Physiol Pharmacol 50(4):373-377.
Lodhi GN, Renner R, Clandinin DR (1969) Studies on the metabolizable energy of rapeseed meal for growing chickens and laying hens. Poult Sci 48(3):964-970.
Loew FM, Doige CE, Manns JG, et al. (1976) Evaluation of dietary rapeseed protein concentrate flours in rats and dogs. Toxicol Appl Pharmacol 35(2):257-267.
Lopez TA, Platonow NS (1975) Alimentacion experimental de bovinos con nabo silvestre (Brassica campestris). Gaceta Vet 37:354-359.
March BE, Biely J, Soong R (1972) Rapeseed meal in the chicken breeder diet. Effects on production, mortality, hatchability, and progeny. Poult Sci 51(5):1589-1596.
March BE, Biely J, Soong R (1975) The effects of rapeseed meal fed during the growing and/or laying periods on mortality and egg production in chickens. Poult Sci 54(6):1875-1882.
March BE, Soong R (1976) Mortality and production characteristics of laying chickens fed high- and low-erucic acid rapeseed oils. Poult Sci 55(4):1557-1560.
McCutcheon JS, Umermura T, Bhatnagar MK, et al. (1976) Cardiopathogenicity of rapeseed oils and oil blends differing in erucic, linoleic, and linolenic acid content. Lipids 11(7):545-552.
Monkiewicz J, Kinal S, Łuczak W (1977) Poziom witaminy A w wątrobie kurczqt rzeźnych a zawartość substancji toksycznych w śrutach rzepakowych. Med Weter 33(2):104-106.
Olomu JM, Robblee AR, Clandinin DR, et al. (1975) Effects of span rapeseed meal on productive performance, egg quality, composition of liver and hearts and incidence of 'fatty livers' in laying hens. Can J Anim Sci 55(1):71-75.
O'Neil JB (1957) Rapeseed oil meal as a vegetable protein supplement in the diet of laying and breeding hens. Poult Sci 36:1146.
Papas A, Ingalls JR, Campbell LD (1979) Studies on the effects of rapeseed meal on thyroid status of cattle, glucosinolate and iodine content of milk and other parameters. J Nutr 109(7):1129-1139.
Pearson AW, Greenwood NM, Butler EJ, et al. (1983) Biochemical changes in layer and broiler chickens when fed on a high-glucosinolate rapeseed meal. Br Poult Sci 24(3):417-427.
Pettit JH, Slinger SJ, Evans EV, et al. (1944) The utilization of sunflower seed oil meal, wheat distillers' dried grains and rapeseed oil meal in poultry rations. Can J Agric Sci 24(5):201-213.
Purves HD (1943) Studies on experimental goitre. IV. The effect of di-iodotyrosine and thyroxine on the goitrogenic action of Brassica seeds. Br J Exp Pathol 24(5):171-173.
Rocquelin G, Sergiel JP, Astorg PO, et al. (1973) Effets des huiles de colza et de canbra sur les lipides et l'anatomie du myocarde du rat: Étude a court terme (de 0 à 60 jours). Ann Biol Anim Biochim Biophys 13(4):587-609.
Rotkiewicz T, Koska J, Kozłowski M, et al. (1981) Über die Eignung nach verschiedenen Verfahren hergestellter Rapsextraktionsschrote für die Broilermast. 2. Zur strumigenen Wirkung von Rapsextraktionsschrot. Nahrung 25(4):371-378.
Seth PC, Clandinin, DR (1973) Effect of including rapeseed meal in the ration of broiler-type chickens on the incidence of perosis and the ineffectiveness of supplemental manganese. Poult Sci 52(3):1158-1160.
Sharpe GL, Larsson KS, Leidén SÅ (1975) Toxicological and teratological studies of a rapeseed protein diet in rats and mice. Nutr Metab 18(5-6):245-257.
Thomas D, Robblee AR, Clandinin DR (1978) Effects of low and high glucosinolate rapeseed meals on productive performance, egg quality, composition of liver and incidence of haemorrhagic liver in laying birds. Br Poult Sci 19(4):449-454.
Timms LM (1983) Forms of leg abnormality observed in male broilers fed on a diet containing 12.5 per cent rapeseed meal. Res Vet Sci 35(2):182-189.
Turner CW (1948) Effect of rapeseed oil meal on the thyroid of the chick. Poult Sci 27:118-120.
Umemura T, Yamashiro S, Bhatnagar MK, et al. (1977) Intravascular fat globules in the heart of pigs fed rapeseed meals. Res Vet Sci 23(1):59-61.
Wiesen B, Kincaid RL, Hillers JK, et al. (1990) The use of rapeseed screenings in diets for lactating cows and subsequent effects on milk yield and composition. J Dairy Sci 73(12):3555-3562.
Witz WM, Carpenter MM, Hayward JW (1950) Nutritional studies with rapeseed meal. Poult Sci 29:786.
Yamashiro S, Bast T (1978) Ultrastructure of livers of broiler chickens fed diets containing rapeseed meal. Res Vet Sci 25(1):21-24.
Yamashiro S, Geissinger HD, Bast T, et al. (1979) Effects of rapeseed products on turkey heart. Anat Histol Embryol 8:187.
Yamashiro S, Umemura T, Bhatnagar MK, et al. (1977) Haemorrhagic liver syndrome of broiler chickens fed diets containing rapeseed products. Res Vet Sci 23(2):179-184.
Yule WJ, McBride RL (1978) Rapeseed meals in broiler diets: Effect on performance and sensory evaluation of carcasses. Br Poult Sci 19(4):543-548.

Note:

Rape is named *Brassica napus* L. var. *napus* or *Brassica rapa* L. subsp. *campestris* (L.) A. R. Clapham in some publications.

BRASSICA RAPA L. subsp. rapa [Brassicaceae]

Synonyms:

Brassica rapa L. subsp. rapifera Metzg.

Common Names:

turnip; white turnip

Citations:

Brakenridge DT (1956) Nitrate poisoning caused by turnips and red root. N Z Vet J 4:165-166.
Debackere M, Hoorens J, Haustraete KH (1966) Vergiftiging door koolzaad - En raapzaadschroot. Vlaams Diergeneesk Tijdschr 35(9-10):393-399.

Kennedy TH, Purves HD (1941) Studies on experimental goitre. I. The effect of Brassica seed diet on rats. Br J Exp Pathol 22(5):241-244.
Muñoz-Bellido FJ, Moyano-Maza JC, Alvarez-Gonzalo M, et al. (2003) Occupational contact dermatitis to turnip (Brassica napa). Allergy 58(11):1198-1199.

Brassica rapa L. subsp. rapifera Metzg. = Brassica rapa L. subsp. rapa

Brassica rutabaga DC. ex H. Lév. = Brassica napus L. napobrassica (L.) Rchb.

Brassica sinapis Vis. = Sinapis arvensis L.

Brazil nut –see– *Bertholletia excelsa* Bonpl.

Brazil rain tree –see– *Brunfelsia australis* Benth.

Brazil rattlebox –see– *Sesbania punicea* (Cav.) Benth.

Brazilian arrowroot –see– *Manihot esculenta* Crantz

Brazilian guarana –see– *Paullinia cupana* Kunth

Brazilian jacaranda –see– *Dalbergia nigra* (Vell.) Allemão ex Benth.

Brazilian pepper tree –see– *Schinus terebinthifolius* Raddi

Brazilian physicnut –see– *Cnidoscolus urens* (L.) Arthur

Brazilian rosebush –see– *Dalbergia nigra* (Vell.) Allemão ex Benth.

Brazilian rosewood –see– *Dalbergia nigra* (Vell.) Allemão ex Benth.

Brazilian walnut –see– *Ocotea porosa* (Nees & Mart.) Barroso

Brechnuss –see– *Strychnos nux-vomica* L.

Brechwurz –see– *Asarum europaeum* L.

breeches flower –see– *Dicentra cucullaria* (L.) Bernh.

Brennessel –see– *Urtica dioica* L.

Brennkraut –see– *Clematis vitalba* L.

bretonica –see– *Melochia tomentosa* L.

brewer's-hop –see– *Humulus lupulus* L.

Breynia officinalis Hemsl. = Breynia vitis-idaea (Burm. f.) C. E. C. Fisch.

BREYNIA VITIS-IDAEA (Burm. f.) C. E. C. Fisch. [Euphorbiaceae]

Synonyms:

Breynia officinalis Hemsl.

Citations:

Lin TJ, Su CC, Lan CK, et al. (2003) Acute poisonings with Breynia officinalis - An outbreak of hepatotoxicity. J Toxicol Clin Toxicol 41(4):513.
Lin TJ, Su CC, Lan CK, et al. (2003) Acute poisonings with Breynia officinalis - An outbreak of hepatotoxicity. J Toxicol Clin Toxicol 41(5):591-594.
Lin TJ, Tsai MS, Chiou NM, et al. (2002) Hepatotoxicity caused by Breynia officinalis. Vet Hum Toxicol 44(2):87-88.

brîcută –see– *Descurainia sophia* (L.) Webb ex Prantl

bright-eyes –see– *Catharanthus roseus* (L.) G. Don

brinjal –see– *Solanum melongena* L.

brionia –see– *Bryonia dioica* Jacq.

bristly foxtail –see– *Setaria pumila* (Poir.) Roem. & Schult.

bristly mallow –see– *Modiola caroliniana* (L.) G. Don

British mandrake –see– *Bryonia dioica* Jacq.

British nettle –see– *Urtica dioica* L.

briuma dandie –see– *Argemone mexicana* L.

broad bean –see– *Vicia faba* L.

broad cocklebur –see– *Xanthium strumarium* L.

broad-leaf endive –see– *Cichorium endivia* L.

broad-leaf laurel –see– *Kalmia latifolia* L.

broad-leaf lupine –see– *Lupinus latifolius* J. Agardh

broad-leaf milkweed –see– *Asclepias eriocarpa* Benth.; *Asclepias latifolia* (Torr.) Raf.; *Asclepias speciosa* Torr.

broad-leaf wart spurge –see– *Euphorbia platyphyllos* L.

broccoli –see– *Brassica oleracea* L. var. italica Plenck

Bromelia ananas L. = Ananas comosus (L.) Merr.

Bromelia comosus L. = Ananas comosus (L.) Merr.

broom –see– *Cytisus scoparius* (L.) Link; *Genista stenopetala* Webb & Berthel.; *Securigera varia* (L.) Lassen; *Templetonia egena* (F. Muell.) Benth.

broom bush –see– *Pimelea trichostachya* Lindl.

broom corn –see– *Sorghum bicolor* (L.) Moench

broom groundsel –see– *Senecio riddellii* Torr. & A. Gray; *Senecio spartioides* Torr. & A. Gray

broom millet –see– *Panicum miliaceum* L.; *Sorghum bicolor* (L.) Moench

broom rape –see– *Orobanche minor* Sm.

broom snakeweed –see– *Gutierrezia sarothrae* (Pursh) Britton & Rusby

broomcorn millet –see– *Panicum miliaceum* L.

broomcorn sorghum –see– *Sorghum bicolor* (L.) Moench

broomtops –see– *Cytisus scoparius* (L.) Link

broomweed –see– *Gutierrezia microcephala* (DC.) A. Gray; *Gutierrezia sarothrae* (Pursh) Britton & Rusby

brown corn –see– *Panicum miliaceum* L.

brown dragon –see– *Arisaema triphyllum* (L.) Schott

brown mustard –see– *Brassica juncea* (L.) Czern.; *Brassica nigra* (L.) W. D. J. Koch

Brown's-larkspur –see– *Delphinium glaucum* S. Watson

browse milk vetch –see– *Astragalus cibarius* E. Sheld.

brugmansia –see– *Brugmansia arborea* (L.) Lagerh.

BRUGMANSIA ARBOREA (L.) Lagerh. [Solanaceae]

Synonyms:

Datura arborea L.; **_Datura cornigera_** Hook.

Common Names:

angel's-tears; angel's-trumpet; angel's-trumpet lily; brugmansia; campana; cornucopia; florifundia; floripondio; maikoa; reina-de-la-noche; saia blanca; tree datura; trombeta blanca; trombeteira; trumpet lily; trumpet vine

Citations:

Belton PA, Gibbons DO (1979) Datura intoxication in West Cornwall. Br Med J 1(6163):585-586.
Ciganda C, Laborde A (2003) Herbal infusions used for induced abortion. J Toxicol Clin Toxicol 41(3):235-239.
Hayman J (1985) Datura poisoning - The angel's trumpet. Pathology 17(3):465-466.
Tokarnia CH, Armién AG, Peixoto PV, et al. (1996) Estudo experimental sobre a toxidez de algumas plantas ornamentais em bovinos. Pesq Vet Bras 16(1):5-20.
Van der Walt SJ, Steyn DG (1943) Recent investigations into the toxicity of plants, etc. XIII. Onderstepoort J Vet Sci Anim Indus 18(1-2):207-224.
Vieira RJ (1976) Intoxicação por folha de "Datura arborea" através da pele. Apresentação de um caso. Rev Assoc Med Bras 22(3):90.
Wallman JD (1963) An odd eye. Med J Aust 1(Feb 3):153-154.

BRUGMANSIA ×CANDIDA Pers. [Solanaceae]

Synonyms:

Datura candida (Pers.) Saff.

Common Names:

angel's-trumpet; borrachero; campana; cornucopia; floripondio; trompetero

Citations:

Finlay P (1998) Anticholinergic poisoning due to Datura candida. Trop Doct 28(3):183-184.
Greene GS, Patterson SG, Warner E (1996) Ingestion of angel's trumpet: An increasingly common source of toxicity. South Med J 89(4):365-369.
Shaw D, Pearn J (1979) Oleander [sic] poisoning. Med J Aust 2(5):267-269.

BRUGMANSIA ×INSIGNIS (Barb. Rodr.) Lockwood ex R. E. Schult. [Solanaceae]

Citations:

Paetzold W, Schneider U, Emrich HM, et al. (1999) Engelstrompeten: Falldarstellung einer drogeninduzierten Psychose durch Brugmansia insignis. Psychiatr Prax 26(3):147-148.

BRUGMANSIA SANGUINEA (Ruiz & Pav.) D. Don [Solanaceae]

Synonyms:

Datura rosei Saff.

Common Names:

puca campuchu; red angel's-trumpet

Citations:

Hudson MJ (1973) Acute atropine poisoning from ingestion of Datura rosei. N Z Med J 77(491):245-248.

BRUGMANSIA SUAVEOLENS (Humb. & Bonpl. ex Willd.) Bercht. & J. Presl [Solanaceae]

Synonyms:

Datura suaveolens Humb. & Bonpl. ex Willd.

Common Names:

angel's-trumpet; änglatrumpet; devil's-trumpet; Engelstrompete; floripon; moonflower

Citations:

Ahremark U, Björkqvist I, Olsson CG (1982) Intoxikation av änglatrumpet. Lakartidningen 79(19):1929-1930.
Anonymous (1984) Datura poisoning from hamburger - Canada. MMWR Morb Mortal Wkly Rep 33(20):282-283.
Chang SS, Wu ML, Deng JF, et al. (1999) Poisoning by Datura leaves used as edible wild vegetables. Vet Hum Toxicol 41(4):242-245.
Francis PD, Clarke CF (1999) Angel trumpet lily poisoning in five adolescents: Clinical findings and management. J Paediatr Child Health 35(1):93-95.
Giess R, Müllges W (1999) Einseitige Mydriasis nach Rückschnitt einer Engelstrompete. Dtsch Med Wochenschr 124(48):1456.
Hall RC, Popkin MK, McHenry LE (1976) Angel's trumpet: A call to the ER. Emerg Med 8(9):241-243.
Hall RC, Popkin MK, McHenry LE (1976) Angel's trumpet may herald psychosis. Med World News 17(18):19.
Hall RC, Popkin MK, McHenry LE (1977) Angel's trumpet psychosis: A central nervous system anticholinergic syndrome. Am J Psychiatry 134(3):312-314.
Havelius U, Asman P (2002) Accidental mydriasis from exposure to angel's trumpet (Datura suaveolens). Acta Ophthalmol (Copenh) 80(3):332-335.
Hornsby P (1969) A case of accidental mydriasis due to the bark of the moon flower. East Afr Med J 46(9):527.
McHenry LE, Hall RC (1978) Angel's trumpet. Lethal and psychogenic aspects. J Fla Med Assoc 65(3):192-196.
Popkin MK (1976) "Angel's trumpet" provides deadly thrills for youth. JAMA 236(3):249.
Rauber Lüthy C, Guirguis M, Meier Abt AS, et al. (1999) Lethal poisoning after ingestion of a tea prepared from the angel's trumpet (Datura suaveolens). J Toxicol Clin Toxicol 37(1):414.
Reichmuth D, Greene G, Matfin G, et al. (1997) A dazed and disoriented man found by the roadside. Hosp Pract (Off Ed) 32(5):37-40.
Smith EA, Meloan CE, Pickell JA, et al. (1991) Scopolamine poisoning from homemade 'moon flower' wine. J Anal Toxicol 15(4):216-219.
Voltz R, Hohlfeld R, Liebler M, et al. (1992) Gardener's mydriasis. Lancet 339(8795):752.

BRUNFELSIA AUSTRALIS Benth. [Solanaceae]

Common Names:

Brazil rain tree; Francissia; morning-noon-night; Paraguayan jasmin; yesterday-today-tomorrow

Citations:

Neilson J, Burren V (1983) Intoxication of two dogs by fruit of Brunfelsia australis. Aust Vet J 60(12):379-380.

Brunfelsia bonodora (Vell.) J. F. Macbr. = Brunfelsia latifolia (Pohl) Benth.

Brunfelsia calycina Benth. = Brunfelsia pauciflora (Cham. & Schltdl.) Benth.

BRUNFELSIA GRANDIFLORA D. Don [Solanaceae]

Citations:

Lloyd HA, Fales HM, Goldman ME, et al. (1985) Brunfelsamidine: A novel convulsant from the medicinal plant Brunfelsia grandiflora. Tetrahedron Lett 26 (22):2623-2624.

BRUNFELSIA LATIFOLIA (Pohl) Benth. [Solanaceae]

Synonyms:

Brunfelsia bonodora (Vell.) J. F. Macbr.

Common Names:

kiss-me-quick; yesterday-today-tomorrow

Citations:

McBarron EJ, De Sarem W (1975) Poisoning of dogs by the fruits of the garden shrub Brunfelsia bonodora. Aust Vet J 51(5):280.

BRUNFELSIA PAUCIFLORA (Cham. & Schltdl.) Benth. [Solanaceae]

Synonyms:

Brunfelsia calycina Benth.

Common Names:

morning-noon-night; yesterday-today; yesterday-today-tomorrow

Citations:

Banton MI, Jowett PL, Renegar KR, et al. (1989) Brunfelsia pauciflora ("Yesterday, to-day and tomorrow") poisoning in a dog. Vet Hum Toxicol 31(5):496-497.

Spainhour CB, Fiske RA, Flory W, et al. (1990) A toxicological investigation of the garden shrub Brunfelsia calycina var. floribunda (yesterday-today-and-tomorrow) in three species. J Vet Diagn Invest 2:3-8.

Tokarnia CH, Gava A, Stolf L, et al. (1991) Intoxicação experimental por Brunfelsia pauciflora (Solanaceae) em bovinos. Pesq Vet Bras 11(1-2):9-12.

Brussels sprouts –see– *Brassica oleracea* L. var. gemmifera Zenker

BRYA EBENUS (L.) DC. [Fabaceae]

Common Names:

Amerikanische Grenadill; cocus; green ebony; grenadil

Citations:

Hausen BM, Bruhn G, Koenig WA (1991) New hydroxyisoflavans as contact sensitizers in cocus wood Brya ebenus DC (Fabaceae). Contact Dermatitis 25(3):149-155.

bryone blanche –see– *Bryonia dioica* Jacq.

Bryonia cretica L. subsp. dioica (Jacq.) Tutin = Bryonia dioica Jacq.

BRYONIA DIOICA Jacq. [Cucurbitaceae]

Synonyms:

Bryonia cretica L. subsp. dioica (Jacq.) Tutin; ***Bryonia lutea*** Bast. ex Ser.

Common Names:

brionia; British mandrake; bryone blanche; bryony; devil's-turnip; English mandrake; Gichtwurz; grapewort; Isle-of-Wight-vine; mandrake; murrainberry; navet-de-diable; nueza blanca; red bryony; Rotbeerige Zaunrübe; Scheißwurz; Schlangenrübe; tetterberry; Teufelsrübe; tuca; white bryony; white wild vine; wild hop

Citations:

Anger AS (1900) Bryony poisoning. Am Vet Rev 23:804-805.

Blackwell WE (1931) Horses poisoned by bryony. Vet Rec 11(9):911-912.

Whur P (1986) White bryony poisoning in a dog. Vet Rec 119(16):411.

Bryonia lutea Bast. ex Ser. = Bryonia dioica Jacq.

BRYONOPSIS LACINIOSA (L.) Naudin [Cucurbitaceae]

Common Names:

native bryony; wild bryony

Citations:

Maiden JH (1896) A fatal case of poisoning by the fruits of a native climbing plant. Bryonia laciniosa, Linn. Agric Gaz New South Wales 7:424-426.

White CT (1924) The native bryony (Bryonia laciniosa). A poisonous plant. Queensland Agric J 22(Dec):442-444.

bryony –see– *Bryonia dioica* Jacq.

Bryophyllum daigremontianum (Raym.-Hamet & H. Perrier) A. Berger = Kalanchoe daigremontiana Raym.-Hamet & H. Perrier

Bryophyllum pinnatum (Lam.) Oken = Kalanchoe pinnata (Lam.) Pers.

Bryophyllum proliferum Bowie ex Hook. = Kalanchoe prolifera (Bowie ex Hook.) Raym.-Hamet

Bryophyllum tubiflorum Harv. = Kalanchoe delagoensis Eckl. & Zeyh.

Buche –see– *Fagus sylvatica* L.

Buchsbaum –see– *Buxus sempervirens* L.

Buchweizen –see– *Fagopyrum esculentum* Moench

buckbean –see– *Menyanthes trifoliata* L.

buckeye –see– *Aesculus flava* Sol.; *Aesculus glabra* Willd.; *Aesculus hippocastanum* L.; *Aesculus pavia* L.

buckle fern –see– *Dryopteris carthusiana* (Vill.) H. P. Fuchs

Buckley's-centaury –see– *Zeltnera calycosa* (Buckley) G. Mans.

buckthorn –see– *Amsinckia intermedia* Fisch. & C. A. Mey.; *Frangula alnus* Mill.; *Karwinskia calderonii* Standl.; *Karwinskia humboldtiana* (Schult.) Zucc.; *Rhamnus cathartica* L.

buckwheat –see– *Fagopyrum esculentum* Moench

Buddha pea –see– *Aeschynomene indica* L.

Buddhist rosary bead –see– *Abrus precatorius* L.

budgeegrass –see– *Cynodon nlemfuensis* Vanderyst; *Cynodon plectostachyus* (K. Schum.) Pilg.

buena moza –see– *Nicotiana glauca* Graham

Buena vista –see– *Codiaeum variegatum* (L.) A. Juss.

buffalo bean –see– *Mucuna pruriens* (L.) DC.

buffalo bur –see– *Solanum rostratum* Dunal

buffelgras –see– *Megathyrsus maximus* (Jacq.) B. K. Simon & S. W. L. Jacobs

buffelgrass –see– *Cenchrus ciliaris* L.

bugbane –see– *Actaea racemosa* L.; *Veratrum viride* Aiton

buis –see– *Buxus sempervirens* L.

buisson ardent –see– *Pyracantha coccinea* M. Roem.

bul rose –see– *Narcissus pseudonarcissus* L.

bulbous buttercup –see– *Ranunculus bulbosus* L.

bull mallow –see– *Malva nicaeensis* All.

bull nettle –see– *Laportea canadensis* (L.) Wedd.; *Solanum aculeatissimum* Jacq.; *Solanum carolinense* L.; *Solanum elaeagnifolium* Cav.

bull pine –see– *Pinus ponderosa* C. Lawson

bull thistle –see– *Silybum marianum* (L.) Gaertn.

bullhead –see– *Tribulus terrestris* L.

bullock bush –see– *Alectryon oleifolius* (Desf.) S. T. Reynolds

bullock's-heart –see– *Annona reticulata* L.

bulls-and-cows –see– *Arum maculatum* L.

BULNESIA SARMIENTOI Lorentz ex Griseb. [Zygophyllaceae]

Common Names:

guaiac; palo santo

Citations:

Williams MC, Rodewijk JC, Olsen JD (1984) Intoxication in cattle, chicks and hamsters from seed of the palo santo tree (Bulnesia sarmientii). Vet Rec 115(25-26):646-648.

bulrush millet –see– *Pennisetum glaucum* (L.) R. Br.

bunchberry –see– *Lantana camara* L.

bunchgrass –see– *Nolina texana* S. Watson

bunk –see– *Conium maculatum* L.

Buphane disticha (L. f.) Herb. = Boophone distictia (L. f.) Herb.

Buphane toxicaria Herb. = Boophone distictia (L. f.) Herb.

bur buttercup –see– *Ceratocephala testiculata* (Crantz) Roth

bur clover –see– *Medicago polymorpha* L.

bur medic –see– *Medicago polymorpha* L.

bur nut –see– *Tribulus terrestris* L.

bur thistle –see– *Xanthium strumarium* L.

bur tree –see– *Sambucus nigra* L.

bur trefoil –see– *Medicago polymorpha* L.

burans –see– *Rhododendron arboreum* Sm.

burdock –see– *Arctium lappa* L.; *Arctium minus* (Hill) Bernh.

Burke's-slangkop –see– *Drimia sanguinea* (Schinz) Jessop

burley tobacco –see– *Nicotiana tabacum* L.

Burma bean –see– *Phaseolus lunatus* L.

Burma lac tree –see– *Gluta usitata* (Wall.) Ding Hou

Burma teak –see– *Tectona grandis* L. f.

Burmese lacquer tree –see– *Gluta usitata* (Wall.) Ding Hou

burning bean –see– *Calia secundiflora* (Ortega) Yakovlev

burningbush –see– *Dictamnus albus* L.; *Bassia scoparia* (L.) A. J. Scott

burnwood –see– *Metopium brownei* (Jacq.) Urb.; *Metopium toxiferum* (L.) Krug & Urb.; *Schinus terebinthifolius* Raddi

burrawang –see– *Macrozamia communis* L. A. S. Johnson; *Macrozamia lucida* L. A. S. Johnson; *Macrozamia spiralis* (Salisb.) Miq.

burrawang palm –see– *Macrozamia spiralis* (Salisb.) Miq.

burro weed –see– *Isocoma plurifolia* (Torr. & A. Gray) Greene

Bursa bursa-pastoris (L.) Britton = Capsella bursa-pastoris (L.) Medik.

BURTTIA PRUNOIDES Baker f. & Exell [Connaraceae]

Citations:

Msengi LM, Mosha RD, Matovelo JA, et al. (1987) The toxicity of Burttia prunoides in rats and goats. Vet Hum Toxicol 29(5):398-400.

Mugera GM, Jiwa SF (1970) Acute Burttia prunoides toxicity in livestock. Bull Epizootic Dis Afr 18(3):253-258.

Musonera NA, Kindele N, Shukla RR (1979) Parasitic pneumonia and toxic hepatitis in cattle from the Kundelungu Ranch - Shaba, Zaire. Bull Anim Health Prod Afr 27:97-100.

burweed –see– *Xanthium strumarium* L.

burweed marsh elder –see– *Cyclachaena xanthiifolia* (Nutt.) Fresen.; *Iva annua* L.

bush tree –see– *Buxus sempervirens* L.

bushi –see– *Aconitum carmichaelii* Debeaux

bushman's-poison –see– *Acokanthera oppositifolia* (Lam.) Codd

buta buta –see– *Cerbera odollam* Gaertn.

BUTEA MONOSPERMA (Lam.) Taub. [Fabaceae]

Common Names:

palas; palash

Citations:

Kulakkattolickal AT (1987) Piscicidal plants of Nepal: Preliminary toxicity screening using grass carp (Ctenopharyngodon idella) fingerlings. J Ethnopharmacol 21(1):1-9.

buti –see– *Lantana camara* L.

butón-de-oro –see– *Ranunculus repens* L.

butter bean –see– *Phaseolus lunatus* L.

butter daisy –see– *Ranunculus acris* L.; *Ranunculus bulbosus* L.; *Verbesina encelioides* (Cav.) Benth. & Hook. f. ex A. Gray

butter print –see– *Abutilon theophrasti* Medik.

butter weed –see– *Senecio glabellus* Poir.

Butterblume –see– *Ranunculus acris* L.; *Taraxacum officinale* F. H. Wigg. aggr.

buttercup –see– *Allamanda cathartica* L.; *Ranunculus acris* L.; *Ranunculus bulbosus* L.; *Ranunculus multifidus* Forssk.; *Ranunculus sceleratus* L.

butterfly banners –see– *Dicentra cucullaria* (L.) Bernh.

butterfly flag –see– *Diplarrhena moraea* Labill.

butterfly milkweed –see– *Asclepias tuberosa* L.

butterfly pea –see– *Clitoria ternatea* L.

butterfly weed –see– *Asclepias tuberosa* L.

button print –see– *Abutilon theophrasti* Medik.

button weed –see– *Abutilon theophrasti* Medik.; *Ixiolaena brevicompta* F. Muell.; *Malva parviflora* L.

buttongrass –see– *Dactyloctenium radulans* (R. Br.) P. Beauv.

BUXUS SEMPERVIRENS L. [Buxaceae]

Common Names:

Abassian boxwood; box; boxwood; Buchsbaum; buis; bush tree; cimisu; dudgeon; English boxwood; European boxwood; evergreen box; Iranian boxwood; Splintbaum; Turkey boxwood

Citations:

Bastien A, Grisvard M, Jean-Blain C, et al. (1973) Intoxication de jeunes bovins par le buis (Buxus sempervirens L.). Bull Soc Sci Vet Med Comp Lyon 75(5):289-290.

Camy G, Leveille JL, Nevers B (1986) Cas clinique: Intoxication de bovins par le buis (Buxus sempervirens). Point Veterinaire 18(97):203-204.

Krüger A, Matschullat G (1970) Buchsbaumvergiftung beim Schwein. Prakt Tierarzt 51:235-236.

Markin LE (1930) Boxwood sensitiveness. J Allergy 1:346-349.

van Soest H, Gotink WM, van der Vooren LJ (1965) Buxusvergiftiging bij varkens en runderen. Tijdschr Diergeneeskd 90(6):387-389.

Byfield fern –see– *Bowenia serrulata* (W. Bull) Chamb.

Byrsocarpus coccineus Schumach. & Thonn. = Rourea coccinea (Schumach. & Thonn.) Hook. f.

C

caapi –see– *Banisteriopsis caapi* (Spruce ex Griseb.) C. V. Morton

cabaret –see– *Asarum europaeum* L.

cabbage –see– *Brassica oleracea* L. var. capitata L.

cabbage lettuce –see– *Lactuca sativa* L. var. capitata L.

cabbagebark –see– *Andira inermis* (W. Wright) Kunth ex DC.

cabinet cherry –see– *Prunus serotina* Ehrh.

cacachilla –see– *Karwinskia humboldtiana* (Schult.) Zucc.; *Karwinskia calderonii* Standl.

cacahuacuahuitl –see– *Theobroma cacao* L.

cacahuil –see– *Anacardium occidentale* L.

cacao –see– *Theobroma cacao* L.

cacatsi –see– *Karwinskia humboldtiana* (Schult.) Zucc.

CACHRYS LIBANOTIS L. [Apiaceae]

Citations:

Ena P, Cerri R, Dessi G, et al. (1991) Phototoxicity due to Cachrys libanotis. Contact Dermatitis 24(7):1-5.

Ena P, Dessi G, Chiarolini F, et al. (1989) Phytophotodermatitis from a Cachrys species. Contact Dermatitis 20(10):144-145.

CADABA APHYLLA (Thunb.) Wild [Capparaceae]

Synonyms:

Cadaba juncea (Sparrm.) Harv. ex Hook. f. nom. illegit.

Common Names:

swartstorm

Citations:

Van der Walt SJ, Steyn DG (1943) Recent investigations into the toxicity of plants, etc. XIII. Onderstepoort J Vet Sci Anim Indus 18(1-2):207-224.

Cadaba juncea (Sparrm.) Harv. ex Hook. f. nom. illegit. = Cadaba aphylla (Thunb.) Wild

cade juniper –see– *Juniperus oxycedrus* L.

cadia-del-perro –see– *Krameria ixine* L.

cadjust –see– *Mucuna pruriens* (L.) DC.

CAESALPINIA GILLIESII (Hook.) D. Dietr. [Fabaceae]

Synonyms:

Poinciana gilliesii Hook.

Common Names:

bird-of-paradise; Mexican bird-of-paradise; mimosa-du-japon; paradise poinciana; poinciana; yellow bird-of-paradise

Citations:

Anonymous (1958) Toxicity studies of Arizona ornamental plants. Ariz Med 15(7):512-514.

Cann HM, Verhulst HL (1958) Bird of Paradise (Caesalpinia gilliesie (Hook), Fam. Leguminosae). Bull Natl Clgh Poison Control Cent 1958(Nov):3-5.

Cann HM, Verhulst HL (1958) Poisonous plants. Bull Natl Clgh Poison Control Cent 1958(Sep):2-5.

Jouglard J, Airaudo CB Richardot R, (1973) Intoxications accidentelles par ingestion de graines de "Mimosa du Japon." Bull Med Leg Toxicol Med 16(1):55-58.

Shoemaker HA (1958) Bird of paradise seed poisoning. J Okla State Med Assoc 51(11):659-660.

cafta –see– *Catha edulis* (Vahl) Forssk. ex Endl.

caixeta –see– *Simarouba amara* Aubl.

CAJANUS CAJAN (L.) Millsp. [Fabaceae]

Common Names:

Bengal gram; dhal arkar; gungo pea; pigeon pea; rahar dal; red gram

Citations:

Kulakkattolickal AT (1987) Piscicidal plants of Nepal: Preliminary toxicity screening using grass carp (Ctenopharyngodon idella) fingerlings. J Ethnopharmacol 21(1):1-9.

Montgomery RD (1964) Observations on the cyanide content and toxicity of tropical pulses. West Indian Med J 13(1):1-11.

cajotillo –see– *Karwinskia humboldtiana* (Schult.) Zucc.

cajú –see– *Anacardium occidentale* L.

cajuil –see– *Anacardium occidentale* L.

caktrio –see– *Tribulus terrestris* L.

calabash gourd –see– *Lagenaria siceraria* (Molina) Standl.

Calabrian pine –see– *Larix decidua* Mill.

caladio –see– *Caladium bicolor* (Aiton) Vent.

caladium –see– *Caladium bicolor* (Aiton) Vent.

CALADIUM BICOLOR (Aiton) Vent. [Araceae]

Common Names:

angel's-wings; caladio; caladium; cananga; capotillo; corazon-de-cabrito; couer saignant; elephant's-ear; fancy-leaf caladium; Heart-of-Jesus; lagrimas-de-Maria; mother-in-law plant; paleta-de-pintor

Citations:

Mrvos R, Krenzelok EP, Jacobsen TD (2001) Toxidromes associated with the most common plant ingestions. Vet Hum Toxicol 43(6):366-369.

Caladium seguinum (Jacq.) Vent. = Dieffenbachia seguine (Jacq.) Schott

Calamity Jane –see– *Echium plantagineum* L.

calamus –see– *Acorus calamus* L.

calbachechem –see– *Metopium brownei* (Jacq.) Urb.

calendula –see– *Calendula officinalis* L.

CALENDULA OFFICINALIS L. [Asteraceae]

Common Names:

calendula; marigold; pot marigold; Ringelblume

Citations:

Reider N, Komericki P, Hausen BM, et al. (2001) The seamy side of natural medicines: Contact sensitization to arnica (Arnica montana L.) and marigold (Calendula officinalis L.). Contact Dermatitis 45(5):269-272.

Caley pea –see– *Lathyrus hirsutus* L.

calfkill –see– *Kalmia angustifolia* L.

calf's-foot –see– *Arum maculatum* L.

CALIA SECUNDIFLORA (Ortega) Yakovlev [Fabaceae]

Synonyms:

Sophora secundiflora (Ortega) Lag. ex DC.

Common Names:

burning bean; coral bean; dry whiskey; Eve's-necklace; frijolillo; frijolito; mescal bean; mescal Bohne; mountain laurel; red bean; Texas mountain laurel; western mountain laurel

Citations:

Boughton IB, Hardy WT (1935) Mescalbean (Sophora secundiflora) poisonous for livestock. Texas Agric Exp Sta Bull #519:18 pp.

Knauer KW, Reagor JC, Bailey EM Jr, et al. (1995) Mescalbean (Sophora secundiflora) toxicity in a dog. Vet Hum Toxicol 37(3):237-239.

Wells SR (1993) Intentional ingestion of the mescal bean (Sophora secundiflora). Vet Hum Toxicol 35(4):330.

calico bush –see– *Kalmia latifolia* L.

California buckeye –see– *Aesculus californica* (Spach) Nutt.

California bur –see– *Xanthium strumarium* L. var. canadense (Mill.) Torr. & A. Gray

California false hellebore –see– *Veratrum californicum* Durand

California fern –see– *Conium maculatum* L.

California filaree –see– *Erodium cicutarium* (L.) L'Her.

California hellebore –see– *Veratrum californicum* Durand

California holly –see– *Heteromeles salicifolia* (C. Presl) Abrams

California live oak –see– *Quercus agrifolia* Née

California milkweed –see– *Asclepias eriocarpa* Benth.

California mugwort –see– *Artemisia vulgaris* L.

California poison ivy –see– *Toxicodendron diversilobum* (Torr. & A. Gray) Greene

California poison oak –see– *Toxicodendron diversilobum* (Torr. & A. Gray) Greene

California poison sumach –see– *Toxicodendron diversilobum* (Torr. & A. Gray) Greene

California poison vine –see– *Toxicodendron diversilobum* (Torr. & A. Gray) Greene

California privet –see– *Ligustrum ovalifolium* Hassk.

California redwood –see– *Sequoia sempervirens* (D. Don) Endl.

California rhododendron –see– *Rhododendron macrophyllum* D. Don ex G. Don

California rosebay –see– *Rhododendron macrophyllum* D. Don ex G. Don

California swamp hellebore –see– *Veratrum californicum* Durand

California toyon –see– *Heteromeles salicifolia* (C. Presl) Abrams

California water hemlock –see– *Cicuta douglasii* (DC.) J. M. Coult. & Rose

California white cedar –see– *Calocedrus decurrens* (Torr.) Florin

calla lily –see– *Zantedeschia aethiopica* (L.) Spreng.

CALLA PALUSTRIS L. [Araceae]

Common Names:

bog arum; copo-de-leite; Drachenkraut; Kalla; Schlangenkraut; Schweinekraut; Schweinsohr; Sumpfdrachenwurz; Sumpfkalla; Sumpfkraut; water arum; water dragon; wild calla

Citations:

Airaksinen M, Peura P, Alaossi Salokangas L, et al. (1986) Toxicity of plant material used as emergency food during famines in Finland. J Ethnopharmacol 18(3):273-296.

Ramón MF, Ballesteros S, Martínez-Arrieta R, et al. (2003) Plant and herb exposures in Spain. J Toxicol Clin Toxicol 41(4):514.

CALLILEPIS LAUREOLA DC. [Asteraceae]

Common Names:

African oxeye daisy; impila; oxeye daisy; wildemagriet

Citations:

Laurens JB, Bekker LC, Steenkamp V, et al. (2001) Gas chromatographic-mass spectrometric confirmation of atractyloside in a patient poisoned with Callilepis laureola. J Chromatogr B Biomed Sci Appl 765(2):127-133.

Seedat YK (1978) Acute renal failure among Blacks and Indians in South Africa. S Afr Med J 54(11):427-431.

Seedat YK, Hitchcock PJ (1971) Acute renal failure from Callilepis laureola. S Afr Med J 45(30):832-833.

Steenkamp PA, Harding NM, van Heerden FR, et al. (2006) Identification of atractyloside by LC-ESI-MS in alleged herbal poisonings. Forensic Sci Int 163(1-2):81-92.
Steenkamp V, Stewart MJ, Zuckerman M (1999) Detection of poisoning by impila (Callilepis laureola) in a mother and child. Hum Exp Toxicol 18(10):594-597.
Stewart MJ, Moar JJ, Mwesigwa J, et al. (2000) Forensic toxicology in urban South Africa. J Toxicol Clin Toxicol 38(4):415-419.
Stewart MJ, Moar JJ, Steenkamp P, et al. (1999) Findings in fatal cases of poisoning attributed to traditional remedies in South Africa. Forensic Sci Int 101(3):177-183.
Wainwright J, Schonland MM (1977) Toxic hepatitis in black patients in Natal. S Afr Med J 51(17):571-573.
Watson AR, Coovadia HM, Bhoola KD (1979) The clinical syndrome of impila (Callilepis laureola) poisoning in children. S Afr Med J 55(8):290-292.

CALLISIA FRAGRANS (Lindl.) Woodson [Commelinaceae]

Common Names:
basket plant; inch plant

Citations:
Lee SE, Mason KV (2006) Immediate hypersensitivity to leaf extracts of Callisia fragrans (inch plant) in a dog. Vet Dermatol 17(1):70-80.

callotrio –see– *Karwinskia humboldtiana* (Schult.) Zucc.

CALOCEDRUS DECURRENS (Torr.) Florin [Cupressaceae]

Synonyms:
Libocedrus decurrens Torr.

Common Names:
California white cedar; incense cedar

Citations:
Calnan CD (1972) Dermatitis from cedar wood pencils. Trans St Johns Hosp Dermatol Soc 58(1):43-47.
Cavagni G, Caffarelli C, Spattini A, et al. (2003) IgE-mediated allergic rhinitis and conjunctivitis caused by Calocedrus decurrens (incense cedar). Allergy 58(11):1201-1202.

CALOPOGONIUM MUCUNOIDES Desv. [Fabaceae]

Citations:
Shenk JS (1976) The meadow vole as an experimental animal. Lab Anim Sci 26(4):664-669.

calotrope –see– *Calotropis procera* (Aiton) W. T. Aiton

CALOTROPIS GIGANTEA (L.) W. T. Aiton [Apocynaceae]

Common Names:
akanda; algodón-de-seda; arka; crown flower; giant milkweed; madar

Citations:
Crawford HE (1958) Crown flower keratoconjunctivitis. Hawaii Med J 17:244-245.
Rao BS, Sarkar HB, Sheshadri HS (1969) Effect of latex of Calotropis gigantea on pregnancy in the albino rat. J Reprod Fertil 38:234.
Rathnasabathy V, Rao KL, Krishnaswamy A, et al. (1949) A preliminary study of the toxicity of Calotropis gigantea. Indian J Med Res 37(4):483-494.
Wong WW (1949) Keratoconjunctivitis due to crownflower. Hawaii Med J 8:339-341.

CALOTROPIS PROCERA (Aiton) W. T. Aiton [Apocynaceae]

Common Names:
ak; akada; algodão-de-seda; ankra; arka; bomubomu; calotrope; crown flower; Dead Sea apple; French jasmine; giant milkweed; king's-crown; madar; milkweed; mudar; rubber tree; Sodom apple; St. Thomas'-bush; swallowwort; wild cotton; wild down

Citations:
Angelo SJ, Misra SS (1978) Hypopyon in a goat (ulceration caused by sap from plants of the genus Calotropis). Indian Vet Med J 2:219-221.
Biedner B, Rothkoff L, Witztum A (1977) Calotropis procera (Sodom apple) latex keratoconjunctivitis. Isr J Med Sci 13(9):914-916.
El Badwi SM, Adam SE, Shigidi MT, et al. (1998) Studies on laticiferous plants: Toxic effects in goats of Calotropis procera latex given by different routes of administration. Dtsch Tierarztl Wochenschr 105(11):425-427.
Faye B (1985) Contribution à l'étude de la toxité de Calotropis procera. Effet d'une alimentaion à base de Calotropis procera sur la mortalité embryonnaire et néonatale chez la souris de laboratoire. Rev Med Vet Pays Trop 38(1):72-75.
Handa F; Sadana JK; Sharma PK (1984) Allergic contact dermatitis due to the plant Calotropis procera. A case report. Indian J Dermatol 29:27-29.
Mahmoud OM, Adam SE, Tartour G (1979) The effects of Calotropis procera on small ruminants. I. Effects of feeding sheep with the plant. J Comp Pathol 89(2):241-250.
Mahmoud OM, Adam SE, Tartour G (1979) The effects of Calotropis procera on small ruminants. II. Effects of administration of the latex to sheep and goats. J Comp Pathol 89(2):251-263.
Maria DL, Panse AD, Kulkarni TP (1973) Injury to the eye by sap of Calotropis procera. East Arch Ophthalmol 1:296-297.
Pahwa R, Chatterjee VC (1988) The toxicity of Indian Calotropis procera R Br latex in the black rat, Rattus rattus Linn. Vet Hum Toxicol 30(4):305-308.
Singh R, Siddiqui MA, Baruah MC (1978) Plant dermatitis in Delhi. Indian J Med Res 68(Oct):650-655.

caltrops –see– *Tribulus terrestris* L.

CALYCANTHUS FLORIDUS L. [Calycanthaceae]

Common names:

Carolina allspice

Citations:

Bradley RE, Jones TJ (1963) Strychnine-like toxicity in cattle. Case report. Southeastern Vet 14(2):40, 71, 73.

camará –see– *Lantana camara* L.

camaray –see– *Dioclea erecta* Hoehne; *Dioclea latifolia* Benth.

camasa –see– *Lagenaria siceraria* (Molina) Standl.

camboatá –see– *Guarea guidonia* (L.) Sleumer

camel bush –see– *Trichodesma incanum* (Bunge) A. DC.

camel melon –see– *Citrullus lanatus* (Thunb.) Matsum. & Nakai

camel poison –see– *Erythrophleum chlorostachys* (F. Muell.) Baill.

CAMELLIA SINENSIS (L.) Kuntze [Theaceae]

Synonyms:

Camellia thea Link

Common Names:

green tea; oolong tea; tea

Citations:

Bonkovsky HL (2006) Hepatotoxicity associated with supplements containing Chinese green tea (Camellia sinensis) Ann Intern Med 144(1):68-71.

Kaiser HE, Bartone JC (1966) The carcinogenic activity of ordinary tea. A preliminary note. J Natl Med Assoc 58:361.

Kamijo Y, Soma K, Asari Y, et al. (1999) Severe rhabdomyolysis following massive ingestion of oolong tea: Caffeine intoxication with coexisting hyponatremia. Vet Hum Toxicol 41(6):381-383.

Murphy KJ (1975) Liver dysfunction and tea eating. Med J Aust 2(11):428-429.

Segi M (1975) Tea-gruel as a possible factor for cancer of the esophagus. Gann 66:199-202.

Camellia thea Link = Camellia sinensis (L.) Kuntze

camiri nut –see– *Aleurites moluccanus* (L.) Willd.

cammock –see– *Hypericum perforatum* L.

camomile –see– *Anthemis cotula* L.; *Chamaemelum nobile* (L.) All.; *Matricaria recutita* L.

camotillo –see– *Solanum tuberosum* L.

campana –see– *Brugmansia ×candida* Pers.; *Brugmansia arborea* (L.) Lagerh.

campanillo amarillo –see– *Thevetia peruviana* (Pers.) K. Schum.

camphor laurel –see– *Cinnamomum camphora* (L.) J. Presl

camphor tree –see– *Cinnamomum camphora* (L.) J. Presl

campo pea –see– *Lathyrus splendens* Kellogg

CAMPTOTHECA ACUMINATA Decne. [Nyssaceae]

Citations:

Cao GR, Gao JX, Duan DX, et al. (1992) Studies on Camptotheca acuminata leaves: Main toxic principle, poisoning, and treatment in goats. In: James LF, Keeler RF, Bailey JM Jr, et al. (eds.) Poisonous plants. Iowa State Univ Press. Ames, Iowa. pp. 506-508.

can-ja-da –see– *Tephrosia purpurea* (L.) Pers.

caña brava –see– *Bambusa vulgaris* Schrad. ex J. C. Wendl.

caña muda –see– *Dieffenbachia seguine* (Jacq.) Schott

Canadian field pea –see– *Pisum sativum* L.

Canadian garlic –see– *Allium canadense* L.

Canadian hemp –see– *Apocynum cannabinum* L.

Canadian juniper –see– *Juniperus communis* L.

Canadian milk vetch –see– *Astragalus canadensis* L.

Canadian poison ivy –see– *Toxicodendron rydbergii* (Small ex Rydb.) Greene

Canadian red cedar –see– *Thuja plicata Donn* ex D. Don

Canadian root –see– *Asclepias tuberosa* L.

Canadian thistle –see– *Cirsium arvense* (L.) Scop.

Canadian wood nettle –see– *Laportea canadensis* (L.) Wedd.

cáñamo silvestre –see– *Galeopsis ladanum* L.

cananga –see– *Caladium bicolor* (Aiton) Vent.

canario –see– *Allamanda cathartica* L.

CANARIUM INDICUM L. [Burseraceae]

Common Names:

nangai

Citations:

Sten E, Stahl Skov P, Andersen SB, et al. (2002) Allergenic components of a novel food, Micronesian nut nangai (Canarium indicum), shows IgE cross-reactivity in pollen allergic patients. Allergy 57(5):398-404.

Canary Island broom –see– *Genista canariensis* L.

Canary Island ivy –see– *Hedera helix* L. subsp. canariensis (Willd.) Cout.

canarygrass –see– *Phalaris aquatica* L.; *Phalaris canariensis* L.; *Phalaris minor* Retz.

CANAVALIA CATHARTICA Thouars [Fabaceae]

Synonyms:

Canavalia virosa (Roxb.) Wight & Arn.

Common Names:

sem

Citations:

Seena S, Sridhar KR (2004) Nutrient composition and biological evaluation of an unconventional legume, Canavalia cathartica of mangroves. Int J Food Sci Nutr 55(8):615-625.

Singh NK, Srivastava PK, Singh DS, et al. (1984) Accidental poisoning by Canavalia virosa. J Assoc Physicians India 32(5):450-451.

CANAVALIA ENSIFORMIS (L.) DC. [Fabaceae]

Common Names:

Chickasaw lima bean; cut-eye bean; giant stock bean; horsebean; jack swort bean; jackbean; Jackbohne; one-eye bean; overlook bean; sword bean; wonder bean

Citations:

Affleck H (1961) Jack bean poisoning in cattle. Rhodesia Agric J 58:21.
Grant G, More LJ, McKenzie NH, et al. (1995) Nutritional and haemagglutination properties of several tropical seeds. J Agric Sci 124:437-445.
Jayne Williams DJ (1973) Influence of dietary jack beans (Canavalia ensiformis) and of concanavalin A on the growth of conventional and gnotobiotic Japanese quail (Coturnix coturnix japonica). Nat New Biol 243(126):150-151.
Ologhobo A, Mosenthin R, Alaka OO (2003) Histological alterations in the internal organs of growing chicks from feeding raw jackbean or limabean seeds. Vet Hum Toxicol 45(1):10-13.
Risso JF, Montilla JJ (1992) Harina de granos de Canavalia ensiformis L (DC) cruda, almacenada en medio alcalino, autoclavada or extruída, en dietas para cerdos en crecimiento. Arch Latinoam Nutr 42(3):268-274.
Shone DK (1961) Toxicity of the jack bean. Rhodesia Agric J 58:18-20.
Tschiersch B (1962) Zur toxischen Wirkung der Jackbohne. Pharmazie 17:621-623.

CANAVALIA ROSEA (Sw.) DC. [Fabaceae]

Common Names:

beach bean

Citations:

Grant G, More LJ, McKenzie NH, et al. (1995) Nutritional and haemagglutination properties of several tropical seeds. J Agric Sci 124:437-445.

Canavalia virosa (Roxb.) Wight & Arn. = Canavalia cathartica Thouars

cancer jalap –see– *Phytolacca americana* L.

cancerroot –see– *Orobanche minor* Sm.; *Phytolacca americana* L.

candelabra aloe –see– *Aloe arborescens* Mill.

candelabra cactus –see– *Euphorbia ingens* E. Mey. ex Boiss.; *Euphorbia lactea* Haw.

candelabra Euphorbia –see– *Euphorbia ingens* E. Mey. ex Boiss.

candelabra flower –see– *Boophone disticha* (L. f.) Herb.

candelabra tree –see– *Euphorbia ingens* E. Mey. ex Boiss.

candelaria –see– *Verbascum thapsus* L.

candleberry –see– *Aleurites moluccanus* (L.) Willd.

candlenut –see– *Aleurites moluccanus* (L.) Willd.

candlenut tree –see– *Vernicia montana* Lour.

candungo –see– *Lagenaria siceraria* (Molina) Standl.

cane reed –see– *Arundo donax* L.

canela-de-veado –see– *Sessea brasiliensis* Toledo

canker rose –see– *Papaver rhoeas* L.

cannabis –see– *Cannabis sativa* L.

Cannabis indica Lam. = Cannabis sativa L.

CANNABIS SATIVA L. [Cannabaceae]

Synonyms:

Cannabis indica Lam.

Common Names:

bhang; cannabis; Cannabisharz; chancre indien; charas; dagga; fimble; gaja; ganga; ganja; grass; Hanf; Haschisch; hashish; hemp; Indian hemp; kif; maconha; marihuana; marijuana; Mary Jane; Mexican dove plant; pot; reefer; siddi; weed

Citations:

Cardassis J (1951) Intoxication des équidés par Cannabis indica. Rec Med Vet Ec Alfort 127(Dec):971-973.
Clarke EG, Greatorex JC, Potter R (1971) Cannabis poisoning in the dog. Vet Rec 88(Jun 26):694.
Debray H, Vidal F, Enjolras M (1987) Intoxication au cannabis chez une enfant de 13 mois. Presse Med 16(36):1807.
Driemeier D (1997) Marijuana (Cannabis sativa) toxicosis in cattle. Vet Hum Toxicol 39(6):351-352.
Freeman GL (1983) Allergic skin test reactivity to marijuana in the southwest. West J Med 138(6):829-831.
Frost RC (1983) Marijuana toxaemia. Vet Rec 112(18):441.
Frye FL (1968) Acute cannabis intoxication in a pup. J Am Vet Med Assoc 152(5):472.
Ghorbal H, Hamouda C, Bousnina M, et al. (2003) Use of plants to induce chemical submission in Tunisia. Vet Hum Toxicol 45(2):91-93.
Gupta BD, Jani CB, Shah PH (2001) Fatal 'bhang' poisoning. Med Sci Law 41(4):349-352.
Henney SN, Coleman MJ (1984) Canine cannabis intoxication. Vet Rec 114(17):436.
Hervás JA, Fiol M, Vidal C, et al. (1987) Intoxicación por ingestión de hachís en niños. Med Clin (Barc) 88:563.
Jain MC, Arora N (1988) Ganja (Cannabis sativa) refuse as cattle feed. Indian J Anim Sci 58(7):865-867.
Janczyk P, Donaldson CW, Gwaltney S (2004) Two hundred and thirteen cases of marijuana toxicoses in dogs. Vet Hum Toxicol 46(1):19-21.
Jones DL (1978) A case of cannabis ingestion. N Z Vet J 26(5):135-136.
Liskow B, Liss JL, Parker CW (1971) Allergy to marihuana. Ann Intern Med 75(4):571-573.
Majmudar V, Azam NA, Finch T (2006) Contact urticaria to Cannabis sativa. Contact Dermatitis 54(2):127.
Pizzi R (2002) Cannabis ingestion in a chinchilla. Vet Rec 150(2):55.
Smith RA (1988) Coma in a ferret after ingestion of cannabis. Vet Hum Toxicol 30(5):486.
Valentine J (1992) Unusual poisoning in a dog. Vet Rec 130:307.
Welshman MD (1986) Doped Dobermann. Vet Rec 119(20):512.

Note:
Marijuana is named *Cannabis sativa* L. subsp. *indica* (Lam.) E. Small & Cronquist in some publications.

Cannabisharz –see– *Cannabis sativa* L.

canning pea –see– *Pisum sativum* L.

canola –see– *Brassica napus* L. var. napus; *Brassica rapa* L. subsp. campestris (L.) A. R. Clapham

Canson kale –see– *Brassica oleracea* L. var. ramosa DC.

Cansons kale –see– *Brassica oleracea* L. var. medullosa Thell.

cantiva –see– *Allamanda cathartica* L.

canudo –see– *Ipomoea carnea* Jacq. subsp. fistulosa (C. Mart. ex Choisy) D. F. Austin

canutillo –see– *Plumbago scandens* L.

cao wu –see– *Aconitum kusnezoffii* Rchb.

caoba –see– *Swietenia mahagoni* (L.) Jacq.

capa-de-la-reina –see– *Clitoria ternatea* L.

Cape jasmine –see– *Gardenia jasminoides* J. Ellis

Cape lilac –see– *Melia azedarach* L.

Cape poison bulb –see– *Boophone disticha* (L. f.) Herb.

Cape slangkop –see– *Ornithoglossum vulgare* B. Nord.

Cape syringa –see– *Melia azedarach* L.

caper bush –see– *Euphorbia lathyris* L.

caper garden spurge –see– *Euphorbia lathyris* L.

caper spurge –see– *Euphorbia lathyris* L.; *Euphorbia peplus* L.

Capeweed –see– *Arctotheca calendula* (L.) Levyns

Capnoides caseana auct. = Corydalis caseana A. Gray

capotillo –see– *Caladium bicolor* (Aiton) Vent.

Capparis persicifolia A. Rich = Capparis tomentosa Lam.

CAPPARIS TOMENTOSA Lam. [Capparaceae]

Synonyms:
Capparis persicifolia A. Rich.

Common Names:
haujeri

Citations:
Ahmed OM, Adam SE (1980) The toxicity of Capparis tomentosa in goats. J Comp Pathol 90(2):187-195.
Ahmed OM, Adam SE, Edds GT (1981) The toxicity of Capparis tomentosa in sheep and calves. Vet Hum Toxicol 23(6):403-409.
Ahmed SA, Amin AE, Adam SE, et al. (1993) By toxic effects of the dried leaves and stem of Capparis tomentosa on Nubian goats. Dtsch Tierarztl Wochenschr 100(5):192-194.
Salih YM, Idris OF, Wahbi AG, et al. (1980) Toxicity of Capparis tomentosa to sheep and goats. Sudan J Vet Res 2:13-21.

CAPSELLA BURSA-PASTORIS (L.) Medik. [Brassicaceae]

Synonyms:
Bursa bursa-pastoris (L.) Britton

Common Names:
bolsa-de-pastor; bourse-à-pasteur; Hirtentäschel; shepherd's-purse

Citations:
Kellerman WA (1895) Poisoning by shepherd's purse. Bot Gaz 20:325-326.
Temperton H (1943) Some nutritional causes of discolourations in the internal contents of eggs. Harper Adams Utility Poultry J 28(3):71-76.

capsicum –see– *Capsicum annuum* L.; *Capsicum frutescens* L.

CAPSICUM ANNUUM L. [Solanaceae]

Synonyms:
Capsicum annuum L. var. conoides (Mill.) Irish

Common Names:
African bird pepper; bell pepper; capsicum; cayenne pepper; cherry pepper; chili; chili pepper; Christmas pepper; green pepper; guinea pepper; hot pepper; jalapeño; ornamental pepper; paprika; pepper; pod pepper; red pepper; Roter Pfeffer; spice pepper; sweet pepper

Citations:
Hoch-Ligeti C (1951) Production of liver tumours by dietary means: Effect of feeding chillies (Capsicum frutescens and annuum (Linn.)) to rats. Acta Unio Int Contra Cancrum 7(3):606-611.
Jones LA, Tandberg D, Troutman WG (1986) A controlled evaluation of household treatments for "chile burns." Vet Hum Toxicol 28(5):486.
Mrvos R, Krenzelok EP, Jacobsen TD (2001) Toxidromes associated with the most common plant ingestions. Vet Hum Toxicol 43(6):366-369.
Niinimäki A (1984) Delayed-type allergy to spices. Contact Dermatitis 11(1):34-40.
Niinimäki A, Hannuksela M (1981) Immediate skin test reactions to spices. Allergy 36(7):487-493.
Sastre J, Olmo M, Novalvos A, et al. (1996) Occupational asthma due to different spices. Allergy 51(2):117-120.
Swami RH, Judge BS, Furbee RB (2003) Accidental ingestion of cayenne pepper sauce requiring prolonged ventilatory support. J Toxicol Clin Toxicol 41(5):653-654.
Weinberg RB (1981) Hunan hand. N Engl J Med 305(17):1020.
Winograd HL (1977) Acute croup in an older child. An unusual toxic origin. Clin Pediatr (Phila) 16(10):884-887.

Note:
Bell pepper is named *Capsicum annuum* L. var. *annuum* in some publications. There is considerable confusion in the plant toxicology literature in the naming of the *Capsicum* peppers.

Capsicum annuum L. var. conoides (Mill.) Irish = Capsicum annuum L.

CAPSICUM FASTIGIATUM Blume [Solanaceae]

Common Names:

chilly

Citations:

Flecker H (1945) Injuries produced by plants in tropical Queensland. Med J Aust 1(Jun 23):636-637.

CAPSICUM FRUTESCENS L. [Solanaceae]

Common Names:

African chili; African pepper; aji caballero; aji picante; bird pepper; capsicum; cayenne pepper; chile; chile pepper; chili bravo; chili pepper; Guinea pepper; hot pepper; Indian goat pepper; jalapeño; Louisiana pepper; pepper; red cayenne; red chili; spur pepper; Tabasco pepper; wild pepper

Citations:

Al Qarawi AA, Adam SE (1999) Effects of red chili (Capsicum frutescens L) on rats. Vet Hum Toxicol 41(5):293-295.

Cohen HS, Colette S (1984) "Tabasco throat." Am Fam Physician 29(6):47.

Diehl AK, Bauer RL (1978) Jalaproctitis. N Engl J Med 299(20):1137-1138.

Hoch-Ligeti C (1951) Production of liver tumours by dietary means: Effect of feeding chillies (Capsicum frutescens and annuum (Linn.)) to rats. Acta Unio Int Contra Cancrum 7(3):606-611.

capulín tullidor –see– *Karwinskia humboldtiana* (Schult.) Zucc.

capulín tullidora –see– *Karwinskia humboldtiana* (Schult.) Zucc.

capulín tullidore –see– *Karwinskia humboldtiana* (Schult.) Zucc.

capulincillo –see– *Karwinskia humboldtiana* (Schult.) Zucc.

caracuha colorado –see– *Plumeria rubra* L.

CARAGANA ARBORESCENS Lam. [Fabaceae]

Common Names:

Erbsenstrauch

Citations:

Krienke EG (1976) Akzidentelle Vergiftungen durch Pflanzen aus der Sicht einer Giftinformationszentrale. Internist (Berl) 17(8):399-410.

caraotas rosadas –see– *Phaseolus vulgaris* L.

CARAPICHEA IPECACUANHA (Brot.) L. Andersson [Rubiaceae]

Common Names:

ipecaagoene; ipecac; ipecacuanha

Citations:

Robertson WO (1979) Syrup of ipecac associated fatality: A case report. Vet Hum Toxicol 21(2):87-89.

caraway –see– *Carum carvi* L.

cardamom –see– *Elettaria cardamomum* (L.) Maton

cardo alazorado –see– *Centaurea solstitialis* L.

cardo asnal –see– *Silybum marianum* (L.) Gaertn.

cardo cuco –see– *Datura ferox* L.

cardo-de-la-liga –see– *Chamaeleon gummifera* (L.) Cass.

cardo-de-Maria –see– *Silybum marianum* (L.) Gaertn.

cardo lechal –see– *Silybum marianum* (L.) Gaertn.

cardo lechero –see– *Centaurea solstitialis* L.

cardo santo –see– *Argemone mexicana* L.

cardoon –see– *Cynara cardunculus* L.

Carduus marianus L. = Silybum marianum (L.) Gaertn.

careless weed –see– *Amaranthus hybridus* L.; *Amaranthus palmeri* S. Watson; *Amaranthus retroflexus* L.; *Cyclachaena xanthiifolia* (Nutt.) Fresen.

CAREX BUEKII Wimm. [Cyperaceae]

Citations:

Rosca V, Pavel A (1969) Observații privind intoxicația cu diferite specii de rogoz (Carex L.) la taurine. Rev Zootehnie Med Vet 19(11):81-85.

CAREX CONTIGUA Hoppe [Cyperaceae]

Citations:

Rosca V, Pavel A (1969) Observații privind intoxicația cu diferite specii de rogoz (Carex L.) la taurine. Rev Zootehnie Med Vet 19(11):81-85.

CAREX VULPINA L. [Cyperaceae]

Citations:

Rosca V, Pavel A (1969) Observații privind intoxicația cu diferite specii de rogoz (Carex L.) la taurine. Rev Zootehnie Med Vet 19(11):81-85.

CAREYA ARBOREA Roxb. [Lecythidaceae]

Citations:

Kulakkattolickal AT (1987) Piscicidal plants of Nepal: Preliminary toxicity screening using grass carp (Ctenopharyngodon idella) fingerlings. J Ethnopharmacol 21(1):1-9.

cariaquito morado –see– *Melochia tomentosa* L.

CARICA PAPAYA L. [Caricaceae]

Common Names:

betek; papaw; papaya; pawpaw

Citations:

Chakraborty P, Ghosh D, Chowdhury I, et al. (2005) Aerobiological and immunochemical studies on Carica papaya L. pollen: An aeroallergen from India. Allergy 60(7):920-926.

Cordero-Moreno R (1973) Etiologic factors in tropical eye diseases. Am J Ophthalmol 75(3):349-364.

Fasal P (1945) Cutaneous diseases in the tropics. A clinical study based on observations in Malaya. Arch Derm Syphilol 51(3):163-171.

Osgood H (1945) Atopic sensitivity to caroid (papain). Report of a case with discussion of some of the properties and uses of the drug. J Allergy 16:245-250.

Carlina gummifera Less. = Chamaeleon gummifera (L.) Cass.

carnation –see– *Dianthus caryophyllus* L.

carnation poppy –see– *Papaver somniferum* L.

carne asada –see– *Andira inermis* (W. Wright) Kunth ex DC.

carob –see– *Ceratonia siliqua* L.

carob bean –see– *Ceratonia siliqua* L.

Carolina canarygrass –see– *Phalaris caroliniana* Walter

Carolina horse nettle –see– *Solanum carolinense* L.

Carolina jasmine –see– *Gelsemium sempervirens* (L.) J. St.-Hil.

Carolina jessamine –see– *Gelsemium sempervirens* (L.) J. St.-Hil.

Carolina love nettle –see– *Solanum pseudocapsicum* L.

Carolina maple –see– *Acer rubrum* L.

Carolina modiola –see– *Modiola caroliniana* (L.) G. Don

Carolina nightshade –see– *Solanum carolinense* L.

Carolina wild woodbine –see– *Gelsemium sempervirens* (L.) J. St.-Hil.

Carolina yellow jessamine –see– *Gelsemium sempervirens* (L.) J. St.-Hil.

carotte-a-moreau –see– *Cicuta douglasii* (DC.) J. M. Coult. & Rose

carpetweed –see– *Drymaria holosteoides* Benth.; *Kallstroemia hirsutissima* Vail ex Small

carrot –see– *Daucus carota* L.

carrot bush –see– *Eremophila deserti* (A. Cunn. ex Benth.) Chinnock

carrot fern –see– *Conium maculatum* L.

carrot weed –see– *Parthenium hysterophorus* L.; *Trachymene glaucifolia* (F. Muell.) Benth.

carrotgrass –see– *Parthenium hysterophorus* L.

CARTHAMUS TINCTORIUS L. [Asteraceae]

Common Names:

hong hua; safflower

Citations:

Chan JC, Chan TY, Chan KL, et al. (1994) Anticholinergic poisoning from Chinese herbal medicines. Aust N Z J Med 24(3):317.

van der Willigen AH, van Joost T, Stolz E, et al. (1987) Contact dermatitis to safflower. Contact Dermatitis 17(11):184-186.

CARUM CARVI L. [Apiaceae]

Common Names:

caraway; krisna jirak

Citations:

Niinimäki A, Hannuksela M (1981) Immediate skin test reactions to spices. Allergy 36(7):487-493.

CARYA ILLINOINENSIS (Wangenh.) K. Koch [Juglandaceae]

Common Names:

pecan

Citations:

Swinny B (1951) Periorbital dermatitis. Ann Allergy 9(6):774-777.

CARYOTA MITIS Lour. [Arecaceae]

Common Names:

besedin palm; clustered fishtail palm; fishtail palm; tufted fishtail palm

Citations:

Snyder DS, Hatfield GM, Lampe KF (1979) Examination of the itch response from the raphides of the fishtail palm Caryota mitis. Toxicol Appl Pharmacol 48(2):287-292.

CARYOTA URENS L. [Arecaceae]

Common Names:

fishtail palm; sago palm; toddy fishtail palm; wine palm

Citations:

Flecker H (1945) Injuries produced by plants in tropical Queensland. Med J Aust 1(Jun 23):636-637.

casca bark –see– *Erythrophleum suaveolens* (Guill. & Perr.) Brenan

cashes –see– *Conium maculatum* L.

cashew –see– *Anacardium occidentale* L.

cashew apple –see– *Anacardium occidentale* L.

cassada –see– *Manihot esculenta* Crantz

cassava –see– *Manihot esculenta* Crantz

cassia –see– *Cinnamomum aromaticum* Nees; *Senna tora* (L.) Roxb.

Cassia acutifolia Delile. = Senna alexandrina Mill.

Cassia didymobotrya Fresen. = Senna didymobotrya (Fresen.) H. S. Irwin & Barneby

Cassia floribunda Cav. = Senna ×floribunda (Cav.) H. S. Irwin & Barneby

Cassia italica (Mill.) Spreng. = Senna italica Mill.

CASSIA JAVANICA L. subsp. nodosa (Buch.-Ham. ex Roxb.) K. Larsen & S. S. Larsen [Fabaceae]

Common Names:

pink-and-white-shower

Citations:

Grant G, More LJ, McKenzie NH, et al. (1995) Nutritional and haemagglutination properties of several tropical seeds. J Agric Sci 124:437-445.

Cassia lindheimeriana Scheele = Senna lindheimeriana (Scheele) H. S. Irwin & Barneby

Cassia obtusifolia L. = Senna obtusifolia (L.) H. S. Irwin & Barneby

Cassia occidentalis L. = Senna occidentalis (L.) Link

Cassia roemeriana Scheele = Senna roemeriana (Scheele) H. S. Irwin & Barneby

Cassia senna L. = Senna alexandrina Mill.

Cassia tora L. = Senna tora (L.) Roxb.

CASSINE BUCHANANII Loes. [Celastraceae]

Synonyms:

Elaeodendron buchananii (Loes.) Loes.; ***Elaeodendron keniense*** Loes.

Citations:

Mugera GM (1970) Toxic and medicinal plants of East Africa. I. Bull Epizootic Dis Afr 18(4):377-387.

Cassuvium pomiferum Lam. = Anacardium occidentale L.

CASTANEA SATIVA Mill. [Fagaceae]

Common Names:

chestnut; Kastanie; sweet chestnut

Citations:

Zapatero L, Baeza ML, Sierra Z, et al. (2005) Anaphylaxis by fruits of the Fagaceae family: Acorn and chestnut. Allergy 60(12):1542.

CASTANOSPERMUM AUSTRALE A. Cunn. & C. Fraser ex Hook. [Fabaceae]

Common Names:

Australian bean tree; Australian black bean tree; bean tree; black bean; Meto; Moreton Bay chestnut

Citations:

Hindmarsh WL, Hart L (1938) Castanospermum australe (black bean, Moreton Bay chestnut). Green seeds poisonous to stock. New South Wales Dep Agric Vet Res Rep #7(1937):109-114.

McKenzie RA, Reichmann KG, Dimmock CK, et al. (1988) The toxicity of Castanospermum australe seeds for cattle. Aust Vet J 65(6):165-167.

Reichmann KG, Twist JO, McKenzie RA (1989) Inhibition of alpha-glucosidase in cattle by Castanospermum australe: An attempted phenocopy of Pompe's disease. Aust Vet J 66(3):86-89.

Shaw D, Pearn J (1979) Oleander [sic] poisoning. Med J Aust 2(5):267-269.

castlebean –see– *Sesbania vesicaria* (Jacq.) Elliott

castor bean –see– *Ricinus communis* L.

casuarina –see– *Casuarina equisetifolia* L.

CASUARINA CRISTATA Miq. [Casuarinaceae]

Synonyms:

Casuarina lepidophloia F. Muell.

Common Names:

Australian pine

Citations:

Zivitz N (1942) Allergy to Australian pine. J Allergy 13:314-316.

CASUARINA EQUISETIFOLIA L. [Casuarinaceae]

Common Names:

Australian pine; beef wood tree; casuarina; horsetail tree; she oak

Citations:

Zivitz N (1942) Allergy to Australian pine. J Allergy 13:314-316.

Casuarina lepidopholia F. Muell. = Casuarina cristata Miq.

cat mint –see– *Nepeta cataria* L.

catalina –see– *Euphorbia pulcherrima* Willd. ex Klotzsch

cataria –see– *Nepeta cataria* L.

catawba rhododendron –see– *Rhododendron catawbiense* Michx.

catayo –see– *Persicaria punctata* (Elliott) Small

catechu –see– *Acacia catechu* (L. f.) Willd.

caterpillar weed –see– *Heliotropium europaeum* L.

CATHA EDULIS (Vahl) Forssk. ex Endl. [Celastraceae]

Synonyms:

Celastrus edulis Vahl

Common Names:

Abyssinian tea; cafta; chat; ciat; flower-of-paradise; gad; gat; inlonge; kat; khat; miraa; ol meraa; qat; tschat

Citations:

Alem A, Shibre T (1997) Khat induced psychosis and its medico-legal implication: A case report. Ethiop Med J 35(2):137-139.

Carothers JC (1945) Miraa as a cause of insanity. East Afr Med J 22(23):4-6.

El Guindy MK (1971) Effects of Catha edulis (khat) chewing on human body. J Egypt Med Assoc 54(4):230-234.

Giannini AJ, Castellani S (1982) A manic-like psychosis due to khat (Catha edulis Forssk.). Clin Toxicol 19(5):455-459.

Heisch RB (1945) A case of poisoning by Catha edulis. East Afr Med J 22(Jan):7-9.

Kennedy JG, Teague J, Rokaw W, et al. (1983) A medical evaluation of the use of qat in North Yemen. Soc Sci Med 17(12):783-793.

Maitai CK (1977) The toxicity of the plant Catha edulis in rats. Toxicon 15(4):363-366.

Nencini P, Ahmed AM, Elmi AS (1986) Subjective effects of khat chewing in humans. Drug Alcohol Depend 18(1):97-105.

Roper JP (1986) The presumed neurotoxic effects of Catha edulis - An exotic plant now available in the United Kingdom. Br J Ophthalmol 70(10):779-781.

CATHARANTHUS PUSILLUS (Murray) G. Don [Apocynaceae]

Synonyms:

Lochnera pusilla (Murray) K. Schum.; ***Vinca pusilla*** Murray

Citations:

Mantramurti ID (1964) Let not Lochnera pussilus poison your livestock. Indian Livestock 2(2):8, 43.

Selva Raj VB, Ganapathy MS (1967) Studies on the toxicity of Lochnera pusilla K. Schum. Indian Vet J 44(10):871-876.

CATHARANTHUS ROSEUS (L.) G. Don [Apocynaceae]

Common Names:

bright-eyes; cayenne jasmine; chichirica; Madagascar periwinkle; old maid; periwinkle; pink periwinkle; rose periwinkle; vinca

Citations:

Deng JF, Lin TJ, Kao WF, et al. (1997) The difficulty in handling poisonings associated with Chinese traditional medicine: A poison control center experience for 1991-1993. Vet Hum Toxicol 39(2):106-114.

Deng JF, Wu ML (2003) Severe bone marrow depression induced by an anti-cancer herb "Catharanthus roseus." J Toxicol Clin Toxicol 41(4):518.

Wu ML, Deng JF, Wu JC, et al. (2004) Severe bone marrow depression induced by an anticancer herb Catharanthus roseus. J Toxicol Clin Toxicol 42(5):667-671.

cathedral bells –see– *Kalanchoe pinnata* (Lam.) Pers.

catmint –see– *Nepeta cataria* L.

catnep –see– *Nepeta cataria* L.

catnip –see– *Nepeta cataria* L.

cat's-ear –see– *Hypochaeris radicata* L.

cat's-eggs –see– *Xanthium strumarium* L.

cat's-head –see– *Tribulus terrestris* L.

cat's-milk –see– *Euphorbia helioscopia* L.

cat's-tail –see– *Equisetum arvense* L.

cat's-tail grass –see– *Equisetum arvense* L.

cat's-valerian –see– *Valeriana officinalis* L.

catuit –see– *Euphorbia tirucalli* L.

CATUNAREGAM SPINOSA (Thunb.) Tirveng. [Rubiaceae]

Citations:

Kulakkattolickal AT (1987) Piscicidal plants of Nepal: Preliminary toxicity screening using grass carp (Ctenopharyngodon idella) fingerlings. J Ethnopharmacol 21(1):1-9.

catweed –see– *Ageratina adenophora* (Spreng.) R. M. King & H. Rob.

catwood –see– *Euonymus europaeus* L.

CAUCALIS PLATYCARPOS L. [Apiaceae]

Synonyms:

Orlaya platycarpos (L.) W. D. J. Koch

Citations:

Sabboykides M (1966) [Poisoning of sheep by the plant Orlaya platycarpa (Umbelliferae).] Hellenike Kteniatrike 9:160-166.

cauliflower –see– *Brassica oleracea* L. var. botrytis L.

caulophylle faux-pigamon –see– *Caulophyllum thalictroides* (L.) Michx.

CAULOPHYLLUM THALICTROIDES (L.) Michx. [Berberidaceae]

Common Names:

blue cohosh; caulophylle faux-pigamon; papooseroot; squawroot

Citations:

Jones TK, Lawson BM (1998) Profound neonatal congestive heart failure caused by maternal consumption of blue cohosh herbal medicine. J Pediatr 132(3 Part 1):550-552.

Rao RB, Hoffman RS (2002) Nicotinic toxicity from tincture of blue cohosh (Caulophyllum thalictroides) used as an abortifacient. Vet Hum Toxicol 44(4):221-222.

Rao RB, Hoffman RS, Desiderio R, et al. (1998) Nicotinic toxicity from tincture of blue cohosh (Caulophyllum thalictroides) used as an abortifacient. J Toxicol Clin Toxicol 36(5):455.

caustic bush –see– *Sarcostemma viminale* (L.) R. Br.

caustic creeper –see– *Euphorbia drummondii* Boiss.

caustic vine –see– *Sarcostemma viminale* (L.) R. Br.

caustic weed –see– *Euphorbia drummondii* Boiss.

caviúna –see– *Dalbergia nigra* (Vell.) Allemão ex Benth.

caviuna vermelho –see– *Machaerium scleroxylon* Tul.

cavolfiore –see– *Brassica oleracea* L. var. botrytis L.

cavolo cappaccio –see– *Brassica oleracea* L. var. capitata L.

cayenne jasmine –see– *Catharanthus roseus* (L.) G. Don

cayenne pepper –see– *Capsicum annuum* L.; *Capsicum frutescens* L.

cayote –see– *Karwinskia humboldtiana* (Schult.) Zucc.

cebil colorado –see– *Anadenanthera colubrina* (Vell.) Brenan var. cebil (Griseb.) Altschul

cebola –see– *Allium cepa* L.

cebolla albarrana –see– *Drimia maritima* (L.) Stearn

cedar-of-Lebanon –see– *Cedrus libani* A. Rich.

cèdre –see– *Thuja occidentalis* L.

cedro prieto –see– *Metopium brownei* (Jacq.) Urb.; *Metopium toxiferum* (L.) Krug & Urb.

cedron –see– *Simaba cedron* Planch.

CEDRUS DEODORA (Roxb. ex D. Don) G. Don [Pinaceae]

Common Names:

deodar cedar

Citations:

Rawat A, Singh A, Singh AB, et al. (2000) Clinical and immunologic evaluation of Cedrus deodora pollen: A new allergen from India. Allergy 55(7):620-626.

CEDRUS LIBANI A. Rich. [Pinaceae]

Common Names:

cedar-of-Lebanon

Citations:

Greenberg M (1972) Respiratory symptoms following brief exposure to cedar of Lebanon (Cedra libani) dust. Clin Allergy 2(3):219-224.

cegadera –see– *Heterophyllaea pustulata* Hook. f.

celandine –see– *Chelidonium majus* L.

celandine poppy –see– *Chelidonium majus* L.

Celastrus edulis Vahl = Catha edulis (Vahl) Forssk. ex Endl.

céleri –see– *Apium graveolens* L.

Celeri graveolens (L.) Britton = Apium graveolens L.

celeriac –see– *Apium graveolens* L.

celery –see– *Apium graveolens* L.

celery-leaf buttercup –see– *Ranunculus sceleratus* L.

celery-leaf crowfoot –see– *Ranunculus sceleratus* L.

celestina azul –see– *Ageratum houstonianum* Mill.

celestrum nigrum –see– *Cestrum nocturnum* L.

celidonia mayor –see– *Chelidonium majus* L.

CELTIS BRIEYI De Wild. [Ulmaceae]

Citations:

Oleffe JA, Sporcq J, Hublet P (1975) Epidemiological study of the wood industry in Belgium. Contact Dermatitis 1(5):315-316.

CENCHRUS CILIARIS L. [Poaceae]

Common Names:

buffelgrass

Citations:

Allen JG, Creeper JH, Forshaw D, et al. (2004) Plant-associated diseases, either new or new to the state, encountered over the last decade (1991-2001) in Western Australia. In: Acamovic T, Stewart CS, Pennycott TW (eds.) Poisonous plants and related toxins. CABI. Wallingford, UK. pp. 540-547.

McKenzie RA, Bell AM, Storie GJ, et al. (1988) Acute oxalate poisoning of sheep by buffelgrass (Cenchrus ciliaris). Aust Vet J 65(1):26.

CENCHRUS INCERTUS M. A. Curtis [Poaceae]

Synonyms:

Cenchrus pauciflorus Benth.

Common Names:

field sandbur; sandbur; southern sandbur

Citations:

Sperry OE, Turk RD, Hoffman GO, et al. (1955) Photosensitization of cattle in Texas. Texas Agric Exp Sta Bull #812:8 pp.

Cenchrus pauciflorus Benth. = Cenchrus incertus M. A. Curtis

CENCHRUS SETIGER Vahl [Poaceae]

Common Names:

birdwoodgrass

Citations:

Allen JG, Creeper JH, Forshaw D, et al. (2004) Plant-associated diseases, either new or new to the state, encountered over the last decade (1991-2001) in Western Australia. In: Acamovic T, Stewart CS, Pennycott TW (eds.) Poisonous plants and related toxins. CABI. Wallingford, UK. pp. 540-547.

Centaurea picris Pall. ex Willd. = Acroptilon repens (L.) DC.

Centaurea repens L. = Acroptilon repens (L.) DC.

CENTAUREA SOLSTITIALIS L. [Asteraceae]

Common Names:

abrepuño; cardo alazorado; cardo lechero; knapweed; St. Barnaby's-thistle; star thistle; yellow bur; yellow centaurea; yellow star thistle

Citations:

Cordy DR (1954) Nigro-pallidal encephalomalacia (chewing disease) in horses on rations high in yellow star thistle. Am Vet Med Assoc Annu Meet 1954:149-154.

Cordy DR (1954) Nigropallidal encephalomalacia in horses associated with ingestion of yellow star thistle. J Neuropathol Exp Neurol 13(2):330-342.

Fowler ME (1965) Nigropallidal encephalomalacia in the horse. J Am Vet Med Assoc 147(6):607-616.

Gard GP, de Sarem WG, Ahrens PJ (1973) Nigropallidal encephalomalacia in horses in New South Wales. Aust Vet J 49(2):107-108.

Martin AA, Yanarella FG, Maurel RA, et al. (1971) Nigropallidal encefalomalacia en equinos provocada por la intoxicacion cronica con "abrepuño" (Centaurea solstitialis L.). Analect Vet 3(1-3):43-53.

Mettler FA, Stern GM (1963) Observations on the toxic effects of yellow star thistle. J Neuropathol Exp Neurol 22(Jan):164-169.

Selfero NA, Merlassino JL, Audisio S (1989) Intoxicacion por abrepuños (Centaurea solstitialis) evaluacion terapeutica. Therios 13(61):42-44.

Young S, Brown WW, Klinger B (1970) Nigropallidal encephalomalacia in horses caused by ingestion of weeds of the genus Centaurea. J Am Vet Med Assoc 157(11):1602-1605.

Centaurium beyrichii (Torr. & A. Gray) B. L. Rob. = Zeltnera beyrichii (Torr. & A. Gray) G. Mans.

Centaurium calycosum (Buckley) Fernald = Zeltnera calycosa (Buckley) G. Mans.

CENTELLA ASIATICA (L.) Urb. [Araliaceae]

Synonyms:

Hydrocotyle asiatica L.

Common Names:

brahmi; fo-ti-tieng; gotu kola; Indian pennywort; madecassol; pennywort; taquip-cohol

Citations:

Eun HC, Lee AY (1985) Contact dermatitis due to madecassol. Contact Dermatitis 13(5):310-313.

Central American walnut –see– *Juglans olanchana* Standl. & L. O. Williams

CENTROSEMA PUBESCENS Benth. [Fabaceae]

Citations:

Shenk JS (1976) The meadow vole as an experimental animal. Lab Anim Sci 26(4):664-669.

century plant –see– *Agave americana* L.; *Agave lechuguilla* Torr.

Cephaelis tomentosa (Aubl.) Vahl = Psychotria poeppigiana Müll. Arg.

CERATOCEPHALA TESTICULATA (Crantz) Roth [Ranunculaceae]

Common Names:

bur buttercup; hornseed buttercup; testiculate buttercup

Citations:

Olsen JD, Anderson TE, Madsen G (1982) Bur buttercup. It will get your sheep if you don't watch out. Utah Sci 43(Spring):10-13.

Olsen JD, Anderson TE, Murphy JC, et al. (1983) Bur buttercup poisoning of sheep. J Am Vet Med Assoc 183(5):538-543.

CERATONIA SILIQUA L. [Fabaceae]

Common Names:

carob; carob bean

Citations:

Silanikove N, Gilboa N, Nir I, et al. (1996) Effect of a daily supplementation of polyethylene glycol on intake and digestion of tannin-containing leaves (Quercus calliprinos, Pistacia lentiscus, and Ceratonia siliqua) by goats. J Agric Food Chem 44:199

CERBERA ODOLLAM Gaertn. [Apocynaceae]

Common Names:

buta buta; famentana; odallum; ordeal tree; pink-eye cerbera; pong pong; samanta; tangena nut; yellow-eye cerbera

Citations:

Gaillard Y, Krishnamoorthy A, Bevalot F (2004) Cerbera odollam: A 'suicide tree' and cause of death in the state of Kerala, India. J Ethnopharmacol 95(2-3):123-126.

Iyer GV, Narendranathan M (1975) A preliminary report on the neurological manifestations of Cerbera odollam poisoning. Indian J Med Res 63(2):312-314.

Kini PM, Pai KN (1965) Cardiotoxic effects of Cerbera odollam. Indian Heart J 17(3):263-270.

Narendranathan M, Das KV, Vijayaraghavan G (1975) Electrocardiographic changes in Cerbera odollam poisoning. J Assoc Physicians India 23(11):757-762.

Narendranathan M, Das KV, Vijayaraghavan G (1975) Prognostic factors in Cerbera odollam poisoning. Indian Heart J 27(4):283-286.

Vijayaraghavan G, Kuruvilla A, Warrier NS, et al. (1974) Cardiovascular effects of cerebrin. Indian Heart J 26(2):79-83.

Cerbera thevetia L. = Thevetia peruviana (Pers.) K. Schum.

CERCOCARPUS MONTANUS Raf. [Rosaceae]

Common Names:

alder-leaf mountain mahogany; hairy mountain mahogany; mountain mahogany

Citations:

Nunez Hernandez G, Wallace JD, Holechek JL, et al. (1991) Condensed tannins and nutrient utilization by lambs and goats fed low-quality diets. J Anim Sci 69(3):1167-1177.

ériman –see– *Monstera deliciosa* Liebm.

CESTRUM AURANTIACUM Lindl. [Solanaceae]

Common Names:

night-blooming jasmine; orange bush; orange cestrum

Citations:

Edgar G (1933) "Orange bush" proved poisonous to stock. Agric Gaz New South Wales 44:785.

Mugera GM, Nderito P (1968) Cestrum poisoning in Kenya livestock. Bull Epizootic Dis Afr 16(4):501-506.

Shone DK (1966) Poisonous plants of Rhodesia. Part I. Rhodesia Zambia Malawi J Agric Res 4:81-94.

Shone DK (1967) Poisonous plants of Rhodesia. Part II. Rhodesia Zambia Malawi J Agric Res 5:29-41.

CESTRUM DIURNUM L. [Solanaceae]

Common Names:

Chinese inkberry; dama-de-dia; day-blooming jasmine; day-blooming jessamine; day cestrum; day jasmine; day jessamine; galan-de-dia; jasmine; jessamine; king-of-the-day; tintan; wild jasmin; wild jasmine

Citations:

Durand R, Figueredo JM, Mendoza E (1999) Intoxication in cattle from Cestrum diurnum. Vet Hum Toxicol 41(1):26-27.

Figueroa V, Sutherland TM (1972) "Muerte súbita" (sudden death) in cattle. 5. The role of toxic plants. Rev Cubana Cienc Agric Eng Ed 6(1):53-59.

Kasali OB, Krook L, Pond WG, et al. (1977) Cestrum diurnum intoxication in normal and hyperparathyroid pigs. Cornell Vet 67(2):190-221.
Krook L, Wasserman RH, McEntee K, et al. (1975) Cestrum diurnum poisoning in Florida cattle. Cornell Vet 65(4):557-575.
Krook L, Wasserman RH, Shively JN, et al. (1975) Hypercalcemia and calcinosis in Florida horses: Implication of the shrub, Cestrum diurnum, as the causative agent. Cornell Vet 65(1):26-56.
Sarkar K, Narbaitz R, Pokrupa R, et al. (1981) The ultrastructure of nephrocalcinosis induced in chicks by Cestrum diurnum leaves. Vet Pathol 18(1):62-70.
Simpson CF, Bruss ML (1979) Ectopic calcification in lambs from feeding the plant Cestrum diurnum. Calcif Tissue Int 29(3):245-250.

CESTRUM LAEVIGATUM Schltdl. [Solanaceae]

Common Names:
coerana; dominguinho; inkberry; wild potato

Citations:
Barros GC, Döbereiner J (1968) Experimentos com Cestrum laevigatum Schlecht. em animais de laboratorio. Pesq Agric Bras 3:307-311.
Döbereiner J, Tokarnia CH, Canella CF (1965) Intoxicação por Cestrum laevigatum Schlecht. em bovinos no Estado do Rio de Janeiro, Brasil. Proc Int Grassland Cong 2:1259-1263.
Döbereiner J, Tokarnia CH, Canella CF (1969) Intoxicação por Cestrum laevigatum Schlecht., a causa de mortandades em bovinos no Estado do Rio de Janeiro. Pesq Agric Bras Vet 4:165-193.
Peixoto PV, Brust LC, Duarte MD, et al. (2000) Cestrum laevigatum poisoning in goats in southeastern Brazil. Vet Hum Toxicol 42(1):13-14.
Thorburn JA (1934) Chase Valley disease. Cestrum laevigatum Schlecht, its toxic effects on ruminants. Onderstepoort J Vet Sci Anim Indus 2(2):667-679.
van der Lugt JJ, Nel PW, Kitching JP (1991) The pathology of Cestrum laevigatum (Schlechtd.) poisoning in cattle. Onderstepoort J Vet Res 58(3):211-221.
van der Lugt JJ, Nel PW, Kitching JP (1992) Experimentally-induced Cestrum laevigatum (Schlechtd.) poisoning in sheep. Onderstepoort J Vet Res 59(2):135-144.

CESTRUM NOCTURNUM L. [Solanaceae]

Common Names:
celestrum nigrum; Chinese inkberry; dama-de-noche; galan-de-noche; huele-de-noche; jasmin-de-nuit; lady-of-the-night; lilas-de-nuit; night-blooming jasmine; night-blooming jessamine; poisonberry

Citations:
Fruthaler GJ (1955) Solanine poisoning. Ochsner Clin Rep 1(2):50-52.

CESTRUM PARQUI L'Her. [Solanaceae]

Common Names:
Chilean cestrum; duraznillo; duraznillo negro; green cestrum; green poisonberry; iodine bush; la parquina; palque; willow-leaf jassamine

Citations:
Guercio V (1966) Su di un episodio da avvelenamento nei suini da «Cestrum parquii». Vet Italiana 17:726-729.
Kudo K, Kelly WR, Oelrichs PB (1985) Experimental poisoning of mice and sheep with Cestrum parqui. In: Seawright AA, Hegarty MP, James LF, et al. (eds.) Plant toxicology. Queensland Poisonous Plant Committee, Yeerongpilly, Queensland. pp. 533-540.
Lopez TA, Spinelli R, Villar JA (1978) Efectos de la dosificacion de Cestrum parqui L'Herit en ovinos y bovinos. Gaceta Vet 40(334):642-650.
Manriquez O, Varas J, Rios JC, et al. (2002) Analysis of 156 cases of plant intoxication received in the toxicologic information center at Catholic University of Chile. Vet Hum Toxicol 44(1):31-32.
McLennan MW, Kelly WR (1984) Cestrum parqui (green cestrum) poisoning in cattle. Aust Vet J 61(9):289-291.
Riet-Correa F, Schild AL, Méndez MC, et al. (1986) Intoxicação por Cestrum parqui (Solanaceae) em bovinos no Rio Grande do Sul. Pesq Vet Bras 6(4):111-115.
Ruiz LF (1930) Tres plantas tóxicas de la flora Argentina. Bol Minist Agric Nac Rep Argentina 29(Mar):45-55.

Ceylon rose –see– *Nerium oleander* L.

cha-de-figo –see– *Ficus carica* L.

chacin –see– *Metopium brownei* (Jacq.) Urb.

chacruna –see– *Psychotria viridis* Ruiz & Pav.

CHAEROPHYLLUM BULBOSUM L. [Apiaceae]

Common Names:
chervil

Citations:
Miedzobrodzki K, Bąk T, Lewandowski L (1963) Zatrucia zwierząt swierząbkiem (Chaerophyllum). Med Weter 19(6):309-310.

chailletia –see– *Dichapetalum cymosum* (Hook.) Engl.

chalice flower –see– *Narcissus pseudonarcissus* L.

CHAMAECYTISUS PROLIFER (L. f.) Link [Fabaceae]

Common Names:
tagasaste; tree lucerne

Citations:
Allen JG, Creeper JH, Forshaw D, et al. (2004) Plant-associated diseases, either new or new to the state, encountered over the last decade (1991-2001) in Western Australia. In: Acamovic T, Stewart CS, Pennycott TW (eds.) Poisonous plants and related toxins. CABI. Wallingford, UK. pp. 540-547.

CHAMAELEON GUMMIFERA (L.) Cass. [Asteraceae]

Synonyms:
Atractylis gummifera L.; ***Carlina gummifera*** Less.

Common Names:
addad; birdlime thistle; cardo-de-la-liga; chameleon leukos; chardon-à-glu; edded; glue chard; glue thistle;

ixia; lion-of-the-earth; mascarida; masticogna; Mediterranean thistle; sticky thistle; Vogelleimdistel; white chameleon

Citations:

Bellimam MA, Karni NE, Bouayoun T, et al. (2001) Study on cases of homicides by poisoning. Vet Hum Toxicol 43(4):244.

Georgiou M, Sianidou L, Hatzis T, et al. (1988) Hepatotoxicity due to Atractylis gummifera L. J Toxicol Clin Toxicol 26(7):487-493.

Guercio V, Randazzo V, Balbo SM (1971) Osservazioni su episodi di avvelenamento da Atractylis gummifera. Veterinaria (Milano) 20(6):311-318.

Hamouda C, Amamou M, Thabet H, et al. (2000) Plant poisonings from herbal medication admitted to a Tunisian toxicologic intensive care unit, 1983-1998. Vet Hum Toxicol 42(3):137-141.

Lemaigre G, Tebbi Z, Galinsky R, et al. (1975) Hépatite fulminante par intoxication due au chardon à glu (Atractylis gummifera-L). Nouv Presse Med 4(40):2865-2868.

Nogue S, Sang P, Botey A, et al. (1992) Insuffisance rénale aigüe due à une intoxication par le chardon à glu (Atractylis gummifera-L). Presse Med 21(3):130.

Tocco L (1922) Sull' avvelenamento per Carlina gummifera. Nota II - Ricerche farmacologiche sul principio attivo della Carlina gummifera (Atractylato di K). Arch Int Pharmacodyn Ther 26:171-186.

CHAMAELIRIUM LUTEUM (L.) A. Gray [Melanthiaceae]

Common Names:

devil's-bit; helonias; unicorn root

Citations:

Moir J (1899) Poisoned by devil's-bit. Vet Rec 11:523-524.

CHAMAEMELUM NOBILE (L.) All. [Asteraceae]

Synonyms:

Anthemis nobilis L.

Common Names:

camomile; chamomile; Hundskamille; Kamille; Roman camomille; sweet camomille

Citations:

Hausen BM (1979) The sensitizing capacity of compositae plants. III. Test results and cross-reactions in compositae-sensitive patients. Dermatologica 159(1):1-11.

CHAMAESYCE HIRTA (L.) Millsp. [Euphorbiaceae]

Synonyms:

Euphorbia hirta L.; ***Euphorbia pilulifera*** L.

Common Names:

asthma herb; Australian asthma herb; Australian snakeweed; dudhi; golandrina; golodrina; grande malnommée; hairy spurge; Jean Robert; milkweed; petite malnommée; Queensland asthma herb; snakeweed; spurge

Citations:

Singh R, Siddiqui MA, Baruah MC (1978) Plant dermatitis in Delhi. Indian J Med Res 68(Oct):650-655.

CHAMAESYCE MACULATA (L.) Small [Euphorbiaceae]

Synonyms:

Euphorbia maculata L.

Common Names:

eyebane; milk purslane; nodding spurge; prostrate spurge; snow-on-the-mountain; spotted spurge

Citations:

Case AA (1957) Photosensitization syndrome in cattle, sheep, and swine. North Am Vet 38(Jun):161-165.

CHAMAESYCE PROSTRATA (Aiton) Small [Euphorbiaceae]

Synonyms:

Euphorbia prostrata Aiton

Common Names:

prostrate euphorbia; prostrate spurge; red caustic creeper

Citations:

Sperry OE, Turk RD, Hoffman GO, et al. (1955) Photosensitization of cattle in Texas. Texas Agric Exp Sta Bull #812:8 pp.

chameleon leukos –see– *Chamaeleon gummifera* (L.) Cass.

chamico –see– *Datura ferox* L.; *Datura stramonium* L.

chamiso hediondo –see– *Artemisia tridentata* Nutt.

chamiza –see– *Atriplex canescens* (Pursh) Nutt.

chamomile –see– *Chamaemelum nobile* (L.) All.; *Matricaria recutita* L.

Champagnerwurz –see– *Veratrum album* L.

chanchanote –see– *Karwinskia humboldtiana* (Schult.) Zucc.

chancre indien –see– *Cannabis sativa* L.

chandan –see– *Santalum album* L.

chandelier plant –see– *Kalanchoe delagoensis* Eckl. & Zeyh.

chanvre-de-Canada –see– *Apocynum cannabinum* L.

chaparral –see– *Larrea tridentata* (DC.) Coville

chaparral death camas –see– *Toxicoscordion fremontii* (Torr.) Rydb.

chaperno –see– *Andira inermis* (W. Wright) Kunth ex DC.

charas –see– *Cannabis sativa* L.

chard –see– *Beta vulgaris* L.

chardon-à-glu –see– *Chamaeleon gummifera* (L.) Cass.

charlock –see– *Sinapis alba* L.; *Sinapis arvensis* L.

chaste tree –see– *Vitex agnus-castus* L.

chat –see– *Catha edulis* (Vahl) Forssk. ex Endl.

chaulai –see– *Amaranthus viridis* L.

chayote –see– *Sechium edule* (Jacq.) Sw.

cheat –see– *Lolium temulentum* L.

chechém –see– *Metopium toxiferum* (L.) Krug & Urb.

chechem negro –see– *Metopium brownei* (Jacq.) Urb.

checkerberry –see– *Gaultheria procumbens* L.

cheeky bugger –see– *Isotoma petraea* F. Muell.

cheesebowl –see– *Papaver rhoeas* L.

cheeseweed –see– *Malva parviflora* L.

CHEILANTHES SIEBERI Kunze [Pteridaceae]

Common Names:

mulga fern; rock fern

Citations:

Clark IA, Dimmock CK (1971) The toxicity of Cheilanthes sieberi to cattle and sheep. Aust Vet J 47(4):149-152.

McKenzie RA (1978) Bovine enzootic haematuria in Queensland. Aust Vet J 54(2):61-64.

Smith BL, Embling PP, Lauren DR, et al. (1989) Carcinogen in rock fern (Cheilanthes sieberi) from New Zealand and Australia. Aust Vet J 66:154-155.

CHEILANTHES TENUIFOLIA (Burm. f.) Sw. [Pteridaceae]

Common Names:

mulga fern; rock fern; rock lip fern

Citations:

Gordon HM (1936) Some field observations on various diseases of sheep. Aust Vet J 12(Feb):28-31.

chekkarmenis –see– *Sauropus androgynus* (L.) Merr.

chekor manis –see– *Sauropus androgynus* (L.) Merr.

chekurmanis –see– *Sauropus androgynus* (L.) Merr.

chela –see– *Ficus amplissima* Sm.

chélidoine –see– *Chelidonium majus* L.

CHELIDONIUM MAJUS L. [Papaveraceae]

Common Names:

Blutkraut; celandine; celandine poppy; celidonia mayor; chélidoine; cock foot; Gelbkraut; great celandine; rock poppy; Schellkraut; Schöllkraut; swallowwort; tetterwort; wartwort; Warzenkraut

Citations:

Benninger J, Schneider HT, Schuppan D, et al. (1999) Acute hepatitis induced by greater celandine (Chelidonium majus). Gastroenterology 117(5):1234-1237.

Koopmann H (1937) Tödliche Schöllkraut-Vergiftung (Chelidonium majus). Sammlung Vergiftungsfallen 8(A682):93-98.

Reeks HC (1903) Poisoning of cattle by common celandine. J Comp Pathol 16:367-371.

Štěrba B, Meissner V (1962) Sušené rostliny s obsahem srdečních glykosidů a alkaloidů jako příčina uhynuté hospodářských zvířat. Veterinarstvi 12:83-86.

Stickel F, Pöschl G, Seitz HK, et al. (2003) Acute hepatitis induced by greater celandine (Chelidonium majus). Scand J Gastroenterol 38(5):565-568.

Strahl S, Ehret V, Dahm HH, et al. (1998) Nekrotisierende Hepatitis nach Einnahme pflanzlicher Heilmittel. Dtsch Med Wochenschr 123(47):1410-1414.

chengkian –see– *Croton tiglium* L.

chenile –see– *Hyoscyamus niger* L.

CHENOPODIUM ALBUM L. [Chenopodiaceae]

Common Names:

fat hen; goosefoot; lamb's-quarter; nabo blanco; pata-de-ganso; quelite cenizo; white goosefoot; wild spinach; witte ganzenrooet

Citations:

Buck WB, Preston KS, Abel M, et al. (1965-1966) Common weeds associated with perirenal edema in swine. Proc Am Coll Vet Toxicol 1965:24-28.

Buck WB, Preston KS, Abel M, et al. (1966) Perirenal edema in swine: A disease caused by common weeds. J Am Vet Med Assoc 148(12):1525-1531.

Çalka Ö, Akdeniz N, Metin A, et al. (2005) Phototoxic dermatitis due to Chenopodium album in a mother and son. Contact Dermatitis 53(1):58-60.

Herweijer CH, Den Houter LF (1970) Vergiftiging bij schapen door witte ganzenvoet (Chenopodium album). Tijdschr Diergeneeskd 95(22):1134-1136.

Herweijer CH, Den Houter LF (1971) Poisoning due to fat hen (Chenopodium album) in sheep. Netherlands J Vet Sci 4(1):52-54.

Ozmen O, Mor F, Unsal A (2003) Nitrate poisoning in cattle fed Chenopodium album hay. Vet Hum Toxicol 45(2):83-84.

Chenopodium ambrosioides L. = Dysphania ambrosioides (L.) Mosyakin & Clemants

Chenopodium atriplicinum (F. Muell.) F. Muell. = Scleroblitum atriplicinum (F. Muell.) Ulbr.

Chenopodium carinatum R. Br. = Dysphania carinata (R. Br.) Mosyakin & Clemants

Chenopodium graveolens Willd. = Dysphania graveolens (Willd.) Mosyakin & Clemants

CHENOPODIUM HYBRIDUM L. [Chenopodiaceae]

Citations:

Ivanov VV (1959) [A case of mass poisoning by Chenopodium hybridum L.] Bot Zh 44(2):198-199.

CHENOPODIUM QUINOA Willd. [Chenopodiaceae]

Common Names:

quinoa; quinua

Citations:

Cheeke PR, Carlsson R (1978) Evaluation of several crops as sources of leaf meal: Composition, effect of drying procedure, and rat growth response. Nutr Rep Int 18(4):465-473.

cherimoya –see– *Annona cherimola* Mill.
cherry –see– *Prunus serotina* Ehrh.; *Prunus virginiana* L.
cherry ballart –see– *Exocarpos cupressiformis* Labill.
cherry bay –see– *Prunus laurocerasus* L.
cherry laurel –see– *Prunus laurocerasus* L.
cherry pepper –see– *Capsicum annuum* L.
cherry pie –see– *Heliotropium europaeum* L.; *Lantana camara* L.
chervil –see– *Chaerophyllum bulbosum* L.
ches tennent –see– *Antiaris toxicaria* Lesch.
chestnut –see– *Castanea sativa* Mill.
cheta mona –see– *Paspalum scrobiculatum* L. var. bispicatum Hack.
Chewing's-fescue –see– *Festuca rubra* L. subsp. fallax (Thuill.) Nyman
chi gurunga –see– *Senecio sceleratus* Schweick.
chibata –see– *Arrabidaea bilabiata* (Sprague) Sandwith
chicalote –see– *Argemone mexicana* L.
chicharron –see– *Metopium brownei* (Jacq.) Urb.
chichicaste –see– *Cnidoscolus urens* (L.) Arthur
chichicaste-de-las-costas –see– *Cnidoscolus urens* (L.) Arthur
chichirica –see– *Catharanthus roseus* (L.) G. Don
chick pea –see– *Cicer arietinum* L.; *Lathyrus hirsutus* L.; *Lathyrus sativus* L.
Chickasaw lima bean –see– *Canavalia ensiformis* (L.) DC.
chickling pea –see– *Lathyrus sativus* L.
chickling vetch –see– *Lathyrus sativus* L.
chickory –see– *Cichorium intybus* L.
chico –see– *Sarcobatus vermiculatus* (Hook.) Torr.
chicory –see– *Cichorium endivia* L.; *Cichorium intybus* L.
chicot –see– *Gymnocladus dioicus* (L.) K. Koch
chiguirera –see– *Persicaria glabra* (Willd.) M. Gómez
chilamatillo –see– *Euphorbia heterophylla* L.
chilbe –see– *Trigonella foenum-graecum* L.
children's-bane –see– *Cicuta maculata* L.
children's-death –see– *Cicuta maculata* L.
chile –see– *Capsicum frutescens* L.
chile-de-perro –see– *Persicaria punctata* (Elliott) Small
chile pepper –see– *Capsicum frutescens* L.
Chilean cestrum –see– *Cestrum parqui* L'Her.
chili –see– *Capsicum annuum* L.
chili bravo –see– *Capsicum frutescens* L.
chili pepper –see– *Capsicum annuum* L.; *Capsicum frutescens* L.
Chillagoe horse poison –see– *Crotalaria aridicola* Domin
chilly –see– *Capsicum fastigiatum* Blume
chilquilla –see– *Pascalia glauca* Ortega
China cockle –see– *Vaccaria hispanica* (Mill.) Rauschert
China rose –see– *Hibiscus rosa-sinensis* L.
China tree –see– *Melia azedarach* L.
Chinaball tree –see– *Melia azedarach* L.
Chinaberry –see– *Melia azedarach* L.
Chinawood oil tree –see– *Vernicia fordii* (Hemsl.) Airy Shaw
Chinese aconite root –see– *Aconitum carmichaelii* Debeaux var. truppelianum (Ulbr.) W. T. Wang & P. K. Hsiao
Chinese bastard anise –see– *Illicium anisatum* L.
Chinese boxthorn –see– *Lycium barbarum* L.
Chinese cabbage –see– *Brassica rapa* L. subsp. campestris (L.) A. R. Clapham
Chinese-cap larkspur –see– *Delphinium glaucum* S. Watson
Chinese cinnamon –see– *Cinnamomum aromaticum* Nees
Chinese ephedra –see– *Ephedra sinica* Stapf
Chinese eupatorium –see– *Eupatorium chinense* L.
Chinese forget-me-not –see– *Cynoglossum officinale* L.
Chinese inkberry –see– *Cestrum diurnum* L.; *Cestrum nocturnum* L.
Chinese juniper –see– *Juniperus chinensis* L.
Chinese jute –see– *Abutilon theophrasti* Medik.
Chinese kidney bean –see– *Wisteria sinensis* (Sims) DC.
Chinese lacquer tree –see– *Toxicodendron vernicifluum* (Stokes) F. A. Barkley
Chinese persimmon –see– *Diospyros kaki* Thunb.
Chinese primrose –see– *Primula obconica* Hance; *Primula praenitens* Ker Gawl.
Chinese rhubarb –see– *Rheum officinale* Baill.
Chinese-rice-paper-plant –see– *Tetrapanax papyrifer* (Hook.) K. Koch
Chinese sacred bamboo –see– *Nandina domestica* Thunb.
Chinese shellac –see– *Toxicodendron vernicifluum* (Stokes) F. A. Barkley
Chinese star anise –see– *Illicium verum* Hook. f.
Chinese sumach –see– *Ailanthus altissima* (Mill.) Swingle
Chinese tallow tree –see– *Triadica sebifera* (L.) Small
Chinese tung oil tree –see– *Vernicia fordii* (Hemsl.) Airy Shaw
Chinese umbrella tree –see– *Melia azedarach* L.
Chinese wisteria –see– *Wisteria sinensis* (Sims) DC.
Chinese yam –see– *Dioscorea polystachya* Turcz.
Chinesewood-oil tree –see– *Vernicia fordii* (Hemsl.) Airy Shaw
Chinesische Glyzinie –see– *Wisteria sinensis* (Sims) DC.
Chinesische Primel –see– *Primula praenitens* Ker Gawl.
chingma –see– *Abutilon theophrasti* Medik.

chinkerinchee –see– *Ornithogalum thyrsoides* Jacq.; *Ornithogalum toxicarium* C. Archer & R. H. Archer

chique chique –see– *Crotalaria pallida* Aiton

CHIRONIA BACCIFERA L. [Gentianaceae]

Common Names:

humansdorp

Citations:

Van der Walt SJ, Steyn DG (1939) Recent investigations into the toxicity of known and unknown poisonous plants in the Union of South Africa. IX. Onderstepoort J Vet Sci Anim Indus 12:335-366.

CHIRONIA TRANSVAALENSIS Gilg [Gentianaceae]

Common Names:

Rhodesian wild gentian; wild gentian

Citations:

Shone DK (1966) Poisonous plants of Rhodesia. Part I. Rhodesia Zambia Malawi J Agric Res 4:81-94.

Shone DK (1967) Poisonous plants of Rhodesia. Part II. Rhodesia Zambia Malawi J Agric Res 5:29-41.

Steyn DG (1933) Recent investigations into the toxicity of known and unknown poisonous plants in the Union of South Africa. Onderstepoort J Vet Sci Anim Indus 1:173-182.

chivalrygrass –see– *Polygonum aviculare* L.

chives –see– *Allium schoenoprasum* L.

CHLORIS GAYANA Kunth [Poaceae]

Common Names:

Rhodesgrass

Citations:

Bargai U (1987) Incidence, etiology and treatment of ovarian follicular cysts in a large dairy herd over a ten-year period. Isr J Vet Med 43(1):56-61.

Chlorophora excelsa (Welw.) Benth. = Milicia excelsa (Welw.) C. C. Berg

CHLOROXYLON SWIETENIA DC. [Rutaceae]

Common Names:

Atlasholz; East Indian satinwood; Ostindisches Satinholz; Ostindisches Seidenholz; satin wood; Satinholz; Westindisches Atlasholz; Westindisches Satinholz; Zitronenholz

Citations:

Cash JT (1911) The dermatitis produced by East Indian satinwood (Chloroxylon swietenia). Br Med J 2:784-790.

chocho –see– *Crotalaria grahamiana* Wight & Arn.; *Sechium edule* (Jacq.) Sw.

chocolate –see– *Theobroma cacao* L.

chocon –see– *Wigandia urens* (Ruiz & Pav.) Kunth var. caracasana (Kunth) D. N. Gibson

chokecherry –see– *Prunus serotina* Ehrh.; *Prunus virginiana* L.

CHONDRILLA JUNCEA L. [Asteraceae]

Common Names:

skeleton weed

Citations:

Anonymous (1954) Investigations into the etiology and control of enzootic (toxaemic) jaundice of sheep. Report of the investigation committee for the year 1952-53. Aust Vet J 30(Jun):182-184.

chongras –see– *Phytolacca americana* L.

chopsui potato –see– *Pachyrhizus erosus* (L.) Urb.

chota mona –see– *Paspalum scrobiculatum* L. var. bispicatum Hack.

chou-de-chien –see– *Mercurialis perennis* L.

chou moellier –see– *Brassica oleracea* L. var. medullosa Thell.

chou moullier –see– *Brassica oleracea* L. var. medullosa Thell.

choux fourrager –see– *Brassica oleracea* L. var. gemmifera Zenker

Christblume –see– *Helleborus niger* L.

Christdorn –see– *Ilex aquifolium* L.

Christmas bells –see– *Kalanchoe daigremontiana* Raym.-Hamet & H. Perrier; *Kalanchoe delagoensis* Eckl. & Zeyh.

Christmas berry –see– *Heteromeles salicifolia* (C. Presl) Abrams; *Lycium barbarum* L.

Christmas candle –see– *Pedilanthus tithymaloides* (L.) Poit.

Christmas cherry –see– *Solanum pseudocapsicum* L.

Christmas flower –see– *Euphorbia pulcherrima* Willd. ex Klotzsch

Christmas holly –see– *Ilex aquifolium* L.; *Ilex opaca* Aiton

Christmas mistletoe –see– *Phoradendron leucarpum* (Raf.) Reveal & M. C. Johnst.

Christmas pepper –see– *Capsicum annuum* L.; *Solanum pseudocapsicum* L.

Christmas poinsettia –see– *Euphorbia pulcherrima* Willd. ex Klotzsch

Christmas rose –see– *Euphorbia pulcherrima* Willd. ex Klotzsch; *Helleborus niger* L.

Christmas star –see– *Euphorbia pulcherrima* Willd. ex Klotzsch

Christmasberry tree –see– *Schinus terebinthifolius* Raddi

Christrose –see– *Helleborus niger* L.

Christ's-thorn –see– *Euphorbia milii* Des Moul.

Christ's-wort –see– *Helleborus niger* L.

Christusdorn –see– *Euphorbia milii* Des Moul.

Christuspalme –see– *Ricinus communis* L.

Christwurz –see– *Helleborus niger* L.

Chrosperma muscaetoxicum (Walt.) Kuntze = Amianthium muscitoxicum (Walter) A Gray

CHROZOPHORA OBLIQUA (Vahl) A. Juss. ex Spreng. [Euphorbiaceae]

Common Names:

rein

Citations:

Adam SE, Al-Redhaiman KN, Al-Qarawi AA (1999) Toxicity of Chrozophora obliqua in rats. Phytother Res 13(7):630-632.

CHROZOPHORA PLICATA (Vahl) A. Juss. ex Spreng. [Euphorbiaceae]

Common Names:

terba

Citations:

Galal M, Adam SE (1988) Experimental Chrozophora plicata poisoning in goats and sheep. Vet Hum Toxicol 30(5):447-452.

chrysanthème matricaire –see– *Tanacetum parthenium* (L.) Sch. Bip.

chrysanthemum –see– *Chrysanthemum ×morifolium* Ramat.

Chrysanthemum leucanthemum L. = Leucanthemum vulgare Lam.

CHRYSANTHEMUM ×MORIFOLIUM Ramat. [Asteraceae]

Common Names:

chrysanthemum; pot mum

Citations:

Bleumink E, Mitchell JC, Nater JP (1973) Contact dermatitis to chrysanthemums. Arch Dermatol 108(2):220-222.

Campolmi P, Sertoli A, Fabbri P, et al. (1978) Alantolactone sensitivity in chrysanthemum contact dermatitis. Contact Dermatitis 4(2):93-102.

Frain-Bell W, Hetherington A, Johnson BE (1979) Contact allergic sensitivity to chrysanthemum and the photosensitivity dermatitis and actinic reticuloid syndrome. Br J Dermatol 101(5):491-501.

Goldstein MB (1931) Dermatitis venenata due to chrysanthemum leaves. JAMA 96(20):1680-1681.

Hausen BM (1979) The sensitizing capacity of compositae plants. III. Test results and cross-reactions in compositae-sensitive patients. Dermatologica 159(1):1-11.

Pilot I (1932) Dermatitis venenata due to chrysanthemums. Report of a case in a florist. J Allergy 3:404-407.

Schmidt RJ, Kingston T (1985) Chrysanthemum dermatitis in South Wales; diagnosis by patch testing with feverfew (Tanacetum parthenium) extract. Contact Dermatitis 13(2):120-121.

Sharma SC, Tanwar RC, Kaur S (1989) Contact dermatitis from chrysanthemums in India. Contact Dermatitis 21(2):69-71.

Chrysanthemum parthenium (L.) Bernh. = Tanacetum parthenium (L.) Sch. Bip.

Chrysanthemum vulgare (L.) Bernh. = Tanacetum vulgare L.

CHRYSOCOMA CILIATA L. [Asteraceae]

Synonyms:

Chrysocoma tenuifolia P. J. Bergius

Common Names:

beesbossie; beeskarroo; bitter bush; bitterbos; bitterbossie; bitterkaroo; bransbossie

Citations:

Steyn DG (1934) The toxicology of plants in South Africa. Onderstepoort J Vet Sci Anim Indus 3(2):359-473.

Van der Vyver FH, Kellerman TS, Bastianello SS, et al. (1985) Valsiekte (falling disease): A nervous disorder in lambs suspected of being caused by the plant Chrysocoma tenuifolia. J S Afr Vet Assoc 56(2):65-68.

Chrysocoma tenuifolia P. J. Bergius = Chrysocoma ciliata L.

chu –see– *Euphorbia royleana* Boiss.

chuanwu –see– *Aconitum carmichaelii* Debeaux

chucho –see– *Nierembergia hippomanica* Miers

chucho-de-la-sierra –see– *Nierembergia hippomanica* Miers

chuen lin –see– *Coptis chinensis* Franch.

chupa –see– *Nicotiana tabacum* L.

churee –see– *Euphorbia royleana* Boiss.

churn staff –see– *Euphorbia helioscopia* L.

chuschu –see– *Nierembergia hippomanica* Miers

chuu-ou-kou –see– *Curcuma longa* L.

ciat –see– *Catha edulis* (Vahl) Forssk. ex Endl.

CICER ARIETINUM L. [Fabaceae]

Common Names:

Bengal gam; Bengal gram; chick pea; coffee pea; garbanzo; Idaho pea; Kichererbse

Citations:

Diaz C, Vivanco F (1949) Studies on cicerism. Bull Inst Med Res Univ Madr 2:95-102.

cicer milk vetch –see– *Astragalus cicer* L.

CICHORIUM ENDIVIA L. [Asteraceae]

Common Names:

broad-leaf endive; chicory; endive

Citations:

Friis B, Hjorth N, Vail JT Jr, et al. (1975) Occupational contact dermatitis from Cichorium (chicory, endive) and Lactuca (lettuce). Contact Dermatitis 1(5):311-313.

Krook G (1977) Occupational dermatitis from Lactuca sativa (lettuce) and Cichorium (endive). Simultaneous occurrence of immediate and delayed allergy as a cause of contact dermatitis. Contact Dermatitis 3(1):27-36.

Vail JT Jr, Mitchell JC (1973) Occupational dermatitis from Cichorium intybus, C. endivia and Lactuca sativa var. longifolia. Contact Dermatol Newsl 14(Aug):413-414.

CICHORIUM INTYBUS L. [Asteraceae]

Synonyms:

Cichorium intybus L. var. sativum (Bisch.) Janch.

Common Names:

blue chicory; blue sailors; chickory; chicory; Hansl-am-Weg; succory; Wegwarte; wild chicory; witloof

Citations:

Bubień Z, Wachnik Z, Żuchowski A (1962) Zatrucia jałówek korzeniami cykorii. Med Weter 18(10):603-605.
Malten KE (1983) Chicory dermatitis from September to April. Contact Dermatitis 9(3):232.
Vail JT Jr, Mitchell JC (1973) Occupational dermatitis from Cichorium intybus, C. endiva and Lactuca sativa var. longifolia. Contact Dermatol Newsl 14(Aug):413-414.
Wachnik Z (1962) Liście cykorii (Cichorium intybus) przyczyna zatrucia świń. Med Weter 18(8):493-495.

Cichorium intybus L. var. sativum (Bisch.) Janch. = Cichorium intybus L.

cicuta –see– *Cicuta douglasii* (DC.) J. M. Coult. & Rose

Cicuta californica A. Gray = Cicuta douglasii (DC.) J. M. Coult. & Rose

CICUTA DOUGLASII (DC.) J. M. Coult. & Rose [Apiaceae]

Synonyms:

Cicuta californica A. Gray; ***Cicuta vagans*** Greene

Common Names:

beaver poison; California water hemlock; carotte-a-moreau; cicuta; cowbane; Douglas'-water hemlock; false parsley; mock eel root; muskrat weed; musquash root; Oregon water hemlock; poison parsnip; snakeroot; snakeweed; spotted cowbane; spotted hemlock; spotted parsley; tuber water hemlock; water hemlock; water parsnip; western water hemlock; wild carrot; wild parsnip

Citations:

Anonymous (1967) Water hemlock in production ponds. Progressive Fish Cult 29:181.
Brodie DA (1901) Poison parsnip in Western Washington. Washington Agric Exp Sta Bull #45:12 pp.
Hedrick UP (1897) A plant that poisons cattle. Cicuta (Cicuta vagans Green.). Oregon Agric Exp Sta Bull #46:12 pp.
Landers D, Seppi K, Blauer W (1985) Seizures and death on a white river float trip. Report of water hemlock poisoning. West J Med 142(5):637-640.
Litovitz TL, Schmitz BF, Holm KC (1989) 1988 Annual report of the American Association of Poison Control Centers National Data Collection System. Am J Emerg Med 7(5):495-545.
Litovitz TL, Schmitz BF, Matyunas N, et al. (1988) 1987 Annual report of the American Association of Poison Control Centers National Data Collection System. Am J Emerg Med 6(5):479-515.
Mutter L (1976) Poisoning by western water hemlock. Can J Public Health 67(5):386.
Panter KE, Baker DC, Kechele PO (1996) Water hemlock (Cicuta douglasii) toxicosis in sheep: Pathologic description and prevention of lesions and death. J Vet Diagn Invest 8(4):474-480.
Withers LM, Cole FR, Nelson RB (1969) Water-hemlock poisoning. N Engl J Med 281(10):566-567.

Cicuta mackenziana Raup = Cicuta virosa L.

CICUTA MACULATA L. [Apiaceae]

Common Names:

American water hemlock; beaver poison; children's-bane; children's-death; cowbane; death-of-man; false parsley; muskrat weed; musquash; musquash poison; musquash root; poison parsley; poison parsnip; poison snakeweed; root poison; snakeweed; spotted cowbane; spotted hemlock; spotted parsley; spotted water hemlock; Wasserchierling; water hemlock; wild carrot; wild hemlock; wild parsnip

Citations:

Anonymous (1899) A case of poisoning - Water hemlock. North Dakota Agric Exp Sta Bull #35:307-310.
Applefeld JJ, Caplan ES (1979) A case of water hemlock poisoning. JACEP 8(10):401-403.
Campbell EW (1966) Plant poisoning Umbelliferae (parsley family). J Maine Med Assoc 57(2):40-42.
Crozier AA (1889) Another death from eating Cicuta maculata. Bot Gaz 14(Jan):17-18.
Egdahl A (1911) A case of poisoning due to eating poison-hemlock (Cicuta maculata). Arch Intern Med 7:348-356.
Fenton WN (1941) Iroquois suicide: A study in the stability of a culture pattern. Smithsonian Bur Am Ethnology Bull #128:80-137.
Gompertz LM (1926) Poisoning with poison hemlock (Cicuta maculata), Report of seventeen cases. JAMA 87(16):1277-1278.
Haggerty DR, Conway JA (1936) Report of poisoning by Cicuta maculata water hemlock. N Y State J Med 36(20):1511-1514.
Hansen AA (1928) Cicuta or water hemlock poisoning. North Am Vet 9(6):34-39.
Heath KB (2001) A fatal case of apparent water hemlock poisoning. Vet Hum Toxicol 43(1):35-36.
Ladd EF (1900) Water hemlock poisoning. North Dakota Agric Exp Sta Bull #44:563-569.
Litovitz TL, Normann SA, Veltri JC (1986) 1985 Annual report of the American Association of Poison Control Centers National Data Collection System. Am J Emerg Med 4(5):427-458.
Majerus TC, Schwartz WK, Oderda GM (1978) Accidental ingestion of water hemlock - Report of a case and review of the literature. Am Assoc Poison Control Centers 1978:34.
Miller MM (1933) Water hemlock poisoning. JAMA 101(11):852-853.
Pammel LH (1895) Poisoning from cowbane (Cicuta maculata L.). Iowa Agric Exp Sta Bull #28:215-228.
Pammel LH (1921) Cowbane. Vet Med 16(Oct):48.

Pammel LH (1928) Cowbane - wild parsnip, Cicuta maculata. North Am Vet 9:25-26.
Quinby CE (1979) Epidemic water hemlock poisoning associated occupationally: Five year follow-up for possible permanent brain damage. Clin Toxicol 15(4):489.
Skidmore LV (1933) Water hemlock (Cicuta maculata L.) poisoning in swine. Vet J 89(2):76-80.
Smith EF (1888) Death from eating Cicuta maculata. Bot Gaz 13(May):128-129.
Starreveld E, Hope CE (1975) Cicutoxin poisoning (water hemlock). Neurology 25(8):730-734.
Stockbridge J (1814) Account of the effects produced by eating a poisonous plant, called Cicuta maculata. New England J Med Surg 3:334-337.
Sweeney K, Gensheimer KF, Knowlton-Field J, et al. (1994) Water hemlock poisoning - Maine, 1992. JAMA 271(19):1475.
Sweeney K, Gensheimer KF, Knowlton-Field J, et al. (1994) Water hemlock poisoning - Maine, 1992. MMWR Morb Mortal Wkly Rep 43(13):229-231.
Undine CA (1938) Poisoning with Cicuta maculata or water hemlock. Minn Med 21:262, 296.

CICUTA MACULATA L. var. angustifolia Hook. [Apiaceae]

Synonyms:

Cicuta occidentalis Greene

Common Names:

cowbane; death-of-man; musquash; peco; poison parsnip; snakeweed; spotted cowbane; spotted parsley; water hemlock; water parsnip; western water hemlock; wild parsnip; Wyoming water hemlock

Citations:

Anonymous (1929) Investigation of stock-poisoning plants. U S Dep Agric Bur Anim Indus Chief Rep 1929:47-48.
Fleming CE (1920) Poisonous range plants. Nevada Agric Exp Sta Annu Rep 1919:39-43.
Fleming CE, Peterson NF, Miller MR, et al. (1920) The poison parsnip or water hemlock (Cicuta occidentalis). A plant deadly to livestock in Nevada. Nevada Agric Exp Sta Bull #100:23 pp.
Marsh CD, Clawson AB, Marsh H (1914) Cicuta, or water hemlock. U S Dep Agric Bull #69:27 pp.
Pammel LH (1921) Western poison cowbane. Vet Med 16(11):33-34.
Stratton MR (1919) Water hemlock poisoning. Colo Med 16(May):104-111.

cicuta maggiore –see– *Conium maculatum* L.

cicuta mayor –see– *Conium maculatum* L.

Cicuta occidentalis Greene = Cicuta maculata L. var. angustifolia Hook.

Cicuta tenuifolia Schrank = Cicuta virosa L.

Cicuta vagans Greene = Cicuta douglasii (DC.) J. M. Coult. & Rose

CICUTA VIROSA L. [Apiaceae]

Synonyms:

Cicuta mackenziana Raup; ***Cicuta tenuifolia*** Schrank

Common Names:

beaver poison; cowbane; cykuta; European water hemlock; five-finger root; Giftwüterich; la ciguë vireuse; poison water hemlock; snakeweed; sprängört; szalej jadowity; Wasserschierling; water hemlock; Waterscheerling; wild carrot

Citations:

Bartel J, Gerber HU (1962) Ein Beitrag zur Vergiftung mit Wasserschierling (Cicuta virosa L.) bei Kindern. Kinderaerztl Prax 12:543-547.
Berndt H (1947) Vergiftungen durch Cicuta virosa L., Wasserschierling. Pharmazie 2:521-523.
Bubien MZ, Magat A, Delatour P (1970) Fréquence des intoxications des animaux en Pologne. Bull Soc Vet Med Comp Lyon 72(2):167-173.
Cederstam R (1976) Sprängörtsförgiftning. Sven Vet 28(24):1114-1116.
Costanza DJ, Hoversten VW (1973) Accidental ingestion of water hemlock. Report of two patients with acute and chronic effects. Calif Med 119(2):78-82.
Czursiedel H (1937) Eine Wasserschierling-Vergiftung (Cicuta virosa). Sammlung Vergiftungsfallen 8(A699):171-172.
Dijkstra RG, Falkena R (1981) Een geval van cicutoxine-intoxicatie bij pony's. Tijdschr Diergeneeskd 106(20):1037-1039.
Knutsen OH, Paszkowski P (1984) New aspects in the treatment of water hemlock poisoning. J Toxicol Clin Toxicol 22(2):157-166.
Kozielec T, Nowotarska T (1973) Zatrucie szalejem jadowitym (Cicuta virosa) sześciorga dzieci. Pediatr Pol 48(12):1519-1522.
Laskowski S, Matyjek J, Koppicz M, et al. (1973) Zatrucie szalejem jadowitym (Cykutą). Pol Arch Med Wewn 50(11):1237-1239.
Laskowski S, Matyjek J, Koppicz M, et al. (1975) Zatrucie szalejem jadowitym (cykutą). Pol Tyg Lek (Wars) 30(12):533-534.
Marcinkowski T, Porawski R (1966) Zatrucie szalejem jadowitym (Cicuta virosa). Wiad Lek 19(2):139-141.
Nilsson NG (1977) Sprängörtsförgiftning - än en gång. Sven Vet 29:725.
Robson P (1965) Water hemlock poisoning. Lancet 2(425):1274-1275.
Schrader A, Schulz O, Volker H, et al. (2001) Aktuelle Vergiftungen durch Pflanzen bei Wiederkäuern in Nord- und Ostdeutschland. Berl Munch Tierarztl Wochenschr 114(5-6):218-221.
Short J (2006) Water hemlock poisoning. Emerg Nurse 14(7):18-19.
Tatar A, Otłowski F (1967) Masowe zatrucie bydła szalejem jadowitym (Cicuta virosa L.). Zycie Weterynaryjne 42(5):137-140.
van Heijst AN, Pikaar SA, van Kesteren RG, et al. (1983) Een vergiftiging door de waterscheeling (Cicuta virosa). Ned Tijdschr Geneeskd 127(53):2411-2413.
Volker H, Schulz O, Albrecht K, et al. (1983) Vergiftungen durch Wasserschierling (Cicuta virosa) bei Mastbullen. Monatsh Veterinarmed 38:11-13.

Vorokhobov LA, Katetnyi VM (1966) [Poisoning by Cicuta virosa in children.] Pediatriia 45(5):80-82.

cicutilla –see– *Parthenium hysterophorus* L.

cila –see– *Drimia maritima* (L.) Stearn

cimicifuga –see– *Actaea racemosa* L.

Cimicifuga racemosa (L.) Nutt. = Actaea racemosa L.

cimisu –see– *Buxus sempervirens* L.

CINNAMOMUM AROMATICUM Nees [Lauraceae]

Common Names:
cassia; Chinese cinnamon; Saigon cinnamon

Citations:
De Benito V, Alzaga R (1999) Occupational allergic contact dermatitis from cassia (Chinese cinnamon) as a flavouring agent in coffee. Contact Dermatitis 40(3):165.

CINNAMOMUM CAMPHORA (L.) J. Presl [Lauraceae]

Common Names:
alcanfor; camphor; camphor laurel; camphor tree; heuile-de-camphre

Citations:
Antman E, Jacob G, Volpe B, et al. (1978) Camphor overdosage: Therapeutic considerations. N Y State J Med 78:896-897.
Craig JO (1953) Poisoning by the volatile oils in childhood. Arch Dis Child 28(142):475-483.
Kopelman R, Mitler S, Kelley R, et al. (1979) Camphor intoxication treated by resin hemoperfusion. JAMA 241(7):727-728.
Phelan WJ 3rd (1976) Camphor poisoning: Over-the-counter dangers. Pediatrics 57(3):428-431.
Seife M, Leon JL (1954) Camphor poisoning following ingestion of nose drops. JAMA 155(12):1059-1060.
Smith JP (1962) These weeds can poison your poultry. Agric Gaz New South Wales 73(Mar):136-142.
Trestrail JH 3rd, Spartz ME (1977) Camphorated and castor oil confusion and its toxic results. Clin Toxicol 11(2):151-158.

CINNAMOMUM TAMALA (Buch.-Ham.) Nees & Eberm. [Lauraceae]

Common Names:
Indian cassia

Citations:
Seetharam KA, Pasricha JS (1987) Condiments and contact dermatitis of the finger-tips. Indian J Dermatol Venereol Leprol 53:325-328.

CINNAMOMUM VERUM J. Presl [Lauraceae]

Synonyms:
Cinnamomum zeylanicum Blume

Common Names:
cinnamon; Zimt

Citations:
Calnan CD (1976) Cinnamon dermatitis from an ointment. Contact Dermatitis 2(3):167-170.
Cummer CL (1940) Dermatitis due to oil of cinnamon. Arch Derm Syphilol 42:674-675.
Epstein FW (1950) Contact dermatitis due to cinnamon. Ohio Med J 46(Jul):659.
Farkas J (1981) Perioral dermatitis from marjoram, bay leaf and cinnamon. Contact Dermatitis 7(2):121.
García-Abujeta JL, de Larramendi HC, Pomares Berna J, et al. (2005) Mud bath dermatitis due to cinnamon oil. Contact Dermatitis 52(4):234.
Goh CL, Ng SK (1988) Bullous contact allergy from cinnamon. Dermatosen 36(6):186-187.
Kern AB (1960) Contact dermatitis from cinnamon. Arch Dermatol 81(Apr):599-600.
Leifer W (1951) Contact dermatitis due to cinnamon. Arch Derm Syphilol 64(1):52-55.
Miller J (1941) Cheilitis from sensitivity to oil of cinnamon present in bubble gum. JAMA 116(2):131-132.
Niinimäki A (1984) Delayed-type allergy to spices. Contact Dermatitis 11(1):34-40.
Perry PA, Dean BS, Krenzelok EP (1990) Cinnamon oil abuse by adolescents. Vet Hum Toxicol 32(2):162-163.
Pilapil VR (1989) Toxic manifestations of cinnamon oil ingestions in a child. Clin Pediatr (Phila) 28(6):276.
Sparks T (1985) Cinnamon oil burn. West J Med 142(6):835.
Tulipan L (1932) Dermatitis from cinnamon. Arch Derm Syphilol 25:921-923.

Cinnamomum zeylanicum Blume = Cinnamomum verum J. Presl

cinnamon –see– *Cinnamomum verum* J. Presl

cinnamon vine –see– *Dioscorea polystachya* Turcz.

cinnamon wood –see– *Sassafras albidum* (Nutt.) Nees

cipolle –see– *Allium cepa* L.

Circassian walnut –see– *Juglans regia* L.

cirri amarillo –see– *Mauria heterophylla* Kunth

CIRSIUM ARVENSE (L.) Scop. [Asteraceae]

Common Names:
Canadian thistle

Citations:
Bergeron JM, Jodoin L (1985) Fiabilité des mesures de poids et d'examens histopathologiques dans les études d'intoxication du campagnol des champs (Microtus pennsylvanicus). Can J Zool 63(4):804-810.

CISSUS ANTARCTICA Vent. [Vitaceae]

Citations:
Hjorth N (1969) Plant dermatitis. Contact Dermatol Newsl 6(Jul):126-127.

CISSUS CUNEIFOLIA Eckl. & Zeyh. [Vitaceae]

Synonyms:

Rhoicissus cuneifolia (Eckl. & Zeyh.) Planch.; ***Rhoicissus tridentata*** (L. f.) Wild & R. B. Drumm. subsp. cuneifolia (Eckl. & Zeyh.) N. R. Urton

Common Names:

wild grape; wilde dreuif

Citations:

Steyn DG, Venter ID (1953) The toxicity of Rhoicissus cuneifolia (E. and Z.) Planch. (wild grape, "wilde druif"). S Afr Med J 27:64-66.

CISSUS QUADRANGULARIS L. [Vitaceae]

Common Names:

vigne-de-Bakel

Citations:

Barakat SE, Adam SE, Maglad MA, et al. (1985) Effects of Cissus quadrangularis on goats and sheep in Sudan. Rev Elev Med Vet Pays Trop 38(2):185-194.

CISTUS CRETICUS L. [Cistaceae]

Citations:

English JS, Cronin E (1988) Allergic contact dermatitis from Cistus creticus. Contact Dermatitis 18(6):123.

CISTUS LADANIFER L. [Cistaceae]

Common Names:

rockrose

Citations:

Ballesteros Moreno E (1965) Nota previa sobre la intoxicación experimental por "Cistus ladaniferus." An Inst Invest Vet 14-15:77-81.

García-González JJ, Crespo V, Barber D, et al. (2001) Cistus ladanifer contact dermatitis. Contact Dermatitis 45(4):238.

CITRULLUS COLOCYNTHIS (L.) Schrad. [Cucurbitaceae]

Synonyms:

Colocynthis vulgaris Schrad.; ***Cucumis colocynthis*** L.

Common Names:

bitter apple; bitter gourd; colocinth; colocynth; coloquinte; coloquíntida; handal; tumin; wild cucumber

Citations:

Al Faraj S (1995) Haemorrhagic colitis induced by Citrullus colocynthis. Ann Trop Med Parasitol 89(6):695-696.

Barri ME, Onsa TO, Elawad AA, et al. (1983) Toxicity of five Sudanese plants to young ruminants. J Comp Pathol 93(4):559-575.

Berrut C, Bisetti A, Widgren S, et al. (1987) Colite pseudomembraneuse causée par l'ingestion de coloquinte. Schweiz Med Wochenschr 117:135-138.

Elawad AA, Abdel Bari EM, Mahmoud OM, et al. (1984) The effect of Citrullus colocynthis on sheep. Vet Hum Toxicol 26(6):481-485.

Gálvez Contreras MD, López Gallardo A, Diez Garcia F, et al. (1996) Intoxicación por coloquíntida, una causa poco frecuente de síndrome diarreico agudo. Med Clin (Barc) 106(15):599.

Hamouda C, Amamou M, Thabet H, et al. (2000) Plant poisonings from herbal medication admitted to a Tunisian toxicologic intensive care unit, 1983-1998. Vet Hum Toxicol 42(3):137-141.

Patrick RL, Willey EN, Fetter BF (1960) Bitter apple (Citrullus colocynthis) poisoning. A discussion of its use as an abortifacient. N C Med J 21(Jan):23-26.

Roe RB (1913) A case of colocynth poisoning. Lancet 1(May 31):1527.

Tidy CM (1868) On poisoning by colocynth (Cucumis colocynthis). Lancet 1(Feb 1):158-159.

CITRULLUS LANATUS (Thunb.) Matsum. & Nakai [Cucurbitaceae]

Synonyms:

Citrullus vulgaris Schrad. ex Eckl. & Zeyh.

Common Names:

bitter melon; camel melon; melon; pie melon; round gourd; tarbuz; watermelon; wild watermelon

Citations:

Sinha SM, Pasricha JS, Sharma RC, et al. (1977) Vegetables responsible for contact dermatitis of the hands. Arch Dermatol 113(6):776-779.

Steyn DG (1950) The toxicity of bitter-tasting cucurbitaceous vegetables - (vegetable marrow, watermelons, etc.) for man. S Afr Med J 24:713-715.

Citrullus vulgaris Schrad. ex Eckl. & Zeyh. = Citrullus lanatus (Thunb.) Matsum. & Nakai

CITRUS AURANTIIFOLIA (Christm.) Swingle [Rutaceae]

Common Names:

lim; lime; Limette; Persian lime; Tahitian lime

Citations:

Cardullo AC, Ruszkowski AM, De Leo VA (1989) Allergic contact dermatitis resulting from sensitivity to citrus peel, geraniol, and citral. J Am Acad Dermatol 21:395-397.

Coffman K, Boyce WT, Hansen RC (1985) Phytophotodermatitis simulating child abuse. Am J Dis Child 139(3):239-240.

Egan CL, Sterling G (1993) Phytophotodermatitis: A visit to Margaritaville. Cutis 51(Jan):41-42.

Field WE, Roe FJ (1965) Tumor promotion in the forestomach epithelium of mice by oral administration of citrus oils. J Natl Cancer Instit 35(5):771-787.

Gross TP, Ratner L, de Rodriguez O, et al. (1987) An outbreak of phototoxic dermatitis due to limes. Am J Epidemiol 125(3):509-514.

Israel E (1985) Outbreak of phototoxic dermatitis from limes - Maryland. MMWR Morb Mortal Wkly Rep 34(30):462-463.

Roe FJ, Peirce WE (1960) Tumor promotion by citrus oils: Tumors of the skin and urethral orifice in mice. J Natl Cancer Instit 24:1389-1402.

Sams WM (1941) Photodynamic action of lime oil (Citrus aurantifolia). Arch Derm Syphilol 44:571-587.
Stokes B (1986) The alarming case of the mysterious plant disease. Chicago Tribune. Sec 5(Aug 7):2.

CITRUS AURANTIUM L. [Rutaceae]

Common Names:
bitter orange; naranjo amarga; petitgrain; Seville orange; sour orange

Citations:
Murray FA (1921) Dermatitis caused by bitter orange. Br Med J 1:739.

CITRUS BERGAMIA Risso & Poit. [Rutaceae]

Common Names:
bergamot; bergamotte; berloque

Citations:
Downing JG (1932) Pigmentation from perfume, "Berlock" dermatitis. N Engl J Med 207:660-662.
Gross P, Robinson LB (1930) Berlock dermatitis: Unusual dermatitis and pigmentation following use of perfume. Arch Derm Syphilol 21:637-641.
Harber LC, Harris H, Leider M, et al. (1964) Berloque dermatitis. Arch Dermatol 90(Dec):572-576.
Oppenheim M (1947) Local sensitization of the skin to Grenz rays by bergamot oil. J Invest Dermatol 8(5):255-262.
Rogin JR, Sheard C (1935) Factors affecting color of skin: Their significance in berlock dermatitis. Arch Derm Syphilol 32:265-283.
Zaynoun S, Hall I, Johnson BE, et al. (1974) A study of bergamot sensitivity. Br J Dermatol 91(Suppl 10):14-15.
Zaynoun ST, Johnson BE, Frain-Bell W (1977) A study of oil of bergamot and its importance as a phototoxic agent. I. Characterization and quantitation of the photoactive component. Br J Dermatol 96(5):475-482.

CITRUS LIMON (L.) Burm. f. [Rutaceae]

Common Names:
Eurecka lemon; lemon; Zitrone

Citations:
Audicana M, Bernaola G (1994) Occupational contact dermatitis from citrus fruits: Lemon essential oil. Contact Dermatitis 31(3):183-185.
Cardullo AC, Ruszkowski AM, De Leo VA (1989) Allergic contact dermatitis resulting from sensitivity to citrus peel, geraniol, and citral. J Am Acad Dermatol 21:395-397.
Fanburg SJ, Kaufman JG (1931) Eczema due to lemon peel. JAMA 97(Aug 8):390.
Horner SG (1931) Dermatitis from oranges and lemons. Lancet 2(Oct 31):961-962.
Naganuma M, Hirose S, Nakayama Y, et al. (1985) A study of the phototoxicity of lemon oil. Arch Dermatol Res 278(1):31-36.
Roe FJ, Peirce WE (1960) Tumor promotion by citrus oils: Tumors of the skin and urethral orifice in mice. J Natl Cancer Instit 24:1389-1402.

CITRUS PARADISI Macfad. [Rutaceae]

Common Names:
grapefruit

Citations:
Beerman H, Fondé GH, Callaway JL (1938) Citrus fruit dermatoses. Arch Derm Syphilol 38:225-234.
Driggers JC, Davis GK, Mehrhof NR (1951) Toxic factor in citrus seed meal. Florida Agric Exp Sta Bull #476:36 pp.
Roe FJ, Peirce WE (1960) Tumor promotion by citrus oils: Tumors of the skin and urethral orifice in mice. J Natl Cancer Instit 24:1389-1402.

CITRUS RETICULATA Blanco [Rutaceae]

Common Names:
mandarin; tangerine

Citations:
Vilaplana J, Romaguera C (2002) Contact dermatitis from the essential oil of tangerine in fragrance. Contact Dermatitis 46(2):108.

CITRUS SINENSIS (L.) Osbeck [Rutaceae]

Common Names:
Apfelsine; orange; shamouti orange; sweet orange

Citations:
Beerman H, Fondé GH, Callaway JL (1938) Citrus fruit dermatoses. Arch Derm Syphilol 38:225-234.
Bendersky G, Lupas JA (1960) Anaphylactoid reaction to ingestion of orange. JAMA 173(3):255-256.
Brun R (1978) Contact dermatitis to orangewood in a manicurist. Contact Dermatitis 4(5):315.
Cardullo AC, Ruszkowski AM, De Leo VA (1989) Allergic contact dermatitis resulting from sensitivity to citrus peel, geraniol, and citral. J Am Acad Dermatol 21:395-397.
Churchill C, Pendleton J, Maddy K, et al. (1986) Outbreak of severe dermatitis among orange pickers - California. MMWR Morb Mortal Wkly Rep 35(29):465-468.
Driggers JC, Davis GK, Mehrhof NR (1951) Toxic factor in citrus seed meal. Florida Agric Exp Sta Bull #476:36 pp.
Field WE, Roe FJ (1965) Tumor promotion in the forestomach epithelium of mice by oral administration of citrus oils. J Natl Cancer Instit 35(5):771-787.
Horner SG (1931) Dermatitis from oranges and lemons. Lancet 2(Oct 31):961-962.
Kesten B, Lyons R (1932) Dermatitis due to contact with orange peel. J Allergy 3(Nov):552-556.
Roe FJ, Peirce WE (1960) Tumor promotion by citrus oils: Tumors of the skin and urethral orifice in mice. J Natl Cancer Instit 24:1389-1402.
Volden G, Krokan H, Kavli G, et al. (1983) Phototoxic and contact toxic reactions of the exocarp of sweet oranges: A common cause of cheilitis? Contact Dermatitis 9(3):201-204.

ciuma rea –see– *Galega officinalis* L.

clavel amarillo –see– *Pascalia glauca* Ortega

clavelitos –see– *Pentalinon luteum* (L.) B. F. Hansen & Wunderlin

cleavers herb –see– *Galium aparine* L.

CLEISTANTHUS COLLINUS (Roxb.) Hook. f. [Phyllanthaceae]

Common Names:

oduvan; oduvanthalai

Citations:

Eswarappa S, Chakraborty AR, Palatty BU, et al. (2003) Cleistanthus collinus poisoning: Case reports and review of the literature. J Toxicol Clin Toxicol 41(4):369-372.

Rao RR (1978) Investigation of leukocytosis and degeneration of skeletal muscles induced by plant toxins. Toxicon 16:143.

Thirumavalavan R (2004) Aggressive potassium correction may halt death in Cleistanthus collinus poisoning. J Toxicol Clin Toxicol 42(5):801.

Thomas K, Dayal AK, Gijsbers A, et al. (1987) Oduvanthalai leaf poisoning. J Assoc Physicians India 35(11):769-771.

Thomas M, Anandan S, Kuruvilla PJ, et al. (2000) Profile of hospital admissions following acute poisoning – Experiences from a major teaching hospital in south India. Adverse Drug React Toxicol Rev 19(4):313-317.

CLEMATIS LIGUSTICIFOLIA Nutt. [Ranunculaceae]

Common Names:

western virgin's-bower

Citations:

Anonymous (1936) Investigations of stock-poisoning plants. U S Dep Agric Bur Anim Indus Chief Rep. 1936:44-45.

CLEMATIS MICROPHYLLA DC. [Ranunculaceae]

Common Names:

half creeper; old-man's-beard; small-leaf clematis

Citations:

Cleland JB (1931) Plants, including fungi, poisonous or otherwise injurious to man in Australia. Series III. Med J Aust 2(Dec 19):775-778.

CLEMATIS VITALBA L. [Ranunculaceae]

Common Names:

berceaux-de-la-vierge; bindwith; biting clematis; Brennkraut; erdei iszalag; hedge vine; herbe-aux-gueux; hierba-de-los-pordioseros; lady's-bower; maiden's-honesty; old man; old-man's-beard; smoke wood; traveler's-joy; virgin's-bower; Waldrebe

Citations:

Moore RH (1971) Poisoning by old man's beard (Clematis vitalba)? Vet Rec 89(21):569-570.

CLEOME SPINOSA Jacq. [Cleomaceae]

Common Names:

Kleopatra nadel; spider flower

Citations:

Szegö L, Maácz J (1968) Dermatitis als Berufskrankheit, hervorgerufen durch die "Kleopatra-Nadel" (Cleome spinosa). Dermatol Wochenschr 154(3):49-56.

CLIDEMIA HIRTA (L.) D. Don [Melastomataceae]

Citations:

Murdiati TB, McSweeney CS, Campbell RS, et al. (1990) Prevention of hydrolysable tannin toxicity in goats fed Clidemia hirta by calcium hydroxide supplementation. J Appl Toxicol 10(5):325-331.

Murdiati TB, McSweeney CS, Campbell RS, et al. (1992) Prevention of hydrolysable tannin toxicity by calcium hydroxide supplementation in goats fed Clidemia hirta. In: James LF, Keeler RF, Bailey JM Jr, et al. (eds.) Poisonous plants. Iowa State Univ Press. Ames, Iowa. pp. 431-435.

climath –see– *Toxicodendron radicans* (L.) Kuntze

climbing buckwheat –see– *Fallopia convolvulus* (L.) Á. Löve

climbing ivy –see– *Toxicodendron radicans* (L.) Kuntze

climbing lily –see– *Gloriosa superba* L.

climbing nightshade –see– *Solanum dulcamara* L.

climbing sumach –see– *Toxicodendron radicans* (L.) Kuntze

CLITORIA TERNATEA L. [Fabaceae]

Common Names:

butterfly pea; capa-de-la-reina; erg elagrab; gokarni; paokeke; wing-leaf clitoria

Citations:

Piala JJ, Madissoo H, Rubin B (1962) Diuretic activity of roots of Clitoria ternatea L. in dogs. Experientia 18(Feb 15):89.

clivers –see– *Galium aparine* L.

CLIVIA MINIATA (Lindl.) Regel [Amaryllidaceae]

Common Names:

kafir lily; klivie

Citations:

Ramón MF, Ballesteros S, Martínez-Arrieta R, et al. (2003) Plant and herb exposures in Spain. J Toxicol Clin Toxicol 41(4):514.

cloak fern –see– *Astrolepis cochisensis* (Goodd.) D. M. Benham & Windham

clotbur –see– *Xanthium spinosum* L.; *Xanthium strumarium* L.; *Xanthium strumarium* L. var. canadense (Mill.) Torr. & A. Gray

cloth-of-gold –see– *Baileya multiradiata* Harv. & A. Gray ex Torr.

clove –see– *Syzygium aromaticum* (L.) Merr. & L. M. Perry

clover broomrape –see– *Orobanche minor* Sm.

clover dodder –see– *Cuscuta epithymum* (L.) L.

clover-leaf poison –see– *Goodia lotifolia* Salisb.

clover tree –see– *Goodia lotifolia* Salisb.

cloves –see– *Syzygium aromaticum* (L.) Merr. & L. M. Perry

cluster bean –see– *Cyamopsis tetragonoloba* (L.) Taub.

cluster-leaf blindgrass –see– *Stypandra imbricata* R. Br.

cluster pine –see– *Pinus pinaster* Aiton

clustered fishtail palm –see– *Caryota mitis* Lour.

c'nenta –see– *Tylecodon ventricosa* (Burm. f.) Toelken

CNEORIDIUM DUMOSUM (Nutt. ex Torr. & A. Gray) Hook. f. ex Baill. [Rutaceae]

Common Names:

berry rue; coast spice bush

Citations:

Tunget CL, Turchen SG, Manoguerra AS, et al. (1993) Cneoridium dumosum exposure resulting in severe phytophotodermatitis. Vet Hum Toxicol 35(4):330.

Tunget CL, Turchen SG, Manoguerra AS, et al. (1994) Sunlight and the plant: A toxic combination: Severe phytophotodermatitis from Cneoridium dumosum. Cutis 54(6):400-402.

CNESTIS CORNICULATA Lam. [Connaraceae]

Citations:

Vickery B, Vickery ML (1974) The toxicity of some members of the Connaraceae family. Br Vet J 130(2):41-43.

CNESTIS FERRUGINEA DC. [Connaraceae]

Citations:

Vickery B, Vickery ML (1974) The toxicity of some members of the Connaraceae family. Br Vet J 130(2):41-43.

CNIDOSCOLUS ANGUSTIDENS Torr. [Euphorbiaceae]

Citations:

Scheman AJ, Conde A (2001) Contact dermatitis from Cnidoscolus angustidens. Contact Dermatitis 45(1):39.

CNIDOSCOLUS URENS (L.) Arthur [Euphorbiaccac]

Synonyms:

Jatropha urens L.

Common Names:

Brazilian physicnut; chichicaste; chichicaste-de-las-costas; güaritoto; mala mujer; nettle; ortiga; ortiga brava; physicnut; pringamoza

Citations:

Lutz O (1914) The poisonous nature of the stinging hairs of Jatropha urens. Am J Pharm 86:527-529.

coada calului –see– *Equisetum arvense* L.

coakum –see– *Phytolacca americana* L.

coal oil brush –see– *Tetradymia glabrata* Torr. & A. Gray

coast buckthorn –see– *Amsinckia intermedia* Fisch. & C. A. Mey.

coast fiddleneck –see– *Amsinckia intermedia* Fisch. & C. A. Mey.

coast manroot –see– *Marah oregonus* (Torr. & S. Watson) J. T. Howell

coast myall –see– *Acacia binervia* (J. C. Wendl.) J. F. Macbr.

coast spice bush –see– *Cneoridium dumosum* (Nutt. ex Torr. & A. Gray) Hook. f. ex Baill.

cobbler's-pegs –see– *Bidens pilosa* L.

coca –see– *Erythroxylum coca* Lam.

cocabolla –see– *Dalbergia retusa* Hemsl.

cocão –see– *Erythroxylum deciduum* A. St.-Hil.

cocco –see– *Colocasia esculenta* (L.) Schott

cochinilla –see– *Metopium brownei* (Jacq.) Urb.

cochinillio –see– *Metopium brownei* (Jacq.) Urb.

Cochlearia armoracia L. = Armoracia rusticana P. Gaertn. et al.

cock foot –see– *Chelidonium majus* L.

cock rose –see– *Papaver rhoeas* L.

cockle –see– *Agrostemma githago* L.; *Vaccaria hispanica* (Mill.) Rauschert

cocklebur –see– *Xanthium sibericum* Patrin; *Xanthium spinosum* L.; *Xanthium strumarium* L.; *Xanthium strumarium* L. var. canadense (Mill.) Torr. & A. Gray

cockroachberry –see– *Solanum aculeatissimum* Jacq.

coco-de-mono –see– *Lecythis ollaria* Loefl.

cocoa –see– *Theobroma cacao* L.

cocobolo –see– *Dalbergia retusa* Hemsl.

coconut –see– *Cocos nucifera* L.

coconut palm –see– *Cocos nucifera* L.

COCOS NUCIFERA L. [Arecaceae]

Common Names:

Bongkrek; coconut; coconut palm; Palmkernen

Citations:

Srinivas CR, Balachandran C, Singh KK (1987) Occupational dermatosis and allergic contact dermatitis in a toddy tapper. Contact Dermatitis 16(5):294-295.

Tella R, Gaig P, Lombardero M, et al. (2003) A case of coconut allergy. Allergy 58(8):825-826.

van Veen AG, Martens WK (1934) Die Giftstoffe der sogenannten Bongkrek-Vergiftungen auf Java. Rec Trav Chim 53:257-268.

Wittczak T, Pas-Wyroslak A, Palczynski C (2005) Occupational allergic conjunctivitis due to coconut fibre dust. Allergy 60(7):970-971.

cocoyam –see– *Colocasia esculenta* (L.) Schott; *Xanthosoma atrovirens* K. Koch & C. D. Bouche; *Xanthosoma sagittifolium* (L.) Schott

cocu –see– *Andira inermis* (W. Wright) Kunth ex DC.

cocus –see– *Brya ebenus* (L.) DC.

CODIAEUM VARIEGATUM (L.) A. Juss. [Euphorbiaceae]

Common Names:

Buena vista; croton; lestun; periqueto; San Francisco; Wunderstrauch

Citations:

Hausen BM, Schulz KH (1977) Occupational contact dermatitis due to croton (Codiaeum variegatum (L.) A. Juss var. pictum (Lodd.) Muell. Arg.). Sensitization by plants of the Euphorbiaceae. Contact Dermatitis 3(6):289-292.

Paulsen E (1998) Occupational dermatitis in Danish gardeners and greenhouse workers (II). Etiological factors. Contact Dermatitis 38(1):14-19.

Schmidt H, Olholm Larsen P (1977) Allergic contact dermatitis from croton (Codiaeum). Contact Dermatitis 3(2):100.

Tafelkruyer J, Van Ketel WG (1976) Sensitivity to Codiaeum variegatum. Contact Dermatitis 2(3):288.

van Ketel WG (1979) Occupational contact dermatitis due to Codiaeum variegatum and possibly to Aeschynanthus pulcher. Dermatosen 27(5):141-142.

coerana –see– *Cestrum laevigatum* Schltdl.

COFFEA ARABICA L. [Rubiaceae]

Common Names:

coffee; coffeebean; Kaffee; Kaffeebohne

Citations:

Bernton HS (1973) On occupational sensitization - A hazard to the coffee industry. JAMA 223(10):1146-1147.

Hammond JC (1944) Dried coffee grounds unsuitable for use in diet of growing chickens. Poult Sci 23(5):454-455.

Karr RM, Lehrer SB, Butcher BT, et al. (1978) Coffee worker's asthma: A clinical appraisal using the radioallergosorbent test. J Allergy Clin Immunol 62(3):143-148.

Kaye M, Freedman SO (1961) Allergy to raw coffee - An occupational disease. Can Med Assoc J 84(Mar 4):469-471.

Lupton ES (1953) Cheilitis due to coffee. Arch Derm Syphilol 68(3):333-334.

Mori H, Hirono I (1977) Effect of coffee on carcinogenicity of cycasin. Br J Cancer 35(3):369-371.

Osterman K, Zetterström O, Johansson SG (1982) Coffee worker's allergy. Allergy 37(5):313-322.

coffee –see– *Coffea arabica* L.

coffee pea –see– *Cicer arietinum* L.

coffee pod –see– *Senna obtusifolia* (L.) H. S. Irwin & Barneby

coffee senna –see– *Senna occidentalis* (L.) Link

coffee tree –see– *Gymnocladus dioicus* (L.) K. Koch; *Sesbania punicea* (Cav.) Benth.

coffeebean –see– *Coffea arabica* L.; *Gymnocladus dioicus* (L.) K. Koch; *Senna occidentalis* (L.) Link; *Senna tora* (L.) Roxb.; *Sesbania drummondii* (Rydb.) Cory; *Sesbania longifolia* DC.; *Sesbania punicea* (Cav.) Benth.; *Sesbania vesicaria* (Jacq.) Elliott

coffeeweed –see– *Senna obtusifolia* (L.) H. S. Irwin & Barneby; *Senna tora* (L.) Roxb.; *Sesbania exaltata* (Raf.) Rydb. ex A. W. Hill; *Sesbania punicea* (Cav.) Benth.; *Sesbania vesicaria* (Jacq.) Elliott

coiron –see– *Festuca argentina* (Speg.) Parodi

cola-de-caballo –see– *Equisetum arvense* L.; *Equisetum fluviatile* L.

cola-de-quirquincho –see– *Huperzia saururus* (Lam.) Trevis.

colchico –see– *Colchicum autumnale* L.

colchicum –see– *Colchicum autumnale* L.

COLCHICUM AUTUMNALE L. [Colchicaceae]

Common Names:

autumn crocus; colchico; colchicum; colchique; fall crocus; fog crocus; Herbstblume; Herbstzeitlose; Lichtblume; meadow crocus; meadow saffron; Michaelmas crocus; mysteria; naked lady; purple crocus; safran-des-prés; son-before-the-father; Wiesensafran; wild saffron; Winterhaube; wonder bulb

Citations:

Adameşteanu I, Adameşteanu C, Salantiu V, et al. (1966) Cercetări anatomoclinice în intoxicația cu Colchicum autumnale la taurine. Rev Zootehnie Med Vet 16(7):58-61.

Brnčić N, Višković I, Perić R, et al. (2001) Accidental plant poisoning with Colchicum autumnale: Report of two cases. Croat Med J 42(6):673-675.

Chareyre S, Meram D, Pulce C, et al. (1989) Acute poisoning of cows by autumnal crocus. Vet Hum Toxicol 31(3):261-262.

Chmir ON, Karmilov VA (1979) [Poisoning with Colchicum autumnale.] Vrach Delo 9:81-82.

Danel VC, Wiart JF, Hardy GA, et al. (2001) Self-poisoning with Colchicum autumnale L. flowers. J Toxicol Clin Toxicol 39(4):409-411.

Flesch F, Krencker E, Mootien E, et al. (2002) Diagnosis wandering in a case of accidental poisoning by Colchicum autumnale. J Toxicol Clin Toxicol 40(3):375-376.

Gabrscek L, Lesnicar G, Krivec B, et al. (2004) Accidental poisoning with autumn crocus. J Toxicol Clin Toxicol 42(1):85-88.

Hitzig WH, Illig R (1958) Colchiumvergiftung bei einem Kleinkind. 1. Klinisches Bild. Helv Paediatr Acta 13(2):117-130.

Jaspersen-Schib R, Theus L, Quirguis-Oeschger M, et al. (1996) Wichtige Pflanzenvergiftungen in der Schweiz 1966-1994. Schweiz Med Wochenschr 126(25):1085-1098.

Kamphues J, Meyer H (1990) Herbstzeitlose (Colchicum autumnale) in Heu und Kolikerkrungen bei Pferden. Tierarztl Prax 18(3):273-275.

Klintschar M, Beham-Schmidt C, Radner H, et al. (1999) Colchicine poisoning by accidental ingestion of meadow saffron (Colchicum autumnale): Pathological and medicolegal aspects. Forensic Sci Int 106(3):191-200.

Lohner E, Gindele HR (1989) Kolchizinvergiftung beim Schwein. Tierarztl Umsch 44(5):314-317.

Mayer H, Wacker R, Dalchow W (1986) Phytotoxikosen durch Kastanien, Oleander, Eicheln und Herbstzeitlose bei verschiedenen Zoo- und Wildtieren. Tierarztl Umsch 41(3):169-178.

Meier PJ, Gossweiler B, Jaspersen-Schib, JR et al. (1992) Vergiftungen mit Arzneimitteln, Haushaltprodukten und Pflanzen in der Kasusistik des Schweizerischen Toxikologischen Informationszentrums. Ther Umsch 49(2):79-85.
Mezger O, Heess W (1932) Herbstzeitlosensamen-Giftmordversuch. Sammlung Vergiftungsfallen 3:47-48.
Panariti E (1996) Meadow saffron (Colchicum autumnale) intoxication in a nomadic Albanian sheep flock. Vet Hum Toxicol 38(3):227-228.
Plackova S, Caganova B (1998) Acute intoxications by mushrooms and plants in Slovakia. J Toxicol Clin Toxicol 36(5):452-453.
Rauber-Lüthy C, Baer W, Rentsch K, et al. (2003) Misdiagnosed fatal meadow saffron poisoning in a toddler. J Toxicol Clin Toxicol 41(5):728.
Sannohe S, Makino Y, Kita T, et al. (2002) Colchicine poisoning resulting from accidental ingestion of meadow saffron (Colchicum autumnale). J Forensic Sci 47(6):1391-1396.
Schrader A, Schulz O, Volker H, et al. (2001) Aktuelle Vergiftungen durch Pflanzen bei Wiederkäuern in Nord- und Ostdeutschland. Berl Munch Tierarztl Wochenschr 114(5-6):218-221.
Schulz O, Hommel H (1975) Herbstzeitlose as Ursache einer Rindervergiftung. Monatsh Veterinarmed 30(9):333-334.
Shergin IK, Tribunskii MP (1971) [Pathomorphologic characteristics of lambs poisoned by meadow saffron.] Veterinariia Moscow 48(5):88-89.
Tovo S (1967) Un caso di avvelenamento mortale da colchico. Minerva Medicoleg 87(6):283-287.
Yamada M, Matsui T, Kobayashi Y, et al. (1999) Supplementary report on experimental autumn crocus (Colchicum autumnale L.) poisoning in cattle: Morphological evidence of apoptosis. J Vet Med Sci 61(7):823-825.
Yamada M, Nakagawa M, Haritani M, et al. (1998) Histopathological study of experimental acute poisoning of cattle by autumn crocus (Colchicum autumnale L.). J Vet Med Sci 60(8):949-952.
Yoneda Y, Hayashi Y, Shiroishi M, et al. (1984) [High performance liquid chromatographic determination of colchicine alkaloid from the tissues and biological fluid of bovine poisoned by the corm of autumn crocus (Colchicum autumnale L.).] Shokuhin Eiseiglku Zasshi 25(5):401-409.
Zoelen GA, Vries I, Meulenbelt J (2002) Colchicine poisoning: Ingestion of Colchicum autumnale flowers. J Toxicol Clin Toxicol 40(3):376-377.

colchique –see– *Colchicum autumnale* L.

cole –see– *Brassica napus* L. var. napus

COLEBROOKEA OPPOSITIFOLIA Sm. [Lamiaceae]

Common Names:

dhursulo

Citations:

Kulakkattolickal AT (1987) Piscicidal plants of Nepal: Preliminary toxicity screening using grass carp (Ctenopharyngodon idella) fingerlings. J Ethnopharmacol 21(1):1-9.

coleus –see– *Solenostemon scutellarioides* (L.) Codd

Coleus blumei Benth. = Solenostemon scutellarioides (L.) Codd

colicweed –see– *Achyranthes aspera* L.; *Dicentra canadensis* (Goldie) Walp.

COLLINSONIA CANADENSIS L. [Lamiaceae]

Common Names:

hardtack, heal all; horse weed; horsebalm; stoneroot

Citations:

Corwin GE (1905) Collinsonia (hardhack) poisoning. Am Vet Rev 29:394.

colocasia –see– *Colocasia esculenta* (L.) Schott

Colocasia antiquorum Schott = Colocasia esculenta (L.) Schott

COLOCASIA ESCULENTA (L.) Schott [Araceae]

Synonyms:

Colocasia antiquorum Schott

Common Names:

alocasia; arvi; cocco; cocoyam; colocasia; dasheen; eddo; eddoe; elephant's-ear; gabi; malanga; suni; taro; taro dasheen; wild taro

Citations:

Mihailidou H, Galanakis E, Paspalaki P, et al. (2002) Pica and the elephant's ear. J Child Neurol 17(11):855-856.
Murillo B, Olivares M, Silva LA, et al. (1981) Valor nutritivo del tuberculo de malanga (Colocasia esculenta) para cerdos y pollos. Arch Latinoam Nutr 31(1):27-43.
Osisiogu IU, Uzo JO, Ugochukwu EN (1974) The irritant effect of cocoyams. Planta Med 26(2):166-169.
Sinha SM, Pasricha JS, Sharma RC, et al. (1977) Vegetables responsible for contact dermatitis of the hands. Arch Dermatol 113(6):776-779.
Tokarnia CH, Armién AG, Peixoto PV, et al. (1996) Estudo experimental sobre a toxidez de algumas plantas ornamentais em bovinos. Pesq Vet Bras 16(1):5-20.

Colocasia indica (Lour.) Kunth = Alocasia macrorrhizos (L.) G. Don

Colocasia macrorrhizos (L.) Schott & Endl. = Alocasia macrorrhizos (L.) G. Don

colocinth –see– *Citrullus colocynthis* (L.) Schrad.

colocynth –see– *Citrullus colocynthis* (L.) Schrad.

Colocynthis vulgaris Schrad. = Citrullus colocynthis (L.) Schrad.

coloquinte –see– *Citrullus colocynthis* (L.) Schrad.

coloquíntida –see– *Citrullus colocynthis* (L.) Schrad.

Colorado bur –see– *Solanum rostratum* Dunal

Colorado loco vetch –see– *Oxytropis lambertii* Pursh

Colorado rubber weed –see– *Hymenoxys richardsonii* (Hook.) Cockerell

colt's-foot –see– *Petasites japonicus* (Siebold & Zucc.) Maxim.; *Tussilago farfara* L.

colt's-tail –see– *Equisetum arvense* L.

Columbia milk vetch –see– *Astragalus miser* Douglas ex Hook. var. serotinus (A. Gray ex J. G. Cooper) Barneby

Columbusgrass –see– *Sorghum ×almum* Parodi

COLUMNEA SPP. [Gesneriaceae]

Common Names:

goldfish plant

Citations:

Paulsen E (1998) Occupational dermatitis in Danish gardeners and greenhouse workers (II). Etiological factors. Contact Dermatitis 38(1):14-19.

colza –see– *Brassica napus* L. var. napus

COMBRETUM LONGISPICATUM (Engl.) Engl. & Diels [Combretaceae]

Citations:

Verdcourt B, Trump EC (1969) Common poisonous plants of East Africa. Collins. London.

COMBRETUM PLATYPETALUM Welw. ex Laws. [Combretaceae]

Common Names:

red wings

Citations:

Shone DK (1966) Poisonous plants of Rhodesia. Part I. Rhodesia Zambia Malawi J Agric Res 4:81-94.

comfrey –see– *Symphytum officinale* L.

comida-de-culebra –see– *Dieffenbachia seguine* (Jacq.) Schott

comigo ninguém pode –see– *Dieffenbachia seguine* (Jacq.) Schott

Commiphora molmol (Engl.) Engl. = Commiphora myrrha (Nees) Engl.

COMMIPHORA MYRRHA (Nees) Engl. [Burseraceae]

Synonyms:

Commiphora molmol (Engl.) Engl.

Common Names:

Arabian myrrh; karam; mirra; myrrh; myrrh gum

Citations:

Al Suwaindan SN, Gad El Rab MO, Al Fakhiry S, et al. (1998) Allergic contact dermatitis from myrrh, a topical herbal medicine used to promote healing. Contact Dermatitis 39(3):137.

Gallo R, Rivara G, Cattarini G, et al. (1999) Allergic contact dermatitis from myrrh. Contact Dermatitis 41(4):230-231.

Omer SA, Adam SE (1999) Toxicity of Commiphora myrrha to goats. Vet Hum Toxicol 41(5):299-301.

COMOCLADIA DENTATA Jacq. [Anacardiaceae]

Common Names:

guao; guao-de-sabana; guao hediondo; guao prieto; guao real; huao; maiden plum

Citations:

Lunin MM (1969) [Burns induced by the tropical plant huao.] Vestn Khir Im I I Grek 102(6):96-100.

comptie –see– *Zamia integrifolia* L. f.

conapi –see– *Deguelia utilis* (A. C. Sm.) A. M. G. Azevedo

concombre d'âne –see– *Ecballium elaterium* (L.) A. Rich.

concombre sauvage –see– *Ecballium elaterium* (L.) A. Rich.

coneflower –see– *Rudbeckia laciniata* L.

confrei –see– *Symphytum officinale* L.

congress weed –see– *Parthenium hysterophorus* L.

congressgrass –see– *Parthenium hysterophorus* L.

CONIUM MACULATUM L. [Apiaceae]

Common Names:

Blutschierling; bunk; California fern; carrot fern; cashes; cicuta maggiore; cicuta mayor; deadly hemlock; Erdschierling; false parsley; fool's-parsley; gefleckter Schierling; grande ciguë; hemlock; herb bennet; Mauerschierling; Mäuseschierling; Nebraska fern; poison fool's-parsley; poison hemlock; poison parsley; poison snakeweed; poisonroot; spotted cowbane; spotted hemlock; spotted parsley; St. Bennet's-herb; stinkweed; wild carrot; wild hemlock; winter fern; wode whistle; Wutschierling

Citations:

Anonymous (1941) Investigations of stock-poisoning plants. U S Dep Agric Bur Anim Indus Chief Rep 1941:62-63.

Anonymous (1951) Unusual case of hemlock poisoning in swine. California Vet 5(2):26.

Biberci E, Altuntas Y, Cobanoglu A, et al. (2002) Acute respiratory arrest following hemlock (Conium maculatum) intoxication. J Toxicol Clin Toxicol 40(4):517-518.

Brenet O, Roy PM, Harry P, et al. (1996) Intoxication à la ciguë: Une évolution parfois bénigne. Presse Med 25(2):82.

Buckingham JL (1936) Poisoning in a pig by hemlock (Conium maculatum). Vet J 92:301-302.

Bunch TD, Panter KE, James LF (1990) Effects of certain poisonous plants on uterine function and fetal development in livestock. J Anim Sci 68(Suppl 1):406.

Copithorne B (1937) Suspected poisoning of goats by hemlock (Conium maculatum). Vet Rec 49(33):1018-1019.

Drummer OH, Roberts AN, Bedford PJ, et al. (1995) Three deaths from hemlock poisoning. Med J Aust 162(11):592-593.

Dyson DA, Wrathall AE (1977) Congenital deformities in pigs possibly associated with exposure to hemlock (Conium maculatum). Vet Rec 100(12):241-242.

Edmonds LD, Selby LA, Case AA (1972) Poisoning and congenital malformations associated with consumption of poison hemlock by sows. J Am Vet Med Assoc 160(9):1319-1324.

Fernandez MC, Ramirez CG, Beamer CL (2002) Human fatality from poison hemlock ingestion. J Toxicol Clin Toxicol 40(5):646.

Foster PF, McFadden R, Trevino R, et al. (2003) Successful transplantation of donor organs from a hemlock poisoning victim. Transplantation 76(5):874-876.

Frank AA, Reed WM (1987) Conium maculatum (poison hemlock) toxicosis in a flock of range turkeys. Avian Dis 31(2):386-388.

Frank BS, Michelson WB, Panter KE, et al. (1995) Ingestion of poison hemlock (Conium maculatum). West J Med 163(6):573-574.

Galey FD, Holstege DM, Fisher EG (1992) Toxicosis in dairy cattle exposed to poison hemlock (Conium maculatum) in hay: Isolation of Conium alkaloids in plants, hay, and urine. J Vet Diagn Invest 4(1):60-64.

Guardigni L, Pasini S, Candoli P, et al. (1990) Il caso clinico. G Clin Med 71(1):65-67.

Gunn A (1881) Cattle poisoned by hemlock. Vet J Ann Comp Pathol 13(Oct):233-235.

Hannam DA (1985) Hemlock (Conium maculatum) poisoning in the pig. Vet Rec 116(12):322.

Jessup DA, Boermans HJ, Kock ND (1986) Toxicosis in tule elk caused by ingestion of poison hemlock. J Am Vet Med Assoc 189(9):1173-1175.

Keeler RF, Balls LD (1978) Teratogenic effects in cattle of Conium maculatum and conium alkaloids and analogs. Clin Toxicol 12(1):49-64.

Keeler RF, Balls LD, Shupe JL, et al. (1980) Teratogenicity and toxicity of coniine in cows, ewes, and mares. Cornell Vet 70(1):19-26.

Lüüs A (1927) Zwei selten Fälle aus der Kinderpraxis. Z Kinderheilkd 45:594-596.

MacDonald H (1937) Hemlock poisoning in horses. Vet Rec 49(38):1211-1212.

Markham K (1985) Hemlock poisoning in piglets. Vet Rec 116(1):27.

McKinney P, Gomez HF, Phillips S, et al. (1992) The fax machine: A new method in plant identification. Vet Hum Toxicol 34(4):353.

Palmer S, Feltham W (1984) Accidental poisoning. Nurs Times 80(19):40-43.

Panter KE, Bunch TD, Keeler RF (1988) Maternal and fetal toxicity of poison hemlock (Conium maculatum) in sheep. Am J Vet Res 49(2):281-283.

Panter KE, Bunch TD, Keeler RF, et al. (1988) Radio ultrasound observations of the fetotoxic effects in sheep from ingestion of Conium maculatum (poison-hemlock). J Toxicol Clin Toxicol 26(3-4):175-187.

Panter KE, Bunch TD, Keeler RF, et al. (1990) Multiple congenital contractures (MCC) and cleft palate induced in goats by ingestion of piperidine alkaloid-containing plants: Reduction in fetal movements as the probable cause. J Toxicol Clin Toxicol 28(1):69-83.

Panter KE, James LF, Keeler RF, et al. (1992) Radio-ultrasound observations of poisonous plant-induced fetotoxicity in livestock. In: James LF, Keeler RF, Bailey JM Jr, et al. (eds.) Poisonous plants. Iowa State Univ Press. Ames, Iowa. pp. 481-488.

Panter KE, Keeler RF, Buck WB (1985) Congenital skeletal malformations induced by maternal ingestion of Conium maculatum (poison hemlock) in newborn pigs. Am J Vet Res 46(10):2064-2066.

Panter KE, Keeler RF, Buck WB (1985) Induction of cleft palate in newborn pigs by maternal ingestion of poison hemlock (Conium maculatum). Am J Vet Res 46(6):1368-1371.

Panter KE, Keeler RF, Buck WB, et al. (1983) Toxicity and teratogenicity of Conium maculatum in swine. Toxicon Suppl 3:333-336.

Panter KE, Keeler RF, Bunch TD, et al. (1990) Congenital skeletal malformations and cleft palate induced in goats by ingestion of Lupinus, Conium and Nicotiana species. Toxicon 28(12):1377-1386.

Penny RH (1953) Hemlock poisoning in cattle. Vet Rec 65(42):699-670.

Plotti GB (1900) Poisoning by virulent hemlock. Am Vet Rev 23:717.

Richter HE (1964) Vergiftung durch Conium maculatum L., Flecken-Schierling. Wien Tierarztl Monatsschr 51:404-407.

Rizzi D, Basile C, Di Maggio A, et al. (1989) Rhabdomyolysis and acute tubular necrosis in coniine (hemlock) poisoning. Lancet 2(Dec 16):1461-1462.

Rizzi D, Basile C, Di Maggio A, et al. (1991) Clinical spectrum of accidental hemlock poisoning. Neurotoxic mani festations, rhabdomyolysis and acute tubular necrosis. Nephrol Dial Transplant 6(12):939-943.

Rizzi D, Introna F Jr, Candela RG, et al. (1988) Rabdiomiolisi tossica e tubulonecrosi nell'avvelenamento da cicuta. Osservazion di quattro casi. Clin Ter 124(3):193-201.

Rubaj B, Zimowski A (1971) Zatrucie nutrii szczwołem plamistym (Conium maculatum L.). Med Weter 27(10):622-623.

Scatizzi A, Di Maggio A, Rizzi D, et al. (1993) Acute renal failure due to tubular necrosis caused by wildfowl-mediated hemlock poisoning. Ren Fail 15(1):93-96.

Short SB, Edwards WC (1989) Accidental Conium maculata poisoning in the rabbit. Vet Hum Toxicol 31(1):54-57.

Tokarnia CH, Döbereiner J, Peixoto PV (1985) Intoxicação experimental por Conium maculatum (Umbeliferae) em bovinos e ovinos. Pesq Vet Bras 5(1):15-25.

Tudor G, Anton E, Diaconescu G (1969) Observații asupra intoxicației cu Conium maculatum (cucuta) la ovine. Rev Zootehnie Med Vet 19(11):74-80.

Widmer WR (1984) Poison hemlock in swine. Vet Med Small Anim Clin 79(3):405-408.

Conophallus konjak Schott = Amorphophallus konjac K. Koch

CONSOLIDA REGALIS Gray [Ranunculaceae]

Synonyms:

Delphinium consolida L.

Common Names:

Ackerrittersporn; Adlerblume; branched larkspur; dolphin flower; Feldrittersporn; field larkspur; king's-consound; knight's-spur; larkspur; mezei szarkaláb; Ottilienkraut; wild larkspur

Citations:

Elphick EE (1931) Sheep poisoned as a result of eating larkspur (Delphinium consolidum). Vet Rec 11(18):512-513.

consuelda –see– *Euphorbia tirucalli* L.

consumption bush –see– *Crotalaria berteroana* DC.

consumption weed –see– *Baccharis halimifolia* L.

conval lily –see– *Convallaria majalis* L.

convallaria –see– *Convallaria majalis* L.

CONVALLARIA MAJALIS L. [Ruscaceae]

Common Names:

conval lily; convallaria; Eischaupe; Jacob's-ladder; lady's-tears; lèrio-do-vale; liljekonvall; lily-of-the-valley; liricon fancy; Maiblume; Maiglöckchen; Maischellchen; májusi gyöngyvirág; May blossom; May lily; Mayflower; mugget; muguet; muguet-de-mai; Nieskraut; our-lady's-tears; Zauke

Citations:

Edgerton PH (1989) Symptoms of digitalis-like toxicity in a family after accidental ingestion of lily of the valley plant. J Emerg Nurs 15:220-223.

Falk W, Weikmann E (1969) Vergiftungen im Kindesalter - Eine Analyse von 642 stationär behandelten kindlichen Vergiftungsfällen. Wien Klin Wochenschr 81(48):867-873.

Haugen S, Bryne E, Falke M, et al. (2001) Grade I-II atrioventricular block following lily-of-the-valley (Convallaria majalis) intake: A report of three cases. J Toxicol Clin Toxicol 39(3):303-304.

Krenzelok EP, Jacobsen TD, Aronis JM (1996) Lily-of-the-valley (Convallaria majalis) exposures: Are the outcomes consistent with the reputation? J Toxicol Clin Toxicol 34:601.

Lamminpaa A, Kinos M (1996) Plant poisonings in children. Hum Exp Toxicol 15(3):245-249.

Moxley RA, Schneider NR, Steinegger DH, et al. (1989) Apparent toxicosis associated with lily-of-the-valley (Convallaria majalis) ingestion in a dog. J Am Vet Med Assoc 195(4):485-487.

Štěrba B, Meissner V (1962) Sušené rostliny s obsahem srdečních glykosidů a alkaloidů jako příčina uhynuté hospodářských zvířat. Veterinarstvi 12:83-86.

convolvulus –see– *Ipomoea muelleri* Benth.

CONVOLVULUS ARVENSIS L. [Convolvulaceae]

Common Names:

Ackerwinde; bindweed; correhuela menor; European bindweed; European morning-glory; field bindweed; laplove; morning-glory; obsession vine

Citations:

Olver H (1872) Supposed poisoning of pigs by convolvulus. The Vet 45(49):727-729.

Pammel LH (1920) European morning glory suspected of being poisonous. Am J Vet Med 15:439.

Schultheiss PC, Knight AP, Traub Dargatz JL, et al. (1995) Toxicity of field bindweed (Convolvulus arvensis) to mice. Vet Hum Toxicol 37(5):452-454.

CONYZA BONARIENSIS (L.) Cronquist [Asteraceae]

Common Names:

fleabane; horseweed

Citations:

Sertoli A, Fabbri P, Campolmi P, et al. (1978) Allergic contact dermatitis to Salvia officinalis, Inula viscosa and Conyza bonariensis. Contact Dermatitis 4(5):314-315.

Conyza coulteri A. Gray = Laennecia coulteri (A. Gray) G. L. Nesom

conyzav –see– *Laennecia coulteri* (A. Gray) G. L. Nesom

cooby –see– *Acacia salicina* Lindl.

Cook tree –see– *Thevetia peruviana* (Pers.) K. Schum.

Cooktown ironwood –see– *Erythrophleum chlorostachys* (F. Muell.) Baill.

Cooktown loquat –see– *Rhodomyrtus macrocarpa* Benth.

coolahgrass –see– *Panicum coloratum* L.

coontie –see– *Zamia integrifolia* L. f.

COOPERIA PEDUNCULATA Herb. [Amaryllidaceae]

Common Names:

evening-star rain lily; giant rain lily; prairie lily; rain lily

Citations:

Casteel SW, Rowe LD, Bailey EM, et al. (1988) Experimentally induced photosensitization in cattle with Cooperia pedunculata. Vet Hum Toxicol 30(2):101-104.

Rowe LD, Norman JO, Corrier DE, et al. (1987) Photosensitization of cattle in southeast Texas: Identification of phototoxic activity associated with Cooperia pedunculata. Am J Vet Res 48(11):1658-1661.

Rowe LD, Rector BS, Bailey EM, et al. (1986) Mammalian photosensitization associated with ingestion of Cooperia pedunculata (rain lily). Proc Am Vet Med Assoc 1986:102.

Cooper's milk vetch –see– *Astragalus miser* Douglas ex Hook. var. serotinus (A. Gray ex J. G. Cooper) Barneby

cop rose –see– *Papaver rhoeas* L.

copa –see– *Schinus terebinthifolius* Raddi

copel tree –see– *Ailanthus altissima* (Mill.) Swingle

copo-de-leite –see– *Calla palustris* L.

copper rose –see– *Papaver rhoeas* L.

copperweed –see– *Oxytenia acerosa* Nutt.

COPTIS CHINENSIS Franch. [Ranunculaceae]

Common Names:

chuen lin

Citations:

Deng JF, Lin TJ, Kao WF, et al. (1997) The difficulty in handling poisonings associated with Chinese traditional medicine: A poison control center experience for 1991-1993. Vet Hum Toxicol 39(2):106-114.

coquelicot –see– *Papaver rhoeas* L.

coquillo –see– *Jatropha curcas* L.

coquito –see– *Jatropha curcas* L.

coração negro –see– *Prunus myrtifolia* (L.) Urb.

coral –see– *Solanum pseudocapsicum* L.

coral bead plant –see– *Abrus precatorius* L.

coral bean –see– *Calia secundiflora* (Ortega) Yakovlev

coral plant –see– *Jatropha multifida* L.; *Jatropha podagrica* Hook.

coral sumach –see– *Metopium toxiferum* (L.) Krug & Urb.; *Schinus terebinthifolius* Raddi

coralberry –see– *Actaea rubra* (Aiton) Willd.

coralillo –see– *Duranta erecta* L.

corazon-de-cabrito –see– *Caladium bicolor* (Aiton) Vent.

CORCHORUS OLITORIUS L. [Malvaceae]

Common Names:

juta; jute

Citations:

Curjel DF, Acton HW (1924) Jute dermatitis. Indian J Med Res 12:257-260.

Johnson SJ (1985) The effects on pigs and chickens of feeding grain contaminated with seeds of Corchorus olitorius. In: Seawright AA, Hegarty MP, James LF, et al. (eds.) Plant toxicology. Queensland Poisonous Plant Committee. Yeerongpilly, Queensland. pp. 165-169.

Johnson SJ, Toleman MA (1982) The toxicity of jute (Corchorus olitorius) seed to pigs. Aust Vet J 58(6):264-265.

McKenzie RA, Callinan RB, Unger DB, et al. (1992) Suspected jute seed (Corchorus olitorius) poisoning of cattle. Aust Vet J 69(5):117-118.

cordatum –see– *Philodendron hederaceum* (Jacq.) Schott

coriander –see– *Coriandrum sativum* L.

CORIANDRUM SATIVUM L. [Apiaceae]

Common Names:

coriander; Koriander

Citations:

Bock SA (1993) Anaphylaxis to coriander: A sleuthing story. J Allergy Clin Immunol 91(6):1232-1233.

Kanerva L, Soini M (2001) Occupational protein contact dermatitis from coriander. Contact Dermatitis 45(6):354-355.

Manzanedo L, Blanco J, Fuentes M, et al. (2004) Anaphylactic reaction in a patient sensitized to coriander seed. Allergy 59(3):362-363.

Stäger J, Wüthrich B, Johansson SG (1991) Spice allergy in celery-sensitive patients. Allergy 46(6):475-478.

Suhonen R, Keskinen H, Björkstén F, et al. (1979) Allergy to coriander. A case report. Allergy 34(5):327-330.

coriaria –see– *Coriaria myrtifolia* L.

CORIARIA ARBOREA Linds. [Coriariaceae]

Common Names:

toot; tree tutu; tupakiki; tutu; Tutubaum

Citations:

Anderson IL (1968) Tutu poisoning in two circus elephants. N Z Vet J 16(8):146-147.

Chilvers CD (1972) Tutu poisoning in an elderly Maori lady. N Z Med J 75(477):85-86.

Clinch PG, Turner JC (1968) Estimation of tutin and hyenanchin in honey. 2. The toxicity of honey samples from test hives during the period 1962-67. N Z J Sci 11:346-351.

Graham JM, Cartridge ME (1961) Tutu poisoning in dogs. N Z Vet J 9:45.

Palmer-Jones T (1947) A recent outbreak of honey poisoning. Part I. Historical and descriptive. N Z J Sci Technol A 29(3):107-114.

Palmer-Jones T (1965) Poisonous honey overseas and in New Zealand. N Z Med J 64(399):631-637.

Palmer-Jones T, White EP (1947) A recent outbreak of honey poisoning. Part VII. Observations on the toxicity and toxin of tutu (Coriaria arborea Lindsay). N Z J Sci Technol A 29(Aug):107-143.

Paterson CR (1947) A recent outbreak of honey poisoning. Part IV. The source of the toxic honey - Field observations. N Z J Sci Technol A 29(Oct):125-129.

CORIARIA MYRTIFOLIA L. [Coriariaceae]

Common Names:

coriaria; emborrachacabras; Gerbermyrte; Gerberstrauch; Lederbaum; masooriberry; Myrtensumach; redoul; roldón; sumach

Citations:

Alonso Casteel P, Moreno Galdó A, Sospedro Martínez E, et al. (1997) Intoxicación grave por Coriaria myrtifolia: A propósito de un caso. An Esp Pediatr 46(1):81-82.

Cahen O, Floras P, Guerineau JM, et al. (1978) Intoxication par les fruits de redoul. Cah Anesthesiol 26(5):693-698.

de Haro L, Pommier P, Tichadou L, et al. (2005) Poisoning by Coriaria myrtifolia Linnaeus: A new case report and review of the literature. Toxicon 46(6):600-603.

Garcia Martin A, Masvidal Aliberch RM, Bofill Bernaldo AM, et al. (1983) Intoxicación por ingesta de Coriaria myrtifolia. Estudio de 25 casos. An Esp Pediatr 19(5):366-370.

Obach R, Plans P (1949) Intoxicacion por la Coriaria myrtifolia. Med Clin (Barc) 13(4):220-222.

Poyen P, Raibaudi R, Bloch C, et al. (1970) Une intoxication aiguë grave par le redoul. Eur J Toxicol 3(6):386-391.

Rimbaud L, Serre H, Passouant P (1943) L'intoxication par le redoul (Coriaria myrtifolia). Gaz Hop 8(Apr):117-118.

Vargas M (1926) Envenenamiento de cinco niños por el roldón (Coriaria myrtifolia). Intensa toxicidad de la coriamirtina. Higiene escolar. El Siglo Medico. pp. 355-359.

CORIARIA RUSCIFOLIA L. [Coriariaceae]

Common Names:

dol diablo; New Zealand toot plant; tutu

Citations:

Manriquez O, Varas J, Rios JC, et al. (2002) Analysis of 156 cases of plant intoxication received in the toxicologic information center at Catholic University of Chile. Vet Hum Toxicol 44(1):31-32.

CORIARIA SINICA Maxim. [Coriariaceae]

Citations:

Gao M, Tang G (1991) [Emergency treatment and care of acute Coriaria sinica Maxim poisoning.] Chung Hua Hu Li Tsa Chih 26(7):295-296.

cork dust –see– *Quercus suber* L.

cork oak –see– *Quercus suber* L.

corkwood –see– *Duboisia myoporoides* R. Br.; *Sesbania grandiflora* (L.) Pers.

corky-fruit water dropwart –see– *Oenanthe pimpinelloides* L.

corn –see– *Zea mays* L.

corn campion –see– *Agrostemma githago* L.

corn chamomile –see– *Anthemis arvensis* L.

corn cockle –see– *Agrostemma githago* L.; *Vaccaria hispanica* (Mill.) Rauschert

corn lily –see– *Veratrum californicum* Durand

corn millet –see– *Panicum miliaceum* L.

corn mustard –see– *Sinapis arvensis* L.

corn poppy –see– *Papaver rhoeas* L.

corn rose –see– *Papaver rhoeas* L.

corn vetch –see– *Securigera varia* (L.) Lassen

corn woundwort –see– *Stachys arvensis* (L.) L.

cornbind –see– *Fallopia convolvulus* (L.) Á. Löve

cornflower –see– *Papaver rhoeas* L.

cornucopia –see– *Brugmansia ×candida* Pers.; *Brugmansia arborea* (L.) Lagerh.

CORNUS OBLONGA Wall. [Cornaceae]

Synonyms:

Cornus paniculata Buch.-Ham. ex D. Don

Common Names:

dogwood; silky dogwood

Citations:

Pammel LH (1925) Dogwood may be poisonous. North Am Vet Nov:62.

Cornus paniculata Buch.-Ham. ex D. Don = Cornus oblonga Wall.

cornwood –see– *Andira inermis* (W. Wright) Kunth ex DC.

coroa-de-cristo –see– *Euphorbia milii* Des Moul.

corona –see– *Mascagnia pubiflora* (A. Juss.) Griseb.

coronilla –see– *Securigera varia* (L.) Lassen

Coronilla varia L. = Securigera varia (L.) Lassen

Coronopus didymus (L.) sm. = *Lepidium didymum* L.

correhuela menor –see– *Convolvulus arvensis* L.

Corsican pine –see– *Larix decidua* Mill.

cortes quillajas –see– *Quillaja saponaria* Molina

CORYDALIS AUREA Willd. [Fumariaceae]

Common Names:

golden corydalis; golden fumeroot; golden smoke; scrambled eggs

Citations:

Smith RA, Lewis D (1990) Apparent Corydalis aurea intoxication of cattle. Vet Hum Toxicol 32(1):63-64.

CORYDALIS CASEANA A. Gray [Fumariaceae]

Synonyms:

Capnoides casaena auct.

Common Names:

fitweed

Citations:

Fleming CE, Miller MR, Vawter LR (1931) The fitweed (Capnoides caseana), a poisonous range plant of the northern Sierra Nevada Mountains. Nevada Agric Exp Sta Bull #121:8-29.

Fleming CE, Miller MR, Vawter LR, et al. (1931) Poisonous range plants. Nevada Agric Exp Sta Annu Rep 1930:12-13.

CORYLUS AVELLANA L. [Betulaceae]

Common Names:

European hazel

Citations:

Skamstrup Hansen K, Ballmer-Weber BK, Lüttkopf D, et al. (2003) Roasted hazelnuts - Allergenic activity evaluated by double-blind, placebo-controlled food challenge. Allergy 58(2):132-138.

Wensing M, Koppelman SJ, Penninks AH, et al. (2001) Hidden hazelnut is a threat to allergic patients. Allergy 56(2):191-192.

corynanthe –see– *Pausinystalia johimbe* (K. Schum.) Pierre ex Beille

Corynanthe johimbe K. Schum. = Pausinystalia johimbe (K. Schum.) Pierre ex Beille

CORYNOCARPUS LAEVIGATUS J. R. Forst. & G. Forst. [Corynocarpaceae]

Common Names:

karaka

Citations:

Bell ME (1974) Toxicology of karaka kernel, karakin and β-nitropropionic acid. N Z J Sci 17:327-334.

Palmer-Jones T, Line LJ (1962) Poisoning of honey bees by nectar from the karaka tree (Corynocarpus laevigata J. R. et G. Forst.). N Z J Agric Res 5(5-6):433-436.

Shaw SD, Billing T (2006) Karaka (Corynocarpus laevigatus) toxicosis in North Island brown kiwi (Apteryx mantelli). Vet Clin North Am Exot Anim Pract 9(3):545-549.

costus –see– *Saussurea costus* (Falc.) Lipsch.

cotieira –see– *Joannesia princeps* Vell.

cotón –see– *Gossypium hirsutum* L.

cotton –see– *Gossypium herbaceum* L.; *Gossypium hirsutum* L.

cotton fireweed –see– *Senecio quadridentatus* Labill.

cotton-leaf physicnut –see– *Jatropha gossypiifolia* L.

cottonbush –see– *Gomphocarpus fruticosa* (L.) W. T. Aiton

cottonweed –see– *Abutilon theophrasti* Medik.

cotyledon –see– *Cotyledon orbiculata* L.

Cotyledon coruscans Haw. = Cotyledon orbiculata L. var. oblonga (Haw.) DC.

COTYLEDON ORBICULATA L. [Crassulaceae]

Common Names:

cotyledon; honde oor; krimpsiektebossie; pig's-ear

Citations:

Anderson LA, Schultz RA, Kellerman TS, et al. (1985) Isolation and characterization of and some observations on poisoning by bufadienolides from Cotyledon orbiculata L. var. orbiculata. Onderstepoort J Vet Res 52(1):21-24.

Terblanche M, Adelaar TF (1965) A note on the toxicity of Cotyledon orbiculata L. J S Afr Vet Assoc 36(4):555-559.

Tustin RC, Thornton DJ, Kleu CB (1984) An outbreak of Cotyledon orbiculata L. poisoning in a flock of Angora goat rams. J S Afr Vet Assoc 55(4):181-184.

Van der Walt SJ, Steyn DG (1939) Recent investigations into the toxicity of known and unknown poisonous plants in the Union of South Africa. IX. Onderstepoort J Vet Sci Anim Indus 12:335-366.

COTYLEDON ORBICULATA L. var. oblonga (Haw.) DC. [Crassulaceae]

Synonyms:

Cotyledon coruscans Haw.

Common Names:

plakkies

Citations:

Van der Walt SJ, Steyn DG (1939) Recent investigations into the toxicity of known and unknown poisonous plants in the Union of South Africa. IX. Onderstepoort J Vet Sci Anim Indus 12:335-366.

couchgrass –see– *Cynodon dactylon* (L.) Pers.

couer saignant –see– *Caladium bicolor* (Aiton) Vent.

COULA EDULIS Baill. [Olacaceae]

Common Names:

ecue; eugomo

Citations:

Dantin-Gallego J, Armayor AF, Riesco J (1952) Some new toxic woods: Some new manifestations of toxicity. Ind Med Surg 21(2):41-46.

Coulter's conyza –see– *Laennecia coulteri* (A. Gray) G. L. Nesom

country walnut –see– *Aleurites moluccanus* (L.) Willd.

countryman's-treacle –see– *Ruta graveolens* L.

COUTOUBEA RAMOSA Aubl. [Gentianaceae]

Citations:

Tokarnia CH, Döbereiner J (1981) Intoxicação experimental por Coutoubea ramosa (Gentianaceae) em bovinos. Pesq Vet Bras 1(2):55-60.

couve-de-Bruxelas –see– *Brassica oleracea* L. var. gemmifera Zenker

cow cabbage –see– *Veratrum californicum* Durand

cow cockle –see– *Agrostemma githago* L.; *Vaccaria hispanica* (Mill.) Rauschert

cow cress –see– *Berula erecta* (Huds.) Coville

cow parsley –see– *Anthriscus sylvestris* (L.) Hoffm.; *Heracleum mantegazzianum* Sommier & Levier

cow parsnip –see– *Heracleum sosnowskyi* Manden.; *Heracleum sphondylium* L.; *Heracleum stevenii* Manden.

cow poison –see– *Delphinium barbeyi* (Huth) Huth

cow sorrel –see– *Rumex acetosella* L.

cow wheat –see– *Scrophularia marilandica* L.

cowage –see– *Mucuna pruriens* (L.) DC.

cowbane –see– *Cicuta douglasii* (DC.) J. M. Coult. & Rose; *Cicuta maculata* L.; *Cicuta maculata* L. var. angustifolia Hook.; *Cicuta virosa* L.

cowbush –see– *Leucaena leucocephala* (Lam.) de Wit

cowgrass –see– *Trifolium pratense* L.

cowhage –see– *Mucuna pruriens* (L.) DC.

cowherb –see– *Vaccaria hispanica* (Mill.) Rauschert

cowitch –see– *Mucuna pruriens* (L.) DC.

cowpea –see– *Vigna unguiculata* (L.) Walp.

cowpen daisy –see– *Verbesina encelioides* (Cav.) Benth. & Hook. f. ex A. Gray

coyote tobacco –see– *Nicotiana attenuata* Torr. ex S. Watson

coyotillo –see– *Karwinskia humboldtiana* (Schult.) Zucc.; *Karwinskia parvifolia* Rose

coyotillo-de-Texas –see– *Karwinskia humboldtiana* (Schult.) Zucc.

coyotio –see– *Karwinskia humboldtiana* (Schult.) Zucc.

crabapple –see– *Malus sylvestris* (L.) Mill.

crabgrass –see– *Digitaria sanguinalis* (L.) Scop.

crab's-eye –see– *Abrus precatorius* L.

crake feet –see– *Hyacinthoides non-scripta* (L.) Chouard ex Rothm.

crambe –see– *Crambe abyssinica* Hochst. ex R. E. Fr.

CRAMBE ABYSSINICA Hochst. ex R. E. Fr. [Brassicaceae]

Common Names:

Abyssinian kale; crambe

Citations:

Lambert JL, Clanton DC, Wolff IA, et al. (1970) Crambe meal protein and hulls in beef cattle rations. J Anim Sci 31(3):601-607.

Van Etten CH, Gagne WE, Robbins DJ, et al. (1969) Biological evaluation of crambe seed meals and derived products by rat feeding. Cereal Chem 46(Mar):145-155.

CRATAEGUS LAEVIGATA (Poir) DC. [Rosaceae]

Common Names:

hawthorn; May tree; Weißdorn; white thorn

Citations:

Smith FP (1940) Eye injuries in agriculture. Trans Ophthalmol Soc U K 60:252-257.

CRATAEGUS MONOGYNA Jacq. [Rosaceae]

Common Names:

hawthorn; single-seed hawthorn; Weißdorn

Citations:

Steinman HK, Lovell CR, Cronin E (1984) Immediate-type hypersensitivity to Crataegus monogyna (hawthorn). Contact Dermatitis 11(5):321.

crazy plant –see– *Scopolia carniolica* Jacq.

crazyweed –see– *Astragalus mollissimus* Torr. var. mollissimus; *Oxytropis lambertii* Pursh

creek lily –see– *Alocasia macrorrhizos* (L.) G. Don

creeping buttercup –see– *Ranunculus repens* L.

creeping caustic –see– *Euphorbia drummondii* Boiss.

creeping Charlie –see– *Glechoma hederacea* L.

creeping Croftonweed –see– *Ageratina riparia* (Regel) R. M. King & H. Rob.

creeping crowfoot –see– *Ranunculus repens* L.

creeping cut-leaf water parsnip –see– *Berula erecta* (Huds.) Coville

creeping fig –see– *Ficus pumila* L.

creeping indigo –see– *Indigofera hendecaphylla* Jacq.; *Indigofera spicata* Forssk.

creeping knapweed –see– *Acroptilon repens* (L.) DC.

creeping lily –see– *Gloriosa superba* L.

creeping mallow –see– *Modiola caroliniana* (L.) G. Don

creeping pimelea –see– *Pimelea prostrata* (J. R. Forst. & G. Forst.) Willd.

creeping spurge –see– *Euphorbia myrsinites* L.

creeping sumach –see– *Toxicodendron radicans* (L.) Kuntze

creeping water parsnip –see– *Berula erecta* (Huds.) Coville

creosote bush –see– *Larrea tridentata* (DC.) Coville

cress-leaf groundsel –see– *Senecio glabellus* Poir.

crested goosefoot –see– *Chenopodium carinatum* R. Br.

crimson clover –see– *Trifolium incarnatum* L.

CROCUS SATIVUS L. [Iridaceae]

Common Names:

azafran; Krokus; saffron; saffron crocus; saffron spice; true saffron

Citations:

Feo F, Martinez J, Martinez A, et al. (1997) Occupational allergy in saffron workers. Allergy 52(6):633-641.

Wüthrich B, Schmid-Grendelmeyer P, Lundberg M (1997) Anaphylaxis to saffron. Allergy 52(4):476-477.

Crofton weed –see– *Ageratina adenophora* (Spreng.) R. M. King & H. Rob.

croix-de-malte –see– *Tribulus terrestris* L.

CROTALARIA ALATA Buch.-Ham. ex D. Don [Fabaceae]

Citations:

Williams MC, Molyneux RJ (1987) Occurrence, concentration, and toxicity of pyrrolizidine alkaloids in Crotalaria seeds. Weed Sci 35(4):476-481.

Crotalaria anagyroides Kunth = Crotalaria micans Link

CROTALARIA ARGYROLOBIOIDES Baker [Fabaceae]

Citations:

Williams MC, Molyneux RJ (1987) Occurrence, concentration, and toxicity of pyrrolizidine alkaloids in Crotalaria seeds. Weed Sci 35(4):476-481.

CROTALARIA ARIDICOLA Domin [Fabaceae]

Common names:

Chillagoe horse poison

Citations:

Cory AH (1920) Report of the chief inspector of stock, Queensland. Annu Rep Dep Agric Stock. 1919-1920:64.

Legg J (1951) Oesophageal and stomach ulceration. Report of the chief inspector of stock, Queensland. Annu Rep Dep Agric Stock. 1950-1951:59.

Newton LG (1962) Oesophageal disease. Report of the chief inspector of stock, Queensland. Annu Rep Dep Agric Stock. 1961-1962:35.

Tucker G (1911) Chillagoe horse disease. Report of the chief inspector of stock, Queensland. Annu Rep Dep Agric Stock. 1910-1911:24.

Tucker G (1912) Chillagoe horse disease and pink-eye. Report of the chief inspector of stock, Queensland. Annu Rep Dep Agric Stock. 1911-1912:32.

White CT (1932) Report of the Government Botanist, Queensland. Annu Rep Dep Agric Stock. 1931-1932:59.

CROTALARIA ASSAMICA Benth. [Fabaceae]

Citations:

Chan MY, Zhao XL, Ogle CW (1989) A comparative study on the hepatic toxicity and metabolism of Crotalaria assamica and Eupatorium species. Am J Chin Med 17(3-4):165-170.

Crotalaria bagamoyoensis Baker f. = Crotalaria laburnoides Klotzsch

CROTALARIA BARKAE Schweinf. [Fabaceae]

Synonyms:

Crotalaria geminiflora Dinter ex Baker f.

Citations:

Jackson JJ, Needham AJ, Lawrence JA (1968) Some recent investigations into Rhodesian toxic plants. Proc Trans Rhodesia Sci Assoc 53:9-12.

Lewis AR, Wilson VJ, Hill RR (1973) Crotalariosis in the common duiker (Sylvicapra grimmia). Rhodesia Vet J 4:29-33.

CROTALARIA BERTEROANA DC. [Fabaceae]

Synonyms:

Crotalaria fulva Roxb.

Common Names:

consumption bush; ragwort; rattlebox; white back; white luck

Citations:

Berry DM, Bras G (1957) Venous occlusion of the liver in crotalaria and senecio poisoning. North Am Vet 38(Nov):323-326, 328.

Bras G, Berry DM, György P (1957) Plants as aetiological factor in veno-occlusive disease of the liver. Lancet 272(May 11):960-962.

Bras G, McLean E (1963) Toxic factors in veno-occlusive disease. Ann N Y Acad Sci 111(Dec 30):392-398.

Lindo V, Bras G (1966) Further investigations on the toxicity of Crotalaria fulva in Jamaica. West Indian Med J 15:34-39.

McLean E, Bras G, György P (1964) Veno-occlusive lesions in livers of rats fed Crotalaria fulva. Br J Exp Pathol 45(Jun):242-247.

CROTALARIA BURKEANA Benth. [Fabaceae]

Common Names:

bottlebrush; klapperbos; rattlebush; stiff sickness bush; stywesiekte bossie; stywesiektebos

Citations:

Steyn DG (1936) Recent investigations into the toxicity of known and unknown poisonous plants in the Union of South Africa. Onderstepoort J Vet Sci Anim Indus 7(1):169-178.

CROTALARIA CRISPATA F. Muell. ex Benth. [Fabaceae]

Citations:

Gardiner MR, Royce R, Bokor A (1965) Studies on Crotalaria crispata, a newly recognised cause of Kimberley horse disease. J Pathol Bact 89(1):43-55.

CROTALARIA DURA J. M. Wood & M. S. Evans [Fabaceae]

Common Names:

jaagsiektebossie; wild lucerne

Citations:

Steyn DG (1937) Recent investigations into the toxicity of known and unknown poisonous plants in the Union of South Africa. VII. Onderstepoort J Vet Sci Anim Indus 9(1):111-124.

Crotalaria fulva Roxb. = Crotalaria berteroana DC.

Crotalaria geminiflora Dinter ex Baker f. = Crotalaria barkae Schweinf.

CROTALARIA GOREENSIS Guill. & Perr. [Fabaceae]

Common Names:

crotalaria; gamba pea; rattlebox

Citations:

Norton JH, O'Rourke PK (1979) Toxicity of Crotalaria goreensis for chickens. Aust Vet J 55(4):173-174.

CROTALARIA GRAHAMIANA Wight & Arn. [Fabaceae]

Common Names:

cho cho

Citations:

Arzt J, Mount ME (1999) Hepatotoxicity associated with pyrrolizidine alkaloid (Crotalaria spp) ingestion in a horse on Easter Island. Vet Hum Toxicol 41(2):96-99.

CROTALARIA JUNCEA L. [Fabaceae]

Common Names:

gergellim bravo; maracá-de-cobra; rattlepod; sanka; sunn hemp

Citations:

Arseculeratne SN, Gunatilaka AA, Panabokke RG (1985) Studies on medicinal plants of Sri Lanka. Part 14. Toxicity of some traditional medicinal herbs. J Ethnopharmacol 13:323-335.

Nobre D, Dagli ML, Haraguchi M (1994) Crotalaria juncea intoxication in horses. Vet Hum Toxicol 36(5):445-448.

Payet M, Camain R, Miege J, et al. (1967) Etude comparée de la toxicité de différentes espèces de Crotalaria. Bull Mem Fac Mixte Med Pharm Dakar 15:266-269.

Steyn DG, van der Walt SJ (1945) Jaagsiekte in horses, and sunn-hemp poisoning in stock. Farming South Africa Jul:1-3.

CROTALARIA LABURNIFOLIA L. [Fabaceae]

Common Names:

bird flower

Citations:

Schoental R, Coady A (1968) The hepatotoxicity of some Ethiopian and East African plants, including some used in traditional medicines. East Afr Med J 45(8):577-580.

CROTALARIA LABURNOIDES Klotzsch [Fabaceae]

Synonyms:

Crotalaria bagamoyoensis Baker f.

Citations:

Heath D, Shaba J, Williams A, et al. (1975) A pulmonary hypertension-producing plant from Tanzania. Thorax 30(4):399-404.

CROTALARIA MAUENSIS Baker f. [Fabaceae]

Citations:

Mugera GM (1970) Toxic and medicinal plants of East Africa. II. Bull Epizootic Dis Afr 18(4):389-403.

CROTALARIA MICANS Link [Fabaceae]

Synonyms:

Crotalaria anagyroides Kunth

Citations:

Tokarnia CH, Döbereiner J (1983) Intoxicação experimental por Crotalaria anagyroides (Leg. Papilionoideae) em bovinos. Pesq Vet Bras 3(4):115-123.

Crotalaria mucronata Desv. = Crotalaria pallida Aiton

CROTALARIA NANA Burm f. [Fabaceae]

Common Names:

jhunjhunia

Citations:

Arora RR, Pyarelal, Ghosh TK, et al. (1981) Epidemiology of veno-occlusive disease in tribal population of Madhya Pradesh and Bihar. J Commun Dis 13(3):147-151.

Shrivastava AB, Kolte GN, Vegad JL, et al. (1985) Haematological study on toxicity of Crotalaria nana in rats. Indian J Anim Sci 55(4):255-258.

Shrivastava AB, Kolte GN, Vegad JL, et al. (1986) Pathological alterations in lungs in Crotalaria nana toxicity in rat. Indian J Anim Sci 56(9):901-906.

Tandon BN, Puri BK, Tandon HD, et al. (1978) Ultra-structure of liver in veno-occlusive disease due to Crotalaria nana Burm. Indian J Med Res 68(Nov):790-797.

CROTALARIA NOVAE-HOLLANDIAE DC. [Fabaceae]

Common Names:

New Holland rattlepod

Citations:

Rose AL, Gardner CA, McConnell JD, et al. (1957) Field and experimental investigation of "walk-about" disease of horses (Kimberly horse disease) in Northern Australia. Crotalaria poisoning in horses. Part II. Aust Vet J 33(Mar):49-62.

CROTALARIA OCHROLEUCA G. Don [Fabaceae]

Common Names:

marejea; sunn hemp

Citations:

Mkiwa FE, Lwoga AB, Mosha RD, et al. (1994) Antinutritional effects of Crotalaria ochroleuca (marejea) in animal feed supplements. Vet Hum Toxicol 36(2):96-100.

Mosha RD, Matovelo JA (1994) Toxicity of Crotalaria ochroleuca seeds in mice. In: Colegate SM, Dorling PR (eds.) Plant-associated toxins. Agricultural, phytochemical and ecological aspects. CABI. New York. pp. 303-306.

CROTALARIA PALLIDA Aiton [Fabaceae]

Synonyms:

Crotalaria mucronata Desv.; ***Crotalaria striata*** DC.

Common Names:

chique chique; crotalaria; gergellim bravo; giant striata crotalaria; guizo-de-cascavel; maracá-de-cobra; rattlebox; rattleweed; smooth crotalaria; streaked rattlepod

Citations:

Arzt J, Mount ME (1999) Hepatotoxicity associated with pyrrolizidine alkaloid (Crotalaria spp) ingestion in a horse on Easter Island. Vet Hum Toxicol 41(2):96-99.

Bierer BW, Vickers CL, Rhodes WH, et al. (1960) Comparison of the toxic effects of Crotalaria spectabilis and Crotalaria giant striata as complete feed contaminants. J Am Vet Med Assoc 136(7):318-322.

Diaz GJ, Roldan LP, Cortes A (2003) Intoxication of Crotalaria pallida seeds to growing broiler chicks. Vet Hum Toxicol 45(4):187-189.

Kelly JW, Barber CW, Pate DD, et al. (1961) Effect of feeding crotalaria seed to young chickens. J Am Vet Med Assoc 139(11):1215-1217.

Laws L (1968) Toxicity of Crotalaria mucronata to sheep. Aust Vet J 44(10):453-455.

Payet M, Camain R, Miege J, et al. (1967) Etude comparée de la toxicité de différentes espèces de Crotalaria. Bull Mem Fac Mixte Med Pharm Dakar 15:266-269.

Smith FH, Osborne JC (1962) Toxic effects of crotalaria seed. Vet Med 57(Mar):234-237.

CROTALARIA POLYSPERMA Kotschy ex Schweinf. [Fabaceae]

Citations:

Andrew HR (1987) An outbreak of laminitic crotalariosis in the Chegutu district. Zimbabwe Vet J 18(1-2):23-26.

CROTALARIA QUINQUEFOLIA L. [Fabaceae]

Citations:

Appelman H, Dirven JG, Kuil H (1962) Giftigheid van Crotalaria quinquefolia als veevoer. Surinaamse, Landbouw 10:67-70.

Williams MC, Molyneux RJ (1987) Occurrence, concentration, and toxicity of pyrrolizidine alkaloids in Crotalaria seeds. Weed Sci 35(4):476-481.

CROTALARIA RECTA Steud. ex A. Rich. [Fabaceae]

Citations:

Schoental R, Coady A (1968) The hepatotoxicity of some Ethiopian and East African plants, including some used in traditional medicines. East Afr Med J 45(8):577-580.

CROTALARIA RETUSA L. [Fabaceae]

Common Names:

earring plant; gergellim bravo; koropo; maracá-de-cobra; rattlebox; rattlepod; rattleweed; shack shack; sonnette; wedge-leaf rattlepod

Citations:

Alfonso HA, Figueredo MA, Rodríguez J, et al. (1986) Intoxicación por Crotalaria retusa y C. spectabilis en pollos y gansos. Primer reporte en Cuba. Rev Salud Anim 8(2):201-202.

Alfonso HA, Sanchez LM, Figeurdo MA, et al. (1993) Intoxication due to Crotalaria retusa and C. spectabilis in chickens and geese. Vet Hum Toxicol 35(6):539.

Emmel MW (1937) The toxicity of Crotalaria retusa L. seeds for the domestic fowl. J Am Vet Med Assoc 91:205-206.

Figueredo M, Rodriguez J, Alfonso HA (1987) Patomorfología de la intoxicación experimental aguda por Crotalaria retusa y C. spectabilis en pollos. Rev Cubana Cienc Vet 18(1-2):63-71.

Hooper PT, Scanlan WA (1977) Crotalaria retusa poisoning of pigs and poultry. Aust Vet J 53(3):109-114.

Nobre VM, Dantas AF, Riet-Correa F, et al. (2005) Acute intoxication by Crotalaria retusa in sheep. Toxicon 45(3):347-352.

Nobre VM, Riet-Correa F, Barbosa Filho JM, et al. (2004) Intoxicação por Crotalaria retusa (Fabaceae) em eqüideos no semi-árido da Paraíba. Pesq Vet Bras 24(3):132-143.

Nobre VM, Riet-Correa F, Dantas AF, et al. (2004) Intoxication by Crotalaria retusa in ruminants and Equidae in the state of Paraíba, northeastern Brazil. In: Acamovic T, Stewart CS, Pennycott TW (eds.) Poisonous plants and related toxins. CABI. Wallingford, UK. pp. 275-279.

Rose AL, Gardner CA, McConnell JD, et al. (1957) Field and experimental investigations of "walk-about" disease of horses (Kimberly horse disease) in Northern Australia. Crotalaria poisoning in horses Part I. Aust Vet J 33(Feb):25-33.

Rose AL, Gardner CA, McConnell JD, et al. (1957) Field and experimental investigation of "walk-about" disease of horses (Kimberly horse disease) in Northern Australia. Crotalaria poisoning in horses. Part II. Aust Vet J 33(Mar):49-62.

Ross AJ (1977) Effects of feeding diets containing Crotalaria retusa L. seed to growing pigs. J Agric Sci (Cambridge) 89(1):101-105.

Ross AJ, Tucker JW (1977) Effects of feeding diets containing Crotalaria retusa L. seed to broiler chickens. J Agric Sci (Cambridge) 89(1):95-99.

Williams MC, Molyneux RJ (1987) Occurrence, concentration, and toxicity of pyrrolizidine alkaloids in Crotalaria seeds. Weed Sci 35(4):476-481.

CROTALARIA SAGITTALIS L. [Fabaceae]

Common Names:

arrow crotalaria; rattlebox; rattlesnake weed; rattleweed; weedy rattlebox; wild pea

Citations:

Gibbons WJ, Hokanson JF, Wiggins AM, et al. (1950) Cirrhosis of the liver in horses. North Am Vet 31(Apr):229-233.

Seibold HR (1957) Crotalaria poisoning in a horse. J Am Vet Med Assoc 130(Apr 15):336.

Stalker M (1884) A new disease among horses - Results of investigation made at the veterinary college at Ames. Am Vet Rev 8:342-345.

CROTALARIA SALTIANA Andrews [Fabaceae]

Common Names:

safari

Citations:

Barri ME, Adam SE (1981) The toxicity of Crotalaria saltiana to calves. J Comp Pathol 91(4):621-627.

Barri ME, Adam SE, Omer OH (1984) Effects of Crotalaria saltiana on Nubian goats. Vet Hum Toxicol 26(6):476-480.

Barri ME, Adam SE, Omer OH (1988) Toxic effects of Crotalaria saltiana in mice. Vet Hum Toxicol 30(5):429-431.

Crotalaria sericea Retz. = Crotalaria spectabilis Roth

CROTALARIA SPECTABILIS Roth [Fabaceae]

Synonyms:

Crotalaria sericea Retz.

Common Names:

gergellim bravo; maracá-de-cobra; rattlebox; rattlepod; rattleweed; showy crotalaria; wild pea

Citations:

Alfonso HA, Figueredo MA, Rodríguez J, et al. (1986) Intoxicación por Crotalaria retusa y C. spectabilis en pollos y gansos. Primer reporte en Cuba. Rev Salud Anim 8(2):201-202.

Alfonso HA, Sanchez LM, Figeurdo MI, et al. (1993) Intoxication due to Crotalaria retusa and C. spectabilis in chickens and geese. Vet Hum Toxicol 35(6):539.

Allen JR (1963) Crotalaria spectabilis toxicity studies in turkeys. Avian Dis 7(3):318-324.

Allen JR, Carstens LA (1970) Clinical signs and pathologic changes in Crotalaria spectabilis-intoxicated rats. Am J Vet Res 31(6):1059-1070.

Allen JR, Carstens LA, Knezevic AL (1965) Crotalaria spectabilis intoxication in rhesus monkeys. Am J Vet Res 26(112):753-757.

Allen JR, Carstens LA, Norback DH, et al. (1970) Ultrastructural and biochemical changes associated with pyrrolizidine-induced hepatic megalocytosis. Cancer Res 30(6):1857-1866.

Allen JR, Childs GR, Cravens WW (1960) Crotalaria spectabilis toxicity in chickens. Proc Soc Exp Biol Med 104(Jul):434-436.

Allen JR, Lalich JJ (1963) Crotalaria spectabilis induced cirrhosis in turkeys. Fed Proc 22:546.

Allen JR, Lalich JJ, Schmittle SC (1963) Crotalaria spectabilis-induced cirrhosis in turkeys. Lab Invest 12(4):512-517.

Becker HN, Bennett WW, Simpson CF (1976) Crotalaria poisoning in swine in Florida (clinical signs, necropsy lesions and microscopic pathology). Proc Int Pig Vet Cong. 1976:S12.

Becker RB, Neal WM, Arnold PT (1941) Crotalaria for forage. Part II. Value of Crotalaria as feed. Florida Agric Exp Sta Bull #361:33-71.
Becker RB, Neal WM, Arnold PT, et al. (1935) A study of the palatability and possible toxicity of 11 species of crotalaria, especially of C. spectabilis Roth. J Agric Res 50(11):911-922.
Bierer BW, Vickers CL, Rhodes WH, et al. (1960) Comparison of the toxic effects of Crotalaria spectabilis and Crotalaria giant striata as complete feed contaminants. J Am Vet Med Assoc 136(7):318-322.
Carlton WW (1967) Crotalaria intoxication in guinea pigs. J Am Vet Med Assoc 151(7):845-855.
Caylor JF, Laurent CK (1961) Effect of level of Crotalaria spectabilis on egg production of White Leghorn hens. Poult Sci 40:818.
Cox DH, Harris DL, Richard TA (1958) Chemical identification of crotalaria poisoning in horses. J Am Vet Med Assoc 133(Oct 15):425-426.
D'Aguanno W, Geiling EM, Davis KJ (1961) Toxicity of Crotalaria spectabilis. Fed Proc 20(1 Pt 1):432.
Emmel MW (1937) The pathology of Crotalaria spectabilis Roth seed poisoning in the domestic fowl. J Am Vet Med Assoc 90(May):627-634.
Emmel MW (1948) Crotalaria poisoning in cattle. J Am Vet Med Assoc 113:164.
Emmel MW, Sanders DA, Henley WW (1935) Additional observations on the toxicity of Crotalaria spectabilis (Roth) for swine. J Am Vet Med Assoc 87:175-176.
Emmel MW, Sanders DA, Henley WW (1935) Crotalaria spectabilis Roth seed poisoning in swine. J Am Vet Med Assoc 86:43-54.
Figueredo M, Rodriguez J, Alfonso HA (1987) Patomorfología de la intoxicación experimental aguda por Crotalaria retusa y C. spectabilis en pollos. Rev Cubana Cienc Vet 18(1-2):63-71.
Gibbons WJ, Durr EH, Cox SA (1953) An outbreak of cirrhosis of the liver in horses. North Am Vet 34(Aug):556-558.
Himes JA, Bruss ML, Simpson CF, et al. (1976) Hypercholeresis in turkeys following the ingestion of Crotalaria spectabilis seeds. Cornell Vet 66(4):551-565.
Kay JM, Gillund TD, Heath D (1967) Mast cells in the lungs of rats fed on Crotalaria spectabilis seeds. Am J Pathol 51(6):1031-1044.
Kay JM, Harris P, Heath D (1967) Pulmonary hypertension produced in rats by ingestion of Crotalaria spectabilis seeds. Thorax 22(2):176-179.
Kay JM, Heath D (1966) Observations on the pulmonary arteries and heart weight of rats fed on Crotalaria spectabilis seeds. J Pathol Bact 92(2):385-394.
Kay JM, Smith P, Heath D (1969) Electron microscopy of Crotalaria pulmonary hypertension. Thorax 24(5):511-526.
Kelly JW, Barber CW, Pate DD, et al. (1961) Effect of feeding crotalaria seed to young chickens. J Am Vet Med Assoc 139(11):1215-1217.
Lalich JJ, Merkow LP (1961) Pulmonary arteritis produced in rats by feeding Crotalaria spectabilis. Lab Invest 10(4):744-750.
Lalich JJ, Merkow LP (1961) Pulmonary arteritis produced in rats by feeding Crotalaria spectabilis. Fed Proc 20:118.
Masugi Y, Oami H, Aihara K, et al. (1965) Renal and pulmonary vascular changes induced by Crotalaria spectabilis in rats. Acta Pathol Jpn 15(4):407-415.
McGrath JP, Duncan JR, Munnell JF (1975) Crotalaria spectabilis toxicity in swine: Characterization of the renal glomerular lesion. J Comp Pathol 85(2):185-194.
Meyrick B, Reid L (1979) Development of pulmonary arterial changes in rats fed Crotalaria spectabilis. Am J Pathol 94(1):37-50.
Payet M, Camain R, Miege J, et al. (1967) Etude comparée de la toxicité de différentes espèces de Crotalaria. Bull Mem Fac Mixte Med Pharm Dakar 15:266-269.
Peckham JC, Sangster LT, Jones OH Jr (1974) Crotalaria spectabilis poisoning in swine. J Am Vet Med Assoc 165(7):633-638.
Piercy PL, Rusoff LL (1946) Crotalaria spectabilis poisoning in Louisiana livestock. J Am Vet Med Assoc 108(827):69-73.
Sanders DA, Shealy AL, Emmel MW (1936) The pathology of Crotalaria spectabilis Roth poisoning in cattle. J Am Vet Med Assoc 89(2):150-156.
Schmittle SC, Richey DJ, Tumlin JT (1959) Toxicity of Crotalaria spectabilis seed in poultry. Poult Sci 38:1244-1245.
Seawright AA, Kelly WR, Hrdlicka J, et al. (1991) Pyrrolizidine alkaloidosis in cattle due to Senecio species in Australia. Vet Rec 129(9):198-199.
Simpson CF, Waldroup PW, Harms RH (1963) Pathologic changes associated with feeding various levels of Crotalaria spectabilis seed to poultry. J Am Vet Med Assoc 142(3):264-271.
Smith FH, Osborne JC (1962) Toxic effects of crotalaria seed. Vet Med 57(Mar):234-237.
Smith P, Heath D (1978) Evagination of vascular smooth muscle cells during the early stages of Crotalaria pulmonary hypertension. J Pathol 124(3):177-183.
Smith P, Kay JM, Heath D (1970) Hypertensive pulmonary vascular disease in rats after prolonged feeding with Crotalaria spectabilis seeds. J Pathol 102(2):97-106.
Souza AC, Hatayde MR, Bechara GH (1997) Aspectos patológicos da intoxicação de suínos por sementes de Crotalaria spectabilis (Fabaceae). Pesq Vet Bras 17(1):12-18.
Stötzer H, Herbst M, Reichl R, et al. (1972) Zur Pathogenese der experimentellen pulmonalen Hypertonie. Modellversuche mit Crotalaria spectabilis an Ratten. Virchows Arch A 356(4):331-342.
Thomas EF (1934) The toxicity of certain species of crotalaria seed for the chicken, quail, turkey and dove. J Am Vet Med Assoc 85(Mar):617-622.
Turner JH, Lalich JJ (1965) Experimental cor pulmonale in the rat. Arch Pathol 79(4):409-418.
Wallace HD, Combs GE (1965) Effect of Crotalaria spectabilis seed on swine. Florida Agric Exp Sta Annu Rep. 1965:84-85.
Wiggins AM (1959) Crotalaria poisoning in cattle. A case report. Auburn Vet 15(Winter):84-85.
Williams MC, Molyneux RJ (1987) Occurrence, concentration, and toxicity of pyrrolizidine alkaloids in Crotalaria seeds. Weed Sci 35(4):476-481.
Young DE (1963) Cardiovascular, hepatic, and renal lesions in rats fed Crotalaria spectabilis. Fed Proc 22:254.

Crotalaria striata DC. = Crotalaria pallida Aiton

CROTALARIA TRICHOTOMA Bojer [Fabaceae]
Synonyms:
Crotalaria usaramoensis Baker f.

Citations:
Payet M, Camain R, Miege J, et al. (1967) Etude comparée de la toxicité de différentes espèces de Crotalaria. Bull Mem Fac Mixte Med Pharm Dakar 15:266-269.

Crotalaria usaramoensis Baker f. = Crotalaria trichotoma Bojer

crotin –see– *Croton tiglium* L.

croton –see– *Codiaeum variegatum* (L.) A. Juss.; *Croton tiglium* L.

CROTON MEGALOBOTRYS Müll. Arg. [Euphorbiaceae]
Citations:
Nyazema NZ (1986) Herbal toxicity in Zimbabwe. Trans R Soc Trop Med Hyg 80(3):448-450.

croton tiglio –see– *Croton tiglium* L.

CROTON TIGLIUM L. [Euphorbiaceae]
Common Names:
ba dau; batu; bois purgatif; bois-de-pavana; bois-des-molluques; chengkian; crotin; croton; croton tiglio; dand; datun; graine-de-moluques; graine-de-tilly; grana tiglio; habbussalatin; jayapala; kamaisa; pa-teou-seou; petit pignon d' Inde; purging croton; tilly; tuba

Citations:
Babu N (1965) Observations on the toxicity of the seed of Croton tiglium Linn. on predatory and weed fishes. Sci Cult 31(6):308-310.

croton weed –see– *Ageratina adenophora* (Spreng.) R. M. King & H. Rob.

crow bells –see– *Hyacinthoides non-scripta* (L.) Chouard ex Rothm.; *Narcissus pseudonarcissus* L.

crow fig –see– *Strychnos nux-vomica* L.

crow leek –see– *Hyacinthoides non-scripta* (L.) Chouard ex Rothm.

crow poison –see– *Amianthium muscitoxicum* (Walter) A. Gray; *Anticlea elegans* (Pursh) Rydb.; *Veratrum viride* Aiton

crowberry –see– *Phytolacca americana* L.

crowflower –see– *Ranunculus acris* L.; *Ranunculus bulbosus* L.

crowfoot –see– *Erodium cicutarium* (L.) L'Her.; *Ranunculus bulbosus* L.; *Ranunculus repens* L.; *Ranunculus sceleratus* L.; *Toxicoscordion venenosum* (S. Watson) Rydb.

crown flower –see– *Calotropis gigantea* (L.) W. T. Aiton; *Calotropis procera* (Aiton) W. T. Aiton

crown-of-the-field –see– *Agrostemma githago* L.

crown-of-thorns –see– *Euphorbia milii* Des Moul.

crownbeard –see– *Verbesina encelioides* (Cav.) Benth. & Hook. f. ex A. Gray

crownvetch –see– *Securigera varia* (L.) Lassen

crozier cycad –see– *Cycas circinalis* L.

crucifixion plant –see– *Euphorbia milii* Des Moul.

cruel vine –see– *Araujia sericifera* Brot.

crunchweed –see– *Sinapis arvensis* L.

crusader's-spears –see– *Drimia maritima* (L.) Stearn

crying tree –see– *Synadenium cupulare* (Boiss.) L. C. Wheeler

CRYPTOLEPIS DECIDUA (Planch. ex Benth.) N. E. Br. [Asclepiadaceae]
Citations:
Van der Walt SJ, Steyn DG (1940) Recent investigations into the toxicity of known and unknown poisonous plants in the Union of South Africa, X. Onderstepoort J Vet Res 15(1-2):261-277.

CRYPTOMERIA JAPONICA (L. f.) D. Don [Cupressaceae]
Common Names:
Japanese cedar

Citations:
Kondo Y, Ipsen H, Løwenstein H, et al. (1997) Comparison of concentrations of Cry j 1 and Cry j 2 in diploid and triploid Japanese cedar (Cryptomeria japonica) pollen extracts. Allergy 52(4):455-459.
Yokozeki H, Satoh T, Katayama I, et al. (2007) Airborne contact dermatitis due to Japanese cedar pollen. Contact Dermatitis 56(7):224-228.

CRYPTOSTEGIA GRANDIFLORA R. Br. [Apocynaceae]
Common Names:
garuda palai; pala; Palay rubber vine; pink allamanda; purple allamanda; rubber vine; ruber-ki-bal

Citations:
Brain C, Fox VE (1994) Suspected cardiac glycoside poisoning in elephants (Loxodonta africana). J S Afr Vet Assoc 65(4):173-174.
Cook DR, Campbell GW, Meldrum AR (1990) Suspected Cryptostegia grandiflora (rubber vine) poisoning in horses. Aust Vet J 67(9):344.
Mathur KS, Dube BK, Kumar P (1964) Cryptostegia grandiflora poisoning simulating digitalis toxicity. J Indian Med Assoc 42(8):381-385.
McGavin MD (1969) Rubber-vine (Cryptostegia grandiflora) toxicity for ruminants. Queensland J Agric Anim Sci 26(1):9-19.

Cryptostemma calendula (L.) Druce = Arctotheca calendula (L.) Levyns

CRYPTOTAENIA JAPONICA Hassk. [Apiaceae]

Common Names:

mitsuba

Citations:

Kanzaki T (1989) Contact dermatitis due to Cryptotaenia japonica Makino. Contact Dermatitis 20(6):60-61.

Cuban physicnut –see– *Jatropha curcas* L.

cubé –see– *Deguelia utilis* (A. C. Sm.) A. M. G. Azevedo

cubé root –see– *Deguelia utilis* (A. C. Sm.) A. M. G. Azevedo

cuca –see– *Erythroxylum coca* Lam.

cuckoo plant –see– *Arisaema triphyllum* (L.) Schott

cuckoopint –see– *Arum maculatum* L.

cucumber –see– *Cucumis sativus* L.

CUCUMIS ACULEATUS Cogn. [Cucurbitaceae]

Citations:

Raymond WD (1947) Notes on a poisonous East African species of wild cucumber (Cucumis aculeatus). East Afr Med J 24(Dec):450-451.

CUCUMIS AFRICANUS L. f. [Cucurbitaceae]

Common Names:

wild cucumber; wild gherkin; wilde Komkommer

Citations:

Raymond WD (1947) Notes on a poisonous East African species of wild cucumber (Cucumis aculeatus). East Afr Med J 24(Dec):450-451.

Steyn DG (1950) The toxicity of bitter-tasting cucurbitaceous vegetables (vegetable marrow, watermelons, etc.) for man. S Afr Med J 24:713-715.

Cucumis colocynthis L. = Citrullus colocynthis (L.) Schrad.

CUCUMIS MELO L. [Cucurbitaceae]

Common Names:

melon

Citations:

García S, Lombardero M, Serra-Baldrich E, et al. (2004) Occupational protein contact dermatitis due to melon. Allergy 59(5):558-559.

CUCUMIS MELO L. subsp. agrestis (Naudin) Pangalo [Cucurbitaceae]

Common names:

ulcardo melon

Citations:

Jubb TF, Creeper JH, McKenzie RA (1995) Poisoning of cattle attributed to Cucumis melo ssp agrestis (ulcardo melon). Aust Vet J 72(7):274-275.

CUCUMIS MYRIOCARPUS Naudin [Cucurbitaceae]

Common Names:

bitter apple; Gifappel; gooseberry melon; paddy melon; prickly cucumber; prickly melon; prickly paddymelon; squash melon; wild cucumber

Citations:

Carter GI (1990) Prickly paddy melon (Cucumis myriocarpus) poisoning of cattle. Aust Vet J 67(7):276.

McKenzie RA, Newman RD, Rayner AC, et al. (1988) Prickly paddy melon (Cucumis myriocarpus) poisoning of cattle. Aust Vet J 65(6):167-170.

Quin JI (1928) The toxicity of Cucumis myriocarpus Naud. S Afr J Sci 25:242-245.

CUCUMIS SATIVUS L. [Cucurbitaceae]

Common Names:

cucumber

Citations:

Edwards EK Jr, Edwards EK (1984) Contact urticaria provoked by pickles. Cutis 33(2):230.

Hjorth N, Roed-Petersen J (1976) Occupational protein contact dermatitis in food handlers. Contact Dermatitis 2(1):28-42.

Sinha SM, Pasricha JS, Sharma RC, et al. (1977) Vegetables responsible for contact dermatitis of the hands. Arch Dermatol 113(6):776-779.

Note:

Cucumber is named *Cucumis sativus* L. var. *sativus* in some publications.

CUCURBITA MAXIMA Duchesne [Cucurbitaceae]

Common Names:

gourd; pumpkin

Citations:

Sinha SM, Pasricha JS, Sharma RC, et al. (1977) Vegetables responsible for contact dermatitis of the hands. Arch Dermatol 113(6):776-779.

CUCURBITA PEPO L. [Cucurbitaceae]

Common Names:

gem squash; marrow; pumpkin; squash; zucchini

Citations:

Herrington ME (1983) Intense bitterness in commercial zucchini. Cucurbit Genetics Coop Rep 6:75-76.

Houston TM (1972) A food illness with a difference. Health People 6(4):8-9.

Steyn DG (1950) The toxicity of bitter-tasting cucurbitaceous vegetables (vegetable marrow, watermelons, etc.) for man. S Afr Med J 24:713-715.

cudweed paper flower –see– *Psilostrophe gnaphalodes* DC.

cuenta-de-oro –see– *Duranta erecta* L.

cuilinbaca –see– *Andira inermis* (W. Wright) Kunth ex DC.

CULLEN CORYLIFOLIUM (L.) Medik. [Fabaceae]

Synonyms:
Psoralea corylifolia L.

Common Names:
boh-gol-zhee; Drüsenklee; Malayan tea; scurfy-pea

Citations:
Maurice PD, Cream JJ (1989) The dangers of herbalism. Br Med J 299(6709):1204.
Nam SW, Baek JT, Lee DS, et al. (2005) A case of acute cholestatic hepatitis associated with the seeds of Psoralea corylifolia (boh-gol-zhee). Clin Toxicol 43(6):589-591.
Perharic L, Shaw D, Colbridge M, et al. (1994) Toxicological problems resulting from exposure to traditional remedies and food supplements. Drug Saf 11(4):284-294.

CULLEN DRUPACEUM (Bunge) C. H. Stirt. [Fabaceae]

Synonyms:
Psoralea drupacea Bunge

Common names:
scurfy pea

Citations:
Arutyunyants SI, Baimuradov TB, Rogozhin PS (1972) [Poisoning of pigs with Psoralea drupacea.] Veterinariia Moscow 49(8):95-96.
Ibadullaev FI, Turdiev I (1973) [Signs of scurfy pea poisoning in swine (experimental data).] Veterinariia Moscow 50(7):92-94.

culver keep –see– *Hyacinthoides non-scripta* (L.) Chouard ex Rothm.

culverkeys –see– *Hyacinthoides non-scripta* (L.) Chouard ex Rothm.

cumin –see– *Cuminum cyminum* L.

CUMINUM CYMINUM L. [Apiaceae]

Common Names:
cumin; kamoon

Citations:
Haroun EM, Mahmoud OM, Adam SE (2002) Effect of feeding Cuminum cyminum fruits, Thymus vulgaris leaves or their mixture to rats. Vet Hum Toxicol 44(2):67-69.
Seetharam KA, Pasricha JS (1987) Condiments and contact dermatitis of the finger-tips. Indian J Dermatol Venereol Leprol 53:325-328.

cungevoi –see– *Alocasia macrorrhizos* (L.) G. Don

cunjeboi –see– *Alocasia macrorrhizos* (L.) G. Don

cunjeboy –see– *Alocasia macrorrhizos* (L.) G. Don

cunjevoi –see– *Alocasia macrorrhizos* (L.) G. Don

cupana –see– *Paullinia cupana* Kunth

CUPRESSUS ARIZONICA Greene [Cupressaceae]

Common Names:
smoothwood

Citations:
Agea E, Bistoni O, Russano A, et al. (2002) The biology of cypress allergy. Allergy 57(10):959-960.
Cortegano I, Civantos E, Aceituno E, et al. (2004) Cloning and expression of a major allergen from Cupressus arizonica pollen, Cup a 3, a PR-5 protein expressed under polluted environment. Allergy 59(5):485-490.

CUPRESSUS MACROCARPA Hartw. ex Gordon [Cupressaceae]

Common Names:
cypress; macrocarpa; macrocarpa pine; Monterey cypress

Citations:
Alexander RE (1943) Abortion and premature calving. N Z J Agr 67:279.
Gilder RP (1943) Effect of macrocarpa cypress on pregnant cows. N Z J Agr 67:278-279.
MacDonald J (1956) Macrocarpa poisoning. N Z Vet J 4:30.
MacLeod AM (1948) Macrocarpa poisoning in dairy herd. Abortions and deaths following eating of green branches. N Z J Agr 76(1):178.
Mason RW (1974) Foetal cerebral leucomalacia associated with Cupressus macrocarpa abortion in cattle. Aust Vet J 50(9):419.
O'Scanaill T (1986) Suspected cypress poisoning in a cow. Ir Vet J 40(9):156.
Sloss V, Brady JW (1983) Abnormal births in cattle following ingestion of Cupressus macrocarpa foliage. Aust Vet J 60(7):223.

CUPRESSUS SEMPERVIRENS L. [Cupressaceae]

Common Names:
cypress; Italian cypress

Citations:
Agea E, Bistoni O, Russano A, et al. (2002) The biology of cypress allergy. Allergy 57(10):959-960.
Dubus JC, Melluso JP, Bodiou AC, et al. (2000) Allergy to cypress pollen. Allergy 55(4):410-411.
Fiorina A, Scordamaglia A, Guerra L, et al. (2002) Prevalence of allergy to Cypress. Allergy 57(9):861-862.
Gould CM (1962) Cypress poisoning. Vet Rec 74(26):743.

curaco aloe –see– *Aloe vera* (L.) Burm. f.

curage –see– *Persicaria hydropiper* (L.) Spach

curamaguey –see– *Pentalinon luteum* (L.) B. F. Hansen & Wunderlin

curcas bean –see– *Jatropha curcas* L.

CURCUMA LONGA L. [Zingiberaceae]

Common Names:
chuu-ou-kou; curcumin; haldi; turmeric

Citations:

Bille N, Larsen JC, Hansen EV, et al. (1985) Subchronic oral toxicity of turmeric oleoresin in pigs. Food Chem Toxicol 23(11):967-973.

Deshpande SS, Lalitha VS, Ingle AD, et al. (1998) Subchronic oral toxicity of turmeric and ethanolic turmeric extract in female mice and rats. Toxicol Lett 95(3):183-193.

Goh CL, Ng SK (1987) Allergic contact dermatitis to Curcuma longa (turmeric). Contact Dermatitis 17(3):186.

Hata M, Sasaki E, Ota M, et al. (1997) Allergic contact dermatitis from curcumin (turmeric). Contact Dermatitis 36(2):107-108.

Ilsley SE, Miller HM, Kamel C (2005) Effects of dietary quillaja saponin and curcumin on the performance and immune status of weaned pigs. J Anim Sci 83(1):82-88.

Kandarkar SV, Sawant SS, Ingle AD, et al. (1998) Subchronic oral hepatotoxicity of turmeric in mice - Histopathological and ultrastructural studies. Indian J Exp Biol 36(7):675-679.

CURCUMA ZEDOARIA (Christm.) Roscoe [Zingiberaceae]

Common Names:

Indian arrowroot; shoti

Citations:

Latif MA, Morris TR, Miah AM, et al. (1979) Toxicity of shoti (Indian arrowroot: Curcuma zedoaria) for rats and chicks. Br J Nutr 41(1):57-63.

curcumin –see– *Curcuma longa* L.

curled dock –see– *Rumex crispus* L.

curly dock –see– *Rumex crispus* L.

currant –see– *Vitis vinifera* L.

currybush –see– *Hypericum revolutum* Vahl

cursed buttercup –see– *Ranunculus sceleratus* L.

cursed crowfoot –see– *Ranunculus sceleratus* L.

Cuscuta arvensis Beyr. ex Engelm. = Cuscuta pentagona Engelm.

CUSCUTA CAMPESTRIS Yunck. [Convolvulaceae]

Common Names:

dodder

Citations:

Morsesyan TB (1971) [Pathological changes in cattle poisoned by the dodder Cuscuta campestris Yuncker.] Biol Zh Armenii 24(7):67-70.

CUSCUTA EPITHYMUM (L.) L. [Convolvulaceae]

Common Names:

alfalfa dodder; clover dodder; dodder; small dodder

Citations:

Pammel LH (1921) Alfalfa dodder. Vet Med 16:48-49.

CUSCUTA PENTAGONA Engelm. [Convolvulaceae]

Synonyms:

Cuscuta arvensis Beyr. ex Engelm.

Common Names:

field dodder

Citations:

Pammel LH (1921) Alfalfa dodder. Vet Med 16:48-49.

custard apple –see– *Annona reticulata* L.; *Annona squamosa* L.; *Asimina triloba* (L.) Dunal

custard day lily –see– *Hemerocallis lilioasphodelus* L.

cut-eye bean –see– *Canavalia ensiformis* (L.) DC.

cut finger –see– *Valeriana officinalis* L.

cut-leaf coneflower –see– *Rudbeckia laciniata* L.

cut-leaf nightshade –see– *Solanum triflorum* Nutt.

cut-leaf philodendron –see– *Monstera deliciosa* Liebm.

cutch –see– *Areca catechu* L.

cutch tree –see– *Acacia catechu* (L. f.) Willd.

Cyamopsis psoraloides (Lam.) DC. = Cyamopsis tetragonoloba (L.) Taub.

CYAMOPSIS TETRAGONOLOBA (L.) Taub. [Fabaceae]

Synonyms:

Cyamopsis psoraloides (Lam.) DC.

Common Names:

cluster bean; guar bean; guar gum; guara

Citations:

Bakshi Y, Creger CR, Couch JR (1964) Studies on guar meal. Poult Sci 43:1302.

Malik MY, Sheikh AA (1967) Studies on the nature of chick-growth inhibiting factor in guar meal. West Pakistan J Agric Res 5(1):116-124.

Malik MY, Sheikh AA (1967) Studies on the use of guar meal in poultry ration. West Pakistan J Agric Res 5(1):107-115.

cycad –see– *Cycas circinalis* L.; *Cycas media* R. Br.; *Cycas revoluta* Thunb.; *Macrozamia riedlei* (Fisch. ex Gaudich.) C. A. Gardner; *Macrozamia spiralis* (Salisb.) Miq.; *Zamia integrifolia* L. f.

CYCAS CIRCINALIS L. [Cycadaceae]

Common Names:

crozier cycad; cycad; fadang; false sago; false sago palm; federico; fern palm; Japanese sago; sago palm; zamia

Citations:

Albretsen JC, Safdar AK, Richardson JA (1998) Cycad palm toxicosis in dogs: 60 cases (1987-1997). J Am Vet Med Assoc 213(1):99-101.

Anderson JL, Hall WT (1964) Neurotoxic effects from cycad leaves. Fed Proc 23:1349.

Anonymous (1965) Cycas circinalis poisoning. Papua New Guinea Dep Agric Annu Rep 1963-1964:18.
Campbell ME, Mickelsen O, Yang MG, et al. (1966) Effects of strain, age and diet on the response of rats to the ingestion of Cycas circinalis. J Nutr 88(1):115-124.
Hoch-Ligeti C, Stutzman E, Arvin JM (1968) Cellular composition during tumor induction in rats by cycad husk. J Natl Cancer Instit 41(2):605-614.
Kurland LT (1988) Amyotrophic lateral sclerosis and Parkinson's disease complex on Guam linked to an environmental neurotoxin. Trends Neurosci 11(2):51-54.
Kurland LT, Mulder DW (1954) Epidemiologic investigations of amyotrophic lateral sclerosis. Neurology 4:355-378.
Laqueur GL (1964) Carcinogenic effects of cycad meal and cycasin, methylazoxymethanol glycoside, in rats and effects of cycasin in germfree rats. Fed Proc 23(6 Part):1386-1388.
Laqueur GL, Mickelsen O, Whiting MG, et al. (1963) Carcinogenic properties of nuts from Cycas circinalis L. indigenous to Guam. J Natl Cancer Instit 31(4):919 951.
Mickelsen O, Campbell E, Yang M, et al. (1964) Studies with cycad. Fed Proc 23(Nov-Dec):1363-1365.
Mugera GM, Whitehair CK, Mickelsen O (1964) Pathology of cycad toxicity in animals. Fed Proc 23:106.
Newberne PM (1976) Biologic effects of plant toxins and aflatoxins in rats. J Natl Cancer Instit 56(3):551-555.
Sanger VL, Yang MG, Mickelsen O (1969) Cycad toxicosis in chickens. J Natl Cancer Instit 43(2):391-395.
Spatz M (1964) Carcinogenic effect of cycad meal in guinea pigs. Fed Proc 23(Nov-Dec):1384-1385.
Spatz M, Laqueur GL (1967) Transplacental induction of tumors in Sprague-Dawley rats with crude cycad material. J Natl Cancer Instit 38(2):233-245.
Spencer PS, Nunn PB, Hugon J, et al. (1987) Guam amyotrophic lateral sclerosis-parkinsonism-dementia linked to a plant excitement neurotoxin. Science 237:517-522.
Whiting MG (1964) Food practices in ALS foci in Japan, the Marianas, and New Guinea. Fed Proc 23(Nov-Dec):1343-1345.
Yang MG, Mickelsen O (1968) Cycad husk from Guam: Its toxicity to rats. Econ Bot 22:149-154.
Yang MG, Mickelsen O, Campbell ME, et al. (1966) Cycad flour used by Guamanians: Effects produced in rats by long-term feeding. J Nutr 90(2):153-156.
Yang MG, Mickelsen O, Laqueur GL, et al. (1966) Irreversibility of damage caused by the ingestion of toxic cycad by rats. Fed Proc 25:661.
Yang MG, Mickelsen O, Sanger VL, et al. (1967) Feeding toxic cycads to rats of various ages. Fed Proc 26:322.
Yang MG, Sanger VL, Mickelsen O, et al. (1968) Carcinogenicity of long-term feeding of cycad husk to rats. Proc Soc Exp Biol Med 127(4):1171-1175.

CYCAS MEDIA R. Br. [Cycadaceae]

Common Names:

cycad; rickety bush; tree zamia; zamia; zamia palm

Citations:

Hooper PT, Best SM, Campbell A (1974) Axonal dystrophy in the spinal cords of cattle consuming the cycad palm, Cycas media. Aust Vet J 50(4):146-149.

CYCAS REVOLUTA Thunb. [Cycadaceae]

Common Names:

Australian nut palm; cycad; false sago palm; Japanese fern palm; leatherleaf palm; sago cycas; sago palm; sotetsu

Citations:

Albretsen JC, Safdar AK, Richardson JA (1998) Cycad palm toxicosis in dogs: 60 cases (1987-1997). J Am Vet Med Assoc 213(1):99-101.
Botha CJ, Naudé TW, Swan GE, et al. (1991) Suspected cycad (Cycas revoluta) intoxication in dogs. J S Afr Vet Assoc 62(4):189-190.
Chang SS, Chan YL, Wu ML, et al. (2004) Acute Cycas seed poisoning in Taiwan. J Toxicol Clin Toxicol 42(1):49-54.
Deng JF, Lin TJ, Kao WF, et al. (1997) The difficulty in handling poisonings associated with Chinese traditional medicine: A poison control center experience for 1991-1993. Vet Hum Toxicol 39(2):106-114.
Iwami O, Niki Y, Watanabe T, et al. (1993) Motor neuron disease on the Kii Peninsula of Japan: Cycad exposure. Neuroepidemiology 12(6):307-312.
Kobayashi A, Tadera K, Yagi F, et al. (1984) [Studies on poisoning of grazing cattle due to ingestion of Cycas revoluta Thunb.] Bull Fac Agric Kagoshima Univ 34:119-129.
Reagor JC, Ray AC, Dubuisson L, et al. (1986) Sago palm (Cycas revoluta) poisoning in the canine. Southwestern Vet 37(1):20.
Spatz M, Laqueur GL (1967) Transplacental induction of tumors in Sprague-Dawley rats with crude cycad material. J Natl Cancer Instit 38(2):233-245.
Whiting MG (1964) Food practices in ALS foci in Japan, the Marianas, and New Guinea. Fed Proc 23(Nov-Dec):1343-1345.
Yasuda N, Kono I, Shimizu T, et al. (1984) [Pathological studies on poisoning of grazing cattle due to ingestion of Cycas revoluta Thunb.] Bull Fac Agric Kagoshima Univ 34:131-136.
Yasuda N, Shimizu T (1998) Cycad poisoning in cattle in Japan: Studies on spontaneous and experimental cases. J Toxicol Sci 23(Suppl 2):126-128.

CYCLACHAENA XANTHIIFOLIA (Nutt.) Fresen. [Asteraceae]

Synonyms:

Iva xanthiifolia Nutt.

Common Names:

burweed marsh elder; careless weed; false ragweed; giant poverty weed; half breed weed; marsh elder

Citations:

Huber HL, Harsh GF (1932) A summer dermatitis caused by a common weed. J Allergy 3:578-582.

cyclamen –see– *Cyclamen persicum* Mill.

CYCLAMEN PERSICUM Mill. [Myrsinaceae]

Common Names:

Alpenveilchen; alpine violet; cyclamen; Persian violet; sowbread; wild cyclamen

Citations:

Hepper FN (2004) Two plant fish-poisons in Lebanon. Vet Hum Toxicol 46(6):338-339.

Jaspersen-Schib R, Theus L, Quirguis-Oeschger M, et al. (1996) Wichtige Pflanzenvergiftungen in der Schweiz 1966-1994. Schweiz Med Wochenschr 126(25):1085-1098.

cykuta –see– *Cicuta virosa* L.

CYMBOPOGON CITRATUS (DC.) Stapf [Poaceae]

Synonyms:

Andropogon citratus DC.

Common Names:

lemongrass

Citations:

Mendelsohn HV (1944) Dermatitis from lemon grass oil (Cymbopogon citratus or Andropogon citratus). Arch Derm Syphilol 50:34-35.

Mendelsohn HV (1946) Lemon grass oil. A primary irritant and sensitizing agent. Arch Derm Syphilol 53:94-98.

CYMBOPOGON NARDUS (L.) Rendle [Poaceae]

Synonyms:

Andropogon nardus L.

Citations:

Keil H (1947) Contact dermatitis due to oil of citronella. J Invest Dermatol 8:327-334.

Lane CG (1922) Dermatitis caused by oil of citronella. Arch Derm Syphilol 5:589-590.

Temple WA, Smith NA, Beasley M (1991) Management of oil of citronella poisoning. J Toxicol Clin Toxicol 29(2):257-262.

CYMOPTERUS IBAPENSIS M. E. Jones [Apiaceae]

Synonyms:

Cymopterus watsonii (J. M. Coult. & Rose) M. E. Jones

Common Names:

desert parsley; spring parsley; wild carrot

Citations:

Binns W, James LF, Brooksby W (1964) Cymopterus watsoni: A photosensitizing plant for sheep. Vet Med Small Anim Clin 59(4):375-379.

Egyed MN, Williams MC (1977) Photosensitizing effects of Cymopterus watsonii and Cymopterus longipes in chickens and turkey poults. Avian Dis 21(4):566-575.

Shlosberg A, Egyed MN (1978) Photosensitization in ducklings induced by seeds of Cymopterus watsonii and C. longipes. Avian Dis 22(4):576-582.

Van Kampen KR, Williams MC, Binns W (1969) Deformities in chickens photosensitized by feeding spring parsley (Cymopterus watsonii). Am J Vet Res 30(9):1663-1665.

Williams MC (1968) Effects of herbicides on the capacity of spring parsley to photosensitize chickens. Weed Sci 16(3):350-352.

Williams MC, Binns W (1968) Experimental photosensitization by spring parsley (Cymopterus watsonii) in chicks. Am J Vet Res 29(1):111-115.

CYMOPTERUS LONGIPES S. Watson [Apiaceae]

Common Names:

sprawling cymopterus

Citations:

Egyed MN, Williams MC (1977) Photosensitizing effects of Cymopterus watsonii and Cymopterus longipes in chickens and turkey poults. Avian Dis 21(4):566-575.

Shlosberg A, Egyed MN (1978) Photosensitization in ducklings induced by seeds of Cymopterus watsonii and C. longipes. Avian Dis 22(4):576-582.

Cymopterus watsonii (J. M. Coult. & Rose) M. E. Jones = Cymopterus ibapensis M. E. Jones

CYNANCHUM OBTUSIFOLIUM L. f. [Asclepiadaceae]

Common Names:

klimop

Citations:

Van der Walt SJ, Steyn DG (1939) Recent investigations into the toxicity of known and unknown poisonous plants in the Union of South Africa. IX. Onderstepoort J Vet Sci Anim Indus 12:335-366.

CYNARA CARDUNCULUS L. [Asteraceae]

Synonyms:

Cynara scolymus L.

Common Names:

artichauts; artichoke; Artischocke; cardoon; globe artichoke; wild artichoke

Citations:

Burry JN, Kuchel R, Reid JG, et al. (1973) Australian bush dermatitis: Compositae dermatitis in South Australia. Med J Aust 1(3):110-116.

Meding B (1983) Allergic contact dermatitis from artichoke, Cynara scolymus. Contact Dermatitis 9(4):314.

Sidi E, Morrill Dobkevitch S, Godechot R (1950) Les dermites provoquees par les artichauts. Presse Med 58:382-383.

Turner T (1980) Compositae dermatitis in South Australia: Contact dermatitis from Ixodia achillaeoides and Cynara cardunculus or the tribulations of a dry flower arranger. Contact Dermatitis 6(6):444.

Cynara scolymus L. = Cynara cardunculus L.

CYNODON AETHIOPICUS Clayton & J. R. Harlan [Poaceae]

Common Names:

Ethiopian dog's-tooth grass; giant stargrass; stargrass

Citations:

Rodel MG (1972) Effects of different grasses on the incidence of neonatal goitre and skeletal deformities in autumn born lambs. Rhodesia Agric J 69:59-60.

Rudert CP, O'Donovan WM (1974) Iodine supplementation of ewes grazing Cynodon aethiopicus Clayton et Harlan. Rhodesia J Agric Res 12(2):141-148.

CYNODON DACTYLON (L.) Pers. [Poaceae]

Synonyms:

Panicum dactylon L.

Common Names:

African stargrass; Bermudagrass; couchgrass; doggrass; doob; doubgrass; grama común; gramilla; gramilla colorada; green couch; Indian couch; Indian doab; pata-de-perdis; quickgrass

Citations:

Allen JG, Creeper JH, Forshaw D, et al. (2004) Plant-associated diseases, either new or new to the state, encountered over the last decade (1991-2001) in Western Australia. In: Acamovic T, Stewart CS, Pennycott TW (eds.) Poisonous plants and related toxins. CABI. Wallingford, UK. pp. 540-547.

Atwood MB (1953) A paralysis of cows of unknown etiology ("Downer cows," "Bermuda grass poisoning"). Georgia Vet 5(1):11-12.

Kidder RW, Beardsley DW, Erwin TC (1961) Photosensitization in cattle grazing frosted common Bermudagrass. Florida Agric Exp Sta Bull #630:21 pp.

Schang PJ, Aramendi MC (1944) Intoxicación de novillos por Cynodon dactylon (L.) Pers. "Bermuda grass," "Gramilla colorada." Gaceta Vet 6(31):264-276.

Singh R, Siddiqui MA, Baruah MC (1978) Plant dermatitis in Delhi. Indian J Med Res 68(Oct):650-655.

Strain GM, Seger CL, Flory W (1982) Toxic Bermuda grass tremor in the goat: An electroencephalographic study. Am J Vet Res 43(1):158-162.

Whitehair CK, Young HC Jr, Gibson ME, et al. (1951) A nervous disturbance in cattle caused by a toxic substance associated with mature Bermuda grass. Oklahoma Agric Exp Sta Misc Publ #MP22:57-59.

Williams GD, Hatkin J, Jones LP (1977) Acute respiratory distress syndrome occurring in Texas pastured cattle. Proc Am Assoc Vet Lab Diagn 20:327-338.

CYNODON NLEMFUENSIS Vanderyst [Poaceae]

Common Names:

African Bermudagrass; African stargrass; budgeegrass; giant stargrass; reuse kweekgrass; stargrass

Citations:

Rodel MG (1972) Effects of different grasses on the incidence of neonatal goitre and skeletal deformities in autumn born lambs. Rhodesia Agric J 69:59-60.

CYNODON PLECTOSTACHYUS (K. Schum.) Pilg. [Poaceae]

Common Names:

budgeegrass; estrella; giant couch; giant stargrass; Naivashagrass; stargrass; summer stargrass

Citations:

Herrington MD, Elliott RC, Brown JF (1971) Diagnosis and treatment of thyroid dysfunction occurring in sheep fed on Cynodon plectostachyus. Rhodesia J Agric Res 9(2):87-93.

CYNOGLOSSUM CAERULEUM Hochst. ex A. DC. [Boraginaceae]

Citations:

Schoental R, Coady A (1968) The hepatotoxicity of some Ethiopian and East African plants, including some used in traditional medicines. East Afr Med J 45(8):577-580.

Note:

This plant name could not be found in the consulted databases.

CYNOGLOSSUM GEOMETRICUM Baker & C. H. Wright [Boraginaceae]

Citations:

Schoental R, Coady A (1968) The hepatotoxicity of some Ethiopian and East African plants, including some used in traditional medicines. East Afr Med J 45(8):577-580.

CYNOGLOSSUM OFFICINALE L. [Boraginaceae]

Common Names:

beggar's-lice; Chinese forget-me-not; dog bur; dog's-tangere; glovewort; gypsy flower; hound's-tongue; Hundszunge; langue-de-chien; sheep lice; sticktight; woolmat

Citations:

Baker DC, Pfister JA, Molyneux RJ, et al. (1991) Cynoglossum officinale toxicity in calves. J Comp Pathol 104(4):403-410.

Baker DC, Smart RA, Ralphs M, et al. (1989) Hound's-tongue (Cynoglossum officinale) poisoning in a calf. J Am Vet Med Assoc 194(7):929-930.

Greatorex JC (1966) Some unusual cases of plant poisoning in animals. Vet Rec 78(21):725-727.

Knight AP, Kimberling CV, Stermitz FR, et al. (1984) Cynoglossum officinale (hound's-tongue) - A cause of pyrrolizidine alkaloid poisoning in horses. J Am Vet Med Assoc 185(6):647-650.

Stegelmeier BL, Gardner DR, James LF, et al. (1996) Pyrrole detection and the pathologic progression of Cynoglossum officinale (houndstongue) poisoning in horses. J Vet Diagn Invest 8(1):81-90.

Stegelmeier BL, Gardner DR, Molyneux RJ, et al. (1994) Cynoglossum officinale (houndstongue) poisoning in horses. Vet Pathol 31:621.

Stegelmeier BL, Gardner DR, Molyneux RJ, et al. (1994) The clinicopathologic changes of Cynoglossum officinale (houndstongue) intoxication in horses. In: Colegate SM,

Dorling PR (eds.) Plant-associated toxins. Agricultural, phytochemical and ecological aspects. CABI. New York. pp. 297-302.
Stein GH (1963) Zur Giftwirkung von Cynoglossum officinale L. auf kleine Nagetiere. Naturwissenschaften 50(16):554.
Zentek J, Aboling S, Kamphues J, et al. (1999) Fallbericht: Tierernährung für Tierärzte - Aktuelle Fälle: Hundszunge (Cynoglossum officinale) im Weide aufwuchs - ein Risiko für die Gesundheit von Pferden. Dtsch Tierarztl Wochenschr 106(11):475-477.

CYPERUS ROTUNDUS L. [Cyperaceae]

Citations:

Aston BC (1909) Researches on poisonous and other plants. N Z Dep Agric Annu Rep 17:178-184.

cypress –see– *Cupressus macrocarpa* Hartw. ex Gordon; *Cupressus sempervirens* L.

cypress spurge –see– *Euphorbia cyparissias* L.

CYPRIPEDIUM ACAULE Aiton [Orchidaceae]

Common Names:

pink lady's-slipper

Citations:

Jennings OE (1907) A note on the poisonous qualities of the yellow ladies' slipper. Plant World 10:189-191.

Cypripedium hirsutum Mill. = Cypripedium reginae Walter

CYPRIPEDIUM REGINAE Walter [Orchidaceae]

Synonyms:

Cypripedium hirsutum Mill.

Common Names:

female nervine; lady's-slipper; large lady's-slipper; nerveroot; showy lady's-slipper; whip-poor-will-shoes; yellow lady's-slipper

Citations:

Burnham SH (1915) Poison oak and soap plant. Am Botanist 21:46-49.

Jennings OE (1907) A note on the poisonous qualities of the yellow ladies' slipper. Plant World 10:189-191.

CYPRIPEDIUM SPECTABILE Salisb. [Orchidaceae]

Common Names:

showy mocassin flower

Citations:

Bacon AE (1902) Some cases of poisoning by Cypripedium spectabile in Vermont. Rhodora 4(May):94-97.

Deane W (1893) Is Cypripedium spectabile poisonous to the touch? Bot Gaz 1893(Apr):143-144.

Jesup HG (1893) Is Cypripedium spectabile poisonous to the touch? Bot Gaz 1893(Apr):142-143.

Cystium diphysum (A. Gray) Rydb. = Astragalus lentiginosus Douglas ex Hook. var diphysus (A. Gray) M. E. Jones

cytise –see– *Laburnum anagyroides* Medik.

Cytisus laburnum L. = Laburnum anagyroides Medik.

Cytisus racemosus Marnock = Genista stenopetala Webb & Berthel.

CYTISUS SCOPARIUS (L.) Link [Fabaceae]

Synonyms:

Sarothamnus scoparius (L.) Wimm. ex W. D. J. Koch

Common Names:

Besenginster; Besentrauch; besom; broom; broomtops; English broom; Galster; Ginster; Irish broom; laburnum; Scotch broom; spartium

Citations:

Perharic L, Shaw D, Colbridge M, et al. (1994) Toxicological problems resulting from exposure to traditional remedies and food supplements. Drug Saf 11(4):284-294.

Czar Pater –see– *Hyacinthus orientalis* L.

Czar Peter –see– *Hyacinthus orientalis* L.

D

dabdabey –see– *Lannea coromandelica* (Houtt.) Merr.

dactyle –see– *Dactylis glomerata* L.

DACTYLIS GLOMERATA L. [Poaceae]

Common Names:

dactyle; Knaulgras; orchardgrass

Citations:

Ćirković TD, Bukilica MN, Gavrović MD, et al. (1999) Physicochemical and immunologic characterization of low-molecular-weight allergoids of Dactylis glomerata pollen proteins. Allergy 54(2):128-134.

DACTYLOCTENIUM RADULANS (R. Br.) P. Beauv. [Poaceae]

Common Names:

buttongrass

Citations:

McKenzie RA, Rayner AC, Thompson GK, et al. (2004) Nitrate-nitrite toxicity in cattle and sheep grazing Dactyloctenium radulans (button grass) in stockyards. Aust Vet J 82(10):630-634.

daffidowndilly –see– *Narcissus pseudonarcissus* L.

daffodil –see– *Narcissus poeticus* L.; *Narcissus pseudonarcissus* L.

daftberry –see– *Atropa belladonna* L.

dagga –see– *Cannabis sativa* L.

dahlbergia –see– *Dalbergia nigra* (Vell.) Allemão ex Benth.; *Dalbergia retusa* Hemsl.

dahlia –see– *Dahlia pinnata* Cav.

DAHLIA PINNATA Cav. [Asteraceae]

Synonyms:

Dahlia variabilis (Willd.) Desf.

Common Names:

dahlia

Citations:

Sharma SC, Kaur S (1989) Airborne contact dermatitis from Compositae plants in northern India. Contact Dermatitis 21(1):1-5.

Dahlia variabilis (Willd.) Desf. = Dahlia pinnata Cav.

dairy pink –see– *Vaccaria hispanica* (Mill.) Rauschert

daisy –see– *Leucanthemum vulgare* Lam.

daisy primrose –see– *Primula praenitens* Ker Gawl.

DALBERGIA LATIFOLIA Roxb. [Fabaceae]

Common Names:

East Indian rosewood; Indian rosewood; Indisches Rosenholz; Ostindisches Palisander; Palisander; rosewood

Citations:

Athavale PN, Shum KW, Gasson P, et al. (2003) Occupational hand dermatitis in a wood turner due to rosewood (Dalbergia latifolia). Contact Dermatitis 48(6):345-346.

Cronin E, Calnan CD (1975) Rosewood knife handle. Contact Dermatitis 1(2):121.

Estlander T, Jolanki R, Alanko K, et al. (2001) Occupational allergic contact dermatitis caused by wood dusts. Contact Dermatitis 44(4):213-217.

Fancalanci S, Giorgini S, Gola M, et al. (1984) Occupational dermatitis in a butcher. Contact Dermatitis 11(5):320-321.

Findley LJ (1972) An unusual case of rosewood dermatitis of the genus Dalbergia (East Indian rosewood). Br J Ind Med 29(3):343-344.

Gallo R, Guarrera M, Hausen BM (1996) Airborne contact dermatitis from East Indian rosewood (Dalbergia latifolia Roxb.). Contact Dermatitis 35(1):60-61.

Hausen BM, Dohn W (1986) Epoxidharz - und Palisanderholzallergie durch Freizeitbeschäftigung. Aktuelle Derm 12(2):47-48.

Hausen BM, Mau HH (1979) Kontaktallergie durch einer Geigen-Kinnkalter aus Palisander. Derm Beruf Umwelt 27(1):18-20.

Haustein UF (1982) Violin chin rest eczema due to East-Indian rosewood (Dalbergia latifolia Roxb.). Contact Dermatitis 8(1):77-78.

DALBERGIA NIGRA (Vell.) Allemão ex Benth. [Fabaceae]

Common Names:

Bahia rosewood; Brazilian jacaranda; Brazilian rosebush; Brazilian rosewood; caviúna; dahlbergia; jacaranda; palissandre bresic; rio palisander; rio rosewood; tree rosewood

Citations:

Estlander T, Jolanki R, Alanko K, et al. (2001) Occupational allergic contact dermatitis caused by wood dusts. Contact Dermatitis 44(4):213-217.

Fisher AA (1986) Erythema multiforme-like eruptions due to exotic woods and ordinary plants: Part I. Cutis 37(2):101-104.

Fisher AA, Bikowski J Jr (1981) Allergic contact dermatitis due to a wooden cross made of Dalbergia nigra. Contact Dermatitis 7(1):45-46.

Hausen BM (1985) Chin rest allergy in a violinist. Contact Dermatitis 12(3):178-180.

Hausen BM, Dohn W (1986) Epoxidharz - und Palisanderholzallergie durch Freizeitbeschäftigung. Aktuelle Derm 12(2):47-48.

Hausen BM, Simatupang MH, Kingreen JC (1972) Untersuchungen zur Überempfindlichkeit gegen Sucupira - und Palisanderholz. Derm Beruf Umwelt 20(1):1-7.

Holst R, Kirby J, Magnusson B (1976) Sensitization to tropical woods giving erythema multiforme-like eruptions. Contact Dermatitis 2(11):295-296.

Schulz KH, Dietrichs HH (1962) Chinone als sensibilisierende Bestandteile von Rio-Palisander- (Dalbergia nigra) und Cocobolo - (Dalbergia retusa) Holz. Allerg Asthma (Leipz) 8(3):125-131.

Tottie M (1938) Jacaranda als Ursache von Dermatitiden. Acta Derm Venereol 19:235-245.

DALBERGIA RETUSA Hemsl. [Fabaceae]

Common Names:

bastard mahogany; cocabolla; cocobolo; dahlbergia; East Indian rosewood

Citations:

Abramowitz EW, Swarts WB (1938) Dermatitis due to cocobolo wood. Arch Derm Syphilol 37:441-443.

Eaton KK (1973) Respiratory allergy to exotic wood dust. Clin Allergy 3(3):307-310.

Freise FW (1936) Vergiftungen durch brasilianische Werkhölzer. I. Sammlung Vergiftungsfallen 7:1-8.

Hausen BM, Münster G (1983) Cocobolo-Holz, ein vergessenes Ekzematogen? Neuere Erkenntnisse über das Hauptallergen des Cocobolo-Holzes (Dalbergia sp.). Dermatosen 31(4):110-117.

Howell JB, Blair DS (1950) Eczema of the hands from wooden-handled objects. Arch Derm Syphilol 62(3):400-404.

Leider M, Schwartzfeld HK (1950) Allergic eczematous contact type dermatitis caused by cocobolo wood (Dalbergia). Arch Derm Syphilol 62(1):125-130.

Levin OL, Behrman HT (1939) Bracelet dermatitis. N Y State J Med 39(Oct 1):1877-1879.

Levin SJ (1941) Cocobolo wood dermatitis: Report of a case due to wooden handles on kitchen knives. J Allergy 12:498-501.

MacKee GM (1913) Dermatitis venenata from coco-bolo wood. J Cutan Dis 31:582-583.

Schulz KH, Dietrichs HH (1962) Chinone als sensibilisierende Bestandteile von Rio-Palisander- (Dalbergia nigra) und Cocobolo - (Dalbergia retusa) Holz. Allerg Asthma (Leipz) 8(3):125-131.

DALECHAMPIA SCANDENS L. [Euphorbiaceae]

Common Names:

bejuco-de-pan; mashasha

Citations:

Reed B, Wilson RC (2001) Voodoo powder as an etiology of leg ulcers. J Am Podiatr Med Assoc 91(6):324-325.

Dallisgrass –see– *Paspalum dilatatum* Poir.

Dalmatian chrysanthemum –see– *Tanacetum cinerariifolium* (Trevir.) Sch. Bip.

Dalmatian sage –see– *Salvia officinalis* L.

dama-de-dia –see– *Cestrum diurnum* L.

dama-de-noche –see– *Cestrum nocturnum* L.

dand –see– *Croton tiglium* L.

dandelion –see– *Hypochaeris radicata* L.; *Taraxacum officinale* F. H. Wigg. aggr.

Dan's-cabbage –see– *Senecio isatideus* DC.; *Senecio retrorsus* DC.

danshen –see– *Salvia miltiorrhiza* Bunge

daphne –see– *Daphne mezereum* L.

DAPHNE MEZEREUM L. [Thymelaeaceae]

Common Names:

Beißbeere; Bergpfeffer; bois gentil; bois joli; daphne; dwarf bay; February daphne; flax olive; Kellerhals; lady laurel; laureola hembra; leño gentil; mezereon; mezereum; mysterious plant; Seidelbast; spurge flax; spurge laurel; spurge olive; Wielandbeere; wild pepper; Wolfsbast; Ziegelbeeren

Citations:

Eagar FS (1887) Case of poisoning by berries of Daphne mezereon. Br Med J 2(Jul 30):239.

Lamminpaa A, Kinos M (1996) Plant poisonings in children. Hum Exp Toxicol 15(3):245-249.

DAPHNIPHYLLUM HUMILE Maxim. ex Franch. & Sav. [Daphniphyllaceae]

Synonyms:

Daphniphyllum macropodum Miq. var. humile (Maxim. ex Franch. & Sav.) K. Rosenthal

Citations:

Sonoda M, Tasaka M, Takahashi K, et al. (1978) [Daphniphyllum plant poisoning in grazing cattle in southern Hokkaido.] J Jpn Vet Med Assoc 31(3):140-144.

Daphniphyllum macropodum Miq. var. humile (Maxim. ex Franch. & Sav.) K. Rosenthal = Daphniphyllum humile Maxim. ex Franch. & Sav.

Darling pea –see– *Swainsona galegifolia* (Andrews) R. Br.; *Swainsona luteola* F. Muell.

darnel –see– *Lolium temulentum* L.

dart poison –see– *Antiaris toxicaria* Lesch.

dasheen –see– *Colocasia esculenta* (L.) Schott

dashin yawo –see– *Acanthospermum hispidum* DC.

date –see– *Phoenix dactylifera* L.

date palm –see– *Phoenix dactylifera* L.

DATISCA GLOMERATA (C. Presl) Baill. [Datiscaceae]

Common Names:

Durango root

Citations:

Galey FD, Johnson BJ, Bruce R, et al. (1990) A case of Datisca glomerata poisoning in beef cattle. Vet Hum Toxicol 32(6):575-576.

Wagnon KA, Hart GH (1945) Durango root (Datisca glomerata) poisoning of range stock. J Am Vet Med Assoc 107(Jul):3-5.

datun –see– *Croton tiglium* L.

datura –see– *Datura inoxia* Mill.; *Datura metel* L.; *Datura stramonium* L.

Datura alba Nees = Datura metel L.

Datura arborea L. = Brugmansia arborea (L.) Lagerh.

Datura candida (Pers.) Saff. = Brugmansia ×candida Pers.

Datura cornigera Hook. = Brugmansia arborea (L.) Lagerh.

Datura fastuosa L. = Datura metel L.

DATURA FEROX L. [Solanaceae]

Common Names:

cardo cuco; chamico; estramonio chino; false castor; fierce thorn apple; long-spine thorn apple; long-spur thorn apple; thorn apple

Citations:

Steenkamp PA, Harding NM, van Heerden FR, et al. (2004) Fatal Datura poisoning: Identification of atropine and scopolamine by high performance liquid chromatography/photodiode array/mass spectrometry. Forensic Sci Int 145(1):31-39.

DATURA INOXIA Mill. [Solanaceae]

Common Names:

apple-of-Peru; datura; dhatura; downy thorn apple; green thorn apple; hairy thorn apple; Indian apple; jimson weed; moonflower; nongue; purple angel's-trumpet; recurved thorn apple; sacred datura

Citations:

De Frates LJ, Hoehns JD, Sakornbut EL, et al. (2005) Antimuscarinic intoxication resulting from the ingestion of moonflower seeds. Ann Pharmacother 39(1):173-176.

Goetz R, Siegel E, Scaglione J, et al. (2003) Suspected moonflower intoxication - Ohio, 2002. MMWR Morb Mortal Wkly Rep 52(33):788-791.

Goetz RJ, Siegel EG, Scaglione JM (2003) Moonflower abuse and intoxication in Northern Ohio. J Toxicol Clin Toxicol 41(5):744.

Hanna JP, Schmidley JW, Braselton WE Jr (1992) Datura delirium. Clin Neuropharmacol 15(2):109-113.

Marciniak J, Sikorski M (1972) Zatrucie alkaloidami bielunia indiańskiego i bielunia dziędzierzawy po spożyciu miodu pszczelego. Pol Tyg Lek (Wars) 27(26):1002-1003.

Meng K, Graetz DK (2004) Moonflower-induced anisocoria. Ann Emerg Med 44(6):665-666.

Ramadan M (2005) Moonflower intoxication in Kansas. Am J Addict 14(4):399-400.

Raman SV, Jacob J (2005) Mydriasis due to Datura inoxia. Emerg Med J 22(4):310-311.

Ramirez M, Rivera E, Ereu C (1999) Fifteen cases of atropine poisoning after honey ingestion. Vet Hum Toxicol 41(1):19-20.

DATURA METEL L. [Solanaceae]

Synonyms:

Datura alba Nees; ***Datura fastuosa*** L.

Common Names:

black datura; datura; devil's-trumpet; downy thorn apple; hairy thorn apple; Hindu datura; jimson weed; ketjubung; metel; mondzo; thorn apple; yangjinhua; zakami

Citations:

But PP (1994) Herbal poisoning caused by adulterants or erroneous substitutes. J Trop Med Hyg 97(6):371-374.

Johnston TF (1972) Datura fastuosa: Its use in Tsongo girls' initiation. Econ Bot 26(4):340-351.

Quek KC, Cheah JS (1974) Poisoning due to ingestion of the seeds of kechubong (Datura fastuosa) for its ganja-like effect in Singapore. J Trop Med Hyg 77(5):111-112.

Schvartsman S, Marcondes E (1965) Intoxicações acidentais agudas na infancia. Revisão de 208 casos. Rev Paul Med 66(1):24-39.

Testa P, Fontanelli G (1988) Intossicazione acuta da alcaloidi atropinosimili della Datura metel. Clin Ter 125(3):203-211.

Zuidema PJ (1985) Vergiftiging door Datura fastuosa (ketjubung). Ned Tijdschr Geneeskd 129(29):1386-1388.

Datura meteloides auct. = Datura wrightii Regel

Datura rosei Saff. = Brugmansia sanguinea (Ruiz & Pav.) D. Don

DATURA STRAMONIUM L. [Solanaceae]

Synonyms:

Datura tatula L.

Common Names:

angel's-trumpet; apple-of-Peru; apple-of-Tolguacha; Asthmador; blou oliebloom; blou stinkblaar; chamico; datura; devil's-apple; devil's-trumpet; devil's-weed; dewtry; dhatura; Doornappel; Dornapfel; estramonio Americano; fireweed; flor-de-la-trompeta; green dragon; herbe-à-sorcier; herbe-des-démoniaques; herbe-du-diable; higuera-del-infierno; higuera loca; Jamestown lily; Jamestown weed; jimson weed; loco seed; mad apple; mad seeds; malpitte; moonflower; olieboon; Peru apple; pigaeble; piggeple; pomme-du-poison; pomme épineuse; purple jimson weed; purple stinkweed; purple stramonium; purple thorn apple; Stechapfel; stink weed; stinkblaar; stinkwort; stramoine; stramonium; teasel bur; thorn apple; tolguacha; Tollkraut; toloache; white-man's-plant; white stramonium

Citations:

Al-Shaikh AM, Sablay ZM (2005) Hallucinogenic plant poisoning in children. Saudi Med J 26(1):118-121.

Alström P, Höjer J (1999) Rituell bärsärkaförgiftning med hallucinogen växt - Behandingsråd vid allvarlig förgiftning med antikolinergika. Lakartidningen 96(50):5612-5614.

Amlo H, Haugeng KL, Wickstrøm E, et al. (1997) Forgiftning med piggeple - Fem tilfeller behandlet med fysostigmin. Tidsskr Nor Laegeforen 117(18):2610-2612.

Anonymous (1925) One dead; three ill from eating jimson weed. Boston Evening Transcript. Sep 12

Arditti J, Spadari M, de Haro L, et al. (2002) Recreational use of jimson weed (Datura stramonium): Ten years experience of the "drug dependence evaluation and information centre" (DDEIC) of Marseilles. J Toxicol Clin Toxicol 40(3):374.

Arena JM (1963) Atropine poisoning: A report of two cases from jimson weed. Clin Pediatr (Phila) 2(4):182-184.

Ballantyne A, Lippiett P, Park J (1976) Herbal cigarettes for kicks. Br Med J 2(6051):1539-1540.

Barnett AH, Jones FW, Williams ER (1977) Acute poisoning with Potter's Asthma Remedy. Br Med J 2(6103):1635.

Barney GH, Wilson BJ (1963) A rare toxicity syndrome in ponies. Vet Med 58:419-421.

Baselga JM, Pigrau C, Martinez Vazquez JM (1985) Datura stramonium: ¿un antiguo alucinógeno en auge? Med Clin (Barc) 84(17):715.

Behrens H, Horn M (1962) Die Verträglichkeit von Stechapfelsamen beim Schwein. Prakt Tierarzt 1962(2):43-44.

Betz P, Janzen J, Roider G, et al. (1991) Psychopathologische Befunde nach oraler Aufnahme von Inhaltstoffen heimischer Nachtschattengewächse. Arch Kriminol 188(5-6):175-182.

Boumba VV, Mitselou A, Vougiouklakis T (2004) Fatal poisoning from ingestion of Datura stramonium seeds. Vet Hum Toxicol 46(2):81-82.

Brunel J (1947) Un cas d'empoisonement grave par des graines de "Datura stramonium." Contrib Inst Bot Univ Montreal 62:31-37.

Bubien MZ, Magat A, Delatour P (1970) Fréquence des intoxications des animaux en Pologne. Bull Soc Vet Med Comp Lyon 72(2):167-173.

Bull-Berg J (1976) Misbruk av piggeplefrø. Tidsskr Nor Laegeforen 96(19-21):1103.

Burkhart KK, Magalski AE, Donovan JW (1999) A retrospective review of the use of activated charcoal and physostigmine in the treatment of jimson weed poisoning. J Toxicol Clin Toxicol 37:389.

Calbo Mayo JM, Barba Romero MA, Broseta Viana L, et al. (2004) Intoxicación familiar accidental por ingesta de Datura stramonium. An Med Interna 21(8):415.

Callahan R, Piccola F, Gensheimer K, et al. (1981) Plant poisonings - New Jersey. MMWR Morb Mortal Wkly Rep 30(6):65-67.

Castañón López L, Martínez Badás JP, Lapeña López de Armentia S, et al. (2000) Intoxicación por Datura stramonium. An Esp Pediatr 53(1):53-55.

Clark JD (2005) The roadside high: Jimson weed toxicity. Air Med J 24(6):234-237.

Clark WL, Jernigan RH, Erwin JW, et al. (1964) Jimson weed poisoning - Tennessee. MMWR Morb Mortal Wkly Rep 13:126-127.

Coremans VP, Lambrecht G, Schepens P, et al. (1994) Anticholinergic intoxication with commercially available thorn apple tea. J Toxicol Clin Toxicol 32(5):589-592.

Crawford L, Friedman, M (1990) The effects of low levels of dietary toxic weed seeds (jimson weed, Datura stramonium and sicklepod, Cassia obtusifolia) on the relative size of rat liver and levels of cytochrome P-450. Toxicol Lett 54(2-3):175-181.

Day EJ, Dilworth BC (1984) Toxicity of jimson weed seed and cocoa shell meal to broilers. Poult Sci 63(3):466-468.

De Witt MS, Swain R, Gibson LB Jr, et al. (1997) The dangers of jimson weed and its abuse by teenagers in the Kanawha Valley of West Virginia. W V Med J 93(4):182-185.

De Young G, Cross EG (1969) Stramonium psychodelia. Can Anaesth Soc J 16(5):429-432.

Dean ES (1963) Self-induced Stramonium intoxication. JAMA 185(11):882.

Di Giacomo JN (1968) Toxic effect of stramonium simulating LSD trip. JAMA 204(3):265-267.

Djordjevic D, Vucinic S, Segrt Z (2004) The experience of national poison control centre in recognition and management of plant poisoning. J Toxicol Clin Toxicol 42(4):509-510.

Dugan GM, Gumbmann MR, Friedman M (1989) Toxicological evaluation of jimson weed (Datura stramonium) seed. Food Chem Toxicol 27(8):501-510.

El Dirdiri NI, Wasfi IA, Adam SE, et al. (1981) Toxicity of Datura stramonium to sheep and goats. Vet Hum Toxicol 23(4):241-246.

Eldor A (1971) [A case of Datura stramonium poisoning.] Harefuah 80(7):386-388.

Engelmeier MP, Finke J (1970) Stechapfelrausch. Pharmako Psychiatrie 3:248-253.

Ertekin V, Selimoğlu MA, Altinkaynak S (2005) A combination of unusual presentations of Datura stramonium intoxication in a child: Rhabdomyolysis and fulminant hepatitis. J Emerg Med 28(2):227-228.

Fama PG (1977) Datura poisoning. N Z Med J 85(Feb 9):108.

Fama PG (1979) Datura poisoning. N Z Med J 90(647):399.

Fanguaf R, Vogt H (1961) Mitteilung über einen Verträglichkeitsversuch mit Stechapfelsamen (Daturasamen) bei legenden Hennen und bei Küken. Arch Geflugelkunde 25:167-171.

Feenaghty DA (1982) Atropine poisoning: Jimsonweed. J Emerg Nurs 8(3):139-141.

Fensbo C, Harbeck C (1979) Datura stramonium anwendt som urtete. Ugeskr Laeger 141(17):1150-1151.

Flunker LK, Damron BL, Sundlof SF (1987) Jimsonweed seed contamination of broiler chick and White Leghorn hen diets. Nutr Rep Int 36(3):551-556.

Forno FJ Jr, Terry RA (1998) Accidental ingestion of jimsonweed by an adolescent. J Am Osteopath Assoc 98(9):502-504.

Friedman M, Dugan G, Gumbmann MR (1990) Composition and toxicology of jimson weed seeds (Datura stramonium). Vet Hum Toxicol 32(Suppl):111.

Garvin JA, Ruh HO (1923) Acute poisoning due to eating the seeds of the jimson weed (Datura stramonium). Arch Pediatr 40:827-831.

Gdyra D, Zwęglińska-Pióro A (1970) Zatrucia Datura stramonium. Pol Tyg Lek (Wars) 25(19):733-735.

Geoffroy H (1964) De certaines intoxications aiguës. Maroc Med 43:603-618.

Ghorbal H, Hamouda C, Bousnina M, et al. (2003) Use of plants to induce chemical submission in Tunisia. Vet Hum Toxicol 45(2):91-93.

Gibson RK (1961) Jimson weed poisoning in children. J Indiana State Med Assoc 54(Jul):1018-1020.
Goldey JA, Dick DA, Porter WL (1966) Cornpicker's pupil. A clinical note regarding mydriasis from jimson weed dust (Stramonium). Ohio State Med J 62(Sep):921.
Goldsmith SR, Frank I, Ungerleider JT (1968) Poisoning from ingestion of a stramonium-belladonna mixture (Flower power gone sour). JAMA 204(2):169-170.
Grandjean EM, de Moerloose P, Zwahlen A (1980) Syndrome atropinique aigu par usage abusif de cigarettes anti-asthmatiques (Datura stramonium). Schweiz Med Wochenschr 110(33):1186-1190.
Guharoy SR, Barajas M (1991) Atropine intoxication from the ingestion and smoking of jimson weed (Datura stramonium). Vet Hum Toxicol 33(6):588-589.
Gururaj AK, Khare CB (1987) Dhatura poisoning: A case report. Med J Malaysia 42(1):68-69.
Hansen AA (1924) Jimson poisoning. Better Crops 2(6):28-29.
Harrison EA, Morgan DH (1976) Abuse of herbal cigarettes containing stramonium. Br Med J 2(6045):1195.
Henson RW, Miller LP, Herron JT (1978) Abuse of thorn apple. Med J Aust 1(5):280.
Hughes JD, Clark JA Jr (1939) Stramonium poisoning, A report of two cases. JAMA 112(24):2500-2502.
Irvine RA, Tang K (1957) Datura poisoning. A case report. West Indian Med J 6(2):126-128.
Jacobziner H, Raybin HW (1961) Fatal salicylate intoxication and stramonium poisoning. N Y State J Med 61:301-303.
Janssens G, De Wilde R (1989) De toxiciteit van Doornappelzaad (Datura stramonium en/of ferox) in slachtvarkensrantsoenen. Vlaams Diergeneesk Tijdschr 58(3):84-86.
Jennings RE (1935) Stramonium poisoning: A review of the literature and report of two cases. J Pediatr 6:657-664.
Jiménez Mejías ME, Fernández A, Montaño Díaz M, et al. (1991) Sindrome anticolinérgico por envenenamiento por Datura stramonium. Med Clin (Barc) 97(6):237.
Johnson RT (1977) Jimson weed toxicity. Clin Med 84(3):14-15.
Jones B (1967) Eating jimsonweed under a tree. Two youngsters survive poison plant. Washington Daily News. Oct 5.
Joubert PH (1990) Poisoning admissions of black South Africans. J Toxicol Clin Toxicol 28(1):85-94.
Kehar ND, Rau KG (1944) Poisoning of livestock by Datura stramonium. Indian J Vet Sci 14:112-114.
King ED Jr (1923) Jimson weed poisoning. J Am Vet Med Assoc 64(Oct):98-99.
Klein-Schwartz W, Oderda GM (1984) Jimsonweed intoxication in adolescents and young adults. Am J Dis Child 138(8):737-739.
Koff M (1966) Poisoning from ingestion of asthma "powders." JAMA 198(9):1034.
Kurzbaum A, Simsolo C, Kvasha L, et al. (2001) Toxic delirium due to Datura stramonium. Isr Med Assoc J 3(7):538-539.
Lanich OK Jr, Ambrus JL (1954) Stramonium poisoning. Pa Med J 57(4):345-348.
Leipold HW, Oehme FW, Cook JE (1973) Congenital arthrogryposis associated with ingestion of jimsonweed by pregnant sows. J Am Vet Med Assoc 162(12):1059-1060.
Levy R (1977) Arrhythmias following physostigmine administration in jimson weed poisoning. J Am Coll Emerg Physicians 6(3):107-108.
Levy R (1977) Jimson seed poisoning: A new hallucinogen on the horizon. J Am Coll Emerg Physicians 6(2):58-61.
Levy R, Cross CJ, Smith MS (1976) Jimson weed poisoning. Ann Intern Med 84(2):223-224.
Litovitz TL, Klein-Schwartz W, Rodgers GC Jr, et al. (2002) 2001 Annual report of the American Association of Poison Control Centers Toxic Exposure Surveillance System. Am J Emerg Med 20(5):391-452.
Locicero R (1955) Intoxicacion por chamico. Sem Med 62(20):983, 986.
Lopez IA (1978) Intoxication by Datura stramonium. Ohio State Med J 74(5):300-301.
Malorny G (1952-1953) Stechapfelsamenvergiftungen nach Genuß von Buchweizenmehlzubereitungen. Sammlung Vergiftungsfallen 14:181-184.
Manriquez O, Varas J, Rios JC, et al. (2002) Analysis of 156 cases of plant intoxication received in the toxicologic information center at Catholic University of Chile. Vet Hum Toxicol 44(1):31-32.
Marciniak J, Sikorski M (1972) Zatrucie alkaloidami bielunia indiańskiego i bielunia dziędzierzawy po spożyciu miodu pszczelego. Pol Tyg Lek (Wars) 27(26):1002-1003.
Martinez Ros C, Adanez Martinez G, Hernandez Ruiperez T, et al. (2001) Collective intoxication caused by Datura stramonium. J Toxicol Clin Toxicol 39(3):299-300.
McNally WD (1915) A case of stramonium poisoning. JAMA 65(Nov 6):1640.
Mebs D, Schmidt K, Raudonat HW, et al. (1986) Stechapfelvergiftung. Dtsch Med Wochenschr 111(19):762.
Meiring PV (1966) Poisoning by Datura stramonium. S Afr Med J 40(14):311-312.
Michalodimitrakis M, Koutselinis A (1984) Discussion of "Datura stramonium: A fatal poisoning." J Forensic Sci 29(4):961-962.
Mikolich JR, Paulson GW, Cross CJ (1975) Acute anticholinergic syndrome due to jimson seed ingestion. Clinical and laboratory observation in six cases. Ann Intern Med 83(3):321-325.
Mitchell JE, Mitchell FN (1955) Jimson weed (Datura stramonium) poisoning in childhood. J Pediatr 47(2):227-230.
Mrvos R, Krenzelok EP, Jacobsen TD (2001) Toxidromes associated with the most common plant ingestions. Vet Hum Toxicol 43(6):366-369.
Nelson PD, Mercer HD, Essig HW, et al. (1982) Jimson weed seed toxicity in cattle. Vet Hum Toxicol 24(5):321-325.
Nogue S, Pujol L, Sanz P, et al. (1995) Datura stramonium poisoning. Identification of tropane alkaloids in urine by gas chromatography-mass spectrometry. J Int Med Res 2(2):132-137.
O'Grady TC, Brown J, Jacamo J (1983) Outbreak of jimson weed abuse among Marine Corps personnel at Camp Pendleton. Mil Med 148(9):732-734.
Onen CL, Othol D, Mbwana SK, et al. (2002) Datura stramonium mass poisoning in Botswana. S Afr Med J 92(3):213-214.
Orr R (1975) Reversal of Datura stramonium delirium with physostigmine: Report of three cases. Anesth Analg 54(1):158.
Perrotta DM, Nickey LN, Raid M, et al. (1995) Jimson weed poisoning - Texas, New York, and California, 1994. JAMA 273(7):532-533.

Perrotta DM, Nickey LN, Raid M, et al. (1995) Jimson weed poisoning -Texas, New York, and California, 1994 MMWR Morb Mortal Wkly Rep 44(3):41-44.
Pinilla Llorente B, Portillo A, Muino-Miguez A, et al. (1992) Intoxicación por Datura stramonium. An Med Interna 9(4):208.
Powers D (1975) Jimson weed intoxication in adolescents. Va Med Monthly 102(12):1051-1053.
Pulewka P (1949) Die Aufklärung einer ungewöhnlichen durch Datura stramonium in Brotmehl hervorgerufenen Massenvergiftung. Klin Wochenschr 27:672-674.
Rannem T (1986) Forgiftning med pigæble. Ugeskr Laeger 148(43):2763.
Robert G, Menichini U (1978) Intossicazione voluttuaria da stramonio. Minerva Med 69(12):763-767.
Robertson MM, Morley JE (1974) Malpitte madness. A report of ten cases. S Afr Med J 48(62):2604-2606.
Roblot F, Montaz L, Delcoustal M, et al. (1995) Intoxication par Datura stramonium: Le diagnostic est clinique, le traitement est symptomatique. Rev Med Interne 16(3):187-190.
Rodgers GC Jr, Von Kanel RL (1993) Conservative treatment of jimsonweed ingestion. Vet Hum Toxicol 35(1):32-33.
Rosen CS, Lechner M (1962) Jimson-weed intoxication. N Engl J Med 267(9):448-450.
Rouquet JP, Bezaury JP, Moron P (1982) A propos de deux épisodes toxicomaniaques par le Datura. Ann Med Psychol (Paris) 140(5):547-550.
Rush B (1973) An account of the effects of the stramonium, or thorn-apple. Clin Pediatr (Phila) 12(1):50-53.
Rwiza HT (1991) Jimson weed food poisoning. An epidemic at Usangi rural government hospital. Trop Geogr Med 43(1-2):85-90.
Salen P, Shih R, Sierzenski P, et al. (2003) Effect of physostigmine and gastric lavage in a Datura stramonium-induced anticholinergic poisoning epidemic. Am J Emerg Med 21(4):316-317.
Sander JF, Berge RD (1955) Jimson weed poisoning, A report of two cases. J Mich State Med Soc 54:691-692.
Satpathy R, Das BB (1979) Accidental poisoning in childhood. J Indian Med Assoc 73(11):190-192.
Savitt DL, Roberts JR, Siegel EG (1986) Anisocoria from jimsonweed. JAMA 255(11):1439-1440.
Schreiber W (1979) Jimson seed intoxication: Recognition and therapy. Mil Med 144(5):329-332.
Schulman ML, Bolton LA (1998) Datura seed intoxication in two horses. J S Afr Vet Assoc 69:27-29.
Schumacher M (1965) A case of atropine alkaloid poisoning. Med J Aust 49(Apr 10):547-548.
Schvartsman S, Marcondes E (1965) Intoxicações acidentais agudas na infancias. Revisão de 208 casos. Rev Paul Med 66(1):24-39.
Shenoy RS (1994) Pitfalls in the treatment of jimsonweed intoxication. Am J Psychiatry 151(9):1396-1397.
Shervette RE 3d, Schydlower M, Lampe RM, et al. (1979) Jimson "loco" weed abuse in adolescents. Pediatrics 63(4):520-523.
Simmat G, Robert R, Gil R, et al. (1983) Tentative d'autolyse par absorption de graines de Datura stramonium. Presse Med 12(38):2399.
Simmons FH (1957) Jimson weed mydriasis in farmers. Am J Ophthalmol 44(1):109-110.
Smidt N, Bieder L, Thomas RG (1978) Datura intoxication. N Z Med J 87(604):61-62.
Snyder J (1995) Cheap high, costly lesson: Jimson weed no cup of tea. Arizona Republic. Nov 9:B1.
Soler-Rodriguez F, Martín A, Garcia-Cambero JP, et al. (2006) Datura stramonium poisoning in horses: A risk factor for colic. Vet Rec 158(4):132-133.
Sportsman LM (1946) Stramonium poisoning. A diagnostic problem with psychiatric implications. J Pediatr 29:345-349.
Steindler R, Langecker H (1937) Ein Fall von Stechapfel-Vergiftung. Sammlung Vergiftungsfallen 8(A686):107-110.
Stiles FC (1951) Stramonium poisoning. J Pediatr 39(3):354-356.
Strobel M, Chevalier J, De Lavarelle B (1991) Coma fébrile avec polynucléose dû à une intoxication par Datura Stramonium. Presse Med 20(43):2214.
Taha SA, Mahdi AH (1984) Datura intoxication in Ridyah. Trans R Soc Trop Med Hyg 78(1):134-135.
Teitelbaum DT (1968) Stramonium poisoning in "Teeny-Boppers." Ann Intern Med 68(1):174-175.
Thabet H, Brahml N, Amamou M, et al. (1999) Datura stramonium poisonings in humans. Vet Hum Toxicol 41(5):320-321.
Thomas J, Gelfand M (1955) Datura poisoning. Cent Afr J Med 1(2):78-79.
Tiongson J, Salen P (1998) Mass ingestion of jimson weed by eleven teenagers. Del Med J 70(11):471-476.
Tostes RA (2002) Accidental Datura stramonium poisoning in a dog. Vet Hum Toxicol 44(1):33-34.
Urich RW, Bowerman DL, Levisky JA, et al. (1982) Datura stramonium: A fatal poisoning. J Forensic Sci 27(4):948-954.
Vanderhoff BT, Mosser KH (1992) Jimson weed toxicity: Management of anticholinergic plant ingestion. Am Fam Physician 46(2):526-530.
Vanmeurs A, Cohen A, Edelbroek P (1992) Atropine poisoning after eating chapattis contaminated with Datura stramonium (thorn apple). Trans R Soc Trop Med Hyg 86(2):221.
Von Kanel RL, Rodgers GC (1992) Conservative treatment of jimsonweed ingestion. Vet Hum Toxicol 34(4):352.
Weintraub S (1960) Stramonium poisoning. Postgrad Med 28(Oct):364-367.
Williams S, Scott P (1984) The toxicity of Datura stramonium (thorn apple) to horses. N Z Vet J 32(4):47.
Worthington TR, Nelson EP, Bryant MJ (1981) Toxicity of thornapple (Datura stramonium L) seeds to the pig. Vet Rec 108(10):208-211.

Datura suaveolens Humb. & Bonpl. ex Willd. = Brugmansia suaveolens (Humb. & Bonpl. ex Willd.) Bercht. & J. Presl

Datura tatula L. = Datura stramonium L.

DATURA WRIGHTII Regel [Solanaceae]

Synonyms:

Datura meteloides auct.

Common Names:

hairy thorn apple; Indian apple; jimson weed; moonflower; recurved thorn apple; sacred datura; Wright's-datura

Citations:
Reader AL 3rd (1977) Mydriasis from Datura wrightii. Am J Ophthalmol 84(2):263-264.

daubentonia –see– *Sesbania drummondii* (Rydb.) Cory; *Sesbania punicea* (Cav.) Benth.

Daubentonia drummondii Rydb. = Sesbania drummondii (Rydb.) Cory

Daubentonia longifolia (Cav.) DC. = Sesbania longifolia DC.

Daubentonia punicea (Cav.) DC. = Sesbania punicea (Cav.) Benth.

DAUCUS CAROTA L. [Apiaceae]

Common Names:
carrot; Karotte; Möhre

Citations:
Fernández-Rivas M, González-Mancebo E, van Leeuwen WA, et al. (2004) Anaphylaxis to raw carrot not linked to pollen allergy. Allergy 59(11):1239-1240.
Klauder JV, Kimmich JM (1956) Sensitization dermatitis to carrots. Arch Dermatol 74(2):149-158.
Meynadier J, Meynadier JM, Guilhou JJ (1982) L'urticaire de contact chez l'atopique. A propos de deux observations. Ann Dermatol Venereol 109(10):871-874.
Muñoz D, Leanizbarrutia I, Lobera T, et al. (1985) Anaphylaxis from contact with carrot. Contact Dermatitis 13(5):345-346.
Peck SM, Spolyar LW, Mason HS (1944) Dermatitis from carrots. Arch Derm Syphilol 49:266-269.
Vickers HR (1941) The carrot as a cause of dermatitis. Br J Dermatol 53:52-57.

Note:
Carrot is named *Daucus carota* L. subsp. *sativus* (Hoffm.) Arcang. var. *sativus* Hoffm. in some publications.

DAVIDSONIA PRURIENS F. Muell. [Unoniaceae]

Common Names:
Davidsonian plum

Citations:
Flecker H (1945) Injuries produced by plants in tropical Queensland. Med J Aust 1(Jun 23):636-637.

Davidsonian plum –see– *Davidsonia pruriens* F. Muell.

day-blooming jasmine –see– *Cestrum diurnum* L.

day-blooming jessamine –see– *Cestrum diurnum* L.

day cestrum –see– *Cestrum diurnum* L.

day jasmine –see– *Cestrum diurnum* L.

day jessamine –see– *Cestrum diurnum* L.

day lily –see– *Hemerocallis dumortieri* C. Morren; *Hemerocallis fulva* (L.) L.; *Hemerocallis minor* Mill.; *Hosta sieboldii* (Paxton) J. W. Ingram

day's-leek –see– *Hyacinthoides non-scripta* (L.) Chouard ex Rothm.

dazao –see– *Ziziphus jujuba* Mill.

dead nettle –see– *Lamium amplexicaule* L.; *Stachys arvensis* (L.) L.

Dead Sea apple –see– *Calotropis procera* (Aiton) W. T. Aiton; *Solanum anguivi* Lam.

dead tongue –see– *Oenanthe crocata* L.

deadly hemlock –see– *Conium maculatum* L.

deadly nightshade –see– *Atropa belladonna* L.; *Solanum dulcamara* L.; *Solanum nigrum* L.

dead-man's-bell –see– *Digitalis purpurea* L.

dead-man's-fingers –see– *Oenanthe crocata* L.

dead-man's-tree –see– *Synadenium cupulare* (Boiss.) L. C. Wheeler

death camas –see– *Amianthium muscitoxicum* (Walter) A. Gray; *Anticlea elegans* (Pursh) Rydb.; *Toxicoscordion fremontii* (Torr.) Rydb.; *Toxicoscordion intermedium* Rydb.; *Toxicoscordion nuttallii* (A. Gray) Rydb.; *Toxicoscordion paniculatum* (Nutt.) Rydb.; *Toxicoscordion venenosum* (S. Watson) Rydb.

death-of-man –see– *Cicuta maculata* L. var. angustifolia Hook.; *Cicuta maculata* L.

death's-herb –see– *Atropa belladonna* L.

deciduous holly –see– *Ilex decidua* Walter

dedalera –see– *Digitalis purpurea* L.

deerwort –see– *Ageratina altissima* (L.) R. M. King & H. Rob.

deerberry –see– *Gaultheria procumbens* L.

deflah –see– *Nerium oleander* L.

DEGUELIA UTILIS (A. C. Sm.) A. M. G. Azevedo [Fabaceae]

Synonyms:
Lonchocarpus nicou auct.

Common Names:
barbasco; bois eniorant; conapi; cubé; cubé root; nicou; real heirri; timbó; timbó legitimo; timbó macaquinho; white hairare

Citations:
Dorne M, Friedman TB (1940) Derris root dermatitis. JAMA 115(Oct 12):1268-1270.
Hansen WH, Davis KJ, Fitzhugh OG (1965) Chronic toxicity of cubé. Toxicol Appl Pharmacol 7(4):535-542.

deli bat bat –see– *Hyoscyamus niger* L.

DELPHINIUM ANDERSONII A. Gray [Ranunculaceae]

Common Names:
Anderson's-larkspur; desert larkspur; low larkspur

Citations:
Fleming CE (1920) Poisonous range plants. Nevada Agric Exp Sta Annu Rep 1919:39-43.

Fleming CE, Miller MR, Vawter LR (1923) The low larkspur (Delphinium andersoni), A plant of the spring range, poisonous to cattle. Nevada Agric Exp Sta Bull #105:22 pp.

DELPHINIUM BARBEYI (Huth) Huth [Ranunculaceae]

Common Names:

Barbey's-larkspur; cow poison; larkspur; peco; poison weed; tall larkspur; western larkspur

Citations:

Anonymous (1933) Investigation of stock-poisoning plants. U S Dep Agric Bur Anim Indus Chief Rep 1933:34-35.
Anonymous (1934) Investigation of stock-poisoning plants. U S Dep Agric Bur Anim Indus Chief Rep 1934:38.
Anonymous (1935) Investigation of stock-poisoning plants. U S Dep Agric Bur Anim Indus Chief Rep 1935:39-40.
Anonymous (1936) Investigations of stock-poisoning plants. U S Dep Agric Bur Anim Indus Chief Rep. 1936:44-45.
Anonymous (1937) Investigations of stock-poisoning plants. U S Dep Agric Bur Anim Indus Chief Rep. 1937:47-48.
Buck WB, Binns W, James L, et al. (1961) Results of feeding of herbicide-treated plants to calves and sheep. J Am Vet Med Assoc 138(6):320-323.
Cronin EH, Nielsen DB, Madson N (1976) Cattle losses, tall larkspur, and their control. J Range Manag 29(5):364-367.
Marsh CD, Clawson AB, Marsh H (1916) Larkspur poisoning of livestock. U S Dep Agric Bull #365:91 pp.
Olsen JD (1977) Unlocking the secrets of larkspur. Utah Sci 38(2):35-38.
Olsen JD (1978) Tall larkspur poisoning in cattle and sheep. J Am Vet Med Assoc 173(6):762-765.
Olsen JD (1982) Mineral supplement can reduce the lethal effect of larkspur poisoning in cattle. Annu Meet Soc Range Manag. p. 13.
Pfister JA, Cheney CD, Gardner DR, et al. (1998) Mineral-salt supplement does not attenuate tall larkspur (Delphinium barbeyi) toxicosis in cattle. J Range Manag 51(5):566-569.
Pfister JA, Panter KE, Manners GD, et al. (1994) Reversal of tall larkspur (Delphinium barbeyi) poisoning in cattle with physostigmine. Vet Hum Toxicol 36(6):511-514.

DELPHINIUM BICOLOR Nutt. [Ranunculaceae]

Common Names:

flathead larkspur; little larkspur; low larkspur; mountain larkspur; purple larkspur

Citations:

Marsh CD, Clawson AB, Marsh H (1916) Larkspur poisoning of livestock. U S Dep Agric Bull #365:91 pp.

Delphinium brownii Rydb. = Delphinium glaucum S. Watson

Delphinium camporum Greene = Delphinium carolinianum Walter subsp. virescens (Nutt.) R. E. Brooks

DELPHINIUM CAROLINIANUM Walter subsp. virescens (Nutt.) R. E. Brooks [Ranunculaceae]

Synonyms:

Delphinium camporum Greene; ***Delphinium virescens*** Nutt.

Common Names:

larkspur; low larkspur; plains larkspur; poison weed; prairie larkspur

Citations:

Crawford AC (1907) The larkspurs as poisonous plants. U S Dep Agric Bur Plant Indus Bull #111(Part 1):12 pp.
Sizelove W, McIlvaim EH, Burrows GE (1992) Possible larkspur intoxications responsible for acute deaths in cattle. Vet Hum Toxicol 34(5):458-459.

Delphinium consolida L. = Consolida regalis Gray

Delphinium cucullatum A. Nelson = Delphinium glaucum S. Watson

DELPHINIUM ELATUM L. [Ranunculaceae]

Common Names:

hoher Rittersporn; larkspur

Citations:

Puschner B, Booth MC, Tor ER, et al. (2002) Diterpenoid alkaloid toxicosis in cattle in the Swiss Alps. Vet Hum Toxicol 44(1):8-10.
Puschner B, Booth MC, Tor ER, et al. (2004) Delphinium alkaloid toxicosis in cattle from Switzerland. In: Acamovic T, Stewart CS, Pennycott TW (eds.) Poisonous plants and related toxins. CABI. Wallingford, UK. pp. 38-43.

DELPHINIUM GLAUCUM S. Watson [Ranunculaceae]

Synonyms:

Delphinium brownii Rydb.; ***Delphinium cucullatum*** A. Nelson

Common Names:

Brown's-larkspur; Chinese-cap larkspur; giant larkspur; Hooker's-larkspur; large larkspur; mountain larkspur; pale larkspur; tall larkspur; tall mountain larkspur; waxy larkspur; western larkspur

Citations:

Marsh CD, Clawson AB, Marsh H (1916) Larkspur poisoning of livestock. U S Dep Agric Bull #365:91 pp.

Delphinium hybridum Stephan ex Willd. = Delphinium schmalhausenii Albov

DELPHINIUM MENZIESII DC. [Ranunculaceae]

Common Names:

field larkspur; little larkspur; low larkspur; Menzies'-larkspur; poison weed; purple larkspur; small larkspur; spring larkspur; western purple larkspur

Citations:

Marsh CD, Clawson AB, Marsh H (1916) Larkspur poisoning of livestock. U S Dep Agric Bull #365:91 pp.

Wilcox EV (1897) Larkspur poisoning of sheep. Montana Agric Exp Sta Bull #15:37-51.

Wilcox EV (1898) Larkspur poisoning of sheep. U S Dep Agric Bur Anim Indus Annu Rep 15:473-479.

Delphinium nelsonii Greene = Delphinium nuttallianum Pritz.

DELPHINIUM NUTTALLIANUM Pritz. [Ranunculaceae]

Synonyms:

Delphinium nelsonii Greene

Common Names:

blue larkspur; dwarf larkspur; low larkspur; meadow larkspur; Nelson's-larkspur; purple larkspur; slim larkspur; upland larkspur

Citations:

Glover GH (1906) Larkspur and other poisonous plants. Colorado Agric Exp Sta Bull #113:24 pp.

Pfister JA, Gardner DR, Stegelmeier BL (2003) Catastrophic cattle loss to low larkspur (Delphinium nuttallianum) in Idaho. Vet Hum Toxicol 45(3):137-139.

DELPHINIUM ×OCCIDENTALE (S. Watson) S. Watson [Ranunculaceae]

Common Names:

duncecap larkspur; poison weed; tall larkspur

Citations:

Anonymous (1934) Investigation of stock-poisoning plants. U S Dep Agric Bur Anim Indus Chief Rep 1934:38.

Anonymous (1935) Investigation of stock-poisoning plants. U S Dep Agric Bur Anim Indus Chief Rep 1935:39-40.

Anonymous (1936) Investigations of stock-poisoning plants. U S Dep Agric Bur Anim Indus Chief Rep. 1936:44-45.

Anonymous (1937) Investigations of stock-poisoning plants. U S Dep Agric Bur Anim Indus Chief Rep. 1937:47-48.

Olsen JD (1983) Relationship of relative total alkaloid concentration and toxicity of duncecap larkspur during growth. J Range Manag 36(5):550-552.

Shupe JL, Balls LD, James LF (1968) Changes in blood serum transaminase associated with lupine and larkspur poisoning in cattle. Cornell Vet 58(1):129-135.

DELPHINIUM ROBUSTUM Rydb. [Ranunculaceae]

Common Names:

robust larkspur

Citations:

Marsh CD, Clawson AB, Marsh H (1916) Larkspur poisoning of livestock. U S Dep Agric Bull #365:91 pp.

DELPHINIUM SCHMALHAUSENII Albov [Ranunculaceae]

Synonyms:

Delphinium hybridum Stephan ex Willd.

Common Names:

delphinium; larkspur

Citations:

Milne JA (1966) A case of delphinium poisoning in rams. N Z Vet J 14(8):127.

DELPHINIUM SCOPULORUM A. Gray [Ranunculaceae]

Common Names:

Rocky Mountain larkspur; tall larkspur; tall mountain larkspur

Citations:

Wilcox EV (1899) Cattle poisoning by the tall larkspur. Montana Agric Exp Sta Bull #22:45-47.

DELPHINIUM TRICORNE Michx. [Ranunculaceae]

Common Names:

dwarf larkspur; low larkspur; staggerweed

Citations:

Hansen AA (1924) The poison plant situation in Indiana. III. Poisonous trees. J Am Vet Med Assoc 66:351-362.

Delphinium virescens Nutt. = Delphinium carolinianum Walter subsp. virescens (Nutt.) R. E. Brooks

DENDROCNIDE MOROIDES (Wedd.) Chew [Urticaceae]

Synonyms:

Laportea moroides Wedd.

Common Names:

Australian nettle; Australische Brennessel; giant nettle; giant stinging tree; gympie bush; gympie gympie; gympie nettle; gympie tree; Nessel

Citations:

Robertson PA, MacFarlane WV (1957) Pain-producing substances from the stinging bush Laportea moroides. Aust J Exp Biol Med Sci 35(4):381-394.

DENDROPANAX TRIFIDUS (Thunb.) Makino [Araliaceae]

Citations:

Oka K, Saito F, Yasuhara T, et al. (1997) The major allergen of Dendropanax trifidus Makino. Contact Dermatitis 36(5):252-255.

Oka K, Saito F, Yasuhara T, et al. (1999) The allergens of Dendropanax trifidus Makino and Fatsia japonica Decne. et Planch. and evaluation of cross-reactions with other plants of the Araliaceae family. Contact Dermatitis 40(4):209-213.

deodar cedar –see– *Cedrus deodora* (Roxb. ex D. Don) G. Don

derris –see– *Paraderris elliptica* (Wall.) Adema

Derris elliptica (Wall.) Benth. = Paraderris elliptica (Wall.) Adema

DERRIS TRIFOLIATA Lour. [Fabaceae]

Common Names:

bagin; sila sila

Citations:

Chakraborty DP, Nandy AC, Philipose MT (1972) Barringtonia acutangula (L.) Gaertn. as a fish poison. Indian J Exp Biol 10(1):78-80.

DESCURAINIA PINNATA (Walter) Britton [Brassicaceae]

Common Names:

tansy mustard

Citations:

Anonymous (1981) Tansy mustard poisoning in a mare. Vet Hum Toxicol 23(3):163.

Pfister JA, Lacey JR, Baker DC, et al. (1990) Is tansy mustard causing photosensitization of cattle in Montana? Rangelands 12(3):170-172.

Staley E (1976) A treatment for tansy mustard poisoning. Bovine Pract 11:35.

Staley E (1976) A treatment for tansy mustard poisoning. Oklahoma Vet 28(2):74.

DESCURAINIA SOPHIA (L.) Webb ex Prantl [Brassicaceae]

Synonyms:

Sisymbrium sophia L.

Common Names:

brîcută; flixweed; tornel; voinicel

Citations:

Contiu I, Rîpeanu M (1971) Îmbolnăviri Ia animale în urma consumului plantei Sisymbrium sofia L. Rev Zootehnie Med Vet 21(8):54-56.

Pfister JA, Lacey JR, Baker DC, et al. (1990) Is tansy mustard causing photosensitization of cattle in Montana? Rangelands 12(3):170-172.

desert baileya –see– *Baileya multiradiata* Harv. & A. Gray ex Torr.

desert crowfoot –see– *Ranunculus cymbalaria* Pursh

desert heliotrope –see– *Phacelia crenulata* Torr. ex S. Watson

desert larkspur –see– *Delphinium andersonii* A. Gray

desert marigold –see– *Baileya multiradiata* Harv. & A. Gray ex Torr.

desert parsley –see– *Cymopterus ibapensis* M. E. Jones

desert poisonbush –see– *Gastrolobium grandiflorum* F. Muell.

desert potato –see– *Jatropha macrorhiza* Benth.

desert rice flower –see– *Pimelea simplex* F. Muell.

desert tea –see– *Ephedra viridis* Coville

desert tobacco –see– *Nicotiana obtusifolia* M. Martens & Galeotti

desert velvet –see– *Psathyrotes annua* (Nutt.) A. Gray

DESMANTHUS LEPTALOBUS Torr. & A. Gray [Fabaceae]

Common Names:

prairie bundleflower; prairie mimosa; slender bundleflower

Citations:

Harvey RB, Rowe LD, Reagor JC (1986) Suspected toxicity of Desmanthus leptalobus (prairie bundleflower, prairie mimosa) to horses. Southwestern Vet 37(2):143-144.

DESMANTHUS VIRGATUS (L.) Willd. [Fabaceae]

Citations:

Figueroa V, Sutherland TM (1972) "Muerte súbita" (sudden death) in cattle. 5. The role of toxic plants. Rev Cubana Cienc Agric Eng Ed 6(1):53-59.

Desmodium canum Schinz & Thell = Desmodium incanum DC.

DESMODIUM INCANUM DC. [Fabaceae]

Synonyms:

Desmodium canum Schinz & Thell; ***Meibomia cana*** S. F. Blake

Citations:

Shenk JS (1976) The meadow vole as an experimental animal. Lab Anim Sci 26(4):664-669.

DETARIUM SENEGALENSE J. F. Gmel. [Fabaceae]

Common Names:

ditakh

Citations:

Berthelot PG, N'Diaye M, Diatta B, et al. (2000) Acute intoxication after ditakh fruit ingestion. Intensive Care Med 26(10):1587.

Burgel QR, Camara P, Collet-Burgel C, et al. (1998) L'intoxication au ditakh: Une intoxication tropicale méconnue - 2 observations. Presse Med 27(30):1528.

Imbert P, Teyssier J (1986) Intoxication aigue par ingestion de ditakh - A propos de 8 observations. Med Trop 46(1):79-83.

Deutsche Kamille –see– *Matricaria recutita* L.

Deutsche Schwertlilie –see– *Iris ×germanica* L.

Deutscher Rhabarber –see– *Rheum ×hybridum* Murray

Deutsches Weidelgras –see– *Lolium perenne* L.

DEVERRA TRIRADIATA Hochst. ex Boiss. [Apiaceae]

Synonyms:

Pituranthos triradiatus (Hochst. ex Boiss.) Asch. & Schweinf

Citations:

Ashkenazy D, Friedman J, Kashman Y, et al. (1984) Photosensitization in ducklings induced by Pituranthos triradiatus. Vet Hum Toxicol 26(2):118-120.

Nyska A, Bar Ishak YR, Ashkenazy D, et al. (1984) Retinal atrophy in albino rats associated with Pituranthos triradiatus (Umbelliferae)-induced photosensitization. Vet Pathol 21(5):551-552.

devil's-apple –see– *Datura stramonium* L.; *Mandragora officinarum* L.; *Podophyllum peltatum* L.; *Solanum aculeatissimum* Jacq.; *Solanum anguivi* Lam.

devil's-backbone –see– *Kalanchoe daigremontiana* Raym.-Hamet & H. Perrier; *Pedilanthus tithymaloides* (L.) Poit.

devil's-bit –see– *Chamaelirium luteum* (L.) A. Gray

devil's-bite –see– *Veratrum viride* Aiton

devil's-dung –see– *Ferula assa-foetida* L.

devil's-ear –see– *Arisaema triphyllum* (L.) Schott

devil's-eye –see– *Hyoscyamus niger* L.

devil's-fig –see– *Argemone mexicana* L.; *Solanum torvum* Sw.

devil's-herb –see– *Mandragora officinarum* L.

devil's-ivy –see– *Epipremnum pinnatum* (L.) Engl.

devil's-lily –see– *Lilium lancifolium* Thunb.

devil's-thorn –see– *Tribulus terrestris* L.

devil's-tongue –see– *Amorphophallus konjac* K. Koch

devil's-trumpet –see– *Brugmansia suaveolens* (Humb. & Bonpl. ex Willd.) Bercht. & J. Presl; *Datura metel* L.; *Datura stramonium* L.

devil's-turnip –see– *Bryonia dioica* Jacq.

devil's-weed –see– *Datura stramonium* L.

dewdrops –see– *Duranta erecta* L.

dewtry –see– *Datura stramonium* L.

dhal arkar –see– *Cajanus cajan* (L.) Millsp.

dharati dhak –see– *Anagallis arvensis* L.

dhatura –see– *Datura inoxia* Mill.; *Datura stramonium* L.

dhobi nut –see– *Semecarpus anacardium* L. f.

dhrek –see– *Melia azedarach* L.

dhurra wheat –see– *Sorghum bicolor* (L.) Moench

dhursulo –see– *Colebrookea oppositifolia* Sm.

Dialiopsis africana Radlk. = Zanha africana (Radlk.) Exell

DIANTHUS CARYOPHYLLUS L. [Caryophyllaceae]

Common Names:

carnation

Citations:

Sánchez-Fernández C, González-Gutiérrez ML, Esteban-López MI, et al. (2004) Occupational asthma caused by carnation (Dianthus caryophyllus) with simultaneous IgE-mediated sensitization to Tetranychus urticae. Allergy 59(1):114-115.

Van Grutten M (1980) Carnation dermatitis in a flower seller. Contact Dermatitis 6(4):289.

Vidal C, Polo F (1998) Occupational allergy caused by Dianthus caryophillus, Gypsophila paniculata, and Lilium longiflorum. Allergy 53(10):995-998.

DICENTRA CANADENSIS (Goldie) Walp. [Fumariaceae]

Synonyms:

Bicuculla canadensis (Goldie) Millsp.; ***Bikukulla canadensis*** (Walp.) Druce

Common Names:

colicweed; Dutchman's-breeches; little blue stagger; staggerweed; trembling stagger; turkey corn; turkey pea

Citations:

Black OF, Eggleston WW, Kelly JW, et al. (1923) Poisonous properties of Bikukulla cucullaria (Dutchman's breeches) and B. canadensis (squirrel corn). J Agric Res 23(2):69-78.

DICENTRA CUCULLARIA (L.) Bernh. [Fumariaceae]

Synonyms:

Bicuculla cucullaria (L.) Millsp.; ***Bikukulla cucullaria*** Millsp.

Common Names:

bleeding heart; breeches flower; butterfly banners; ear drops; kitten breeches; little staggerweed; pearl harlequin; soldier's-cap; squirrel corn; staggerweed; white hearts; wild bleeding heart

Citations:

Black OF, Eggleston WW, Kelly JW, et al. (1923) Poisonous properties of Bikukulla cucullaria (Dutchman's breeches) and B. canadensis (squirrel corn). J Agric Res 23(2):69-78.

Hansen AA (1924) Dutchman's breeches. A recent addition to the poison plants. Better Crops 2(4):26-27, 41.

Dicentra spectabilis (L.) Lem. = Lamprocapnos spectabilis (L.) Fukuhara

DICHAPETALUM BARTERI Engl. [Dichapetalaceae]

Common Names:

akwuosa; goat killer

Citations:

Adaudi AO (1975) Dichapetalum barteri poisoning in goats. Trop Anim Health Prod 7(1):56-57.

Nwude N, Parsons LE, Adaudi AO (1977) Acute toxicity of the leaves and extracts of Dichapetalum barteri (Engl.) in mice, rabbits and goats. Toxicology 7(1):23-29.

DICHAPETALUM CYMOSUM (Hook.) Engl. [Dichapetalaceae]

Synonyms:

Dichapetalum venenatum Engl. & Gilg

Common Names:

blaargif; blinkblaar; chailletia; gifblaar; giftblaar; giftblad; makaon; makoeu; poison leaf

Citations:

Basson PA (1988) Acute and chronic cardiomyopathies caused by monofluoroacetic acid in giftblaar (Dichapetalum cymosum) and cardiac glycosides in slangkop (Urginea sanguinea). J S Afr Vet Assoc 59(2):100.

Dunphy JT (1906) Report of experiments carried out to observe effects of certain poisonous plants on sheep and goats. Transvaal Agric J 4(14):315-323.

Egyed MN (1986) The therapy of experimental poisoning in sheep caused by a plant containing monofluoroacetic acid, Dichapetalum cymosum. Isr J Vet Med 42(1):54.

Egyed MN, Schultz RA (1986) The efficacy of acetamide for the treatment of experimental Dichapetalum cymosum (gifblaar) poisoning in sheep. Onderstepoort J Vet Res 53(4):231-234.

Schultz RA, Kellerman TS, Van den Berg H (2005) The role of fluorescence polarization immuno-assay in the diagnosis of plant-induced cardiac glycoside poisoning livestock in South Africa. Onderstepoort J Vet Res 72(3):189-201.

Steyn DG (1934) Plant poisoning in stock and the development of tolerance. Onderstepoort J Vet Sci Anim Indus 3(1):119-123.

Steyn DG (1934) Recent investigations into the toxicity of known and unknown poisonous plants in the Union of South Africa. Onderstepoort J Vet Sci Anim Indus 3(1):125-130.

Van der Walt SJ, Steyn DG (1946) Recent investigations into the toxicity of plants, etc. XV. Onderstepoort J Vet Sci Anim Indus 21(1):45-55.

DICHAPETALUM MICHELSONII Hauman [Dichapetalaceae]

Common Names:

umutambasha

Citations:

Thienpont D, Vandervelden M (1961) Dichapetalum michelsonii Hauman. Novelle plante toxique pour bétail du Ruanda-Burundi. Rev Elev Med Vet Pays Trop 14(2):209-211.

DICHAPETALUM MOSSAMBICENSE (Klotzsch) Engl. [Dichapetalaceae]

Citations:

Msami HM (1999) An outbreak of suspected poisoning of cattle by Dichapetalum sp. in Tanzania. Trop Anim Health Prod 31(1):1-7.

DICHAPETALUM RUHLANDII Engl. [Dichapetalaceae]

Citations:

Kamau JA, Gachuhi DM, Gyrd-Hansen N, et al. (1978) A study of the toxicity of Dichapetalum ruhlandii (Ludi). Indian Vet J 55(8):626-630.

DICHAPETALUM STUHLMANII Engl. [Dichapetalaceae]

Common Names:

karati

Citations:

Msami HM (1999) An outbreak of suspected poisoning of cattle by Dichapetalum sp. in Tanzania. Trop Anim Health Prod 31(1):1-7.

DICHAPETALUM TOXICARIUM (G. Don) Baill. [Dichapetalaceae]

Common Names:

rat's-bane

Citations:

Van Dijk C, Vickery B, Vickery ML (1974) The toxicity of Dichapetalum toxicarium for Ndama cattle. Trop Anim Health Prod 6(2):117-118.

Dichapetalum venenatum Engl. & Gilg = Dichapetalum cymosum (Hook.) Engl.

DICHROCEPHALA CHRYSANTHEMIFOLIA (Blume) DC. [Asteraceae]

Citations:

Mugera GM (1970) Toxic and medicinal plants of East Africa. II. Bull Epizootic Dis Afr 18(4):389-403.

Dichternarzisse –see– *Narcissus poeticus* L.

DICTAMNUS ALBUS L. [Rutaceae]

Common Names:

burning bush; Diptam; dittamo; dittany; fraxinella; gas plant; lemon-scented gas plant

Citations:

Cummer CL, Dexter R (1937) Dermatitis caused by Dictamnus albus (gas plant). JAMA 109(7):495-497.

Engel S, Horn K (1972) Phytodermatosen durch Dictamnus alba, Sanicula europaea und Phyllodendron consanguineum. Dermatol Monatsschr 158(1):22-27.

Henderson JA, Des Groseilliers JP (1984) Gas plant (Dictamnus albus) phytophotodermatitis simulating poison ivy. Can Med Assoc J 130(7):889-891.

Knüchel M, Luderschmidt C (1986) Bullöse phototoxische Kontaktdermatitis durch Dictamnus albus. <<Brennender Busch>> der Bibel? Dtsch Med Wochenschr 111(38):1445-1447.

Marchi AG, Trevisan G, Kokelj F, et al. (1982) Fitofotodermatite da dittamo - Segnalazione di 49 casi. Minerva Pediatr 34(17):691-694.

Moller H (1978) Phototoxicity of Dictamnus alba. Contact Dermatitis 4(5):264-269.
Suhonen R (1977) Phytophotodermatitis: An experimental study using the chamber method. Contact Dermatitis 3(3):127-132.

DICTAMNUS DASYCARPUS Turcz. [Rutaceae]

Citations:

Stekhun FI, Kyrnakov BA (1962) [Dermatitis caused by Dictamnus dasycarpus.] Vestn Dermatol Venerol 36(1):67-70.

dictamo –see– *Pedilanthus tithymaloides* (L.) Poit.

Didiscus glaucifolius F. Muell. = Trachymene glaucifolia (F. Muell.) Benth.

dieffenbachia –see– *Dieffenbachia seguine* (Jacq.) Schott

Dieffenbachia amoena hort. = Dieffenbachia seguine (Jacq.) Schott

DIEFFENBACHIA BAUSEI Regel [Araceae]

Common Names:

dumb cane

Citations:

Der Marderosian AH, Giller FB, Roia FC Jr (1976) Phytochemical and toxicological screening of household ornamental plants potentially toxic to humans. I. J Toxicol Environ Health 1(6):939-953.

Dieffenbachia exotica hort. = Dieffenbachia seguine (Jacq.) Schott

Dieffenbachia maculata (G. Lodd.) G. Don = Dieffenbachia seguine (Jacq.) Schott

Dieffenbachia picta Schott = Dieffenbachia seguine (Jacq.) Schott

DIEFFENBACHIA PITTIERII Engl. & K. Krause [Araceae]

Citations:

Cumpston KL, Vogel SN, Leikin JB, et al. (2003) Acute airway compromise after brief exposure to a Dieffenbachia plant. J Emerg Med 25(4):391-397.

DIEFFENBACHIA SEGUINE (Jacq.) Schott [Araceae]

Synonyms:

Caladium seguinum (Jacq.) Vent.; ***Dieffenbachia amoena*** hort.; ***Dieffenbachia exotica*** hort.; ***Dieffenbachia maculata*** (G. Lodd.) G. Don; ***Dieffenbachia picta*** Schott

Common Names:

American arum; aninga para; arum; arum-of-the-Antilles; caña muda; comida-de-culebra; comigo-ninguém-pode; dieffenbachia; dumb cane; leopard lily; lluvia-del-plata; mother-in-law-plant; mother-in-law's-tongue; oto-de-lagarto; poison arum; prickblad; Schierling Caladium; Schweigrohr; spotted dumbcane

Citations:

Arai M, Stauber E, Shropshire CM (1992) Evaluation of selected plants for their toxic effects on canaries. J Am Vet Med Assoc 200(9):1329-1331.
Bogdanowski T, Rubisz-Brzezińska J (1984) Dieffenbachia picta - Zmiany toksyczno - alergiczne na błonie śluzowej jamy ustnej i języka. Przegl Dermatol 71(1):61-63.
Corazza M, Romani I, Poli F, et al. (1998) Irritant contact dermatitis due to Dieffenbachia spp. J Eur Acad Dermatol Venereol 10(1):87-89.
Costa de Pasquale R, Ragusa S, Circosta C, et al. (1984) Investigations on Dieffenbachia amoena Gentil. I. Endocrine effects and contraceptive activity. J Ethnopharmacol 12(3):293-303.
Der Marderosian AH, Giller FB, Roia FC Jr (1976) Phytochemical and toxicological screening of household ornamental plants potentially toxic to humans. I. J Toxicol Environ Health 1(6):939-953.
Drach G, Maloney WH (1963) Toxicity of the common houseplant Dieffenbachia. Report of a case. JAMA 184(13):1047-1048.
Ellis W, Barfort P, Mastman GJ (1973) Keratoconjunctivitis with corneal crystals caused by the Dieffenbachia plant. Am J Ophthalmol 76(1):143-147.
Faivre M, Barral C (1974) Un cas d'intoxication par Dieffenbachia picta. Nouv Presse Med 3(20):1313-1314.
Flecker H (1945) Injuries produced by plants in tropical Queensland. Med J Aust 1(Jun 23):636-637.
Fochtman FW, Manno JE, Winek CL, et al. (1969) Toxicity of the genus Dieffenbachia. Toxicol Appl Pharmacol 15(1):38-45.
Krenzelok EP, Jacobsen TD, Aronis JM (1996) A review of 96,659 Dieffenbachia and Philodendron exposures. J Toxicol Clin Toxicol 34:601.
Kuballa B, Lugnier AA, Anton R (1981) Study of Dieffenbachia-induced edema in mouse and rat hindpaw: Respective role of oxalate needles and trypsin-like protease. Toxicol Appl Pharmacol 58(3):444-451
Ladeira AM, Andrade SO, Sawaya P (1975) Studies on Dieffenbachia picta Schott: Toxic effects in guinea pigs. Toxicol Appl Pharmacol 34(3):363-373.
Lim KH (1977) External eye allergy from sap of Dieffenbachia picta. Singapore Med J 18(3):176-177.
Loretti AP, Ilha MR (2003) Accidental fatal poisoning of a dog by Dieffenbachia picta (dumb cane). Vet Hum Toxicol 45(5):233-239.
Louhija A, Luomanmäki K (1968) Kirjovehkan aiheuttama myrkytys. Duodecim 84(3):219-220.
Madaus G, Koch FE (1941) Tierexperimentelle Studien zur Frage der medikamentösen Sterilisierung (durch Caladium seguinum [Dieffenbachia seguina]). Z Gesamte Exp Med 109:68-87.
Manno JE, Fochtman FW, Winek CL, et al. (1967) Toxicity of plants of the genus Dieffenbachia. Toxicol Appl Pharmacol 10:405-406.
Martin H, Martin C (1977) Accident caustique buccopharyngé provoqué par le Dieffenbachia picta. J Fr Otorhinolaryngol 26(9):715-719.
Mrvos R, Krenzelok EP, Jacobsen TD (2001) Toxidromes associated with the most common plant ingestions. Vet Hum Toxicol 43(6):366-369.

Napier A (1912) Irritant poisoning from chewing the shoots and leaves of Dieffenbachia seguina. Lancet 182(Apr 20):1058.

Neves L, Carneiro CM, Pereira NA (1988) Estudo do mecanismo tóxico em Dieffenbachia picta. Acta Amazonica 18(1-2 Suppl):171-174.

Occhioni P, Rizzini CT (1958) Ação tóxica de duas sp. de Dieffenbachia. Rev Bras Med 15(Jan):10-16.

Pedaci L, Krenzelok EP, Jacobsen TD, et al (1999 Dieffenbachia species exposure: An evidence-based assessment of symptom presentation. Vet Hum Toxicol 41:335-338.

Riede B (1971) Augenverletzung mit dem Saft der Pflanze "Dieffenbachia seguine." Dtsch Gesundheitsw 26(2):73-76.

Rizzini CT, Occhioni P (1957) Ação tóxica das Dieffenbachia picta e D. seguine. Rodriguesia 32:5-19.

Wiese M, Kruszewska S, Kolacinski Z (1996) Acute poisoning with Dieffenbachia picta. Vet Hum Toxicol 38(5):356-358.

diente-de-ajo –see– *Allium sativum* L.

Digitale pourprée –see– *Digitalis purpurea* L.

digitalis –see– *Digitalis lanata* Ehrh.; *Digitalis purpurea* L.

DIGITALIS LANATA Ehrh. [Plantaginaceae]

Common Names:

digitalis; Fingerhut; foxglove; Grecian foxglove; white foxglove; Wollfingerhut; yellow foxglove

Citations:

Barnikol H, Hofmann W (1973) Digitalisvergiftung beim Schwein. Tierarztl Umsch 28(12):612-614, 616.

Lo Vecchio F, Seby MV, Johnson D (1998) Digitalis poisoning following the ingestion of an herbal dietary supplement. J Toxicol Clin Toxicol 36(5):457.

Slifman NR, Obermeyer WR, Musser SM, et al. (1998) Contamination of botanical supplements by Digitalis lanata. N Engl J Med 339(12):806-811.

DIGITALIS PURPUREA L. [Plantaginaceae]

Common Names:

bloody finger; dead-man's-bell; dedalera; digitale pourprée; digitalis; fairy bells; fairy caps; fairy fingers; fairy gloves; fairy thimbles; finger flower; Fingerhut; flap dock; folk's-glove; foxes glofa; foxglove; lady's-thimble; lion's-mouth; lusmore; pop dock; purple digitalis; purple foxglove; revebjelle; Roter Fingerhut; thimbles; throatwort; Waldschelle

Citations:

Arai M, Stauber E, Shropshire CM (1992) Evaluation of selected plants for their toxic effects on canaries. J Am Vet Med Assoc 200(9):1329-1331.

Bain RJ (1985) Accidental digitalis poisoning due to drinking herbal tea. Br Med J 290(6482):1624.

Brauner GJ, Greene MH (1972) Digitalis allergy: Digoxin-induced vasculitis. Cutis 10(Oct):441-445.

Brustbauer R, Wenisch C (1997) Bradykardes Vorhofflimmern nach Genuß von Kräutertee. Dtsch Med Wochenschr 122(3):930-932.

Carmichael MA (1987) Suspected foxglove poisoning. Vet Rec 120(15):375.

Corrigall W, Moody RR, Forbes JC (1978) Foxglove (Digitalis purpurea) poisoning in farmed red deer (Cervus elaphus). Vet Rec 102(6):119-122.

Der Marderosian AH, Giller FB, Roia FC Jr (1976) Phytochemical and toxicological screening of household ornamental plants potentially toxic to humans. I. J Toxicol Environ Health 1(6):939-953.

Dickstein ES, Kunkel FW (1980) Foxglove tea poisoning. Am J Med 69(1):167-169.

Jowett NI (2002) Foxglove poisoning. Hosp Med 63(12):758-759.

Lacassie E, Marquet P, Martin-Dupont S, et al. (2000) A non-fatal case of intoxication with foxglove, documented by means of liquid chromatography-electrospray-mass spectrometry. J Forensic Sci 45(5):1154-1158.

MacLean A (1966) Suspected foxglove poisoning in sheep. Vet Rec 79(25):817-818.

Newman LS, Feinberg MW, LeWine HE (2004) A bitter tale. N Engl J Med 351(6):594-599.

Omvik P (1981) Revebjelleforgiftning. Tidsskr Nor Laegeforen 101(15):949-950.

Parker WH (1951) Foxglove (Digitalis purpurea) poisoning in turkeys. Vet Rec 63(24):416.

Porter R, Schultz D, Robertson WO (1999) Alternative medicine toxicology: Digitalis poisoning. J Toxicol Clin Toxicol 37(5):617.

Simpkiss M, Holt D (1983) Digitalis poisoning due to the accidental ingestion of foxglove leaves. Ther Drug Monit 5(2):217.

Stillman AE, Huxtable RJ, Fox DW, et al. (1977) Poisoning associated with herbal teas - Arizona, Washington. MMWR Morb Mortal Wkly Rep 26(32):257-259.

Thierry S, Blot F, Lachérade JC, et al. (2000) Poisoning with foxglove extract: Favorable evolution without Fab fragments. Intensive Care Med 26(10):1586.

Thomas DL, Quick MP, Morgan RP (1987) Suspected foxglove (Digitalis purpurea) poisoning in a dairy cow. Vet Rec 120(13):300-301.

Zimowski A (1973) Zatrucie nutrii naparstnicq purpurowq (Digitalis purpurea). Med Weter 29(4):226.

DIGITARIA SANGUINALIS (L.) Scop. [Poaceae]

Synonyms:

Syntherisma sanguinalis (L.) Dulac

Common Names:

crabgrass; summergrass

Citations:

Pickens EM, Welch HF, Shivers CC (1920) A disease of cattle due to crab grass (Digitaria sanguinalis). Cornell Vet 10(1):8-16.

Diholcos bisulcatus (Hook.) Rydb. = Astragalus bisulcatus (Hook.) A. Gray

Diholcos micranthus Rydb. = Astragalus bisulcatus (Hook.) A. Gray

dikkopbos –see– *Hertia pallens* (DC.) Kuntze

dill –see– *Anethum graveolens* L.

DIMORPHANDRA GARDERIANA Tul. [Fabaceae]

Common Names:

fava d'anta

Citations:

Döbereiner J, Tokarnia CH (1985) Giftpflanzenbedingte Nierenschädigungen bei Rindern in Brasilien. Dtsch Tierarztl Wochenschr 92(10):411-415.

Döbereiner J, Tokarnia CH, Gava A, et al. (1985) Intoxicação experimental em bovinos pela fava de Dimorphandra gardneriana (Leg. Caesalpinoideae). Pesq Vet Bras 5(2):47-51.

DIMORPHANDRA MOLLIS Benth. [Fabaceae]

Common Names:

faveira

Citations:

Döbereiner J, Tokarnia CH (1985) Giftpflanzenbedingte Nierenschädigungen bei Rindern in Brasilien. Dtsch Tierarztl Wochenschr 92(10):411-415.

Menezes Filho JA (1985) Intoxicação experimental pela fava de Dimorphandra mollis Benth. (Leg. Caesalpinoideae) em bovinos. Pesq Vet Bras 5(3):93-96.

Santos FC (1979) Intoxicação experimental em bovinos pela "Faveira"- Dimorphandra mollis Benth. Arq Esc Vet Univ Fed Minas Gerais 31:476-477.

Santos FC, Couto ES, Santos HL (1974) Intoxicação experimental em bovinos pela "Faveira"- Dimorphandra mollis Benth. Arq Esc Vet Univ Fed Minas Gerais 26(3):319-329.

Tokarnia CH, Döbereiner J (1967) Intoxicação experimental pela fava da "Faveira" (Dimorphandra mollis Benth.) em bovinos. Pesq Agric Bras 2:367-373.

DIMORPHOTHECA CUNEATA (Thunb.) Less. [Asteraceae]

Citations:

Marais JS, Rimington C (1934) Isolation of the poisonous principle of Dimorphotheca cuneata Less. Onderstepoort J Vet Sci Anim Indus 3(1):111-117.

DIOCLEA ERECTA Hoehne [Fabaceae]

Common Names:

camaray; mucunan; tatu; tehetu; toti

Citations:

Pinto NR, Baruzzi RG (1991) Male pubertal seclusion and risk of death in Indians from Alto Xingu, central Brazil. Hum Biol 63(6):821-834.

DIOCLEA GRANDIFLORA Mart. ex Benth. [Fabaceae]

Citations:

Grant G, More LJ, McKenzie NH, et al. (1995) Nutritional and haemagglutination properties of several tropical seeds. J Agric Sci 124:437-445.

DIOCLEA LATIFOLIA Benth. [Fabaceae]

Common Names:

camaray; mucunan; tatu; tehetu; toti

Citations:

Pinto NR, Baruzzi RG (1991) Male pubertal seclusion and risk of death in Indians from Alto Xingu, central Brazil. Hum Biol 63(6):821-834.

DIOCLEA SCLEROCARPA Ducke [Fabaceae]

Citations:

Grant G, More LJ, McKenzie NH, et al. (1995) Nutritional and haemagglutination properties of several tropical seeds. J Agric Sci 124:437-445.

DIOSCOREA ALATA L. [Dioscoreaceae]

Common Names:

renta yam; white yam; wing-stalk yam

Citations:

Anonymous (1992) Toxic hypoglycemic syndrome: Jamaica, 1989-1991. MMWR Morb Mortal Wkly Rep 41(4):53-55.

Lawrence AW (1977) Cases of poisoning presenting in the Casualty Department of the Cornwall Regional Hospital, Jamaica. West Indian Med J 26(4):211-215.

Dioscorea batatas Decne. = Dioscorea polystachya Turcz.

DIOSCOREA COMMUNIS (L.) Caddick & Wilkin [Dioscoreaceae]

Synonyms:

Tamus communis L.

Common Names:

black bryony; black-eye root; Isle-of-Wight-vine; lady's-seal; mandrake; murrainberry; nueza negra; oxberry; Schmerwurz; Schwarze Zaunrübe; tamier

Citations:

Schmidt RJ, Moult SP (1983) The dermatitic properties of black bryony (Tamus communis L.). Contact Dermatitis 9(5):390-396.

DIOSCOREA CONVOLVULACEA Schltdl. & Cham. [Dioscoreaceae]

Synonyms:

Dioscorea hirsuta M. Martens & Galeotti

Common Names:

gadong; tuber yam

Citations:

Pruis GW (1941) Gadoenvergiftiging? Genesk Tijdschr Nederl Indie 81(16):864-873.

Dioscorea hirsuta M. Martens & Galeotti = Dioscorea convolvulacea Schltdl. & Cham.

DIOSCOREA POLYSTACHYA Turcz. [Dioscoreaceae]

Synonyms:

Dioscorea batatas Decne.

Common Names:

Chinese yam; cinnamon vine

Citations:

Kubo Y, Nonaka S, Yoshida H (1988) Allergic contact dermatitis from Dioscorea batatas Decaisne. Contact Dermatitis 18(5):111-112.

DIOSCOREA QUARTINIANA A. Rich. [Dioscoreaceae]

Citations:

Steyn DG (1965) An investigation into cases of suspected poisoning in Africans in northern Rhodesia. S Afr Med J 39(16):344-350.

DIOSCOREA SIMULANS Prain & Burkill [Dioscoreaceae]

Citations:

Zhang YG, Huang GZ (1988) Poisoning by toxic plants in China. Report of 19 autopsy cases. Am J Forensic Med Pathol 9(4):313-319.

DIOSPYROS KAKI Thunb. [Ebenaceae]

Common Names:

Chinese persimmon; Japanese date plum; Japanese persimmon; kaki; persimmon

Citations:

Benharroch D, Krugliak P, Porath A, et al. (1993) Pathogenetic aspects of persimmon bezoars, A case-control retrospective study. J Clin Gastroenterol 17(2):149-152.

Chisholm EM, Leong HT, Chung SC, et al. (1992) Phytobezoar: An uncommon cause of small bowel obstruction. Ann R Coll Surg Engl 74:342-344.

Choi SO, Kang JS (1988) Gastrointestinal phytobezoars in childhood. J Pediatr Surg 23(4):338-341.

Chont LK (1942) Phytobezoar and its formation in vitro. Radiology 38(Jan):14-21.

Krausz MM, Moriel EZ, Ayalon A, et al. (1986) Surgical aspects of gastrointestinal persimmon phytobezoar treatment. Am J Surg 152(5):526-530.

Morey DA, Means RL, Hirsley EL (1955) Diospyrobezoar in the postgastrectomy stomach. Arch Surg 71(6):946-948.

Moriel EZ, Ayalon A, Eid A, et al. (1983) An unusually high incidence of gastrointestinal obstruction by persimmon bezoars in Israeli patients after ulcer surgery. Gastroenterology 84:752-755.

Prandini M, Marchesi S (1999) Anaphylaxis to persimmon. Allergy 54(8):897.

Verstandig AG, Klin B, Bloom RA, et al. (1989) Small bowel phytobezoars: Detection with radiography. Radiology 172:705-707.

DIOSPYROS LYCIOIDES Desf. [Ebenaceae]

Synonyms:

Royena decidua Burch.

Common Names:

bloubos

Citations:

Van der Walt SJ (1944) Recent investigations into the toxicity of plants, etc. in the Union of South Africa. Onderstepoort J Vet Sci Anim Indus 20(1):75-83.

Van der Walt SJ, Steyn DG (1946) Recent investigations into the toxicity of plants, etc. XV. Onderstepoort J Vet Sci Anim Indus 21(1):45-55.

Diospyros macassar A. Chev. = Diospyros Rumphii Bakh.

DIOSPYROS RUMPHII Bakh. [Ebenaceae]

Synonyms:

Diospyros macassar A. Chev.

Common Names:

Macasszar; Makassarholz

Citations:

Buschke A, Joseph A (1927) Ueber Hautentzündung, hervorgerufen durch Makassarholz, mit Berücksichtigung gewerbehygienischer Fragen. Dtsch Med Wochenschr 23(Sep 23):1641-1642.

DIOSPYROS TEXANA Scheele [Ebenaceae]

Common Names:

black persimmon; Mexican persimmon; Texas persimmon

Citations:

Chont LK (1942) Phytobezoar and its formation in vitro. Radiology 38(Jan):14-21.

DIOSPYROS VIRGINIANA L. [Ebenaceae]

Common Names:

eastern persimmon; persimmon; possum wood; wild persimmon

Citations:

Chont LK (1942) Phytobezoar and its formation in vitro. Radiology 38(Jan):14-21.

Cummings CA, Copedge KJ, Confer AW (1997) Equine gastric impaction, ulceration, and perforation due to persimmon (Diospyros virginiana) ingestion. J Vet Diagn Invest 9:311-313.

Honnas CM, Schumacher J (1985) Primary gastric impaction in a pony. J Am Vet Med Assoc 187:501-502.

Kellam LL, Johnson PJ, Kramer J, et al. (2000) Gastric impaction and obstruction of the small intestine associated with persimmon phytobezoar in a horse. J Am Vet Med Assoc 216:1279-1281.

Morgan SE, Bellamy J (1994) Persimmon colic in a mare. Equine Pract 16:8-10.

Wilson, RB, Scruggs DW (1992) Duodenal obstruction associated with persimmon fruit ingestion by two horses. Equine Vet Sci 12:26-27.

Dipcadi glaucum (Burch. ex Ker Gawl.) Baker = Ornithogalum magnum (Baker) J. C. Manning & Goldblatt

DIPLARRHENA MORAEA Labill. [Iridaceae]

Common Names:

butterfly flag; native lily; white lily

Citations:

Seddon HR, Carne HR (1926) Diplarrhena moraea, Labill. A plant harmful to stock. New South Wales Dep Agric Sci Bull #26:44-45.

DIPLAZIUM ESCULENTUM (Retz.) Sw. [Dryopteridaceae]

Citations:

Devi S, Yasmeem, Singh J, et al. (1989) Patch testing animals to allergenic fern spores. J Toxicol Cutaneous Ocul Toxicol 8(2):167-172.

Prasad B (1977) Role of fern Diplazium esculentum in chronic bovine haematuria. Pantnagar J Res 2:265.

Prasad B, Joshi HC, Choudhuri PC (1977) Role of fern Diplazium esculentum in chronic bovine haematuria. I. Some clinical and biochemical studies. Pantnagar J Res 2:69-73.

DIPLOCYCLOS PALMATUS (L.) C. Jeffrey [Cucurbitaceae]

Common Names:

native bryony; wild bryony

Citations:

Mugera GM (1970) Toxic and medicinal plants of East Africa. II. Bull Epizootic Dis Afr 18(4):389-403.

DIPLOTAXIS ERUCOIDES DC. [Brassicaceae]

Common Names:

white wall rocket

Citations:

García-Ortega P, Bartolomé B, Enrique E, et al. (2001) Allergy to Diplotaxis erucoides pollen: Occupational sensitization and cross-reactivity with other common pollens. Allergy 56(7):679-683.

Diptam –see– *Dictamnus albus* L.

DIPTERYX ODORATA (Aubl.) Willd. [Fabaceae]

Common Names:

fava tonka; fève Tonka; tonga bean; tonka bean

Citations:

Hogan RP 3rd (1983) Hemorrhagic diathesis caused by drinking an herbal tea. JAMA 249(19):2679-2680.

dishwatergrass –see– *Polygonum aviculare* L.

DISTEMONANTHUS BENTHAMIANUS Baill. [Fabaceae]

Common Names:

ayan; movingui

Citations:

Morgan JW, Thomson J (1967) Ayan dermatitis. Br J Ind Med 24(2):156-158.

ditakh –see– *Detarium senegalense* J. F. Gmel.

DITAXIS DESERTORUM (Müll. Arg.) Pax & K. Hoffm. [Euphorbiaceae]

Citations:

Tokarnia CH, Chagas BR, Chagas AD, et al. (1997) Anemia hemolítica causada por Ditaxis desertorum (Euphorbiaceae) em bovinos. Pesq Vet Bras 17(3-4):112-116.

ditch crowfoot –see– *Ranunculus sceleratus* L.

ditchbur –see– *Xanthium strumarium* L.

dittamo –see– *Dictamnus albus* L.

dittany –see– *Dictamnus albus* L.

DITTRICHIA GRAVEOLENS (L.) Greuter [Asteraceae]

Synonyms:

Inula graveolens (L.) Desf.

Common Names:

stinkwort

Citations:

MacPherson J (1932) Plants injurious to man in Australia. Med J Aust 1(Feb 20):281.

Setchell BP, McInnes P, Christie DG (1964) Poisoning of sheep with anthelmintic doses of carbon tetrachloride. IV. Effect of stinkwort (Inula graveolens). Aust Vet J 40(Jan):30-31.

DITTRICHIA VISCOSA (L.) Greuter [Asteraceae]

Synonyms:

Inula viscosa (L.) Aiton

Common Names:

sticky elecampane

Citations:

Pinedo JM, Gonzalez de Canales F, Hinojosa JL, et al. (1987) Contact dermatitis to sesquiterpene lactones in Inula viscosa Aiton. Contact Dermatitis 17(5):322-323.

Sertoli A, Fabbri P, Campolmi P, et al. (1978) Allergic contact dermatitis to Salvia officinalis, Inula viscosa and Conyza bonariensis. Contact Dermatitis 4(5):314-315.

djaring –see– *Archidendron jiringa* (Jack) I. C. Nielsen

djengkol –see– *Archidendron jiringa* (Jack) I. C. Nielsen

djenkol –see– *Archidendron jiringa* (Jack) I. C. Nielsen

djenkol bean –see– *Archidendron jiringa* (Jack) I. C. Nielsen

Djenkolbohne –see– *Archidendron jiringa* (Jack) I. C. Nielsen

djering –see– *Archidendron jiringa* (Jack) I. C. Nielsen

dock –see– *Rumex acetosella* L.

doctor gum –see– *Metopium toxiferum* (L.) Krug & Urb.; *Schinus terebinthifolius* Raddi

dodder –see– *Cuscuta campestris* Yunck.; *Cuscuta epithymum* (L.) L.

doddies –see– *Jacobaea vulgaris* Gaertn.

DODONAEA VISCOSA Jacq. [Sapindaceae]

Citations:

Colodel EM, Traverso SD, Seitz AL, et al. (2003) Spontaneous poisoning by Dodonaea viscosa (Sapindaceae) in cattle. Vet Hum Toxicol 45(3):147-148.

DOELLINGERIA UMBELLATA (Mill.) Nees [Asteraceae]

Synonyms:

Aster umbellatus Mill.

Common Names:

umbellate aster

Citations:

Bergeron JM, Jodoin L (1985) Fiabilité des mesures de poids et d'examens histopathologiques dans les études d'intoxication du campagnol des champs (Microtus pennsylvanicus). Can J Zool 63(4):804-810.

dog apple –see– *Asimina triloba* (L.) Dunal

dog bur –see– *Cynoglossum officinale* L.

dog camomile –see– *Anthemis cotula* L.

dog chamomile –see– *Anthemis cotula* L.

dog daisy –see– *Anthemis cotula* L.

dog fennel –see– *Anthemis cotula* L.

dog flea weed –see– *Parthenium hysterophorus* L.

dog leek –see– *Hyacinthoides non-scripta* (L.) Chouard ex Rothm.

dog mercury –see– *Mercurialis perennis* L.

dog parsley –see– *Aethusa cynapium* L.

dog poison –see– *Aethusa cynapium* L.

dogbane –see– *Apocynum cannabinum* L.

dog's-tangere –see– *Cynoglossum officinale* L.

dogweed –see– *Eremophila deserti* (A. Cunn. ex Benth.) Chinnock

dogwood –see– *Cornus oblonga* Wall.; *Euonymus europaeus* L.; *Solanum dulcamara* L.

dohány –see– *Nicotiana tabacum* L.

dol diablo –see– *Coriaria ruscifolia* L.

Dolichos axillaris E Mey. = Macrotyloma axillare (E. Mey.) Verdc.

Dolichos lablab L. = Lablab purpureus (L.) Sweet

Dolichos pruriens L. = Mucuna pruriens (L.) DC.

doll's-eyes –see– *Actaea rubra* (Aiton) Willd.

dolphin flower –see– *Consolida regalis* Gray

dome-d'onze-heures –see– *Ornithogalum umbellatum* L.

dominguinho –see– *Cestrum laevigatum* Schltdl.

Don Juan –see– *Nicotiana glauca* Graham

donkey tail –see– *Euphorbia myrsinites* L.

doob –see– *Cynodon dactylon* (L.) Pers.

doolan –see– *Acacia salicina* Lindl.

doom bark –see– *Erythrophleum suaveolens* (Guill. & Perr.) Brenan

Doornappel –see– *Datura stramonium* L.

doorweed –see– *Polygonum aviculare* L.

dooryard knotweed –see– *Polygonum aviculare* L.

dopnit –see– *Wikstroemia ovata* C. A. Mey.

Dorfpflanze –see– *Anthemis cotula* L.

dormilión –see– *Rudbeckia laciniata* L.

Dornapfel –see– *Datura stramonium* L.

Dornfarn –see– *Dryopteris carthusiana* (Vill.) H. P. Fuchs

doss cabbage –see– *Senecio latifolius* DC.

doubgrass –see– *Cynodon dactylon* (L.) Pers.

double bean –see– *Phaseolus lunatus* L.; *Vicia faba* L.

douce amère –see– *Solanum dulcamara* L.

Douglas'-spur lupine –see– *Lupinus argenteus* Pursh var. argenteus

Douglas'-water hemlock –see– *Cicuta douglasii* (DC.) J. M. Coult. & Rose

downy sartwellia –see– *Sartwellia flaveriae* A. Gray

downy spurge –see– *Euphorbia pilosa* L.

downy thorn apple –see– *Datura inoxia* Mill.; *Datura metel* L.

drabok –see– *Lolium temulentum* L.

DRACAENA FRAGRANS (L.) Ker Gawl [Ruscaceae]

Citations:

Ramón MF, Ballesteros S, Martínez-Arrieta R, et al. (2003) Plant and herb exposures in Spain. J Toxicol Clin Toxicol 41(4):514.

Drachenkraut –see– *Calla palustris* L.

dragon arum –see– *Pinellia ternata* (Thunb.) Makino

dragon bone tree –see– *Euphorbia lactea* Haw.

dragon bones –see– *Euphorbia lactea* Haw.

dragon turnip –see– *Arisaema triphyllum* (L.) Schott

dragonroot –see– *Arisaema triphyllum* (L.) Schott

drake –see– *Lolium temulentum* L.

dravanti –see– *Jatropha curcas* L.

DREGEA VOLUBILIS (L. f.) Benth. ex Hook. f. [Asclepiadaceae]

Common Names:

anguna

Citations:

Tennekoon KH, Jeevathayaparan S, Kurukulasooriya AP, et al. (1991) Possible hepatotoxicity of Nigella sativa seeds and Dregea volubilis leaves. J Ethnopharmacol 31:283-289.

drilgras –see– *Equisetum ramosissimum* Desf.

Drimia alta R. A. Dyer = Drimia robusta Baker

DRIMIA ALTISSIMA (L. f.) Ker Gawl. [Hyacinthaceae]

Synonyms:

Urginea altissima (L. f.) Baker

Common Names:

maerman; slangkop

Citations:

Shone DK (1966) Poisonous plants of Rhodesia. Part I. Rhodesia Zambia Malawi J Agric Res 4:81-94.

Shone DK (1967) Poisonous plants of Rhodesia. Part II. Rhodesia Zambia Malawi J Agric Res 5:29-41.

Steyn DG (1936) Recent investigations into the toxicity of known and unknown poisonous plants in the Union of South Africa. Onderstepoort J Vet Sci Anim Indus 7(1):169-178.

DRIMIA DEPRESSA (Baker) Jessop [Hyacinthaceae]

Synonyms:

Urginea capitata (Hook.) Baker

Common Names:

berg slangkop

Citations:

Mitchell DT, Canhan AS, Bayer AJ (1934) Urginea capitata Baker - the berg slangkop. Its toxic effect on ruminants. Onderstepoort J Vet Sci Anim Indus 2(2):681-689.

DRIMIA MACROCENTRA (Baker) Jessop [Hyacinthaceae]

Synonyms:

Urginea macrocentra Baker

Common Names:

Natal slangkop

Citations:

Van der Walt SJ, Steyn DG (1943) Recent investigations into the toxicity of plants, etc. XIII. Onderstepoort J Vet Sci Anim Indus 18(1-2):207-224.

DRIMIA MARITIMA (L.) Stearn [Hyacinthaceae]

Synonyms:

Urginea maritima (L.) Baker

Common Names:

albariã; cebolla albarrana; cila; crusader's-spears; Mäasezwiebel; medicinal squill; Mediterranean squill; Meerzweibel; mouse onion; onion-de-mer; red squill; scille maritime; scille officinale; sea onion; sea squill; squill; Steinheil; white squill

Citations:

Hamouda C, Amamou M, Thabet H, et al. (2000) Plant poisonings from herbal medication admitted to a Tunisian toxicologic intensive care unit, 1983-1998. Vet Hum Toxicol 42(3):137-141.

Leblanc FJ, Lee CO (1939) A study of the toxic properties of red squill. J Am Pharm Assoc. Mar:151-154.

Nagle AC (1948) Red squill poisoning in a dog. J Am Vet Med Assoc 112:139.

Rietz JH, Moore EN (1943) Red squill (raticide) poisoning in swine. J Am Vet Med Assoc 102(791):120-121.

Tuncok Y, Kozan O, Cavdar C, et al. (1995) Urginea maritima (squill) toxicity. J Toxicol Clin Toxicol 33(1):83-86.

Ward JC, Barber CW, Garlough FE, et al. (1937) Susceptibility of hogs to red squill. J Am Pharm Assoc 26(2):137-139.

DRIMIA PHYSODES (Jacq.) Jessop [Hyacinthaceae]

Synonyms:

Urginea physodes (Jacq.) Baker

Common Names:

slangkop

Citations:

Nel PW, Schultz RA, Jordaan P, et al. (1987) Cardiac glycoside poisoning in sheep caused by Urginea physodes (Jacq.) Bak. and the isolated physodine A. Onderstepoort J Vet Res 54(4):641-644.

DRIMIA ROBUSTA Baker [Hyacinthaceae]

Synonyms:

Drimia alta R. A. Dyer

Citations:

Van der Walt SJ, Steyn DG (1943) Recent investigations into the toxicity of plants, etc. XIII. Onderstepoort J Vet Sci Anim Indus 18(1-2):207-224.

DRIMIA SANGUINEA (Schinz) Jessop [Hyacinthaceae]

Synonyms:

Urginea burkei Baker; ***Urginea sanguinea*** Schinz

Common Names:

blueweed; Burke's-slangkop; slangkop; Transvaal slangkop

Citations:

Basson PA (1988) Acute and chronic cardiomyopathies caused by monofluoroacetic acid in giftblaar (Dichapetalum cymosum) and cardiac glycosides in slangkop (Urginea sanguinea). J S Afr Vet Assoc 59(2):100.

Dunphy JT (1906) Report of experiments carried out to observe effects of certain poisonous plants on sheep and goats. Transvaal Agric J 4(14):315-323.

Joubert JP, Schultz RA (1982) The treatment of Urginea sanguinea Schinz poisoning in sheep with activated charcoal and potassium chloride. J S Afr Vet Assoc 53(1):25-28.

McVann A, Havlik I, Joubert PH, et al. (1992) Cardiac glycoside poisoning involved in deaths from traditional medicines. S Afr Med J 81(3):139-141.

Schultz RA, Kellerman TS, Van den Berg H (2005) The role of fluorescence polarization immuno-assay in the diagnosis of plant-induced cardiac glycoside poisoning livestock in South Africa. Onderstepoort J Vet Res 72(3):189-201.

dronkgras –see– *Equisetum ramosissimum* Desf.

DROSANTHEMUM FLORIBUNDUM (Haw.) Schwantes [Aizoaceae]

Synonyms:

Drosanthemum hispidum (L.) Schwantes

Common Names:

vygie

Citations:

Van der Walt SJ, Steyn DG (1939) Recent investigations into the toxicity of known and unknown poisonous plants in the Union of South Africa. IX. Onderstepoort J Vet Sci Anim Indus 12:335-366.

Drosanthemum hispidum (L.) Schwantes = Drosanthemum floribundum (Haw.) Schwantes

Drosselkirschbaum –see– *Frangula alnus* Mill.

drumhead cabbage –see– *Brassica oleracea* L. var. *capitata* L.

Drummond's-rattlebush –see– *Sesbania drummondii* (Rydb.) Cory

Drummond's-sesbane –see– *Sesbania drummondii* (Rydb.) Cory

drumstick tree –see– *Moringa oleifera* Lam.

drunk –see– *Lolium temulentum* L.

Drüsenklee –see– *Cullen corylifolium* (L.) Medik.

dry whiskey –see– *Calia secundiflora* (Ortega) Yakovlev

drymaria –see– *Drymaria pachyphylla* Wooton & Standl.

DRYMARIA ARENARIOIDES Humb. & Bonpl. ex Schult. [Caryophyllaceae]

Common Names:

alfombrilla; drymary

Citations:

Dollahite JW (1959) Toxicity of Drymaria arenarioides for cattle, sheep, and goats. J Am Vet Med Assoc 135(Jul 15):125-127.

Dollahite JW, Anthony WV (1956) Toxicity of Drymaria arenarioides for cattle, sheep and goats. Texas Agric Exp Sta Prog Rep #1911:4 pp.

Larios F, Jabalera J (1983) Toxicidad de alfombrilla (Drymaria arenarioides) em bovinos alimentados experimentalmente. Tecnica Pecuaria en Mexico 44:86-91.

Larios MV, Javalera J (1976) Alteraciones anatomopatologicas, hemograma, y quimica sanguinea en el intoxicacion experimental aguda por alfombrilla (Drymaria arenarioides) en bovinos. Tecnica Pecuaria en Mexico 30:111.

Ocampo MV (1972) Investigaciones biomédicas de los efectos tóxicos y abortivos de las saponinas de Drymaria arenarioides, en ratonas gestantes. Veterinaria (Mexico) 3(4):94-97.

Williams MC (1978) Toxicity of saponins in alfombrilla (Drymaria arenarioides). J Range Manag 31(3):182-184.

DRYMARIA HOLOSTEOIDES Benth. [Caryophyllaceae]

Common Names:

carpet weed; drymary; thick-leaf drymary

Citations:

Pammel LH (1922) Carpet weed suspected of being poisonous. Vet Med 17:36.

DRYMARIA PACHYPHYLLA Wooton & Standl. [Caryophyllaceae]

Common Names:

drymaria; drymary; inkweed; thick-leaf drymary

Citations:

Lantow JL (1929) The poisoning of livestock by Drymaria pachyphylla. New Mexico Agric Exp Sta Bull #173:13 pp.

Mathews FP (1933) The toxicity of Drymaria pachyphylla for cattle, sheep and goats. J Am Vet Med Assoc 83:255-260.

Mathews FP (1938) Poisonous plants in the Davis Mountains area. Texas Agric Exp Sta Annu Rep 51:13-14.

drymary –see– *Drymaria arenarioides* Humb. & Bonpl. ex Schult.; *Drymaria holosteoides* Benth.; *Drymaria pachyphylla* Wooton & Standl.

DRYOPTERIS AFFINIS (Lowe) Fraser-Jenk. subsp. borreri (Newman) Fraser-Jenk. [Dryopteridaceae]

Synonyms:

Dryopteris borreri Newman

Common names:

trusty male fern

Citations:

MacLeod NS, Greig A, Bonn JM, et al. (1978) Poisoning in cattle associated with Dryopteris filix-mas and D borreri. Vet Rec 102(11):239-240.

Dryopteris borreri Newman = Dryopteris affinis (Lowe) Fraser-Jenk. subsp. borreri (Newman) Fraser-Jenk.

DRYOPTERIS CARTHUSIANA (Vill.) H. P. Fuchs [Dryopteridaceae]

Common Names:

buckle fern; Dornfarn

Citations:

Murray V (1966) Suspected poisoning by common buckle fern (Dryopteris family). Ir Vet J 20:122-124.

DRYOPTERIS FILIX-MAS (L.) Schott [Dryopteridaceae]

Common Names:

aspidium; Farnkraut; felce maschia; filis mas; filix mas; fougère mâle; helecho macho; male fern; Wurmfarn

Citations:

Balaščák J, Gdovin T, Stiko M (1968) Über das auftreten von Haematuria vesicalis bovis chronica in der Slowakei. Folia Vet Kosice 12(2):111.

Edgar JT, Thin IM (1968) Plant poisoning involving male fern. Vet Rec 82:33-34.

Greiner H (1952-1954) Wurmfarnvergiftung. Sammlung Vergiftungsfallen 14:124-125.

Hall MC (1914) Unusual case of fatal poisoning from administration of male-fern as vermifuge. JAMA 63(3):242-243.

Houtzagers JJ (1965) Nierinsufficiëntie na toediening van filix mas. Tijdschr Geneeskd 109(4):178-179.

MacLeod NS, Greig A, Bonn JM, et al. (1978) Poisoning in cattle associated with Dryopteris filix-mas and D borreri. Vet Rec 102(11):239-240.

Mitchell GB, Wain EB (1983) Suspected male fern poisoning in cattle. Vet Rec 113(8):188.

Rosen ES, Edgar JT, Smith JL (1969) Male fern retro-bulbar neuropathy in cattle. Ophthalmol Soc UK 89:285-299.

Rosen ES, Edgar JT, Smith JL (1970) Male fern retro-bulbar neuropathy in cattle. J Small Anim Pract 10(11):619-625.

Smyth PJ (1968) An outbreak of male fern (Dryopteris filix-mas) poisoning. Ir Vet J 22:69-71.

Stryczek J (1984) Paproć przyczynq dyskwalifikacji jelit bydła. Med Weter 40(8):488-489.

Valtonen EJ, Takki S (1968) Acute hepatocellular damage caused by oleoresin of the male fern in the rat: An electron microscope study. Acta Pharmacol Toxicol (Copenh) 26(2):169-176.

DRYOPTERIS JUXTAPOSITA H. Christ [Dryopteridaceae]

Citations:

Gounalan S, Somvanshi R, Kataria M, et al. (1999) Effect of bracken (Pteridium aquilinum) and dryopteris (Dryopteris juxtaposita) fern toxicity in laboratory rabbits. Indian J Exp Biol 37(10):980-985.

Kataria M, Somvanshi R, Dash S (1998) Biochemical and histological changes in blood, erythrocytes and tissue of rats on feeding Dryopteris juxtaposita fern. Indian J Exp Biol 36(5):510-513.

dubbelljie –see– *Tribulus terrestris* L.

DUBOISIA MYOPOROIDES R. Br. [Solanaceae]

Common Names:

corkwood

Citations:

Barnes R, Jones T, Simpson G, et al. (1971) An unusual occupational hazard. Med J Aust 2(20):1018-1020.

Biersteker K, Koning C, van Lith GH, et al. (1970) An air-borne epidemic of one-sided mydriasis. Arch Environ Health 20(3):410-411.

Pearn J (1981) Corked up: Clinical hyoscine poisoning with alkaloids of the native corkwood, Duboisia. Med J Aust 2(8):422-423.

duckbush –see– *Gomphocarpus fruticosa* (L.) W. T. Aiton

duckretter –see– *Veratrum viride* Aiton

duck's-foot –see– *Podophyllum peltatum* L.

dudgeon –see– *Buxus sempervirens* L.

dudhi –see– *Chamaesyce hirta* (L.) Millsp.

Dugaldia hoopesii (A. Gray) Rydb. = Hymenoxys hoopesii (A. Gray) Bierner

Duke-of-Argyll's-tea-tree –see– *Lycium barbarum* L.

dumb cane –see– *Dieffenbachia bausei* Regel; *Dieffenbachia seguine* (Jacq.) Schott

dunal –see– *Asimina triloba* (L.) Dunal

duncecap larkspur –see– *Delphinium barbeyi* (Huth) Huth

Durand's-false hellebore –see– *Veratrum californicum* Durand

Durango root –see– *Datisca glomerata* (C. Presl) Baill.

duranta –see– *Duranta erecta* L.

DURANTA ERECTA L. [Verbenaceae]

Synonyms:

Duranta repens L.

Common Names:

adonis morado; coralillo; cuenta-de-oro; dewdrops; duranta; espina-de-palma; garbancillo; geisha girl; golden dewdrop; pigeonberry; sky flower; yellow hat tree

Citations:

Scanlan SN, Eagles DA, Vacher NE, et al. (2006) Duranta erecta poisoning in nine dogs and a cat. Aust Vet J 84(10):367-370.

Duranta repens L. = Duranta erecta L.

duras milk blanco –see– *Solanum glaucophyllum* Desf.

duraznillo –see– *Cestrum parqui* L'Her.; *Solanum rostratum* Dunal

duraznillo blanco –see– *Solanum glaucophyllum* Desf.; *Xanthium strumarium* L. var. canadense (Mill.) Torr. & A. Gray

duraznillo negro –see– *Cestrum parqui* L'Her.

durian –see– *Durio zibethinus* L.

DURIO ZIBETHINUS L. [Malvacaceae]

Common Names:

durian; house mango; nanas

Citations:

Fasal P (1945) Cutaneous diseases in the tropics. A clinical study based on observations in Malaya. Arch Derm Syphilol 51(3):163-171.

Olivieri J, Quiliquini-Chambard AM, Hauser C (2002) Allergy to durian. Allergy 57(3):263.

durra –see– *Sorghum bicolor* (L.) Moench

Dürrwurz –see– *Inula conyzae* (Griess.) Meikle

duscle –see– *Solanum nigrum* L.

Dutch clover –see– *Trifolium repens* L.

Dutch hyacinth –see– *Hyacinthus orientalis* L.

Dutchman's-breeches –see– *Dicentra canadensis* (Goldie) Walp.; *Thamnosma texana* (A. Gray) Torr.

duwweljie –see– *Tribulus terrestris* L.

dwale –see– *Atropa belladonna* L.

dwarf bay –see– *Daphne mezereum* L.

dwarf Darling pea –see– *Swainsona luteola* F. Muell.

dwarf elder –see– *Sambucus nigra* L.

dwarf juniper –see– *Juniperus communis* L.

dwarf larkspur –see– *Delphinium nuttallianum* Pritz.; *Delphinium tricorne* Michx.

dwarf laurel –see– *Kalmia angustifolia* L.

dwarf milkweed –see– *Asclepias verticillata* L.

dwarf nettle –see– *Urtica dioica* L.

dwarf poison milkweed –see– *Asclepias pumila* (A. Gray) Vail

dwarf sheep laurel –see– *Kalmia angustifolia* L.

dwarf silky oak –see– *Grevillea banksii* R. Br.

dwarf sumach –see– *Rhus typhina* L.

dwarf yellow day lily –see– *Hemerocallis minor* Mill.

dwayberry –see– *Atropa belladonna* L.

DYERA COSTULATA (Miq.) Hook. f. [Apocynaceae]

Common Names:

jelutong

Citations:

Siregar RS (1975) Occupational dermatoses among foresters. Contact Dermatitis 1(1):33-37.

DYERA LOWII Hook. f. [Apocynaceae]

Citations:

Siregar RS (1975) Occupational dermatoses among foresters. Contact Dermatitis 1(1):33-37.

Dysosma pleiantha (Hance) Woodson = Podophyllum pleianthum Hance

DYSPHANIA AMBROSIOIDES (L.) Mosyakin & Clemants [Chenopodiaceae]

Synonyms:

Chenopodium ambrosioides L.

Common Names:

ambroisie; American wormseed; American wormwood; epazote; epazote-de-comer; hipazote; Jerusalem oak; Mexican goosefoot; Mexican tea; pasote; pazote; sagrado; Spanish tea; stinkweed; wormseed; wormseed goosefoot

Citations:

Anonymous (1934) Investigation of stock-poisoning plants. U S Dep Agric Bur Anim Indus Chief Rep 1934:38.

Farquharson J (1938) Deafness due to toxicity of oil of Chenopodium. J Am Vet Med Assoc 46(6):329.

Guirola L, Garcia G, Torrealba A, et al. (1992) Acute renal failure from the ingestion of toxic plants. Vet Hum Toxicol 34(6):548.

Kinsley AT (1936) Chenopodium intoxication in swine. Vet Med 31(Jun):253.

Nyazema NZ (1986) Herbal toxicity in Zimbabwe. Trans R Soc Trop Med Hyg 80(3):448-450.

Pammel LH (1923) Worm seed or Mexican tea poisonous to geese. Vet Med 18:736.

DYSPHANIA CARINATA (R. Br.) Mosyakin & Clemants [Chenopodiaceae]

Synonyms:

Chenopodium carinatum R. Br.

Common Names:

boggabri; crested goosefoot; green crumbweed; keel crumbweed; keel goosefoot; ridged goosefoot; scented goosefoot

Citations:

Anonymous (1929) The poison plants committee. J CSIRO Aust 2:40-48.

Seddon HR (1932) Chenopodium carinatum. ("crested goosefoot" or boggabri.") Poisonous to stock under certain circumstances. Agric Gaz New South Wales 43:763-764.

DYSPHANIA GRAVEOLENS (Willd.) Mosyakin & Clemants [Chenopodiaceae]

Synonyms:

Chenopodium graveolens Willd.

Common names:

epazote

Citations:

Montoya-Cabrera MA, Escalante-Galindo P, Meckes-Fisher M, et al. (1996) Envenenmiento mortal causado por el aceite de epazote, Chenopodium graveolens. Gac Med Mex 132(4):433-437.

E

eagle fern –see– *Pteridium aquilinum* (L.) Kuhn; *Pteridium aquilinum* (L.) Kuhn var. latiusculum (Desf.) Underw. ex A. Heller

ear drops –see– *Dicentra cucullaria* (L.) Bernh.

Earle's locoweed –see– *Astragalus mollissimus* Torr. var. earlei (Greene ex Rydb.) Tidestr.

earring plant –see– *Crotalaria retusa* L.

earth gall –see– *Veratrum viride* Aiton

earthnut –see– *Arachis hypogaea* L.

East Coast rengas –see– *Gluta renghas* L.

East Indian ginger –see– *Zingiber officinale* Roscoe

East Indian rosewood –see– *Dalbergia latifolia* Roxb.; *Dalbergia retusa* Hemsl.

East Indian satinwood –see– *Chloroxylon swietenia* DC.

Easter flower –see– *Euphorbia pulcherrima* Willd. ex Klotzsch

Easter lily –see– *Lilium longiflorum* Thunb.

Easter rose –see– *Helleborus niger* L.

eastern baccharis –see– *Baccharis halimifolia* L.

eastern bitterweed –see– *Helenium amarum* (Raf.) H. Rock

eastern black nightshade –see– *Solanum ptycanthum* Dunal

eastern chokecherry –see– *Prunus virginiana* L.

eastern mistletoe –see– *Phoradendron leucarpum* (Raf.) Reveal & M. C. Johnst.

eastern oakleaf –see– *Toxicodendron diversilobum* (Torr. & A. Gray) Greene

eastern oakleaf poison ivy –see– *Toxicodendron pubescens* Mill.

eastern persimmon –see– *Diospyros virginiana* L.

eastern poison oak –see– *Toxicodendron pubescens* Mill.

eastern whorled milkweed –see– *Asclepias verticillata* L.

easy tree philodendron –see– *Philodendron bipinnatifidum* Schott ex Endl.

Ebenbaum –see– *Taxus baccata* L.

ECBALLIUM ELATERIUM (L.) A. Rich. [Cucurbitaceae]

Common Names:

bitter apple; concombre d'âne; concombre sauvage; elaterium; momordique; Spritzgurke; squirting cucumber; wild cucumber

Citations:

Raikhlin B, Bentur Y (1999) Ecballium elaterium - Medicine or poison? J Toxicol Clin Toxicol 37:415.

Raikhlin B, Bentur Y (2000) Ecballium elaterium (squirting cucumber) - Remedy or poison? J Toxicol Clin Toxicol 38(3):305-308.

Satar S, Gokel Y, Toprak N, et al. (2001) Life-threatening uvular angioedema caused by Ecballium elaterium. Eur J Emerg Med 8(4):337-339.

Vlachos P, Kanitsakis NN, Kokonas N (1994) Fatal cardiac and renal failure due to Ecballium elaterium (squirting cucumber). J Toxicol Clin Toxicol 32(6):737-738.

ECHINOCHLOA COLONA (L.) Link [Poaceae]

Citations:

Sperry OE, Turk RD, Hoffman GO, et al. (1955) Photosensitization of cattle in Texas. Texas Agric Exp Sta Bull #812:8 pp.

ECHINOCHLOA FRUMENTACEA Link [Poaceae]

Common Names:

Japanese barnyard millet; Japanese millet; samak; sanwa millet

Citations:

Allen JG, Creeper JH, Forshaw D, et al. (2004) Plant-associated diseases, either new or new to the state, encountered over the last decade (1991-2001) in Western Australia. In: Acamovic T, Stewart CS, Pennycott TW (eds.) Poisonous plants and related toxins. CABI. Wallingford, UK. pp. 540-547.

Echinopanax elatus Nakai = Oplopanax elatus (Nakai) Nakai

ECHINOPOGON OVATUS (G. Forst.) P. Beauv. [Poaceae]

Common Names:

hedgehoggrass; rough beardedgrass

Citations:

Seddon HR, Carne HR (1926) Staggers in stock due to rough-bearded grass (Echinopogon ovatus). New South Wales Dep Agric Sci Bull #26:34-40.

echium –see– *Echium plantagineum* L.; *Echium vulgare* L.

Echium lycopsis auct. = Echium plantagineum L.

ECHIUM PLANTAGINEUM L. [Boraginaceae]

Synonyms:

Echium lycopsis auct.

Common Names:

Calamity Jane; echium; Lady Campbell's-weed; Paterson's-curse; purple bugloss; purple viper's-bugloss; Riverina bluebell; Salvation Jane; viper's-bugloss

Citations:

Culvenor CC, Jago MV, Peterson JE, et al. (1984) Toxicity of Echium plantagineum (Paterson's curse). I. Marginal toxic effects in Merino wethers from long-term feeding. Aust J Agric Res 35(2):293-304.

Méndez MC, Riet-Correa F, Schild AL, et al. (1985) Intoxicação por Echium plantagineum (Boraginaceae) em bovinos no Rio Grande do Sul. Pesq Vet Bras 5(2):57-64.

Peterson JE, Jago MV (1984) Toxicity of Echium plantagineum (Paterson's curse). II. Pyrrolizidine alkaloid poisoning in rats. Aust J Agric Res 35(2):305-315.

Schild AC, Motta AC, Riet-Correa F, et al. (2004) Photosensitization in cattle in southern Brazil. In: Acamovic T, Stewart CS, Pennycott TW (eds.) Poisonous plants and related toxins. CABI. Wallingford, UK. pp. 162-166.

Seaman JT (1978) Pyrrolizidine alkaloid poisoning of horses. Aust Vet J 54(3):150.

Seaman JT (1985) Hepatogenous chronic copper poisoning in sheep associated with grazing Echium plantagineum. Aust Vet J 62(7):247-248.

Seaman JT (1987) Pyrrolizidine alkaloid poisoning of sheep in New South Wales. Aust Vet J 64(6):164-167.

Seaman JT, Dellow JJ (1987) A survey of landholder attitudes to Echium plantagineum L. in three shires in the central west of New South Wales. J Aust Inst Agric Sci 53(4):296-301.

Seaman JT, Dixon RJ (1989) Investigations into the toxicity of Echium plantagineum in sheep. 2. Pen feeding experiments. Aust Vet J 66(9):286-292.

Seaman JT, Turvey WS, Ottaway SJ, et al. (1989) Investigations into the toxicity of Echium plantagineum in sheep. 1. Field grazing experiments. Aust Vet J 66(9):279-285.

Sharrock AG (1969) Pyrrolizidine alkaloid poisoning in a horse in New South Wales. Aust Vet J 45(8):388.

St. George-Grambauer TD, Rac R (1962) Hepatogenous chronic copper poisoning in sheep in South Australia due to the consumption of Echium plantagineum L. (Salvation Jane). Aust Vet J 38(May):288-293.

ECHIUM VULGARE L. [Boraginaceae]

Common Names:

blue devil; blue thistle; blue-weed; echium; viper's-bugloss

Citations:

Moyano MR, García A, Rueda A, et al. (2006) Echium vulgare and Senecio vulgaris poisoning in fighting bulls. J Vet Med A Physiol Pathol Clin Med 53(1):24-25.

Echte Kamille –see– *Matricaria recutita* L.

Echter Eisenhut –see– *Aconitum napellus* L.

Echter Steinklee –see– *Melilotus officinalis* Lam.

Echter Sturmhut –see– *Aconitum napellus* L.

Echtes Mahagoni –see– *Swietenia macrophylla* King; *Swietenia mahagoni* (L.) Jacq.

ecue –see– *Coula edulis* Baill.

edded –see– *Chamaeleon gummifera* (L.) Cass.

eddo –see– *Colocasia esculenta* (L.) Schott

eddoe –see– *Colocasia esculenta* (L.) Schott

Edelteak –see– *Madhuca longifolia* (L.) J. F. Macbr.

Edelwicke –see– *Lathyrus odoratus* L.

Efeu –see– *Hedera helix* L.

Efeutute –see– *Epipremnum pinnatum* (L.) Engl.

eggplant –see– *Solanum melongena* L.

egynyári szélfü –see– *Mercurialis annua* L.

Egyptian clover –see– *Trifolium alexandrinum* L.

Eibe –see– *Taxus baccata* L.

Eibenbaum –see– *Taxus baccata* L.

Eibennadeln –see– *Taxus baccata* L.

Eichel –see– *Quercus robur* L.

Eichenbaum –see– *Quercus robur* L.

EICHHORNIA CRASSIPES (Mart.) Solms [Pontederiaceae]

Common Names:

water hyacinth

Citations:

Babu NS (1987) Studies on the pathology of ovine and caprine kidneys with special reference to the effect of feeding water hyacinth to sheep. Indian J Vet Pathol 11:86-87.

Babu NS, Paliwal OP, Charan K, et al. (1988) Effects of water hyacinth feeding in sheep with special reference to renal lesion. Indian J Vet Pathol 12:33-36.

Singh KP (1987) Studies on the pathology of ovine and caprine liver with particular reference to feeding of water hyacinth in sheep. Indian J Vet Pathol 11:92-93.

Einjährige Bingelkraut –see– *Mercurialis annua* L.

Eischaupe –see– *Convallaria majalis* L.

Eisenhut –see– *Aconitum napellus* L.

Eisensäbelholz –see– *Mesua ferrea* L.

el halouk –see– *Orobanche minor* Sm.

ELAEAGNUS ANGUSTIFOLIA L. [Elaeagnaceae]

Common Names:

árbol-del-paraíso; matapolilla; oleaster; olivo-de-Bohemia; Russian olive; silverberry

Citations:

Sastre J, Lluch-Bernal M, Bustillo AM, et al. (2004) Allergenicity and cross-reactivity of Russian olive pollen (Elaeagnus angustifolia). Allergy 59(11):1118-1186.

Elaeodendron buchananii (Loes.) Loes. = Cassine buchananii Loes.

Elaeodendron keniense Loes. = Cassine buchananii Loes.

elaterium –see– *Ecballium elaterium* (L.) A. Rich.

elderberry –see– *Sambucus cerulea* Raf.; *Sambucus nigra* L.

eléboro blanco –see– *Veratrum album* L.

eléboro fetido –see– *Helleborus foetidus* L.

elecampane –see– *Inula helenium* L.

elegant death camas –see– *Anticlea elegans* (Pursh) Rydb.

elephant grass –see– *Pennisetum purpureum* Schumach.

elephant's-ear –see– *Alocasia macrorrhizos* (L.) G. Don; *Caladium bicolor* (Aiton) Vent.; *Colocasia esculenta* (L.) Schott; *Epipremnum pinnatum* (L.) Engl.; *Mandragora officinarum* L.; *Podophyllum peltatum* L.; *Xanthosoma sagittifolium* (L.) Schott

elephant's-scratch wort –see– *Mucuna pruriens* (L.) DC.

ELETTARIA CARDAMOMUM (L.) Maton [Zingiberaceae]

Synonyms:

Amomum cardamomum L.

Common Names:

cardamom; lesser cardamom; small cardamon

Citations:

Hjorth N, Roed-Petersen J (1976) Occupational protein contact dermatitis in food handlers. Contact Dermatitis 2(1):28-42.

Mobacken H, Fregert S (1975) Allergic contact dermatitis from cardamon. Contact Dermatitis 1(3):175-176.

Niinimäki A (1984) Delayed-type allergy to spices. Contact Dermatitis 11(1):34-40.

Seetharam KA, Pasricha JS (1987) Condiments and contact dermatitis of the finger-tips. Indian J Dermatol Venereol Leprol 53:325-328.

eleuthero ginseng –see– *Eleutherococcus senticosus* (Rupr. & Maxim.) Maxim.

ELEUTHEROCOCCUS SENTICOSUS (Rupr. & Maxim.) Maxim. [Araliaceae]

Common Names:

eleuthero ginseng; Siberian ginseng

Citations:

Koren G, Randor S, Martin S, et al. (1990) Maternal ginseng use associated with neonatal androgenization. JAMA 264(22):2866.

Siegel RK (1980) Ginseng and high blood pressure. JAMA 243(1):32.

Ellangowan –see– *Eremophila deserti* (A. Cunn. ex Benth.) Chinnock

Ellangowan poison bush –see– *Eremophila deserti* (A. Cunn. ex Benth.) Chinnock

ELLIOTTIA PANICULATA (Siebold & Zucc.) Benth. & Hook. f. [Ericaceae]

Synonyms:

Tripetaleia paniculata Siebold & Zucc.

Common Names:

hotsutsuji

Citations:

Tokuda Y, Sumita E (1925) Studies on poisonous honey in Japan. I. On the source of poisonous honey. Bee World 7(1):4-5.

Elsebaum –see– *Prunus padus* L.

EMBELIA RIBES Burm. f. [Myrsinaceae]

Common Names:

embilia; enkoko; umkoko; vidanga

Citations:

Kholkute SD, Kekare MB, Jathar VS, et al. (1978) Antifertility effects of Embelia ribes Burm. Indian J Exp Biol 16(10):1035-1037.

Low G, Rogers LJ, Brumley SP, et al. (1985) Visual deficits and retinotoxicity caused by the naturally occurring anthelmintics, Embelia ribes and Hagenia abyssinica. Toxicol Appl Pharmacol 81(2):220-230.

Purandare TV, Kholkute SD, Gurjar A, et al. (1979) Semen analysis and hormonal levels in bonnet macaques administered Embelia ribes berries, an indigenous plant having contraceptive activity. Indian J Exp Biol 17(9):935-936.

embilia –see– *Embelia ribes* Burm. f.

Emblica officinalis Gaertn. = Phyllanthus emblica L.

emborrachacabras –see– *Coriaria myrtifolia* L.

emetico –see– *Jatropha multifida* L.

Emory's-locoweed –see– *Astragalus emoryanus* (Rydb.) Cory

Emory's-milk vetch –see– *Astragalus emoryanus* (Rydb.) Cory

emu bush –see– *Eremophila latrobei* F. Muell.

ENCELIA CALIFORNICA Nutt. [Asteraceae]

Citations:

Webber HJ (1932) Dermatitis produced by Encelia californica Nutt. Science 76(1971):324-325.

ENCEPHALARTOS HILDEBRANDTII A. Braun & C. D. Bouché [Zamiaceae]

Citations:

Mugera GM (1969) Induction of kidney tumours in the rat by feeding Encephalartos hildebrandtii for short periods. Br J Cancer 23(4):755-756.

Mugera GM, Nderito P (1968) Toxic properties of Encephalartos hildebrandtii. East Afr Med J 45(12):732-741.

Mugera GM, Nderito P (1968) Tumours of the liver, kidney, and lungs in rats fed Encephalartos hildebrandtii. Br J Cancer 22(3):563-568.

ENCEPHALARTOS HORRIDUS (Jacq.) Lehm. [Zamiaceae]

Citations:

Van der Walt SJ (1944) Recent investigations into the toxicity of plants, etc. in the Union of South Africa. Onderstepoort J Vet Sci Anim Indus 20(1):75-83.

ENCEPHALARTOS LAEVIFOLIUS Stapf & Burtt Davy [Zamiaceae]

Citations:

Tustin RC (1974) Toxicity and carcinogenicity of some South African cycad (Encephalartos) species. S Afr Med J 48(57):2369-2373.

ENCEPHALARTOS LANATUS Stapf & Burtt Davy [Zamiaceae]

Citations:

Tustin RC (1983) Notes on the toxicity and carcinogenicity of some South African cycad species with special reference to that of Encephalartos lanatus. J S Afr Vet Assoc 54(1):33-42.

ENCEPHALARTOS LEBOMBOENSIS I. Verd. [Zamiaceae]

Citations:

Tustin RC (1974) Toxicity and carcinogenicity of some South African cycad (Encephalartos) species. S Afr Med J 48(57):2369-2373.

ENCEPHALARTOS LEHMANNII Lehm. [Zamiaceae]

Citations:

Van der Walt SJ, Steyn DG (1946) Recent investigations into the toxicity of plants, etc. XV. Onderstepoort J Vet Sci Anim Indus 21(1):45-55.

ENCEPHALARTOS UMBELUZIENSIS R. A. Dyer [Zamiaceae]

Citations:

Tustin RC (1974) Toxicity and carcinogenicity of some South African cycad (Encephalartos) species. S Afr Med J 48(57):2369-2373.

ENCEPHALARTOS VILLOSUS Lem. [Zamiaceae]

Citations:

Tustin RC (1974) Toxicity and carcinogenicity of some South African cycad (Encephalartos) species. S Afr Med J 48(57):2369-2373.

enchanter's-herb –see– *Verbena officinalis* L.

endive –see– *Cichorium endivia* L.

Endymion nonscripta (L.) Garcke = Hyacinthoides non-scripta (L.) Chouard ex Rothm.

Engelstrompete –see– *Brugmansia suaveolens* (Humb. & Bonpl. ex Willd.) Bercht. & J. Presl

Engelwurz –see– *Angelica archangelica* L.

Engleman ivy –see– *Parthenocissus quinquefolia* (L.) Planch.

English bean –see– *Vicia faba* L.

English bluebell –see– *Hyacinthoides non-scripta* (L.) Chouard ex Rothm.

English boxwood –see– *Buxus sempervirens* L.

English broom –see– *Cytisus scoparius* (L.) Link

English daisy –see– *Bellis perennis* L.

English holly –see– *Ilex aquifolium* L.; *Ilex opaca* Aiton

English ivy –see– *Hedera helix* L.

English laurel –see– *Prunus laurocerasus* L.

English mandrake –see– *Bryonia dioica* Jacq.

English oak –see– *Quercus robur* L.

English plantage –see– *Plantago lanceolata* L.

English plantain –see– *Plantago lanceolata* L.

English ryegrass –see– *Lolium perenne* L.

English walnut –see– *Juglans regia* L.

English yew –see– *Taxus baccata* L.

enkoko –see– *Embelia ribes* Burm. f.

ENTADA PHASEOLOIDES (L.) Merr. [Fabaceae]

Common Names:

balugo; gila bean; godogo; matchbox bean

Citations:

Grant G, More LJ, McKenzie NH, et al. (1995) Nutritional and haemagglutination properties of several tropical seeds. J Agric Sci 124:437-445.

ENTANDROPHRAGMA ANGOLENSE (Welw.) C. DC. [Meliaceae]

Citations:

Oleffe JA, Sporcq J, Hublet P (1975) Epidemiological study of the wood industry in Belgium. Contact Dermatitis 1(5):315-316.

ENTANDROPHRAGMA CANDOLLEI Harms [Meliaceae]

Citations:

Oleffe JA, Sporcq J, Hublet P (1975) Epidemiological study of the wood industry in Belgium. Contact Dermatitis 1(5):315-316.

ENTANDROPHRAGMA CYLINDRICUM (Sprague) Sprague [Meliaceae]

Common Name:

sapele

Citations:

Álvarez-Cuesta C, Gala Ortiz G, Rodríguez Díaz E, et al. (2004) Occupational asthma and IgE-mediated contact dermatitis from sapele wood. Contact Dermatitis 51(2):88-89.

Correia O, Barros MA, Mesquita-Guimarães J, (1992) Airborne contact dermatitis from the woods Acacia melanoxylon and Entandrophragma cylindricum. Contact Dermatitis 27(5):343-344.

Estlander T, Jolanki R, Alanko K, et al. (2001) Occupational allergic contact dermatitis caused by wood dusts. Contact Dermatitis 44(4):213-217.

Halpin B (1961) Toxicity in fowl due to hardwood litter. Vet Rec 73(18):454-455.

ENTANDROPHRAGMA UTILE (Dawe & Sprague) Sprague [Meliaceae]

Citations:

Oleffe JA, Sporcq J, Hublet P (1975) Epidemiological study of the wood industry in Belgium. Contact Dermatitis 1(5):315-316.

ENTEROLOBIUM CONTORTISILIQUUM (Vell.) Morong [Fabaceae]

Common Names:

tamboril; timbaúba

Citations:

Grecco FB, Dantas AF, Riet-Correa F, et al. (2002) Cattle intoxication from Enterolobium contortisiliquum pods. Vet Hum Toxicol 44(3):160-162.

Marques DC, Santos HL, Couto ES, et al. (1974) Intoxicação experimental pelo tamboril Enterolobium contortisiliquum Vell. Morong em bovinos. Arq Esc Vet Univ Fed Minas Gerais 26(3):283-286.

Schild AC, Motta AC, Riet-Correa F, et al. (2004) Photosensitization in cattle in southern Brazil. In: Acamovic T, Stewart CS, Pennycott TW (eds.) Poisonous plants and related toxins. CABI. Wallingford, UK. pp. 162-166.

Tokarnia CH, Canella CF, Döbereiner J (1960) Intoxicação experimental pela fava da "Timbaúba" (Enterolobium contortisiliquum (Vell.) Morong) em bovinos. Arq Inst Biol Anim (Rio Janeiro) 3:73-81.

Tokarnia CH, Döbereiner J, Dutra IS, et al. (1999) Experimentos em bovinos com as favas de Enterolobium contortisiliquum e E. timbouva para verificar propriedades fotosensibilizantes e/ou abortivas. Pesq Vet Bras 19(1):39-45.

ENTEROLOBIUM CYCLOCARPUM (Jacq.) Griseb. [Fabaceae]

Citations:

Grant G, More LJ, McKenzie NH, et al. (1995) Nutritional and haemagglutination properties of several tropical seeds. J Agric Sci 124:437-445.

ENTEROLOBIUM ELLIPTICUM Benth. [Fabaceae]

Citations:

Andrade SO, Lee JM, Sartori C, et al. (1970) Estudos sôbre os efeitos tóxicos do Enterolobium ellipticum Benth (Leguminosae). Arq Inst Biol (Sao Paulo) 37(Suppl 1):38.

ENTEROLOBIUM GUMMIFERUM (Mart.) J. F. Macbr. [Fabaceae]

Common Names:

tamboril-do-campo

Citations:

Deutsch J, Döbereiner J, Tokarnia CH (1965) Fotossensibilidade hepatogénica em bovinos na intoxicação pela fava de Enterolobium gummiferum (Mart.) Macbr. Proc Int Grassland Cong 2:1279-1282.

ENTEROLOBIUM TIMBOUVA Mart. [Fabaceae]

Common Names:

orelha-de-preto; timborá

Citations:

Tokarnia CH, Döbereiner J, Dutra IS, et al. (1999) Experimentos em bovinos com as favas de Enterolobium contortisiliquum e E. timbouva para verificar propriedades fotosensibilizantes e/ou abortivas. Pesq Vet Bras 19(1):39-45.

entjiebos –see– *Hertia cluytiifolia* (DC.) Kuntze

Epaltes alata (Sond.) Steetz = Epaltes gariepina (DC.) Steetz

EPALTES GARIEPINA (DC.) Steetz [Asteraceae]

Synonyms:

Epaltes alata (Sond.) Steetz

Citations:

Steyn DG (1934) Recent investigations into the toxicity of known and unknown poisonous plants in the Union of South Africa. Onderstepoort J Vet Sci Anim Indus 3(1):1205-130.

Van der Walt SJ, Steyn DG (1943) Recent investigations into the toxicity of plants, etc. XIII. Onderstepoort J Vet Sci Anim Indus 18(1-2):207-224.

epazote –see– *Dysphania ambrosioides* (L.) Mosyakin & Clemants; *Dysphania graveolens* (Willd.) Mosyakin & Clemants

epazote-de-comer –see– *Dysphania ambrosioides* (L.) Mosyakin & Clemants

epena –see– *Virola calophylloidea* Markgr.

ephedra –see– *Ephedra sinica* Stapf

EPHEDRA SINICA Stapf [Ephedraceae]

Common Names:

Chinese ephedra; ephedra; guarana; ma huang

Citations:

Liu Y, Wolf LR, Zhu W (1997) Epidemiology of adult poisoning at China Medical University. J Toxicol Clin Toxicol 35(2):175-180.

Mrvos R, Krenzelok EP, Jacobsen TD (2001) Toxidromes associated with the most common plant ingestions. Vet Hum Toxicol 43(6):366-369.

Ooms TG, Khan SA, Means C (2001) Suspected caffeine and ephedrine toxicosis resulting from ingestion of an herbal supplement containing guarana and ma huang in dogs: 47 cases (1997-1999). J Am Vet Med Assoc 218(2):225-229.

Pace S (1996) Ma huang food supplement toxicity in two adolescents. J Toxicol Clin Toxicol 34:598.

Palmer M, Haller C, McKinney P, et al. (1998) A gap in the safety net: A multi-center prospective study of herbals and other dietary supplements. J Toxicol Clin Toxicol 36(5):454.

Palmer M, Haller C, McKinney P, et al. (1999) Botanicals and other dietary supplements: Adverse events by age. J Toxicol Clin Toxicol 37(5):609.

Perrotta DM, Goody G, Culmo C, (1996) Adverse events associated with ephedrine-containing products - Texas, December 1993 - September 1995. MMWR Morb Mortal Wkly Rep 45(32):689-692.

Sawyers B, LoVecchio F (2002) Transient ischemic attack associated with metabolife use. J Toxicol Clin Toxicol 40(5):644-645.

Sharma AN, Cinci PM, Hoffman RS (2000) Ma huang-induced myocardial infarction with normal coronary arteries. J Toxicol Clin Toxicol 38(5):521.

Zaacks SM, Klein L, Tan CD, et al. (1999) Hypersensitivity myocarditis associated with ephedra use. J Toxicol Clin Toxicol 37(4):485-489.

EPHEDRA VIRIDIS Coville [Ephedraceae]

Common Names:

American ephedra; desert tea; joint fir; Mexican tea; Mormon tea; popotillo; teamster's-tea; whorehouse tea

Citations:

Keeler RF (1989) Investigation of maternal and embryo/fetal toxicity of Ephedra viridis and Ephedra nevadensis in sheep and cattle. J Range Manag 42(1):31-35.

Epheu –see– *Hedera helix* L.

Epheublättrige Gundelrebe –see– *Glechoma hederacea* L.

epile oleander –see– *Thevetia peruviana* (Pers.) K. Schum.

EPIMEDIUM GRANDIFLORUM C. Morren [Berberidaceae]

Citations:

Tuckler V, Peck C, Nesbitt C, et al. (2003) Intractable priapism associated with herbal stimulants. J Toxicol Clin Toxicol 41(5):671-672.

Epipremnum aureum (Lindenl. & André) G. S. Bunting = Epipremnum pinnatum (L.) Engl.

EPIPREMNUM PINNATUM (L.) Engl. [Araceae]

Synonyms:

Epipremnum aureum (Linden & André) G. S. Bunting; ***Scindapsus aureus*** (Linden & André) Engl.

Common Names:

amapalo amarillo; devil's-ivy; Efeutute; elephant's-ear; golden Ceylon creeper; golden hunter's-robe; golden pothos; hunter's-robe; ivy arum; malanga; malanga trepadora; marble queen; pothos; Solomon Island ivy; taro vine; variegated philodendron

Citations:

Krenzelok EP, Mrvos R, Jacobsen TD (2002) Contrary to the literature, vomiting is not a common manifestation with plant exposures. Vet Hum Toxicol 44(5):298-300.

Mobacken H (1975) Allergic plant dermatitis from Scindapsus aureus. Contact Dermatitis 1(1):60-61.

Mrvos R, Krenzelok EP, Jacobsen TD (2001) Toxidromes associated with the most common plant ingestions. Vet Hum Toxicol 43(6):366-369.

Paulsen E (1998) Occupational dermatitis in Danish gardeners and greenhouse workers (II). Etiological factors. Contact Dermatitis 38(1):14-19.

eppich –see– *Hedera helix* L.

EQUISETUM ARVENSE L. [Equisetaceae]

Common Names:

barba ursului; bottlebrush; cat's-tail; cat's-tail grass; coada calului; cola-de-caballo; colt's-tail; false horsetail; field horsetail; foxtail rush; horsepipe; horsetail; joint rush; Kleiner Schachtelhalm; mare's-tail; meadow pine; pine top; pinegrass; prêle; scouring rush; shavegrass; snakegrass

Citations:

Agustìn-Ubide MP, Martinez-Cócera C, Alonso-Llamazares A, et al. (2004) Diagnostic approach to anaphylaxis by carrot, related vegetables and horsetail (Equisetum arvense) in a homemaker. Allergy 59(7):786-787.

Bocoş EI (1971) Observații asupra intoxicației cu Equisetum la animale. Rev Zootehnie Med Vet 21(8):57-59.

Bubien MZ, Magat A, Delatour P (1970) Fréquence des intoxications des animaux en Pologne. Bull Soc Vet Med Comp Lyon 72(2):167-173.

Henderson JA, Evans EV, McIntosh RA (1952) The antithiamine action of Equisetum. J Am Vet Med Assoc 120(903):375-378.

Hudson R (1924) Poisoning by horsetail (Equisetum arvense). Vet J 80:40.

Jones LR (1901) Are our native horsetails or ferns poisonous? Proc Soc Promot Agric Sci 22:70-74.

Maeda H, Miyamoto K, Sano T (1997) Occurrence of dermatitis in rats fed a cholesterol diet containing field horsetail (Equisetum arvense L.). J Nutr Sci Vitaminol (Tokyo) 43(5):553-563.

Pammel LH (1919) Poisoning from horsetail. Am J Vet Med 14:245-246.

Rich FA (1902) Equisetum poisoning. Am Vet Rev 26:944-954.

Rich FA, Jones LR (1902) A poisonous plant. The common horsetail (Equisetum arvense). Vermont Agric Exp Sta Bull #95:187-190.

Sudan BJ (1985) Seborrhoeic dermatitis induced by nicotine of horsetails (Equisetum arvense L.). Contact Dermatitis 13(3):201-202.

EQUISETUM FLUVIATILE L. [Equisetaceae]

Common Names:

cola-de-caballo; horsetail; swamp horsetail; water horsetail

Citations:

Klebesadel LJ, Mitchell WW (1964) An outbreak of horse poisoning from swamp horsetail (Equisetum fluviatile L.). J Range Manag 17(6):333-334.

EQUISETUM PALUSTRE L. [Equisetaceae]

Common Names:

erba cavalina; gli equiseti; herbe-aux-grenouilles; horsetail; marsh horsetail; marsh weed; mocsári zsurló; myrsnelle; prêle commune; queue-de-cheval; Sumpfschachtelhalm

Citations:

Butkus J, Čepulevičius M (1972) Apsinuodijimas asiūkliais ar babezielioze? Trudy Litovskii Vet Akademii 10:181-184.

Kamphues J (1990) Verweigerung der Aufnahme von Heu bei Zuchtbullen dedingt durch eine Kontamination mit Sumpfschachtelhalm (Equisetum palustre). Tierarztl Prax 18(4):349-351.

Khoteev VS, Chevguz FK (1970) [Large outbreak of Equisetum palustre poisoning in horses.] Veterinariia Moscow 47(8):68-69.

Lindt S (1959) Über eine Schachtelhalmvergiftung bei Kälbern. Schweiz Arch Tierheilkd 101:461-464.

Richter HE (1961) Neuerlicher Vergiftungsfall durch Equisetum palustre L., Sumpfschachtelhalm. Wien Tierarztl Monatsschr 48(9):761-762.

EQUISETUM RAMOSISSIMUM Desf. [Equisetaceae]

Common Names:

drilgras; dronkgras; horsetail; mare's-tail; perde stert

Citations:

Steyn DG (1933) Recent investigations into the toxicity of known and unknown poisonous plants in the Union of South Africa. Onderstepoort J Vet Sci Anim Indus 1:173-182.

EQUISETUM TELMATEIA Ehrh. [Equisetaceae]

Common Names:

giant horsetail

Citations:

Miró O, Pedrol E, Nogué S, et al. (1996) Hiponatremia e hipopotasemia graves inducidas por el consumo de Equisetum telmateia. Med Clin (Barc) 106(6):639.

eramusu –see– *Hemidesmus indicus* (L.) W. T. Aiton

erand –see– *Ricinus communis* L.

erba cavalina –see– *Equisetum palustre* L.

Erbse –see– *Lathyrus odoratus* L.; *Pisum sativum* L.

Erbsenstrauch –see– *Caragana arborescens* Lam.

Erdapfel –see– *Solanum tuberosum* L.

Erdbeere –see– *Fragaria* ×*ananassa* Duchesne ex Rozier

erdei iszalag –see– *Clematis vitalba* L.

erdei szélfü –see– *Mercurialis perennis* L.

Erdepheu –see– *Glechoma hederacea* L.

Erdklee –see– *Trifolium subterraneum* L.

Erdnuß –see– *Arachis hypogaea* L.

Erdschierling –see– *Conium maculatum* L.

EREMOPHILA DESERTI (A. Cunn. ex Benth.) Chinnock [Scrophulariaceae]

Synonyms:

Myoporum deserti A. Cunn. ex Benth.

Common Names:

carrot bush; dogweed; Ellangowan; Ellangowan poison bush; ngaione; pencil bush; poison bush; turkey bush

Citations:

Allen JG, Seawright AA (1973) The effect of prior treatment with phenobarbitone, dicophane (DDT) and β-diethylaminoethyl phenylpropyl acetate (SKF 525A) on experimental intoxication of sheep with the plant Myoporum deserti Cunn. Res Vet Sci 15(2):167-179.

Johnstone IL, Allen GH (1944) Observations on the poisoning of sheep by Myoporum deserti (turkey-brush or Ellangowan-brush). Aust Vet J 20:227-230.

EREMOPHILA MACULATA (Ker Gawl.) F. Muell. [Scrophulariaceae]

Common Names:

fuchsia bush; native fuchsia; spotted emubush; spotted fuchsia

Citations:

Finnemore H (1931) The poisoning of stock on the Georgina River. The native fuchsia and gidgea. J CSIRO Aust 4:220-224.

Ramsay AA, Henry M (1929) Rosewood (Heterodendrum oleaefolium) and native fuchsia (Eremophila maculata) two poisonous plants. Agric Gaz New South Wales 40:834-837.

Seddon HR, King RO (1930) The fatal dose for sheep of cyanogenetic plants containing sambunigrin or prunasin. J CSIRO Aust 3:14-24.

erg elagrab –see– *Clitoria ternatea* L.

ERIOCHLOA CONTRACTA Hitchc. [Poaceae]

Common Names:

prairie cupgrass

Citations:

Sperry OE, Turk RD, Hoffman GO, et al. (1955) Photosensitization of cattle in Texas. Texas Agric Exp Sta Bull #812:8 pp.

ERIOBOTRYA JAPONICA (Thunb.) Lindl. [Rosaceae]

Common Names:

Japanese plum; loquat; loquat plum

Citations:

Weber MA, Garner M (2002) Cyanide toxicosis in Asian small-clawed otters (Amblonyx cinereus) secondary to ingestion of loquat (Eriobotrya japonica). J Zoo Wildl Med 33(2):145-146.

ERODIUM CICUTARIUM (L.) L'Her. [Geraniaceae]

Common Names:

alfilaria; alfilerillo; California filaree; pingrass; redstem filaree; stork's-bill

Citations:

Ford GE (1965) Photosensitivity due to Erodium spp. Aust Vet J 41(Feb):56.

EROPHACA BAETICA (L.) Boiss. [Fabaceae] [Fabaceae]

Synonyms:

Astragalus lusitanicus Lam.

Common Names:

garbancillo; garbanzuelos

Citations:

Abdennebi EH, Quazzani N, Jossang A, et al. (2001) Inhibition of glycosidases by Astragalus lusitanicus and correlation with toxicity. Vet Hum Toxicol 43(5):266-269.

Abdennebi EH, Quazzani NE, Lamnaouer D (1998) Clinical and analytical studies of sheep dosed with various preparations of Astragalus lusitanicus. Vet Hum Toxicol 40(6):327-331.

El Hamidi M, Leipold HW (1989) Poisoning of sheep by Astragalus lusitanicus in Morocco: Field and experimental studies. Zentralbl Veterinarmed A 36(2):115-121.

Garcia Roman A, Moyano Salvago MR, Soler Rodriguez F, et al. (1987) Physiopathologic changes in lambs fed with Astragalus lusitanicus Lam. Vet Hum Toxicol 29(5):387-389.

Moyano Salvago R, Gázquez Ortíz A, García Román A, et al. (1989) Alteraciones anatomopatológicas producidas en la intoxicación experimental por Astragalus lusitanicus Lam. Medicina Vet 6(9):491-494.

Soler F, Pedrera JD, Roncero V, et al. (1991) Description of a field case of Astragalus lusitanicus intoxication in goat. Acta Vet Scand Suppl 87:351-353.

Soler Rodriguez F, García Román A, Moyano Salvago MR, et al. (1990) Effect of Astragalus lusitanicus Lam in experimental subchronic neurotoxicity of lambs. Vet Hum Toxicol 32(6):551-554.

ers –see– *Vicia ervilia* (L.) Willd.

ers ervillier –see– *Vicia ervilia* (L.) Willd.

erun –see– *Erythrophleum suaveolens* (Guill. & Perr.) Brenan

erva-de-passarinho –see– *Phoradendron piperoides* (Kunth) Trel.

erva-de-rato –see– *Palicourea marcgravii* A. St.-Hil.; *Psychotria poeppigiana* Müll. Arg.

ervilhaca –see– *Vicia sativa* L.

Ervum ervilia L. = Vicia ervilia (L.) Willd.

Ervum lens L. = Lens culinaris Medik.

ERYCIBE HENRYI Prain [Convolvulaceae]

Common names:

ting-kung-teng

Citations:

Lin CC, Chen JC (2002) Medicinal herb Erycibe henri Prain ("ting kung teng") resulting in acute cholinergic syndrome. J Toxicol Clin Toxicol 40(2):185-187.

ERYCIBE OBTUSIFOLIA Benth. [Convolvulaceae]

Common Names:

lei-gong-teng

Citations:

Hsu HY, Lin CC, Chen JY, et al. (1998) Toxic effects of Erycibe obtusifolia, a Chinese medicinal herb, in mice. J Ethnopharmacol 62(2):101-105.

ERYTHROPHLEUM AFRICANUM (Welw. ex Benth.) Harms [Fabaceae]

Common Names:

ordeal tree; samburu

Citations:

Nwude N, Chineme CN (1981) Toxic effects of the leaves of Erythrophleum africanum Harms in sheep. Bull Anim Health Prod Afr 29(4):349-354.

ERYTHROPHLEUM CHLOROSTACHYS (F. Muell.) Baill. [Fabaceae]

Synonyms:

Erythrophleum laboucheri F. Muell. ex Benth.

Common Names:

black bean; camel poison; Cooktown ironwood; ironweed; ironwood; northern ironwood

Citations:

Petrie JM, Priestly H (1921) Physiological activity of Erythrophloeum laboucherii. Proc Linn Soc N S W 46:340-348.

Erythrophleum guineense G. Don = Erythrophleum suaveolens (Guill. & Perr.) Brenan

Erythrophleum laboucheri F. Muell. ex Benth. = Erythrophleum chlorostachys (F. Muell.) Baill.

ERYTHROPHLEUM SUAVEOLENS (Guill. & Perr.) Brenan [Fabaceae]

Synonyms:

Erythrophleum guineense G. Don

Common Names:

casca bark; doom bark; erun; marcona bark; ordeal bark; ordeal tree; red-water bark; sasswood; sassy bark; saucy bark

Citations:

Akinlade NO (1967) A case of sasswood poisoning. West Afr Med J 16(2):63-64.

Nwude N, Chineme CN (1980) Investigations into the toxicity of the leaves of Erythrophleum guineense Don. in sheep. Res Vet Sci 28(1):112-115.

ERYTHROPHLEUM SUCCIRUBRUM Gagnep. [Fabaceae]

Common Names:

phan saat; saat; tria

Citations:

Echeverria P, Taylor DN, Bodhidatta L, et al. (1986) Deaths following ingestion of a cardiotoxic plant in Kampuchean children in Thailand. Southeast Asian J Trop Med Public Health 17(4):601-603.

ERYTHROXYLUM ARGENTINUM O. E. Schulz [Erythroxylaceae]

Citations:

Barros RR, Teixeira FR, Oliveira FN, et al. (2004) Poisoning in sheep from the ingestion of fruits of Erythroxylum argentinum. Vet Hum Toxicol 46(4):173-175.

ERYTHROXYLUM COCA Lam. [Erythroxylaceae]

Common Names:

coca; cuca; spadic bush

Citations:

Valentine JL, Fremming BD, Chappell RH, et al. (1988) Effects of extended feeding of decocainized Erythroxylon coca leaves on growth and selected organs in rats and rabbits. Hum Toxicol 7:21-26.

ERYTHROXYLUM DECIDUUM A. St.-Hil. [Erythroxylaceae]

Common Names:

cocão

Citations:

Colodel EM, Seitz AL, Schmitz M, et al. (2004) Intoxicação por Erythroxylum deciduum (Erythroxylaceae) em ovinos. Pesq Vet Bras 24(3):165-168.

ESENBECKIA LEIOCARPA Engl. [Rutaceae]

Common Names:

Guarantã

Citations:

Freise FW (1936) Vergiftungen durch brasilianische Werkhölzer. II. Jacareúba-Holz und Seidenholz. Sammlung Vergiftungsfallen 7(C33):61-72.

esparto –see– *Macrochloa tenacissima* (Loefl. ex L.) Kunth

esparto grass –see– *Nassella tenuissima* (Trin.) Barkworth

espigon –see– *Tribulus terrestris* L.

espina colorado –see– *Solanum sisymbriifolium* Lam.

espina-de-palma –see– *Duranta erecta* L.

espinafre –see– *Spinacia oleracea* L.

espirradeira –see– *Nerium oleander* L.

Essigbaum –see– *Rhus typhina* L.

estafiate –see– *Artemisia ludoviciana* Nutt.

estramonio Americano –see– *Datura stramonium* L.

estramonio chino –see– *Datura ferox* L.

estrella –see– *Cynodon plectostachyus* (K. Schum.) Pilg.

estrella federal –see– *Euphorbia pulcherrima* Willd. ex Klotzsch

esula redonda –see– *Euphorbia peplus* L.

esura –see– *Agelaea pentagyna* (Lam.) Baill.

Ethiopian dog's-toothgrass –see– *Cynodon aethiopicus* Clayton & J. R. Harlan

EUCALYPTUS CAMALDULENSIS Dehnh. [Myrtaceae]

Common Names:

red gum

Citations:

Hughes RJ (1973) Abnormal yolk color and mottling caused by dietary tannic acid and 'tannins'. J Poultry Sci 52(5):1784-1786.

EUCALYPTUS CLADOCALYX F. Muell. [Myrtaceae]

Common Names:

sugar gum

Citations:

Anonymous (1929) The poison plants committee. J CSIRO Aust 2:40-48.

Reece RL, Handson P (1982) Observations on the accidental poisoning of birds by organophosphate insecticides and other toxic substances. Vet Rec 111(20):453-455.

Steyn DG (1948) The toxicity of various species of Eucalyptus trees (blue gum, blougom, bloekom). J S Afr Vet Assoc 19(Mar):25-29.

Webber JJ, Roycroft CR, Callinan JD (1985) Cyanide poisoning of goats from sugar gums (Eucalyptus cladocalyx). Aust Vet J 62(1):28.

EUCALYPTUS GLOBULUS Labill. [Myrtaceae]

Common Names:

blue gum eucalyptus; gum tree

Citations:

Krenzelok EP, Mrvos R, Jacobsen TD (2002) Contrary to the literature, vomiting is not a common manifestation with plant exposures. Vet Hum Toxicol 44(5):298-300.

Mrvos R, Krenzelok EP, Jacobsen TD (2001) Toxidromes associated with the most common plant ingestions. Vet Hum Toxicol 43(6):366-369.

Ramón MF, Ballesteros S, Martínez-Arrieta R, et al. (2003) Plant and herb exposures in Spain. J Toxicol Clin Toxicol 41(4):514.

EUCALYPTUS SALIGNA Sm. [Myrtaceae]

Citations:

Holst R, Kirby J, Magnusson B (1976) Sensitization to tropical woods giving erythema multiforme-like eruptions. Contact Dermatitis 2(11):295-296.

EUCOMIS UNDULATA Aiton [Hyacinthaceae]

Citations:

Van der Walt SJ (1944) Recent investigations into the toxicity of plants, etc. in the Union of South Africa. Onderstepoort J Vet Sci Anim Indus 20(1):75-83.

Eugenia caryophyllata Thunb. = Syzygium aromaticum (L.) Merr. & L. M. Perry

EUGENIA MYRTIFOLIA Salisb. [Myrtaceae]

Citations:

Anderson CR (1944) Contact dermatitis from alfalfa and bur clover. Arch Derm Syphilol 50:201.

Eugenia smithii Poir. = Acmena smithii (Poir.) Merr. & L. M. Perry

eugomo –see– *Coula edulis* Baill.

EUONYMUS EUROPAEUS L. [Celastraceae]

Common Names:

ananbeam; bonnet-de-prêtre; catwood; dogwood; European euonymus; European spindle tree; fusain; gatter; louse berry tree; pegwood; Pfaffenhütchen; pigwood; prickwood; skewerwood; Spillbaum; Spindelbaum; spindle tree

Citations:

Hermkes L (1941) Eine seltene Vergiftung mit den Früchten des Spindelbaumes (Pfaffenhütchen). Munch Med Wochenschr 88(37):1011-1012.

Herold DA, Wahl R, Maasch HJ, et al. (1991) Occupational wood-dust sensitivity from Euonymus europaeus (spindle tree) and investigation of cross reactivity between E. e. and Artemisia vulgaris pollen (mugwort). Allergy 46(3):186-190.

Eupatorium adenophorum Spreng. = Ageratina adenophora (Spreng.) R. M. King & H. Rob.

Eupatorium ageratoides L. f. = Ageratina altissima (L.) R. M. King & H. Rob.

EUPATORIUM ALTISSIMUM L. [Asteraceae]

Common Names:

thoroughwort

Citations:

Brier AJ (1939-1940) Contact dermatitis from thurowort (Eupatorium altissimum). J Allergy 11:402-406.

EUPATORIUM CHINENSE L. [Asteraceae]

Common Names:

Chinese eupatorium; lan tsao

Citations:

Pak C, Read BE (1937) Chinese eupatorium. Chin J Physiol 12(3):263-274.

Eupatorium glandulosum Michx. = Ageratina adenophora (Spreng.) R. M. King & H. Rob.

Eupatorium riparium Regel = Ageratina riparia (Regel) R. M. King & H. Rob.

Eupatorium rugosum Houtt. = Ageratina altissima (L.) R. M. King & H. Rob.

Eupatorium urticaefolium Reichard = Ageratina altissima (L.) R. M. King & H. Rob.

Eupatorium wrightii A. Gray = Ageratina wrightii (A. Gray) R. M. King & H. Rob.

EUPHORBIA ABYSSINICA J. F. Gmel. [Euphorbiaceae]

Citations:

Abebe W (1992) Adverse effects of traditional drug preparations. J Ethnopharmacol 36:93-94.

EUPHORBIA ANTIQUORUM L. [Euphorbiaceae]

Synonyms:

Euphorbia trigona Haw.

Common Names:

fleshy spurge; Malayan spurge; milkbush; spurge cactus

Citations:

Hallén L (1986) Sprättivägplanta och high chapparall - Modekrukväxter med risker. Lakartidningen 83(34):2756.

Euphorbia biglandulosa Desf. = Euphorbia rigida M. Bieb.

EUPHORBIA CAERULESCENS Haw. [Euphorbiaceae]

Common Names:

blue euphorbia; sweet noors

Citations:

Evans FJ (1978) The irritant toxins of blue euphorbia (Euphorbia caerulescens Haw.). Toxicon 16(1):51-57.

EUPHORBIA CHARACIAS L. [Euphorbiaceae]

Common Names:

phlomos; tántago-de-Valencia; tárago mayor

Citations:

Eke T, Al-Husainy S, Raynor MK (2000) The spectrum of ocular inflammation caused by euphorbia plant sap. Arch Ophthalmol 118(1):13-16.

EUPHORBIA COOPERI N. E. Br. ex A. Berger [Euphorbiaceae]

Citations:

Evans FJ (1978) The irritant toxins of blue euphorbia (Euphorbia caerulescens Haw.). Toxicon 16(1):51-57.

EUPHORBIA CYPARISSIAS L. [Euphorbiaceae]

Common Names:

cypress spurge; farkas kutyatej; Glelskraut; Krötengraß; lechetrezna; spurge; Warzengraß; Zypressen-Wolfsmilch

Citations:

Asilian A, Faghihi G (2004) Severe irritant contact dermatitis from cypress spurge. Contact Dermatitis 51(1):37-39.

Lisch K (1980) Die Wirkung des Milchsaftes von Euphorbiazeen auf das Auge. Klin Monatsbl Augenheilkd 176(3):469-471.
Ritter S (1985) Vergiftungen durch Pflanzen. Dtsch Apoth Ztg 125(37):1834-1836.
Rössler G (1985) Keratoconjunctivitis Euphorbiae. Klin Monatsbl Augenheilkd 186(5):380-381.

Euphorbia dallachyana Baill. = Euphorbia drummondii Boiss.

EUPHORBIA DENDROIDES L. [Euphorbiaceae]

Common Names:
titimalo

Citations:
Bioman P, Savir H (1974) [Chemical ocular injury by Euphorbia dendroides.] Harefuah 87(Oct):308-309.

EUPHORBIA DRUMMONDII Boiss. [Euphorbiaceae]

Synonyms:
Euphorbia dallachyana Baill.

Common Names:
balsam; caustic creeper; caustic weed; creeping caustic; mat balsam; mat spurge; milk spurge; milkweed; onat spurge; poison weed; red soldier

Citations:
Anonymous (1929) The poison plants committee. J CSIRO Aust 2:40-48.
Seddon HR (1927) Euphorbia drummondii, "milk weed," a plant poisonous to sheep. J CSIRO Aust 1:268-273.
Seddon HR (1928) Milk weed (Euphorbia drummondii) proved poisonous to sheep. Agric Gaz New South Wales 39:777-782.

EUPHORBIA ESULA L.

Common names:
leafy spurge

Citations:
Heemstra JM, Kronberg SL, Neiger RD, et al. (1999) Behavioral, nutritional, and toxicological responses of cattle to ensiled leafy spurge. J Anim Sci 77(3):600-610.
Johnson A, Peake RW (1960) Effect of selective grazing by sheep on the control of leafy spurge (Euphorbia esula L.). J Range Manag 12:192-195.

EUPHORBIA FRANCKIANA A. Berger [Euphorbiaceae]

Citations:
Evans FJ (1978) The irritant toxins of blue euphorbia (Euphorbia caerulescens Haw.). Toxicon 16(1):51-57.

EUPHORBIA FULGENS Karw. ex Klotzsch [Euphorbiaceae]

Common Names:
scarlet plume

Citations:
Hausen BM, Ketels-Harken H, Schulz KH (1976) Berufsbedingte Inhalationsallergie durch Pollen von Euphorbia fulgens Karw. Dtsch Med Wochenschr 101(15):567-570.

Euphorbia geniculata Ortega = Euphorbia heterophylla L.

EUPHORBIA GERARDIANA Jacq. [Euphorbiaceae]

Common Names:
Wolfsmilch

Citations:
Geßner O (1936) Gewerbliche Dermatitis durch Wolfsmilcharten (Euphorbiaceen). Sammlung Vergiftungsfallen 7(A651):217-218.

EUPHORBIA GRANDICORNIS Goebel ex N. E. Br. [Euphorbiaceae]

Citations:
Likes K, Chavez M, Picchioni A, et al. (undated) Keratoconjunctivitis and contact dermatitis from exposure to the sap from Euphorbia spp. Unpublished manuscript from the Arizona Poison Control and Drug Information Center, Tucson.

EUPHORBIA GRANTII Oliv. [Euphorbiaceae]

Citations:
Likes K, Chavez M, Picchioni A, et al. (undated) Keratoconjunctivitis and contact dermatitis from exposure to the sap from Euphorbia spp. Unpublished manuscript from the Arizona Poison Control and Drug Information Center, Tucson.

EUPHORBIA HELIOSCOPIA L. [Euphorbiaceae]

Common Names:
cat's-milk; churn staff; Hundswolfkraut; ibby dalegrass; irby dalegrass; lecheruela; lechetrezna; little good; little goody; mamona; Milchkraut; Sonnenwolfsmilch; spurge; sun spurge; wartwort; Wolfsmilch; wortgrass

Citations:
Cleland JB (1931) Plants, including fungi, poisonous or otherwise injurious to man in Australia. Series III. Med J Aust 2(Dec 19):775-778.
Guggenheim I (1926) Bindehaut- und Hornhautentzündung durch Saft der Euphorbia helioscopia (Wolfsmilch). Klin Monatsbl Augenheilkd 77:521-523.
Rössler G (1985) Keratoconjunctivitis Euphorbiae. Klin Monatsbl Augenheilkd 186(5):380-381.
Vlachos P, Poulos L, Koutselinis A, et al. (1978) Euphorbia poisoning (case reports). Environ Biol Med 6:104.

EUPHORBIA HERMENTIANA Lem. [Euphorbiaceae]

Common Names:
milk bush

Citations:

Worobec SM, Hickey TA, Kinghorn AD, et al. (1981) Irritant contact dermatitis from an ornamental Euphorbia. Contact Dermatitis 7(1):19-22.

EUPHORBIA HETEROPHYLLA L. [Euphorbiaceae]

Synonyms:

Euphorbia geniculata Ortega

Common Names:

chilamatillo; fiddler's-spurge; golondrina; green euphorb; hierba-del-duende; Japanese poinsettia; Mexican fire plant; painted leaf; painted spurge; wild poinsettia

Citations:

Deodikar GB, Thakar CV, Phadke RP, et al. (1958) Poisoning of honeybees foraging on Euphorbia geniculata. Bee World 39(5):118-120.

Euphorbia hirta L. = Chamaesyce hirta (L.) Millsp.

EUPHORBIA INGENS E. Mey. ex Boiss. [Euphorbiaceae]

Common Names:

candelabra cactus; candelabra Euphorbia; candelabra tree; kankerbos; naboom

Citations:

Nyazema NZ (1986) Herbal toxicity in Zimbabwe. Trans R Soc Trop Med Hyg 80(3):448-450.

EUPHORBIA LACTEA Haw. [Euphorbiaceae]

Common Names:

candelabra cactus; dragon bone tree; dragon bones; false cactus; hatrack cactus; milkstripe euphorbia; monkey puzzle; mottled spurge

Citations:

Cordero-Moreno R (1973) Etiologic factors in tropical eye diseases. Am J Ophthalmol 75(3):349-364.

Crowder JI, Sexton RR (1964) Keratoconjunctivitis resulting from the sap of candelebra cactus and the pencil tree. Arch Ophthalmol 72(Oct):476-484.

EUPHORBIA LATHYRIS L. [Euphorbiaceae]

Common Names:

antigopher plant; caper bush; caper garden spurge; caper spurge; garden spurge; gopher plant; Kreuzblättrige Wolfsmilch; Maulwurfskraut; milkweed; mole plant; myrtle spurge; Pillenbaum; sassy jack; Springkraut; Springwolfsmilch; springwort; tártago; wild caper; wolf's-milk

Citations:

Antcliff RJ (1994) Euphorbia lathyris latex keratoconjunctivitis. Eye 8(Part 6):696-698.

Dietze U, Heydenreich A (1982) Augenschädigung durch eine zur Wühlmausbekämpfung eigensetzte Wolfsmilchpflanze (Euphorbia lathyris). Fol Ophthalmol 7:261-264.

Frohn A, Frohn C, Steuhl KP, et al. (1993) Wolfsmilchverätzung. Ophthalmologe 90(1):58-61.

Geidel K (1962) Klinische Beobachtung und tierexperimentelle Untersuchungen über die Wirkung von Saft der Euphorbia lathyris (Springwolfsmilch) am Auge. Klin Monatsbl Augenheilkd 141(Sep):374-379.

Pullar EM (1939) Studies on five suspected poisonous plants. Aust Vet J 15:19-23.

EUPHORBIA LEDIENII A. Berger [Euphorbiaceae]

Citations:

Evans FJ (1978) The irritant toxins of blue euphorbia (Euphorbia coerulescens Haw.). Toxicon 16(1):51-57.

EUPHORBIA LEUCONEURA Boiss. [Euphorbiaceae]

Common Names:

borsteuphorbia

Citations:

Hallén L (1986) Sprättivägplanta och high chapparall - Modekrukväxter med risker. Lakartidningen 83(34):2756.

Euphorbia maculata L. = Chamaesyce maculata (L.) Small

EUPHORBIA MARGINATA Pursh [Euphorbiaceae]

Common Names:

ghostweed; lepadena; ostromlecza paskowany; snow-on-the-mountain; variegated spurge; white margin spurge

Citations:

Agersborg G (1882) Poisoning in a cow by Euphorbia marginata. Am Vet Rev 6:353.

Eke T, Al-Husainy S, Raynor MK (2000) The spectrum of ocular inflammation caused by euphorbia plant sap. Arch Ophthalmol 118(1):13-16.

Pinedo JM, Saavedra V, Gonzalez-de-Canales F, et al. (1985) Irritant dermatitis due to Euphorbia marginata. Contact Dermatitis 13(1):44.

Szafran L (1967) Ostre zapalenie spojówek i rogówki wywołane sokiem ostromlecza paskowanego (Euphorbia marginata). Pol Tyg Lek (Wars) 22(18):671-672.

EUPHORBIA MATABELENSIS Pax [Euphorbiaceae]

Citations:

Gundidza M, Sorg B, Hecker E (1993) A skin irritant principle from Euphorbia matabelensis Pax. J Ethnopharmacol 39(3):209-212.

EUPHORBIA MAURITANICA L. [Euphorbiaceae]

Common Names:

geelmelkbos

Citations:

Terblanche M, Adelaar TF, van Straten AM (1966) Euphorbia mauritanica L. as a poisonous plant in South Africa. J S Afr Vet Assoc 37(3):311-315.

EUPHORBIA MILII Des Moul. [Euphorbiaceae]

Common Names:

Christ's-thorn; Christusdorn; coroa-de-cristo; crown-of-thorns; crucifixion plant

Citations:

Eke T, Al-Husainy S, Raynor MK (2000) The spectrum of ocular inflammation caused by euphorbia plant sap. Arch Ophthalmol 118(1):13-16.

McKinney P, Gomez HF, Phillips S, et al. (1992) The fax machine: A new method in plant identification. Vet Hum Toxicol 34(4):353.

EUPHORBIA MYRSINITES L. [Euphorbiaceae]

Common Names:

creeping spurge; donkey tail; spurge

Citations:

Lisch K (1980) Die Wirkung des Milchsaftes von Euphorbiazeen auf das Auge. Klin Monatsbl Augenheilkd 176(3):469-471.

Sierputowska M, Sliwinska M (1968) Dermatitis toxica bullosa phytogenes wywołane przez Euphorbia myrsinites. Przegl Dermatol 55(6):787-790.

Spoerke DG, Temple AR (1979) Dermatitis after exposure to a garden plant (Euphorbia myrsinites). Am J Dis Child 133(1):28-29.

EUPHORBIA PAGANORUM A. Chev. [Euphorbiaceae]

Citations:

Croizat L (1935) Poisonous Euphorbiae. Desert Plant Life 7:130.

EUPHORBIA PALUSTRIS L. [Euphorbiaceae]

Common Names:

Sumpfwolfsmilch

Citations:

Eke T, Al-Husainy S, Raynor MK (2000) The spectrum of ocular inflammation caused by euphorbia plant sap. Arch Ophthalmol 118(1):13-16.

EUPHORBIA PEPLUS L. [Euphorbiaceae]

Common Names:

caper spurge; esula redonda; Gartenwolfsmilch; lechetrezna; milkweed; mole plant; petty spurge; Stierkraut; wartweed; Wolfsmilch

Citations:

Biedner BZ, Sachs U, Witztum A (1981) Euphorbia peplus latex keratoconjunctivitis. Ann Ophthalmol 13(6):739-740.

Calnan CD (1975) Petty spurge (Euphorbia peplus L.). Contact Dermatitis 1(2):128.

Eke T (1994) Acute kerato-uveitis associated with topical self-administration of the sap of the petty spurge (Euphorbia peplus). Eye 8(Part 6):694-696.

Hartmann K (1940) Augenschädigung durch den Saft der Euphorbia peplus (Wolfsmilch). Klin Monatsbl Augenheilkd 104:324-326.

Nawito M, Ahmed YF, Zayed SM (1998) Dietary cancer risk from conditional cancerogens in produce of livestock fed on species of spurge (Euphorbiaceae). 2. Pathophysiological investigations in lactating goats fed on the skin irritant herb Euphorbia peplus and in their milk-raised kids. J Cancer Res Clin Oncol 124(3-4):179-185.

Seddon HR (1929) Feeding experiments with Euphorbia peplus ("petty spurge"). New South Wales Dep Agric Sci Bull 33:99-101.

Zayed SM, Farghaly M, Taha H (1998) Dietary cancer risk from conditional cancerogens in produce of livestock fed on species of spurge (Euphorbiaceae). 3. Milk of lactating goats fed on the skin irritant herb Euphorbia peplus is polluted by tumor promoters of the ingenane diterpene ester type. J Cancer Res Clin Oncol 124(6):301-306.

EUPHORBIA PILOSA L. [Euphorbiaceae]

Common Names:

downy spurge; hairy spurge; Wolfsmilch

Citations:

Geßner O (1936) Gewerbliche Dermatitis durch Wolfmilcharten (Euphorbiaceen). Sammlung Vergiftungsfallen 7(A651):217-218.

Euphorbia pilulifera L. = Chamaesyce hirta (L.) Millsp.

EUPHORBIA PLATYPHYLLOS L. [Euphorbiaceae]

Common Names:

broad-leaf wart spurge

Citations:

Eke T, Al-Husainy S, Raynor MK (2000) The spectrum of ocular inflammation caused by euphorbia plant sap. Arch Ophthalmol 118(1):13-16.

Euphorbia prostrata Aiton = Chamaesyce prostrata (Aiton) Small

EUPHORBIA PULCHERRIMA Willd. ex Klotzsch [Euphorbiaceae]

Synonyms:

Poinsettia pulcherrima (Willd. ex Klotzsch) Graham

Common Names:

bandera; beteta; catalina; Christmas flower; Christmas poinsettia; Christmas rose; Christmas star; Easter flower; estrella federal; flor-de-fuego; flor-de-navidad; flor-de-nochebuena; flor-de-pascua; flor-de-Santa-Caterina; lobster flower; paño-de-holanda; papagallo; pastora; poinsettia; Weihnachtsstern

Citations:

Anonymous (1980) Tabulations of 1977 case reports. Bull Natl Clgh Poison Control Cent 24(6):1-4.

D'Arcy WG (1974) Severe contact dermatitis from poinsettia. Arch Dermatol 109(6):909-910.

Der Marderosian AH, Giller FB, Roia FC Jr (1976) Phytochemical and toxicological screening of household ornamental plants potentially toxic to humans. I. J Toxicol Environ Health 1(6):939-953.
Edwards N (1983) Local toxicity from a poinsettia plant: A case report. J Pediatr 102(3):404-405.
Gock K, Schlatter C, Jenny E (1974) Akute Hundevergiftungen. Analyse der im Schweizerischen Toxikologischen Informationszentrum registrierten Fälle. Schweiz Arch Tierheilkd 116(11):565-585.
Ibáñez MD, Fernández-Nieto M, Martínez J, et al. (2004) Asthma induced by latex from 'Christmas flower' (Euphorbia pulcherrima). Allergy 59(10):1127-1128.
Klug S, Saleem G, Honcharuk L, et al. (1990) Toxicity potential of poinsettia, is the plant really toxic? Am Assoc Poison Control Cent 1990:112.
Krenzelok EP, Jacobsen TD, Aronis JM (1996) Poinsettia exposures have good outcomes. Just as we thought. Am J Emerg Med 14(7):671-674.
Krenzelok EP, Jacobsen TD, Aronis M (1998) The most prevalent symptoms associated with common plant exposures. J Toxicol Clin Toxicol 36(5):453.
Krenzelok EP, Mrvos R, Jacobsen TD (2002) Contrary to the literature, vomiting is not a common manifestation with plant exposures. Vet Hum Toxicol 44(5):298-300.
Massmanian A (1998) Contact dermatitis due to Euphorbia pulcherrima Willd, simulating a phototoxic reaction. Contact Dermatitis 38(2):113-114.
Mrvos R, Krenzelok EP, Jacobsen TD (2001) Toxidromes associated with the most common plant ingestions. Vet Hum Toxicol 43(6):366-369.
Paulsen E (1998) Occupational dermatitis in Danish gardeners and greenhouse workers (II). Etiological factors. Contact Dermatitis 38(1):14-19.
Runyon R (1980) Toxicity of fresh poinsettia (Euphorbia pulcherrima) to Sprague-Dawley rats. Clin Toxicol 16:167-173.
Santucci B, Picardo M, Cristaudo A (1985) Contact dermatitis from Euphorbia pulcherrima. Contact Dermatitis 12(5):285-286.
Tokarnia CH, Armién AG, Peixoto PV, et al. (1996) Estudo experimental sobre a toxidez de algumas plantas ornamentais em bovinos. Pesq Vet Bras 16(1):5-20.

EUPHORBIA RIGIDA M. Bieb. [Euphorbiaceae]

Synonyms:

Euphorbia biglandulosa Desf.

Citations:

Frohn A, Frohn C, Steuhl KP, et al. (1993) Wolfsmilchverätzung. Ophthalmologe 90(1):58-61.
Lisch K (1980) Die Wirkung des Milchsaftes von Euphorbiazeen auf das Auge. Klin Monatsbl Augenheilkd 176(3):469-471.
Smirnow LD, Efremov AI (1970) [Dermatitis due to Euphorbia rigida M.B.] Vestn Dermatol Venerol 44(6):67-69.

EUPHORBIA ROBBIAE Turrill [Euphorbiaceae]

Citations:

Eke T, Al-Husainy S, Raynor MK (2000) The spectrum of ocular inflammation caused by euphorbia plant sap. Arch Ophthalmol 118(1):13-16.

EUPHORBIA ROYLEANA Boiss. [Euphorbiaceae]

Common Names:

chu; churee; tordanda

Citations:

Sofat BK, Sood GC, Chandel RD, et al. (1972) Euphorbia royleana latex keratitis. Am J Ophthalmol 74(4):634-637.
Sood GC, Sofat BK, Chandel RD (1971) Injury to the eye by the sap of Euphorbia royleana. Br J Ophthalmol 55(12):856-857.

EUPHORBIA SALICIFOLIA Host [Euphorbiaceae]

Common Names:

Wolfsmilch

Citations:

Geßner O (1936) Gewerbliche Dermatitis durch Wolfmilcharten (Euphorbiaceen). Sammlung Vergiftungsfallen 7(A651):217-218.

EUPHORBIA TIRUCALLI L. [Euphorbiaceae]

Common Names:

bali bali; catuit; consuelda; finger tree; gaton; Geraldton carnation weed; Indian spurge tree; Indian tree; Malabar tree; milk bush; milk hedge; milk tree; minyara; monkey fiddle; naked lady; pencil cactus; pencil tree; rubber euphorbia; sapu; solsoldong; soro soro; sosueldo; spurge tree; suel-da-con-suelda; suerda-consuerda; utupa

Citations:

Cordero-Moreno R (1973) Etiologic factors in tropical eye diseases. Am J Ophthalmol 75(3):349-364.
Crowder JI, Sexton RR (1964) Keratoconjunctivitis resulting from the sap of candelebra cactus and the pencil tree. Arch Ophthalmol 72(Oct):476-484.
Uichanco JB, Aquino AR (1962) Euphorbia tirucalli as a systemic poison and its possible use for euthanasia. Philippine J Vet Med 1:161-167.

Euphorbia tithymaloides L. = Pedilanthus tithymaloides (L.) Poit.

Euphorbia trigona Haw. = Euphorbia antiquorum L.

EUPHORBIA TRIANGULARIS Desf. ex A. Berger [Euphorbiaceae]

Citations:

Gschwendt M, Hecker E (1974) Über die Wirkstoffe der Euphorbiaceen II. Hautreizende und cocarcinogene Faktoren aus Euphorbia triangularis Desf. Z Krebsforsch 81:193-210.

Euphorbia viminale L. = Sarcostemma viminale (L.) R. Br.

Eurecka lemon –see– *Citrus limon* (L.) Burm. f.

Europäische Haselwurz –see– *Asarum europaeum* L.

European aconite –see– *Aconitum napellus* L.

European beech –see– *Fagus sylvatica* L.

European bindweed –see– *Convolvulus arvensis* L.

European bittersweet –see– *Solanum dulcamara* L.

European boxwood –see– *Buxus sempervirens* L.

European cherry laurel –see– *Prunus laurocerasus* L.

European dog mercury –see– *Mercurialis perennis* L.

European elder –see– *Sambucus nigra* L.

European euonymus –see– *Euonymus europaeus* L.

European hazel –see– *Corylus avellana* L.

European heliotrope –see– *Heliotropium europaeum* L.

European hop –see– *Humulus lupulus* L.

European horse chestnut –see– *Aesculus hippocastanum* L.

European laurel –see– *Laurus nobilis* L.

European mandrake –see– *Mandragora officinarum* L.

European May Day tree –see– *Prunus padus* L.

European mistletoe –see– *Viscum album* L.

European morning-glory –see– *Convolvulus arvensis* L.

European nightshade –see– *Solanum dulcamara* L.

European oak –see– *Quercus robur* L.

European pennyroyal –see– *Mentha pulegium* L.

European pine –see– *Pinus pinaster* Aiton

European privet –see– *Ligustrum vulgare* L.

European spindle tree –see– *Euonymus europaeus* L.

European vervain –see– *Verbena officinalis* L.

European water hemlock –see– *Cicuta virosa* L.

European white hellebore –see– *Veratrum album* L.

European white lupin –see– *Lupinus albus* L.

European yew –see– *Taxus baccata* L.

EUXYLOPHORA PARAENSIS Huber [Rutaceae]

Common Names:

Páo amarello; Páo setim; Perobas

Citations:

Freise FW (1936) Vergiftungen durch brasilianische Werkhölzer. II. Jacareúba-Holz und Seidenholz. Sammlung Vergiftungsfallen 7(C33):61-72.

evening-star rain lily –see– *Cooperia pedunculata* Herb.

evening trumpet flower –see– *Gelsemium sempervirens* (L.) J. St.-Hil.

evening trumpet vine –see– *Gelsemium sempervirens* (L.) J. St.-Hil.

evergreen box –see– *Buxus sempervirens* L.

evergreen holm oak –see– *Quercus ilex* L.

everlasting pea –see– *Lathyrus hirsutus* L.; *Lathyrus latifolius* L.; *Lathyrus pusillus* Elliott; *Lathyrus sylvestris* L.

everlasting thorn –see– *Pyracantha coccinea* M. Roem.

Eve's-necklace –see– *Calia secundiflora* (Ortega) Yakovlev

ewe-aran –see– *Spigelia anthelmia* L.

ewon agogo –see– *Lantana camara* L.

EXACUM AFFINE Balf. f. [Gentianaceae]

Citations:

Paulsen E (1998) Occupational dermatitis in Danish gardeners and greenhouse workers (II). Etiological factors. Contact Dermatitis 38(1):14-19.

EXCOECARIA PARVIFOLIA Müll. Arg. [Euphorbiaceae]

Common Names:

blind-your-eyes; gutta percha tree; poison tree

Citations:

Logan DA (1925) Gutta percha tree (Excoecaria parvifolia) and its effects on sheep. Queensland Agric J 24(Sep 1):232-233.

exile oleander –see– *Thevetia peruviana* (Pers.) K. Schum.

exile tree –see– *Thevetia peruviana* (Pers.) K. Schum.

eyebane –see– *Chamaesyce maculata* (L.) Small

ezonegi –see– *Allium schoenoprasum* L.

F

faba bean –see– *Vicia faba* L.

fabismo –see– *Vicia faba* L.

fadang –see– *Cycas circinalis* L.

FADOGIA HOMBLEI De Wild. [Rubiaceae]

Synonyms:

Fadogia monticola Robyns

Common Names:

wild date

Citations:

Adelaar TF, Terblanche M, Smit JD (1966) A report on negative experiments with ferric chloride as a prophylactic agent against gousiekte. J S Afr Vet Assoc 37(2):199-201.

Fourie N, Schultz RA, Prozesky L, et al. (1989) Clinical pathological changes in gousiekte, a plant-induced cardiotoxicosis of ruminants. Onderstepoort J Vet Res 56(1):73-80.

Hurter LR, Naudé TW, Adelaar TF, et al. (1972) Ingestion of the plant Fadogia monticola Robyns as an additional cause of gousiekte in ruminants. Onderstepoort J Vet Res 39(1):71-81.

Prozesky L, Bastianello SS, Fourie N, et al. (2005) A study of the pathology and pathogenesis of the myocardial lesions in gousiekte, a plant-induced cardiotoxicosis of ruminants. Onderstepoort J Vet Res 72(3):219-230.

Fadogia monticola Robyns = Fadogia homblei De Wild.

Fagara flava (Vahl) Krug & Urb. = Zanthoxylum flavum Vahl

FAGARA HEITZII Aubrév. & Pellegr. [Rutaceae]

Common Names:

linoncillo; olon

Citations:

Condé-Salazar L, Guimaraens D, Romero LV, et al. (1987) Allergic contact dermatitis to olonwood. Contact Dermatitis 16(4):231-232.

fagopyrisme –see– *Fagopyrum esculentum* Moench

FAGOPYRUM ESCULENTUM Moench [Polygonaceae]

Synonyms:

Fagopyrum sagittatum Gilib.; ***Polygonum fagopyrum*** L.

Common Names:

Buchweizen; buckwheat; fagopyrisme; hainkorn; Heidegrütze; Heidekorn; trigo sarracero; wild buckwheat

Citations:

Bichlmaier H (1912) Experimentelle Untersuchungen über Buchweizenerkranckung. Monatsh Prakt Tierheil 23:305-318.

Blumstein GI (1935) Buckwheat sensitivity. J Allergy 7:74-79.

Bruce EA (1917) Fagopyrismus (buckwheat poisoning) and similar affections. J Am Vet Med Assoc 52(Nov):189-194.

Collet P, Henry E (1952) Photosensibilisation du Dindon par le Sarrazin. Bull Soc Vet Lyon 54-55(1):437-442.

Göhte CJ, Wieslander G, Ancker K, et al. (1983) Buckwheat allergy: Health food, an inhalation health risk. Allergy 38(3):155-159.

Horesh AJ (1972) Buckwheat sensitivity in children. Ann Allergy 30(12):685-689.

Kubo H, Fujimoto H, Nakayama H (1938) Über die Leberveranderungen durch die Zufuhr von Buchweizen, gewöhnlicher Nahrungsmittel, bei Kaninchen, Ratten und Mausen. (I. Mitteilung). Acta Pathol Jpn 28:325-336.

Lee SY, Lee KS, Hong CH, et al. (2001) Three cases of childhood nocturnal asthma due to buckwheat allergy. Allergy 56(8):763-766.

Park JW, Kang DB, Kim CW, et al. (2000) Identification and characterization of the major allergens of buckwheat. Allergy 55(11):1035-1041.

Priouzeau MM (1942) Fagopyrisme chez les bovidés. Rec Med Vet Ec Alfort 118:160-168.

Schiffner R, Przybilla B, Burgdorff T, et al. (2001) Anaphylaxis to buckwheat. Allergy 56(10).1020-1021.

Sheard C, Caylor HD, Schlotthauer C (1928) Photosensitization of animals after the ingestion of buckwheat. J Exp Med 47:1013-1028.

Valdivieso R, Moneo I, Pola J, et al. (1989) Occupational asthma and contact urticaria caused by buckwheat flour. Ann Allergy 63(2):149-152.

Fagopyrum sagittatum Gilib. = Fagopyrum esculentum Moench

FAGUS GRANDIFOLIA Ehrh. [Fagaceae]

Common Names:

American beech; beechnut

Citations:

Morton JF (1979) Plant tannins and esophageal cancer. In: Deichmann WB (ed.) Toxicology and occupational medicine. Elsevier. New York. pp. 129-137.

FAGUS SYLVATICA L. [Fagaceae]

Common Names:

beech; beechnut; Buche; European beech; Rotbuche

Citations:

Hayes MJ, Turner M (1990) Beechmast poisoning. Vet Rec 127(20):508.

Morton JF (1979) Plant tannins and esophageal cancer. In: Deichmann WB (ed.) Toxicology and occupational medicine. Elsevier. New York. pp. 129-137.

Räsänen L, Jolanki R, Estlander T, et al. (1998) Occupational contact allergy from beechwood. Contact Dermatitis 38(1):55.

Wilkens WM, Cranwell MP (1990) Beechmast poisoning in ponies. Vet Rec 127(17):435.

FAIDHERBIA ALBIDA (Delile) A. Chev. [Fabaceae]

Citations:

Steyn DG (1965) An investigation into cases of suspected poisoning in Africans in northern Rhodesia. S Afr Med J 39(16):344-350.

fairy bells –see– *Digitalis purpurea* L.

fairy caps –see– *Digitalis purpurea* L.

fairy fingers –see– *Digitalis purpurea* L.

fairy gloves –see– *Digitalis purpurea* L.

fairy thimbles –see– *Digitalis purpurea* L.

fakete nadálytö –see– *Symphytum officinale* L.

falaris –see– *Phalaris aquatica* L.

fall crocus –see– *Colchicum autumnale* L.

fall panicgrass –see– *Panicum dichotomiflorum* Michx.

fall panicum –see– *Panicum dichotomiflorum* Michx.

fall poison –see– *Ageratina altissima* (L.) R. M. King & H. Rob.; *Amianthium muscitoxicum* (Walter) A. Gray

Fallkraut –see– *Arnica montana* L.

FALLOPIA CONVOLVULUS (L.) Á. Löve [Polygonaceae]

Synonyms:

Polygonum convolvulus L.

Common Names:

black bindweed; climbing buckwheat; cornbind; ivy bindweed; knot bindweed; Windenknöterich

Citations:

Salgues R (1961) Le genre Polygonum L. (Polygonacées) Etudes chimiques et toxicologiques. Les faits hématologiques de l'empoisonnement expérimental. Qualitas Plant Mat Veg 8:367-395.

Wilsdorf G, Heckel R, Juhrke J (1991) Orientierende toxikologische Untersuchungen an Wistar-Ratten zur Bewertung von Windenknöterich - (Polygonum convolvulus) samen. Arch Exp Veterinarmed 45(1):123-129.

FALLOPIA MULTIFLORA (Thunb.) Haraldson [Polygonaceae]

Synonyms:

Polygonum multiflorum Thunb.

Common Names:

ho-shou-wu; shou-wu-pian; shouwu

Citations:

Battinelli L, Daniele C, Mazzanti G, et al. (2004) New case of acute hepatitis following the consumption of shou wu pian, a Chinese herbal product derived from Polygonum multiflorum. Ann Intern Med 140(7):E589-E590.

But PP, Tomlinson B, Lee KL (1996) Hepatitis related to the Chinese medicine shou-wu-pian manufactured from Polygonum multiflorum. Vet Hum Toxicol 38(4):280-282.

Chan TY, Chan JC, Tomlinson B, et al. (1994) Poisoning by Chinese herbal medicines in Hong Kong: A hospital-based study. Vet Hum Toxicol 36(6):546-547.

Panis B, Wong DR, Hooymans PM, et al. (2005) Recurrent toxic hepatitis in a Caucasian girl related to the use of shou-wu-pian, a Chinese herbal preparation. J Pediatr Gastroenterol Nutr 41(2):256-258.

FALLOPIA SACHALINENSIS (F. Schmidt) Ronse Decr. [Polygonaceae]

Citations:

Lamminpaa A, Kinos M (1996) Plant poisonings in children. Hum Exp Toxicol 15(3):245-249.

falsa visnaga –see– *Ammi majus* L.

Falsche Akazie –see– *Robinia pseudoacacia* L.

falsche Efparfette –see– *Securigera varia* (L.) Lassen

Falscher Kalmus –see– *Iris pseudoacorus* L.

false acacia –see– *Robinia pseudoacacia* L.

false angustura bark –see– *Strychnos ignatii* P. J. Bergius

false banana –see– *Asimina triloba* (L.) Dunal

false cactus –see– *Euphorbia lactea* Haw.

false castor –see– *Datura ferox* L.

false chamomile –see– *Matricaria recutita* L.

false coffeebean –see– *Gymnocladus dioicus* (L.) K. Koch

false dandelion –see– *Hypochaeris radicata* L.

false ebony –see– *Laburnum anagyroides* Medik.

false heliotrope –see– *Phacelia crenulata* Torr. ex S. Watson

false hellebore –see– *Veratrum album* L.; *Veratrum californicum* Durand; *Veratrum japonicum* (Baker) Loes.; *Veratrum viride* Aiton

false horsetail –see– *Equisetum arvense* L.

false indigo –see– *Baptisia australis* (L.) R. Br.

false jessamine –see– *Gelsemium sempervirens* (L.) J. St.-Hil.

false lupine –see– *Thermopsis montana* Nutt.

false mistletoe –see– *Phoradendron leucarpum* (Raf.) Reveal & M. C. Johnst.

false parsley –see– *Aethusa cynapium* L.; *Cicuta douglasii* (DC.) J. M. Coult. & Rose; *Cicuta maculata* L.; *Conium maculatum* L.

false pennyroyal –see– *Hedeoma pulegioides* (L.) Pers.

false pepper –see– *Schinus terebinthifolius* Raddi

false peyote –see– *Ariocarpus retusus* Scheidw.

false poinciana –see– *Sesbania drummondii* (Rydb.) Cory; *Sesbania punicea* (Cav.) Benth.

false ragweed –see– *Cyclachaena xanthiifolia* (Nutt.) Fresen.; *Iva angustifolia* Nutt. ex DC.; *Parthenium hysterophorus* L.

false sago palm –see– *Cycas circinalis* L.; *Cycas revoluta* Thunb.; *Cycas circinalis* L.

false sunflower –see– *Helenium autumnale* L.; *Helenium microcephalum* DC.

false sycamore –see– *Melia azedarach* L.

false tragacanth –see– *Sterculia urens* Roxb.

false upas tree –see– *Strychnos ignatii* P. J. Bergius

false white cedar –see– *Thuja occidentalis* L.

false white hellebore –see– *Veratrum californicum* Durand

false wormwood –see– *Parthenium hysterophorus* L.

falso perejil –see– *Aethusa cynapium* L.

famentana –see– *Cerbera odollam* Gaertn.

fancy-leaf caladium –see– *Caladium bicolor* (Aiton) Vent.

fanweed –see– *Thlaspi arvense* L.

farfara –see– *Tussilago farfara* L.

FARFUGIUM JAPONICUM (L.) Kitam. [Asteraceae]

Common Names:
tsuwabuki

Citations:
Hirono I, Ueno I, Aiso S, et al. (1983) Carcinogenic activity of Farfugium japonicum and Senecio cannabifolius. Cancer Lett 20(2):191-198.

farkas kutyatej –see– *Euphorbia cyparissias* L.

farkasölö sisakvirág –see– *Aconitum lycoctonum* L. subsp. vulparia (Rchb.) Nyman

Farn –see– *Pteridium aquilinum* (L.) Kuhn

Farnkraut –see– *Dryopteris filix-mas* (L.) Schott; *Pteridium aquilinum* (L.) Kuhn

fat hen –see– *Chenopodium album* L.

fatarita –see– *Sorghum bicolor* (L.) Moench

fatberry –see– *Rhodotypos scandens* (Thunb.) Makino

FATSIA JAPONICA (Thunb.) Decne. & Planch. [Araliaceae]

Common Names:
aralia; Aralie

Citations:
Oka K, Saito F, Yasuhara T, et al. (1999) The allergens of Dendropanax trifidus Makino and Fatsia japonica Decne. et Planch. and evaluation of cross-reactions with other plants of the Araliaceae family. Contact Dermatitis 40(4):209-213.

Faulbaum –see– *Frangula alnus* Mill.; *Prunus padus* L.

faux acajou –see– *Blighia sapida* K. D. Koenig

faux gui américain –see– *Phoradendron leucarpum* (Raf.) Reveal & M. C. Johnst.

faux thuya –see– *Thuja occidentalis* L.

fava –see– *Vicia faba* L.

fava bean –see– *Vicia faba* L.

fava d'anta –see– *Dimorphandra garderiana* Tul.

fava tonka –see– *Dipteryx odorata* (Aubl.) Willd.

Favabohne –see– *Vicia faba* L.

faveira –see– *Dimorphandra mollis* Benth.

favismo –see– *Vicia faba* L.

feathergrass –see– *Stipa pennata* L.

February daphne –see– *Daphne mezereum* L.

fedegoso –see– *Senna occidentalis* (L.) Link

federico –see– *Cycas circinalis* L.

fehér zászpa –see– *Veratrum album* L.

Feige –see– *Ficus carica* L.

Feigenbaum –see– *Ficus carica* L.

fekete bodza –see– *Sambucus nigra* L.

felce maschia –see– *Dryopteris filix-mas* (L.) Schott

Feldmohn –see– *Papaver rhoeas* L.

Feldrittersporn –see– *Consolida regalis* Gray

felonwood –see– *Solanum dulcamara* L.

felonwort mortal –see– *Solanum dulcamara* L.

female nervine –see– *Cypripedium reginae* Walter

fennel –see– *Foeniculum vulgare* Mill.

Fensterblatt –see– *Monstera deliciosa* Liebm.

fenugreek –see– *Trigonella foenum-graecum* L.

fern palm –see– *Cycas circinalis* L.

Ferula abyssinica A. Rich. = Ferula communis L.

FERULA ASSA-FOETIDA L. [Apiaceae]

Common Names:
asafetida; asafoetida; devil's-dung; Stinkasant

Citations:
Bottei EM, DuMont CM, Otten EJ (1999) Methemoglobinemia caused by cutaneously absorbed asafoetida. J Toxicol Clin Toxicol 41(5):620-621.

Seetharam KA, Pasricha JS (1987) Condiments and contact dermatitis of the finger-tips. Indian J Dermatol Venereol Leprol 53:325-328.

FERULA COMMUNIS L. [Apiaceae]

Synonyms:
Ferula abyssinica A. Rich.

Common Names:
fessoukh; giant fennel

Citations:

Aragno M, Tagliapietra S, Nano GM, et al. (1988) Experimental studies on the toxicity of Ferula communis in the rat. Res Commun Chem Pathol Pharmacol 59(3):399-402.

Colella G, Marocchio L, Palermo D, et al. (1987) La tossicita' di Ferula communis in Puglia. Rilievi ematologici, chimico-tossicologici ed anatomo-istopatologici. Acta Med Vet (Napoli) 33(3):237-250.

Corticelli B, Deiana S (1957) Sul comportamento elettroforetico dei protidi serici e plasmatici del coniglio intossicato con Ferula communis. Boll Soc Ital Biol Sper 33(5):625-628.

Corticelli B, Deiana S, Palmas G (1957) Azione protettiva ed azione antiemorragica della vitamina K1 nella intossicazione da Ferula communis. Boll Soc Ital Biol Sper 33(5):629-631.

Gil MC, Gomez L, Roy TJ, et al. (2002) Testicular and epididymal changes in rams following intoxication by Ferula communis. Vet Rec 150(1):24-25.

Infante Miranda F (1965) Intoxicación por Ferula communis L. Arch Zootecnia 14(53):1-30.

Shlosberg A, Egyed MN (1985) Experimental Ferula communis (giant fennel) toxicosis in sheep. Zentralbl Veterinarmed A 32(10):778-784.

Shlosberg A, Egyed MN (1986) Stability of prothrombin times in sheep dosed with natural and synthetic anticoagulants. Res Vet Sci 40(2):141-143.

Tligui N, El Haouzi M, El Himer H (1998) Vitamin K1 therapy of Ferula communis variety brevifolia toxicosis in sheep. In: Garland T, Barr AC (eds.) Toxic plants and other natural toxicants. CABI. New York. pp. 159-164.

FERULA HERMONIS Boiss. [Apiaceae]

Common Names:

zallouh

Citations:

El Thaher TS, Matalka KZ, Taha HA, et al. (2001) Ferula hermonis 'zallouh' and enhancing erectile function in rats: Efficacy and toxicity study. Int J Impot Res 13(4):247-251.

fescue –see– *Festuca arundinacea* Schreb.

fessoukh –see– *Ferula communis* L.

FESTUCA ARGENTINA (Speg.) Parodi [Poaceae]

Common Names:

coiron; pampasgrass

Citations:

Jones FS, Arnold JF (1917) Staggers in sheep in Patagonia. J Exp Med 26(6):805-823.

Uzal FA, Woodman MP, Giraudo CG, et al. (1996) An attempt to reproduce 'mal seco' in horses by feeding them Festuca argentina. Vet Rec 139(3):68-70.

FESTUCA ARUNDINACEA Schreb. [Poaceae]

Synonyms:

Festuca elatior L. subsp. arundinacea (Schreb.) Celak.

Common Names:

alta fescue; fescue; fétaque roseau; reed fescue; tall fescue; tall meadow fescue; William's-grass

Citations:

Anas K, Cross DL, Poling R, et al. (1998) A survey concerning the equine fescue toxicosis malady. J Equine Vet Sci 18(10):631-637.

Ashley G (1958) Fescue poisoning of cattle on Florida muck land. J Am Vet Med Assoc 132:493-494.

Breuhaus BA (2003) Thyroid function in mature horses ingesting endophyte-infected fescue seed. J Am Vet Med Assoc 223(4):340-345.

Browning R Jr (2004) Effects of endophyte-infected tall fescue on indicators of thermal status and growth in Hereford and Senepol steers. J Anim Sci 82(2):634-643.

Dougherty CT, Lauriault LM, Bradley NW, et al. (1991) Induction of tall fescue toxicosis in heat-stressed cattle and its alleviation with thiamin. J Anim Sci 69(3):1008-1018.

Farnell DR, Futrell MC, Watson VH, et al. (1975) Field studies on etiology and control of fescue toxicosis. J Environ Quality 4(1):120-122.

Franco DA, Taylor LA (1971) Gangrenous fescue lameness in a Charolais heifer. Vet Med Small Anim Clin 66(10):1012-1015.

Jackson JA Jr, Hemken RW, Boling JA, et al. (1984) Summer fescue toxicity in dairy steers fed tall fescue seed. J Anim Sci 58(5):1057-1061.

Jensen R, Deem AW, Knaus D (1956) Fescue lameness in cattle. I. Experimental production of the disease. Am J Vet Res 17(63):196-201.

Jones KL, King SS, Griswold KE, et al. (2003) Domperidone can ameliorate deleterious reproductive effects and reduced weight gain associated with fescue toxicosis in heifers. J Anim Sci 81(10):2568-2574.

Oliver JW, Schultze AE, Rohrbach BW, et al. (2000) Alterations in hemograms and serum biochemical analytes of steers after prolonged consumption of endophyte-infected tall fescue. J Anim Sci 78(4):1029-1035.

Parish JA, McCann MA, Watson RH, et al. (2003) Use of nonergot alkaloid-producing endophytes for alleviating tall fescue toxicosis in sheep. J Anim Sci 81(5):1316-1322.

Poppenga RH, Mostrom MS, Haschek WM, et al. (1984) Mare agalactia, placental thickening, and high foal mortality associated with the grazing of tall fescue: A case report. Proc Am Assoc Vet Lab Diagn 27:325-336.

Pulsford MF (1950) A note on lameness in cattle grazing on tall meadow fescue (Festuca arundinacea) in South Australia. Aust Vet J 26:87-88.

Shlosberg A, Hermes JC, Smith Wood L, et al. (2004) The effect of feeding broilers tall fescuegrass (Arundinacea) seeds containing the endophyte toxin ergovaline on ascites syndrome. In: Acamovic T, Stewart CS, Pennycott TW (eds.) Poisonous plants and related toxins. CABI. Wallingford, UK. pp. 558-563.

Watson DF, Rooney JR, Hoag WG (1957) Fescue foot lameness in cattle - Some observations on the disease in Virginia. J Am Vet Med Assoc 130(Mar 1):217-219.

Wolfe BA, Bush M, Monfort SL, et al. (1998) Abdominal lipomatosis attributed to tall fescue toxicosis in deer. J Am Vet Med Assoc 213(12):1783-1786.

Festuca elatior subsp. arundinacea (Schreb.) Celak. = Festuca arundinacea Schreb.

FESTUCA RUBRA L. subsp. fallax (Thuill.) Nyman [Poaceae]

Synonyms:

Festuca rubra L. var. commutata Gaudin

Common Names:

Chewing's-fescue; red fescue

Citations:

Haag JR (1954) Toxicity of nematode infested Chewings fescue seed. Science 102(2651):406-407.

Shaw JN, Muth OH (1949) Some types of forage poisoning in Oregon cattle and sheep. J Am Vet Med Assoc 114(866):315-317.

Festuca rubra L. var. commutata Gaudin = Festuca rubra L. subsp. fallax (Thuill.) Nyman

fétaque roseau –see– *Festuca arundinacea* Schreb.

fetid buckeye –see– *Aesculus glabra* Willd.

fetid nightshade –see– *Hyoscyamus niger* L.

fetid sterculia –see– *Sterculia foetida* L.

fetterbush –see– *Pieris floribunda* (Pursh) Benth. & Hook. f.

fèves –see– *Vicia faba* L.

Feuerblume –see– *Papaver rhoeas* L.

Feuerbohne –see– *Phaseolus coccineus* L.

Feuerdorn –see– *Pyracantha coccinea* M. Roem.

feuilles-de-houx –see– *Berberis aquifolium* Pursh

fève-du-loup –see– *Lupinus albus* L.

fève Tonka –see– *Dipteryx odorata* (Aubl.) Willd.

fever twig –see– *Solanum dulcamara* L.

feverfew –see– *Parthenium hysterophorus* L.; *Tanacetum cinerariifolium* (Trevir.) Sch. Bip.; *Tanacetum parthenium* (L.) Sch. Bip.

Fichte –see– *Picea abies* (L.) H. Karst.

fico –see– *Ficus carica* L.

Ficus –see– *Ficus carica* L.

FICUS AMPLISSIMA Sm. [Moraceae]

Synonyms:

Ficus tsiela Roxb.

Common Names:

chela

Citations:

Nair ND, Valsala KV, Ramachandran KM, et al. (1985) Experimental studies on chela leaf (Ficus tsiela Roxb.) poisoning in calves. Kerala J Vet Sci 16(2):94-99.

Rajan A, Nair ND, Valsala KV, et al. (1986) Pathology of a nervous disorder in cattle caused by the toxicity of the leaves of the tree Ficus tsiela Roxb. Indian Vet J 63(3):184-186.

FICUS BENJAMINA L. [Moraceae]

Common Names:

Benjamin tree; ficus maqui; fig; Java willow; small-leaf rubber plant; weeping ficus; weeping fig

Citations:

Axelsson G, Skedinger M, Zetterström O (1985) Allergy to weeping fig - A new occupational disease. Allergy 40(6):461-464.

Axelsson IG, Johansson SG, Zetterström O (1987) A new indoor allergen from a common non-flowering plant. Allergy 42(8):604-611.

Axelsson IG, Johansson SG, Zetterström O (1987) Occupational allergy to weeping fig in plant keepers. Allergy 42(3):161-167.

Kanerva L, Estlander T, Petman L, et al. (2001) Occupational allergic contact urticaria to yucca (Yucca aloifolia), weeping fig (Ficus benjamina), and spathe flower (Spathiphyllum wallisii). Allergy 56(10):1008-1011.

Mrvos R, Krenzelok EP, Jacobsen TD (2001) Toxidromes associated with the most common plant ingestions. Vet Hum Toxicol 43(6):366-369.

FICUS CARICA L. [Moraceae]

Common Names:

cha-de-figo; Feige; Feigenbaum; fico; Ficus; fig; figues

Citations:

Andreichuk IE (1984) [Fig dermatitis.] Vestn Dermatol Venerol 4(Apr):67-68.

Gandolfo M, Baeza M, De Barrio M (2001) Anaphylaxis after eating figs. Allergy 56(5):462-463.

Goitre M, Bedello PG, Cane D, et al. (1984) Fitofotodermatite da fico. G Ital Dermatol Venereol 119(6):435-436.

Ippen H (1982) Phototoxische Reaktion auf Feigen. Hautarzt 33(6):337-339.

Kitchevatz MM (1934) Etiologie et pathogénèse de la dermatite des figues. Bull Soc Fr Dermatol Syphiligr 41:1751-1759.

Kuske H (1939) Experimentelle Untersuchungen zur Photosensibilisierung der Haut durch pflanzliche Wirkstoffe. 1. Mitteilung. Lichtsensibilisierung durch Furocumarine als Ursache verschiedener phytogener Dermatosen. Arch Derm Syphilol 178:112-123.

Lembo G, Lo Presti M, Balato N (1985) Phytophotodermatitis due to Ficus carica. Photodermatol 2(2):119-120.

Londoño F, Rueda LA (1968) Fotodermatitis por "Ficus carica" en gestantes. Med Cutanea 11(6):573-580.

FICUS CORDATA Thunb. subsp. salicifolia (Vahl) C. C. Berg [Moraceae]

Citations:

Myburgh JG, Fournie N, Van der Lugt JJ, et al. (1994) A nervous disorder in cattle, caused by the plants Ficus ingens var. ingens and Ficus cordata subsp. salicifolia. Onderstepoort J Vet Res 61(2):171-176.

FICUS INGENS (Miq.) Miq. [Moraceae]

Citations:

Myburgh JG, Fournie N, Van der Lugt JJ, et al. (1994) A nervous disorder in cattle, caused by the plants Ficus ingens var. ingens and Ficus cordata subsp. salicifolia. Onderstepoort J Vet Res 61(2):171-176.

FICUS MACROPHYLLA Desf. ex Pers. [Moraceae]

Citations:

Schultz RA, Kellerman TS, Van den Berg H (2005) The role of fluorescence polarization immuno-assay in the diagnosis of plant-induced cardiac glycoside poisoning livestock in South Africa. Onderstepoort J Vet Res 72(3):189-201.

ficus maqui –see– *Ficus benjamina* L.

FICUS PUMILA L. [Moraceae]

Common Names:

creeping fig; wild fig

Citations:

English PB, Grey LP (1943) Sap dermatitis and conjunctivitis caused by the wild fig (Ficus pumila). Med J Aust 1(Jun 26):578-579.

Ficus tsiela Roxb. = Ficus amplissima Sm.

fiddle flower –see– *Pedilanthus tithymaloides* (L.) Poit.

fiddleneck –see– *Amsinckia intermedia* Fisch. & C. A. Mey.

fiddler's-spurge –see– *Euphorbia heterophylla* L.

Fieberklee –see– *Menyanthes trifoliata* L.

field balm –see– *Glechoma hederacea* L.

field bean –see– *Lablab purpureus* (L.) Sweet; *Phaseolus vulgaris* L.; *Vicia faba* L.

field bindweed –see– *Convolvulus arvensis* L.

field dodder –see– *Cuscuta pentagona* Engelm.

field hedge nettle –see– *Stachys arvensis* (L.) L.

field horsetail –see– *Equisetum arvense* L.

field larkspur –see– *Consolida regalis* Gray; *Delphinium menziesii* DC.

field mustard –see– *Sinapis arvensis* L.

field pea –see– *Pisum sativum* L.

field pennycress –see– *Thlaspi arvense* L.

field poppy –see– *Papaver rhoeas* L.

field sandbur –see– *Cenchrus incertus* M. A. Curtis

field soapwort –see– *Vaccaria hispanica* (Mill.) Rauschert

field sorrel –see– *Rumex acetosella* L.

field woundwort –see– *Stachys arvensis* (L.) L.

fierce thorn apple –see– *Datura ferox* L.

fiery thorn –see– *Pyracantha coccinea* M. Roem.

Fiesenkerbel –see– *Heracleum mantegazzianum* Sommier & Levier

fig –see– *Ficus benjamina* L.; *Ficus carica* L.

figuera infernal –see– *Ricinus communis* L.

figues –see– *Ficus carica* L.

filis mas –see– *Dryopteris filix-mas* (L.) Schott

filix mas –see– *Dryopteris filix-mas* (L.) Schott

fimble –see– *Cannabis sativa* L.

fine-leaf sneezeweed –see– *Helenium amarum* (Raf.) H. Rock

fine-leaf water dropwort –see– *Oenanthe aquatica* (L.) Poir.

finger cherry –see– *Rhodomyrtus macrocarpa* Benth.

finger cherry loquat –see– *Rhodomyrtus macrocarpa* Benth.

finger flower –see– *Digitalis purpurea* L.

finger tree –see– *Euphorbia tirucalli* L.

finger weed –see– *Amsinckia intermedia* Fisch. & C. A. Mey.

Fingerhut –see– *Digitalis lanata* Ehrh.; *Digitalis purpurea* L.

fir club moss –see– *Huperzia selago* (L.) Bernh. ex Schrank & Mart.

fireball –see– *Bassia scoparia* (L.) A. J. Scott

firecracker plant –see– *Aesculus pavia* L.

firesticks –see– *Gomphocarpus fruticosa* (L.) W. T. Aiton

firethorn –see– *Pyracantha coccinea* M. Roem.

fireweed –see– *Bassia scoparia* (L.) A. J. Scott; *Datura stramonium* L.; *Gaillardia pulchella* Foug.; *Senecio linearifolius* A. Rich; *Senecio madagascariensis* Poir.

fireweed fiddleneck –see– *Amsinckia intermedia* Fisch. & C. A. Mey.

fisa –see– *Blighia sapida* K. D. Koenig

Fischerkappe –see– *Aconitum napellus* L.

fishtail palm –see– *Caryota mitis* Lour.; *Caryota urens* L.

fitweed –see– *Corydalis caseana* A. Gray

FITZROYA CUPRESSOIDES (Molina) I. M. Johnst. [Cupressaceae]

Synonyms:

Fitzroya patagonica Hook. f. ex Lindl.

Common Names:

alerce; l'alerce

Citations:

Oleffe JA, Dedeken H, Sporcq J (1975) Occupational dermatitis from alerce (Fitzroya cupressoides). Contact Dermatitis 1(5):319.

Oleffe J, Dedeken H, Sporcq J, et al. (1975) Allergie cutanée professionnelle à l'alerce (Fitzroya cupressoides). Berufsdermatosen 23(6):196-207.

Oleffe JA, Sporcq J, Hublet P (1975) Epidemiological study of the wood industry in Belgium. Contact Dermatitis 1(5):315-316.

Fitzroya patagonica Hook. f. ex Lindl. = Fitzroya cupressoides (Molina) I. M. Johnst.

five-finger root –see– *Cicuta virosa* L.

five-hook bassia –see– *Bassia hyssopifolia* (Pall.) Kuntze

Flachs –see– *Linum usitatissimum* L.

flag iris –see– *Iris ×germanica* L.

flag lily –see– *Iris ×germanica* L.

flame lily –see– *Gloriosa superba* L.

flame tree –see– *Sterculia foetida* L.

Flanders poppy –see– *Papaver rhoeas* L.

flannel mullein –see– *Verbascum thapsus* L.

flap dock –see– *Digitalis purpurea* L.

flat billy button –see– *Ixiolaena brevicompta* F. Muell.

flat cedar –see– *Thuja plicata Donn* ex D. Don

flat pea –see– *Lathyrus sylvestris* L.

flathead larkspur –see– *Delphinium bicolor* Nutt.

flatweed –see– *Hypochaeris radicata* L.

flax –see– *Linum rigidum* Pursh; *Linum usitatissimum* L.

flax olive –see– *Daphne mezereum* L.

flaxweed –see– *Pimelea simplex* F. Muell.; *Pimelea trichostachya* Lindl.

fleabane –see– *Conyza bonariensis* (L.) Cronquist

Fleischkraut –see– *Armoracia rusticana* P. Gaertn. et al.

fleshy spurge –see– *Euphorbia antiquorum* L.

fleur-de-lis –see– *Iris ×germanica* L.

fleur-de-luce –see– *Iris pseudoacorus* L.

Flieder –see– *Sambucus nigra* L.

Fliegenkraut –see– *Artemisia vulgaris* L.

Flinders poppy –see– *Pimelea decora* Domin

flixweed –see– *Descurainia sophia* (L.) Webb ex Prantl

Flökraut –see– *Artemisia vulgaris* L.; *Persicaria maculosa* Gray

floppers –see– *Kalanchoe pinnata* (Lam.) Pers.

flor-de-barbero –see– *Allamanda cathartica* L.

flor-de-coral –see– *Jatropha multifida* L.

flor-de fuego –see– *Euphorbia pulcherrima* Willd. ex Klotzsch

flor-de-la-pascua –see– *Euphorbia pulcherrima* Willd. ex Klotzsch

flor-de-la-trompeta –see– *Datura stramonium* L.

flor-de-mayo –see– *Plumeria rubra* L.

flor-de-navidad –see– *Euphorbia pulcherrima* Willd. ex Klotzsch

flor-de-nochebuena –see– *Euphorbia pulcherrima* Willd. ex Klotzsch

flor-de-pascua –see– *Euphorbia pulcherrima* Willd. ex Klotzsch

flor-de-Santa-Caterina –see– *Euphorbia pulcherrima* Willd. ex Klotzsch

flor-de-sapo –see– *Pascalia glauca* Ortega

Florentine iris –see– *Iris ×germanica* L. nothovar. florentina Dykes

florestina –see– *Florestina tripteris* DC.

FLORESTINA TRIPTERIS DC. [Asteraceae]

Common Names:

florestina; sticky florestin; sticky palafoxia

Citations:

Boughton IB, Hardy WT (1939) Hard yellow livers of sheep and cattle and feeding trials of suspected plants. Texas Agric Exp Sta Annu Rep 52:238-240.

Florida arrowroot –see– *Zamia integrifolia* L. f.

Florida coffee –see– *Senna occidentalis* (L.) Link

Florida coontie –see– *Zamia integrifolia* L. f.

Florida holly –see– *Schinus terebinthifolius* Raddi

Florida poison tree –see– *Metopium toxiferum* (L.) Krug & Urb.

florifundia –see– *Brugmansia arborea* (L.) Lagerh.

floripon –see– *Brugmansia suaveolens* (Humb. & Bonpl. ex Willd.) Bercht. & J. Presl

floripondio –see– *Brugmansia ×candida* Pers.; *Brugmansia arborea* (L.) Lagerh.

floss flower –see– *Ageratum conyzoides* L.; *Ageratum houstonianum* Mill.

FLOURENSIA CERNUA DC. [Asteraceae]

Common Names:

blackbrush; hojase; tarbush; varnish bush

Citations:

Dollahite JW, Allen TJ (1975) The toxicity of the fruit of Flourensia cernua (tarbush) (blackbrush). Southwestern Vet 28(2):113-117.

Fredrickson E, Thilsted J, Estell R, et al. (1994) Effect of chronic ingestion of tarbush (Flourensia cernua) on ewe lambs. Vet Hum Toxicol 35(5):409-415.

Hailey TL, Thomas JW, Robinson RM (1966) Pronghorn die-off in Trans-Pecos Texas. J Wildl Manage 30(3):488-496.

Mathews FP (1941) Poisonous plants in the Davis Mountains. Texas Agric Exp Sta Annu Rep 54:93.

Mathews FP (1944) The toxicity of the ripe fruit of blackbrush or tarbrush (Flourensia cernua) for sheep and goats. Texas Agric Exp Sta Bull #664:5-16.

Rick E , Tellez M, Fredrickson E, et al. (2001) Extracts of Flourensia cernua reduce consumption of alfalfa pellets by sheep. J Chem Ecol 27(11):2275-2285.

flower-of-paradise –see– *Catha edulis* (Vahl) Forssk. ex Endl.

flux weed –see– *Hypericum punctatum* Lam.

fly poison –see– *Amianthium muscitoxicum* (Walter) A. Gray

flying saucers –see– *Ipomoea tricolor* Cav.; *Ipomoea violacea* L.

fo-ti-tieng –see– *Centella asiatica* (L.) Urb.

fo tzu –see– *Aconitum carmichaelii* Debeaux

fodder beet –see– *Beta vulgaris* L.

fodder radish –see– *Raphanus sativus* L.

FOENICULUM VULGARE Mill. [Apiaceae]

Common Names:

anis hinono; fennel

Citations:

Levy SB (1948) Bronchial asthma due to ingestion of fennel and fennel seed. Ann Allergy 6:415-416.

Seetharam KA, Pasricha JS (1987) Condiments and contact dermatitis of the finger-tips. Indian J Dermatol Venereol Leprol 53:325-328.

foenugreek –see– *Trigonella foenum-graecum* L.

fog crocus –see– *Colchicum autumnale* L.

foirolle –see– *Mercurialis annua* L.

foironde –see– *Mercurialis annua* L.

folk's-glove –see– *Digitalis purpurea* L.

fool's-cicely –see– *Aethusa cynapium* L.

fool's-parsley –see– *Aethusa cynapium* L.; *Conium maculatum* L.

foothill death camas –see– *Toxicoscordion paniculatum* (Nutt.) Rydb.

forest grasstree –see– *Xanthorrhoea media* R. Br.

forest milk vetch –see– *Astragalus miser* Douglas ex Hook. var. hylophilus (Rydb.) Barneby

forest primrose –see– *Hypericum revolutum* Vahl

forest vetchling –see– *Lathyrus sylvestris* L.

fougère aigle –see– *Pteridium aquilinum* (L.) Kuhn

fougère grand aigle –see– *Pteridium aquilinum* (L.) Kuhn

fougère mâle –see– *Dryopteris filix-mas* (L.) Schott

four-seed vetch –see– *Vicia tetrasperma* (L.) Schreb.

four-wing milk vetch –see– *Astragalus tetrapterus* A. Gray

four-wing saltbush –see– *Atriplex canescens* (Pursh) Nutt.

foxes glofa –see– *Digitalis purpurea* L.

foxglove –see– *Digitalis lanata* Ehrh.; *Digitalis purpurea* L.

foxtail –see– *Hordeum jubatum* L.

foxtail barley –see– *Hordeum jubatum* L.

foxtail millet –see– *Setaria pumila* (Poir.) Roem. & Schult. subsp. pumila

foxtail rush –see– *Equisetum arvense* L.

foxtailgrass –see– *Setaria pumila* (Poir.) Roem. & Schult. subsp. pumila

FRAGARIA ×ANANASSA Duchesne ex Rozier [Rosaceae]

Common Names:

Erdbeere; strawberry; wild strawberry

Citations:

Grattan CE, Harman RR (1985) Contact urticaria to strawberry. Contact Dermatitis 13(3):191-192.

Francisco Alvarez –see– *Luehea divaricata* Mart.

Francissia –see– *Brunfelsia australis* Benth.

frangipani –see– *Plumeria rubra* L.

frangula –see– *Frangula alnus* Mill.

FRANGULA ALNUS Mill. [Rhamnaceae]

Synonyms:

Rhamnus frangula L.

Common Names:

alder buckthorn; aulne noir; black alder; black dogwood; bourdaine; buckthorn; Drosselkirschbaum; Faulbaum; frangula; glossy buckthorn; Pulverholz; trollhegg; vuilboom; Wegdorn

Citations:

van den Dikkenberg MI, Holtkamp BM (1987) Vuilboomintoxicatie bij paarden. Tijdschr Diergeneeskd 112(6):340-341.

Französisher Spinat –see– *Rumex acetosa* L.

fraxinella –see– *Dictamnus albus* L.

FRAXINUS EXCELSIOR L. [Oleaceae]

Common Names:

ash tree

Citations:

Fernández-Rivas M, Pérez-Carral C, Senent CJ (1997) Occupational asthma and rhinitis caused by ash (Fraxinus excelsior) wood dust. Allergy 52(2):196-199.

Hemmer W, Focke M, Wantke F, et al. (2000) Ash (Fraxinus excelsior)-pollen allergy in central Europe: Specific role of pollen panallergens and the major allergen of ash pollen, Fra e 1. Allergy 55(10):923-930.

freckled milk vetch –see– *Astragalus lentiginosus* Douglas ex Hook.; *Astragalus lentiginosus* Douglas ex Hook. var. diphysus (A. Gray) M. E. Jones

French bean –see– *Phaseolus vulgaris* L.

French honeysuckle –see– *Galega officinalis* L.

French hydrangea –see– *Hydrangea macrophylla* (Thunb.) Ser.

French jasmine –see– *Calotropis procera* (Aiton) W. T. Aiton

French lilac –see– *Galega officinalis* L.

French marigold –see– *Tagetes patula* L.

French maritime pine tree –see– *Pinus pinaster* Aiton

French millet –see– *Panicum miliaceum* L.

French physicnut –see– *Jatropha multifida* L.

Frenchberry –see– *Rhamnus cathartica* L.

Frenchweed –see– *Thlaspi arvense* L.

freshwater mangrove –see– *Barringtonia acutangula* (L.) Gaertn.

friar's-cap –see– *Aconitum napellus* L.

friar's-cowl –see– *Arum maculatum* L.

frijolillo –see– *Calia secundiflora* (Ortega) Yakovlev

frijolito –see– *Calia secundiflora* (Ortega) Yakovlev

fringe-leaf paspalum –see– *Paspalum setaceum* Michx. var. ciliatifolium (Michx.) Vasey

Froschkraut –see– *Ranunculus sceleratus* L.

fruit-of-heaven –see– *Mangifera indica* L.

fruit salad plant –see– *Monstera deliciosa* Liebm.

fu tse –see– *Aconitum carmichaelii* Debeaux

fuchsia bush –see– *Eremophila maculata* (Ker Gawl.) F. Muell.

fuh –see– *Onychium contiguum* Wall. ex C. Hope

FURCRAEA CABUYA Trel. [Agavaceae]

Citations:

Robledo Clavijo A, Cano Puerta G (1963) Agavosis, un sindrome provocado por el contacto con fibras de las plantas de la familia Agavaceae. Estudio clinico y farmacológico. Antioquia Medica 13(6):405-448.

fusain –see– *Euonymus europaeus* L.

Futtererbse –see– *Pisum sativum* L.

Futterrübe –see– *Beta vulgaris* L.

Futterwicke –see– *Vicia sativa* L.

G

gabi –see– *Colocasia esculenta* (L.) Schott

gad –see– *Catha edulis* (Vahl) Forssk. ex Endl.

gadalin kura –see– *Kalanchoe lanceolata* (Forssk.) Pers.

gadon machiji –see– *Trianthema portulacastrum* L.

gadong –see–*Dioscorea convolvulacea* Schltdl. & Cham.

gaillardia –see– *Gaillardia pulchella* Foug. var. picta (Sweet) A. Gray

GAILLARDIA ARISTATA Pursh [Asteraceae]

Citations:

Burry JN (1980) Dermatitis from Gaillardia aristata: Compositae dermatitis in South Australia. Contact Dermatitis 6(2):157.

Gaillardia picta D. Don = Gaillardia pulchella Foug. var. picta (Sweet) A. Gray

GAILLARDIA PULCHELLA Foug. var. picta (Sweet) A. Gray [Asteraceae]

Synonyms:

Gaillardia picta D. Don

Common Names:

gaillardia; Kokardenblume

Citations:

Zschunke E (1955) Allergische Dermatitis durch Gaillardia picta (Kokardenblume). Dermatol Wochenschr 132(51):1321-1323.

gaja –see– *Cannabis sativa* L.

galan-de-dia –see– *Cestrum diurnum* L.

galan-de-noche –see– *Cestrum nocturnum* L.

galega –see– *Galega officinalis* L.

GALEGA OFFICINALIS L. [Fabaceae]

Common Names:

American sanfoin; ciuma rea; French honeysuckle; French lilac; galega; Geißraute; goat's-rue; lavanése-rue-de-chèvre; scrintioare cosita

Citations:

Durieux M (1968) Intoxication de moutons par <<Galega officinalis>>. Bull Soc Vet Pratique 52(6):383-385.

Gresham AC, Booth K (1991) Poisoning of sheep by goat's rue. Vet Rec 129(9):197-198.

Keeler RF, Baker DC, Evans JO (1988) Individual animal susceptibility and its relationship to induced adaptation or tolerance in sheep to Galega officinalis L. Vet Hum Toxicol 30(5):420-423.

Keeler RF, Baker DC, Panter KE (1992) Concentration of galegine in Verbesina encelioides and Galega officinalis and the toxic and pathologic effects induced by the plants. J Environ Pathol Toxicol Oncol 11(2):75-81.

Keeler RF, Johnson AE, Stuart LD, et al. (1986) Toxicosis from and possible adaptation to Galega officinalis in sheep and the relationship to Verbesina encelioides toxicosis. Vet Hum Toxicol 28(4):309-315.

Le Souef HD (1950) Poisoning of sheep by goat's rue. N Z J Agr 80:580.

Predoiu I (1966) O intoxicație la ovine cu fîn de Galega officinalis. Rev Zootehnie Med Vet 16(6):73-75.

Puyt JD, Faliu L, Keck G, et al. (1981) Fatal poisoning of sheep by Galega officinalis (French honeysuckle). Vet Hum Toxicol 23(6):410-411.

Williams MC (1978) Toxicological investigations on goatsrue. Proc West Soc Weed Sci 31:156-157.

GALENIA AFRICANA L. [Aizoaceae]

Common Names:

geelbos; kraalbos

Citations:

Van der Lugt JJ, Fourie N, Schultz RA (1988) Galenia africana L. (Aizoaceae) poisoning in sheep. J S Afr Vet Assoc 59(2):100.

Van der Lugt JJ, Schultz RA, Fourie N, et al. (1992) Galenia africana L. poisoning in sheep and goats: Hepatic and cardiac changes. Onderstepoort J Vet Res 59(4):323-333.

GALENIA PUBESCENS (Eckl. & Zeyh.) Druce [Aizoaceae]

Common Names:

green galenia

Citations:

Williams MC (1979) Toxicological investigations on Galenia pubescens. Weed Sci 27(5):506-508.

GALEOPSIS LADANUM L. [Lamiaceae]

Common Names:

cáñamo silvestre; red hemp nettle

Citations:

Aparicio R, Onate JM, Arizcun A, et al. (1999) Rabdomiólisis epidémica por ingestión de codornices. Estudio clínico, epidemiológico y experimental. Medica Clinica 112(4):143-146.

GALINSOGA PARVIFLORA Cav. [Asteraceae]

Citations:

Hausen BM (1979) The sensitizing capacity of compositae plants. III. Test results and cross-reactions in compositae-sensitive patients. Dermatologica 159(1):1-11.

GALIUM APARINE L. [Rubiaceae]

Common Names:

cleavers herb; clivers; goosegrass

Citations:

Januszewski J, Lewandowski L, Mazurkiewicz M (1988) Wpływ skarmiania paszy zanieczyszczonej nasionami przytulii czepnej (Galium aparine L.) na stan zdrowotny kurczqt typu nieśnego. Med Weter 44(6):365-367.

GALIUM ODORATUM (L.) Scop. [Rubiaceae]

Common Names:

sweet woodruff; szagos müge; Waldmeister; woodruff

Citations:

Hogan RP 3rd (1983) Hemorrhagic diathesis caused by drinking an herbal tea. JAMA 249(19):2679-2680.

Ridker PM (1989) Health hazards of unusual herbal teas. Am Fam Physician 39(5):153-156.

galletagrass –see– *Pleuraphis rigida* Thurb.

gallinazo –see– *Lablab purpureus* (L.) Sweet

gallita bush –see– *Karwinskia humboldtiana* (Schult.) Zucc.

galnebaer –see– *Atropa belladonna* L.

galster –see– *Cytisus scoparius* (L.) Link

gamba pea –see– *Crotalaria goreensis* Guill. & Perr.

Gambel's-oak –see– *Quercus gambelii* Nutt.

gamboge –see– *Garcinia hanburyi* Hook. f.

gan-di-huang –see– *Rehmanni glutinosa* (Gaertn.) Steud.

ganga –see– *Cannabis sativa* L.

ganja –see– *Cannabis sativa* L.

Gänseblume –see– *Leucanthemum vulgare* Lam.

ganskweek –see– *Lasiospermum bipinnatum* (Thunb.) Druce

garbancillo –see– *Astragalus bergii* Hieron.; *Astragalus mollissimus* Torr. var. earlei (Greene ex Rydb.) Tidestr.; *Astragalus wootonii* E. Sheld.; *Duranta erecta* L.; *Erophaca baetica* (L.) Boiss.

garbanzo –see– *Cicer arietinum* L.

garbanzuelos –see– *Erophaca baetica* (L.) Boiss.

GARCINIA HANBURYI Hook. f. [Clusiaceae]

Common Names:

gamboge

Citations:

Kong L, Ye D, Wang S, et al. (1996) [Acute toxicity and anti-inflammatory effect of processed products of gamboge.] Chung Kuo Chung Yao Tsa Chih 21(4):214-216, 255.

garden heliotrope –see– *Valeriana officinalis* L.

garden huckleberry –see– *Solanum nigrum* L.

garden nightshade –see– *Solanum nigrum* L.

garden orache –see– *Atriplex hortensis* L.

garden pea –see– *Pisum sativum* L.

garden sage –see– *Salvia officinalis* L.

garden sorrel –see– *Rumex acetosa* L.

garden spurge –see– *Euphorbia lathyris* L.

gardener's-garters –see– *Phalaris arundinacea* L.

gardenia –see– *Gardenia jasminoides* J. Ellis

GARDENIA JASMINOIDES J. Ellis [Rubiaceae]

Common Names:

Cape jasmine; gardenia; sokujikoh

Citations:

Chung MS, Hong CB, Yeh CH, et al. (1976) [Gardenia poisoning in pigs. II. Clinical and clinico-pathological observations.] J Chin Soc Vet Sci 2(1):30-34.

Hong CB, Chung MS, Yang CH, et al. (1976) [Gardenia poisoning in pigs. I. Reproduction of a black discoloration of the viscera in animals with the seed-pods of Gardenia jasminoides Ellis. f. grandiflora Makino.] J Chin Soc Vet Sci 2(1):26-29.

Kubo Y, Nonaka S, Yoshida H (1990) Allergic contact dermatitis from gardenia fruit. Contact Dermatitis 22(3):65-67.

garget –see– *Phytolacca americana* L.

gari –see– *Manihot esculenta* Crantz

garlic –see– *Allium sativum* L.

garlic guineahen weed –see– *Petiveria alliacea* L.

Garry's-oak –see– *Quercus garryana* Douglas ex Hook.

Gartenbohne –see– *Phaseolus coccineus* L.; *Phaseolus vulgaris* L.

Gartenmelde –see– *Atriplex hortensis* L.

Gartenraute –see– *Ruta graveolens* L.

Gartenschierling –see– *Aethusa cynapium* L.

Gartenwolfsmilch –see– *Euphorbia peplus* L.

garuda palai –see– *Cryptostegia grandiflora* R. Br.

gas plant –see– *Dictamnus albus* L.

GASTROLOBIUM GRANDIFLORUM F. Muell. [Fabaceae]

Common Names:

Australian poison bush; desert poisonbush; heart-leaf; heart-leaf poison bush; poison bush; wallflower poison; York Road poison

Citations:

Campbell GW, Kingston GC (1992) Heart leaf poison bush (Gastrolobium grandiflorum) toxicity in a western Queensland sheep flock. Vet Hum Toxicol 34(3):265-266.

Mahoney DF (1956) Heart-leaf poisoning of cattle. Queensland Agric J 82(Jun 1):361-362

gat –see– *Catha edulis* (Vahl) Forssk. ex Endl.

gaton –see– *Euphorbia tirucalli* L.

gatter –see– *Euonymus europaeus* L.

gauge plant –see– *Senecio integerrimus* Nutt.

GAULTHERIA PROCUMBENS L. [Ericaceae]

Common Names:

boxberry; checkerberry; deerberry; mountain tea; partridgeberry; Scheinbeere; teaberry; winterberry; wintergreen

Citations:

Craig JO (1953) Poisoning by the volatile oils in childhood. Arch Dis Child 28(142):475-483.

Hofman M, Diaz JE, Martella C (1996) Massive oil of wintergreen overdose from oriental herbal medication. J Toxicol Clin Toxicol 34:596-597.

Litovitz TL, Normann SA, Veltri JC (1986) 1985 Annual report of the American Association of Poison Control Centers National Data Collection System. Am J Emerg Med 4(5):427-458.

Litovitz TL, Schmitz BF, Bailey KM (1990) 1989 Annual report of the American Association of Poison Control Centers National Data Collection System. Am J Emerg Med 8(5):394-442.

gautan kura –see– *Solanum incanum* L.

geel tulp –see– *Moraea pallida* (Baker) Goldblatt

geelbos –see– *Galenia africana* L.

geelmelkbos –see– *Euphorbia mauritanica* L.

Gefleckter Aronstab –see– *Arum maculatum* L.

Gefleckter Schierling –see– *Conium maculatum* L.

GEIGERIA ASPERA Harv. [Asteraceae]

Common Names:

vermeerbos; vermeerbosie; vomiting bush

Citations:

Steyn DG (1936) Recent investigations into the toxicity of known and unknown poisonous plants in the Union of South Africa. Onderstepoort J Vet Sci Anim Indus 7(1):169-178.

GEIGERIA BURKEI Harv. [Asteraceae]

Citations:

Botha CJ, Gous TA, Penrith ML, et al. (1997) Vermeersiekte caused by Geigeria burkei Harv. subsp. burkei var. hirtella Merxm. in the northern province of South Africa. J S Afr Vet Assoc 68(3):97-101.

GEIGERIA FILIFOLIA Mattf. [Asteraceae]

Common Names:

vermeerbos

Citations:

Joubert JP (1983) Attempted prevention and treatment of Geigeria filifolia Mattf. poisoning (vermeersiekte) in sheep. J S Afr Vet Assoc 54(4):255-258.

GEIGERIA ORNATIVA O. Hoffm. [Asteraceae]

Citations:

Pienaar JG, Kriek NP, Naudé TW, et al. (1973) Lesions in sheep skeletal and oesophageal muscle in vermeersiekte (Geigeria ornativa O. Hoffm. poisoning). Onderstepoort J Vet Res 40(3):127-137.

Van der Lugt JJ, Van Heerden J (1993) Experimental vermeersiekte (Geigeria ornativa O. Hoffm. poisoning) in sheep. II. Histological and ultrastructural lesions. J S Afr Vet Assoc 64(2):82-88.

Van Heerden J, Van der Lugt JJ, Durante E, et al. (1993) Experimental vermeersiekte (Geigeria O. Hoffm. poisoning) in sheep. I. An evaluation of diagnostic aids and an assessment of the preventive effect of ethoxyquin. J S Afr Vet Assoc 64(2):76-81.

geisha girl –see– *Duranta erecta* L.

Geißklee –see– *Laburnum anagyroides* Medik.

Geißraute –see– *Galega officinalis* L.

Gelbe Alpenrose –see– *Rhododendron aureum* Georgi

Gelbe Lupine –see– *Lupinus luteus* L.

Gelbe Narzisse –see– *Narcissus pseudonarcissus* L.

Gelbe Schneerose –see– *Rhododendron aureum* Georgi

Gelbe Schwertlilie –see– *Iris ×germanica* L.; *Iris pseudoacorus* L.

Gelbe Wolfsbohne –see– *Lupinus luteus* L.

Gelber Eisenhut –see– *Aconitum lycoctonum* L. subsp. vulparia (Rchb.) Nyman

Gelber Jasmin –see– *Gelsemium sempervirens* (L.) J. St.-Hil.

Gelber Sturmhut –see– *Aconitum lycoctonum* L. subsp. vulparia (Rchb.) Nyman

Gelbkraut –see– *Chelidonium majus* L.

gelsemium –see– *Gelsemium elegans* (Gardner & Champ.) Benth.; *Gelsemium sempervirens* (L.) J. St.-Hil.

GELSEMIUM ELEGANS (Gardner & Champ.) Benth. [Gelsemiaceae]

Common Names:

gelsemium

Citations:

Chao TC, Lo DS, Bloodworth BC (1992) Common poisons in Singapore – Past and present. Med Sci Law 32(2):139-147.

Zhang YG, Huang GZ (1988) Poisoning by toxic plants in China. Report of 19 autopsy cases. Am J Forensic Med Pathol 9(4):313-319.

GELSEMIUM SEMPERVIRENS (L.) J. St.-Hil. [Gelsemiaceae]

Common Names:

Carolina jasmine; Carolina jessamine; Carolina wild woodbine; Carolina yellow jessamine; evening trumpet flower; evening trumpet vine; false jessamine; Gelber Jasmin; gelsemium; jasmine; jasmin-de-Caroline; jessamine; wild jessamine; woodbine; yellow jasmine; yellow jessamine

Citations:

Blaw ME, Adkisson MA, Levin D, et al. (1979) Poisoning with Carolina jessamine (Gelsemium sempervirens [L.] Ait.). J Pediatr 94(6):998-1001.

Craig FR, Williamson H (1963) Poultry sickness traced to yellow jessamine. North Carolina Res Farming 22(1):3.
Freeman JN (1873) Gelsemium. Lancet 2(Sep 27):475.
Ringer S, Murrell W (1876) On Gelsemium sempervirens. Action on man. Lancet 1(May 6):661-663.
Ringer S, Murrell W (1878) On Gelsemium sempervirens. Lancet 1(Jun 22):892-894.
Thompson LJ, Frazier K, Stiver S, et al. (2002) Multiple animal intoxications associated with Carolina jessamine (Gelsemium sempervirens) ingestions. Vet Hum Toxicol 44(5):272-273.
Williamson JH, Craig FR, Barber CW, et al. (1964) Some effects of feeding Gelsemium sempervirens (yellow jessamine) to young chickens and turkeys. Avian Dis 8(2):183-190.

gem squash –see– *Cucurbita pepo* L.

genda –see– *Tagetes patula* L.

genièvre –see– *Juniperus communis* L.

GENISTA CANARIENSIS L. [Fabaceae]

Common Names:

Canary Island broom

Citations:

Fadiman J (1965) Genista canariensis: A minor psychedelic. Econ Bot 19:383.

GENISTA STENOPETALA Webb & Berthel. [Fabaceae]

Synonyms:

Cytisus racemosus Marnock

Common Names:

broom

Citations:

Meier-Abt PJ (2002) Swiss Toxicologial Information Centre Annual Report 2002. www.toxi.ch.

genkol bean –see– *Archidendron jiringa* (Jack) I. C. Nielsen

gentian –see– *Gentiana lutea* L.

GENTIANA LUTEA L. [Gentianaceae]

Common Names:

bitterwort; gentian; gentiane jaune; grande gentiane

Citations:

Perharic L, Shaw D, Colbridge M, et al. (1994) Toxicological problems resulting from exposure to traditional remedies and food supplements. Drug Saf 11(4):284-294.

gentiane jaune –see– *Gentiana lutea* L.

Geoffroea inermis W. Wright = Andira inermis (W. Wright) Kunth ex DC.

Geoffroea jamaicensis W. Wright = Andira inermis (W. Wright) Kunth ex DC.

Geraldton carnation weed –see– *Euphorbia tirucalli* L.

geranium –see– *Pelargonium ×domesticum* L. H. Bailey

Gerbermyrte –see– *Coriaria myrtifolia* L.

Gerberstrauch –see– *Coriaria myrtifolia* L.

gergellim bravo –see– *Crotalaria spectabilis* Roth; *Crotalaria juncea* L.; *Crotalaria pallida* Aiton; *Crotalaria retusa* L.

German camomille –see– *Matricaria recutita* L.; *Matricaria recutita* L.

German primrose –see– *Primula obconica* Hance

germander –see– *Teucrium chamaedrys* L.

germandrée petit chêne –see– *Teucrium chamaedrys* L.

Gerste –see– *Hordeum vulgare* L.

gesse pourpre –see– *Lathyrus clymenum* L.

ghostweed –see– *Euphorbia marginata* Pursh

ghwanobos –see– *Senecio ilicifolius* L.

giant arbor vitae –see– *Thuja plicata* Donn ex D. Don

giant buttercup –see– *Ranunculus acris* L.

giant couch –see– *Cynodon plectostachyus* (K. Schum.) Pilg.

giant cow parsnip –see– *Heracleum mantegazzianum* Sommier & Levier

giant elephant's-ear –see– *Alocasia macrorrhizos* (L.) G. Don

giant fennel –see– *Ferula communis* L.

giant hogweed –see– *Heracleum mantegazzianum* Sommier & Levier

giant horsetail –see– *Equisetum telmateia* Ehrh.

giant larkspur –see– *Delphinium glaucum* S. Watson

giant milkweed –see– *Calotropis gigantea* (L.) W. T. Aiton; *Calotropis procera* (Aiton) W. T. Aiton

giant nettle –see– *Dendrocnide moroides* (Wedd.) Chew; *Urtica ferox* G. Forst.

giant panicum –see– *Urochloa mutica* (Forssk.) T. Q. Nguyen

giant parsnip –see– *Heracleum mantegazzianum* Sommier & Levier

giant pigweed –see– *Trianthema portulacastrum* L.

giant poverty weed –see– *Cyclachaena xanthiifolia* (Nutt.) Fresen.

giant ragweed –see– *Ambrosia trifida* L.

giant rain lily –see– *Cooperia pedunculata* Herb.

giant Russian hogweed –see– *Heracleum mantegazzianum* Sommier & Levier

giant stargrass –see– *Cynodon aethiopicus* Clayton & J. R. Harlan; *Cynodon nlemfuensis* Vanderyst; *Cynodon plectostachyus* (K. Schum.) Pilg.

giant stinging tree –see– *Dendrocnide moroides* (Wedd.) Chew

giant stock bean –see– *Canavalia ensiformis* (L.) DC.

giant striata crotalaria –see– *Crotalaria pallida* Aiton

giant taro –see– *Alocasia macrorrhizos* (L.) G. Don

gibata –see– *Arrabidaea bilabiata* (Sprague) Sandwith

Gichtrose –see– *Rhododendron aureum* Georgi

Gichtwurz –see– *Bryonia dioica* Jacq.

gidee gidee –see– *Abrus precatorius* L.

gidyea tree –see– *Acacia georginae* F. M. Bailey

Gifappel –see– *Cucumis myriocarpus* Naudin; *Pachystigma pygmaeum* (Schltr.) Robyns

gifblaar –see– *Dichapetalum cymosum* (Hook.) Engl.

gifbloom –see– *Boophone disticha* (L. f.) Herb.

gifbol –see– *Boophone disticha* (L. f.) Herb.

gifboom –see– *Acokanthera oppositifolia* (Lam.) Codd

gifbossie –see– *Thesium namaquense* Schltr.

Giftbaum –see– *Toxicodendron radicans* (L.) Kuntze

Giftbeere –see– *Nicandra physalodes* (L.) Gaertn.

giftblad –see– *Dichapetalum cymosum* (Hook.) Engl.

Giftefeu –see– *Toxicodendron radicans* (L.) Kuntze

Gifteiche –see– *Toxicodendron radicans* (L.) Kuntze

Gifthahnenfuß –see– *Ranunculus sceleratus* L.

Giftige Rebendolde –see– *Oenanthe crocata* L.

Giftprimel –see– *Primula obconica* Hance

Giftsumach –see– *Toxicodendron pubescens* Mill.; *Toxicodendron radicans* (L.) Kuntze

Giftwicke –see– *Securigera varia* (L.) Lassen

Giftwüterich –see– *Cicuta virosa* L.

gila bean –see– *Entada phaseoloides* (L.) Merr.

gilisztaüzö varádics –see– *Tanacetum vulgare* L.

gill-over-the-ground –see– *Glechoma hederacea* L.

gin nan –see– *Ginkgo biloba* L.

gingali –see– *Sesamum indicum* L.

ginger –see– *Zingiber officinale* Roscoe

gingibre –see– *Zingiber officinale* Roscoe

ginigawat –see– *Megathyrsus maximus* (Jacq.) B. K. Simon & S. W. L. Jacobs

ginkgo –see– *Ginkgo biloba* L.

GINKGO BILOBA L. [Ginkgoaceae]

Common Names:

ginkgo; Ginkgobaum; ginko; ginkyo; gin nan; kew tree; maidenhair tree; silver apricot

Citations:

Becker LE, Skipworth GB (1975) Ginkgo-tree dermatitis, stomatitis, and proctitis. JAMA 231(11):1162-1163.

Bilbert GJ (1997) Ginkgo biloba. Neurology 48(4):1137.

Bolus M (1939) Dermatitis venenata due to ginkgo berries. Arch Derm Syphilol 39:530.

Castelli D, Colin L, Camel E, et al. (1998) Pretreatment of skin with a Ginkgo biloba extract/sodium carboxymethyl-β-1,3-glucan formulation appears to inhibit the elicitation of allergic contact dermatitis in man. Contact Dermatitis 38(3):123-126.

Fujisawa M, Hori Y, Nakajima M, et al. (2002) Gas chromatography-mass spectrometry analysis of 4-O-methylpyridoxine (MPN) in the serum of patients with ginkgo seed poisoning. J Anal Toxicol 26(3):138-143.

Hori Y, Fujisawa M, Shimada K, et al. (2004) Rapid analysis of 4-O-methylpyridoxine in the serum of patients with Ginkgo biloba seed poisoning by ion-pair high-performance liquid chromatography. Biol Pharm Bull 27(4):486-491.

Kajiyama Y, Fujii K, Takeuchi H, et al. (2002) Ginkgo seed poisoning. Pediatrics 109(2):325-327.

Nakamura T (1985) Gingko tree dermatitis. Contact Dermatitis 12(5):281-282.

Rowin J, Lewis SL (1996) Spontaneous bilateral subdural hematomas associated with chronic Ginkgo biloba ingestion. Neurology 46(Jun):1775-1776.

Saito J (1929) Klinische und experimentelle Untersuchung der durch Ginkgofrucht verursachten Dermatitis und ihrer hautreizenden Bestandteile. Jpn J Dermatol Urol 29:105-129.

Saito J (1930) Klinische und experimentelle Untersuchung der durch Ginkgofrucht verursachten Dermatitis und ihrer hautreizenden Bestandteile. Tohoku J Exp Med 16:385-412.

Sowers WF, Weary PE, Collins OD, et al. (1965) Ginkgo-tree dermatitis. Arch Dermatol 91(May):452-456.

Starr AM (1913) Poisoning by Ginkgo. Bot Gaz 55:251.

Tabayashi T (1927) [Experimental study of dermatitis owing to fruits of the Gingko.] Nippon Hifuka Gakkai Zasshi 27(A):962-963.

Tomb RR, Foussereau J, Sell Y (1988) Mini-epidemic of contact dermatitis from ginkgo tree fruit (Ginkgo biloba L.). Contact Dermatitis 19(8):281-283.

Vale S (1998) Subarachnoid haemorrhage associated with Ginkgo biloba. Lancet 352(9121):36.

Vollmer H, Halter K (1938) Dermatitis durch Früchte von Ginkgo biloba. Sammlung Vergiftungsfallen 9(A721):17-20.

Vollmer H, Halter K (1939) Dermatitis durch Früchte von Ginkgo biloba. Z Haut Geschlechtskr 62:208.

Yagi M, Wada K, Sakata M, et al. (1993) [Studies on the constituents of edible and medicinal plants. IV. Determination of 4-O-methylpyridoxine in serum of the patient with gin-nan food poisoning.] J Pharm Soc Jpn 113(8):596-599.

Ginkgobaum –see– *Ginkgo biloba* L.

ginko –see– *Ginkgo biloba* L.

ginkyo –see– *Ginkgo biloba* L.

ginseng –see– *Panax ginseng* C. A. Mey.

ginster –see– *Cytisus scoparius* (L.) Link

giraffe thorn –see– *Acacia ×giraffae* Willd.

glabrous hemp –see– *Apocynum cannabinum* L.

glacier ivy –see– *Hedera helix* L.

glands –see– *Quercus robur* L.

GLANDULARIA ELEGANS (Kunth) Umber [Verbenaceae]

Synonyms:

Verbena elegans Kunth

Citations:

Potter PC, Mather S, Lockey P, et al. (1995) Immediate and delayed contact hypersensitivity to verbena plants. Contact Dermatitis 33(5):343-346.

GLANDULARIA ×HYBRIDA (hort. ex Groenl. & Rümpler) G. L. Nesom & Pruski [Verbenaceae]

Synonyms:

Verbena ×hybrida hort. ex Groenl. & Rümpler

Citations:

Potter PC, Mather S, Lockey P, et al. (1995) Immediate and delayed contact hypersensitivity to verbena plants. Contact Dermatitis 33(5):343-346.

Glanzpeterlein –see– *Aethusa cynapium* L.

Glanzpetersilie –see– *Aethusa cynapium* L.

glaucous zygadenus –see– *Anticlea elegans* (Pursh) Rydb.

GLECHOMA HEDERACEA L. [Lamiaceae]

Synonyms:

Nepeta hederacea (L.) Trev.

Common Names:

alefoot; alehoof; creeping Charlie; Epheublättrige Gündelrebe; Erdepheu; field balm; gill-over-the-ground; ground ivy; kerek repkény; lierre tenestre; run-away robin

Citations:

Bergeron JM, Jodoin L (1985) Fiabilité des mesures de poids et d'examens histopathologiques dans les études d'intoxication du campagnol des champs (Microtus pennsylvanicus). Can J Zool 63(4):804-810.

Hazslinszky B (1935) Ueber Vergiftung bei Pferden durch Glechoma. Dtsch Tierarztl Wochenschr 43(Nov 9):708-709.

Nicolau A, Bârză H, Duca H, et al. (1956) Vergiftungen durch die Pflanze Glechoma hederacea beim Pferde. Monatsh Veterinarmed 11(20):534-538.

Glelskraut –see– *Euphorbia cyparissias* L.

gli equiseti –see– *Equisetum palustre* L.

GLINUS OPPOSITIFOLIUS (L.) Aug. DC. [Molluginaceae]

Common Names:

kaipajia

Citations:

Alikutty KM (1975) A field report on the occurrence of certain cases of weed toxicity in cattle. Kerala J Vet Sci 6(1-2):149-150.

Alikutty KM, Aleyas NM (1978) Experimental weed (Glinus oppositifolius) toxicity in cattle. Kerala J Vet Sci 9(1):145-148.

GLIRICIDIA MACULATA (Kunth) Kunth ex Walp. [Fabaceae]

Citations:

Van der Walt SJ, Steyn DG (1943) Recent investigations into the toxicity of plants, etc. XIII. Onderstepoort J Vet Sci Anim Indus 18(1-2):207-224.

GLIRICIDIA SEPIUM (Jacq.) Kunth ex Walp. [Fabaceae]

Common Names:

lengua-de-perico; madera negra; madre; madre-de-cacao; madriado; mata ratón; mother-of-cocoa; piñon amoroso; piñon florido; raton; sangre-de-drago

Citations:

Glander KE (1977) Poison in a monkey's Garden of Eden. Nat Hist 86(Mar):35-41.

globe artichoke –see– *Cynara cardunculus* L.

GLOCHIDION HEYNEANUM (Wight & Arn.) Wight [Phyllanthaceae]

Synonyms:

Glochidion velutinum Wight.

Common Names:

lattighat

Citations:

Kulakkattolickal AT (1987) Piscicidal plants of Nepal: Preliminary toxicity screening using grass carp (Ctenopharyngodon idella) fingerlings. J Ethnopharmacol 21(1):1-9.

Glochidion velutinum Wight = Glochidion heyneanum (Wight & Arn.) Wight

Glockenbilsenkraut –see– *Scopolia carniolica* Jacq.

gloriosa lily –see– *Gloriosa superba* L.

Gloriosa rothschildiana O'Brien = Gloriosa superba L.

Gloriosa simplex auct. = Gloriosa superba L.

GLORIOSA SUPERBA L. [Colchicaceae]

Synonyms:

Gloriosa rothschildiana O'Brien; ***Gloriosa simplex*** auct.

Common Names:

climbing lily; creeping lily; flame lily; gloriosa lily; kar-ihari; Malabar glory lily; mora; Rothschild's-glory lily; superb lily; Turk's-cap

Citations:

Aleem HM (1992) Gloriosa superba poisoning. J Assoc Physicians India 40(8):541-542.

Angunawela RM, Fernando HA (1971) Acute ascending polyneuropathy and dermatitis following poisoning by tubers of Gloriosa superba. Ceylon Med J 16(4):233-235.

Fernando R (2002) The national poisons information centre in Sri Lanka: The first ten years. J Toxicol Clin Toxicol 40(5):551-555.

Fernando R, Fernando DN (1990) Poisoning with plants and mushrooms in Sri Lanka: A retrospective hospital based study. Vet Hum Toxicol 32(6):579-581.
Gooneratne BW (1966) Massive generalized alopecia after poisoning by Gloriosa superba. Br Med J 1(5494):1023-1024.
Jose J, Ravindran M (1988) A rare case of poisoning by Gloriosa superba. J Assoc Physicians India 36(7):451-452.
Lucas GN (1997) Plant poisoning: A hospital-based study in Sri Lanka. Indian J Pediatr 64(4):495-502.
Mendis S (1989) Colchicine cardiotoxicity following ingestion of Gloriosa superba tubers. Postgrad Med J 65(768):752-755.
Nagaratnam N, De Silva DP, De Silva N (1973) Colchicine poisoning following ingestion of Gloriosa superba tubers. Trop Geogr Med 25(1):15-17.
Senanayake N, Karalliedde L (1988) Patterns of acute poisoning in a medical unit in central Sri Lanka. Forensic Sci Int 36:101-104.
Senewiratne B, Thambipillai S (1974) Pattern of poisoning in a developing agricultural country. Br J Prev Soc Med 28(1):32-36.

glory lily –see– *Gloriosa superba* L.

glossy buckthorn –see– *Frangula alnus* Mill.

glottidium –see– *Sesbania vesicaria* (Jacq.) Elliott

Glottidium vesicarium (Jacq.) R. M. Harper = Sesbania vesicaria (Jacq.) Elliott

glovewort –see– *Cynoglossum officinale* L.

glue chard –see– *Chamaeleon gummifera* (L.) Cass.

glue thistle –see– *Chamaeleon gummifera* (L.) Cass.

GLUTA RENGHAS L. [Anacardiaceae]

Synonyms:

Stagmaria vermiciflua Jack

Common Names:

ape nut; black varnish tree; East Coast rengas; Javanese lac tree; jitong; rengas; renghas

Citations:

Fasal P (1945) Cutaneous diseases in the tropics. A clinical study based on observations in Malaya. Arch Derm Syphilol 51(3):163-171.
Grevenstuk A (1937) Vergiftung durch Renghas und durch Japanlack. Sammlung Vergiftungsfallen 8(A664):35-38.

GLUTA USITATA (Wall.) Ding Hou [Anacardiaceae]

Synonyms:

Melanorrhoea usitata Wall.

Common Names:

Burma lac tree; Burmese lacquer tree

Citations:

Kullavanijaya P, Ophaswongse S (1997) A study of dermatitis in the lacquerware industry. Contact Dermatitis 36(5):244-246.

Glyceria aquatica (L.) Wahlb. = Glyceria maxima (Hartm.) Holmb.

GLYCERIA GRANDIS S. Watson [Poaceae]

Common Names:

American mannagrass; tall mannagrass

Citations:

Puls R, Newschwander FP, Greenway JA (1978) Cyanide poisoning from Glyceria grandis S. Wats. ex Gray (tall mannagrass) in a British Columbia beef herd. Can Vet J 19(9):264-265.

GLYCERIA MAXIMA (Hartm.) Holmb. [Poaceae]

Synonyms:

Glyceria aquatica (L.) Wahlb.; ***Poa aquatica*** L.

Common Names:

reed sweetgrass; vizi harmatkása; water gromwells

Citations:

Barton NJ, McOrist S, McQueen DS, et al. (1983) Poisoning of cattle by Glyceria maxima. Aust Vet J 60(7):220-221.
Fritzsch R (1966) Beitrag zur Futterintoxikation beim Rind durch Giftstoffe obligater Futterpflanzen. Monatsh Veterinarmed 21(9):327-331.
Guilhon J (1988) L'enseignement de la botanique et la mort subite au pré. Bull Acad Vet Fr 61(3):267-271.
Rainey JW (1946) Hydrocyanic acid poisoning from Poa aquatica. Aust Vet J 22:187.
Sharman JR (1967) Cyanide poisoning of cattle grazing "reed sweet-grass." N Z Vet J 15:7-8.
Shulakov MG, Vorob Eva FD (1977) [Toxicity of hay containing Glyceria maxima (reed sweet-grass).] Veterinariia Moscow 54(8):103-105.

glicine –see– *Wisteria sinensis* (Sims) DC.

glycine chinensis –see– *Wisteria sinensis* (Sims) DC.

GLYCINE MAX (L.) Merr. [Fabaceae]

Common Names:

soja; Sojabohne; soy; soya bean; soybean

Citations:

Almquist HJ, Merritt JB (1952) Effect of raw soybean meal on growth of the chick. Proc Soc Exp Biol Med 79:277-279.
Almquist HJ, Merritt JB (1952) Effect of soybean antitrypsin on growth of the chick. Arch Biochem 35(2):352-354.
Anderson DW, Howard HW (1959) Feeding of soybean products and development of goiter. Pediatrics 24(5):854-855.
Avanzi CF (1972) Effetti tireostatici della soia. Agric Ital 72(4):293-296.
Balloun SL, Johnson EL (1953) Anticoagulant properties of unheated soybean meal in chick diets. Arch Biochem 42(2):355-359.
Booth AN, Robbins DJ, Ribelin WE, et al. (1960) Effect of raw soybean meal and amino acids on pancreatic hypertrophy in rats. Proc Soc Exp Biol Med 104:681-683.

Booth AN, Robbins DJ, Ribelin WE, et al. (1964) Prolonged pancreatic hypertrophy and reversibility in rats fed raw soybean meal. Proc Soc Exp Biol Med 116(Aug-Sep):1067-1069.

Bornstein S, Lipstein B (1963) The influence of age of chicks on their sensitivity to raw soybean oil meal. Poult Sci 42:61-70.

Bray DJ (1964) Pancreatic hypertrophy in laying pullets induced by unheated soybean meal. Poult Sci 43(2):382-384.

Carlson CW, Saxena HC, Jensen LS, et al. (1964) Rachitogenic activity of soybean fractions. J Nutr 82(Apr):507-511.

Clarke E, Wiseman J (2004) Effect of varying trypsin inhibitor activity of full fat soya on nutritional value for broiler chicks. In: Acamovic T, Stewart CS, Pennycott TW (eds.) Poisonous plants and related toxins. CABI. Wallingford, UK. pp. 512-519.

Collins JL, Beaty BF (1980) Heat inactivation of trypsin inhibitor in fresh green soybeans and physiological responses of rats fed the beans. J Food Sci 45(3):542-546.

Cross RF (1953) Observations on the bovine hemorrhagic disease caused by trichloroethylene-processed soybean oil meal. J Am Vet Med Assoc 123(Feb):103-105.

Finzi A, Romboli I (1972) Cecità congenita in pulcini in seguito a somministrazione protratta di soia cruda a galline in deposizione. Rivista Ital Sost Grasse 49:252-253.

Garlich JD, Nesheim MC (1965) Effect of sodium taurocholate on fat malabsorption induced by feeding unheated soybean proteins. Proc Soc Exp Biol Med 118(Apr):1022-1025.

Garlich JD, Nesheim MC (1966) Relationship of fractions of soybeans and a crystalline soybean trypsin inhibitor to the effects of feeding unheated soybean meal to chicks. J Nutr 88(1):100-110.

Goldberg A, Guggenheim K (1964) Effect of antibiotics on pancreatic enzymes of rats fed soybean flour. Arch Biochem Biophys 108(Nov):250-254.

Grant G, Dorward PM, Buchan W (1995) Consumption of diets containing raw soya beans (Glycine max), kidney beans (Phaseolus vulgaris), cowpeas (Vigna unguiculata) or lupin seeds (Lupinus angustifolius) by rats up to 700 days: Effects on body composition and organ weights. Br J Nutr 73(1):17-29.

Grant G, Dorward PM, Pusztai A (1993) Pancreatic enlargement is evident in rats fed diets containing raw soybeans (Glycine max) or cowpeas (Vigna unguiculata) for 800 days but not in those fed diets based on kidney beans (Phaseolus vulgaris) or lupinseed (Lupinus angustifolius). J Nutr 123(12):2207-2215.

Holm GC, Eveleth DF, Dinusson WE (1953) Trichloroethylene soybean meal poisoning in sheep. J Am Vet Med Assoc 122(914):380-382.

Hove EL, King S, Hill GD (1978) Composition, protein quality, and toxins of seeds of the grain legumes Glycine max, Lupinus spp., Phaseolus spp., Pisum sativum, and Vicia faba. N Z J Agric Res 21:457-462.

Hydovitz JD (1960) Occurrence of goiter in an infant on a soy diet. N Engl J Med 262(7):351-353.

Kimura S, Suwa J, Ito M, et al. (1979) Experimental studies on the role of defatted soybean in the development of malignant goiter. In: Miller et al. (eds.) Naturally occurring carcinogens, mutagens and modulators of carcinogenesis. Jpn Sci Soc Press. Tokyo. pp. 101-110.

Kwong E, Barnes RH (1963) Effect of soybean trypsin inhibitor on methionine and cystine utilization. J Nutr 81(4):392-398.

Lecks HI, Baker D (1965) Pseudotumor cerebri - An allergic phenomenon? A discussion of 17 cases including two of infants manifesting pseudotumor while receiving soybean feedings. Clin Pediatr (Phila) 4(1):32-37.

Maggio P, Monsó E, Baltasar M, et al. (2003) Occupational asthma caused by soybean hull: A workplace equivalent to epidemic asthma. Allergy 58(4):350-351.

McCarrison R (1934) The goitrogenic action of soya-bean and ground-nut. Indian J Med Res 21:179-181.

Moss MH (1969) Hypoprothrombinemic bleeding in a young infant associated with a soy protein formula. Am J Dis Child 117(5):540-542.

Nitsan Z, Alumot E (1965) The effect of raw soybean levels in the diet on the proteolytic activity and pancreatic hypertrophy in the growing chick. Poult Sci 44(5):1210-1214.

Patten JR, Richards EA, Pope H 2nd (1971) The effect of raw soybean on the pancreas of adult dogs. Proc Soc Exp Biol Med 137(1):59-63.

Pubols MH, Saxena HC, McGinnis J (1964) Pancreatic enzyme levels in chicks fed unheated soybean meal. Proc Soc Exp Biol Med 117(3):713-717.

Rackis JJ, McGee JE, Gumbmann MR, et al. (1979) Effects of soy proteins containing trypsin inhibitors in long term feeding studies in rats. J Am Oil Chem Soc 56(3):162-168.

Rackis JJ, Smith AK, Nash AM, et al. (1963) Feeding studies on soybeans. Growth and pancreatic hypertrophy in rats fed soybean meal fractions. Cereal Chem 40(Sep):531-538.

Romboli I, Finzi A (1974) Indagine sul fattore antitiroideo della soia. Riv Zoot Vet 2:123-137.

Salman AJ, Pubols MH, McGinnis J (1968) Chemical and microscopic nature of pancreata from chicks fed unheated soybean meal. Proc Soc Exp Biol Med 128(1):258-261.

Saxena HC, Jensen LS, McGinnis J (1961) Growth inhibition by raw soybean meal for chicks and turkey poults. Poult Sci 40:1452-1453.

Saxena HC, Jensen LS, McGinnis J, et al. (1963) Histophysiological studies on chick pancreas as influenced by feeding raw soybean meal. Proc Soc Exp Biol Med 112(Feb):390-393.

Setchell KD, Gosselin SJ, Welsh MB, et al. (1987) Dietary estrogens - A probable cause of infertility and liver disease in captive cheetahs. Gastroenterology 93(2):225-233.

Sharpless GR (1938) A new goiter-producing diet for the rat. Proc Soc Exp Biol Med 38:166-168.

Sharpless GR, Pearsons J, Prato GS (1939) Production of goiter in rats with raw and with treated soy bean flour. J Nutr 17(6):545-555.

Stewart GH, Lawrence JA (1978) An outbreak of hepatosis dietetica in Rhodesia. Rhodesia Vet J 8:80-86.

Suwa J, Koyanagi T, Kimura S (1979) Studies on soybean factors which produced goiter in rats. J Nutr Sci Vitaminol (Tokyo) 25(4):309-315.

Taramarcaz P, Hauser C, Eigenmann PA (2001) Soy anaphylaxis. Allergy 56(8):792.

Tudor RJ, Dayan AD (1987) Comparative subacute effects of dietary raw soya flour on the pancreas of three species, the marmoset, mouse and rat. Food Chem Toxicol 25(10):739-745.

Wilgus HS Jr, Gassner FX, Patton AR, et al. (1941) The goitrogenicity of soybeans. J Nutr 22:43-52.

Glycine wightii (Wight & Arn.) Verdc. = Neonotonia wightii (Wight & Arn.) J. A. Lackey

GLYCYRRHIZA GLABRA L. [Fabaceae]

Common Names:

Lakritze; licorice; licorise; liksidsrod; liquorice; réglisse

Citations:

Barrella M, Lauria G, Quatrale R, et al. (1997) Hypokaliemic rhabdomyolysis associated with liquorice ingestion: Report of an atypical case. Ital J Neurol Sci 18(4):217-220.

Cartier A, Malo JL, Labrecque M (2002) Occupational asthma due to liquorice roots. Allergy 57(9):863.

Cataldo F, Di Stefano P, Violante M, et al. (1997) Pseudoiperaldosteronismo secondario ad intossicazione da liquirizia associato a gastrite emorragica. Pediatr Med Chir 19(3):219-221.

Chamberlain JJ, Abolnik IZ (1997) Pulmonary edema following a licorice binge. West J Med 167(3):184-185.

Delcroix C, Poncin E, Pourrat O, et al. (1985) Tubulopathie proximale au cours d'une intoxication par la réglisse. Presse Med 14(46):2346-2347.

Famularo G, Corsi FM, Giacanelli M (1999) Iatrogenic worsening of hypokalemia and neuromuscular paralysis associated with the use of glucose solutions for potassium replacement in a young woman with licorice intoxication and furosemide abuse. Acad Emerg Med 6(9):960-964.

Minaire Y, Baltassat P, Gouezo F, et al. (1964) A propos de deux cas d'intoxication par le réglisse. Ann Pharm Fr 22(3):245.

Morell F, Orriols R, Lucas A, et al. (1983) Seudohiperaldosteronismo por ingestión de regaliz. Med Clin (Barc) 80(1):49-50.

Rausch Stroomann JG (1968) Reversible severe hypertension due to licorice ingestion. N Engl J Med 279(Sep 12):606.

Yoshida S, Takayama Y (2003) Licorice-induced hypokalemia as a treatable cause of dropped head syndrome. Clin Neurol Neurosurg 105(4):286-287.

Glyzinie –see– *Wisteria sinensis* (Sims) DC.

gnawed canarygrass –see– *Phalaris paradoxa* L.

GNIDIA ANTHYLLOIDES Gilg [Thymelaeaceae]

Synonyms:

Lasiosiphon anthylloides (L. f.) Meisn.

Citations:

Alexander R (1928) Lasiosiphon anthylloides as a poisonous plant. Rep Director Vet Educ Res South Africa 13:233-240.

GNIDIA BURCHELLII Gilg [Thymelaeaceae]

Synonyms:

Lasiosiphon burchellii Meisn.

Common Names:

harpuisbar; harpuisbos

Citations:

Terblanche M, Pieterse MJ, Adelaar TF, et al. (1966) Further studies of the toxicology of Lasiosiphon burchellii Meisn. J S Afr Vet Assoc 37(2):185-189.

GNIDIA KRAUSSIANA Meisn. [Thymelaeaceae]

Synonyms:

Lasiosiphon kraussianus (Meisn.) Burtt Davy

Common Names:

yellow heads

Citations:

Nwude N, Parsons LE (1978) Haematological and biochemical changes in Lasiosiphon kraussianus (Meisn.) poisoning in cattle. Bull Anim Health Prod Afr 26(1):73-77.

Nwude N, Parsons LE (1978) Toxicity of Lasiosiphon kraussianus (Meisn.) in livestock. Bull Anim Health Prod Afr 26(1):67-72.

Nyazema NZ (1986) Herbal toxicity in Zimbabwe. Trans R Soc Trop Med Hyg 80(3):448-450.

GNIDIA LATIFOLIA (Oliv.) Gilg [Thymelaeaceae]

Synonyms:

Lasiosiphon latifolius (Oliv.) Brenan

Citations:

Kiptoon JC, Mugera GM, Waiyaki PG (1982) Haematological and biochemical changes in cattle poisoned by Gnidia latifolia syn. Lasiosiphon latifolius (Thymelaeaceae). Toxicology 25(2-3):129-139.

Mugera GM, Kiptoon JC, Waiyaki PG (1982) Studies on the toxicity of Gnidia latifolia (Meisn) in cattle. Bull Anim Health Prod Afr 30(3):251-256.

GNIDIA POLYCEPHALA Gilg [Thymelaeaceae]

Common names:

januariebos

Citations:

Grant R, Booker HH, Basson PA, et al. (1988) Cardiomyopathies caused by januariebos (Gnidia polycephala) and avocado (Persea americana) leaves. J S Afr Vet Assoc 59(2):101.

goabean –see– *Psophocarpus tetragonolobus* (L.) DC.

goat head –see– *Tribulus terrestris* L.

goat killer –see– *Dichapetalum barteri* Engl.

goat nut –see– *Simmondsia chinensis* (Link) C. K. Schneid.

goat weed –see– *Ageratum conyzoides* L.; *Hypericum perforatum* L.; *Tribulus terrestris* L.

goat's-rue –see– *Galega officinalis* L.

gobernadora –see– *Larrea tridentata* (DC.) Coville

gobu –see– *Stipa sibirica* (L.) Lam.

godogo –see– *Entada phaseoloides* (L.) Merr.

gogu –see– *Stipa sibirica* (L.) Lam.

gokharu –see– *Tribulus terrestris* L.

gokarni –see– *Clitoria ternatea* L.

golandrina –see– *Chamaesyce hirta* (L.) Millsp.

gold-dust plant –see– *Aucuba japonica* Thunb.

gold-leaf plant –see– *Aucuba japonica* Thunb.

golden banner –see– *Thermopsis montana* Nutt.

golden Ceylon creeper –see– *Epipremnum pinnatum* (L.) Engl.

golden chain –see– *Laburnum anagyroides* Medik.

golden corn –see– *Semecarpus anacardium* L. f.

golden corydalis –see– *Corydalis aurea* Willd.

golden crownbeard –see– *Verbesina encelioides* (Cav.) Benth. & Hook. f. ex A. Gray

golden dewdrop –see– *Duranta erecta* L.

golden feather –see– *Tanacetum cinerariifolium* (Trevir.) Sch. Bip.

golden fumeroot –see– *Corydalis aurea* Willd.

golden germander –see– *Teucrium polium* L.

golden glow –see– *Rudbeckia laciniata* L.

golden hunter's-robe –see– *Epipremnum pinnatum* (L.) Engl.

golden oats –see– *Trisetum flavescens* (L.) P. Beauv.

golden pothos –see– *Epipremnum pinnatum* (L.) Engl.

golden privet –see– *Ligustrum ovalifolium* Hassk.

golden rail –see– *Laburnum anagyroides* Medik.

golden rain –see– *Laburnum anagyroides* Medik.

golden smoke –see– *Corydalis aurea* Willd.

golden tip –see– *Goodia lotifolia* Salisb.

golden trumpet –see– *Allamanda cathartica* L.

goldenrod –see– *Ageratina altissima* (L.) R. M. King & H. Rob.; *Solidago speciosa* Nutt.; *Solidago spectabilis* (D. C. Eaton) A. Gray

goldfish plant –see– *Columnea spp.*

Goldhafer –see– *Trisetum flavescens* (L.) P. Beauv.

Goldregen –see– *Laburnum anagyroides* Medik.

golio –see– *Malva parviflora* L.

golodrina –see– *Chamaesyce hirta* (L.) Millsp.

golondrina –see– *Euphorbia heterophylla* L.

GOMPHOCARPUS FRUTICOSA (L.) W. T. Aiton [Apocynaceae]

Synonyms:

Asclepias florida N. E. Br.; ***Asclepias fruticosa*** L.; ***Asclepias phillipsiae*** N. E. Br.

Common Names:

balloon cotton; blaasoppies; bladder cottonbush; cottonbush; duckbush; firesticks; melkbos; milkweed; narrow-leaf cottonbush; swan plant; tontelbos; wild cotton; wild cottonbush; wildekapok

Citations:

Schultz RA, Kellerman TS, Van den Berg H (2005) The role of fluorescence polarization immuno-assay in the diagnosis of plant-induced cardiac glycoside poisoning livestock in South Africa. Onderstepoort J Vet Res 72(3):189-201.

GOMPHRENA CELOSIOIDES Mart. [Amaranthaceae]

Common Names:

gomphrena weed; soft khaki weed; white eye

Citations:

Newton LG (1952) Gomphrena celosioides - A plant causing ataxia in horses. Aust Vet J 28(Jun):151-154.

Gomphrena decumbens Jacq. = Gomphrena serrata L.

GOMPHRENA SERRATA L. [Amaranthaceae]

Synonyms:

Gomphrena decumbens Jacq.

Common Names:

gomphrena weed

Citations:

Clay AL (1944) Gomphrena poisoning. Aust Vet J 20(2):211-212.

gomphrena weed –see– *Gomphrena celosioides* Mart.; *Gomphrena serrata* L.

gonde –see– *Oenanthe palustris* (Chiov.) C. Norman

GONIOLIMON TATARICUM (L.) Boiss. [Plumbaginaceae]

Synonyms:

Limonium tataricum (L.) Mill.

Common Names:

sea lavender

Citations:

Quirce S, Garcia Figueroa B, Olaguibel JM, et al. (1993) Occupational asthma and contact urticaria from dried flowers of Limonium tataricum. Allergy 48(4):285-290.

GONYSTYLUS BANCANUS (Miq.) Kurz [Thymelaeaceae]

Common Names:

ramin

Citations:

Beck MH, Roberts MM (1982) A case of ramin wood sensitivity. Contact Dermatitis 8(1):74-75.

Bruynzeel DP, de Haan P (1987) Sensitivity to ramin wood. Contact Dermatitis 17(12):318-319.

Hinojosa M, Losada E, Moneo I, et al. (1986) Occupational asthma caused by African maple (obeche) and ramin: Evidence of cross reactivity between these two woods. Clin Allergy 16(2):145-153.

Howie AD, Boyd G, Moran F (1976) Pulmonary hypersensitivity to ramin (Gonystylus bancanus). Thorax 31:585-587.

GOODIA LOTIFOLIA Salisb. [Fabaceae]

Common Names:

black scours; booroo molie; clover-leaf poison; clover tree; golden tip

Citations:

Maiden JH (1895) Is Goodia poisonous to stock? Agric Gaz New South Wales 6:306-307.

GOODIA MEDICAGINEA F. Muell. [Fabaceae]

Common Names:

black scours; small golden tip

Citations:

Maiden JH (1895) Is Goodia poisonous to stock? Agric Gaz New South Wales 6:306-307.

gooseberry –see– *Nicandra physalodes* (L.) Gaertn.

gooseberry melon –see– *Cucumis myriocarpus* Naudin

goosefoot –see– *Chenopodium album* L.

goosegrass –see– *Galium aparine* L.; *Triglochin maritima* L.

gopher plant –see– *Euphorbia lathyris* L.

gordolobo –see– *Senecio flaccidus* Less. var. flaccidus; *Verbascum thapsus* L.

gordolobo yerba –see– *Senecio flaccidus* Less. var. flaccidus; *Verbascum thapsus* L.

gosmore –see– *Hypochaeris radicata* L.

GOSSWEILERODENDRON BALSAMIFERUM (Vermoesen) Harms [Fabaceae]

Common Names:

agba

Citations:

Oleffe JA, Sporcq J, Hublet P (1975) Epidemiological study of the wood industry in Belgium. Contact Dermatitis 1(5):315-316.

GOSSYPIUM HIRSUTUM L. [Malvaceae]

Common Names:

algodón; cotón; cotton; ngodoncillo; upland cotton; wild cotton

Citations:

Akhmedov MN, Arutyunyants SI, Lavrova AD, et al. (1975) [Toxicity of cottonseed meal for poultry.] Veterinariia Moscow 52(11):80-83.

Ambož M (1964) Případ ortravy bavlníkovými pokrutinami. Veterinarstvi 14:41-42.

Ambrose AM, Robbins DJ (1951) Studies on the chronic oral toxicity of cottonseed meal and cottonseed pigment glands. J Nutr 43:357-370.

Binns HR (1938) Cottonseed poisoning of pigs. J Comp Pathol Exp Ther 51:296-304.

Boyd EM, Boulanger MA (1969) Acute oral toxicity of cottonseed oil. Toxicol Appl Pharmacol 14:432-438.

Boyd EM, Krijnen CJ (1971) Intolerance to cottonseed oil in rats fed a low protein diet. Food Cosmet Toxicol 9(3):389-394.

Brown GT (1929) Cotton seed and kapok sensitization. JAMA 93(5):370-374.

Carruth FE (1917) Methods for approximating the relative toxicity of cottonseed products. J Biol Chem 32:87-90.

Coppock CE, West JW, Moya JR, et al. (1985) Effects of amount of whole cottonseed on intake, digestibility, and physiological responses of dairy cows. J Dairy Sci 68(9):2248-2258.

Cummins KA, Russell RW (1985) Effects of feeding whole cottonseed to lactating dairy cows on glucose and palmitate metabolism. J Dairy Sci 68(8):2009-2015.

De Peters EJ, Taylor SJ, Franke AA, et al. (1985) Effects of feeding whole cottonseed on composition of milk. J Dairy Sci 68(4):897-902.

Dinwiddie RR (1903) Pig feeding experiments with cotton seed meal. Arkansas Agric Exp Sta Bull #76:133-148.

East NE, Anderson M, Lowenstine LJ (1994) Apparent gossypol-induced toxicosis in adult dairy goats. J Am Vet Med Assoc 204(4):642-643.

Evans RJ, Bauer DH, Vaghefi SB, et al. (1969) Influence of feeding cottonseed oil to laying hens on the low density lipoproteins of their eggs. J Nutr 99(4):485-490.

Fitzsimmons RC, Newcombe M, Moul IE (1989) The long-term effects of feeding ground and whole cottonseed to laying hens. Can J Anim Sci 69(2):425-431.

Fulmer E (1905) Tests in feeding cotton seed products. Part III. Effect of feeding cottonseed meal upon the health of animals. Washington Agric Exp Sta Bull #67:28-42.

Gallup WD (1926) Eliminating the toxicity of cottonseed meal. J Dairy Sci 9(4):359-372.

Gao H, Yang ZS, Jin SX (1985) [Primary observations on distal renal tubule acidosis in 177 cases caused by gossypol intoxication.] Chung Hua Nei Ko Tsa Chih 24(7):419-421, 447.

Goldberg SA, Maynard LA (1922-) Studies of cottonseed poisoning. I. The pathological tissue changes resulting from continuous feed of cottonseed meal. J Am Vet Med Assoc 62:450-457.

Graham GG, Morales E, Acevedo G, et al. (1970) Dietary protein quality in infants and children. III. Prolonged feeding of cottonseed flour. Am J Clin Nutr 23(2):165-169.

Hale F, Lyman CM (1962) Effective utilization of cottonseed meal in swine rations. J Anim Sci 21:998.

Hale WH, Lambeth C, Theurer B, et al. (1969) Digestibility and utilization of cottonseed hulls by cattle. J Anim Sci 29(5):773-776.

Halverson JO, Sherwood FW (1930) Investigations in the feeding of cottonseed meal to cattle. North Carolina Agric Exp Sta Tech Bull #39:158 pp.

Haschek WM, Beasley VR, Buck WB, et al. (1989) Cottonseed meal (gossypol) toxicosis in a swine herd. J Am Vet Med Assoc 195(5):613-615.

Hawkins GE, Cummins KA, Silverio M, et al. (1985) Physiological effects of whole cottonseed in the diet of lactating dairy cows. J Dairy Sci 68(10):2608-2614.

Hendricks JD, Sinnhuber RO, Loveland PM, et al. (1980) Hepatocarcinogenicity of glandless cottonseeds and cottonseed oil to rainbow trout (Salmo gairdnerii). Science 208(4441):309-311.

Heywang BW, Bird HR (1952) Failure of several amino compounds and of sardine meal to inactivate gossypol in diets of breeding chickens. Poult Sci 31:805-809.

Heywang BW, Bird HR (1955) Relationship between the weight of chicks and levels of dietary free gossypol supplied by different cottonseed products. Poult Sci 34(1):1239-1247.

Heywang BW, Denton CA, Bird HR (1949) The effect of the dietary level of cottonseed meal on hatchability. Poult Sci 28:610-617.

Heywang BW, Kemmerer AR (1966) Effect of gossypol source and level on chick growth. Poult Sci 45:1429-1430.

Hotis RP, Woodward TE (1935) Heavy cottonseed meal feeding in relation to udder troubles in dairy cows. U S Dep Agric Tech Bull #473:15 pp.

Hudson LM, Kerr LA, Maslin WR (1988) Gossypol toxicosis in a herd of beef calves. J Am Vet Med Assoc 192(9):1303-1305.

Jarrett JA (1983) Too much cottonseed led to poisoning. Hoards Dairyman 128(8):568-570.

Kemmerer AR, Heywang BW, Vavich MG, et al. (1965) Effect of cottonseed oil on egg hatchability. Poult Sci 44(5):1315-1318.

Kramer RY, Garner DL, Ericson SA, et al. (1991) The effect of cottonseed components on testicular development in pubescent rams. Vet Hum Toxicol 33(1):11-16.

Leighton RE, Anthony WB, Huff JS, et al. (1953) Relation of breed and free gossypol levels to cottonseed meal toxicity in dairy calves. J Dairy Sci 36:601-602.

Lindsey TO, Hawkins GE, Guthrie LD (1980) Physiological responses of lactating cows to gossypol from cottonseed meal rations. J Dairy Sci 63(4):562-573.

Lorenz FW (1939) Egg deterioration due to ingestion by hens of malvaceous materials. Poult Sci 18:295-300.

Macy IG, Mendel LB (1920) Comparative studies on the physiological value and toxicity of cotton seed and some of its products. J Pharmacol Exp Ther 16:345-390.

Naber EC, Morgan CL (1957) Studies on a hatchability depressing factor in cottonseed oil. Poult Sci 36:429-431.

Nagalakshmi D, Sharma AK, Sastry VR (2000) Pathological lesions in lambs fed raw or processed cottonseed meal. Vet Res Commun 24(5):349-359.

Neser JA, Grimbeek PJ, Langford MJ, et al. (1988) Gossipol poisoning in pigs. J S Afr Vet Assoc 59(2):104.

Orgad Klopfer U, Adler H (1986) Gossypol poisoning in calves. Isr J Vet Med 42(1):16-18.

Osborne TB, Mendel LB (1917) The use of cotton seed as food. J Biol Chem 29(2):289-317.

Patton CS, Legendre AM, Gompf RE, et al. (1985) Heart failure caused by gossypol poisoning in two dogs. J Am Vet Med Assoc 187(6):625-627.

Pérez-Buriel J, Watts AB (1973) Avances sobre la inactivacion del gosipol en la harina de algodon. Agron Trop (Ven) 23(3):323-331.

Ringrose RC, Morgan CL, Lease EJ (1941) The effect of cottonseed oil on the hatchability of eggs. Poult Sci 20:57-61.

Rogers PA, Henaghan TP, Wheeler B (1975) Gossypol poisoning in young calves. Ir Vet J 29(1):9-13.

Rojas SW, Scott ML (1969) Factors affecting the nutritive value of cottonseed meal as a protein source in chick diets. Poult Sci 48(3):819-835.

Rommel GM, Vedder EB (1915) Beriberi and cottonseed poisoning in pigs. J Agric Res 5(11):489-493.

Schaible PJ, Moore LA, Moore JM (1933) Gossypol a cause of discoloration in egg yolks from hens fed cottonseed meal. Poult Sci 12:334.

Schwartze EW, Alsberg CL (1924) Relation between toxicity of cottonseed and its gossypol content. J Agric Res 28(2):173-189.

Smith FH (1963) Isolation of gossypol from tissue of porcine livers. J Am Oil Chem Soc 40(2):60-61.

Smith FH, Clawson AJ (1970) The effects of dietary gossypol on animals. J Am Oil Chem Soc 47(11):443-447.

Taub SJ (1934) Cottonseed allergy. JAMA 103(5):334-335.

Tone JN, Jensen DR (1970) Effect of ingested gossypol on the growth performance of rats. Experientia 26(9):970-971.

Tone JN, Jensen DR (1976) The accumulation pattern of ingested gossypol in selected organs of rats. Experientia 32(3):369-371.

Wells CA, Ewing PV (1916) Cottonseed meal as an incomplete food. J Biol Chem 27(1):15-25.

West JL (1939) Lesions of gossypol poisoning in dogs fed cottonseed meal. Vet Med 34(10):603.

Wilde JK (1944) Post-mortem lesions noted in pigs used in cottonseed feeding trials. Vet J 100:209-213.

Withers WA, Carruth FE (1918) Comparative toxicity of cottonseed products. J Agric Res 14(10):425-452.

Withers WA, Carruth FE (1918) Gossypol, the toxic substance in cottonseed. J Agric Res 12(2):83-102.

Note:

Cottonseed in most toxicity reports is probably from *Gossypium hirsutum* L. but some may come from other *Gossypium* species.

gotu kola –see– *Centella asiatica* (L.) Urb.

goudronnier –see– *Semecarpus ater* (G. Forst.) Vieill.

gourd –see– *Cucurbita maxima* Duchesne; *Lagenaria siceraria* (Molina) Standl.

gousiektebossie –see– *Pachystigma pygmaeum* (Schltr.) Robyns

gout stalk –see– *Jatropha podagrica* Hook.

graceful narrowgrass –see– *Triglochin maritima* L.

grain sorghum –see– *Sorghum bicolor* (L.) Moench

graine-de-moluques –see– *Croton tiglium* L.

graine-de-tilly –see– *Croton tiglium* L.

graines-d'arrow-root-de-Floride –see– *Zamia integrifolia* L. f.

graines-de-Ricin –see– *Ricinus communis* L.

grama común –see– *Cynodon dactylon* (L.) Pers.

gramilla –see– *Cynodon dactylon* (L.) Pers.

gramilla colorada –see– *Cynodon dactylon* (L.) Pers.

grana tiglio –see– *Croton tiglium* L.

grand trèfle rouge –see– *Trifolium pratense* L.

grande berce –see– *Heracleum mantegazzianum* Sommier & Levier

grande ciguë –see– *Conium maculatum* L.

grande consoude –see– *Symphytum officinale* L.

grande gentiane –see– *Gentiana lutea* L.

grande malnommée –see– *Chamaesyce hirta* (L.) Millsp.

grande ortie –see– *Urtica dioica* L.

grandiúva –see– *Trema micrantha* (L.) Blume

grape –see– *Vitis labrusca* L.; *Vitis vinifera* L.

grapewort –see– *Bryonia dioica* Jacq.

grapefruit –see– *Citrus paradisi* Macfad.

Graserich –see– *Allium sativum* L.

grass –see– *Cannabis sativa* L.

grass-leaf day lily –see– *Hemerocallis minor* Mill.

grass pea –see– *Lathyrus sativus* L.

grass vetchling –see– *Lathyrus nissolia* L.

grassland lupin –see– *Lupinus arbustus* Douglas ex Lindl.

grassland lupine –see– *Lupinus arbustus* Douglas ex Lindl.; *Lupinus argenteus* Pursh Lupinus; *argenteus* Pursh var. holosericeus (Nutt.) Barneby; *Lupinus formosus* Greene

grassy death camas –see– *Toxicoscordion nuttallii* (A. Gray) Rydb.; *Toxicoscordion venenosum* (S. Watson) Rydb.

graveyard flower –see– *Plumeria rubra* L.

graveyard plumeria –see– *Plumeria rubra* L.

gray fly poison –see– *Amianthium muscitoxicum* (Walter) A. Gray

gray fuchsia bush –see– *Eremophila latrobei* F. Muell.

gray oak –see– *Quercus grisea* Liebm.; *Quercus rubra* L.

gray Swainsona pea –see– *Swainsona canescens* (Benth.) F. Muell.

GRAYIA SPINOSA (Hook.) Moq. [Chenopodiaceae]

Common Names:

hop sage

Citations:

Fleming CE, Miller MR, Vawter LR, et al. (1931) Poisonous range plants. Nevada Agric Exp Sta Annu Rep #1928:21-22.

greasewood –see– *Larrea tridentata* (DC.) Coville; *Sarcobatus vermiculatus* (Hook.) Torr.

Great Basin lupine –see– *Lupinus argenteus* Pursh var. holosericeus (Nutt.) Barneby

great burdock –see– *Arctium lappa* L.

great celandine –see– *Chelidonium majus* L.

great laurel –see– *Kalmia latifolia* L.

great millet –see– *Sorghum bicolor* (L.) Moench

great morel –see– *Atropa belladonna* L.

great mullein –see– *Verbascum thapsus* L.

great nettle –see– *Urtica dioica* L.

Great Plains whorled milkweed –see– *Asclepias pumila* (A. Gray) Vail

great ragweed –see– *Ambrosia trifida* L.

greater ammi –see– *Ammi majus* L.

greater periwinkle –see– *Vinca major* L.

Grecian foxglove –see– *Digitalis lanata* Ehrh.

Grecian laurel –see– *Laurus nobilis* L.

Greek hay seed –see– *Trigonella foenum-graecum* L.

green amaranth –see– *Amaranthus viridis* L.

green amaranthus –see– *Amaranthus retroflexus* L.

green arum –see– *Alocasia macrorrhizos* (L.) G. Don

green bean –see– *Phaseolus vulgaris* L.

green bristle grass –see– *Setaria italica* (L.) P. Beauv.

green canola –see– *Brassica napus* L. var. napus

green cestrum –see– *Cestrum parqui* L'Her.

green couch –see– *Cynodon dactylon* (L.) Pers.

green crumbweed –see– *Chenopodium carinatum* R. Br.

green dragon –see– *Datura stramonium* L.

green ebony –see– *Brya ebenus* (L.) DC.

green euphorb –see– *Euphorbia heterophylla* L.

green galenia –see– *Galenia pubescens* (Eckl. & Zeyh.) Druce

green hellebore –see– *Helleborus viridis* L.; *Veratrum album* L.; *Veratrum viride* Aiton

green milkweed –see– *Asclepias viridis* Walter

green panic –see– *Megathyrsus maximus* (Jacq.) B. K. Simon & S. W. L. Jacobs

green pea –see– *Pisum sativum* L.

green pepper –see– *Capsicum annuum* L.

green pigweed –see– *Amaranthus hybridus* L.; *Amaranthus retroflexus* L.

green poisonberry –see– *Cestrum parqui* L'Her.

green potato –see– *Solanum tuberosum* L.

Green River milk vetch –see– *Astragalus pubentissimus* Torr. & A. Gray

green saure –see– *Rumex acetosa* L.

green sorrel –see– *Rumex acetosa* L.

green tea –see– *Camellia sinensis* (L.) Kuntze

green thorn apple –see– *Datura inoxia* Mill.

green timber milk vetch –see– *Astragalus miser* Douglas ex Hook. var. hylophilus (Rydb.) Barneby

green-head coneflower –see– *Rudbeckia laciniata* L.

green-stem paper flower –see– *Psilostrophe sparsiflora* (A. Gray) A. Nelson

Greiskraut –see– *Jacobaea vulgaris* Gaertn.

grenadil –see– *Brya ebenus* (L.) DC.

grermandrée-petit-chêne –see– *Teucrium chamaedrys* L.

GREVILLEA BANKSII R. Br. [Proteaceae]

Common Names:

Bank's-grevillea; dwarf silky oak; kahili flower; rain flower; red silky oak; silky oak tree

Citations:

Arnold HL (1941) Kahili flower (Grevillea banksii) dermatitis - A preliminary report. Hawaii Med J 1(1):15-18.

Arnold HL Jr (1942) Dermatitis due to the blossom of Grevillea banksii. A newly recognized and common entity in Hawaii. Arch Derm Syphilol 45(6):1037-1051.

GREVILLEA JUNIPERINA R. Br. [Proteaceae]

Citations:

Apted J (1988) Acute contact urticaria from Grevillea juniperina. Contact Dermatitis 18(2):126.

GREVILLEA ROBUSTA A. Cunn. ex R. Br. [Proteaceae]

Common Names:

Australian silk oak; silky oak; silver oak

Citations:

Hoffman TE, Hausen BM, Adams RM (1985) Allergic contact dermatitis to "silver oak" wooden arm bracelets. J Am Acad Dermatol 13(5 Part 1):778-779.

May SB (1960) Dermatitis due to Grevillea robusta (Australian silk oak). Report of a case. Arch Dermatol 82(Dec):1006.

gringging –see– *Melia azedarach* L.

Große Brennessel –see– *Urtica dioica* L.

Große Klette –see– *Arctium lappa* L.

Großer Sauerampfer –see– *Rumex acetosa* L.

ground cherry –see– *Physalis longifolia* Nutt.; *Solanum aculeatissimum* Jacq.

ground hemlock –see– *Taxus baccata* L.

ground ivy –see– *Glechoma hederacea* L.; *Modiola caroliniana* (L.) G. Don

ground lemon –see– *Podophyllum peltatum* L.

groundnut –see– *Arachis hypogaea* L.

groundsel –see– *Jacobaea vulgaris* Gaertn.; *Senecio eremophilus* Richardson var. kingii Greenm.; *Senecio flaccidus* Less. var. flaccidus; *Senecio latifolius* DC.; *Senecio riddellii* Torr. & A. Gray; *Senecio spartioides* Torr. & A. Gray

groundsel bush –see– *Baccharis glomeruliflora* Pers.; *Baccharis halimifolia* L.

groundsel tree –see– *Baccharis halimifolia* L.

Grüne Bohne –see– *Phaseolus vulgaris* L.

Grüne Christwurz –see– *Helleborus viridis* L.

Grüne Nieswurz –see– *Helleborus viridis* L.

Grüner Salat –see– *Lactuca sativa* L.

Grünmais –see– *Zea mays* L. subsp. mays

guacis –see– *Leucaena leucocephala* (Lam.) de Wit

guaiac –see– *Bulnesia sarmienti* Lorentz ex Griseb.

guajillo –see– *Acacia berlandieri* Benth.

guanábana –see– *Annona muricata* L.

guang-fang-ji –see– *Aristolochia fangchi* Y. C. Wu ex L. D. Chow & S. M. Hwang

guango –see– *Samanea saman* (Jacq.) Merr.

guano bos –see– *Senecio ilicifolius* L.

guao –see– *Comocladia dentata* Jacq.; *Metopium brownei* (Jacq.) Urb.

guao-de-costa –see– *Metopium brownei* (Jacq.) Urb.; *Metopium toxiferum* (L.) Krug & Urb.

guao-de-sabana –see– *Comocladia dentata* Jacq.

guao hediondo –see– *Comocladia dentata* Jacq.

guao prieto –see– *Comocladia dentata* Jacq.

guao real –see– *Comocladia dentata* Jacq.

guar bean –see– *Cyamopsis tetragonoloba* (L.) Taub.

guar gum –see– *Cyamopsis tetragonoloba* (L.) Taub.

guara –see– *Cyamopsis tetragonoloba* (L.) Taub.

guarana –see– *Ephedra sinica* Stapf; *Paullinia cupana* Kunth

guarantã –see– *Esenbeckia leiocarpa* Engl.

GUAREA GUIDONIA (L.) Sleumer [Meliaceae]

Synonyms:

Guarea rusbyi (Britton) Rusby; ***Guarea trichilioides*** L.

Common Names:

camboatá

Citations:

Saad AD, Linardi MC (1970) Aspectos histopatológicos da intoxicação de cobaias (Cavia porcellus) pela Guarea trichilioides L. (Meliaceae). Arq Inst Biol (Sao Paulo) 37(4):239-250.

Guarea rusbyi (Britton) Rusby = Guarea guidonia (L.) Sleumer

Guarea trichilioides L. = Guarea guidonia (L.) Sleumer

güaritoto –see– *Cnidoscolus urens* (L.) Arthur

Guatamalan rhubarb –see– *Jatropha podagrica* Hook.

guava –see– *Psidium guajava* L.

guaya –see– *Lathyrus sativus* L.

guaycur –see– *Limonium brasiliense* (Boiss.) Kuntze

guayiga –see– *Zamia debilis* L. f.; *Zamia integrifolia* L. f.

guayule –see– *Parthenium argentatum* A. Gray

guijiu –see– *Podophyllum hexandrum* Royle

Guinae corn –see– *Sorghum bicolor* (L.) Moench

Guinea pepper –see– *Capsicum annuum* L.; *Capsicum frutescens* L.

Guineagrass –see– *Megathyrsus maximus* (Jacq.) B. K. Simon & S. W. L. Jacobs

guineahen weed –see– *Petiveria alliacea* L.

guizo-de-cascavel –see– *Crotalaria pallida* Aiton

gulf stargrass –see– *Brachyachne convergens* (F. Muell.) Stapf

gum acacia –see– *Acacia senegal* (L.) Willd.

gum arabic –see– *Acacia senegal* (L.) Willd.

gum karaya –see– *Sterculia urens* Roxb.

gum tree –see– *Eucalyptus globulus* Labill.

gumai –see– *Stipa sibirica* (L.) Lam.

gumbo –see– *Abelmoschus esculentus* (L.) Moench

gungo pea –see– *Cajanus cajan* (L.) Millsp.

gur ghas –see– *Stipa sibirica* (L.) Lam.

GUTIERREZIA MICROCEPHALA (DC.) A. Gray [Asteraceae]

Common Names:

broomweed; perennial broomweed; perennial threadleaf snakeweed; slinkweed; small-head matchbush; snakeweed; sticky snakeweed; stinkweed; threadleaf broomweed; threadleaf snakeweed; turpentine weed

Citations:

Dollahite JW, Allen TJ (1959) Feeding perennial broomweed to cattle, swine, sheep, goats, rabbits, guinea pigs and chickens. Texas Agric Exp Sta Prog Rep #2105:6 pp.

Dollahite JW, Anthony WV (1955) Experimental production of premature calves and retained placentas by feeding perennial broomweed. Texas Agric Exp Sta Prog Rep #1825:6 pp.

Dollahite JW, Anthony WV (1956) Experimental production of abortion, premature calves and retained placentas by feeding a species of perennial broomweed. Texas Agric Exp Sta Prog Rep #1884:3 pp.

Dollahite JW, Anthony WV (1957) Experimental production of abortion, premature calves and retained placentas by feeding a species of perennial broomweed. Southwestern Vet 10(Winter):128-131.

Dollahite JW, Anthony WV (1957) Poisoning of cattle with Gutierrezia microcephala, a perennial broomweed. J Am Vet Med Assoc 130(Jun 15):525-530.

Edrington TS, Flores-Rodriguez GI, Smith GS (1990) Effects of ingested snakeweed (Gutierrezia microcephala) herbage on size of testes and seminal vesicals and on semen quality in rats. Proc West Sec Am Soc Anim Sci 41:88-90.

Edrington TS, Flores-Rodriguez GI, Smith GS, et al. (1993) Effect of ingested snakeweed (Gutierrezia microcephala) foliage on reproduction, semen quality, and serum clinical profiles of male rats. J Anim Sci 71(6):1520-1525.

Edrington TS, Smith GS, Reynolds PR, et al. (1991) Comparative toxicity of snakeweed (Gutierrezia microcephala) collected from two sites. Proc West Sec Am Soc Anim Sci 42:173-175.

Edrington TS, Smith GS, Ross TT, et al. (1992) Toxicity of new growth and whole plant foliage of snakeweed (Gutierrezia microcephala) as changed by drying. Proc West Sec Am Soc Anim Sci 43:414-417.

Edrington TS, Smith GS, Ross TT, et al. (1993) Embryonic mortality in Sprague-Dawley rats induced by snakeweed (Gutierrezia microcephala). J Anim Sci 71(8):2193-2198.

Edrington TS, Smith GS, Samford MD, et al. (1991) Ingested snakeweed (Gutierrezia microcephala) foliage related to embryonic and fetal mortality of albino rats. Proc West Sec Am Soc Anim Sci 42:12-14.

Flores-Rodriguez GI, Smith GS, McDaniel KC (1989) Effects of ingested snakeweed (Gutierrezia microcephala) herbage on reproduction, serum progesterone, and blood constituents of female albino rats. Proc West Sec Am Soc Anim Sci 40:217-221.

Mathews FP (1936) The toxicity of broomweed (Gutierrezia microcephala) for sheep, cattle and goats. J Am Vet Med Assoc 88:55-61.

Mathews FP, Schmidt H (1934) Miscellaneous poisonous plants. Texas Agric Exp Sta Annu Rep 47:12-13.

Oetting BC, Ross TT, Walraven K, et al. (1990) Effects of ingested snakeweed herbage on estrual activity, blood progesterone, and serum clinical profiles of fine-wool ewes. Proc West Sec Am Soc Anim Sci 41:23-26.

Smith GS, Ross TT, Hallford DM, et al. (1994) Toxicology of snakeweeds (Gutierrezia microcephala, G. sarothrae). J Anim Sci 72(Suppl 2):102.

Staley EC, Smith GS, Greenberg JA (1995) Effects on reproduction in female offspring from Sprague-Dawley rats fed 10% snakeweed (Gutierrezia microcephala) throughout pregnancy and concurrent treatment with safflower oil. Vet Hum Toxicol 37(5):440-442.

Staley EC, Smith GS, Greenberg JA (1996) Decreased reproductive effects from snakeweed (Gutierrezia microcephala) in Sprague-Dawley rats with increased dietary snakeweed consumption. Vet Hum Toxicol 38(4):259-264.

Williams JL, Campos D, Ross TT, et al. (1992) Snakeweed (Gutierrezia spp.) toxicosis in beef heifers. Proc West Sec Am Soc Anim Sci 43:67-69.

GUTIERREZIA SAROTHRAE (Pursh) Britton & Rusby [Asteraceae]

Common Names:

broom snakeweed; broomweed; hierba-de-San Nicolas; hierba-de-vibora; matchbrush; matchweed; perennial broomweed; perennial snakeweed; slinkweed; snakeweed; turpentine broomweed; turpentine weed; yellow weed

Citations:

Oetting BC, Ross TT, Walraven K, et al. (1990) Effects of ingested snakeweed herbage on estrual activity, blood progesterone, and serum clinical profiles of fine-wool ewes. Proc West Sec Am Soc Anim Sci 41:23-26.

Smith GS, Ross TT, Hallford DM, et al. (1994) Toxicology of snakeweeds (Gutierrezia microcephala, G. sarothrae). J Anim Sci 72(Suppl 2):102.

gutka –see– *Nicotiana tabacum* L.

gutta percha tree –see– *Excoecaria parvifolia* Müll. Arg.

gwanja kusa –see– *Blighia sapida* K. D. Koenig

GYMNOCLADUS DIOICUS (L.) K. Koch [Fabaceae]

Common Names:

American coffeebean; American coffeeberry; bonduc; chicot; coffee tree; coffeebean; false coffeebean; Kentucky coffee tree; Kentucky coffeebean; Kentucky mahogany; nicker tree; stump tree

Citations:

Hansen AA (1928) Toxic trees. North Am Vet 9(10):49-53.

Troxel MT, Poppenga RH (2005) Kentucky coffee tree intoxication in a dog with cerebellovestibular signs, bradycardia, hypotension, and hypoglycemia. J Vet Intern Med 19(4):599-601.

GYMNOSPORIA SPINOSA (Blanco) Merr. & Rolfe [Celastraceae]

Common Names:

kankera

Citations:

Singh AN, Pachalag SV, Rathore BL, et al. (1973) The effect of kankera (Gymnosporia spinosa Forssk) leaves on health of sheep. Indian Vet J 50(7):722-723.

gympie bush –see– *Dendrocnide moroides* (Wedd.) Chew

gympie gympie –see– *Dendrocnide moroides* (Wedd.) Chew

gympie nettle –see– *Dendrocnide moroides* (Wedd.) Chew

gympie tree –see– *Dendrocnide moroides* (Wedd.) Chew

GYPSOPHILA PANICULATA L. [Caryophyllaceae]

Common Names:

baby's-breath; tall gysophyl

Citations:

Twiggs JT, Yunginger JW, Agarwal MK, et al. (1982) Occupational asthma in a florist caused by the dried plant, baby's breath. J Allergy Clin Immunol 69(5):474-477.

Vidal C, Polo F (1998) Occupational allergy caused by Dianthus caryophillus, Gypsophila paniculata, and Lilium longiflorum. Allergy 53(10):995-998.

Gypsophila vaccaria Sibth. & Sm. = Vaccaria hispanica (Mill.) Rauschert

gypsy flower –see– *Cynoglossum officinale* L.

H

habb-el-meluk –see– *Jatropha curcas* L.

habbussalatin –see– *Croton tiglium* L.

habet-el-arus –see– *Abrus precatorius* L.

Haemanthus multiflorus Martyn = Scadoxus multiflorus (Martyn) Raf.

Hafelkraut –see– *Asarum europaeum* L.

Hafelmünch –see– *Asarum europaeum* L.

Hafelührlein –see– *Asarum europaeum* L.

Hafer –see– *Avena sativa* L.

Hagedorn –see– *Securigera varia* (L.) Lassen

Hagenburger's-ivy –see– *Hedera helix* L. subsp. canariensis (Willd.) Cout.

HAGENIA ABYSSINICA (Bruce) J. F. Gmel. [Rosaceae]

Common Names:

kosso; kousso

Citations:

Abebe W (1992) Adverse effects of traditional drug preparations. J Ethnopharmacol 36:93-94.

Low G, Rogers LJ, Brumley SP, et al. (1985) Visual deficits and retinotoxicity caused by the naturally occurring anthelmintics, Embelia ribes and Hagenia abyssinica. Toxicol Appl Pharmacol 81(2):220-230.

Hahnenfuß –see– *Ranunculus bulbosus* L.

hai yu –see– *Alocasia macrorrhizos* (L.) G. Don

Hain Sauerklee –see– *Oxalis acetosella* L.

Hainkorn –see– *Fagopyrum esculentum* Moench

hair bell –see– *Hyacinthoides non-scripta* (L.) Chouard ex Rothm.

hairy caltrop –see– *Kallstroemia hirsutissima* Vail ex Small

hairy cat's-ear –see– *Hypochaeris radicata* L.

hairy indigo –see– *Indigofera hirsuta* L.

hairy lupine –see– *Lupinus argenteus* Pursh

hairy medic –see– *Medicago polymorpha* L.

hairy mountain mahogany –see– *Cercocarpus montanus* Raf.

hairy nightshade –see– *Solanum sarrachoides* Sendtn.

hairy spurge –see– *Chamaesyce hirta* (L.) Millsp.; *Euphorbia pilosa* L.

hairy thorn apple –see– *Datura inoxia* Mill.; *Datura metel* L.; *Datura wrightii* Regel

hairy vetch –see– *Vicia hirsuta* (L.) Gray; *Vicia villosa* Roth

hairy vetchling –see– *Lathyrus hirsutus* L.

hairy wild lettuce –see– *Hypochaeris radicata* L.

HAKEA DRUPACEA (C. F. Gaertn.) Roem. & Schult. [Proteaceae]

Citations:

Apted J (1988) Acute contact urticaria from Hakea suaveolens. Contact Dermatitis 18(8):126.

haldi –see– *Curcuma longa* L.

HALDINA CORDIFOLIA (Roxb.) Ridsdale [Rubiaceae]

Synonyms:

Adina cordifolia (Roxb.) Hook. f. ex Brandis

Common Names:

karam

Citations:

Kulakkattolickal AT (1987) Piscicidal plants of Nepal: Preliminary toxicity screening using grass carp (Ctenopharyngodon idella) fingerlings. J Ethnopharmacol 21(1):1-9.

HALESIA TETRAPTERA J. Ellis [Styracaceae]

Common Names:

silver bill; snow drop

Citations:

Anonymous (1906) Poisoning of horses by "Snow Drop." Am Vet Rev 30:298.

half breed weed –see– *Cyclachaena xanthiifolia* (Nutt.) Fresen.

half creeper –see– *Clematis microphylla* DC.

halogeton –see– *Halogeton glomeratus* (M. Bieb.) C. A. Mey.

HALOGETON GLOMERATUS (M. Bieb.) C. A. Mey. [Chenopodiaceae]

Common Names:

barilla; halogeton

Citations:

Anderson WA, Huffman WT (1957) Halogeton poisoning in a ewe. J Am Vet Med Assoc 130(Apr 15):330-331.

Cook CW, Stoddart LA (1953) The halogeton problem in Utah. Utah Agric Exp Sta Bull #364:1-44.

Fleming CE, Miller MR, Vawter LR, et al. (1948) Poisonous range plants - Halogeton glomeratus. A desert range plant poisonous to sheep in Nevada. Nevada Agric Exp Sta Annu Rep 1947:16-18.

James LF (1968) Serum electrolyte, acid-base balance, and enzyme changes in acute Halogeton glomeratus poisoning in sheep. Can J Comp Med Vet Sci 32(4):539-543.

James LF (1970) Locomotor disturbance of cattle grazing Halogeton glomeratus. J Am Vet Med Assoc 156(9):1310-1312.

James LF, Johnson AE (1970) Prevention of fatal Halogeton glomeratus poisoning in sheep. J Am Vet Med Assoc 157(4):437-442.

Lincoln SD, Black B (1980) Halogeton poisoning in range cattle. J Am Vet Med Assoc 176(8):717-718.

Littledike ET, James LF, Cook H (1976) Oxalate (Halogeton) poisoning of sheep: Certain physiopathologic changes. Am J Vet Res 37(6):661-666.

Shupe JL, James LF (1969) Additional physiopathologic changes in Halogeton glomeratus (oxalate) poisoning in sheep. Cornell Vet 59(1):41-55.

Van Kampen KR, James LF (1969) Acute halogeton poisoning of sheep: Pathogenesis of lesions. Am J Vet Res 30(10):1779-1783.

handal –see– *Citrullus colocynthis* (L.) Schrad.

hanf –see– *Cannabis sativa* L.

hangon so –see– *Senecio cannabifolius* Less.

Hansl-am-Weg –see– *Cichorium intybus* L.

Haplopappus heterophyllus (A. Gray) S. F. Blake = Isocoma plurifolia (Torr. & A. Gray) Greene

hard heads –see– *Acroptilon repens* (L.) DC.

hardhack –see– *Collinsonia canadensis* L.

Hardinggrass –see– *Phalaris aquatica* L.

harebell –see– *Hyacinthoides non-scripta* (L.) Chouard ex Rothm.

harebur –see– *Arctium lappa* L.

haricot bean –see– *Phaseolus vulgaris* L.

haricot-de-lima –see– *Phaseolus lunatus* L.

haricot-du-Cap –see– *Phaseolus lunatus* L.

harmal –see– *Peganum harmala* L.; *Rhazya stricta* Decne.

harmel –see– *Peganum harmala* L.; *Rhazya stricta* Decne.

harmel peganum –see– *Peganum harmala* L.

harpuis bos –see– *Gnidia burchellii* Gilg

harpuisbar –see– *Gnidia burchellii* Gilg

harro –see– *Terminalia chebula* Retz.

Hartheu –see– *Hypericum perforatum* L.

Hartriegel –see– *Ligustrum vulgare* L.

hart's-thorn –see– *Rhamnus cathartica* L.

Haschisch –see– *Cannabis sativa* L.

Haselwurz –see– *Asarum europaeum* L.

Hasenöhrlein –see– *Asarum europaeum* L.

hashish –see– *Cannabis sativa* L.

hathisunda –see– *Heliotropium ellipticum* Ledeb.

hatrack cactus –see– *Euphorbia lactea* Haw.

haujeri –see– *Capparis tomentosa* Lam.

haute bois –see– *Sambucus nigra* L.

heavenwood –see– *Ailanthus altissima* (Mill.) Swingle

Hawaiian baby woodrose –see– *Argyreia nervosa* (Burm. f.) Bojer

hawthorn –see– *Crataegus laevigata* (Poir) DC.; *Crataegus monogyna* Jacq.

Hayden's-poison vetch –see– *Astragalus bisulcatus* (Hook.) A. Gray

hazelwort –see– *Asarum europaeum* L.

hazelwurz –see– *Asarum europaeum* L.

he broom –see– *Laburnum anagyroides* Medik.

headache –see– *Papaver rhoeas* L.

heal all –see– *Collinsonia canadensis* L.

heart ivy –see– *Hedera helix* L.; *Schefflera actinophylla* (Endl.) Harms

heart-leaf –see– *Gastrolobium grandiflorum* F. Muell.

heart-leaf cocklebur –see– *Xanthium strumarium* L.

heart-leaf ivy –see– *Philodendron hederaceum* (Jacq.) Schott

heart-leaf nettle –see– *Urtica chamaedryoides* Pursh

heart-leaf philodendron –see– *Philodendron hederaceum* (Jacq.) Schott

heart-leaf poison bush –see– *Gastrolobium grandiflorum* F. Muell.

Heart-of-Jesus –see– *Caladium bicolor* (Aiton) Vent.

heart's-ease –see– *Persicaria maculosa* Gray

heartweed –see– *Persicaria maculosa* Gray

heavenly bamboo –see– *Nandina domestica* Thunb.

heavenly blue –see– *Ipomoea tricolor* Cav.; *Ipomoea violacea* L.

Hecke –see– *Prunus laurocerasus* L.

hedegar –see– *Silybum marianum* (L.) Gaertn.

HEDEOMA PULEGIOIDES (L.) Pers. [Lamiaceae]

Common Names:

American pennyroyal; false pennyroyal; hédéome; huile d'hédéome; mosquito plant; North American pennyroyal; pennyroyal; poleo chino; squaw mint

Citations:

Anderson IB, Mullen WH, Meeker JE, et al. (1996) Pennyroyal toxicity: Measurement of toxic metabolite levels in two cases and review of the literature. Ann Intern Med 124(8):726-734.

Early DF (1961) Pennyroyal: A rare cause of epilepsy. Lancet 2(Sep 9):580-581.

Sudekum M, Poppenga RH, Raju N, et al. (1992) Pennyroyal oil toxicosis in a dog. J Am Vet Med Assoc 200(6):817-818.

hédéome –see– *Hedeoma pulegioides* (L.) Pers.

Hedera canariensis Willd. = Hedera helix L. subsp. canariensis (Willd.) Cout.

HEDERA HELIX L. [Araliaceae]

Common Names:

Efeu; English ivy; Epheu; Eppich; glacier ivy; heart ivy; hiedra; ivy; lierre; lierre grimpant; needlepoint ivy; poet's-ivy; ripple ivy

Citations:

Boyle J, Harman RM (1985) Contact dermatitis to Hedera helix (common ivy). Contact Dermatitis 12(2):111-112.
Brömel J, Zettl K (1986) Efeuvergiftung bei Rehwild. Prakt Tierarzt 67(11):967-968.
García M, Fernández E, Navarro JA, et al. (1995) Allergic contact dermatitis from Hedera helix L. Contact Dermatitis 33(9):133-134.
Goldman L, Preston RH, Muegel HR (1956) Dermatitis venenata from English Ivy (Hedera helix). Arch Dermatol 74(3):311-312.
Hausen BM, Brohan J, Konig WA, et al. (1987) Allergic and irritant contact dermatitis from falcarinol and didehydrofalcarinol in common ivy (Hedera helix L.). Contact Dermatitis 17(7):1-9.
Highman WJ (1924) The pathogenesis of dermatitis, including eczema. A case of English ivy poisoning. Arch Derm Syphilol 9:344-354.
Jøhnke H, Bjarnason B (1994) Allergisk kontakdermatitis over for efeu (Hedera helix L.). Ugeskr Laeger 156(25):3778-3779.
Mahe-Quinio M, Rossinyol G, Foucaud A (1975) Empoisonnement mortel de poulets par les graines de lierre. Plantes Med Phytotherap 9(3):182-186.
Mitchell JC (1981) Allergic contact dermatitis from Hedera helix and Brassaia actinophylla (Araliaceae). Contact Dermatitis 7(3):158-159.
Paulsen E (1998) Occupational dermatitis in Danish gardeners and greenhouse workers (II). Etiological factors. Contact Dermatitis 38(1):14-19.
Ramón MF, Ballesteros S, Martínez-Arrieta R, et al. (2003) Plant and herb exposures in Spain. J Toxicol Clin Toxicol 41(4):514.
Roed-Petersen J (1975) Allergic contact hypersensitivity to ivy (Hedera helix). Contact Dermatitis 1(1):57.
Rynes SE (1949) House ivy dermatitis. Ann Allergy 7:62-64.
Sánchez-Pérez J, Córdoba S, Hausen BM, et al. (1998) Allergic contact dermatitis from common ivy confirmed with stored allergens. Contact Dermatitis 39(5):259-260.
Turton PH (1925) Poisoning by ivy. Br Med J 2:294.
Yesudian PD, Franks A (2002) Contact dermatitis from Hedera helix in a husband and wife. Contact Dermatitis 46(2):125-126.

HEDERA HELIX L. subsp. canariensis (Willd.) Cout. [Araliaceae]

Synonyms:

Hedera canariensis Willd.

Common Names:

Algerian ivy; Canary Island ivy; Hagenburger's-ivy; ivy; Madeira ivy; variegated ivy

Citations:

Calnan CD (1981) Dermatitis from ivy (Hedera canariensis variegata). Contact Dermatitis 7(2):124-125.
Dorsey CS (1957) Contact dermatitis from Algerian ivy. Arch Dermatol 75(5):671-675.
Dorsey CS (1959) Algerian ivy dermatitis. A California disease. Calif Med 90(2):155-159.
Hambly EM, Wilkinson DS (1978) Sensitivity to variegated ivy (Hedera canariensis). Contact Dermatitis 4(4):239-240.
Massmanian A, Valcuende Cavero F, Ramirez Bosca A, et al. (1988) Contact dermatitis from variegated ivy (Hedera helix subsp. canariensis Willd.). Contact Dermatitis 18(7):247-248.

Hederich –see– *Raphanus raphanistrum* L.
hedge apple –see– *Maclura pomifera* (Raf.) C. K. Schneid.
hedge nettle –see– *Lamium amplexicaule* L.
hedge plant –see– *Ligustrum vulgare* L.
hedge vine –see– *Clematis vitalba* L.
hedgehoggrass –see– *Echinopogon ovatus* (G. Forst.) P. Beauv.
hediondilla –see– *Larrea tridentata* (DC.) Coville
hedlock –see– *Sinapis arvensis* L.
Hedrich –see– *Raphanus sativus* L.
Heidegrütze –see– *Fagopyrum esculentum* Moench
Heidekorn –see– *Fagopyrum esculentum* Moench
Heidewacholder –see– *Juniperus communis* L.
Heilkraut –see– *Heracleum sphondylium* L.
heléboro –see– *Helleborus foetidus* L.
helecho –see– *Pteridium aquilinum* (L.) Kuhn
helecho macho –see– *Dryopteris filix-mas* (L.) Schott

HELENIUM AMARUM (Raf.) H. Rock [Asteraceae]

Synonyms:

Helenium tenuifolium Nutt.

Common Names:

American sneezeweed; bitter sneezeweed; bitterweed; eastern bitterweed; fine-leaf sneezeweed; sneezeweed; southeastern bitterweed; Spanish daisy

Citations:

Dollahite JW, Rowe LD, Kim HL, et al. (1973) Toxicity of Helenium amarum (bitter sneezeweed) to sheep. Southwestern Vet 26(Winter):135-137.

HELENIUM AUTUMNALE L. [Asteraceae]

Common Names:

autumn sneezeweed; bitter sneezeweed; bitterweed; false sunflower; narrow-leaf sneezeweed; oxeye; poison sneezeweed; sneezeweed; sneezewort; staggerweed; staggerwort; swamp sunflower; yellow oxeye; yellow star

Citations:

Anonymous (1896) Poisonous plants. U S Dep Agric Annu Rep 1896:99.

Pammel LH (1917) Young sneezeweed poisonous. Am J Vet Med 12:461-462.
Phares DL (1889) Helenium autumnale. Mississippi Agric Exp Sta Bull #9(12):11-14.

Helenium hoopesii A. Gray = Hymenoxys hoopesii (A. Gray) Bierner

HELENIUM INTEGRIFOLIUM (Kunth) Benth. & Hook. f. ex Hemsl. [Asteraceae]

Citations:

de Aluja AS, Paasch L (1973) Intoxicación de borregos con la planta Helenium integrifolium. Veterinaria (Mexico) 4(3):214-222.

HELENIUM MICROCEPHALUM DC. [Asteraceae]

Common Names:

false sunflower; small-head sneezeweed; sneezeweed

Citations:

Boughton IB, Hardy WT (1939) Hard yellow livers of sheep and cattle and feeding trials of suspected plants. Texas Agric Exp Sta Annu Rep 52:238-240.

Boughton IB, Hardy WT (1940) Feeding trials of suspected plants. Texas Agric Exp Sta Annu Rep 53:236-237.

Dollahite JW, Hardy WT, Henson JB (1964) Toxicity of Helenium microcephalum (smallhead sneezeweed). J Am Vet Med Assoc 145(7):694-696.

HELENIUM QUADRIDENTATUM Labill. [Asteraceae]

Common Names:

rosilla

Citations:

Alfonso HA, Figueredo MA, Rodriguez J, et al. (1986) Toxicidad de Helenium quadridentatum en terneros y ratones albino macho. Primer reporte en Cuba. Rev Salud Anim 8(3):295-296.

Helenium tenuifolium Nutt. = Helenium amarum (Raf.) H. Rock

HELIANTHUS ANNUUS L. [Asteraceae]

Common Names:

annual sunflower; Sonnenblume; sunflower; wild artichoke; wild sunflower

Citations:

Atis S, Tutluoglu B, Sahin K, et al. (2002) Sensitization to sunflower pollen and lung functions in sunflower processing workers. Allergy 57(1):35-39.

Hausen BM (1979) The sensitizing capacity of compositae plants. III. Test results and cross-reactions in compositae-sensitive patients. Dermatologica 159(1):1-11.

Vandenplas O, Van der Borght T, Delwiche JP (1998) Occupational asthma caused by sunflower-seed dust. Allergy 53(9):907-908.

HELICHRYSUM ARGYROSPHAERUM DC. [Asteraceae]

Citations:

Basson PA, Kellerman TS, Albl P, et al. (1975) Blindness and encephalopathy caused by Helichrysum argyrosphaerum DC. (Compositae) in sheep and cattle. Onderstepoort J Vet Res 42(4):135-147.

HELICHRYSUM BLANDOWSKIANUM Steetz ex Sond. [Asteraceae]

Common Names:

woolly everlasting; woolly everlasting daisy

Citations:

McAuliffe PR, White WE (1976) "Woolly everlasting daisy" (Helichrysum blandowskianum) toxicity in cattle and sheep. Aust Vet J 52(8):366-368.

HELICHRYSUM CEPHALOIDEUM DC. [Asteraceae]

Citations:

Van der Walt SJ, Steyn DG (1940) Recent investigations into the toxicity of known and unknown poisonous plants in the Union of South Africa, X. Onderstepoort J Vet Res 15(1-2):261-277.

Helichrysum diosmifolium (Vent.) Sweet = Ozothamnus diosmifolius (Vent.) DC.

heliotrope –see– *Heliotropium europaeum* L.; *Heliotropium lasiocarpum* Fisch. & C. A. Mey.

HELIOTROPIUM AEGYPTIACUM Lehm. [Boraginaceae]

Synonyms:

Heliotropium cinerascens Steud. ex DC. & A. DC.

Citations:

Schoental R, Coady A (1968) The hepatotoxicity of some Ethiopian and East African plants, including some used in traditional medicines. East Afr Med J 45(8):577-580.

HELIOTROPIUM AMPLEXICAULE Vahl [Boraginaceae]

Common Names:

blue heliotrope; violet heliotrope; wild verbena

Citations:

Ketterer PJ, Glover PE, Smith LW (1987) Blue heliotrope (Heliotropium amplexicaule) poisoning in cattle. Aust Vet J 64(4):115-117.

Heliotropium cinerascens Steud. ex DC. & A. DC. = Heliotropium aegyptiacum Lehm.

HELIOTROPIUM CIRCINATUM Griseb. [Boraginaceae]

Citations:

Eroksuz H, Eroksuz Y, Ozer H, et al. (2003) Toxicity of dietary Heliotropium circinatum to rats. Vet Hum Toxicol 45(4):198-201.

HELIOTROPIUM DOLOSUM De Not. [Boraginaceae]

Citations:

Eroksuz H, Eroksuz Y, Ozer H, et al. (2001) Toxicity of dietary Heliotropium dolosum seed to mice. Vet Hum Toxicol 43(3):152-155.

Eroksuz H, Eroksuz Y, Ozer H, et al. (2001) Toxicity of dietary Heliotropium dolosum seeds to broiler chickens. Vet Hum Toxicol 43(6):334-338.

Eroksuz H, Eroksuz Y, Ozer H, et al. (2002) Toxicity of dietary Heliotropium dolosum seed to Japanese quail. Vet Hum Toxicol 44(5):264-268.

Heliotropium eichwaldii Steud. = Heliotropium ellipticum Ledeb.

HELIOTROPIUM ELLIPTICUM Ledeb. [Boraginaceae]

Synonyms:

Heliotropium eichwaldii Steud.

Common Names:

hathisunda; khalibui

Citations:

Datta DV, Khuroo MS, Mattocks AR, et al. (1978) Herbal medicines and veno-occlusive disease in India. Postgrad Med J 54(634):511-515.

Datta DV, Khuroo MS, Mattocks AR, et al. (1978) Veno-occlusive disease of liver due to heliotropium plant, used as medicinal herb (report of 6 cases with review of literature). J Assoc Physicians India 26(5):383-393.

HELIOTROPIUM EUROPAEUM L. [Boraginaceae]

Common Names:

caterpillar weed; cherry pie; European heliotrope; heliotrope; herbe-aux-verrues; herbe-de-St. Fiacre; potato weed; tournesol; toursole; wandary curse; wild heliotrope

Citations:

Anonymous (1951) Investigations into the etiology and control of enzootic (toxaemic) jaundice of sheep. Report of the investigation committee for the year 1950-51. Aust Vet J 27(Aug):203-206.

Anonymous (1954) Investigations into the etiology and control of enzootic (toxaemic) jaundice of sheep. Report of the investigation committee for the year 1952-53. Aust Vet J 30(Jun):182-184.

Bull LB, Dick AT, Keast JC, et al. (1956) An experimental investigation of the hepatotoxic and other effects on sheep of consumption of Heliotropium europaeum L.: Heliotrope poisoning of sheep. Aust J Agric Res 7:281-332.

Bull LB, Rogers ES, Keast JC, et al. (1961) Heliotropium poisoning in cattle. Aust Vet J 37(Feb):37-43.

Edgar G, Albiston HE, Bull LB (1949) Investigations into the etiology and control of enzootic (toxaemic) jaundice of sheep. Aust Vet J 25(Sep):202-208.

Gaul KL, Gallagher PF, Reyes D, et al. (1994) Poisoning of pigs and poultry by stock feed contaminated with heliotrope seed. In: Colegate SM, Dorling PR (eds.) Plant-associated toxins. Agricultural, phytochemical and ecological aspects. CABI. New York. pp. 137-142.

Gregory TS (1963) Trial of cobalt administration for the prevention of Heliotropium poisoning. Aust Vet J 39(Feb):64.

Harper PA, Walker KH, Krahenbuhl RE, et al. (1985) Pyrrolizidine alkaloid poisoning in calves due to contamination of straw by Heliotropium europeum. Aust Vet J 62(11):382-383.

Harris DJ, Nowara G (1995) The characteristics and causes of sheep losses in the Victorian Mallee. Aust Vet J 72(9):331-340.

Hill BD, Gaul KL, Noble JW (1997) Poisoning of feedlot cattle by seeds of Heliotropium europaeum. Aust Vet J 75(5):360-361.

Howell JM, Deol HS, Dorling PR, et al. (1991) Experimental copper and heliotrope intoxication in sheep: Morphological changes. J Comp Pathol 105(1):49-74.

Jones RT, Drummond GR, Chatham RO (1981) Heliotropium europaeum poisoning of pigs. Aust Vet J 57(8):396.

Kinnaird PJ, Chan P, Leaver DD (1968) Heliotrope poisoning in cattle. Aust Vet J 44(1):39.

Lanigan GW, Payne AL, Peterson JE (1978) Antimethanogenic drugs and Heliotropium europaeum poisoning in penned sheep. Aust J Agric Res 29:1281-1292.

Lanigan GW, Peterson JE (1979) Bromosulphophthalein clearance rates in sheep with pyrrolizidine liver damage. Aust Vet J 55(5):220-224.

McKenna CT, Orchard HE (1949) Heliotrope poisoning in sheep. J Agric South Aust 52:436-437.

McLennan MW, Dodson ME, Rac R (1972) Heliotropium poisoning in cattle. Aust Vet J 48(8):480.

Pass DA, Hogg GG, Russell RG, et al. (1979) Poisoning of chickens and ducks by pyrrolizidine alkaloids of Heliotropium europaeum. Aust Vet J 55(6):284-288.

Peterson JE, Payne AL, Culvenor CC (1992) Heliotropium europaeum poisoning of sheep with low liver copper concentrations and the preventive efficacy of cobalt and antimethanogen. Aust Vet J 69(3):51-56.

HELIOTROPIUM INDICUM L. [Boraginaceae]

Citations:

van Weeren PR, Morales JA, Rodriguez LL, et al. (1999) Mortality supposedly due to intoxication by pyrrolizidine alkaloids from Heliotropium indicum in a horse population in Costa Rica: A case report. Vet Q 21(2):59-62.

HELIOTROPIUM LASIOCARPUM Fisch. & C. A. Mey. [Boraginaceae]

Common Names:

heliotrope

Citations:

Chauvin P, Dillon JC, Moren A (1994) Épidémie d'intoxication alimentaire à l'héliotrope. Tadjikistan, Novembre 1992 - Mars 1993. Sante 4(4):263-268.

Chauvin P, Dillon JC, Moren A, et al. (1993) Heliotrope poisoning in Tadjikistan. Lancet 341:1663.

Culvenor CC, Edgar JA, Smith LW, et al. (1986) Heliotropium lasiocarpum Fisch and Mey identified as cause of veno-occlusive disease due to a herbal tea. Lancet 1:978.

Levin GS, Mal'tsev VM, Kalugina VI, et al. (1970) [Reproduction of chronic affection of the liver in dogs by feeding seeds of Heliotropium.] Patol Fiziol Eksp Ter 14(6):41-45.

HELIOTROPIUM OVALIFOLIUM Forssk. [Boraginaceae]

Citations:

Abu Damir H, Adam SE, Tartour G (1982) The effects of Heliotropium ovalifolium on goats and sheep. Br Vet J 138(6):463-472.

Creeper JH, Mitchell AA, Jubb TF, et al. (1999) Pyrrolizidine alkaloid poisoning of horses grazing a native heliotrope (Heliotropium ovalifolium). Aust Vet J 77(6):401-402.

HELIOTROPIUM SCOTTEAE Rendle [Boraginaceae]

Citations:

Wahome WM, Muchiri DJ, Mugera GM (1994) An acute toxicity study of Heliotropium scotteae Rendle in mice. Vet Hum Toxicol 38(4):295-297.

HELIOTROPIUM SUPINUM L. [Boraginaceae]

Citations:

Schoental R, Coady A (1968) The hepatotoxicity of some Ethiopian and East African plants, including some used in traditional medicines. East Afr Med J 45(8):577-580.

Wiltjer JC, Walker CE (1974) Rectal prolapse in cattle associated with pyrrolizidine alkaloid poisoning. Aust Vet J 50(12):579-580.

HELIOTROPIUM TERNATUM Vahl [Boraginaceae]

Common Names:

sali

Citations:

Dunham LJ, Sheets RH, Morton JF (1974) Proliferative lesions in cheek pouch and esophagus of hamsters treated with plants from Curacao, Netherland Antilles. J Natl Cancer Instit 53(5):1259-1269.

HELIPTERUM CHARSLEYAE F. Muell. [Asteraceae]

Citations:

Burry JN, Kuchel R, Reid JG, et al. (1973) Australian bush dermatitis: Compositae dermatitis in South Australia. Med J Aust 1(3):110-116.

hellebore –see– *Helleborus foetidus* L.; *Helleborus niger* L.; *Veratrum album* L.; *Veratrum californicum* Durand; *Veratrum viride* Aiton

hellebore blanc –see– *Veratrum album* L.

hellébore noir –see– *Helleborus niger* L.

HELLEBORUS FOETIDUS L. [Ranunculaceae]

Common Names:

bar foot; Bärenfuß; bear's-foot; eléboro fetido; heléboro; hellebore; hierba ballestera; hierba-de-los-ballesteros; hierba llavera; ox heel; pie-de-grifo; setterwort; Stinkende Nieswurz; stinking hellebore; stinkwort; wilde Christwurz

Citations:

Holliman A, Milton D (1990) Helleborus foetidus poisoning of cattle. Vet Rec 127(13):339-340.

HELLEBORUS NIGER L. [Ranunculaceae]

Common Names:

bear's-foot; black hellebore; Christblume; Christmas rose; Christrose; Christ's-wort; Christwurz; Easter rose; hellebore; hellébore noir; Nieswurz; rose-de-nöel; Schneerose; Schwarze Neiswurz

Citations:

Fürth E (1905) Ueber eine Vergiftung mit Helleborus niger. Med Klin 1(14):330-331.

HELLEBORUS VIRIDIS L. [Ranunculaceae]

Common Names:

Bärenfuß; bastard hellebore; bear's-foot; boar's-foot; green hellebore; Grüne Christwurz; Grüne Nieswurz; Nieswurz

Citations:

Berselli L (1936) Contributo alla casistica degli avvelenamenti da elleboro verde nei bovini. Nuova Vet 14(7):21-22.

Bossi M, Brambilla G, Cavalli A, et al. (1981) Aritmia minacciosa da non comune intossicazione alimentaire. G Ital Cardiol 11(12):2254-2257.

Johnson CT, Routledge JK (1971) Suspected Helleborus viridis poisoning of cattle. Vet Rec 89(7):202.

Hellerkraut –see– *Thlaspi arvense* L.

helmet flower –see– *Aconitum napellus* L.; *Scutellaria lateriflora* L.

helonias –see– *Chamaelirium luteum* (L.) A. Gray

HEMEROCALLIS DUMORTIERI C. Morren [Hemerocallidaceae]

Common Names:

day lily

Citations:

Hadley RM, Richardson JA, Gwaltney-Brant SM (2003) A retrospective study of daylily toxicosis in cats. Vet Hum Toxicol 45(1):38-39.

Hemerocallis flava (L.) L. = Hemerocallis lilioasphodelus L.

Hemerocallis graminea Andrews = Hemerocallis minor Mill.

HEMEROCALLIS LILIOASPHODELUS L. [Hemerocallidaceae]

Synonyms:

Hemerocallis flava (L.) L.

Common Names:

custard day lily; day lily; orange day lily; pale day lily; tawny day lily; tawny lily; lemon day lily; tall yellow day lily

Citations:

Der Marderosian AH, Giller FB, Roia FC Jr (1976) Phytochemical and toxicological screening of household ornamental plants potentially toxic to humans. I. J Toxicol Environ Health 1(6):939-953.

HEMEROCALLIS MINOR Mill. [Hemerocallidaceae]

Synonyms:

Hemerocallis graminea Andrews

Common Names:

day lily; dwarf yellow day lily; grass-leaf day lily

Citations:

Chen HT, Jing PM, Yan XY (1987) [Poisoning by roots of Hemerocallis minor Mill. in sheep: Changes in the optic tract and pathogenesis.] Acta Vet Zootech Sin 18(4):256-261.

Hadley RM, Richardson JA, Gwaltney-Brant SM (2003) A retrospective study of daylily toxicosis in cats. Vet Hum Toxicol 45(1):38-39.

Hemerocallis sieboldii Paxton = Hosta sieboldii (Paxton) J. W. Ingram

HEMIDESMUS INDICUS (L.) W. T. Aiton [Apocynaceae]

Common Names:

anantamul; eramusu; Indian sarsaparilla

Citations:

Arseculeratne SN, Gunatilaka AA, Panabokke RG (1985) Studies on medicinal plants of Sri Lanka. Part 14. Toxicity of some traditional medicinal herbs. J Ethnopharmacol 13:323-335.

HEMIPOGON SETACEUS Decne. [Asclepiadaceae]

Citations:

Andrade SO, Camargo WV, Fernandes N (1963) II. Investigações sôbre plantas tóxicas no Estado de São Paulo. Arq Inst Biol (Sao Paulo) 30(Oct):189-203.

hemlock –see– *Conium maculatum* L.

hemlock dropwort –see– *Oenanthe crocata* L.

hemlock water dropwort –see– *Oenanthe crocata* L.

hemlock water parsnip –see– *Sium suave* Walter

hemp –see– *Cannabis sativa* L.

hemp agrimony –see– *Ageratina adenophora* (Spreng.) R. M. King & H. Rob.

hemp dogbane –see– *Apocynum cannabinum* L.

hemp sesbania –see– *Sesbania exaltata* (Raf.) Rydb. ex A. W. Hill

henbane –see– *Hyoscyamus niger* L.

henbit –see– *Lamium amplexicaule* L.

henbit dead-nettle –see– *Lamium amplexicaule* L.

henna –see– *Lawsonia inermis* L.

hennebane –see– *Hyoscyamus niger* L.

HERACLEUM DULCE Fisch. [Apiaceae]

Citations:

Efremov IA (1961) [Photodermatitis caused by Heracleum dulce.] Vestn Dermatol Venerol 35(3):64-66.

Heracleum giganteum auct. = Heracleum stevenii Manden.

Heracleum laciniatum auct. = Heracleum stevenii Manden.

HERACLEUM MANTEGAZZIANUM Sommier & Levier [Apiaceae]

Common Names:

Bärenklau; cow parsley; Fiesenkerbel; giant cow parsnip; giant hogweed; giant parsnip; giant Russian hogweed; grande berce; Herkuleskraut; Herkulesstaude; hogweed; Mantegazzis Bärenklau; Riesenbärenklau; Riesenherkulestaude; wild parsnip; wild rhubarb

Citations:

Alderfligel C (1974) Cas d'intoxication par la berce geante. Bull Med Leg Toxicol Med 17(4):278-279.

Andrews AH, Giles CJ, Thomsett LR (1985) Suspected poisoning of a goat by giant hogweed. Vet Rec 116(8):205-207.

Camm E, Buck HW, Mitchell JC (1976) Phytophotodermatitis from Heracleum mantegazzianum. Contact Dermatitis 2(2):68-72.

Dolowy WC (1996) Giant hogweed photodermatitis in two dogs in Bellevue, Washington. J Am Vet Med Assoc 209(4):722.

Drever JC, Hunter JA (1970) Giant hogweed dermatitis. Scott Med J 15(9):315-319.

Francis H (1970) Man against the giant hogweed. Lancet 2(666):269.

Harwood DG (1985) Giant hogweed and ducklings. Vet Rec 116(11):300.

Hintermann J (1967) Dermatose chez une chienne due a <<Heracleum mantegazzianum Somm. et Levier>>. Schweiz Arch Tierheilkd 109(12):654-656.

Jones JG, Russell DG (1968) Giant hogweed dermatitis. Practitioner 200(199):704-706.

Kosenow W, Kneiknecht A, Schürmann C (1983) Phytophototoxische Dermatitis durch Riesenbarenklau bzw. Herkulesstaude (Heracleum mantegazzianum). Pädiat Prax 29:115-123.

Miescher G, Burckhardt W (1937) Herakleumdermatitis. Schweiz Med Wochenschr 67:82-83.

Prinz L, Köstler H (1976) Ein Bericht über 3 Fälle von toxischer Phytophotodermatitis durch Heracleum mantegazzianum (Riesenherkulesstaude). Dermatol Monatsschr 162(11):881-886.
Sauter C (2001) Heracléum mantegázzianum: Photosensibilisierung. Schweiz Rundsch Med Prax 90(1-2):73-74.
Schulz KH, Spier HW (1951) Pflanzendermatitis infolge Photosensibilisierung durch Bärenklau (Heracleum mantegazzianum). Hautarzt 2:77-78.
Smellie JH (1968) Giant hogweed. Br Med J 3(5610):123.
Tiedemann A, Schultze H (1987) Bullöse Fotodermatitis nach Kontakt mit Stengelsaft des Riesenbärenklau (Heracleum mantegazzianum). Z Arztl Fortbild (Jena) 81(5):235-236.

HERACLEUM SOSNOWSKYI Manden. [Apiaceae]

Common Names:

cow parsnip; Sosnovsky's-cow parsnip

Citations:

Maksakova GP (1978) [A case of contact dermatitis caused by Sosnovsky's-cow parsnip.] Vestn Dermatol Venerol 8:48-49.

HERACLEUM SPHONDYLIUM L. [Apiaceae]

Common Names:

cow parsnip; Heilkraut; hogweed; Wiesenbärenklau

Citations:

Kunkel O (1954) Dermatitis durch Heracleum sphondylium. Med Welt 31-32:1050.
Michon P, Dornier R, Metz J, et al. (1956) Dermatite purpurique d'origine végétale par Heracleum sphondylium. Rev Med Nancy 81(Nov):924-929.

HERACLEUM STEVENII Manden. [Apiaceae]

Synonyms:

Heracleum giganteum auct.; ***Heracleum laciniatum*** auct.

Common Names:

cow parsnip; tromso palm

Citations:

Kavli G, Midelfart K, Raa J, et al. (1983) Phototoxicity from furocoumarins (psoralens) of Heracleum laciniatum in a patient with vitiligo. Action spectrum studies on bergapten, pimpinellin, angelicin and sphondin. Contact Dermatitis 9(5):364-366.
Kavli G, Volden G, Midelfart K, et al. (1983) In vivo and in vitro phototoxicity of different parts of Heracleum laciniatum. Contact Dermatitis 9(4):269-273.
Kavli G, Volden G, Midelfart K, et al. (1983) Phototoxicity of Heracleum laciniatum. Case reports and experimental studies. Contact Dermatitis 9(1):27-32.
Kavli G, Volden G, Raa J (1982) Accidental induction of photocontact allergy to Heracleum laciniatum. Acta Derm Venereol 62(5):435-438.
Pátková V (1984) Dermatitida po bolševníku. Cesk Pediatr 39(2):118-119.

herb bennett –see– *Conium maculatum* L.; *Valeriana officinalis* L.

herb John –see– *Hypericum perforatum* L.

herb mercury –see– *Mercurialis annua* L.; *Mercurialis perennis* L.

herb-of-grace –see– *Ruta graveolens* L.; *Verbena officinalis* L.

herb-of-repentance –see– *Ruta graveolens* L.

herb-of-the-cross –see– *Verbena officinalis* L.

herbe-à-sorcier –see– *Datura stramonium* L.

herbe-aux-grenouilles –see– *Equisetum palustre* L.

herbe-aux-gueux –see– *Clematis vitalba* L.

herbe-aux-magiciens –see– *Mandragora officinarum* L.

herbe-aux-verrues –see– *Heliotropium europaeum* L.

herbe-de-bicho –see– *Persicaria punctata* (Elliott) Small

herbe-de-grace –see– *Ruta graveolens* L.

herbe-de-St. Fiacre –see– *Heliotropium europaeum* L.

herbe-de-St. Innocent –see– *Persicaria hydropiper* (L.) Spach

Herbe-de-St. Jacques –see– *Jacobaea vulgaris* Gaertn.

herbe-des-démoniaques –see– *Datura stramonium* L.

herbe-du-diable –see– *Datura stramonium* L.

herbe-du-siège –see– *Scrophularia auriculata* L.

Herbstblume –see– *Colchicum autumnale* L.

Herbstzeitlose –see– *Colchicum autumnale* L.

Herkuleskraut –see– *Heracleum mantegazzianum* Sommier & Levier

Herkulesstaude –see– *Heracleum mantegazzianum* Sommier & Levier

HERTIA CLUYTIIFOLIA (DC.) Kuntze [Asteraceae]

Synonyms:

Othonna cluytiifolia (DC.) Sch. Bip

Common Names:

entjiebos

Citations:

Van der Walt SJ, Steyn DG (1943) Recent investigations into the toxicity of plants, etc. XIII. Onderstepoort J Vet Sci Anim Indus 18(1-2):207-224.

HERTIA PALLENS (DC.) Kuntze [Asteraceae]

Synonyms:

Othonna pallens DC.

Common Names:

dikkopbos; springbokbos; springbokbossie; Vaalbos

Citations:

Prozesky L, Kellerman TS, Jordaan P, et al. (1985) An ovine hepatotoxicosis caused by the plant Hertia pallens (DC.) Kuntze (Asteraceae). Onderstepoort J Vet Res 52(4):233-238.

Steyn DG (1937) Recent investigations into the toxicity of known and unknown poisonous plants in the Union of South Africa. VII. Onderstepoort J Vet Sci Anim Indus 9(1):111-124.

Van der Walt SJ, Steyn DG (1943) Recent investigations into the toxicity of plants, etc. XIII. Onderstepoort J Vet Sci Anim Indus 18(1-2):207-224.

Heterodendrum oleifolium Desf. = Alectryon oleifolius (Desf.) S. T. Reynolds

Heteromeles arbutifolia M. Roem = Heteromeles salicifolia (C. Presl) Abrams

HETEROMELES SALICIFOLIA (C. Presl) Abrams [Rosaceae]

Synonyms:

Heteromeles arbutifolia M. Roem

Common Names:

California holly; California toyon; Christmas berry; tollon; toyon

Citations:

Tegzes JH, Puschner B, Melton LA (2003) Cyanide toxicosis in goats after ingestion of California holly (Heteromeles arbutifolia). J Vet Diagn Invest 15(5):478-480.

HETEROPHYLLAEA PUSTULATA Hook. f. [Rubiaceae]

Common Names:

cegadera

Citations:

Hansen EW, Martiarena CA (1967) Contribución al estudio de la toxicidad de Heterophyllaea pustulata Hook "Cegadera" en el ganado dermatitis. Queratoconjunctivitis tóxica experimental en especies animales receptivas. Rev Inv Ag 4th Ser 4(7):81-113.

heuile-de-camphre –see– *Cinnamomum camphora* (L.) J. Presl

heuvo vegetal –see– *Blighia sapida* K. D. Koenig

HEVEA BRASILIENSIS (Willd. ex A. Juss.) Müll. Arg. [Euphorbiaceae]

Common Names:

rubber

Citations:

Rihs HP, Dumont B, Rozynek P, et al. (2003) Molecular cloning, purification, and IgE-binding of a recombinant class I chitinase from Hevea brasiliensis leaves (rHev b 11.0102). Allergy 58(3):246-251.

Hexham scent –see– *Melilotus indicus* (L.) All.

hibbert bean –see– *Phaseolus lunatus* L.

Hibiscus esculentus L. = Abelmoschus esculentus (L.) Moench

HIBISCUS ROSA-SINENSIS L. [Malvaceae]

Common Names:

China rose; japa; shoe flower

Citations:

Paulsen E (1998) Occupational dermatitis in Danish gardeners and greenhouse workers (II). Etiological factors. Contact Dermatitis 38(1):14-19.

Hicks yew –see– *Taxus ×media* Rehder

hiedra –see– *Hedera helix* L.; *Toxicodendron diversilobum* (Torr. & A. Gray) Greene

hierba ballestera –see– *Helleborus foetidus* L.; *Veratrum album* L.

hierba-de-cancer –see– *Acalypha indica* L.

hierba-de-la-India –see– *Ipomoea carnea* Jacq.

hierba-de-los-ballesteros –see– *Helleborus foetidus* L.

hierba-de-los-pordioseros –see– *Clematis vitalba* L.

hierba-de-pasmo –see– *Baccharis pteronioides* DC.

hierba-de-San Nicolas –see– *Gutierrezia sarothrae* (Pursh) Britton & Rusby

hierba-de-santiago –see– *Jacobaea vulgaris* Gaertn.

hierba-de-vibora –see– *Gutierrezia sarothrae* (Pursh) Britton & Rusby

hierba-de-Zorro –see– *Lantana camara* L.

hierba-del-duende –see– *Euphorbia heterophylla* L.

hierba-des-pescado –see– *Plumbago scandens* L.

hierba llavera –see– *Helleborus foetidus* L.

hierba mora –see– *Solanum nigrum* L.

hierbe-de-la-flecha –see– *Sapium biloculare* (S. Watson) Pax

hierbe mala –see– *Sapium biloculare* (S. Watson) Pax

high laurel –see– *Kalmia latifolia* L.

higuera-del-infierno –see– *Datura stramonium* L.

higuera loca –see– *Datura stramonium* L.

higuerilla –see– *Ricinus communis* L.

hikamas –see– *Pachyrhizus erosus* (L.) Urb.

Hilaria rigida (Thurb.) Benth. ex Scribn. = Pleuraphis rigida Thurb.

hills-of-snow –see– *Hydrangea macrophylla* (Thunb.) Ser.

Himmelschwertel –see– *Iris ×germanica* L.; *Iris pseudoacorus* L.

hinchador –see– *Toxicodendron striatum* (Ruiz & Pav.) Kuntze

hinchahuevos –see– *Toxicodendron striatum* (Ruiz & Pav.) Kuntze

Hindu datura –see– *Datura metel* L.

hipazote –see– *Dysphania ambrosioides* (L.) Mosyakin & Clemants

HIPPOMANE MANCINELLA L. [Euphorbiaceae]

Common Names:

arbol-de-la-muerte; beach apple; le mancenilier; mancenillier; manchineal; manchineel; manzanilla; Manzanillabaum; manzanillo; manzanillo-de-la-costa; manzanillo-de-playa; pinipiniche; shore apple

Citations:

Botterel F, Brun S, Bourée P (2000) Dermite du mancenillier. Presse Med 29(2):81.

Bygbjerg IC, Johansen HK (1991) Manzinellaforgiftning kompliceret med streptokokpharyngytis og impetigo. Ugeskr Laeger 154(1):27-28.

Caddy DJ (1894) Poisoning by manchineel. Lancet 2(Dec 22):1478-1479.

Chareyre S, Meram D, Descotes J (1991) Intoxication par le mancenillier. A propos d'un cas. J Toxicol Clin Exp 11(1):59-61.

Grana PC (1946) Conjunctivitis and dermatitis due to "Beach Apple." Report of thirteen cases. Arch Ophthalmol 35:421-422.

Lowe NJ (1974) Primary irritant dermatitis from the manchineel tree (Hippomane mancinella) in the eastern Caribbean. J R Nav Med Serv 60:133-136.

Mauze J, Arnaud G (1953) A propos des accidents provoques par le mancenilier des Antilles. Bull Soc Pathol Exot 46(4):496-498.

Pitts JF, Barker NH, Gibbons DC, et al. (1993) Manchineel keratoconjunctivitis. Br J Ophthalmol 77(5):284-288.

Satulsky EM (1943) Dermatitis venenata caused by the manzanillo tree. Arch Derm Syphilol 47(36):36-38.

Satulsky EM, Wirts CA (1943) Dermatitis venenata caused by the manzanillo tree. Further observations and report of sixty cases. Arch Derm Syphilol 47:797-798.

Snow JS, Harley RD (1944) Dermatitis venenata and keratoconjunctivitis caused by the manzanillo tree. Arch Derm Syphilol 49:236-239.

Hirschdorn –see– *Rhamnus cathartica* L.

HIRSCHFELDIA INCANA (L.) Lagr.- Foss. [Brassicaceae]

Common Names:

mustard

Citations:

Lanson RK, Abdulla A (1963) Effects of feeding mustard seed to immature chickens and laying hens. Poult Sci 42:1283-1284.

Hirtentäschel –see– *Capsella bursa-pastoris* (L.) Medik.

HISTIOPTERIS INCISA (Thunb.) J. Sm. [Dennstaedtiaceae]

Common names:

bat's-wing fern; water fern

Citations:

Saito K, Nagao T, Takatsuki S, et al. (1990) The sesquiterpenoid carcinogen of bracken fern, and some analogues, from the Pteridaceae. Phytochemistry 29:1475-1479.

ho-shou-wu –see– *Fallopia multiflora* (Thunb.) Haraldson

hoary alyssum –see– *Berteroa incana* (L.) DC.

hoary cress –see– *Lepidium draba* L. var. draba

hoary false alyssum –see– *Berteroa incana* (L.) DC.

hoary nettle –see– *Urtica dioica* L.

hog apple –see– *Podophyllum peltatum* L.

hog brake –see– *Pteridium aquilinum* (L.) Kuhn

hog gum –see– *Metopium toxiferum* (L.) Krug & Urb.

hog millet –see– *Panicum miliaceum* L.

hogbane –see– *Hyoscyamus niger* L.

hogbean –see– *Hyoscyamus niger* L.

hogweed –see– *Amaranthus spinosus* L.; *Ambrosia artemisiifolia* L.; *Heracleum mantegazzianum* Sommier & Levier; *Heracleum sphondylium* L.

Hoher Rittersporn –see– *Delphinium elatum* L.

hoja-de-San Pablo –see– *Wigandia urens* (Ruiz & Pav.) Kunth var. caracasana (Kunth) D. N. Gibson

hojase –see– *Flourensia cernua* DC.

hoko –see– *Phytolacca dodecandra* L'Her.

Holcus halepense L. = Sorghum halepense (L.) Pers.

Holcus sorghum L. = Sorghum bicolor (L.) Moench

Holcus sudanensis (Piper) L. H. Bailey = Sorghum ×drummondii (Steud.) Millsp. & Chase

Holder –see– *Sambucus nigra* L.

HOLIGARNA FERRUGINEA Marchand [Anacardiaceae]

Citations:

Kulkarni SB, Srinivas CR, Krupashankar DS, et al. (1987) Chemical characterization and clinical studies on the allergenic principles of Holigarna ferruginea March (Anacardiaceae). Indian Drugs 24(4):180-182.

Srinivas CR, Kulkarni SB, Menon SK, et al. (1987) Allergenic agent in contact dermatitis from Holigarna ferruginea. Contact Dermatitis 17(4):219-222.

holly –see– *Ilex aquifolium* L.; *Ilex opaca* Aiton

hollyhock –see– *Alcea rosea* L.

holm oak –see– *Quercus ilex* L.

HOLOCALYX BALANSAE Micheli [Fabaceae]

Synonyms:

Holocalyx glaziovii Taub. ex Glaz.

Common Names:

alecrim; alecrim-de-campinas

Citations:

Bicudo PL (1978) Intoxicação experimental de bovinos pelo "Alecrim de Campinas," Holocalyx glaziovii, Taub. Arq Esc Vet Univ Fed Minas Gerais 30:369-370.

Souza JA, Martins EO, Zezza Neto L (1972) Ação da hidroxocobalamina na intoxicação aguda do coelho pelo Holocalix balansae Mich. Rev Fac Med Vet Zootec Univ Sao Paulo 9(1):159-164.

Souza JA, Muelier SB, Martins EO (1971) Ação da hidroxocobalamina na intoxicação aguda do coelho pelo Holocalix balansae, Mich. Rev Fac Med Vet Sao Paulo 8(3):721-727.

Holocalyx glaziovii Taub. ex Glaz. = Holocalyx balansae Micheli

holy herb –see– *Verbena officinalis* L.

holy thistle –see– *Silybum marianum* (L.) Gaertn.

holy tree –see– *Melia azedarach* L.

Homalobus campestris Torr. & A. Gray = Astragalus convallarius Greene

Homalobus clementis Rydb. = Astragalus tenellus Pursh

Homalobus multiflorus (Pursh) Torr. & A. Gray = Astragalus tenellus Pursh

Homalobus stipitatus Rydb. = Astragalus tenellus Pursh

Homalobus strigulosus Rydb. = Astragalus tenellus Pursh

Homeria glauca (Wood & Evans) N. E. Br. = Moraea pallida (Baker) Goldblatt

Homeria miniata (Andrews) Sweet = Moraea miniata Andrews

Homeria pallida Baker = Moraea pallida (Baker) Goldblatt

Homeria pura N. E. Br. = Moraea pallida (Baker) Goldblatt

hondala –see– *Adenia hondala* (Gaertn.) W. J. de Wilde

honde oor –see– *Cotyledon orbiculata* L.

Honduran mahogany –see– *Swietenia macrophylla* King

Honduran walnut –see– *Metopium brownei* (Jacq.) Urb.

honey mesquite –see– *Prosopis glandulosa* Torr.; *Prosopis juliflora* (Sw.) DC.

honey-of-Trebizond –see– *Rhododendron ponticum* L.

honey plant –see– *Scrophularia marilandica* L.

hong hua –see– *Carthamus tinctorius* L.

hoodwort –see– *Scutellaria lateriflora* L.

Hooker's-larkspur –see– *Delphinium glaucum* S. Watson

hop –see– *Humulus lupulus* L.

hop sage –see– *Grayia spinosa* (Hook.) Moq.

hop tree –see– *Ptelea trifoliata* L. subsp. angustifolia (Benth.) V. L. Bailey

Hopfen –see– *Humulus lupulus* L.

hophorn bean –see– *Ostrya virginiana* (Mill.) K. Koch

hops –see– *Humulus lupulus* L.

HORDEUM JUBATUM L. [Poaceae]

Common Names:

foxtail; foxtail barley; skunkgrass; squirreltail; squirreltail barley; squirreltailgrass; ticklegrass; wild barley; wild foxtail

Citations:

Fleming CE, Peterson NF (1919) Don't feed foxtail hay to lambing ewes! Nevada Agric Exp Sta Bull #97:18 pp.

HORDEUM VULGARE L. [Poaceae]

Common Names:

barley; Gerste

Citations:

Chinn WT (1963) Pneumonia in barley-fed cattle. Vet Rec 75(9):256-257.

Cronin E (1979) Contact dermatitis from barley dust. Contact Dermatitis 5(3):196.

Harris AH (1962) Apparent hazards of high barley feeding to cattle. Vet Rec 74(49):1434.

Hickey TF (1963) Mortality in barley-fed cattle. Vet Rec 75(8):210-211.

Horrox MA (1962) Mortality in barley-fed cattle. Vet Rec 74(52):1551.

Martin B (1963) Mortality in barley-fed cattle. Vet Rec 75(3):78-79.

Nisbet A, Beddows AC, Jones NA (1963) Over-feeding of barley to sheep. Vet Rec 75(6):152.

Nolan FJ (1962) Mortality in barley-fed cattle. Vet Rec 74(48):1376-1377.

Pereira F, Rafael M, Lacerda MH (1998) Contact dermatitis from barley. Contact Dermatitis 39(5):261-262.

Phillips GD (1963) Better British beef and barley feed. Vet Rec 75(22):578-579.

Preston TR (1963) Symposium: Recent innovations in calf-rearing. II. Barley-beef production. Vet Rec 75(51):1399-1402.

Rix JC (1966) Poisoning from excess intake of moist stored barley. Vet Rec 78(16):574.

Simonsson A, Bjorklund NE (1978) Some effects of the fineness of ground barley on gastric lesions and gastric contents in growing pigs. Swedish J Agric Res 8:97-106.

Solomons B (1971) Sensitization to oats and barley. Contact Dermatol Newsl 10(Jul):231.

Thompson AG (1925) Barley itch. Br Med J 1:71.

Wills WK (1909) Barley itch. Br J Dermatol 21:249-252.

Note:

Barley is named *Hordeum vulgare* L. subsp. *vulgare* in some publications.

horehound –see– *Marrubium vulgare* L.

Hornklee –see– *Lotus corniculatus* L.

hornseed buttercup –see– *Ceratocephala testiculata* (Crantz) Roth

Horn's-milk vetch –see– *Astragalus hornii* A. Gray

horse apple –see– *Maclura pomifera* (Raf.) C. K. Schneid.

horse balm –see– *Collinsonia canadensis* L.

horse chestnut –see– *Aesculus californica* (Spach) Nutt.; *Aesculus glabra* Willd.; *Aesculus hippocastanum* L.; *Aesculus pavia* L.

horse gam –see– *Lablab purpureus* (L.) Sweet

horse mango –see– *Mangifera foetida* Lour.

horse millet –see– *Pennisetum glaucum* (L.) R. Br.

horse mint –see– *Mentha ×piperita* L. nothosubsp. citrata (Ehrh.) Briq.

horse nettle –see– *Solanum carolinense* L.; *Solanum dimidiatum* Raf.

horse pipe –see– *Equisetum arvense* L.

horse purslane –see– *Trianthema portulacastrum* L.

horse sorrel –see– *Rumex acetosella* L.

horse tamarind –see– *Leucaena leucocephala* (Lam.) de Wit

horsebane –see– *Oenanthe aquatica* (L.) Poir.

horsebean –see– *Canavalia ensiformis* (L.) DC.; *Lablab purpureus* (L.) Sweet; *Lupinus leucophyllus* Douglas ex Lindl.; *Strychnos ignatii* P. J. Bergius; *Vicia faba* L.

horsebrush –see– *Tetradymia canescens* DC.; *Tetradymia glabrata* Torr. & A. Gray

horseradish –see– *Armoracia rusticana* P. Gaertn. et al.

horseradish tree –see– *Moringa oleifera* Lam.

horsetail –see– *Equisetum arvense* L.; *Equisetum fluviatile* L.; *Equisetum palustre* L.; *Equisetum ramosissimum* Desf.

horsetail milkweed –see– *Asclepias subverticillata* (A. Gray) Vail; *Asclepias verticillata* L.

horsetail tree –see– *Casuarina equisetifolia* L.

horseweed –see– *Conyza bonariensis* (L.) Cronquist; *Collinsonia canadensis* L.

hortensia –see– *Hydrangea macrophylla* (Thunb.) Ser.

HOSLUNDIA OPPOSITA Vahl [Lamiaceae]

Citations:

Verdcourt B, Trump EC (1969) Common poisonous plants of East Africa. Collins. London.

HOSTA SIEBOLDII (Paxton) J. W. Ingram [Aguacaceae]

Synonyms:

Hemerocallis sieboldii Paxton

Common Names:

day lily

Citations:

Hadley RM, Richardson JA, Gwaltney-Brant SM (2003) A retrospective study of daylily toxicosis in cats. Vet Hum Toxicol 45(1):38-39.

hot pepper –see– *Capsicum annuum* L.; *Capsicum frutescens* L.

hotsutsuji –see– *Elliottia paniculata* (Siebold & Zucc.) Benth. & Hook. f.

Hottentot poison bush –see– *Acokanthera oppositifolia* (Lam.) Codd; *Acokanthera schimperi* (A. DC.) Oliv.

houndberry –see– *Solanum nigrum* L.

hound's-tongue –see– *Cynoglossum officinale* L.

houx –see– *Ilex aquifolium* L.

hoya –see– *Hoya carnosa* (L. f.) R. Br.

HOYA AUSTRALIS R. Br. ex J. Traill [Apocynaceae]

Common Names:

wax flower

Citations:

Legg J, White CT (1939) Hoya australis (wax flower): A native plant poisonous to stock. Queensland Agric J 51:17-19.

Legg J, White CT (1939) Hoya australis: A plant poisonous to stock. Aust Vet J 15:34-36.

HOYA CARNOSA (L. f.) R. Br. [Apocynaceae]

Common Names:

hoya; wax flower; waxplant

Citations:

Kuliszkiewicz-Janus M, Klinger M (1987) Przełom hemolityczny powikłany ostrą niewydolnością nerek u heterozygotycznej kobiety z niedoborem dehydrogenazy glukozo-6-fosforamowej wywołany nadwrazliwością na woskownicę mięsistą (Hoya carnosa). Pol Arch Med Wewn 78(1):43-47.

Kuliszkiewicz-Janus M, Tyran W, Szajerka G (1992) Haemolytic crises caused by Hoya carnosa in a patient with G6PD deficiency. Acta Haematol Pol 23(1):63-67.

Rothe A (1986) Hoya carnosa - Is it allergenic? Contact Dermatitis 14(4):250-252.

hualtaco tree –see– *Loxopterygium huasango* Spruce ex Engl.

huao –see– *Comocladia dentata* Jacq.; *Metopium brownei* (Jacq.) Urb.; *Metopium toxiferum* (L.) Krug & Urb.

huasango tree –see– *Loxopterygium huasango* Spruce ex Engl.

hubam –see– *Melilotus albus* Medik.

huckleberry –see– *Solanum nigrum* L.

huele-de-noche –see– *Cestrum nocturnum* L.

huevil –see– *Vestia foetida* (Ruiz & Pav.) Hoffmanns.; *Vestia lycioides* Willd.

Huflattich –see– *Tussilago farfara* L.

hug melose –see– *Achyranthes aspera* L.

Hühnertod –see– *Hyoscyamus niger* L.

huile-de-menthe-pouliot –see– *Mentha pulegium* L.

huile-de-persil-sauvage –see– *Petroselinum crispum* (Mill.) Nyman ex A. W. Hill

huile-de-sassafras-américain –see– *Sassafras albidum* (Nutt.) Nees

huile-de-sauge –see– *Salvia officinalis* L.

huile d'hédéome –see– *Hedeoma pulegioides* (L.) Pers.

huilihuiste –see– *Karwinskia calderonii* Standl.

Hülse –see– *Ilex aquifolium* L.

hulst –see– *Ilex aquifolium* L.

humansdorp –see– *Chironia baccifera* L.

HUMULUS LUPULUS L. [Cannabaceae]

Common Names:

brewer's-hop; European hop; hop; Hopfen; hops

Citations:

Cookson JS, Lawton A (1953) Hop dermatitis in Herefordshire. Br Med J 2(4832):376-379.

Duncan KL, Hare WR, Buck WB (1997) Malignant hyperthermia-like reaction secondary to ingestion of hops in five dogs. J Am Vet Med Assoc 210(1):51-54.

Estrada JL, Gozalo F, Cecchini C, et al. (2002) Contact urticaria from hops (Humulus lupulus) in a patient with previous urticaria–angioedema from peanut, chestnut and banana. Contact Dermatitis 46(2):127.

Galbraith SN (1924) Occupational diseases of hop-picking. Lancet 2:885.

Newmark FM (1978) Hops allergy and terpene sensitivity: An occupational disease. Ann Allergy 41(5):311-312.

O'Donovan WJ (1924) Hop dermatitis. Lancet 2:597-598.

Raith L, Jager K (1984) Hop allergy. Contact Dermatitis 11(1):53.

Smithies BM (1929) The occupational diseases of hop-picking. Lancet 2:494-495.

Streich CJ (1924) Hop dermatitis. Lancet 2:727.

Hundsdill –see– *Aethusa cynapium* L.

Hundskamille –see– *Anthemis cotula* L.; *Chamaemelum nobile* (L.) All.

Hundskraut –see– *Mercurialis perennis* L.

Hundspetersilie –see– *Aethusa cynapium* L.

Hundswolfkraut –see– *Euphorbia helioscopia* L.

Hundszunge –see– *Cynoglossum officinale* L.

hunter's-robe –see– *Epipremnum pinnatum* (L.) Engl.

HUPERZIA SAURURUS (Lam.) Trevis. [Lycopodiaceae]

Synonyms:

Lycopodium saururus Lam.

Common Names:

cola-de-quirquincho

Citations:

Ciganda C, Laborde A (2001) Herbal infusions used for induced abortion. J Toxicol Clin Toxicol 39(3):318-319.

Ciganda C, Laborde A (2003) Herbal infusions used for induced abortion. J Toxicol Clin Toxicol 41(3):235-239.

HUPERZIA SELAGO (L.) Bernh. ex Schrank & Mart. [Lycopodiaceae]

Synonyms:

Lycopodium selago L.

Common Names:

fir club moss

Citations:

Felgenhauer N, Zilker T, Worek F, et al. (2000) Intoxication with huperzine A, a potent anticholinesterase found in the fir club moss. J Toxicol Clin Toxicol 38(7):803-808.

Laborde A, Ciganda C (1998) Poisoning by herbal infusions ingested as abortifacient agents. J Toxicol Clin Toxicol 36(5):454-455.

HUPERZIA SERRATA (Thunb.) Trevis. [Lycopodiaceae]

Synonyms:

Lycopodium serratum Thunb.

Common Names:

jin-bu-huan

Citations:

Picciotto A, Campo N, Brizzolara R, et al. (1998) Chronic hepatitis induced by jin bu huan. J Hepatol 28(1):165-167.

Woolf GM, Petrovic LM, Rojter SE, et al. (1994) Acute hepatitis associated with the Chinese herbal product jin bu huan. Ann Intern Med 121(10):729-735.

HURA CREPITANS L. [Euphorbiaceae]

Common Names:

assacu; javillo; molinillo; monkey dinner bell; monkey pistol; possumwood; sandbox; tronador; West Indian sandbox tree

Citations:

Swaddiwudhipong W, Kunasol P, Sangwanloy O, et al. (1989) Foodborne disease outbreaks of chemical etiology in Thailand, 1981-1987. Southeast Asian J Trop Med Public Health 20(1):125-132.

hurab ekhowsa –see– *Acanthospermum hispidum* DC.

husk tomato –see– *Physalis longifolia* Nutt.

hyacinth –see– *Hyacinthus orientalis* L.

hyacinth bean –see– *Lablab purpureus* (L.) Sweet

HYACINTHOIDES NON-SCRIPTA (L.) Chouard ex Rothm. [Hyacinthaceae]

Synonyms:

Endymion non-scriptus (L.) Garcke; ***Scilla non-scripta*** (L.) Hoffmanns. & Link

Common Names:

blue bells-of-England; blue bottle; bluebell; crake feet; crow bells; crow leek; culver keep; culverkeys; day's-leek; dog leek; English bluebell; hair bell; harebell; wild hyacinth

Citations:

Thursby-Pelham RH (1967) Suspected Scilla non-scripta (bluebell) poisoning in cattle. Vet Rec 80(24):709-710.

HYACINTHUS ORIENTALIS L. [Hyacinthaceae]

Common Names:

Anna Maria; Czar Pater; Czar Peter; Dutch hyacinth; hyacinth; Hyazinthe; oriental hyacinth; Roman hyacinth

Citations:

Lamminpaa A, Estlander J, Jolanki K, et al. (1996) Occupational allergic contact dermatitis caused by decorative plants. Contact Dermatitis 34(5):330-335.

Muñoz D, Urrutia I, Leanizbarrutia I, et al. (1989) Contact dermatitis from plants in a geriatric nurse. Contact Dermatitis 20(13):227-228.

Piirilä P, Hannu T, Keskinen H, et al. (1998) Occupational asthma to hyacinth. Allergy 53(3):328-329.

van der Werff PJ (1959) Occupational diseases among workers in the bulb industries. Acta Allergol 14:338-355.

van Dongen K (1945) Twee doodelijke vergiftigingen door het eten van hyacinthenbollen. Ned Tijdschr Geneeskd 89:223-225.

Hyazinthe –see– *Hyacinthus orientalis* L.

hybrid vetch –see– *Vicia villosa* Roth subsp. varia (Host) Corb.

Hybridklee –see– *Trifolium hybridum* L.

hydrangea –see– *Hydrangea macrophylla* (Thunb.) Ser.

Hydrangea hortensia Siebold = Hydrangea macrophylla (Thunb.) Ser.

HYDRANGEA MACROPHYLLA (Thunb.) Ser. [Hydrangaceae]

Synonyms:

Hydrangea hortensia Siebold

Common Names:

hills-of-snow; French hydrangea; hortensia; hydrangea; Popo-hau; sevenbark hydrangea

Citations:

Avenel-Audran M, Hausen BM, le Sellin J, et al. (2000) Allergic contact dermatitis from hydrangea – is it so rare? Contact Dermatitis 43(4):189-191.

Bruce EA (1920) Hydrangea poisoning. J Am Vet Med Assoc 58:313-315.

Bruynzeel DP (1986) Allergic contact dermatitis to hydrangea. Contact Dermatitis 14(2):128.

De Rooij J, Bruynzeel DP, Rustemeyer T (2006) Occupational allergic contact dermatitis from hydrangea. Contact Dermatitis 54(1):65-66.

Kuligowski ME, Chang A, Leemreize JH (1992) Allergic contact hand dermatitis from hydrangea: Report of a 10th case. Contact Dermatitis 26(4):269-270.

Note:

Hydrangea is named *Hydrangea macrophylla* (Thunb.) Ser. subsp. *macrophylla* in some publications.

Hydrocotyle asiatica L. = Centella asiatica (L.) Urb.

HYLOTELEPHIUM SPECTABILE (Boreau) H. Ohba [Crassulaceae]

Synonyms:

Sedum spectabile Boreau

Citations:

Végh E (1985) Dísznövény okozta mérgezés esete sertéseken. Magyar Allator Lapja 40(8):505-506.

Hymenocyclus smithii L. Bolus = Malephora smithii (L. Bolus) H. E. K. Hartmann

Hymenoxys floribunda (A. Gray) Cockerell = Hymenoxys richardsonii (Hook.) Cockerell

HYMENOXYS HOOPESII (A. Gray) Bierner [Asteraceae]

Synonyms:

Dugaldia hoopesii (A. Gray) Rydb.; ***Helenium hoopesii*** A. Gray

Common Names:

mountain sneezeweed; orange sneezeweed; sneezeweed; sunflower; western sneezeweed; yellow weed; yerba-del-lobo

Citations:

Buck WB, Binns W, James L, et al. (1961) Results of feeding of herbicide-treated plants to calves and sheep. J Am Vet Med Assoc 138(6):320-323.

Buck WB, James L, Binns W (1961) Changes in serum transaminase activities associated with plant and mineral toxicity in sheep and cattle. Proc Am Coll Vet Toxicol 1961:13-24.

Buck WB, James LF, Binns W (1961) Changes in serum transaminase activities associated with plant and mineral toxicity in sheep and cattle. Cornell Vet 51(Oct):568-585.

Marsh CD (1916) Cause of spewing sickness of sheep. U S Dep Agric Bur Anim Indus Doc. Oct 25:1-4.

Marsh CD, Clawson AB, Couch JF, et al. (1921) Western sneezeweed (Helenium hoopesii) as a poisonous plant. U S Dep Agric Bull #947:46 pp.

HYMENOXYS LEMMONII (Greene) Cockerell [Asteraceae]

Common Names:

alkali hymenoxys; Lemmon's-hymenoxys

Citations:

Fleming CE, Miller MR, Vawter LR, et al. (1934) Poisonous plants. Nevada Agric Exp Sta Annu Rep 1933:10-13.

HYMENOXYS ODORATA DC. [Asteraceae]

Synonyms:

Actinea odorata (DC.) Kuntze

Common Names:

annual bitterweed; bitter actinea; bitter rubberweed; bittersweet; bitterweed; bitterweed actinea; limonillo; rubberweed; western bitterweed

Citations:

Anonymous (1931) Investigation of stock-poisoning plants. U S Dep Agric Bur Anim Indus Chief Rep 1931:56-57.

Anonymous (1932) Investigation of stock-poisoning plants. U S Dep Agric Bur Anim Indus Chief Rep 1932:44.

Anonymous (1933) Investigation of stock-poisoning plants. U S Dep Agric Bur Anim Indus Chief Rep 1933:34-35.

Boughton IB, Hardy WT (1937) Toxicity of bitterweed (Actinea odorata) for sheep. Texas Agric Exp Sta Bull #552:15 pp.

Bridges GW, Bailey EM Jr, Camp BJ (1980) Prevention of bitterweed intoxication of sheep. Vet Hum Toxicol 22(2):87-90.
Calhoun MC, Ueckert DN, Livingston CW Jr, et al. (1981) Effects of bitterweed (Hymenoxys odorata) on voluntary feed intake and serum constituents of sheep. Am J Vet Res 42(10):1713-1717.
Calhoun MC, Ueckert DN, Livingston CW Jr, et al. (1982) Effect of 2,4-D on hymenoxon concentration and toxicity of bitterweed (Hymenoxys odorata) force-fed to sheep. J Range Manag 35(4):489-492.
Clawson AB (1931) "Bitterweed" poisoning of livestock. U S Dep Agric Mimeo Rep. Apr:1-3.
Clawson AB (1931) A preliminary report on the poisonous effects of bitter rubber weed (Actinea odorata) on sheep. J Agric Res 43(8):693-701.
Hardy WT, Cory VL, Schmidt H, et al. (1931) Bitterweed poisoning in sheep. Texas Agric Exp Sta Bull #433:18 pp.
Kim HL, Anderson AC, Herrig BW, et al. (1982) Protective effects of antioxidants on bitterweed (Hymenoxys odorata DC) toxicity in sheep. Am J Vet Res 43(11):1945-1950.
Kim HL, Anderson AC, Terry MK, et al. (1981) Protective effect of butylated hydroxyanisole on acute hymenoxon and bitterweed poisoning. Res Commun Chem Pathol Pharmacol 33(2):365-368.
Poage GW 3rd, Scott CB, Bisson MG, et al. (2000) Activated charcoal attenuates bitterweed toxicosis in sheep. J Range Manag 53(1):73-78.
Steel EG, Witzel DA, Blanks A (1976) Acquired coagulation factor X activity deficiency connected with Hymenoxys odorata DC (Compositae), bitterweed poisoning in sheep. Am J Vet Res 37(12):1383-1386.
Taylor CA, Ralphs MH (1992) Reducing livestock losses from poisonous plants through grazing management. J Range Manag 45(1):9-12.
Terry MK, Kim HL, Corrier DE, et al. (1981) The acute oral toxicity of hymenoxon in sheep. Res Commun Chem Pathol Pharmacol 31(1):181-184.
Witzel DA, Jones LP, Ivie GW (1977) Pathology of subacute bitterweed (Hymenoxys odorata) poisoning in sheep. Vet Pathol 14(1):73-78.
Witzel DA, Rowe LD, Clark DE (1974) Physiopathologic studies on acute Hymenoxys odorata (bitterweed) poisoning in sheep. Am J Vet Res 35(7):931-934.

HYMENOXYS RICHARDSONII (Hook.) Cockerell [Asteraceae]

Synonyms:

Actinea richardsonii (Hook.) Kuntze; ***Hymenoxys floribunda*** (A. Gray) Cockerell

Common Names:

Colorado rubber weed; pingree; pingue; rubberweed

Citations:

Aanes WA (1961) Pingue (Hymenoxys richardsonii) poisoning in sheep. Am J Vet Res 22:47-52.
Anonymous (1936) Investigations of stock-poisoning plants. U S Dep Agric Bur Anim Indus Chief Rep. 1936:44-45.
Pammel LH (1919) Pingue poisoning in Arizona. Am J Vet Med 14:32.

hyoscyamus –see– *Hyoscyamus niger* L.

HYOSCYAMUS ALBUS L. [Solanaceae]

Common Names:

saccharine blanche; white henbane; white saccharin

Citations:

Selmi H, Hezmiri H, Tabarki B, et al. (1996) Intoxication aigue par la saccharine blanche. A propos de 5 observations pédiatriques. Tunis Med 74(1):45-47.

HYOSCYAMUS NIGER L. [Solanaceae]

Common Names:

beleño negro; Bilsen; Bilsenbohne; Bilsenkraut; Bilsensee; black henbane; Bolmört; chenile; deli-bat-bat; devil's-eye; fetid nightshade; henbane; hennebane; hogbane; hogbean; Hühnertod; hyoscyamus; infidel opium; insane root; jusquiame; jusquiame noire; poison tobacco; Saukraut; Schwarzes Bilsenkraut; stinking nightshade; stinky nightshade; Teufelsauge; Zahnkraut

Citations:

Betz P, Janzen J, Roider G, et al. (1991) Psychopathologische Befunde nach oraler Aufnahme von Inhaltstoffen heimischer Nachtschattengewächse. Arch Kriminol 188(5-6):175-182.
Bubien MZ, Magat A, Delatour P (1970) Fréquence des intoxications des animaux en Pologne. Bull Soc Vet Med Comp Lyon 72(2):167-173.
Daneshvar S, Mirhossaini ME, Balali M (1992) Hyoscyamus poisoning in Mashhad. Toxicon 30:501.
Graev M, Fallani M (1960) Avvelenamento collettivo da ingestione da Hyoscyamus niger (Con un caso mortale). Minerva Medicoleg 80(Nov-Dec):225-230.
Kurkcuoglu M (1970) Henbane (Hyoscyamus niger) poisonings in the vicinity of Erzurum. Turk J Pediatr 12(1):48-56.
Nagy L, Sipos I (1980) A beléndek, mint kábítószerpótló. Morphol Igazsagugyi Orv Sz 20(4):312.
Sands JM, Sands R (1976) Henbane chewing. Med J Aust 2(2):55, 58.
Spoerke DG, Hall AH, Dodson CD, et al. (1987) Mystery root ingestion. J Emerg Med 5(5):385-388.
Tugrul L (1985) Abuse of henbane by children in Turkey. Bull Narc 37(2-3):75-78.
Welsby JR (1903) Henbane poisoning. Vet Rec 16(794):181.

HYOSCYAMUS RETICULATUS L. [Solanaceae]

Citations:

Tugrul L (1985) Abuse of henbane by children in Turkey. Bull Narc 37(2-3):75-78.

hypericum –see– *Hypericum perforatum* L.

HYPERICUM AETHIOPICUM Thunb. var glaucescens Sonder [Hypericaceae]

Citations:

Quin JI (1933) Studies on the photosensitisation of animals in South Africa. 3. The photodynamic action of Hypericum ethiopicum var. glaucescens Sond. and Hypericum leucoptychodes (Syn. H. lanceolatum Lam.). Onderstepoort J Vet Sci Anim Indus 1(2):491-496

Note:

This plant name could not be found in the databases searched.

Hypericum crispum L. = Hypericum triquetrifolium Turra

Hypericum lanceolatum auct. = Hypericum revolutum Vahl

Hypericum leucoptychodes Steud. ex A. Rich. = Hypericum revolutum Vahl

Hypericum maculatum Walter = Hypericum punctatum Lam.

HYPERICUM PERFORATUM L. [Hypericaceae]

Common Names:

amber; aran; cammock; goat weed; Hartheu; herb John; hypericum; Johanniskraut; Klamath weed; lyukaslevelü orbáncfü; millepertuis; penny John; rosin rose; St. John's-wort; Tipton weed

Citations:

Anonymous (1929) The poison plants committee. J CSIRO Aust 2:40-48.

Araya OS, Ford EJ (1981) An investigation of the type of photosensitization caused by the ingestion of St John's Wort (Hypericum perforatum) by calves. J Comp Pathol 91(1):135-141.

Bourke CA (2000) Sunlight associated hyperthermia as a consistent and rapidly developing clinical sign in sheep intoxicated by St John's wort (Hypericum perforatum). Aust Vet J 78(7):483-488.

Bourke CA (2003) The effect of shade, shearing and wool type in the protection of Merino sheep from Hypericum perforatum (St John's wort) poisoning. Aust Vet J 81(8):494-498.

Bourke CA, White JG (2004) Reassessment of the toxicity of Hypericum perforatum (St John's wort) for cattle. Aust Vet J 82(11):707-710.

Brown TM (2000) Acute St. John's wort toxicity. Am J Emerg Med 18(2):231-232.

Cunningham IJ (1947) Photosensitivity diseases in New Zealand. V. Photosensitisation by St. John's wort (Hypericum perforatum). N Z J Sci Technol A 29:207-213.

Dodd S (1920) St. John's wort and its action on live stock. J Comp Pathol 33:105-114.

Dodd S (1920) St. John's wort and its effects on livestock. Agric Gaz New South Wales 31:265-272.

Henry M (1922) Feeding and contact experiments with St. Johnswort. Agric Gaz New South Wales 33:205-207.

Kako MD, Al-Sultan II, Saleem AN (1993) Studies of sheep experimentally poisoned with Hypericum perforatum. Vet Hum Toxicol 35(4):298-300.

Kümper H (1989) Hypericismus bei Schafen. Tierarztl Prax 17(3):257-261.

Marsh CD, Clawson AB (1930) Toxic effect of St. Johnswort (Hypericum perforatum) on cattle and sheep. U S Dep Agric Tech Bull #202:23 pp.

Pace N (1942) The etiology of hypericism, A photosensitivity produced by St. Johnswort. Am J Physiol 136:650-656.

Palmer M, Haller C, McKinney P, et al. (1999) Botanicals and other dietary supplements: Adverse events by age. J Toxicol Clin Toxicol 37(5):609.

Rogers TB (1914) On the action of St. John's wort as a sensitizing agent for non-pigmented skin. Am Vet Rev 46:145-162.

Schrader A, Schulz O, Völker H, et al. (2001) Aktuelle Vergiftungen durch Pflanzen bei Wiederkäuern in Nord- und Ostdeutschland. Berl Munch Tierarztl Wochenschr 114(5-6):218-221.

Seddon HR, Belschner HG (1929) The effect of immature St. John's Wort (Hypericum perforatum) on sheep. J CSIRO Aust 2(4):229-231.

Seddon HR, White HC (1928) Some observations on the toxic principle of St. Johnswort (Hypericum perforatum). New South Wales Dep Agric Vet Res Rep 5:106-111.

Waksman JC, Heard K, Jolliff H, et al. (2000) Serotonin syndrome associated with the use of St. John's wort (Hypericum perforatum) and paroxetine. J Toxicol Clin Toxicol 38(5):521.

HYPERICUM PUNCTATUM Lam. [Hypericaceae]

Synonyms:

Hypericum maculatum Walter

Common Names:

flux weed; spotted St. John's-wort; St. John's-wort

Citations:

Vind R (1957) Equine photodermatitis. J Am Vet Med Assoc 131(Dec 1):529.

HYPERICUM REVOLUTUM Vahl [Hypericaceae]

Synonyms:

Hypericum lanceolatum auct.; ***Hypericum leucoptychodes*** Steud. ex A. Rich.

Common Names:

currybush; forest primrose; kerriebos; St. John's-wort

Citations:

Quin JI (1933) Studies on the photosensitisation of animals in South Africa. 3. The photodynamic action of Hypericum ethiopicum var. glaucescens Sond. and Hypericum leucoptychodes (Syn. H. lanceolatum Lam.). Onderstepoort J Vet Sci Anim Indus 1(2):491-496

HYPERICUM TRIQUETRIFOLIUM Turra [Hypericaceae]

Synonyms:

Hypericum crispum L.

Common Names:

St. John's-wort

Citations:

Bale S (1978) Poisoning of sheep, goats and cows by the weed Hypericum triquetrifolium. Refu Vet 35:36-37.

Lang E, Sendil C (1971) Hypericum crispum, die Ursache einer be Pflanzenfressern in der Türkei beobachteten Lichtkrankheit. Tierarztl Umsch 26:73-76.

Ray G (1914) Note sur les effets toxiques du Millepertuis à feuilles crispées (Hypericum crispum). Rec Med Vet Ec Alfort 68:39-42.

hypo –see– *Antiaris toxicaria* Lesch.

HYPOCHAERIS RADICATA L. [Asteraceae]

Common Names:

cat's-ear; dandelion; false dandelion; flatweed; gosmore; hairy cat's-ear; hairy wild lettuce; rough cat's-ear; spotted cat's-ear; summer dandelion

Citations:

Cahill JI, Goulden BE, Pearce HG (1985) A review and some observations on stringhalt. N Z Vet J 33(7):101-104.

Galey FD, Hullinger PJ, McCaskill J (1991) Outbreaks of stringhalt in Northern California. Vet Hum Toxicol 33(2):176-177.

Gay CC, Fransen S, Richards J, et al. (1993) Hypochaeris-associated stringhalt in North America. Equine Vet J 25(5):456-457.

Huntington PJ, Jeffcott LB, Friend SC, et al. (1989) Australian stringhalt - Epidemiological, clinical and neurological investigations. Equine Vet J 21(4):266-273.

Torre F (2005) Clinical diagnosis and results of surgical treatment of 13 cases of acquired bilateral stringhalt (1991-2003). Equine Vet J 37(2):181-183.

hyssop –see– *Hyssopus officinalis* L.

HYSSOPUS OFFICINALIS L. [Lamiaceae]

Common Names:

hyssop

Citations:

Millet Y, Jouglard J, Steinmetz MD, et al. (1981) Toxicity of some essential plant oils. Clinical and experimental study. Clin Toxicol 18(12):1485-1498.

Millet Y, Tognetti P, Lavaire-Pierlovisi M, et al. (1979) Étude expérimentale des propriétés toxiques convulsivantes des essences de sauge et d'hysope du commerce. Rev Electroencephalogr Neurophysiol Clin 9(1):12-18.

I

ibby dalegrass –see– *Euphorbia helioscopia* L.

Ibe –see– *Taxus baccata* L.

ibogaine –see– *Tabernanthe iboga* Baill.

Iboza riparia (Hochst.) N. E. Br. = Tetradenia riparia (Hochst.) Codd

iceberg lettuce –see– *Lactuca sativa* L. var. capitata L.

Iceland poppy –see– *Papaver nudicaule* L.

Idaho pea –see– *Cicer arietinum* L.

If –see– *Taxus baccata* L.

Ignatius bean –see– *Strychnos ignatii* P. J. Bergius

ILEX AQUIFOLIUM L. [Aquifoliaceae]

Common Names:

acebo; Beinved; Christdorn; Christmas holly; English holly; holly; houx; Hülse; hulst; kristkorn; Stecheiche; Stechhulsen; Stechpalme; Walddistelstrauch

Citations:

Mrvos R, Krenzelok EP, Jacobsen TD (2001) Toxidromes associated with the most common plant ingestions. Vet Hum Toxicol 43(6):366-369.

Ramón MF, Ballesteros S, Martínez-Arrieta R, et al. (2003) Plant and herb exposures in Spain. J Toxicol Clin Toxicol 41(4):514.

ILEX DECIDUA Walter [Aquifoliaceae]

Common Names:

deciduous holly; meadow holly; possum haw; swamp holly; winterberry

Citations:

Anonymous (1936) Investigations of stock-poisoning plants. U S Dep Agric Bur Anim Indus Chief Rep. 1936:44-45.

ILEX MYRTIFOLIA Walter [Aquifoliaceae]

Common Names:

myrtle-leaf holly

Citations:

Pence M, Frazier KS, Hawkins L, et al. (2001) The potential toxicity of Ilex myrtifolia in beef cattle. Vet Hum Toxicol 43(3):172-174.

ILEX OPACA Aiton [Aquifoliaceae]

Common Names:

American holly; Christmas holly; English holly; holly; white holly

Citations:

Rodrigues TD, Johnson PN, Jeffrey LP (1984) Holly berry ingestion: Case report. Vet Hum Toxicol 26(2):157-158.

ILEX PARAGUARIENSIS A. St.-Hil. [Aquifoliaceae]

Common Names:

maté; Paraguayan tea; South American tea; yerba maté

Citations:

Anonymous (1995) Anticholinergic poisoning associated with an herbal tea - New York City, 1994. MMWR Morb Mortal Wkly Rep 44(11):193-195.

De Annuntis GJ, Fill S, Meggs WJ, et al. (1994) Anticholinergic poisoning from Paraguay tea: A need for further regulation. Vet Hum Toxicol 36(4):359.

McGee JO, Patrick RS, Wood CB, et al. (1976) A case of veno-occlusive disease of the liver in Britain associated with herbal tea consumption. J Clin Pathol 29(9):788-794.

ILLICIUM ANISATUM L. [Illiciaceae]

Synonyms:

Illicium religiosum Siebold & Zucc.

Common Names:

anis; bastard anise; Chinese bastard anise; Japanese anise; Japanese star anise; man t'sao; poison bay; shu-mang-tsao; star anise

Citations:

Read BE, Kiang PC (1927) Bastard anise poisoning and its antidotal measures. Chin J Physiol 1:15-21.

Wijnands-Kleukers AP, Johanns ES, Vries I, et al. (2002) Intensive search for the cause of an epidemic of seizures after drinking herbal tea (star anise toxicity). J Toxicol Clin Toxicol 40(3):263-264.

Illicium religiosum Siebold & Zucc. = Illicium anisatum L.

ILLICIUM VERUM Hook. f. [Illiciaceae]

Common Names:

anès estrella; Chinese star anise; star anise

Citations:

Deng JF, Lin TJ, Kao WF, et al. (1997) The difficulty in handling poisonings associated with Chinese traditional medicine: A poison control center experience for 1991-1993. Vet Hum Toxicol 39(2):106-114.

Ize Ludlow D, Ragone S, Bernstein JN, et al. (2004) Chemical composition of Chinese star anise (Illicium verum) and neurotoxicity in infants. JAMA 291(5):562-563.

Martínez-Arrieta R, Ballesteros S (2001) A case of severe consequences after mislabeling herbal preparations. J Toxicol Clin Toxicol 39(5):548.

Ramón MF, Ballesteros S, Martínez-Arrieta R, et al. (2003) Plant and herb exposures in Spain. J Toxicol Clin Toxicol 41(4):514.

Ramón MF, Martínez-Arrieta R, Ballesteros S, et al. (2002) Star anise toxicity. An outbreak of pediatric poisoning. J Toxicol Clin Toxicol 40(3):328.

Rudzki E, Grzywa Z (1976) Sensitizing and irritating properties of star anise oil. Contact Dermatitis 2(5):305-308.

imperial ginseng –see– *Panax ginseng* C. A. Mey.

impila –see– *Callilepis laureola* DC.

Inca lily –see– *Alstroemeria ligtu* L.

incense cedar –see– *Calocedrus decurrens* (Torr.) Florin

inch plant –see– *Callisia fragrans* (Lindl.) Woodson

Indian acalypha –see– *Acalypha indica* L.

Indian aconite –see– *Aconitum ferox* Wall. ex Ser.

Indian aloe –see– *Aloe vera* (L.) Burm. f.

Indian apple –see– *Datura inoxia* Mill.; *Datura wrightii* Regel; *Podophyllum peltatum* L.

Indian arrowroot –see– *Curcuma zedoaria* (Christm.) Roscoe

Indian bead –see– *Abrus precatorius* L.

Indian cassia –see– *Cinnamomum tamala* (Buch.-Ham.) Nees & Eberm.

Indian clover –see– *Melilotus indicus* (L.) All.

Indian corn –see– *Veratrum californicum* Durand; *Zea mays* L.

Indian couch –see– *Cynodon dactylon* (L.) Pers.

Indian cress –see– *Tropaeolum majus* L.

Indian doab –see– *Cynodon dactylon* (L.) Pers.

Indian fig –see– *Opuntia ficus-indica* (L.) Mill.

Indian fig cactus –see– *Opuntia ficus-indica* (L.) Mill.

Indian goat pepper –see– *Capsicum frutescens* L.

Indian gum –see– *Sterculia urens* Roxb.

Indian hellebore –see– *Veratrum viride* Aiton

Indian hemp –see– *Apocynum cannabinum* L.; *Cannabis sativa* L.

Indian horse chestnut –see– *Aesculus indica* (Wall. ex Cambess.) Hook.

Indian ink tree –see– *Semecarpus anacardium* L. f.

Indian jack-in-the-pulpit –see– *Arisaema triphyllum* (L.) Schott

Indian laurel –see– *Terminalia elliptica* Willd.

Indian licorice –see– *Abrus precatorius* L.

Indian lilac –see– *Melia azedarach* L.

Indian mallow –see– *Abutilon theophrasti* Medik.

Indian mandrake –see– *Podophyllum hexandrum* Royle

Indian marking nut –see– *Semecarpus anacardium* L. f.

Indian melilot –see– *Melilotus indicus* (L.) All.

Indian milkweed –see– *Asclepias eriocarpa* Benth.

Indian millet –see– *Pennisetum glaucum* (L.) R. Br.; *Sorghum bicolor* (L.) Moench

Indian mustard –see– *Brassica juncea* (L.) Czern.

Indian mutters –see– *Lathyrus sativus* L.

Indian nosy –see– *Asclepias tuberosa* L.

Indian oak –see– *Barringtonia acutangula* (L.) Gaertn.

Indian pea –see– *Lathyrus sativus* L.

Indian pennywort –see– *Centella asiatica* (L.) Urb.

Indian podophyllum –see– *Podophyllum hexandrum* Royle

Indian poke –see– *Veratrum viride* Aiton

Indian red –see– *Sanguinaria canadensis* L.

Indian rosewood –see– *Dalbergia latifolia* Roxb.

Indian sanicle –see– *Ageratina altissima* (L.) R. M. King & H. Rob.

Indian sarsaparilla –see– *Hemidesmus indicus* (L.) W. T. Aiton

Indian senna –see– *Senna alexandrina* Mill.

Indian snakeroot –see– *Rauvolfia serpentina* (L.) Benth. ex Kurz

Indian spurge tree –see– *Euphorbia tirucalli* L.

Indian teak –see– *Tectona grandis* L. f.

Indian tragacanth –see– *Sterculia urens* Roxb.

Indian tree –see– *Euphorbia tirucalli* L.

Indian turnip –see– *Arisaema triphyllum* (L.) Schott

Indian uncus –see– *Veratrum viride* Aiton

Indian walnut –see– *Aleurites moluccanus* (L.) Willd.

INDIGASTRUM PARVIFLORUM (B. Heyne ex Wight & Arn.) Schrire [Fabaceae]

Citations:

Aylward JH, Court RD, Haydock KP, et al. (1987) Indigofera species with agronomic potential in the tropics. Rat toxicity studies. Aust J Agric Res 38(1):177-186.

indigo –see– *Indigofera australis* Willd.; *Indigofera suffruticosa* Mill.; *Indigofera tinctoria* L.

indigo plant –see– *Swainsona galegifolia* (Andrews) R. Br.

INDIGOFERA ALTERNANS DC. [Fabaceae]

Citations:

Aylward JH, Court RD, Haydock KP, et al. (1987) Indigofera species with agronomic potential in the tropics. Rat toxicity studies. Aust J Agric Res 38(1):177-186.

INDIGOFERA ANTUNESIANA Harms [Fabaceae]

Citations:

Aylward JH, Court RD, Haydock KP, et al. (1987) Indigofera species with agronomic potential in the tropics. Rat toxicity studies. Aust J Agric Res 38(1):177-186.

INDIGOFERA ARRECTA Hochst. ex A. Rich [Fabaceae]

Common Names:

Java indigo

Citations:

Aylward JH, Court RD, Haydock KP, et al. (1987) Indigofera species with agronomic potential in the tropics. Rat toxicity studies. Aust J Agric Res 38(1):177-186.

Rosenberg MM, Zoebisch OC (1952) A chick test for toxicity in forage legumes. Agron J 44:315-318.

INDIGOFERA ASTRAGALINA DC. [Fabaceae]

Citations:

Aylward JH, Court RD, Haydock KP, et al. (1987) Indigofera species with agronomic potential in the tropics. Rat toxicity studies. Aust J Agric Res 38(1):177-186.

INDIGOFERA AUSTRALIS Willd. [Fabaceae]

Common Names:

Australian indigo; indigo; native indigo; southern indigo; wild indigo

Citations:

Anonymous (1929) The poison plants committee. J CSIRO Aust 2:40-48.

INDIGOFERA CRYPTANTHA Benth. ex Harv. [Fabaceae]

Citations:

Aylward JH, Court RD, Haydock KP, et al. (1987) Indigofera species with agronomic potential in the tropics. Rat toxicity studies. Aust J Agric Res 38(1):177-186.

Indigofera dominii H. Eichler = Indigofera linnaei Ali

Indigofera endecaphylla Jacq. ex Poir., orth. var. = Indigofera hendecaphylla Jacq.

Indigofera enneaphylla L. = Indigofera linnaei Ali

INDIGOFERA HENDECAPHYLLA Jacq. [Fabaceae]

Synonyms:

Indigofera endecaphylla Jacq. ex Poir., orth. var.

Common Names:

creeping indigo; prostrate indigo; trailing indigo

Citations:

Britten EJ, Matsumoto H, Palafox AL (1959) Comparative toxic effects of 3-nitropropionic acid, sodium nitrite and Indigofera endecaphylla on chicks. Agron J 51:462-464.

Emmel MW, Ritchey GE (1941) The toxicity of Indigofera endecaphylla Jacq. for rabbits. J Am Soc Agron 33:675-677.

Hutton EM, Windrum GM, Kratzing CC (1958) Studies on the toxicity of Indigofera endecaphylla. I. Toxicity for rabbits. J Nutr 64(3):321-337.

Hutton EM, Windrum GM, Kratzing CC (1958) Studies on the toxicity of Indigofera endecaphylla. II. Toxicity for mice. J Nutr 65(3):429-440.

Jeganathan P (1953) Toxic effects of feeding Indigofera endecaphylla (Jacq) to calves. Ceylon Vet J 1(Dec):83-85.

Nordfeldt S, Henke LA, Morita K, et al. (1952) Feeding tests with Indigofera endecaphylla Jacq. (creeping Indigo) and some observations on its poisonous effects on domestic animals. Hawaii Agric Exp Sta Tech Bull #15:23 pp.

Nordfeldt S, Younge OR (1949) Toxicity of creeping Indigo to livestock. Hawaii Agric Exp Sta Prog Note #55:2 pp.

Payne WJ, Naidu RK (1955) The toxicity of creeping indigo (Indigofera endecaphylla). Agric J Fiji 26:1-3.

Rosenberg MM, Palafox AL (1950) The effect of creeping indigo (Indigofera endecaphylla) when fed to growing chickens. World Poultry Sci J 6:284-291.

Rosenberg MM, Palafox AL (1951) The effect of creeping indigo (Indigofera endecaphylla) on laying chickens. World Poultry Sci J 7:9-15.

Rosenberg MM, Zoebisch OC (1952) A chick test for toxicity in forage legumes. Agron J 44:315-318.

Yelf JD (1959) The toxicity of creeping indigo in Fiji. Fiji Agric J 29:9.

INDIGOFERA HIRSUTA L. [Fabaceae]

Common Names:

hairy indigo; wild indigo

Citations:

Aylward JH, Court RD, Haydock KP, et al. (1987) Indigofera species with agronomic potential in the tropics. Rat toxicity studies. Aust J Agric Res 38(1):177-186.

Rosenberg MM, Zoebisch OC (1952) A chick test for toxicity in forage legumes. Agron J 44:315-318.

INDIGOFERA HOCHSTETTERI Baker [Fabaceae]

Citations:

Suliman HB, Wasfi IA, Tartour G, et al. (1983) The effects of Indigofera hochstetteri on goats. Rev Elev Med Vet Pays Trop 36(4):393-402.

INDIGOFERA LINNAEI Ali [Fabaceae]

Synonyms:

Indigofera dominii H. Eichler; ***Indigofera enneaphylla*** L.

Common Names:

bakario; bird's-foot indigo; Birdsville indigo; nine-leaf indigo

Citations:

Anonymous (1950) Birdsville disease of horses. Aust Vet J 26(Sep):261.

Bell AT, Everist SL (1951) Indigofera enneaphylla: A plant toxic to horses (Birdsville disease). Aust Vet J 27(Aug):185-188.

Bell AT, Hall WT (1952) Birdsville disease of horses. Feeding trials with Indigofera enneaphylla. Aust Vet J 28(Jun):141-144.

Hegarty MP, Kelly WR, McEwan D, et al. (1988) Hepatotoxicity to dogs of horse meat contaminated with indospicine. Aust Vet J 65(11):337-340.

Murray LF, Moore T, Sharman IM (1965) The toxicity of Indigofera enneaphylla L. in rats. Aust J Agric Res 16(4):713-720.

Rose AL, Banks AW, McConnell JD (1951) Birdsville disease in the Northern Territory. Aust Vet J 27(8):189-196.

Indigofera mucronata Spreng. ex DC = Indigofera trita L. f. subsp. scabra (Roth) de Kort & G. Thijsse

INDIGOFERA NIGRITANA Hook. f. [Fabaceae]

Citations:

Aylward JH, Court RD, Haydock KP, et al. (1987) Indigofera species with agronomic potential in the tropics. Rat toxicity studies. Aust J Agric Res 38(1):177-186.

Indigofera patens Eckl. & Zeyh. = Indigofera sessilifolia DC.

INDIGOFERA SCHIMPERI Jaub. & Spach [Fabaceae]

Citations:

Aylward JH, Court RD, Haydock KP, et al. (1987) Indigofera species with agronomic potential in the tropics. Rat toxicity studies. Aust J Agric Res 38(1):177-186.

INDIGOFERA SESSILIFOLIA DC. [Fabaceae]

Synonyms:

Indigofera patens Eckl. & Zeyh.

Citations:

Van der Walt SJ, Steyn DG (1939) Recent investigations into the toxicity of known and unknown poisonous plants in the Union of South Africa. IX. Onderstepoort J Vet Sci Anim Indus 12:335-366.

INDIGOFERA SPICATA Forssk. [Fabaceae]

Common Names:

creeping indigo; trailing indigo

Citations:

Aylward JH, Court RD, Haydock KP, et al. (1987) Indigofera species with agronomic potential in the tropics. Rat toxicity studies. Aust J Agric Res 38(1):177-186.

Bindon BM, Lamond DR (1966) Examination of tropical legumes for deleterious effects on animal reproduction. Proc Aust Soc Anim Prod 6:109-116.

Britten EJ, Palafox AL, Frodyma MM, et al. (1963) Level of 3-nitropropanoic acid in relation to toxicity of Indigofera spicata in chicks. Crop Sci 3(5):415-416.

Christie GS, Wilson M, Hegarty MP (1975) Effects on the liver in the rat of ingestion of Indigofera spicata, a legume containing an inhibitor of arginine metabolism. J Pathol 117(4):195-205.

INDIGOFERA SUFFRUTICOSA Mill. [Fabaceae]

Common Names:

anil indigo; anileira; indigo; timbe mirim

Citations:

Aylward JH, Court RD, Haydock KP, et al. (1987) Indigofera species with agronomic potential in the tropics. Rat toxicity studies. Aust J Agric Res 38(1):177-186.

Barbosa Neto JD, Oliveira CM, Peixoto PV, et al. (2001) Anemia hemolítica causda por Indigofera suffruticosa (Leg. Papilionoideae) em bovinos. Pesq Vet Bras 21(1):18-22.

Indigofera teysmannii Miq. = Indigofera zollingeriana Miq.

INDIGOFERA TINCTORIA L. [Fabaceae]

Common Names:

indigo

Citations:

Aylward JH, Court RD, Haydock KP, et al. (1987) Indigofera species with agronomic potential in the tropics. Rat toxicity studies. Aust J Agric Res 38(1):177-186.

INDIGOFERA TRITA L. f. [Fabaceae]

Common Names:

tipsywood

Citations:

Aylward JH, Court RD, Haydock KP, et al. (1987) Indigofera species with agronomic potential in the tropics. Rat toxicity studies. Aust J Agric Res 38(1):177-186.

INDIGOFERA TRITA L. f. subsp. scabra (Roth) de Kort & G. Thijsse [Fabaceae]

Synonyms:

Indigofera mucronata Spreng. ex DC.

Citations:

Aylward JH, Court RD, Haydock KP, et al. (1987) Indigofera species with agronomic potential in the tropics. Rat toxicity studies. Aust J Agric Res 38(1):177-186.

INDIGOFERA TRUXILLENSIS Kunth [Fabaceae]

Citations:

Carvalho FS (1985) Aspectos clínicos, laboratoriais e anátomo-histológicos na intoxicação experimental pela Indigofera truxillensis H.B.K. em bovinos. Arq Bras Med Vet Zootecnia 37(5):519-523.

Carvalho FS, Pessoa JM, Souza R, et al. (1986) Níveis sangüíneos de TGO, TGP, fósforo, câlcio, uréia e dextrose na intoxicação experimental de ovinos pela Indigofera truxillensis H.B.K. Arq Bras Med Vet Zootecnia 38(2):137-146.

INDIGOFERA ZOLLINGERIANA Miq. [Fabaceae]

Synonyms:

Indigofera teysmannii Miq.

Citations:

Krishna L, Vaid J, Singh B (1986) Pathological study on Indigofera teysmanni toxicity in sheep. Indian J Comp Microbiol Immunol Infect Dis 7(1):14-17.

Singh B, Negi SS, Vaid J, et al. (1985) Palatability, voluntary intake and nutritive value of Indigofera teysmanni in sheep. Cheiron 14(6):315-318.

Indisches Rosenholz –see– *Dalbergia latifolia* Roxb.

Indochinese lacquer –see– *Toxicodendron succedaneum* (L.) Kuntze

indoor linden –see– *Sparrmannia africana* L. f.

infidel opium –see– *Hyoscyamus niger* L.

Inkalilie –see– *Alstroemeria ligtu* L.

Inkarnatklee –see– *Trifolium incarnatum* L.

inkberry –see– *Cestrum laevigatum* Schltdl.; *Phytolacca americana* L.

inkweed –see– *Drymaria pachyphylla* Wooton & Standl.

inland pigweed –see– *Portulaca oleracea* L.

inlonge –see– *Catha edulis* (Vahl) Forssk. ex Endl.

insane root –see– *Hyoscyamus niger* L.

insect flower –see– *Tanacetum cinerariifolium* (Trevir.) Sch. Bip.

interior live oak –see– *Quercus wislizeni* DC.

INULA BRITANNICA L. [Asteraceae]

Citations:

Hegyi E (1967) Plant dermatitis (Inula britannica L.). Contact Dermatol Newsl 2(Jul):4.

INULA CONYZAE (Griess.) Meikle [Asteraceae]

Common Names:

Dürrwurz

Citations:

Reinboth W (1967) Vergiftungen durch Inula conyza (Dürrwurz) bei Rindern. Monatsh Veterinarmed 22(15):611-612.

Ulbrich M, Lorenz H, Rittenbach P, et al. (1966) Inula conyza (Dürrwurz) als Ursache einer Massenvergiftung bei Rindern. Monatsh Veterinarmed 21(23):896-902.

Vockrodt H (1973) Vergiftungen landwirtschaftlicher Nutztiere durch Schadpflanzen und deren Vorbeuge. Monatsh Veterinarmed 28(2):59-62.

Inula graveolens (L.) Desf. = Dittrichia graveolens (L.) Greuter

INULA HELENIUM L. [Asteraceae]

Common Names:

alant; elecampane

Citations:

Lamminpaa A, Estlander J, Jolanki K, et al. (1996) Occupational allergic contact dermatitis caused by decorative plants. Contact Dermatitis 34(5):330-335.

Mateo MP, Velasco M, Miquel FJ, et al. (1995) Erythema-multiforme-like eruption following contact dermatitis from sesquiterpene lactones in herbal medicine. Contact Dermatitis 33(6):449-450.

Piankova ZP, Nugmanova ML (1975) [Dermatitis caused by elecampane.] Vestn Dermatol Venerol 12:52-54.

Inula viscosa (L.) Aiton = Dittrichia viscosa (L.) Greuter

iodine bush –see– *Cestrum parqui* L'Her.

Ioxylon pomiferum Raf. = Maclura pomifera (Raf.) C. K. Schneid.

Ipé –see– *Tabebuia serratifolia* (Vahl) G. Nicholson

ipecaagoene –see– *Carapichea ipecacuanha* (Brot.) L. Andersson

ipecac –see– *Carapichea ipecacuanha* (Brot.) L. Andersson

ipecacuanha –see– *Carapichea ipecacuanha* (Brot.) L. Andersson

ipil ipil –see– *Leucaena leucocephala* (Lam.) de Wit

ipoh akar –see– *Strychnos ignatii* P. J. Bergius

ipoh tree –see– *Antiaris toxicaria* Lesch.

IPOMOEA ASARIFOLIA (Desr.) Roem. & Schult. [Convolvulaceae]

Common Names:

salsa

Citations:

Barbosa JD, Oliveira CM, Duarte MD, et al. (2005) Intoxicações experimental e natural por Ipomoea asarifolia (Convolvulaceae) em búfalos e outros ruminantes. Pesq Vet Bras 25(4):231-234.

Döbereiner J, Tokarnia CH, Canella CF (1960) Intoxicação experimental pela "salsa" (Ipomoea asarifolia R. et Schult.) em ruminantes. Arq Inst Biol Anim (Rio Janeiro) 3:39-57.

Medeiros RM, Barbosa RC, Riet-Correa F, et al. (2003) Tremorgenic syndrome in goats caused by Ipomoea asarifolia in northeastern Brazil. Toxicon 41:933-935.

IPOMOEA BATATAS (L.) Lam. [Convolvulaceae]

Common Names:

Süßkartoffel; sweet potato

Citations:

Hill BD, Wright HF (1992) Acute interstitial pneumonia in cattle associated with consumption of mould-damaged sweet potatoes (Ipomoea batatas). Aust Vet J 69(2):36-37.

Medeiros RM, Simoes SV, Tabosa IM, et al. (2001) Bovine atypical interstitial pneumonia associated with the ingestion of damaged sweet potatoes (Ipomoea batatas) in northeastern Brazil. Vet Hum Toxicol 43(4):205-207.

Peckham JC, Mitchell FE, Jones OH Jr, et al. (1972) Atypical interstitial pneumonia in cattle fed moldy sweet potatoes. J Am Vet Med Assoc 160(2):169-172.

Rogers GM, Poore MH, Ferko BL, et al. (1999) Dental wear and growth performance in steers fed sweetpotato cannery waste. J Am Vet Med Assoc 214(5):681-686.

Stephens C (1941) Black rot sweet potato poisoning in cattle. J Am Vet Med Assoc 98(769):317.

Note:

Sweet potato is named *Ipomoea batatas* (L.) Lam. var. *batatas* in some publications.

IPOMOEA CARNEA Jacq. [Convolvulaceae]

Common Names:

awier; baros; besharam; hierba-de-la-India; palo-santo-de-castilla; shrubby morning-glory; tree morning-glory

Citations:

Adam SE, Tartour G, Obeid HM, et al. (1973) Effects of Ipomoea carnea on the liver and on serum enzymes in young ruminants. J Comp Pathol 83(4):531-542.

Damir HA, Adam SE, Tartour G (1987) The effects of Ipomoea carnea on goats and sheep. Vet Hum Toxicol 29(4):316-319.

de Balogh KK, Dimande AP, van der Lugt JJ, et al. (1998) Ipomoea carnea: The cause of a lysosomal storage disease in goats in Mozambique. In: Garland T, Barr AC (eds.) Toxic plants and other natural toxicants. CABI. New York. pp. 428-434.

de Balogh KK, Dimande AP, van der Lugt JJ, et al. (1999) A lysosomal storage disease induced by Ipomoea carnea in goats in Mozambique. J Vet Diagn Invest 11(3):266-273.

Hueza IM, Dagli SL, Górniak SL (2003) Toxic effects of prenatal Ipomoea carnea administration to rats. Vet Hum Toxicol 45(6):298-302.

Hueza IM, Fonseca ES, Paulino CA, et al. (2004) Ipomoea carnea induced enhanced macrophage activity (phagocytosis and peroxide production) in immunocompetent rats. In: Acamovic T, Stewart CS, Pennycott TW (eds.) Poisonous plants and related toxins. CABI. Wallingford, UK. pp. 316-321.

Idris OF, Tartour G, Adam SE, et al. (1973) Toxicity to goats of Ipomoea carnea. Trop Anim Health Prod 5(2):119-123.

Misra A, Misra SN (1965) Study on the toxic effect of Ipomoea carnea in sheep, goats and cattle. Indian Vet J 42(9):703-707.

Schumaher-Henrique B, Górniak SL, Dagli ML, et al. (2003) The clinical, biochemical, haematological and pathological effects of long-term administration of Ipomoea carnea to growing goats. Vet Res Commun 27(4):311-319.

Schumaher-Henrique B, Górniak SL, Dagli ML, et al. (2004) Evaluation of Ipomoea carnea toxicity to growing goats: Clinical, biochemical, haematological and pathological alterations. In: Acamovic T, Stewart CS, Pennycott TW (eds.) Poisonous plants and related toxins. CABI. Wallingford, UK. pp. 90-97.

Srilatha CH, Naidu NR, Rao PR (1997) Pathology of Ipomoea carnea toxicity in goats. Indian J Anim Sci 67(3):253-254.

Tartour G, Adam SE, Obeid HM, et al. (1974) Development of anaemia in goats fed with Ipomoea carnea. Br Vet J 130(3):271-279.

Tartour G, Adam SE, Obeid HM, et al. (1974) Serum iron, total iron-binding capacity and bilirubin concentrations in young ruminants fed Ipomoea carnea. Acta Vet (Beogr) 24(6):261-268.

Tartour G, Obeid HM, Adam SE, et al. (1973) Haematological changes in sheep and calves following prolonged oral administration of Ipomoea carnea. Trop Anim Health Prod 5(4):284-292.

Tirkey K, Yadava KP, Jha GJ, et al. (1987) Effect of feeding of Ipomoea carnea leaves on goats. Indian J Anim Sci 57(8):863-866.

Zakir MD, Vadlamudi VP, More PR (1989) Some blood biochemical changes in Ipomoea carnea toxicity in Osmanabadi goats. J Maharashtra Agric Univ 14(1):126-127.

IPOMOEA CARNEA Jacq. subsp. fistulosa (Mart. ex Choisy) D. F. Austin [Convolvulaceae]

Synonyms:

Ipomoea fistulosa Mart. ex Choisy

Common Names:

canudo; mata-bode; morning-glory; shrubby morning-glory

Citations:

Tokarnia CH, Döbereiner J, Canella CF (1960) Estudo experimental sôbre a toxidex do "Canudo" (Ipomoea fistulosa Mart.) em ruminantes. Arq Inst Biol Anim (Rio Janeiro) 3:59-71.

Ipomoea fistulosa Mart. ex Choisy = Ipomoea carnea Jacq. subsp. fistulosa (Mart. ex Choisy) D. F. Austin

IPOMOEA HEDERACEA Jacq. [Convolvulaceae]

Common Names:

ivy-leaf morning-glory; morning-glory

Citations:

Dugan GM, Gumbmann MR (1990) Toxicological evaluation of morning glory seed: Subchronic 90-day feeding study. Food Chem Toxicol 28(8):553-559.

IPOMOEA LACUNOSA L. [Convolvulaceae]

Common Names:

white morning-glory

Citations:

Dugan GM, Gumbmann MR (1990) Toxicological evaluation of morning glory seed: Subchronic 90-day feeding study. Food Chem Toxicol 28(8):553-559.

IPOMOEA MUELLERI Benth. [Convolvulaceae]

Common Names:

convolvulus; lamb poison; morning-glory

Citations:

Gardiner MR, Royce R, Oldroyd B (1965) Ipomoea muelleri intoxication of sheep in Western Australia. Br Vet J 121:272-277.

IPOMOEA RIEDELII Meissn. [Convolvulaceae]

Citations:

Barbosa RC, Riet-Correa F, Medeiros RM, et al. (2006) Intoxication by Ipomoea sericophylla and Ipomoea riedelii in goats in the state of Paraíba, Northeastern Brazil. Toxicon 47(4):371-379.

IPOMOEA SERICOPHYLLA Meisn. [Convolvulaceae]

Citations:

Barbosa RC, Riet-Correa F, Medeiros RM, et al. (2006) Intoxication by Ipomoea sericophylla and Ipomoea riedelii in goats in the state of Paraíba, Northeastern Brazil. Toxicon 47(4):371-379.

IPOMOEA TRICOLOR Cav. [Convolvulaceae]

Common Names:

badoh negro; blue morning-glory; blue water; flying saucers; heavenly-blue; morning-glory; summer skies; wedding bells; woodrose

Citations:
Ingram AL Jr (1964) Morning glory seed reaction. JAMA 190(13):1133-1134.

IPOMOEA VIOLACEA L. [Convolvulaceae]

Common Names:
badoh; blue-star morning-glory; flying saucers; heavenly-blue; morning-glory; pearly gates; pearly gates morning-glory; summer skies; wedding bells

Citations:
Cohen S (1964) Suicide following morning glory seed ingestion. Am J Psychiatry 120(Apr):1024-1025.
Fink PJ, Goldman MJ, Lyons I (1966) Morning glory seed psychosis. Arch Gen Psychiatry 15:209-213.

Iranian boxwood –see– *Buxus sempervirens* L.

irby dalegrass –see– *Euphorbia helioscopia* L.

iris desmarais –see– *Iris pseudoacorus* L.

IRIS ×GERMANICA L. [Iridaceae]

Common Names:
Deutsche Schwertlilie; flag iris; flag lily; fleur-de-lis; Gelbe Schwertlilie; Himmelschwertel; Vielchenwurzel

Citations:
Mrvos R, Krenzelok EP, Jacobsen TD (2001) Toxidromes associated with the most common plant ingestions. Vet Hum Toxicol 43(6):366-369.
Ramón MF, Ballesteros S, Martínez-Arrieta R, et al. (2003) Plant and herb exposures in Spain. J Toxicol Clin Toxicol 41(4):514.

IRIS ×GERMANICA L. nothovar. florentina Dykes [Iridaceae]

Common Names:
Florentine iris; orris root

Citations:
Winter GR (1948) Allergic manifestations caused by the use of a dentifrice containing orris root powder. J Periodont 19:108.

iris jaune –see– *Iris pseudoacorus* L.

IRIS PSEUDOACORUS L. [Iridaceae]

Common Names:
acoro bastardo; Blaue Ilge; Falscher Kalmus; fleur-de-luce; Gelbe Schwertlilie; Himmelsschwertel; iris desmarais; iris jaune; Jacob's-sword; lirio amarillo; Teichlilie; Veilchenwurzel; Wasserschwertlilie; water flag; yellow flag; yellow iris; yellow water iris

Citations:
Boddie GF (1947) Toxicological problems in veterinary medicine. Vet Rec 59(37):470-486.
Calnan CD (1970) Iris pseudacorus L. Contact Dermatol Newsl 8(Aug):171.

Irish broom –see– *Cytisus scoparius* (L.) Link

Irish potato –see– *Solanum tuberosum* L.

Irish yew –see– *Taxus baccata* L.

iroko –see– *Milicia excelsa* (Welw.) C. C. Berg

iroko teak –see– *Milicia excelsa* (Welw.) C. C. Berg

iron oak –see– *Quercus douglasii* Hook. & Arn.; *Quercus stellata* Wangenh.

ironweed –see– *Ambrosia trifida* L.; *Amsinckia intermedia* Fisch. & C. A. Mey.; *Erythrophleum chlorostachys* (F. Muell.) Baill.

ironwood –see– *Erythrophleum chlorostachys* (F. Muell.) Baill.

ironwood tree –see– *Machaerium scleroxylon* Tul.

Ishin –see– *Blighia sapida* K. D. Koenig

isin –see– *Blighia sapida* K. D. Koenig

Isle-of-Wight-vine –see– *Bryonia dioica* Jacq.; *Dioscorea communis* (L.) Caddick & Wilkin

Isocoma heterophylla (A. Gray) Greene = Isocoma plurifolia (Torr. & A. Gray) Greene

ISOCOMA PLURIFOLIA (Torr. & A. Gray) Greene [Asteraceae]

Synonyms:
Aplopappus heterophyllus (A. Gray) S. F. Blake; ***Bigelowia coronopifolia*** A. Gray; ***Haplopappus heterophyllus*** (A. Gray) S. F. Blake; ***Isocoma heterophylla*** (A. Gray) Greene; ***Isocoma wrightii*** (A. Gray) Rydb.

Common Names:
alkali weed, burro weed; jimmy weed; rayless goldenrod; sick weed; woody aster

Citations:
Couch JF (1941) Trembles (milk sickness) produced by toxic butter. Vet Med 36(5):244-245.
Jordan EO, Harris NM (1908) The cause of milksickness or trembles. JAMA 50(21):1665-1673.
Marsh CD, Roe GC (1921) The "alkali disease" of livestock in the Pecos Valley. U S Dep Agric Circ #180:8 pp.
Marsh CD, Roe GC, Clawson AB (1926) Rayless goldenrod (Aplopappus heterophyllus) as a poisonous plant. U S Dep Agric Bull #1391:24 pp.

Isocoma wrightii (A. Gray) Rydb. = Isocoma plurifolia (Torr. & A. Gray) Greene

ISOTOMA PETRAEA F. Muell. [Campanulaceae]

Common Names:
cheeky bugger; rock isotome

Citations:
Higginson AR (1957) Further notes on plants injurious to man. South Australia Naturalist 32(Dec):26.

ISOTROPIS CUNEIFOLIA (Sm.) Domin [Fabaceae]

Common Names:

lambpoison; pea blossom poison

Citations:

Gardiner MR, Royce RD (1967) Poisoning of sheep and cattle in Western Australia due to species of Isotropis (Papilionaceae). Aust J Agric Res 18:505-513.

ISOTROPIS DRUMMONDII Meisn. [Fabaceae]

Citations:

Gardiner MR, Royce RD (1967) Poisoning of sheep and cattle in Western Australia due to species of Isotropis (Papilionaceae). Aust J Agric Res 18:505-513.

ISOTROPIS FORRESTII F. Muell. [Fabaceae]

Citations:

Colegate SM, Dorling PR, Huxtable CR, et al. (1992) Nephrotoxicity of Isotropis forrestii: Pathology of intoxication and the structural determination of (+)-forrestine, a neurotoxic alkaloid isolated from the plant. In: James LF, Keeler RF, Bailey JM Jr, et al. (eds.) Poisonous plants. Iowa State Univ Press. Ames, Iowa. pp. 414-417.

Cooper TB, Huxtable CR, Vogel P (1986) The nephrotoxicity of Isotropis forrestii in sheep. Aust Vet J 63(6):178-182.

Italian bean –see– *Vicia faba* L.

Italian clover –see– *Trifolium incarnatum* L.

Italian cocklebur –see– *Xanthium strumarium* L. var. canadense (Mill.) Torr. & A. Gray

Italian cypress –see– *Cupressus sempervirens* L.; *Lolium multiflorum* Lam.

Italian ryegrass –see– *Lolium perenne* L.

itamo –see– *Pedilanthus tithymaloides* (L.) Poit.

itchweed –see– *Veratrum viride* Aiton

itsit –see– *Trianthema portulacastrum* L.

IVA ANGUSTIFOLIA Nutt. ex DC. [Asteraceae]

Common Names:

false ragweed; marsh elder; narrow-leaf marsh elder; narrow-leaf sumpweed

Citations:

Murphy MJ, Reagor JC, Ray AC, et al. (1983) Bovine abortion associated with ingestion of Iva angustifolia (narrowleaf sumpweed). Proc Am Assoc Vet Lab Diagn 26:161-166.

Smith WA, Prince HE, Cole ML (1941-1942) Contact dermatitis from the narrow leaf marsh elder (Iva angustifolia). J Allergy 13:371-379.

IVA ANNUA L. [Asteraceae]

Synonyms:

Iva ciliata Willd.

Common Names:

burweed marsh elder; marsh elder; rough marsh elder

Citations:

Brunsting LA, Williams DH (1936) Ragweed (contact) dermatitis. JAMA 106(18):1533-1535.

Burrows GE, Schwab RP, Stein LE, et al. (1998) Comparison of the reproductive effects of Baptisia australis, Iva annua and Sophora nuttalliana in rats. In: Garland T, Barr AC (eds.) Toxic plants and other natural toxicants. CABI. New York. pp. 297-302.

Iva ciliata Willd. = Iva annua L.

IVA MICROCEPHALA Nutt. [Asteraceae]

Common Names:

marsh elder

Citations:

Williams O, Speairs R, Boggs HW (1960) Hypersensitivity to marsh elder: Difficulties encountered in patch testing. J La State Med Soc 112(Jun):216-219.

Iva xanthiifolia Nutt. = Cyclachaena xanthiifolia (Nutt.) Fresen.

ivraie –see– *Lolium perenne* L.

ivray –see– *Lolium temulentum* L.

ivy –see– *Hedera helix* L.; *Hedera helix* L. subsp. canariensis (Willd.) Cout.

ivy arum –see– *Epipremnum pinnatum* (L.) Engl.

ivy bindweed –see– *Fallopia convolvulus* (L.) Á. Löve

ivy bush –see– *Kalmia latifolia* L.

ivy-leaf morning-glory –see– *Ipomoea hederacea* Jacq.

ivywood –see– *Kalmia latifolia* L.

ixia –see– *Chamaeleon gummifera* (L.) Cass.

IXIOLAENA BREVICOMPTA F. Muell. [Asteraceae]

Common Names:

button weed; flat billy button

Citations:

Walker KH, Thompson DR, Seaman JT (1980) Suspected poisoning of sheep by Ixiolaena brevicompta. Aust Vet J 56(2):64-66.

IXODIA ACHILLIOIDES R. Br. [Asteraceae]

Citations:

Turner T (1980) Compositae dermatitis in South Australia: Contact dermatitis from Ixodia achillaeoides and Cynara cardunculus or the tribulations of a dry flower arranger. Contact Dermatitis 6(6):444.

J

jaagsiektebossie –see– *Crotalaria dura* J. M. Wood & M. S. Evans

jacaranda –see– *Dalbergia nigra* (Vell.) Allemão ex Benth.

jack-in-the-pulpit –see– *Arisaema triphyllum* (L.) Schott; *Pinellia ternata* (Thunb.) Makino

jack swort bean –see– *Canavalia ensiformis* (L.) DC.

jackbean –see– *Canavalia ensiformis* (L.) DC.

Jackbohne –see– *Canavalia ensiformis* (L.) DC.

JACOBAEA AQUATICA (Hill) P. Gaertn. et al. var. aquatica [Asteraceae]

Synonyms:

Senecio aquaticus Hill; ***Senecio barbariefolius*** Rchb.

Common Names:

marsh ragwort

Citations:

Dybing O, Erichsen S (1959) Liver changes in rats after administration of Senecio aquaticus. Acta Pathol Microbiol Scand 47:1-8.

Evans WC, Evans ET (1949) Poisoning of farm animals by the marsh ragwort (Senecio aquaticus Huds.). Nature 164(4157):30-31.

JACOBAEA AQUATICA (Hill) P. Gaertn. et al. var. erratica (Bertol.) Pelser & Meijden [Asteraceae]

Synonyms:

Senecio erraticus Bertol.; ***Senecio erraticus*** Bertol. subsp. barbareifolius (Fiori) Beger

Common Names:

ragwort

Citations:

Araya O, Fuentealba IC (1990) Chronic hepato-toxicity of Senecio erraticus in calves from two 50-day feeding periods in consecutive years. Vet Hum Toxicol 32(6):555-557.

Araya O, Hernandez JR, Espinoza AE, et al. (1983) Serum changes and histologic liver lesions due to experimental ingestion of ragwort (Senecio erraticus) in sheep. Vet Hum Toxicol 25(1):4-7.

Araya O, Illanes O, Wittwer F (1986) Seneciosis en novillos despues de la exposicion natural a la ingestion de Senecio erraticus. Vet Argent 3(21):62, 64-67.

Sova Z, Vaněk J, Jicha J (1974) [Experimental production of liver fibrosis in horses by feeding them on groundsel (Senecio erraticus subsp. barbaraeifolius).] Sbornik Pro Ekon Fak Cesk Budejovicich Biol Rada 12:123-140.

Vaněk J (1958) Vergiftung mit Kreuzkraut (Senecio) als Ursache der Ždárer Pferdeseuche. Schweiz Z Pathol Bakteriol 21(4):821-848.

JACOBAEA RAPHANIFOLIA (Wall. ex DC.) B. Nord. [Asteraceae]

Synonyms:

Senecio diversifolius Wall. ex DC.

Citations:

Winter H, Seawright AA, Hrdlicka J, et al. (1993) Pyrrolizidine alkaloid poisoning of yaks: Diagnosis of pyrrolizidine alkaloid exposure by the demonstration of sulphur-conjugated pyrrolic metabolites of the alkaloid in circulating haemoglobin. Aust Vet J 70(8):312-313.

JACOBAEA VULGARIS Gaertn. [Asteraceae]

Synonyms:

Senecio jacobaea L.

Common Names:

benweed; doddies; Greiskraut; groundsel; Herbe-de-St. Jacques; hierba-de-santiago; jacobée; Jacobskreuzkraut; Jacobskruiskruid; Jakobskraut; ragwort; seggrum; staggerwort; stinking Willy; suzon; tansy ragwort

Citations:

Betty RW, Markson LM (1954) Liver biopsy in the diagnosis of ragwort (Senecio jacobea) poisoning in a herd of cattle. Vet Rec 66(28):398-400.

Brodrick T (1997) Ragwort poisoning in a pedigree cow. Vet Rec 141(18):480.

Brüggemann RJ (2003) Jacobskruiskruid. Tijdschr Diergeneeskd 128:123.

Burns J (1972) The heart and pulmonary arteries in rats on Senecio jacobaea. J Pathol 106(3):187-194.

Cheeke PR, Garman GR (1973) Protein effect on tansy ragwort toxicity in rats. J Anim Sci 37(1):276.

Cheeke PR, Garman GR (1974) Influence of dietary protein and sulfur amino acid levels on the toxicity of Senecio jacobaea (tansy ragwort) to rats. Nutr Rep Int 9(3):197-207.

Cheeke PR, Pierson-Goeger ML (1983) Toxicity of Senecio jacobaea and pyrrolizidine alkaloids in various laboratory animals and avian species. Toxicol Lett 18(3):343-349.

Cheeke PR, Schmitz JA, Lassen ED, et al. (1985) Effects of dietary supplementation with ethoxyquin, magnesium oxide, methionine hydroxy analog, and B vitamins on tansy ragwort (Senecio jacobaea) toxicosis in beef cattle. Am J Vet Res 46(10):2179-2183.

Cockburn RS, Eaton G, Hudson JR, et al. (1955) Acute poisoning of cattle by common ragwort (Senecio jacobaea L.). Vet Rec 67(Aug 20):640.

Craig AM, Pearson EG, Meyer C, et al. (1991) Clinicopathologic studies of tansy ragwort toxicosis in ponies: Sequential serum and histopathological changes. J Equine Vet Sci 11(5):261-271.

Craig AM, Pearson EG, Meyer C, et al. (1991) Serum liver enzyme and histopathologic changes in calves with chronic and chronic-delayed Senecio jacobaea toxicosis. Am J Vet Res 52(12):1969-1978.

Dean RE, Winward AH (1974) An investigation into the possibility of tansy ragwort poisoning of black-tailed deer. J Wildl Dis 10(2):166-169.

Dewes HF, Lowe MD (1985) Haemolytic crisis associated with ragwort poisoning and rail chewing in two thoroughbred fillies. N Z Vet J 33(9):159-160.

Ford EJ, Ritchie HE, Thorpe E (1968) Serum changes following the feeding of ragwort (Senecio jacobea) to calves. J Comp Pathol 78(2):207-218.

Garrett BJ, Holtan DW, Cheeke PR, et al. (1984) Effects of dietary supplementation with butylated hydroxyanisole, cysteine, and vitamins B on tansy ragwort (Senecio jacobaea) toxicosis in ponies. Am J Vet Res 45(3):459-464.

Giles CJ (1983) Outbreak of ragwort (Senecio jacobaea) poisoning in horses. Equine Vet J 15(3):248-250.

Goeger DE, Cheeke PR, Buhler DR (1979) Effect of dietary tansy ragwort (Senecio jacobaea) on dairy goats and toxicity of their milk. J Anim Sci 49(Suppl 1):370.

Goeger DE, Cheeke PR, Ramsdell HS, et al. (1983) Comparison of the toxicities of Senecio jacobaea, Senecio vulgaris and Senecio glabellus in rats. Toxicol Lett 15(1):19-23.

Goeger DE, Cheeke PR, Schmitz JA, et al. (1982) Toxicity of tansy ragwort (Senecio jacobaea) to goats. Am J Vet Res 43(2):252-254.

Gopinath C, Ford EJ (1977) The effect of ragwort (Senecio jacobaea) on the liver of the domestic fowl (Gallus domesticus): A histopathological and enzyme histochemical study. Br Poult Sci 18(2):137-141.

Harding JD, Lewis G, Done JT, et al. (1964) Experimental poisoning by Senecio jacobaea in pigs. Pathol Vet 1:204-220.

Hooper PT (1974) The pathology of Senecio jacobaea poisoning of mice. J Pathol 113(4):227-230.

Johnson AE (1978) Tolerance of cattle to tansy ragwort (Senecio jacobaea). Am J Vet Res 39(9):1542-1544.

Johnson AE (1982) Failure of mineral-vitamin supplements to prevent tansy ragwort (Senecio jacobaea) toxicosis in cattle. Am J Vet Res 43(4):718-723.

Johnson AE, Smart RA (1983) Effects on cattle and their calves of tansy ragwort (Senecio jacobaea) fed in early gestation. Am J Vet Res 44(7):1215-1219.

Lanux-Van Gorder V (2000) Tansy ragwort poisoning in a horse in southern Ontario. Can Vet J 41(5):409-410.

Markson LM (1960) The pathogenesis of the hepatic lesion in calves poisoned experimentally with Senecio jacobaea. Proc R Soc Med 53(Apr):283-284.

Miranda CL, Buhler DR, Cheeke PR (1979) Toxicity of tansy ragwort in rats. Toxicol Appl Pharmacol 48:A192.

Miranda CL, Cheeke PR, Schmitz JA, et al. (1980) Toxicity of Senecio jacobaea (tansy ragwort) in rats. Toxicol Appl Pharmacol 56(3):432-442.

Palfrey GD, MacLean KS, Langille WM (1967) Correlation between incidence of ragwort (Senecio jacobaea L.) poisoning and lack of mineral in cattle. Weed Res 7(2):171-175.

Pearson EG (1977) Clinical manifestations of tansy ragwort poisoning. Mod Vet Pract 58(5):421-424.

Standley HP (1924) Poisoning by ragwort (Senecio jacobaea). Vet J 80:35-36.

Swick RA, Cheeke PR, Goeger DE, et al. (1982) Effect of dietary Senecio jacobaea and injected Senecio alkaloids and monocrotaline on guinea pigs. J Anim Sci 55(6):1411-1416.

Swick RA, Cheeke PR, Miranda CL, et al. (1984) The effect of consumption of the pyrrolizidine alkaloid-containing plant Senecio jacobaea on iron and copper metabolism in the rat. J Environ Pharmacol Toxicol 5(4-5):59-69.

Thorpe E, Ford EJ (1968) Development of hepatic lesions in calves fed with ragwort (Senecio jacobaea). J Comp Pathol 78(2):195-205.

van Wuijckhuise L, Mars MH (2003) Antwoord auteurs: Nogmaals Jacobskruiskruidvergiftiging. Tijdschr Diergeneeskd 128(4):123-124.

Vockrodt H (1973) Vergiftungen landwirtschaftlicher Nutztiere durch Schadpflanzen und deren Vorbeuge. Monatsh Veterinarmed 28(2):59-62.

Vos JH, Geerts AA, Borgers JW, et al. (2002) Jacobskruiskruid: Bedrieglijke schoonheid. Vergifting met Senecio jacobaea. Tijdschr Diergeneeskd 127(24):753-756.

White RD, Swick RA, Cheeke PR (1984) Effects of dietary copper and molybdenum on tansy ragwort (Senecio jacobaea) toxicity in sheep. Am J Vet Res 45(1):159-161.

jacobée –see– *Jacobaea vulgaris* Gaertn.

Jacob's-ladder –see– *Convallaria majalis* L.; *Pedilanthus tithymaloides* (L.) Poit.

Jacob's-sword –see– *Iris pseudoacorus* L.

Jakobskraut –see– *Jacobaea vulgaris* Gaertn.

Jacobskreuzkraut –see– *Jacobaea vulgaris* Gaertn.

Jacobskruikruid –see– *Jacobaea vulgaris* Gaertn.

jaggery –see– *Saccharum officinarum* L.

jalap –see– *Phytolacca americana* L.

jalapa –see– *Allamanda cathartica* L.

jalapeño –see– *Capsicum annuum* L.; *Capsicum frutescens* L.

Jamaican apple –see– *Annona reticulata* L.

Jamaican sumach –see– *Metopium toxiferum* (L.) Krug & Urb.

Jamestown lily –see– *Datura stramonium* L.

Jamestown weed –see– *Datura stramonium* L.

jamuna –see– *Syzygium cumini* (L.) Skeels

janca tree –see– *Metopium toxiferum* (L.) Krug & Urb.

januariebos –see– *Gnidia polycephala* Gilg

japa –see– *Hibiscus rosa-sinensis* L.

Japanese anise –see– *Illicium anisatum* L.

Japanese aucuba –see– *Aucuba japonica* Thunb.

Japanese barnyard millet –see– *Echinochloa frumentacea* Link

Japanese black pine –see– *Pinus thunbergii* Parl.

Japanese cedar –see– *Cryptomeria japonica* (L. f.) D. Don

Japanese daisy –see– *Tanacetum cinerariifolium* (Trevir.) Sch. Bip.

Japanese date plum –see– *Diospyros kaki* Thunb.

Japanese fern palm –see– *Cycas revoluta* Thunb.

Japanese lacquer tree –see– *Toxicodendron vernicifluum* (Stokes) F. A. Barkley

Japanese laurel –see– *Aucuba japonica* Thunb.

Japanese millet –see– *Echinochloa frumentacea* Link

Japanese pepper –see– *Schinus terebinthifolius* Raddi

Japanese persimmon –see– *Diospyros kaki* Thunb.

Japanese pieris –see– *Pieris japonica* (Thunb.) D. Don ex G. Don

Japanese plum –see– *Eriobotrya japonica* (Thunb.) Lindl.

Japanese poinsettia –see– *Euphorbia heterophylla* L.

Japanese sago –see– *Cycas circinalis* L.

Japanese shellac –see– *Toxicodendron vernicifluum* (Stokes) F. A. Barkley

Japanese show lily –see– *Lilium lancifolium* Thunb.

Japanese star anise –see– *Illicium anisatum* L.

Japanese sumach –see– *Toxicodendron vernicifluum* (Stokes) F. A. Barkley

Japanese tallow tree –see– *Triadica sebifera* (L.) Small

Japanese varnish –see– *Toxicodendron vernicifluum* (Stokes) F. A. Barkley

Japanese wax tree –see– *Toxicodendron succedaneum* (L.) Kuntze

Japanese wisteria –see– *Wisteria floribunda* (Willd.) DC.

Japanese yew –see– *Taxus cuspidata* Siebold & Zucc.

Japanische Lackbaum –see– *Toxicodendron vernicifluum* (Stokes) F. A. Barkley

Japanische Primel –see– *Primula obconica* Hance

Japanlack –see– *Toxicodendron vernicifluum* (Stokes) F. A. Barkley

Japansk idegran –see– *Taxus cuspidata* Siebold & Zucc.

jarak blanda –see– *Jatropha curcas* L.

jaramago blanco –see– *Raphanus raphanistrum* L.

jaranjinha –see– *Polygala klotzschii* Chodat

jasmin –see– *Jasminum officinale* L.; *Jasminum sambac* (L.) Aiton

jasmin amarillo –see– *Allamanda cathartica* L.

jasmin-de-Caroline –see– *Gelsemium sempervirens* (L.) J. St.-Hil.

jasmin-de-nuit –see– *Cestrum nocturnum* L.

jasmin frances –see– *Moringa oleifera* Lam.

jasmine –see– *Cestrum diurnum* L.; *Gelsemium sempervirens* (L.) J. St.-Hil.; *Jasminum officinale* L.

JASMINUM ANGULARE Vahl [Oleaceae]

Citations:

Van der Walt SJ, Steyn DG (1943) Recent investigations into the toxicity of plants, etc. XIII. Onderstepoort J Vet Sci Anim Indus 18(1-2):207-224.

JASMINUM OFFICINALE L. [Oleaceae]

Common Names:

jasmin; jasmine; poet's-jasmine; white jasmine; yasmin

Citations:

Bedi BM (1971) Jasmine flower - Contact dermatitis (report of a case). Indian J Dermatol 16(3):61-62.

JASMINUM SAMBAC (L.) Aiton [Oleaceae]

Common Names:

Arabian jasmine; jasmin

Citations:

Singh R, Siddiqui MA, Baruah MC (1978) Plant dermatitis in Delhi. Indian J Med Res 68(Oct):650-655.

Jatropha aceroides (Pax & K. Hoffm.) Hutch. = Jatropha lobata (Forssk.) Müll. Arg. subsp. aceroides Pax & K. Hoffm.

JATROPHA CURCAS L. [Euphorbiaceae]

Common Names:

American purging nut; arbol santo; Barbados purging nut; big-purge nut; bini-da-zugu; black vomit nut; botuje; coquillo; coquito; Cuban physicnut; curcas bean; dravanti; habb-el-meluk; jarak blanda; kukai haole; physicnut; pinhão-de-purga; piñoncillo; purge nut; purgeerboontjie; purging nut; ramjada; Schwarze Brechnuß; tártago; tempate; tuba; tuba tuba; wild castor seed

Citations:

Abdu Aguye I, Sannusi A, Alafiya Tayo RA, et al. (1986) Acute toxicity studies with Jatropha curcas L. Hum Toxicol 5(4):269-274.

Adam SE (1974) Toxic effects of Jatropha curcas in mice. Toxicology 2(1):67-76.

Adam SE, Magzoub M (1975) Toxicity of Jatropha curcas (Euphorbiaceae) for goats. Toxicology 4(3):347-354.

Ahmed OM, Adam SE (1979) Effects of Jatropha curcas on calves. Vet Pathol 16(4):476-482.

Ahmed OM, Adam SE (1979) Toxicity of Jatropha curcas in sheep and goats. Res Vet Sci 27(1):89-96.

el Badwi SM, Adam SE, Hapke HJ (1992) Toxic effects of low levels of dietary Jatropha curcas seed on brown Hisex chicks. Vet Hum Toxicol 34(2):112-115.

el Badwi SM, Mousa HM, Adam SE, et al. (1992) Response of brown Hisex chicks to low levels of Jatropha curcas, Ricinus communis or their mixture. Vet Hum Toxicol 34(4):304-306.

Ho RK (1960) Acute poisoning from the ingestion of seeds of Jatropha curcas. Report of five cases. Hawaii Med J 19(4):421-423.

Joubert PH, Brown JM, Hay IT, et al. (1984) Acute poisoning with Jatropha curcas (purging nut tree) in children. S Afr Med J 65(18):729-730.

Liberalino AA, Bambirra EA, Moraes Santos T, et al. (1988) Jatropha curcas L. seeds: Chemical analysis and toxicity. Arq Biol Tecnologia 31(4):539-550.

Makkar HP, Becker K (1998) Jatropha curcas toxicity: Identification of toxic principle(s). In: Garland T, Barr AC (eds.) Toxic plants and other natural toxicants. CABI. New York. pp. 554-558.
Mampane KJ, Joubert PH, Hay IT (1987) Jatropha curcas: Use as a traditional tswana medicine and its role as a cause of acute poisoning. Phytother Res 1(1):50-51.
Panigrahi S, Francis BJ, Cano LA, et al. (1984) Toxicity of Jatropha curcas seeds from Mexico to rats and mice. Nutr Rep Int 29(5):1089-1099.
Randall JA (1914) Twenty-two cases of poisoning by the seed of Jatropha curcas. U S Nav Med Bull 8:290-291.
Taufa T (1973) Poisoning of school children with physic nuts. Papua New Guinea Med J 16(1):60-61.

JATROPHA DIOICA Sessé [Euphorbiaceae]

Common Names:

leather plant; leatherstem; rubber plant; rubber stem; sangre-de-drago

Citations:

Boughton IB, Hardy WT (1939) Hard yellow livers of sheep and cattle and feeding trials of suspected plants. Texas Agric Exp Sta Annu Rep 52:238-240.

JATROPHA GLAUCA Griseb. [Euphorbiaceae]

Citations:

Barri ME, Onsa TO, Elawad AA, et al. (1983) Toxicity of five Sudanese plants to young ruminants. J Comp Pathol 93(4):559-575.

JATROPHA GOSSYPIIFOLIA L. [Euphorbiaceae]

Common Names:

bellyache plant; bini-da-zugu; botuje papa; cottonleaf physicnut; physicnut; tuá tuá; wild cassava; wild physicnut

Citations:

Guirola L, Garcia G, Torrealba A, et al. (1992) Acute renal failure from the ingestion of toxic plants. Vet Hum Toxicol 34(6):548.

JATROPHA LOBATA (Forssk.) Müll. Arg. subsp. aceroides Pax & K. Hoffm. [Euphorbiaceae]

Synonyms:

Jatropha aceroides (Pax & K. Hoffm.) Hutch.

Citations:

Barri ME, Onsa TO, Elawad AA, et al. (1983) Toxicity of five Sudanese plants to young ruminants. J Comp Pathol 93(4):559-575.

JATROPHA MACRORHIZA Benth. [Euphorbiaceae]

Common Names:

desert potato; jicamilla

Citations:

Consroe PF, Glow DE (1975) Clinical toxicology of the "Desert potato": Two case reports of acute Jatropha macrorhiza root ingestion. Ariz Med 32(6):475-477.

Jatropha manihot L. = Manihot esculenta Crantz

JATROPHA MULTIFIDA L. [Euphorbiaceae]

Common Names:

arbre corail; avellana-purgante-de-Santo Domingo; botuje; bouquet corail; coral plant; emetico; flor-de-coral; French physicnut; médicinier bâtard; médicinier d'espagne; noisetier purgatif; petit médicinier; physicnut; pinhoen; pino-de-coral; piñol-de-cumaña; purgante-de-España; purging nut; Santa Ana; small physicnut; tartago

Citations:

Baruffa G (1964) Avvelenamento da semi di Jatropha multifida (L). Arch Ital Sci Med Trop Parassitol 45(May):261-264.
Fawcett NP (1978) Pediatric facets of poisonous plants. J Fla Med Assoc 65(3):199-204.
Flecker H (1945) Injuries produced by plants in tropical Queensland. Med J Aust 1(Jun 23):636-637.
Levin Y, Sherer Y, Bibi H, et al. (2000) Rare Jatropha multifida intoxication in two children. J Emerg Med 19(2):173-175.
Swaddiwudhipong W, Kunasol P, Sangwanloy O, et al. (1989) Foodborne disease outbreaks of chemical etiology in Thailand, 1981-1987. Southeast Asian J Trop Med Public Health 20(1):125-132.

JATROPHA PODAGRICA Hook. [Euphorbiaceae]

Common Names:

coral plant; gout stalk; Guatamalan rhubarb; physicnut; podagrica

Citations:

Swaddiwudhipong W, Kunasol P, Sangwanloy O, et al. (1989) Foodborne disease outbreaks of chemical etiology in Thailand, 1981-1987. Southeast Asian J Trop Med Public Health 20(1):125-132.

Jatropha urens L. = Cnidoscolus urens (L.) Arthur

Java bean –see– *Phaseolus lunatus* L.

Java indigo –see– *Indigofera arrecta* Hochst. ex A. Rich

Java olive –see– *Sterculia foetida* L.

Java teak –see– *Tectona grandis* L. f.

Java willow –see– *Ficus benjamina* L.

Javabohne –see– *Phaseolus lunatus* L.

Javanese lac tree –see– *Gluta renghas* L.

Javanischer Giftbaum –see– *Antiaris toxicaria* Lesch.

javillo –see– *Hura crepitans* L.

jayapala –see– *Croton tiglium* L.

Jean Robert –see– *Chamaesyce hirta* (L.) Millsp.

jelutong –see– *Dyera costulata* (Miq.) Hook. f.

jengkol bean –see– *Archidendron jiringa* (Jack) I. C. Nielsen

jenkol –see– *Archidendron jiringa* (Jack) I. C. Nielsen

jenkol bean –see– *Archidendron jiringa* (Jack) I. C. Nielsen jequirity bean –see– *Abrus precatorius* L.

jequirity pea –see– *Abrus precatorius* L.

jering –see– *Archidendron jiringa* (Jack) I. C. Nielsen

Jerusalem berry –see– *Solanum pseudocapsicum* L.

Jerusalem cherry –see– *Solanum pseudocapsicum* L.

Jerusalem oak –see– *Dysphania ambrosioides* (L.) Mosyakin & Clemants

Jerusalemkirsche –see– *Solanum pseudocapsicum* L.

jessamine –see– *Cestrum diurnum* L.; *Gelsemium sempervirens* (L.) J. St.-Hil.

jetbead –see– *Rhodotypos scandens* (Thunb.) Makino

jetberry bush –see– *Rhodotypos scandens* (Thunb.) Makino

Jewbush –see– *Pedilanthus tithymaloides* (L.) Poit.

jhunjhunia –see– *Crotalaria nana* Burm f.

jicama –see– *Pachyrhizus erosus* (L.) Urb.

jicamilla –see– *Jatropha macrorhiza* Benth.

jimãozinho –see– *Polygala klotzschii* Chodat

jimbey –see– *Leucaena leucocephala* (Lam.) de Wit

jimmy fern –see– *Astrolepis cochisensis* (Goodd.) D. M. Benham & Windham

jimmy weed –see– *Isocoma plurifolia* (Torr. & A. Gray) Greene

jimson weed –see– *Datura inoxia* Mill.; *Datura metel* L.; *Datura stramonium* L.; *Datura wrightii* Regel

jin-bu-huan –see– *Polygala chinensis* L.; *Huperzia serrata* (Thunb.) Trevis.

jitong –see– *Gluta renghas* L.

jivanti –see– *Leptadenia reticulata* (Retz.) Wight & Arn.

jo jo –see– *Soliva sessilis* Ruiz & Pav.

joá-preto –see– *Solanum fastigiatum* Willd.

Joan Silverpin –see– *Papaver somniferum* L.

JOANNESIA PRINCEPS Vell. [Euphorbiaceae]

Common Names:

anda; anda assu; cotieira

Citations:

Nishioka SA, Escalante RD (1997) Poisoning by the ingestion of seeds of the fruit of the "cotieira" (Joannesia princeps). Rev Paul Med 115(1):1366-1367.

jocote marañon –see– *Anacardium occidentale* L.

joggemscholtzbossie –see– *Pteronia pallens* L. f.

Johanniskraut –see– *Hypericum perforatum* L.

Johnsongrass –see– *Sorghum halepense* (L.) Pers.

joint charlock –see– *Raphanus raphanistrum* L.

joint fir –see– *Ephedra viridis* Coville

joint rush –see– *Equisetum arvense* L.; *Juncus holoschoenus* R. Br.

jojoba –see– *Simmondsia chinensis* (Link) C. K. Schneid.

Jonesiella recedens Rydb. = Astragalus praelongus E. Sheld.

jonquil –see– *Narcissus jonquilla* L.

jorro jorro –see– *Thevetia peruviana* (Pers.) K. Schum.

Joseph's-coat –see– *Perilla frutescens* (L.) Britton

jowar –see– *Sorghum bicolor* (L.) Moench

juar –see– *Sorghum bicolor* (L.) Moench

juglans –see– *Juglans nigra* L.

JUGLANS NIGRA L. [Juglandaceae]

Common Names:

black walnut; juglans; walnut

Citations:

Eaton SA, Allen D, Eades SC, et al. (1995) Digital Starling forces and hemodynamics during early laminitis induced by an aqueous extract of black walnut (Juglans nigra) in horses. Am J Vet Res 56(10):1338-1344.

Galey FD, Twardock AR, Goetz TE, et al. (1990) Gamma scintigraphic analysis of the distribution of perfusion of blood in the equine foot during black walnut (Juglans nigra)-induced laminitis. Am J Vet Res 51(4):688-695.

McConnico R, Stokes AM, Eades SC, et al. (2005) Investigation of the effect of black walnut extract on in vitro transport and structure of equine colonic mucosa. Am J Vet Res 66(3):443-449.

Peterson DE (1984) Equine laminitis associated with black walnut toxicity. Minnesota Vet 24(1):38-43.

Ralston SL, Rich VA (1983) Black walnut toxicosis in horses. J Am Vet Med Assoc 183(10):1095.

Siegel JM (1954) Dermatitis due to black walnut juice. Arch Derm Syphilol 70(4):511-513.

Uhlinger C (1989) Black walnut toxicosis in ten horses. J Am Vet Med Assoc 195(3):343-344.

JUGLANS OLANCHANA Standl. & L. O. Williams [Juglandaceae]

Common Names:

Central American walnut

Citations:

Bush RK, Clayton D (1983) Asthma due to Central American walnut (Juglans olanchana) dust. Clin Allergy 13(4):389-394.

JUGLANS REGIA L. [Juglandaceae]

Common Names:

Circassian walnut; English walnut; Madeira walnut; nogal; Persian walnut; Walnuß

Citations:

Bonamonte D, Foti C, Angelini G (2001) Hyperpigmentation and contact dermatitis due to Juglans regia. Contact Dermatitis 44(2):101-102.

Mendonca C, Madan V, Austin S, et al. (2005) Occupational contact urticaria from walnut associated with hand eczema. Contact Dermatitis 53(3):173-174.

Neri I, Bianchi F, Giacomini F, et al. (2006) Acute irritant contact dermatitis due to Juglans regia. Contact Dermatitis 55(1):62-63.

juhsóska –see– *Rumex acetosella* L.

jujube –see– *Ziziphus jujuba* Mill.

jumbee bead –see– *Abrus precatorius* L.

jumbey –see– *Leucaena leucocephala* (Lam.) de Wit

JUNCUS HOLOSCHOENUS R. Br. [Juncaceae]

Common Names:

joint rush

Citations:

Albiston HE (1937) The joint leaf rush (Juncus holoschoenus): A cyanogenetic plant. Aust Vet J 13(Oct):200.

JUNELLIA LIGUSTRINA (Lag.) Moldenke [Verbenaceae]

Synonyms:

Lippia ligustrina (Lag.) Britton

Common Names:

whitebrush

Citations:

Mathews FP (1941) Poisonous plants in the Davis Mountains. Texas Agric Exp Sta Annu Rep 54:93.

Mathews FP (1942) Whitebrush (Lippia ligustrina) poisoning in horses. J Am Vet Med Assoc 101(784):35-38.

juniper –see– *Juniperus communis* L.; *Juniperus osteosperma* (Torr.) Little; *Juniperus oxycedrus* L.

JUNIPERUS CHINENSIS L. [Cupressaceae]

Common Names:

Chinese juniper

Citations:

Dooms-Goossens A, Maertens M, van Lint L, et al. (1984) Colophony-induced sensitivity to Juniperus chinensis L. 'Hetzii'? Contact Dermatitis 10(3):185-187.

JUNIPERUS COMMUNIS L. [Cupressaceae]

Common Names:

Canadian juniper; dwarf juniper; genièvre; Heidewacholder; juniper; közöbséges boróka; Kranbaum; little juniper; Machandelbaum; mineur; mountain juniper; pétron; vulgaire; Wacholder

Citations:

Gardner DR, Panter KE, James LF, et al. (1998) Abortifacient effects of lodgepole pine (Pinus contorta) and common juniper (Juniperus communis) on cattle. Vet Hum Toxicol 40(5):260-263.

JUNIPERUS MONOSPERMA (Engelm.) Sarg. [Cupressaceae]

Common Names:

one-seed juniper

Citations:

Holechek JL, Munshikpu AV, Saiwana L, et al. (1990) Influences of six shrub diets varying in phenol content on intake and nitrogen retention by goats. Trop Grasslands 24(2):93-98.

JUNIPERUS OSTEOSPERMA (Torr.) Little [Cupressaceae]

Common Names:

juniper; little Utah juniper; Utah juniper

Citations:

Johnson AE, James LF, Spillett J (1976) The abortifacient and toxic effects of big sagebrush (Artemisia tridentata) and juniper (Juniperus osteosperma) on domestic sheep. J Range Manag 29(4):278-280.

JUNIPERUS OXYCEDRUS L. [Cupressaceae]

Common Names:

cade juniper; juniper

Citations:

Grolnick M (1938) Dermatitis due to haemorrhoidal ointment containing Krameria and oil of cade. JAMA 110(13):951-953.

Lord LW (1933) Oil of cadeberry. A little known drug that is valuable in certain dermatoses. Arch Derm Syphilol 28:29-31.

Nelson T (1932) Contact dermatitis to oil of cade. J Allergy 3:319.

Rahmani H, Leonhardt S, Beladdale D, et al. (2004) Severe acute lung oedema after rectal enema with cade oil. J Toxicol Clin Toxicol 42(4):487.

JUNIPERUS PINCHOTII Sudw. [Cupressaceae]

Common Names:

red-berry juniper

Citations:

Straka E, Scott CB, Taylor CA Jr, et al. (2004) Biological control of the toxic shrub juniper. In: Acamovic T, Stewart CS, Pennycott TW (eds.) Poisonous plants and related toxins. CABI. Wallingford, UK. pp. 436-442.

Juno's-tears –see– *Verbena officinalis* L.

jusquiame –see– *Hyoscyamus niger* L.

jusquiame noire –see– *Hyoscyamus niger* L.

juta –see– *Corchorus capsularis* L.

jute –see– *Corchorus capsularis* L.

K

Kaffee –see– *Coffea arabica* L.

Kaffeebohne –see– *Coffea arabica* L.

kafferkoring –see– *Sorghum bicolor* (L.) Moench

kafir corn –see– *Sorghum bicolor* (L.) Moench

kafir lily –see– *Clivia miniata* (Lindl.) Regel

kafir onion –see– *Boophone disticha* (L. f.) Herb.

kahili flower –see– *Grevillea banksii* R. Br.

kaipajua –see– *Glinus oppositifolius* (L.) Aug. DC.

kaki –see– *Diospyros kaki* Thunb.

kalanchoe –see– *Kalanchoe daigremontiana* Raym.-Hamet & H. Perrier; *Kalanchoe delagoensis* Eckl. & Zeyh.; *Kalanchoe pinnata* (Lam.) Pers.; *Kalanchoe prolifera* (Bowie ex Hook.) Raym.-Hamet

KALANCHOE DAIGREMONTIANA Raym.-Hamet & H. Perrier [Crassulaceae]

Synonyms:

Bryophyllum daigremontianum (Raym.-Hamet & H. Perrier) A. Berger

Common Names:

air plant; Christmas bells; devil's-backbone; kalanchoe; Mexican hat plant; mission bells; mother-of-millions; mother-of-thousands

Citations:

McKenzie RA, Dunster PJ (1986) Hearts and flowers: Bryophyllum poisoning of cattle. Aust Vet J 63(7):222-227.

McKenzie RA, Franke FP, Dunster PJ (1987) The toxicity to cattle and bufadienolide content of six Bryophyllum species. Aust Vet J 64(10):298-301.

Williams MC, Smith MC (1984) Toxicity of Kalanchoe spp to chicks. Am J Vet Res 45(3):543-546.

KALANCHOE DELAGOENSIS Eckl. & Zeyh. [Crassulaceae]

Synonyms:

Bryophyllum tubiflorum Harv.; ***Kalanchoe tubiflora*** (Harv.) Raym.-Hamet

Common Names:

chandelier plant; Christmas bells; kalanchoe; mission bells; mother-of-millions

Citations:

McKenzie RA, Dunster PJ (1986) Hearts and flowers: Bryophyllum poisoning of cattle. Aust Vet J 63(7):222-227.

McKenzie RA, Dunster PJ (1987) Curing experimental Bryophyllum tubiflorum poisoning of cattle with activated carbon, electrolyte replacement solution and antiarrhythmic drugs. Aust Vet J 64(7):211-214.

Williams MC, Smith MC (1984) Toxicity of Kalanchoe spp to chicks. Am J Vet Res 45(3):543-546.

KALANCHOE FEDTSCHENKOI Raym.-Hamet & H. Perrier [Crassulaceae]

Common Names:

lavender scallops; South American air plant

Citations:

McKenzie R (1987) Mother-of-millions. In: Covacevich et al. (eds.) Toxic plants and animals. A guide for Australia. pp. 51-55.

Williams MC, Smith MC (1984) Toxicity of Kalanchoe spp to chicks. Am J Vet Res 45(3):543-546.

KALANCHOE INTEGRA (Medik.) Kuntze [Crassulaceae]

Citations:

Varma RK (1986) Antidotal treatment of experimental Kalanchoe integra poisoning. Indian J Anim Sci 56(4):411.

Varma RK, Garg BD, Kharole MU, et al. (1981) Chronic toxicity studies on Kalanchoe integra in sheep. Indian J Anim Sci 51(5):522-526.

KALANCHOE LANCEOLATA (Forssk.) Pers. [Crassulaceae]

Common Names:

gadalin kura

Citations:

Anderson LA, Schultz RA, Joubert JP, et al. (1983) Krimpsiekte and acute cardiac glycoside poisoning in sheep caused by bufadienolides from the plant Kalanchoe lanceolata Forsk. Onderstepoort J Vet Res 50(4):295-300.

Masvingwe C, Mavenyengwa M (1997) Kalanchoe lanceolata poisoning in Brahman cattle in Zimbabwe: The first field outbreak. J S Afr Vet Assoc 68(1):18-20.

Schultz RA, Kellerman TS, Van den Berg H (2005) The role of fluorescence polarization immuno-assay in the diagnosis of plant-induced cardiac glycoside poisoning livestock in South Africa. Onderstepoort J Vet Res 72(3):189-201.

KALANCHOE PINNATA (Lam.) Pers. [Crassulaceae]

Synonyms:

Bryophyllum pinnatum (Lam.) Oken

Common Names:

air plant; cathedral bells; floppers; kalanchoe; life plant; live forever; live leaf; Mexican love plant; resurrection plant; sinalo todo

Citations:

McKenzie R (1987) Mother-of-millions. In: Covacevich et al. (eds.) Toxic plants and animals. A guide for Australia. pp. 51-55.

McKenzie RA, Dunster PJ (1986) Hearts and flowers: Bryophyllum poisoning of cattle. Aust Vet J 63(7):222-227.

Reppas GP (1995) Byophyllum pinnatum poisoning of cattle. Aust Vet J 72(11):425-427.

KALANCHOE PROLIFERA (Bowie ex Hook.) Raym.-Hamet [Crassulaceae]

Synonyms:

Bryophyllum proliferum Bowie ex Hook.

Common Names:

kalanchoe

Citations:

McKenzie R (1987) Mother-of-millions. In: Covacevich et al. (eds.) Toxic plants and animals. A guide for Australia. pp. 51-55.

McKenzie RA, Dunster PJ (1986) Hearts and flowers: Bryophyllum poisoning of cattle. Aust Vet J 63(7):222-227.

Shone DK (1966) Poisonous plants of Rhodesia. Part I. Rhodesia Zambia Malawi J Agric Res 4:81-94.

Shone DK (1967) Poisonous plants of Rhodesia. Part II. Rhodesia Zambia Malawi J Agric Res 5:29-41.

Kalanchoe tubiflora (Harv.) Raym.-Hamet = Kalanchoe delagoensis Eckl. & Zeyh.

kale –see– *Brassica oleracea* L. var. viridis L.

kali mona –see– *Paspalum scrobiculatum* L. var. bispicatum Hack.

kaliphool –see– *Salvia coccinea* Buc'hoz ex Etl.

Kalla –see– *Calla palustris* L.

KALLSTROEMIA HIRSUTISSIMA Vail ex Small [Zygophyllaceae]

Common Names:

carpetweed; hairy caltrop

Citations:

Mathews FP (1944) The toxicity of Kallstroemia hirsutissima (carpet weed) for cattle, sheep, and goats. J Am Vet Med Assoc 105(810):152-155.

Wilson RD, Witzel DA, Verlander JM (1982) Somatosensory-evoked response of ataxic Angora goats in suspected haloxon-delayed neurotoxicity. Am J Vet Res 43(12):2224-2226.

Kallstroemia intermedia Rydb. = Kallstroemia parviflora Norton

KALLSTROEMIA PARVIFLORA Norton [Zygophyllaceae]

Synonyms:

Kallstroemia intermedia Rydb.

Common Names:

warty caltrop

Citations:

Dollahite JW (1975) Toxicity of Kallstroemia parviflora (warty caltrop) to sheep, goats and rabbits. Southwestern Vet 28(2):135-139.

KALMIA ANGUSTIFOLIA L. [Ericaceae]

Common Names:

Berglorbeer; calfkill; dwarf laurel; dwarf sheep laurel; kidkill; lamb laurel; lambkill; lambkill Kalmia; laurel; Lorbeerose; low laurel; narrow-leaf laurel; sheep laurel; sheep poison; sheep poison laurel; sheepkill; small laurel; spoonwood; spoonwood ivy; wicky; wintergreen

Citations:

Marsh CD, Clawson AB (1930) Mountain-laurel (Kalmia latifolia) and sheep laurel (Kalmia angustifolia) as stock-poisoning plants. U S Dep Agric Tech Bull #219:22 pp.

Pritchard WR (1956) Laurel (Kalmia angustifolia) poisoning of sheep. North Am Vet 37(Jun):461-462.

Wasserman B (1959) Sheep laurel poisoning in the cat. A case report. J Am Vet Med Assoc 135(Dec 1):569.

KALMIA LATIFOLIA L. [Ericaceae]

Common Names:

American laurel; big ivy; broad-leaf laurel; calico bush; great laurel; high laurel; ivy bush; ivywood; laurel; mountain ivy; mountain laurel; poison laurel; rose laurel; round-leaf laurel; spoon hunt; spoon hutch; spoonwood; wicky; wintergreen; wood laurel

Citations:

Anonymous (1922) The problem of poisonous range plants. Nevada Agric Exp Sta Annu Rep 1921:10-12.

Anonymous (1969) Plant toxin poisoned the honey. New Sci (1956) 41(Jan 30):218.

Crawford AC (1908) Mountain laurel, a poisonous plant. U S Dep Agric Bur Plant Indus Bull #121(2):21-35.

Marsh CD, Clawson AB (1930) Mountain-laurel (Kalmia latifolia) and sheep laurel (Kalmia angustifolia) as stock-poisoning plants. U S Dep Agric Tech Bull #219:22 pp.

Rusby HH (1902) The poisonous properties of mountain laurel. Druggists Circ Chem Gaz 46(Feb):27.

KALMIA MICROPHYLLA (Hook.) A. Heller [Ericaceae]

Common Names:

alpine kalmia; alpine laurel; bog laurel; pale laurel; small-leaf laurel; western laurel; western swamp kalmia; western swamp laurel

Citations:

Clawson AB (1933) Alpine kalmia (Kalmia microphylla) as a stock-poisoning plant. U S Dep Agric Tech Bull #391:9 pp.

Fleming CE (1920) Poisonous range plants. Nevada Agric Exp Sta Annu Rep 1919:39-43.

Kalmus –see– *Acorus calamus* L.

kalo nut –see– *Vernicia fordii* (Hemsl.) Airy Shaw
kaluduru –see– *Nigella sativa* L.
kamaisa –see– *Croton tiglium* L.
kamala –see– *Mallotus philippensis* (Lam.) Müll. Arg.
Kamalindo –see– *Tamarindus indica* L.
kambala –see– *Milicia excelsa* (Welw.) C. C. Berg
kambala teak –see– *Milicia excelsa* (Welw.) C. C. Berg
kameelboom –see– *Acacia* ×*giraffae* Willd.
kameeldoring –see– *Acacia* ×*giraffae* Willd.
Kamille –see– *Chamaemelum nobile* (L.) All.; *Matricaria recutita* L.
kamoon –see– *Cuminum cyminum* L.
kan-to-ka –see– *Tussilago farfara* L.
Kanadische Rebe –see– *Parthenocissus quinquefolia* (L.) Planch.
kandlaarbos –see– *Tylecodon wallichii* (Harv.) Toelken
kaner –see– *Nerium oleander* L.
kanir –see– *Thevetia peruviana* (Pers.) K. Schum.
kanja –see– *Millettia pinnata* (L.) Panigrahi
kankera –see– *Gymnosporia spinosa* (Blanco) Merr. & Rolfe
kankerbos –see– *Euphorbia ingens* E. Mey. ex Boiss.
Kansas thistle –see– *Solanum rostratum* Dunal
Kapuziner Bohne –see– *Phaseolus coccineus* L.
Kapuzinerkresse –see– *Tropaeolum majus* L.
karaka –see– *Corynocarpus laevigatus* J. R. Forst. & G. Forst.
karam –see– *Commiphora myrrha* (Nees) Engl.; *Haldina cordifolia* (Roxb.) Ridsdale
karanja –see– *Milletia pinnata* (L.) Panigrahi
karar –see– *Terminalia chebula* Retz.
karara –see– *Mucuna pruriens* (L.) DC.
karati –see– *Dichapetalum stuhlmanii* Engl.
karaya gum –see– *Sterculia urens* Roxb.
karayu gum –see– *Sterculia urens* Roxb.
karela –see– *Momordica charantia* L.
karihari –see– *Gloriosa superba* L.
Karotte –see– *Daucus carota* L.
Kartoffel –see– *Solanum tuberosum* L.

KARWINSKIA CALDERONII Standl. [Rhamnaceae]

Common Names:

buckthorn; cacachilla; huilihuiste; tullidora

Citations:

Ascherio A, Bermudez CS, Garcia D (1992) Outbreak of buckthorn paralysis in Nicaragua. J Trop Pediatr 38(2):87-89.

KARWINSKIA HUMBOLDTIANA (Schult.) Zucc. [Rhamnaceae]

Common Names:

buckthorn; cacachilla; cacatsi; cajotillo; callotrio; capulín tullidor; capulín tullidore; capulincillo; cayote; chanchanote; coyotillo; coyotillo-de-Texas; coyotio; gallita bush; la tullidora; margarita; palo negrito; pimientillo; riventdore; tanglefoot; tullidora; wild cherry

Citations:

Arreola Nava ME, Vázquez Castellaños JL, González Castañeda ME (2000) Factores geográficos en la epidemiología de la intoxicación por Karwinskia (tullidora) en Mexico. Cad Saude Publica 16(1):255-260.

Bermúdez-De Rocha MV, Lozano-Meléndez FE, Tamez Rodríguez VA, et al. (1995) Frecuencia da intoxicación con Karwinskia humboldtiana en México. Salud Publica Mex 37(1):57-62.

Bermúdez MV, Gonzalez Spencer D, Guerrero M, et al. (1986) Experimental intoxication with fruit and purified toxins of buckthorn (Karwinskia humboldtiana). Toxicon 24(11-12):1091-1097.

Bermúdez MV, Martinez FJ, Salazar ME, et al. (1992) Experimental acute intoxication with ripe fruit of Karwinskia humboldtiana (tullidora) in rat, guinea-pig, hamster, and dog. Toxicon 30(11):1493-1496.

Bustamante-Sarabia J, Olvera-Rabiela JE, Correa Nieto-Canedo L (1978) Intoxication fatal por tullidora (Karwinskia humboldtiana). Comunicacion de un caso. Gac Med Mex 114(5):241-244.

Calderon-Gonzalez R, Rizzi-Hernandez H (1967) Buckthorn polyneuropathy. N Engl J Med 277(2):69-71.

Carrada-Bravo T, López-Leal H, Vázquez-Arias G, et al. (1983) Brote epidémico de polirradiculoneuritis por tullidora Karwinskia humboldtiana. Bol Med Hosp Infant Mex 40(3):139-147.

Charlton KM, Claborn LD, Pierce KR (1971) A neuropathy in goats caused by experimental coyotillo (Karwinskia humboldtiana) poisoning: Clinical and neurophysiologic studies. Am J Vet Res 32(9):1381-1389.

Charlton KM, Pierce KR (1969) Peripheral neuropathy in experimental coyotillo poisoning in goats. Tex Rep Biol Med 27(2):389-399.

Charlton KM, Pierce KR (1970) A neuropathy in goats caused by experimental coyotillo (Karwinskia humboldtiana) poisoning. 2. Lesions in the peripheral nervous system: Teased fiber and acid phosphatase studies. Pathol Vet 7(5):385-407.

Charlton KM, Pierce KR (1970) A neuropathy in goats caused by experimental coyotillo (Karwinskia humboldtiana) poisoning. 3. Distribution of lesions in peripheral nerves. Pathol Vet 7(5):408-419.

Charlton KM, Pierce KR (1970) A neuropathy in goats caused by experimental coyotillo (Karwinskia humboldtiana) poisoning. 4. Light and electron microscopic lesions in peripheral nerves. Pathol Vet 7(5):420-434.

Charlton KM, Pierce KR, Storts RW, et al. (1970) A neuropathy in goats caused by experimental coyotillo (Karwinskia humboldtiana) poisoning. 5. Lesions in the central nervous system. Pathol Vet 7(5):435-447.

Del Pozo EC (1965) Los effectos paralizantes de la "tullidora" estudios clinícos y experimentales. Gac Med Mex 95(2):179-182.

Dewan ML, Henson JB, Dollahite JW, et al. (1965) Toxic myodegeneration in goats produced by feeding mature fruits from the coyotillo plant (Karwinskia humboldtiana). Am J Pathol 46(2):215-226.
Escobar Izquierdo A, Nieto D (1965) Aspectos neuropatologicos de la intoxicación con Karwinskia humboldtiana intoxication. Estudio experimental. Gac Med Mex 95(2):163-177.
Espinola Cantón JJ (1972) Estudio preliminar, con el microscopio electrónico, de las alteraciones estructurales de las células hepáticas, renales, musculares y en las vainas de mielina de ratones intoxicados con Karwinskia humboldtiana. Veterinaria 3(1):12-17.
Flores Otero G, Cueva J, Muñoz Martinez EJ, et al. (1987) Spectrophotometric and chromatographic detection of Karwinskia humboldtiana (tullidora) toxin in rat serum after tullidora ingestion. Toxicon 25(4):419-426.
Jarmillo-Juárez F, Ortiz GG, Vázquez ML, et al. (1995) Renal failure during acute toxicity produced by tullidora ingestion (Karwinskia humboldtiana). Gen Pharmacol 26(3):649-653.
Jaramillo-Juárez F, Rodríguez-Vázquez ML, Muñoz-Martínez J, et al. (2005) The ATP levels in kidneys and blood are mainly decreased by acute ingestion of tullidora (Karwinskia humboldtiana). Toxicon 46(1):99-103.
Marsh CD, Clawson AB, Roe GC (1928) Coyotillo (Karwinskia humboldtiana) as a poisonous plant. U S Dep Agric Tech Bull #29:26 pp.
Martinez HR, Bermudez MV, Rangel Guerra RA, et al. (1998) Clinical diagnosis in Karwinskia humboldtiana polyneuropathy. J Neurol Sci 154(1):49-54.
Muñoz-Martínez EJ, Cueva J, Joseph-Nathan P (1983) Denervation caused by tullidora (Karwinskia humboldtiana). Neuropathol Appl Neurobiol 9(2):121-134.
Ortiz GG, Gonzalez Burgos I, Feria Velasco A (1992) Structural study of the acute effect of Karwinskia humboldtiana on cerebral motor cortex, hippocampus, and caudate nucleus of the rat. Gen Pharmacol 23(3):543-547.
Padron Puyou F (1951) Estudio clinico-experimental de la paralisis por Karwinskia humboldtiana ("Tullidora") en niños. Gac Med Mex 81(2-3-4):299-311.
Puértolas Márquez MA, Nava Jiménez O, Medina Lopéz HA, et al. (1984) Polirradiculoneuritis por Karwinskia humboldtiana. Informe de seis casos. Rev Med Inst Mex Seguro Soc 22(1):25-27.

KARWINSKIA JOHNSTONII R. Fernández [Rhamnaceae]

Citations:

Arreola Nava ME, Vázquez Castellanos JL, González Castaneda ME (2000) Factores geográficos en la epidemiología de la intoxicación por Karwinskia (tullidora) en Mexico. Cad Saude Publica 16(1):255-260.

KARWINSKIA MOLLIS Schltdl. [Rhamnaceae]

Citations:

Arreola Nava ME, Vázquez Castellanos JL, González Castaneda ME (2000) Factores geográficos en la epidemiología de la intoxicación por Karwinskia (tullidora) en Mexico. Cad Saude Publica 16(1):255-260.

KARWINSKIA PARVIFOLIA Rose [Rhamnaceae]

Common Names:

coyotillo

Citations:

Arreola Nava ME, Vázquez Castellanos JL, González Castaneda ME (2000) Factores geográficos en la epidemiología de la intoxicación por Karwinskia (tullidora) en Mexico. Cad Saude Publica 16(1):255-260.

KARWINSKIA RZEDOWSKII R. Fernández [Rhamnaceae]

Citations:

Arreola Nava ME, Vázquez Castellanos JL, González Castaneda ME (2000) Factores geográficos en la epidemiología de la intoxicación por Karwinskia (tullidora) en Mexico. Cad Saude Publica 16(1):255-260.

Kastanie –see– *Aesculus glabra* Willd.; *Castanea sativa* Mill.

kat –see– *Catha edulis* (Vahl) Forssk. ex Endl.

kateli –see– *Argemone mexicana* L.

katzimelk –see– *Prosopis glandulosa* Torr.

kaunch –see– *Mucuna pruriens* (L.) DC.

kava –see– *Piper methysticum* G. Forst.

kava kava –see– *Piper methysticum* G. Forst.

kavaanuo bos –see– *Senecio ilicifolius* L.

kawa –see– *Piper methysticum* G. Forst.

KEDROSTIS NANA Cogn. [Cucurbitaceae]

Citations:

Van der Walt SJ, Steyn DG (1940) Recent investigations into the toxicity of known and unknown poisonous plants in the Union of South Africa, X. Onderstepoort J Vet Res 15(1-2):261-277.

keel crumbweed –see– *Chenopodium carinatum* R. Br.

keel goosefoot –see– *Chenopodium carinatum* R. Br.

kejaat –see– *Pterocarpus angolensis* DC.

keladi –see– *Alocasia longiloba* Miq.

keladi chandek –see– *Alocasia longiloba* Miq.

Kellerhals –see– *Daphne mezereum* L.

Kellogg's-spur lupine –see– *Lupinus argenteus* Pursh var. heteranthus (S. Watson) Barneby

Kelsey's-milk vetch –see– *Astragalus atropubescens* J. M. Coult. & Fisch.

Kentish balsam –see– *Mercurialis perennis* L.

Kentucky coffee tree –see– *Gymnocladus dioicus* (L.) K. Koch

Kentucky coffeebean –see– *Gymnocladus dioicus* (L.) K. Koch

Kentucky mahogany –see– *Gymnocladus dioicus* (L.) K. Koch

kerek repkény –see– *Glechoma hederacea* L.

Kermesbeere –see– *Phytolacca americana* L.

kerriebos –see– *Hypericum revolutum* Vahl

kesari dal –see– *Lathyrus sativus* L.

ketjubung –see– *Datura metel* L.

kew –see– *Piper methysticum* G. Forst.

kew tree –see– *Ginkgo biloba* L.

khair –see– *Acacia catechu* (L. f.) Willd.

khalibui –see– *Heliotropium ellipticum* Ledeb.

khandai –see– *Melilotus albus* Medik.

khat –see– *Catha edulis* (Vahl) Forssk. ex Endl.

khaya –see– *Khaya anthotheca* (Welw.) C. DC.

KHAYA ANTHOTHECA (Welw.) C. DC. [Meliaceae]

Synonyms:

Khaya nyasica Stapf ex Baker f.

Common Names:

African mahogany; khaya

Citations:

Estlander T, Jolanki R, Alanko K, et al. (2001) Occupational allergic contact dermatitis caused by wood dusts. Contact Dermatitis 44(4):213-217.

Morgan JW, Orsler RJ, Wilkinson DS (1968) Dermatitis due to the wood dusts of Khaya anthotheca and Machaerium scleroxylon. Br J Ind Med 25(2):119-125.

Morgan JW, Wilkinson DS (1965) Sensitization to Khaya anthotheca. Nature 207(5001):1101.

Wilkinson DS (1968) Wood dermatitis. Sensitization to Khaya anthotheca (African mahogany). Br J Dermatol 80(3):195-196.

Wilkinson DS (1971) Tests with different species of Khaya woods. Contact Dermatol Newsl 9(Jan):216.

KHAYA EURYPHYLLA Harms [Meliaceae]

Citations:

Wilkinson DS (1971) Tests with different species of Khaya woods. Contact Dermatol Newsl 9(Jan):216.

KHAYA IVORENSIS A. Chev. [Meliaceae]

Synonyms:

Khaya klainei Pierre ex Pellegr.

Common Names:

African mahogany; Afrikanisches Mahagoni; samanguilla

Citations:

Dantin-Gallego J, Armayor AF, Riesco J (1952) Some new toxic woods: Some new manifestations of toxicity. Ind Med Surg 21(2):41-46.

Estlander T, Jolanki R, Alanko K, et al. (2001) Occupational allergic contact dermatitis caused by wood dusts. Contact Dermatitis 44(4):213-217.

Wilkinson DS (1968) Wood dermatitis. Sensitization to Khaya anthotheca (African mahogany). Br J Dermatol 80(3):195-196.

Wilkinson DS (1971) Tests with different species of Khaya woods. Contact Dermatol Newsl 9(Jan):216.

Khaya klainei Pierre ex Pellegr. = Khaya ivorensis A. Chev.

KHAYA MADAGASCARIENSIS Jum. & H. Perrier [Meliaceae]

Citations:

Wilkinson DS (1971) Tests with different species of Khaya woods. Contact Dermatol Newsl 9(Jan):216.

Khaya nyasica Stapf ex Baker f. = Khaya anthotheca (Welw.) C. DC.

KHAYA SENEGALENSIS (Desr.) A. Juss. [Meliaceae]

Common Names:

Afrikanisches Mahagoni; Mahagoni

Citations:

Dantin-Gallego J, Armayor AF, Riesco J (1952) Some new toxic woods: Some new manifestations of toxicity. Ind Med Surg 21(2):41-46.

khella –see– *Ammi visnaga* (L.) Lam.

khesara dahl –see– *Lathyrus sativus* L.

khesari –see– *Lathyrus sativus* L.

khesari dahl –see– *Lathyrus sativus* L.

khesari del –see– *Lathyrus sativus* L.

khilla –see– *Ammi visnaga* (L.) Lam.

khillala –see– *Ammi visnaga* (L.) Lam.

Kiawe bean –see– *Prosopis juliflora* (Sw.) DC.

Kichererbse –see– *Cicer arietinum* L.; *Lathyrus sativus* L.

kidkill –see– *Kalmia angustifolia* L.

kidney bean –see– *Phaseolus vulgaris* L.

Kiefer –see– *Pinus sylvestris* L.

kif –see– *Cannabis sativa* L.

kikar –see– *Acacia jacquemontii* Benth.; *Acacia leucophloea* (Roxb.) Willd.

kikuyagrass –see– *Pennisetum clandestinum* Hochst. ex Chiov.

Kikuyugrass –see– *Pennisetum clandestinum* Hochst. ex Chiov.

kimbar mahalba –see– *Rourea coccinea* (Schumach. & Thonn.) Hook. f.

Kimberley couch –see– *Brachyachne convergens* (F. Muell.) Stapf

King Edward daisy –see– *Leucanthemum vulgare* Lam.

King Island melilot –see– *Melilotus indicus* (L.) All.

king-of-fruits –see– *Mangifera indica* L.

king-of-the-day –see– *Cestrum diurnum* L.

kings-and-queens –see– *Arum maculatum* L.

king's-consound –see– *Consolida regalis* Gray

king's-crown –see– *Calotropis procera* (Aiton) W. T. Aiton

Kirschlorbeer –see– *Prunus laurocerasus* L.

kiss-me-quick –see– *Brunfelsia latifolia* (Pohl) Benth.

kissieblaaren –see– *Malva parviflora* L.

kitten-breeches –see– *Dicentra cucullaria* (L.) Bernh.

Kittie McWanie –see– *Asclepias curassavica* L.

kiwach –see– *Mucuna pruriens* (L.) DC.

kiwachi –see– *Mucuna pruriens* (L.) DC.

kiwash –see– *Mucuna pruriens* (L.) DC.

kiwi –see– *Actinidia chinensis* Planch.; *Actinidia deliciosa* (A. Chev.) C. F. Liang & A. R. Ferguson

klaaslouwbossie –see– *Athanasia trifurcata* (L.) L.

Klamath weed –see– *Hypericum perforatum* L.

klapperbos –see– *Crotalaria burkeana* Benth.

Klappermohn –see– *Papaver rhoeas* L.

Klatschmohn –see– *Papaver rhoeas* L.

Klatschrose –see– *Papaver rhoeas* L.

Kleebaum –see– *Laburnum anagyroides* Medik.

kleiner Ampfer –see– *Rumex acetosella* L.

Kleingrass –see– *Panicum coloratum* L.

kleiner Sauerampfer –see– *Rumex acetosella* L.

kleiner Schachtelhalm –see– *Equisetum arvense* L.

Kleopatra nadel –see– *Cleome spinosa* Jacq.

klesari dal –see– *Lathyrus sativus* L.

Klettenhaare –see– *Arctium lappa* L.

klimop –see– *Cynanchum obtusifolium* L. f.

Klivie –see– *Clivia miniata* (Lindl.) Regel

Knallerbsenstrauch –see– *Symphoricarpos albus* (L.) S. F. Blake

knapweed –see– *Centaurea solstitialis* L.

Knaulgras –see– *Dactylis glomerata* L.

knight's-spur –see– *Consolida regalis* Gray

knitbone –see– *Symphytum officinale* L.

Knoblauch –see– *Allium sativum* L.

Knobleig –see– *Allium sativum* L.

Knoflack –see– *Allium sativum* L.

Knorpelmöhre –see– *Ammi majus* L.

Knospen –see– *Tussilago farfara* L.

knot bindweed –see– *Fallopia convolvulus* (L.) Á. Löve

knotweed –see– *Polygonum aviculare* L.

koa haole –see– *Leucaena leucocephala* (Lam.) de Wit

kochia –see– *Bassia scoparia* (L.) A. J. Scott

Kochia scoparia (L.) Schrad. = Bassia scoparia (L.) A. J. Scott

kocsányos tölgy –see– *Quercus robur* L.

Kohl –see– *Brassica oleracea* L. var. capitata L.

koibos –see– *Tetragonia schenkii* (Schinz) Engl.

Kokardenblume –see– *Gaillardia pulchella* Foug. var. picta (Sweet) A. Gray

konjac –see– *Amorphophallus konjac* K. Koch

konti –see– *Zamia integrifolia* L. f.

Kopfsalat –see– *Lactuca sativa* L. var. capitata L.

Korallenstrauch –see– *Solanum pseudocapsicum* L.

Korean ginseng –see– *Panax ginseng* C. A. Mey.

Korean lespedeza –see– *Kummerowia stipulacea* (Maxim.) Makino

Korean pine –see– *Pinus koraiensis* Siebold & Zucc.

Korean red ginseng –see– *Panax ginseng* C. A. Mey.

Koriander –see– *Coriandrum sativum* L.

Kornrade –see– *Agrostemma githago* L.

koro –see– *Mauria heterophylla* Kunth

koropo –see– *Crotalaria retusa* L.

kosso –see– *Hagenia abyssinica* (Bruce) J. F. Gmel.

koto –see– *Pterygota bequaertii* De Wild.; *Pterygota macrocarpa* K. Schum.

kousso –see– *Hagenia abyssinica* (Bruce) J. F. Gmel.

kouterbossie –see– *Athanasia trifurcata* (L.) L.

kovanna bos –see– *Senecio burchellii* DC.; *Senecio ilicifolius* L.

kowhai –see– *Sophora microphylla* Aiton

közöbséges boróka –see– *Juniperus communis* L.

kraalbos –see– *Galenia africana* L.

kra niang –see– *Archidendron jiringa* (Jack) I. C. Nielsen

krakos –see– *Archidendron jiringa* (Jack) I. C. Nielsen

KRAMERIA IXINE L. [Krameriaceae]

Common Names:

cadia-del-perro

Citations:

Dunham LJ, Sheets RH, Morton JF (1974) Proliferative lesions in cheek pouch and esophagus of hamsters treated with plants from Curacao, Netherland Antilles. J Natl Cancer Instit 53(5):1259-1269.

Morton JF (1979) Plant tannins and esophageal cancer. In: Deichmann WB (ed.) Toxicology and occupational medicine. Elsevier. New York. pp. 129-137.

KRAMERIA LAPPACEA (Dombey) Burdet & B. B. Simpson [Krameriaceae]

Common Names:

Peruvian rhatany; red rhatany; rhatanhia; rhatania

Citations:

Grolnick M (1938) Studies in contact dermatitis. III. Active sensitization with Krameria in man. J Invest Dermatol 1:179-189.

Kranbaum –see– *Juniperus communis* L.

Krätzraute –see– *Ruta graveolens* L.

Kren –see– *Armoracia rusticana* P. Gaertn. et al.

Kreuzblättrige Wolfsmilch –see– *Euphorbia lathyris* L.

Kreuzdorn –see– *Rhamnus cathartica* L.

Kreuzkraut –see– *Senecio vernalis* Waldst. & Kit.

Kriechender Hahnenfuß –see– *Ranunculus repens* L.

krimpsiektebos –see– *Tylecodon wallichii* (Harv.) Toelken

krimpsiektebossie –see– *Cotyledon orbiculata* L.

krisna jirak –see– *Carum carvi* L.

kristkorn –see– *Ilex aquifolium* L.

kristpalme –see– *Ricinus communis* L.

Krokus –see– *Crocus sativus* L.

Kronenschütchen –see– *Securigera varia* (L.) Lassen

Kronwicke –see– *Securigera varia* (L.) Lassen

Krötengraß –see– *Euphorbia cyparissias* L.

Krötenpeterlein –see– *Aethusa cynapium* L.

Kruiskruid –see– *Senecio vulgaris* L.

kuan-yin-lien –see– *Alocasia macrorrhizos* (L.) G. Don

kubabul –see– *Leucaena leucocephala* (Lam.) de Wit

kukai haole –see– *Jatropha curcas* L.

kukui nut –see– *Aleurites moluccanus* (L.) Willd.

KUMMEROWIA STIPULACEA (Maxim.) Makino [Fabaceae]

Synonyms:

Lespedeza stipulacea Maxim.

Common Names:

Korean lespedeza; lespedeza

Citations:

Muhrer ME, Gentry RF (1948) A hemorrhagic factor in moldy Lespedeza hay. Missouri Agric Exp Sta Res Bull #429:11 pp.

kunde –see– *Vigna unguiculata* (L.) Walp.

kundur laut –see– *Sarcolobus globosus* Wall.

kurrajong –see– *Brachychiton populneus* (Schott & Endl.) R. Br.

kusth –see– *Saussurea costus* (Falc.) Lipsch.

L

la ciguë vireuse –see– *Cicuta virosa* L.

la-nielle-des-blés –see– *Agrostemma githago* L.

la parquina –see– *Cestrum parqui* L'Her.

la tullidora –see– *Karwinskia humboldtiana* (Schult.) Zucc.

la yedra –see– *Toxicodendron diversilobum* (Torr. & A. Gray) Greene

lablab –see– *Lablab purpureus* (L.) Sweet

LABLAB PURPUREUS (L.) Sweet [Fabaceae]

Synonyms:

Dolichos lablab L.

Common Names:

field bean; gallinazo; horse gam; horsebean; hyacinth bean; lablab

Citations:

Jaffé WG (1950) Estudio sobre la inhibición del crecimeniento de ratas causado por algunas semillas de leguminosas. Acta Cient Venez 1(2):62-64.

Jaffé WG (1950) La toxicidad de las caraotas crudas para conejos. Acta Cient Venez 1(1):16-17.

labriform milkweed –see– *Asclepias labriformis* M. E. Jones

laburnum –see– *Cytisus scoparius* (L.) Link; *Laburnum anagyroides* Medik.

LABURNUM ALPINUM (Mill.) J. Presl [Fabaceae]

Common Names:

alpine golden-chain; Scotch laburnum; Scottish laburnum

Citations:

Mitchell RG (1951) Laburnum poisoning in children. Report on ten cases. Lancet 261(2):57-58.

LABURNUM ANAGYROIDES Medik. [Fabaceae]

Synonyms:

Cytisus laburnum L.; ***Laburnum vulgare*** J. Presl

Common Names:

bean tree; bean trefoil; Bohnenbaum; Bohnenstrauch; cytise; false ebony; Geißklee; golden chain; golden rail; golden rain; Goldregen; he broom; Kleebaum; laburnum; rain tree

Citations:

Auchterlonie L (1948) Laburnum poisoning. Vet Rec 60(48):633-634.

Biggs MG (1883) Case of poisoning by laburnum. Br Med J 1(Jun 9):1117-1118.

Bramley A, Goulding R (1981) Laburnum "poisoning." Br Med J 283(6301):1220-1221.

Bramley A, Goulding R (1982) Laburnum "poisoning." Br Med J 284(6309):116.

Clarke ML, Clarke EG, King T (1971) Fatal laburnum poisoning in a dog. Vet Rec 88(7):199-200.

Furet Y, Ernouf D, Brechot JF, et al. (1986) Intoxication collective aux fleurs de cytise. Presse Med 15(23):1103-1104.

Honegger RE, Furrer J (1975) Einige bemerkenswerte Todesfälle bei Reptilien. Salamandra 11(3-4):179-181.

Keeler RF, Baker DC (1990) Myopathy in cattle induced by alkaloid extracts from Thermopsis montana, Laburnum anagyroides, and a Lupinus sp. J Comp Pathol 103:169-182.

Leyland A (1981) Laburnum (Cytisus laburnum) poisoning in two dogs. Vet Rec 109(13):287.

Mitchell RG (1951) Laburnum poisoning in children. Report on ten cases. Lancet 261(2):57-58.

Mořkovský O, Kučera J (1980) Hromadná otrava dětí v kolektivním zařízení semeny štědřence. Cesk Pediatr 35(5-6):284-285.

Richards HG, Stephens A (1970) A fatal case of laburnum seed poisoning. Med Sci Law 10(4):260-266.

Roberts AM (1877) A case of Laburnum poisoning. Lancet 2(Sep 1):341.

Vance J (1877) Poisoning by Laburnum seeds. Lancet 2(Sep 15):414.

Wheelhouse CG (1870) Laburnum poisoning. Br Med J 1(Jan 22):79.

Laburnum vulgare J. Presl = Laburnum anagyroides Medik.

lace flower –see– *Ammi majus* L.

lace-tree Philodendron –see– *Philodendron bipinnatifidum* Schott ex Endl.

lacquer –see– *Toxicodendron vernicifluum* (Stokes) F. A. Barkley

lacquer tree –see– *Toxicodendron vernicifluum* (Stokes) F. A. Barkley

LACTUCA SATIVA L. [Asteraceae]

Common Names:

Grüner Salat; lettuce; Salat; wild lettuce

Citations:

Friis B, Hjorth N, Vail JT Jr, et al. (1975) Occupational contact dermatitis from Cichorium (chicory, endive) and Lactuca (lettuce). Contact Dermatitis 1(5):311-313.

Gregory M (1997) Lettuce as a suspected cause of narcosis in a duckling. Vet Rec 141(12):316.

Helander I (1984) Contact dermatitis to lettuce. Contact Dermatitis 11(4):249.

Krook G (1977) Occupational dermatitis from Lactuca sativa (lettuce) and Cichorium (endive). Simultaneous occurrence of immediate and delayed allergy as a cause of contact dermatitis. Contact Dermatitis 3(1):27-36.
Mitchell D, Beck MH, Hausen BM (1989) Contact sensitivity to lettuce in a chef. Contact Dermatitis 20(9):398-399.
Rinkel HJ, Balyeat RM (1932) Occupational dermatitis due to lettuce. JAMA 98(2):137-138.

LACTUCA SATIVA L. var. capitata L. [Asteraceae]

Common Names:

cabbage lettuce; iceberg lettuce; Kopfsalat

Citations:

Hausen BM (1979) The sensitizing capacity of compositae plants. III. Test results and cross-reactions in compositae-sensitive patients. Dermatologica 159(1):1-11.

LACTUCA SATIVA L. var. longifolia Lam. [Asteraceae]

Common Names:

romaine lettuce

Citations:

Vail JT Jr, Mitchell JC (1973) Occupational dermatitis from Cichorium intybus, C. endiva and Lactuca sativa var. longifolia. Contact Dermatol Newsl 14(Aug):413-414.

ladino clover –see– *Trifolium repens* L.

Ladinoklee –see– *Trifolium repens* L.

Lady Campbell's-weed –see– *Echium plantagineum* L.

lady laurel –see– *Daphne mezereum* L.

lady-of-the-night –see– *Cestrum nocturnum* L.

lady's-bower –see– *Clematis vitalba* L.

lady's-finger –see– *Abelmoschus esculentus* (L.) Moench

lady's-seal –see– *Dioscorea communis* (L.) Caddick & Wilkin

lady's-slipper –see– *Cypripedium reginae* Walter; *Paphiopedilum haynaldianum* (Rchb. f.) Stein

lady's-sorrel –see– *Oxalis acetosella* L.

lady's-tears –see– *Convallaria majalis* L.

lady's-thimble –see– *Digitalis purpurea* L.

lady's-thistle –see– *Silybum marianum* (L.) Gaertn.

lady's-thumb –see– *Persicaria maculosa* Gray

LAENNECIA COULTERI (A. Gray) G. L. Nesom [Asteraceae]

Synonyms:

Conyza coulteri A. Gray

Common Names:

conyza; Coulter's-conyza

Citations:

Boughton IB, Hardy WT (1941) Feeding trials of suspected plants. Texas Agric Exp Sta Annu Rep 54:159.

LAGENARIA SICERARIA (Molina) Standl. [Cucurbitaceae]

Common Names:

amargo; bottle gourd; calabash gourd; camasa; cand-ungo; caracho; gourd; maracos; marimbas; marimbo; tula-de-mate

Citations:

Barri ME, Onsa TO, Elawad AA, et al. (1983) Toxicity of five Sudanese plants to young ruminants. J Comp Pathol 93(4):559-575.

LAGERSTROEMIA PARVIFLORA Roxb. [Lythraceae]

Common Names:

bodhangero

Citations:

Kulakkattolickal AT (1987) Piscicidal plants of Nepal: Preliminary toxicity screening using grass carp (Ctenopharyngodon idella) fingerlings. J Ethnopharmacol 21(1):1-9.

lagrimas-de-Maria –see– *Caladium bicolor* (Aiton) Vent.

laguana –see– *Annona muricata* L.

lakh –see– *Lathyrus sativus* L.

lakh dahl –see– *Lathyrus sativus* L.

lakhori –see– *Lathyrus sativus* L.

Lakritze –see– *Glycyrrhiza glabra* L.

lal burass –see– *Rhododendron arboreum* Sm.

lal champa –see– *Plumeria rubra* L.

l'alerce –see– *Fitzroya cupressoides* (Molina) I. M. Johnst.

lamb-in-a-pulpit –see– *Arum maculatum* L.

lamb laurel –see– *Kalmia angustifolia* L.

Lambert's-crazyweed –see– *Oxytropis lambertii* Pursh

Lambert's-locoweed –see– *Oxytropis lambertii* Pursh

Lambert's-point locoweed –see– *Oxytropis lambertii* Pursh

lambkill –see– *Kalmia angustifolia* L.

lambkill Kalmia –see– *Kalmia angustifolia* L.

lambpoison –see– *Ipomoea muelleri* Benth.; *Isotropis cuneifolia* (Sm.) Domin

lamb's-quarter –see– *Chenopodium album* L.

lamb's-tongue –see– *Scleroblitum atriplicinum* (F. Muell.) Ulbr.

lamb's-tongue groundsel –see– *Senecio integerrimus* Nutt.

LAMIUM AMPLEXICAULE L. [Lamiaceae]

Common Names:

dead nettle; hedge nettle; henbit; henbit dead nettle;

Citations:

Dodd S, Henry M (1921) Staggers or shivers in live stock. Agric Gaz New South Wales 32:327-329.
White CT (1921) Two plants poisonous to stock. Queensland Agric J New Series 16(Sep):194-196.

LAMPROCAPNOS SPECTABILIS (L.) Fukuhara [Fumariaceae]

Synonyms:

Dicentra spectabilis (L.) Lem.

Common Names:

bleeding heart

Citations:

Harville CH (1933) Contact dermatitis due to a common plant. J Allergy 4:527-529.

lamtoro –see– *Leucaena leucocephala* (Lam.) de Wit

lan tsao –see– *Eupatorium chinense* L.

lance-leaf sage –see– *Salvia reflexa* Hornem.

land cress –see– *Lepedium didymum* L.

lang –see– *Lathyrus sativus* L.

lang du –see– *Alocasia macrorrhizos* (L.) G. Don

langue-de-chien –see– *Cynoglossum officinale* L.

langwort –see– *Veratrum album* L.

LANNEA COROMANDELICA (Houtt.) Merr. [Anacardiaceae]

Common Names:

dabdabey

Citations:

Kulakkattolickal AT (1987) Piscicidal plants of Nepal: Preliminary toxicity screening using grass carp (Ctenopharyngodon idella) fingerlings. J Ethnopharmacol 21(1):1-9.

lanoline bush –see– *Zieria smithii* Andrews

lansina –see– *Ricinus communis* L.

lantana –see– *Lantana camara* L.

LANTANA CAMARA L. [Verbenaceae]

Synonyms:

Lantana tiliifolia Cham.

Common Names:

bunchberry; buti; camará; cherry pie; ewon agogo; hierba-de-Zorro; lantana; large-leaf lantana; latina; panch; pasarín; racha; red sage; San Rafaelito; shrub verbena; tickberry; wild sage; yellow sage

Citations:

Agarwala ON, Negi SS, Mahadevan V (1962) Serum bilirubin and icteric index values in cattle and sheep in experimental Lantana poisoning. Curr Sci 31(12):506-507.

Aguilera R, Zaldívar V, Margolles E, et al. (1986) Indicadores del funcionamiento hepático en la intoxicación experimental aguda de vacas por Lantana camara. Rev Salud Anim 8(3):211-220.

Black H, Carter RG (1985) Lantana poisoning of cattle and sheep in New Zealand. N Z Vet J 33(8):136-137.

Brito MF, Tokarnia CH, Döbereiner J (2004) A toxidez de diversas lantanas para bovinos e ovinos no Brasil. Pesq Vet Bras 24(3):153-159.

de Aluja AS (1970) Lantana camara poisoning in cattle in Mexico. Vet Rec 86(21):628.

de Aluja AS, Sanz R, Espinosa F (1970) El mal de playa - Intoxicación del ganado bovino con Lantana camara. Veterinaria (Mexico) 1(4):7-13.

de Aluja AS, Skewes HR (1971) Further investigation regarding the toxicity of members of the genus Lantana in Mexico. Proc World Vet Cong 1:327-331.

Dhillon KS, Paul BS (1971) Clinical studies of Lantana camara (L.) poisoning in buffalo calves, with special reference to its effect on rumen motility. Indian J Anim Sci 41(10):945-948.

Dhillon KS, Paul BS, Garg BD (1970) Some haematological aspects in Lantana camara poisoning in buffalo calves. J Res Punjab Agric Univ 7(2):262-266.

Dhillon SS, Singh B (1965) Pathology of Lantana camara poisoning in buffalo calves. J Res Punjab Agric Univ 2(1):39-42.

Dwivedi SK, Shivnani GA, Joshi HC (1971) Clinical and biochemical studies in Lantana poisoning in ruminants. Indian J Anim Sci 41(10):948-953.

Fourie N, Van der Lugt JJ, Newsholme SJ, et al. (1987) Acute Lantana camara toxicity in cattle. J S Afr Vet Assoc 58(4):173-178.

Frisch JE, O'Neill CJ, Burrow HM (1984) The incidence and effect of poisoning with Lantana camara in different cattle breeds. J Agric Sci 102(1):191-195.

Ganai GN, Jha GJ (1991) Immunosuppression due to chronic Lantana camara L. toxicity in sheep. Indian J Exp Biol 29(8):762-766.

Gemmell RT, Pass MA (1978) Effect of ingestion of Lantana camara L. on the ultrastructure of the liver cells of the sheep. J Anat 126:630.

Gopinath C, Ford EJ (1969) The effect of Lantana camara on the liver of sheep. J Pathol 99(1):75-85.

Hari R, Shivnani GA, Joshi HC (1973) Efficacy of certain drug treatments in lantana poisoning in buffalo calves in relation to biochemical changes. Indian J Anim Sci 43(9):829-833.

Hari R, Shivnani GA, Joshi HC (1973) Therapeutic efficacy in lantana poisoning in buffalo calves in relation to clinical and haematological studies. Indian Vet J 50(8):764-770.

Hunt S, McCosker PJ (1970) Observations on serum adenosine deaminase activity in experimentally produced liver diseases of cattle and sheep: Yellow-wood, lantana, carbon tetrachloride and chronic copper poisoning. Br Vet J 126(2):74-81.

Ide A, Tutt CL (1998) Acute Lantana camara poisoning in a Boer goat kid. J S Afr Vet Assoc 69(1):30-32.

Johnson JH, Jensen JM (1998) Hepatotoxicity and secondary photosensitization in a red kangaroo (Megaleia rufus) due to ingestion of Lantana camara. J Zoo Wildl Med 29(2):203-207.

Kariuki DP, McGrane J (1979) Lantana poisoning in cattle - Case report. Kenya Vet 3(1):23.

Lal M, Kalra DB (1960) Lantana poisoning in domesticated animals. Indian Vet J 37:263-269.

Lin SC, Wu YH, Tsai JF, et al. (1985) [Studies on the hepatogenous photosensitization of cattle in Taiwan. IV. Experimental Lantana camara poisoning in white Taiwan native hybrid goats.] Taiwan J Vet Med Anim Husbandry 45:81-91.

McDonald PA (1955) Cases of lantana poisoning seen. Georgia Vet 7(3):18-19.

McSweeney CS, Pass MA (1983) Effect of lantana on the composition of the extracellular fluid of sheep. Vet Hum Toxicol 25(5):330-334.
McSweeney CS, Pass MA (1983) The mechanism of ruminal stasis in lantana-poisoned sheep. Q J Exp Physiol 68(3):301-313.
McSweeney CS, Pass MA, Henry P (1983) Changes in rumen contents associated with lantana poisoning of sheep. Comp Biochem Physiol 75C(2):361-367.
McSweeney CS, Stewart C, Pass MA (1985) Treatment of Lantana poisoning of cattle and sheep. In: Seawright AA, Hegarty MP, James LF, et al. (eds.) Plant toxicology. Queensland Poisonous Plant Committee. Yeerongpilly. Queensland. pp. 61-69.
Morton JF (1994) Lantana, or red sage (Lantana camara L. [Verbenaceae]), notorious weed and popular garden flower: Some cases of poisoning in Florida. Econ Bot 48(3):259-270.
Obwolo MJ, Odiawo GO, Goedegebuure SA (1990) Clinicopathological features of experimental low-dose Lantana camara poisoning in indigenous Zimbabwean goats. Zimbabwe Vet J 21(1):1-7.
Pass MA, Gemmell RT, Heath TJ (1978) Effect of Lantana on the ultrastructure of the liver of sheep. Toxicol Appl Pharmacol 43(3):589-596.
Pass MA, Heath T (1977) Gallbladder paralysis in sheep during lantana poisoning. J Comp Pathol 87(2):301-306.
Pass MA, Heath TJ (1978) The effect of Lantana camara on intestinal motility in sheep. J Comp Pathol 88(1):149-156.
Pass MA, Seawright AA, Heath T (1976) Effect of ingestion of Lantana camara L. on bile formation in sheep. Biochem Pharmacol 25(18):2101-2102.
Pass MA, Stewart C (1984) Administration of activated charcoal for the treatment of lantana poisoning of sheep and cattle. J Appl Toxicol 4(5):267-269.
Sastry MS, Singh YP (1979) Experimental lantana poisoning and its treatment in livestock. Indian Vet J 56(12):1007-1012.
Seawright AA (1963) Studies on experimental intoxication of sheep with Lantana camara. Aust Vet J 39(Sep):340-344.
Seawright AA (1964) Studies on the pathology of experimental lantana (Lantana camara L.) poisoning of sheep. Pathol Vet 1:504-529.
Seawright AA (1965) Electron microscopic observations of the hepatocytes of sheep in lantana poisoning. Pathol Vet 2:175-196.
Seawright AA, Allen JG (1972) Pathology of the liver and kidney in lantana poisoning of cattle. Aust Vet J 48(6):323-331.
Sharma OP, Dawra RK, Krishna L, et al. (1988) Toxicity of Lantana (Lantana camara L.) leaves and isolated toxins to rabbits. Vet Hum Toxicol 30(3):214-218.
Sharma OP, Dawra RK, Makkar HP (1982) Effect of Lantana camara toxicity on lipid peroxidation in guinea pig tissues. Res Commun Chem Pathol Pharmacol 38(1):153-156.
Sharma OP, Makkar HP, Dawra RK (1982) Biochemical changes in hepatic microsomes of guinea-pig under lantana toxicity. Xenobiotica 12(4):265-269.
Sharma OP, Makkar HP, Dawra RK (1983) Effect of lantana toxicity on lysosomal and cytosol enzymes in guinea pig liver. Toxicol Lett 16(1-2):41-45.
Sharma OP, Makkar HP, Dawra RK, et al. (1981) Hepatic and renal toxicity of lantana in the guinea pig. Toxicol Lett 7(4-5):347-351.
Sharma OP, Makkar HP, Dawra RK, et al. (1982) Changes in blood constituents of guinea pigs in lantana toxicity. Toxicol Lett 11(1-2):73-76.
Sharma OP, Makkar HP, Pal RN, et al. (1980) Lantadene A content and toxicity of the lantana plant (Lantana camara, Linn.) to guinea pigs. Toxicon 18(4):485-488.
Sharma OP, Makkar HP, Pal RN, et al. (1981) Fragility of erythrocytes in animals affected by lantana poisoning. Clin Toxicol 18(1):25-35.
Sharma OP, Vaid J, Sharma PD (1991) Comparison of lantadenes content and toxicity of different taxa of the lantana plant. J Chem Ecol 17(11):2283-2291.
Shone DK (1959) Lantana poisoning of cattle. Rhodesia Agric J 56:238-239.
Silva FM, Couto ES (1971) Intoxicação experimental de bovinos pela Lantana camara no Estado de Pernambuco. Arq Esc Vet Univ Fed Minas Gerais 23:77-89.
Srivastava SM, Sinha SP (1980) An outbreak of Lantana poisoning in sheep. Indian Vet Med J 4(3):136-137.
Tokarnia CH, Armién AG, Barros SS, et al. (1999) Estudos complementares sobre a toxidez de Lantana camara (Verbenaceae) em bovinos. Pesq Vet Bras 19(3-4):128-132.
Tokarnia CH, Döbereiner J, Lazzari AA, et al. (1984) Intoxicação por Lantana spp. (Verbenaceae) em bovinos nos Estados de Mato Grosso e Rio de Janeiro. Pesq Vet Bras 4(4):129-141.
Uppal RP (1970) The pharmacotoxicological investigations of Lantana camara Linn. in sheep. Haryana Vet 9(2):91-92.
Uppal RP, Paul BS (1978) Assessment of hepatic dysfunction in experimental Lantana poisoning in sheep. Indian Vet J 55(Oct):798-802.
Verhulst HL, Page LA (1962) Lantana. Bull Natl Clgh Poison Control Cent 1962:6.
Wolfson SL (1963) Acute poisoning in children from ingestion of the green fruit of Lantana camara. Am Assoc Poison Control Cent 1963:4.
Wolfson SL, Solomons TW (1964) Poisoning by fruit of Lantana camara. An acute syndrome observed in children following ingestion of the green fruit. Am J Dis Child 107:173-176.
Yadava JM, Verma NS (1978) An outbreak of Lantana poisoning in domesticated animals. Indian Vet Med J 2:1-9.

LANTANA GLUTINOSA Poepp. [Verbenaceae]

Citations:

Riet-Correa F, Mendéz MC, Schild AL, et al. (1984) Intoxicação por Lantana glutinosa (Verbenaceae) em bovinos no Estado de Santa Catarina. Pesq Vet Bras 4(4):147-153.

Lantana tiliifolia Cham. = Lantana camara L.

lao-hu-yu –see– *Alocasia macrorrhizos* (L.) G. Don

lapacho –see– *Tabebuia impetiginosa* (Mart. ex DC.) Standl.

laplove –see– *Convolvulus arvensis* L.

LAPORTEA CANADENSIS (L.) Wedd. [Urticaceae]

Synonyms:

Urticastrum divaricatum (L.) Kuntze

Common Names:
bull nettle; Canadian wood nettle; nettle; stinging nettle; wood nettle

Citations:

Masias MA, Positano RG (1990) Urticaceae poisoning. J Am Podiatr Med Assoc 80(11):613-616.

Laportea moroides Wedd. = Dendrocnide moroides (Wedd.) Chew

lappa –see– *Arctium lappa* L.

Lappa major Gaertn. = Arctium lappa L.

Lappa vulgaris Hill = Arctium lappa L.

larch –see– *Larix decidua* Mill.

larch pine –see– *Larix decidua* Mill.

larcha –see– *Citrus aurantium* L.

large lady's-slipper –see– *Cypripedium reginae* Walter

large larkspur –see– *Delphinium glaucum* S. Watson

large-leaf lantana –see– *Lantana camara* L.

large poison bride bush –see– *Pavetta schumanniana* F. Hoffm. ex K. Schum.

large watergrass –see– *Paspalum dilatatum* Poir.

LARIX DECIDUA Mill. [Pinaceae]

Common Names:
Calabrian pine; Corsican pine; larch; larch pine

Citations:

Karlberg AT, Lidén C (1985) Clinical experience and patch testing using colophony (rosin) from different sources. Br J Dermatol 113(4):475-481.

larkspur –see– *Consolida regalis* Gray; *Delphinium barbeyi* (Huth) Huth; *Delphinium carolinianum* Walter subsp. virescens (Nutt.) R. E. Brooks; *Delphinium elatum* L.; *Delphinium glaucum* S. Watson; *Delphinium menziesii* DC.; *Delphinium schmalhausenii* Albov

LARREA TRIDENTATA (DC.) Coville [Zygophyllaceae]

Common Names:
chaparral; creosote bush; gobernadora; greasewood; hediondilla

Citations:

Alderman S, Kailas S, Goldfarb S (1994) Cholestatic hepatitis after ingestion of chaparral leaf: Confirmation by endoscopic retrograde cholangiopancreatography and liver biopsy. J Clin Gastroenterol 19(3):242-247.

Anonymous (1992) Chaparral-induced toxic hepatitis - California and Texas, 1992. JAMA 268(23):3295, 3298.

Batchelor WB, Heathcote J, Wanless IR (1995) Chaparral-induced hepatic injury. Am J Gastroenterol 90(5):831-833.

Blanc PD, Trainor WD, Lim DT (1986) Herbal tea asthma. Br J Ind Med 43(2):137-138.

Caldwell SH, Feeley JW, Wieboldt TF, et al. (1994) Acute hepatitis with use of over-the-counter herbal remedies. Va Med Q 121(1):31-33.

Clark F, Reed DR (1992) Chaparral-induced toxic hepatitis - California and Texas, 1992. MMWR Morb Mortal Wkly Rep 41(43):812-814.

Estes JD, Stolpman D, Olyaei A, et al. (2003) High prevalence of potentially hepatotoxic herbal supplement use in patients with fulminant hepatic failure. Arch Surg 138(8):852-858.

Gordon DW, Rosenthal G, Hart J, et al. (1995) Chaparral ingestion. The broadening spectrum of liver injury caused by herbal medications. JAMA 273(6):489-490.

Holechek JL, Munshikpu AV, Saiwana L, et al. (1990) Influences of six shrub diets varying in phenol content on intake and nitrogen retention by goats. Trop Grasslands 24(2):93-98.

Katz M, Saibil F (1990) Herbal hepatitis: Subacute hepatic necrosis secondary to chaparral leaf. J Clin Gastroenterol 12(2):203-206.

Kauma H, Koskela R, Mäkisalo H, et al. (2004) Toxic acute hepatitis and hepatic fibrosis after consumption of chaparral tablets. Scand J Gastroenterol 39(11):1168-1171.

Leonforte JF (1986) Contact dermatitis from Larrea (creosote bush). J Am Acad Dermatol 14:202-207.

Shasky DR (1986) Contact dermatitis from Larrea tridentata (creosote bush). J Am Acad Dermatol 15:302.

Sheikh NM, Philen RM, Love LA (1997) Chaparral-associated hepatotoxicity. Arch Intern Med 157(8):913-919.

Smart CR, Hogle HH, Robins RK, et al. (1969) Interesting observation on nordihydroguaiaretic acid (NSC-4291; NDGA) and a patient with malignant melanoma. A preliminary report. Cancer Chemother Rep 53(2):147-151.

Smith LM (1937) Dermatitis caused by creosote bush. J Allergy 8:187-188.

Lasiosiphon anthylloides (L. f.) Meisn. = Gnidia anthylloides Gilg

Lasiosiphon burchellii Meisn. = Gnidia burchellii Gilg

Lasiosiphon kraussianus (Meisn.) Burtt Davy = Gnidia kraussiana Meisn.

Lasiosiphon latifolius (Oliv.) Brenan = Gnidia latifolia (Oliv.) Gilg

LASIOSPERMUM BIPINNATUM (Thunb.) Druce [Asteraceae]

Common Names:
ganskweek

Citations:

Adelaar TF, Terblanche M, Smit JD, et al. (1964) A hitherto unknown poisonous plant: Lasiospermum bipinnatum (Thunb.) Druce. Preliminary communication. J S Afr Vet Assoc 35(1):11-16.

Fair AE, Tustin RC, Adelaar TF (1970) Poisoning of cattle by ganskweek (Lasiospermum bipinnatum (Thunb.) Druce). J S Afr Vet Assoc 41(3):231-232.

Kellerman TS, Basson PA, Naudé TW, et al. (1973) Photosensitivity in South Africa. 1. A comparative study of Asaemia axillaris (Thunb.) Harv. ex Jackson and Lasiospermum bipinnatum (Thunb.) Druce poisoning in sheep. Onderstepoort J Vet Res 40(3):115-126.

Penrith ML, Van Vollenhoven E (1994) Pulmonary and hepatic lesions associated with suspected ganskweek (Lasiospermum bipinnatum) poisoning in cattle. J S Afr Vet Assoc 65(3):122-124.

Thornton DJ (1977) Ganskweek (Lasiospermum bipinnatum) poisoning in cattle. J S Afr Vet Assoc 48(3):210-211.

Williams MC (1990) The pathology of experimental Lasiospermum bipinnatum (Thunb.) Druce (Asteraceae) poisoning in sheep. I. Hepatic lesions. Onderstepoort J Vet Res 57(4):249-261.

Williams MC (1990) The pathology of experimental Lasiospermum bipinnatum (Thunb.) Druce (Asteraceae) poisoning in sheep. II. Pulmonary and miscellaneous lesions. Onderstepoort J Vet Res 57(4):263-268.

late locoweed –see– *Astragalus miser* Douglas ex Hook. var. serotinus (A. Gray ex J. G. Cooper) Barneby

late milk vetch –see– *Astragalus miser* Douglas ex Hook. var. serotinus (A. Gray ex J. G. Cooper) Barneby

late yellow locoweed –see– *Oxytropis campestris* (L.) DC. var. spicata Hook.

LATHYRUS APHACA L. [Fabaceae]

Common Names:

yellow flower pea; yellow vetchling

Citations:

Lewis HB, Fajans RS, Esterer MB, et al. (1948) The nutritive values of some legumes, Lathyrism in the rat. The sweet pea (Lathyrus odoratus), Lathyrus sativus, Lathyrus cicera and some other species of Lathyrus. J Nutr 36:537-559.

LATHYRUS CLYMENUM L. [Fabaceae]

Common Names:

gesse pourpre; Spanish vetch; Spanish vetchling

Citations:

Schribaux E (1895) Nombreux empoisonnements causés par la gesse pourpre. Agric Pratique 59(2):52-55.

LATHYRUS HIRSUTUS L. [Fabaceae]

Common Names:

Caley pea; chick pea; everlasting pea; hairy vetchling; red vetch; rough pea; singletary pea; vetchling; wild winter pea

Citations:

Burrows GE, Tate LH, Tripp ML, et al. (1993) Suspected intoxications due to Lathyrus. Vet Hum Toxicol 35(3):262-263.

Lewis HB, Fajans RS, Esterer MB, et al. (1948) The nutritive values of some legumes, Lathyrism in the rat. The sweet pea (Lathyrus odoratus), Lathyrus sativus, Lathyrus cicera and some other species of Lathyrus. J Nutr 36:537-559.

Sugg RS, Simms BT, Baker KG (1944) Studies of toxicity of wild winter peas (Lathyrus hirsutus) for cattle. Vet Med 39(4):308-311.

Turney DM, Copeland DH, Salmon WD (1943) Lathyrism in relation to the use of Caley peas (Lathyrus hirsutus) for livestock. Alabama Agric Exp Sta Annu Rep 54-55:18.

Turney DM, Salmon WD, Copeland DH (1944) Lathyrism in relation to the use of Lathyrus hirsutus for livestock. Alabama Agric Exp Sta Annu Rep 56-57:18-19.

LATHYRUS LATIFOLIUS L. [Fabaceae]

Common Names:

everlasting pea; perennial pea; perennial sweet pea

Citations:

Burrows GE, Tate LH, Tripp ML, et al. (1993) Suspected intoxications due to Lathyrus. Vet Hum Toxicol 35(3):262-263.

Lewis HB, Schulert AR (1949) Experimental lathyrism in the white rat and mouse. Proc Soc Exp Biol Med 71:440-441.

Schulert AR, Lewis HB (1952) Experimental lathyrism. Proc Soc Exp Biol Med 81(1):86-89.

LATHYRUS NISSOLIA L. [Fabaceae]

Common Names:

grass vetchling

Citations:

Greatorex JC (1966) Some unusual cases of plant poisoning in animals. Vet Rec 78(21):725-727.

LATHYRUS ODORATUS L. [Fabaceae]

Common Names:

annual sweet pea; Edelwicke; Erbse; Platterbse; sweet pea

Citations:

Abramovich A, Devoto FC (1969) Comparison entre fentes palatines provoquées par hypervitaminose A et Lathyrus odoratus. C R Soc Biol 162(10):1853-1854.

Bachhuber TE, Lalich JJ (1954) Production of dissecting aneurysms in rats fed Lathyrus odoratus. Science 120(Oct 29):712-713.

Bachhuber TE, Lalich JJ (1955) Effect of sweet pea meal on the rat aorta. Arch Pathol 59(2):247-253.

Chang CY, Witschi E, Ponseti IV (1954) Teratogenic development in Xenopus larvae caused by sweet pea seeds (Lathyrus odoratus) and their extracts. Anat Rec 120:816.

Chang CY, Witschi E, Ponseti IV (1955) Teratogenic effects of Lathyrus odoratus seeds on development and regeneration of vertebrate limbs. Proc Soc Exp Biol Med 90(1):45-50.

Churchill DW, Gelfant S, Lalich JJ, et al. (1955) Alterations in the polysaccharides and elastic fibers in the aortas of rats fed toxic Lathyrus factor. Lab Invest 4(1):1-8.

Dasler W (1954) Incisor ash versus femur ash in sweet pea lathyrism (odoratism). J Nutr 54:397-402.

Dasler W (1954) Observations on odoratism (sweet pea lathyrism) in the rat. J Nutr 53:105-113.

Dasler W (1954) Partial protection against odoratism (sweet pea lathyrism) by diets high in gelatin or casein. Proc Soc Exp Biol Med 85:485-488.

Dasler W, Milliser RV (1957) Experimental lathyrism in mice fed diets containing sweet peas or ß-aminopropionitrile. Proc Soc Exp Biol Med 96:171-174.

Gardner AF (1960) Morphologic and histochemical studies of skeletal lesions in rats fed sweet pea (Lathyrus odoratus) seeds. Am J Vet Res 21(Mar):298-305.

Geiger BJ, Steenbock H, Parsons HT (1933) Lathyrism in the rat. J Nutr 6(5):427-442.
Geiger BJ, Steenbock H, Parsons HT (1976) Lathyrism in the rat. Nutr Rev 34(8):240-241.
Gillman T (1958) Mast cell increases after calciferol intoxication and in experimental odoratism. Acta Haematol 19(3):179-186.
Grant RA, Hathorn M, Gillman T (1960) Aortic, serum, connective tissue and osseous chemistry in lathyritic rats. Nature 186(4719):164-165.
Lewis HB, Esterer MB (1943) Experimental lathyrism in the white rat. Proc Soc Exp Biol Med 53:263-264.
Lewis HB, Fajans RS, Esterer MB, et al. (1948) The nutritive values of some legumes, Lathyrism in the rat. The sweet pea (Lathyrus odoratus), Lathyrus sativus, Lathyrus cicera and some other species of lathyrus. J Nutr 36:537-559.
Okuyama S (1963) [Biochemical, histological and roentgenological studies on skeletal lesions in rats fed with Lathyrus odoratus (sweet pea) seeds.] Fukushima Med J 11:77-104.
Ponseti IV, Baird WA (1952) Scoliosis and dissecting aneurysm of the aorta in rats fed with Lathyrus odoratus seeds. Am J Pathol 28(6):1059-1077.
Ponseti IV, Shepard RS (1954) Lesions of mesodermal tissues in rats fed Lathyrus odoratus seeds. Fed Proc 13(Mar):473.
Ponseti IV, Wawzonek S, Shepard RS, et al. (1956) Further studies on lathyrism in the rat. Proc Soc Exp Biol Med 92(2):366-369.
Raharjo YC, Cheeke PR, Arscott GH (1988) Effects of dietary butylated hydroxyanisole and cysteine on toxicity of Lathyrus odoratus to broiler and Japanese quail chicks. Poult Sci 67(1):153-155.
Robinson JJ, Bast TH (1934) Bone changes due to lathyrism in rats. Anat Rec 59(3):283-295.
Ruth EB (1954) Lathyrism in the rat: A gross study of skeletal lesions. Anat Rec 118:406-407.
Schulert AR, Lewis HB (1952) Experimental lathyrism. Proc Soc Exp Biol Med 81(1):86-89.
Selye H (1957) Über die humorale Beeinflussung des experimentellen Lathyrismus. Naunyn Schmiedebergs Arch Exp Pathol Pharmakol 230(2):155-160.
Stamler FW (1955) Reproduction in rats fed Lathyrus peas or aminonitriles. Proc Soc Exp Biol Med 90(1):294-298.
Steffek AJ, Verrusio AC, Watkins CA (1972) Cleft palate in rodents after maternal treatment with various lathyrogenic agents. Teratology 5(1):33-40.
Vivanco F, Jimenez Diaz C (1951) Further studies on the toxic effects of the leguminous proteins (leguminism). Lathyrus odoratus (lathyrism). Bull Inst Med Res Univ Madr 4(1):1-13.
Walker DG (1957) Elastic fiber alterations in rats treated with Lathyrus odoratus. Histopathologic study of elastic cartilage and of elastic fibers in arteries and membranes, with special reference to the occurrence of extra-aortic dissecting aneurysms. Arch Pathol 64(4):434-435.
Walker DG, Wirtschafter ZT (1956) Histopathogenesis of aortic aneurysms in the Lathyrus-fed rat. Arch Pathol 61(2):125-135.
Walker DG, Wirtschafter ZT (1956) Resorption of embryos in rats on Lathyrus odoratus diet. J Nutr 58:147-159.

LATHYRUS PUSILLUS Elliott [Fabaceae]

Common Names:

everlasting pea; singletary pea; vetchling

Citations:

Lee JG (1950) Experimental lathyrism produced by feeding singletary pea (Lathyrus pusillus) seed. J Nutr 40:587-594.
Lee JG, Dupuy HP, Rolfs HE (1956) Dietary protein and the development of rat lathyrism. J Nutr 58:433-442.

LATHYRUS SATIVUS L. [Fabaceae]

Common Names:

almortas; bitter vetch; chick pea; chickling pea; chickling vetch; grass pea; guaya; Indian mutters; Indian pea; kesari dal; khesara dahl; khesari; khesari dahl; khesari del; kichererbse; klesari dal; lakh; lakh dahl; lakhori; lang; matra; pulse; Saatplatterbse; sweet pea; teora dahl; teswa; tochina; vetchling; white vetch

Citations:

Bhagvat K (1946) Toxic effects in guinea pigs of diet containing large proportion of Lathyrus sativus. Indian J Med Res 34:299-304.
Cohn DF, Streifler M (1983) Intoxication by the chickling pea (Lathyrus sativus): Nervous system and skeletal findings. Arch Toxikol Suppl 6:190-193.
Dwivedi MP, Prasad BG (1964) An epidemiological study of lathyrism in the district of Rewa, Madhya Pradesh. Indian J Med Res 52(1):81-116.
Ganapathy KT, Dwivedi MP, Nagrajan V, et al. (1963) Experiments on chicks fed on Lathyrus sativus. Indian J Med Res 51(5):865-870.
Gebreab T, Gabriel ZW, Maffi M, et al. (1978) Neurolathyrism: A review and a report of an epidemic. Ethiop Med J 16(1):1-11.
Getahun H, Lambein F, Van der Stuyft P (2002) ABO blood groups, grass pea preparation, and neurolathyrism in Ethiopia. Trans R Soc Trop Med Hyg 96(6):700-703.
Getahun H, Lambein F, Vanhoorne M (2002) Neurolathyrism in Ethiopia: Assessment and comparison of knowledge and attitude of health workers and rural inhabitants. Soc Sci Med 54(10):1513-1524.
Getahun H, Lambein F, Vanhoorne M, et al. (2002) Pattern and associated factors of the neurolathyrism epidemic in Ethiopia. Trop Med Int Health 7(2):118-124.
Getahun T, Lambein F, Vanhoorne M, et al. (2003) Food-aid cereals to reduce neurolathyrism related to grass-pea preparations during famine. Lancet 362(9398):1808-1810.
Getahun H, Mekonnen A, Tekle Haimanot R, et al. (1999) Epidemic of neurolathyrism in Ethiopia. Lancet 354(9175):306-307.
Haimanot RT, Kidane Y, Wuhib E, et al. (1990) Lathyrism in rural northwestern Ethiopia: A highly prevalent neurotoxic disorder. Int J Epidemiol 19(3):664-672.
Haque A, Hossain M, Khan JK, et al. (1994) New findings and symptomatic treatment for neurolathyrism, a motor neuron disease occurring in north west Bangladesh. Paraplegia 32(3):193-195.
Haque A, Hossain M, Lambein F, et al. (1997) Evidence of osteolathyrism among patients suffering from neurolathyrism in Bangladesh. Nat Toxins 5:43-46.

Hirano A, Llena JF, Streifler M, et al. (1976) Anterior horn cell changes in a case of neurolathyrism. Acta Neuropathol (Berl) 35(4):277-283.
Hugon J, Ludolph A, Roy DN, et al. (1988) Studies on the etiology and pathogenesis of motor neuron diseases. II. Clinical and electrophysiologic features of pyramidal dysfunction in macaques fed Lathyrus sativus and IDPN. Neurology 38(3):435-442.
Kulkarni SW, Attal HC, Choubey BS (1977) An epidemiologic study of lathyrism in Amgaon block, Bhandara district. Indian J Med Res 66(4):602-610.
Labella V, Rizza ML, Alfano F, et al. (1997) Dietary consumption of Lathyrus sativus seeds induces behavioral changes in the rat. Environ Res 74(1):61-66.
Lewis HB, Fajans RS, Esterer MB, et al. (1948) The nutritive values of some legumes, Lathyrism in the rat. The sweet pea (Lathyrus odoratus), Lathyrus sativus, Lathyrus cicera and some other species of lathyrus. J Nutr 36:537-559.
Maru M, Getahun A, Hoshna S (1988) Prevalence of paralytic poliomyelitis in rural and urban populations in Ethiopia: Report of a house-to-house survey. Am J Trop Med Hyg 38(3):633-635.
Menzies DW, Mills KW (1957) The aortic and skeletal lesions of lathyrism in rats on a diet of sweet pea. J Pathol Bact 73:223-237.
Misra UK, Sharma VP (1994) Peripheral and central conduction studies in neurolathyrism. J Neurol Neurosurg Psychiatry 57:572-577.
Panda NC, Mitra A, Kedary S (1972) Experimental lathyrism in chicks. Indian J Anim Sci 42(11):949-951.
Pratap Rudra MP, Singh MR, Junaid MA, et al. (2004) Metabolism of dietary ODAP in humans may be responsible for the low incidence of neurolathyrism. Clin Biochem 37(4):318-322.
Rotter RG, Marquardt RR, Low RK, et al. (1990) Influence of autoclaving on the effects of Lathyrus sativus fed to chicks. Can J Anim Sci 70(2):739-741.
Shrivastava KK, Sarasa Bharati R, Arora MM (1982) Rare postmortem findings in a case of human lathyrism. Indian J Pathol Microbiol 25(3):225-228.
Steyn DG (1933) Lathyrus sativus L. (chickling vetch; khesari; Indian pea) as a stock food. Onderstepoort J Vet Sci Anim Indus 1(1):163-171.
Streifler M, Cohn DF (1981) Chronic central nervous system toxicity of the chickling pea (Lathyrus sativus). Clin Toxicol 18(12):1513-1517.
Striefler M, Cohn DF, Hirano A, et al. (1977) The central nervous system in a case of neurolathyrism. Neurology 27(12):1176-1178.
Toledano A, Jiménez-Castellanos, López Aydillo NR (1965) Efecto de la harina de almortas (Lathyrus sativus) en pupilla frente a los ratones blancos por via oral. Rev Clin Esp 97(5):328-333.
Valdivieso R, Quirce S, Sainz T (1988) Bronchial asthma caused by Lathyrus sativus flour. Allergy 43(7):536-539.
Voelcker JA (1925) Lathyrus poisoning in the horse. Br Vet J 81:134-135.
Weintroub S, Cohen DF, Salama R, et al. (1980) Skeletal findings in human neurolathyrism. Is there a human osteolathyrism? Eur Neurol 19(2):121-127.
Zagami V (1931) Sugli effetti dell'alimentazione esclusiva con semi di «Lathyrus sativus L.» nei ratti albini. Atti Acad Naz Lincei 14:218.

LATHYRUS SPHAERICUS Retz. [Fabaceae]

Citations:

Lewis HB, Fajans RS, Esterer MB, et al. (1948) The nutritive values of some legumes, Lathyrism in the rat. The sweet pea (Lathyrus odoratus), Lathyrus sativus, Lathyrus cicera and some other species of lathyrus. J Nutr 36:537-559.

LATHYRUS SPLENDENS Kellogg [Fabaceae]

Common Names:

campo pea; pride-of-California

Citations:

Schulert AR, Lewis HB (1952) Experimental lathyrism. Proc Soc Exp Biol Med 81(1):86-89.

Lathyrus strictus Nutt. = Lathyrus vestitus Nutt. subsp. alefeldii (T. G. White) Broich

LATHYRUS SYLVESTRIS L. [Fabaceae]

Common Names:

everlasting pea; flat pea; forest vetchling; narrow-leaf everlasting pea; vetch; wood pea

Citations:

Daniel TW, Wolberg FB, Miller VL, et al. (1946) Chemical composition and digestibility of flat pea forage in three states of maturity. J Anim Sci 5:80-86.
Huang TC, Cunha TJ, Ham WE (1950) The deleterious effects of flat pea seed for rats. Am J Vet Res 11(39):217-220.
Lewis HB, Fajans RS, Esterer MB, et al. (1948) The nutritive values of some legumes, Lathyrism in the rat. The sweet pea (Lathyrus odoratus), Lathyrus sativus, Lathyrus cicera and some other species of lathyrus. J Nutr 36:537-559.
Rasmussen MA, Allison MJ, Foster JG (1993) Flatpea intoxication in sheep and indications of ruminal adaptation. Vet Hum Toxicol 35(2):123-127.
Rowe LD, Ivie GW, DeLoach JR, et al. (1993) The toxic effects of mature flatpea (Lathyrus sylvestris L cv lathco) on sheep. Vet Hum Toxicol 35(2):127-133.

LATHYRUS TINGITANUS L. [Fabaceae]

Common Names:

Platterbse; Tangier pea

Citations:

Griebel C (1950) Erkrankungen durch Bohnenflocken (Phaseolus vulgaris L.) und Platterbsen (Lathyrus tingitanus L.). Z Lebensm Unters Forsch 90:191-197.
Lewis HB, Fajans RS, Esterer MB, et al. (1948) The nutritive values of some legumes, Lathyrism in the rat. The sweet pea (Lathyrus odoratus), Lathyrus sativus, Lathyrus cicera and some other species of lathyrus. J Nutr 36:537-559.

LATHYRUS VESTITUS Nutt. subsp. alefeldii (T. G. White) Broich [Fabaceae]

Synonyms:

Lathyrus strictus Nutt.

Citations:

Schulert AR, Lewis HB (1952) Experimental lathyrism. Proc Soc Exp Biol Med 81(1):86-89.

latina –see– *Lantana camara* L.

lattighat –see– *Glochidion heyneanum* (Wight & Arn.) Wight

Lauchpflanze –see– *Allium porrum* L.

laureal real –see– *Prunus laurocerasus* L.

laurel –see– *Kalmia angustifolia* L.; *Kalmia latifolia* L.; *Laurus nobilis* L.; *Prunus laurocerasus* L.

laurel blanco –see– *Nerium oleander* L.

laurel cherry –see– *Prunus laurocerasus* L.

laurel colorado –see– *Nerium oleander* L.

laurel rosa –see– *Nerium oleander* L.

laurel rosado –see– *Nerium oleander* L.

laurel rose –see– *Nerium oleander* L.

laureola hembra –see– *Daphne mezereum* L.

laurier rose –see– *Nerium oleander* L.

laurierkers –see– *Prunus laurocerasus* L.

LAURUS NOBILIS L. [Lauraceae]

Common Names:

bay; bay laurel; European laurel; Grecian laurel; laurel; Lorbeer; sweet bay

Citations:

Farkas J (1981) Perioral dermatitis from marjoram, bay leaf and cinnamon. Contact Dermatitis 7(2):121.

Foussereau J, Muller JC, Benezra C (1975) Contact allergy to Frullania and Laurus nobilis: Cross-sensitization and chemical structure of the allergens. Contact Dermatitis 1(4):223-230.

Hausen BM (1985) Lorbeer Allergie. Dtsch Med Wochenschr 110(6):634-638.

Özden MG, Öztaş P, Öztaş MO, et al. (2001) Allergic contact dermatitis from Laurus nobilis (laurel) oil. Contact Dermatitis 45(3):178.

LAVANDULA ANGUSTIFOLIA Mill. [Lamiaceae]

Common Names:

lavender

Citations:

Brandão FM (1986) Occupational allergy to lavender oil. Contact Dermatitis 15(4):249-250.

Coulson IH, Ali Khan AS (1999) Facial 'pillow' dermatitis due to lavender oil allergy. Contact Dermatitis 41(2):111.

Sugiura M, Hayakawa R, Kato Y, et al. (2000) Results of patch testing with lavender oil in Japan. Contact Dermatitis 43(3):157-160.

lavender –see– *Lavandula angustifolia* Mill.

lavender scallops –see– *Kalanchoe fedtschenkoi* Raym.-Hamet & H. Perrier

lavanése-rue-de-chèvre –see– *Galega officinalis* L.

LAWSONIA INERMIS L. [Lythraceae]

Common Names:

henna

Citations:

Cronin E (1980) Immediate-type hypersensitivity to henna. Contact Dermatitis 5(3):198-199.

Nigam PK, Saxena AK (1988) Allergic contact dermatitis from henna. Contact Dermatitis 18(4):55-56.

Pasricha JS, Gupta R, Panjwani S (1980) Contact dermatitis to henna (Lawsonia). Contact Dermatitis 6(15):288-289.

Singh R, Siddiqui MA, Baruah MC (1978) Plant dermatitis in Delhi. Indian J Med Res 68(Oct):650-655.

le chones –see– *Asclepias latifolia* (Torr.) Raf.

le mancenilier –see– *Hippomane mancinella* L.

lead tree –see– *Leucaena leucocephala* (Lam.) de Wit

leadwort –see– *Plumbago scandens* L.

leaf flower –see– *Phyllanthus abnormis* Baill.

leaf mustard –see– *Brassica juncea* (L.) Czern.

leafless milkweed –see– *Sarcostemma viminale* (L.) R. Br.

leafy spurge –see– *Euphorbia esula* L.

leather plant –see– *Jatropha dioica* Sessé

leatherleaf fern –see– *Rumohra adiantiformis* (G. Forst.) Ching

leatherleaf palm –see– *Cycas revoluta* Thunb.

leatherstem –see– *Jatropha dioica* Sessé

lebbeck tree –see– *Albizia lebbeck* (L.) Benth.

Lebensbaum –see– *Platycladus orientalis* (L.) Franco; *Thuja occidentalis* L.

Leberstockkraut –see– *Levisticum officinale* W. D. J. Koch

leche-de-gallina –see– *Ornithogalum umbellatum* L.

lecheguilla –see– *Agave lechuguilla* Torr.

lecheruela –see– *Euphorbia helioscopia* L.

lechetrezna –see– *Euphorbia cyparissias* L.; *Euphorbia helioscopia* L.; *Euphorbia peplus* L.

lechuguilla –see– *Agave lechuguilla* Torr.

Lecythis elliptica Kunth = Lecythis minor Jacq.

LECYTHIS MINOR Jacq. [Lecythidaceae]

Synonyms:

Lecythis elliptica Kunth

Common Names:

sapucaia nut

Citations:

Dickson JD (1969) Notes on hair and nail loss after ingesting sapucaia nuts (Lecythis elliptica). Econ Bot 23:133-134.

LECYTHIS OLLARIA Loefl. [Lecythidaceae]

Common Names:

coco-de-mono; monkey nut; monkey pod; monkey's-coconut; sapucacia

Citations:

Kerdel-Vegas F (1964) Generalized hair loss due to the ingestion of "Coco de Mono" (Lecythis ollaria). J Invest Dermatol 42(Jan):91-94.

Kerdel-Vegas F, Aronow L (1966) Epilating effect and cytotoxic principle of Lecythis ollaria. Dermatol Iber Lat Am 1:57-69.

LEDEBOURIA COOPERI (Hook. f.) Jessop [Hyacinthaceae]

Synonyms:

Scilla cooperi Hook. f.

Common Names:

squill

Citations:

Crossley A, Gelfand M (1959) Poisoning by Scilla cooperi: A description of a case. Cent Afr J Med 5(10):537-539.

Steyn DG (1933) Recent investigations into the toxicity of known and unknown poisonous plants in the Union of South Africa. Onderstepoort J Vet Sci Anim Indus 1:173-182.

Lederbaum –see– *Coriaria myrtifolia* L.

Lederfarn –see– *Rumohra adiantiformis* (G. Forst.) Ching

Lederkraut –see– *Asarum europaeum* L.

Ledum columbianum Piper = Rhododendron ×columbianum (Piper) Harmaja

leek –see– *Allium porrum* L.

lei-gong-teng –see– *Erycibe obtusifolia* Benth.; *Tripterygium wilfordii* Hook. f.

Lein –see– *Linum usitatissimum* L.

Lemmon's-hymenoxys –see– *Hymenoxys lemmonii* (Greene) Cockerell

lemon –see– *Citrus limon* (L.) Burm. f.

lemon day lily –see– *Hemerocallis lilioasphodelus* L.

lemon-scented gas plant –see– *Dictamnus albus* L.

lemon verbena –see– *Aloysia citrodora* Palau

lemongrass –see– *Cymbopogon citratus* (DC.) Stapf

len –see– *Linum usitatissimum* L.

lengua-de-perico –see– *Gliricidia sepium* (Jacq.) Kunth ex Walp.

lengua-de-vaca –see– *Rumex crispus* L.

leño gentil –see– *Daphne mezereum* L.

LENS CULINARIS Medik. [Fabaceae]

Common Names:

lentil; Linse

Citations:

Ibáñez Sandín D, Martínez San Ireneo M, Marañón Lizana F, et al. (1999) Specific IgE determinations to crude and boiled lentil (Lens culinaris) extracts in lentil-sensitive children and controls. Allergy 54(11):1209-1214.

Lent lily –see– *Narcissus pseudonarcissus* L.

Lent rose –see– *Narcissus pseudonarcissus* L.

lentil –see– *Lens culinaris* Medik.

lentille bâtarde –see– *Vicia ervilia* (L.) Willd.

lentille ers –see– *Vicia ervilia* (L.) Willd.

LEONTODON AUTUMNALIS L. [Asteraceae]

Citations:

Hausen BM (1979) The sensitizing capacity of compositae plants. III. Test results and cross-reactions in compositae-sensitive patients. Dermatologica 159(1):1-11.

leopard lily –see– *Dieffenbachia seguine* (Jacq.) Schott

leopard palm –see– *Amorphophallus konjac* K. Koch

leopardbane –see– *Arnica montana* L.

lepadena –see– *Euphorbia marginata* Pursh

LEPIDIUM DIDYMUM L. [Brassicaceae]

Common Names:

land cress; lesser swine's-cress

Citations:

Park RJ (1965) Benzyl thiocyanate taint in the milk of dairy cattle ingesting Coronopus didymus Sm. Nature 207:640.

LEPIDIUM DRABA L. [Brassicaceae]

Common Names:

hoary cress; whitetop

Citations:

Fleming CE, Miller MR, Vawter LR, et al. (1934) Poisonous plants. Nevada Agric Exp Sta Annu Rep 1933:10-13.

LEPIDOZAMIA PEROFFSKYANA Regel [Zamiaceae]

Citations:

Cook RW, Gill PA, Boulton JG, et al. (2003) Hepatic lesions in bovine cycad toxicosis. Proc Annu Conf Aust Soc Vet Pathol pp. 27-28.

Gobé GC (1994) Apoptosis in brain and gut tissue of mice fed a seed preparation of the cycad Lepidozamia peroffskyana. Biochem Biophys Res Commun 205(1):327-333.

Gobé GC, Pound AW (1985) Toxic properties of the Australian cycad, Lepidozamia peroffskyana. Queensland Agric J 111(5):261-262.

LEPTADENIA RETICULATA (Retz.) Wight & Arn. [Apocynaceae]

Common names:

jivanti

Citations:

Anjaria JV, Gupta I (1970) Preliminary observations on toxic effects of Leptadenia reticulata (jivanti) and leptaden. Gujvet 4(1):16-18.

Leptopus decaisnei (Benth.) Pojark. = Andrachne decaisnei Benth.

lèrio-do-vale –see– *Convallaria majalis* L.

lesoma –see– *Boophone disticha* (L. f.) Herb.

lespedeza –see– *Kummerowia stipulacea* (Maxim.) Makino

Lespedeza stipulacea Maxim. = Kummerowia stipulacea (Maxim.) Makino

lesser broomrape –see– *Orobanche minor* Sm.

lesser burdock –see– *Arctium minus* (Hill) Bernh.

lesser cardamom –see– *Elettaria cardamomum* (L.) Maton

lesser celandine –see– *Chelidonium majus* L.

lesser hemlock –see– *Aethusa cynapium* L.

lesser loosestrife –see– *Lythrum hyssopifolia* L.

lesser rushy milk vetch –see– *Astragalus convallarius* Greene

lesser swine's-cress –see– *Lepidium didymum* L.

lestun –see– *Codiaeum variegatum* (L.) A. Juss.

lettuce –see– *Lactuca sativa* L.

leucaena –see– *Leucaena leucocephala* (Lam.) de Wit

Leucaena glauca auct. = Leucaena leucocephala (Lam.) de Wit

LEUCAENA LEUCOCEPHALA (Lam.) de Wit [Fabaceae]

Synonyms:

Leucaena glauca auct.

Common Names:

cowbush; guacis; horse tamarind; ipil ipil; jimbey; jumbey; koa haole; kubabul; lamtoro; lead tree; leucaena; subabul; tan tan tree; tangantangan; vi vi; virvi; West Indian lead tree; white popinac; wild tamarind

Citations:

Acamovic T, D'Mello JP (1994) Influence of Leucaena seed and leaf meal diets on young chicks. In: Colegate SM, Dorling PR (eds.) Plant-associated toxins. Agricultural, phytochemical and ecological aspects. CABI. New York. pp. 189-194.

Akpokodje JU, Otesile EB (1987) Leucaena leucocephala toxicity in Ndama cattle. Bull Anim Health Prod Afr 35(1):77-78.

Almeida AP, Kommers GD, Nogueira AP, et al. (2006) Avaliação do efeito tóxico de Leucaena leucocephala (Leg. Mimosoideae) em ovinos. Pesq Vet Bras 26(3):190-194.

Anderson RC, Anderson TJ, Nisbet DJ, et al. (2001) Drought associated poisoning of cattle in South Texas by the high quality forage legume Leucaena leucocephala. Vet Hum Toxicol 43(2):95-96.

Bindon BM, Lamond DR (1966) Examination of tropical legumes for deleterious effects on animal reproduction. Proc Aust Soc Anim Prod 6:109-116.

Blunt CG, Jones RJ (1977) Steer liveweight gains in relation to the proportion of time on Leucaena leucocephala pastures. Trop Grasslands 11(2):159-164.

Compère R (1959) Etude toxicologique du Leucaena glauca chez les bovins. Bull Agric Congo Belg Ruanda Urundi 50:1311-1320.

Damseaux J (1956) Etude de trois légumineuses fourragères introduites au Congo Belge en vue de l'alimentation du bétail. Bull Agric Congo Belg 47(1):93-111.

D'Mello JP, Thomas D (1978) The nutritive value of dried leucaena leaf meal from Malawi: Studies with young chicks. Trop Agric 55:45-50.

Donaldson LE, Hamilton RI, Lambourne LJ, et al. (1970) Assessing Leucaena leucocephala for deleterious effects in cattle and sheep. Proc Int Grassland Cong. pp. 780-782.

Falvey L (1976) The effects of Leucaena leucocephala on cattle in the Northern Territory. Aust Vet J 52(5):243.

Franzolin Neto R, Velloso L (1986) Leucaena leucocephala (Lam.) de Wit en raçõeas para ovinos. 2. Toxicidade. Rev Soc Bras Zootec 15(5):415-424.

Gloria LA, Gerpacio AL, Aglibut FB, et al. (1966) Leucaena glauca Benth for poultry and livestock. III. Protein and energy levels and minerals in minimizing toxic effects of mimosine in chick rations. Philippine Agriculturist 50:235-246.

Hamilton RI, Donaldson LE, Lambourne LJ (1968) Enlarged thyroid glands in calves born to heifers fed a sole diet of Leucaena leucocephala. Aust Vet J 44(Oct):484.

Hamilton RI, Donaldson LE, Lambourne LJ (1971) Leucaena leucocephala as a feed for dairy cows: Direct effect on reproduction and residual effect on the calf and lactation. Aust J Agric Res 22:681-692.

Hathcock JN, Labadan MM, Mateo JP (1975) Effects of dietary protein level on toxicity of Leucaena leucocephala to chicks. Nutr Rep Int 11(1):55-62.

Hegarty MP, Schinckel PG, Court RD (1964) Reaction of sheep to the consumption of Leucaena glauca Benth. and to its toxic principle mimosine. Aust J Agric Res 15:153-165.

Holmes JH (1979) Toxicity of Leucaena leucocephala. I. Equal toxic effects of two Leucaena strains on two breeds of tropical cattle. Papua New Guinea Agric J 30(4):65-69.

Holmes JH (1981) Toxicity of Leucaena leucocephala for steers in the wet tropics. Trop Anim Health Prod 13(2):94-100.

Holmes JH, Humphrey JD, Walton EA, et al. (1981) Cataracts, goitre and infertility in cattle grazed on an exclusive diet of Leucaena leucocephala. Aust Vet J 57(6):257-261.

Iwanaga II, Otagaki KK, Wayman O (1957) Dehydrated koa haole (Leucaena glauca) in rations for growing and fattening swine. Proc West Sec Am Soc Anim Prod 8(25):1-5.

Jones RJ, Blunt CG, Holmes JH (1976) Enlarged thyroid glands in cattle grazing Leucaena pastures. Trop Grasslands 10:113-116.

Jones RJ, Blunt CG, Nurnberg BI (1978) Toxicity of Leucaena leucocephala. The effect of iodine and mineral supplements on penned steers fed a sole diet of Leucaena. Aust Vet J 54(8):387-392.

Jones RJ, Hegarty MP (1984) The effect of different proportions of Leucaena leucocephala in the diet of cattle on growth, feed intake, thyroid function and urinary excretion of 3-hydroxy-4(1H)-pyridone. Aust J Agric Res 35(2):317-325.

Joshi HS (1968) The effect of feeding Leucaena leucocephala (Lam.) De Wit on reproduction in rats. Aust J Agric Res 19(2):341-352.

Labadan MM (1969) The effects of various treatments and additives on the feeding value of ipil-ipil leaf meal in poultry. Philippine Agriculturist 53:392-401.
Labadan MM, Abilay TA, Alejar AS, et al. (1969) The effects of feeding high levels of ipil-ipil (Leucaena leucocephala) leaf meal on comb and testes growth of single comb White Leghorn cockerels. Philippine Agriculturist 53:402-410.
Letts GA (1963) Leucaena glauca and ruminants. Aust Vet J 39(Jul):287-288.
Little DA, Hamilton RI (1971) Leucaena leucocephala and thyroid function of newborn lambs. Aust Vet J 47(9):457-458.
Malynicz G (1974) The effect of adding Leucaena leucocephala meal to commercial rations for growing pigs. Papua New Guinea Agric J 25:12-14.
Martinez MA, Seifert HS (1991) Untersuchungen zur Giftigkeit von L. leucocephala für Ziegen im Nord-Osten Mexikos. Berl Munch Tierarztl Wochenschr 104(8):257-262.
Montagna W, Yun JS (1963) The effects of the seeds of Leucaena glauca on the hair follicles of the mouse. J Invest Dermatol 40:325-332.
Prasad J (1988) Clinico-pathological aspects of experimental Leucaena toxicity in lambs. Indian J Anim Sci 58(10):1181-1182.
Prasad J (1989) A note on toxic effects of Leucaena leucocephala in goats: A clinical study. Indian J Vet Med 9(2):151-152.
Prasad J (1989) Toxic syndrome in calves given a Leucaena mixed forage diet. Indian J Vet Med 9(2):149-150.
Prasad J, Paliwal OP (1989) Pathological changes in experimentally induced Leucaena toxicity in lambs. Indian Vet J 66(8):711-714.
Ram JJ, Atreja PP, Chopra RC, et al. (1994) Mimosine degradation in calves fed a sole diet of Leucaena leucocephala in India. Trop Anim Health Prod 26(4):199-206.
Singh N, Singh SV, Sinha NK, et al. (1989) Ovine wool shedding syndrome in subabool toxicity. Indian J Vet Med 9(1):67-68.
Szyszka M, Meulen U (1984) Der „acceptable daily intake" für Mimosin bei landwirtschaftlichen Nutztieren und seine Bedeutung für die Einsetzbarkeit von Leucaena leucocephala in der Tierernährung. Dtsch Tierarztl Wochenschr 91(7-8):260-262.
Vohradsky F (1972) Observations on influence of feeding horse tamarind (Leucaena glauca Benth.) on the health of cattle in Ghana. Ghana J Agric Sci 5(Part 2):153-156.
Wayman O, Iwanaga II, Hugh WI (1970) Fetal resorption in swine caused by Leucaena leucocephala (Lam.) de Wit. in the diet. J Anim Sci 30(4):583-588.

LEUCANTHEMUM ×SUPERBUM (Bergmans ex J. W. Ingram) D. H. Kent [Asteraceae]

Common Names:

Shasta daisy

Citations:

Burry JN, Kuchel R, Reid JG, et al. (1973) Australian bush dermatitis: Compositae dermatitis in South Australia. Med J Aust 1(3):110-116.

LEUCANTHEMUM VULGARE Lam. [Asteraceae]

Synonyms:

Chrysanthemum leucanthemum L.

Common Names:

daisy; Gänseblume; King Edward daisy; Margarite; marguerite daisy; Orakelblume; oxeye daisy; pyrethrum; whiteweed

Citations:

Kren O (1925) Ein Beitrag zur Pflanzendermatitis. Arch Derm Syphilol 149:96-98.

LEUCOTHOE DAVISIAE Torr. ex A. Gray [Ericaceae]

Common Names:

beach laurel; black laurel; mountain laurel; Sierra laurel

Citations:

Fowler ME (1985) Plant poisoning in two pack llamas. California Vet 39(3):17-20.

LEUCOTHOE GRAYANA Maxim. [Ericaceae]

Citations:

Kohanawa M, Shoya S, Ikeda K, et al. (1970) Experimental Leucothoe grayana poisoning in goats. Natl Inst Anim Health Q (Tokyo) 10(3):160-170.

lever wood –see– *Ostrya virginiana* (Mill.) K. Koch

LEVISTICUM OFFICINALE W. D. J. Koch [Apiaceae]

Common Names:

Badekraut; Leberstockkraut; Liebstöckel; lovage; Saukraut

Citations:

Aswood-Smith MJ, Ceska O, Teoman A, et al. (1992) Photosensitivity from harvesting lovage (Levisticum officinale). Contact Dermatitis 26(5):356-357.
Calnan CD (1969) Lovage sensitivity. Contact Dermatol Newsl 5(Feb):99.

Libesafpel –see– *Lycopersicon esculentum* Mill.

Libocedrus decurrens Torr. = Calocedrus decurrens (Torr.) Florin

licheta –see– *Agrostemma githago* L.

Lichtblume –see– *Colchicum autumnale* L.

licorice –see– *Glycyrrhiza glabra* L.

licorise –see– *Glycyrrhiza glabra* L.

Liebstöckel –see– *Levisticum officinale* W. D. J. Koch

lierre –see– *Hedera helix* L.

lierre grimpant –see– *Hedera helix* L.

lierre tenestre –see– *Glechoma hederacea* L.

l'if –see– *Taxus baccata* L.

life plant –see– *Kalanchoe pinnata* (Lam.) Pers.

light-yellow sophora –see– *Sophora flavescens* Aiton

Liguster –see– *Ligustrum vulgare* L.

ligustrum –see– *Ligustrum vulgare* L.

Ligustrum amurense Carrière = Ligustrum obtusifolium Siebold & Zucc. subsp. suave (Kitag.) Kitag.

LIGUSTRUM OBTUSIFOLIUM Siebold & Zucc. subsp. suave (Kitag.) Kitag. [Oleaceae]

Synonyms:

Ligustrum amurense Carrière

Common Names:

Amur River privet; northern privet; privet

Citations:

Kerr LA, Kelch WJ (1999) Fatal privet (Ligustrum amurease) toxicosis in Tennessee cows. Vet Hum Toxicol 41(6):391-392.

LIGUSTRUM OVALIFOLIUM Hassk. [Oleaceae]

Common Names:

California privet; golden privet; oval-leaf privet; privet

Citations:

Parkinson SC (1986) Suspected privet poisoning. Vet Rec 119(19):483-484.

LIGUSTRUM VULGARE L. [Oleaceae]

Common Names:

aligustre; Beinholz; European privet; Hartriegel; hedge plant; Liguster; ligustrum; prim; primaryvet; privet; Rainweide; Tintenbeerstrauch; Tintenbeertraube; troène; wax-leaf privet; Zaunriegel

Citations:

Arai M, Stauber E, Shropshire CM (1992) Evaluation of selected plants for their toxic effects on canaries. J Am Vet Med Assoc 200(9):1329-1331.

Geßner O (1943-1944) Tödliche Vergiftung durch Früchte des Ligusterstrauches (Ligustrum vulgare L.) bei einem 5-jährigen Kinde. Sammlung Vergiftungsfallen 13:1-2.

Kozlov VA, Guliaeva TN (1983) [Poisoning by the fruit of the privet (Ligustrum vulgare).] Sud Med Ekspert 26(3):56-57.

Turner TW (1904) Some interesting cases. Vet Rec 17:319-320.

liksidsrod –see– *Glycyrrhiza glabra* L.

lilas-de-nuit –see– *Cestrum nocturnum* L.

LILIUM LANCIFOLIUM Thunb. [Liliaceae]

Synonyms:

Lilium tigrinum Ker Gawl.

Common Names:

devil's-lily; Japanese show lily; tiger lily

Citations:

Gathings AM (1991) Acute renal failure in a cat, was it the tiger lily? Napinet Rep 3(2):2.

Langston CE (2002) Acute renal failure caused by lily ingestion in six cats. J Am Vet Med Assoc 220(1):49-52.

LILIUM LONGIFLORUM Thunb. [Liliaceae]

Common Names:

Easter lily; trumpet lily

Citations:

Hall JO (1990) Are Easter lilies toxic to cats? Napinet Rep 3(2):1.

Lahti A (1986) Contact urticaria and respiratory symptoms from tulips and lilies. Contact Dermatitis 14(5):317-319.

Langston CE (2002) Acute renal failure caused by lily ingestion in six cats. J Am Vet Med Assoc 220(1):49-52.

Piirilä P, Kanerva L, Alanko K, et al. (1999) Occupational IgE-mediated asthma, rhinoconjunctivitis, and contact urticaria caused by Easter lily (Lilium longiflorum) and tulip. Allergy 54(3):273-277.

Vidal C, Polo F (1998) Occupational allergy caused by Dianthus caryophillus, Gypsophila paniculata, and Lilium longiflorum. Allergy 53(10):995-998.

Volmer PA (2002) How dangerous are winter and spring holiday plants to pets? Vet Med (Dec):879-884.

Lilium tigrinum Ker Gawl. = Lilium lancifolium Thunb.

liljekonvall –see– *Convallaria majalis* L.

lily-of-the-Nile –see– *Zantedeschia aethiopica* (L.) Spreng.

lily-of-the-valley –see– *Convallaria majalis* L.

lily-of-the-valley bush –see– *Pieris japonica* (Thunb.) D. Don ex G. Don

lily pilly –see– *Acmena smithii* (Poir.) Merr. & L. M. Perry

lilygrass –see– *Arum maculatum* L.

lim –see– *Citrus aurantiifolia* (Christm.) Swingle

lima bean –see– *Phaseolus lunatus* L.

limber pine –see– *Pinus contorta* Douglas ex Loudon

lime –see– *Citrus aurantiifolia* (Christm.) Swingle; *Tilia ×europaea* L.

Limette –see– *Citrus aurantiifolia* (Christm.) Swingle

limonillo –see– *Hymenoxys odorata* DC.

LIMONIUM BRASILIENSE (Boiss.) Kuntze [Plumbaginaceae]

Synonyms:

Statice brasiliensis Boiss.

Common Names:

guaycur

Citations:

Ciganda C, Laborde A (2001) Herbal infusions used for induced abortion. J Toxicol Clin Toxicol 39(3):318-319.

Limonium tataricum (L.) Mill. = Goniolimon tataricum (L.) Boiss.

lin –see– *Linum usitatissimum* L.

linden –see– *Tilia cordata* Mill.; *Tilia ×europaea* L.

Lindheimer's-senna –see– *Senna lindheimeriana* (Scheele) H. S. Irwin & Barneby

linen flax –see– *Linum usitatissimum* L.

ling lang –see– *Sophora flavescens* Aiton

linho –see– *Linum usitatissimum* L.

linoncillo –see– *Fagara heitzii* Aubrév. & Pellegr.

Linse –see– *Lens culinaris* Medik.

linseed –see– *Linum usitatissimum* L.

Linsen –see– *Linum usitatissimum* L.

LINUM NEOMEXICANUM Greene [Linaceae]

Common Names:

yellow pine flax

Citations:

Eggleston WW, Black OF, Kelly JW (1930) Linum neomexicanum (yellow pine flax) and one of its poisonous constituents. J Agric Res 41(10):715-718.

LINUM RIGIDUM Pursh [Linaceae]

Common Names:

flax; stiff-stem flax; stiff-stem yellow flax; yellow flax

Citations:

Anonymous (1896) Poisonous plants. U S Dep Agric Annu Rep 1896:99.

LINUM USITATISSIMUM L. [Linaceae]

Common Names:

Flachs; flax; Lein; Leinsamen; len; lin; linen flax; linho; linseed; Linsen; Saatlein

Citations:

Auld SJ (1912) The formation of prussic acid from linseed cake and other feeding stuffs. J Bd Agric 19:446-460.

Bakhiet AO, Adam SE (1995) Response of Bovans chicks to low dietary levels of Linum usitatissimum seeds. Vet Hum Toxicol 37(6):534-535.

Barnes MH (1931) Linseed dermatitis. J Indus Hyg 13(2):49-55.

Blood DC, Steel JD (1944) Observations on the cyanogenetic properties of linseed "nuts." II. Clinical syndrome in sheep. Aust Vet J 20:338-343.

Brioux C, Richart A (1928) L'acide cyanhydrique des tourteaux de lin. Toxicité de certains de ces tourteaux. Bull Acad Vet Fr I New Series:134-146.

Care AD (1954) Goitrogenic activity in linseed. N Z J Sci Technol A 36(Dec):321-327.

Care AD (1954) Goitrogenic properties of linseed. Nature 173(4395):172-173.

Debackere M, Hoorens J, Haustraete KH (1966) Vergiftiging door koolzaad - En raapzaadschroot. Vlaams Diergeneesk Tijdschr 35(9-10):393-399.

Dunne GT (1924) Poisoning in calves by nascent hydrocyanic acid evolved by cake in solution. Vet J 80:40-42.

Enge EH (1933) Flax straw poisoning in the horse. Vet Med 28(Apr):161.

Franklin MC, Reid RL (1944) Observations of the cyanogenetic properties of linseed "nuts." I. Feeding trials with sheep. Aust Vet J 20(Dec):332-337.

Kratzer FH, Williams DE (1948) The effect of pyridoxine upon growth of chicks fed linseed oil meal. Poult Sci 27:671.

Kratzer FH, Williams DE (1948) The improvement of linseed oil meal for chick feeding by the addition of synthetic vitamins. Poult Sci 27:236-238.

Kratzer FH, Williams DE, Marshall B, et al. (1954) Some properties of the chick growth inhibitor in linseed oil meal. J Nutr 52:555-563.

Ladd EF, Johnson AK (1911) Are flaxseed screenings poisonous to stock? North Dakota Agric Exp Sta Special Bull #1:316-317.

Lezaun A, Fraj J, Colás C, et al. (1998) Anaphylaxis from linseed. Allergy 53(1):105-106.

MacGregor HI, McGinnis J (1948) Toxicity of linseed meal for chicks. Poult Sci 27:141-145.

Montgomerie RF (1924) Hydrocyanic acid generated from linseed cake meal. A case of poisoning in calves. Vet J 80:311-314.

Nicholson D (1927) Cutaneous reactions to flax and linen in five cases of bronchial asthma and two of hayfever with poor results from specific treatment. Can Med Assoc J 17:552-554.

Perrot M (1928) Intoxication des moutons par des tourteaux de lin. Rec Med Vet Ec Alfort 104:15-18.

Phipps WR, Martini MC, Lampe JW, et al. (1993) Effect of flax seed ingestion on the menstrual cycle. J Clin Endocrinol Metab 77(5):1215-1219.

Potteau B, Cluzan R (1966) Incidences nutritionnelles et toxicologiques de l'ingestion d'huile de lin chauffée. I. Effets généraux et action sur l'utilisation des protéines de la ration. Ann Biol Anim Biochim Biophys 6(1):47-64.

Potteau B, Leclerc J (1966) Incidences nutritionnelles et toxicologiques de l'ingestion d'huile de lin chauffée. II. Effet sur l'efficacité biologique du calcium alimentaire. Ann Biol Anim Biochim Biophys 6(1):65-71.

Tou JC, Chen J, Thompson LU (1999) Dose, timing, and duration of flaxseed exposure affect reproductive indices and sex hormone levels in rats. J Toxicol Environ Health A 56:555-570.

Vokoun FJ (1927) Linseed oil dermatitis. JAMA 89(1):20-21.

lion-of-the-earth –see– *Chamaeleon gummifera* (L.) Cass.

lion's-mouth –see– *Digitalis purpurea* L.

Lippia citriodora Kunth = Aloysia citrodora Palau

Lippia ligustrina (Lag.) Britton = Junellia ligustrina (Lag.) Moldenke

LIPPIA PRETORIENSIS H. Pearson [Verbenaceae]

Citations:

Quin JI (1933) Studies on the photosensitization of animals in South Africa. V. The toxicity of Lippia rehmannii (Pears) and Lippia pretoriensis (Pears). Onderstepoort J Vet Sci Anim Indus 1(2):501-504.

LIPPIA REHMANNII H. Pearson [Verbenaceae]

Citations:

Quin JI (1933) Studies on the photosensitization of animals in South Africa. V. The toxicity of Lippia rehmannii (Pears) and Lippia pretoriensis (Pears). Onderstepoort J Vet Sci Anim Indus 1(2):501-504.

liquorice –see– *Glycyrrhiza glabra* L.

liricon fancy –see– *Convallaria majalis* L.

lirio amarillo –see– *Iris pseudoacorus* L.

lirio cala –see– *Zantedeschia aethiopica* (L.) Spreng.

LITHRAEA BRASILIENSIS Marchand [Anacardiaceae]

Common Names:

aroeira

Citations:

Oliveira Lima A (1953) Über das antigene Verhalten der Ölharze einiger Gattungen der Familie Anacardiaceae. Int Arch Allergy 4:169-174.

Santos OL, Filgueira AL (1994) "Aroeira"-induced photosensitization. Int J Dermatol 33(3):222.

LITHRAEA CAUSTICA (Molina) Hook. & Arn. [Anacardiaceae]

Common Names:

litre tree

Citations:

Manriquez O, Varas J, Rios JC, et al. (2002) Analysis of 156 cases of plant intoxication received in the toxicologic information center at Catholic University of Chile. Vet Hum Toxicol 44(1):31-32.

Oliveira Lima A (1953) Über das antigene Verhalten der Ölharze einiger Gattungen der Familie Anacardiaceae. Int Arch Allergy 4:169-174.

LITHRAEA MOLLEOIDES (Vell.) Engl. [Anacardiaceae]

Citations:

Oliveira Lima A (1953) Über das antigene Verhalten der Ölharze einiger Gattungen der Familie Anacardiaceae. Int Arch Allergy 4:169-174.

litre tree –see– *Lithraea caustica* (Molina) Hook. & Arn.

little blue stagger –see– *Dicentra canadensis* (Goldie) Walp.

little-flower lupine –see– *Lupinus argenteus* Pursh var. holosericeus (Nutt.) Barneby

little good –see– *Euphorbia helioscopia* L.

little goody –see– *Euphorbia helioscopia* L.

little juniper –see– *Juniperus communis* L.

little larkspur –see– *Delphinium bicolor* Nutt.; *Delphinium menziesii* DC.

little lettuce –see– *Agave lechuguilla* Torr.

little mallow –see– *Malva parviflora* L.

little peach –see– *Solanum glaucophyllum* Desf.

little-seed Canarygrass –see– *Phalaris minor* Retz.

little staggerweed –see– *Dicentra cucullaria* (L.) Bernh.

little Utah juniper –see– *Juniperus osteosperma* (Torr.) Little

live forever –see– *Kalanchoe pinnata* (Lam.) Pers.

live leaf –see– *Kalanchoe pinnata* (Lam.) Pers.

live oak –see– *Quercus virginiana* Mill.

l'ivraie –see– *Lolium temulentum* L.

lluvia-del-plata –see– *Dieffenbachia seguine* (Jacq.) Schott

Loasa vulcanica André = Nasa triphylla (Juss.) Weigand

lobelia –see– *Lobelia berlandieri* A. DC.

LOBELIA BERLANDIERI A. DC. [Campanulaceae]

Common Names:

Berlandier lobelia; lobelia

Citations:

Dollahite JW, Allen TJ (1962) Poisoning of cattle, sheep, and goats with Lobezla [Lobelia] and Centarium species. Southwestern Vet 15(Winter):126-130.

Lopez R, Martinez-Burnes J, Vargas G, et al. (1994) Taxonomical, clinical and pathological findings in moradilla (lobelia-like) poisoning in sheep. Vet Hum Toxicol 36(3):195-198.

lobster flower –see– *Euphorbia pulcherrima* Willd. ex Klotzsch

Lochnera pusilla (Murray) K. Schum. = Catharanthus pusillus (Murray) G. Don

loco seed –see– *Datura stramonium* L.

locoweed –see– *Astragalus mollissimus* Torr. var. mollissimus; *Astragalus pubentissimus* Torr. & A. Gray; *Oxytropis lambertii* Pursh; *Oxytropis sericea* Nutt.

locust tree –see– *Robinia pseudoacacia* L.

lodgepole pine –see– *Pinus contorta* Douglas ex Loudon

løg –see– *Allium cepa* L.

lolch –see– *Lolium temulentum* L.

LOLIUM MULTIFLORUM Lam. [Poaceae]

Common Names:

annual ryegrass; Italian ryegrass; Welches Weidelgras

Citations:

Norris RT, Richards IS, Petterson DS (1981) Treatment of ovine annual ryegrass toxicity with chlordiazepoxide: A field evaluation. Aust Vet J 57(6):302-303.

Richards IS, Petterson DS, Purcell DA (1979) Treatment of ovine annual ryegrass toxicity with chlordiazepoxide. Aust Vet J 55(6):282-283.

LOLIUM PERENNE L. [Poaceae]

Common Names:

Deutsches Weidelgras; English ryegrass; Italian ryegrass; ivraie; perennial ryegrass; raigras perenne; ryegrass

Citations:

Blythe LL, Andreason CB, Pearson EG, et al. (1998) Atypical pneumonia associated with ryegrass staggers in calves. In: Garland T, Barr AC (eds.) Toxic plants and other natural toxicants. CABI. New York. pp. 69-72.

Clegg FG, Watson WA (1960) Ryegrass staggers in sheep. Vet Rec 72(36):731-733.

Galey FD, Tracy ML, Craigmill AL, et al. (1991) Staggers induced by consumption of perennial ryegrass in cattle and sheep from northern California. J Am Vet Med Assoc 199(4):466-470.

Keogh RG (1973) Induction and prevention of ryegrass staggers in grazing sheep. N Z J Exp Agr 1(1):55-57.

Mackintosh CG, Orr MB, Gallagher RT, et al. (1982) Ryegrass staggers in Canadian wapiti deer. N Z Vet J 30:106-107.

Mitchell PJ, McCaughan CJ (1992) Perennial ryegrass staggers in fallow deer (Dama dama). Aust Vet J 69(10):258-259.

Pearson EG, Andreasen CB, Blythe LL, et al. (1996) Atypical pneumonia associated with ryegrass staggers in calves. J Am Vet Med Assoc 209(6):1137-1142.

Peterson AJ, Bass JJ, Byford MJ (1978) Decreased plasma testosterone concentrations in rams affected by ryegrass staggers. Res Vet Sci 25(3):266-268.

Wheatley WM (1998) Management, environmental and livestock interactions impact on perennial ryegrass/Neotyphodium/livestock associations. In: Garland T, Barr AC (eds.) Toxic plants and other natural toxicants. CABI. New York. pp. 45-48.

LOLIUM RIGIDUM Gaudin [Poaceae]

Common Names:

annual ryegrass; Wimmera ryegrass

Citations:

Berry PH, Cook RD, McHowell J, et al. (1976) Lesions in sheep and guinea pigs pen fed parasitised annual ryegrass (Lolium rigidum). Aust Vet J 52(11):540-541.

Berry PH, Howell JM, Cook RD (1980) Morphological changes in the central nervous system of sheep affected with experimental annual ryegrass (Lolium rigidum) toxicity. J Comp Pathol 90(4):603-617.

Berry PH, Howell JM, Cook RD, et al. (1980) Central nervous system changes in sheep and cattle affected with natural or experimental annual ryegrass toxicity. Aust Vet J 56(8):402-403.

Berry PH, Richards RB, Howell JM, et al. (1982) Hepatic damage in sheep fed annual ryegrass, Lolium rigidum, parasitised by Anguina agrostis and Corynebacterium rathayi. Res Vet Sci 32(2):148-156.

Berry PH, Wise JL (1975) Wimmera rye grass toxicity in Western Australia. Aust Vet J 51(11):525-530.

Creeper JH, Vale W, Walsh R (1996) Annual ryegrass toxicosis in horses. Aust Vet J 74(6):465-467.

Gwynn R, Hadlow AJ (1971) Toxicity syndrome in sheep grazing Wimmera ryegrass in Western Australia. Aust Vet J 47(Aug):408.

Lanigan GW, Payne AL, Frahn JL (1976) Origin of toxicity in parasitised annual ryegrass (Lolium rigidum). Aust Vet J 52(5):244-246.

McIntosh GH, Thomas MR (1967) Toxicity of parasitised Wimmera ryegrass, Lolium rigidum, for sheep and cattle. Aust Vet J 43(9):349-353.

LOLIUM TEMULENTUM L. [Poaceae]

Common Names:

bearded darnel; cheat; darnel; drabok; drake; drunk; ivray; l'ivraie; lolch; poison darnel; poison ryegrass; Schwindelhafer; sturdy rye; tares; Taumelhafer; Taumelkorn; Taumelloch; Tollkorn; Tollkraut; winter ryegrass

Citations:

Tait J (1842) A case of poisoning in pigs. Veterinarian 15:212-213.

Urbain A, Nouvel J (1939) Cas d'intoxication alimentaire dus à l'ivraie observés chez des animaux sauvages en captivité. Bull Acad Vet Fr 12:77-82.

London plane tree –see– *Platanus ×acerifolia* (Aiton) Willd.

long-spine thorn apple –see– *Datura ferox* L.

long-spur lupine –see– *Lupinus arbustus* Douglas ex Lindl.

long-spur thorn apple –see– *Datura ferox* L.

LONICERA TATARICA L. [Caprifoliaceae]

Common Names:

Tatarian honeysuckle; Tatarische Heckenkirsche; Tatarisches Geißblatt

Citations:

Lamminpaa A, Kinos M (1996) Plant poisonings in children. Hum Exp Toxicol 15(3):245-249.

Schurno A (1958) Beobachtung einer Vergiftung durch Beeren der tatarischen Heckenkirsche. Kinderaerztl Prax 26(8):357-360.

loogbos –see– *Psilocaulon absimile* N. E. Br.

look –see– *Allium sativum* L.

LOPHIRA ALATA Banks ex C. F. Gaertn. [Ochnaceae]

Synonyms:

Lophira procera A. Chev.

Common Names:

azobe tree

Citations:

Anonymous (1977) Paralysis posterior bij zeugen en Azobézaagsel. Tijdschr Diergeneeskd 102(9):582.

Lophira procera A. Chev. = Lophira alata Banks ex C. F. Gaertn.

LOPHOLAENA CORIIFOLIA (Sond.) Phill. & C. A. Sm. [Asteraceae]

Synonyms:

Lopholaena randii S. Moore

Citations:

Quin JI (1933) Studies on the photosensitization of animals in South Africa. 4. The toxicity of Lopholaena coriifolia (Harv.) Phill. & C. A. Sm. (= L. randii sp. Moore). Onderstepoort J Vet Sci Anim Indus 1(2):497-499.

Lopholaena randii S. Moore = Lopholaena coriifolia (Sond.) Phill. & C. A. Sm.

LOPHOPETALUM DUBIUM M. A. Lawson [Celastraceae]

Common Names:

perupok

Citations:

Tanaka S, Matsumoto Y, Tamada Y (2003) Allergic contact dermatitis due to perupok wood. Contact Dermatitis 48(5):273.

LOPHOPETALUM FLORIBUNDUM Wight [Celastraceae]

Common Names:

perupok

Citations:

Tanaka S, Matsumoto Y, Tamada Y (2003) Allergic contact dermatitis due to perupok wood. Contact Dermatitis 48(5):273.

LOPHOPHORA WILLIAMSII (Lem. ex Salm-Dyck) J. M. Coult. [Cactaceae]

Common Names:

anhalonium; mescal; mescal bean; mescal button; Peyolylkaktus; peyote; peyote cactus; peyote muscal; the-bad-seed

Citations:

Blum K, Futterman SL, Pascarosa P (1977) Peyote, a potential ethnopharmacologic agent for alcoholism and other drug dependencies: Possible biochemical rationale. Clin Toxicol 11(4):459-472.

loquat –see– *Eriobotrya japonica* (Thunb.) Lindl.; *Rhodomyrtus macrocarpa* Benth.

loquat plum –see– *Eriobotrya japonica* (Thunb.) Lindl.

Lorbeer –see– *Laurus nobilis* L.

Lorbeerkirsche –see– *Prunus laurocerasus* L.

Lorbeerose –see– *Kalmia angustifolia* L.; *Nerium oleander* L.

lords-and-ladies –see– *Arum maculatum* L.; *Zantedeschia aethiopica* (L.) Spreng.

lorier bol –see– *Nerium oleander* L.

LOTUS CORNICULATUS L. [Fabaceae]

Common Names:

bird's-foot trefoil; Hornklee; Pantoffeln; szarvas kerep; Taubenkröpferl

Citations:

Dougherty RW, Christensen RB (1953) In vivo absorption studies of hydrocyanic acid of plant juice origin. Cornell Vet 43(3):481-486.

Stafford KJ, West DM, Alley MR, et al. (1995) Suspected photosensitisation in lambs grazing birdsfoot trefoil (Lotus corniculatus). N Z Vet J 43:114-117.

louco –see– *Plumbago scandens* L.

Louisiana pepper –see– *Capsicum frutescens* L.

Louisiana wormwood –see– *Artemisia ludoviciana* Nutt.

louseberry tree –see– *Euonymus europaeus* L.

lovage –see– *Levisticum officinale* W. D. J. Koch

love apple –see– *Lycopersicon esculentum* Mill.; *Solanum aculeatissimum* Jacq.; *Solanum carolinense* L.; *Solanum pseudocapsicum* L.

love bean –see– *Abrus precatorius* L.

love pea –see– *Abrus precatorius* L.

low arrowgrass –see– *Triglochin maritima* L.

low larkspur –see– *Delphinium andersonii* A. Gray; *Delphinium bicolor* Nutt.; *Delphinium carolinianum* Walter subsp. virescens (Nutt.) R. E. Brooks; *Delphinium menziesii* DC.; *Delphinium nuttallianum* Pritz.; *Delphinium tricorne* Michx.

low laurel –see– *Kalmia angustifolia* L.

low ragweed –see– *Ambrosia artemisiifolia* L.

low senna –see– *Senna tora* (L.) Roxb.

low-whorled milkweed –see– *Asclepias pumila* (A. Gray) Vail

Löwenzahn –see– *Taraxacum officinale* F. H. Wigg. aggr.

LOXOPTERYGIUM HUASANGO Spruce ex Engl. [Anacardiaceae]

Common Names:

hualtaco tree; huasango tree

Citations:

Paulson GA (1941-1943) Mango and hualtaco dermatitis. Med Bull Standard Oil Co 5:197-200.

lucerne –see– *Medicago sativa* L.

lucky bean –see– *Abrus precatorius* L.

lucky nut –see– *Thevetia peruviana* (Pers.) K. Schum.

LUEHEA DIVARICATA Mart. [Malvaceae]

Common Names:

Francisco Alvarez

Citations:

Ciganda C, Laborde A (2001) Herbal infusions used for induced abortion. J Toxicol Clin Toxicol 39(3):318-319.

Ciganda C, Laborde A (2003) Herbal infusions used for induced abortion. J Toxicol Clin Toxicol 41(3):235-239.

LUFFA ACUTANGULA (L.) Roxb. [Cucurbitaceae]

Common Names:

kali-tori; ridge gourd; smooth luffa

Citations:

Sinha SM, Pasricha JS, Sharma RC, et al. (1977) Vegetables responsible for contact dermatitis of the hands. Arch Dermatol 113(6):776-779.

lumbán –see– *Aleurites moluccanus* (L.) Willd.

lumbang –see– *Aleurites moluccanus* (L.) Willd.

lumbang nut –see– *Reutealis trisperma* (Blanco) Airy Shaw

lunara lupine –see– *Lupinus formosus* Greene

lupin –see– *Lupinus albus* L.; *Lupinus angustifolius* L.

lupin blanc –see– *Lupinus albus* L.

lupin bleu –see– *Lupinus angustifolius* L.

lupin jaune –see– *Lupinus luteus* L.

lupine –see– *Lupinus albus* L.; *Lupinus albus* L. var. albus; *Lupinus argenteus* Pursh; *Lupinus diffusus* Nutt.; *Lupinus leucophyllus* Douglas ex Lindl.

LUPINUS ALBUS L. [Fabaceae]

Common Names:

European white lupin; fève-du-loup; lupin; lupin blanc; lupine; pois-de-loup; sweet lupin; white lupin; white lupine

Citations:

Agid Y, Pertuiset B, Dubois B (1988) Motoneuron disease as manifestation of lupin seed toxicity. Lancet 1(Jun 11):1347.

Anonymous (1934) Investigation of stock-poisoning plants. U S Dep Agric Bur Anim Indus Chief Rep 1934:38.

Castanon JI, Perez-Lanzac, J (1990) Substitution of fixed amounts of soybean meal for field beans (Vicia faba), sweet lupines (Lupinus albus), cull peas (Pisum sativum) and vetches (Vicia sativa) in diets for high performance laying Leghorn hens. Br Poult Sci 31(1):173-184.

Cubillos V, Cubillos A, Guerra J (1982) Estudio histopatológico en hígados de Gallus gallus alimentados con semilla de lupin (altramuz) dulce (L. albus var. multolupa). Zentralbl Veterinarmed A 29(3):215-222.

Guillaume J, Chenieux JC, Rideau M (1979) Feeding value of Lupinus albus L. in chicken diets (with emphasis on the role of alkaloids). Nutr Rep Int 20(1):57-65.

Hove EL, King S, Hill GD (1978) Composition, protein quality, and toxins of seeds of the grain legumes Glycine max, Lupinus spp., Phaseolus spp., Pisum sativum, and Vicia faba. N Z J Agric Res 21:457-462.

Litkey J, Dailey MW (2007) Anticholinergic toxicity associated with the ingestion of lupini beans. Am J Emerg Med 25(2):215-217.

Perez-Escamilla R, Vohra P, Klasing K (1988) Lupins (Lupinus albus var. ultra) as a replacement for soybean meal in diets for growing chickens and turkey poults. Nutr Rep Int 38(3):583-593.

Ruiz LP Jr, White SF, Hove EL (1977) The alkaloid content of sweet lupin seed used in feeding trials on pigs and rats. Anim Feed Sci Technol 2(1):59-66.

Watkins BA, Manning B, Al-Athari AK (1988) The effects of Lupinus albus cultivar ultra on broiler performance. Nutr Rep Int 38(1):173-181.

Wittenburg H, Nehring, K (1965) Untersuchungen über die Wirkung reiner Lupinenalkaloide auf den tierischen Organismus. Die Wirkung von Lupanin auf Ratten. Pharmazie 20:156-158.

LUPINUS ALBUS L. var. albus [Fabaceae]

Synonyms:

Lupinus termis Forssk.

Common Names:

lupine; turmus

Citations:

Lavy R (1964) Thrombocytopenic purpura due to Lupinus termis bean. J Allergy 35(5):386-389.

Tannous RI, Nayfeh SN (1969) Effect of feeding lupine seeds on spermatogenesis in the rat. Aust J Biol Sci 22(4):1071-1075.

Tannous RI, Shadarevian S, Cowan JW (1968) Rat studies on quality of protein and growth-inhibiting action of alkaloids of lupine (Lupinus termis). J Nutr 94(2):161-165.

LUPINUS ANGUSTIFOLIUS L. [Fabaceae]

Synonyms:

Lupinus varius L.

Common Names:

bitter lupin; Blaue Lupine; blue lupin; lupin; lupin bleu; narrow-leaf lupin; narrow-leaf lupine; New Zealand blue lupine; New Zealand lupin; Western Australian blue lupin; wild lupin

Citations:

Allen JG (1981) An evaluation of lupinosis in cattle in Western Australia. Aust Vet J 57(5):212-215.

Allen JG, Tudor GD, Petterson DS (1998) The feeding of lupin grain can cause rumen acidosis and rumenitis. In: Garland T, Barr AC (eds.) Toxic plants and other natural toxicants. CABI. New York. pp. 143-148.

Arnold GW, Hill JL, Maller RA, et al. (1976) Comparison of lupin varieties for nutritive value as dry standing feed for weaner sheep and for incidence of lupinosis. Aust J Agric Res 27(3):423-435.

Brash AG (1943) Lupin poisoning of sheep. N Z J Agr 67:83-84.

Gardiner MR (1965) Mineral metabolism in sheep lupinosis. I. Iron and cobalt. J Comp Pathol 75(4):397-408.

Gardiner MR (1965) The pathology of lupinosis of sheep. Gross- and histo-pathology. Pathol Vet 2(5):417-445.

Gardiner MR (1967) Cattle lupinosis. A clinical and pathological study. J Comp Pathol 77(1):63-69.

Godfrey NW, Mercy AR, Emms Y, et al. (1984) Tolerance of growing pigs to lupine (Lupinus angustifolius) alkaloids. Aust J Exp Agric 25(4):791-795.

Grant G, Dorward PM, Buchan W (1995) Consumption of diets containing raw soya beans (Glycine max), kidney beans (Phaseolus vulgaris), cowpeas (Vigna unguiculata)

or lupin seeds (Lupinus angustifolius) by rats up to 700 days: Effects on body composition and organ weights. Br J Nutr 73(1):17-29.

Grant G, Dorward PM, Pusztai A (1993) Pancreatic enlargement is evident in rats fed diets containing raw soybeans (Glycine max) or cowpeas (Vigna unguiculata) for 800 days but not in those fed diets based on kidney beans (Phaseolus vulgaris) or lupinseed (Lupinus angustifolius). J Nutr 123(12):2207-2215.

Karasinski D, Bednarczyk M, Peretiatkowicz M, et al. (1988) The influence of alkaloids in seeds of Lupinus angustifolius on the growth and some meat features of ducks. Bull Pol Acad Sci Biol Sci 36(10-12):215-224.

Marsh CD, Clawson AB, Marsh H (1916) Lupines as poisonous plants. U S Dep Agric Bull #405:45 pp.

Olver MD, Jonker A (1997) Effect of sweet, bitter and soaked micronised bitter lupins on broiler performance. Br Poult Sci 38(2):203-208.

Olver MD, Jonker A (1998) Effects of sweet, bitter and soaked micronised bitter lupins on duckling performance. Br Poult Sci 39(5):622-626.

LUPINUS ARBUSTUS Douglas ex Lindl. [Fabaceae]

Synonyms:

Lupinus laxiflorus auct.

Common Names:

grassland lupin; grassland lupine; long-spur lupine; spur lupine

Citations:

Clawson AB (1931) Two lupines shown to be poisonous to livestock. U S Dep Agric Official Record 10(Feb26):71.

Panter KE, Gardner DR, Molyneux RJ (1998) Teratogenic and fetotoxic effects of two piperidine alkaloid-containing lupines (L. formosus and L. arbustus) in cows. J Nat Toxins 7(2):131-140.

Panter KE, Gardner DR, Shea RE, et al. (1998) Toxic and teratogenic piperidine alkaloids from Lupinus, Conium and Nicotiana species. In: Garland T, Barr AC (eds.) Toxic plants and other natural toxicants. CABI. New York. pp. 345-350.

Panter KE, Keeler RF (1991) Toxicologic and teratogenic comparison of Lupinus formosus, L. arbustus and L. caudatus in goats. J Anim Sci 69(Suppl 1):393.

LUPINUS ARGENTEUS Pursh [Fabaceae]

Common Names:

blue bean; grassland lupine; hairy lupine; lupine; silver lupine; silvery bean; silvery lupine; tailcup lupine; wild bean

Citations:

James LF, Binns W, Shupe JL (1968) Blood changes in cattle and sheep fed lupine. Am J Vet Res 29(3):557-560.

Marsh CD, Clawson AB, Marsh H (1916) Lupines as poisonous plants. U S Dep Agric Bull #405:45 pp.

Panter KE, Mayland HF, Gardner DR, et al. (2001) Beef cattle losses after grazing Lupinus argenteus (silvery lupine). Vet Hum Toxicol 43(5):279-282.

LUPINUS ARGENTEUS Pursh var. argenteus [Fabaceae]

Synonyms:

Lupinus laxiflorus Douglas ex Lindl.

Common Names:

Douglas'-spur lupine

Citations:

Anonymous (1929) Investigation of stock-poisoning plants. U S Dep Agric Bur Anim Indus Chief Rep 1929:47-48.

LUPINUS ARGENTEUS Pursh var. heteranthus (S. Watson) Barneby [Fabaceae]

Common Names:

Kellogg's-spur lupine; tailcup lupine; tailspur lupine

Citations:

Anonymous (1929) Investigation of stock-poisoning plants. U S Dep Agric Bur Anim Indus Chief Rep 1929:47-48.

Anonymous (1932) Investigation of stock-poisoning plants. U S Dep Agric Bur Anim Indus Chief Rep 1932:44.

Anonymous (1933) Investigation of stock-poisoning plants. U S Dep Agric Bur Anim Indus Chief Rep 1933:34-35.

Binns W, James LF, Keeler RF, et al. (1968) Effects of teratogenic agents in range plants. Cancer Res 28(11):2323-2326.

Gardner DR, Panter KE (1993) Comparison of blood plasma alkaloid levels in cattle, sheep, and goats fed Lupinus caudatus. J Nat Toxins 2(1):1-11.

Gardner DR, Panter KE (1994) Anagyrine and ammodendrine alkaloid levels in the blood of cattle, sheep and goats fed teratogenic lupin species. In: Colegate SM, Dorling PR (eds.) Plant-associated toxins. Agricultural, phytochemical and ecological aspects. CABI. New York. pp. 173-177.

Keeler RF, Binns W, James LF, et al. (1969) Preliminary investigation of the relationship between bovine congenital lathyrism induced by aminoacetonitrile and the lupine induced crooked calf disease. Can J Comp Med Vet Sci 33(1):89-92.

Panter KE, James LF, Keeler RF, et al. (1992) Radio-ultrasound observations of poisonous plant-induced fetotoxicity in livestock. In: James LF, Keeler RF, Bailey JM Jr, et al. (eds.) Poisonous plants. Iowa State Univ Press. Ames, Iowa. pp. 481-488.

Shupe JL, Balls LD, James LF (1968) Changes in blood serum transaminase associated with lupine and larkspur poisoning in cattle. Cornell Vet 58(1):129-135.

Shupe JL, Binns W, James LF, et al. (1967) Crooked calf syndrome, A plant-induced congenital deformity. Zuchthygiene 2:145-152.

Shupe JL, Binns W, James LF, et al. (1968) A congenital deformity in calves induced by the maternal consumption of lupin. Aust J Agric Res 19(2):335-340.

LUPINUS ARGENTEUS Pursh var. holosericeus (Nutt.) Barneby [Fabaceae]

Synonyms:

Lupinus holosericeus Nutt.

Common Names:
grassland lupine; Great Basin lupine; little-flower lupine; silky lupine; spur lupine; tailcup lupine

Citations:
Binns W, James LF (1961) A congenital deformity in calves, similar to "crooked calf disease," has been experimentally produced by feeding heifers lupine and lead. Proc Am Soc Anim Prod Western Sect 12(66):1-3.
Binns W, James LF, Shupe JL, et al. (1962) Crooked calf disease produced experimentally. Utah Farm Home Sci 23(2):35-37.

LUPINUS ARGENTEUS Pursh var. rubricaulis (Greene) S. L. Welsh [Fabaceae]

Common Names:
mountain silvery lupine

Citations:
Anonymous (1934) Investigation of stock-poisoning plants. U S Dep Agric Bur Anim Indus Chief Rep 1934:38.

LUPINUS COSENTINII Guss. [Fabaceae]

Common Names:
sandhill lupin; sandplain lupin; West Australian blue lupin

Citations:
Arnold GW, Hill JL, Maller RA, et al. (1976) Comparison of lupin varieties for nutritive value as dry standing feed for weaner sheep and for incidence of lupinosis. Aust J Agric Res 27(3):423-435.
Croker KP, Allen JG, Gittins SP, et al. (1998) The development of lupinosis in weaner sheep grazed on sandplain lupins. In: Garland T, Barr AC (eds.) Toxic plants and other natural toxicants. CABI. New York. pp. 459-463.
Hawkins CD (1994) "Acorn" calves and retained placentae following grazing on sandplain lupins (Lupinus cosentinii). In: Colegate SM, Dorling PR (eds.) Plant-associated toxins. Agricultural, phytochemical and ecological aspects. CABI. New York. 351-356.

Lupinus cruckshanksii Hook. = Lupinus mutabilis Sweet

Lupinus cumulicola Small = Lupinus diffusus Nutt.

Lupinus cyanus Rydb. = Lupinus leucophyllus Douglas ex Lindl.

LUPINUS DIFFUSUS Nutt. [Fabaceae]

Synonyms:
Lupinus cumulicola Small

Common Names:
lupine; sandhill lupine; sky-blue lupine

Citations:
Duncan WH, Piercy PL, Starling RJ (1955) Toxicological studies of southeastern plants. I. Leguminosae. Econ Bot 9(3):243-255.

LUPINUS DIGITATUS Forssk. [Fabaceae]

Common Names:
Western Australian blue lupin

Citations:
Gardiner MR (1967) The role of copper in the pathogenesis of subacute and chronic lupinosis of sheep. Aust Vet J 43(Jul):243-248.

LUPINUS FORMOSUS Greene [Fabaceae]

Common Names:
grassland lupine; lunara lupine; summer lupine

Citations:
Gardner DR, Panter KE (1994) Ammodendrine and related piperidine alkaloid levels in the blood plasma of cattle, sheep and goats fed Lupinus formosus. J Nat Toxins 3(2):107-116.
Gardner DR, Panter KE (1994) Anagyrine and ammodendrine alkaloid levels in the blood of cattle, sheep and goats fed teratogenic lupin species. In: Colegate SM, Dorling PR (eds.) Plant-associated toxins. Agricultural, phytochemical and ecological aspects. CABI. New York. pp. 173-177.
Keeler RF, Panter KE (1989) Piperidine alkaloid composition and relation to crooked calf disease-inducing potential of Lupinus formosus. Teratology 40(5):423-432.
Keeler RF, Panter KE (1992) Induction of crooked calf disease by the piperidine alkaloid-containing plant Lupinus formosus. In: James LF, Keeler RF, Bailey JM Jr, et al. (eds.) Poisonous plants. Iowa State Univ Press. Ames, Iowa. pp. 239-244.
Panter KE, Gardner DR, Molyneux RJ (1994) Comparison of toxic and teratogenic effects of Lupinus formosus, L. arbustus and L. caudatus in goats. J Nat Toxins 3(2):83-93.
Panter KE, Gardner DR, Molyneux RJ (1998) Teratogenic and fetotoxic effects of two piperidine alkaloid-containing lupines (L. formosus and L. arbustus) in cows. J Nat Toxins 7(2):131-140.
Panter KE, Keeler RF (1991) Toxicologic and teratogenic comparison of Lupinus formosus, L. arbustus and L. caudatus in goats. J Anim Sci 69(Suppl 1):393.

Lupinus holosericeus Nutt. = Lupinus argenteus Pursh var. holosericeus (Nutt.) Barneby

LUPINUS LATIFOLIUS J. Agardh [Fabaceae]

Common Names:
broad-leaf lupine

Citations:
Craigmill AL, Crosby D, Kilgore W (1983) The transfer of teratogenic lupine alkaloids to human beings through milk. J Am Vet Med Assoc 183:351.
Kilgore WW, Crosby DG, Craigmill AL, et al. (1981) Toxic plants as possible human teratogens. Calif Agric 35(Nov-Dec):6.

Lupinus laxiflorus auct. = Lupinus arbustus Douglas ex Lindl.

Lupinus laxiflorus Douglas ex Lindl. = Lupinus argenteus Pursh var. argenteus

LUPINUS LEUCOPHYLLUS Douglas ex Lindl. [Fabaceae]

Synonyms:

Lupinus cyaneus Rydb.

Common Names:

blue pea; horsebean; lupine; poison lupine; velvet lupine; western lupine; white-leaf lupine; wild pea; woolly-leaf lupine; woolly lupine

Citations:

Anonymous (1933) Investigation of stock-poisoning plants. U S Dep Agric Bur Anim Indus Chief Rep 1933:34-35.

Anonymous (1936) Investigations of stock-poisoning plants. U S Dep Agric Bur Anim Indus Chief Rep. 1933:44-45.

Knowles AD (1915) Lupinosis of horses and the treatment. J Am Vet Med Assoc 48:286-303.

Marsh CD, Clawson AB, Marsh H (1916) Lupines as poisonous plants. U S Dep Agric Bull #405:45 pp.

Lupinus leucopsis J. Agardh = Lupinus sericeus Pursh

LUPINUS LUTEUS L. [Fabaceae]

Common Names:

Gelbe Lupine; Gelbe Wolfsbohne; lupin jaune; yellow lupin; yellow lupine

Citations:

Arnold GW, Hill JL, Maller RA, et al. (1976) Comparison of lupin varieties for nutritive value as dry standing feed for weaner sheep and for incidence of lupinosis. Aust J Agric Res 27(3):423-435.

Gardiner MR (1965) The pathology of lupinosis of sheep. Gross- and histo-pathology. Pathol Vet 2(5):417-445.

Soler Rodriguez F, Miguez Santiyan MP, Pedrera Zamorano JD, et al. (1991) An outbreak of lupinosis in sheep. Vet Hum Toxicol 33(5):492-494.

Turska R (1977) Pathological changes of bile canaliculi of chicken fed with lupine. Acta Med Pol 18(4):363-364.

Turska R, Jamroz D, Czarna Z, et al. (1976) Ultrastructural changes in hepatocytes of chicken fed with lupine. Anat Histol Embryol 5:96-97.

LUPINUS MUTABILIS Sweet [Fabaceae]

Synonyms:

Lupinus cruckshanksii Hook.

Common Names:

tallhue

Citations:

Hove EL, King S, Hill GD (1978) Composition, protein quality, and toxins of seeds of the grain legumes Glycine max, Lupinus spp., Phaseolus spp., Pisum sativum, and Vicia faba. N Z J Agric Res 21:457-462.

Ruiz MA, Sotelo A (2001) Chemical composition, nutritive value, and toxicology evaluation of Mexican wild lupins. J Agric Food Chem 49(11):5336-5339.

Lupinus ornatus Douglas = Lupinus sericeus Pursh

LUPINUS REFLEXUS Rose [Fabaceae]

Citations:

Ruiz MA, Sotelo A (2001) Chemical composition, nutritive value, and toxicology evaluation of Mexican wild lupins. J Agric Food Chem 49(11):5336-5339.

LUPINUS ROTUNDIFLORUS M. E. Jones [Fabaceae]

Citations:

Ruiz MA, Sotelo A (2001) Chemical composition, nutritive value, and toxicology evaluation of Mexican wild lupins. J Agric Food Chem 49(11):5336-5339.

LUPINUS SERICEUS Pursh [Fabaceae]

Synonyms:

Lupinus leucopsis J. Agardh; ***Lupinus ornatus*** Douglas

Common Names:

Big Bend lupine; Pursh's-silky lupine; silky lupine; silvery lupine; sink lupine

Citations:

Anonymous (1935) Investigation of stock-poisoning plants. U S Dep Agric Bur Anim Indus Chief Rep 1935:39-40.

Binns W, James LF (1961) A congenital deformity in calves, similar to "crooked calf disease," has been experimentally produced by feeding heifers lupine and lead. Proc Am Soc Anim Prod Western Sect 12(66):1-3.

Binns W, James LF, Keeler RF, et al. (1968) Effects of teratogenic agents in range plants. Cancer Res 28(11):2323-2326.

Binns W, James LF, Shupe JL, et al. (1962) Crooked calf disease produced experimentally. Utah Farm Home Sci 23(2):35-37.

James LF, Binns W, Shupe JL (1968) Blood changes in cattle and sheep fed lupine. Am J Vet Res 29(3):557-560.

Marsh CD, Clawson AB, Marsh H (1916) Lupines as poisonous plants. U S Dep Agric Bull #405:45 pp.

Nelson SB (1906) Feeding wild plants to sheep. Washington Agric Exp Sta Bull #73:64 pp.

Shupe JL, Balls LD, James LF (1968) Changes in blood serum transaminase associated with lupine and larkspur poisoning in cattle. Cornell Vet 58(1):129-135.

Shupe JL, Binns W, James LF, et al. (1967) Crooked calf syndrome, A plant-induced congenital deformity. Zuchthygiene 2:145-152.

Shupe JL, Binns W, James LF, et al. (1967) Lupine, a cause of crooked calf disease. J Am Vet Med Assoc 151(2):198-203.

Shupe JL, Binns W, James LF, et al. (1968) A congenital deformity in calves induced by the maternal consumption of lupin. Aust J Agric Res 19(2):335-340.

LUPINUS SIMULANS Rose [Fabaceae]

Citations:

Ruiz MA, Sotelo A (2001) Chemical composition, nutritive value, and toxicology evaluation of Mexican wild lupins. J Agric Food Chem 49(11):5336-5339.

LUPINUS SPLENDENS Rose [Fabaceae]

Citations:

Ruiz MA, Sotelo A (2001) Chemical composition, nutritive value, and toxicology evaluation of Mexican wild lupins. J Agric Food Chem 49(11):5336-5339.

LUPINUS SULPHUREUS Douglas ex Hook. [Fabaceae]

Common Names:

sulfur lupine; yellow lupine

Citations:

Panter KE, Gardner DR, Gay CC, et al. (1997) Observations of Lupinus sulphureus-induced "crooked calf disease." J Range Manag 50:587-592.

Lupinus termis Forssk. = Lupinus albus L. var. albus

Lupinus varius L. = Lupinus angustifolius L.

lusmore –see– *Digitalis purpurea* L.

luya –see– *Zingiber officinale* Roscoe

luzerne –see– *Medicago sativa* L.

Lychnis githago (L.) Scop. = Agrostemma githago L.

LYCIUM BARBARUM L. [Solanaceae]

Synonyms:

Lycium halimifolium Mill.; ***Lycium vulgare*** Dunal

Common Names:

Bocksdorn; boxthorn; Chinese boxthorn; Christmas berry; Duke-of-Argyll's-tea-tree; marriage vine; matrimony vine; morali; Teufelszwirn

Citations:

Hansen AA (1927) Stock poisoning by plants in the nightshade family. J Am Vet Med Assoc 71:221-227.

Lycium halimifolium Mill. = Lycium barbarum L.

Lycium vulgare Dunal = Lycium barbarum L.

LYCOPERSICON ESCULENTUM Mill. [Solanaceae]

Common Names:

Libesafpel; love apple; Tomate; tomaten Düsche; tomates halentano; tomato

Citations:

Hjorth N, Roed-Petersen J (1976) Occupational protein contact dermatitis in food handlers. Contact Dermatitis 2(1):28-42.

Lain ES (1918) Dermatitis Lycopersicum esculentum (tomato plant). JAMA 71(14):1114-1117.

Sinha SM, Pasricha JS, Sharma RC, et al. (1977) Vegetables responsible for contact dermatitis of the hands. Arch Dermatol 113(6):776-779.

Note:

Tomato is named *Lycopersicon esculentum* Mill. var. *esculentum* in some publications.

LYCOPODIUM CLAVATUM L. [Lycopodiaceae]

Citations:

Salén EB (1951) Lycopodium allergy. Acta Allergol 4:308-319.

Lycopodium saururus Lam. = Huperzia saurus (Lam.) Trevis.

Lycopodium selago L. = Huperzia selago (L.) Bernh. ex Schrank & Mart.

Lycopodium serratum Thunb. = Huperzia serrata (Thunb.) Trevis.

LYGODESMIA TEXANA (Torr. & A. Gray) Greene ex Small [Asteraceae]

Common Names:

Texas skeleton plant

Citations:

Sperry OE, Turk RD, Hoffman GO, et al. (1955) Photosensitization of cattle in Texas. Texas Agric Exp Sta Bull #812:8 pp.

lyngwort –see– *Veratrum album* L.

LYTHRUM HYSSOPIFOLIA L. [Lythraceae]

Common Names:

lesser loosestrife

Citations:

Allen JG, Creeper JH, Forshaw D, et al. (2004) Plant-associated diseases, either new or new to the state, encountered over the last decade (1991-2001) in Western Australia. In: Acamovic T, Stewart CS, Pennycott TW (eds.) Poisonous plants and related toxins. CABI. Wallingford, UK. pp. 540-547.

Glastonbury JR, Walker RI, Links IJ, et al. (1991) Nephrosis in sheep associated with the ingestion of Lythrum hyssopifolia L., "lesser loosestrife." Proc Annu Conf Aust Soc Vet Path pp. 65-66.

lyukaslevelü orbáncfü –see– *Hypericum perforatum* L.

M

ma huang –see– *Ephedra sinica* Stapf

ma niang –see– *Archidendron jiringa* (Jack) I. C. Nielsen

ma-qian-zi –see– *Strychnos nux-vomica* L.

Mäasezwiebel –see– *Drimia maritima* (L.) Stearn

mabola persimmon –see– *Prunus myrtifolia* (L.) Urb.

maca colorada –see– *Andira inermis* (W. Wright) Kunth ex DC.

MACADAMIA INTEGRIFOLIA Maiden & Betche [Proteaceae]

Common Names:

Australian bush nut; bopple nut; macadamia nut; popple nut; Queensland nut

Citations:

Hansen SR, Buck WB, Meerdink G, et al. (2000) Weakness, tremors, and depression associated with macadamia nuts in dogs. Vet Hum Toxicol 42(1):18-21.

Hornfeldt CS, Borys DJ (1985) Review of veterinary cases received by the Hennepin Poison Center in 1984. Vet Hum Toxicol 27(6):525-528.

Lerch M, Egger C, Bircher AJ (2005) Allergic reactions to macadamia nut. Allergy 60(1):130-131.

McKenzie RA, Purvis-Smith GR, Allan SJ, et al. (2000) Macadamia nut poisoning of dogs. Aust Vet Pract 30:6-9.

macadamia nut –see– *Macadamia integrifolia* Maiden & Betche; *Macadamia tetraphylla* L. A. S. Johnson

MACADAMIA TETRAPHYLLA L. A. S. Johnson [Proteaceae]

Common Names:

Australian bush nut; bopple nut; macadamia nut; popple nut; Queensland nut

Citations:

Hansen SR, Buck WB, Meerdink G, et al. (2000) Weakness, tremors, and depression associated with macadamia nuts in dogs. Vet Hum Toxicol 42(1):18-21.

Lerch M, Egger C, Bircher AJ (2005) Allergic reactions to macadamia nut. Allergy 60(1):130-131.

macallo –see– *Andira inermis* (W. Wright) Kunth ex DC.

macasszar –see– *Diospyros macassar* A. Chev.

macaw tree –see– *Sesbania punicea* (Cav.) Benth.

macca –see– *Zea mays* L.

mace –see– *Myristica fragrans* Houtt.

MACHAERIUM SCLEROXYLON Tul. [Fabaceae]

Common Names:

caviuna vermelha; ironwood tree; pao ferro; pao ferro rosewood; Santos Palisander; Santos rosewood

Citations:

Condé-Salazar L, Garcia Diez A, Rafeensperger F, et al. (1980) Contact allergy to the Brazilian rosewood substitute Machaerium scleroxylon Tul. (pao ferro). Contact Dermatitis 6(13):246-250.

Estlander T, Jolanki R, Alanko K, et al. (2001) Occupational allergic contact dermatitis caused by wood dusts. Contact Dermatitis 44(4):213-217.

Hausen BM (1982) Häufigkeit und Bedeutung toxischer und allergischer Kontaktdermatitiden durch Machaerium scleroxylum Tul. (pao ferro), einem Ersatzholz für Palisander (Dalbergia nigra All.). Hautarzt 33(6):321-328.

Hausen BM, Simatupang MH, Kingreen JC (1972) Untersuchungen zur Überempfindlichkeit gegen Sucupira - und Palisanderholz. Derm Beruf Umwelt 20(1):1-7.

Hjorth N (1974) Occupational dermatitis from pao ferro (synonyms: Caviuna vermelha, Machaerium scleroxylon). Contact Dermatol Newsl 16(Aug):473, 521.

Holst R, Kirby J, Magnusson B (1976) Sensitization to tropical woods giving erythema multiforme-like eruptions. Contact Dermatitis 2(11):295-296.

Irvine C, Reynolds A, Finlay AY (1988) Erythema multiforme-like reaction to "rosewood." Contact Dermatitis 19(9):224-225.

Morgan JW, Orsler RJ, Wilkinson DS (1968) Dermatitis due to the wood dusts of Khaya anthotheca and Machaerium scleroxylon. Br J Ind Med 25(2):119-125.

Roed-Petersen J, Menné T, Nielsen KM, et al. (1987) Is it possible to work with pao ferro (Machaerium scleroxylon Tul.)? Arch Dermatol Res 279(Suppl):S108-S110.

Rojas-Hijazo B, Lezaun A, Hausen BM, et al. (2007) Airborne contact dermatitis in gaitas (flageolets) constructors after exposure to sawdust of caviuna. Contact Dermatitis 56(9):274-277.

Machandelbaum –see– *Juniperus communis* L.

machandeuse –see– *Metopium toxiferum* (L.) Krug & Urb.

machanoise –see– *Metopium toxiferum* (L.) Krug & Urb.

machelikoane –see– *Acalypha indica* L.

machineel –see– *Hippomane mancinella* L.

MACLURA POMIFERA (Raf.) C. K. Schneid. [Moraceae]

Synonyms:

Ioxylon pomiferum Raf.

Common Names:

bodark; bois d'arc; bowwood; hedge apple; horse apple; Osage apple; Osage orange

Citations:

Pullar EM (1939) Studies on five suspected poisonous plants. Aust Vet J 15:19-23.

Schur A (1932) Dermatitis venenata. Report of a case due to the Osage orange. Arch Dermatol 26:312-313.

maconha –see– *Cannabis sativa* L.

macrocarpa pine –see– *Cupressus macrocarpa* Hartw. ex Gordon

MACROCHLOA TENACISSIMA (Loefl. ex L.) Kunth [Poaceae]

Common Names:

esparto

Citations:

Zamarron C, Del Campo F, Paredos C (1992) Extrinsic allergic alveolitis to exposure to esparto dust. J Intern Med 232(2):177-179.

MACROTYLOMA AXILLARE (E. Mey.) Verdc. [Fabaceae]

Synonyms:

Dolichos axillaris E. Mey.

Citations:

Shenk JS (1976) The meadow vole as an experimental animal. Lab Anim Sci 26(4):664-669.

macrotys –see– *Actaea racemosa* L.

MACROZAMIA COMMUNIS L. A. S. Johnson [Zamiaceae]

Common Names:

burrawang; wild pineapple; zamia

Citations:

Healy PJ (1969) Studies on poisoning by Macrozamia communis. I. Biochemical disturbances in the liver. Biochem Pharmacol 18(1):85-92.

Healy PJ (1969) Studies on poisoning by Macrozamia communis. II. Effects of modifying agents upon changes induced in the liver by Macrozamia communis. Biochem Pharmacol 18(1):93-99.

MACROZAMIA DOUGLASII W. Hill ex F. M. Bailey [Zamiaceae]

Citations:

Hall WT (1954) Zamia staggers in cattle. Queensland Agric J 79:173-177.

Hall WT (1957) Toxicity of the leaves of Macrozamia spp. for cattle. Queensland J Agric Sci 14(2):41-52.

MACROZAMIA LUCIDA L. A. S. Johnson [Zamiaceae]

Common Names:

burrawang; wild pineapple; Zamia

Citations:

Hall WT, McGavin MD (1968) Clinical and neuropathological changes in cattle eating the leaves of Macrozamia lucida or Bowenia serrulata (Family Zamiaceae). Pathol Vet 5(1):26-34.

MACROZAMIA MIQUELII (F. Muell.) A. DC. [Zamiaceae]

Common Names:

wild pineapple; zamia; zamia palm

Citations:

Turner F (1893) The zamia palm (Macrozamia miquelii, F. v. M.) and its relation to the disease known as rickets in cattle. Agric Gaz New South Wales 4:158-161.

MACROZAMIA PAULI-GUILIELMI W. Hill & F. Muell. [Zamiaceae]

Common Names:

zamia

Citations:

Hall WT (1954) Zamia staggers in cattle. Queensland Agric J 79:173-177.

Hall WT (1957) Toxicity of the leaves of Macrozamia spp. for cattle. Queensland J Agric Sci 14(2):41-52.

MACROZAMIA RIEDLEI (Fisch. ex Gaudich.) C. A. Gardner [Zamiaceae]

Common Names:

cycad; zamia; zamia palm

Citations:

Gabbedy BJ, Meyer EP, Dickson J (1975) Zamia palm (Macrozamia riedlei) poisoning of sheep. Aust Vet J 51(6):303-305.

Gardiner MR (1970) Chronic ovine hepatosis following feeding of Macrozamia riedlei nuts. Aust J Agric Res 21(3):519-526.

MACROZAMIA SPIRALIS (Salisb.) Miq. [Zamiaceae]

Synonyms:

Zamia fraseri Van Houtte ex Regel

Common Names:

burrawang; burrawang palm; cycad; wild pineapple; zamia; zamia palm

Citations:

Edwards HH (1894) The disease known as "rickets" or "wobbles." J Agric West Aust 1(18):225-234.

Hall WT (1954) Zamia staggers in cattle. Queensland Agric J 79:173-177.

Hall WT (1957) Toxicity of the leaves of Macrozamia spp. for cattle. Queensland J Agric Sci 14(2):41-52.

Seddon HR, Belschner HG (1930) Poisoning of sheep by the seeds of burrawang (Macrozamia spiralis). Agric Gaz New South Wales 41:451-457.

mad apple –see– *Datura stramonium* L.

mad seeds –see– *Datura stramonium* L.

Madagascar periwinkle –see– *Catharanthus roseus* (L.) G. Don

madar –see– *Calotropis gigantea* (L.) W. T. Aiton; *Calotropis procera* (Aiton) W. T. Aiton

madársóska –see– *Oxalis acetosella* L.

madderwort –see– *Artemisia absinthium* L.

madecassol –see– *Centella asiatica* (L.) Urb.

Madeira ivy –see– *Hedera helix* L. subsp. canariensis (Willd.) Cout.

Madeira walnut –see– *Juglans regia* L.

Madeira winter cherry –see– *Solanum pseudocapsicum* L.

madera negra –see– *Gliricidia sepium* (Jacq.) Kunth ex Walp.

MADHUCA LONGIFOLIA (L.) J. F. Macbr. [Sapotaceae]

Common Names:

Edelteak; ilpa oil plant; mahua; Moah; Moah Holz

Citations:

Arseculeratne SN, Gunatilaka AA, Panabokke RG (1985) Studies on medicinal plants of Sri Lanka. Part 14. Toxicity of some traditional medicinal herbs. J Ethnopharmacol 13:323-335.

madre –see– *Gliricidia sepium* (Jacq.) Kunth ex Walp.

madre-de-cacao –see– *Gliricidia sepium* (Jacq.) Kunth ex Walp.

madriado –see– *Gliricidia sepium* (Jacq.) Kunth ex Walp.

madweed –see– *Scutellaria lateriflora* L.

maerman –see– *Drimia altissima* (L. f.) Ker Gawl.

MAESA LANCEOLATA Forssk. [Maesaceae]

Common Names:

ol odoa

Citations:

Mugera GM (1970) Toxic and medicinal plants of East Africa. II. Bull Epizootic Dis Afr 18(4):389-403.

Magenkraut –see– *Artemisia absinthium* L.

Magot –see– *Sapium biloculare* (S. Watson) Pax

maguey –see– *Agave americana* L.

Mahagoni –see– *Khaya senegalensis* (Desr.) A. Juss.

mahogany –see– *Swietenia macrophylla*; *Swietenia mahagoni* (L.) Jacq.; *Triplochiton scleroxylon* K. Schum.

mahonia –see– *Berberis aquifolium* Pursh

Mahonia aquifolium (Pursh) Nutt. = Berberis aquifolium Pursh

Mahonie –see– *Berberis aquifolium* Pursh

Mährrettig –see– *Armoracia rusticana* P. Gaertn. et al.

mahua –see– *Madhuca longifolia* (L.) J. F. Macbr.

mai-meu-dong-tang –see– *Ophiopogon japonicus* (Thunb.) Ker Gawl.

MAIANTHEMUM BIFOLIUM (L.) F. W. Schmidt [Ruscaceae]

Common Names:

Bittekonvall; Maiblom; May lily; Schattenblume

Citations:

Lamminpaa A, Kinos M (1996) Plant poisonings in children. Hum Exp Toxicol 15(3):245-249.

Maiblom –see– *Maianthemum bifolium* (L.) F. W. Schmidt

Maiblume –see– *Convallaria majalis* L.

maiden apple –see– *Momordica charantia* L.

maiden plum –see– *Comocladia dentata* Jacq.

maidenhair tree –see– *Ginkgo biloba* L.; *Moringa oleifera* Lam.

maiden's-honesty –see– *Clematis vitalba* L.

Maiglöckchen –see– *Convallaria majalis* L.

maigoya –see– *Plectranthus barbatus* Andrews

maikoa –see– *Brugmansia arborea* (L.) Lagerh.

Maischellchen –see– *Convallaria majalis* L.

maize –see– *Zea mays* L.

Majoran –see– *Origanum majorana* L.

májusi gyöngyvirág –see– *Convallaria majalis* L.

makaon –see– *Dichapetalum cymosum* (Hook.) Engl.

makoi –see– *Solanum nigrum* L.

makoré –see– *Tieghemella heckelii* Pierre ex A. Chev.

mala mujer –see– *Cnidoscolus urens* (L.) Arthur; *Solanum rostratum* Dunal; *Toxicodendron striatum* (Ruiz & Pav.) Kuntze

Malabar glory lily –see– *Gloriosa superba* L.

Malabar tree –see– *Euphorbia tirucalli* L.

malanga –see– *Colocasia esculenta* (L.) Schott; *Epipremnum pinnatum* (L.) Engl.; *Xanthosoma sagittifolium* (L.) Schott

malanga trepadora –see– *Epipremnum pinnatum* (L.) Engl.

malasampaga –see– *Wikstroemia ovata* C. A. Mey.

Malayan arrow poison –see– *Antiaris toxicaria* Lesch.

Malayan spurge –see– *Euphorbia antiquorum* L.

Malayan tea –see– *Cullen corylifolium* (L.) Medik.

male fern –see– *Dryopteris filix-mas* (L.) Schott

MALEPHORA SMITHII (L. Bolus) H. E. K. Hartmann [Aizoaceae]

Synonyms:

Hymenocyclus smithii L. Bolus

Common Names:

vygie

Citations:

Van der Walt SJ, Steyn DG (1939) Recent investigations into the toxicity of known and unknown poisonous plants in the Union of South Africa. IX. Onderstepoort J Vet Sci Anim Indus 12:335-366.

malkop ui –see– *Ornithogalum magnum* (Baker) J. C. Manning & Goldblatt

MALLOTUS PHILIPPENSIS (Lam.) Müll. Arg. [Euphorbiaceae]

Synonyms:

Rottlera tinctoria Roxb.

Common Names:

kamala; monkey face tree

Citations:

Kulakkattolickal AT (1987) Piscicidal plants of Nepal: Preliminary toxicity screening using grass carp (Ctenopharyngodon idella) fingerlings. J Ethnopharmacol 21(1):1-9.

mallow –see– *Malva parviflora* L.

malpitte –see– *Datura stramonium* L.

maluie –see– *Ornithogalum magnum* (Baker) J. C. Manning & Goldblatt

MALUS DOMESTICA Borkh. [Rosaceae]

Common Names:

Apfel; apple

Citations:

Duboucher C, Escamilla R, Rocchiccioli F, et al. (1986) Pulmonary lipogranulomatosis due to excessive consumption of apples. Chest 90(4):611-612.

Hansen AA (1928) Stock poisoning plants. Hydrocyanic poisoning. North Am Vet 9(5):24-27.

Kammerer M, Sachot E, Blanchot D (2001) Ethanol toxicosis from the ingestion of rotten apples by a dog. Vet Hum Toxicol 43(6):349-350.

Merrill SD (1952) Apple poisoning in dairy cows. Vet Med 47(Oct):405-406.

Meynadier J, Meynadier JM, Guilhou JJ (1982) L'urticaire de contact chez l'atopique. A propos de deux observations. Ann Dermatol Venereol 109(10):871-874.

Skamstrup Hansen K, Vestergaard H, Stahl Skov P, et al. (2001) Double-blind, placebo-controlled food challenge with apple. Allergy 56(2):109-117.

Teli AA, Chauhan HV, Gupta BS, et al. (1986) Acidosis in ewes caused by feeding of damaged apple (Malus sylvestris) diet. Indian Vet J 63(Jul):591-593.

MALUS SYLVESTRIS (L.) Mill. [Rosaceae]

Common Names:

crabapple

Citations:

Shaw JM (1986) Suspected cyanide poisoning in two goats caused by ingestion of crab apple leaves and fruits. Vet Rec 119(10):242-243.

malva –see– *Malva pusilla* Sm.

Malva borealis Wallman = Malva pusilla Sm.

MALVA NICAEENSIS All. [Malvaceae]

Common Names:

bull mallow

Citations:

Shlosberg A, Egyed MN, Efron Y (1975) Malva nicaeensis - A potentially poisonous plant for cattle. Refu Vet 32(4):155-156.

MALVA PARVIFLORA L. [Malvaceae]

Common Names:

alkali mallow; button weed; cheeseweed; golio; kissieblaaren; little mallow; mallow; marshmallow; small-flower mallow

Citations:

Allen JG, Creeper JH, Forshaw D, et al. (2004) Plant-associated diseases, either new or new to the state, encountered over the last decade (1991-2001) in Western Australia. In: Acamovic T, Stewart CS, Pennycott TW (eds.) Poisonous plants and related toxins. CABI. Wallingford, UK. pp. 540-547.

Anonymous (2005) Marshmallow toxicity in young sheep. South Australian state report. Animal Health Surveillance 10 (4):14.

Dodd S, Henry M (1921) Staggers or shivers in live stock. Agric Gaz New South Wales 32:327-329.

Lorenz FW (1939) Egg deterioration due to ingestion by hens of malvaceous materials. Poult Sci 18:295-300.

Main DC, Butler AR (2006) Probable Malva parviflora (small flowered mallow) intoxication in sheep in Western Australia. Aust Vet J 84(4):134-135.

MALVA PUSILLA Sm. [Malvaceae]

Synonyms:

Malva borealis Wallman

Common Names:

malva; round-leaf mallow; running mallow

Citations:

Hester JH (1906) The injurious effects of malva plant. Am Vet Rev 30:106-108.

mamona –see– *Euphorbia helioscopia* L.; *Ricinus communis* L.

man-in-the-ground –see– *Marah oregonus* (Torr. & S. Watson) J. T. Howell

man-better-man –see– *Achyranthes aspera* L.

man t'sao –see– *Illicium anisatum* L.

mancenillier –see– *Hippomane mancinella* L.; *Metopium toxiferum* (L.) Krug & Urb.

manchineal –see– *Hippomane mancinella* L.

mandarin –see– *Citrus reticulata* Blanco

Mandelbaum –see– *Prunus dulcis* (Mill.) D. A. Webb

mandioc –see– *Manihot esculenta* Crantz

mandioca –see– *Manihot esculenta* Crantz

mandragora –see– *Mandragora officinarum* L.

Mandragora autumnalis Bertol. = Mandragora officinarum L.

MANDRAGORA OFFICINARUM L. [Solanaceae]

Synonyms:

Mandragora autumnalis Bertol.

Common Names:

Alraunwurzel; bidh-el-ghoul; devil's-apple; devil's-herb; elephant's-ear; European mandrake; herbe-aux-magiciens; mandragora; Mandragore d'Europe; mandrake; medicinal mandrake; Satan's-apple

Citations:

De Salvo R, Sinardi AU, Santamaria LB, et al. (1980) Su un raro caso di intossicazione acuta da mandragora. Criteri diagnostici e terapeutici. Minerva Anestesiol 46(12):1265-1272.

Ghorbal H, Hamouda C, Bousnina M, et al. (2003) Use of plants to induce chemical submission in Tunisia. Vet Hum Toxicol 45(2):91-93.

Jiménez Mejías ME, Montaño Díaz M, López Pardo F, et al. (1990) Intoxicación atropinica por Mandragora autumnalis. Descripción de quince casos. Med Clin (Barc) 95(18):689-692.

Piccillo GA, Miele L, Mondati E, et al. (2006) Anticholinergic syndrome due to 'Devil's herb': When risks come from the ancient time. Int J Clin Pract 60(4):492-494.

Piccillo GA, Mondati EG, Moro PA (2002) Six clinical cases of Mandragora autumnalis poisoning: Diagnosis and treatment. Eur J Emerg Med 9(4):342-347.

Vlachos P, Poulos L (1982) A case of mandrake poisoning. J Toxicol Clin Toxicol 19(5):521-522.

Mandragore d'Amerique –see– *Podophyllum peltatum* L.

Mandragore d'Europe –see– *Mandragora officinarum* L.

mandrake –see– *Arum maculatum* L.; *Bryonia dioica* Jacq.; *Dioscorea communis* (L.) Caddick & Wilkin; *Mandragora officinarum* L.; *Pinellia ternata* (Thunb.) Makino; *Podophyllum peltatum* L.

manga –see– *Mangifera indica* L.

manganillo-de-cerro –see– *Toxicodendron striatum* (Ruiz & Pav.) Kuntze

mangel –see– *Beta vulgaris* L.

mangel wurzel –see– *Beta vulgaris* L.

mangga –see– *Mangifera indica* L.

manggaboom –see– *Mangifera indica* L.

manggom –see– *Mangifera indica* L.

MANGIFERA CAESIA Jack [Anacardiaceae]

Common Names:

binjai

Citations:

Fasal P (1945) Cutaneous diseases in the tropics. A clinical study based on observations in Malaya. Arch Derm Syphilol 51(3):163-171.

Landor JV (1943) Dermatitis venenata caused by smoke. Br J Dermatol 55:17-19.

MANGIFERA FOETIDA Lour. [Anacardiaceae]

Common Names:

horse mango

Citations:

Fasal P (1945) Cutaneous diseases in the tropics. A clinical study based on observations in Malaya. Arch Derm Syphilol 51(3):163-171.

MANGIFERA INDICA L. [Anacardiaceae]

Common Names:

aam; apple-of-the-tropics; fruit-of-heaven; king-of-fruits; manga; mangga; manggaboom; manggom; mango; mangot; mangotine; mangue; manguier; pied mango; skin mango

Citations:

Brown A, Brown FR (1940) Mango dermatitis. J Allergy 12:310-311.

Dang RW, Bell DB 2nd (1967) Anaphylactic reaction to the ingestion of mango. Case report. Hawaii Med J 27(2):149-150.

Fasal P (1945) Cutaneous diseases in the tropics. A clinical study based on observations in Malaya. Arch Derm Syphilol 51(3):163-171.

Geller M (1989) Poison ivy, mangoes, cashews, and dermatitis. Ann Intern Med 110(12):1036-1037.

Geroso AM, Elpern DJ (1992) Some observations on mango and mokihana dermatitis from Hawaii. Contact Dermatitis 26(5):346-347.

Goldberg LC (1954) Mango dermatitis. JAMA 156(Nov 6):954.

Goldstein N (1968) The ubiquitous urushiols. Contact dermatitis from mango, poison ivy, and other "poison" plants. Cutis 4(6):679-685.

Hammerstein CP (1959) Mango dermatitis experiences. Proc Florida Mango Forum. pp. 14-16.

Hershko K, Weinberg I, Ingber A (2005) Exploring the mango-poison ivy connection: The riddle of discriminative plant dermatitis. Contact Dermatitis 52(1):3-5.

Keil H, Wasserman D, Dawson CR (1946) Mango dermatitis and its relationship to poison ivy hypersensitivity. Ann Allergy 4(Jul-Aug):268-281.

Kirby-Smith JL (1938) Mango dermatitis. Am J Trop Med 18:373-384.

Lindenbaum S (1962) [Allergic reactions to Mangifera indica (mango).] Harefuah 62:422-423.

Paschke A, Kinder H, Zunker K, et al. (2001) Characterization of cross-reacting allergens in mango fruit. Allergy 56(3):237-242.

Paulson GA (1941-1943) Mango and hualtaco dermatitis. Med Bull Standard Oil Co 5:197-200.

Rubin JM, Shapiro J, Muehlbauer P, et al. (1965) Shock reaction following ingestion of mango. JAMA 193(5):397-398.

Simmons JS, Bolin ZE (1921) Dermatitis venenata produced by an irritant present in the stem sap of the mango (Mangifera indica L.). Am J Trop Med 1:351-374.

Singh R, Siddiqui MA, Baruah MC (1978) Plant dermatitis in Delhi. Indian J Med Res 68(Oct):650-655.

Tucker MO, Swan CR (1998) Images in clinical medicine: The mango-poison ivy connection. N Engl J Med 339(4):235.
Weinstein S, Bassiri-Tehrani S, Cohen DE (2004) Allergic contact dermatitis to mango flesh. Int J Dermatol 43(3):195-196.
Zakon SJ (1939) Contact dermatitis due to mango. JAMA 113(Nov 11):1808.

mango –see– *Mangifera indica* L.

mangold –see– *Beta vulgaris* L.

Mangold Rübe –see– *Beta vulgaris* L.

mangot –see– *Mangifera indica* L.

mangotine –see– *Mangifera indica* L.

mangue –see– *Mangifera indica* L.

manguier –see– *Mangifera indica* L.

maniçoba –see– *Manihot carthagenensis* (Jacq.) Müll. Arg. subsp. glaziovii (Müll. Arg.) Allem

maniçoba-do-ceará –see– *Manihot carthagenensis* (Jacq.) Müll. Arg. subsp. glaziovii (Müll. Arg.) Allem

manihot –see– *Manihot esculenta* Crantz

Manihot aipi Pohl = Manihot esculenta Crantz

MANIHOT CARTHAGENENSIS (Jacq.) Müll. Arg. subsp. glaziovii (Müll. Arg.) Allem [Euphorbiaceae]

Synonyms:

Manihot glaziovii Müll. Arg.

Common Names:

maniçoba; maniçoba-do-ceará

Citations:

Amorim SL, Medeiros RM, Riet-Correa F (2005) Intoxicação experimental por Manihot glaziovii (Euphorbiaceae) em caprinos. Pesq Vet Bras 25(3):179-187.
Canella CF, Döbereiner J, Tokarnia CH (1968) Intoxicação experimental pela "maniçoba" (Manihot glaziovii Muell. Arg.) em bovinos. Pesq Agric Bras 3:347-350.
Tokarnia CH, Peixoto PV, Brito MF, et al. (1999) Estudos experimentais com plantas cianogênicas em bovinos. Pesq Vet Bras 19(2):84-90.

MANIHOT ESCULENTA Crantz [Euphorbiaceae]

Synonyms:

Jatropha manihot L.; ***Manihot aipi*** Pohl; ***Manihot utilissima*** Pohl

Common Names:

akpu; bitter cassava; Brazilian arrowroot; cassada; cassava; gari; mandioc; mandioca; manihot; manioc; Maniok; simaltarul; sweet cassava; tapioca; ubi kayu; yuca

Citations:

Abuye C, Kelbessa U, Wolde Gebriel S (1998) Health effects of cassava consumption in south Ethiopia. East Afr Med J 75(3):166-170.
Adewusi SR, Akindahunsi AA (1994) Cassava processing, consumption, and cyanide toxicity. J Toxicol Environ Health 43(1):13-23.
Akanji AO, Famuyiwa OO (1993) The effects of chronic cassava consumption, cyanide intoxication and protein malnutrition on glucose tolerance in growing rats. Br J Nutr 69(1):269-276.
Akintonwa A (1996) Acute cyanide poisoning attributed to Manihot esculenta. J Toxicol Clin Toxicol 34:599.
Akintonwa A, Tunwashe OL (1992) Fatal cyanide poisoning from cassava-based meal. Hum Exp Toxicol 11(1):47-49.
Ariffin WA, Choo KE, Karnaneedi S (1992) Cassava (ubi kayu) poisoning in children. Med J Malaysia 47(3):231-234.
Assan R, Boukersi H, Clauser E (1984) Cassava pancreatitis in western Europe. Lancet 2(8414):1278.
Ballinas Díaz J, Larios Saldaña A, Cruz Mondragón C, et al. (1997) Elaboración de una harina integral de yuca (M. esculenta Crantz) para alimentación de pollitos de engorde. 2. Evaluación de una harina integral de yuca en pollitos de engorde. Arch Latinoam Nutr 47(4):387-390.
Banea M, Tylleskär T, Rosling H (1997) Konzo and ebola in Bandundu region of Zaire. Lancet 349(9052):621.
Banea-Mayambu JP, Tylleskär T, Gitebo N, et al. (1997) Geographical and seasonal association between linamarin and cyanide exposure from cassava and the upper motor neurone disease konzo in former Zaire. Trop Med Int Health 2(12):1143-1151.
Bourdoux P, Delange F, Gerard M, et al. (1978) Evidence that cassava ingestion increases thiocyanate formation: A possible etiologic factor in endemic goiter. J Clin Endocrinol Metab 46(4):613-621.
Cheok SS (1978) Acute cassava poisoning in children in Sarawak. Trop Doct 8(Jul):99-101.
Clark A (1936) Report on the effects of certain poisons contained in food-plants of West Africa upon the health of the native races. J Trop Med Hyg 39(23-24):269-295.
Cliff J, Coutinho J (1995) Acute intoxication from newly-introduced cassava during drought in Mozambique. Trop Doct 25(4):193.
Cliff J, Nicala D (1997) Long-term follow-up of konzo patients. Trans R Soc Trop Med Hyg 91(4):447-449.
Cliff J, Nicala D, Saute F, et al. (1997) Konzo associated with war in Mozambique. Trop Med Int Health 2(11):1068-1074.
Dawood MY (1969) Acute tapioca poisoning in a child. J Singapore Paediat Soc 11(2):154-158.
Ekpechi OL (1967) Pathogenesis of endemic goitre in eastern Nigeria. Br J Nutr 21(3):537-545.
Ekpechi OL, Dimitriadou A, Fraser R (1966) Goitrogenic activity of cassava (a staple Nigerian food). Nature 210(5041):1137-1138.
Espinoza OB, Perez M, Ramirez MS (1992) Bitter cassava poisoning in eight children: A case report. Vet Hum Toxicol 34(1):65.
Gaitan E, Cooksey RC, Legan J, et al. (1994) Antithyroid effects in vivo and in vitro of babassu and mandioca: A staple food in goiter areas of Brazil. Eur J Endocrinol 131(21):138-144.
Geldof AA, Becking JL, de Vries CD, et al. (1992) Histopathological changes in rat pancreas after fasting and cassava feeding. In Vivo 6(5):545-551.

Jackson LC (1988) Possible adaptation to serum thiocyanate overload associated with chronic sublethal dietary cyanide ingestion. Hum Biol 60(4):615-622.
Kamalu BP (1993) Pathological changes in growing dogs fed on a balanced cassava (Manihot esculenta Crantz) diet. Br J Nutr 69(3):921-934.
Mathangi DC, Mohan V, Namasivayam A (1999) Effect of cassava on motor co-ordination and neurotransmitter level in the albino rat. Food Chem Toxicol 37(1):57-60.
Mozambique Ministry of Health (1984) Mantakassa: An epidemic of spastic paraparesis associated with chronic cyanide intoxication in a cassava staple area of Mozambique. 2. Nutritional factors and hydrocyanic acid content of cassava products. Bull World Health Organ 62(3):485-492.
Narendranathan M, Sharma KN, Sosamma PI (1989) Serum rhodanese in goitre and calcific pancreatitis of tropics. J Assoc Physicians India 37(10):648-649.
Ngudi DD, Kuo YH, Lambein F (2003) Cassava cyanogens and free amino acids in raw and cooked leaves. Food Chem Toxicol 41(8):1193-1197.
Njoh J (1990) Tropical ataxic neuropathy in Liberians. Trop Geogr Med 42(1):92-94.
Okafor PN, Okorowkwo CO, Maduagwu EN (2002) Occupational and dietary exposures of humans to cyanide poisoning from large-scale cassava processing and ingestion of cassava foods. Food Chem Toxicol 40(7):1001-1005.
Oluwole OS, Onabolu AO, Cotgreave IA, et al. (2002) Low prevalence of ataxic polyneuropathy in a community with high exposure to cyanide from cassava foods. J Neurol 249(8):1034-1040.
Onabolu A, Bokanga M, Tylleskär T, et al. (2001) High cassava production and low dietary cyanide exposure in mid-west Nigeria. Public Health Nutr 4(1):3-9.
Osman BA, Ng ML, Bakar AA, et al. (1993) The effect of cassava leave intake on thyroid hormone and urinary iodine. East Afr Med J 70(5):314-315.
Osuntokun BO (1968) An ataxic neuropathy in Nigeria. Brain 91(2):215-248.
Osuntokun BO (1972) Chronic cyanide neurotoxicity and neuropathy in Nigerians. Plant Foods Hum Nutr 2(3-4):215-266.
Panigrahi S, Rickard J, O'Brien GM, et al. (1992) Effects of different rates of drying cassava root on its toxicity to broiler chicks. Br Poult Sci 33(5):1025-1041.
Sandhyamani S (1991) Cardiovascular effects in bonnet monkeys (Macca radiata) of a cassava-based protein-deficient diet. Vet Hum Toxicol 33(5):429-430.
Sandhyamani S (1992) Vasculopathic and cardiomyopathic changes induced by low-protein high-carbohydrate tapioca based diet in bonnet monkey. Vasculopathic and cardiomyopathic changes in induced malnutrition. Am J Cardiovas Pathol 4(1):41-50.
Sezi CL (1996) Effects of cassava diet on Cercopithecus aethiops livers: A case for cassava as the cause of both tropical splenomegaly syndrome (TSS) and endomyocardial fibrosis (EMF). East Afr Med J 73(5 Suppl):S24-S28.
Singh JD (1981) The teratogenic effects of dietary cassava on the pregnant albino rat: A preliminary report. Teratology 24:289-291.
Sreeja VG, Leelamma S (1996) Effect of protein supplemented cassava diet in rats. Indian J Biochem Biophys 33(2):149-151.
Sreeja VG, Leelamma S (1998) Hyperglycemic effect of low protein cassava diet. Indian J Exp Biol 36(3):308-310.
Tewe OO, Maner JH (1981) Performance and pathophysiological changes in pregnant pigs fed cassava diets containing different levels of cyanide. Res Vet Sci 30(2):147-151.
Tewe OO, Maner JH, Gomez G (1977) Influence of cassava diets on placental thiocyanate transfer, tissue rhodanese activity and performance of rats during gestation. J Sci Food Agric 28(8):750-756.
Tylleskär T, Banea M, Bikangi N, et al. (1991) Epidemiological evidence from Zaire for a dietary etiology of konzo, an upper motor neuron disease. Bull World Health Organ 69(5):581-589.
Tylleskär T, Banea M, Bikangi N, et al. (1992) Cassava cyanogens and konzo, an upper motorneuron disease found in Africa. Lancet 339(8787):208-211.
Tylleskär T, Banea M, Bikangi N, et al. (1995) Dietary determinants of a non-progressive spastic paraparesis (Konzo): A case-referent study in a high incidence area of Zaire. Int J Epidemiol 24(5):949-956.
Tylleskär T, Howlett WP, Rwiza HT, et al. (1993) Konzo: A distinct disease entity with selective upper motor neuron damage. J Neurol Neurosurg Psychiatry 56(6):638-643.
Tylleskär T, Legue FD, Peterson S, et al. (1994) Konzo in the Central African Republic. Neurology 44(5):959-961.
Van Heijst AN, Maes RA, Mtanda AT, et al. (1994) Chronic cyanide poisoning in relation to blindness and tropical neuropathy. J Toxicol Clin Toxicol 32(5):549-556.
Vannasaeng S, Vichayanrat A, Nitiyanant W, et al. (1982) Diabetes mellitus in the tropics: A case with pancreatic calcification and chronic cassava toxicity. J Med Assoc Thai 65(6):330-332.

Manihot glaziovii Müll. Arg. = Manihot carthagenensis (Jacq.) Müll. Arg. subsp. glaziovii (Müll. Arg.) Allem

MANIHOT TRIPARTITA (Spreng.) Müll. Arg. [Euphorbiaceae]

Citations:

Fernandes NS, Nazário W, Camargo WV (1972) Manihot tripartita, nova espécie de mandioca tóxica para bovinos. Ocorrência clínica no município de Luciara (MT). Biologico 38(6):163-167.

Manihot utilissima Pohl = Manihot esculenta Crantz

manioc –see– *Manihot esculenta* Crantz

Maniok –see– *Manihot esculenta* Crantz

mansonia –see– *Mansonia altissima* (A. Chev.) A. Chev.

MANSONIA ALTISSIMA (A. Chev.) A. Chev. [Malvaceae]

Synonyms:

Sterculiacea altissima

Common Names:

Afrikanische Schwarznuß; bété; mansonia

Citations:

Bertocchi D, Dovadola E (1967) Una forma di difterite dei suinetti provocata da truciolo di «Mansonia altissima». Nuova Vet 43:459-462.

Bertschinger HU, Lott-Stolz G (1970) Erkrangungen bei Schweinen durch Holzabfälle von Mansonia altissima. Schweiz Arch Tierheilkd 112(12):641-651.

Bourne LB (1956) Dermatitis from mansonia wood. Br J Ind Med 13(1):55-58.

Horner S, Wigley JE (1936) A case of dermatitis venenata due to mansonia wood (Sterculiacea altissima). Br J Dermatol 48:26-28.

Oleffe JA, Sporcq J, Hublet P (1975) Epidemiological study of the wood industry in Belgium. Contact Dermatitis 1(5):315-316.

Salamone L, di Blasi S, Coniglio L (1969) Rilievi sulla patologia da legno di mansonia. Folia Medica 52:427-449.

Vacirca G, Agosti M, Crotti S (1970) Osservazioni clinicosperimentali sull'intossicazione da Mansonia altissima nel vitello. Atti Soc Ital Sci Vet 24:461-463.

Ventruoli M, Quadri E (1969) Intossicazione da «Mansonia» nel cane. Contributo clinico e spermitale. Nuova Vet 45:71-81.

Note:

The synonym *Sterculiacea altissima* could not be found in the databases searched.

Mantegazzis Bärenklau –see– *Heracleum mantegazzianum* Sommier & Levier

many-ray Baileya –see– *Baileya multiradiata* Harv. & A. Gray ex Torr.

manzanilla –see– *Hippomane mancinella* L.

Manzanillabaum –see– *Hippomane mancinella* L.

manzanillo –see– *Hippomane mancinella* L.

manzanillo-de-la-costa –see– *Hippomane mancinella* L.

manzanillo-de-playa –see– *Hippomane mancinella* L.

maracá-de-cobra –see– *Crotalaria juncea* L.; *Crotalaria pallida* Aiton; *Crotalaria retusa* L.; *Crotalaria spectabilis* Roth

MARAH OREGONUS (Torr. & S. Watson) J. T. Howell [Cucurbitaceae]

Synonyms:

Micrampelis oregona (Torr. & A. Gray) Greene

Common Names:

coast manroot; man-in-the-ground; old man; wild cucumber

Citations:

Hugelmeyer CD, Putnam TS, Burton BT, et al. (1987) Fatal poisoning due to ingestion of wild cucumber seed tea. Vet Hum Toxicol 29(6):462-463.

Litovitz TL, Martin TG, Schmitz B (1987) 1986 Annual report of the American Association of Poison Control Centers National Data Collection System. Am J Emerg Med 5(6):405-445.

maranga calalu –see– *Moringa oleifera* Lam.

maranguey –see– *Zamia integrifolia* L. f.; *Zamia portoricensis* Urb.

marañón –see– *Anacardium occidentale* L.

Maranta galanga L. = Alpinia galanga (L.) Sw.

marble queen –see– *Epipremnum pinnatum* (L.) Engl.

marcona bark –see– *Erythrophleum suaveolens* (Guill. & Perr.) Brenan

marejea –see– *Crotalaria ochroleuca* G. Don

Märek –see– *Armoracia rusticana* P. Gaertn. et al.

mare's-tail –see– *Equisetum arvense* L.; *Equisetum ramosissimum* Desf.

margarita –see– *Karwinskia humboldtiana* (Schult.) Zucc.

Margarite –see– *Leucanthemum vulgare* Lam.

margosa tree –see– *Azadirachta indica* A. Juss.

marguerite daisy –see– *Argyranthemum frutescens* (L.) Sch. Bip.; *Leucanthemum vulgare* Lam.

MARGYRICARPUS PINNATUS (Lam.) Kuntze [Rosaceae]

Common Names:

yerba-de-la-perdiz

Citations:

Ciganda C, Laborde A (2001) Herbal infusions used for induced abortion. J Toxicol Clin Toxicol 39(3):318-319.

Laborde A, Ciganda C (1998) Poisoning by herbal infusions ingested as abortifacient agents. J Toxicol Clin Toxicol 36(5):454-455.

marian thistle –see– *Silybum marianum* (L.) Gaertn.

marigold –see– *Calendula officinalis* L.; *Tagetes erecta* L.

marigold paper flower –see– *Psilostrophe tagetina* (Nutt.) Greene

marihuana –see– *Cannabis sativa* L.

marijuana –see– *Cannabis sativa* L.

Maris Kestrel kale –see– *Brassica oleracea* L. var. medullosa Thell.

marjoram –see– *Origanum majorana* L.

markery –see– *Toxicodendron radicans* (L.) Kuntze

marking nut –see– *Semecarpus anacardium* L. f.; *Semecarpus australiensis* Engl.

Markstammkohl –see– *Brassica oleracea* L. var. medullosa Thell.

markweed –see– *Toxicodendron radicans* (L.) Kuntze

marmeleiro bravo –see– *Prunus myrtifolia* (L.) Urb.

marriage vine –see– *Lycium barbarum* L.; *Solanum dulcamara* L.

marronier d'Inde –see– *Aesculus hippocastanum* L.

marrow –see– *Cucurbita pepo* L.

marrow stem kale –see– *Brassica oleracea* L. var. medullosa Thell.

marrube blanc –see– *Marrubium vulgare* L.

MARRUBIUM VULGARE L. [Lamiaceae]

Common Names:

horehound; marrube blanc; white horehound

Citations:

Kchouk M, Chadli A (1963) Sur les propriétés abortives du marrube blanc (Marrubium vulgare L.). Arch Inst Pasteur Tunis 40(Jun):129-132.

MARSDENIA ROSTRATA R. Br. [Asclepiadaceae]

Common Names:

milk vine; poison vine

Citations:

Seddon HR, Carne HR (1925) A vine poisonous to stock (Marsdenia rostrata). Agric Gaz New South Wales 36:99-103.

Seddon HR, Carne HR (1925) Marsdenia rostrata: A vine poisonous to stock. New South Wales Dep Agric Sci Bull #24:34-43.

marsh arrowgrass –see– *Triglochin maritima* L.

marsh buttercup –see– *Ranunculus cymbalaria* Pursh

marsh crowfoot –see– *Ranunculus sceleratus* L.

marsh elder –see– *Cyclachaena xanthiifolia* (Nutt.) Fresen.; *Iva angustifolia* Nutt. ex DC.; *Iva annua* L.; *Iva microcephala* Nutt.

marsh horsetail –see– *Equisetum palustre* L.

marsh pepper –see– *Persicaria hydropiper* (L.) Spach

marsh ragwort –see– *Jacobaea aquatica* (Hill) P. Gaertn. et al. var. aquatica

marsh trefoil –see– *Menyanthes trifoliata* L.

marsh turnip –see– *Arisaema triphyllum* (L.) Schott

marsh weed –see– *Equisetum palustre* L.

marshmallow –see– *Malva parviflora* L.

MARSILEA DRUMMONDII A. Braun [Marsileaceae]

Common Names:

nardoo fern

Citations:

Anonymous (1976) Nardoo poisoning. Agric Gaz New South Wales 87:44.

Pritchard D, Eggleston GW, Macadam JF (1978) Nardoo fern and polioencephalomalacia. Aust Vet J 54(4):204-205.

marton d'Inde –see– *Aesculus hippocastanum* L.

marupá –see– *Simarouba amara* Aubl.

Maruta cotula DC. = Anthemis cotula L.

Mary Jane –see– *Cannabis sativa* L.

MASCAGNIA PUBIFLORA (A. Juss.) Griseb. [Malpighiaceae]

Common Names:

corona; timbó

Citations:

Döbereiner J, Gava A, Consorte LB, et al. (1986) Intoxicação experimental por Mascagnia pubiflora [Malpighiaceae] em coelhos. Pesq Vet Bras 6(2):51-57.

Fernandes NS, Macruz R (1964) Toxicidade de "corona" - Mascagnia pubiflora (Juss) Griseb. [Malpighiaceae]. Arq Inst Biol (Sao Paulo) 31(1):1-4.

Santos FC, Fischer P, Jardim EC (1976) Intoxicação experimental em bovinos por "Timbó" Mascagnia pubiflora. An Esc Agron Vet 6(1):97-103.

Tokarnia CH, Döbereiner J (1973) Intoxicação por Mascagnia pubiflora em bovinos no Estado de Mato Grosso. Pesq Agric Bras Vet 8(6):61-68.

MASCAGNIA RIGIDA (A. Juss.) Griseb. [Malpighiaceae]

Common Names:

tingui

Citations:

Batatinha MJ, Spinosa HS, Bernardi MM (1988) Toxic effects of Mascagnia rigida in laboratory animals. Vet Hum Toxicol 30(3):259.

Medeiros RM, Neto SA, Barbosa RC, et al. (2002) Sudden bovine death from Mascagnia rigida in northeastern Brazil. Vet Hum Toxicol 44(5):286-288.

Santos HL (1975) Aspectos clínicos, laboratoriais e anátomo-histopatológicos, na intoxicação experimental de bovinos pela Mascagnia rigida (Juss) Gr. Arq Esc Vet Univ Fed Minas Gerais 27:398-399.

Tokarnia CH, Canella CF, Döbereiner J (1961) Intoxicação por um "tingui" (Mascagnia rigida Griseb.) em bovinos no nordeaste do Brasil. Arq Inst Biol Anim (Rio Janeiro) 4:203-215.

Tokarnia CH, Döbereiner J, Canella CF (1987) Intoxicação experimental por Mascagnia rigida (Malpighiaceae) em coelhos. Pesq Vet Bras 7(1):11-16.

Tokarnia CH, Döbereiner J, Peixoto PV (1985) Intoxicação por Mascagnia aff. rigida (Malpighiaceae) em bovinos no norte do Espírito Santo. Pesq Vet Bras 5(3):77-91.

Tokarnia CH, Peixoto PV, Döbereiner J (1985) Intoxicação experimental por Mascagnia aff. rigida (Malpighiaceae) em coelhos. Pesq Vet Bras 5(4):121-128.

mascarida –see– *Chamaeleon gummifera* (L.) Cass.

mashasha –see– *Dalechampia scandens* L.

masooriberry –see– *Coriaria myrtifolia* L.

mastic –see– *Pistacia lentiscus* L.

mastic gum –see– *Pistacia lentiscus* L.

mastich –see– *Pistacia lentiscus* L.

mastiche –see– *Pistacia lentiscus* L.

masticogna –see– *Chamaeleon gummifera* (L.) Cass.

mastis –see– *Pistacia lentiscus* L.

mastix –see– *Pistacia lentiscus* L.

mat balsam –see– *Euphorbia drummondii* Boiss.

mat spurge –see– *Euphorbia drummondii* Boiss.

mata bode –see– *Ipomoea carnea* Jacq. subsp. fistulosa Mart. ex Choisy) D. F. Austin mata pasto –see– *Senna occidentalis* (L.) Link

mata ratón –see– *Gliricidia sepium* (Jacq.) Kunth ex Walp.

matalahuga aniseed –see– *Pimpinella anisum* L.

matapolilla –see– *Elaeagnus angustifolia* L.

matar –see– *Pisum sativum* L.

matchbox bean –see– *Entada phaseoloides* (L.) Merr.

matchbrush –see– *Gutierrezia sarothrae* (Pursh) Britton & Rusby

matchweed –see– *Gutierrezia sarothrae* (Pursh) Britton & Rusby

maté –see– *Ilex paraguariensis* A. St.-Hil.

matra –see– *Lathyrus sativus* L.

Matricaria chamomilla auct. = Matricaria recutita L.

MATRICARIA NIGELLIFOLIA DC. [Asteraceae]

Common Names:

rivierals; staggers weed; stootsiektebossie; water kerwal

Citations:

Newsholme SJ, Kellerman TS, Welman WG (1984) Pathology of a nervous disorder (pushing disease or "stootsiekte") in cattle caused by the plant Matricaria nigellifolia DC. (Asteraceae). Onderstepoort J Vet Res 51(2):119-127.

MATRICARIA RECUTITA L. [Asteraceae]

Synonyms:

Matricaria chamomilla auct.

Common Names:

camomile; chamomile; Deutsche Kamille; Echte Kamille; false chamomile; German camomille; German chamomile; Kamille

Citations:

Casterline CL, Bonita L (1980) Allergy to chamomile tea. JAMA 244(4):330-331.

Hausen BM (1979) The sensitizing capacity of compositae plants. III. Test results and cross-reactions in compositae-sensitive patients. Dermatologica 159(1):1-11.

Hausen BM, Schulz KH (1978) Polyvalente Kontaktallergie bei einer Floristin. Derm Beruf Umwelt 26(5):175-176.

Lipsitz DJ (1984) Herbal teas and water intoxication in a young child. J Fam Pract 18(6):933-937.

Lundh K, Hindsén M, Gruvberger B, et al. (2006) Contact allergy to herbal teas derived from Asteraceae plants. Contact Dermatitis 54(4):196-201.

Rodríguez-Serna M, Sánchez-Motilla JM, Ramón R, et al. (1998) Allergic and systemic contact dermatitis from Matricaria chamomilla tea. Contact Dermatitis 39(4):192-193.

Rudzki E, Rapiejko P, Rebandel P (2003) Occupational contact dermatitis, with asthma and rhinitis, from camomile in a cosmetician also with contact urticaria from both camomile and lime flowers. Contact Dermatitis 49(3):162.

Rycroft RJ (2003) Recurrent facial dermatitis from chamomile tea. Contact Dermatitis 48(4):229.

van Ketel WG (1982) Allergy to Matricaria chamomilla. Contact Dermatitis 8(2):143.

van Ketel WG (1987) Allergy to Matricaria chamomilla. Contact Dermatitis 16(1):50-51.

matrimony vine –see– *Lycium barbarum* L.

MATTEUCCIA STRUTHIOPTERIS (L.) Tod. [Dryopteridaceae]

Common Names:

ostrich fern

Citations:

Bills D, Arias L, Constantine P, et al. (1994) Ostrich fern poisoning - New York and Western Canada, 1994. JAMA 273(12):912-913.

Bills D, Arias L, Constantine P, et al. (1994) Ostrich fern poisoning - New York and Western Canada, 1994. MMWR Morb Mortal Wkly Rep 43(37):677, 683, 684.

Bruneau A, Lummis W, Ramsay D (2000) Food poisoning associated with the ingestion of fiddleheads - Quebec 1999. Can Commun Dis Rep 26(20):165-170.

Morgan P, Morton T, Iverson F, et al. (1994) Ostrich fern poisoning - Western Canada and New York, 1994. Can Commun Dis Rep 20(18):160-163.

Mauerschierling –see– *Conium maculatum* L.

Maulwurfskraut –see– *Euphorbia lathyris* L.

MAURIA HETEROPHYLLA Kunth [Anacardiaceae]

Synonyms:

Mauria puberula Tul.

Common Names:

cirri amarillo; koro; pepeo tree

Citations:

Hurtado I, Medina JD, Dao L, et al. (1982) Studies on the skin-sensitizing properties of the "pepeo" tree, Mauria puberula (Anacardiaceae). J Am Acad Dermatol 7(3):341-345.

Mauria puberula Tul. = Mauria heterophylla Kunth

Mäuseschierling –see– *Conium maculatum* L.

May apple –see– *Podophyllum peltatum* L.

May blossom –see– *Convallaria majalis* L.

May lily –see– *Convallaria majalis* L.; *Maianthemum bifolium* (L.) F. W. Schmidt

May tree –see– *Crataegus laevigata* (Poir) DC.

Mayflower –see– *Convallaria majalis* L.

Maygrass –see– *Phalaris caroliniana* Walter

Maypop –see– *Passiflora incarnata* L.

mays –see– *Zea mays* L.

Mayweed –see– *Anthemis cotula* L.

meadow buttercup –see– *Ranunculus acris* L.

meadow cone –see– *Rudbeckia laciniata* L.

meadow crocus –see– *Colchicum autumnale* L.

meadow death camas –see– *Anticlea elegans* (Pursh) Rydb.; *Toxicoscordion venenosum* (S. Watson) Rydb.

meadow fern –see– *Onoclea sensibilis* L.

meadow garlic –see– *Allium canadense* L.

meadow holly –see– *Ilex decidua* Walter

meadow larkspur –see– *Delphinium nuttallianum* Pritz.

meadow leek –see– *Allium canadense* L.

meadow maple –see– *Acer rubrum* L.

meadow pine –see– *Equisetum arvense* L.

meadow poke –see– *Veratrum viride* Aiton

meadow ranunculus –see– *Ranunculus acris* L.

meadow saffron –see– *Colchicum autumnale* L.

meadow sorrel –see– *Rumex acetosa* L.; *Rumex acetosella* L.

meadow sweet –see– *Ammi majus* L.

meadow turnip –see– *Arisaema triphyllum* (L.) Schott

mealie –see– *Zea mays* L.

mealy rosettes –see– *Psathyrotes annua* (Nutt.) A. Gray

medic –see– *Medicago polymorpha* L.

Medicago denticulata Willd. = Medicago polymorpha L.

Medicago hispida Gaertn. = Medicago polymorpha L.

MEDICAGO POLYMORPHA L. [Fabaceae]

Synonyms:

Medicago denticulata Willd.; ***Medicago hispida*** Gaertn.

Common Names:

bur clover; bur medic; bur trefoil; hairy medic; medic; trefoil

Citations:

Anderson CR (1944) Contact dermatitis from alfalfa and bur clover. Arch Derm Syphilol 50:201.

Bull LB, MacIndoe RH (1926) Photosensitization in sheep: Trefoil dermatitis. Aust Vet J 2(Sep):85-91.

Byrne KV (1937) Dermatitis in white pigs due to photosensitization. Aust Vet J 13(Apr):74-75.

Byrne KV (1937) Dermatitis of white pigs associated with lucerne and trefoil grazing. Agric Gaz New South Wales 48(Apr 1):214.

Dodd S (1916) Trefoil dermatitis or the sensitization of unpigmented skin to the sun's rays by the ingestion of trefoil. J Comp Pathol Ther 29:47-62.

Donaldson LE (1983) Clover disease in two Mississippi cattle herds. J Am Vet Med Assoc 182(4):412-413.

MEDICAGO SATIVA L. [Fabaceae]

Common Names:

alfalfa; lucerne; luzerne; razha; trefoil

Citations:

Adler JH, Trainin D (1962) The apparent effect of alfalfa on the reproductive performance of dairy cattle. Proc Int Cong Anim Reprod 3:451-456.

Allen JG, Creeper JH, Forshaw D, et al. (2004) Plant-associated diseases, either new or new to the state, encountered over the last decade (1991-2001) in Western Australia. In: Acamovic T, Stewart CS, Pennycott TW (eds.) Poisonous plants and related toxins. CABI. Wallingford, UK. pp. 540-547.

Anderson CR (1944) Contact dermatitis from alfalfa and bur clover. Arch Derm Syphilol 50:201.

Bornstein S, Adler JH (1963) The oestrogenic effect of alfalfa meal on the growing chicken. Refu Vet 20:184-175.

Bozhilov K (1961) [Dermatitis in cows caused by fresh lucerne.] Veterinarna Sbirka 58(6):11-12.

Byrne KV (1937) Dermatitis of white pigs associated with lucerne and trefoil grazing. Agric Gaz New South Wales 48(Apr 1):214.

Casteel SW, Rotinghaus GE, Johnson GC, et al. (1994) Hepatotoxicosis in cattle induced by consumption of alfalfa-grass hay. In: Colegate SM, Dorling PR (eds.) Plant-associated toxins. Agricultural, phytochemical and ecological aspects. CABI. New York. pp. 307-312.

Coop IE (1977) Depression of lambing percentage from mating on lucerne. Proc NZ Soc Anim Prod 37:149-151.

Coop IE, Clark VR (1960) The reproductive performance of ewes mated on lucerne. N Z J Agric Res 3(Dec):922-933.

Gumbrell RC, Jagusch KT (1973) "Red gut" syndrome in lambs grazing lucerne. N Z Vet J 21(8):178-179.

Hall JW, Walker I, Majak W (1994) Evaluation of two supplements for the prevention of alfalfa bloat. Can Vet J 35(11):702-705.

House JK, George LW, Oslund KL, et al. (1996) Primary photosensitization related to ingestion of alfalfa silage by cattle. J Am Vet Med Assoc 209(9):1604-1607.

Jagusch KT, Gumbrell RC, Dellow DW (1976) Red gut in lamb lucerne grazing trials at Lincoln. Proc NZ Soc Anim Prod 36:190-197.

Jagusch KT, Gumbrell RC, Mobley MC, et al. (1977) Effect of salt blocks and roughage as supplements to grazing lucerne on growth rate of lambs and incidence of deaths from 'red gut'. N Z J Exp Agr 5(1):19-22.

Kaufman WH (1954) Alfalfa seed dermatitis. JAMA 155(12):1058-1059.

Kienholz EW, Jensen LS, McGinnis J (1962) Evidence for chick growth inhibitors in several legume seeds. Poult Sci 41:367-371.

Krause E (1969) Zum Vorkommen östrogenwirksamer Stoffe in Futterpflanzen. Arch Exp Veterinarmed 23(3):481-487.

Lotan E, Adler JH (1966) Early effects of excessive alfalfa feeding on bovine fertility. Refu Vet 23(1):112-110.

Marschang F (1969) Massenvergiftung bei Schweinen nach Verfütterung von Luzernemehl, das mit kumarinhaltigen Pflanzen verunreinigt war. Monatsh Veterinarmed 24(9):340-344.

Monlux AW, Glenn BL, Panciera RJ, et al. (1963) Bovine hepatogenous photosensitivity associated with the feeding of alfalfa hay. J Am Vet Med Assoc 142(9):989-994.

Natscheff B, Kristanoff T, Ibrischimoff N, et al. (1960) Über den Luzerneausschlag bei Kühen und Kälbern. Berl Munch Tierarztl Wochenschr 73(14):263-265.

Roberts JL, Hayashi JA (1983) Exacerbation of SLE associated with alfalfa ingestion. N Engl J Med 308(22):1361.

Scales GH, Moss RA, Kelly RW (1977) Reproductive performance of ewes mated on lucerne. Proc NZ Soc Anim Prod 37:152-157.
Stifel FB, Vetter RL (1967) Effects of feed intake upon bloating patterns in lambs fed alfalfa soilage. J Anim Sci 26(2):385-388.
Sudarić F, Nevjestić A, Rukavina L, et al. (1976) Trovanje dikumarinom teladi hranjenih djetelinom slaboc kvaliteta. Veterinarski Glasnik 30(11):905-910.
Vetter O, Wujanz G, Priboth W (1968) Über Schadwirkungen bei Milchkühen nach Aufnahme hämolysinhaltiger Luzerne. Monatsh Veterinarmed 23(22):871-877.

medicinal aloe –see– *Aloe arborescens* Mill.; *Aloe vera* (L.) Burm. f.

medicinal leopardbane –see– *Arnica montana* L.

medicinal mandrake –see– *Mandragora officinarum* L.

medicinal rhubarb –see– *Rheum officinale* Baill.

medicinal squill –see– *Drimia maritima* (L.) Stearn

medicinal valerian –see– *Valeriana officinalis* L.

médicinier bâtard –see– *Jatropha multifida* L.

médicinier d'espagne –see– *Jatropha multifida* L.

Mediterranean aloe –see– *Aloe vera* (L.) Burm. f.

Mediterranean canarygrass –see– *Phalaris minor* Retz.

Mediterranean squill –see– *Drimia maritima* (L.) Stearn

Mediterranean thistle –see– *Chamaeleon gummifera* (L.) Cass.

Meerrettich –see– *Armoracia rusticana* P. Gaertn. et al.

Meerzweibel –see– *Drimia maritima* (L.) Stearn

MEGATHYRSUS MAXIMUS (Jacq.) B. K. Simon & S. W. L. Jacobs [Poaceae]

Synonyms:

Panicum maximum Jacq.

Common Names:

blousaadgras; buffelgras; ginigawat; green panic; Guineagrass

Citations:

Döbereiner J, Rosa IV, Lazzari AA (1976) "Cara inchada" (doença peridentária) em bezerros mantidos em pastos de Panicum maximum. Pesq Agric Bras Vet 11(9):43-47.

Mehldorn –see– *Securigera varia* (L.) Lassen

Meibomia cana S. F. Blake = Desmodium incanum DC.

meiran –see– *Origanum majorana* L.

mejorana –see– *Salvia coccinea* Buc'hoz ex Etl.

melaleuca –see– *Melaleuca alternifolia* (Maiden & Betche) Cheel

MELALEUCA ALTERNIFOLIA (Maiden & Betche) Cheel [Myrtaceae]

Common Names:

Australian tea tree; melaleuca; tea oil tree

Citations:

Bischoff K, Guale F (1998) Australian tea tree (Melaleuca alternifolia) oil poisoning in three purebred cats. J Vet Diagn Invest 10(2):208-210.
Elliott C (1993) Tea tree oil poisoning. Med J Aust 159:830-831.
Hornfeldt CS (1993) Melaleuca oil poisoning: A case report. Vet Hum Toxicol 35(4):329.
Jacobs MR, Hornfeldt CS (1994) Melaleuca oil poisoning. J Toxicol Clin Toxicol 32(4):461-464.
Kaluzienski M (2000) Partial paralysis and altered behavior in dogs treated with melaleuca oil. J Toxicol Clin Toxicol 38(5):518-519.
Knight TE, Hausen BM (1994) Melaleuca oil (tea tree oil) dermatitis. J Am Acad Dermatol 30(3):423-427.
Palmer M, Haller C, McKinney P, et al. (1999) Botanicals and other dietary supplements: Adverse events by age. J Toxicol Clin Toxicol 37(5):609.
Veien NK, Rosner K, Skovgaard GL (2004) Is tea tree oil an important contact allergen? Contact Dermatitis 50(6):378-379.
Villar D, Knight MJ, Hansen SR, et al. (1994) Toxicity of melaleuca oil and related essential oils applied topically on dogs and cats. Vet Hum Toxicol 36(2):139-142.

Melampyrum cristatum L. = Scrophularia marilandica L.

Melanorrhoea usitata Wall. = Gluta usitata (Wall.) Ding Hou

MELIA AZEDARACH L. [Meliaceae]

Common Names:

African lilac tree; arbol-de-quitasol; bakkain; bead tree; bessiboom; Cape lilac; Cape syringa; China tree; Chinaball tree; Chinaberry; Chinese umbrella tree; dhrek; false sycamore; gringging; holy tree; Indian lilac; paraíso; paraíso vegetal; Persian lilac; pride-of-China; pride-of-India; sering; syringa; Texas umbrella tree; umbrella tree; white cedar

Citations:

Bahri S, Sani Y, Hooper PT (1992) Myodegeneration in rats fed Melia azedarach. Aust Vet J 69(2):33.
Brown JH (1939) China berry poisoning in hogs. J Am Vet Med Assoc 95(Jul):448.
Carratala RE (1939) Intoxicacion mortal por frutos de Melia azedarach L. Rev Asoc Med Argent 53(May 15):338-340.
Hare WR, Schutzman H, Lee BR, et al. (1997) Chinaberry poisoning in two dogs. J Am Vet Med Assoc 210(11): 1638-1640.
Hothi DS, Balwant Singh, Kwatra MS, et al. (1976) A note on the comparative toxicity of Melia azedarach (Dhrek) berries to piglets, buffalo-calves, rabbits and fowls. J Res Punjab Agric Univ 13:232-234.
Kiat TK (1969) Melia azedarach poisoning. Singapore Med J 10(1):24-28.
Kwatra MS, Singh B, Hothi DS, et al. (1974) Poisoning by Melia azedarach in pigs. Vet Rec 95(18):421.

Méndez MC, Aragao M, Elias F, et al. (2002) Experimental intoxication by the leaves of Melia azedarach (Meliaceae) in cattle. Pesq Vet Bras 22(1):19-24.
Méndez MC, Elias F, Aragao M, et al. (2002) Intoxication of cattle by the fruits of Melia azedarach. Vet Hum Toxicol 44(3):145-148.
Pammel LH (1921) China tree poisonous. Vet Med 16(10):47.
Seddon HR (1931) Toxicity of Melia azedarach, "white cedar." Med J Aust 1(Jun 27):778.
Steyn DG, Rindl M (1929) Preliminary report on the toxicity of the fruit of Melia azedarach (syringa berries). Trans R Soc South Africa 17(part 4):295-308.
White CT (1920) The white cedar (Melia azedarach var australasica): A plant poisonous to pigs. Queensland Agric J 14:146-147.

MELICOPE ANISATA (H. Mann) T. G. Hartley & B. C. Stone [Rutaceae]

Synonyms:

Pelea anisata H. Mann

Common Names:

mokihana

Citations:

Elpern DJ, Mitchell JC (1984) Phytophotodermatitis from mokihana fruits (Pelea anisata H. Mann, fam. Rutaceae) in Hawaiian lei. Contact Dermatitis 10(4):224-226.
Geroso AM, Elpern DJ (1992) Some observations on mango and mokihana dermatitis from Hawaii. Contact Dermatitis 26(5):346-347.

MELIENTHA SUAVIS Pierre [Opiliaceae]

Common Names:

pak wan

Citations:

Kerr AF (1931) Poisoning by pak wan (Melientha suavis) in Siam. Trans R Soc Trop Med Hyg 25(2):141-143.

melilot –see– *Melilotus officinalis* Lam.

meliloto –see– *Melilotus albus* Medik.

melilotus –see– *Melilotus indicus* (L.) All.

MELILOTUS ALBUS Medik. [Fabaceae]

Common Names:

Arctic sweet clover; Bokhara Klee; hubam; khandai; meliloto; méliot blanc; Steinklee; sweet clover; trébol dulce; trébol locoweed; Weißer Steinklee; white melilot; white sweet clover

Citations:

Cannon CY, Greenwood D (1930) Effect of diet of sweet clover on the calcium in the blood serum. J Dairy Sci 13:424-431.
Dotta U, Ferraudo M (1969) Particolari aspetti clinici di un episodio de intossicazione alimentaire da meliloto nei bovini. Atti Soc Ital Buiatria 1:153-166.
Meads EB, Taylor PA, Pallister WA (1964) An unusual outbreak of sweet clover poisoning in cattle. Can Vet J 5(3):65-71.
Pra AR (1984) Intoxicacion por henificacion de Melilotus alba (trebol dulce). Vet Argent 1(8):800, 802-806.
Smith WK, Brink RA (1938) Relation of bitterness to the toxic principle in sweetclover. J Agric Res 57(2):145-154.
Wiesner E, Rex JO, Wiesner B (1968) Gerinnungsuntersuchungen an gesunden Rindern nach Verfütterung von Steinklee. 2. Mittleilung: Fütterungsversuche. Monatsh Veterinarmed 23(20):788-791.

MELILOTUS INDICUS (L.) All. [Fabaceae]

Common Names:

Hexham scent; Indian clover; Indian melilot; King Island melilot; melilotus; soup clover; sour clover; sour sweet clover; sweet clover

Citations:

Wignall WN, Banks AW, Hackett E, et al. (1961) Dicoumarol poisoning of cattle and sheep in South Australia. Aust Vet J 37(Dec):456-459.

MELILOTUS OFFICINALIS (L.) Lam. [Fabaceae]

Synonyms:

Trifolium officinale L.

Common Names:

Echter Steinklee; melilot; méliot jaune; orvosi somkóró; Steinklee; sweet clover; trevo doce amnelo; yellow sweet clover

Citations:

Edmondson P, Loeffler DG, Burrows GE (1988) Subcutaneous swelling in a cow. Vet Hum Toxicol 30(3):265.
Wiesner E, Rex JO, Wiesner B (1965) Gerinnungsuntersuchungen an gesunden Rindern nach Verfütterung von Steinklee. 1. Mitteilung: Normalwerte. Monatsh Veterinarmed 20(11-12):469-475.

méliot blanc –see– *Melilotus albus* Medik.

méliot jaune –see– *Melilotus officinalis* Lam.

melkbos –see– *Gomphocarpus fruticosa* (L.) W. T. Aiton

melktou –see– *Sarcostemma viminale* (L.) R. Br.

MELOCHIA PYRAMIDATA L. [Sterculiaceae]

Citations:

Palmer AC, Woodham CB (1975) Derrengue, a paralysis of cattle in El Salvador ascribed to ingestion of Melochia pyramidata. Vet Rec 96(25):547-548.

MELOCHIA TOMENTOSA L. [Sterculiaceae]

Common Names:

basora corra; bretonica; cariaquito morado

Citations:

Morton JF (1979) Plant tannins and esophageal cancer. In: Deichmann WB (ed.) Toxicology and occupational medicine. Elsevier. New York. pp. 129-137.

melon –see– *Citrullus lanatus* (Thunb.) Matsum. & Nakai; *Cucumis melo* L.

memory root –see– *Arisaema triphyllum* (L.) Schott

Menispermum acutum Thunb. = Sinomenium acutum (Thunb.) Rehder & E. H. Wilson

Mentha citrata Ehrh. = Mentha ×piperita L. nothosubsp. citrata (Ehrh.) Briq.

MENTHA ×PIPERITA L. nothosubsp. citrata (Ehrh.) Briq. [Lamiaceae]

Synonyms:

Mentha citrata Ehrh.

Common Names:

bergamot mint; horse mint; mint; orange mint

Citations:

Sams WM (1940) Occupational dermatitis due to mint. Report of two cases. Arch Derm Syphilol 41:503-505.

MENTHA ×PIPERITA L. nothosubsp. piperita [Lamiaceae]

Common Names:

peppermint; Pfefferminze; yerba buena

Citations:

Eickholt TH, Box RH (1965) Toxicities of peppermint and Pycnanthemum albescens oils, fam. Labiateae. J Pharm Sci 54(7):1071-7072.

Lipsitz DJ (1984) Herbal teas and water intoxication in a young child. J Fam Pract 18(6):933-937.

Parys BT (1983) Chemical burns resulting from contact with peppermint oil mar; a case report. Burns 9(5):374-375.

mentha pouliot –see– *Mentha pulegium* L.

MENTHA PULEGIUM L. [Lamiaceae]

Common Names:

European pennyroyal; huile-de-menthe-pouliot; mentha pouliot; mosquito plant; pennyroyal; póleo; squaw mint

Citations:

Anderson IB, Mullen WH, Meeker JE, et al. (1996) Pennyroyal toxicity: Measurement of toxic metabolite levels in two cases and review of the literature. Ann Intern Med 124(8):726-734.

Bakerink JA, Gospe JM Jr, Dimand RJ, et al. (1996) Multiple organ failure after ingestion of pennyroyal oil from herbal tea in two infants. Pediatrics 98(5):944-947.

Cienki JJ, Connolly HV, Nelson SA, et al. (1994) Hepatoxicity and death after treatment with Broncodin herbal tea. Vet Hum Toxicol 36(4):359.

Ciganda C, Laborde A (2001) Herbal infusions used for induced abortion. J Toxicol Clin Toxicol 39(3):318-319.

Ciganda C, Laborde A (2003) Herbal infusions used for induced abortion. J Toxicol Clin Toxicol 41(3):235-239.

Early DF (1961) Pennyroyal: A rare cause of epilepsy. Lancet 2(Sep 9):580-581.

Laborde A, Ciganda C (1998) Poisoning by herbal infusions ingested as abortifacient agents. J Toxicol Clin Toxicol 36(5):454-455.

Litovitz TL, Felberg L, Soloway RA, et al. (1995) 1994 Annual report of the American Association of Poison Control Centers Toxic Exposure Surveillance System. Am J Emerg Med 12(5):551-597.

Litovitz TL, Felberg L, White S, et al. (1996) 1995 Annual report of the American Association of Poison Control Centers Toxic Exposure Surveillance System. Am J Emerg Med 14(5):487-537.

McCormick MA, Manoguerra AS (1988) Toxicity of pennyroyal oil: A case report and review. Vet Hum Toxicol 30(4):347.

Roé E, Serra-Baldrich E, Dalmau J, et al. (2005) Mentha pulegium contact dermatitis. Contact Dermatitis 53(6):355.

Sudekum M, Poppenga RH, Raju N, et al. (1992) Pennyroyal oil toxicosis in a dog. J Am Vet Med Assoc 200(6):817-818.

Sullivan JB Jr, Rumack BH, Thomas H Jr, et al. (1979) Pennyroyal oil poisoning and hepatotoxicity. JAMA 242(26):2873-2874.

Sullivan JB, Rumack BH, Thomas H Jr, et al. (1979) Pennyroyal oil: Review of toxicity and report of a fatal case with massive hepatic necrosis. Clin Toxicol 15(4):489.

Vallance WB (1955) Pennyroyal poisoning. A fatal case. Lancet 269(Oct 22):850-851.

MENTHA SATUREIOIDES R. Br. [Lamiaceae]

Common names:

native pennyroyal, squeejit

Citations:

White CT (1934) Annual Report of the Government Botanist, Queensland 1933-34.

White CT (1936) Annual Report of the Government Botanist, Queensland 1935-36.

MENTHA SPICATA L. [Lamiaceae]

Synonyms:

Mentha viridis (L.) L.

Common Names:

spearmint

Citations:

Akdogan M, Kilinc I, Oncu M, et al. (2003) Investigation of biochemical and histopathological effects of Mentha piperita L. and Mentha spicata L. on kidney tissue in rats. Hum Exp Toxicol 22:213-219.

Bonamonte D, Mundo L, Daddabbo M, et al. (2001) Allergic contact dermatitis from Mentha spicata (spearmint). Contact Dermatitis 45(5):298.

Clayton R, Orton D (2004) Contact allergy to spearmint oil in a patient with oral lichen planus. Contact Dermatitis 51(5-6):314-315.

Skrebova N, Brocks K, Karlsmark T (1998) Allergic contact cheilitis from spearmint oil. Contact Dermatitis 39(1):35-36.

Tomson N, Murdoch S, Finch TM (2004) The dangers of making mint sauce. Contact Dermatitis 51(2):92-93.

Mentha viridis (L.) L. = Mentha spicata L.

MENYANTHES TRIFOLIATA L. [Menyanthaceae]

Common Names:

Bitterklee; bitterroot; bogbean; buckbean; Fieberklee; marsh trefoil; water shamrock

Citations:

Airaksinen M, Peura P, Alaossi Salokangas L, et al. (1986) Toxicity of plant material used as emergency food during famines in Finland. J Ethnopharmacol 18(3):273-296.

Menzies'-larkspur –see– *Delphinium menziesii* DC.

mercurial perenne –see– *Mercurialis perennis* L.

mercuriale-du-bois –see– *Mercurialis perennis* L.

MERCURIALIS ANNUA L. [Euphorbiaceae]

Common Names:

annual mercury; baron's-mercury; Bingelkraut; boys-and-girls; egynyári szélfü; Einjährige Bingelkraut; foirolle; foironde; herb mercury; mercury; Schutt Bingelkraut; vignette

Citations:

Bizzeti M, De Lucia PG, Corazza M, et al. (1987) Intossicazione da Mercurialis annua nella pecora. Ann Fac Med Vet Pisa 60:165-175.

Bokori J, Kovács F, Haraszti E (1955) Lovak egynyáriszélfü (Mercurialis annua) mérgezése. Magyar Allator Lapja 10:191-196.

Durieux J, Durieux M (1973) Quelques intoxications vegetales. Bull Mens Soc Vet Prat Fr 57(2):71-76.

Guilhon J (1988) L'enseignement de la botanique et la mort subite au pré. Bull Acad Vet Fr 61(3):267-271.

Landau M, Egyed MN, Flesh D (1973) Mercurialis annua poisoning in housed sheep. Refu Vet 30(3-4):131-135.

Polidori F, Maggi M (1954) Problemi di alimentazione degli animali su alcuni casi di avvelenamento da «Mercurialis annua» nei bovini. Ricerche sperimentali sulla differente tossicita' della pianta allo stato fresco e dopo essiccamento. Nuova Vet 30:146-150.

Sendil C (1978) Koyun ve buzağilarda deneysel Mercurialis annua zehirlenmesinde klinik bulgular ve sağitim üzerinde araştirmalar. Vet Fakul Dergisi Ankara Univ 25:480-499.

Senf W, Seffner W (1965) Beitrag zur Bingelkrautvergiftung beim Schaf. Monatsh Veterinarmed 20(15):622-625.

Welchman DB, Gibbens JC, Giles N, et al. (1995) Suspected annual mercury (Mercurialis annua) poisoning of lambs grazing fallow arable land. Vet Rec 137(23):592-593.

MERCURIALIS PERENNIS L. [Euphorbiaceae]

Common Names:

Ausdauerndes Bingelkraut; chou-de-chien; dog's-mercury; erdei szélfü; European dog mercury; herb mercury; Hundskraut; Kentish balsam; mercurial perenne; mercuriale-du-bois; mercury; Stinkerich; Waldbingelkraut; wild spinach

Citations:

Baker JR, Faull WB (1968) Dog's mercury (Mercurialis perennis L.) poisoning in sheep. Vet Rec 82:485-489.

Bismarck R, Floehr W (1974) Über Bingelkrautvergiftungen in einer weidenden Kuhherde. Dtsch Tierarztl Wochenschr 81(18):433-434.

Eaton G (1941) A series of cases of poisoning in cattle. Vet Rec 53(10):145-146.

Mortensen OS (1999) Ikke alt grønt er sundt! Ugeskr Laeger 161(16):2384.

Rugman F, Meecham J, Edmondson J (1983) Mercurialis perennis (dog's mercury) poisoning: A case of mistaken identity. Br Med J (Clin Res Ed) 287(6409):1924.

Watson PJ (1998) Suspected dog's mercury (Mercurialis perennis) poisoning in cattle. Vet Rec 142(5):116-117.

mercury –see– *Mercurialis annua* L.; *Mercurialis perennis* L.; *Toxicodendron radicans* (L.) Kuntze

MERREMIA TUBEROSA (L.) Rendle [Convolvulaceae]

Common Names:

wood rose

Citations:

Hruby K (2002) Poisoning from natural drugs rediscovered for recreational abuse. J Toxicol Clin Toxicol 40(3):276-277.

MERWILLA PLUMBEA (Lindl.) Speta [Liliaceae]

Synonyms:

Scilla natalensis Planch.

Citations:

Van der Walt SJ, Steyn DG (1943) Recent investigations into the toxicity of plants, etc. XIII. Onderstepoort J Vet Sci Anim Indus 18(1-2):207 224.

mescal –see– *Lophophora williamsii* (Lem. ex Salm-Dyck) J. M. Coult.

mescal bean –see– *Calia secundiflora* (Ortega) Yakovlev; *Lophophora williamsii* (Lem. ex Salm-Dyck) J. M. Coult.

mescal Bohne –see– *Calia secundiflora* (Ortega) Yakovlev

mescal button –see– *Lophophora williamsii* (Lem. ex Salm-Dyck) J. M. Coult.

MESEMBRYANTHEMUM NODIFLORUM L. [Azioaceae]

Common Names:

slender ice plant

Citations:

Jacob RH, Peet RL (1989) Acute oxalate toxicity of sheep associated with slender iceplant (Mesembryanthemum nodiflorum). Aust Vet J 66(3):91-92.

mesic milk vetch –see– *Astragalus diversifolius* A. Gray

MESONA CHINENSIS Benth. [Lamiaceae]

Citations:

Goh CL (1988) Occupational dermatitis from Mesonia chinensis. Contact Dermatitis 18(7):113.

mesquite –see– *Prosopis glandulosa* Torr.; *Prosopis juliflora* (Sw.) DC.

mesquite bean –see– *Prosopis juliflora* (Sw.) DC.

MESUA FERREA L. [Clusiaceae]

Common Names:

Eisensäbelholz; Tagayasan

Citations:

Iwakawa K (1911) Über das entzündungserregende Pulver des japanischen Nutzholzes "Tagayasan." Naunyn Schmiedebergs Arch Exp Pathol Pharmakol 65:315-324.

meswak –see– *Salvadora persica* L.

metel –see– *Datura metel* L.

methi –see– *Trigonella foenum-graecum* L.

methika –see– *Trigonella foenum-graecum* L.

meto –see– *Castanospermum australe* A. Cunn. & C. Fraser ex Hook.

METOPIUM BROWNEI (Jacq.) Urb. [Anacardiaceae]

Synonyms:

Rhus metopium L.

Common Names:

black poisonwood; bois mulatre; boxchechem; burnwood; calbachechem; cedro prieto; chacin; chechem negro; chicharron; cochinilla; cochinillio; guao; guao-de-costa; Honduran walnut; huao; palo-de-rosa

Citations:

Lunin MM (1969) [Burns induced by the tropical plant huao.] Vestn Khir Im I I Grek 102(6):96-100.

METOPIUM TOXIFERUM (L.) Krug & Urb. [Anacardiaceae]

Common Names:

almendron; bearwood; black poisonwood; burnwood; cedro prieto; chechém; coral sumach; doctor gum; Florida poison tree; guao-de-costa; hog gum; huao; Jamaican sumach; janca tree; machandeuse; machanoise; mancenillier; mountain manchineel; papayo; poison tree; poison weed; poisonwood

Citations:

Grabham M (1899) Acute dermatitis caused by handling Rhus. Br Med J 1(May 5):1140.

Jackson WP (1946) Plant dermatitis in the Bahamas. Br Med J 2(Aug 31):298.

Lunin MM (1969) [Burns induced by the tropical plant huao.] Vestn Khir Im I I Grek 102(6):96-100.

Wahlberg JE, Lovell CR (1996) Poison wood dermatitis in a future prime minister. Contact Dermatitis 34(5):363.

Mexican bird-of-paradise –see– *Caesalpinia gilliesii* (Hook.) D. Dietr.

Mexican bird seed –see– *Parthenium hysterophorus* L.

Mexican breadfruit –see– *Monstera deliciosa* Liebm.

Mexican devil –see– *Ageratina adenophora* (Spreng.) R. M. King & H. Rob.

Mexican dove plant –see– *Cannabis sativa* L.

Mexican elder –see– *Sambucus mexicana* C. Presl ex DC.

Mexican fire plant –see– *Euphorbia heterophylla* L.

Mexican fireweed –see– *Bassia scoparia* (L.) A. J. Scott

Mexican goosefoot –see– *Dysphania ambrosioides* (L.) Mosyakin & Clemants

Mexican hat plant –see– *Kalanchoe daigremontiana* Raym.-Hamet & H. Perrier

Mexican jumping bean –see– *Sapium biloculare* (S. Watson) Pax

Mexican love plant –see– *Kalanchoe pinnata* (Lam.) Pers.

Mexican marigold –see– *Tagetes minuta* L.

Mexican persimmon –see– *Diospyros texana* Scheele

Mexican poppy –see– *Argemone mexicana* L.

Mexican prickly poppy –see– *Argemone mexicana* L.

Mexican rubber plant –see– *Parthenium argentatum* A. Gray

Mexican sandbur –see– *Tribulus terrestris* L.

Mexican sleepygrass –see– *Achnatherum robustum* (Vasey) Barkworth

Mexican tea –see– *Dysphania ambrosioides* (L.) Mosyakin & Clemants; *Ephedra viridis* Coville

Mexican weed –see– *Ricinus communis* L.

Mexican whorled milkweed –see– *Asclepias mexicana* Cav.

mezei sóska –see– *Rumex acetosa* L.

mezei szarkaláb –see– *Consolida regalis* Gray

mezereon –see– *Daphne mezereum* L.

mezereum –see– *Daphne mezereum* L.

Michaelmas crocus –see– *Colchicum autumnale* L.

Michaux's-milk vetch –see– Astragalus michauxii (Kuntze) F. J. Herm.

Micrampelis oregona (Torr. & A. Gray) Greene = Marah oregonus (Torr. & S. Watson) J. T. Howell

Milchkraut –see– *Euphorbia helioscopia* L.

milfoil –see– *Achillea millefolium* L.

MILICIA EXCELSA (Welw.) C. C. Berg [Moraceae]

Synonyms:

Chlorophora excelsa (Welw.) Benth.

Common Names:
abang; African teak; iroko; iroko teak; kambala; kambala teak; odum

Citations:
Azofra J, Olaguibel JM (1989) Occupational asthma caused by iroko wood. Allergy 44(2):156-158.
Beer WE (1970) Sensitivity to iroko wood in a wood machinist. Contact Dermatol Newsl 7(Jan):159.
Davidson JM (1941) Toxic effects of iroko. An African wood. Lancet 1:38-39.
Fernández de Corres L, Corrales JL, Muñoz D, et al. (1984) Dermatitis alérgicas de contacto por plantas. Allergol Immunopathol (Madr) 12(4):313-319.
Hinnen U, Willa-Craps C, Elsner P (1995) Allergic contact dermatitis from iroko and pine wood dust. Contact Dermatitis 33(6):428.
Oleffe JA, Sporcq J, Hublet P (1975) Epidemiological study of the wood industry in Belgium. Contact Dermatitis 1(5):315-316.
Ordman D (1949) Bronchial asthma caused by the inhalation of wood dust. Ann Allergy 7:492-496,505.
Schulz KH (1957) Allergische Kontaktdermatitis durch exotische Hölzer, insbesondere durch Kambala-Teakholz. Berufsdermatosen 5:238-244.
Stingeni L, Mariotti M, Lisi P (1998) Airborne allergic contact dermatitis from iroko (Chlorophora excelsa). Contact Dermatitis 38(5):287-288.
Thienemann K (1941) Kambala-Teakholz-Dermatosen. Arch Derm Syphilol 182:551-570.
Whiting DA (1971) Plant dermatitis in the southern Transvaal. S Afr Med J 45(7):163-167.
Wilkinson DS (1969) Sensitivity to iroko wood in a boatbuilder. Contact Dermatol Newsl 6(Jul):142.

MILIUSA VELUTINA Hook.f. & Thomson [Annonaceae]

Citations:
Kulakkattolickal AT (1987) Piscicidal plants of Nepal: Preliminary toxicity screening using grass carp (Ctenopharyngodon idella) fingerlings. J Ethnopharmacol 21(1):1-9.

milk bush –see– *Euphorbia antiquorum* L.; *Euphorbia hermentiana* Lem.; *Euphorbia tirucalli* L.; *Sarcostemma viminale* (L.) R. Br.

milk hedge –see– *Euphorbia tirucalli* L.

milk purslane –see– *Chamaesyce maculata* (L.) Small

milk spurge –see– *Euphorbia drummondii* Boiss.

milk thistle –see– *Silybum marianum* (L.) Gaertn.

milk tree –see– *Euphorbia tirucalli* L.; *Thevetia peruviana* (Pers.) K. Schum.

milk vine –see– *Marsdenia rostrata* R. Br.

milkstripe euphorbia –see– *Euphorbia lactea* Haw.

milkweed –see– *Apocynum cannabinum* L.; *Asclepias curassavica* L.; *Asclepias latifolia* (Torr.) Raf.; *Calotropis procera* (Aiton) W. T. Aiton; *Chamaesyce hirta* (L.) Millsp.; *Euphorbia drummondii* Boiss.; *Euphorbia lathyris* L.; *Euphorbia peplus* L.; *Gomphocarpus fruticosa* (L.) W. T. Aiton; *Pratia erecta* Gaudich.

milkwort –see– *Polygala sanguinea* L.

millepertuis –see– *Hypericum perforatum* L.

millet –see– *Panicum miliaceum* L.; *Sorghum bicolor* (L.) Moench

MILLETTIA LAURENTII De Wild. [Fabaceae]

Common Names:
wengé

Citations:
Oleffe JA, Sporcq J, Hublet P (1975) Epidemiological study of the wood industry in Belgium. Contact Dermatitis 1(5):315-316.

MILLETTIA PACHYCARPA Benth. [Fabaceae]

Citations:
Chakraborty DP, Nandy AC, Philipose MT (1972) Barringtonia acutangula (L.) Gaertn. as a fish poison. Indian J Exp Biol 10(1):78-80.

MILLETTIA PINNATA (L.) Panigrahi [Fabaceae]

Synonyms:
Pongamia glabra Vent.

Common Names:
kanja; karanja

Citations:
Natanam R, Kadirvel R, Ravi R (1989) The toxic effects of karanja (Pongamia glabra Vent) oil and cake on growth and feed efficiency in broiler chicks. Anim Feed Sci Technol 27(1-2):95-100.

MILLETTIA STUHLMANNII Taub. [Fabaceae]

Common Names:
panga panga; partridge

Citations:
Oleffe JA, Sporcq J, Hublet P (1975) Epidemiological study of the wood industry in Belgium. Contact Dermatitis 1(5):315-316.
Ordman D (1949) Bronchial asthma caused by the inhalation of wood dust. Ann Allergy 7:492-496,505.

milo –see– *Sorghum bicolor* (L.) Moench

mimosa –see– *Acacia ×giraffae* Willd.; *Albizia julibrissin* Durazz.

mimosa-du-japon –see– *Caesalpinia gilliesii* (Hook.) D. Dietr.

Mimosa sirissa Roxb. = Albizia lebbeck (L.) Benth.

mineur –see– *Juniperus communis* L.

mingwort –see– *Artemisia absinthium* L.

minnie minnies –see– *Abrus precatorius* L.

mint –see– *Mentha ×piperita* L. nothosubsp. citrata (Ehrh.) Briq.

mint plant –see– *Perilla frutescens* (L.) Britton

mintweed –see– *Salvia reflexa* Hornem.; *Stachys arvensis* (L.) L.

minyara –see– *Euphorbia tirucalli* L.

mio mio –see– *Baccharis coridifolia* DC.

mioga –see– *Zingiber mioga* (Thunb.) Roscoe

miraa –see– *Catha edulis* (Vahl) Forssk. ex Endl.

mirra –see– *Commiphora myrrha* (Nees) Engl.

mission bells –see– *Kalanchoe daigremontiana* Raym.-Hamet & H. Perrier; *Kalanchoe delagoensis* Eckl. & Zeyh.

mist flower –see– *Ageratina riparia* (Regel) R. M. King & H. Rob.

Mistel –see– *Viscum album* L.

mistletoe –see– *Phoradendron leucarpum* (Raf.) Reveal & M. C. Johnst.; *Phoradendron piperoides* (Kunth) Trel.; *Viscum album* L.

mitsuba –see– *Cryptotaenia japonica* Hassk.

mnyala –see– *Euphorbia tirucalli* L.

Moah –see– *Madhuca longifolia* (L.) J. F. Macbr.

mock eel root –see– *Cicuta douglasii* (DC.) J. M. Coult. & Rose

mocsári zsurló –see– *Equisetum palustre* L.

Modecca palmata Lam. = Adenia hondala (Gaertn.) W. J. de Wilde

MODIOLA CAROLINIANA (L.) G. Don [Malvaceae]

Synonyms:

Modiola multifida Moench

Common Names:

bristly mallow; Carolina modiola; creeping mallow; ground ivy; red mallow; red willow; wheel mallow

Citations:

King ED Jr (1920) Poisonous plants of the south. J Am Vet Med Assoc 57:302-313.

Modiola multifida Moench = Modiola caroliniana (L.) G. Don

Möhre –see– *Daucus carota* L.

mokihana –see– *Melicope anisata* (H. Mann) T. G. Hartley & B. C. Stone

mole bean –see– *Ricinus communis* L.; *Sesbania vesicaria* (Jacq.) Elliott

mole plant –see– *Euphorbia lathyris* L.; *Euphorbia peplus* L.

molinillo –see– *Hura crepitans* L.

Molten disease plant –see– *Senecio retrorsus* DC.

momchina –see– *Triadica sebifera* (L.) Small

MOMORDICA CHARANTIA L. [Cucurbitaceae]

Common Names:

African cucumber; balsam pear; balsamino; bitter cucumber; bitter gourd; bitter melon; karela; maiden apple; wild balsam apple; wild cucumber

Citations:

Foley RH (1976) Acute poisoning in a puppy caused by the balsam pear (Momordica charantia). Vet Med Small Anim Clin 71(6):761-762.

Tennekoon KH, Jeevathayaparan S, Angunawala P, et al. (1994) Effect of Momordica charantia on key hepatic enzymes. J Ethnopharmacol 44:93-97.

MOMORDICA DIOICA Roxb. ex Willd. [Cucurbitaceae]

Common Names:

bitter gourd

Citations:

Sinha SM, Pasricha JS, Sharma RC, et al. (1977) Vegetables responsible for contact dermatitis of the hands. Arch Dermatol 113(6):776-779.

momordique –see– *Ecballium elaterium* (L.) A. Rich.

monagrass –see– *Paspalum scrobiculatum* L. var. bispicatum Hack.

Mondbohne –see– *Phaseolus lunatus* L.

mondzo –see– *Datura metel* L.

monim –see– *Asimina triloba* (L.) Dunal

monkey dinner bell –see– *Hura crepitans* L.

monkey face tree –see– *Mallotus philippensis* (Lam.) Müll. Arg.

monkey fiddle –see– *Euphorbia tirucalli* L.

monkey nut –see– *Arachis hypogaea* L.; *Lecythis ollaria* Loefl.

monkey pistol –see– *Hura crepitans* L.

monkey pod –see– *Lecythis ollaria* Loefl.

monkey puzzle –see– *Euphorbia lactea* Haw.

monkey's-coconut –see– *Lecythis ollaria* Loefl.

monk's-cowl –see– *Aconitum napellus* L.

monkshood –see– *Aconitum lycoctonum* L. subsp. vulparia (Rchb.) Nyman; *Aconitum napellus* L.

monstera –see– *Monstera deliciosa* Liebm.

MONSTERA DELICIOSA Liebm. [Araceae]

Common Names:

cériman; cut-leaf philodendron; Fensterblatt; fruit salad plant; Mexican breadfruit; monstera; piñanona; split-leaf philodendron; Swiss-cheese plant; windowleaf

Citations:

Der Marderosian AH, Giller FB, Roia FC Jr (1976) Phytochemical and toxicological screening of household ornamental plants potentially toxic to humans. I. J Toxicol Environ Health 1(6):939-953.

Whiting DA (1971) Plant dermatitis in the southern Transvaal. S Afr Med J 45(7):163-167.

MONTANOA TOMENTOSA Cerv. [Asteraceae]

Common Names:

yucuyahui; zoapatle

Citations:

Montoya-Cabrera MA, Simental-Toba A, Sánchez-Rodríguez S, et al. (1998) Depresión cardiorrespiratoria en ocho recién nacidos cuyas madres ingirieron infusiones de yucuyahui (zoapatle-Montanoa tomentosa) durante el trabajo de parto. Gac Med Mex 134(5):611-615.

Monterey cypress –see– *Cupressus macrocarpa* Hartw. ex Gordon

Monterey pine –see– *Pinus radiata* D. Don

moonflower –see– *Brugmansia suaveolens* (Humb. & Bonpl. ex Willd.) Bercht. & J. Presl; *Datura inoxia* Mill.; *Datura stramonium* L.; *Datura wrightii* Regel

Moorbeere –see– *Vaccinium uliginosum* L.

mora –see– *Gloriosa superba* L.

MORAEA ERICI-ROSENII R. E. Fr. [Iridaceae]

Citations:

Shone DK (1966) Poisonous plants of Rhodesia. Part I. Rhodesia Zambia Malawi J Agric Res 4:81-94.

Shone DK (1967) Poisonous plants of Rhodesia. Part II. Rhodesia Zambia Malawi J Agric Res 5:29-41.

MORAEA MINIATA Andrews [Iridaceae]

Synonyms:

Homeria miniata (Andrews) Sweet

Common Names:

tulp; two-leaf Cape-tulip

Citations:

Filmer JF (1926) Cape tulip (Homeria miniata). A poison plant. J Agric West Aust 3:240-243.

MORAEA PALLIDA (Baker) Goldblatt [Iridaceae]

Synonyms:

Homeria glauca (Wood & Evans) N. E. Br.; ***Homeria pallida*** Baker; ***Homeria pura*** N. E. Br.

Common Names:

geel tulp; Natal yellow tulp; Transvaal yellow tulp; tulp; yellow tulip; yellow tulp

Citations:

Awan GE, Schultz RA, Kellerman TS, et al. (1995) In vivo effects of a novel calcium antagonist (R56865) against induced epoxyscillirosidin and tulp poisoning in sheep. Onderstepoort J Vet Res 62(3):163-166.

Button C, Mulders MS (1984) Further physiopathological features of experimental Homeria glauca (Wood and Evans) N. E. Br. poisoning in Merino sheep. Onderstepoort J Vet Res 51(2):95-96.

Button C, Reyers F, Meltzer DG, et al. (1983) Some physiopathological features of experimental Homeria glauca (Wood & Evans) N. E. Br. poisoning in Merino sheep. Onderstepoort J Vet Res 50(3):191-196.

Schultz RA, Kellerman TS, Van den Berg H (2005) The role of fluorescence polarization immuno-assay in the diagnosis of plant-induced cardiac glycoside poisoning livestock in South Africa. Onderstepoort J Vet Res 72(3):189-201.

Snyman LD, Schultz RA, Joubert JP, et al. (2003) Conditioned feed aversion as a means to prevent tulp (Homeria pallida) poisoning in cattle. Onderstepoort J Vet Res 70(1):43-48.

MORAEA POLYSTACHYA (Thunb.) Ker Gawl. [Iridaceae]

Common Names:

blue tulp; tulip

Citations:

Joubert JP, Schultz RA (1982) The minimal effective dose of activated charcoal in the treatment of sheep poisoned with the cardiac glycoside containing plant Moraea polystachya (Thunb) Ker Gawl. J S Afr Vet Assoc 53(4):265-266.

Joubert JP, Schultz RA (1982) The treatment of Moraea polystachya (Thunb) Ker-Gawl (cardiac glycoside) poisoning in sheep and cattle with activated charcoal and potassium chloride. J S Afr Vet Assoc 53(4):249-253.

Schultz RA, Kellerman TS, Van den Berg H (2005) The role of fluorescence polarization immuno-assay in the diagnosis of plant-induced cardiac glycoside poisoning livestock in South Africa. Onderstepoort J Vet Res 72(3):189-201.

MORAEA SETACEA Ker Gawl. [Iridaceae]

Common Names:

blou tulp; blue tulip; bokuintjie; tulp

Citations:

Van der Walt SJ, Steyn DG (1940) Recent investigations into the toxicity of known and unknown poisonous plants in the Union of South Africa, X. Onderstepoort J Vet Res 15(1-2):261-277.

MORAEA SPATHULATA (L. f.) Klatt [Iridaceae]

Common Names:

yellow tulp

Citations:

Van der Walt SJ, Steyn DG (1939) Recent investigations into the toxicity of known and unknown poisonous plants in the Union of South Africa. IX. Onderstepoort J Vet Sci Anim Indus 12:335-366.

MORAEA THOMSONII Baker [Iridaceae]

Synonyms:

Moraea trita N. E. Br. var. foliata N. E. Br.

Common Names:

tulp

Citations:

Van der Walt SJ, Steyn DG (1946) Recent investigations into the toxicity of plants, etc. XV. Onderstepoort J Vet Sci Anim Indus 21(1):45-55.

Moraea trita N. E. Br. var. foliata N. E. Br. = Moraea thomsonii Baker

morali –see– *Lycium barbarum* L.

morella douce-amère –see– *Solanum dulcamara* L.

morelle furieuse –see– *Atropa belladonna* L.

morelle noire –see– *Solanum nigrum* L.

morelle noire-de-l'est –see– *Solanum ptycanthum* Dunal

morenita –see– *Bassia scoparia* (L.) A. J. Scott

Moreton Bay chestnut –see– *Castanospermum australe* A. Cunn. & C. Fraser ex Hook.

morgeline –see– *Anagallis arvensis* L.

MORINDA CITRIFOLIA L. [Rubiaceae]

Common Names:

noni

Citations:

Mueller BA, Scott MK, Sowinski KM, et al. (2000) Noni juice (Morinda citrifolia): Hidden potential for hyperkalemia? Am J Kidney Dis 35(2):310-312.

Stadlbauer V, Fickert P, Lackner C, et al. (2005) Hepatotoxicity of noni juice: Report of two cases. World J Gastroenterol 11(30):4758-4760.

MORINGA CONCANENSIS Nimmo [Moringaceae]

Citations:

Shukla S, Mathur R, Prakash AO (1988) Anti-implantation efficacy of Moringa oleifera Lam. and Moringa concanensis Nimmo in rats. Int J Crude Drug Res 26(Mar):29-32.

MORINGA OLEIFERA Lam. [Moringaceae]

Common Names:

angela; arbol-de-las-perlas; ben; drumstick tree; horseradish tree; jasmin frances; maidenhair tree; maranga calalu; paraiso; paraíso blanco; paraiso-de-España; paraiso frances; paraist estranjaro; perlas-del-oriente; sahajana

Citations:

Grant G, More LJ, McKenzie NH, et al. (1995) Nutritional and haemagglutination properties of several tropical seeds. J Agric Sci 124:437-445.

Shukla S, Mathur R, Prakash AO (1988) Anti-implantation efficacy of Moringa oleifera Lam. and Moringa concanensis Nimmo in rats. Int J Crude Drug Res 26(Mar):29-32.

Mormon tea –see– *Ephedra viridis* Coville

morning-glory –see– *Convolvulus arvensis* L.; *Ipomoea carnea* Jacq. subsp. fistulosa (Mart. ex Choisy) D. F. Austin; *Ipomoea hederacea* Jacq.; *Ipomoea muelleri* Benth.; *Ipomoea tricolor* cav.; *Ipomoea violacea* L.; *Turbina corymbosa* (L.) Raf.

morning-noon-night –see– *Brunfelsia australis* Benth.; *Brunfelsia pauciflora* (Cham. & Schltdl.) Benth.

MORUS ALBA L. [Moraceae]

Common Names:

Weißer Maulbeerbaum

Citations:

Navarro AM, Orta JC, Sanchez MC, et al. (1997) Primary sensitization to Morus alba. Allergy 52(11):1144-1145.

MORUS NIGRA L. [Moraceae]

Common Names:

black mulberry; mulberry; Schwarzer Maulbeerbaum

Citations:

James T (1961) Mulberry sickness. S Afr Med J 35:1086.

mosquito plant –see– *Hedeoma pulegioides* (L.) Pers.; *Mentha pulegium* L.

mostarda blanca –see– *Sinapis alba* L.

mostaza campestre –see– *Sinapis arvensis* L.

mostaza silvestre –see– *Sinapis arvensis* L.

moth mullein –see– *Verbascum thapsus* L.

moth plant –see– *Araujia sericifera* Brot.

mother-in-law plant –see– *Caladium bicolor* (Aiton) Vent.; *Dieffenbachia seguine* (Jacq.) Schott

mother-in-law's-tongue –see– *Dieffenbachia seguine* (Jacq.) Schott; *Sansevieria trifasciata* Prain

mother-of-cocoa –see– *Gliricidia sepium* (Jacq.) Kunth ex Walp.

mother-of-millions –see– *Kalanchoe daigremontiana* Raym.-Hamet & H. Perrier; *Kalanchoe delagoensis* Eckl. & Zeyh.

mother-of-thousands –see– *Kalanchoe daigremontiana* Raym.-Hamet & H. Perrier

mottled spurge –see– *Euphorbia lactea* Haw.

moulmain teak –see– *Tectona grandis* L. f.

mountain alkanet –see– *Arnica montana* L.

mountain death camas –see– *Anticlea elegans* (Pursh) Rydb.

mountain fetterbush –see– *Pieris floribunda* (Pursh) Benth. & Hook. f.

mountain grape –see– *Berberis aquifolium* Pursh

mountain graperoot –see– *Berberis aquifolium* Pursh

mountain ivy –see– *Kalmia latifolia* L.

mountain juniper –see– *Juniperus communis* L.

mountain larkspur –see– *Delphinium bicolor* Nutt.; *Delphinium glaucum* S. Watson

mountain laurel –see– *Calia secundiflora* (Ortega) Yakovlev; *Kalmia latifolia* L.; *Leucothoe davisiae* Torr. ex A. Gray

mountain mahogany –see– *Cercocarpus montanus* Raf.

mountain manchineel –see– *Metopium toxiferum* (L.) Krug & Urb.; *Schinus terebinthifolius* Raddi

mountain orach –see– *Atriplex hortensis* L.

mountain pieris –see– *Pieris floribunda* (Pursh) Benth. & Hook. f.

mountain pink –see– *Zeltnera beyrichii* (Torr. & A. Gray) G. Mans.; *Zeltnera calycosa* (Buckley) G. Mans.

mountain rosebay –see– *Rhododendron catawbiense* Michx.

mountain silvery lupine –see– *Lupinus argenteus* Pursh var. rubricaulis (Greene) S. L. Welsh

mountain sneezeweed –see– *Hymenoxys hoopesii* (A. Gray) Bierner

mountain sorrel –see– *Rumex acetosella* L.

mountain tea –see– *Gaultheria procumbens* L.

mountain thermopsis –see– *Thermopsis montana* Nutt.

mountain tobacco –see– *Arnica montana* L.

mountain white oak –see– *Quercus douglasii* Hook. & Arn.

mouron-des-champs –see– *Anagallis arvensis* L.

mouse onion –see– *Drimia maritima* (L.) Stearn

moutarde blanche –see– *Sinapis alba* L.

moutarde-des-champs –see– *Sinapis arvensis* L.

moutarde jaune –see– *Sinapis alba* L.

moutarde noire –see– *Brassica nigra* (L.) W. D. J. Koch

movingui –see– *Distemonanthus benthamianus* Baill.

mu oil tree –see– *Vernicia montana* Lour.

mucuna –see– *Mucuna pruriens* (L.) DC.

MUCUNA PRURIENS (L.) DC. [Fabaceae]

Synonyms:

Dolichos pruriens L.; ***Stizolobium pruriens*** (L.) Medik.

Common Names:

buffalo bean; cadjust; cowage; cowhage; cowitch; elephant's-scratch wort; karara; kaunch; kiwach; kiwachi; kiwash; mucuna; pica pica; pois-a-gratter; stinging bean; velvet bean; vine gungo pea; voodoo bean

Citations:

Broadbent JL (1953) Observations on itching produced by cowhage, and on the part played by histamine as a mediator in itch sensation. Br J Pharmacol 8(3):263-270.

Fairbrothers D, Kirby E, Lester RM, et al. (1985) Mucuna pruriens-associated pruritus - New Jersey. MMWR Morb Mortal Wkly Rep 34(48):732-734.

Fairbrothers D, Kirby E, Lester RM, et al. (1986) Mucuna pruriens-associated pruritus - New Jersey. JAMA 255(3):313.

Harms RH, Simpson CF, Waldroup PW (1961) Influence of feeding various levels of velvet beans to chicks and laying hens. J Nutr 75(Sep):127-131.

Miller ER, Massengale ON, Barnes MA (1925) Some effects resulting from eating velvet beans. J Am Pharm Assoc 14:1113-1114.

Shelley WB, Arthur RP (1955) Studies on cowhage (Mucuna pruriens) and its pruritogenic proteinase, mucunain. Arch Dermatol 72(5):399-406.

Note:

Velvet bean is named *Mucuna pruriens* (L.) DC. var. *utilis* (Wall. ex Wight) Baker ex Burck in some publications.

mucunan –see– *Dioclea erecta* Hoehne; *Dioclea latifolia* Benth.

mudar –see– *Calotropis procera* (Aiton) W. T. Aiton

mugget –see– *Convallaria majalis* L.

muguet –see– *Convallaria majalis* L.

muguet-de-mai –see– *Convallaria majalis* L.

mugwort –see– *Artemisia absinthium* L.; *Artemisia vulgaris* L.; *Parthenium hysterophorus* L.

mulberry –see– *Morus nigra* L.

mulga fern –see– *Cheilanthes sieberi* Kunze; *Cheilanthes tenuifolia* (Burm. f.) Sw.

mullein –see– *Verbascum thapsus* L.

mullein pink –see– *Agrostemma githago* L.

murajes –see– *Anagallis arvensis* L.

murrainberry –see– *Bryonia dioica* Jacq.; *Dioscorea communis* (L.) Caddick & Wilkin

MUSA ×PARADISIACA L. [Musaceae]

Synonyms:

Musa ×sapientum L.

Common Names:

banana; Banane; plantain

Citations:

Cordero-Moreno R (1973) Etiologic factors in tropical eye diseases. Am J Ophthalmol 75(3):349-364.

McKinney B, Crawford MA (1965) Fibrosis in guinea-pig heart produced by plantain diet. Lancet 2(418):880-882.

Meynadier J, Meynadier JM, Guilhou JJ (1982) L'urticaire de contact chez l'atopique. A propos de deux observations. Ann Dermatol Venereol 109(10):871-874.

Tokuyama K, Takei K, Arakawa H, et al. (1997) Banana allergy in infants. Allergy 52(3):350-351.

Musa ×sapientum L = Musa ×paradisiaca L.

Muskatnuß –see– *Myristica fragrans* Houtt.

muskrat weed –see– *Cicuta douglasii* (DC.) J. M. Coult. & Rose; *Cicuta maculata* L.

musquash –see– *Cicuta maculata* L.; *Cicuta maculata* L. var. angustifolia Hook.

musquash poison –see– *Cicuta maculata* L.

musquashroot –see– *Cicuta douglasii* (DC.) J. M. Coult. & Rose; *Cicuta maculata* L.

mustard –see– *Brassica nigra* (L.) W. D. J. Koch; *Hirschfeldia incana* (L.) Lagr.- Foss.; *Sinapis alba* L.; *Sinapis arvensis* L.

mustard tree –see– *Nicotiana glauca* Graham

Mutterkraut –see– *Tanacetum parthenium* (L.) Sch. Bip.

myall –see– *Acacia pendula* A. Cunn. ex G. Donn

myoporum –see– *Myoporum laetum* G. Forst.

MYOPORUM ACUMINATUM R. Br. [Scrophulariaceae]

Common Names:

boobialla; strychnine bush; water bush

Citations:

Legg J, White CT (1941) Myoporum acuminatum (strychnine bush). A plant poisonous to stock. Queensland Agric J 56:124-125.

Legg J, White CT (1941) Myoporum acuminatum: A plant poisonous to stock. Aust Vet J 17:104-105.

Myoporum deserti A. Cunn. ex Benth. = Eremophila deserti (A. Cunn. ex Benth.) Chinnock

MYOPORUM INSULARE R. Br. [Scrophulariaceae]

Common Names:

boobyalla; Tasmanian ngaio

Citations:

Jerrett IV, Chinnock RJ (1983) Outbreaks of photosensitisation and deaths in cattle due to Myoporum aff. insulare R. Br. toxicity. Aust Vet J 60(6):183-186.

MYOPORUM LAETUM G. Forst. [Scrophulariaceae]

Common Names:

myoporum; ngaio

Citations:

Bonel-Raposo JB, Méndez MC, Andrade GB, et al. (1998) Experimental intoxication by Myoporum laetum in cattle. Vet Hum Toxicol 40(5):275-277.

Bonel-Raposo JB, Méndez MC, Riet-Correa F, et al. (1998) Experimental intoxication by Myoporum laetum in sheep. Vet Hum Toxicol 40(3):132-135.

Bonel-Raposo J, Driemeier D, Barros SS, et al. (2003) Evolução das lesões histológicas e ultra-estruturais no fígado de ovinos e bovinos experimentalmente intoxicados por Myoporum laetum. Pesq Vet Bras 23(4):149-155.

Denz FA, Hanger WG (1961) The liver toxin in Myoporum laetum. J Pathol Bact 81(Jan):91-99.

Hankin TH (1940) Pigs poisoned by eating ngaio leaves. N Z J Agr 61:360-361.

McCann PJ (1946) Ngaio poisoning of cattle. N Z J Agr 72:139-140.

Webster WM (1926) Two recent cases of plant poisoning among stock. N Z J Agr 33:102-105.

MYOPORUM TETRANDRUM (Labill.) Domin [Scrophulariaceae]

Common Names:

boobialla

Citations:

Allen JG, Seawright AA, Hrdlicka J (1978) The toxicity of Myoporum tetrandrum (Boobialla) and myoporaceous furanoid essential oils for ruminants. Aust Vet J 54(6):287-292.

myristica –see– *Myristica fragrans* Houtt.

MYRISTICA FRAGRANS Houtt. [Myristicaceae]

Common Names:

mace; Muskatnuß; myristica; nutmeg; nutmet; nux moshata

Citations:

Abernethy MK, Becker LB (1992) Acute nutmeg intoxication. Am J Emerg Med 10(5):429.

Ahmad A, Thompson HS (1975) Nutmeg mydriasis. JAMA 234(3):274.

Åkesson HO, Wålinder J (1965) Nutmeg intoxication. Lancet 1(Jun 12):1271-1272.

Bartlett BF (1911) Nutmeg poisoning. Br Med J. Aug 5:269.

Beck R (1914) Zwei Fälle von Muskatnussvergiftung. Munch Med Wochenschr 61(16):878.

Dale HH (1909) Note on nutmeg-poisoning. Proc R Soc Med Therap Pharm Sect 2 Part 1(4):69-73.

Demetriades AK, Wallman PD, McGuiness A, et al. (2005) Low cost, high risk: Accidental nutmeg intoxication. Emerg Med J 22(3):223-225.

Dinakar HS (1978) Nutmeg abuse. Am J Psychiatry 135(12):1571.

Dodge WT (1887) Nutmeg poisoning. Med Rec 32(Nov 12):624.

Faguet RA, Rowland KF (1978) "Spice cabinet" intoxication. Am J Psychiatry 135(7):860-861.

Forrester MB (2005) Nutmeg intoxication in Texas, 1998-2004. Hum Exp Toxicol 24(11):563-566.

Fras I, Friedman JJ (1969) Hallucinogenic effects of nutmeg in adolescent. N Y State J Med 69(Feb 1):463-465.

Green RC Jr (1959) Nutmeg poisoning. JAMA 171(10):1342-1344.

Green RC Jr (1959) Nutmeg poisoning. Va Med Monthly 86(Oct):586-590.

Hamilton J (1906) Nutmeg poisoning. Br Med J Oct 6:900.

Hentschel H, Greyer H, Stein U (2000) Ingestion of nutmeg (Myristica fragrans). J Toxicol Clin Toxicol 38(2):234.

Hruby K (2002) Poisoning from natural drugs rediscovered for recreational abuse. J Toxicol Clin Toxicol 40(3):276-277.

Jacobziner H, Raybin HW (1965) Poisonings due to mace (nutmeg), furniture polish, and lead. N Y State J Med 65(Sep 1):2270-2279.

Johnson J (1906) Nutmeg poisoning. Br Med J 2(Oct 13):984.

Löhrer F, Kaiser R (1999) Biogene Suchtmittel. Neue Konsumgewohnheiten bei jungen Abhängigen? Nervenarzt 70(11):1029-1033.

McCord JA, Jervey LP (1962) Nutmeg (myristicin) poisoning. J S C Med Assoc 58:436-439.

Mendelsohn G (1907) Zwei Fälle von Vergiftung mit Muskatnuß. Dtsch Med Wochenschr 33:2001.

Messiha FS, Zaki NN (1984) Behavioral and histological adverse reactions of nutmeg. Vet Hum Toxicol 26(5):410.

Painter JC, Shanor SP, Winek CL (1971) Nutmeg poisoning - A case report. Clin Toxicol 4(1):1-4.

Panayotopoulos DJ, Chisholm DD (1970) Hallucinogenic effect of nutmeg. Br Med J 1(5698):754.

Payne RB (1963) Nutmeg intoxication. N Engl J Med 269(1):36-38.

Pitter RA (1902) A case of nutmeg poisoning. Lancet 1(Apr 12):1035.

Pytte M, Rygnestad T (1998) Muskatnøtt - Mer enn et krydder. Tidsskr Nor Laegeforen 118(28):4346-4347.

Reekie JS (1909) Nutmeg poisoning. JAMA 52(Jan 2):62.
Sangalli BC, Chiang W (2000) Toxicology of nutmeg abuse. J Toxicol Clin Toxicol 38(6):671-678.
Sangalli BC, Enser B, De Tino M (1999) Recipe for nutmeg (Myristica fragrans) abuse. J Toxicol Clin Toxicol 37(5):618.
Simpson TG (1895) Case of poisoning by nutmeg. Lancet 1(Jan 19):150.
Smith SM (1902) Nutmeg poisoning. Lancet 1(Jun 21):1798.
Weiss G (1960) Hallucinogenic and narcotic-like effects of powdered myristica (nutmeg). Psychiatr Q 34(Apr):346-356.
Wilkinson AN (1906) Poisoning by nutmeg. Br Med J 1(Mar 3):539.
Wilkinson KD (1911) Nutmeg poisoning. Br Med J 1(Apr 29):993.

myrobalan –see– *Terminalia chebula* Retz.

MYROXYLON BALSAMUM (L.) Harms [Fabaceae]

Common Names:

balsam-of-tolu; tolu balsam

Citations:

Bjarnason B, Flosadóttir E, Fischer T (2000) Assessment of balsam of Peru patch tests. Contact Dermatitis 42(6):326-329.
Calnan CD (1975) Active sensitization to para and balsam of Peru. Contact Dermatitis 1(2):126-127.
Cummer CL (1927) Dermatitis produced by balsam of Peru. Arch Derm Syphilol 16:44-50.
Forsbeck M, Skog E (1977) Immediate reactions to patch tests with balsam of Peru. Contact Dermatitis 3(4):201-205.
Fregert S, Rorsman H (1963) Simultaneous hypersensitivity to balsam of pine and to balsam of Peru. Acta Derm Venereol 42:21-22.
Mitchell JC (1975) Patch testing with some components of balsam of Peru. Contact Dermatitis 1(6):391-392.
Tanaka S, Matsumoto Y, Dlova N, et al. (2004) Immediate contact reactions to fragrance mix constituents and Myroxylon pereirae resin. Contact Dermatitis 51(1):20-21.
Temesvári E, Soos G, Podányi B, et al. (1978) Contact urticaria provoked by balsam of Peru. Contact Dermatitis 4(2):65-68.
Trattner A, David M (2003) Patch testing with fine fragrances: Comparison with fragrance mix, balsam of Peru and a fragrance series. Contact Dermatitis 49(6):287-289.
Veien NK, Hattel T, Justesen O, et al. (1983) Oral challenge with balsam of Peru in patients with eczema: A preliminary study. Contact Dermatitis 9(1):75-76.
Veien NK, Hattel T, Justesen O, et al. (1985) Reduction of intake of balsams in patients sensitive to balsam of Peru. Contact Dermatitis 13(5):270-273.

Note:

Balsam-of-Peru is named *Myroxylon balsamum* (L.) Harms var. *pereirae* (Royle) Harms in some publications.

myrrh –see– *Commiphora myrrha* (Nees) Engl.
myrrh gum –see– *Commiphora myrrha* (Nees) Engl.
myrsnelle –see– *Equisetum palustre* L.
Myrtensumach –see– *Coriaria myrtifolia* L.
myrtle laurelcherry –see– *Prunus myrtifolia* (L.) Urb.
myrtle-leaf holly –see– *Ilex myrtifolia* Walter
myrtle spurge –see– *Euphorbia lathyris* L.
mysteria –see– *Colchicum autumnale* L.
mysterious plant –see– *Daphne mezereum* L.
mysterygrass –see– *Toxicoscordion venenosum* (S. Watson) Rydb.

N

nabo blanco –see– *Chenopodium album* L.

nabo-del-diablo –see– *Oenanthe crocata* L.

nabo silvestre –see– *Brassica rapa* L. subsp. campestris (L.) A. R. Clapham

naboom –see– *Euphorbia ingens* E. Mey. ex Boiss.

nai habarala –see– *Alocasia cucullata* (Lour.) G. Don

Naivashagrass –see– *Cynodon plectostachyus* (K. Schum.) Pilg.

naked lady –see– *Colchicum autumnale* L.; *Euphorbia tirucalli* L.

NAMA PARRYI A. Gray [Hydrophyllaceae]

Citations:

Saunders GF (1903) Poisonous effects of a California shrub. Plant World 6:245-246.

nanas –see– *Durio zibethinus* L.

Nancy –see– *Vaccaria hispanica* (Mill.) Rauschert

nandina –see– *Nandina domestica* Thunb.

NANDINA DOMESTICA Thunb. [Berberidaceae]

Common Names:

Chinese sacred bamboo; heavenly bamboo; nandina; sacred bamboo

Citations:

Bradley M, Neiman LJ, Burrows GE (1988) Seizures in a puppy. Vet Hum Toxicol 30(2):121.

nangai –see– *Canarium indicum* L.

nap-at-noon –see– *Ornithogalum umbellatum* L.

napiergrass –see– *Pennisetum purpureum* Schumach.

naranjillo –see– *Solanum bonariense* L.

naranjo amarga –see– *Citrus aurantium* L.

narcisco –see– *Nerium oleander* L.

narcisse –see– *Narcissus poeticus* L.

narcisse-des-poées –see– *Narcissus poeticus* L.

narcissus –see– *Narcissus poeticus* L.; *Narcissus pseudonarcissus* L.

NARCISSUS JONQUILLA L. [Amaryllidaceae]

Common Names:

jonquil

Citations:

Stryker GV (1936) Contact dermatitis caused by the jonquil (Narcissus jonquilla). J Ind Hyg Toxicol 18:462-465.

Vigneau CH, Tsao J, Chamaillard C, et al. (1982) Accidental absorption of daffodils (Narcissus jonquilla): Two common intoxications. Vet Hum Toxicol 24(Suppl):133-135.

NARCISSUS POETICUS L. [Amaryllidaceae]

Common Names:

daffodil; Dichternarzisse; narcisse; narcisse-des-poées; narcissus; Narzisse; poet's-daffodil; poet's-narcissus; Sternblume; white sun lily; whitsun lily

Citations:

Derbes VJ (1942) Occupational, seasonal hay fever and asthma due to narcissus bulbs. South Med J 35(10):912-913.

Jacobziner H, Raybin HW (1962) Poisonings due to mistaken identities of products. N Y State J Med 62:3460-3461.

van der Werff PJ (1959) Occupational diseases among workers in the bulb industries. Acta Allergol 14:338-355.

NARCISSUS PSEUDONARCISSUS L. [Amaryllidaceae]

Common Names:

averil; bell rose; bul rose; chalice flower; crow bells; daffidowndilly; daffodil; Gelbe Narzisse; Lent lily; Lent rose; narcissus; Narzisse; Osterglocke; paciencia; trumpet narcissus

Citations:

Bleumink E, Nater JP (1974) Contact dermatitis in a gardener caused by daffodils. Derm Beruf Umwelt 22(3):123-126.

Gonçalo S, Freitas JD, Sousa I (1987) Contact dermatitis and respiratory symptoms from Narcissus pseudonarcissus. Contact Dermatitis 16(10):115-116.

Hausen BM, Oestmann G (1988) Untersuchungen über die Häufigkeit berufsbedingter allergischer Hauterkrangungen auf einem Blumengroßmarkt. Derm Beruf Umwelt 36(4):117-124.

Litovitz TL, Fahey BA (1982) Please don't eat the daffodils. N Engl J Med 306(9):547.

Mrvos R, Krenzelok EP, Jacobsen TD (2001) Toxidromes associated with the most common plant ingestions. Vet Hum Toxicol 43(6):366-369.

Muñoz D, Urrutia I, Leanizbarrutia I, et al. (1989) Contact dermatitis from plants in a geriatric nurse. Contact Dermatitis 20(13):227-228.

Saxon-Buri S (2004) Daffodil toxicosis in an adult cat. Can Vet J 45(3):248-250.

Walsh D (1910) Investigation of a dermatitis amongst flower-pickers in the Scilly Islands, the so-called "lily rash." Br Med J 2(Sep 24):854-856.

Wilson T (1924) Poisoning caused by eating daffodil bulbs. Missouri Bot Garden Bull 12:52.

Wilson T (1924) The common daffodil (Narcissus pseudonarcissus) as a poison. Pharm J Pharmacist 112(Feb 9):141-142.

nardoo fern –see– *Marsilea drummondii* A. Braun

Narrawa bur –see– *Solanum cinereum* R. Br.

narrow dock –see– *Rumex crispus* L.

narrow-leaf butterweed –see– *Senecio spartioides* Torr. & A. Gray

narrow-leaf cottonbush –see– *Gomphocarpus fruticosa* (L.) W. T. Aiton

narrow-leaf everlasting pea –see– *Lathyrus sylvestris* L.

narrow-leaf laurel –see– *Kalmia angustifolia* L.

narrow-leaf lupin –see– *Lupinus angustifolius* L.

narrow-leaf lupine –see– *Lupinus angustifolius* L.

narrow-leaf marsh elder –see– *Iva angustifolia* Nutt. ex DC.

narrow-leaf milkweed –see– *Asclepias mexicana* Cav.

narrow-leaf sage –see– *Salvia reflexa* Hornem.

narrow-leaf sneezeweed –see– *Helenium autumnale* L.

narrow-leaf sumpweed –see– *Iva angustifolia* Nutt. ex DC.

narrow-leaf water parsnip –see– *Berula erecta* (Huds.) Coville

narrow-leaf whorled milkweed –see– *Asclepias mexicana* Cav.

NARTHECIUM ASIATICUM Maxim. [Nartheciaceae]

Citations:

Suzuki K, Kobayashi M, Ito A, et al. (1985) Narthecium asiaticum Maxim. poisoning of grazing cattle: Observations on spontaneous and experimental cases. Cornell Vet 75(2):348-365.

NARTHECIUM OSSIFRAGUM (L.) Huds. [Nartheciaceae]

Common Names:

bog asphodel

Citations:

Dishington IW, Laksesvela B (1976) Alveldsykens etiologi belyst ved BSP-test. Nord Vet Med 28(11):547-549.

Ender F (1955) Undersøkelser over alveldsykens etiologi. Nord Vet Med 7:329-377.

Flåøyen A (1991) A difference in susceptibility of two breeds of sheep to the "Alveld toxin." Vet Res Commun 15(6):455-457.

Flåøyen A, Binde M, Bratberg B, et al. (1995) Nephrotoxicity of Narthecium ossifragum in cattle in Norway. Vet Rec 137(11):259-263.

Flåøyen A, Bratberg B, Fjølstad M, et al. (1994) A pasture-related nephrotoxicosis of cattle in Norway: Clinical signs and pathological findings. In: Colegate SM, Dorling PR (eds.) Plant-associated toxins. Agricultural, phytochemical and ecological aspects. CABI. New York. pp. 557-560.

Flåøyen A, Bratberg B, Frøslie A, et al. (1995) Nephrotoxicity and hepatotoxicity in calves apparently caused by experimental feeding with Narthecium ossifragum. Vet Res Commun 19(1):63-73.

Flåøyen A, Bratberg B, Frøslie A, et al. (1997) Further studies on the presence, qualities and effects of the toxic principles from Narthecium ossifagum plants. Vet Res Commun 21(2):137-148.

Flåøyen A, Bratberg B, Grønstol H (1995) Nephrotoxicity in lambs apparently caused by experimental feeding with Narthecium ossifragum. Vet Res Commun 19(1):75-79.

Flåøyen A, Handeland K, Stuve G, et al. (1999) Experimental Narthecium ossifragum nephrotoxicity in cervids from Norway. J Wildl Dis 35(1):24-30.

Flåøyen A, Johansen J, Olsen J (1995) Narthecium ossifragum associated photosensitization in sheep in the Faroe Islands. Acta Vet Scand 36(2):277-278.

Laksesvela B, Dishington IW (1983) Bog asphodel (Narthecium ossifragum) as a cause of photosensitisation in lambs in Norway. Vet Rec 112(16):375-378.

Malone FE, Kennedy S, Reilly GA, et al. (1992) Bog asphodel (Narthecium ossifragum) poisoning in cattle. Vet Rec 131(5):100-103.

Malone FE, Kennedy S, Reilly GA, et al. (1998) Bog asphodel (Narthecium ossifragum) poisoning in cattle. In: Garland T, Barr AC (eds.) Toxic plants and other natural toxicants. CABI. New York. pp. 564-568.

Scheie E, Ryste EV, Flåøyen A (2003) Measurement of phylloeryththrin (phytoporthyrin) in plasma or serum and skin from sheep photosensitised after ingestion of Narthecium ossifragum. N Z Vet J 51(3):99-103.

Wisløff T, Flåøyen A, Ottesen N, et al. (2003) Narthecium ossifragum (L.) Huds. causes kidney damage in goats: Morphologic and functional effects. Vet Pathol 40(3):317-327.

Narzisse –see– *Narcissus poeticus* L.; *Narcissus pseudonarcissus* L.

NASA TRIPHYLLA (Juss.) Weigend [Loasaceae]

Synonyms:

Loasa vulcanica André

Citations:

Mitchell JC, Rook AJ (1979) Botanical dermatology. Plants and plant products injurious to the skin. Greengrass. Vancouver.

nasgaal –see– *Solanum nigrum* L.

NASSELLA TENUISSIMA (Trin.) Barkworth [Poaceae]

Synonyms:

Stipa tenacissima Trin.

Common names:

esparto grass

Citations:

Baz G, Hinojosa M, Quirce S, et al. (1999) Occupational asthma caused by esparto grass (Stipa tenacissima) fibers. Allergy 54(1):86-87.

nastergal –see– *Solanum nigrum* L.

nasturium –see– *Tropaeolum majus* L.

Nasturtium armoracia (L.) Fr. = Armoracia rusticana P. Gaertn. et al.

Natal camel thorn –see– *Acacia lasiopetala* Oliv.

Natal cherry –see– *Solanum pseudocapsicum* L.

Natal round yellow bean –see– *Phaseolus vulgaris* L.

Natal slangkop –see– *Drimia macrocentra* (Baker) Jessop

Natal yellow tulp –see– *Moraea pallida* (Baker) Goldblatt

native bryony –see– *Bryonopsis laciniosa* (L.) Naudin; *Diplocyclos palmatus* (L.) C. Jeffrey

native couchgrass –see– *Brachyachne convergens* (F. Muell.) Stapf

native fuchsia –see– *Eremophila latrobei* F. Muell.; *Eremophila maculata* (Ker Gawl.) F. Muell.

native indigo –see– *Indigofera australis* Willd.

native lily –see– *Diplarrhena moraea* Labill.

native loquat –see– *Rhodomyrtus macrocarpa* Benth.

native parsnip –see– *Trachymene glaucifolia* (F. Muell.) Benth.

native tobacco –see– *Nicotiana suaveolens* Lehm.

naughty-man's-cherry –see– *Atropa belladonna* L.

navet-de-diable –see– *Bryonia dioica* Jacq.

navy bean –see– *Phaseolus vulgaris* L.

nea –see– *Strophanthus gratus* (Wall. & Hook.) Baill.

Nebraska fern –see– *Conium maculatum* L.

needle bur –see– *Amaranthus spinosus* L.

needlepoint ivy –see– *Hedera helix* L.

neem –see– *Azadirachta indica* A. Juss.

neempathi –see– *Azadirachta indica* A. Juss.

negroés olive tree –see– *Terminalia chebula* Retz.

neguillon –see– *Agrostemma githago* L.

Nelson's-larkspur –see– *Delphinium nuttallianum* Pritz.

nenta –see– *Tylecodon wallichii* (Harv.) Toelken

NEOBASSIA PROCERIFLORA (F. Muell.) A. J. Scott [Chenopodiaceae]

Synonyms:

Threlkeldia proceriflora F. Muell.

Common Names:

soda bush

Citations:

Legg J, Francis WD (1939) Threlkeldia proceriflora: A plant poisonous to stock. Aust Vet J 15:168-171.

Mathams RH, Sutherland AK (1952) The oxalate content of some Queensland pasture plants. Queensland J Agric Sci 9:317-334.

NEONOTONIA WIGHTII (Wight & Arn.) J. A. Lackey [Chenopodiaceae]

Synonyms:

Glycine wightii (Wight & Arn.) Verdc.

Citations:

Shenk JS (1976) The meadow vole as an experimental animal. Lab Anim Sci 26(4):664-669.

NEORAUTANENIA CORIACEA C. A. Sm. [Fabaceae]

Citations:

Steyn DG (1962) The toxicity of (A) Neorautanenia ficifolia (Benth.) C. A. Smith and (B) Neorautanenia coriacea C. A. Smith. S Afr Med J 36(Nov 3):922-923.

NEORAUTANENIA FICIFOLIUS (Benth.) C. A. Sm. [Fabaceae]

Citations:

Steyn DG (1962) The toxicity of (A) Neorautanenia ficifolia (Benth.) C. A. Smith and (B) Neorautanenia coriacea C. A. Smith. S Afr Med J 36(Nov 3):922-923.

Nepaul aconite –see– *Aconitum ferox* Wall. ex Ser.

NEPETA CATARIA L. [Lamiaceae]

Common Names:

cataria; catmint; catnep; catnip

Citations:

Bailey RL (1970) Tropical plants prove fatal. Mod Vet Pract 51(6):42, 46.

Hatch RC (1972) Effect of drugs on catnip (Nepeta cataria)-induced pleasure behavior in cats. Am J Vet Res 33(1):143-155.

Jackson B, Reed A (1969) Catnip and alteration of consciousness. JAMA 207(7):1349-1350.

Massoco CO, Silva MR, Gorniak SL, et al. (1995) Behavioral effects of acute and long-term administration of catnip (Nepeta cataria) in mice. Vet Hum Toxicol 37(6):530-533.

Osterhoudt KC, Lee SK, Callahan JM, et al. (1997) Catnip and the alteration of human consciousness. Vet Hum Toxicol 39(6):373-375.

Nepeta hederacea (L.) Trev. = Glechoma hederacea L.

NEPHROLEPIS EXALTATA (L.) Schott [Nephrolepidaceae]

Citations:

Stoof TJ, Bruynzeel DP (1989) Contact allergy to Nephrolepsis ferns. Contact Dermatitis 20(8):234-235.

nerium –see– *Nerium oleander* L.

Nerium indicum Mill. = Nerium oleander L.

Nerium odorum Aiton = Nerium oleander L.

NERIUM OLEANDER L. [Apocynaceae]

Synonyms:

Nerium indicum Mill.; ***Nerium odorum*** Aiton

Common Names:

adelfa; Ceylon rose; deflah; espirradeira; kaner; laurel blanco; laurel colorado; laurel rosa; laurel rosado; laurel rose; laurier rose; Lorbeerose; lorier bol; narcisco; nerium; oleander; pink oleander; rosa laurel; rose laurel; rosebay; Rosenlorbeer; South Sea rose; yee tho

Citations:

Adam SE, Al-Yahya MA, Al-Farhan AH (2002) Toxicity of Nerium oleander and Rhazya stricta in Najdi sheep: Hematologic and clinicopathologic alterations. Am J Chin Med 30(2-3):255-262.

Alfonso HA, Sanchez LM, Merino N, et al. (1991) Intoxicacion por Nerium oleander en ocas. Rev Salud Anim 13:183-184.

Alfonso HA, Sanchez LM, Merino N, et al. (1994) Intoxication due to Nerium oleander in geese. Vet Hum Toxicol 36(1):47.

Anonymous (1971) Oleander poisoning in equines. J R Army Vet Corps 42:8-9.

Apted J (1983) Oleander dermatitis. Contact Dermatitis 9(4):321.

Arai M, Stauber E, Shropshire CM (1992) Evaluation of selected plants for their toxic effects on canaries. J Am Vet Med Assoc 200(9):1329-1331.

Arao T, Fuke C, Takaesu H, et al. (2002) Simultaneous determination of cardenolides by sonic spray ionization liquid chromatography-ion trap mass spectrometry - A fatal case of oleander poisoning. J Anal Toxicol 26(4):222-227.

Bárdosi Z (1963) Leanderlevél okozta mérgezések. Magyar Allator Lapja 18(9):361.

Bartell S, Anchor A (1996) Use of digoxin-specific antibody fragments in suicidal oleander toxicity. J Toxicol Clin Toxicol 34:600.

Benaissa L (1966) Hémolyse aiguë par inhalation prolongée de fleurs de lauriers-rose. Tunis Med 44(2):115-119.

Burton LE, Picchioni AL, Chin L (1965) Dipotassium edetate as an antidote in poisoning from oleander and its chief glycoside, oleandrin. Arch Int Pharmacodyn Ther 158(1):202-211.

Der Marderosian AH, Giller FB, Roia FC Jr (1976) Phytochemical and toxicological screening of household ornamental plants potentially toxic to humans. I. J Toxicol Environ Health 1(6):939-953.

Driggers DA, Solbrig R, Steiner JF, et al. (1989) Acute oleander poisoning. A suicide attempt in a geriatric patient. West J Med 151(Dec):660-662.

Duraković Z, Duraković A, Duraković S (1996) Oleander poisoning treated by resin haemoperfusion. J Indian Med Assoc 94(4):149-150.

Duraković Z, Plavšić F, Duraković S, et al. (1985) Otrovanje oleanderom. Arh Hig Rada Toksikol 36(4):387-393.

Eliakis CE, Eliakis EC, Coutselinis AS, et al. (1961) Discussion expérimentale sur la toxicité des feuilles du laurier-rose. Symtômes, altérations anatomopathologiques et méthodes d'identification de ses glucosides. Ann Med Leg Criminol Police Sci Toxicol 41(Jul-Aug):367-388.

Galey FD, Holstege DM, Johnson BJ, et al. (1998) Toxicity and diagnosis of oleander (Nerium oleander) poisoning in livestock. In: Garland T, Barr AC (eds.) Toxic plants and other natural toxicants. CABI. New York. pp. 215-219.

Galey FD, Holstege DM, Plumlee KH, et al. (1996) Diagnosis of oleander poisoning in livestock. J Vet Diagn Invest 8(3):358-364.

Graeme KA, Lo Vecchio FA, Selden BS, et al. (1998) Cardiotoxicity from ingestion of unprocessed Nerium oleander leaves treated with FAB fragments. J Toxicol Clin Toxicol 36(5):457.

Haynes BE, Bessen HA, Wightman WD (1985) Oleander tea: Herbal draught of death. Ann Emerg Med 14(4):350-353.

Hughes KJ, Dart AJ, Hodgson DR (2002) Suspected Nerium oleander (Oleander) poisoning in a horse. Aust Vet J 80(7):412-415.

Kaojarern S, Sukhupunyarak S, Mokkhavesa C (1986) Oleander yee tho poisoning. J Med Assoc Thai 69:108-112.

Khasigian P, Everson G, Bellinghausen R, et al. (1998) Poisoning following oleander smoke inhalation. J Toxicol Clin Toxicol 36(5):456-457.

Kirsch M (1997) Akute Glykosidintoxikation durch Aufnahme von Oleanderblättern (Nerium oleander) beim einem Meerschweinchen. Tierarztl Prax 25(4):398-400.

Krumholz B (1951) [Poisoning of cattle by oleander leaves.] Refu Vet 8:41-43,54-55.

Le Couteur DG, Fisher AA (2002) Chronic and criminal administration of Nerium oleander. J Toxicol Clin Toxicol 40(4):523-524.

Liu SK (1957) The pathology of oleander poisoning in cattle. Mem Coll Agric Taiwan Univ 5:75-82.

Livingston ML (1976) Case of oleander poisoning. Florida Vet J 6(2):18-19.

Maham M, Rezakhani A (1994) Treatment of oleander poisoning in cattle and donkeys. In: Colegate SM, Dorling PR (eds.) Plant-associated toxins. Agricultural, phytochemical and ecological aspects. CABI. New York. pp. 538-540.

Mahin L, Marzou A, Huart A (1984) A case report of Nerium oleander poisoning in cattle. Vet Hum Toxicol 26(4):303-304.

Marchán Carranza E, Gijón Rodríguez J, Márquez Moreno MD, et al. (1995) Intoxicación digitálica por hojas de Nerium oleander (adelfa). Rev Clin Esp 195(7):516.

Marri R (1937) Eine Vergiftung mit Nerium oleander. Sammlung Vergiftungsfallen 8(A702):177-180.

Marri R (1939) Oleanderblätter-Vergiftung. Sammlung Vergiftungsfallen 10(A808):145-148.

Mayer H, Wacker R, Dalchow W (1986) Phytotoxikosen durch Kastanien, Oleander, Eicheln und Herbstzeitlose bei verschiedenen Zoo- und Wildtieren. Tierarztl Umsch 41(3):169-178.

Meyer HP, van der Linden WJ, van der Linde-Sipman JS (1993) Een geval van oleanderintoxicatie bij de kat. Tijdschr Diergeneeskd 118(13):436-438.

Miller R (1973) Oleander poisoning in a two-toed sloth. J Zoo Anim Med 4(2):14.

Miller RM (1976) Oleander poisoning in a sloth. J Zoo Anim Med 7(4):11.

Minardi I, Lunelli IA, Capovilla LC, et al. (1982) Intoxicação acidental de bovinos por espirradeira (Nerium oleander L.). Rev Set Cienc Agrar 4:89-92.

Monzani V, Rovellini A, Schinco G, et al. (1997) Acute oleander poisoning after a self-prepared tisane. J Toxicol Clin Toxicol 35(6):667-668.

Nishioka SA, Resende ES (1995) Transitory complete atrioventricular block associated to ingestion of Nerium oleander. Rev Assoc Med Bras 41(1):60-62.

Oryan A, Maham M, Rezakhani A, et al. (1996) Morphological studies on experimental oleander poisoning in cattle. Zentralbl Veterinarmed A 43(10):625-634.

Osterloh J, Herold S, Pond S (1982) Oleander interference in the digoxin radioimmunoassay in a fatal ingestion. JAMA 247(11):1596-1597.
Pearn J (1987) Oleander poisoning. In: Covacevich et al. (eds.) Toxic plants and animals. A guide for Australia. pp. 37-49.
Ramón MF, Ballesteros S, Martínez-Arrieta R, et al. (2003) Plant and herb exposures in Spain. J Toxicol Clin Toxicol 41(4):514.
Ratigan WJ (1921-1922) Oleander poisoning in a bear. J Am Vet Med Assoc 60:96-97.
Reagor JC (1985) Increased oleander poisoning after extensive freezes in south/southeast Texas. Southwestern Vet 36(2):95.
Refsdal AO (1970) Forgitning hos hest med Nerium oleander. Norsk Veterinaertidsskrift 82(7-8):403-404.
Rezakhani A, Maham M (1992) Oleander poisoning in cattle of the Fars Province, Iran. Vet Hum Toxicol 34(6):549.
Rezakhani A, Maham M (1994) Cardiac manifestations of oleander poisoning in cattle and donkeys. In: Colegate SM, Dorling PR (eds.) Plant-associated toxins. Agricultural, phytochemical and ecological aspects. CABI. New York. pp. 534-537.
Romano GA, Mombelli G (1990) Intoxikation mit Oleanderblättern. Schweiz Med Wochenschr 120(16):596-597.
Safadi R, Levy I, Amitai Y, et al. (1995) Beneficial effect of digoxin-specific Fab antibody fragments in oleander intoxication. Arch Intern Med 155(19):2121-2125.
Schwartz WL, Bay WW, Dollahite JW, et al. (1974) Toxicity of Nerium oleander in the monkey (Cebus apella). Vet Pathol 11(3):259-277.
Shahidullah M, Sobhan MA, Rahman KA, et al. (1987) Study of haematological changes in acute Nerium indicum poisoning in guineapig (Clavia cobaya). Bangladesh Vet J 21(3-4):1-7.
Shlosberg A, Ohad DG, Bellaiche M, et al. (1998) Monitoring of physiological and pathological changes in turkey poults fed leaves of potentially cardiomyotoxic Nerium oleander and Persea americana. In: Garland T, Barr AC (eds.) Toxic plants and other natural toxicants. CABI. New York. pp. 131-136.
Shoshan A (1951) [Nerium oleander is a poisonous plant.] Refu Vet 8:43-44, 54.
Shropshire CM, Stauber E, Arai A (1992) Evaluation of selected plants for acute toxicosis in budgerigars. J Am Vet Med Assoc 200(7):936-939.
Shumaik GM, Wu AW, Ping AC (1988) Oleander poisoning: Treatment with digoxin-specific Fab antibody fragments. Ann Emerg Med 17(7):732-735.
Singh R, Siddiqui MA, Baruah MC (1978) Plant dermatitis in Delhi. Indian J Med Res 68(Oct):650-655.
Smith PA, Aldridge BM, Kittleson MD (2003) Oleander toxicosis in a donkey. J Vet Intern Med 17(1):111-114.
Soto-Blanco B, Fontenele-Neto JD, Silva DM, et al. (2006) Acute cattle intoxication from Nerium oleander pods. Trop Anim Health Prod 38(6):451-454.
Spevak L, Šoć M (1975) Dva slučaja trovanja čajem od oleanderovog lišća. Arh Hig Rada Toksikol 26(2):147-150.
Szabuniewicz M, Schwartz WL, McCrady JD, et al. (1972) Experimental oleander poisoning and treatment. Southwestern Vet 25(2):105-114.
Thimmiah K (1972) Nerium poisoning in cattle. Indian Vet J 49(9):942-946.
Tracqui A, Kintz P, Branche F, et al. (1998) Confirmation of oleander poisoning by HPLC/MS. Int J Legal Med 111(1):32-34.
Trautvetter E, Kasbohm C, Werner J (1969) Oleandervergiftung mit respiratorisch gekoppeltem AV-Block bei einem Hund. Berl Munch Tierarztl Wochenschr 82(16):306-308.
Vermunt J (1987) Oleander - Decorative and very poisonous. N Z Vet J 35(8):138-139.
Wilson FW (1909) Oleander poisoning in live stock. Arizona Agric Exp Sta Bull #59:381-397.

nerprun –see– *Rhamnus cathartica* L.

nerveroot –see– *Cypripedium reginae* Walter

Nessel –see– *Dendrocnide moroides* (Wedd.) Chew

nettle –see– *Cnidoscolus urens* (L.) Arthur; *Laportea canadensis* (L.) Wedd.; *Urtica chamaedryoides* Pursh; *Urtica dioica* L.

nettle tree –see– *Urtica ferox* G. Forst.

nettleberry –see– *Atropa belladonna* L.

Nevada goldenrod –see– *Solidago spectabilis* (D. C. Eaton) A. Gray

New Holland rattlepod –see– *Crotalaria novae-hollandiae* DC.

New York ironweed –see– *Vernonia noveboracensis* (L.) Michx.

New Zealand blue lupine –see– *Lupinus angustifolius* L.

New Zealand lupin –see– *Lupinus angustifolius* L.

New Zealand stinging nettle –see– *Urtica ferox* G. Forst.

New Zealand toot plant –see– *Coriaria ruscifolia* L.

ngaio –see– *Myoporum laetum* G. Forst.

ngaione –see– *Eremophila deserti* (A. Cunn. ex Benth.) Chinnock

ngodoncillo –see– *Gossypium hirsutum* L.

niang –see– *Archidendron jiringa* (Jack) I. C. Nielsen

niang yai –see– *Archidendron jiringa* (Jack) I. C. Nielsen

NICANDRA PHYSALODES (L.) Gaertn. [Solanaceae]

Common Names:

apple-of-Peru; Giftbeere; gooseberry; shoofly plant; wild gooseberry

Citations:

Cohen RD (1970) Bloat in sheep grazing wild gooseberry, Nicandra physaloides. Aust Vet J 46(11):559.

nicker tree –see– *Gymnocladus dioicus* (L.) K. Koch

NICOTIANA ATTENUATA Torr. ex S. Watson [Solanaceae]

Common Names:

coyote tobacco; wild tobacco

Citations:

Fleming CE, Miller MR, Vawter LR, et al. (1934) Poisonous plants. Nevada Agric Exp Sta Annu Rep 1933:10-13.

Marsh CD, Clawson AB, Roe GC (1927) Wild tobaccos (Nicotiana trigonophylla Dunal and Nicotiana attenuata Torrey) as stock-poisoning plants. U S Dep Agric Tech Bull #22:22 pp.

NICOTIANA GLAUCA Graham [Solanaceae]

Common Names:

buena moza; Don Juan; mustard tree; smooth tobacco; tobacco; tree tobacco; wild tobacco; wild tree tobacco; wildetabak

Citations:

Bunch TD, Panter KE, James LF (1990) Effects of certain poisonous plants on uterine function and fetal development in livestock. J Anim Sci 68(Suppl 1):406.

Castorena JL, Garriott JC, Barnhardt FE, et al. (1987) A fatal poisoning from Nicotiana glauca. J Toxicol Clin Toxicol 25(5):429-435.

Crowe MW, Keeler RF (1985) Comparative clinical and pathologic aspects of Nicotiana glauca and Nicotiana tabacum teratogenicity in swine. In: Seawright AA, Hegarty MP, James LF, et al. (eds.) Plant toxicology. Queensland Poisonous Plant Committee. Yeerongpilly, Queensland. pp. 317-323.

Curry S, Bond R, Kunkel D (1988) Acute nicotine poisonings after ingestions of tree tobacco. Vet Hum Toxicol 30(4):369.

Keeler RF (1979) Congenital defects in calves from maternal ingestion of Nicotiana glauca of high anabasine content. Clin Toxicol 15(4):417-426.

Keeler RF, Balls LD, Panter K (1981) Teratogenic effects of Nicotiana glauca and concentration of anabasine, the suspect teratogen in plant parts. Cornell Vet 71(1):47-53.

Keeler RF, Crowe MW (1983) Congenital deformities in swine induced by wild tree tobacco, Nicotiana glauca. J Toxicol Clin Toxicol 20(1):47-58.

Keeler RF, Crowe MW (1984) Teratogenicity and toxicity of wild tree tobacco, Nicotiana glauca in sheep. Cornell Vet 74(1):50-59.

Keeler RF, Shupe JL, Crowe MW, et al. (1981) Nicotiana glauca-induced congenital deformities in calves: Clinical and pathologic aspects. Am J Vet Res 42(7):1231-1234.

Manoguerra AS, Freeman D (1983) Acute poisoning from the ingestion of Nicotiana glauca. J Toxicol Clin Toxicol 19(8):861-864.

Mellick LB, Makowski T, Mellick GA, et al. (1999) Neuromuscular blockade after ingestion of tree tobacco (Nicotiana glauca). Ann Emerg Med 34(1):101-104.

Mizrachi N, Levy S, Goren Z (2000) Fatal poisoning from Nicotiana glauca leaves: Identification of anabasine by gas-chromatography/mass spectrometry. J Forensic Sci 45(3):736-741.

Panter KE, Bunch TD, Keeler RF, et al. (1990) Multiple congenital contractures (MCC) and cleft palate induced in goats by ingestion of piperidine alkaloid-containing plants: Reduction in fetal movements as the probable cause. J Toxicol Clin Toxicol 28(1):69-83.

Panter KE, James LF, Keeler RF, et al. (1992) Radio-ultrasound observations of poisonous plant-induced fetotoxicity in livestock. In: James LF, Keeler RF, Bailey JM Jr, et al. (eds.) Poisonous plants. Iowa State Univ Press. Ames, Iowa. pp. 481-488.

Panter KE, Keeler RF (1992) Induction of cleft palate in goats by Nicotiana glauca during a narrow gestational period and the relation to reduction in fetal movement. J Nat Toxins 1(1):25-32.

Panter KE, Keeler RF, Bunch TD, et al. (1990) Congenital skeletal malformations and cleft palate induced in goats by ingestion of Lupinus, Conium and Nicotiana species. Toxicon 28(12):1377-1386.

Panter KE, Weinzweig J, Gardner DR, et al. (2000) Comparison of cleft palate induction by Nicotiana glauca in goats and sheep. Teratology 61(3):203-210.

Plumlee KH, Holstege DM, Blanchard PC, et al. (1993) Nicotiana glauca toxicosis of cattle. J Vet Diagn Invest 5(3):498-499.

Steenkamp PA, van Heerden FR, van Wyk BE (2002) Accidental fatal poisoning by Nicotiana glauca: Identification of anabasine by high performance liquid chromatography/photodiode array/mass spectrometry. Forensic Sci Int 127(3):208-217.

NICOTIANA OBTUSIFOLIA M. Martens & Galeotti [Solanaceae]

Synonyms:

Nicotiana trigonophylla Dunal

Common Names:

desert tobacco; tabaquillo; wild tobacco

Citations:

Marsh CD, Clawson AB, Roe GC (1927) Wild tobaccos (Nicotiana trigonophylla Dunal and Nicotiana attenuata Torrey) as stock-poisoning plants. U S Dep Agric Tech Bull #22:22 pp.

NICOTIANA SUAVEOLENS Lehm. [Solanaceae]

Common Names:

native tobacco

Citations:

Seddon HR, McGrath TT (1933) Nicotiana suaveolens (native tobacco) proved poisonous to stock. New South Wales Dep Agric Vet Res Rep 6:119-121.

NICOTIANA TABACUM L. [Solanaceae]

Common Names:

burley tobacco; chupa; dohány; Gutka; pan masala; punche; shammah; tabac; tabaco; Tabak; tobacci; tobacco

Citations:

Arcury TA, Quandt SA, Garcia DI, et al. (2002) A clinic-based, case-control comparison of green tobacco sickness among minority farm workers: Clues for prevention. South Med J 95(9):1008-1011.

Babu S, Bhat RV, Kumar PU, et al. (1996) A comparative clinico-pathological study of oral submucous fibrosis in habitual chewers of pan masala and betelquid. J Toxicol Clin Toxicol 34(3):317-322.

Bele-Binda, Mohobo E (1975) Un cas d'intoxication aiguë au tabac par lavement. Ann Anesthesiol Fr 16(2):97-100.
Bleasdale R (1906) Tobacco poisoning in a child. Br Med J 1(May 19):1155-1156.
Borys DJ, Setzer SC, Ling LJ (1987) CNS depression in an infant after the ingestion of tobacco: A case report. Vet Hum Toxicol 29(6):477.
Bryden PA, McKnight RH, Spiller HA (2001) Pediatric nicotine exposures from ingestion of spittoon contents. J Toxicol Clin Toxicol 39(5):524.
Carl P, Crawford M, Ravlo O (1984) Tobaksforgiftning hos børn. Ugeskr Laeger 146(15):1160-1161.
Crowe MW (1969) Skeletal anomalies in pigs associated with tobacco. Mod Vet Pract 50(13):54-55.
Crowe MW (1972) A study of the teratogenic capability of tobacco (Nicotiana tobacum [tabacum]) and those chemicals commonly applied to the growing plant. Kentucky Univ Tobacco Health Res Inst Conf Rep 3:256-266.
Crowe MW (1973) A study of the teratogenic capability of tobacco (Nicotiana tobacum [tabacum]) and those chemicals commonly applied to the growing plant. Kentucky Univ Tobacco Health Res Inst Conf Rep 4:198-202.
Crowe MW (1978) Tobacco - a cause of congenital arthrogryposis. In: Keeler RF, Van Kampen KR, James LF, et al. (eds.) Effects of poisonous plants on livestock. Academic Press. New York. pp. 419-427.
Crowe MW, Keeler RF (1985) Comparative clinical and pathologic aspects of Nicotiana glauca and Nicotiana tabacum teratogenicity in swine. In: Seawright AA, Hegarty MP, James LF, et al. (eds.) Plant toxicology. Queensland Poisonous Plant Committee. Yeerongpilly, Queensland. pp. 317-323.
Crowe MW, Pike HT (1973) Congenital arthrogryposis associated with ingestion of tobacco stalks by pregnant sows. J Am Vet Med Assoc 162(6):453-455.
Crowe MW, Swerczek TW (1974) Congenital arthrogryposis in offspring of sows fed tobacco (Nicotiana tabacum). Am J Vet Res 35(8):1071-1073.
Dahl B, Caravati EM (2003) Intracerebral hemorrhage associated with ingestion of tobacco snuff. J Toxicol Clin Toxicol 41(5):673.
Edmonson WD (1996) Green tobacco sickness (bradycardia in a young farmer). J Tenn Med Assoc 89(3):85-86.
Garcia Estrada H, Fischman CM (1977) An unusual case of nicotine poisoning. Clin Toxicol 10(4):391-393.
Garvin EC (1913) A case of fatal poisoning by tobacco. Cleveland Med J 12:130.
Gehlbach SH, Perry LD, Williams WA, et al. (1975) Nicotine absorption by workers harvesting green tobacco. Lancet 1(Mar 1):478-480.
Gehlbach SH, Williams WA, Perry LD, et al. (1974) Green-tobacco sickness. An illness of tobacco harvesters. JAMA 229(14):1880-1883.
Ghosh SK, Gokani VN, Parikh JR, et al. (1987) Protection against "green symptoms" from tobacco in Indian harvesters: A preliminary intervention study. Arch Environ Health 41(2):121-124.
Ghosh SK, Parikh JR, Gokani VN, et al. (1980) Studies on occupational health problems in agricultural tobacco workers. J Soc Occup Med 30(3):113-117.
Johnson DW, Wilkins V (1996) Ingestion of tobacco products by children: A prospective study. J Toxicol Clin Toxicol 34:600-601.
Kaplan B (1968) Acute nicotine poisoning in a dog. Vet Med Small Anim Clin 63(Nov):1033-1034.
McGee D, Picciotti M, Spevack T (1991) Two year review of tobacco ingestions. Vet Hum Toxicol 33(4):370.
McKnight RH, Levine EJ, Rodgers GC Jr (1994) Detection of green tobacco sickness by a regional poison center. Vet Hum Toxicol 36(6):505-510.
McKnight RH, Spiller HA (2005) Green tobacco sickness in children and adolescents. Public Health Rep 120(6):602-605.
Menges RW, Selby LA, Marienfeld CJ, et al. (1970) A tobacco related epidemic of congenital limb deformities in swine. Environ Res 3(4):285-302.
Misumi J, Koyama W, Miura J (1983) [2 cases of "green tobacco disease" among tobacco harvesters and percutaneous nicotine absorption in rats.] Jpn J Indus Health 25(1):3-9.
Nakamura T (1984) Tobacco dermatitis in Japanese harvesters. Contact Dermatitis 10(5):310.
Oberst BB, McIntyre RA (1953) Acute nicotine poisoning. Case report. Pediatrics 11:338-340.
Ohyama K, Uehara T, Nohara T, et al. (1986) [Contact dermatitis due to tobacco leaves.] Nippon Hifuka Gakkai Zasshi 96(14):1677-1680.
Parikh JR, Gokani VN, Doctor PB, et al. (2005) Acute and chronic health effects due to green tobacco exposure in agricultural workers. Am J Ind Med 47(6):494-499.
Pecegueiro M (1988) Airborne contact dermatitis to tobacco. Contact Dermatitis 17(1):50-51.
Poniecka H (1990) Rośliny jako przyczyna alergii kontaktowej w materiale kliniki dermatologicznej am w białymstoku. Przegl Dermatol 77(4):262-265.
Ross MP, Revolinski D, Taurman L, et al. (1994) Detection of a peadiatric occupational poisoning in Kentucky. Vet Hum Toxicol 36(4):360.
Samitz MH, Mori P, Long CF (1939) Dermatological hazards in the cigar industry. Ind Med Surg 18(10):434-439.
Satpathy R, Das BB (1979) Accidental poisoning in childhood. J Indian Med Assoc 73(11):190-192.
Saxena K, Scheman A (1985) Suicide plan by nicotine poisoning: A review of nicotine toxicity. Vet Hum Toxicol 27(6):495-497.
Smolinske SC, Spoerke DG, Spiller SK, et al. (1988) Cigarette and nicotine chewing gum toxicity in children. Hum Toxicol 7:27-31.
Trapé-Cardoso M, Bracker A, Dauser D, et al. (2005) Cotinine levels and green tobacco sickness among shade tobacco workers. J Agromedicine 10(2):27-37.
Trapé-Cardoso M, Bracker A, Grey M, et al. (2003) Shade tobacco and green tobacco sickness in Connecticut. J Occup Environ Med 45(6):656-661.
Vero F, Genovese S (1941) Occupational dermatitis in cigar makers due to contact with tobacco leaves. Arch Derm Syphilol 43:257-263.
Vig MM (1990) Nicotine poisoning in a dog. Vet Hum Toxicol 32(6):573-575.
Weizenecker R, Deal WB (1970) Tobacco cropper's sickness. J Fla Med Assoc 57(12):13-14.
Willis HW (1937) Acute nicotine poisoning. Report of a case in a child. J Pediatr 10:65-68.

Nicotiana trigonophylla Dunal = Nicotiana obtusifolia M. Martens & Galeotti nicou –see– *Deguelia utilis* (A. C. Sm.) A. M. G. Azevedo

NIDORELLA FOETIDA (L.) DC. [Asteraceae]

Citations:

Schneider DJ, Green JR, Collett MG (1987) Ovine hepatogenous photosensitivity caused by the plant Nidorella foetida (Thunb.) DC. (Asteraceae). Onderstepoort J Vet Res 54(1):53-57.

nielle –see– *Agrostemma githago* L.

NIEREMBERGIA HIPPOMANICA Miers [Solanaceae]

Common Names:

chucho; chucho-de-la-sierra; chuschu

Citations:

Botha CJ, Schultz RA, van der Lugt JJ, et al. (1999) Neurotoxicity in calves induced by the plant, Nierembergia hippomanica Miers var. violacea Millan in South Africa. Onderstepoort J Vet Res 66(3):237-244.

NIEREMBERGIA VEITCHII Hook. [Solanaceae]

Citations:

Riet-Correa F, Schild AL, Méndez MC, et al. (1987) Enzootic calcinosis in sheep caused by the ingestion of Nierembergia veitchii (Solanaceae). Pesq Vet Bras 7(3):85-95.

Nieskraut –see– *Convallaria majalis* L.

Niespulver –see– *Veratrum album* L.

Nieswurz –see– *Helleborus niger* L.; *Helleborus viridis* L.; *Veratrum album* L.

NIGELLA SATIVA L. [Ranunculaceae]

Common Names:

black cumin; kaluduru

Citations:

Tennekoon KH, Jeevathayaparan S, Kurukulasooriya AP, et al. (1991) Possible hepatotoxicity of Nigella sativa seeds and Dregea volubilis leaves. J Ethnopharmacol 31:283-289.

Zedlitz S, Kaufmann R, Boehncke WH (2002) Allergic contact dermatitis from black cumin (Nigella sativa) oil-containing ointment. Contact Dermatitis 46(3):188.

night-blooming jasmine –see– *Cestrum aurantiacum* Lindl.; *Cestrum nocturnum* L.

night-blooming jessamine –see– *Cestrum nocturnum* L.

nightingales –see– *Arum maculatum* L.

nightshade –see– *Pentalinon luteum* (L.) B. F. Hansen & Wunderlin; *Solanum carolinense* L.; *Solanum dulcamara* L.; *Solanum glaucophyllum* Desf.; *Solanum nigrum* L.; *Solanum pseudocapsicum* L.; *Solanum triflorum* Nutt.

nim tree –see– *Azadirachta indica* A. Juss.

nimb –see– *Azadirachta indica* A. Juss.

nine-leaf indigo –see– *Indigofera linnaei* Ali

ningpo varnish –see– *Toxicodendron vernicifluum* (Stokes) F. A. Barkley

nilphulia –see– *Ageratum conyzoides* L.

nipple fruit –see– *Solanum pseudocapsicum* L.

Noah Holz –see– *Madhuca longifolia* (L.) J. F. Macbr.

nodding blue lily –see– *Stypandra glauca* R. Br.

nodding spurge –see– *Chamaesyce maculata* (L.) Small

noeniebossie –see– *Boscia foetida* Schinz

nogal –see– *Juglans regia* L.

nogal-de-la-India –see– *Aleurites moluccanus* (L.) Willd.

noisetier purgatif –see– *Jatropha multifida* L.

noix d'acajou –see– *Anacardium occidentale* L.

noix d'arec –see– *Areca catechu* L.

noix vomique –see– *Strychnos nux-vomica* L.

nole –see– *Semecarpus ater* (G. Forst.) Vieill.

NOLINA MICROCARPA S. Watson [Ruscaceae]

Common Names:

beargrass; sacahuista; sacahuiste; small-seed nolina

Citations:

Rankins DL Jr, Smith GS, Ross TT (1986) Rat study of beargrass (sacahuiste) toxicity. Proc West Sec Am Soc Anim Sci 37:224-226.

Rankins DL Jr, Smith GS, Ross TT, et al. (1988) Nolina microcarpa toxicosis in sheep. Proc West Sec Am Soc Anim Sci 39:218-221.

Rankins DL Jr, Smith GS, Ross TT, et al. (1993) Characterization of toxicosis in sheep dosed with blossoms of sacahuiste (Nolina microcarpa). J Anim Sci 71(9):2489-2498.

Samford MD, Ross TT, Edrington TS, et al. (1991) Toxicity of beargrass blossoms diminished during storage. Proc West Sec Am Soc Anim Sci 42:54-57.

Smith GS, Zornes ML, Schemnitz SD, et al. (1992) Toxicity of seeds of Nolina microcarpa (sacahuiste, beargrass) to chukar partridge (Alectoris chukar). Proc West Sec Am Soc Anim Sci 43:394-396.

NOLINA TEXANA S. Watson [Ruscaceae]

Common Names:

beargrass; bunchgrass; sacahuista; sacahuiste; Texas sacahuista; Texas sacahuiste

Citations:

Anonymous (1933) Investigation of stock-poisoning plants. U S Dep Agric Bur Anim Indus Chief Rep 1933:34-35.

Mathews FP (1932) Miscellaneous poisonous plants. Texas Agric Exp Sta Annu Rep 45:11-12.

Mathews FP (1938) Poisonous plants in the Davis Mountains area. Texas Agric Exp Sta Annu Rep 51:13-14.

Mathews FP (1940) Poisoning in sheep and goats by sacahuiste (Nolina texana) buds and blooms. Texas Agric Exp Sta Bull #585:5-19.

NOLLETIA GARIEPINA (DC.) Mattf. [Asteraceae]

Citations:

du Plessis EC, Joubert JP, Prozesky L, et al. (2004) Nephrotic syndrome in cattle caused by the shrub Nolletia gariepina in the Kalahari sandveld of southern Africa. In: Acamovic T, Stewart CS, Pennycott TW (eds.) Poisonous plants and related toxins. CABI. Wallingford, UK. pp. 76-78.

Meintjes RA, Botha CJ, Prozesky L (2005) Toxicity, pathophysiology and pathology in sheep following dosing of the nephrotoxic plant Nolletia gariepina (DC) Mattf. Onderstepoort J Vet Res 72(1):39-53.

NOLTEA AFRICANA (L.) Endl. [Rhamnaceae]

Common Names:

seepbossie

Citations:

Van der Walt SJ, Steyn DG (1939) Recent investigations into the toxicity of known and unknown poisonous plants in the Union of South Africa. IX. Onderstepoort J Vet Sci Anim Indus 12:335-366.

nonclimbing poison ivy –see– *Toxicodendron rydbergii* (Small ex Rydb.) Greene

nongue –see– *Datura inoxia* Mill.

noni –see– *Morinda citrifolia* L.

Noogoora bur –see– *Xanthium strumarium* L.; *Xanthium strumarium* L. var. glabratum (DC.) Cronquist

North American pennyroyal –see– *Hedeoma pulegioides* (L.) Pers.

northern grasstree –see– *Xanthorrhoea media* R. Br.

northern ironwood –see– *Erythrophleum chlorostachys* (F. Muell.) Baill.

northern poison ivy –see– *Toxicodendron rydbergii* (Small ex Rydb.) Greene

northern privet –see– *Ligustrum obtusifolium* Siebold & Zucc. subsp. suave (Kitag.) Kitag.

northern red oak –see– *Quercus rubra* L.

nosebleed –see– *Achillea millefolium* L.

novet –see– *Brassica rapa* L. subsp. campestris (L.) A. R. Clapham

Nubian senna –see– *Senna alexandrina* Mill.

nueza blanca –see– *Bryonia dioica* Jacq.

nueza negra –see– *Dioscorea communis* (L.) Caddick & Wilkin

nuez-de-India –see– *Aleurites moluccanus* (L.) Willd.

NUPHAR LUTEA (L.) Sm. [Nymphaeaceae]

Citations:

Airaksinen M, Peura P, Alaossi Salokangas L, et al. (1986) Toxicity of plant material used as emergency food during famines in Finland. J Ethnopharmacol 18(3):273-296.

nutmeg –see– *Myristica fragrans* Houtt.

nutmet –see– *Myristica fragrans* Houtt.

Nuttall's-death camas –see– *Toxicoscordion nuttallii* (A. Gray) Rydb.

nux moshata –see– *Myristica fragrans* Houtt.

nux vomica –see– *Strychnos nux-vomica* L.

nyári hérics –see– *Adonis aestivalis* L.

NYMPHAEA ODORATA Aiton subsp. tuberosa (Paine) Wiersema & Hellq. [Nymphaeaceae]

Synonyms:

Nymphaea tuberosa Paine

Common Names:

water lily; white water lily

Citations:

Hansen AA (1930) Indiana plants injurious to livestock. Purdue Agric Exp Sta Circ #175:38 pp.

Nymphaea tuberosa Paine = Nymphaea odorata Aiton subsp. tuberosa (Paine) Wiersema & Hellq.

O

oak-leaf poison ivy –see– *Toxicodendron diversilobum* (Torr. & A. Gray) Greene; *Toxicodendron pubescens* Mill.

oakbrush –see– *Quercus gambelii* Nutt.

oatgrass –see– *Trisetum flavescens* (L.) P. Beauv.

oats –see– *Avena sativa* L.

obeche –see– *Triplochiton scleroxylon* K. Schum.

obsession vine –see– *Convolvulus arvensis* L.

OCOTEA POROSA (Nees & Mart.) Barroso [Lauraceae]

Common Names:

Brazilian walnut

Citations:

Schwartz L (1931) Dermatitis venenata due to contact with Brazilian walnut wood. Public Health Rep 46(Aug 14):1938-1943.

octopus –see– *Aloe arborescens* Mill.

octopus tree –see– *Schefflera actinophylla* (Endl.) Harms

odallum –see– *Cerbera odollam* Gaertn.

odum –see– *Milicia excelsa* (Welw.) C. C. Berg

oduvan –see– *Cleistanthus collinus* (Roxb.) Hook. f.

oduvanthalai –see– *Cleistanthus collinus* (Roxb.) Hook. f.

oenanthe –see– *Oenanthe aquatica* (L.) Poir.; *Oenanthe crocata* L.

OENANTHE AQUATICA (L.) Poir. [Apiaceae]

Common Names:

fine-leaf water dropwort; horsebane; oenanthe; rizi mételykóró

Citations:

Bubien MZ, Magat A, Delatour P (1970) Fréquence des intoxications des animaux en Pologne. Bull Soc Vet Med Comp Lyon 72(2):167-173.

OENANTHE CROCATA L. [Apiaceae]

Common Names:

belder root; ben dock; dead-man's-fingers; dead tongue; giftige Rebendolde; hemlock dropwort; hemlock water dropwort; nabo-del-diablo; oenanthe; perreil; Safranrebendolde; water dropwort; water dropwort hemlock

Citations:

Anger JP, Anger F, Chauvel Y, et al. (1976) Intoxication mortellepar Oenanthe safranée (Oenanthe crocata). Eur J Toxicol Environ Hyg 9(2):119-125.

Ball MJ, Flather ML, Forfar JC (1987) Hemlock water dropwort poisoning. Postgrad Med J 63:363-365.

Boyle JG (1958) Water dropwort poisoning. State Vet J 13(Sep):17-18.

Downs C, Phillips J, Ranger A, et al. (2002) A hemlock water dropwort curry: A case of multiple poisoning. Emerg Med J 19(5):472-473.

Fitzgerald P, Moss N, O'Mahony S, et al. (1987) Accidental hemlock poisoning. Br Med J 295(6613):1657.

Guilhon J (1988) L'enseignement de la botanique et la mort subite au pré. Bull Acad Vet Fr 61(3):267-271.

King LA, Lewis MJ, Parry D, et al. (1985) Identification of oenanthotoxin and related compounds in hemlock water dropwort poisoning. Hum Toxicol 4(4):355-364.

Marras G (1965) Su di un caso di avvelenamento acuto mortale da «Oenanthe crocata». Minerva Medicoleg 85(3):83-85.

McGrath J (1937) Notes on a case of poisoning by Oenanthe crocata (water dropwort). Ir J Med Sci Jul:309-311.

Milne J (1945) Fatal poisoning in a horse by Oenanthe crocata. Vet Rec 57(3):30.

Mitchell MI, Routledge PA (1977) Poisoning by hemlock water dropwort. Lancet 1(8008):423-424.

Mitchell MI, Routledge PA (1978) Hemlock water dropwort poisoning: A review. Clin Toxicol 12(4):417-426.

Northall FS, Dauncey EA, Butler JM (2003) An overview of plant and fungal poisonings in the UK, and some interesting cases. J Toxicol Clin Toxicol 41(4):518-519.

O'Connor TM (1966) Treatment for poisoning by Oenanthe crocata. Vet Rec 79(5):157-158.

O'Mahony S, Fitzgerald P, Whelton MJ (1987) Poisoning by hemlock water dropwort. Ir J Med Sci 156(8):241.

Pallares JM, Saban J, Bouza C, et al. (1985) Reversible autonomic dysfunction in Oenanthe crocata poisoning evaluated by simple bedside tests. Hum Toxicol 4:521-526.

Swinscow D (1953) Accidental poisoning of young children. Arch Dis Child 28(137):26-29.

Thomas WE (1937) Water dropwort poisoning. Br Med J 1(May 15):1045-1046.

Wilson AL, Johnston WG, McCusker HB, et al. (1958) Hemlock water dropwort (Oenanthe crocata) poisoning in cattle. Vet Rec 70:587-590.

OENANTHE PALUSTRIS (Chiov.) C. Norman [Apiaceae]

Synonyms:

Stephanorossia palustris Chiov.

Common Names:

gonde

Citations:

Mugera GM (1970) Toxic and medicinal plants of East Africa. I. Bull Epizootic Dis Afr 18(4):377-387.

OENANTHE PIMPINELLOIDES L. [Apiaceae]

Common Names:

corky-fruit water dropwart

Citations:

Forsyth AA (1966) Treatment for poisoning by Oenanthe crocata. Vet Rec 79(2):55.

OENANTHE SILAIFOLIA M. Bieb. [Apiaceae]

Citations:

Aspiotis N, Lavrentiades G, Andreou C (1960) Etude pharmacodynamique de la plante toxique Oenanthe silaifolia. Bull Acad Vet Fr 33(1):75-80.

oficial-de-sala –see– *Asclepias curassavica* L.

Ohio buckeye –see– *Aesculus glabra* Willd.

okpu –see– *Blighia sapida* K. D. Koenig

okra –see– *Abelmoschus esculentus* (L.) Moench

ol dewo –see– *Sarcostemma viminale* (L.) R. Br.

ol meraa –see– *Catha edulis* (Vahl) Forssk. ex Endl.

ol odoa –see– *Maesa lanceolata* Forssk.

old maid –see– *Catharanthus roseus* (L.) G. Don

old maid's-pink –see– *Agrostemma githago* L.

old man –see– *Clematis vitalba* L.; *Marah oregonus* (Torr. & S. Watson) J. T. Howell

old-man-in-the-spring –see– *Senecio vulgaris* L.

old-man rosemary –see– *Rosmarinus officinalis* L.

old-man sagebrush –see– *Artemisia filifolia* Torr.

old-man's-beard –see– *Clematis microphylla* DC.; *Clematis vitalba* L.

old woman –see– *Artemisia absinthium* L.

OLEA EUROPAEA L. [Oleaceae]

Common Names:

olive

Citations:

Azofra J (2004) Olive allergy. Allergy 59(5):559.

Carnés Sánchez J, Iraola VM, Sastre J, et al. (2002) Allergenicity and immunochemical characterization of six varieties of Olea europaea. Allergy 57(4):313-318.

Hausen BM, Rothenborg HW (1981) Allergic contact dermatitis caused by olive wood jewelry. Arch Dermatol 117(11):732-734.

Pajarón MJ, Vila L, Prieto I, et al. (1997) Cross-reactivity of Olea europaea with other Oleaceae species in allergic rhinitis and bronchial asthma. Allergy 52(8):829-835.

Quiralte J, Florido F, Arias de Saavedra JM, et al. (2002) Olive allergen-specific IgE responses in patients with Olea europaea pollinosis. Allergy 57(Suppl 71):47-52.

van Joost T, Smitt JH, van Ketel WG (1981) Sensitization to olive oil (Olea europeae). Contact Dermatitis 7(6):309-310.

Williams J, Roberts H, Tate B (2007) Contact urticaria to olives. Contact Dermatitis 56(1):52-53.

Williams JD, Tate BJ (2006) Occupational allergic contact dermatitis from olive oil. Contact Dermatitis 55(4):251-252.

oleander –see– *Nerium oleander* L.; *Thevetia peruviana* (Pers.) K. Schum.

OLEARIA AXILLARIS F. Muell. [Asteraceae]

Citations:

Burry JN, Kuchel R, Reid JG, et al. (1973) Australian bush dermatitis: Compositae dermatitis in South Australia. Med J Aust 1(3):110-116.

oleaster –see– *Elaeagnus angustifolia* L.

olieblom –see– *Ricinus communis* L.

olieboon –see– *Datura stramonium* L.

olive –see– *Olea europaea* L.

olivo-de-Bohemia –see– *Elaeagnus angustifolia* L.

oloiuqui –see– *Turbina corymbosa* (L.) Raf.

ololiuqui –see– *Turbina corymbosa* (L.) Raf.

olomiojo –see– *Thevetia peruviana* (Pers.) K. Schum.

olon –see– *Fagara heitzii* Aubrév. & Pellegr.

onat spurge –see– *Euphorbia drummondii* Boiss.

one-eye bean –see– *Canavalia ensiformis* (L.) DC.

one-seed juniper –see– *Juniperus monosperma* (Engelm.) Sarg.

one-stem butterweed –see– *Senecio integerrimus* Nutt.

onehunga weed –see– *Soliva sessilis* Ruiz & Pav.

ongaonga –see– *Urtica ferox* G. Forst.

onion –see– *Allium cepa* L.

onion-de-mer –see– *Drimia maritima* (L.) Stearn

ONOCLEA SENSIBILIS L. [Dryopteridaceae]

Common Names:

meadow fern; sensitive fern

Citations:

Waller EF, Prince FS, Hodgdon AR, et al. (1944) Sensitive-fern poisoning of horses. New Hampshire Agric Sta Tech Bull #83:8 pp.

ONYCHIUM CONTIGUUM Wall. ex C. Hope [Pteridiaceae]

Common Names:

fuh

Citations:

Dawra RK, Sharma OP, Somvanshi R (2001) A preliminary study on the carcinogenicity of the common fern Onychium contiguum. Vet Res Commun 25(5):413-420.

oolong tea –see– *Camellia sinensis* (L.) Kuntze

OPHIOPOGON JAPONICUS (Thunb.) Ker Gawl. [Ruscaceae]

Common Names:

mai-meu-dong-tang; ophiopogonics

Citations:

Mochitomi Y, Inoue A, Kawabata H, et al. (1998) Stevens-Johnson syndrome caused by a health drink (Eberu) containing ophiopogonis tuber. J Dermatol 25(10):662-665.

ophiopogonics –see– *Ophiopogon japonicus* (Thunb.) Ker Gawl.

opium poppy –see– *Papaver somniferum* L.

OPLOPANAX ELATUS (Nakai) Nakai [Araliaceae]

Synonyms:

Echinopanax elatus Nakai

Citations:

Lopatin AI, Kolesnikova NP (1975) [On skin lesions caused by the use of Echinopanax elatum Nakai.] Vestn Dermatol Venerol 12:83-85.

OPUNTIA FICUS-INDICA (L.) Mill. [Cactaceae]

Common Names:

Indian fig; Indian fig cactus; prickly pear; sabbara; sabra

Citations:

Shanon J, Sagher F (1956) Sabra dermatitis. An occupational dermatitis due to prickly pear handling simulating scabies. Arch Dermatol 74(3):269-275.

Shanon Y, Sagher F (1965) Sabra dermatitis. Dermatite professionale causata da contatto con Fico d'India che assomiglia alla scabbia. Dermatol Int 4:125-127.

Yoon HJ, Won CH, Moon SE (2004) Allergic contact dermatitis due to Opuntia ficus-indica var. saboten. Contact Dermatitis 51(5-6):311-312.

OPUNTIA ROBUSTA H. L. Wendl. ex Pfeiff. [Cactaceae]

Citations:

Mitchell JC, Rook AJ (1979) Botanical Dermatology. Plants and plant products injurious to the skin. Greengrass. Vancouver.

orache –see– *Atriplex hortensis* L.; *Atriplex patula* L.

Orakelblume –see– *Leucanthemum vulgare* Lam.

orange –see– *Citrus sinensis* (L.) Osbeck

orange apocynum –see– *Asclepias tuberosa* L.

orange cestrum –see– *Cestrum aurantiacum* Lindl.

orange day lily –see– *Hemerocallis lilioasphodelus* L.

orange milkweed –see– *Asclepias tuberosa* L.

orange mint –see– *Mentha ×piperita* L. nothosubsp. citrata (Ehrh.) Briq.

orange sneezeweed –see– *Hymenoxys hoopesii* (A. Gray) Bierner

orange swallowwort –see– *Asclepias tuberosa* L.

orangeroot –see– *Asclepias tuberosa* L.

orashe –see– *Atriplex hortensis* L.

Orbignya phalerata Mart. = Attalea speciosa Mart. ex Spreng.

orchardgrass –see– *Dactylis glomerata* L.

ordeal bark –see– *Erythrophleum suaveolens* (Guill. & Perr.) Brenan

ordeal nut –see– *Strychnos nux-vomica* L.

ordeal tree –see– *Cerbera odollam* Gaertn.; *Erythrophleum africanum* (Welw. ex Benth.) Harms; *Erythrophleum suaveolens* (Guill. & Perr.) Brenan

oregano –see– *Origanum vulgare* L.

Oregon grape –see– *Berberis aquifolium* Pursh

Oregon holly grape –see– *Berberis aquifolium* Pursh

Oregon maple –see– *Acer macrophyllum* Pursh

Oregon oak –see– *Quercus garryana* Douglas ex Hook.

Oregon water hemlock –see– *Cicuta douglasii* (DC.) J. M. Coult. & Rose

Oregon white oak –see– *Quercus garryana* Douglas ex Hook.

oreille d'homme –see– *Asarum europaeum* L.

orelha-de-preto –see– *Enterolobium timbouva* Mart.

Orelia grandiflora Aubl. = Allamanda cathartica L.

oriental arbor vitae –see– *Platycladus orientalis* (L.) Franco

oriental cashew –see– *Semecarpus anacardium* L. f.

oriental ginseng –see– *Panax ginseng* C. A. Mey.

oriental hyacinth –see– *Hyacinthus orientalis* L.

oriental lacquer tree –see– *Toxicodendron vernicifluum* (Stokes) F. A. Barkley

oriental mustard –see– *Brassica juncea* (L.) Czern.

oriental poppy –see– *Papaver somniferum* L.

oriental sumach –see– *Toxicodendron succedaneum* (L.) Kuntze

oriental wax tree –see– *Toxicodendron succedaneum* (L.) Kuntze

ORIGANUM COMPACTUM Benth. [Lamiaceae]

Citations:

Verrept H, Schillemans L (1994) Ziektegedrag van Marokkaanse migranten met vakantie in Marokko. Ned Tijdschr Geneeskd 138(7):337-339.

ORIGANUM MAJORANA L. [Lamiaceae]

Common Names:

Majoran; marjoram; Meiran; Wurstkraut

Citations:

Farkas J (1981) Perioral dermatitis from marjoram, bay leaf and cinnamon. Contact Dermatitis 7(2):121.

ORIGANUM VULGARE L. [Lamiaceae]

Common Names:

oregano

Citations:

Benito M, Jorro G, Morales C, et al. (1996) Labiatae allergy: Systemic reactions due to oregano and thyme. Ann Allergy Asthma Immunol 76(5):416-418.

Ciganda C, Laborde A (2001) Herbal infusions used for induced abortion. J Toxicol Clin Toxicol 39(3):318-319.

oris root –see– *Iris ×germanica* L. nothovar. florentina Dykes

Orlaya platycarpos (L.) W. D. J. Koch = Caucalis platycarpos L.

ornamental French bean –see– *Phaseolus vulgaris* L.

ornamental pepper –see– *Capsicum annuum* L.

Ornithogalum caudatum Jacq. = Ornithogalum longebracteatum Jacq.

ORNITHOGALUM LONGEBRACTEATUM Jacq. [Hyacinthaceae]

Synonyms:

Ornithogalum caudatum Jacq.

Citations:

Mugera GM (1970) Toxic and medicinal plants of East Africa. II. Bull Epizootic Dis Afr 18(4):389-403.

Pohl RW (1965) Contact dermatitis from the juice of Ornithogalum caudatum. Toxicon 3(2):167-168.

ORNITHOGALUM MAGNUM (Baker) J. C. Manning & Goldblatt [Hyacinthaceae]

Synonyms:

Dipcadi glaucum (Burch. ex Ker Gawl.) Baker

Common Names:

malkop ui; maluie; poison onion; wild onion; wilde ui

Citations:

Steyn DG (1936) Recent investigations into the toxicity of known and unknown poisonous plants in the Union of South Africa. Onderstepoort J Vet Sci Anim Indus 7(1):169-178.

ORNITHOGALUM NANODES F. M. Leight. [Hyacinthaceae]

Citations:

Bamhare C (1998) Suspected cardiac glycoside intoxication in sheep and goats in Namibia due to Ornithogalum nanodes (Leighton). Onderstepoort J Vet Res 65(1):25-30.

ORNITHOGALUM THYRSOIDES Jacq. [Hyacinthaceae]

Common Names:

chinkerinchee; star-of-Bethlehem; viooltje; wonder flower; zuidewindlelie

Citations:

Allen JG, Creeper JH, Forshaw D, et al. (2004) Plant-associated diseases, either new or new to the state, encountered over the last decade (1991-2001) in Western Australia. In: Acamovic T, Stewart CS, Pennycott TW (eds.) Poisonous plants and related toxins. CABI. Wallingford, UK. pp. 540-547.

Breukink HJ (1963) Ornithogalosis of Zuidewindlelie - vergiftiging bij schapen. Tijdschr Diergeneeskd 88(10):653-658.

ORNITHOGALUM TOXICARIUM C. Archer & R. H. Archer [Hyacinthaceae]

Common Names:

chinkerinchee

Citations:

Botha CJ, Schultz RA, van der Lugt JJ, et al. (2000) A krimpsiekte-like syndrome in small stock poisoned by Ornithogalum toxicarium Archer & Archer. J S Afr Vet Assoc 71(1):6-9.

ORNITHOGALUM UMBELLATUM L. [Hyacinthaceae]

Common Names:

dome-d'onze-heures; leche-de-gallina; nap-at-noon; snowdrop; star flower; star-of-Bethlehem; summer snowflake

Citations:

Hansen AA (1924) The poison plant situation in Indiana. III. Poisonous trees. J Am Vet Med Assoc 66:351-362.

Hansen AA (1930) Indiana plants injurious to livestock. Purdue Agric Exp Sta Circ #175:38 pp.

Ornithoglossum glaucum Salisb. = Ornithoglossum vulgare B. Nord.

ORNITHOGLOSSUM VULGARE B. Nord. [Liliaceae]

Synonyms:

Ornithoglossum glaucum Salisb.

Common Names:

Cape slangkop

Citations:

Steyn DG (1933) Recent investigations into the toxicity of known and unknown poisonous plants in the Union of South Africa. Onderstepoort J Vet Sci Anim Indus 1:173-182.

OROBANCHE MINOR Sm. [Orobanchaceae]

Common Names:

beech drops; broom rape; cancerroot; clover broomrape; el halouk; lesser broomrape; zobb elkeh

Citations:

Kamel SH (1956) Étude chimique et toxicologique d'une plante égyptienne: Orobanche minor Sutton. Rev Elev Med Vet Pays Trop 9:43-48.

oromerillo –see– *Baccharis coridifolia* DC.

orris root –see– *Iris ×germanica* L. nothovar. florentina Dykes

ortiga –see– *Cnidoscolus urens* (L.) Arthur; *Wigandia urens* (Ruiz & Pav.) Kunth var. caracasana (Kunth) D. N. Gibson

ortiga brava –see– *Cnidoscolus urens* (L.) Arthur

ortiguilla –see– *Urtica chamaedryoides* Pursh

orvosi somkóró –see– *Melilotus officinalis* Lam.

ORYZA SATIVA L. [Poaceae]

Common Names:

Reis; rice

Citations:

Alderson HE, Rawlins AG (1925) Rice workers' dermatitis. Cal West Med 23(1):42-45.

González-Mendiola R, Martín-García C, Carnés J, et al. (2003) Asthma induced by the inhalation of vapours during the process of boiling rice. Allergy 58(11):1202 1203.

Joseph T (1968) Paddy grain injury of the eye. Br J Ophthalmol 52(2):191-197.

Nakamura T (1983) Contact dermatitis to Oryza. Contact Dermatitis 9(1):80.

Orhan F, Sekerel BE (2003) A case of isolated rice allergy. Allergy 58(5):456-457.

van den Hoogenband HM, van Ketel WG (1983) Allergy to rice. Contact Dermatitis 9(6):527-528.

Wüthrich B, Scheitlin T, Ballmer-Weber B (2002) Isolated allergy to rice. Allergy 57(3):263-264.

Yamakawa Y, Ohsuna H, Aihara M, et al. (2001) Contact urticaria from rice. Contact Dermatitis 44(2):91-93.

Yang KL, Tschou TK, T'ang CC, et al. (1965) A study on dermatitis in rice farmers. Chin Med J 84(3):143-160.

os gras –see– *Panicum schinzii* Hack.

Osage apple –see– *Maclura pomifera* (Raf.) C. K. Schneid.

Osage orange –see– *Maclura pomifera* (Raf.) C. K. Schneid.

oseille-de-bûcheron –see– *Oxalis acetosella* L.

oseille longue –see– *Rumex acetosa* L.

Osterglocke –see– *Narcissus pseudonarcissus* L.

Ostindisches Palisander –see– *Dalbergia latifolia* Roxb.

Ostindisches Satinholz –see– *Chloroxylon swietenia* DC.

Ostindisches Seidenholz –see– *Chloroxylon swietenia* DC.

ostrich fern –see– *Matteuccia struthiopteris* (L.) Tod.

ostromlecza paskowany –see– *Euphorbia marginata* Pursh

OSTRYA VIRGINIANA (Mill.) K. Koch [Betulaceae]

Common Names:

hophorn bean; lever wood

Citations:

Meeham J (1889) Poisonous plants. Garden Forest 2(57):154.

otaheite walnut –see– *Aleurites moluccanus* (L.) Willd.

Othonna cluytiifolia (DC.) Sch. Bip = Hertia cluytiifolia (DC.). Kuntze

Othonna pallens DC. = Hertia pallens (DC.) Kuntze

oto-de-lagarto –see– *Dieffenbachia seguine* (Jacq.) Schott

Ottilienkraut –see– *Consolida regalis* Gray

ouabaia –see– *Strophanthus gratus* (Wall. & Hook.) Baill.

ouabain –see– *Acokanthera oppositifolia* (Lam.) Codd; *Acokanthera schimperi* (A. DC.) Oliv.

oumiedos –see– *Boscia foetida* Schinz

our lady's-tears –see– *Convallaria majalis* L.

our lady's-thistle –see– *Silybum marianum* (L.) Gaertn.

oval-leaf privet –see– *Ligustrum ovalifolium* Hassk.

oven –see– *Afzelia africana* Sm. ex Pers.

overlook bean –see– *Canavalia ensiformis* (L.) DC.

ox heel –see– *Helleborus foetidus* L.

OXALIS ACETOSELLA L. [Oxalidaceae]

Synonyms:

Oxalis montana Raf.

Common Names:

acederilla; alleluia; hain Sauerklee; lady's-sorrel; madársóska; oseille-de-bûcheron; pain-de-coucou; sarelle; sauerblec; sourgrass; trébol acedo; trèfle aigrelet; wood sorrel

Citations:

Lamminpaa A, Kinos M (1996) Plant poisonings in children. Hum Exp Toxicol 15(3):245-249.

Volmer PA (2002) How dangerous are winter and spring holiday plants to pets? Vet Med (Dec):879-884.

Oxalis cernua Thunb. = Oxalis pes-caprae L.

Oxalis montana Raf. = Oxalis acetosella L.

OXALIS PES-CAPRAE L. [Oxalidaceae]

Synonyms:

Oxalis cernua Thunb.

Common Names:

Bermuda buttercup; Bermuda oxalis; sorrel; soursob; wood sorrel

Citations:

Bull LB (1929) Poisoning of sheep by soursobs (Oxalis cernua): Chronic oxalic acid poisoning. Aust Vet J 5(Jun):60-69.

Cosseddu AM (1967) Rilievi analitici sul latte prodotto da bovine alimentate con "Oxalis cernua Thunb." Agric Ital 67:3-11.

Di Domizio G (1962) Nefropatia ossalica in bovini alimentati con «Oxalis cernua Thumb». II. Riproduzione sperimentale dell'ossalopatia. Arch Vet Ital 13(2):97-126.

Lai P, Cosseddu AM (1967) Bovine alimentate con «Oxalis cernua»: Contenuto in acido ossalico del latte e stato sanitario dei vitelli allattati. Arch Vet Ital 18(3):171-189.

Manunta G, Floris B (1984) Ossalosi alimentare sperimentale degli ovini: Quadro elettroforetico e riserva alcalina ematici. Atti Soc Ital Sci Vet 38(Sep):224-227.
Manunta G, Naitana S (1984) Ossalosi alimentare degli ovini: Variazioni ormono-ematochimiche ed urinarie. Atti Soc Ital Sci Vet 38(Sep):221-224.
Manunta G, Naitana S, Floris B, et al. (1985) Ormone paratiroideo, tirocalciotonina, calcio, fosforo, magnesio, zinco, rame, fosfatasi alcalina, cloruri ematici e modificazioni ossee in ovini con ossalidosi sperimentale alimentare. Clin Vet (Milano) 108(3):185-206.
Manunta G, Naitana S, Floris B, et al. (1985) Quadro elettroforetico ematico, stato alcalotico, deposizione renale di sali di calcio e loro lisi in ovine alimentati con acetosella o con mais. Clin Vet (Milano) 108(3):207-218.

oxberry –see– *Dioscorea communis* (L.) Caddick & Wilkin

oxeye –see– *Helenium autumnale* L.

oxeye daisy –see– *Callilepis laureola* DC.; *Leucanthemum vulgare* Lam.

oxkiller –see– *Boophone disticha* (L. f.) Herb.

oxytenia –see– *Oxytenia acerosa* Nutt.

OXYTENIA ACEROSA Nutt. [Asteraceae]

Common Names:

copperweed; oxytenia

Citations:

Anonymous (1936) Investigations of stock-poisoning plants. U S Dep Agric Bur Anim Indus Chief Rep. 1936:44-45.
Anonymous (1937) Investigations of stock-poisoning plants. U S Dep Agric Bur Anim Indus Chief Rep. 1937:47-48.
Anonymous (1939) Investigations of stock-poisoning plants. U S Dep Agric Bur Anim Indus Chief Rep. 1939:59-60.
Schwartz L, Warren LH (1940) Dermatitis caused by contact with copperweed (Oxytenia acerosa). J Allergy 12:63-68.
Thorp F Jr, Durrell LW, Harshfield GS, et al. (1940) Oxytenia found to be poisonous to livestock. Colorado Agric Exp Sta Farm Bull #2(2):18-19.
Thorp F Jr, Harshfield GS, Durrell LW, et al. (1940) Oxytenia acerosa - A plant poisonous to livestock. J Am Vet Med Assoc 96(Jan):97.

Oxytropis albiflora (A. Nelson) K. Schum. = Oxytropis sericea Nutt.

OXYTROPIS CAMPESTRIS (L.) DC. var. spicata Hook. [Fabaceae]

Synonyms:

Aragallus spicatus (Hook.) Rydb.; ***Oxytropis campestris*** (L.) DC. var. gracilis (A. Nelson) Barneby

Common Names:

late yellow locoweed; white locoweed; woolly locoweed

Citations:

Harries WN, Baker FP, Johnston A (1972) Case report. An outbreak of locoweed poisoning in horses in southwestern Alberta. Can Vet J 13(6):141-145.

Oxytropis campestris (L.) DC. var. gracilis (A. Nelson) Barneby = Oxytropis campestris (L.) DC. var. spicata Hook.

OXYTROPIS LAMBERTII Pursh [Fabaceae]

Synonyms:

Aragallus lambertii (Pursh) Greene; ***Spiesia lambertii*** (Pursh) Kuntze

Common Names:

Colorado loco vetch; crazyweed; Lambert's-crazyweed; Lambert's-locoweed; Lambert's-point locoweed; locoweed; pink locoweed; point locoweed; purple locoweed; rattleweed; stemless locoweed; white locoweed

Citations:

Anonymous (1907) Studies of poisonous plants. U S Dep Agric Annu Rep 1906:46, 207-209.
Anonymous (1909) Poisonous-plant investigations. U S Dep Agric Annu Rep 1908:57-58, 306-308, 430.
Crawford AC (1908) Laboratory work on loco-weed investigations. U S Dep Agric Bur Plant Indus Bull #121(Part 3):39-40.
Klench JP (1888) Rattleweed or loco-disease. Am Vet Rev 12:395-402.
Marsh CD (1908) Results of loco-weed investigations in the field. U S Dep Agric Bur Plant Indus Bull #121(Part 3):37-38.
Marsh CD (1909) The locoweed disease of the plains. U S Dep Agric Bur Anim Indus Bull #112:130 pp.
Marsh CD (1919) The locoweed disease. U S Dep Agric Farmers Bull #1054:24 pp.
McEachran W (1889) The loco disease. Colorado Agric Exp Sta Annu Rep 1889:78-79.
Peters AT, Sturdevant LB (1908) Loco weed poisoning in horses. Nebraska Agric Exp Sta Annu Rep 21:74-107.
Stalker M (1886) The "Loco" plant and its effect on animals. U S Dep Agric Bur Anim Indus Annu Rep 3:271-276.

OXYTROPIS OCHROCEPHALA Bunge [Fabaceae]

Citations:

Cao GR, Duan DX, Li SJ, et al. (1990) [A study on the toxicity of Oxytropis ochrocephala in goats.] Acta Vet Zootech Sin 21(1):80-86.
Cao GR, Li SJ, Duan DX, et al. (1992) The toxic principle of Chinese locoweeds (Oxytropis and Astragalus): Toxicity in goats. In: James LF, Keeler RF, Bailey JM Jr, et al. (eds.) Poisonous plants. Iowa State Univ Press. Ames, Iowa. pp. 117-121.

OXYTROPIS PUBERULA Boriss. [Fabaceae]

Citations:

Moldagulov MA (1989) [Poisoning of animals by Oxytropis puberula, a plant of Kazakhstan.] Veterinariia Moscow 1:56-57.

OXYTROPIS SERICEA Nutt. [Fabaceae]

Synonyms:

Aragallus sericeus (Nutt. ex Torr. & A. Gray) Greene; ***Oxytropis albiflora*** (A. Nelson) K. Schum.

Common Names:
locoweed; point locoweed; rattleweed; silky crazyweed; white locoweed; white-point locoweed

Citations:

Bachman SE, Galyean ML, Smith GS, et al. (1992) Early aspects of locoweed toxicosis and evaluation of a mineral supplement or clinoptilolite as dietary treatments. J Anim Sci 70(10):3125-3132.
Dugarte-Stavanja M, Smith GS, Edrington TS, et al. (1997) Failure of dietary bentonite clay, Silent Herder mineral supplement, or parenteral Banamine to alleviate locoweed toxicosis in rats. J Anim Sci 75(7):1867-1875.
James LF, Hartley WJ, Nielsen D, et al. (1986) Locoweed (Oxytropis sericea) poisoning and congestive heart failure in cattle. J Am Vet Med Assoc 189(12):1549-1556.
James LF, Hartley WJ, Panter KE, et al. (1985) Relationship of locoweed to congestive right heart failure in cattle. In: Seawright AA, Hegarty MP, James LF, et al. (eds.) Plant toxicology. Queensland Poisonous Plant Committee, Yeerongpilly. Queensland. pp. 438-445.
James LF, Hartley WJ, Van Kampen KR, et al. (1983) Relationship between ingestion of the locoweed Oxytropis sericea and congestive right-sided heart failure in cattle. Am J Vet Res 44(2):254-259.
James LF, Nielsen DB, Panter KE (1983) Too much heart: Locoweed and heart failure in cattle. Utah Sci 44(4):118-119.
James LF, Panter KE, Broquist HP, et al. (1991) Swainsonine-induced high mountain disease in calves. Vet Hum Toxicol 33(3):217-219.
James LF, Shupe JL, Binns W, et al. (1967) Abortive and teratogenic effects of locoweed on sheep and cattle. Am J Vet Res 28(126):1379-1388.
James LF, Van Kampen KR, Johnson AE (1970) Physiopathologic changes in locoweed poisoning of livestock. Am J Vet Res 31(4):663-672.
Ortiz AR, Hallford DM, Galyean ML, et al. (1997) Effects of locoweed (Oxytropis sericea) on growth, reproduction, and serum hormone profile in young rams. J Anim Sci 75(12):3229-3234.
Panter KE, James LF, Nielson D, et al. (1988) The relationship of Oxytropis sericea (green and dry) and Astragalus lentiginosus with high mountain disease in cattle. Vet Hum Toxicol 30(4):318-323.
Panter KE, Ralphs MH, James LF, et al. (1999) Effects of locoweed (Oxytropis sericea) on reproduction in cows with a history of locoweed consumption. Vet Hum Toxicol 41(5):282-286.
Pfister JA, Astorga JB, Panter KE, et al. (2006) Maternal ingestion of locoweed. I. Effects on ewe-lamb bonding and behaviour. Small Ruminant Res 65:51-63.
Pfister J, Davidson T, Panter K, et al. (2006) Maternal ingestion of locoweed. III. Effects on lamb behaviour at birth. Small Ruminant Res 65(1-2):70-78.
Pfister JA, Stegelmeier BL, Cheney CD, et al. (1996) Operant analysis of chronic locoweed intoxication in sheep. J Anim Sci 74 (11): 2622-2632.
Pulsipher GD, Galyean ML, Hallford DM, et al. (1994) Effects of graded levels of bentonite on serum clinical profiles, metabolic hormones, and serum swainsonine concentrations in lambs fed locoweed (Oxytropis sericea). J Anim Sci 72(6):1561-1569.
Richards JB, Hallford DM, Duff GC, et al. (1999) Serum luteinizing hormone, testosterone, and thyroxine and growth responses of ram lambs fed locoweed (Oxytropis sericea) and treated with vitamin E/selenium. Theriogenology 52(6):1055-1066.
Stegelmeier BL, James LF, Gardner DR, et al. (2005) Locoweed (Oxytropis sericea)-induced lesions in mule deer (Odocoileius hemionus). Vet Pathol 42(5):566-578.
Stegelmeier BL, James LF, Hall JO, et al. (2001) Neurologic disease in range goats associated with Oxytropis sericea (locoweed) poisoning and water deprivation. Vet Hum Toxicol 43(5):302-304.
Stegelmeier BL, James LF, Panter KE, et al. (1995) Serum swainsonine concentration and α-mannosidase activity in cattle and sheep ingesting Oxytropis sericea and Astragalus lentiginosus (locoweeds). Am J Vet Res 56(2):149-154.
Stegelmeier BL, James LF, Panter KE, et al. (1999) Dose response of sheep poisoned with locoweed (Oxytropis sericea). J Vet Diagn Invest 11:448-456.
Stegelmeier BL, James LF, Panter KE, et al. (2004) The clinical and morphological changes of intermittent locoweed (Oxytropis sericea) poisoning in sheep. In: Acamovic T, Stewart CS, Pennycott TW (eds.) Poisonous plants and related toxins. CABI. Wallingford, UK. pp. 431-435.
Taylor JB, Strickland J, May T, et al. (2000) Effect of subacute swainsonine (locoweed; Oxytropis sericea) consumption or immunocompetence and serum constituents of sheep in a nutrient-restricted state. Vet Hum Toxicol 42(4):199-204.

Ozark milkweed –see– *Asclepias viridis* Walter

OZOTHAMNUS DIOSMIFOLIUS (Vent.) DC. [Asteraceae]

Synonyms:
Helichrysum diosmifolium (Vent.) Sweet

Common Names:
sage bush

Citations:

McMahon R, Freeman S (1986) Allergic contact dermatitis to Helichrysum diosmifolium. Australas J Dermatol 27(3):138-140.

P

pa-teou-seou –see– *Croton tiglium* L.; *Triadica sebifera* (L.) Small

Pachystigma pygmaeum (Schltr.) Robyns = Vangueria pygmae Schltr.

Pachystigma thamnus Robyns = Vangueria thamnus (Robyns) Lantz

PACHYRHIZUS EROSUS (L.) Urb. [Fabaceae]

Common Names:

chopsui potato; hikamas; jicama; yam bean

Citations:

Amarasingham RD, Lee H (1969) A review of poisoning cases examined by the Department of Chemistry, Malaysia, from 1963 to 1967. Med J Malaysia 23(3):220-227.

Hung YM, Hung SY, Olson KR, et al. (2007) Yam bean seed poisoning mimicking cyanide intoxication. Intern Med J 37(2):130-132.

Zhang YG, Huang GZ (1988) Poisoning by toxic plants in China. Report of 19 autopsy cases. Am J Forensic Med Pathol 9(4):313-319.

paciencia –see– *Narcissus pseudonarcissus* L.

Pacific Labrador tea –see– *Rhododendron ×columbianum* (Piper) Harmaja

Pacific onion –see– *Allium validum* S. Watson

Pacific poison oak –see– *Toxicodendron diversilobum* (Torr. & A. Gray) Greene

Pacific rhododendron –see– *Rhododendron macrophyllum* D. Don ex G. Don

Pacific yew –see– *Taxus brevifolia* Nutt.

packalacca –see– *Phytolacca dioica* L.

PACKERA GLABELLA (Poir.) C. Jeffrey [Asteraceae]

Synonyms:

Senecio glabellus Poir.

Common Names:

bitterweed; butter weed; cress-leaf groundsel; yellowtop

Citations:

Goeger DE, Cheeke PR, Ramsdell HS, et al. (1983) Comparison of the toxicities of Senecio jacobaea, Senecio vulgaris and Senecio glabellus in rats. Toxicol Lett 15(1):19-23.

PACKERA PLATTENSIS (Nutt.) W. A. Weber & Á. Löve [Asteraceae]

Synonyms:

Senecio plattensis Nutt.

Common Names:

Platte groundsel; prairie groundsel; prairie ragwort; squaw weed

Citations:

Lilley CW (1980) Pyrrolizidine alkaloid toxicosis Senecio plattensis. Equine Pract 2(2):6.

PACKERA SANGUISORBAE (DC.) C. Jeffrey [Asteraceae]

Citations:

Rosiles MR, Paasch LH (1982) Megalocitosis hepática en ovinos. Veterinaria (Mexico) 13:151-153.

paddy melon –see– *Cucumis myriocarpus* Naudin

padauk –see– *Pterocarpus dalbergioides* DC.; *Pterocarpus soyauxii* Taub.

pagoda tree –see– *Plumeria rubra* L.

pain-de-coucou –see– *Oxalis acetosella* L.

painted leaf –see– *Euphorbia heterophylla* L.

painted spurge –see– *Euphorbia heterophylla* L.

pak wan –see– *Melientha suavis* Pierre

pala –see– *Cryptostegia grandiflora* R. Br.

palas –see– *Butea monosperma* (Lam.) Taub.

palash –see– *Butea monosperma* (Lam.) Taub.

Palay rubber vine –see– *Cryptostegia grandiflora* R. Br.

pale day lily –see– *Hemerocallis lilioasphodelus* L.

pale larkspur –see– *Delphinium glaucum* S. Watson

pale laurel –see– *Kalmia microphylla* (Hook.) A. Heller

paleta-de-pintor –see– *Caladium bicolor* (Aiton) Vent.

PALICOUREA AENEOFUSCA (Müll. Arg.) Standl. [Rubiaceae]

Citations:

Tokarnia CH, Döbereiner J, Couceiro JE, et al. (1983) Intoxicação por Palicourea aeneofusca (Rubiaceae), a causa de "mortes súbitas" em bovinos na zona da mata de Pernambuco. Pesq Vet Bras 3(3):75-79.

PALICOUREA GRANDIFLORA (Kunth) Standl. [Rubiaceae]

Citations:

Döbereiner J, Tokarnia CH (1982) Intoxicação experimental por Palicourea grandiflora (Rubiaceae) em coelhos. Pesq Vet Bras 2(3):121-124.

PALICOUREA JURUANA Krause [Rubiaceae]

Citations:

Tokarnia CH, Döbereiner J (1982) Intoxicação experimental por Palicourea juruana (Rubiaceae) em bovinos e coelhos. Pesq Vet Bras 2(1):17-26.

PALICOUREA MARCGRAVII A. St.-Hil. [Rubiaceae]

Common Names:

erva-de-rato

Citations:

Barbosa JD, Oliveira CM, Tokarnia CH, et al. (2003) Comparação da sensibilidade de bovinos e búfalos à intoxicação por Palicourea marcgravii (Rubiaceae). Pesq Vet Bras 23(4):167-172.

Camargo W (1962) Uma nova "erva-de-rato" tóxico para bovinos Palicourea barbiflora (?), Rubiaceae, comparação com a Palicourea marcgravii var. pubescens e com a Psychotria officinalis, Rubiaceae. Arq Inst Biol (Sao Paulo) 29(1):1-11.

Costa MV, Pessoa JM, Nascimento EF, et al. (1984) Lesões em bovinos intoxicados pela Palicourea marcgravii St. Hill. Arq Bras Med Vet Zootecnia 36(5):571-580.

Döbereiner J, Tokarnia CH (1959) Intoxicação de bovinos pela "erva de rato" (Palicourea marcgravii St. Hil.) no Vale do Itapicuru, Maranhão. Arq Inst Biol Anim (Rio Janeiro) 2:83-91.

Górniak SL, Palermo-Neto J, Spinosa HS (1989) Effects of Palicourea marcgravii on laboratory rodents. J Ethnopharmacol 25(1989):221-226.

Gorniak SL, Palermo-Neto J, Spinosa HS (1993) Effect of CNS depressant drugs on acute intoxication from Palicourea marcgravii St Hill in rats. Vet Hum Toxicol 35(1):19-21.

Gorniak SL, Palermo-Neto J, Spinosa HS (1994) Effects of acetamide on experimentally-induced Palicourea marcgravii (St Hill) poisoning in rats. Vet Hum Toxicol 36(2):101-102.

Peixoto PV, Tokarnia CH, Döbereiner J, et al. (1987) Intoxicação experimental por Palicourea marcgravii (Rubiaceae) em coelhos. Pesq Vet Bras 7(4):117-129.

Tokarnia CH, Döbereiner J (1986) Intoxicação por Palicourea marcgravii (Rubiaceae) em bovinos no Brasil. Pesq Vet Bras 6(3):73-92.

Tokarnia CH, Peixoto PV, Döbereiner J (1986) Intoxicação experimental por Palicourea marcgravii (Rubiaceae) em ovinos. Pesq Vet Bras 6(4):121-131.

Tokarnia CH, Peixoto PV, Döbereiner J (1991) Intoxicação experimental por Palicourea marcgravii (Rubiaceae) em caprinos. Pesq Vet Bras 11(3-4):65-70.

Palicourea longipedunculata Gardner = Psychotria longipedunculata (Gardner) Müll. Arg.

Palisander –see– *Dalbergia latifolia* Roxb.

palissandre bresic –see– *Dalbergia nigra* (Vell.) Allemão ex Benth.

Palliser poison vetch –see– *Astragalus miser* Douglas ex Hook. var. serotinus (A. Gray ex J. G. Cooper) Barneby

palma christi –see– *Ricinus communis* L.

palma-de-Cristo –see– *Ricinus communis* L.

Palmer's-amaranth –see– *Amaranthus palmeri* S. Watson

Palmer's-pigweed –see– *Amaranthus palmeri* S. Watson

Palmkernen –see– *Cocos nucifera* L.

palmyra palm –see– *Borassus flabellifer* L.

palo-de-compadre –see– *Toxicodendron striatum* (Ruiz & Pav.) Kuntze

palo-de-la-cruz –see– *Plumeria rubra* L.

palo-de-rosa –see– *Metopium brownei* (Jacq.) Urb.

palo-de-San Pablo –see– *Wigandia urens* (Ruiz & Pav.) Kunth var. caracasana (Kunth) D. N. Gibson

palo-de-sarna –see– *Toxicodendron striatum* (Ruiz & Pav.) Kuntze

palo-de-viruela –see– *Toxicodendron striatum* (Ruiz & Pav.) Kuntze

palo negrito –see– *Karwinskia humboldtiana* (Schult.) Zucc.

palo santo –see– *Bulnesia sarmienti* Lorentz ex Griseb.

palo-santo-de-castilla –see– *Ipomoea carnea* Jacq.

palo sarno –see– *Toxicodendron striatum* (Ruiz & Pav.) Kuntze

palque –see– *Cestrum parqui* L'Her.

palu-de-cashupete –see– *Anacardium occidentale* L.

pampagrass –see– *Festuca argentina* (Speg.) Parodi

pan masala –see– *Nicotiana tabacum* L.

PANAX GINSENG C. A. Mey. [Araliaceae]

Common Names:

all heal; American ginseng; Asian ginseng; ginseng; imperial ginseng; Korean ginseng; Korean red ginseng; oriental ginseng

Citations:

Palmer M, Haller C, McKinney P, et al. (1999) Botanicals and other dietary supplements: Adverse events by age. J Toxicol Clin Toxicol 37(5):609.

Siegel RK (1979) Ginseng abuse syndrome. Problems with the panacea. JAMA 241(15):1614-1615.

Siegel RK (1980) Ginseng and high blood pressure. JAMA 243(1):32.

panch –see– *Lantana camara* L.

panga panga –see– *Millettia stuhlmannii* Taub.

panicgrass –see– *Panicum coloratum* L.; *Panicum schinzii* Hack.

panic millet –see– *Panicum miliaceum* L.

panicle amaranth –see– *Amaranthus cruentus* L.

panicle death camas –see– *Toxicoscordion paniculatum* (Nutt.) Rydb.

PANICUM COLORATUM L. [Poaceae]

Common Names:

coolahgrass; Kleingrass; panicgrass

Citations:

Bridges CH, Camp BJ, Livingston CW, et al. (1987) Kleingrass (Panicum coloratum L.) poisoning in sheep. Vet Pathol 24(6):525-531.

Cornick JL, Carter GK, Bridges CH (1988) Kleingrass-associated hepatotoxicosis in horses. J Am Vet Med Assoc 193(8):932-935.

Muchiri DJ, Bridges CH, Ueckert DN, et al. (1980) Photosensitization of sheep on kleingrass pasture. J Am Vet Med Assoc 177(Aug 15):353-354.

PANICUM COLORATUM L. var. makarikariense Gooss. [Poaceae]

Common names:

bambatsi grass

Citations:

Regnault TR (1990) Secondary photosensitisation of sheep grazing bambatsi grass (Panicum coloratum var makarikariense). Aust Vet J 67(11):419.

Panicum dactylon L. = Cynodon dactylon (L.) Pers.

PANICUM DICHOTOMIFLORUM Michx. [Poaceae]

Common Names:

fall panicgrass; fall panicum; smooth witchgrass

Citations:

Miles CO, Munday SC, Holland PT, et al. (1991) Identification of sapogenin glucuronide in the bile of sheep affected by Panicum dichotomiflorum toxicosis. N Z Vet J 39:150-152.

Miles CO, Wilkins AL, Munday SC, et al. (1992) Identification of the calcium salt of epismilagenin β-D-glucuronide in the bile crystals of sheep affected by Panicum dichotomiflorum and Panicum schinizii toxicoses. J Agric Food Chem 40:1606-1609.

Panicum laevifolium Hack. = Panicum schinzii Hack.

Panicum maximum Jacq. = Megathyrsus maximus (Jacq.) B. K. Simon & S. W. L. Jacobs

PANICUM MILIACEUM L. [Poaceae]

Common Names:

broom millet; broomcorn millet; brown corn; corn millet; French millet; hog millet; millet; panic millet; proso millet; Russian millet; sonwa millet; white millet

Citations:

Allen JG, Creeper JH, Forshaw D, et al. (2004) Plant-associated diseases, either new or new to the state, encountered over the last decade (1991-2001) in Western Australia. In: Acamovic T, Stewart CS, Pennycott TW (eds.) Poisonous plants and related toxins. CABI. Wallingford, UK. pp. 540-547.

Bajmócy E, Glávits R (1988) Növendék juhok zöles legeltetésével kapscolatos fotoszenzibilizációs megbetegedése. Magyar Allator Lapja 43(8):471-477.

Bohle B, Hirt W, Nachbargauer P, et al. (2003) Allergy to millet: Another risk for atopic bird keepers. Allergy 58(4):325-328.

Hinebauch TD (1893) Millet disease. J Comp Med Vet Arch 14:313-322.

Hinebauch TD (1894) Millet disease in horses. Vet J Ann Comp Pathol 38:1-10.

Olivieri J, Hauser C (1998) Anaphylaxis to millet. Allergy 53(1):109-110.

Stuck BA, Blum A, Klimek L, et al. (2001) Millet, a potentially life-threatening allergen. Allergy 56(4):350.

PANICUM SCHINZII Hack. [Poaceae]

Synonyms:

Panicum laevifolium Hack.

Common Names:

blousaad; os gras; panicgrass; sweetgrass

Citations:

Lancaster MJ, Vit I, Lyford RL (1991) Analysis of the crystals from sheep grazing Panicum schinzii (sweet grass). Aust Vet J 68(8):281.

Miles CO, Wilkins AL, Munday SC, et al. (1992) Identification of the calcium salt of epismilagenin β-D-glucuronide in the bile crystals of sheep affected by Panicum dichotomiflorum and Panicum schinizii toxicoses. J Agric Food Chem 40:1606-1609.

paño-de-holanda –see– *Euphorbia pulcherrima* Willd. ex Klotzsch

Pantoffeln –see– *Lotus corniculatus* L.

panu habarala –see– *Alocasia cucullata* (Lour.) G. Don

páo amarello –see– *Euxylophora paraensis* Huber

páo branco –see– *Zanthoxylum flavum* Vahl

pao d'arco –see– *Tabebuia impetiginosa* (Mart. ex DC.) Standl.

pao ferro –see– *Machaerium scleroxylon* Tul.

pao ferro rosewood –see– *Machaerium scleroxylon* Tul.

páo setim –see– *Euxylophora paraensis* Huber

paokeke –see– *Clitoria ternatea* L.

papa –see– *Solanum tuberosum* L.

papagallo –see– *Euphorbia pulcherrima* Willd. ex Klotzsch

PAPAVER NUDICAULE L. [Papaveraceae]

Common Names:

Arctic poppy; Iceland poppy

Citations:

Anonymous (1960) Poisoning with Iceland poppies (Papaver nudicaule). N Z Dep Agric Annu Rep 1960:190.

Malmanche ID (1970) Suspected Papaver nudicaule (Iceland poppy) poisoning in two horses. N Z Vet J 18(5):96-97.
McLennan GC (1929) Poisoning of sheep by ingestion of Iceland poppies (Papaver nudicaule). Aust Vet J 5(Sep):117.
McLennan GC (1930) Iceland poppy poisoning. Aust Vet J 6(Mar):40.
Terblanche M, Adelaar TF (1964) A note on the toxicity of Papaver nudicaule L. (Iceland poppy). J S Afr Vet Assoc 35(3):383-384.

PAPAVER RHOEAS L. [Papaveraceae]

Common Names:

African rose; blind-eyes; canker rose; cheesebowl; cock rose; cop rose; copper rose; coquelicot; corn poppy; corn rose; cornflower; Feldmohn; Feuerblume; field poppy; Flanders poppy; headache; Klappermohn; Klatschmohn; Klatschrose; pavot coq; red poppy; redweed

Citations:

Barbe ME (1882) Effects of the Papaver rhoeas (coquelicot) upon two cows. Am Vet Rev 5:513-514.
Trasbot ML (1888) Narcotism simulating typhoid fever - Produced by Papaver rhoeas. Am Vet Rev 12:83-84.

PAPAVER SOMNIFERUM L. [Papaveraceae]

Common Names:

balewort; carnation poppy; Joan Silverpin; opium poppy; oriental poppy; peony poppy; Saatmohn; Schlafmohn; white poppy

Citations:

Chao TC, Lo DS, Bloodworth BC (1992) Common poisons in Singapore - Past and present. Med Sci Law 32(2):139-147.
Dutta AK, Seth A, Goyal PK, et al. (1998) Poisoning in children: Indian scenario. Indian J Pediatr 65(3):365-370.
Ekins BR, Rubio EL, Carson RS, et al. (1986) Hmong opiate folk remedy toxicity in two infants. Vet Hum Toxicol 28(5):481.
Gupta S, Govil YC, Misra PK, et al. (1998) Trends in poisoning in children: Experience at a large referral teaching hospital. Natl Med J India 11(4):166-168.
Lagneau F, Gallard P (1946) Intoxication de Bovins par d'œillette. Rec Med Vet Ec Alfort 122:310-313.
Lézy (1946) Intoxication de Bovins par des capsules d'œillette. Rec Med Vet Ec Alfort 122:23-24.
Odendaal JS (1986) Verdagte opiumvergiftiging in twee jong honde. J S Afr Vet Assoc 57(Jun):113-114.
Rubio EL, Ekins BR, Singh PD, et al. (1987) Hmong opiate folk remedy toxicity in three infants. Vet Hum Toxicol 29(4):323-325.

papaw –see– *Asimina triloba* (L.) Dunal; *Carica papaya* L.

papaya –see– *Carica papaya* L.

papayo –see– *Metopium toxiferum* (L.) Krug & Urb.

paper flower –see– *Baileya multiradiata* Harv. & A. Gray ex Torr.; *Psilostrophe gnaphalodes* DC.; *Psilostrophe tagetina* (Nutt.) Greene

PAPHIOPEDILUM HAYNALDIANUM (Rchb. f.) Stein [Orchidaceae]

Common Names:

lady's-slipper

Citations:

Hausen BM (1980) Allergic contact dermatitis to quinones in Paphiopedilum haynaldianum (Orchidaceae). Arch Dermatol 116(3):327-328.

papooseroot –see– *Caulophyllum thalictroides* (L.) Michx.

paprika –see– *Capsicum annuum* L.

para nut –see– *Bertholletia excelsa* Bonpl.

paragrass –see– *Urochloa mutica* (Forssk.) T. Q. Nguyen

PARADERRIS ELLIPTICA (Wall.) Adema [Fabaceae]

Synonyms:

Derris elliptica (Wall.) Benth.

Common Names:

derris; tuba root

Citations:

Ambrose AM, Haag HB (1938) Toxicological studies of derris. Indus Engineer Chem 30(5):592-595.
Dorne M, Friedman TB (1940) Derris root dermatitis. JAMA 115(Oct 12):1268-1270.
Haag HB, Taliaferro I (1940) Toxicologic studies on cubé. J Pharmacol Exp Ther 69:13-20.
Kulakkattolickal AT (1987) Piscicidal plants of Nepal: Preliminary toxicity screening using grass carp (Ctenopharyngodon idella) fingerlings. J Ethnopharmacol 21(1):1-9.
Rickard PP, Cox PA (1986) Use of derris as a fish poison in Guadalcanal, Solomon Islands. Econ Bot 40(4):479-484.

paradise poinciana –see– *Caesalpinia gilliesii* (Hook.) D. Dietr.

paradise tree –see– *Ailanthus altissima* (Mill.) Swingle

paradoxagrass –see– *Phalaris paradoxa* L.

Paraguayan jasmin –see– *Brunfelsia australis* Benth.

Paraguayan tea –see– *Ilex paraguariensis* A. St.-Hil.

paraíso –see– *Melia azedarach* L.; *Moringa oleifera* Lam.

paraíso blanco –see– *Moringa oleifera* Lam.

paraíso-de-España –see– *Moringa oleifera* Lam.

paraíso frances –see– *Moringa oleifera* Lam.

paraíso vegetal –see– *Melia azedarach* L.

paraist estranjaro –see– *Moringa oleifera* Lam.

parietaria –see– *Parietaria judaica* L.

PARIETARIA JUDAICA L. [Urticaceae]

Common Names:

parietaria; pellitory

Citations:

Andri L, Senna GE, Betteli C, et al. (1992) Local nasal immunotherapy in allergic rhinitis to Parietaria. A double-blind study. Allergy 47(4 Pt 1):318-323.
Bass D, Baldo BA (1984) Parietaria as a cause of asthma. Med J Aust 114:511.
Costa MA, Duro G, Izzo V, et al. (2000) The IgE-binding epitopes of rPar j 2, a major allergen of Parietaria judaica pollen, are heterogeneously recognized among allergic subjects. Allergy 55(3):246-250.
Kaufman HS (1990) Parietaria: An unrecognized cause of respiratory allergy in the United States. Ann Allergy 64(3):293-296.
Ortolani C, Pastorello EA, Incorvaia C, et al. (1994) A double-blind, placebo-controlled study of immunotherapy with an alginate-conjugated extract of Parietaria judaica in patients with Parietaria hay fever. Allergy 49(1):13-21.
Pajano GB, Passalacqua G, Vita D, et al. (2004) Sublingual immunotherapy abrogates seasonal bronchial hyperresponsiveness in children with Parietaria-induced respiratory allergy: A randomized controlled trial. Allergy 59(8):883-887.
Purello-D'Ambrosio F, Gangemi S, Isola S, et al. (1999) Sublingual immunotherapy: A double-blind, placebo-controlled trial with Parietaria judaica extract standardized in mass units in patients with rhinoconjunctivitis, asthma, or both. Allergy 54(9):968-973.

parlor ivy –see– *Philodendron hederaceum* (Jacq.) Schott

parlour primula –see– *Primula obconica* Hance

paroba d'água –see– *Sessea brasiliensis* Toledo

parrot flower –see– *Sesbania grandiflora* (L.) Pers.

Parry's-aster –see– *Xylorhiza glabriuscula* Nutt.

parsley –see– *Petroselinum crispum* (Mill.) Nyman ex A. W. Hill

parsnip –see– *Pastinaca sativa* L.; *Trachymene ochracea* L. A. S. Johnson

parson-in-the-pulpit –see– *Arisaema triphyllum* (L.) Schott

parthenium –see– *Parthenium hysterophorus* L.

PARTHENIUM ARGENTATUM A. Gray [Asteraceae]

Common Names:

guayule; Mexican rubber plant; rubber plant

Citations:

Smith LM, Hughes RP (1938) Dermatitis caused by Mexican rubber plant. Arch Derm Syphilol 38:780.

PARTHENIUM HYSTEROPHORUS L. [Asteraceae]

Common Names:

bastard feverfew; carrot weed; carrotgrass; cicutilla; congress weed; congressgrass; dog-flea weed; false ragweed; false wormwood; feverfew; Mexican bird seed; mugwort; parthenium; poundcake bush; Santa Maria feverfew; scourge-of-India; white broomweed; whitehead; wild feverfew; wild wormwood

Citations:

Ahmed MN, Rao PR, Mahendar M, et al. (1988) Experimental introduction of acute toxicity in buffalo calves by feeding Parthenium hysterophorus Linn. Indian J Anim Sci 58(6):731-734.
Ahmed NN, Rao PR, Mahendar M, et al. (1988) A study on changes in blood chemistry in Parthenium toxicity in buffalo calves. Cheiron 17(2):57-60.
Bhutani LK, Rao DS (1978) Photocontact dermatitis caused by Parthenium hysterophorus. Dermatologica 157(4):206-209.
Dutta RK, Babu BV (1985) A study of parthenium dermatitis. Indian J Dermatol 30(1):1-6.
Fisher AA (1979) Contact dermatitis due to American parthenium weed in India. Cutis 23(1):20-21, 32, 36.
French SW (1930) A case of skin sensitivity to Parthenium hysterophorus. Mil Surg 66:673-676.
Handa S, Sahoo B, Sharma VK (2001) Oral hyposensitization in patients with contact dermatitis from Parthenium hysterophorus. Contact Dermatitis 44(5):279-282.
Kahn IS, Grothaus EM (1936) Parthenium hysterophorus: Antigenic properties, respiratory and cutaneous. Tex State J Med 32(Aug):284-288.
Mahajan VK, Sharma NL, Sharma RC (2004) Parthenium dermatitis: Is it a systemic contact dermatitis or an airborne contact dermatitis? Contact Dermatitis 51(5-6):231-234.
More PR, Vadlamudi VP, Qureshi MI (1982) Note on the toxicity of Parthenium hysterophorus in livestock. Indian J Anim Res 52(6):456-457.
More PR, Vadlamudi VP, Qureshi MI (1982) Toxicity of Parthenium hysterophorus Linn. in bovines: Changes in some biochemical constituents of blood. Indian Vet J 59(7):515-517.
Narasimhan TR, Ananth M, Swamy MN, et al. (1977) Toxicity of Parthenium hysterophorus L. to cattle and buffaloes. Experientia 33(10):1358-1359.
Narasimhan TR, Ananth M, Swamy MN, et al. (1980) Toxicity of Parthenium hysterophorus: Partheniosis in cattle and buffaloes. Indian J Anim Sci 50(2):173-178.
Ogden HD (1957) Diagnosis and treatment of Parthenium dermatitis. J La State Med Soc 109(10):378-379.
Rao PV, Mangala A, Rao BS, et al. (1977) Clinical and immunological studies on persons exposed to Parthenium hysterophorus L. Experientia 33(10):1387-1388.
Sahoo B, Handa S, Kumar B (2001) Eczema herpeticum in parthenium dermatitis. Contact Dermatitis 44(2):106-107.
Sharma VK, Sethuraman G, Bhat R (2005) Evolution of clinical pattern of parthenium dermatitis: A study of 74 cases. Contact Dermatitis 53(2):84-88.
Sharma VK, Sethuraman G, Tejasvi T (2004) Comparison of patch test contact sensitivity to acetone and aqueous extracts of Parthenium hysterophorus in patients with airborne contact dermatitis. Contact Dermatitis 50(4):230-232.
Siddiqui MA, Singh R, Sharma RC (1978) Contact dermatitis due to Parthenium hysterophorus. Indian J Med Res 68(Sep):481-484.

Singh KK, Srinivas CR, Balachandran C, et al. (1987) Parthenium dermatitis sparing vitiliginous skin. Contact Dermatitis 16(9):174.
Singh R, Siddiqui MA, Baruah MC (1978) Plant dermatitis in Delhi. Indian J Med Res 68(Oct):650-655.
Srinivas CR, Balachandran C, Shenoi SD, et al. (1991) Azathioprine in the treatment of parthenium dermatitis. Br J Dermatol 124(4):394-395.
Srinivas CR, Krupashankar DS, Singh KK, et al. (1988) Oral hyposensitization in parthenium dermatitis. Contact Dermatitis 18(6):242-243.
Verma KK, Sirka CS, Ramam M, et al. (2002) Parthenium dermatitis presenting as photosensitive lichenoid eruption. A new clinical variant. Contact Dermatitis 46(5):286-289.
Wedner HJ, Wilson P, Lewis WH (1989) Allergic reactivity to Parthenium hysterophorus pollen: An ELISA study of 582 sera from the United States Gulf Coast. J Allergy Clin Immunol 84(Aug):263-271.

PARTHENOCISSUS QUINQUEFOLIA (L.) Planch. [Vitaceae]

Synonyms:
Ampelopsis quinquefolia (L.) Michx.

Common Names:
American ivy; Amerikanischer Efeu; Engleman ivy; Kanadische Rebe; Virginia creeper; Virginia ivy; woodbine

Citations:
Bernays HL (1876) Poisoning by Virginia creeper. Br Med J 2(Jul 1):32.
Bernays HL (1876) Poisoning by Virginia creeper. Lancet 2(Jul 15):105.
Shropshire CM, Stauber E, Arai A (1992) Evaluation of selected plants for acute toxicosis in budgerigars. J Am Vet Med Assoc 200(7):936-939.
Warren LE (1923) Poisoning from Virginia creeper. Hygeia 1(7):459-460.

partidge –see– *Andira inermis* (W. Wright) Kunth ex DC.; *Millettia stuhlmannii* Taub.

partridgeberry–see– *Gaultheria procumbens* L.

pas-d'Ane-commun –see– *Tussilago farfara* L.

pasarín –see– *Lantana camara* L.

PASCALIA GLAUCA Ortega [Asteraceae]

Synonyms:
Wedelia glauca (Ortega) O. Hoffm. ex Hicken

Common Names:
chilquilla; clavel amarillo; flor-de-sapo; seca tierra; sunchillo; suncho; yerba-de-sapo; yuyo-de-sapo

Citations:
Collazo L, Riet-Correa F (1996) Experimental intoxication of sheep and cattle with Wedelia glauca. Vet Hum Toxicol 38(3):200-203.
Giusti L (1934-1936) Experiencias sobre la acción fisiológica de la Wedelia glauca y del Astragalus bergii. Rev Argentina Agron 1:223-228.
Morán BL, Kosic CV (1965) Intoxicacion de cerdos por Wedelia glauca. Gaceta Vet 27:461-462.

pasote –see– *Dysphania ambrosioides* (L.) Mosyakin & Clemants

paspalum –see– *Paspalum dilatatum* Poir.

Paspalum ciliatifolium Michx. = Paspalum setaceum Michx. var. ciliatifolium (Michx.) Vasey

Paspalum commersonii Lam. = Paspalum scrobiculatum L. var. bispicatum Hack.

PASPALUM DILATATUM Poir. [Poaceae]

Common Names:
Bahiagrass; Dallisgrass; large watergrass; paspalum

Citations:
Hayes WF (1948) Probable Dallis grass poisoning. Texas Vet Bull #10(1):5-6.
Horne CH (1948) Dallis grass poisoning. Texas Vet Bull 10(3):1.

PASPALUM NOTATUM Flüggé [Poaceae]

Common Names:
Bahiagrass

Citations:
Davies JM, Bright ML, Rolland JM, et al. (2005) Bahia grass pollen specific IgE is common in seasonal rhinitis patients but has limited cross-reactivity with ryegrass. Allergy 60(2):251-255.

PASPALUM SCROBICULATUM L. var. bispicatum Hack. [Poaceae]

Synonyms:
Paspalum commersonii Lam.

Common Names:
cheta mona; chota mona; kali mona; monagrass

Citations:
Mia B (1960) Probable "Mona" grass Paspalum commersonii poisoning. Indian Vet J 37:31-34.

PASPALUM SETACEUM Michx. var. ciliatifolium (Michx.) Vasey [Poaceae]

Synonyms:
Paspalum ciliatifolium Michx.

Common Names:
fringe-leaf paspalum

Citations:
Sperry OE, Turk RD, Hoffman GO, et al. (1955) Photosensitization of cattle in Texas. Texas Agric Exp Sta Bull #812:8 pp.

pasqueflower –see– *Anemone patens* L.

PASSIFLORA ADENOPODA DC. [Passifloraceae]

Citations:

Saenz JA, Nassar M (1972) Toxic effect of the fruit of Passiflora adenopoda DC. on humans: Phytochemical determination. Rev Biol Trop 20(1):137-140.

Passiflora alba Link & Otto = Passiflora subpeltata Ortega

PASSIFLORA INCARNATA L. [Passifloraceae]

Common Names:

apricot vine; Maypop; passionflower; passionvine; purple passionflower

Citations:

Fisher AA, Purcell P, Le Couteur DG (2000) Toxicity of Passiflora incarnata L. J Toxicol Clin Toxicol 38(1):63-66.

Solbakken AM, Rørbakken G, Gunderson T (1997) Naturmedisin som rusmiddel. Tidsskr Nor Laegeforen 117(8):1140-1141.

PASSIFLORA SUBPELTATA Ortega [Passifloraceae]

Synonyms:

Passiflora alba Link & Otto

Common Names:

white passion vine; wild passion fruit; wild passionvine

Citations:

Anonymous (1929) The poison plants committee. J CSIRO Aust 2:40-48.

Dodd S (1910) Report on experiments with the wild passionflower vine in connection with the death of cattle in the Beaudesert district (Queensland). J Comp Pathol Ther 23:205-212.

passion vine –see– *Passiflora incarnata* L.

passionflower –see– *Passiflora incarnata* L.

PASTINACA SATIVA L. [Apiaceae]

Common Names:

parsnip; Pastinak; wild parsnip

Citations:

Aberer W (1992) Occupational dermatitis from organically grown parsnip (Pastinaca sativa L). Contact Dermatitis 26(1):62.

Belisario JC (1952) Parsnip dermatitis in the tropics under active service conditions. Aust J Dermatol 1(3):183-193.

Campbell AN, Cooper CE, Dahl MG (1982) "Non-accidental injury" and wild parsnips. Br Med J 284(6317):708.

Fernández de Corres L, Corrales JL, Muñoz D, et al. (1984) Dermatitis alérgicas de contacto por plantas. Allergol Immunopathol (Madr) 12(4):313-319.

Jensen T, Hansen KG (1939) Active spectral range for phytogenic photodermatosis produced by Pastinaca sativa (dermatitis bullosa striata pratensis, Oppenheim). Arch Derm Syphilol 40:566-577.

Pedersen NB, Arlés UB (1997) Phototoxic reaction to parsnip and UVA sunbed. Contact Dermatitis 39(2):97.

Picardo M, Cristaudo A, de Luca C, et al. (1986) Contact dermatitis to Pastinaca sativa. Contact Dermatitis 15(6):98-99.

Poniecka H (1990) Rośliny jako przyczyna alergii kontaktowej w materiale kliniki dermatologicznej am w białymstoku. Przegl Dermatol 77(4):262-265.

Starck V (1945) Percutaneous photosensitization due to handling of parsnips. Acta Derm Venereol 25:179-188.

Note:

Parsnip is named *Pastinaca sativa* L. subsp. *sativa* in some publications.

Pastinak –see– *Pastinaca sativa* L.

pastora –see– *Euphorbia pulcherrima* Willd. ex Klotzsch

pata-de-ganso –see– *Chenopodium album* L.

pata-de-perdis –see– *Cynodon dactylon* (L.) Pers.

Paternostererbse –see– *Abrus precatorius* L.

Paterson's-curse –see– *Echium plantagineum* L.

PAULLINIA CUPANA Kunth [Sapindaceae]

Common Names:

Brazilian guarana; cupana; guarana

Citations:

Ooms TG, Khan SA, Means C (2001) Suspected caffeine and ephedrine toxicosis resulting from ingestion of an herbal supplement containing guarana and ma huang in dogs: 47 cases (1997-1999). J Am Vet Med Assoc 218(2):225-229.

Sawyers B, LoVecchio F (2002) Transient ischemic attack associated with metabolife use. J Toxicol Clin Toxicol 40(5):644-645.

PAUSINYSTALIA JOHIMBE (K. Schum.) Pierre ex Beille [Rubiaceae]

Synonyms:

Corynanthe johimbe K. Schum.

Common Names:

African yohimbine; corynanthe; yo yo; yohimbé

Citations:

Linden CH, Vellman WP, Rumack B (1985) Yohimbine: A new street drug. Ann Emerg Med 14(10):1002-1004.

Palmer M, Haller C, McKinney P, et al. (1998) A gap in the safety net: A multi-center prospective study of herbals and other dietary supplements. J Toxicol Clin Toxicol 36(5):454.

PAVETTA HARBORII S. Moore [Rubiaceae]

Common Names:

small poison bride bush

Citations:

Adelaar TF, Terblanche M, Smit JD (1966) A report on negative experiments with ferric chloride as a prophylactic agent against gousiekte. J S Afr Vet Assoc 37(2):199-201.

Pretorius PJ, Terblanche M (1967) A preliminary study on the symptomatology and cardiodynamics of gousiekte in sheep and goats. J S Afr Vet Assoc 38(1):29-53.
Pretorius PJ, Terblanche M, Van der Walt JD, et al. (1973) Cardiac failure in ruminants caused by gousiekte. Recent Adv Stud Cardiac Struct Metab 2:385-397.
Prozesky L, Bastianello SS, Fourie N, et al. (2005) A study of the pathology and pathogenesis of the myocardial lesions in gousiekte, a plant-induced cardiotoxicosis of ruminants. Onderstepoort J Vet Res 72(3):219-230.

PAVETTA SCHUMANNIANA F. Hoffm. ex K. Schum. [Rubiaceae]

Common Names:
large poison bride bush

Citations:
Jackson JJ, Needham AJ, Lawrence JA (1968) Some recent investigations into Rhodesian toxic plants. Proc Trans Rhodesia Sci Assoc 53:9-12.

Pavia glabra Spach = Aesculus glabra Willd.

pavot coq –see– *Papaver rhoeas* L.

pawpaw –see– *Asimina triloba* (L.) Dunal; *Carica papaya* L.

pazote –see– *Dysphania ambrosioides* (L.) Mosyakin & Clemants

pea –see– *Pisum sativum* L.

pea blossom poison –see– *Isotropis cuneifolia* (Sm.) Domin

pea bush –see– *Tephrosia purpurea* (L.) Pers.

peach –see– *Prunus persica* (L.) Batsch

peach-leaf poison bush –see– *Trema tomentosa* (Roxb.) H. Hara

peachwort –see– *Persicaria maculosa* Gray

peanut –see– *Arachis hypogaea* L.

pear –see– *Pyrus communis* L.

pearl harlequin –see– *Dicentra cucullaria* (L.) Bernh.

pearl millet –see– *Pennisetum glaucum* (L.) R. Br.

pearly gates –see– *Ipomoea violacea* L.

peas –see– *Pisum sativum* L.

peavine –see– *Astragalus emoryanus* (Rydb.) Cory

peavine clover –see– *Trifolium pratense* L.

pecan –see– *Carya illinoinensis* (Wangenh.) K. Koch

peco –see– *Cicuta maculata* L. var. angustifolia Hook.; *Delphinium barbeyi* (Huth) Huth

PEDILANTHUS TITHYMALOIDES (L.) Poit. [Euphorbiaceae]

Synonyms:
Euphorbia tithymaloides L.

Common Names:
bejuco-de-estrella; bitamo; Christmas candle; devil's-backbone; dictamo; fiddle flower; itamo; Jacob's-ladder; Jewbush; pie-de-niño; pinipini; poison halipan; redbird cactus; ribbon cactus; slipper flower; slipper plant; zig-zag plant

Citations:
Lim TK, Soepadmo E (1984) Eye injury from plant sap of Pedilanthus tithymaloides Poit. Singapore Med J 25(6):412-419.

pedunculate oak –see– *Quercus robur* L.

peganum –see– *Peganum harmala* L.

PEGANUM HARMALA L. [Nitraraceae]

Common Names:
African rue; harmal; harmel peganum; peganum; Syrian rue; wild rue

Citations:
Anonymous (1936) Investigations of stock-poisoning plants. U S Dep Agric Bur Anim Indus Chief Rep. 1936:44-45.
Anonymous (1937) Investigations of stock-poisoning plants. U S Dep Agric Bur Anim Indus Chief Rep. 1937:47-48.
Anonymous (1942) Investigations of stock-poisoning plants. U S Dep Agric Bur Anim Indus Chief Rep 1942:29.
Ben Salah N, Amamou M, Jerbi Z, et al. (1986) Un cas de surdosage en Peganum harmala L. J Toxicol Clin Exp 6(5):319-322.
Black WL, Parker KW (1936) Toxicity tests on African rue (Peganum harmala L.). New Mexico Agric Exp Sta Bull #240:14 pp.
Boughton IB, Hardy WT (1939) Hard yellow livers of sheep and cattle and feeding trials of suspected plants. Texas Agric Exp Sta Annu Rep 52:238-240.
Hamouda C, Amamou M, Thabet H, et al. (2000) Plant poisonings from herbal medication admitted to a Tunisian toxicologic intensive care unit, 1983-1998. Vet Hum Toxicol 42(3):137-141.
Mathews FP (1941) Poisonous plants in the Davis Mountains. Texas Agric Exp Sta Annu Rep 54:93.
Moran EA, Couch JF, Clawson AB (1940) Peganum harmala, a poisonous plant in the Southwest. Vet Med 35(4):234-235.

pegwood –see– *Euonymus europaeus* L.

PELARGONIUM ×DOMESTICUM L. H. Bailey [Geraniaceae]

Common Names:
geranium

Citations:
Hjorth N (1969) Plant dermatitis. Contact Dermatol Newsl 6(Jul):126-127.
Oehme FW (1978) The hazard of plant toxicities to the human population. In: Keeler RF, Van Kampen KR, James LF, et al. (eds.) Effects of poisonous plants on livestock. Academic Press. New York. pp. 67-80.

PELARGONIUM GRAVEOLENS L'Her. [Geraniaceae]

Citations:

Hjorth N (1969) Plant dermatitis. Contact Dermatol Newsl 6(Jul):126-127.

PELARGONIUM ×HORTORUM L. H. Bailey [Geraniaceae]

Citations:

Hjorth N (1969) Plant dermatitis. Contact Dermatol Newsl 6(Jul):126-127.

Pelea anisata H. Mann = Melicope anisata (H. Mann) T. G. Hartley & B. C. Stone

pellitory –see– *Parietaria judaica* L.; *Tanacetum cinerariifolium* (Trevir.) Sch. Bip.

pencil bush –see– *Eremophila deserti* (A. Cunn. ex Benth.) Chinnock

pencil cactus –see– *Euphorbia tirucalli* L.

pencil caustic –see– *Sarcostemma viminale* (L.) R. Br.

pencil tree –see– *Euphorbia tirucalli* L.

Pennisetum americanum (L.) Leeke = Pennisetum glaucum (L.) R. Br.

PENNISETUM CLANDESTINUM Hochst. ex Chiov. [Poaceae]

Common Names:

kikuyugrass

Citations:

Bryson RW, Newsholme SJ (1978) Kikuyu grass poisoning of cattle in Natal. J S Afr Vet Assoc 49(1):19-21.

Busch J, Harris GH, Coup MR, et al. (1969) Acute ruminal indigestion, alkalosis and death of cattle grazing kikuyu. N Z Vet J 17(9):182-183.

Cordes DO, Coup MR, Harris GH, et al. (1969) Acute ruminal indigestion, alkalosis and death of cattle grazing Kikuyu grass. N Z Vet J 17(5):77-81.

Gabbedy BJ, Gwynn R, Hopkinson WI, et al. (1974) Kikuyu poisoning of cattle in Western Australia. Aust Vet J 50(8):369-370.

Martinovich D, Mortimer PH, di Menna ME (1972) Similarities between so-called kikuyu poisoning of cattle and two experimental mycotoxicoses. N Z Vet J 20(4):57-58.

Martinovich D, Smith B (1972) Kikuyu poisoning in sheep. N Z Vet J 20(9):169.

Martinovich D, Smith B (1973) Kikuyu poisoning of cattle. 1. Clinical and pathological findings. N Z Vet J 21(4):55-63.

Peet RL, Dickson J, Hare M (1990) Kikuyu poisoning in goats and sheep. Aust Vet J 67(6):229-230.

Smith B, Martinovich D (1973) Kikuyu poisoning of cattle. 2. Epizootiological aspects. N Z Vet J 21(5):85-89.

Van Heerden J, Williams MC, Van Rensburg IB, et al. (1978) An outbreak of "kikuyu poisoning" in western Transvaal. J S Afr Vet Assoc 49(1):27-30.

Wells AW (1958) Nitrite poisoning in pigs on kikuyu (Pennisetum clandestinum). Aust Vet J 34(Nov):402.

Wong PT, Roth IJ, Jackson AR (1987) Kikuyu poisoning of cattle in New South Wales and its relationship to pasture fungi on kikuyu grass (Pennisetum clandestinum). Aust Vet J 64(8):229-232.

PENNISETUM GLAUCUM (L.) R. Br. [Poaceae]

Synonyms:

Pennisetum americanum (L.) Leeke; ***Pennisetum typhoides*** (Burm. f.) Stapf & C. E. Hubb.

Common Names:

bajra; bulrush millet; horse millet; Indian millet; pearl millet; spiked millet

Citations:

Gadir WS, Adam SE (2000) Effects of pearl millet (Pennisetum typhoides), and fermented and processed fermented millet on Nubian goats. Vet Hum Toxicol 42(3):133-136.

Klopfenstein CF, Hoseney RC, Leipold HW (1983) Further studies on the goitrogenic effects of pearl millet diets. Nutr Rep Int 28(5):1137-1144.

PENNISETUM PURPUREUM Schumach. [Poaceae]

Common Names:

elephant grass; Napiergrass

Citations:

Dhillon KS, Paul BS, Bajwa RS, et al. (1971) A preliminary report on a peculiar type of Napier grass (Pennisetum purpureum, 'Pusa Giant') poisoning in buffalo calves. Indian J Anim Sci 41(11):1034-1036.

Seiler RJ, Omar AR, Salim N (1979) Nitrate poisoning in cattle fed napier grass (Pennisetum purpureum). Kajian Veterinar 11:10-13.

Pennisetum typhoides (Burm. f.) Stapf & C. E. Hubb. = Pennisetum glaucum (L.) R. Br

penny John –see– *Hypericum perforatum* L.

pennycress –see– *Thlaspi arvense* L.

pennycress mustard –see– *Thlaspi arvense* L.

pennyroyal –see– *Hedeoma pulegioides* (L.) Pers.; *Mentha pulegium* L.

pennywort –see– *Centella asiatica* (L.) Urb.

PENTALINON LUTEUM (L.) B. F. Hansen & Wunderlin [Apocynaceae]

Synonyms:

Urechites lutea (L.) Britton

Common Names:

bejuco marrullero; clavelitos; curamaguey; nightshade; wild allamanda; yellow nightshade

Citations:

Infante JF, Perdomo J, Merino N, et al. (1984) Estudio morfopatologico en bovinos intoxicados experimentale con bejuco marrullero (Urechites lutea) en la provincia Gramma. Rev Salud Anim 6(2):193-204.

Joa R, Merino N, Marrero E, et al. (1985) Estudio anatomopatológico en bovinos intoxicados experimentalmente con glicósidos aislados de Urechites lutea (L) Britton. Rev Cubana Cienc Vet 16(1):41-52.
Marrero E, Fernandez O, Pompa A, et al. (1984) Alteraciones electrocardiográficas en terneros intoxicados con glicósidos de Urechites lutea (L) Britton. Rev Cubana Cienc Vet 15(2):179-189.
Marrero Faz E (1996) Urechites lutea (L) Britton toxicity in cattle. Vet Hum Toxicol 38(4):313-314.

peonia –see– *Abrus precatorius* L.

peony poppy –see– *Papaver somniferum* L.

pepeo tree –see– *Mauria heterophylla* Kunth

PEPEROMIA OBTUSIFOLIA (L.) A. Dietr. [Piperaceae]

Common Names:

American rubber plant; baby rubber plant; pepper face

Citations:

Der Marderosian AH, Giller FB, Roia FC Jr (1976) Phytochemical and toxicological screening of household ornamental plants potentially toxic to humans. I. J Toxicol Environ Health 1(6):939-953.

pepper –see– *Capsicum annuum* L.; *Capsicum frutescens* L.; *Piper nigrum* L.

pepper face –see– *Peperomia obtusifolia* (L.) A. Dietr.

pepper plant –see– *Piper methysticum* G. Forst.

pepper tree –see– *Schinus terebinthifolius* Raddi

pepper turnip –see– *Arisaema triphyllum* (L.) Schott

peppermint –see– *Mentha ×piperita* L. nothosubsp. piperita

perde stert –see– *Equisetum ramosissimum* Desf.

perennial broomweed –see– *Gutierrezia microcephala* (DC.) A. Gray; *Gutierrezia sarothrae* (Pursh) Britton & Rusby

perennial nettle –see– *Urtica dioica* L.

perennial pea –see– *Lathyrus latifolius* L.

perennial ryegrass –see– *Lolium perenne* L.

perennial snakeweed –see– *Gutierrezia sarothrae* (Pursh) Britton & Rusby

perennial sweet pea –see– *Lathyrus latifolius* L.

perennial threadleaf snakeweed –see– *Gutierrezia microcephala* (DC.) A. Gray

PERGULARIA GARIEPENSIS N. E. Br. [Asclepiadaceae]

Citations:

Steyn DG (1937) Recent investigations into the toxicity of known and unknown poisonous plants in the Union of South Africa. VII. Onderstepoort J Vet Sci Anim Indus 9(1):111-124.

PERICOPSIS ELATA (Harms) Meeuwen [Fabaceae]

Synonyms:

Afrormosia elata Harms

Common Names:

afrormosia

Citations:

Oleffe JA, Sporcq J, Hublet P (1975) Epidemiological study of the wood industry in Belgium. Contact Dermatitis 1(5):315-316.

PERILLA FRUTESCENS (L.) Britton [Lamiaceae]

Common Names:

beefsteak plant; blueweed; Joseph's-coat; mint plant; perilla mint; purple mint plant; rattlesnake weed; wild coleus

Citations:

Kerr LA, Burrows GE (1985) Perilla frutescens toxicity in cattle. Toxicon 23:31.
Kerr LA, Johnson BJ, Burrows GE (1986) Intoxication of cattle by Perilla frutescens (purple mint). Vet Hum Toxicol 28(5):412-416.
Linnabary RD, Warren J, Wilson BJ, et al. (1978) Acute bovine pulmonary emphysema produced by Perilla frutescens. Mod Vet Pract 59(9):684-686.

perilla mint –see– *Perilla frutescens* (L.) Britton

periqueto –see– *Codiaeum variegatum* (L.) A. Juss.

periwinkle –see– *Catharanthus roseus* (L.) G. Don; *Vicia villosa* Roth subsp. varia (Host) Corb.; *Vinca major* L.

perlas-del-oriente –see– *Moringa oleifera* Lam.

peroba-da-campos –see– *Aspidosperma gomezianum* A. DC.

peroba d'aqua –see– *Sessea brasiliensis* Toledo

Perobas –see– *Euxylophora paraensis* Huber

perreil –see– *Oenanthe crocata* L.

PERSEA AMERICANA Mill. var. americana [Lauraceae]

Common Names:

aguacate; alligator pear; avocado; avocado pear

Citations:

Appleman D (1944) Preliminary report on toxicity of avocado leaves. Calif Avocado Soc Yearb 1944:37.
Bee JG (1968) Suspected avocado poisoning of a horse. Mod Vet Pract 49(11):52-53.
Buoro IB, Nyamwange SB, Chai D, et al. (1994) Putative avocado toxicity in two dogs. Onderstepoort J Vet Res 61(1):107-109.
Burger WP, Naudé TW, van Rensburg IB, et al. (1994) Avocado (Persea americana) poisoning in ostriches. In: Colegate SM, Dorling PR (eds.) Plant-associated toxins. Agricultural, phytochemical and ecological aspects. CABI. New York. pp. 546-551.

Craigmill AL, Seawright AA, Mattila T, et al. (1989) Pathological changes in the mammary gland and biochemical changes in milk of the goat following oral dosing with leaf of the avocado (Persea americana). Aust Vet J 66(7):206-211.
Grant R, Basson PA, Booker HH, et al. (1991) Cardiomyopathy caused by avocado (Persea americana Mill) leaves. J S Afr Vet Assoc 62(1):21-22.
Grant R, Booker HH, Basson PA, et al. (1988) Cardiomyopathies caused by januariebos (Gnidia polycephala) and avocado (Persea americana) leaves. J S Afr Vet Assoc 59(2):101.
Hargis AM, Stauber E, Casteel S, et al. (1989) Avocado (Persea americana) intoxication in caged birds. J Am Vet Med Assoc 194(1):64-66.
McKenzie RA, Brown OP (1991) Avocado (Persea americana) poisoning of horses. Aust Vet J 68(2):77-78.
Sani Y, Atwell RB, Seawright AA (1991) The cardiotoxicity of avocado leaves. Aust Vet J 68(4):150-151.
Sani Y, Seawright AA, Ng JC, et al. (1994) The toxicity of avocado leaves (Persea americana) for the heart and lactating mammary gland of the mouse. In: Colegate SM, Dorling PR (eds.) Plant-associated toxins. Agricultural, phytochemical and ecological aspects. CABI. New York. pp. 552-556.
Shlosberg A, Ohad DG, Bellaiche M, et al. (1998) Monitoring of physiological and pathological changes in turkey poults fed leaves of potentially cardiomyotoxic Nerium oleander and Persea americana. In: Garland T, Barr AC (eds.) Toxic plants and other natural toxicants. CABI. New York. pp. 131-136.
Shropshire CM, Stauber E, Arai A (1992) Evaluation of selected plants for acute toxicosis in budgerigars. J Am Vet Med Assoc 200(7):936-939.
Stadler P, Van Rensburg IB, Naudé TE (1991) Suspected avocado (Persea americana) poisoning in goats. J S Afr Vet Assoc 62(4):186-188.
Valeri H, Gimeno F (1953) Estudio fito-quimico toxicologico de los frutos de aguacate (Persea americana - C. Bauhin, Pinax 441, 1623). Rev Med Vet y Parasit Caracas 12(1 & 4):131-165.

Persea americana Mill. var. guatemalensis (L. O. Williams) Scora = Persea americana Mill. var. nubigena (L. O. Williams) L. E. Kopp

PERSEA AMERICANA Mill. var. nubigena (L. O. Williams) L. E. Kopp [Lauraceae]

Synonyms:

Persea americana Mill. var. guatemalensis (L. O. Williams) Scora

Citations:

Burger WP, Naudé TW, Van Rensburg IB, et al. (1994) Cardiomyopathy in ostriches (Struthio camelus) due to avocado (Persea americana var. guatemalensis) intoxication. J S Afr Vet Assoc 65(3):113-118.
Craigmill AL, Eide RN, Shultz TA, et al. (1984) Toxicity of avocado (Persea americana (Guatamalan var)) leaves: Review and preliminary report. Vet Hum Toxicol 26(5):381-383.

Persian lilac –see– *Melia azedarach* L.

Persian lime –see– *Citrus aurantiifolia* (Christm.) Swingle

Persian violet –see– *Cyclamen persicum* Mill.

Persian walnut –see– *Juglans regia* L.

persicaria –see– *Persicaria maculosa* Gray

PERSICARIA AMPHIBIA (L.) Delarbre [Polygonaceae]

Synonyms:

Polygonum amphibium L.

Citations:

Salgues R (1961) Le genre Polygonum L. (Polygonacées) Etudes chimiques et toxicologiques. Les faits hématologiques de l'empoisonnement expérimental. Qualitas Plant Mat Veg 8:367-395.

PERSICARIA GLABRA (Willd.) M. Gómez [Polygonaceae]

Synonyms:

Polygonum glabrum Willd.

Common Names:

chiguirera

Citations:

Kulakkattolickal AT (1987) Piscicidal plants of Nepal: Preliminary toxicity screening using grass carp (Ctenopharyngodon idella) fingerlings. J Ethnopharmacol 21(1):1-9.

PERSICARIA HYDROPIPER (L.) Spach [Polygonaceae]

Synonyms:

Polygonum hydropiper L.

Common Names:

biting persicaria; curage; herbe-de-St. Innocent; marsh pepper; Pfeffer Knöterich; pirae; poivre d'eau; renonée âcre; smartweed; Wasserpfeffer; water pepper; water smartweed

Citations:

Kulakkattolickal AT (1987) Piscicidal plants of Nepal: Preliminary toxicity screening using grass carp (Ctenopharyngodon idella) fingerlings. J Ethnopharmacol 21(1):1-9.
Salgues R (1961) Le genre Polygonum L. (Polygonacées) Etudes chimiques et toxicologiques. Les faits hématologiques de l'empoisonnement expérimental. Qualitas Plant Mat Veg 8:367-395.

PERSICARIA MACULOSA Gray [Polygonaceae]

Synonyms:

Polygonum persicaria L.

Common Names:

Flöhkraut; heart's-ease; heartweed; lady's-thumb; peachwort; persicaria; Pfirsichblättriger Knöterich; redshank; spotted smartweed; willow weed

Citations:

Salgues R (1961) Le genre Polygonum L. (Polygonacées) Etudes chimiques et toxicologiques. Les faits hématologiques de l'empoisonnement expérimental. Qualitas Plant Mat Veg 8:367-395.

PERSICARIA NEPALENSIS (Meisn.) H. Gross [Polygonaceae]

Synonyms:

Polygonum nepalense Meisn.

Common Names:

pirae

Citations:

Kulakkattolickal AT (1987) Piscicidal plants of Nepal: Preliminary toxicity screening using grass carp (Ctenopharyngodon idella) fingerlings. J Ethnopharmacol 21(1):1-9.

PERSICARIA PUNCTATA (Elliott) Small [Polygonaceae]

Synonyms:

Polygonum punctatum Elliott

Common Names:

catayo; chile-de-perro; herbe-de-bicho; poor-man's-pepper; smartweed; tabaquillo-de-los-rios; water smartweed; yerba-del-diablo; yerba picanta

Citations:

Salgues R (1961) Le genre Polygonum L. (Polygonacées) Etudes chimiques et toxicologiques. Les faits fématologiques de l'empoisonnement expérimental. Qualitas Plant Mat Veg 8:367-395.

persil sauvage –see– *Petroselinum crispum* (Mill.) Nyman ex A. W. Hill

persimmon –see– *Diospyros kaki* Thunb.; *Diospyros virginiana* L.

Peru apple –see– *Datura stramonium* L.

perupok –see– *Lophopetalum dubium* M. Laws.; *Lophopetalum floribundum* Wight

Peruvian bark –see– *Alstonia constricta* F. Muell.

Peruvian creeper –see– *Araujia sericifera* Brot.

Peruvian rhatany –see– *Krameria lappacea* (Dombey) Burdet & B. B. Simpson

pessegueiro bravo –see– *Prunus myrtifolia* (L.) Urb.

petasites –see– *Petasites japonicus* (Siebold & Zucc.) Maxim.

PETASITES JAPONICUS (Siebold & Zucc.) Maxim. [Asteraceae]

Common Names:

colt's-foot; petasites

Citations:

Fushimi K, Kato K, Kato T, et al. (1978) Carcinogenicity of flower stalks of Petasites japonicus Maxim. in mice and Syrian golden hamsters. Toxicol Lett 1(1978):201-294.

Hirono I, Shimizu M, Fushimi K, et al. (1973) Carcinogenic activity of Petasites japonicus Maxim., a kind of coltsfoot. Gann 64(Oct):527-528.

Petersilie –see– *Petroselinum crispum* (Mill.) Nyman ex A. W. Hill

Petersilienschierling –see– *Aethusa cynapium* L.

petit médicinier –see– *Jatropha multifida* L.

petit pignon d' Inde –see– *Croton tiglium* L.

petite ciguë –see– *Aethusa cynapium* L.

petite malnommée –see– *Chamaesyce hirta* (L.) Millsp.

petitgrain –see– *Citrus aurantium* L.

PETIVERIA ALLIACEA L. [Phytolaccaceae]

Common Names:

anamú; garlic guineahen weed; guineahen weed

Citations:

Nuñez Bello V, Vanegas Díaz LA, Torres Gamez JE, et al. (1983) Caquexia muscular distrofica y su relacion clinico-patologica con la neurotoxicidad retardada. Rev Inst Colombiano Agropecuario 18(4):345-353.

pétron –see– *Juniperus communis* L.

PETROSELINUM CRISPUM (Mill.) Nyman ex A. W. Hill [Apiaceae]

Synonyms:

Apium petroselinum L.; ***Petroselinum hortense*** auct.; ***Petroselinum sativum*** Hoffm., nom. nud.

Common Names:

huile-de-persil-sauvage; parsley; persil sauvage; Petersilie

Citations:

Arai M, Stauber E, Shropshire CM (1992) Evaluation of selected plants for their toxic effects on canaries. J Am Vet Med Assoc 200(9):1329-1331.

Ciganda C, Laborde A (2001) Herbal infusions used for induced abortion. J Toxicol Clin Toxicol 39(3):318-319.

Ciganda C, Laborde A (2003) Herbal infusions used for induced abortion. J Toxicol Clin Toxicol 41(3):235-239.

Laborde A, Ciganda C (1998) Poisoning by herbal infusions ingested as abortifacient agents. J Toxicol Clin Toxicol 36(5):454-455.

Mele V (1968) Sulla intossicazione de prezzemolo usato come mezzo abortivo. Folia Med (Napoli) 51(8):601-613.

Perelman B, Kuttin ES (1988) Parsley-induced photosensitivity in ostriches and ducks. Avian Pathol 17(1):183-192.

Stransky L, Tsankov N (1980) Contact dermatitis from parsley (Petroselinum). Contact Dermatitis 6(17):233-234.

Petroselinum hortense auct. = Petroselinum crispum (Mill.) Nyman ex A. W. Hill

Petroselinum sativum Hoffm., nom. nud. = Petroselinum crispum (Mill.) Nyman ex A. W. Hill

petty morel –see– *Solanum nigrum* L.

petty spurge –see– *Euphorbia peplus* L.

Peucedanum graveolens (L.) Benth. & Hook. f. = Anethum graveolens L.

PEUMUS BOLDUS Molina [Monimiaceae]

Common Names:

boldo; boldus

Citations:

Piscaglia F, Leoni S, Venturi A, et al. (2005) Caution in the use of boldo in herbal laxatives: A case of hepatotoxicity. Scand J Gastroenterol 40(2):236-239.

Peyolylkaktus –see– *Lophophora williamsii* (Lem. ex Salm-Dyck) J. M. Coult.

peyote –see– *Lophophora williamsii* (Lem. ex Salm-Dyck) J. M. Coult.

peyote muscal –see– *Lophophora williamsii* (Lem. ex Salm-Dyck) J. M. Coult.

Pfaffenhütchen –see– *Euonymus europaeus* L.

Pfeffer Knöterich –see– *Persicaria hydropiper* (L.) Spach

Pfefferminze –see– *Mentha ×piperita* L. nothosubsp. piperita

Pfefferwurzel –see– *Armoracia rusticana* P. Gaertn. et al.

Pferdebohne –see– *Vicia faba* L.

Pferdekastanie –see– *Aesculus hippocastanum* L.

Pfirsich –see– *Prunus persica* (L.) Batsch

Pfirsichblättriger Knöterich –see– *Persicaria maculosa* Gray

Pflaume –see– *Prunus domestica* L.

Phaca canadensis MacMill. = Astragalus canadensis L.

PHACELIA BRACHYLOBA (Benth.) A. Gray [Boraginaceae]

Citations:

Munz PA (1932) Dermatitis produced by Phacelia (Hydrophyllaceae). Science 76(1965):194.

PHACELIA CAMPANULARIA A. Gray [Boraginaceae]

Citations:

Munz PA (1932) Dermatitis produced by Phacelia (Hydrophyllaceae). Science 76(1965):194.

PHACELIA CRENULATA Torr. ex S. Watson [Boraginaceae]

Common Names:

desert heliotrope; false heliotrope; rama zorilla; scorpion weed; skunk bush

Citations:

Berry CZ, Shapiro SI, Dahlen RF (1962) Dermatitis venenata from Phacelia crenulata. Arch Dermatol 85(Jun):737-739.

Munz PA (1932) Dermatitis produced by Phacelia (Hydrophyllaceae). Science 76(1965):194.

PHACELIA GRANDIFLORA (Benth.) A. Gray [Boraginaceae]

Citations:

Munz PA (1932) Dermatitis produced by Phacelia (Hydrophyllaceae). Science 76(1965):194.

PHACELIA MINOR (Harv.) Thell. [Boraginaceae]

Citations:

Munz PA (1932) Dermatitis produced by Phacelia (Hydrophyllaceae). Science 76(1965):194.

PHACELIA PEDICELLATA A. Gray [Boraginaceae]

Citations:

Munz PA (1932) Dermatitis produced by Phacelia (Hydrophyllaceae). Science 76(1965):194.

phalaris –see– *Phalaris aquatica* L.

PHALARIS ANGUSTA Nees ex Trin. [Poaceae]

Common Names:

aveia louca; aveia-de-sangue; timothy canarygrass

Citations:

Gava A, Sousa RS, de Deus MS, et al. (1999) Phalaris angusta (Gramineae) como causa de enfermidade neurológica em bovinos no Estado de Santa Catarina. Pesq Vet Bras 19(1):35-38.

Odriozola E, Campero C, Lopez T, et al. (1991) Neuropathological effects and deaths of cattle and sheep in Argentina from Phalaris angusta. Vet Hum Toxicol 33(5):465-467.

Sousa RS, Irigoyen LF (1999) Intoxicação experimental por Phalaris angusta (Gramineae) em bovinos. Pesq Vet Bras 19(3-4):116-122.

PHALARIS AQUATICA L. [Poaceae]

Synonyms:

Phalaris tuberosa L.

Common Names:

Australian phalaris; canarygrass; falaris; Hardinggrass; phalaris; ronphagrass; Toowoomba canarygrass

Citations:

Bourke CA, Carrigan MJ, Seaman JT, et al. (1987) Delayed development of clinical signs in sheep affected by Phalaris aquatica staggers. Aust Vet J 64(1):31-32.

Bourke CA, Colegate SM, Rendell D, et al. (2005) Peracute ammonia toxicity: A consideration in the pathogenesis of Phalaris aquatica 'Polioencephalomalacia-like sudden death' poisoning of sheep and cattle. Aust Vet J 83(3):168-171.

Bourke CA, Rendell D, Colegate SM (2003) Clinical observations and differentiation of the peracute Phalaris aquatica poisoning syndrome in sheep known as 'polioencephalomalacia-like sudden death'. Aust Vet J 81(11):698-700.

Del Potro DH, Odriozola ER, Odeon A, et al. (1984) Intoxicacion de ovinos con falaris. Vet Argent 1(8):763-766.

Gaggino OP, Carrillo BJ (1965) Tembleque del falaris (Phalaris staggers). IV. Su observación en terneros de tambo. Idia 206:31-32.
Gaggino OP, Carrillo BJ, Frontera AR (1963) "Phalaris staggers" su observacion en el sudeste de la provincia de Buenos Aires. Gaceta Vet 25(150):51-56.
Gaggino OP, Carrillo BJ, Frontera AR (1963) "Tembleque del falaris" en el sudeste de la provincia de Buenos Aires. Idia 182(Feb):10-14.
Gallagher CH, Koch JH, Hoffman H (1967) Deaths of ruminants grazing Phalaris tuberosa in Australia. Aust Vet J 43(Nov):495-500.
Gallagher CH, Koch JH, Moore RM, et al. (1964) Toxicity of Phalaris tuberosa for sheep. Nature 204(4958):542-545.
Kennedy DJ, Cregan PD, Glastonbury JR, et al. (1986) Poisoning of cattle grazing of a low-alkaloid cultivar of Phalaris aquatica, Sirolan. Aust Vet J 63(3):88-89.
Kerr DR (1972) Rapid death of cattle grazing recently irrigated Phalaris tuberosa. Aust Vet J 48(7):421.
Lee HJ, Kuchel RE (1953) The aetiology of Phalaris staggers in sheep. 1. Preliminary observations on the preventative role of cobalt. Aust J Agric Res 4:88-99.
Lee HJ, Kuchel RE, Trowbridge RF (1956) The aetiology of Phalaris staggers in sheep. II. The toxicity to sheep of three types of pasture containing Phalaris tuberosa. Aust J Agric Res 7:333-344.
Le Souef HD (1948) Poisoning of sheep by Phalaris tuberosa. Vet J 24(Jan):12-13.
McDonald IW (1942) A "staggers" syndrome in sheep and cattle associated with grazing on Phalaris tuberosa. Aust Vet J 18(Oct):182-189.
Milne JA (1955) The occurrence of Phalaris staggers in sheep. N Z Vet J 3:119-121.
Monteseirín J, Pérez-Formoso JL, Sánchez-Hernández MC, et al. (2002) Occupational contact dermatitis from canary-grass seed. Contact Dermatitis 47(4):247.
Moore RM, Arnold GW, Hutchings RJ, et al. (1961) Poisoning of Merino sheep on Phalaris tuberosa L. pastures. Aust J Sci 24:88-89.
Moore RM, Hutchings RJ (1967) Mortalities among sheep grazing Phalaris tuberosa. Aust J Exp Agric Anim Husb 7(24):17-21.
Southcott WH (1956) Observations on Phalaris staggers in sheep. Aust Vet J 32(9):225-228.
Watson ER (1956) An outbreak of Phalaris staggers in sheep at Kojonuj, Western Australia. J Aust Inst Agric Sci 22:209-211.

PHALARIS ARUNDINACEA L. [Poaceae]

Common Names:

gardener's-garters; reed canarygrass; reed phalaris; ribbongrass; strandrøyr

Citations:

Bergeron JM, Jodoin L (1985) Fiabilité des mesures de poids et d'examens histopathologiques dans les études d'intoxication du campagnol des champs (Microtus pennsylvanicus). Can J Zool 63(4):804-810.
Marten GC, Jordan RM, Hovin AW (1981) Improved lamb performance associated with breeding for alkaloid reduction in reed canarygrass. Crop Sci 21(Mar-Apr):295-298.
Simpson BH, Jolly RD, Thomas SH (1969) Phalaris arundinacea as a cause of deaths and inco-ordination in sheep. N Z Vet J 17(12):240-244.
Ulvund MJ (1985) Chronic poisoning in a lamb grazing Phalaris arundinacea. Acta Vet Scand 26(2):286-288.
Ulvund MJ (1985) Incoordination and hyperexcitability in a lamb grazing Phalaris arundinacea in southwestern Norway. Nord Vet Med 37(4):259-260.

PHALARIS CANARIENSIS L. [Poaceae]

Common Names:

birdseedgrass; canarygrass

Citations:

East NE, Higgins RJ (1988) Canary grass (Phalaris sp) toxicosis in sheep in California. J Am Vet Med Assoc 192(5):667-669.

PHALARIS CAROLINIANA Walter [Poaceae]

Common Names:

Carolina canarygrass; Maygrass; southern canarygrass

Citations:

Nicholson SS, Olcott BM, Usenik EA, et al. (1989) Delayed Phalaris grass toxicosis in sheep and cattle. J Am Vet Med Assoc 195(3):345-346.

PHALARIS COERULESCENS Desf. [Poaceae]

Common Names:

blue canarygrass

Citations:

Colegate SM, Anderton N, Edgar J, et al. (1999) Suspected blue canary grass (Phalaris coerulescens) poisoning of horses. Aust Vet J 77(8):537-538.

PHALARIS MINOR Retz. [Poaceae]

Common Names:

little-seed Canarygrass; Mediterranean canarygrass; wild canarygrass

Citations:

Mendel VE, Crenshaw GL, Baker NF, et al. (1969) Staggers in pastured cattle. J Am Vet Med Assoc 154(7):769-772.
Van Halderen A, Green JR, Schneider DJ (1990) An outbreak of suspected Phalaris staggers in sheep in the Western Cape province. J S Afr Vet Assoc 61(1):39-40.

PHALARIS PARADOXA L. [Poaceae]

Common Names:

gnawed canarygrass; paradoxagrass

Citations:

Bourke CA, Colegate SM, Slattery S, et al. (2003) Suspected Phalaris paradoxa (paradoxa grass) poisoning in horses. Aust Vet J 81(10):635-637.

Phalaris tuberosa L. = Phalaris aquatica L.

phan saat –see– *Erythrophleum succirubrum* Gagnep.

PHASEOLUS ACUTIFOLIUS A. Gray [Fabaceae]

Common Names:

white tepary bean

Citations:

Grant G, More LJ, McKenzie NH, et al. (1995) Nutritional and haemagglutination properties of several tropical seeds. J Agric Sci 124:437-445.

PHASEOLUS COCCINEUS L. [Fabaceae]

Common Names:

Bohne; Feuerbohne; Gartenbohne; Kapuziner Bohne; runner bean; scarlet runner bean; seven-years'-bean; Turkische Bohne

Citations:

Faschingbauer H, Kofler L (1929) Ueber Giftwirkung von rohen Bohnen und Bohnenkeimlingen. Wien Klin Wochenschr 42(33):1069-1072.

Grant G, More LJ, McKenzie NH, et al. (1995) Nutritional and haemagglutination properties of several tropical seeds. J Agric Sci 124:437-445.

Haidvogl M, Fritsch G, Grubbauer HM (1979) Vergiftung durch rohe Gartenbohnen (Phaseolus vulgaris und Phaseolus coccineus) im Kindesalter. Paediat Paedol 14(3):293-296.

Narayan S, Sansom JE (2002) Contact urticaria from runner bean (Phaseolus coccineus). Contact Dermatitis 47(4):243.

PHASEOLUS LUNATUS L. [Fabaceae]

Common Names:

Burma bean; butter bean; double bean; haricot-de-lima; haricot-du-cap; hibbert bean; Java bean; Javabohne; lima bean; Mondbohne; Rangoon bean; Rangoonbohne; Sieva bean; sugar bean; white Indian bean; wild lima bean; wild sieva bean

Citations:

Anonymous (1941) Investigations of stock-poisoning plants. U S Dep Agric Bur Anim Indus Chief Rep 1941:62-63.

Dammann C, Behrens MD (1906) Massenvergiftungen von Pferden, Rindern, und Schweinen durch blausäurenhaltige Bohnen. Dtsch Tierarztl Wochenschr 14(1):1-4.

Hove EL, King S, Hill GD (1978) Composition, protein quality, and toxins of seeds of the grain legumes Glycine max, Lupinus spp., Phaseolus spp., Pisum sativum, and Vicia faba. N Z J Agric Res 21:457-462.

Jaffé WG (1950) Estudio sobre la inhibición del crecimeniento de ratas causado por algunas semillas de leguminosas. Acta Cient Venez 1(2):62-64.

Montgomery RD (1964) Observations on the cyanide content and toxicity of tropical pulses. West Indian Med J 13(1):1-11.

Mosselman G (1906) Empoissonement de bêtes bovines par les graines de haricot de Lima (Phaseolus lunatus) et recherches sur la toxicité de cette plante comestible. Ann Med Vet 55:141-153.

Ologhobo A, Mosenthin R, Alaka OO (2003) Histological alterations in the internal organs of growing chicks from feeding raw jackbean or limabean seeds. Vet Hum Toxicol 45(1):10-13.

Robertson A, Wijnne AJ (1905) Blauwzuurvergifting na gebruik van kratokboonen. Pharm Weekbl 42(19):393-399.

PHASEOLUS VULGARIS L. [Fabaceae]

Common Names:

bean; black bean; black kidney bean; Bohne; caraotas rosadas; field bean; French bean; Gartenbohne; green bean; Grüne Bohne; haricot bean; kidney bean; Natal round yellow bean; navy bean; ornamental French bean; pinto bean; red bean; red kidney bean; red Mexican bean; red navy bean; Speisebohnen; string bean; wax bean; white bean; Van Zyl sugar bean; Van Zyl suikerboontjie; yellow wax bean

Citations:

Begbie R, King TP (1981) The nutritional toxicity of the seed lectins of Phaseolus vulgaris in the pig. Biochem Soc Trans 9:145.

Bulavenko LS (1965) [A case of poisoning by seeds of the ornamental French bean.] Gig Sanit 30(10):90-91.

Carmalt J, Rosel KP, Burns T, et al. (2003) Suspected white kidney bean (Phaseolus vulgaris) toxicity in horses and cattle. Aust Vet J 81(11):674-676.

Desai ID (1966) Effect of kidney beans (Phaseolus vulgaris) on plasma tocopherol-level and its relation to nutritional muscular dystrophy in the chick. Nature 209(5025):810.

Grant G, Dorward PM, Buchan W (1995) Consumption of diets containing raw soya beans (Glycine max), kidney beans (Phaseolus vulgaris), cowpeas (Vigna unguiculata) or lupin seeds (Lupinus angustifolius) by rats up to 700 days: Effects on body composition and organ weights. Br J Nutr 73(1):17-29.

Grant G, Dorward PM, Pusztai A (1993) Pancreatic enlargement is evident in rats fed diets containing raw soybeans (Glycine max) or cowpeas (Vigna unguiculata) for 800 days but not in those fed diets based on kidney beans (Phaseolus vulgaris) or lupinseed (Lupinus angustifolius). J Nutr 123(12):2207-2215.

Grant G, More LJ, McKenzie NH, et al. (1995) Nutritional and haemagglutination properties of several tropical seeds. J Agric Sci 124:437-445.

Greer F, Brewer AC, Pusztai A (1985) Effect of kidney bean (Phaseolus vulgaris) toxin on tissue weight and composition and some metabolic functions of rats. Br J Nutr 54(1):95-103.

Griebel C (1950) Erkrankungen durch Bohnenflocken (Phaseolus vulgaris L.) und Platterbsen (Lathyrus tingitanus L.). Z Lebensm Unters Forsch 90:191-197.

Haidvogl M, Fritsch G, Grubbauer HM (1979) Vergiftung durch rohe Gartenbohnen (Phaseolus vulgaris und Phaseolus coccineus) im Kindesalter. Paediat Paedol 14(3):293-296.

Hintz HF, Hogue DE, Krook L (1967) Toxicity of red kidney beans (Phaseolus vulgaris) in the rat. J Nutr 93(1):77-86.

Hove EL, King S, Hill GD (1978) Composition, protein quality, and toxins of seeds of the grain legumes Glycine max, Lupinus spp., Phaseolus spp., Pisum sativum, and Vicia faba. N Z J Agric Res 21:457-462.

Jaffé WG (1949) Toxicity of raw kidney-beans. Experientia 5:81.

Jaffé WG (1950) Estudio sobre la inhibición del crecimeniento de ratas causado por algunas semillas de leguminosas. Acta Cient Venez 1(2):62-64.

Jaffé WG (1950) La toxicidad de las caraotas crudas para conejos. Acta Cient Venez 1(1):16-17.

Jayne-Williams DJ, Burgess CD (1974) Further observations on the toxicity of navy beans (Phaseolus vulgaris) for Japanese quail (Coturnix coturnix japonica). J Appl Bact 37(1):149-169.

Jayne-Williams DJ, Hewitt D (1972) The relationship between the intestinal microflora and the effects of diets containing raw navy beans (Phaseolus vulgaris) on the growth of Japanese quail (Coturnix coturnix japonica). J Appl Bact 35(2):331-344.

Kakade ML, Keahey KK, Whitehair CK, et al. (1965) Morphological changes in rats fed navy beans. Proc Soc Exp Biol Med 119(4):934-937.

Kerk P (1973) Diarree bij varkens door voerbonen (Phaseolus species). Tijdschr Diergeneeskd 98:1230-1233.

King TP, Begbie R, Cadenhead A (1983) Nutritional toxicity of raw kidney beans in pigs. Immunocytochemical and cytopathological studies on the gut and the pancreas. J Sci Food Agric 34(12):1404-1412.

Konishi T, Ichijo S (1970) [Clinical studies on bean-hulls poisoning in horses. I. Clinical and biochemical observations in spontaneous cases.] Res Bull Obihiro Zootech Univ 6:242-257.

Konishi T, Ichijo S (1970) [Clinical studies on bean-hulls poisoning of horse. III. Experimental feeding of poisonous bean-hull in rabbit.] Res Bull Obihiro Zootech Univ 6:275-282.

Konishi T, Ichijo S (1970) [Clinical studies on poisoning by bean pods in horses. II. Clinical and biochemical observations in experimental cases.] Res Bull Obihiro Zootech Univ 6:258-273.

Loosli JK, Krook L, Hintz HF, et al. (1964) The nature of kidney bean (Phaseolus vulgaris) toxicity in the rat. Fed Proc 23:500.

Montgomery RD (1964) Observations on the cyanide content and toxicity of tropical pulses. West Indian Med J 13(1):1-11.

Nehring K, Laube W (1962) Untersuchungen über Schadensfälle nach Verfütterung von Speisebohnen (Phaseolus vulgaris) an Rinder und Schweine. Monatsh Veterinarmed 17:888-890.

Noah ND, Bender AE, Reaidi GB, et al. (1980) Food poisoning from raw kidney beans. Br Med J 281(6234):236-237.

Paleček F (1969) Pulmonary damage in rats fed by beans (Phaseolus vulgaris). Experientia 25(3):285.

Pusztai A (1980) Nutritional toxicity of the kidney bean (Phaseolus vulgaris). Anim Nutr Allied Sci Annu Rep Studies 36:110-118.

Rattray EA, Palmer R, Pusztai A (1974) Toxicity of kidney beans (Phaseolus vulgaris L.) to conventional and gnotobiotic rats. J Sci Food Agric 25(8):1035-1040.

Rodhouse JC, Haugh CA, Roberts D, et al. (1990) Red kidney bean poisoning in the UK: An analysis of 50 suspected incidents between 1976 and 1989. Epidemiol Infect 105(3):485-491.

Toro F, Benshimol AL, Gonzalez Elorriaga M, et al. (1992) Spleen and thymus histology and proliferative response of splenic cells in rats fed raw and cooked Phaseolus vulgaris beans. Arch Latinoam Nutr 42(4):395-402.

Ueda A, Ono T, Yamigawa S (1967) [Neurohistopathological studies on bean-hulls poisoning of horses.] Res Bull Obihiro Zootech Univ 5(2):149-154.

Untawale GG, McGinnis J (1979) Effect of rye and levels of raw and autoclaved beans (Phaseolus vulgaris) on adhesion of microflora to the intestinal mucosa. Poult Sci 58(4):928-933.

Wagh PV, Klaustermeier DF, Waibel PE, et al. (1963) Nutritive value of red kidney beans (Phaseolus vulgaris) for chicks. J Nutr 80:191-195.

Note:

Bean is named *Phaseolus vulgaris* L. var. *vulgaris* in some publications.

pheasant's-eye –see– *Adonis annua* L.; *Adonis microcarpa* DC.

philodendron –see– *Philodendron bipinnatifidum* Schott ex Endl.; *Philodendron hederaceum* (Jacq.) Schott

PHILODENDRON BIPINNATIFIDUM Schott ex Endl. [Araceae]

Synonyms:

Philodendron selloum K. Koch

Common Names:

easy tree philodendron; lace-tree Philodendron; philodendron; saddleleaf

Citations:

Paulsen E (1998) Occupational dermatitis in Danish gardeners and greenhouse workers (II). Etiological factors. Contact Dermatitis 38(1):14-19.

PHILODENDRON CONSANGUINEUM Schott [Araceae]

Citations:

Engel S, Horn K (1972) Phytodermatosen durch Dictamnus alba, Sanicula europaea und Phyllodendron consanguineum. Dermatol Monatsschr 158(1):22-27.

Philodendron cordatum hort. = Philodendron hederaceum (Jacq.) Schott

PHILODENDRON HEDERACEUM (Jacq.) Schott [Araceae]

Synonyms:

Philodendron cordatum hort.; ***Philodendron scandens*** K. Koch & Sello; ***Philodendron scandens*** K. Koch & Sello subsp. oxycardium (Schott) G. S. Bunting

Common Names:

cordatum; heart-leaf ivy; heart-leaf philodendron; parlor ivy; philodendron; sweetheart plant

Citations:

Ayres S Jr, Ayres S 3rd (1958) Philodendron as a cause of contact dermatitis. Arch Dermatol 78(3):330-333.

Dorsey C (1958) Philodendron dermatitis. Calif Med 88(4):329-330.

Fernández de Corres L, Corrales JL, Muñoz D, et al. (1984) Dermatitis alérgicas de contacto por plantas. Allergol Immunopathol (Madr) 12(4):313-319.

Hammershøy O, Verdich J (1980) Allergic contact dermatitis from Philodendron scandens Koch et Sello subsp oxycardium (Schott) Bunting ("Philodendron scandens cordatum"). Contact Dermatitis 6(2):95-99.
Harris JH (1942) Dermatitis of the eyelids due to philodendron (Scandens cardatum) plants. Arch Derm Syphilol 45:1066-1068.
Knight TE (1991) Philodendron-induced dermatitis: Report of cases and review of the literature. Cutis 48(5):375-378.
Mrvos R, Krenzelok EP, Jacobsen TD (2001) Toxidromes associated with the most common plant ingestions. Vet Hum Toxicol 43(6):366-369.

PHILODENDRON LACERUM (Jacq.) Schott [Araceae]

Citations:

Knight TE (1991) Philodendron-induced dermatitis: Report of cases and review of the literature. Cutis 48(5):375-378.

Philodendron scandens K. Koch & Sello = Philodendron hederaceum (Jacq.) Schott

Philodendron scandens K. Koch & Sello subsp. oxycardium (Schott) G. S. Bunting = Philodendron hederaceum (Jacq.) Schott

Philodendron selloum K. Koch = Philodendron bipinnatifidum Schott ex Endl.

PHLEUM PRATENSE L. [Poaceae]

Common Names:

timothy

Citations:

Bergeron JM, Jodoin L (1985) Fiabilité des mesures de poids et d'examens histopathologiques dans les études d'intoxication du campagnol des champs (Microtus pennsylvanicus). Can J Zool 63(4):804-810.
Mitchell JH, Mitchell WF (1945) Season dermatitis due to the albumin fraction of timothy pollen. J Allergy 16:48-50.

phlomos –see– *Euphorbia characias* L.

PHOENIX DACTYLIFERA L. [Arecaceae]

Common Names:

date palm

Citations:

Gonzalo MA, Moneo I, Ventas P, et al. (1997) Immediate hypersensitivity reaction to date. Allergy 52(5):598-599.
Kwaasi AA, Harfi HA, Parhar RS, et al. (1999) Allergy to date fruits: Characterization of antigens and allergens of fruits of the date palm (Phoenix dactylifera L.). Allergy 54(12):1270-1277.
Kwaasi AA, Harfi HA, Parhar RS, et al. (2002) Cross-reactivities between date palm (Phoenix dactylifera L.) polypeptides and foods implicated in the oral allergy syndrome. Allergy 57(6):508-518.

Phoradendron (Pursh) flavescens Nutt. ex A. Gray = Phoradendron leucarpum (Raf.) Reveal & M. C. Johnst.

Phoradendron latifolium Griseb. = Phoradendron piperoides (Kunth) Trel.

PHORADENDRON LEUCARPUM (Raf.) Reveal & M. C. Johnst. [Santalaceae]

Synonyms:

Phoradendron flavescens (Pursh) Nutt. ex A. Gray; ***Phoradendron serotinum*** (Raf.) M. C. Johnst.; ***Phoradendron tomentosum*** (DC.) A. Gray

Common Names:

American mistletoe; Christmas mistletoe; eastern mistletoe; false mistletoe; faux gui américain; misteltoe; Texas mistletoe

Citations:

Anonymous (1984) Boy sickened by mistletoe. Associated Press. Dec 16
Buster OC (1887-1888) Poisoning by mistletoe (Viscum flavescens). Texas Courier Record Med 5:218-219.
Hall AH, Spoerke DG, Rumack BH (1986) Assessing mistletoe toxicity. Ann Emerg Med 15(11):1320-1323.
Hanzlik PJ, French WO (1924) The pharmacology of Phoradendron flavescens (American mistletoe). J Pharmacol Exp Ther 23(4):269-306.
Krenzelok EP, Jacobsen TD, Aronis J (1997) American mistletoe exposure. Am J Emerg Med 15(5):516-520.
Moore HW (1963) Mistletoe poisoning. A review of the available literature, and the report of a case of probable fatal poisoning. J S C Med Assoc 59(8):269-271.
Spiller HA, Willias DB, Gorman SE, et al. (1996) Retrospective study of mistletoe ingestion. J Toxicol Clin Toxicol 34(4):405-408.

PHORADENDRON PIPEROIDES (Kunth) Trel. [Santalaceae]

Synonyms:

Phoradendron latifolium Griseb.

Common Names:

erva-de-passarinho; mistletoe

Citations:

Queiroz-Neto A, Hubner DV, Cortez DA, et al. (1985) Avaliação da toxicidade do chá de erva-de-passarinho (Phoradendron latifolium). Ars Vet 1(1):9-16.

Phoradendron serotinum (Raf.) M. C. Johnst. = Phoradendron leucarpum (Raf.) Reveal & M. C. Johnst.

Phoradendron tomentosum (DC.) A. Gray = Phoradendron leucarpum (Raf.) Reveal & M. C. Johnst.

PHYLLANTHUS ABNORMIS Baill. [Phyllanthaceae]

Common Names:

abnormal leaf flower; abnormal leg flower; leaf flower; spurge

Citations:

Mathews FP (1941) Poisonous plants in the Davis Mountains. Texas Agric Exp Sta Annu Rep 54:93.

Mathews FP (1945) The toxicity of a spurge (Phyllanthus abnormis) for cattle, sheep, and goats. Cornell Vet 35(Oct):336-346.

PHYLLANTHUS EMBLICA L. [Phyllanthaceae]

Synonyms:

Emblica officinalis Gaertn.

Common Names:

amila

Citations:

Kulakkattolickal AT (1987) Piscicidal plants of Nepal: Preliminary toxicity screening using grass carp (Ctenopharyngodon idella) fingerlings. J Ethnopharmacol 21(1):1-9.

PHYSALIS LONGIFOLIA Nutt. [Solanaceae]

Common Names:

ground cherry; husk tomato

Citations:

Pammel LH (1919) Ground cherry leaves supposed to be poisonous. Am J Vet Med 14:606.

physicnut –see– *Cnidoscolus urens* (L.) Arthur; *Jatropha curcas* L.; *Jatropha gossypiifolia* L.; *Jatropha multifida* L.; *Jatropha podagrica* Hook.

phytolacca –see– *Phytolacca americana* L.

PHYTOLACCA ACINOSA Roxb. [Phytolaccaceae]

Synonyms:

Phytolacca esculenta Van Houtte

Citations:

Ota T, Shiraishi H, Uemura A, et al. (1967) [Poisoning by the plant Phytolacca esculenta van Houtte.] J Jpn Vet Med Assoc 20:291-292.

PHYTOLACCA AMERICANA L. [Phytolaccaceae]

Synonyms:

Phytolacca decandra L.; ***Phytolacca rigida*** Small

Common Names:

American cancer; American nightshade; cancer jalap; cancerroot; chongras; coakum; crowberry; garget; inkberry; jalap; Kermesbeere; phytolacca; phytolaque rouge; pigeonberry; pocan; pocan bush; poke; pokeberry; pokeroot; pokesalad; pokeweed; red ink plant; red pokeweed; redweed; Scharlachbeere; skoke; Virginia poke

Citations:

Barker BE, Farnes P, La Marche PH (1966) Peripheral blood plasmacytosis following systemic exposure to Phytolacca americana (pokeweed). Pediatrics 38(3):490-493.

Barker BE, Farnes P, La Marche PH (1967) Haematological effects of pokeweed. Lancet 1(Feb 25):437.

Barnett BD (1975) Toxicity of pokeberries (fruit of Phytolacca americana Large) for turkey poults. Poult Sci 54(4):1215-1217.

Brooker J, Obar C, Courtemanche L (2001) A fatality from Phytolacca americana (pokeweed) root ingestion. J Toxicol Clin Toxicol 39(5):549-550.

Callahan R, Piccola F, Gensheimer K, et al. (1981) Plant poisonings - New Jersey. MMWR Morb Mortal Wkly Rep 30(6):65-67.

Collins HL, McClung M, Spyker DA (1980) Poke - Poison or poultice. Vet Hum Toxicol 22:364.

Edwards N, Rodgers GC (1982) Pokeberry pancake breakfast - or - It's gonna be a great day! Vet Hum Toxicol 24(Suppl):135-137.

French C (1900) Pokeroot poisoning. New York Med J 72(Oct 13):653-654.

Guthrie A (1887) Poisoning by poke root. JAMA 9:125.

Hamilton RJ, Shih RD, Hoffman RS (1995) Mobitz type I heart block after pokeweed ingestion. Vet Hum Toxicol 37(1):66-67.

Jaeckle KA, Freeman FR (1981) Pokeweed poisoning. South Med J 74(5):639-640.

Kingsbury JM, Hillman RB (1965) Pokeweed (Phytolacca) poisoning in a dairy herd. Cornell Vet 55(4):534-538.

Krenzelok EP, Mrvos R, Jacobsen TD (2002) Contrary to the literature, vomiting is not a common manifestation with plant exposures. Vet Hum Toxicol 44(5):298-300.

Lawrence RA (1990) The clinical effect of pokeweed root ingestion upon 32 adults. Am Assoc Poison Control Centers 1990:116.

Lewis WH, Smith PR (1979) Poke root herbal tea poisoning. JAMA 242(25):2759-2760.

Matyunas NJ, Rodgers GC Jr (1985) Evaluation of pokeberry ingestion in children. Vet Hum Toxicol 28(4):298.

Mrvos R, Krenzelok EP, Jacobsen TD (2001) Toxidromes associated with the most common plant ingestions. Vet Hum Toxicol 43(6):366-369.

Musgrove WC (1858) Poisoning by Phytolaccoe radix (poke root) - Four cases. Southern Med Surg J 14(Apr):230-231.

Patterson FD (1929) Pokeweed causes heavy losses in swine herd. Vet Med 24(Mar):114.

Peixoto PV, Wouters F, Lemos RA, et al. (1997) Phytolacca decandra poisoning in sheep in southern Brazil. Vet Hum Toxicol 39(5):302-303.

Roberge R, Brader E, Martin ML, et al. (1986) The root of evil - Pokeweed intoxication. Ann Emerg Med 15(4):470-473.

Rodgers GC (1984) Pokeroot poisoning: Report of two cases and a review of literature. Vet Hum Toxicol 26(Suppl 2):52-53.

Smith GW, Constable PD (2002) Suspected pokeweed toxicity in a Boer goat. Vet Hum Toxicol 44(6):351-353.

Stein ZL (1979) Pokeweed-induced gastroenteritis. Am J Hosp Pharm 36(10):1303.

White GR (1902) Phytolacca poisoning in cattle. Am Vet Rev 26:339-340.

Phytolacca decandra L. = Phytolacca americana L.

PHYTOLACCA DIOICA L. [Phytolaccaceae]

Common Names:

bella sombra; packalacca; umbu

Citations:

Storie GJ, McKenzie RA, Fraser IR (1992) Suspected packalacca (Phytolacca dioica) poisoning of cattle and chickens. Aust Vet J 69(1):21-22.

PHYTOLACCA DODECANDRA L'Her. [Phytolaccaceae]

Common Names:

hoko

Citations:

Ephraim M, Alemshet WA, Tesfaye L, et al. (1979) Studies on the toxicity of Phytolacca dodecandra to Ethiopian Highland sheep. Bull Anim Health Prod Afr 27:79-86.

Mettam RW (1939) Poisoning by Phytolacca dodecandra L'Herit. Family: Phytolaccaceae. Vet J 95:135-138.

Mugera GM (1970) Phytolacca dodecandra L'Herit toxicity in livestock in Kenya. Bull Epizootic Dis Afr 18(1):41-43.

Phytolacca esculenta Van Houtte = Phytolacca acinosa Roxb.

Phytolacca rigida Small = Phytolacca americana L.

phytolaque rouge –see– *Phytolacca americana* L.

pica pica –see– *Mucuna pruriens* (L.) DC.

PICEA ABIES (L.) H. Karst. [Pinaceae]

Synonyms:

Picea excelsa (Lam.) Link

Common Names:

balsam-of-spruce; Fichte

Citations:

Fregert S, Rorsman H (1963) Hypersensitivity to balsam of pine and spruce. Arch Dermatol 87(Jun):693-695.

Picea excelsa (Lam.) Link = Picea abies (L.) H. Karst.

pickaback plant –see– *Tolmiea menziesii* Torr. & A. Gray

PICRADENIOPSIS OPPOSITIFOLIA (Nutt.) Rydb. [Asteraceae]

Synonyms:

Bahia oppositifolia (Nutt.) DC.

Common Names:

Bahia; plains Bahia

Citations:

Deem AW, Thorp F Jr, Durrell LW (1939) Range plant newly found to be poisonous. Science 89(May 12):435.

pie-de-grifo –see– *Helleborus foetidus* L.

pie-de-niño –see– *Pedilanthus tithymaloides* (L.) Poit.

picry –see– *Toxicodendron radicans* (L.) Kuntze

pie melon –see– *Citrullus lanatus* (Thunb.) Matsum. & Nakai

pie plant –see– *Rheum* ×*hybridum* Murray

pied-de-veau –see– *Arum maculatum* L.

pied mango –see– *Mangifera indica* L.

pieris –see– *Pieris japonica* (Thunb.) D. Don ex G. Don

PIERIS FLORIBUNDA (Pursh) Benth. & Hook. f. [Ericaceae]

Synonyms:

Andromeda floribunda Pursh

Common Names:

fetterbush; mountain fetterbush; mountain pieris

Citations:

Taylor H (1918) Poisoning by Andromeda floribunda. Vet Rec 30(1545):335-336.

PIERIS FORMOSA (Wall.) D. Don [Ericaceae]

Citations:

Hollands RD, Hughes MC (1986) Pieris formosanum poisoning in the goat. Vet Rec 118(14):407-408.

PIERIS JAPONICA (Thunb.) D. Don ex G. Don [Ericaceae]

Common Names:

Japanese pieris; lily-of-the-valley bush; pieris

Citations:

Brahm E, Buntenkötter S, Simanowski W (1973) Vergiftungen von wiederkäuenden Paarhufern durch Ericaceen im Dormunder Tierpark. In: Ippen R et al. (eds.) Erkrank der Zoot XV. Int Symp, Kolmarden Akad Verlag. pp. 125-130.

Gillis WT, Ball W Sr, Ball W Jr (1961) Pieris poisoning in zoo animals. Michigan State Univ Vet 22(1):24-26.

Koller LD, Helfer DH (1978) Adiaspiromycosis in the lungs of a goat. J Am Vet Med Assoc 173(1):80-81.

Pizzi R, Goodman G, Gunn-Moore D, et al. (2005) Pieris japonica intoxication in an African spurred tortoise (Geochelone sulcata). Vet Rec 156(15):487-488.

Plumlee KH, Van Alstine WG, Sullivan JM (1992) Japanese pieris toxicosis of goats. J Vet Diagn Invest 4(3):363-364.

Power SB, O'Donnell PG, Quirk EG (1991) Pieris poisoning in sheep. Vet Rec 128(25):599-600.

Smith MC (1978) Japanese pieris poisoning in the goat. J Am Vet Med Assoc 173(1):78-79.

Smith MC (1979) Fetal mummification in a goat due to Japanese pieris (Pieris japonica) poisoning. Cornell Vet 69(1):85-87.

Visser IJ, van den Hoven R, Vos JH, et al. (1988) Pieris japonica (pieris)-intoxicatie bij twee geiten. Tijdschr Diergeneeskd 113(4):185-189.

pig nut –see– *Simmondsia chinensis* (Link) C. K. Schneid.

pigaeble –see– *Datura stramonium* L.

pigeon pea –see– *Cajanus cajan* (L.) Millsp.

pigeonberry –see– *Duranta erecta* L.; *Phytolacca americana* L.

pigeongrass –see– *Setaria pumila* (Poir.) Roem. & Schult.; *Verbena officinalis* L.

piggeple –see– *Datura stramonium* L.

piggyback plant –see– *Tolmiea menziesii* (Pursh) Torr. & A. Gray

pignola –see– *Pinus pinaster* Aiton

pignon –see– *Pinus pinaster* Aiton

pig's-ear –see– *Cotyledon orbiculata* L.

pigweed –see– *Amaranthus hybridus* L.; *Amaranthus palmeri* S. Watson; *Amaranthus quitensis* Kunth; *Amaranthus retroflexus* L.

pigwood –see– *Euonymus europaeus* L.

pila dhatura –see– *Argemone mexicana* L.

pila kaner –see– *Thevetia peruviana* (Pers.) K. Schum.

Pillenbaum –see– *Euphorbia lathyris* L.

Pimelea altior F. Muell. = Pimelea latifolia R. Br. subsp. altior (F. Muell.) Threlfall

Pimelea continua J. M. Black = Pimelea simplex F. Muell. subsp. continua (J. M. Black) Threlfall

PIMELEA DECORA Domin [Thymelaeaceae]

Common Names:

Flinders poppy; pimelea poppy

Citations:

Hill MW (1970) Toxicity of Pimelea decora in horses. Aust Vet J 46(6):287-289.

PIMELEA FLAVA R. Br. [Thymelaeaceae]

Common Names:

yellow riceflower

Citations:

Cleland JB (1931) Plants, including fungi, poisonous or otherwise injurious to man in Australia. Series III. Med J Aust 2(Dec 19):775-778.

PIMELEA LATIFOLIA R. Br. subsp. altior (F. Muell.) Threlfall [Thymelaeaceae]

Synonyms:

Pimelea altior F. Muell.

Citations:

Rogers RJ, Roberts KH (1976) Pimelea altior poisoning of cattle. Aust Vet J 52(4):193-194.

PIMELEA NEOANGELICA Threlfall [Thymelaeaceae]

Citations:

Storie GJ, Norman JL, McKenzie RA (1986) Pimelea neoanglica poisoning of cattle. Aust Vet J 63(7):235.

PIMELEA PAUCIFLORA R. Br. [Thymelaeaceae]

Common Names:

poison pimelea; scanty rice flower; scrub kurrajong

Citations:

Anonymous (1929) The poison plants committee. J CSIRO Aust 2:40-48.

Seddon HR, Hindmarsh WL, McGrath TT (1933) Pimelea pauciflora (scrub kurrajong), proved to be toxic for sheep. New South Wales Dep Agric Vet Res Rep #6(Part 3):122-125.

pimelea poppy –see– *Pimelea decora* Domin

PIMELEA PROSTRATA (J. R. Forst. & G. Forst.) Willd. [Thymelaeaceae]

Common Names:

creeping pimelea; prostrate pimelea; Strathmore weed

Citations:

Webster WM (1926) Two recent cases of plant poisoning among stock. N Z J Agr 33:102-105.

PIMELEA SIMPLEX F. Muell. [Thymelaeaceae]

Common Names:

desert rice flower; flaxweed; poverty weed

Citations:

McClure TJ, Farrow BR (1971) Chronic poisoning of cattle by desert rice flower (Pimelea simplex) and its resemblance to St. George disease as seen in north-western New South Wales. Aust Vet J 47(3):100-102.

Roberts HB, Healy PJ (1971) Pimelea simplex and St. George disease of cattle. Aust Vet J 47(3):123-124.

PIMELEA SIMPLEX F. Muell. subsp. continua (J. M. Black) Threlfall [Thymelaeaceae]

Synonyms:

Pimelea continua J. M. Black

Citations:

Kelly WR (1975) The pathology and haematological changes in experimental Pimelea spp. poisoning in cattle ("St. George disease"). Aust Vet J 51(5):233-243.

Kelly WR, Seawright AA (1978) Pimelea spp. poisoning of cattle. In: Keeler RF, Van Kampen KR, James LF, et al. (eds.) Effects of poisonous plants on livestock. Academic Press. New York. pp. 293-300.

PIMELEA TRICHOSTACHYA Lindl. [Thymelaeaceae]

Common Names:

bootlace plant; Borgia's-bouquet; broom bush; flaxweed; poverty weed; spiked riceflower; wild flax

Citations:

Clark IA (1971) A note on the pathogenesis of St. George disease of cattle. Aust Vet J 47(6):285-286.

Clark IA (1973) The pathogenesis of St George disease of cattle. Res Vet Sci 14(3):341-349.

Kelly WR (1975) 59Fe utilization and excretion in anaemia of cattle caused by Pimelea trichostachya intoxication. Aust Vet J 51(11):504-510.

Kelly WR (1975) The pathology and haematological changes in experimental Pimelea spp. poisoning in cattle ("St. George disease"). Aust Vet J 51(5):233-243.

Kelly WR, Seawright AA (1978) Pimelea spp. poisoning of cattle. In: Keeler RF, Van Kampen KR, James LF, et al. (eds.) Effects of poisonous plants on livestock. Academic Press. New York. pp. 293-300.

PIMENTA DIOICA (L.) Merr. [Myrtaceae]

Common Names:

allspice

Citations:

Niinimäki A (1984) Delayed-type allergy to spices. Contact Dermatitis 11(1):34-40.

pimienta-de-Brasil –see– *Schinus terebinthifolius* Raddi

pimientillo –see– *Karwinskia humboldtiana* (Schult.) Zucc.

pimpernel –see– *Anagallis arvensis* L.

PIMPINELLA ANISUM L. [Apiaceae]

Common Names:

anis; matalahuga aniseed

Citations:

González-Gutiérrez ML, Sánchez-Fernández C, Esteban-López MI, et al. (2000) Allergy to anis. Allergy 55(2):195-196.

Loveman AB (1938) Report of a case of sensitivity of the mucous membranes and the skin to oil of anise. Arch Derm Syphilol 37:70-81.

Tuckler V, Peck C, Nesbitt C, et al. (2002) Seizure in an infant from aniseed oil toxicity. J Toxicol Clin Toxicol 40(5):689.

pin cherry –see– *Prunus serotina* Ehrh.

pinang wang –see– *Areca catechu* L.

piñanona –see– *Monstera deliciosa* Liebm.

pine top –see– *Equisetum arvense* L.

pineapple –see– *Ananas comosus* (L.) Merr.

pinegrass –see– *Equisetum arvense* L.

PINELLIA TERNATA (Thunb.) Makino [Araceae]

Common Names:

dragon arum; Jack-in-the-pulpit; mandrake

Citations:

Zhang ZZ (1983) [Experimental study on "18 against compatibilities" of Chinese traditional medicine: Acute toxicity of Aconitum carmichaeli and Pinellia ternata.] Bull Chin Materia Medica 8(4):33-34.

pingrass –see– *Erodium cicutarium* (L.) L'Her.

pingree –see– *Hymenoxys richardsonii* (Hook.) Cockerell

pingue –see– *Hymenoxys richardsonii* (Hook.) Cockerell

pinhão-de-purga –see– *Jatropha curcas* L.

pinhoen –see– *Jatropha multifida* L.

pinipini –see– *Pedilanthus tithymaloides* (L.) Poit.

pinipiniche –see– *Hippomane mancinella* L.

pink allamanda –see– *Cryptostegia grandiflora* R. Br.

pink-and-white shower –see– *Cassia javanica* L. subsp. nodosa (Buch.-Ham. ex Roxb.) K. Larsen & S. S. Larsen

pink cockle –see– *Vaccaria hispanica* (Mill.) Rauschert

pink-eye cerbera –see– *Cerbera odollam* Gaertn.

pink lady's-slipper –see– *Cypripedium acaule* Aiton

pink locoweed –see– *Oxytropis lambertii* Pursh

pink oleander –see– *Nerium oleander* L.

pink peppercorn –see– *Schinus terebinthifolius* Raddi

pink periwinkle –see– *Catharanthus roseus* (L.) G. Don

pink siris –see– *Albizia julibrissin* Durazz.

pino-de-coral –see– *Jatropha multifida* L.

pinoccios –see– *Pinus pinaster* Aiton

piñol-de-cumaña –see– *Jatropha multifida* L.

piñon amoroso –see– *Gliricidia sepium* (Jacq.) Kunth ex Walp.

piñon florido –see– *Gliricidia sepium* (Jacq.) Kunth ex Walp.

piñon nut –see– *Pinus pinaster* Aiton

piñoncillo –see– *Jatropha curcas* L.

pinto bean –see– *Phaseolus vulgaris* L.

PINUS CONTORTA Douglas ex Loudon [Pinaceae]

Common Names:

limber pine; lodgepole pine; shore pine

Citations:

Gardner DR, Panter KE, James LF, et al. (1998) Abortifacient effects of lodgepole pine (Pinus contorta) and common juniper (Juniperus communis) on cattle. Vet Hum Toxicol 40(5):260-263.

PINUS CUBENSIS Griseb. [Pinaceae]

Citations:

Hernandez J, Aguilera JM, Rodriguez F, et al. (1979) Abortos en vacas provocados por la ingestion de hojas o agujas de pino (Pinus cubensis). Parte II. Cienc Tec Agric Vet 1(1-2):55-71.

Hernandez JA, Rodriguez F, Gutierrez E, et al. (1979) Ensayo biologico sobre el efecto del pino (Pinus cubensis) como efecto causal de abortos en ratonas y cobayas gestantes. Cienc Tec Agric Vet 1(1-2):7-12.

PINUS KORAIENSIS Siebold & Zucc. [Pinaceae]

Common Names:

Korean pine

Citations:

Kim IH, Choi KC, An BS, et al. (2003) Effect on abortion of feeding Korean pine needles to pregnant Korean native cows. Can J Vet Res 67(3):194-197.

PINUS PINASTER Aiton [Pinaceae]

Common Names:

cluster pine; European pine; French maritime pine; pignola; pignon; pinoccios; piñon nut; star pine

Citations:

de las Marinas D, Vila L, Sanz ML (1998) Allergy to pine nuts. Allergy 53(2):220-222.

García-Menaya JM, Gonzalo-Garijo MA, Moneo I, et al. (2000) A 17-kDa allergen detected in pine nuts. Allergy 55(3):291-293.

Roux N, Hogendijk S, Hauser C (1998) Severe anaphylaxis to pine nuts. Allergy 53(2):213-214.

PINUS PONDEROSA C. Lawson [Pinaceae]

Common Names:

blackjack pine; bull pine; ponderosa pine; western yellow pine; yellow pine

Citations:

Adams DC, Pfister JA, Short RE, et al. (1992) Pine needle effects on in vivo and in vitro digestibility of crested wheatgrass. J Range Manag 45(3):249-253.

Anderson CK, Lozano EA (1977) Pine needle toxicity in pregnant mice. Cornell Vet 67(2):229-235.

Call JW, James LF (1978) Pine needle abortion in cattle. In: Keeler RF, Van Kampen KR, James LF, et al. (eds.) Effects of poisonous plants on livestock. Academic Press. New York. pp. 587-590.

Chow FH, Hanson KJ, Hamar DW, et al. (1972) Reproductive failure of mice caused by pine needle ingestion. J Reprod Fertil 30(1):169-172.

Christenson LK, Short RE, Ford SP (1992) Effects of ingestion of ponderosa pine needles by late-pregnant cows on uterine blood flow and steroid secretion. J Anim Sci 70(2):531-537.

Christenson LK, Short RE, Rosazza JP, et al. (1992) Specific effects of blood plasma from beef cows fed pine needles during late pregnancy on increasing tone of caruncular arteries in vitro. J Anim Sci 70(2):525-530.

Ford SP, Christenson LK, Rosazza JP, et al. (1992) Effects of ponderosa pine needle ingestion on uterine vascular function in late-gestation beef cows. J Anim Sci 70(5):1609-1614.

James LF, Call JW, Stevenson AH (1977) Experimentally induced pine needle abortion in range cattle. Cornell Vet 67(2):294-299.

James LF, Short RE, Panter KE, et al. (1989) Pine needle abortion in cattle: A review and report of 1973-1984 research. Cornell Vet 79(1):39-52.

Jensen R, Pier AC, Kaltenbach CC, et al. (1989) Evaluation of histopathologic and physiologic changes in cows having premature births after consuming ponderosa pine needles. Am J Vet Res 50(2):285-289.

MacDonald MA (1952) Pine needle abortion in range beef cattle. J Range Manag 5:150-155.

Neff TE, Adams CJ, Jackson LL (1979) Fetal death in mice induced by ponderosa pine needles. Fed Proc 38:1189.

Neff TE, Adams CJ, Jackson LL (1982) Pathological effects of pine needle ingestion in pregnant mice. Cornell Vet 72(2):128-136.

Panter KE, James LF, Molyneux RJ (1992) Ponderosa pine needle-induced parturition in cattle. J Anim Sci 70(5):1604-1608.

Panter KE, James LF, Molyneux RJ, et al. (1990) Premature bovine parturition induced by ponderosa pine: Effects of pine needles, bark and branch tips. Cornell Vet 80(4):329-338.

Phelps DA, Short RE, Panter KE, et al. (1987) Effects of presence of CL on pine needle-induced abortion in cattle and of alternative treatments for retained placenta. J Anim Sci 65(Suppl 1):415-416.

Short RE, Bellows RA, Staigmiller RB, et al. (1994) Pine needle abortion in cattle: Effects of diet variables on consumption of pine needles and parturition response. J Anim Sci 72(4):805-810.

Short RE, Ford SP, Grings EE, et al. (1995) Abortifacient response and plasma vasoconstrictive activity after feeding needles from ponderosa pine trees to cattle and sheep. J Anim Sci 73(7):2102-2104.

Short RE, Ford SP, Rosazza JP, et al. (1996) Effects of feeding pine needle components to late pregnant cattle. Proc West Sec Am Soc Anim Sci 47:193-195.

Short RE, James LF, Panter KE, et al. (1992) Effects of feeding ponderosa pine needles during pregnancy - Comparative studies with bison, cattle, goats, and sheep. J Anim Sci 70(11):3498-3504.

Stegelmeier BL, Gardner DR, James LF, et al. (1994) The toxic and abortifacient effects of ponderosa pine. Vet Pathol 31:607.

Stegelmeier BL, Gardner DR, James LF, et al. (1996) The toxic and abortifacient effects of ponderosa pine. Vet Pathol 33(1):22-28.

Stuart LD, James LF, Panter KE, et al. (1989) Pine needle abortion in cattle: Pathological observations. Cornell Vet 79(1):61-69.

PINUS RADIATA D. Don [Pinaceae]

Common Names:

Monterey pine; radiata pine

Citations:

Burry JN (1976) Contact dermatitis from radiata pine. Contact Dermatitis 2(12):262-263.

Burry JN (1977) Environmental dermatitis. Contact dermatitis from Pinus radiata. Med J Aust 2(1):13-14.

Knowles RL, Dewes HF (1980) Pinus radiata implicated in abortion. N Z Vet J 28(Mar 28):103.

PINUS SYLVESTRIS L. [Pinaceae]

Common Names:

balsam-of-pine; Kiefer

Citations:

Beer WE (1970) Sensitivity to pine in a joiner's mate. Contact Dermatol Newsl 8(Aug):191.

Bonfiglio JF, Hintze KL, Sigell LT (1987) A multi-center evaluation of the incidence and severity of pine-oil ingestions. Vet Hum Toxicol 29(6):481.

Fregert S, Rorsman H (1963) Hypersensitivity to balsam of pine and spruce. Arch Dermatol 87(Jun):693-695.

Fregert S, Rorsman H (1963) Simultaneous hypersensitivity to balsam of pine and to balsam of Peru. Acta Derm Venereol 42:21-22.

PINUS THUNBERGII Parl. [Pinaceae]

Common Names:

Japanese black pine

Citations:

Nakamura T (1986) Contact dermatitis to Japanese black pine. Contact Dermatitis 14(5):317.

PIPER METHYSTICUM G. Forst. [Piperaceae]

Common Names:

awa; kava; kava kava; kawa; kew; pepper plant

Citations:

Ballesteros S, Adán S, Ramón MF, et al. (2001) Severe adverse effect associated with kava-kava. J Toxicol Clin Toxicol 39(3):312.

Brauer RB, Stangl M, Stewart JR, et al. (2003) Acute liver failure after administration of herbal tranquilizer kava-kava (Piper methysticum). J Clin Psychiatry 64(2):216-218.

Clough AR, Bailie RS, Currie B (2003) Liver function test abnormalities in users of aqueous kava extracts. J Toxicol Clin Toxicol 41(6):821-829.

Garner LF, Klinger JD (1985) Some visual effects caused by the beverage kava. J Ethnopharmacol 13(3):307-311.

Humberston CL, Akhtar J, Krenzelok EP (2001) Acute hepatitis induced by kava kava, an herbal product derived from Piper methysticum. J Toxicol Clin Toxicol 39(5):549.

Humberston CL, Akhtar J, Krenzelok EP (2003) Acute hepatitis induced by kava kava. J Toxicol Clin Toxicol 41(2):109-113.

Leung N (2004) Acute urinary retention secondary to kava ingestion. Emerg Med Australas 16(1):94.

Mathews JD, Riley MD, Fejo L, et al. (1988) Effect of heavy usage of kava on physical health: Summary of a pilot survey in an aboriginal community. Med J Aust 148(Jun 6):548-555.

Palmer M, Haller C, McKinney P, et al. (1999) Botanicals and other dietary supplements: Adverse events by age. J Toxicol Clin Toxicol 37(5):609.

Schulze J, Raasch W, Siegers CP (2003) Toxicity of kava pyrones, drug safety and precautions - A case study. Phytomedicine 10(Suppl 4):68-73.

Stickel F, Baumüller HM, Seitz K, et al. (2003) Hepatitis induced by kava (Piper methysticum rhizoma). J Hepatol 39(1):62-67.

Strahl S, Ehret V, Dahm HH, et al. (1998) Nekrotisierende Hepatitis nach Einnahme pflanzlicher Heilmittel. Dtsch Med Wochenschr 123(47):1410-1414.

PIPER NIGRUM L. [Piperaceae]

Common Names:

black pepper; pepper; table pepper; white pepper

Citations:

Cohle SD (1986) Homicidal asphyxia by pepper aspiration. J Forensic Sci 31(4):1475-1478.

Cohle SD, Trestrail JD 3rd, Graham MA, et al. (1988) Fatal pepper aspiration. Am J Dis Child 142(6):633-636.

Mrvos R, Krenzelok EP, Jacobsen TD (2001) Toxidromes associated with the most common plant ingestions. Vet Hum Toxicol 43(6):366-369.

Sheahan K, Page DV, Kemper T, et al. (1988) Childhood sudden death secondary to accidental aspiration of black pepper. Am J Forensic Med Pathol 9(1):51-53.

Stäger J, Wüthrich B, Johansson SG (1991) Spice allergy in celery-sensitive patients. Allergy 46(6):475-478.

pipiltzinzintli –see– *Salvia divinorum* Epling & Jativa

Piptadenia macrocarpa Benth. = Anadenanthera colubrina (Vell.) Brenan var. cebil (Griseb.) Altschul

PIPTADENIA VIRIDIFLORA (Kunth) Benth. [Fabaceae]

Citations:

Brito MF, Franca TN, Oliveira KD, et al. (2000) Estudos experimentais em coelhos com plantas cianogênicas. Pesq Vet Bras 20(2):65-70.

Tokarnia CH, Peixoto PV, Brito MF, et al. (1999) Estudos experimentais com plantas cianogênicas em bovinos. Pesq Vet Bras 19(2):84-90.

PIQUERIA TRINERVIA Cav. [Asteraceae]

Citations:

Langley WD (1937) Sensitivity to piqueria pollen with case report. J Allergy 9:60-61.

Piquiá marfim –see– *Balfourodendron riedelianum* (Engl.) Engl.

pirae –see– *Persicaria hydropiper* (L.) Spach; Polygonum nepalense

PISONIA ALBA Span. [Nyctaginaceae]

Citations:

Fernando R, Fernando DN (1990) Poisoning with plants and mushrooms in Sri Lanka: A retrospective hospital based study. Vet Hum Toxicol 32(6):579-581.

pistachio nut –see– *Pistacia vera* L.

PISTACIA LENTISCUS L. [Anacardiaceae]

Common Names:

mastic; mastic gum; mastich; mastiche; mastis; mastix

Citations:

Keynan N, Tamir R, Waisel Y, et al. (1997) Allergenicity of the pollen of pistacia. Allergy 52(3):323-330.

Silanikove N, Gilboa N, Nir I, et al. (1996) Effect of a daily supplementation of polyethylene glycol on intake and digestion of tannin-containing leaves (Quercus calliprinos, Pistacia lentiscus, and Ceratonia siliqua) by goats. J Agric Food Chem 44:199-205.

PISTACIA TEREBINTHUS L. subsp. palaestina (Boiss.) Engl. [Anacardiaceae]

Citations:

Keynan N, Tamir R, Waisel Y, et al. (1997) Allergenicity of the pollen of pistacia. Allergy 52(3):323-330.

PISTACIA VERA L. [Anacardiaceae]

Common Names:

pistachio nut

Citations:

Keynan N, Tamir R, Waisel Y, et al. (1997) Allergenicity of the pollen of pistacia. Allergy 52(3):323-330.

Liccardi G, Russo M, Mistrello G, et al. (1999) Sensitization to pistachio is common in Parietaria allergy. Allergy 54(6):643-645.

PISUM SATIVUM L. [Fabaceae]

Common Names:

Alaska pea; Austrian pea; black pea; Canadian field pea; Erbse; field pea; Futtererbse; green pea; matar; pea; split pea

Citations:

Anonymous (1963) Peavine silage paralysis. Feedstuffs 35(Jun 22):24.

Castanon JI, Perez-Lanzac, J (1990) Substitution of fixed amounts of soybean meal for field beans (Vicia faba), sweet lupines (Lupinus albus), cull peas (Pisum sativum) and vetches (Vicia sativa) in diets for high performance laying Leghorn hens. Br Poult Sci 31(1):173-180.

Clarke EG, Humphreys DJ (1971) Toxic factors in pea haulm silage effluent - The factors toxic to fish. J Sci Food Agric 22(4):205-207.

Hove EL, King S, Hill GD (1978) Composition, protein quality, and toxins of seeds of the grain legumes Glycine max, Lupinus spp., Phaseolus spp., Pisum sativum, and Vicia faba. N Z J Agric Res 21:457-462.

Kienholz EW, Jensen LS, McGinnis J (1959) Improvement of nutritional value of peas by cooking. Proc Soc Exp Biol Med 102:35-38.

Kienholz EW, Jensen LS, McGinnis J (1962) Evidence for chick growth inhibitors in several legume seeds. Poult Sci 41:367-371.

Martinez JA, Esparza ML, Larralde J (1995) Immunological changes in growing mice fed on diets containing casein or peas (Pisum sativum var. belinda) as the source of protein. Br J Nutr 73(1):87-97.

Reardon CJ, McKenzie RA (2002) Pea mania: Deranged behaviour in cattle grazing a pea crop (Pisum sativum var arvense). Aust Vet J 80(10):617-619.

Salmon WD, Sewell WE (1936) Lameness in hogs produced by Austrian pea (Pisum arvense) forage. Alabama Agric Exp Sta Annu Rep #47:17-18.

Whiting F, Connell R, Plummer PJ, et al. (1957) Incoordination (cerebellar ataxia) among lambs from ewes fed peavine silage. Can J Comp Med Vet Sci 21(3):77-84.

Note:

Pea is named *Pisum sativum* L. subsp. *sativum* var. *arvense* (L.) Poir. in some publications.

Pithecellobium jiringa (Jack) Prain = Archidendron jiringa (Jack) I. C. Nielsen

Pithecellobium lobatum Benth. = Archidendron jiringa (Jack) I. C. Nielsen

Pithecolobium saman Benth. = Samanea saman (Jacq.) Merr.

pitis pitis –see–*Sarcolobus globosus* Wall.

Pituranthos triradiatus (Hochst. ex Boiss.) Asch. & Schweinf. = Deverra triradiata Hochst. ex Boiss.

plains Bahia –see– *Picradeniopsis oppositifolia* (Nutt.) Rydb.

plains larkspur –see– *Delphinium carolinianum* Walter subsp. virescens (Nutt.) R. E. Brooks

plains whorled milkweed –see– *Asclepias pumila* (A. Gray) Vail

plakkies –see– *Cotyledon orbiculata* L. var. oblonga (Haw.) DC.; *Tylecodon ventricosa* (Burm. f.) Toelken; *Tylecodon wallichii* (Harv.) Toelken

plane tree –see– *Platanus orientalis* L.

PLANTAGO LANCEOLATA L. [Plantaginaceae]

Common Names:

English plantage; English plantain; llantén menor; plantain

Citations:

Anastasov G (1972) [Cardiotoxic effect of medicinal herbs containing glucosides (intoxication from lanceolate plantain - Plantago lanceolata, family Plantaginaceae).] Vutr Boles 11(4):112-116.

Asero R, Mistrello G, Roncarolo D, et al. (2000) Detection of allergens in plantain (Plantago lanceolata) pollen. Allergy 55(11):1059-1062.

plantain –see– *Musa ×paradisiaca* L.; *Plantago lanceolata* L.

PLATANUS ×ACERIFOLIA (Aiton) Willd. [Platanaceae]

Common Names:

London plane tree

Citations:

Asturias JA, Ibarrola I, Bartolomé B, et al. (2002) Purification and characterization of Pla a 1, a major allergen from Platanus acerifolia pollen. Allergy 57(3):221-227.

Enrique E, Cisteró-Bahima A, Bartolomé B, et al. (2002) Platanus acerifolia pollinosis and food allergy. Allergy 57(4):351-356.

PLATANUS ORIENTALIS L. [Platanaceae]

Common Names:

plane tree

Citations:

Ross AF, Mitchell JC (1974) Respiratory irritation by leaf hair of the tree Platanus. Ann Allergy 32(2):94-97.

Platte groundsel –see– *Senecio plattensis* Nutt.

Platterbse –see– *Lathyrus odoratus* L.; *Lathyrus tingitanus* L.

PLATYCLADUS ORIENTALIS (L.) Franco [Cupressaceae]

Synonyms:

Thuja orientalis L.

Common Names:

Abendlandischer Lebensbaum; Lebensbaum; oriental arbor vitae

Citations:

Zubiri A, Obras-Loscertales JM (1970) Caso curioso de dermitis por contacto. Actas Dermosifiliogr 61(9):277-280.

PLECTRANTHUS BARBATUS Andrews [Lamiaceae]

Common Names:

maigoya

Citations:

Owili DM (1977) Perianal dermatitis in Kenya due to Plectranthus barbatus leaves (maigoya leaves). East Afr Med J 54(10):571-573.

PLEURAPHIS RIGIDA Thurb. [Poaceae]

Synonyms:

Hilaria rigida (Thurb.) Benth. ex Scribn.

Common Names:

big galleta; galletagrass

Citations:

Quortrup ER, McFarland RJ (1956) Animal losses involving noxious weeds in San Diego County. California Vet 9(5).14-17.

pleurisy root –see– *Asclepias tuberosa* L.

plum –see– *Prunus domestica* L.

plumajillo –see– *Achillea millefolium* L.

PLUMBAGO SCANDENS L. [Plumbaginaceae]

Common Names:

borico; canutillo; hierba-des-pescado; leadwort; louco

Citations:

Medeiros RM, Barbosa RC, Lima EF, et al. (2001) Intoxication of goats by Plumbago scandens in northeastern Brazil. Vet Hum Toxicol 43(3):167 169.

Tokarnia CH, Döbereiner J (1982) Intoxicação experimental por Plumbago scandens (Plumbaginaceae) em bovinos. Pesq Vet Bras 2(3):105-112.

plumeria –see– *Plumeria rubra* L.

PLUMERIA RUBRA L. [Apocynaceae]

Common Names:

caracuha colorado; flor-de-mayo; frangipani; graveyard flower; graveyard plumeria; lal champa; pagoda tree; palo-de-la-cruz; plumeria; rose-pink frangipani; temple flower; temple tree

Citations:

Gunawardana VK, Goonasekera MM, Gunaherath GM, et al. (1998) Embryotoxic effect of Plumeria rubra. In: Garland T, Barr AC (eds.) Toxic plants and other natural toxicants. CABI. New York. pp. 317-322.

Poa aquatica L. = Glyceria maxima (Hartm.) Holmb.

pocan bush –see– *Phytolacca americana* L.

pod peppe –see– *Capsicum annuum* L.

podagrica –see– *Jatropha podagrica* Hook.

podophyllum –see– *Podophyllum peltatum* L.

Podophyllum emodi Wall. ex Honigberger = Podophyllum hexandrum Royle

PODOPHYLLUM HEXANDRUM Royle [Berberidaceae]

Synonyms:

Podophyllum emodi Wall. ex Honigberger

Common Names:

guijiu; Indian mandrake; Indian podophyllum; tao-er-qi

Citations:

But PP (1994) Herbal poisoning caused by adulterants or erroneous substitutes. J Trop Med Hyg 97(6):371-374.

But PP, Tomlinson B, Cheung KO, et al. (1996) Adulterants of herbal products can cause poisoning. Br Med J 313(7049):117.

Ng TH, Chan YW, Yu YL, et al. (1991) Encephalopathy and neuropathy following ingestion of a Chinese herbal broth containing podophyllin. J Neurol Sci 101:107-113.

PODOPHYLLUM PELTATUM L. [Berberidaceae]

Common Names:

American mandrake; American May apple; bajiaolian; behen; devil's-apple; duck's-foot; elephant's-ear; ground lemon; hog apple; Indian apple; Mandragore d'Amerique; mandrake; May apple; podophyllum; raccoonberry; umbrella leaf; vegetable calomel; vegetable mercury; wild buck foot; wild jalap; wild lemon; wild mandrake

Citations:

Balucani M, Zellers DD (1964) Podophyllum resin poisoning with complete recovery. JAMA 189(8):639-640.

Campbell AN (1980) Accidental poisoning with podophyllin. Lancet 1(Jan 26):206-207.

Cassidy DE, Drewry J, Fanning JP (1982) Podophyllum toxicity: A report of a fatal case and review of the literature. J Toxicol Clin Toxicol 19(1):35-44.

Chamberlain MJ, Reynolds AL, Yeoman WB (1972) Toxic effect of podophyllum application in pregnancy. Br Med J 3(5823):391-392.

Chang LW, Yang CM, Chen CF, et al. (1992) Experimental podophyllotoxin (bajiaolian) poisoning: II. Effects on the liver, intestine, kidney, pancreas and testis. Biomed Environ Sci 5(4):293-302.

Clark AN, Parsonage MJ (1957) A case of podophyllum poisoning with involvement of the nervous system. Br Med J 2(5054):1155-1157.

Dobb GJ, Edis RH (1984) Coma and neuropathy after ingestion of herbal laxative containing podophyllin. Med J Aust 140(8):495-496.

Fenton WN (1941) Iroquois suicide: A study in the stability of a culture pattern. Smithsonian Bur Am Ethnology Bull #128:80-137.

Filley CM, Graff-Radford NR, Lacey JR, et al. (1982) Neurologic manifestations of podophyllin toxicity. Neurology 32(3):308-311.

Frasca T, Brett AS, Yoo SD (1997) Mandrake toxicity: A case of mistaken identity. Arch Intern Med 157(17):2007-2009.

Holdright DR, Jahangiri M (1990) Accidental poisoning with podophyllin. Hum Exp Toxicol 9:55-56.
Juurlink DN, Sellens C, Thompson M, et al. (1999) Danger in the doctor's office: Two cases of severe neurologic sequelae after ingestion of podophyllin. J Toxicol Clin Toxicol 37(5):620.
Kaymakçalan S (1964) Fatal poisoning with podophyllum resin. JAMA 190(Nov 9):558.
McFarland MF 3rd, McFarland J (1981) Accidental ingestion of Podophyllum. Clin Toxicol 18(8):973-977.
McIntosh RA (1928) May apple poisoning in a cow. Ontario Vet Coll Rep 1927(29):18-20.
Moher LM, Mauer SA (1979) Podophyllin toxicity: Case report and literature review. J Fam Pract 9(2):237-240.
Nelson LM (1953) Use of podophyllin (Podophyllum resin) in dermatology. Arch Derm Syphilol 67(5):488-495.
O'Donovan WJ (1935) Dermatitis due to Podophyllum resin. Br J Dermatol 47:13-21.
Rate RG, Leche J, Chervenak C (1979) Podophyllin toxicity. Ann Intern Med 90(4):723.

PODOPHYLLUM PLEIANTHUM Hance [Berberidaceae]

Synonyms:
Dysosma pleiantha (Hance) Woodson

Common Names:
baijiaolian

Citations:

Deng JF, Lin TJ, Kao WF, et al. (1997) The difficulty in handling poisonings associated with Chinese traditional medicine: A poison control center experience for 1991-1993. Vet Hum Toxicol 39(2):106-114.
Kao WF, Deng JF (1992) Bajiaolian intoxication - A dangerous toxic effect in herbal therapeutics. Toxicon 30:524.
Kao WF, Hung DZ, Tsai WJ, et al. (1992) Podophyllotoxin intoxication: Toxic effect of Bajiaolian in herbal therapeutics. Hum Exp Toxicol 11(6):480-487.

poet's-daffodil –see– *Narcissus poeticus* L.

poet's-ivy –see– *Hedera helix* L.

poet's-jasmine –see– *Jasminum officinale* L.

poet's-narcissus –see– *Narcissus poeticus* L.

pohon upas –see– *Antiaris toxicaria* Lesch.

pohun ipoh –see– *Antiaris toxicaria* Lesch.

poinciana –see– *Caesalpinia gilliesii* (Hook.) D. Dietr.

Poinciana gilliesii Hook. = Caesalpinia gilliesii (Hook.) D. Dietr.

poinsettia –see– *Euphorbia pulcherrima* Willd. ex Klotzsch

Poinsettia pulcherrima (Willd. ex Klotzsch) Graham = Euphorbia pulcherrima Willd. ex Klotzsch

point locoweed –see– *Oxytropis lambertii* Pursh; *Oxytropis sericea* Nutt.

pois-a-gratter –see– *Mucuna pruriens* (L.) DC.

pois-de-loup –see– *Lupinus albus* L.

poison arum –see– *Dieffenbachia seguine* (Jacq.) Schott

poison ash –see– *Toxicodendron radicans* (L.) Kuntze; *Toxicodendron vernix* (L.) Kuntze

poison bay –see– *Illicium anisatum* L.

poison bean –see– *Sesbania drummondii* (Rydb.) Cory; *Thermopsis montana* Nutt.

poison blackcherry –see– *Atropa belladonna* L.

poison bush –see– *Eremophila deserti* (A. Cunn. ex Benth.) Chinnock; *Gastrolobium grandiflorum* F. Muell.; *Thesium namaquense* Schltr.

poison camas –see– *Toxicoscordion nuttallii* (A. Gray) Rydb.; *Toxicoscordion venenosum* (S. Watson) Rydb.

poison chickweed –see– *Anagallis arvensis* L.

poison creeper –see– *Toxicodendron radicans* (L.) Kuntze

poison darnel –see– *Lolium temulentum* L.

poison dogwood –see– *Toxicodendron vernix* (L.) Kuntze

poison elder –see– *Toxicodendron vernix* (L.) Kuntze

poison flower –see– *Solanum dulcamara* L.

poison fool's-parsley –see– *Conium maculatum* L.

poison halipan –see– *Pedilanthus tithymaloides* (L.) Poit.

poison hemlock –see– *Conium maculatum* L.

poison ivy –see– *Toxicodendron diversilobum* (Torr. & A. Gray) Greene; *Toxicodendron pubescens* Mill.; *Toxicodendron radicans* (L.) Kuntze; *Toxicodendron radicans* (L.) *Kuntze* subsp. negundo (Greene) Gillis; *Toxicodendron rydbergii* (Small ex Rydb.) Greene; *Toxicodendron vernix* (L.) Kuntze

poison lac –see– *Toxicodendron vernicifluum* (Stokes) F. A. Barkley

poison laurel –see– *Kalmia latifolia* L.

poison leaf –see– *Dichapetalum cymosum* (Hook.) Engl.

poison lily –see– *Veratrum viride* Aiton

poison lupine –see– *Lupinus leucophyllus* Douglas ex Lindl.

poison mercury –see– *Toxicodendron radicans* (L.) Kuntze

poison nightshade –see– *Solanum dulcamara* L.

poison nut –see– *Strychnos nux-vomica* L.

poison oak –see– *Metopium toxiferum* (L.) Krug & Urb.; *Toxicodendron pubescens* Mill.; *Toxicodendron rydbergii* (Small ex Rydb.) Greene; *Toxicodendron vernix* (L.) Kuntze

poison onion –see– *Ornithogalum magnum* (Baker) J. C. Manning & Goldblatt

poison parsley –see– *Cicuta maculata* L.; *Conium maculatum* L.

poison parsnip –see– *Cicuta douglasii* (DC.) J. M. Coult. & Rose; *Cicuta maculata* L.; *Cicuta maculata* L. var. angustifolia Hook.

poison peach –see– *Trema tomentosa* (Roxb.) H. Hara

poison pimelea –see– *Pimelea pauciflora* R. Br.

poison pod albizia –see– *Albizia versicolor* Welw. ex Oliv.

poison primrose –see– *Primula obconica* Hance

poison primula –see– *Primula obconica* Hance

poison ragwort –see– *Senecio isatideus* DC.

poison root –see– *Conium maculatum* L.

poison ryegrass –see– *Lolium temulentum* L.

poison sedge –see– *Schoenus asperocarpus* F. Muell.

poison sego –see– *Toxicoscordion nuttallii* (A. Gray) Rydb.; *Toxicoscordion paniculatum* (Nutt.) Rydb.; *Toxicoscordion venenosum* (S. Watson) Rydb.

poison snakeweed –see– *Cicuta maculata* L.; *Conium maculatum* L.

poison sneezeweed –see– *Helenium autumnale* L.

poison suckleya –see– *Suckleya suckleyana* (Torr.) Rydb.

poison sumach –see– *Toxicodendron radicans* (L.) Kuntze; *Toxicodendron striatum* (Ruiz & Pav.) Kuntze; *Toxicodendron vernix* (L.) Kuntze

poison swamp sumach –see– *Toxicodendron vernix* (L.) Kuntze

poison tobacco –see– *Hyoscyamus niger* L.

poison tree –see– *Excoecaria parvifolia* Müll. Arg.; *Metopium toxiferum* (L.) Krug & Urb.; *Toxicodendron vernix* (L.) Kuntze

poison vetch –see– *Astragalus bisulcatus* (Hook.) A. Gray; *Astragalus michauxii* (Kuntze) F. J. Herm.; *Astragalus miser* Douglas ex Hook. var. oblongifolius (Rydb.) Cronquist; *Astragalus tetrapterus* A. Gray

poison vine –see– *Marsdenia rostrata* R. Br.; *Toxicodendron radicans* (L.) Kuntze

poison water hemlock –see– *Cicuta virosa* L.

poison weed –see– *Anagallis arvensis* L.; *Delphinium barbeyi* (Huth) Huth; *Delphinium carolinianum* Walter subsp. virescens (Nutt.) R. E. Brooks; *Delphinium menziesii* DC.; *Delphinium ×occidentale* (S. Watson) S. Watson; *Euphorbia drummondii* Boiss.; *Metopium toxiferum* (L.) Krug & Urb.; *Toxicodendron vernix* (L.) Kuntze

poison wild onion –see– *Toxicoscordion venenosum* (S. Watson) Rydb.

poisonberry –see– *Cestrum nocturnum* L.; *Solanum dulcamara* L.; *Solanum nigrum* L.

poisongrass –see– *Toxicoscordion venenosum* (S. Watson) Rydb.

poisonwood –see– *Toxicodendron vernix* (L.) Kuntze; *Metopium toxiferum* (L.) Krug & Urb.; *Schinus terebinthifolius* Raddi

poivre d'eau –see– *Persicaria hydropiper* (L.) Spach

poke –see– *Phytolacca americana* L.

pokeberry –see– *Phytolacca americana* L.

pokeroot –see– *Phytolacca americana* L.

pokesalad –see– *Phytolacca americana* L.

pokeweed –see– *Phytolacca americana* L.

pokok ipoh –see– *Antiaris toxicaria* Lesch.

póleo –see– *Mentha pulegium* L.

poleo chino –see– *Hedeoma pulegioides* (L.) Pers.

POLYCARPON PROSTRATUM (Forssk.) Asch. & Schweinf. [Caryophyllaceae]

Citations:

Alikutty KM (1975) A field report on the occurrence of certain cases of weed toxicity in cattle. Kerala J Vet Sci 6(1-2):149-150.

POLYGALA CHINENSIS L. [Polygalaceae]

Common Names:

jin-bu-huan

Citations:

Horowitz RS, Dart RC, Hurlbut K, et al. (1994) Acute hepatitis associated with a Chinese herbal product, jin bu huan (JBH). Vet Hum Toxicol 36(4):359.

Horowitz RS, Feldhaus K, Dart RC, et al. (1996) The clinical spectrum of jin bu huan toxicity. Arch Intern Med 156(8):899-903.

Palmer M, Haller C, McKinney P, et al. (1999) Botanicals and other dietary supplements: Adverse events by age. J Toxicol Clin Toxicol 37(5):609.

POLYGALA KLOTZSCHII Chodat [Polygalaceae]

Common Names:

jaranjinha; jimãozinho

Citations:

Camargo WV, Mengato W, Platzech F, et al. (1968) Intoxicação de bovinos pela Polygala klotzchii, Chodat, na região do Pontal (Estado de São Paulo). Biologico 34:221-223.

Rezende AM, Döbereiner J, Tokarnia CH (1981) Intoxicação experimental por Polygala klotzchii (Polygalaceae) em coelhos. Pesq Vet Bras 1(2):61-64.

Temperini JA, Retz L, Cappellaro CF (1977) Estudo experimental de toxicidade da Polygala klotzchii Chodat. Biologico 43:194-198.

Tokarnia CH, Döbereiner J, Canella CF (1976) Intoxicação experimental por Polygala klotzchii em bovinos. Pesq Agric Bras Vet 11(9):73-86.

POLYGALA SANGUINEA L. [Polygalaceae]

Common Names:

milkwort

Citations:

Pammel LH (1929) Horses and milkwort. Vet Med 24(Dec):514.

POLYGONATUM MULTIFLORUM (L.) All. [Ruscaceae]

Common Names:

Solomon's-seal; Vielblütige Weißwurz

Citations:

Baxter CP (1983) Solomon's seal poisoning in a dog. Vet Rec 113(11):247-248.

POLYGONUM ACUMINATUM Kunth [Polygonaceae]

Citations:

Salgues R (1961) Le genre Polygonum L. (Polygonacées) Etudes chimiques et toxicologiques. Les faits hématologiques de l'empoisonnement expérimental. Qualitas Plant Mat Veg 8:367-395.

Polygonum amphibium L. = Persicaria amphibia (L.) Delarbre

POLYGONUM AVICULARE L. [Polygonaceae]

Common Names:

chivalrygrass; dishwatergrass; dooryard knotweed; doorweed; knotgrass; knotweed; prostrate knotweed; wireweed

Citations:

Knight PR (1979) Suspected nitrite toxicity in horses associated with the ingestion of wireweed (Polygonum aviculare). Aust Vet Pract 9(3):175-177.

Salgues R (1961) Le genre Polygonum L. (Polygonacées) Etudes chimiques et toxicologiques. Les faits hématologiques de l'empoisonnement expérimental. Qualitas Plant Mat Veg 8:367-395.

Polygonum bistorta L. = Bistorta officinalis Delarbre

Polygonum convolvulus L. = Fallopia convolvulus (L.) Á. Löve

Polygonum fagopyrum L. = Fagopyrum esculentum Moench

Polygonum glabrum Willd. = Persicaria glabra (Willd.) M. Gómez

Polygonum hydropiper L. = Persicaria hydropiper (L.) Spach

POLYGONUM MITE Schrank [Polygonaceae]

Citations:

Salgues R (1961) Le genre Polygonum L. (Polygonacées) Etudes chimiques et toxicologiques. Les faits hématologiques de l'empoisonnement expérimental. Qualitas Plant Mat Veg 8:367-395.

Polygonum multiflorum Thunb. = Fallopia multiflora (Thunb.) Haraldson

Polygonum nepalense Meisn. = Persicaria nepalensis (Meisn.) H. Gross

Polygonum persicaria L. = Persicaria maculosa Gray

Polygonum punctatum Elliott = Persicaria punctata (Elliott) Small

pomegranate –see– *Punica granatum* L.

pomme d'acajou –see– *Semecarpus ater* (G. Forst.) Vieill.

pomme-du-poison –see– *Datura stramonium* L.

pomme épineuse –see– *Datura stramonium* L.

pommier cajou –see– *Anacardium occidentale* L.

pommier d'amour –see– *Solanum pseudocapsicum* L.

ponderosa pine –see– *Pinus ponderosa* C. Lawson

pong pong –see– *Cerbera odollam* Gaertn.

Pongamia glabra Vent. = Milletia pinnata (L.) Panigrahi

ponso –see– *Tanacetum vulgare* L.

Pontic rhododendron –see– *Rhododendron ponticum* L.

poolroot –see– *Ageratina altissima* (L.) R. M. King & H. Rob.

poolwort –see– *Ageratina altissima* (L.) R. M. King & H. Rob.

poor-man's-alfalfa –see– *Bassia scoparia* (L.) A. J. Scott

poor-man's-banana –see– *Asimina triloba* (L.) Dunal

poor-man's-liquid amber –see– *Toxicodendron radicans* (L.) Kuntze

poor-man's-pepper –see– *Persicaria punctata* (Elliott) Small

poor-man's-weatherglass –see– *Anagallis arvensis* L.

pop dock –see– *Digitalis purpurea* L.

popcorn tree –see– *Triadica sebifera* (L.) Small

popolo –see– *Solanum anguivi* Lam.

popotillo –see– *Ephedra viridis* Coville

popple nut –see– *Macadamia integrifolia* Maiden & Betche; *Macadamia tetraphylla* L. A. S. Johnson

poppy weed –see– *Argemone mexicana* L.

Portland sago –see– *Arum maculatum* L.

Portland starchroot –see– *Arum maculatum* L.

PORTULACA GRANDIFLORA Hook. [Portulacaceae]

Common Names:

rock moss

Citations:

Deng JF, Lin TJ, Kao WF, et al. (1997) The difficulty in handling poisonings associated with Chinese traditional medicine: A poison control center experience for 1991-1993. Vet Hum Toxicol 39(2):106-114.

PORTULACA OLERACEA L. [Portulacaceae]

Common Names:

inland pigweed; purslane; pusley

Citations:

Deng JF, Lin TJ, Kao WF, et al. (1997) The difficulty in handling poisonings associated with Chinese traditional medicine: A poison control center experience for 1991-1993. Vet Hum Toxicol 39(2):106-114.

Mathams RH, Sutherland AK (1952) The oxalate content of some Queensland pasture plants. Queensland J Agric Sci 9:317-334.

possum haw –see– *Ilex decidua* Walter

possumwood –see– *Diospyros virginiana* L.; *Hura crepitans* L.

post locust –see– *Robinia pseudoacacia* L.

post oak –see– *Quercus stellata* Wangenh.

pot –see– *Cannabis sativa* L.

pot marigold –see– *Calendula officinalis* L.

pot mum –see– *Chrysanthemum ×morifolium* Ramat.

potato –see– *Solanum tuberosum* L.

potato bush –see– *Solanum esuriale* Lindl.

potato weed –see– *Heliotropium europaeum* L.; *Solanum dimidiatum* Raf.; *Solanum esuriale* Lindl.

pothos –see– *Epipremnum pinnatum* (L.) Engl.; *Scindapsus pictus* Hassk.

poundcake bush –see– *Parthenium hysterophorus* L.

poverty weed –see– *Pimelea simplex* F. Muell.; *Pimelea trichostachya* Lindl.

powderpuff lily –see– *Scadoxus multiflorus* (Martyn) Raf. subsp. multiflorus

prairie bundleflower –see– *Desmanthus leptalobus* Torr. & A. Gray

prairie crocus –see– *Anemone patens* L.

prairie cupgrass –see– *Erichloa contracta* Hitchc.

prairie groundsel –see– *Senecio plattensis* Nutt.

prairie larkspur –see– *Delphinium carolinianum* Walter subsp. virescens (Nutt.) R. E. Brooks

prairie lily –see– *Cooperia pedunculata* Herb.

prairie mimosa –see– *Desmanthus leptalobus* Torr. & A. Gray

prairie ragwort –see– *Senecio plattensis* Nutt.

prairie sage –see– *Artemisia ludoviciana* Nutt.

PRATIA CONCOLOR (R. Br.) Druce [Campanulaceae]

Synonyms:

Pratia erecta Gaudich.

Common Names:

milkweed

Citations:

McCarthy JH (1921) Poisonous "milk weed" (Pratia erecta). Queensland Agric J. Sep:196.

Pratia erecta Gaudich. = Pratia concolor (R. Br.) Druce

prayer bead –see– *Abrus precatorius* L.

prayer bean –see– *Abrus precatorius* L.

precatory bean –see– *Abrus precatorius* L.

precatory pea –see– *Abrus precatorius* L.

prêle –see– *Equisetum arvense* L.

prêle commune –see– *Equisetum palustre* L.

prickblad –see– *Dieffenbachia seguine* (Jacq.) Schott

prickly cucumber –see– *Cucumis myriocarpus* Naudin

prickly melon –see– *Cucumis myriocarpus* Naudin

prickly nightshade –see– *Solanum rostratum* Dunal

prickly paddymelon –see– *Cucumis myriocarpus* Naudin

prickly pear –see– *Opuntia ficus-indica* (L.) Mill.

prickly poppy –see– *Argemone mexicana* L.

prickly saltwort –see– *Salsola tragus* L.

prickly yellow poppy –see– *Argemone mexicana* L.

prickwood –see– *Euonymus europaeus* L.

pride-of-California –see– *Lathyrus splendens* Kellogg

pride-of-China –see– *Melia azedarach* L.

pride-of-India –see– *Melia azedarach* L.

priest's-pentle –see– *Arisaema triphyllum* (L.) Schott

prim –see– *Ligustrum vulgare* L.

primaryvet –see– *Ligustrum vulgare* L.

Primel –see– *Primula obconica* Hance; *Primula veris* L.

primrose –see– *Primula auricula* L.; *Primula malacoides* Franch.; *Primula obconica* Hance; *Primula vulgaris* Huds.

primula –see– *Primula obconica* Hance

PRIMULA ALLIONII Loisel. [Primulaceae]

Citations:

Aplin CG, Lovell CR (2001) Contact dermatitis due to hardy primula species and their cultivars. Contact Dermatitis 44(1):23-29.

PRIMULA AURICULA L. [Primulaceae]

Common Names:

auricula; bear's-ears; primrose

Citations:

Aplin CG, Lovell CR (2001) Contact dermatitis due to hardy primula species and their cultivars. Contact Dermatitis 44(1):23-29.

PRIMULA FARINOSA L. [Primulaceae]

Common Names:

wild primrose

Citations:

Sharpe HA (1912) Primula dermatitis. Its occurrence in rural districts. JAMA 59(24):2148-2149.

PRIMULA FORRESTII Balf. f. [Primulaceae]

Citations:

Aplin CG, Lovell CR (2001) Contact dermatitis due to hardy primula species and their cultivars. Contact Dermatitis 44(1):23-29.

PRIMULA MALACOIDES Franch. [Primulaceae]

Common Names:

primrose

Citations:

Paulsen E (1998) Occupational dermatitis in Danish gardeners and greenhouse workers (II). Etiological factors. Contact Dermatitis 38(1):14-19.

Whiting DA (1971) Plant dermatitis in the southern Transvaal. S Afr Med J 45(7):163-167.

PRIMULA MARGINATA Curtis [Primulaceae]

Citations:

Aplin CG, Lovell CR (2001) Contact dermatitis due to hardy primula species and their cultivars. Contact Dermatitis 44(1):23-29.

PRIMULA OBCONICA Hance [Primulaceae]

Common Names:

Chinese primrose; German primrose; Giftprimel; Japanische Primel; parlour primula; poison primrose; poison primula; Primel; primrose; primula

Citations:

Apted J (1988) Contact dermatitis and Primula obconica. Med J Aust 149(Oct 17):452.

Bandmann HJ, Breit R, Fregert S (1973) Kontaktallergie gegenüber Primula obconica Häufigkeit - Aktive Sensibilisierung bei der Epicutantestung. Hautarzt 24(6):240-243.

Barton EA (1899) On the effects of Primula obconica on the skin. Lancet 1(45):1717.

Dooms-Goossens A, Biesemans G, Vandaele M, et al. (1989) Primula dermatitis: More than one allergen? Contact Dermatitis 21(14):122-124.

Epstein E (1990) Primula contact dermatitis: An easily overlooked diagnosis. Cutis 45(6):411-416.

Fernández de Corres L, Leanizbarrutia I, Muñoz D (1987) Contact dermatitis from Primula obconica Hance. Contact Dermatitis 16(7):195-197.

Fernández de Corrés L, Leanizbarrutia I, Muñoz D, et al. (1987) Contact dermatitis from a neighbour's primula. Contact Dermatitis 16(8):234-235.

Fregert S, Hjorth N, Schulz KH (1968) Patch testing with synthetic primin in persons sensitive to Primula obconica. Arch Dermatol 98(2):144-147.

Gallo R, Sorbara S, Rongioletti F (2005) Contact erythema multiforme from Primula obconica. Contact Dermatitis 53(6):351-352.

Ingber A (1991) Primula photodermatitis in Israel. Contact Dermatitis 25(4):265-266.

Lapière K, Matthieu L, Meuleman L, et al. (2001) Primula dermatitis mimicking lichen planus. Contact Dermatitis 44(3):199.

Mitchell JN, Rook AJ (1977) Primula dermatitis. Contact Dermatitis 3(6):286.

Mowad CM (1998) Routine testing for Primula obconica: Is it useful in the United states? Am J Contact Dermat 9(4):231-233.

Nakamura T (1983) Primula dermatitis in Japan. Contact Dermatitis 9(4):328-329.

Paulsen E, Christensen LP, Andersen KE (2006) Miconidin and miconidin methyl ether from Primula obconica Hance: New allergens in an old sensitizer. Contact Dermatitis 55(4):203-209.

Poniecka H (1990) Rośliny jako przyczyna alergii kontaktowej w materiale kliniki dermatologicznej am w białymstoku. Przegl Dermatol 77(4):262-265.

Rook A, Wilson HT (1965) Primula dermatitis. Br Med J 1(5429):220-222.

Rytand DA, Burnham DK, Cox AJ Jr (1948) Periarteritis nodosa following the dermatitis of poison oak and of primrose. Stanford Med Bull 6(2):319-323.

Sweet EA (1907) Poisoning by Primula obconica. JAMA 49(4):329.

Virgili A, Corazza M (1991) Unusual primin dermatitis. Contact Dermatitis 24(1):63-64.

Yasuda H, Kumakiri M, Miura Y, et al. (1983) [Primula dermatitis] Hokkaido Igaku Zasshi 58(6):617-621.

Primula officinalis (L.) Hill = Primula veris L.

PRIMULA PRAENITENS Ker Gawl. [Primulaceae]

Synonyms:

Primula sinensis Sabine ex Lindl.

Common Names:

Chinese primrose; daisy primrose

Citations:

Densten JC (1903) Primula sinensis (daisy primrose) poisoning mistaken for eczema. Med Summary Philadelphia 25:227.

Montgomery DW, Culver GD (1915) Dermatitis caused by primula poisoning. Cal State J Med 13(8):306-307.

Primula sinensis Sabine ex Lindl. = Primula praenitens Ker Gawl.

PRIMULA VERIS L. [Primulaceae]

Synonyms:

Primula officinalis (L.) Hill

Common Names:

Primel; Schlüsselblume; Wiesenschlüsselblume

Citations:

Geßner O (1936) Dermatitis durch deutsche Primeln (Primula officinalis L.). Sammlung Vergiftungsfallen 7(A650):215-216.

PRIMULA VULGARIS Huds. [Primulaceae]

Common Names:

primrose

Citations:

Aplin CG, Lovell CR (2001) Contact dermatitis due to hardy primula species and their cultivars. Contact Dermatitis 44(1):23-29.

Prince-of-Wales-feather –see– *Amaranthus retroflexus* L.

pringamoza –see– *Cnidoscolus urens* (L.) Arthur; *Wigandia urens* (Ruiz & Pav.) Kunth var. caracasana (Kunth) D. N. Gibson

privet –see– *Ligustrum obtusifolium* Siebold & Zucc. subsp. suave (Kitag.) Kitag.; *Ligustrum ovalifolium* Hassk.; *Ligustrum vulgare* L.

proso millet –see– *Panicum miliaceum* L.

PROSOPIS ALBA Griseb. [Fabaceae]

Citations:

Lyon CK, Gumbmann MR, Becker R (1988) Value of mesquite leaves as forage. J Sci Food Agric 44(2):111-117.

PROSOPIS ARTICULATA S. Watson [Fabaceae]

Citations:

Lyon CK, Gumbmann MR, Becker R (1988) Value of mesquite leaves as forage. J Sci Food Agric 44(2):111-117.

PROSOPIS CHILENSIS (Molina) Stuntz [Fabaceae]

Citations:

Lyon CK, Gumbmann MR, Becker R (1988) Value of mesquite leaves as forage. J Sci Food Agric 44(2):111-117.

PROSOPIS GLANDULOSA Torr. [Fabaceae]

Common Names:

algarroba; honey mesquite; katzimelk; mesquite; Texas mesquite

Citations:

Holechek JL, Munshikpu AV, Saiwana L, et al. (1990) Influences of six shrub diets varying in phenol content on intake and nitrogen retention by goats. Trop Grasslands 24(2):93-98.

Washburn KE, Breshears ME, Ritchey JW, et al. (2002) Honey mesquite toxicosis in a goat. J Am Vet Med Assoc 220(12):1837-1839.

PROSOPIS JULIFLORA (Sw.) DC. [Fabaceae]

Common Names:

algarrobo; honey mesquite; Kiawe bean; mesquite; mesquite bean

Citations:

Dollahite JW (1964) Management of the disease produced in cattle on an unbalanced diet of mesquite beans. Southwestern Vet 17(Summer):293-295.

Dollahite JW, Anthony WV (1957) Malnutrition in cattle on an unbalanced diet of mesquite beans. Cattleman Oct:30, 31, 48-52.

Fernández Baca S, Vallenas A, Novoa C, et al. (1967) Estudio experimental de la 'coquera' en caprinos. Rev Fac Med Vet Univ Nacional Lima 18-20:131-159.

Fox EC (1941) Mesquite wood dermatitis. Arch Derm Syphilol 44:1098-1100.

Seifert HS, Beller KA (1969) Blausäurevergiftung beim Rind, verursacht durch Abweiden von Zuckerrohr (Saccharum officinarum) und Zufütterung von Früchten des Algarrobaumes (Prosopis juliflora). Berl Munch Tierarztl Wochenschr 82(5):88-91.

Sivasankar N, Bjat AG, Patil SA (1991) A note on honey mesquite Prosopis juliflora as a piscicide. Indian Forester 117(2):151-152.

Stewart CD (1940) Dermatitis due to mesquite wood. Report of a case. Arch Derm Syphilol 42:937.

Tabosa IM, Riet-Correa F, Barros SS, et al. (2006) Neurohistologic and ultrastructural lesions in cattle experimentally intoxicated with the plant Prosopis juliflora. Vet Pathol 43(5):695-701.

Tabosa IM, Souza JC, Graca DL, et al. (2000) Neuronal vacuolation of the trigeminal nuclei in goats caused by ingestion of Prosopis juliflora pods (mesquite beans). Vet Hum Toxicol 42(3):155-158.

PROSOPIS NIGRA (Griseb.) Hieron. [Fabaceae]

Citations:

Lyon CK, Gumbmann MR, Becker R (1988) Value of mesquite leaves as forage. J Sci Food Agric 44(2):111-117.

PROSOPIS VELUTINA Wooton [Fabaceae]

Citations:

Lyon CK, Gumbmann MR, Becker R (1988) Value of mesquite leaves as forage. J Sci Food Agric 44(2):111-117.

prostrate amaranth –see– *Amaranthus blitoides* S. Watson

prostrate euphorbia –see– *Chamaesyce prostrata* (Aiton) Small

prostrate indigo –see– *Indigofera hendecaphylla* Jacq.

prostrate knotweed –see– *Polygonum aviculare* L.

prostrate pigweed –see– *Amaranthus blitoides* S. Watson

prostrate pimelia –see– *Pimelea prostrata* (J. R. Forst. & G. Forst.) Willd.

prostrate spurge –see– *Chamaesyce maculata* (L.) Small; *Chamaesyce prostrata* (Aiton) Small

prune –see– *Prunus domestica* L.

pruneaux –see– *Prunus domestica* L.

Prunus amygdalus Batsch –see– *Prunus dulcis* (Mill.) D. A. Webb

PRUNUS ARMENIACA L. [Rosaceae]

Common Names:

apricot; Aprikose

Citations:

Anonymous (1972) Cyanide poisoning from the ingestion of apricot kernels. California Morbidity 34(Sep 1):1.

Göransson K (1981) Contact urticaria to apricot stone. Contact Dermatitis 7(5):282.

Johnson H, Rich G (1975) Cyanide poisoning from apricot kernels. California Morbidity #51(Dec 26):1.

Lasch EE, El Shawa R (1981) Multiple cases of cyanide poisoning by apricot kernels in children from Gaza. Pediatrics 68(1):5-7.

Miller KW, Anderson JL, Stoewsand GS (1981) Amygdalin metabolism and effect on reproduction of rats fed apricot kernels. J Toxicol Environ Health 7:457-467.

Rubino MJ, Davidoff F (1979) Cyanide poisoning from apricot seeds. JAMA 241(4):359.

Rubino MJ, Davidoff F, Baselt R, et al. (1979) Cyanide poisoning from apricot seeds: Case report and review of the literature. Clin Toxicol 15(4):489.

Sayre JW, Kaymakcalan S (1964) Cyanide poisoning from apricot seeds among children in central Turkey. N Engl J Med 270(21):1113-1115.

Suchard JR, Wallace KL, Gerkin RD (1998) Acute cyanide toxicity caused by apricot kernel ingestion. Ann Emerg Med 32(6):742-744.

Townsend WA (1975) Cyanide poisoning from ingestion of apricot kernels. California Morbidity 45(Nov 14):1.

Wallace KL, Gerkin R, Mitchell RB (1996) Acute cyanide poisoning due to apricot kernel ingestion. J Toxicol Clin Toxicol 34:599.

PRUNUS COCOMILIA Ten. [Rosaceae]

Synonyms:

Prunus pseudoarmeniaca Heldr. & Sartori

Common Names:

wild apricot; zerdali

Citations:

Sayre JW, Kaymakcalan S (1964) Cyanide poisoning from apricot seeds among children in central Turkey. N Engl J Med 270(21):1113-1115.

Prunus demissa (Nutt.) D. Dietr. = Prunus virginiana L.

PRUNUS DOMESTICA L. [Rosaceae]

Common Names:

Pflaume; plum; prune; pruneaux

Citations:

Austvoll A (1954) Plumstones in pig swill. Vet Rec 66:681.

Cardassis J, Giannakoulas D (1961) [Hydrocyanic acid poisoning from plum stones in pigs]. Hellenike Kteniatrike 4:136-141.

Jaworski K, Głodek S (1965) Zatrucia śliwkami u krów i trzody chlewnej. Med Weter 21(1):44.

PRUNUS DULCIS (Mill.) D. A. Webb [Rosaceae]

Synonyms:

Prunus amygdalus Batsch; ***Prunus dulcis*** (Mill.) D. A. Webb var. amara (DC.) Buchheim

Common Names:

almond; amandier; bitter almond; Bitter Mandel; Mandelbaum; wild bitter almond

Citations:

Pack WK, Raudonat HW, Schmidt K (1972) Über ein todliche Blausäurevergiftung nach dem Genuß bitterer Mandeln (Prunus amygdalus). Z Rechtsmed 70(1):53-54.

Shragg TA, Albertson TE, Fisher CJ Jr (1982) Cyanide poisoning after bitter almond ingestion. West J Med 136(1):65-69.

Tholhuysen LJ (1960) Dodelijke vergiftiging van twee runderen door bittere amandelen. Tijdschr Diergeneeskd 85:1243-1244.

Prunus dulcis (Mill.) D. A. Webb var. amara (DC.) Buchheim = Prunus dulcis (Mill.) D. A. Webb

PRUNUS LAUROCERASUS L. [Rosaceae]

Common Names:

black cherry; cherry bay; cherry laurel; English laurel; European cherry laurel; Hecke; Kirschlorbeer; laureal real; laurel; laurel cherry; laurierkers; Lorbeerkirsche

Citations:

Gill DA, McGregor P (1928) Laurel poisoning in stock. A case of deaths among calves. N Z J Agr 37(Dec 20):407.

Robb W, Campbell D (1941) Poisoning of sheep by the consumption of laurel leaves. Vet Rec 53(7):93-95.

Wilson DR, Gordon WJ (1941) Laurel poisoning in sheep. Vet Rec 53(7):95-97.

Wollgarten B (1997) Der klinische Fall. Tierarztl Prax 25(3):223-225.

Prunus melanocarpa (A. Nelson) Rydb. = Prunus virginiana L.

PRUNUS MYRTIFOLIA (L.) Urb. [Rosaceae]

Synonyms:

Prunus sphaerocarpa Sw.

Common Names:

black heart; coração negro; mabola persimmon; marmeleiro bravo; myrtle laurelcherry; pessegueiro bravo

Citations:

Saad AD, Camargo WV (1967) Intoxicação cianídrica em animais domésticos. O «pessegueiro bravo» (Prunus sphaerocarpa Sw.) planta cianogênica da média Mojiana, como responsável por mortes de bovinos e caprinos. Biologico 33:211-220.

Saad AD, Camargo WV (1970) Intoxicação em animais domésticos pelo Brumus [Prunus] sphaerocarpa. Arq Inst Biol (Sao Paulo) 37(Suppl 1):38.

PRUNUS PADUS L. [Rosaceae]

Common Names:

Ahlkirsche; bird cherry; Elsebaum; European May Day tree; Faulbaum; Traubenkirsche

Citations:

Hofstee AW (1989) Acute vergiftigingen bij dieren in Nederland over de periode 1985-1988. Tijdschr Diergeneeskd 114(22):1154-1158.

Sargison ND, Williamson DS, Duncan JR, et al. (1996) Prunus padus (bird cherry) poisoning in cattle. Vet Rec 138(8):188.

PRUNUS PERSICA (L.) Batsch [Rosaceae]

Common Names:

peach; Pfirsich

Citations:

Carrasquer Moya C, Peláez Hernández A, Durá Calatayud M, et al. (2002) Allergy to peach. Allergy 57(8):756-757.

Cuesta-Herranz J, Lázaro M, de las Heras M, et al. (1998) Peach allergy pattern: Experience in 70 patients. Allergy 53(1):78-82.

Hansen AA (1928) Stock poisoning plants. Hydrocyanic poisoning. North Am Vet 9(5):24-27.

Prunus pseudoarmeniaca Heldr. & Sartori = Prunus cocomilia Ten.

PRUNUS SEROTINA Ehrh. [Rosaceae]

Common Names:

black cherry; black wild cherry; cabinet cherry; cherry; chokecherry; pin cherry; rum cherry; Späte Traubenkirsche; whiskey cherry; wild beach cherry; wild black cherry; wild cherry

Citations:

Enge EH (1939) Wild-cherry poisoning in pigs. J Am Vet Med Assoc 94:123-124.
Gough JF (1995) Black cherry poisoning in an Angora goat. Can Vet J 36(1):45.
Hansen AA (1924) Prussic acid poisoning in livestock. A real problem. Better Crops 2(3):26-27,38.
Radi ZA, Styer EL, Thompson LJ (2004) Prunus spp. intoxication in ruminants: A case in a goat and diagnosis by identification of leaf fragments in rumen contents. J Vet Diagn Invest 16(6):593-599.
Selby LA, Menges RW, Houser EC, et al. (1971) Outbreak of swine malformations associated with the wild black cherry, Prunus serotina. Arch Environ Health 22(4):496-501.

Prunus sphaerocarpa Sw. = Prunus myrtifolia (L.) Urb.

PRUNUS SPINOSA L. [Rosaceae]

Common Names:

blackthorn; Schlehe; Schwarzdorn; sloe

Citations:

Buhr AJ (1960) The thorn in the flesh. Lancet 1:309-310.

PRUNUS VIRGINIANA L. [Rosaceae]

Synonyms:

Prunus demissa (Nutt.) D. Dietr.; ***Prunus melanocarpa*** (A. Nelson) Rydb.; ***Prunus virginiana*** L. var. demissa (Nutt.) Torr.; ***Prunus virginiana*** L. var. melanocarpa (A. Nelson) Sarg.

Common Names:

black cherry; chokecherry; eastern chokecherry; western chokecherry; wild cherry; wild chokecherry

Citations:

Anonymous (1935) Investigation of stock-poisoning plants. U S Dep Agric Bur Anim Indus Chief Rep 1935:39-40.
Anonymous (1936) Investigations of stock-poisoning plants. U S Dep Agric Bur Anim Indus Chief Rep. 1936:44-45.
Fleming CE, Dill R (1928) The poisoning of sheep on mountain grazing ranges in Nevada by the western chokecherry (Prunus demissa). Nevada Agric Exp Sta Bull #110:14 pp.
Fleming CE, Miller MR, Vawter LR (1926) The common choke-cherry (Prunus demissa) as a plant poisonous to sheep and cattle. Nevada Agric Exp Sta Bull #109:8-30.
Fleming CE, Miller MR, Vawter LR, et al. (1931) Poisonous range plants. Nevada Agric Exp Sta Annu Rep #1928:21-22.
Jackson T (1995) Cyanide poisoning in two donkeys. Vet Hum Toxicol 37(6):567-568.
Ogilvie S (1955) Chokecherry toxic to an antelope. J Mammol 36:146.
Pijoan M (1942) Cyanide poisoning from choke cherry seed. Am J Med Sci 204:550-553.
Stauffer VD (1970) Hydrocyanic acid poisoning from choke cherry leaves. J Am Vet Med Assoc 157(10):1324.
Stauffer VD (1981) Poison caused by ingestion of chokecherry leaves in a dog. Vet Med Small Anim Clin 76(11):1573.

psathyrotes –see– *Psathyrotes annua* (Nutt.) A. Gray

PSATHYROTES ANNUA (Nutt.) A. Gray [Asteraceae]

Common Names:

desert velvet; mealy rosettes; psathyrotes; velvet brittlestem; velvet rosettes

Citations:

Binns W, James LF, Shupe JL (1962) Psathyrotes annua, poisonous plant for sheep. Vet Med 57(Jun):509-511.
Buck WB, James LF, Binns W (1961) Changes in serum transaminase activities associated with plant and mineral toxicity in sheep and cattle. Proc Am Coll Vet Toxicol 1961:13-24.
Buck WB, James LF, Binns W (1961) Changes in serum transaminase activities associated with plant and mineral toxicity in sheep and cattle. Cornell Vet 51(Oct):568-585.

Pseudocalymma elegans (Vell.) Kuhlm. = Arrabidaea elegans (Vell.) A. H. Gentry

PSIDIUM GUAJAVA L. [Myrtaceae]

Common Names:

guava

Citations:

Obi M, Miyazaki Y, Yokozeki H, et al. (2001) Allergic contact dermatitis due to guava tea. Contact Dermatitis 44(2):116-117.

PSILOCAULON ABSIMILE N. E. Br. [Aizoaceae]

Common Names:

asbos; loogbos

Citations:

Rimington C, Steyn DG (1933) Psilocaulon absimile NE Br as a stock poison. I. Determination of oxalic, malic, tartaric acids, etc. Onderstepoort J Vet Sci Anim Indus 1(2):439-455.

psilostrophe –see– *Psilostrophe tagetina* (Nutt.) Greene

PSILOSTROPHE GNAPHALODES DC. [Asteraceae]

Common Names:

cudweed paper flower; paper flower

Citations:

Anonymous (1933) Investigation of stock-poisoning plants. U S Dep Agric Bur Anim Indus Chief Rep 1933:34-35.

Mathews FP (1934) Psilostrophe tagetinae and Psilostrophe gnaphalodes, two plants poisonous to sheep and cattle on the ranges of the Southwest. Texas Agric Exp Sta Bull #500:5-13.

PSILOSTROPHE SPARSIFLORA (A. Gray) A. Nelson [Asteraceae]

Common Names:

green-stem paper flower

Citations:

Anonymous (1940) Investigations of stock-poisoning plants. U S Dep Agric Bur Anim Indus Chief Rep 1940:62.

Anonymous (1941) Investigations of stock-poisoning plants. U S Dep Agric Bur Anim Indus Chief Rep 1941:62-63.

PSILOSTROPHE TAGETINA (Nutt.) Greene [Asteraceae]

Common Names:

marigold paper flower; paper flower; psilostrophe; woolly paper flower

Citations:

Anonymous (1933) Investigation of stock-poisoning plants. U S Dep Agric Bur Anim Indus Chief Rep 1933:34-35.

Mathews FP (1934) Psilostrophe tagetinae and Psilostrophe gnaphalodes, Two plants poisonous to sheep and cattle on the ranges of the Southwest. Texas Agric Exp Sta Bull #500:5-13.

PSOPHOCARPUS TETRAGONOLOBUS (L.) DC. [Fabaceae]

Common Names:

Goa bean; winged bean

Citations:

Lovell CR, Rycroft RJ (1984) Contact urticaria from winged bean (Psophocarpus tetragonolobus). Contact Dermatitis 10(5):314-315.

Psoralea corylifolia L. = Cullen corylifolium (L.) Medik.

Psoralea drupacea Bunge = Cullen drupaceum (Bunge) C. H. Stirt.

PSYCHOTRIA BARBIFLORA DC. [Rubiaceae]

Citations:

Andrade SO, Camargo WV, Fernandes N (1963) II. Investigações sôbre plantas tóxicas no Estado de São Paulo. Arq Inst Biol (Sao Paulo) 30(Oct):189-203.

Camargo W (1962) Uma nova "erva-de-rato" tóxico para bovinos Palicourea barbiflora (?), Rubiaceae, comparação com a Palicourea marcgravii var. pubescens e com a Psychotria officinalis, Rubiaceae. Arq Inst Biol (Sao Paulo) 29(1):1-11.

PSYCHOTRIA CARTHAGINENSIS Jacq. [Rubiaceae]

Citations:

Flores FA, Lewis WH (1978) Drinking the South American hallucinogenic ayahuasca. Econ Bot 32(Apr-Jun):154-156.

PSYCHOTRIA LONGIPEDUNCULATA (Gardner) Müll. Arg. [Rubiaceae]

Synonyms:

Palicourea longipedunculata Muell.

Citations:

Mello EM, Sampaio Fernandes JS (1940) Contribuição ao estudo das plantas tóxicas Brasileiras. Ministerio Agricultura. pp. 1-50.

PSYCHOTRIA POEPPIGIANA Müll. Arg. [Rubiaceae]

Synonyms:

Cephaelis tomentosa (Aubl.) Vahl

Common names:

erva de rato

Citations:

Nazario W (1973) Cephaelis tomentosa: Uma nova erva de rato, tóxica para bovinos - Ocorrência clinica no município de Luciara (Mt). Atualidades Veterinarias 2(9):37-38.

PSYCHOTRIA VIRIDIS Ruiz & Pav. [Rubiaceae]

Common Names:

chacruna

Citations:

Flores FA, Lewis WH (1978) Drinking the South American hallucinogenic ayahuasca. Econ Bot 32(Apr-Jun):154-156.

Ptelea angustifolia Benth. = Ptelea trifoliata L. subsp. angustifolia (Benth.) V. L. Bailey

PTELEA TRIFOLIATA L. subsp. angustifolia (Benth.) V. L. Bailey [Rutaceae]

Synonyms:

Ptelea angustifolia Benth.

Common Names:

hop tree

Citations:

Muenscher WC, Brown BI (1944) Dermatitis and photosensitization produced by Ptelea angustifolia. Madrono 7(6):184-188.

PTERIDIUM AQUILINUM (L.) Kuhn [Dennstaedtiaceae]

Synonyms:

Pteridium aquilinum (L.) Kuhn var. esculentum (G. Forst.) Kuhn; ***Pteridium aquilinum*** (L.) Kuhn var. latiusculum (Desf.) Underw. ex A. Heller; ***Pteridium***

esculentum (G. Forst.) Cockayne; ***Pteridium latiusculum*** (Desf.) Hieron. ex R. E. Fr.; ***Pteris aquilina*** L.

Common Names:

Adlerfarn; bracken fern; brake fern; braken fern; eagle fern; Farn; Farnkraut; fougère aigle; fougère grand aigle; helecho; hog brake; samambaia

Citations:

Almond N (1894) Fern poisoning. J Comp Pathol Ther 7:165-167.

Alonso Amelot ME, Avendano M (2001) Possible association between gastric cancer and bracken fern in Venezuela: An epidemiologic study. Int J Cancer 91(2):252-259.

Andrade SO, Ladeira AM, Shirose I, et al. (1977) Efeitos tóxicos da "samambaia" (Pteridium aquilinum), na alimentação de cobaias. Arq Inst Biol (Sao Paulo) 44(4):195-207.

Andrade SO, Sunohara Y (1970) Atividade antitiamínica do Pteridium aquilinum. Arq Inst Biol (Sao Paulo) 37(1):39.

Barnett KC, Blakemore WF, Mason J (1972) Bracken retinopathy in sheep. Trans Ophthalmol Soc U K 92:741-744.

Barnett KC, Watson WA (1970) Bright blindness in sheep. A primary retinopathy due to feeding bracken (Pteris aquilina). Res Vet Sci 11(3):289-290.

Blakeway J (1924) Bracken poisoning in pigs. Vet J 80:89-90.

Brough RJ (1963) 'Out of season' bracken poisoning. Vet Rec 75(7):181-182.

Bryan GT, Pamukcu AM (1979) Bracken fern (BF), a natural urinary bladder carcinogen. In: Deichmann WB (ed.) Toxicology and occupational medicine. Elsevier. New York. pp. 229-232.

Campo MS, Jarrett WF, O'Neil W, et al. (1994) Latent papillomavirus infection in cattle. Res Vet Sci 56(2):151-157.

Carpenter KJ, Phillipson AT, Thomson W (1950) Experiments with dried bracken (Pteris aquilina). Br Vet J 106:292-308.

Chang MC, Feng TK (1964) [Bracken poisoning of dairy cattle.] Acta Vet Zootech Sin 7(1):69-78.

Cordy DR (1952) The pathology of experimental bracken poisoning in rats. Cornell Vet 42(1):108-117.

Cordy DR (1957) Experimental production and control of bracken fern poisoning in rats. J Am Vet Med Assoc 130(Apr 15):333-334.

Costa AJ (1973) Quadro clínico anátomopatológico da intoxicação aguda por samambaia nos bovinos (Pteridium aquilinum). Atualidades Veterinarias 1(6):22-23.

Craig JF, Davies GO (1940) Some observations on bracken poisoning. Vet Rec 52(27):499.

Dalton RG (1964) The effects of batyl alcohol in the haematology of cattle poisoned with bracken. Vet Rec 76(15):411-416.

Derflinger ER (1944) Fern allergy. Fort Dodge Biochem Rev 15(2):3-5, 16.

Diniz JM, Basile JR, Camargo NJ (1984) Intoxicação natural de asininos por Pteridium aquilinum (L.) Kuhn no Brasil. Arq Bras Med Vet Zootecnia 36(5):515-522.

Döbereiner J, Tokarnia CH, Canella CF (1967) Ocorrência da hematúria enzoótica e de carcinomas epidermóides no trato digestivo superior em bovinos no Brasil. Pesq Agric Bras 2:489-504.

Dzuvić A (1969) Veränderungen an Blutgefäßen der Harnblase bei experimenteller Haematuria vesicalis bovis. Dtsch Tierarztl Wochenschr 76(14):368, 371-373.

Edwards BL (1983) Poisoning by Pteridium aquilinum in pregnant sows. Vet Rec 112(May 7):459-460.

Evans ET, Evans WC (1949) The effect of the inclusion of bracken (Pteris aquilina) in the diets of rats. Biochem J Proc 44:ix.

Evans ET, Evans WC, Roberts HE (1951) Studies on bracken poisoning in the horse. Br Vet J 107:364-371.

Evans ET, Evans WC, Roberts HE (1951) Studies on bracken poisoning in the horse. II. Br Vet J 107:399-411.

Evans IA (1979) Bracken carcinogenicity. Res Vet Sci 26(3):339-348.

Evans IA, Humphreys DJ, Goulden L, et al. (1963) Effects of bracken rhizomes on the pig. J Comp Pathol 73(Jul):229-243.

Evans IA, Mainwaring-Burt R, Allport BR, et al. (1972) Methods of study and the effects of the naturally occurring bracken fern carcinogen in a nonhuman primate, Macaca fascicularis. In: Goldsmith EI et al. (eds.) Medical primatology. Basel 111:241-249.

Evans IA, Mason J (1965) Carcinogenic activity of bracken. Nature 208(5013):913-914.

Evans WC, Evans ET (1949) Studies on the biochemistry of pasture plants. 3. The effects of inclusion of bracken (Pteris aquilina) in the diets of rats, and the problem of bracken poisoning in farm animals. Br Vet J 105:175-186.

Evans WC, Evans ET, Hughes LE (1954) Studies on bracken poisoning in cattle. I. Br Vet J 110:295-306.

Evans WC, Evans ET, Hughes LE (1954) Studies on bracken poisoning in cattle. II. 1950 bracken poisoning experiments (Lluest farm). Br Vet J 110:365-380.

Evans WC, Evans ET, Hughes LE (1954) Studies on bracken poisoning in cattle. III. Field outbreaks of bovine bracken poisoning. Br Vet J 110:426-442.

Evans WC, Evans IA, Axford RF, et al. (1961) Studies on bracken poisoning in cattle. VII. The toxicity of bracken rhizomes. Vet Rec 73(35):852-853.

Evans WC, Evans IA, Edwards CM, et al. (1957) Bracken poisoning of cattle - Therapeutic treatment. Biochem J 65:6P.

Evans WC, Evans IA, Humphreys DJ, et al. (1975) Induction of thiamine deficiency in sheep, with lesions similar to those of cerebrocortical necrosis. J Comp Pathol 85(2):253-267.

Evans WC, Widdop B, Harding JD (1972) Experimental poisoning by bracken rhizomes in pigs. Vet Rec 90(17):471-475.

Fletcher JM (1944) Bracken poisoning in cattle. Vet Rec 56(49):478.

Foggie A (1951) Suspected bracken poisoning in sheep. Vet Rec 63(13):242.

Fox HJ, France WH (1954) Bracken poisoning in cattle. Vet Rec 66(Nov 13):711-712.

France WH (1955) Bracken poisoning in cattle. Vet Rec 67(Oct 22):802-804.

Fushimi K, Kato T, Hirono I (1973) [Influence of bracken meal on pregnant mice.] Acta Sch Med Univ Gifu 21:448-452.

Galpin OP, Whitaker CJ, Whitaker R, et al. (1990) Gastric cancer in Gwynedd. Possible links with bracken. Br J Cancer 61(5):737-740.

Gava A, Silva Neves D, Gava D, et al. (2002) Bracken fern (Pteridium aquilinum) poisoning in cattle in southern Brazil. Vet Hum Toxicol 44(6):362-365.

Gdovin T, Sokol J (1968) Zur Etiologie der Haematuria Vesicalis Bovis Chronica in der Slowakei. Folia Vet Kosice 12(2):105-106.

Gerenutti M, Spinosa HS, Bernardi MM (1992) Effects of bracken fern (Pteridium aquilinum L Kuhn) feeding during the development of female rats and their offspring. Vet Hum Toxicol 34(4):307-310.

Gleeson LN (1944) Suspected bracken poisoning. Vet Rec 56(41):374-375.

Gorišek J, Maržan B (1965) Veränderungen des Blutbildes und der Blutgerinnung bei mit Adlerfarn (Pteridium aquilinum) vergifteten Kälbern. Wien Tierarztl Monatsschr 52(5):530-538.

Goto M, Itakura C (1973) [Pathological studies on an acute hemorrhagic disease in pastured cattle in Tottori Prefecture. Cases occurring in 1967.] Jpn J Vet Sci 35(2):115-122.

Gounalan S, Somvanshi R, Kataria M, et al. (1999) Effect of bracken (Pteridium aquilinum) and dryopteris (Dryopteris juxtaposita) fern toxicity in laboratory rabbits. Indian J Exp Biol 37(10):980-985.

Gregorović V, Skušek F, Šenk L (1962) Adlerfarn (Pteris aquilina) als Ursache einer Massenvergiftung in einem Kälbermastgroßbetrieb. Dtsch Tierarztl Wochenschr 69(12):327-329.

Guilhon J, Obray, J, Queinnec G (1955) Reproduction expériment du syndrome hémorrhagique des jeunes bovidés bretons par ingestion de fougére aigle de la region Pansienne. Bull Acad Vet Fr 28:457-462.

Haag JR, Weswig PH, Freed AM (1947) Antithiamine activity of bracken fern. Fed Proc 6:408-409.

Hadwen S (1917) So-called staggers in horses caused by the ingestion of Pteris aquilina, the common bracken. J Am Vet Med Assoc 50:702-704.

Hadwen S, Bruce EA (1920) The poisoning of horses by the common bracken (Pteris aquilinum L.). Vet J 76:98-109.

Hadwen S, Bruce EA (1933) The poisoning of horses by the common bracken (Pteris aquilina L.). Vet J 89:120-128.

Hagan WA (1925) Braken poisoning of cattle. Cornell Vet 15(Jul):326-332.

Hagan WA (1926) Bracken poisoning of cattle. New York State Vet Coll Rep 29:146-151.

Hagan WA, Zeissig A (1927) Experimental bracken poisoning of cattle. Cornell Vet 17(Apr):194-208.

Harding JD (1972) Bracken poisoning in pigs. Agriculture 79(7):313-314.

Heath GB, Wood B (1958) Bracken poisoning in cattle. J Comp Pathol 68(2):201-212.

Hirono I (1986) Human carcinogenic risk in the use of bracken fern. Proc Int Sympos Princess Takamatsu Cancer Res Fund 16:139-145.

Hirono I, Aiso S, Hosaka S, et al. (1983) Induction of mammary cancer in CD rats fed bracken diet. Carcinogenesis 4(7):885-887.

Hirono I, Aiso S, Yamaji T, et al. (1984) Hyperplastic nodules of the liver induced in rats fed bracken diet. Cancer Lett 22(2):151-155.

Hirono I, Fushimi K, Mori H, et al. (1973) Comparative study of carcinogenic activity in each part of bracken. J Natl Cancer Instit 50(5):1367-1371.

Hirono I, Ito M, Yagyu S, et al. (1993) Reproduction of progressive retinal degeneration (bright blindness) in sheep by administration of ptaquiloside contained in bracken. J Vet Med Sci 55(6):979-983.

Hirono I, Shibuya C, Fushimi K, et al. (1970) Studies on carcinogenic properties of bracken, Pteridium aquilinum. J Natl Cancer Instit 45(1):179-188.

Hirono I, Shibuya C, Shimizu M, et al. (1972) Carcinogenic activity of processed bracken used as human food. J Natl Cancer Instit 48:1245-1250.

Hopkins NC (1987) Enzootic haematuria in Nepal. Trop Anim Health Prod 19(3):159-164.

Hosaka S, Nagayama H, Hirono I, et al. (1983) Enhanced activity of ornithine decarboxylase of the ileum in rats by bracken fern (Pteridium aquilinum). Br J Cancer 48(2):311-314.

Howell RM, Evans IA (1967) Chromatographic characteristics of fibrinogen and seromucoid in bovine bracken poisoning. J Comp Pathol 77(2):117-128.

Jarrett WF (1978) Transformation of warts to malignancy in alimentary carcinoma in cattle. Bull Cancer 65(2):191-194.

Jarrett WF (1982) Bracken and cancer. Proc R Soc Edinburgh B 81(1-2):79-83.

Jarrett WF, McNeil PE, Grimshaw WT, et al. (1978) High incidence area of cattle cancer with a possible interaction between an environmental carcinogen and a papilloma virus. Nature 274(5668):215-217.

Kato T (1978) [Experimental studies on the carcinogenicity of bracken fern (Pteridium aquilinum) in the rat.] Gifu Daigaku Igakubu Kiyo 26(2):318-333.

Kelleway RA, Geovjian L (1978) Acute bracken fern poisoning in a 14-month-old horse. Vet Med Small Anim Clin 73(3):295-296.

Kitahara T (1974) [Experimental studies on bracken poisoning in cattle.] Bull Nippon Vet Zootech Coll 23(Nov):88-107.

Konishi T, Ichijo S, Ogawa S, et al. (1971) [Clinical and pathological studies on bracken poisoning of cattle in Tokashi District.] Res Bull Obihiro Zootech Univ 7(Series 1):216-239.

Langham RF (1957) Bracken fern poisoning in a cow. J Am Vet Med Assoc 130(Apr 15):334-335.

Langley HR (1944) Bracken poisoning in cattle. Vet Rec 56(52):518.

Lee JM, Andrade SO, Camargo WV, et al. (1966) Hematúria em bovinos no Estado de São Paulo. Arq Inst Biol (Sao Paulo) 33(2):27-37.

Lynch J (1935) Bracken poisoning. Vet Rec 15(35):1067-1068.

Maeda T (1975) [Fundamental studies on the aetiology of haematuria vesicalis bovis. I. The induction of carcinomas and haemorrhages of the urinary bladder in guinea pigs by feeding bracken fern (Pteridium aquilinum).] Bull Fac Agric 27:79-88.

Maeda T (1978) [Studies on chronic bovine hematuria vesicalis due to tumors. V. Etiological investigation.] J Jpn Vet Med Assoc 31(5):277-282.

Marrero E, Bulnes C, Sánchez LM, et al. (2001) Pteridium aquilinum (bracken fern) toxicity in cattle in the humid Chaco of Tarija, Bolivia. Vet Hum Toxicol 43(3):156-158.

Marrero E, Bulnes C, Sánchez LM, et al. (2004) Chronic toxicity in cattle due to Pteridium aquilinum (bracken fern) in Tarija Dept., Bolivia: An interdisciplinary investigation. In: Acamovic T, Stewart CS, Pennycott TW (eds.) Poisonous plants and related toxins. CABI. Wallingford, UK. pp. 248-252.

Mason J, Barnett KC, Blakemore WF, et al. (1973) Some biochemical observations on a primary retinal degeneration in sheep. Exp Eye Res 15(1):51-60.
McCrea CT, Head KW (1981) Sheep tumours in north east Yorkshire. II. Experimental induction of tumours. Br Vet J 137(1):21-30.
McKenzie RA (1978) Bovine enzootic haematuria in Queensland. Aust Vet J 54(2):61-64.
McLauchlan D (1951) Bracken poisoning. Vet Rec 63(13):241-242.
Miura S, Ohshima K (1961) [A pathological study on the first cases of bracken poisoning in cows in Japan.] Jpn J Vet Sci 23:347-353.
Moon FE, Raafat MA (1951) Some biochemical aspects of bracken 'poisoning' in the ruminant animal. I. Vitamin factors. J Sci Food Agric 2(3):228-240.
Moon FE, Raafat MA (1951) The experimental production of bracken "poisoning" in sheep. J Comp Pathol 61:88-100.
Moreira-Souto MA, Kommers GD, Barros CS, et al. (2006) Neoplasias do trato alimentar superior de bovinos associadas ao consumo espontâneo de samambaia (Pteridium aquilina). Pesq Vet Bras 26(2):112-122.
Mueller SB, Madureira FR, Alencar Filho RA, et al. (1975) Tentativa de reprodução experimental da hematúria enzoótica em bovinos pela administracação de samambaia. Arq Inst Biol (Sao Paulo) 42:203-211.
Mugera GM, Nderito P, Sorheim AO (1969) The pathology of urinary bladder tumours in Kenya Zebu cattle. J Comp Pathol 79(2):251-254.
Naftalin JM, Cushnie GH (1951) Pathology of bracken poisoning. Vet Rec 63(18):332.
Naftalin JM, Cushnie GH (1954) Experimental bracken poisoning in calves. J Comp Pathol 64(1):75-86.
Naftalin JM, Cushnie GH (1954) Pathology of bracken poisoning in cattle. J Comp Pathol 64(1):54-74.
Naftalin JM, Cushnie GH (1956) Haematology of experimental bracken poisoning of cattle. I. Changes in blood and bone marrow. II. Attempts to modify the course of the bone marrow damage. J Comp Pathol Exp Ther 66(4):354-372.
Nandi SN (1982) Bovine haematuria of Indian hills (Himalayas): A haemorrhagic syndrome, due to development of transition cell carcinomas of the urinary bladder, associated with the consumption of Himalayan weeds and ferns, as cheap fodder. Dtsch Tierarztl Wochenschr 89(12):479-482.
Nivelle A, Monsallier G (1985) Cas cliniques: Intoxication par la fougère grand aigle chez les bovins. Point Veterinaire 17(90):355-357.
Odawara T, Uchino T, Mitani S, et al. (1975) [Observations on some findings in the rumen juice of experimental poisoned cattle by bracken.] J Jpn Vet Med Assoc 28(5):227-232.
O'Moore LB (1949) The treatment with Vitamin B1 of bracken staggers in the bovine. Vet Rec 61(46):768.
Oyamada T, Yoshikawa T (1987) Histopathogenesis of intestinal tumors induced by oral administration of bracken fern, Pteridium aquilinum in rats. Nippon Juigaku Zasshi 49(4):687-696.
Pamukcu AM, Ertürk E, Price JM, et al. (1972) Lymphatic leukemia and pulmonary tumors in female Swiss mice fed bracken fern (Pteris aquilina). Cancer Res 32(7):1442-1445.
Pamukcu AM, Ertürk E, Yalçiner S, et al. (1976) Carcinogenic activity of milk from bracken fern fed cows. Proc Am Assoc Cancer Res 17:14.
Pamukcu AM, Ertürk E, Yalçiner S, et al. (1976) Histogenesis of urinary bladder cancer induced in rats by bracken fern. Invest Urol 14(3):213-218.
Pamukcu AM, Ertürk E, Yalçiner S, et al. (1978) Carcinogenic and mutagenic activities of milk from cows fed bracken fern (Pteridium aquilinum). Cancer Res 38(6):1556-1560.
Pamukcu AM, Göksoy SK, Price JM (1967) Urinary bladder neoplasms induced by feeding bracken fern (Pteris aquilina) to cows. Cancer Res 27(1):917-924.
Pamukcu AM, Milli U, Bryan GT (1981) Protective effect of nicotinamide on bracken fern induced carcinogenicity in rats. Nutr Cancer 3(2):86-93.
Pamukcu AM, Price JM (1969) Induction of intestinal and urinary bladder cancer in rats by feeding bracken fern (Pteris aquilina). J Natl Cancer Instit 43(1):275-281.
Pamukcu AM, Price JM, Bryan GT (1976) Naturally occurring and braken-fern-induced bovine urinary bladder tumors. Clinical and morphological characteristics. Vet Pathol 13(2):110-122.
Pamukcu AM, Yalçiner S, Hatcher JF, et al. (1980) Quercetin, a rat intestinal and bladder carcinogen present in bracken fern (Pteridium aquilinum). Cancer Res 40(10):3468-3472.
Pamukcu AM, Yalçiner S, Price JM, et al. (1970) Effects of the coadministration of thiamine on the incidence of urinary bladder carcinomas in rats fed bracken fern. Cancer Res 30(11):2671-2674.
Parker WH, McCrea CT (1965) Bracken (Pteris aquilina) poisoning of sheep in the North York Moors. Vet Rec 77(30):861-865.
Perkins LE (1950) Experiments on bracken poisoning. Br Vet J 106(10):377-385.
Phillipson AT, Reid RS (1954) Studies of the toxicity of bracken (Pteris aquilina). J Comp Pathol 64(13):243-259.
Philp RB, Gowdey CW (1967) Anaemia, thrombocytopenia and reduced platelet adhesiveness in rats fed bracken fern, and protective effects of batyl alcohol. Nature 216(5114):498-499.
Pinto C, Januário T, Geraldes M, et al. (2004) Bovine enzootic haematuria on Sao Miguel Islands - Azores. In: Acamovic T, Stewart CS, Pennycott TW (eds.) Poisonous plants and related toxins. CABI. Wallingford, UK. pp. 564-574.
Pirie HM (1973) Unusual occurrence of squamous carcinoma of the upper alimentary tract in cattle in Britain. Res Vet Sci 15(1):135-138.
Price JM, Pamukcu AM (1968) The induction of neoplasms of the urinary bladder of the cow and the small intestine of the rat by feeding bracken fern (Pteris aquilina). Cancer Res 28(11):2247-2251.
Rajendran MP, Chennakesavalu M, Rao CV, et al. (1983) Experimental production of enzootic bovine haematuria with bracken fern. Indian Vet J 60:173-178.
Rao DS, Joshi HC, Kumar M (1987) Immune status of calves suffering from bracken-fern-induced haematuria. Indian J Anim Sci 57(11):1207-1208.
Rao DS, Joshi HC, Kumar M (1988) Biochemical findings in bracken fern toxicity in calves. Int J Anim Sci 3(1):33-37.
Rao DS, Joshi HC, Kumar M (1988) Haemogram in bracken fern (Pteris aquilina) (Pteridium aquilinum) poisoning in rats. Int J Anim Sci 3(1):39-43.
Rao DS, Joshi HC, Kumar M (1988) Leucine amino peptidase and gamma glutamyl transpeptidase activity in bracken-fern-induced haematuria in calves and rats. Indian J Anim Sci 58(5):544-547.

Rim GM, Lee HI (1972) [Pathologic studies of experimental bracken poisoning in mice.] Korean J Vet Res 12(2):145-151.
Roberts HE, Evans ET, Evans WC (1949) The production of "bracken staggers" in the horse and its treatment by Vitamin B1 therapy. Vet Rec 61(35):549-550.
Rosas H, Quintero SO, Gómez J, et al. (1974) Rats and bovine toxic lesions by Pteridium aquilinum. J Anim Sci 38(1):231.
Rosenberger G (1965) Längere Aufnahme von Adlerfarn (Pteris aquilina) - Die Ursache der chronischen vesikalen Haematurie des Rindes. Wien Tierarztl Monatsschr 52(5):415-421.
Rosenberger G, Heeschen W (1960) Adlerfarn (Pteris aquilina) - die Ursache des sog. Stallrotes der Rinder (Haematuria vesicalis bovis chronica). Dtsch Tierarztl Wochenschr 67(8):201-208.
Samaddar J (1973) Role of Pteris aquilina in enzootic bovine haematuria. Indian J Anim Sci 43(6):510-514.
Santos RC, Brasileiro-Filho G, Hojo ES (1987) Induction of tumors in rats by bracken fern (Pteridium aquilinum) from Ouro Preto (Minas Gerais, Brazil). Braz J Med Biol Res 20(1):73-77.
Schacham P, Philp RB, Gowdey CW (1970) Antihematopoietic and carcinogenic effects of bracken fern (Pteridium aquilinum) in rats. Am J Vet Res 31(1):191-197.
Scholl E (1962) Über Farnvergiftung beim Rind. Schweiz Arch Tierheilkd 104:323-328.
Shearer GD (1945) Some observations on the poisonous properties of bracken (Pteris aquilina). J Comp Pathol 55:301-307.
Singh RP, Joshi HC, Kumar M (1987) Experimental bracken fern toxicity in calves: Changes in blood and urine. Indian J Vet Med 7(2):96-100.
Skerman KD, Newton LG (1952) Bracken fern poisoning of cattle. Queensland Agric J 74(3):163-167.
Smith BL, Embling PP, Agnew MP, et al. (1988) Carcinogenicity of bracken fern (Pteridium esculentum) in New Zealand. N Z Vet J 36(2):56-58.
Sofrenovic D, Stamatovic S, Bratanovic U, et al. (1965) Die pathomorphologischen Veränderungen der durch Fütterung mit Adlerfarn (Pteris aquilina) hervorgerufenen vesikalen Hämaturie beim Rind. Dtsch Tierarztl Wochenschr 72(18):409-413.
Stamatović S, Bratanović U, Sofrenović D (1965) Das klinische Bild der durch Verfütterung von Adlerfarn (Pteris aquilina) experimentell hervorgerufenen Haematuria Vesicalis der Rinder. Wien Tierarztl Monatsschr 52(6):589-596.
Stockman S (1917) Bracken poisoning in cattle in Great Britain. J Comp Pathol Exp Ther 30:311-316.
Stockman S (1922) Bracken poisoning in cattle in Great Britain. J Comp Pathol Exp Ther 35:273-275.
Sunderman FM (1987) Bracken poisoning in sheep. Aust Vet J 64(1):25-26.
Swann HC, Barrowman JC (1959) Bracken poisoning in pigs. Vet Rec 71(23):493.
Tokarnia CH, Döbereiner J, Canella CF (1967) Ocorrência da intoxicação aguda pela "samambaia" (Pteridium aquilinum (L.) Kuhn) em bovinos no Brasil. Pesq Agric Bras 2:329-336.
Tomlinson CJ (1983) Bracken poisoning/PGE. Goat Vet Soc J 4(2):43-44.
Tustin RC, Adelaar TF, Meldal-Johnsen CM (1968) Bracken poisoning in cattle in the Natal midlands. J S Afr Vet Assoc 39(3):91-99.
Twomey DF, Holt GJ, Reid HW (2002) Malignant catarrhal fever in cattle with suspected bracken poisoning. Vet Rec 151(16):486-487.
Ushijima J, Matsukawa K, Yuasa A, et al. (1983) Toxicities of bracken fern in guinea pigs. Jpn J Vet Sci 45(5):593-602.
Ushimaru Y (1978) [Studies for the detection of a bracken carcinogen.] Gifu Daigaku Igakubu Kiyo 26(3):368-380.
Wagnon KA (1959) A study of bracken fern poisoning of cattle on a California forest range. J Range Manag 12:249-255.
Wang YD, Xu LR, Wen LJ, et al. (1984) [Studies on experimental bovine bracken poisoning.] Acta Vet Zootech Sin 15(4):235-239.
Watson WA, Terlecki S, Patterson DS, et al. (1972) Experimentally produced progressive retinal degeneration (bright blindness) in sheep. Br Vet J 128(9):457-469.
Wells HE (1949) Bracken poisoning. Agriculture 56:204-205.
Wilson AL, Robertson JM (1961) The toxicity of bracken rhizomes. Vet Rec 73(41):1010.
Winterhalter M (1973) Otrovanje krave sa bujadi (Pteris aquilina). Veterinarski Glasnik 27(9):683-685.
Xenos E, Stoilis E (1969) [Bracken-poisoning in cattle in Macedonia.] Hellenike Kteniatrike 12(4):166-171.
Yamane O, Hayashi T, Sako S (1975) Studies on blood coagulation disorders in domestic animals. Thrombelastograms [sic] of normal cattle and cattle affected with bracken poisoning. Jpn J Vet Sci 37(6):577-583.
Yamane O, Hayashi T, Sako S, et al. (1975) [Experimental studies on bracken poisoning of cattle. Clinical findings and blood figures.] Bull Fac Agric Tottori Univ 27:68-78.
Yamane O, Hayashi T, Sako S, et al. (1975) [Haematological studies on grazing cattle with petechiae on the visible mucous membranes.] J Jpn Vet Med Assoc 28(10):516-521.
Yamane O, Hayashi T, Sako S, et al. (1975) [Studies on haemorrhagic diathesis in experimental bovine bracken poisoning.] J Jpn Vet Med Assoc 28(5):219-223.
Yamane O, Hayashi T, Sako S, et al. (1975) Studies on hemorrhagic diathesis of experimental bovine bracken poisoning. I. Detection of circulating anticoagulants. Jpn J Vet Sci 37(4):335-340.
Yamane O, Hayashi T, Sako S, et al. (1975) Studies on hemorrhagic diathesis of experimental bovine bracken poisoning. II. Heparin-like substance level in blood. Jpn J Vet Sci 37(4):341-347.
Yasuda Y, Kihara T, Nishimura H (1974) Embryotoxic effects of feeding bracken fern (Pteridium aquilinum) to pregnant mice. Toxicol Appl Pharmacol 28(2):264-268.
Yoshikawa T, Oyamada T, Yoshikawa H, et al. (1981) Histopathogenesis of bracken fern-induced experimental tumor of urinary bladder. Nippon Juigaku Zasshi 43(6):875-885.
Yunoki K, Hayashi T, Morita N (1972) Induction of intestinal tumors in rats with feeding of bracken fern. Acta Med Univ Kagoshima 14:249-254.

Pteridium aquilinum (L.) Kuhn var. esculentum (G. Forst.) Kuhn = Pteridium aquilinum (L.) Kuhn

Pteridium aquilinum (L.) Kuhn var. latiusculum (Desf.) Underw. ex A. Heller = Pteridium aquilinum (L.) Kuhn

Pteridium esculentum (G. Forst.) Cockayne = Pteridium aquilinum (L.) Kuhn

Pteridium latiusculum (Desf.) Hieron. ex R. E. Fr. = Pteridium aquilinum (L.) Kuhn

Pteris aquilina L. = Pteridium aquilinum (L.) Kuhn

PTERIS ENSIFORMIS Burm. f. [Pteridaceae]

Citations:

de Cock PA, Vorwerk H, Bruynzeel DP (1998) Hand dermatitis caused by ferns. Contact Dermatitis 39(6):324.

PTEROCARPUS ANGOLENSIS DC. [Fabaceae]

Common Names:

kejaat

Citations:

Ordman D (1949) Bronchial asthma caused by the inhalation of wood dust. Ann Allergy 7:492-496, 505.

PTEROCARPUS DALBERGIOIDES DC. [Fabaceae]

Common Names:

padauk

Citations:

Mehta AJ, Statham BN (2007) Allergic contact dermatitis to purpleheart and padauk wood (Pterocarpus dalbergioides). Contact Dermatitis 56(5):245.

Pterocarpus michelii Britton = Pterocarpus santalinoides L'Hér. ex DC.

PTEROCARPUS SANTALINOIDES L'Hér. ex DC. [Fabaceae]

Synonyms:

Pterocarpus michelii Britton

Citations:

Pinto NR, Baruzzi RG (1991) Male pubertal seclusion and risk of death in Indians from Alto Xingu, central Brazil. Hum Biol 63(6):821-834.

PTEROCARPUS SOYAUXII Taub. [Fabaceae]

Common Names:

padouk

Citations:

Kiec-Swierczynska M, Krecisz B, Swierczynska-Machura D, et al. (2004) Occupational allergic contact dermatits caused by padauk wood (Pterocarpus soyauxii Taub.). Contact Dermatitis 50(6):384-385.

PTERONIA PALLENS L. f. [Asteraceae]

Common Names:

aasvoëbossie; joggemscholtzbossie; stolsbossie; witgatbossie

Citations:

Prozesky L, Kellerman TS, Welman WG (1986) An ovine hepatotoxicosis caused by the plant Pteronia pallens (Asteraceae) L. f. Onderstepoort J Vet Res 53(1):9-12.

PTEROSPERMUM ACERIFOLIUM (L.) Willd. [Malvaceae]

Citations:

Kulakkattolickal AT (1987) Piscicidal plants of Nepal: Preliminary toxicity screening using grass carp (Ctenopharyngodon idella) fingerlings. J Ethnopharmacol 21(1):1-9.

PTERYGOTA BEQUAERTII De Wild. [Sterculiaceae]

Common Names:

koto

Citations:

Estlander T, Jolanki R, Alanko K, et al. (2001) Occupational allergic contact dermatitis caused by wood dusts. Contact Dermatitis 44(4):213-217.

PTERYGOTA MACROCARPA K. Schum. [Sterculaceae]

Common Names:

koto

Citations:

Estlander T, Jolanki R, Alanko K, et al. (2001) Occupational allergic contact dermatitis caused by wood dusts. Contact Dermatitis 44(4):213-217.

puca campuchu –see– *Brugmansia sanguinea* (Ruiz & Pav.) D. Don

puccoon –see– *Sanguinaria canadensis* L.

pulsatilla –see– *Anemone patens* L.

Pulsatilla hirsutissima Britton = Anemone patens L.

Pulsatilla patens (L.) Mill. = Anemone patens L.

pulse –see– *Lathyrus sativus* L.

pulse milk vetch –see– *Astragalus tenellus* Pursh

Pulverholz –see– *Frangula alnus* Mill.

pumpkin –see– *Cucurbita maxima* Duchesne; *Cucurbita pepo* L. subsp. pepo

punche –see– *Nicotiana tabacum* L.

punchon –see– *Verbascum thapsus* L.

puncture vine –see– *Tribulus terrestris* L.

puncture weed –see– *Tribulus terrestris* L.

PUNICA GRANATUM L. [Lythraceae]

Common Names:

pomegranate

Citations:

Gaig P, Bartolomé B, Lleonart R, et al. (1999) Allergy to pomegranate (Punica granatum). Allergy 54(3):287-288.

Igea JM, Cuesta J, Cuevas M, et al. (1991) Adverse reaction to pomegranate ingestion. Allergy 46(6):472-474.
Valsecchi R, Reseghetti A, Leghissa P, et al. (1998) Immediate contact hypersensitivity to pomegranate. Contact Dermatitis 38(1):44-45.

puppetroot –see– *Veratrum viride* Aiton

purgante-de-España –see– *Jatropha multifida* L.

purge nut –see– *Jatropha curcas* L.

purgeerboontjie –see– *Jatropha curcas* L.

purging buckthorn –see– *Rhamnus cathartica* L.

purging croton –see– *Croton tiglium* L.

purging nut –see– *Jatropha curcas* L.; *Jatropha multifida* L.

purple allamanda –see– *Cryptostegia grandiflora* R. Br.

purple angel's-trumpet –see– *Datura inoxia* Mill.

purple bugloss –see– *Echium plantagineum* L.

purple clover –see– *Trifolium pratense* L.

purple cockle –see– *Agrostemma githago* L.

purple crocus –see– *Colchicum autumnale* L.

purple digitalis –see– *Digitalis purpurea* L.

purple foxglove –see– *Digitalis purpurea* L.

purple goosefoot –see– *Scleroblitum atriplicinum* (F. Muell.) Ulbr.

purple ivy –see– *Rhododendron catawbiense* Michx.

purple jimson weed –see– *Datura stramonium* L.

purple larkspur –see– *Delphinium bicolor* Nutt.; *Delphinium menziesii* DC.; *Delphinium nuttallianum* Pritz.

purple laurel –see– *Rhododendron catawbiense* Michx.

purple-laurel rhododendron –see– *Rhododendron catawbiense* Michx.

purple locoweed –see– *Astragalus mollissimus* Torr. var. mollissimus; *Oxytropis lambertii* Pursh

purple mint plant –see– *Perilla frutescens* (L.) Britton

purple passionflower –see– *Passiflora incarnata* L.

purple rattlebox –see– *Sesbania punicea* (Cav.) Benth.

purple rattlebush –see– *Sesbania punicea* (Cav.) Benth.

purple rhododendron –see– *Rhododendron catawbiense* Michx.

purple sesban –see– *Sesbania punicea* (Cav.) Benth.

purple sesbane –see– *Sesbania punicea* (Cav.) Benth.

purple sesbania –see– *Sesbania punicea* (Cav.) Benth.

purple stinkweed –see– *Datura stramonium* L.

purple stramonium –see– *Datura stramonium* L.

purple thorn apple –see– *Datura stramonium* L.

purple viper's-bugloss –see– *Echium plantagineum* L.

Pursh's-silky lupine –see– *Lupinus sericeus* Pursh

purslane –see– *Portulaca oleracea* L.

pusley –see– *Portulaca oleracea* L.

PYCNANTHEMUM ALBESCENS Torr. & A. Gray [Lamiaceae]

Common Names:

albescens

Citations:

Eickholt TH, Box RH (1965) Toxicities of peppermint and Pycnanthemum albescens oils, fam. Labiateae. J Pharm Sci 54(7):1071-7072.

PYGMAEOTHAMNUS CHAMAEDENDRON Robyns [Rubiaceae]

Citations:

Steyn DG (1933) Recent investigations into the toxicity of known and unknown poisonous plants in the Union of South Africa. Onderstepoort J Vet Sci Anim Indus 1:173-182.

pyracantha –see– *Pyracantha coccinea* M. Roem.

PYRACANTHA COCCINEA M. Roem. [Rosaceae]

Common Names:

buisson ardent; everlasting thorn; Feuerdorn; firethorn; pyracantha; vuudoorn

Citations:

Anonymous (1980) Tabulations of 1977 case reports. Bull Natl Clgh Poison Control Cent 24(6):1-4.
Spoerke DG, Temple AR (1978) One year's experience with potential plant poisonings reported to the Intermountain Regional Poison Control Center. Vet Hum Toxicol 20(2):85-89.

pyrethrum –see– *Leucanthemum vulgare* Lam.; *Tanacetum cinerariifolium* (Trevir.) Sch. Bip.

Pyrethrum cinerariifolium Trevir. = Tanacetum cinerariifolium (Trevir.) Sch. Bip.

PYRUS COMMUNIS L. [Rosaceae]

Common Names:

Birne; pear

Citations:

Krauskopf J (1978) Fytodermatózy způsobené rostlinami růžovitými. Cesk Dermatol 53(4):253-258.

Q

qat –see– *Catha edulis* (Vahl) Forssk. ex Endl.

Quaker bonnet –see– *Scutellaria lateriflora* L.

Queen Anne's-lace –see– *Ammi majus* L.

Queensland asthma herb –see– *Chamaesyce hirta* (L.) Millsp.

Queensland maple –see– *Alocasia macrorrhizos* (L.) G. Don

Queensland nut –see– *Macadamia integrifolia* Maiden & Betche; *Macadamia tetraphylla* L. A. S. Johnson

Queensland umbrella tree –see– *Schefflera actinophylla* (Endl.) Harms

queen's-nightshade –see– *Solanum esuriale* Lindl.

queen's-root –see– *Stillingia treculiana* (Müll. Arg.) I. M. Johnst.

queen's-umbrella tree –see– *Schefflera actinophylla* (Endl.) Harms

quelite –see– *Amaranthus retroflexus* L.

quelite cenizo –see– *Chenopodium album* L.

quelite espinoso –see– *Amaranthus spinosus* L.

quelite morado –see– *Amaranthus hybridus* L.

quena –see– *Solanum esuriale* Lindl.

QUERCUS AGRIFOLIA Née [Fagaceae]

Common Names:

California live oak

Citations:

Kinde H (1988) A fatal case of oak poisoning in a double-wattled cassowary (Casuarius casuarius). Avian Dis 32(4):849-851.

QUERCUS ALBA L. [Fagaceae]

Common Names:

white oak

Citations:

Chung-MacCoubrey AL, Hagerman AE, Kirkpatrick RL (1997) Effects of tannins on digestion and detoxification activity in gray squirrels (Sciurus carolinensis). Physiol Zool 70(3):270-277.

Cockrill JM, Beasley JN (1979) Renal damage to cattle during acorn poisoning. Vet Med Small Anim Clin 74(1):82-85.

Kradel DC, Cowan RL (1977) Fall deaths of numerous deer may be caused by acorn toxin. Sci Agric Pennsylvania 25(1):2.

Quercus breviloba (Torr.) Sarg. = Quercus sinuata Walter var. breviloba (Torr.) C. H. Müll.

Quercus calliprinos Webb = Quercus coccifera L.

QUERCUS COCCIFERA L. [Fagaceae]

Synonyms:

Quercus calliprinos Webb

Citations:

Silanikove N, Gilboa N, Nir I, et al. (1996) Effect of a daily supplementation of polyethylene glycol on intake and digestion of tannin-containing leaves (Quercus calliprinos, Pistacia lentiscus, and Ceratonia siliqua) by goats. J Agric Food Chem 44:199-205.

Yeruham I, Avidar Y, Perl S, et al. (1998) Probable toxicosis in cattle in Israel caused by the oak Quercus calliprinos. Vet Hum Toxicol 40(6):336-340.

QUERCUS DOUGLASII Hook. & Arn. [Fagaceae]

Common Names:

blue oak; iron oak; mountain white oak

Citations:

Ostrowski SR, Smith BP, Spier SJ, et al. (1989) Compensatory weight gain in steers recovered from oak bud toxicosis. J Am Vet Med Assoc 195(4):481-484.

Plumlee KH, Johnson B, Galey FD (1998) Comparison of disease in calves dosed orally with oak or commercial tannic acid. J Vet Diagn Invest 10(3):263-267.

Plumlee KH, Johnson B, Galey FD (1998) Disease in cattle dosed orally with oak or tannic acid. In: Garland T, Barr AC (eds.) Toxic plants and other natural toxicants. CABI. New York. pp. 549-553.

Spier SJ, Smith BP, Seawright AA, et al. (1987) Oak toxicosis in cattle in northern California: Clinical and pathologic findings. J Am Vet Med Assoc 191(8):958-964.

QUERCUS GAMBELII Nutt. [Fagaceae]

Common Names:

Gambel's-oak; oakbrush; Rocky Mountain white oak; scrub oak; Utah white oak

Citations:

Marsh CD, Clawson AB, Marsh H (1919) Oak-leaf poisoning of domestic animals. U S Dep Agric Bull #767:36 pp.

Nastis AS, Malechek JC (1981) Digestion and utilization of nutrients in oak browse by goats. J Anim Sci 53(2):283-290.

QUERCUS GARRYANA Douglas ex Hook. [Fagaceae]

Common Names:

Garry's-oak; Oregon oak; Oregon white oak; western white oak

Citations:

Kasari TR, Pearson EG, Hultgren BD (1986) Oak (Quercus garryana) poisoning of range cattle in Southern Oregon. Compend Cont Educ Pract Vet 8(9):F17-F24.

QUERCUS GRISEA Liebm. [Fagaceae]

Common Names:

gray oak

Citations:

Holechek JL, Munshikpu AV, Saiwana L, et al. (1990) Influences of six shrub diets varying in phenol content on intake and nitrogen retention by goats. Trop Grasslands 24(2):93-98.

QUERCUS HAVARDII Rydb. [Fagaceae]

Common Names:

sand shin oak; sand shinnery oak; shin oak; shinnery oak

Citations:

Dollahite JW (1961) Shin oak (Quercus harvardi) poisoning in cattle. Southwestern Vet 14(Spring):198-201.

Dollahite JW, Housholder GT, Camp BJ (1963) Calcium hydroxide, a possible antidote for shin oak (Quercus havardi) poisoning in cattle. Southwestern Vet 16(Winter):115-117.

Dollahite JW, Housholder GT, Camp BJ (1966) Oak poisoning in livestock. Texas Agric Exp Sta Bull #1049:8 pp.

Dollahite JW, Pigeon RF, Camp BJ (1962) The toxicity of gallic acid, pyrogallol, tannic acid, and Quercus havardi in the rabbit. Am J Vet Res 23(Nov):1264-1267.

Householder GT, Dollahite JW (1963) Some clinical biochemical changes in the blood serum of calves fed Quercus havardi. Southwestern Vet 16(Winter):107-113.

Marsh CD, Clawson AB, Marsh H (1919) Oak-leaf poisoning of domestic animals. U S Dep Agric Bull #767:36 pp.

Smith HA (1959) The diagnosis of oak poisoning. Southwestern Vet 13(Fall):34-36.

QUERCUS ILEX L. [Fagaceae]

Common Names:

evergreen holm oak; holm oak

Citations:

Vega A, Dominguez C, Cosmes P, et al. (1998) Anaphylactic reaction to ingestion of Quercus ilex acorn nut. Clin Exp Allergy 28(6):739-742.

Zapatero L, Baeza ML, Sierra Z, et al. (2005) Anaphylaxis by fruits of the Fagaceae family: Acorn and chestnut. Allergy 60(12):1542.

Quercus incana Roxb. = Quercus leucotrichophora A. Camus

QUERCUS LEUCOTRICHOPHORA A. Camus [Fagaceae]

Synonyms:

Quercus incana Roxb.

Common Names:

ban; bluejack oak; sandjack oak; turkey oak; upland willow oak

Citations:

Garg SK, Makkar HP, Nagal KB, et al. (1992) Oak (Quercus incana) leaf poisoning in cattle. Vet Hum Toxicol 34(2):161-164.

Kaushal JR, Gill RS, Negi SS (1971) Utilisation of oak (Quercus incana) kernels in poultry mashes. Indian Vet J 48(4):398-407.

Lohan OP, Lall D, Vaid J, et al. (1983) Utilization of oak tree (Quercus incana) fodder in cattle rations and fate of oak-leaf tannins in the ruminant system. Indian J Anim Sci 53(10):1057-1063.

QUERCUS LOBATA Née [Fagaceae]

Common Names:

valley oak

Citations:

Fowler ME, Richards WP (1965) Acorn poisoning in a cow and a sheep. J Am Vet Med Assoc 147(11):1215-1220.

Quercus pedunculata Ehrh. = Quercus robur L.

QUERCUS ROBUR L. [Fagaceae]

Synonyms:

Quercus pedunculata Ehrh.

Common Names:

Eichel; English oak; European oak; glands; kocsányos tölgy; pedunculate oak; September oak; Sommereiche; Stieleiche

Citations:

Cedervall A, Johansson HE, Jönsson L (1973) Acorn poisoning in cattle. Nord Vet Med 25(12):639-644.

Clarke EG, Cotchin E (1956) A note on the toxicity of the acorn. Br Vet J 112:135-139.

Commergnat J (1973) Intoxication des bovins par les glands. Bull Mens Soc Vet Prat Fr 57(7):371-374.

Dixon PM, McPherson EA, Rowland AC, et al. (1979) Acorn poisoning in cattle. Vet Rec 104(13):284-285.

Jerrett IV, Haynes MP, McCausland IP (1981) Acorn poisoning in calves. Aust Vet Assoc Yearbook. pp. 193-194.

Neser JA, Coetzer JA, Boomker J, et al. (1982) Oak (Quercus rubor) poisoning in cattle. J S Afr Vet Assoc 53(3):151-155.

Stöber M, Ziegler HP, von Benten K (1974) Beitrag zur Eichelvergiftung des Rindes - Krankheitfälle im Herbst 1973. Dtsch Tierarztl Wochenschr 81(7):155-161.

Stöber M, Ziegler HP, von Benten K (1976) Acorn poisoning in cattle. Bovine Pract 11(Nov):36-41.

Tudor G (1971) Observații asupra intoxicației cu ghindă verde la ovine. Rev Zootehnie Med Vet 21(8):60-62.

QUERCUS RUBRA L. [Fagaceae]

Synonyms:

Quercus rubra L. var. borealis (F. Michx.) Farw.

Common Names:

gray oak; northern red oak; red oak

Citations:

Chung-MacCoubrey AL, Hagerman AE, Kirkpatrick RL (1997) Effects of tannins on digestion and detoxification activity in gray squirrels (Sciurus carolinensis). Physiol Zool 70(3):270-277.

Cockrill JM, Beasley JN (1979) Renal damage to cattle during acorn poisoning. Vet Med Small Anim Clin 74(1):82-85.

Duncan CS (1961) Oak leaf poisoning in two horses. Cornell Vet 51(Jan):159-162.

Kradel DC, Cowan RL (1977) Fall deaths of numerous deer may be caused by acorn toxin. Sci Agric Pennsylvania 25(1):2.

Quercus rubra L. var. borealis (F. Michx.) Farw. = Quercus rubra L.

QUERCUS SINUATA Walter var. breviloba (Torr.) C. H. Müll. [Fagaceae]

Synonyms:

Quercus breviloba (Torr.) Sarg.

Common Names:

shin oak

Citations:

Pammel LH (1917) A further discussion of oak poisoning. Am J Vet Med 12:462.

Pammel LH (1917) More cases of poisoning from oak. Am J Vet Med 12:712.

Smith HA (1959) The diagnosis of oak poisoning. Southwestern Vet 13(Fall):34-36.

QUERCUS STELLATA Wangenh. [Fagaceae]

Common Names:

iron oak; post oak

Citations:

Dollahite JW (1961) Shin oak (Quercus harvardi) poisoning in cattle. Southwestern Vet 14(Spring):198-201.

Dollahite JW, Housholder GT, Camp BJ (1966) Effect of calcium hydroxide on the toxicity of post oak (Quercus stellata) in calves. J Am Vet Med Assoc 148(8):908-912.

Dollahite JW, Housholder GT, Camp BJ (1966) Oak poisoning in livestock. Texas Agric Exp Sta Bull #1049:8 pp.

QUERCUS SUBER L. [Fagaceae]

Common Names:

cork dust; cork oak

Citations:

Cancella LC (1963) Suberosis: A pneumoconiosis due to cork dust - The present stage of the problem. Ind Med Surg 32(2):435-445.

QUERCUS VIRGINIANA Mill. [Fagaceae]

Common Names:

live oak

Citations:

Taylor CA, Ralphs MH (1992) Reducing livestock losses from poisonous plants through grazing management. J Range Manag 45(1):9-12.

QUERCUS WISLIZENI DC. [Fagaceae]

Common Names:

interior live oak

Citations:

Wilson AD, Weir WC, Torell DT (1971) Evaluation of chamise (Adenostoma fasiculatum) and interior live oak (Quercus wislizenii) as feed for sheep. J Anim Sci 32(5):1042-1045.

queue-de-cheval –see– *Equisetum palustre* L.

quickgrass –see– *Cynodon dactylon* (L.) Pers.

quillaia –see– *Quillaja saponaria* Molina

quillaja –see– *Quillaja saponaria* Molina

quillaja bark –see– *Quillaja saponaria* Molina

QUILLAJA SAPONARIA Molina [Rosaceae]

Common Names:

cortes quillajas; quillaia; quillaja; soapbark tree; soapwort

Citations:

Bradley R, Harchelroad F (1994) Root beer as a cause for thrombocytopenia. Vet Hum Toxicol 36(4):357.

Ilsley SE, Miller HM, Kamel C (2005) Effects of dietary quillaja saponin and curcumin on the performance and immune status of weaned pigs. J Anim Sci 83(1):82-88.

Raghuprasad PK, Brooks SM, Litwin A, et al. (1980) Quillaja bark (soapbark)-induced asthma. J Allergy Clin Immunol 65(4):285-287.

Sieben P, Rørdam AM, Thomsen ES (1982) Forgiftning med nysepulver. Ugeskr Laeger 144(24):1780.

quinine bush –see– *Alstonia constricta* F. Muell.

quinoa –see– *Chenopodium quinoa* Willd.

quinua –see– *Chenopodium quinoa* Willd.

R

rabaniza –see– *Raphanus raphanistrum* L.

rábano magistro –see– *Armoracia rusticana* P. Gaertn. et al.

rábano ragisco –see– *Armoracia rusticana* P. Gaertn. et al.

rábano rusticano –see– *Armoracia rusticana* P. Gaertn. et al.

rábano silvestre –see– *Armoracia rusticana* P. Gaertn. et al.; *Raphanus raphanistrum* L.

rabbitbrush –see– *Psilostrophe sparsiflora* (A. Gray) A. Nelson

rabo-de-gato –see– *Achyranthes aspera* L.

raccoonberry –see– *Podophyllum peltatum* L.

racha –see– *Lantana camara* L.

racine-de-mahonia –see– *Berberis aquifolium* Pursh

radiata pine -see- *Pinus radiata* D. Don

Radicula armoracia (L.) B. L. Rob. = Armoracia rusticana P. Gaertn. et al.

radish –see– *Raphanus sativus* L.

ragweed –see– *Jacobaea vulgaris* Gaertn.

ragwort –see– *Crotalaria berteroana* DC.; *Jacobaea aquatica* (Hill) P. Gaertn. et al. var. erratica (Bertol.) Pelser & Meijden; *Jacobaea vulgaris* Gaertn.; *Senecio alpinus* (L.) Scop.; *Senecio brasiliensis* (Spreng.) Less.; *Senecio isatideus* DC.; *Senecio latifolius* DC.; *Senecio retrorsus* DC.; *Senecio ruwenzoriensis* S. Moore; *Senecio sceleratus* Schweick.

rahar dal –see– *Cajanus cajan* (L.) Millsp.

rai' dore –see– *Senna occidentalis* (L.) Link

raigras perenne –see– *Lolium perenne* L.

rain flower –see– *Grevillea banksii* R. Br.

rain lily –see– *Cooperia pedunculata* Herb.

rain tree –see– *Laburnum anagyroides* Medik.; *Samanea saman* (Jacq.) Merr.

rainberry –see– *Rhamnus cathartica* L.

rainbow-leaf –see– *Smodingium argutum* E. Mey. ex Sond.

Rainfarn –see– *Tanacetum vulgare* L.

rainfort sauvage –see– *Raphanus raphanistrum* L.

Rainweide –see– *Ligustrum vulgare* L.

raisin –see– *Vitis vinifera* L.

raiz forte –see– *Armoracia rusticana* P. Gaertn. et al.

rajma –see– *Acacia leucophloea* (Roxb.) Willd.

ral tree –see– *Semecarpus anacardium* L. f.

Ral'dore –see– *Senna occidentalis* (L.) Link

rama zorilla –see– *Phacelia crenulata* Torr. ex S. Watson

ramin –see– *Gonystylus bancanus* (Miq.) Kurz

ramjada –see– *Jatropha curcas* L.

rancher's-fiddleneck –see– *Amsinckia intermedia* Fisch. & C. A. Mey.

ranger's-buttons –see– *Sphenosciadium capitellatum* A. Gray

rangiora –see– *Brachyglottis repanda* J. R. Forst. & G. Forst.

Rangoon bean –see– *Phaseolus lunatus* L.

Rangoon teak –see– *Tectona grandis* L. f.

Rangoonbohne –see– *Phaseolus lunatus* L.

ransoms –see– *Allium ursinum* L.

RANUNCULUS ACRIS L. [Ranunculaceae]

Common Names:

acrid buttercup; blister plant; bouton d'or; butter daisy; Butterblume; buttercup; crowflower; giant buttercup; Hahnenfuß; meadow buttercup; meadow ranunculus; réti boglárka; tall buttercup; tall crowfoot; tall field buttercup; upright meadow crowfoot; yellow gowan

Citations:

Harshberger JW (1894) An additional poisonous plant. Bot Gaz 19(322):159.

Harshberger JW (1894) Ranunculus acris, an additional poisonous plant. Garden Forest 7(322):170.

Hidiroglou M, Knutti HJ (1963) The effects of green tall buttercup in roughage on the growth and health of beef cattle and sheep. Can J Anim Sci 43(Jun):68-71.

Pammel LH (1929) Buttercup or crow-foot poisoning. Vet Med 24(Dec):540-541.

Winters JB (1976) Severe urticarial reaction in a dog following ingestion of tall field buttercup. Vet Med Small Anim Clin 71(3):307.

RANUNCULUS BULBOSUS L. [Ranunculaceae]

Common Names:

bulbous buttercup; butter daisy; buttercup; crowflower; crowfoot; Hahnenfuß; St. Anthony's-turnip

Citations:

Gunning OV (1949) Suspected buttercup poisoning in a Jersey cow. Br Vet J 105:393.

Kelch WJ, Kerr LA, Adair HS, et al. (1992) Suspected buttercup (Ranunculus bulbosus) toxicosis with secondary photosensitization in a Charolais heifer. Vet Hum Toxicol 34(3):238-239.

Pammel LH (1929) Buttercup or crow-foot poisoning. Vet Med 24(Dec):540-541.

RANUNCULUS CYMBALARIA Pursh [Ranunculaceae]

Common Names:

desert crowfoot; marsh buttercup; seaside buttercup; shore buttercup

Citations:

Fleming CE (1920) Poisonous range plants. Nevada Agric Exp Sta Annu Rep 1919:39-43.

RANUNCULUS DAMASCENUS Boiss. & Gaill. [Ranunculaceae]

Citations:

Metın A, Çalka Ö, Behçet L, et al. (2001) Phytodermatitis from Ranunculus damascenus. Contact Dermatitis 44(3):183.

RANUNCULUS MULTIFIDUS Forssk. [Ranunculaceae]

Synonyms:

Ranunculus pinnatus Poir.; ***Ranunculus pubescens*** Thunb.

Common Names:

buttercup

Citations:

Shone DK (1966) Poisonous plants of Rhodesia. Part I. Rhodesia Zambia Malawi J Agric Res 4:81-94.

Shone DK (1967) Poisonous plants of Rhodesia. Part II. Rhodesia Zambia Malawi J Agric Res 5:29-41.

Ranunculus pinnatus Poir. = Ranunculus multifidus Forssk.

Ranunculus pubescens Thunb. = Ranunculus multifidus Forssk.

RANUNCULUS REPENS L. [Ranunculaceae]

Common Names:

butón-de-oro; creeping buttercup; creeping crowfoot; crowfoot; kriechender Hahnenfuß

Citations:

Morales H (1989) Abortos en una lechería da la VIII Región de Chile atribuidos al consumo de botón de oro (Ranunculus repens, L.). Arch Med Vet 21(2):163-166.

RANUNCULUS SCELERATUS L. [Ranunculaceae]

Common Names:

blister buttercup; buttercup; celery-leaf buttercup; celery-leaf crowfoot; crowfoot; cursed buttercup; cursed crowfoot; ditch crowfoot; Froschkraut; Gifthahnenfuß; marsh crowfoot; water crowfoot

Citations:

Miedzobrodzki K (1971) Przypadek zatrucia krów jaskrem (Ranunculus sceleratus). Med Weter 27(7):411-412.

rape –see– *Brassica napus* L. var. napus; *Brassica rapa* L. subsp. campestris (L.) A. R. Clapham

rape kale –see– *Brassica napus* L. var. pabularia (DC.) Rchb.

Raphanus niger Mill. = Raphanus sativus L.

RAPHANUS RAPHANISTRUM L. [Brassicaceae]

Common Names:

Hederich; jaramago blanco; joint charlock; rabaniza; rábano silvestre; rainfort sauvage; ravenelle; Rettich; white charlock; wild mustard; wild radish

Citations:

Hale OM, Utley PR (1985) Effects of feeding wheat contaminated with wild radish (Raphanus raphanistrum) to growing pigs. J Anim Sci 61(5):1172-1177.

Trouche (1936) Intoxication d'un troupeau d'agnelles par suite d'une absorption abondante de ravenelles. Rev Med Vet 88:682-683.

RAPHANUS SATIVUS L. [Brassicaceae]

Synonyms:

Raphanus niger Mill.

Common Names:

Ackerrettich; fodder radish; Hederich; radish; white radish

Citations:

El Sayed F, Manzur F, Marguery MC, et al. (1995) Urticarial manifestations due to Raphanus niger. Contact Dermatitis 32(4):241.

Mitchell JC, Jordan WP (1974) Allergic contact dermatitis from the radish, Raphanus sativus. Br J Dermatol 91(2):183-189.

Sinha SM, Pasricha JS, Sharma RC, et al. (1977) Vegetables responsible for contact dermatitis of the hands. Arch Dermatol 113(6):776-779.

Raps –see– *Brassica napus* L. var. napus

rapsfromel –see– *Brassica napus* L. var. napus

RAPUTIA ALBA (Mart.) Engl. [Rutaceae]

Citations:

Freise FW (1936) Vergiftungen durch brasilianische Werkhölzer. II. Jacareúba-Holz und Seidenholz. Sammlung Vergiftungsfallen 7(C33):61-72.

rastrero –see– *Trifolium repens* L.

raton –see– *Gliricidia sepium* (Jacq.) Kunth ex Walp.

rat's-bane –see– *Dichapetalum toxicarium* (G. Don) Baill.

ratti –see– *Abrus precatorius* L.

rattlebox –see– *Crotalaria berteroana* DC.; *Crotalaria goreensis* Guill. & Perr.; *Crotalaria pallida* Aiton; *Crotalaria retusa* L.; *Crotalaria sagittalis* L.; *Crotalaria spectabilis* Roth; *Sesbania drummondii* (Rydb.) Cory; *Sesbania punicea* (Cav.) Benth.

rattlebush –see– *Crotalaria burkeana* Benth.; *Sesbania drummondii* (Rydb.) Cory; *Sesbania punicea* (Cav.) Benth.

rattlepod –see– *Crotalaria juncea* L.; *Crotalaria retusa* L.; *Crotalaria spectabilis* Roth rattleroot –see– *Actaea racemosa* L.

rattlesnake weed –see– *Crotalaria sagittalis* L.

rattleweed –see– *Astragalus lentiginosus* Douglas ex Hook. var. diphysus (A. Gray) M. E. Jones; *Astragalus wootonii* E. Sheld.; *Crotalaria pallida* Aiton; *Crotalaria retusa* L.; *Crotalaria sagittalis* L.; *Crotalaria spectabilis* Roth; *Oxytropis lambertii* Pursh; *Oxytropis sericea* Nutt.

raunja –see– *Acacia leucophloea* (Roxb.) Willd.

Rauschbeere –see– *Vaccinium uliginosum* L.

Raute –see– *Ruta graveolens* L.

RAUVOLFIA SERPENTINA (L.) Benth. ex Kurz [Apocynaceae]

Common Names:

Indian snakeroot; rauwolfia; serpentwood; snakeroot

Citations:

Hager W, Friedrich KH, Wink K, et al. (1968) Suizidversuch mit Ajmalin. Dtsch Med Wochenschr 93(38):1809-1812.

Müller FJ (1969) Orale Intoxikation einer Erwachsenen mit Gilurytmal. Med Klin 64(31):1389-1393.

Rogos R (1967) Suizidversuch mit Ajmalin. Z Gesamte Inn Med 22(14):432-435.

RAUVOLFIA VOMITORIA Afzel. [Apocynaceae]

Common Names:

Akanta

Citations:

Jornod JC, Barrelet JA (1965) Suicidal attempt by overdosage of ajmaline. Am Heart J 70(5):719-720.

rauwolfia –see– *Rauvolfia serpentina* (L.) Benth. ex Kurz

ravenelle –see– *Raphanus raphanistrum* L.; *Sinapis arvensis* L.

rayless goldenrod –see– *Isocoma plurifolia* (Torr. & A. Gray) Greene

razha –see– *Medicago sativa* L.

real heirri –see– *Deguelia utilis* (A. C. Sm.) A. M. G. Azevedo

Rebelwurz –see– *Asarum europaeum* L.

Rebhuhnholz –see– *Andira inermis* (W. Wright) Kunth ex DC.

recurved thorn apple –see– *Datura inoxia* Mill.; *Datura wrightii* Regel

red angel's-trumpet –see– *Brugmansia sanguinea* (Ruiz & Pav.) D. Don

red baneberry –see– *Actaea rubra* (Aiton) Willd.

red bead vine –see– *Abrus precatorius* L.

red bean –see– *Calia secundiflora* (Ortega) Yakovlev; *Phaseolus vulgaris* L.

red-berry juniper –see– *Juniperus pinchotii* Sudw.

red bryony –see– *Bryonia dioica* Jacq.

red buckeye –see– *Aesculus pavia* L.

red bur –see– *Sclerolaena anisacanthoides* (F. Muell.) Domin

red caustic creeper –see– *Chamaesyce prostrata* (Aiton) Small

red cayenne –see– *Capsicum frutescens* L.

red cedar –see– *Thuja occidentalis* L.; *Thuja plicata Donn* ex D. Don

red chickweed –see– *Anagallis arvensis* L.

red chili –see– *Capsicum frutescens* L.

red clover –see– *Trifolium pratense* L.

red cole –see– *Armoracia rusticana* P. Gaertn. et al.

red cotton –see– *Asclepias curassavica* L.

red darling pea –see– *Swainsona galegifolia* (Andrews) R. Br.

red fescue –see– *Festuca rubra* L. subsp. fallax (Thuill.) Nyman

red ginger –see– *Zingiber officinale* Roscoe

red gram –see– *Cajanus cajan* (L.) Millsp.

red gum –see– *Eucalyptus camaldulensis* Dehnh.

red heart-of-bullock bush –see– *Alectryon oleifolius* (Desf.) S. T. Reynolds

red hemp nettle –see– *Galeopsis ladanum* L.

red ink plant –see– *Phytolacca americana* L.

red kidney bean –see– *Phaseolus vulgaris* L.

red lac –see– *Toxicodendron succedaneum* (L.) Kuntze

red laurel –see– *Rhododendron catawbiense* Michx.

red mallow –see– *Modiola caroliniana* (L.) G. Don

red maple –see– *Acer rubrum* L.

red Mexican bean –see– *Phaseolus vulgaris* L.

red navy bean –see– *Phaseolus vulgaris* L.

red oak –see– *Quercus rubra* L.

red orache –see– *Atriplex rosea* L.

red paperbark tree –see– *Albizia tanganyicensis* Baker f.

red pepper –see– *Capsicum annuum* L.

red pimpernel –see– *Anagallis arvensis* L.

red pokeweed –see– *Phytolacca americana* L.

red poppy –see– *Papaver rhoeas* L.

red puccoon –see– *Sanguinaria canadensis* L.

red rhatany –see– *Krameria lappacea* (Dombey) Burdet & B. B. Simpson

red sage –see– *Lantana camara* L.; *Salvia coccinea* Buc'hoz ex Etl.

red salvia –see– *Salvia coccinea* Buc'hoz ex Etl.

red silky oak –see– *Grevillea banksii* R. Br.

red soldier –see– *Euphorbia drummondii* Boiss.

red sorrel –see– *Rumex acetosella* L.

red squill –see– *Drimia maritima* (L.) Stearn

red-stem filaree –see– *Erodium cicutarium* (L.) L'Her.

red-stem peavine –see– *Astragalus emoryanus* (Rydb.) Cory

red vetch –see– *Lathyrus hirsutus* L.

red-water bark –see– *Erythrophleum suaveolens* (Guill. & Perr.) Brenan

red willow –see– *Modiola caroliniana* (L.) G. Don

red wings –see– *Combretum platypetalum* Welw. ex Laws.

redbird cactus –see– *Pedilanthus tithymaloides* (L.) Poit.

redhead cottonbush –see– *Asclepias curassavica* L.

redoul –see– *Coriaria myrtifolia* L.

redroot –see– *Amaranthus hybridus* L.; *Amaranthus palmeri* S. Watson; *Amaranthus retroflexus* L.; *Sanguinaria canadensis* L.

redroot pigweed –see– *Amaranthus retroflexus* L.

redscale –see– *Atriplex rosea* L.

redshank –see– *Amaranthus cruentus* L.; *Persicaria maculosa* Gray

redtop –see– *Asclepias curassavica* L.

redweed –see– *Papaver rhoeas* L.; *Phytolacca americana* L.

redwood –see– *Sequoia sempervirens* (D. Don) Endl.

reed canarygrass –see– *Phalaris arundinacea* L.

reed fescue –see– *Festuca arundinacea* Schreb.

reed phalaris –see– *Phalaris arundinacea* L.

reed sweetgrass –see– *Glyceria maxima* (Hartm.) Holmb.

reefer –see– *Cannabis sativa* L.

reflexed amaranthus –see– *Amaranthus retroflexus* L.

réglisse –see– *Glycyrrhiza glabra* L.

REHMANNI GLUTINOSA (Gaertn.) Steud. [Phrymaceae]

Common Names:

gan-di-huang

Citations:

Chan JC, Chan TY, Chan KL, et al. (1994) Anticholinergic poisoning from Chinese herbal medicines. Aust N Z J Med 24(3):317.

Reichblütige Glyzinie –see– *Wisteria floribunda* (Willd.) DC.

rein –see– *Chrozophora obliqua* (Vahl) A. Juss. ex Spreng.

reina-de-la-noche –see– *Brugmansia arborea* (L.) Lagerh.

Reis –see– *Oryza sativa* L.

Reiterkappe –see– *Aconitum napellus* L.

remolacha –see– *Beta vulgaris* L.

rengas –see– *Gluta renghas* L.

renghas –see– *Gluta renghas* L.

renonée âcre –see– *Persicaria hydropiper* (L.) Spach

renta yam –see– *Dioscorea alata* L.

repolho –see– *Brassica oleracea* L. var. capitata L.

Rere –see– *Senna occidentalis* (L.) Link

resurrection plant –see– *Kalanchoe pinnata* (Lam.) Pers.

retama –see– *Thevetia peruviana* (Pers.) K. Schum.

réti boglárka –see– *Ranunculus acris* L.

Rettich –see– *Raphanus raphanistrum* L.

reunja –see– *Acacia leucophloea* (Roxb.) Willd.

reuse kweekgrass –see– *Cynodon nlemfuensis* Vanderyst

REUTEALIS TRISPERMA (Blanco) Airy Shaw [Euphorbiaceae]

Synonyms:

Aleurites trisperma Blanco

Common Names:

banucalad nut; lumbang nut; soft lumbang

Citations:

Emmel MW (1947) The toxic principle of the species Aleurites. J Am Vet Med Assoc 111(Nov):386-387.

Emmel MW (1947) The toxic principle of the tung tree. Florida Agric Exp Sta Bull #431:35 pp.

revebjelle –see– *Digitalis purpurea* L.

reverchonia –see– *Reverchonia arenaria* A. Gray

REVERCHONIA ARENARIA A. Gray [Phyllanthaceae]

Common Names:

reverchonia; sand reverchonia; spurge

Citations:

Anonymous (1942) Investigations of stock-poisoning plants. U S Dep Agric Bur Anim Indus Chief Rep 1942:29.

Rhabarber –see– *Rheum ×hybridum* Murray

RHAMNUS CATHARTICA L. [Rhamnaceae]

Common Names:

buckthorn; Frenchberry; hart's-thorn; Hirschdorn; Kreuzdorn; purging buckthorn; rainberry; Rhine thorn; Rhineberry; way thorn

Citations:

Anderson T, Konracki S, Lyman DO, et al. (1978) Diarrhea from herbal tea - New York, Pennsylvania. MMWR Morb Mortal Wkly Rep 27(8):248-249.

Banach K (1980) Ostre zatrucia antrazwiazkami spowodowane spożyciem owoców szakłaku pospolitego. Wiad Lek 33(5):405-408.

Calderon-Gonzalez R, Rizzi-Hernadez H (1967) Buckthorn polyneuropathy. N Eng J Med 277:69-71.

Lichtensteiger CA, Johnston NA, Beasley VR (1997) Rhamnus cathartica (buckthorn) hepatocellular toxicity in mice. Toxicol Pathol 25(5):449-452.

Rhamnus frangula L. = Frangula alnus Mill.

rhapontic rhubarb –see– *Rheum ×hybridum* Murray

RHAPONTICUM REPENS (L.) Hidalgo [Asteraceae]

Synonyms:

Acroptilon repens (L.) DC.; ***Centarurea repens*** L.; ***Centaurea picris*** Pall. ex Willd.

Common Names:

creeping knapweed; hard heads; Russian centaurea; Russian knapweed; Russian star thistle; Turkestan thistle

Citations:

Dil'bazi GI (1974) [Poisoning of buffaloes by hay heavily contaminated with Centaurea picris.] Veterinariia Moscow 51(2):106-107.

Farrell RK, Sande RD, Lincoln SD (1971) Nigropallidal encephalomalacia in a horse. J Am Vet Med Assoc 158(7):1201-1204.

Larson KA, Young S (1970) Nigropallidal encephalomalacia in horses in Colorado. J Am Vet Med Assoc 156(5):626-628.

Young S, Brown WW, Klinger B (1970) Clinical and pathologic aspects of Russian knapweed poisoning in horses (nigropallidal encephalomalacia). J Am Vet Med Assoc 156(May 1):1219.

Young S, Brown WW, Klinger B (1970) Nigropallidal encephalomalacia in horses caused by ingestion of weeds of the genus Centaurea. J Am Vet Med Assoc 157(11):1602-1605.

Young S, Brown WW, Klinger B (1970) Nigropallidal encephalomalacia in horses fed Russian knapweed (Centaurea repens L). Am J Vet Res 31(8):1393-1404.

Rhapontikrhabarber –see– *Rheum ×hybridum* Murray

rhatanhia –see– *Krameria lappacea* (Dombey) Burdet & B. B. Simpson

rhatania –see– *Krameria lappacea* (Dombey) Burdet & B. B. Simpson

RHAZYA STRICTA Decne. [Apocynaceae]

Common Names:

harmal; harmel

Citations:

Adam SE (1998) Toxicity of Rhazya stricta to sheep. Vet Hum Toxicol 40(2):68-69.

Adam SE (1999) Experimental Rhazya stricta toxicosis in rats. Vet Hum Toxicol 41(1):5-8.

Adam SE, Al-Yahya MA, Al-Farhan AH (2002) Toxicity of Nerium oleander and Rhazya stricta in Najdi sheep: Hematologic and clinicopathologic alterations. Am J Chin Med 30(2-3):255-262.

RHEUM ×HYBRIDUM Murray [Polygonaceae]

Synonyms:

Rheum rhaponticum auct.

Common Names:

Deutscher Rhabarber; pie plant; Rhabarber; rhapontic rhubarb; Rhapontikrhabarber; rhubarb; rhubarbe; ruibarbo; water plant; wine plant

Citations:

Anonymous (1917) Death from eating rhubarb leaves. JAMA 68(25):1928.

Anonymous (1917) Poisoning by Rhubarb leaves. Lancet 1(Jun 2):847-848.

Anonymous (1917) Poisoning from rhubarb leaves. JAMA 68(26):847.

Arena JM (1972) Rhubarb hyperphagia hazards. JAMA 219(5):626.

Benson HW (1919) Poisoning from eating canned rhubarb stems. JAMA 73(Oct 11):1152.

Fühner H (1926) Deutscher Rhabarber. Munch Med Wochenschr 73:400-402.

Maillart (1917) Un cas d'empoisonnement par des feuilles de rhubarbe. Rev Med Suisse Romande 37:344-348, 468-471.

Naudin L (1932) Intoxication d'une chèvre par le limbe de feuilles de rhubarbe. Rec Med Vet Ec Alfort 108:91-92.

Robb HF, Sippy JJ (1919) Death from rhubarb leaves due to oxalic acid poisoning. JAMA 73(8):627-628.

Spoerke DG, Temple AR (1978) One year's experience with potential plant poisonings reported to the Intermountain Regional Poison Control Center. Vet Hum Toxicol 20(2):85-89.

Streicher E (1964) Akutes Nierenversagen und Ikterus nach einer Vergiftung mit Rhaberberblättern. Dtsch Med Wochenschr 89(50):2379-2381.

Tallqvist H, Väänänen I (1960) Death of a child from oxalic acid poisoning due to eating rhubarb leaves. Ann Paediatr Fenn 6(2):1-4.

RHEUM OFFICINALE Baill. [Polygonaceae]

Common Names:

Chinese rhubarb; medicinal rhubarb; rhubarb

Citations:

Liu WK, Ng TB, Chan WY, et al. (1996) Toxic effects of Chinese medicinal herbs on pregnant mice. Toxicologist 30(1 Part 2):192.

Rheum rhaponticum auct. = Rheum ×hybridum Murray

rheumatism root –see– *Apocynum cannabinum* L.

rheumatism weed –see– *Apocynum cannabinum* L.

Rhine thorn –see– *Rhamnus cathartica* L.

Rhineberry –see– *Rhamnus cathartica* L.

Rhodesgrass –see– *Chloris gayana* Kunth

Rhodesian ragwort –see– *Senecio latifolius* DC.; *Senecio sceleratus* Schweick.

Rhodesian wild gentian –see– *Chironia transvaalensis* Gilg

Rhodesierholz –see– *Santalum album* L.

rhododendron –see– *Rhododendron indicum* (L.) Sweet; *Rhododendron macrophyllum* D. Don ex G. Don; *Rhododendron polifolium* Franch; *Rhododendron ponticum* L.; *Rhododendron simsii* Planch.

RHODODENDRON ARBOREUM Sm. [Ericaceae]

Common Names:

burans; lal burass

Citations:

Carey FM, Lewis JJ, MacGregor JL, et al. (1959) Pharmacological and chemical observations on some toxic nectars. J Pharm Pharmacol 11:269T-274T.

RHODODENDRON AUREUM Georgi [Ericaceae]

Common Names:

Gelbe Alpenrose; Gelbe Schneerose; Gichtrose; Siberian snow rose

Citations:

Igumnow AK (1967) [Rhododendron dermatitis in Transbaikalia.] Vestn Dermatol Venerol 41(4):70-71.

RHODODENDRON BARBATUM Wall. ex G. Don [Ericaceae]

Citations:

Carey FM, Lewis JJ, MacGregor JL, et al. (1959) Pharmacological and chemical observations on some toxic nectars. J Pharm Pharmacol 11:269T-274T.

Purohit K (1960) Rhododendron poisoning in animals. Indian Vet J 37:631-633.

RHODODENDRON BRACHYCARPUM D. Don ex G. Don [Ericaceae]

Citations:

Hikino H, Ohizumi Y, Konno C, et al. (1979) Subchronic toxicity of ericaceous toxins and Rhododendron leaves. Chem Pharm Bull (Tokyo) 27(4):874-879.

Rhododendron californicum Hook. = Rhododendron macrophyllum D. Don ex G. Don

RHODODENDRON CAMPANULATUM D. Don [Ericaceae]

Citations:

Purohit K (1960) Rhododendron poisoning in animals. Indian Vet J 37:631-633.

RHODODENDRON CATAWBIENSE Michx. [Ericaceae]

Common Names:

catawba rhododendron; mountain rosebay; purple ivy; purple laurel; purple rhododendron; red laurel; rosebay; rosebay laurel

Citations:

Brahm E, Buntenkötter S, Simanowski W (1973) Vergiftungen von wiederkäuenden Paarhufern durch Ericaceen im dormunder Tierpark. In: Ippen R et al. (eds.) Erkrank der Zoot XV. Int Symp, Kolmarden Akad Verlag. pp. 125-130.

RHODODENDRON ×COLUMBIANUM (Piper) Harmaja [Ericaceae]

Synonyms:

Ledum columbianum Piper

Common Names:

Pacific Labrador tea; swamp laurel

Citations:

Eastwood A (1933) Ledum as a poisonous plant. Leaflet Western Bot 1(5):43.

RHODODENDRON DEGRONIANUM Carrière [Ericaceae]

Synonyms:

Rhododendron metternickii Siebold & Zucc. var. pentamerum Maxim.

Citations:

Hikino H, Ohizumi Y, Konno C, et al. (1979) Subchronic toxicity of ericaceous toxins and Rhododendron leaves. Chem Pharm Bull (Tokyo) 27(4):874-879.

RHODODENDRON FABERI HEMSL.subsp. prattii (Franch.) D. F. Chamb. [Ericaceae]

Synonyms:

Rhododendron prattii Franch.

Citations:

Carey FM, Lewis JJ, MacGregor JL, et al. (1959) Pharmacological and chemical observations on some toxic nectars. J Pharm Pharmacol 11:269T-274T.

RHODODENDRON FULVUM Balf. f. & W. W. Sm. [Ericaceae]

Citations:

Carey FM, Lewis JJ, MacGregor JL, et al. (1959) Pharmacological and chemical observations on some toxic nectars. J Pharm Pharmacol 11:269T-274T.

RHODODENDRON INDICUM (L.) Sweet [Ericaceae]

Common Names:

azalea; rhododendron

Citations:

Puschner B, Holstege DM, Lamberski N, et al. (2001) Grayanotoxin poisoning in three goats. J Am Vet Med Assoc 218(4):573-575.

Tokarnia CH, Armién AG, Peixoto PV, et al. (1996) Estudo experimental sobre a toxidez de algumas plantas ornamentais em bovinos. Pesq Vet Bras 16(1):5-20.

RHODODENDRON LEDIFOLIUM G. Don [Ericaceae]

Citations:

Tokarnia CH, Armién AG, Peixoto PV, et al. (1996) Estudo experimental sobre a toxidez de algumas plantas ornamentais em bovinos. Pesq Vet Bras 16(1):5-20.

RHODODENDRON LUTEUM Sweet [Ericaceae]

Common Names:

pontic azalea

Citations:

Sütilüpinar N, Mat A, Satganoğlu Y (1993) Poisoning by toxic honey in Turkey. Arch Toxikol 67(2):148-150.

RHODODENDRON MACABEANUM G. Watt ex Balf. f. [Ericaceae]

Citations:

Carey FM, Lewis JJ, MacGregor JL, et al. (1959) Pharmacological and chemical observations on some toxic nectars. J Pharm Pharmacol 11:269T-274T.

RHODODENDRON MACROPHYLLUM D. Don ex G. Don [Ericaceae]

Synonyms:

Rhododendron californicum Hook.

Common Names:

California rhododendron; California rosebay; Pacific rhododendron; rhododendron; western rhododendron

Citations:

Casteel S, Wagstaff DJ (1989) Rhododendron macrophyllum poisoning in a group of goats and sheep. Vet Hum Toxicol 31(2):176-177.

Rhododendron metternickii Siebold & Zucc. var. pentamerum Maxim. = Rhododendron degronianum Carrière

RHODODENDRON MOLLE (Blume) G. Don [Ericaceae]

Citations:

Chan JC, Chan TY, Chan KL, et al. (1994) Anticholinergic poisoning from Chinese herbal medicines. Aust N Z J Med 24(3):317.

Chan TY, Chan JC, Tomlinson B, et al. (1994) Poisoning by Chinese herbal medicines in Hong Kong: A hospital-based study. Vet Hum Toxicol 36(6):546-547.

RHODODENDRON NIVEUM Hook. f. [Ericaceae]

Citations:

Carey FM, Lewis JJ, MacGregor JL, et al. (1959) Pharmacological and chemical observations on some toxic nectars. J Pharm Pharmacol 11:269T-274T.

RHODODENDRON POLIFOLIUM Franch [Ericaceae]

Common Names:

rhododendron

Citations:

Bubien MZ, Magat A, Delatour P (1970) Fréquence des intoxications des animaux en Pologne. Bull Soc Vet Med Comp Lyon 72(2):167-173.

RHODODENDRON PONTICUM L. [Ericaceae]

Common Names:

Asiatische Alpenrose; honey-of-Trebizond; Pontic rhododendron; rhododendron

Citations:

Biberoğlu K, Biberoğlu S, Komsuoğlu B (1988) Transient Wolff-Parkinson-White syndrome during honey intoxication. Isr J Med Sci 24(4-5):253-254.

Biberoğlu S, Biberoğlu K, Komsuoğlu B (1988) Mad honey. JAMA 259(13):1943.

Black DH (1991) Rhododendron poisoning in sheep. Vet Rec 128(15):363-364.

Frape D, Ward A (1993) Suspected rhododendron poisoning in dogs. Vet Rec 132(20):515-516.

Gossinger H, Hruby K, Haubenstock A, et al. (1983) Cardiac arrhythmias in a patient with grayanotoxin-honey poisoning. Vet Hum Toxicol 25(5):328-329.

Higgins RJ, Hannam DA, Humphreys DJ, et al. (1985) Rhododendron poisoning in sheep. Vet Rec 116(11):294-295.

Matschullat G (1974) Rhododendronvergiftung bei Schafen. Prakt Tierarzt 55(11):624, 626.

Onat FY, Yegen BC, Lawrence R, et al. (1991) Mad honey poisoning in man and rat. Rev Environ Health 9(1):3-10.

Sütilüpinar N, Mat A, Satganoğlu Y (1993) Poisoning by toxic honey in Turkey. Arch Toxikol 67(2):148-150.

Rhododendron prattii Franch. = Rhododendron faberi Hemsl. subsp. prattii (Franch.) D. F. Chamb.

RHODODENDRON SIMSII Planch. [Ericaceae]

Common Names:

azalea; rhododendron

Citations:

Rose A, Pitchford W, Monin T, et al. (1988) Acute weakness and death in a cat. Vet Hum Toxicol 30(4):334-335.

RHODODENDRON SINOGRANDE Balf. f. & W. W. Sm. [Ericaceae]

Citations:

Carey FM, Lewis JJ, MacGregor JL, et al. (1959) Pharmacological and chemical observations on some toxic nectars. J Pharm Pharmacol 11:269T-274T.

RHODODENDRON THOMSONII Hook. f. [Ericaceae]

Citations:

Carey FM, Lewis JJ, MacGregor JL, et al. (1959) Pharmacological and chemical observations on some toxic nectars. J Pharm Pharmacol 11:269T-274T.

RHODOMYRTUS MACROCARPA Benth. [Myrtaceae]

Common Names:

Australian finger cherry; Cooktown loquat; finger cherry; finger cherry loquat; native loquat; wannakai

Citations:

D'Ombrain A (1944) Sudden blindness after eating finger cherries (Rhodomyrtus macrocarpa). Med J Aust 2(Sep 2):263.

English FP, Bennett Y (1989) Blindness and the finger cherry tree. Med J Aust 151:112-113.

Flecker H (1944) Sudden blindness after eating "finger cherries" (Rhodomyrtus macrocarpa). Med J Aust 1944(Aug 19):183-185.
Lucas CE (1944) Sudden blindness after eating finger cherries (Rhodomyrtus macrocarpa). Aust Med J. Sep 2:263.

RHODOTYPOS SCANDENS (Thunb.) Makino [Rosaceae]

Common Names:
fatberry; jetbead; jetberry bush

Citations:
Rascoff H, Wasser S (1953) Poisoning in a child simulating diabetic coma. Report of a case. JAMA 152(12):1134-1135.

Rhoicissus cuneifolia (Eckl. & Zeyh.) Planch. = Cissus cuneifolia Eckl. & Zeyh.

Rhoicissus tridentata (L. f.) Wild & R. B. Drumm. subsp. cuneifolia (Eckl. & Zeyh.) N. R. Urton = Cissus cuneifolia Eckl. & Zeyh.

rhubarb –see– *Rheum officinale* Baill.; *Rheum ×hybridum* Murray

rhubarbe –see– *Rheum ×hybridum* Murray

rhus –see– *Toxicodendron succedaneum* (L.) Kuntze

RHUS CHINENSIS Mill. [Anacardiaceae]

Synonyms:
Rhus semialata Murray

Citations:
Biberstein H (1929) Über Hautreaktionen bei Applikation von verschiedenen Rhusarten. Klin Wochenschr 8:99-102.

RHUS CORIARIA L. [Anacardiaceae]

Common Names:
sumach

Citations:
Pilgram RE, Fleagle GS (1970) Indian sandal strap dermatitis. JAMA 211(8):1378.

RHUS CRENATA Thunb. [Anacardiaceae]

Citations:
Biberstein H (1929) Über Hautreaktionen bei Applikation von verschiedenen Rhusarten. Klin Wochenschr 8:99-102.

Rhus diversiloba Torr. & A. Gray = Toxicodendron diversilobum (Torr. & A. Gray) Greene

Rhus metopium L. = Metopium brownei (Jacq.) Urb.

Rhus radicans L. = Toxicodendron radicans (L.) Kuntze

Rhus semialata Murray = Rhus chinensis Mill.

Rhus striata Ruiz & Pav. = Toxicodendron striatum (Ruiz & Pav.) Kuntze

Rhus succedanea L. = Toxicodendron succedaneum (L.) Kuntze

Rhus toxicodendron L. = Toxicodendron pubescens Mill.

rhus tree –see– *Toxicodendron succedaneum* (L.) Kuntze

RHUS TRICHOCARPA Miq. [Anacardiaceae]

Citations:
Franklin JJ (1952) Another plant (Rhus trichocarpa) to be labelled poisonous. Plants Gardens 8:271.

RHUS TYPHINA L. [Anacardiaceae]

Common Names:
antler tree; dwarf sumach; Essigbaum; staghorn sumach; velvet sumach; Virginia sumach

Citations:
Biberstein H (1929) Über Hautreaktionen bei Applikation von verschiedenen Rhusarten. Klin Wochenschr 8:99-102.

Rhus vernicifera DC. = Toxicodendron vernicifluum (Stokes) F. A. Barkley

Rhus verniciflua Stokes = Toxicodendron vernicifluum (Stokes) F. A. Barkley

Rhus vernix L. = Toxicodendron vernix (L.) Kuntze

RHUS VIRENS Lindh. ex A. Gray [Anacardiaceae]

Common Names:
sumach

Citations:
Mathews FP, Schmidt H (1934) Miscellaneous poisonous plants. Texas Agric Exp Sta Annu Rep 47:12-13.

ribbon cactus –see– *Pedilanthus tithymaloides* (L.) Poit.

ribbongrass –see– *Phalaris arundinacea* L.

rice –see– *Oryza sativa* L.

richweed –see– *Ageratina altissima* (L.) R. M. King & H. Rob.

ricin –see– *Ricinus communis* L.

ricino –see– *Ricinus communis* L.

RICINUS COMMUNIS L. [Euphorbiaceae]

Common Names:
African coffee tree; amerikansk olieplante; arandi; castor bean; Christuspalme; erand; figuera infernal; graines-de-Ricin; higuerilla; kristpalme; lansina; mamona; Mexican weed; mole bean; olieblom; Palma Christi; palma-de-Cristo; ricin; ricino; rizin; Rizinus; tangan tangan; wonder tree; Wonderbohne; wonderboon; Wunderbaum; zurma

Citations:
Abdüladir-Lütfi A (1935) Tödliche Intoxikation durch die Samen der Rizinuspflanze. Dtsch Med Wochenschr 61(Mar 15):416-417.
Albretsen JC, Gwaltney Brant SM, Khan SA (2000) Evaluation of castor bean toxicosis in dogs: 98 cases. J Am Anim Hosp Assoc 36(3):229-233.

Anderson KE, Nielsen R (1984) Lipstick dermatitis to castor oil. Contact Dermatitis 11(4):253-254.

Anderson TS (1948) Castor poisoning in Ayrshire cattle. Vet Rec 60:28.

Anonymous (1982) Castor-oil bean toxicity. Vet Rec 111:172.

Aplin PJ, Eliseo T (1997) Ingestion of castor oil plant seeds. Med J Aust 167(5):260-261.

Aslani MR, Maleki M, Mohri M, et al. (2007) Castor bean (Ricinus communis) toxicosis in a sheep flock. Toxicon 49(3):400-406.

Astolfi E, Polack NR (1961) Intoxicación accidental por ingestión de semillas de ricino. Arch Argent Pediatr 57(Oct):337-339.

Belzunegui T, Charles AB, Hernández R, et al. (1988) Intoxicación por ingestión de semillas de ricino. A propósito de un caso. Med Clin (Barc) 90(17):716-717.

Benesi FJ (1979) Influência do farelo de mamona (Ricinus communis L.) destoxicado sobre o proteinograma sangüíneo e desempho de suínos. Arq Esc Vet Univ Minas Gerais Belol Horizonte. pp. 541-542.

Bernton HS (1945) Castor bean sensitiveness. Case report with discussion of principles. South Med J 38:670-677.

Bispham WN (1903) Report of cases of poisoning by fruit of Ricinus communis. Am J Med Sci 126:319-321.

Brito MF, Tokarnia CH (1996) Intoxicação experimental pelas sementes trituradas de Ricinus communis (Euphorbiaceae) em coelhos. Pesq Vet Bras 16(4):1-7.

Bubien MZ, Magat A, Delatour P (1970) Fréquence des intoxications des animaux en Pologne. Bull Soc Vet Med Comp Lyon 72(2):167-173.

Challoner KR, McCarron MM (1990) Castor bean intoxication. Ann Emerg Med 19(10):1177-1183.

Clemens E (1963) Über Toxizität und Verträglichkeit von Ricinusextraktionsschrot bei verscheidenen Tierarten. Landwirtschaftliche Forsch Sonderheft 17:202-211.

Cooper WC, Perone VB, Scheel LD, et al. (1964) Occupational hazards from castor bean pomace: Tests for toxicity. Am Ind Hyg Assoc J 25(Sep-Oct):431-438.

Davison AG, Britton MG, Forrester JA, et al. (1983) Asthma in merchant seamen and laboratory workers caused by allergy to castor beans: Analysis of allergens. Clin Allergy 13(6):553-561.

Döbereiner J, Tokarnia CH, Canella CF (1981) Experimental poisoning of cattle by the pericarp of the fruit of Ricinus communis. Pesq Vet Bras 1(3):95-97.

El Badwi SM, Adam SE, Hapke HJ (1992) Experimental Ricinus communis poisoning in chicks. Phytother Res 6:205-208.

El Badwi SM, Mousa HM, Adam SE, et al. (1992) Response of brown Hisex chicks to low levels of Jatropha curcas, Ricinus communis or their mixture. Vet Hum Toxicol 34(4):304-306.

El Mauhoub M, Khalifa MM, Jaswal OB, et al. (1983) Ricin syndrome: A possible new teratogenic syndrome associated with ingestion of castor oil seed in early pregnancy: A case report. Ann Trop Paediatr 3(2):57-61.

Fernando R, Fernando DN (1990) Poisoning with plants and mushrooms in Sri Lanka: A retrospective hospital based study. Vet Hum Toxicol 32(6):579-581.

Figley KD, Rawling FF (1950) Castor bean: An industrial hazard as a contaminant of green coffee dust and used burlap bags. J Allergy 21:545-553.

Fox MW (1961) Castor seed residue poisoning in dairy cattle. Vet Rec 73(36):885-886.

Fuller G, Walker HG Jr, Mottola AC, et al. (1971) Potential for detoxified castor meal. J Am Oil Chem Soc 48:616-618.

Geary T (1950) Castor bean poisoning. Vet Rec 62(32):472-473.

Geoffroy H (1964) De certaines intoxications aiguës. Maroc Med 43:603-618.

Gullan AG (1905) Acute poisoning by a single castor-oil seed. Br Med J 1:988-989.

Hamouda C, Amamou M, Thabet H, et al. (2000) Plant poisonings from herbal medication admitted to a Tunisian toxicologic intensive care unit, 1983-1998. Vet Hum Toxicol 42(3):137-141.

Hebanowski M (1964) Przypadek ostrej niedomogi nerek w przebiegu zatrucia nasionami rącznika. Pol Tyg Lek (Wars) 19(31):1204-1205.

Hemphill RE (1981) Visual hallucinations and psychosis associated with the leaf of the castor oil tree. S Afr Med J 59(4):96.

Ingle VN, Kale VG, Talwalkar YB (1966) Accidental poisoning in children with particular reference to castor beans. Indian J Pediatr 33(223):237-240.

Jacobziner H, Raybin HW (1960) Briefs on accidental chemical poisonings in New York City. Incidents reported from out-of-town. N Y State J Med 60(Jun 15):1979-1983.

Jensen WI, Allen JP (1981) Naturally occurring and experimentally induced castor bean (Ricinus communis) poisoning in ducks. Avian Dis 25(1):184-194.

Johnson J (1959) Report of castor bean poisoning in Alameda County. California State Dep Agric Field Lett #52:3.

Joubert PH (1990) Poisoning admissions of black South Africans. J Toxicol Clin Toxicol 28(1):85-94.

Kanerva L, Estlander T, Jolanski R (1990) Long-lasting contact urticaria from castor bean. J Am Acad Dermatol 23(2):351-355.

Karolini T, Żarnowska-Ćwiertka W (1965) Przyoadek zatrucia pokarmowego nasionami rącznika. Przegl Epidemiol 19(2):272-273.

Kaszás T, Papp G (1960) Ricinussamen - Vergiftung von Schulkindern. Arch Toxikol 18(3):145-150.

Katsnel'son IB, Besser VL, Ionov IT, et al. (1960) [Poisoning with the seed of Ricinus communis (clinical-experimental observations)] Sov Med 24:131-135.

Kinamore PA, Jaeger RW, de Castro FJ, et al. (1980) Abrus and Ricinus ingestion: Management of three cases. Clin Toxicol 17(3):401-405.

Kingma DJ (1971) Ricine-intoxicatie door het kauwen op één wonderboon. Ned Tijdschr Geneeskd 115(28):1190-1191.

Koch LA, Caplan J (1942) Castor bean poisoning. Am J Dis Child 64:485-486.

Kopferschmitt J, Flesch F, Lugnier A, et al. (1983) Acute voluntary intoxication by ricin. Hum Toxicol 2(2):239-342.

Kraszewska Z, Świtlik I, Stalewski R, et al. (1965) Ostre zatrucie nasionami drzewa rącznikowego. Pol Tyg Lek (Wars) 20(8):279-281.

Krieger-Huber S (1980) Rizin-Vergiftungen mit tödlichem Ausgang bei Hunden nach Aufnahme des biologischen Naturdüngers "Oscorna animalin." Kleintierpraxis 25:281-286.

Layton LL (1977) Castorbean allergens in tissues of catfish reared on a diet containing castorbean pomace. J Sci Food Agric 28(5):399-404.

Lensch J (1966) Rizinvergiftung beim Rind. Tierarztl Umsch 21:21-22.

Lindenbaum SE (1966) Case report: Pollinosis due to Ricinus communis or castor bean plant. Ann Allergy 24(1):23-25.

Lockey SD Jr, Dunkelberger L (1968) Anaphylaxis from an Indian necklace. JAMA 206(13):2900-2901.

Lucas GN (1997) Plant poisoning: A hospital-based study in Sri Lanka. Indian J Pediatr 64(4):495-502.

Malizia E, Sarcinelli L, Andreucci G (1977) Ricinus poisoning: A familiar epidemy. Acta Pharmacol Toxicol 41(Suppl 2):351-361.

Maretić Z (1980) Otrovanje sjemenkama ricinusa. Arh Hig Rada Toksikol 31(3):251-257.

McCunn J, Andrew H, Clough GW (1945) Castor-bean poisoning in horses. Vet J 101:136-138.

Meldrum WP (1900) Poisoning by castor oil seeds. Br Med J 1(Feb 10):317.

Mel'nik IL, Koltyn EM (1973) [Castor pomaces, a cause of disease in newborn calves.] Veterinariia Moscow 49(1):98-99.

Metz G, Bocher D, Metz J (2001) IgE-mediated allergy to castor bean dust in a landscape gardener. Contact Dermatitis 44(6):367.

Moore J (1924) Poisoning from castor beans. Vet J 80:75-76.

Navarro-Rouimi R, Charpin D (1999) Anaphylactic reaction to castor bean seeds. Allergy 54(10):1117.

Palatnick W, Tenenbein M (1997) Hepatotoxicity due to castor bean ingestion. J Toxicol Clin Toxicol 35:528.

Palatnick W, Tenenbein M (2000) Hepatotoxicity from castor bean ingestion in a child. J Toxicol Clin Toxicol 38(1):67-69.

Pammel LH (1920) Castor seed poisonous. Am J Vet Med 15:171-172.

Pammel LH (1921) Castor bean poisoning. Vet Med 16(3):45.

Pammel LH (1921) Castor oil plant - poisonous. Vet Med 16(12):49.

Panzani R, Johansson SG (1986) Results of skin test and RAST in allergy to a clinically potent allergen (castor bean). Clin Allergy 16(3):259-266.

Panzani R, Layton LL (1963) Allergy to the dust of Ricinus communis (castor bean): Clinical studies upon human beings and passively sensitized monkeys. Int Arch Allergy Appl Immunol 22(6):350-368.

Peck EF (1942) Castor seed poisoning in a camel - A note on gastric lavage. Vet Rec 54:184.

Pevny I (1979) Ricinusschrot-Allergie. Derm Beruf Umwelt 27(6):159-162.

Popescu IG, Păun R, Molner C (1965) Die spezifische hyposensibilisierende Behandlung des beruflichen Rizinusasthmas. Allerg Asthma (Leipz) 11(5):239-244.

Purushotham NP, Raghavan GV, Rao MS, et al. (1985) Pathology of thyroid gland of sheep fed on castor bean meal (Ricinus communis). Indian J Vet Pathol 9:70-73.

Purushotham NP, Raghavan GV, Rao MS, et al. (1985) Studies on the pathology of experimental feeding of castor bean meal (Ricinus communis) in sheep. Indian Vet J 62(2):116-118.

Ramakrishnan S, Balasubramanian K, Madhavan M (1972) Biochemical and pathological studies on castor seed poisoning. J Assoc Physicians India 20(10):781-784.

Ramón MF, Ballesteros S, Martínez-Arrieta R, et al. (2003) Plant and herb exposures in Spain. J Toxicol Clin Toxicol 41(4):514.

Ratner B, Gruehl HL (1929) Respiratory anaphylaxis (asthma) and ricin poisoning induced with castor bean dust. Am J Hyg 10:236-244.

Rauber A, Heard J (1985) Castor bean toxicity re-examined: A new perspective. Vet Hum Toxicol 27(6):498-502.

Rauber AP, Heard JH (1985) The great castor bean panic. Vet Hum Toxicol 28:308.

Righi V (1961) Avvelenamento in bovini da farina di semi di ricino erroneamente mescolati a semi di fava. Veterinaria Italiana 12:893-895.

Ritter S (1985) Vergiftungen durch Pflanzen. Dtsch Apoth Ztg 125(37):1834-1836.

Robbins WJ (1923) A case of supersensitiveness to the poisonous action of the castor bean. Science 58(1503):305-306.

Sai S (1983) Lipstick dermatitis caused by castor oil. Contact Dermatitis 9(1):75.

Saragea M, Vladutiu A, Negur T, et al. (1966) Antibodies to Ricinus communis (castor bean) in a rural community. Ann Allergy 24(4):179-182.

Satpathy R, Das BB (1979) Accidental poisoning in childhood. J Indian Med Assoc 73(11):190-192.

Shone DK (1966) Poisonous plants of Rhodesia. Part I. Rhodesia Zambia Malawi J Agric Res 4:81-94.

Shrivastava HP, Prasad MC, Sadagopan VR (1988) Hepatorenal changes in chicks fed diets containing raw and treated castor meals. Indian J Vet Pathol 12:49-53.

Smith FJ (1886) Poisoned by castor beans. Am Vet Rev 10:367.

Snodgrass WR (1982) Castor bean ingestion. Lawrence Rev Natural Products 3(8):32.

Soto Blanco B, Sinhorini IL, Górniak SL, et al. (2002) Ricinus communis cake poisoning in a dog. Vet Hum Toxicol 44(3):155-156.

Spyker DA, Sauer K, Kell SO, et al. (1982) A castor bean poisoning and a widely available bioassay for ricin. Vet Hum Toxicol 24(4):293.

Steffan MA (1987) Anaphylaktischer Schock auf Rizinusbohnenschrot - Kasuistischer Beitrag. Derm Beruf Umwelt 35(5):177-178.

Stewart MJ, Moar JJ, Mwesigwa J, et al. (2000) Forensic toxicology in urban South Africa. J Toxicol Clin Toxicol 38(4):415-419.

Stewart MJ, Moar JJ, Steenkamp P, et al. (1999) Findings in fatal cases of poisoning attributed to traditional remedies in South Africa. Forensic Sci Int 101(3):177-183.

Targosz D, Winnik L, Szkolnicka B (2002) Suicidal poisoning with castor bean (Ricinus communis) extract injected subcutaneously - Case report. J Toxicol Clin Toxicol 40(3):398.

Thorpe SC, Kemeny DM, Panzani R, et al. (1987) The relationship between total serum IgE and castor bean-specific IgE antibodies in castor bean-sensitive patients from Marseilles. Int Arch Allergy Appl Immunol 82:456-460.

Thorpe SC, Kemeny DM, Panzani R, et al. (1988) Allergy to castor bean. I. Its relationship to sensitization to common inhalant allergens (atopy). J Allergy Clin Immunol 82:62-66.

Tokarnia CH, Canella CF, Döbereiner J, et al. (1967) Experimentos com plantas suspeitas de serem tóxicas realizados em bovinos no Estado do Rio De Janeiro, que resultaram negativos ou em perturbações leves passageiras. Pesq Agric Bras 2:343-351.
Tokarnia CH, Döbereiner J, Canella CF (1975) Intoxicação experimental em bovinos pelas folhas de Ricinus communis. Pesq Agric Bras Vet 10(8):1-7.
Topping MD, Henderson RT, Luczynska CM, et al. (1982) Castor bean allergy among workers in the felt industry. Allergy 37(8):603-608.
Topping MD, Tyrer FH, Lowing RK (1981) Castor bean allergy in the upholstery department of a furniture factory. Br J Ind Med 38(3):293-296.
Vilhjalmsdottir L, Fisher H (1971) Castor bean meal as a protein source for chickens: Detoxification and determination of limiting amino acids. J Nutr 101(9):1185-1192.
Vincent P (1924) Castor oil seed poisoning in cattle. Vet J 80:89.
Vinther S, Matzen P (1983) Forgiftning med Ricinus communis. Ugeskr Laeger 145(20):1546-1547.
Vroege D (1971) Ricine-intoxicatie door het kauwen op één wonderboon. Ned Tijdschr Geneeskd 115(38):1580-1581.
Wedin GP, Neal JS, Everson GW, et al. (1985) Castor bean poisoning: Two case reports. Vet Hum Toxicol 28(4):299.
Wedin GP, Neal JS, Everson GW, et al. (1986) Castor bean poisoning. Am J Emerg Med 4(3):259-261.
Westermarck H (1952) Ricinushaltiga linfrökakor, som orsak till dödsfall och produktionsminskning på nöt. Nord Vet Med 4(Oct):1005-1009.
Wilcox FP (1955) Poisoning - Castor bean. Los Angeles County Livestock Dep Annu Rep 1954-1955:38-41.
Wolfle J, Kowalewski S (1995) Epidemiology of ingestions in a regional poison control center over twenty years. Vet Hum Toxicol 37(4):367-368.
Wolfromm R, Guibert L, Rivolier J, et al. (1967) Épidémic d'allergie au de Dieppe (1963-1965). Presse Med 75(43):2157-2160.
Zerbst GH (1944) Unusual hazard in a fertilizer factory. Indus Med 13:552.
Zifroni A (1985) [Castor bean poisoning.] Harefuah 108(2):102-103, 110.

rickety bush –see– *Cycas media* R. Br.

Riddell's-groundsel –see– *Senecio riddellii* Torr. & A. Gray

ridge goosefoot –see– *Chenopodium carinatum* R. Br.

ridge gourd –see– *Luffa acutangula* (L.) Roxb.

RIEDELIELLA GRACILIFLORA Harms [Fabaceae]

Citations:

Dagli ML, Perrone EA, Haraguchi M, et al. (1995) The acute toxicity of Riedeliella graciflora in laboratory animals. Vet Hum Toxicol 37(6):544-546.
Górniak SL, Dagli ML, Perrone EA, et al. (1995) The acute toxicity of Riedeliella graciflora in calves. Vet Hum Toxicol 37(5):447-448.

Riesenbärenklau –see– *Heracleum mantegazzianum* Sommier & Levier

Riesenherkulestaude –see– *Heracleum mantegazzianum* Sommier & Levier

Ringelblume –see– *Calendula officinalis* L.; *Taraxacum officinale* F. H. Wigg. aggr.

rio palisander –see– *Dalbergia nigra* (Vell.) Allemão ex Benth.

rio rosewood –see– *Dalbergia nigra* (Vell.) Allemão ex Benth.

ripple ivy –see– *Hedera helix* L.

Rivea corymbosa (L.) Hallier f. = Turbina corymbosa (L.) Raf.

riventdore –see– *Karwinskia humboldtiana* (Schult.) Zucc.

river myall –see– *Acacia binervia* (J. C. Wendl.) J. F. Macbr.

Riverina bluebell –see– *Echium plantagineum* L.

rivierals –see– *Matricaria nigellifolia* DC.

rizi mételykóró –see– *Oenanthe aquatica* (L.) Poir.

rizin –see– *Ricinus communis* L.

Rizinus –see– *Ricinus communis* L.

robin –see– *Robinia pseudoacacia* L.

robinia –see– *Robinia pseudoacacia* L.

ROBINIA PSEUDOACACIA L. [Fabaceae]

Common Names:

acacia falsa; akác; Akazie; black acacia; black locust; Falsche Akazie; false acacia; locust tree; post locust; robin; robinia; Robinie; robinier; Scheinakazie; shipmast locust; Silberregen; Weiße Robinie; white locust; yellow locust

Citations:

Arai M, Stauber E, Shropshire CM (1992) Evaluation of selected plants for their toxic effects on canaries. J Am Vet Med Assoc 200(9):1329-1331.
Artero Sivera A, Arnedo Pena A, Pastor Cubo A (1989) Estudio clinicoepidemiologico de un envenenamiento accidental por Robinia pseudoacacia L. en escolares. An Esp Pediatr 30(3):191-194.
Barnes MF (1921) Black locust poisoning of chickens. J Am Vet Med Assoc 59:370-372.
Breazile JE (1959-1960) Black locust (Robin) poisoning in cattle. Vet Scope (Missouri Univ) 9(1):12, 20.
Costa Bou X, Soler I Ros JM, Seculi Palacios JL (1990) Intoxicación por Robinia pseudoacacia. An Esp Pediatr 32(1):68-69.
Emery ZT (1887) Report of thirty-two cases of poisoning by locust bark. New York Med J 45:92.
Filandrinos DT, Sioris LJ (1992) Transient elevation of liver function tests following ingestion of black locust seeds. Vet Hum Toxicol 34(4):351.
Gardiner WW (1903) Poisoned from eating locust-tree bark (Robinia pseudoacacia, false acacia). Am Vet Rev 27(Oct):599-600.
Hansen AA (1924) Robitin - A potent plant poison. Better Crops 2(2):22-23, 44.

Hansen AA (1924) The poison plant situation in Indiana. III. Poisonous trees. J Am Vet Med Assoc 66:351-362.
Hansen AA (1928) Toxic trees. North Am Vet 9(10):49-53.
Hopper DW (1999) False acacia poisoning in horses. Vet Rec 145(4):115.
Hui A, Marraffa JM, Stork CM (2004) A rare ingestion of the black locust tree. J Toxicol Clin Toxicol 42(1):93-95.
Hui A, Stork CM (2002) A rare intoxication of black locust tree bark. J Toxicol Clin Toxicol 40(5):618-619.
Keller H, Dewitz W (1969) Vergiftungen bei 9 Pferden durch Rinde der "Falschen Akazie" (Robinia pseudoacacia). Dtsch Tierarztl Wochenschr 76(5):115-117.
Landolt G, Feige K, Schoberl M (1997) Vergiftung bei Pferden durch die Rinde der «Falschen Akazie» (Robinia pseudoacacia). Schweiz Arch Tierheilkd 139(8):363-366.
Pammel LH (1927) The toxicology of black locust. North Am Vet 8(Jan):41-43.
Shropshire CM, Stauber E, Arai A (1992) Evaluation of selected plants for acute toxicosis in budgerigars. J Am Vet Med Assoc 200(7):936-939.
Tasaki B, Tanaka U (1918) On the toxic constituents in the bark of Robinia pseudoacacia L. J Coll Agric Imperial Inst Tokyo 3(5):337-356.
Thursby Pelham RH (1999) False acacia poisoning in horses. Vet Rec 145(5):148.
Waldron CA (1908) Poisoning from locust bark. Am Vet Rev 33:456-459.

Robinie –see– *Robinia pseudoacacia* L.

robinier –see– *Robinia pseudoacacia* L.

robust larkspur –see– *Delphinium robustum* Rydb.

robust needlegrass –see– *Achnatherum robustum* (Vasey) Barkworth

rock centaury –see– *Zeltnera beyrichii* (Torr. & A. Gray) G. Mans.

rock fern –see– *Cheilanthes sieberi* Kunze; *Cheilanthes tenuifolia* (Burm. f.) Sw.

rock isotome –see– *Isotoma petraea* F. Muell.

rock lip fern –see– *Cheilanthes tenuifolia* (Burm. f.) Sw.

rock moss –see– *Portulaca grandiflora* Hook.

rock poppy –see– *Chelidonium majus* L.

rockrose –see– *Cistus ladanifer* L.

Rocky Mountain larkspur –see– *Delphinium scopulorum* A. Gray

Rocky Mountain sage –see– *Salvia reflexa* Hornem.

Rocky Mountain white oak –see– *Quercus gambelii* Nutt.

ROEPERA AMMOPHILA (F. Muell.) Beier & Thulin [Zygophyllaceae]

Synonyms:

Zygophyllum ammophilum F. Muell.

Common Names:

sand twinleaf; twinleaf

Citations:

Chippendale G, Humble A, Siebert B (1964) Zygophyllum ammophilum and presumed nitrate poisoning in cattle. Aust Vet J 40(Jun):241.

ROEPERA FOETIDA (Schrad. & J. C. Wendl.) Beier & Thulin [Zygophyllaceae]

Synonyms:

Zygophyllum foetidum Schrad. & J. C. Wendl.

Common Names:

skuimbos

Citations:

Steyn DG (1934) Recent investigations into the toxicity of known and unknown poisonous plants in the Union of South Africa. Onderstepoort J Vet Sci Anim Indus 3(1):125-130.

Roggen –see– *Secale cereale* L.

roldón –see– *Coriaria myrtifolia* L.

roly poly –see– *Salsola tragus* L.

romaine lettuce –see– *Lactuca sativa* L. var. longifolia Lam.

Roman camomille –see– *Chamaemelum nobile* (L.) All.

Roman hyacinth –see– *Hyacinthus orientalis* L.

Roman wormwood –see– *Artemisia pontica* L.

romerillo –see– *Artemisia ludoviciana* Nutt.; *Baccharis coridifolia* DC.

ROMULEA BULBOCODIUM (L.) Sebast. & Mauri [Iridaceae]

Common names:

onion grass, Guildford grass

Citations:

Eckel JF (1965) A sheep breeding problem in the Inverleigh district. Victorian Vet Proc 23:42-43.
Galvin J (1998) Suspected onion grass toxicity. Quarterly Report for 1 April to 30 June 1998, Victoria (Agriculture Victoria). Anim Health Surveillance Q3 (2):12.
Gorrie CJ (1962) Ovine abortion in Victoria. Aust Vet J 38:138-142.

ronphagrass –see– *Phalaris aquatica* L.

roode muur –see– *Anagallis arvensis* L.

rooibessie –see– *Solanum kwebense* N. E. Br.

rooimuur –see– *Anagallis arvensis* L.

rooisuikerblom –see– *Tylecodon grandiflorus* (Burm. f.) Toelken

root poison –see– *Cicuta maculata* L.

Rorippa armoracia (L.) Hitchc. = Armoracia rusticana P. Gaertn. et al.

rosa laurel –see– *Nerium oleander* L.

rosary bean –see– *Abrus precatorius* L.

rose campion –see– *Agrostemma githago* L.

rose-de-nöel –see– *Helleborus niger* L.

rose laurel –see– *Kalmia latifolia* L.; *Nerium oleander* L.

rose periwinkle –see– *Catharanthus roseus* (L.) G. Don

rose-pink frangipani –see– *Plumeria rubra* L.

rose silkweed –see– *Asclepias incarnata* L.

rosebay –see– *Nerium oleander* L.

rosebay laurel –see– *Rhododendron catawbiense* Michx.

rosemary –see– *Rosmarinus officinalis* L.

Rosenkohl –see– *Brassica oleracea* L. var. gemmifera Zenker

Rosenlorbeer –see– *Nerium oleander* L.

Rosenmelde –see– *Atriplex rosea* L.

rosewood –see– *Alectryon oleifolius* (Desf.) S. T. Reynolds; *Dalbergia latifolia* Roxb.

rosilla –see– *Helenium quadridentatum* Labill.

rosin rose –see– *Hypericum perforatum* L.

rosita –see– *Zeltnera calycosa* (Buckley) G. Mans.

Rosmarin –see– *Rosmarinus officinalis* L.

ROSMARINUS OFFICINALIS L. [Lamiaceae]

Common Names:

old-man rosemary; rosemary; Rosmarin; vervions

Citations:

Armisén M, Rodríguez V, Vidal C (2003) Photoaggravated allergic contact dermatitis due to Rosmarinus officinalis cross-reactive with Thymus vulgaris. Contact Dermatitis 48(1):52-53.

González-Mahave I, Lobesa T, Del Pozo MD, et al. (2006) Rosemary contact dermatitis and cross-reactivity with other labiate plants. Contact Dermatitis 54(4):210-212.

Guin JD (2001) Rosemary cheilitis: One to remember. Contact Dermatitis 45(1):63.

Martínez-González MC, Goday Buján JJ, Martínex Gómez W, et al. (2007) Concomitant allergic contact dermatitis due to Rosmarinus officinalis (rosemary) and Thymus vulgaris (thyme). Contact Dermatitis 56(1):49-50.

Serra E, Vila A, Peramiquel L, et al. (2005) Allergic contact dermatitis due to rosemary. Contact Dermatitis 53(3):179-180.

Roßkastanie –see– *Aesculus hippocastanum* L.

rosy crown vetch –see– *Securigera varia* (L.) Lassen

Rotbeerige Zaunrübe –see– *Bryonia dioica* Jacq.

Rotbuche –see– *Fagus sylvatica* L.

rote Bete –see– *Beta vulgaris* L.

Roteibe –see– *Taxus baccata* L.

Roter Fingerhut –see– *Digitalis purpurea* L.

Roter Pfeffer –see– *Capsicum annuum* L.

Roth vetch –see– *Vicia villosa* Roth

Rothschild's-glory lily –see– *Gloriosa superba* L.

Rotklee –see– *Trifolium pratense* L.

Rottlera tinctoria Roxb. = Mallotus philippensis (Lam.) Müll. Arg.

rough beardedgrass –see– *Echinopogon ovatus* (G. Forst.) P. Beauv.

rough cat's-ear –see– *Hypochaeris radicata* L.

rough marsh elder –see– *Iva annua* L.

rough pea –see– *Lathyrus hirsutus* L.

rough pigweed –see– *Amaranthus retroflexus* L.

round gourd –see– *Citrullus lanatus* (Thunb.) Matsum. & Nakai

round-leaf laurel –see– *Kalmia latifolia* L.

round-leaf mallow –see– *Malva pusilla* Sm.

ROUREA COCCINEA (Schumach. & Thonn.) Hook. f. [Connaraceae]

Synonyms:

Byrsocarpus coccineus Schumach. & Thonn.

Common Names:

kimbar mahalba

Citations:

Amégée Y (1983) Le tournis à Byrsocarpus des petits ruminants au Sud du Togo. Rev Elev Med Vet Pays Trop 36(1):27-31.

Vickery B, Vickery ML (1974) The toxicity of some members of the Connaraceae family. Br Vet J 130(2):41-43.

royal poinciana –see– *Sesbania punicea* (Cav.) Benth.

Royena decidua Burch. = Diospyros lycioides Desf.

rubber euphorbia –see– *Euphorbia tirucalli* L.

rubber hevea –see– *Hevea brasiliensis* (Willd. ex A. Juss.) Müll. Arg.

rubber plant –see– *Jatropha dioica* Sessé; *Parthenium argentatum* A. Gray

rubber stem –see– *Jatropha dioica* Sessé

rubber tree –see– *Calotropis procera* (Aiton) W. T. Aiton

rubber vine –see– *Araujia sericifera* Brot.; *Cryptostegia grandiflora* R. Br.

rubberroot –see– *Asclepias tuberosa* L.

rubberweed –see– *Hymenoxys odorata* DC.; *Hymenoxys richardsonii* (Hook.) Cockerell

Rübe –see– *Beta vulgaris* L.

ruber-ki-bal –see– *Cryptostegia grandiflora* R. Br.

RUBIA PEREGRINA L. [Rubiaceae]

Citations:

Castelain M, Ducombs G (1988) Contact dermatitis from madder. Contact Dermatitis 19(10):228-229.

ruda –see– *Ruta chalepensis* L.; *Ruta graveolens* L.

ruda-del-monte –see– *Thamnosma texana* (A. Gray) Torr.

RUDBECKIA LACINIATA L. [Asteraceae]

Common Names:

coneflower; cut-leaf coneflower; dormilión; golden glow; green-head coneflower; meadow cone; tall coneflower; thimble weed

Citations:

Gates FC (1930) Botanical notes, 1928-1929. Trans Kans Acad Sci 33:26-28.

Pammel LH (1922) Golden glow. Vet Med 17(6):296.

Skidmore LV, Peterson NF (1932) Observations on the toxicity of golden glow (Rudbeckia lacinata) to swine and other animals. J Am Vet Med Assoc 81(Nov):655-662.

rue –see– *Ruta chalepensis* L.; *Ruta graveolens* L.; *Ruta montana* Mill.

rue-des-jardin –see– *Ruta graveolens* L.

ruibarbillo –see– *Rumex acetosella* L.

ruibarbo –see– *Rheum ×hybridum* Murray

rum cherry –see– *Prunus serotina* Ehrh.

RUMEX ACETOSA L. [Polygonaceae]

Common Names:

acedera común; Französisher Spinat; garden sorrel; green saure; green sorrel; Großer Sauerampfer; meadow sorrel; mezei sóska; oseille longue; Sauerampfer; sharp dock; sheep's-sorrel; sorrel dock; sorrel grass; sour dock; sour leek; sour sorrel; sourgrass; surelle; surete; tall sorrel; vinagrera; vinette

Citations:

Coward TG (1949) Acute, fatal poisoning in sheep due to ingestion of common sorrel (Rumex acetosa). Vet Rec 61(46):765-766.

Vollmer H (1939) Sauerampfer - Vergiftung eines Kindes. Sammlung Vergiftungsfallen 10(A816):175-178.

RUMEX ACETOSELLA L. [Polygonaceae]

Common Names:

cow sorrel; dock; field sorrel; horse sorrel; juhsóska; kleiner Ampfer; kleiner Sauerampfer; meadow sorrel; mountain sorrel; red sorrel; ruibarbillo; sand sorrel; sheep sorrel; sorrel; sour dock; sour weed; sourgrass; wood sorrel

Citations:

Schrader A, Schulz O, Volker H, et al. (2001) Aktuelle Vergiftungen durch Pflanzen bei Wiederkäuern in Nord- und Ostdeutschland. Berl Munch Tierarztl Wochenschr 114(5-6):218-221.

RUMEX CRISPUS L. [Polygonaceae]

Common Names:

acedera; curled dock; curly dock; lengua-de-vaca; narrow dock; sorrel; sour dock; yellow dock

Citations:

Farré M, Xirgu J, Salgado A, et al. (1989) Fatal oxalic acid poisoning from sorrel soup. Lancet 2(8678):1524.

Panciera RJ, Martin T, Burrows GE, et al. (1990) Acute oxalate poisoning attributable to ingestion of curly dock (Rumex crispus) in sheep. J Am Vet Med Assoc 196(12):1981-1984.

Reig R, Blanche C, Fontarnau R, et al. (1990) Fatal poisoning by Rumex crispus (curled dock): Pathological findings and application of scanning electron microscopy. Vet Hum Toxicol 32(5):468-470.

RUMEX VENOSUS Pursh [Polygonaceae]

Common Names:

sour green; veined dock; wild begonia; wild hydrangea; wing sorrel

Citations:

Dickie CW, Hamann MH, Carroll WD, et al. (1978) Oxalate (Rumex venosus) poisoning in cattle. J Am Vet Med Assoc 173(1):73-74.

RUMOHRA ADIANTIFORMIS (G. Forst.) Ching [Dryopteridaceae]

Synonyms:

Arachniodes adiantiformis (Forst) Tindale

Common Names:

leatherleaf fern; Lederfarn

Citations:

Hausen BM, Schulz KH (1978) Occupational allergic contact dermatitis due to leatherleaf fern Arachniodes adiantiformis (Forst) Tindale. Br J Dermatol 98(3):325-329.

Note:

The synonym *Arachniodes adiantiformis* (Forst) Tindale could not be found in the databases searched.

runaway robin –see– *Glechoma hederacea* L.

Runkelrübe –see– *Beta vulgaris* L.

runner bean –see– *Phaseolus coccineus* L.

running mallow –see– *Malva pusilla* Sm.

running-myrtle –see– *Vinca major* L.

rush –see– *Equisetum arvense* L.

Russian centaurea –see– *Acroptilon repens* (L.) DC.

Russian comfrey –see– *Symphytum officinale* L.

Russian knapweed –see– *Acroptilon repens* (L.) DC.

Russian millet –see– *Panicum miliaceum* L.

Russian olive –see– *Elaeagnus angustifolia* L.

Russian-sickle milk vetch –see– *Astragalus falcatus* Lam.

Russian star thistle –see– *Acroptilon repens* (L.) DC.

Russian thistle –see– *Salsola tragus* L.

rustic treacle –see– *Allium sativum* L.

Ruta bracteosa DC. = Ruta chalepensis L.

RUTA CHALEPENSIS L. [Rutaceae]

Synonyms:

Ruta bracteosa DC.

Common Names:
ruda; rue

Citations:
Brener S, Friedman J (1985) Phytophotodermatitis induced by Ruta chalepensis L. Contact Dermatitis 12(4):230-232.
Ciganda C, Laborde A (2001) Herbal infusions used for induced abortion. J Toxicol Clin Toxicol 39(3):318-319.
Ciganda C, Laborde A (2003) Herbal infusions used for induced abortion. J Toxicol Clin Toxicol 41(3):235-239.
Gonçalo S, Correia C, Couto JS, et al. (1989) Contact and photocontact dermatitis from Ruta chalepensis. Contact Dermatitis 21(16):200-201.
Laborde A, Ciganda C (1998) Poisoning by herbal infusions ingested as abortifacient agents. J Toxicol Clin Toxicol 36(5):454-455.
Valsecchi R, Cassina GP, Migliori M, et al. (1985) Tego dermatitis. Contact Dermatitis 12(4):230-232.

RUTA CORSICA DC. [Rutaceae]

Citations:
Ena P, Camarda I (1990) Phytophotodermatitis from Ruta corsica. Contact Dermatitis 22(4):63.

RUTA GRAVEOLENS L. [Rutaceae]

Common Names:
arruda; ave grace; countryman's-treacle; Gartenraute; herb-of-grace; herb-of-repentance; herbe-de-grace; Raute; ruda; rue; rue-des-jardin; sadab; sitab; vulgarmente ruda; Weinkraut; Weinraute; Wiesenraute

Citations:
Ciganda C, Laborde A (2001) Herbal infusions used for induced abortion. J Toxicol Clin Toxicol 39(3):318-319.
Ciganda C, Laborde A (2003) Herbal infusions used for induced abortion. J Toxicol Clin Toxicol 41(3):235-239.
El Agraa SE, El Badwi SM, Adam SE (2002) Preliminary observations on experimental Ruta graveolens toxicosis in Nubian goats. Trop Anim Health Prod 34(4):271-281.
Gandhi M, Lal R, Sankaranarayanan A, et al. (1991) Postcoital antifertility activity of Ruta graveolens in female rats and hamsters. J Ethnopharmacol 34:49-59.
Gawkrodger DJ, Savin JA (1983) Phytodermatitis due to common rue (Ruta graveolens). Contact Dermatitis 9(3):224.
Hanslian L, Kadlec K (1968) Bulosní dermatitida po kontaktu s routou obecnou. Cesk Farm 17(6):293-295.
Heskel NS, Amon RB, Storrs FJ, et al. (1983) Phytophotodermatitis due to Ruta graveolens. Contact Dermatitis 9(4):278-280.
Sharma AN, Nelson LS, Hoffman RS (2001) Ruta graveolens as an ethnic abortifacient. J Toxicol Clin Toxicol 39(3):312-313.
Szegö L, Dolinay V (1954) Durch "Ruta graveolens" verursachte massenhafte Dermatitis Bullosa als Berufsschadigung. Dermatol Wochenschr 130(45):1180-1185.
Szegö L, Tamas-Fekete M (1955) Untersuchungen über die Rolle der physikalischen Factoren bei durch Ruta graveolens verursachten Dermatitis. Berufsdermatosen 3:84-88.
Wessner D, Hofmann H, Ring J (1999) Phytophotodermatitis due to Ruta graveolens applied as protection against evil spells. Contact Dermatitis 41(4):232.

RUTA MONTANA Mill. [Rutaceae]

Common Names:
rue

Citations:
Hamouda C, Amamou M, Thabet H, et al. (2000) Plant poisonings from herbal medication admitted to a Tunisian toxicologic intensive care unit, 1983-1998. Vet Hum Toxicol 42(3):137-141.
Ortiz-Frutos FJ, Sanchez B, Garcia B, et al. (1995) Photocontact dermatitis from rue (Ruta montana L.). Contact Dermatitis 33(7):284.

rutabaga –see– *Brassica napus* L. var. napobrassica (L.) Rchb.

RUTIDOSIS HELICHRYSOIDES DC. [Asteraceae]

Citations:
Burry JN, Kuchel R, Reid JG, et al. (1973) Australian bush dermatitis: Compositae dermatitis in South Australia. Med J Aust 1(3):110-116.

Rydberg's-poison ivy –see– *Toxicodendron rydbergii* (Small ex Rydb.) Greene

Rydberg's-weedy milk vetch –see– *Astragalus miser* Douglas ex Hook. var. oblongifolius (Rydb.) Cronquist

rye –see– *Secale cereale* L.

ryegrass –see– *Lolium perenne* L.

S

saat –see– *Erythrophleum succirubrum* Gagnep.

Saatlein –see– *Linum usitatissimum* L.

Saatmohn –see– *Papaver somniferum* L.

Saatplatterbse –see– *Lathyrus sativus* L.

Saatwicke –see– *Vicia sativa* L.

sabbara –see– *Opuntia ficus-indica* (L.) Mill.

sabra –see– *Opuntia ficus-indica* (L.) Mill.

sacahuista –see– *Nolina microcarpa* S. Watson; *Nolina texana* S. Watson

sacahuiste –see– *Nolina microcarpa* S. Watson; *Nolina texana* S. Watson

saccharine blanche –see– *Hyoscyamus albus* L.

SACCHARUM OFFICINARUM L. [Poaceae]

Common Names:

jaggery; sugar cane; Zuckerrohr

Citations:

Clark A (1936) Report on the effects of certain poisons contained in food-plants of West Africa upon the health of the native races. J Trop Med Hyg 39(23-24):269-295.

Lehman CW (1976) Sugar cane smoke, an allergenic agent. Hawaii Med J 35(11):336-339.

Seetharam KA, Pasricha JS (1987) Condiments and contact dermatitis of the finger-tips. Indian J Dermatol Venereol Leprol 53:325-328.

Seifert HS, Beller KA (1969) Blausäurevergiftung beim Rind, verursacht durch Abweiden von Zuckerrohr (Saccharum officinarum) und Zufütterung von Früchten des Algarrobaumes (Prosopis juliflora). Berl Munch Tierarztl Wochenschr 82(5):88-91.

Valarezo S, Preston TR (1973) Effects of two molasses diets and one cereal diet on the incidence of pendulous crop in turkeys. Cuban J Agric Sci 7(2):219-226.

sacha sandia –see– *Xanthium strumarium* L. var. canadense (Mill.) Torr. & A. Gray

sacred bamboo –see– *Nandina domestica* Thunb.

sacred datura –see– *Datura inoxia* Mill.; *Datura wrightii* Regel

sadab –see– *Ruta graveolens* L.

saddleleaf –see– *Philodendron bipinnatifidum* Schott ex Endl.

Sadenbaum –see– *Platycladus orientalis* (L.) Franco

safari –see– *Crotalaria saltiana* Andrews

safed kikar –see– *Acacia leucophloea* (Roxb.) Willd.

safflower -see- *Carthamnus tinctorius* L.

saffron –see– *Crocus sativus* L.

saffron crocus –see– *Crocus sativus* L.

saffron spice –see– *Crocus sativus* L.

safran-des-prés –see– *Colchicum autumnale* L.

Safranrebendolde –see– *Oenanthe crocata* L.

sage –see– *Salvia officinalis* L.

sage bush –see– *Ozothamnus diosmifolius* (Vent.) DC.

sagebrush –see– *Artemisia tridentata* Nutt.

sagewort –see– *Artemisia ludoviciana* Nutt.

sago cycas –see– *Cycas revoluta* Thunb.

sago palm –see– *Caryota urens* L.; *Cycas circinalis* L.; *Cycas revoluta* Thunb.

sagrado –see– *Dysphania ambrosioides* (L.) Mosyakin & Clemants

sahajana –see– *Moringa oleifera* Lam.

saia blanca –see– *Brugmansia arborea* (L.) Lagerh.

Saigon cinnamon –see– *Cinnamomum aromaticum* Nees

SAINTPAULIA IONANTHA H. Wendl. [Gesneriaceae]

Common Names:

African violet

Citations:

Anonymous (1980) Tabulations of 1977 case reports. Bull Natl Clgh Poison Control Cent 24(6):1-4.

Temesvari E (1982) Occupational contact sensitivity from Sainpaulia ionantha. Contact Dermatitis 8(6):426.

sal –see– *Shorea robusta* C. F. Gaertn.

salad mustard –see– *Sinapis alba* L.

salago –see– *Wikstroemia ovata* C. A. Mey.

Salat –see– *Lactuca sativa* L.

Salbei –see– *Salvia officinalis* L.

sali –see– *Heliotropium ternatum* Vahl

SALIX ALBA L. [Salicaceae]

Common Names:

white willow; willow

Citations:

Chivato T, Juan F, Montoro A, et al. (1996) Anaphylaxis induced by ingestion of a pollen compound. J Invest Allergol Clin Immunol 6(3):208-209.

SALIX VIMINALIS L. [Salicaceae]

Citations:

Gonçalo S, Sousa I, Moreno A, et al. (1986) Occupational dermatitis from Salix viminalis. Contact Dermatitis 14(3):188-189.

Sally wattle –see– *Acacia binervia* (J. C. Wendl.) J. F. Macbr.

saloop –see– *Sassafras albidum* (Nutt.) Nees

salsa –see– *Ipomoea asarifolia* (Desr.) Roem. & Schult.

Salsola kali L. = Salsola tragus L.

Salsola pestifer A. Nelson = Salsola tragus L.

SALSOLA TRAGUS L. [Chenopodiaceae]

Synonyms:

Salsola kali L.; ***Salsola pestifer*** A. Nelson

Common Names:

prickly saltwort; roly poly; Russian thistle; saltwort; soft roly poly; tumbleweed

Citations:

Carnés J, Fernández-Caldas E, Marina A, et al. (2003) Immunochemical characterization of Russian thistle (Salsola kali) pollen extracts. Purification of the allergen Sal k 1. Allergy 58(11):1152-1156.

Mathams RH, Sutherland AK (1952) The oxalate content of some Queensland pasture plants. Queensland J Agric Sci 9:317-334.

Powell RF, Smith EB (1978) Tumbleweed dermatitis. Arch Dermatol 114(5):751-754.

Salsola tuberculata (Fenzl ex Moq.) Schinz var. tomentosa C. A. Smith ex Aellen = Salsola tuberculatiformis Botsch.

SALSOLA TUBERCULATIFORMIS Botsch. [Chenopodiaceae]

Synonyms:

Salsola tuberculata (Fenzl ex Moq.) Schinz var. tomentosa C. A. Smith ex Aellen

Citations:

Basson PA, Morgenthal JC, Bilbrough RB, et al. (1969) "Grootlamsiekte," a specific syndrome of prolonged gestation in sheep caused by a shrub, Salsola tuberculata (Fenzl ex Moq) Schinz var. tomentosa C. A. Smith ex Aellen. Onderstepoort J Vet Res 36(1):59-103.

Joubert JP, Basson PA, Lucks HJ, et al. (1972) "Grootlamsiekte," a specific syndrome of prolonged gestation in sheep: Further investigations. Onderstepoort J Vet Res 39(1):59-70.

salt groundsel –see– *Baccharis halimifolia* L.

saltwort –see– *Salsola tragus* L.

SALVADORA PERSICA L. [Salvadoraceae]

Common Names:

meswak

Citations:

Harfi HA, Lundberg M (1997) Meswak, a novel allergen. Allergy 52(4):474-475.

Salvation Jane –see– *Echium plantagineum* L.

salvia –see– *Salvia officinalis* L.

SALVIA COCCINEA Buc'hoz ex Etl. [Lamiaceae]

Common Names:

kaliphool; mejorana; red sage; red salvia; Texas sage; tropical sage

Citations:

Hindmarsh WL (1938) Salvia coccinea. A garden escape poisonous to stock. New South Wales Dep Agric Vet Res Rep #7:118-119.

Pal B, Varshneya C, Prasad B, et al. (1997) Nitrate bearing Salvia coccinea poisoning in Gaddi goats in Kangra Valley. Indian J Anim Sci 67(1):37-38.

Varshneya C, Kishtwaria RS, Prasad B (1995) Kali phool (Salvia coccinea) poisoning in Gaddi goats. Indian Vet J 72(Jun):616-618.

SALVIA DIVINORUM Epling & Játiva [Lamiaceae]

Common Names:

pipiltzinzintli; Ska Maria Pastora

Citations:

Arditti J, Bourdon JH, de Haro L, et al. (2003) Salvia divinorum, a new abuse potential plant. J Toxicol Clin Toxicol 41(4):516.

Valdés LJ 3rd, Díaz JL, Paul AG (1983) Ethnopharmacology of Ska Maria Pastora (Salvia divinorum Epling and Jativa-M.). J Ethnopharmacol 7(3):287-312.

SALVIA MILTIORRHIZA Bunge [Lamiaceae]

Common Names:

danshen

Citations:

Yu CM, Chan JC, Sanderson JE (1997) Chinese herbs and warfarin potentiation by 'danshen'. J Intern Med 241(4):337-339.

SALVIA OFFICINALIS L. [Lamiaceae]

Common Names:

Dalmatian sage; garden sage; huile-de-sauge; sage; Salbei; salvia

Citations:

Millet Y, Jouglard J, Steinmetz MD, et al. (1981) Toxicity of some essential plant oils. Clinical and experimental study. Clin Toxicol 18(12):1485-1498.

Sertoli A, Fabbri P, Campolmi P, et al. (1978) Allergic contact dermatitis to Salvia officinalis, Inula viscosa and Conyza bonariensis. Contact Dermatitis 4(5):314-315.

Tong T, Schnier AB, Williams SR, et al. (2003) Sage tea related convulsions in a pediatric unit. J Toxicol Clin Toxicol 41(5):727-728.

SALVIA REFLEXA Hornem. [Lamiaceae]

Common Names:

annual sage; blue sage; lance-leaf sage; mintweed; narrow-leaf sage; Rocky Mountain sage; wild mint

Citations:

Smith JP (1962) These weeds can poison your poultry. Agric Gaz New South Wales 73(Mar):136-142.

Williams CH, Hines HJ (1940) The toxic properties of Salvia reflexa. Aust Vet J 16:14-20.

SALVIA ROSMARINOIDES A. St.-Hil. ex Benth. [Lamiaceae]

Citations:

Andrade SO, Camargo WV, Fernandes N (1963) II. Investigações sôbre plantas tóxicas no Estado de São Paulo. Arq Inst Biol (Sao Paulo) 30(Oct):189-203.

samak –see– *Echinochloa frumentacea* Link

samambaia –see– *Pteridium aquilinum* (L.) Kuhn

SAMANEA SAMAN (Jacq.) Merr. [Fabaceae]

Synonyms:

Pithecolobium saman Benth.

Common Names:

guango; rain tree; zamang

Citations:

Carter GR, Arnold RN (1948) An azoturia-like condition in mares eating guango (Pithecolobium saman, Benth) pods in Jamaica. Can J Comp Med Vet Sci 12(9):255-259.

samanguilla –see– *Khaya ivorensis* A. Chev.

samanta –see– *Cerbera odollam* Gaertn.

samba –see– *Triplochiton scleroxylon* K. Schum.

SAMBUCUS CERULEA Raf. [Adoxaceae]

Synonyms:

Sambucus mexicana auct.

Common Names:

blue elderberry; elderberry

Citations:

Kunitz S, Melton RJ, Updyke T, et al. (1984) Outbreak of suspected cyanide poisoning from elderberry juice. California Morbidity 7(Feb 24):1.

Kunitz S, Melton RJ, Updyke T, et al. (1984) Poisoning from elderberry juice - California. MMWR Morb Mortal Wkly Rep 33(13):173-174.

Sambucus mexicana auct. = Sambucus cerulea Raf.

SAMBUCUS MEXICANA C. Presl ex DC. [Adoxaceae]

Common Names:

blueberry elderberry; Mexican elderberry; sauco; tapiro

Citations:

Anonymous (1984) Poisoning from elderberry juice. JAMA 251(16):2075.

SAMBUCUS NIGRA L. [Adoxaceae]

Common Names:

Aalhorn; arn tree; barr tree; black elderberry; boon tree; boor tree; booty tree; bourtree; bur tree; dwarf elderberry; elderberry; European elderberry; fekete bodza; Flieder; haute bois; Holder; Schwarzer Holunder; sureau noire

Citations:

Anonymous (1895) Poisonous and medicinal plants. U S Dep Agric Annu Rep 1894:165-166.

Anonymous (1980) Tabulations of 1977 case reports. Bull Natl Clgh Poison Control Cent 24(6):1-4.

SAMBUCUS RACEMOSA L. [Adoxaceae]

Citations:

Lamminpaa A, Kinos M (1996) Plant poisonings in children. Hum Exp Toxicol 15(3):245-249.

samburu –see– *Erythrophleum africanum* (Welw. ex Benth.) Harms

San Francisco –see– *Codiaeum variegatum* (L.) A. Juss.

San Jose –see– *Allamanda cathartica* L.

San Jose amarillo –see– *Allamanda cathartica* L.

San Pablo –see– *Wigandia urens* (Ruiz & Pav.) Kunth var. caracasana (Kunth) D. N. Gibson

San Rafaelito –see– *Lantana camara* L.

sand briar –see– *Solanum carolinense* L.

sand bur –see– *Solanum rostratum* Dunal

sand groundsel –see– *Senecio riddellii* Torr. & A. Gray

sand jack oak –see– *Quercus incana* W. Bartram

sand reverchonia –see– *Reverchonia arenaria* A. Gray

sand sagebrush –see– *Artemisia filifolia* Torr.

sand shin oak –see– *Quercus havardii* Rydb.

sand shinnery oak –see– *Quercus havardii* Rydb.

sand sorrel –see– *Rumex acetosella* L.

sand twinleaf –see– *Roepera ammophila* (F. Muell.) Beier & Thulin

sandalwood –see– *Santalum album* L.

sandbox –see– *Hura crepitans* L.

sandbur –see– *Cenchrus incertus* M. A. Curtis

sandfly zieria –see– *Zieria smithii* Andrews

sandhill lupin –see– *Lupinus cosentinii* Guss.

sandhill lupine –see– *Lupinus diffusus* Nutt.

sandplain lupin –see– *Lupinus cosentinii* Guss.

sandsage –see– *Artemisia filifolia* Torr.

sangre-de-drago –see– *Gliricidia sepium* (Jacq.) Kunth ex Walp.; *Jatropha dioica* Sessé

Sanguinaire-du-Canada –see– *Sanguinaria canadensis* L.

sanguinaria –see– *Sanguinaria canadensis* L.

SANGUINARIA CANADENSIS L. [Papaveraceae]

Common Names:

bloodroot; bloodwort; Indian red; puccoon; red puccoon; redroot; sanguinaria; Sanguinaire-du-Canada

Citations:

Downey W (1907) An investigation of the properties of the Sanguinaria canadensis; or puccoon. Bull Lloyd Lib #9(5):1-34.

SANICULA EUROPAEA L. [Apiaceae]

Citations:

Engel S, Horn K (1972) Phytodermatosen durch Dictamnus alba, Sanicula europaea und Phyllodendron consanguineum. Dermatol Monatsschr 158(1):22-27.

sanka –see– *Crotalaria juncea* L.

SANSEVIERIA HYACINTHOIDES (L.) Druce [Ruscaceae]

Synonyms:

Sansevieria thyrsiflora Thunb.

Citations:

Der Marderosian AH, Giller FB, Roia FC Jr (1976) Phytochemical and toxicological screening of household ornamental plants potentially toxic to humans. I. J Toxicol Environ Health 1(6):939-953.

Sansevieria thyrsiflora Thunb. = Sansevieria hyacinthoides (L.) Druce

SANSEVIERIA TRIFASCIATA Prain [Ruscaceae]

Common Names:

mother-in-law's-tongue; snake plant

Citations:

Rose A, Pitchford W, Monin T, et al. (1988) Acute weakness and death in a cat. Vet Hum Toxicol 30(4):334-335.

SANSEVIERIA TRIFASCIATA Prain var. laurenti (De Wild.) N. E. Br. [Ruscaceae]

Common Names:

yellow-leaf mother-in-law's-tongue

Citations:

Der Marderosian AH, Giller FB, Roia FC Jr (1976) Phytochemical and toxicological screening of household ornamental plants potentially toxic to humans. I. J Toxicol Environ Health 1(6):939-953.

Santa Ana –see– *Jatropha multifida* L.

Santa Juane –see– *Abrus precatorius* L.

Santa Maria feverfew –see– *Parthenium hysterophorus* L.

SANTALUM ALBUM L. [Santalaceae]

Common Names:

chandan; Rhodesierholz; sandalwood; white sandalwood

Citations:

Starke JC (1967) Photoallergy to sandalwood oil. Arch Derm Syphilol 96(1):62-63.

Santos Palisander –see– *Machaerium scleroxylon* Tul.

Santos rosewood –see– *Machaerium scleroxylon* Tul.

sanve –see– *Sinapis arvensis* L.

sanwa millet –see– *Echinochloa frumentacea* Link

sapele –see– *Entandrophragma cylindricum* (Sprague) Sprague

SAPIUM BILOCULARE (S. Watson) Pax [Euphorbiaceae]

Common Names:

hierbe-de-la-flecha; hierbe mala; Magot; Mexican jumping bean; yerba-de-la-flecha

Citations:

Bradley CE (1956) Yerba de la fleche - Arrow and fish poison of the American Southwest. Econ Bot 10:362-366.

Sapium sebiferum (L.) Roxb. = Triadica sebifera (L.) Small

sapu –see– *Euphorbia tirucalli* L.

sapucacia –see– *Lecythis ollaria* Loefl.

sapucaia nut –see– *Lecythis minor* Jacq.

sara –see– *Zea mays* L.

SARCOBATUS VERMICULATUS (Hook.) Torr. [Sarcobataceae]

Common Names:

black greasewood; chico; greasewood

Citations:

Fleming CE, Miller MR, Vawter LR (1928) The greasewood (Sarcobatus vermiculatus), a range plant poisonous to sheep. Nevada Agric Exp Sta Bull #115:22 pp.

Fleming CE, Miller MR, Vawter LR, et al. (1931) Poisonous range plants. Nevada Agric Exp Sta Annu Rep #1928:21-22.

Marsh CD, Clawson AB, Couch JF (1923) Greasewood as a poisonous plant. U S Dep Agric Circ #279:4 pp.

SARCOCEPHALUS DIDERRICHII De Wild. [Rubiaceae]

Common Names:

acajou-jaune-du-Congo; afrikanischer Buchsbaum; bois-gris-jaunàtre; west African boxwood; yellow grey box

Citations:

Harvey RJ (1906) The physiological properites of 'West African Boxwood'. Biochem J 1:39-53.

SARCOLOBUS GLOBOSUS Wall. [Apocynaceae]

Common Names:

banok; kundur laut; pitis pitis

Citations:

Amarasingham RD, Lee H (1969) A review of poisoning cases examined by the Department of Chemistry, Malaysia, from 1963 to 1967. Med J Malaysia 23(3):220-227.

Arokiasamy M (1968) Toxicity of Sarcolobus globosus as observed in a cat. Malaysian Vet J 4(3):196-199.

Millard AH, Orr W (1958) Observations on the toxicity of Sarcolobus globosus, Wall. Malaysian Vet J 2(1):24-30.

SARCOSTEMMA VIMINALE (L.) R. Br. [Apocynaceae]

Synonyms:

Euphorbia viminale L.

Common Names:

caustic bush; caustic vine; leafless milkweed; melktou; milk bush; ol dewo; pencil caustic; spantou melkbos

Citations:

Gilruth JA (1931) Caustic vine (Sarcostemma australe) as a poison plant. J CSIRO Aust 4:58-60.

Gilruth JA, Murnane D (1931) Sarcostemma australe (caustic vine): A plant that is poisonous to stock. J CSIRO Aust 4:225-231.

Philip JR, Jackson JJ, Shone DK (1958) Sarcostemma viminale poisoning in sheep and cattle. J S Afr Vet Assoc 29(4):319-320.

Terblanche M, van Straten AM (1966) Further studies on Sarcostemma viminale R. Br. poisoning. J S Afr Vet Assoc 37(3):317-319.

sarelle –see– *Oxalis acetosella* L.

Sarothamnus scoparius (L.) Wimm. ex W. D. J. Koch = Cytisus scoparius (L.) Link

sarson –see– *Brassica juncea* (L.) Czern.

sartwellia –see– *Sartwellia flaveriae* A. Gray

SARTWELLIA FLAVERIAE A. Gray [Asteraceae]

Common Names:

downy sartwellia; sartwellia; thread-leaf sartwellia

Citations:

Mathews FP (1940) The toxicity of Sartwellia flaveriae to goats. J Agric Res 61(4):287-292.

Mathews FP, Schmidt H (1934) Miscellaneous poisonous plants. Texas Agric Exp Sta Annu Rep 47:12-13.

Mathews FP, Schmidt H (1935) Miscellaneous poisonous plants. Texas Agric Exp Sta Annu Rep 48:16-17.

Saskatoon serviceberry –see– *Amelanchier alnifolia* (Nutt.) Nutt. ex M. Roem.

sassafras –see– *Sassafras albidum* (Nutt.) Nees

SASSAFRAS ALBIDUM (Nutt.) Nees [Lauraceae]

Common Names:

ague tree; American sassafras; cinnamonwood; huile-de-sassafras américain; saloop; sassafras; saxifras

Citations:

Craig JO (1953) Poisoning by the volatile oils in childhood. Arch Dis Child 28(142):475-483.

Grande GA, Dannewitz SR (1987) Symptomatic sassafras oil ingestion. Vet Hum Toxicol 29(6):447.

sasswood –see– *Erythrophleum suaveolens* (Guill. & Perr.) Brenan

sassy bark –see– *Erythrophleum suaveolens* (Guill. & Perr.) Brenan

sassy jack –see– *Euphorbia lathyris* L.

Satan's-apple –see– *Mandragora officinarum* L.

satayanashi –see– *Argemone mexicana* L.

Satinholz –see– *Chloroxylon swietenia* DC.

satinwood –see– *Chloroxylon swietenia* DC.; *Zanthoxylum flavum* Vahl

satyanasi –see– *Argemone mexicana* L.

Saubohne –see– *Vicia faba* L.

sauco –see– *Sambucus mexicana* C. Presl ex DC.

saucy bark –see– *Erythrophleum suaveolens* (Guill. & Perr.) Brenan

Sauerampfer –see– *Rumex acetosa* L.

sauerblec –see– *Oxalis acetosella* L.

Saukraut –see– *Hyoscyamus niger* L.; *Levisticum officinale* W. D. J. Koch

Sauropus albicans Blume = Sauropus androgynus (L.) Merr.

SAUROPUS ANDROGYNUS (L.) Merr. [Phyllanthaceae]

Synonyms:

Sauropus albicans Blume

Common Names:

asin asin; chekkarmenis; chekor manis; chekurmanis

Citations:

Bender AE, Ismail KS (1976) Nutritive value and toxicity of a Malaysian food, Sauropus albicans. Proc Nutr Soc 32(2):79A-80A.

Chang H, Wang JS, Tseng HH, et al. (1997) Histopathological study of Sauropus androgynus-associated constrictive bronchiolitis: A new cause of constrictive bronchiolitis obliterans. Am J Surg Pathol 21(1):35-42.

Chang YL (1998) Segmental necrosis of small bronchi after prolonged intakes of Sauropus androgynus in Taiwan. Am J Respir Crit Care Med 157(2):594-598.

Chang YL, Chen JS, Wu HD, et al. (2000) Retransplantation of contralateral lung in a patient with Sauropus androgynus-induced bronchobronchiolitis obliterans. Transplant Proc 32(7):2432-2434.

Chen CW, Hsiue TR, Chen KW, et al. (1996) Increased IL-5 and IL-10 transcription in bronchial cells after Sauropus androgynus ingestion. J Formos Med Assoc 95(9):699-702.
Deng JF (1996) Outbreak of obstructive ventilatory impairment associated with consumption of Sauropus androgynus vegetable. J Toxicol Clin Toxicol 34:600.
Ger LP, Chiang AA, Lai RS, et al. (1997) Association of Sauropus androgynus and bronchiolitis obliterans syndrome: A hospital-based case-control study. Am J Epidemiol 145(9):842-849.
Higenbottam TW (1997) Bronchiolitis obliterans following the ingestion of an Asian shrub leaf. Thorax 52(Suppl 3):S68-S72.
Hsiue TR, Guo YL, Chen KW, et al. (1998) Dose-response relationship and irreversible obstructive ventilatory defect in patients with consumption of Sauropus androgynus. Chest 113(1):71-76.
Hsu H, Chang H, Su J, et al. (1998) Lung transplantation in Sauropus androgynus consumption patients in Taiwan. Transplant Proc 30(7):3393-3394.
Lin TJ, Lu CC, Chen KW, et al. (1996) Outbreak of obstructive ventilatory impairment associated with consumption of Sauropus androgyous vegetable. J Toxicol Clin Toxicol 34(1):1-8.
Oonakahara K, Matsuyama W, Higashimoto I, et al. (2005) Outbreak of bronchiolitis obliterans associated with consumption of Sauropus androgynus in Japan - Alert of food-associated pulmonary disorders from Japan. Respiration 72(2):221.
Wang J, Tseng HH, Lai RS, et al. (2000) Sauropus androgynus-constrictive obliterative bronchitis/bronchiolitis - Histopathological study of pneumonectomy and biopsy specimens with emphasis on the inflammatory process and disease progression. Histopath 37(5):402-410.
Wu CL, Hsu WH, Chiang CD (1998) The effect of large-dose prednisolone on patients with obstructive lung disease associated with consuming Sauropus androgynus. Chung Hua I Hsueh Tsa Chih 61(1):34-38.
Wu CL, Hsu WH, Chiang CD, et al. (1997) Lung injury related to consuming Sauropus androgynus vegetable. J Toxicol Clin Toxicol 35(3):241-248.
Yang CF, Wu MT, Chiang AA, et al. (1997) Correlation of high-resolution CT and pulmonary function in bronchiolitis obliterans: A study based on 24 patients associated with consumption of Sauropus androgynus. AJR Am J Roentgenol 168(4):1045-1050.

SAUSSUREA COSTUS (Falc.) Lipsch. [Asteraceae]

Common Names:
costus; kusth

Citations:
Maibach HI, Mitchell JC (1975) Costus absolute (Saussurea): Predictive assay for allergic contact sensitization in guinea-pigs. Contact Dermatitis 1(3):184.

saxifras –see– *Sassafras albidum* (Nutt.) Nees

SCADOXUS MULTIFLORUS (Martyn) Raf. [Amaryllidaceae]

Synonyms:
Haemanthus multiflorus Martyn

Common Names:
blood lily; powderpuff lily

Citations:
Mugera GM (1970) Toxic and medicinal plants of East Africa. II. Bull Epizootic Dis Afr 18(4):389-403.

scanty rice flower –see– *Pimelea pauciflora* R. Br.
scarlet buckeye –see– *Aesculus pavia* L.
scarlet maple –see– *Acer rubrum* L.
scarlet milkweed –see– *Asclepias curassavica* L.
scarlet pimpernel –see– *Anagallis arvensis* L.
scarlet plume –see– *Euphorbia fulgens* Karw. ex Klotzsch
scarlet rhus –see– *Toxicodendron succedaneum* (L.) Kuntze
scarlet runner bean –see– *Phaseolus coccineus* L.
scarlet sumach –see– *Toxicodendron succedaneum* (L.) Kuntze
scarlet wistaria tree –see– *Sesbania grandiflora* (L.) Pers.
scarletberry –see– *Solanum dulcamara* L.
scented goosefoot –see– *Chenopodium carinatum* R. Br.
scentless chamomile –see– *Anthemis arvensis* L.
Schafgarbe –see– *Achillea millefolium* L.
Schaflinse –see– *Securigera varia* (L.) Lassen
Scharlachbeere –see– *Phytolacca americana* L.
Schattenblume –see– *Maianthemum bifolium* (L.) F. W. Schmidt
schefflera –see– *Schefflera actinophylla* (Endl.) Harms

SCHEFFLERA ACTINOPHYLLA (Endl.) Harms [Araliaceae]

Synonyms:
Brassaia actinophylla Endl.

Common Names:
Australian ivy palm; Australian umbrella tree; heart ivy; octopus tree; queens umbrella tree; Queensland umbrella tree; ripple ivy; schefflera; umbrella tree

Citations:
Grob M, Wüthrich B (1998) Occupational allergy to the umbrella tree (Schefflera). Allergy 53(10):1008-1009.
Mitchell JC (1981) Allergic contact dermatitis from Hedera helix and Brassaia actinophylla (Araliaceae). Contact Dermatitis 7(3):158-159.
Stowe CM, Fangmann G (1975) Schefflera toxicosis in a dog. J Am Vet Med Assoc 167(1):74.

SCHEFFLERA ARBORICOLA (Hayata) Merr. [Araliaceae]

Citations:
Paulsen E (1998) Occupational dermatitis in Danish gardeners and greenhouse workers (II). Etiological factors. Contact Dermatitis 38(1):14-19.

SCHEFFLERA KWANGSIENSIS Merr. ex H. L. Li [Araliaceae]

Synonyms:

Schefflera volkensii (Harms) Harms

Citations:

Calnan CD (1981) Dermatitis from Schefflera. Contact Dermatitis 7(6):341.

Schefflera volkensii (Harms) Harms = Schefflera kwangsiensis Merr. ex H. L. Li

Scheinakazie –see– *Robinia pseudoacacia* L.

Scheinbeere –see– *Gaultheria procumbens* L.

Scheißwurz –see– *Bryonia dioica* Jacq.

Schellkraut –see– *Chelidonium majus* L.

Schierling Caladium –see– *Dieffenbachia seguine* (Jacq.) Schott

SCHINUS TEREBINTHIFOLIUS Raddi [Anacardiaceae]

Common Names:

aroeira; Brazilian pepper tree; burnwood; Christmasberry tree; copa; coral sumach; doctor gum; false pepper; Florida holly; Japanese pepper; mountain manchineel; pepper tree; pimienta-de-Brasil; pink peppercorn; poisonwood

Citations:

Anonymous (1978) Berry banquet leaves 115 birds dead in South Dade. Miami Herald. Mar 2:C1.

Britt JO Jr, Howard EB (1984) Pepper tree berry poisoning in a flock of wild cedar waxwings. Avian Exotic Pract 1(3):11-13.

Catalano PN (1984) Mango sap and poison ivy dermatitis. J Am Acad Dermatol 10(3):522.

SCHIZOCARPHUS NERVOSUS (Burch.) Van der Merwe [Hyacinthaceae]

Synonyms:

Scilla rigidifolia Kunth.

Citations:

Van der Walt SJ (1944) Recent investigations into the toxicity of plants, etc. in the Union of South Africa. Onderstepoort J Vet Sci Anim Indus 20(1):75-83.

Van der Walt SJ, Steyn DG (1946) Recent investigations into the toxicity of plants, etc. XV. Onderstepoort J Vet Sci Anim Indus 21(1):45-55.

Schlafmohn –see– *Papaver somniferum* L.

Schlangenkraut –see– *Calla palustris* L.

Schlangenrübe –see– *Bryonia dioica* Jacq.

Schlehe –see– *Prunus spinosa* L.

Schlüsselblume –see– *Primula veris* L.

Schmerwurz –see– *Dioscorea communis* (L.) Caddick & Wilkin

Schneeball –see– *Viburnum rhitidophyllum* Hemsl.

Schneebeere –see– *Symphoricarpos albus* (L.) S. F. Blake

Schneerose –see– *Helleborus niger* L.

SCHOENUS ASPEROCARPUS F. Muell. [Cyperaceae]

Common Names:

poison sedge

Citations:

Colegate SM, Dorling PR, Huxtable CR (1994) Poison sedges (Schoenus spp.): Toxic effects and identification of the toxic principle. In: Colegate SM, Dorling PR (eds.) Plant-associated toxins. Agricultural, phytochemical and ecological aspects. CABI. New York. pp. 275-280.

Nairn ME, Aplin TE, Petterson DS, et al. (1971) Poison sedge can kill stock. J Agric West Aust 12:45-47.

Royce RD (1952) Poison sedge and cyanogenesis. J Agric Western Australia 3rd Series 1(Jul-Aug):497-500.

Schöllkraut –see– *Chelidonium majus* L.

Schutt Bingelkraut –see– *Mercurialis annua* L.

Schwarzdorn –see– *Prunus spinosa* L.

Schwarze Brechnuß –see– *Jatropha curcas* L.

schwarze Neiswurz –see– *Helleborus niger* L.

schwarze Zaunrübe –see– *Dioscorea communis* (L.) Caddick & Wilkin

Schwarzer Holunder –see– *Sambucus nigra* L.

Schwarzer Maulbeerbaum –see– *Morus nigra* L.

Schwarzer Nachtshatten –see– *Solanum nigrum* L.

Schwarzer Senf –see– *Brassica nigra* (L.) W. D. J. Koch

Schwarze Bilsenkraut –see– *Hyoscyamus niger* L.

Schwarzkraut –see– *Asarum europaeum* L.

Schwarzwurz –see– *Asarum europaeum* L.

Schweden Klee –see– *Trifolium hybridum* L.

Schweigrohr –see– *Dieffenbachia seguine* (Jacq.) Schott

Schweinekraut –see– *Calla palustris* L.

Schweinsohr –see– *Calla palustris* L.

Schwindelhafer –see– *Lolium temulentum* L.

Scilla cooperi Hook. f. = Ledebouria cooperi (Hook. f.) Jessop

Scilla natalensis Planch. = Merwilla plumbea (Lindl.) Speta

Scilla nonscripta (L.) Hoffmanns. & Link = Hyacinthoides non-scripta (L.) Chouard ex Rothm.

Scilla nutans Sm. = Hyacinthoides non-scripta (L.) Chouard ex Rothm.

Scilla rigidifolia Kunth = Schizocarphus nervosus (Burch.) Van der Merwe

scille maritime –see– *Drimia maritima* (L.) Stearn

scille officinale –see– *Drimia maritima* (L.) Stearn

Scindapsus aureus (Linden & André) Engl. = Epipremnum pinnatum (L.) Engl.

SCINDAPSUS PICTUS Hassk. [Araceae]

Common Names:

pothos

Citations:

Ramón MF, Ballesteros S, Martínez-Arrieta R, et al. (2003) Plant and herb exposures in Spain. J Toxicol Clin Toxicol 41(4):514.

SCLEROBLITUM ATRIPLICINUM (F. Muell.) Ulbr. [Chenopodiaceae]

Synonyms:

Chenopodium atriplicinum (F. Muell.) F. Muell.

Common Names:

lamb's-tongue; purple goosefoot

Citations:

King RO (1937) Chenopodium atriplicinum (lamb's tongue) a plant toxic for sheep in immature stages of growth. New South Wales Dep Agric Vet Res Rep 7:95-97.

Seddon HR, Belschner HG (1931) Mortalities in sheep associated with grazing on young pasture plants, with special reference to Chenopodium atriplicinus. Aust Vet J 7:68-70.

Seddon HR, Belschner HG (1931) Mortalities in sheep grazing on young plants of Chenopodium atriplicinum. Agric Gaz New South Wales 42:865-866.

SCLEROLAENA ANISACANTHOIDES (F. Muell.) Domin [Chenopodiaceae]

Synonyms:

Bassia echinopsila (F. Muell.) F. Muell.

Common Names:

red bur

Citations:

Mathams RH, Sutherland AK (1952) The oxalate content of some Queensland pasture plants. Queensland J Agric Sci 9:317-334.

scopolia –see– *Scopolia carniolica* Jacq.

SCOPOLIA CARNIOLICA Jacq. [Solanaceae]

Common Names:

crazy plant; Glockenbilsenkraut; scopolia; Tollkraut

Citations:

Cuculić M, Kalodera Z, Sindik J, et al. (1988) Obiteljsko trovanje korijenom bijelog buna (Scopolia carniolica Jacq.). Arh Hig Rada Toksikol 39(3):345-348.

Hatziisaak T, Weber A (1998) Hexentee. Schweiz Rundsch Med Prax 87(49):1705-1708.

scorpion weed –see– *Phacelia crenulata* Torr. ex S. Watson

Scotch broom –see– *Cytisus scoparius* (L.) Link

Scotch laburnum –see– *Laburnum alpinum* (Mill.) J. Presl

Scottish laburnum –see– *Laburnum alpinum* (Mill.) J. Presl

scourge-of-India –see– *Parthenium hysterophorus* L.

scouring rush –see– *Equisetum arvense* L.

scrambled eggs –see– *Corydalis aurea* Willd.

scrintioare cosita –see– *Galega officinalis* L.

Scrophularia aquatica auct. = Scrophularia auriculata L.

SCROPHULARIA AURICULATA L. [Scrophulariaceae]

Synonyms:

Scrophularia aquatica auct.

Common Names:

betoine aquatique; herbe-du-siège; tigerwort; water betony; water figwort; water scrophularia

Citations:

Ewart RH (1937) Poisoning in young cattle by a member of Scrophulariaceae. Vet Rec 49:1514.

Runciman T (1885) Paralysis in ewes caused by eating "Scrophularia aquatica" (greater water, tiger-wort, water betony). Am Vet Rev 9:79-80.

SCROPHULARIA MARILANDICA L. [Scrophulariaceae]

Synonyms:

Melampyrum cristatum L.

Common Names:

cow wheat; honey plant; Simpson money plant

Citations:

Lazareva DN, Karimova SG, Zarudy FS, et al. (1967) [On the neurologic effect of cow wheat (Melampyrum cristatus L.).] Farmakol Toksikol 29:175-177.

scrub kurrajong –see– *Pimelea pauciflora* R. Br.

scrub oak –see– *Quercus gambelii* Nutt.

scullcap –see– *Scutellaria lateriflora* L.

scurfy pea –see– *Cullen corylifolium* (L.) Medik.; *Cullen drupaceum* (Bunge) C. H. Stirt.

scutellaria –see– *Scutellaria lateriflora* L.

SCUTELLARIA LATERIFLORA L. [Lamiaceae]

Common Names:

helmet flower; hoodwort; madweed; Quaker bonnet; scullcap; scutellaria; skullcap

Citations:

Estes JD, Stolpman D, Olyaei A, et al. (2003) High prevalence of potentially hepatotoxic herbal supplement use in patients with fulminant hepatic failure. Arch Surg 138(8):852-858.

sea arrowgrass –see– *Triglochin maritima* L.

sea lavender –see– *Goniolimon tataricum* (L.) Boiss.

sea myrtle –see– *Baccharis halimifolia* L.

sea onion –see– *Drimia maritima* (L.) Stearn

sea purslane –see– *Trianthema portulacastrum* L.

sea squill –see– *Drimia maritima* (L.) Stearn

seaside arrowgrass –see– *Triglochin maritima* L.

seaside buttercup –see– *Ranunculus cymbalaria* Pursh

sebipira –see– *Bowdichia nitida* Spruce ex Benth.

seca tierra –see– *Pascalia glauca* Ortega

SECALE CEREALE L. [Poaceae]

Common Names:

annual rye; Roggen; rye

Citations:

Karasek E (1961) Intoxikationen durch ausgekeimten Roggen in einer Schafherde. Monatsh Veterinarmed 16:265-266.

MacAuliffe T, Pietraszek A, McGinnis J (1976) Variable rachitogenic effects of grain and alleviation by extraction or supplementation with Vitamin D, fat and antibiotics. Poult Sci 55(6):2142-2147.

MacAuliffe T, Zaviezo D, McGinnis J (1979) Effect of gamma irradiation, fractionation, and penicillin supplementation on the rachitogenic activity of rye for chicks. Poult Sci 58(2):329-332.

Proudfoot FG, Hulan HW (1986) The nutritive value of ground rye as a feed ingredient for adult Leghorn hens. Can J Anim Sci 66(1):311-315.

Untawale GG, McGinnis J (1979) Effect of rye and levels of raw and autoclaved beans (Phaseolus vulgaris) on adhesion of microflora to the intestinal mucosa. Poult Sci 58(4):928-933.

SECHIUM EDULE (Jacq.) Sw. [Cucurbitaceae]

Common Names:

chayote; chocho

Citations:

Jensen LP, Lai AR (1986) Chayote (Sechium edule) causing hypokalemia in pregnancy. Am J Obstet Gynecol 155(5):1048-1049.

SECURIGERA VARIA (L.) Lassen [Fabaceae]

Synonyms:

Coronilla varia L.

Common Names:

axwort; broom; cornvetch; coronilla; crownvetch; Falsche Efparfette; Giftwicke; Giftwide; Hagedorn; Kronenschütchen; Kronwicke; Mehldorn; rosy crown vetch; Schaflinse

Citations:

Gustine DL, Shenk JS, Wangsness PJ, et al. (1973) Isolation and characterization of crownvetch constituents deleterious to nonruminants. Proc Fescue Toxicity Conf. Lexington, Kentucky. pp. 140-144.

Salyi G, Sztojkov V, Hilbert Miklovica M (1988) A nutria tarka koronafürt (Coronilla varia L.) okozta mérgezése. Magyar Allator Lapja 43(5):313-316.

Shenk JS (1976) The meadow vole as an experimental animal. Lab Anim Sci 26(4):664-669.

Shenk JS, Wangsness PJ, Leach RM, et al. (1976) Relationship between β-nitropropionic acid content of crownvetch and toxicity in nonruminant animals. J Anim Sci 42(3):616-621.

Zuyu L, Xueqin F, Zihua Z, et al. (1987) [The effects of feeding diets containing Astragalus adsurgens Pall and Coronilla varia L. on broiler chickens.] Acta Vet Zootech Sin 18(3):157-162.

sedano –see– *Apium graveolens* L.

Sedum spectabile Boreau = Hylotelephium spectabile (Boreau) H. Ohba

seepbossie –see– *Noltea africana* (L.) Endl.

seeroogblom –see– *Boophone disticha* (L. f.) Herb.

seggrum –see– *Jacobaea vulgaris* Gaertn.

Seidelbast –see– *Daphne mezereum* L.

selino –see– *Apium graveolens* L.

Sellerie –see– *Apium graveolens* L.

selyemkóró –see– *Asclepias syriaca* L.

sem –see– *Canavalia cathartica* Thouars

SEMECARPUS ANACARDIUM L. f. [Anacardiaceae]

Synonyms:

Anacardium orientale auct. ex Steud.

Common Names:

bella gutta tree; bibba nut; bibha; bichi nut; black nut; dhobi nut; golden corn; Indian ink tree; Indian marking nut; marking nut; oriental cashew; ral tree

Citations:

Cade A, Nelson CS (1996) Semecarpus anacardium-induced facial oedema. Br J Dermatol 135(2):338-339.

Fitz-Hugh T Jr, Livingood CS, Rogers AM (1943) Dhobie mark dermatitis. U S Army Med Dep Bull #69:55-58.

Fox H (1921) Dermatitis venenata from oriental cashew nut. Report of a case. Arch Derm Syphilol 3:202-203.

Goldsmith NR (1943) Dermatitis from Semecarpus anacardium (bhilawanol or the marking nut). Spread by contaminated mail. JAMA 123(Sep 4):27.

Livingood CS, Rogers AM, Fitz-Hugh T Jr (1943) Dhobie mark dermatitis. JAMA 123(1):23-26.

Matthai TP, Date A (1979) Renal cortical necrosis following exposure to sap of the marking-nut tree (Semecarpus anacardium). Am J Trop Med Hyg 28(4):773-774.

Shankar DS (1992) Contact urticaria induced by Semecarpus anacardium. Contact Dermatitis 26(3):200.

Waud SP, Fein H (1943) Dermatitis venenata caused by the ink from the bichi nut. Bull U S Army Med Dep #69(Oct):59-60.

SEMECARPUS ATER (G. Forst.) Vieill. [Anacardiaceae]

Common Names:

acajou; goudronnier; nole; pomme d'acajou

Citations:

Hitch JM (1944) Dermatitis venenata caused by Semecarpus atra. U S Nav Med Bull 42(5):1111-1115.

SEMECARPUS AUSTRALIENSIS Engl. [Anacardiaceae]

Common Names:

marking nut; tar tree

Citations:

Flecker H (1945) Injuries produced by plants in tropical Queensland. Med J Aust 1(Jun 23):636-637.

Seminole bread –see– *Zamia integrifolia* L. f.

sempervivum –see– *Aloe vera* (L.) Burm. f.

SENECIO ABYSSINICUS Sch. Bip. ex A. Rich. [Asteraceae]

Citations:

Williams AO, Schoental R (1970) Hepatotoxicity of Senecio abyssinicus. Experimental and ultrastructural studies. Trop Geogr Med 22(2):201-210.

SENECIO ALPINUS (L.) Scop. [Asteraceae]

Common Names:

Alpenkreuzkraut; ragwort

Citations:

Braun U, Linggi T, Pospischil A (1999) Ultrasonographic findings in three cows with chronic ragwort (Senecio alpinus) poisoning. Vet Rec 144(5):122-126.

Pohlenz J, Lüthy J, Minder HP, et al. (1980) Enzootische Leberzirrhose beim Rind, verursacht durch Pyrrolizidinalkaloide nach Aufnahme von Senecio alpinus (Alpenkreuzkraut). Schweiz Arch Tierheilkd 122(4):183-194.

Stunzi H (1982) Vergiftungen bei Kühen nach Aufnahme von Alpenkreuzkraut. Mitt Schweiz Landwirtsch 30:35-36.

Senecio ambrosioides Rydb. = Senecio eremophilus Richardson var. kingii Greenm.

Senecio aquaticus Hill = Jacobaea aquatica (Hill) P. Gaertn. et al. var. aquatica

Senecio australis Willd. = Senecio linearifolius A. Rich

Senecio barbareaefolius Rchb. = Jacobaea aquatica (Hill) P. Gaertn. et al. var. aquatica

SENECIO BRASILIENSIS (Spreng.) Less. [Asteraceae]

Common Names:

ragwort

Citations:

Barros CS, Castilhos LM, Santos MN, et al. (1987) Liver biopsy in ragwort poisoning. Vet Rec 121(16):382.

Barros CS, Driemeier D, Pilati C, et al. (1992) Senecio spp poisoning in cattle in Southern Brazil. Vet Hum Toxicol 34(3):241-246.

Barros CS, Metzdorf LL, Peixoto PV (1987) Ocorrência de surtos de intoxicação por Senecio spp. (Compositae) em bovinos no Rio Grande do Sul. Pesq Vet Bras 7(4):101-107.

Barros CS, Metzdorf LL, Santos MN, et al. (1989) Intoxicação experimental por Senecio brasiliensis (Compositae) em ovinos. Pesq Vet Bras 9(3-4):55-67.

Gava A, Barros CS (1997) Senecio spp. poisoning of horses in Southern Brazil. Pesq Vet Bras 17(1):36-40.

Ilha MR, Loretti AP, Barros SS, et al. (2001) Intoxicação espontânea por Senecio brasiliensis (Asteraceae) em ovinos no Rio Grande do Sul. Pesq Vet Bras 21(3):123-138.

Karam FS, Soares MP, Haraguchi M, et al. (2004) Aspectos epidemiológicos da seneciose na região sul do Rio Grande do Sul. Pesq Vet Bras 24(4):191-198.

Méndez MC, Riet-Correa F, Schild AL (1987) Intoxicação por Senecio spp. (Compositae) em bovinos no Rio Grande do Sul. Pesq Vet Bras 7(2):51-56.

Nazario W, Portugal MA, Fancelli MI (1988) Considerações sobre o papel do Senecio brasiliensis, Lessing, Acompanhamento de acidente tóxico em bovinos, ocorrido em São Paulo. Pesq Agric Bras 23(5):537-542.

Tokarnia CH, Döbereiner J (1984) Intoxicação experimental por Senecio brasiliensis (Compositae) em bovinos. Pesq Vet Bras 4(2):39-65.

SENECIO BURCHELLII DC. [Asteraceae]

Common Names:

kovanna bos; sprinkaanbos

Citations:

Chase WH (1904) The Molteno cattle disease. Agric J Cape Good Hope 25:675-678.

Robertson W (1906) Cirrhosis of the liver in stock in Cape Colony, produced by two species of Senecio (Senecio burchelli and Senecio latifolius). J Comp Pathol Ther 19(2):97-110.

Willmot FC, Robertson GW (1920) Senecio disease, or cirrhosis of the liver due to senecio poisoning. Lancet 2(Oct 23):848-849.

SENECIO CANNABIFOLIUS Less. [Asteraceae]

Common Names:

hangon so

Citations:

Hirono I, Ueno I, Aiso S, et al. (1983) Carcinogenic activity of Farfugium japonicum and Senecio cannabifolius. Cancer Lett 20(2):191-198.

SENECIO CISPLATINUS Cabrera [Asteraceae]

Citations:

Méndez MC, Riet-Correa F, Schild AL (1987) Intoxicação por Senecio spp. (Compositae) em bovinos no Rio Grande do Sul. Pesq Vet Bras 7(2):51-56.

Schild AC, Motta AC, Riet-Correa F, et al. (2004) Photosensitization in cattle in southern Brazil. In: Acamovic T, Stewart CS, Pennycott TW (eds.) Poisonous plants and related toxins. CABI. Wallingford, UK. pp. 162-166.

SENECIO CUNNINGHAMII DC. [Asteraceae]

Citations:

Wiltjer JC, Walker CE (1974) Rectal prolapse in cattle associated with pyrrolizidine alkaloid poisoning. Aust Vet J 50(12):579-580.

SENECIO DESFONTAINEI Druce [Asteraceae]

Citations:

El Sergany MA, Morsy AA, Dessouky MI, et al. (1984) Studies on blood cytology, serum enzymes and the histopathology of the liver in chicks fed dried Senecio desfontainei. Arch Geflugelkunde 48(3):77-80.

Senecio diversifolius Wall. ex DC. = Jacobaea raphanifolia (Wall. ex DC.) B. Nord.

Senecio douglasii DC. var. longilobus (Benth.) L. D. Benson = Senecio flaccidus Less. var. flaccidus

SENECIO EREMOPHILUS Richardson var. kingii Greenm. [Asteraceae]

Synonyms:

Senecio ambrosioides Rydb.

Common Names:

groundsel

Citations:

Anonymous (1942) Investigations of stock-poisoning plants. U S Dep Agric Bur Anim Indus Chief Rep 1942:29.

Senecio erraticus Bertol. = Jacobaea aquatica (Hill) P. Gaertn. et al. var. erratica (Bertol.) Pelser & Meijden

Senecio erraticus Bertol. subsp. barbareifolius (Fiori) Beger = Jacobaea aquatica (Hill) P. Gaertn. et al. var. erratica (Bertol.) Pelser & Meijden

SENECIO FLACCIDUS Less. var. flaccidus [Asteraceae]

Synonyms:

Senecio douglasii DC. var. longilobus (Benth.) L. D. Benson; ***Senecio longilobus*** Benth.

Common Names:

gordolobo yerba; groundsel; threadleaf groundsel; woolly groundsel

Citations:

Anonymous (1933) Investigation of stock-poisoning plants. U S Dep Agric Bur Anim Indus Chief Rep 1933:34-35.

Boughton IB, Hardy WT (1939) Hard yellow livers of sheep and cattle and feeding trials of suspected plants. Texas Agric Exp Sta Annu Rep 52:238-240.

Clawson AB (1933) The American groundsels species of Senecio as stock poisoning plants. Vet Med 28(Mar):105-110.

Dollahite JW (1972) The use of sheep and goats to control Senecio poisoning in cattle. Southwestern Vet 25(3):223-226.

Fox DW, Hart MC, Bergeson PS, et al. (1978) Pyrrolizidine (senecio) intoxication mimicking Reye Syndrome. J Pediatr 93(6):980-982.

Harris PN, Chen KK (1970) Development of hepatic tumours in rats following ingestion of Senecio longilobus. Cancer Res 30(12):2881-2886.

Henderson FG, Harris PN, Chen KK (1954) Hepatotoxic action of Senecio longilobus. J Pharmacol Exp Ther 110:26.

Johnson AE, Molyneux RJ (1984) Toxicity of threadleaf groundsel (Senecio douglasii var longilobus) to cattle. Am J Vet Res 45(1):26-31.

Mathews FP (1932) Miscellaneous poisonous plants. Texas Agric Exp Sta Annu Rep 45:11-12.

Mathews FP (1933) Poisoning of cattle by species of groundsel (Senecio longilobus Benth., and Senecio riddellii Torr. and Gray). Texas Agric Exp Sta Bull #481:5-20.

Mathews FP, Schmidt H (1934) Miscellaneous poisonous plants. Texas Agric Exp Sta Annu Rep 47:12-13.

Mathews FP, Schmidt H (1935) Miscellaneous poisonous plants. Texas Agric Exp Sta Annu Rep 48:16-17.

Molyneux RJ, Johnson AE, Stuart LD (1988) Delayed manifestation of Senecio-induced pyrrolizidine alkaloidosis in cattle: Case reports. Vet Hum Toxicol 30(3):201-205.

Stillman AE, Huxtable R, Consroe P, et al. (1977) Hepatic veno-occlusive disease due to pyrrolizidine (Senecio) poisoning in Arizona. Gastroenterology 73(2):349-352.

Stillman AE, Huxtable RJ, Fox DW, et al. (1977) Poisoning associated with herbal teas - Arizona, Washington. MMWR Morb Mortal Wkly Rep 26(32):257-259.

Stillman AE, Huxtable RJ, Fox D, et al. (1977) Pyrrolizidine (Senecio) poisoning in Arizona: Severe liver damage due to herbal teas. Ariz Med 34(8):545-546.

Vardiman PH (1953) The bromsulfalein liver function test and biopsy of the liver in the diagnosis of senecio poisoning in cattle. Am J Vet Res 14(51):175-178.

Senecio gigas Vatke = Solanecio gigas (Vatke) C. Jeffrey

Senecio glabellus Poir. = Packera glabella (Poir.) C. Jeffrey

SENECIO GLUTINOSUS Thunb. [Asteraceae]

Citations:

Steyn DG (1937) Recent investigations into the toxicity of known and unknown poisonous plants in the Union of South Africa. VII. Onderstepoort J Vet Sci Anim Indus 9(1):111-124.

SENECIO HETEROTRICHUS DC. [Asteraceae]

Citations:

Karam FS, Soares MP, Haraguchi M, et al. (2004) Aspectos epidemiológicos da seneciose na região sul do Rio Grande do Sul. Pesq Vet Bras 24(4):191-198.

Méndez MC, Riet-Correa F, Schild AL (1987) Intoxicação por Senecio spp. (Compositae) em bovinos no Rio Grande do Sul. Pesq Vet Bras 7(2):51-56.

Schild AC, Motta AC, Riet-Correa F, et al. (2004) Photosensitization in cattle in southern Brazil. In: Acamovic T, Stewart CS, Pennycott TW (eds.) Poisonous plants and related toxins. CABI. Wallingford, UK. pp. 162-166.

SENECIO ILICIFOLIUS L. [Asteraceae]

Common Names:

ghwanobos; guano bos; kavaanuo bos; Kovanna bos; sprinkaanbos

Citations:

Willmot FC, Robertson GW (1920) Senecio disease, or cirrhosis of the liver due to senecio poisoning. Lancet 2(Oct 23):848-849.

SENECIO INTEGERRIMUS Nutt. [Asteraceae]

Common Names:

gauge plant; lamb's-tongue groundsel; one-stem butterweed; western groundsel; wet-the-bed

Citations:

Clawson AB (1933) The American groundsels species of Senecio as stock poisoning plants. Vet Med 28(Mar):105-110.

SENECIO ISATIDEUS DC. [Asteraceae]

Common Names:

Dan's-cabbage; poison ragwort; ragwort

Citations:

Van der Walt SJ (1944) Recent investigations into the toxicity of plants, etc. in the Union of South Africa. Onderstepoort J Vet Sci Anim Indus 20(1):75-83.

Senecio jacobaea L. = Jacobaea vulgaris Gaertn.

SENECIO JUNCEUS (DC.) Harv. [Asteraceae]

Citations:

Lyford CL, Vergara GG, Moeller DD (1976) Hepatic veno-occlusive disease originating in Ecuador. Gastroenterology 70(1):105-108.

SENECIO LATIFOLIUS DC. [Asteraceae]

Common Names:

doss cabbage; groundsel; ragwort; Rhodesian ragwort

Citations:

Burrows GE, Morton DJ, Basudde CD (1990) The effects of feeding cabbage (Brassica oleracea) and Rhodesian ragwort (Senecio latifolius) on the disposition of antipyrine and acetaminophen in sheep. Toxicon 28(6):603.

Robertson W (1906) Cirrhosis of the liver in stock in Cape Colony, produced by two species of Senecio (Senecio burchelli and Senecio latifolius). J Comp Pathol Ther 19(2):97-110.

SENECIO LAUTUS G. Forst. ex Willd. [Asteraceae]

Common Names:

fireweed

Citations:

Kirkland PD, Moore RE, Walker KH, et al. (1982) Deaths in cattle associated with Senecio lautus consumption. Aust Vet J 59(2):64.

Noble JW, Crossley J, Hill BD, et al. (1994) Pyrrolizidine alkaloids of cattle associated with Senecio lautus. Aust Vet J 71(7):196-200.

SENECIO LEPTOLOBUS DC. [Asteraceae]

Citations:

Karam FS, Soares MP, Haraguchi M, et al. (2004) Aspectos epidemiológicos da seneciose na região sul do Rio Grande do Sul. Pesq Vet Bras 24(4):191-198.

Méndez MC, Riet-Correa F, Schild AL (1987) Intoxicação por Senecio spp. (Compositae) em bovinos no Rio Grande do Sul. Pesq Vet Bras 7(2):51-56.

Schild AC, Motta AC, Riet-Correa F, et al. (2004) Photosensitization in cattle in southern Brazil. In: Acamovic T, Stewart CS, Pennycott TW (eds.) Poisonous plants and related toxins. CABI. Wallingford, UK. pp. 162-166.

SENECIO LINEARIFOLIUS A. Rich [Asteraceae]

Synonyms:

Senecio australis Willd.

Common Names:

fireweed

Citations:

Dickinson CG (1930) Waratah horse disease. Aust Vet J 6(Jun):65-67.

Senecio longilobus Benth. = Senecio flaccidus Less. var. flaccidus

SENECIO MADAGASCARIENSIS Poir. [Asteraceae]

Common Names:

fireweed

Citations:

Small AC, Kelly WR, Seawright AA, et al. (1993) Pyrrolizidine alkaloidosis in a two month old foal. Zentralbl Veterinarmed A 40(3):213-218.

SENECIO MOOREI R. E. Fr. [Asteraceae]

Citations:

Kamau JA, Mugera GM (1975) Induction of liver tumours in rats following feeding with Senecio moorei for a long period. Bull Anim Health Prod Afr 23:249-263.

Mugera GM (1970) Toxic and medicinal plants of East Africa. II. Bull Epizootic Dis Afr 18(4):389-403.

SENECIO MYRIOCEPHALUS Sch. Bip. ex A. Rich. [Asteraceae]

Citations:

Schoental R, Coady A (1968) The hepatotoxicity of some Ethiopian and East African plants, including some used in traditional medicines. East Afr Med J 45(8):577-580.

SENECIO OXYPHYLLUS DC. [Asteraceae]

Citations:

Barros CS, Driemeier D, Pilati C, et al. (1992) Senecio spp poisoning in cattle in Southern Brazil. Vet Hum Toxicol 34(3):241-246.

Driemeier D, Barros CS (1992) Intoxicação experimental por Senecio oxyphyllus (Compositae) em bovinos. Pesq Vet Bras 12(1-2):33-42.

Gava A, Barros CS (1997) Senecio spp. poisoning of horses in Southern Brazil. Pesq Vet Bras 17(1):36-40.

Karam FS, Soares MP, Haraguchi M, et al. (2004) Aspectos epidemiológicos da seneciose na região sul do Rio Grande do Sul. Pesq Vet Bras 24(4):191-198.

Schild AC, Motta AC, Riet-Correa F, et al. (2004) Photosensitization in cattle in southern Brazil. In: Acamovic T, Stewart CS, Pennycott TW (eds.) Poisonous plants and related toxins. CABI. Wallingford, UK. pp. 162-166.

Senecio plattensis Nutt. = Packera plattensis (Nutt.) W. A. Weber & Á. Löve

SENECIO QUADRIDENTATUS Labill. [Asteraceae]

Common Names:

cotton fireweed

Citations:

Dickson J, Hill R (1977) Cotton fireweed - Potential poison. J Agric West Aust 18(3):109-110.

Kater JC (1965) Cotton fireweed (Senecio quadridentata) poisoning in cattle. New South Wales Vet Inspector 29:45-47.

SENECIO RETRORSUS DC. [Asteraceae]

Common Names:

Dan's-cabbage; Molten disease plant; ragwort

Citations:

Van der Walt SJ, Steyn DG (1939) Recent investigations into the toxicity of known and unknown poisonous plants in the Union of South Africa. IX. Onderstepoort J Vet Sci Anim Indus 12:335-366.

SENECIO RIDDELLII Torr. & A. Gray [Asteraceae]

Common Names:

broom groundsel; groundsel; Riddell's-groundsel; sand groundsel; threadleaf groundsel; woolly groundsel

Citations:

Carpenter PT (1938) Acute senecio poisoning. U S Army Vet Bull 32:32-41.

Hill KR, Martin HM (1958) Hepatic veno-occlusive disease and megalocytosis in senecio poisoning in horses. Further investigations of "the walking disease" of northwestern Nebraska. Br Vet J 114:345-350.

Johnson AE, Molyneux RJ, Stuart LD (1985) Toxicity of Riddell's groundsel (Senecio riddellii) to cattle. Am J Vet Res 46(3):577-582.

Mathews FP (1932) Miscellaneous poisonous plants. Texas Agric Exp Sta Annu Rep 45:11-12.

Mathews FP (1933) Poisoning of cattle by species of groundsel (Senecio longilobus Benth., and Senecio riddellii Torr. and Gray). Texas Agric Exp Sta Bull #481:5-20.

Mathews FP, Schmidt H (1934) Miscellaneous poisonous plants. Texas Agric Exp Sta Annu Rep 47:12-13.

Mathews FP, Schmidt H (1935) Miscellaneous poisonous plants. Texas Agric Exp Sta Annu Rep 48:16-17.

Molyneux RJ, Johnson AE, Stuart LD (1988) Delayed manifestation of Senecio-induced pyrrolizidine alkaloidosis in cattle: Case reports. Vet Hum Toxicol 30(3):201-205.

Rosenfeld I, Beath OA (1947) Pharmacological action of Senecio riddellii. J Am Pharm Assoc 36(11):331-334.

Van Es L, Cantwell LR, Martin HM, et al. (1929) On the nature and cause of "the Walking Disease" of northwestern Nebraska. Nebraska Agric Exp Sta Res Bull #43:47 pp.

Vardiman PH (1952) Experimental feeding of senecio silage to calves. J Am Vet Med Assoc 121(Nov):397-400.

Vardiman PH (1953) The bromsulfalein liver function test and biopsy of the liver in the diagnosis of senecio poisoning in cattle. Am J Vet Res 14(51):175-178.

SENECIO RUWENZORIENSIS S. Moore [Asteraceae]

Common Names:

ragwort

Citations:

Thorold PW, Sapiro ML (1953) Preliminary note on the alkaloids and toxicity of Senecio ruwenzoriensis S. Moore. East Afr Agric J 19(Oct):103.

SENECIO SCELERATUS Schweick. [Asteraceae]

Common Names:

chi gurunga; ragwort; Rhodesian ragwort

Citations:

Shone DK (1966) Poisonous plants of Rhodesia. Part I. Rhodesia Zambia Malawi J Agric Res 4:81-94.

SENECIO SCHIMPERI Sch. Bip. ex A. Rich. [Asteraceae]

Citations:

Schoental R, Coady A (1968) The hepatotoxicity of some Ethiopian and East African plants, including some used in traditional medicines. East Afr Med J 45(8):577-580.

SENECIO SCHULTZII Hochst. ex A. Rich. [Asteraceae]

Citations:

Schoental R, Coady A (1968) The hepatotoxicity of some Ethiopian and East African plants, including some used in traditional medicines. East Afr Med J 45(8):577-580.

SENECIO SELLOI (Spreng.) DC. [Asteraceae]

Citations:

Barros CS, Metzdorf LL, Peixoto PV (1987) Ocorrência de surtos de intoxicação por Senecio spp. (Compositae) em bovinos no Rio Grande do Sul. Pesq Vet Bras 7(4):101-107.

Karam FS, Soares MP, Haraguchi M, et al. (2004) Aspectos epidemiológicos da seneciose na região sul do Rio Grande do Sul. Pesq Vet Bras 24(4):191-198.

Méndez MC, Riet-Correa F, Schild AL (1987) Intoxicação por Senecio spp. (Compositae) em bovinos no Rio Grande do Sul. Pesq Vet Bras 7(2):51-56.

Odriozola E, Campero C, Casaro A, et al. (1994) Pyrrolizidine alkaloidosis in Argentinian cattle caused by Senecio selloi. Vet Hum Toxicol 36(3):205-208.

Schild AC, Motta AC, Riet-Correa F, et al. (2004) Photosensitization in cattle in southern Brazil. In: Acamovic T, Stewart CS, Pennycott TW (eds.) Poisonous plants and related toxins. CABI. Wallingford, UK. pp. 162-166.

SENECIO SPARTIOIDES Torr. & A. Gray [Asteraceae]

Common Names:

broom groundsel; groundsel; narrow-leaf butterweed

Citations:

Anonymous (1933) Investigation of stock-poisoning plants. U S Dep Agric Bur Anim Indus Chief Rep 1933:34-35.

Anonymous (1937) Investigations of stock-poisoning plants. U S Dep Agric Bur Anim Indus Chief Rep. 1937:47-48.

Clawson AB (1933) The American groundsels species of Senecio as stock poisoning plants. Vet Med 28(Mar):105-110.

SENECIO SUBALPINUS Koch [Asteraceae]

Citations:

Smith RA, Panariti E (1995) Intoxication of Albanian cattle after ingestion of Senecio subalpinus. Vet Hum Toxicol 37(5):478-479.

SENECIO TWEEDIEI Hook. & Arn. [Asteraceae]

Citations:

Mendez MD, Riet-Correa F (1993) Intoxication by Senecio tweediei in cattle in Southern Brazil. Vet Hum Toxicol 35(1):55.

Schild AC, Motta AC, Riet-Correa F, et al. (2004) Photosensitization in cattle in southern Brazil. In: Acamovic T, Stewart CS, Pennycott TW (eds.) Poisonous plants and related toxins. CABI. Wallingford, UK. pp. 162-166.

SENECIO VERNALIS Waldst. & Kit. [Asteraceae]

Common Names:

Kreuzkraut

Citations:

Eroksuz H, Eroksuz Y, Ozer H, et al. (2003) Toxicity of Senecio vernalis to laying hens and evaluation of residues in eggs. Vet Hum Toxicol 45(2):76-80.

Hippchen C, Entzeroth R, Röder E, et al. (1986) Experimentelle Untersuchungen zur Lebertoxizität von Senecioalkaloiden aus Senecio vernalis an Ziegen. Prakt Tierarzt 67(4):322-324.

Hupka E (1955) Ist die Verfütterung von Senecio (Kreuzkraut) als Ursache der Leberverhurtung des Pferdes Anzusehen? Dtsch Tierarztl Wochenschr 62(1-2):1-3.

SENECIO VULGARIS L. [Asteraceae]

Common Names:

groundsel; Kreuzkraut; Kruiskruid; old-man-in-the-spring; Séneçon commun; simson

Citations:

Buckmaster GW, Cheeke PR, Shull LR (1976) Pyrrolizidine alkaloid poisoning in rats: Protective effects of dietary cysteine. J Anim Sci 43(2):464-473.

Fowler ME (1968) Pyrrolizidine alkaloid poisoning in calves. J Am Vet Med Assoc 152(8):1131-1137.

Goeger DE, Cheeke PR, Ramsdell HS, et al. (1983) Comparison of the toxicities of Senecio jacobaea, Senecio vulgaris and Senecio glabellus in rats. Toxicol Lett 15(1):19-23.

Gulick BA, Liu IK, Qualls CW Jr, et al. (1980) Effect of pyrrolizidine alkaloid-induced hepatic disease on plasma amino acid patterns in the horse. Am J Vet Res 41(11):1894-1898.

Lessard P, Wilson WD, Olander HJ, et al. (1986) Clinicopathologic study of horses surviving pyrrolizidine alkaloid (Senecio vulgaris) toxicosis. Am J Vet Res 47(8):1776-1780.

Mendel VE, Witt MR, Gitchell BS, et al. (1988) Pyrrolizidine alkaloid-induced liver disease in horses: An early diagnosis. Am J Vet Res 49(4):572-578.

Moyano MR, García A, Rueda A, et al. (2006) Echium vulgare and Senecio vulgaris poisoning in fighting bulls. J Vet Med A Physiol Pathol Clin Med 53(1):24-25.

Sedlmeier H, Dahme E, Schiefer B (1963) Veränderungen an der Rattenleber bei chronischer Seneciovergiftung im Vergleich zu den durch Buttergelb hervorgerufenen Lebergeschwülsten. Zentralbl Veterinarmed A 10(1):26-38.

séné-du-Sénégal –see– *Senna italica* Mill.

Séneçon commun –see– *Senecio vulgaris* L.

Senf –see– *Brassica nigra* (L.) W. D. J. Koch

senna –see– *Senna alexandrina* Mill.; *Senna obtusifolia* (L.) H. S. Irwin & Barneby; *Senna occidentalis* (L.) Link

SENNA ALEXANDRINA Mill. [Fabaceae]

Synonyms:

Cassia acutifolia Delile; ***Cassia senna*** L.

Common Names:

Alexandrian senna; Indian senna; Nubian senna; senna; senna mekka; te-de-sena; Tinnevelly senna

Citations:

El Sayed NY, Abdelbari EM, Mahmoud OM, et al. (1983) The toxicity of Cassia senna to Nubian goats. Vet Q 5(2):80-85.

Omer SA, Ibrahim FH, Khalid SA, et al. (1992) Toxicological interactions of Abrus precatorius and Cassia senna in the diet of Lohmann broiler chicks. Vet Hum Toxicol 34(4):310-313.

Rizzo M, Vanderperren B, Hantson P (2003) Acute liver and kidney injury after abuse of senna alkaloids. J Toxicol Clin Toxicol 41(4):514-515.

Smyth D, Gallo A, McGuigan M (2000) A prospective case series of pediatric senna poisoning. J Toxicol Clin Toxicol 38(5):519.

Spiller HA, Winter ML, Weber JA, et al. (2002) Skin breakdown and blisters from senna containing laxative ingestion in young children. J Toxicol Clin Toxicol 40(5):609.

senna bean –see– *Senna occidentalis* (L.) Link; *Sesbania drummondii* (Rydb.) Cory

SENNA DIDYMOBOTRYA (Fresen.) H. S. Irwin & Barneby [Fabaceae]

Synonyms:

Cassia didymobotrya Fresen.

Common Names:

wild senna

Citations:

Mugera GM (1970) Toxic and medicinal plants of East Africa. II. Bull Epizootic Dis Afr 18(4):389-403.

SENNA ×FLORIBUNDA (Cav.) H. S. Irwin & Barneby [Fabaceae]

Synonyms:

Cassia floribunda Cav.

Common Names:

arsenic bush; smooth senna

Citations:

Mugera GM (1970) Toxic and medicinal plants of East Africa. II. Bull Epizootic Dis Afr 18(4):389-403.

SENNA ITALICA Mill. [Fabaceae]

Synonyms:

Cassia italica (Mill.) Spreng.

Common Names:

séné-du-Sénégal; wild senna

Citations:

Galal M, Adam SE, Maglad MA, et al. (1985) The effects of Cassia italica on goats and sheep. Acta Vet (Beogr) 35(3):163-174.

SENNA LINDHEIMERIANA (Scheele) H. S. Irwin & Barneby [Fabaceae]

Synonyms:

Cassia lindheimeriana Scheele

Common Names:

Lindheimer's-senna

Citations:

Boughton IB, Hardy WT (1939) Hard yellow livers of sheep and cattle and feeding trials of suspected plants. Texas Agric Exp Sta Annu Rep 52:238-240.

senna mekka –see– *Senna alexandrina* Mill.

SENNA OBTUSIFOLIA (L.) H. S. Irwin & Barneby [Fabaceae]

Synonyms:

Cassia obtusifolia L.

Common Names:

coffee pod; coffee weed; senna; sicklepod senna

Citations:

Charles OW (1976) Coffee weed toxicity in animals. Proc Georgia Conf Feed Indust 1976:67-76.

Charles OW, Muller HD (1975) Coffee weed (Cassia obtusifolia) toxicity response of laying hens. Poult Sci 54:1745.

Crawford L, Friedman, M (1990) The effects of low levels of dietary toxic weed seeds (jimson weed, Datura stramonium and sicklepod, Cassia obtusifolia) on the relative size of rat liver and levels of cytochrome P-450. Toxicol Lett 54(2-3):175-181.

Dugan GM, Gumbmann MR (1990) Toxicological evaluation of sicklepod and black nightshade seeds in short-term feeding studies in rats. Food Chem Toxicol 28(2):101-107.

Flory W, Spainhour GB Jr, Colvin B, et al. (1992) The toxicologic investigation of a feed grain contaminated with seeds of the plant species Cassia. J Vet Diagn Invest 4(1):65-69.

Flunker LK, Damron BL, Sundlof SF, et al. (1989) Response of White Leghorn hens to various dietary levels of Cassia obtusifolia and nutrient fortification as a means of alleviating depressed performance. Poult Sci 68(7):909-913.

Hebert CD, Flory W (1983) Determination of the oral toxicity of Cassia obtusifolia seeds in chickens. Vet Hum Toxicol 25(3):164-166.

Henson JB, Dollahite JW, Bridges CH, et al. (1965) Myodegeneration in cattle grazing Cassia species. J Am Vet Med Assoc 147(2):142-145.

McCormack JE, Neisler WE (1980) Cassia obtusifolia (sicklepod) toxicity in a dairy herd. Vet Med Small Anim Clin 75(12):1849-1851.

Nicholson SS, Thornton JT, Rimes AJ Jr (1977) Toxic myopathy in dairy cattle caused by Cassia obtusifolia in greenchop. Bovine Pract 12(Nov):120.

Page RK, Vezey S, Charles OW, et al. (1977) Effects on feed consumption and egg production of coffee bean seed (Cassia obtusifolia) fed to White Leghorn hens. Avian Dis 21(1):90-96.

Putnam MR, Boosinger T, Spano J, et al. (1988) Evaluation of Cassia obtusifolia (sicklepod) seed consumption in Holstein calves. Vet Hum Toxicol 30(4):316-318.

Reddy B, Charles OW (1977) The toxicity of coffee weed (Cassia obtusifolia) when fed to young chicks and laying hens. Poult Sci 56:1357.

Voss KA, Brennecke LH (1991) Toxicological and hematological effects of sicklepod (Cassia obtusifolia) seeds in Sprague-Dawley rats: A subchronic feeding study. Toxicon 29(11):1329-1336.

SENNA OCCIDENTALIS (L.) Link [Fabaceae]

Synonyms:

Cassia occidentalis L.

Common Names:

antbush; coffee senna; coffeebean; fedegoso; Florida coffee; mata pasto; rai 'dore; Ral'dore; Rere; senna; stinkingweed; stinkingwood; styptic weed; wild coffee

Citations:

Barros CS, Ilha MR, Bezerra PS Jr, et al. (1999) Intoxicação por Senna occidentalis (Leg. Caesalpinoideae) em bovinos em pastoreio. Pesq Vet Bras 19(2):68-70.

Barros CS, Pilati C, Andujar MB, et al. (1990) Intoxicação por Cassia occidentalis (Leg. Caes.) em bovinos. Pesq Vet Bras 10(3-4):47-58.

Barth AT, Kommers GD, Salles MS, et al. (1994) Coffee senna (Senna occidentalis) poisoning in cattle in Brazil. Vet Hum Toxicol 36(6):541-545.

Brocq-Rousseu, Bruère P (1925) Accidents mortels sur des chevaux, dus a la graine de Cassia accidentalis L. C R Seanc Soc Biol 92:555-557.

Calore EE (1998) Toxic peripheral neuropathy of chicks fed Senna occidentalis seeds. Ecotoxicol Environ Saf 39(1):27-30.

Calore EE, Calore NM, Weg R, et al. (1999) The lysosomal enzymes acid phosphatase and cathepsin D in rats intoxicated with Senna occidentalis seeds. J Submicrosc Cytol Pathol 31(2):259-264.

Calore EE, Cavaliere MJ, Haraguchi M, et al. (1997) Experimental mitochondrial myopathy induced by chronic intoxication by Senna occidentalis seeds. J Neurol Sci 146(1):1-6.

Cavaliere MJ, Górniak SL, Dagli M, et al. (1997) Mitochondrial myopathy in Senna occidentalis-seed-fed chicken. Ecotoxicol Environ Saf 37(2):181-185.

Colvin BM, Harrison LR, Sangster LT, et al. (1986) Cassia occidentalis toxicosis in growing pigs. J Am Vet Med Assoc 189(4):423-426.

Haraguchi M, Calore EE, Dagli ML, et al. (1998) Muscle atrophy induced in broiler chicks by parts of Senna occidentalis seeds. Vet Res Commun 22(4):265-271.

Haraguchi M, Raspantini PC, Dagli ML, et al. (2004) Effect of Senna occidentalis seed in laying hens: Evaluation of the toxicity and egg production. In: Acamovic T, Stewart CS, Pennycott TW (eds.) Poisonous plants and related toxins. CABI. Wallingford, UK. pp. 269-274.

Henson JB, Dollahite JW (1966) Toxic myodegeneration in calves produced by experimental Cassia occidentalis intoxication. Am J Vet Res 27(119):947-949.

Hueza IM, Latorre AO, Raspantini PC, et al. (2007) Effect of Senna occidentalis seeds on immunity in broiler chickens. J Vet Med A Physiol Pathol Clin Med 54(4):179-185.

Irigoyen LF, Graça DL, Barros CS (1991) Intoxicação experimental por Cassia (Leg. Caes.) em eqüinos. Pesq Vet Bras 11(1-2):35-44.

Marrero Faz E, Bulnes Goicochea C, Perez Ruano M (1998) Cassia occidentalis occidentalis toxicosis in heifers. Vet Hum Toxicol 40(5):307.

Martin BW, Terry MK, Bridges CH, et al. (1981) Toxicity of Cassia occidentalis in the horse. Vet Hum Toxicol 23(6):416-417.

Martins E, Martins VM, Riet-Correa F, et al. (1986) Intoxicação por Cassia occidentalis (Leguminoseae) em suínos. Pesq Vet Bras 6(2):35-38.

Mercer HD, Neal FC, Himes JA, et al. (1967) Cassia occidentalis toxicosis in cattle. J Am Vet Med Assoc 151(6):735-741.

Moussu R (1925) L'intoxication par les graines de Cassia occidentalis L. est due a une toxalbumine. C R Soc Biol 92:862-863.

Neal FC, Plummer CB (1967) Cassia occidentalis poisoning in cattle. Florida Agric Exp Sta Annu Rep pg. 207.

O'Hara PJ, Pierce KR (1974) A toxic cardiomyopathy caused by Cassia occidentalis. I. Morphologic studies in poisoned rabbits. Vet Pathol 11(2):97-109.

O'Hara PJ, Pierce KR (1974) A toxic cardiomyopathy caused by Cassia occidentalis. II. Biochemical studies in poisoned rabbits. Vet Pathol 11(2):110-124.

O'Hara PJ, Pierce KR, Read WK (1969) Degenerative myopathy associated with ingestion of Cassia occidentalis L.: Clinical and pathologic features of the experimentally induced disease. Am J Vet Res 30(12):2173-2180.

O'Hara PJ, Pierce KR, Read WK (1970) Effects of vitamin E and selenium on Cassia occidentalis intoxication in cattle. Am J Vet Res 31(12):2151-2156.

Rogers RJ, Gibson J, Reichmann KG (1979) The toxicity of Cassia occidentalis for cattle. Aust Vet J 55(9):408-412.

Schmitz DG, Denton JH (1977) Senna bean toxicity in cattle. Southwestern Vet 30(2):165-170.

Silva TC, Gorniak SL, Oloris SC, et al. (2003) Effects of Senna occidentalis on chick bursa of Fabricius. Avian Pathol 32(6):633-637.

Simpson CF, Damron BL, Harms RH (1971) Toxic myopathy of chicks fed Cassia occidentalis seeds. Avian Dis 15(2):284-290.

Suliman HB, Shommein AM (1986) Toxic effect of the roasted and unroasted beans of Cassia occidentalis in goats. Vet Hum Toxicol 28(1):6-11.

Suliman HB, Wasfi IA, Adam SE (1982) The toxicity of Cassia occidentalis to goats. Vet Hum Toxicol 24(5):326-330.

Tasaka AC, Calore EE, Cavalier MJ, et al. (1998) Experimental poisoning in rabbits fed with Senna occidentalis seeds. In: Garland T, Barr AC (eds.) Toxic plants and other natural toxicants. CABI. New York. pp. 527-530.

Tasaka AC, Sinhorini IL, Dagli ML, et al. (2004) Perinatal study of Senna occidentalis intoxication in rabbits. In: Acamovic T, Stewart CS, Pennycott TW (eds.) Poisonous plants and related toxins. CABI. Wallingford, UK. pp. 459-464.

Torres WL, Nakano M, Nobre D, et al. (1971) Intoxicação em aves ocasionada por Cassia occidentalis L. Biologico 37(8):204-208.

SENNA ROEMERIANA (Scheele) H. S. Irwin & Barneby [Fabaceae]

Synonyms:

Cassia roemeriana Scheele

Common Names:

twin-leaf senna; two-leaf senna

Citations:

Rowe LD, Corrier DE, Reagor JC, et al. (1985) Hepatotoxicosis and toxic myodegeneration in goats produced by Cassia roemeriana (twoleaf senna). Am Vet Med Assoc Annu Meet 1985:115.

Rowe LD, Corrier DE, Reagor JC, et al. (1987) Experimentally induced Cassia roemeriana poisoning in cattle and goats. Am J Vet Res 48(6):992-997.

SENNA TORA (L.) Roxb. [Fabaceae]

Synonyms:

Cassia tora L.

Common Names:

cassia; coffeebean; coffeeweed; low senna; sickle senna

Citations:

Grant G, More LJ, McKenzie NH, et al. (1995) Nutritional and haemagglutination properties of several tropical seeds. J Agric Sci 124:437-445.

Sood AK, Sharma M, Katoch BS (1990) Improvement in the nutritive value of Indian horse-chestnut kernel and cassia seeds by chemical treatments for feeding of chicks. Indian J Anim Sci 60(11):1364-1369.

sennie bean –see– *Sesbania drummondii* (Rydb.) Cory

sensitive fern –see– *Onoclea sensibilis* L.

September oak –see– *Quercus robur* L.

SEQUOIA SEMPERVIRENS (D. Don) Endl. [Cupressaceae]

Common Names:

California redwood; redwood

Citations:

Chan-Yeung M, Abboud R (1976) Occupational asthma due to California redwood (Sequoia sempervirens) dusts. Am Rev Respir Dis 114(7):1027-1031.
Cohen HI, Merigan TC, Kosek JC, et al. (1967) Sequoiosis: A granulomatous pneumonitis associated with redwood sawdust inhalation. Am J Med 43(5):785-794.
do Pico GA (1978) Asthma due to dust from redwood (Sequoia sempervirens). Chest 73(3):424-425.

sering –see– *Melia azedarach* L.

Seriphidium tridentatum (Nutt.) W. A. Weber = Artemisia tridentata Nutt.

serpentwood –see– *Rauvolfia serpentina* (L.) Benth. ex Kurz

sesame –see– *Sesamum indicum* L.

SESAMUM INDICUM L. [Pedaliaceae]

Synonyms:

Sesamum orientale L.

Common Names:

gingali; sesame; til

Citations:

Levy Y, Danon YL (2001) Allergy to sesame seed in infants. Allergy 56(2):193-194.
Neering H, Vitányi BE, Malten KE, et al. (1975) Allergens in sesame oil contact dermatitis. Acta Derm Venereol 55(1):31-34.
Torsney PJ (1964) Hypersensitivity to sesame seed. J Allergy 35(6):514-519.
van Dijk E, Neering H, Vitányi BE (1973) Contact hypersensitivity to sesame oil in patients with leg ulcers and eczema. Acta Dermatovener 53(2):133-135.

Sesamum orientale L. = Sesamum indicum L.

sesban –see– *Sesbania exaltata* (Raf.) Rydb. ex A. W. Hill; *Sesbania grandiflora* (L.) Pers.; *Sesbania sesban* (L.) Merr.; *Sesbania vesicaria* (Jacq.) Elliott

sesbane –see– *Sesbania punicea* (Cav.) Benth.

sesbania –see– *Sesbania exaltata* (Raf.) Rydb. ex A. W. Hill

SESBANIA CANNABINA (Retz.) Pers. [Fabaceae]

Common Names:

sesbania pea

Citations:

Willimas KC, Daniels LJ (1985) The tolerance of pigs to the contamination of grain with seeds of Sesbania cannabina. In: Seawright AA, Hegarty MP, James LF, et al. (eds.) Plant toxicology. Queensland Poisonous Plant Committee. Yeerongpilly, Queensland. pp. 157-164.

Sesbania cavanillesii S. Watson = Sesbania longifolia DC.

SESBANIA DRUMMONDII (Rydb.) Cory [Fabaceae]

Synonyms:

Daubentonia drumondii Rydb.

Common Names:

coffeebean; daubentonia; Drummond's-rattlebush; Drummond's-sesbane; false poinciana; poisonbean; rattlebox; rattlebush; sennabean; sennie bean

Citations:

Flory W, Hebert CD (1984) Determination of the oral toxicity of Sesbania drummondii seeds in chickens. Am J Vet Res 45(5):955-958.
Venugopalan CS, Flory W, Tucker TA, et al. (1984) Assessment of smooth muscle function in Sesbania drummondii toxicosis in Gallus domesticus. Am J Vet Res 45(4):764-768.

SESBANIA EXALTATA (Raf.) Rydb. ex A. W. Hill [Fabaceae]

Synonyms:

Sesbania macrocarpa Muhl., nom. nud.

Common Names:

bequilla; coffeeweed; hemp sesbania; sesban; sesbania

Citations:

Flunker LK, Damron BL, Wilson HR (1991) Research note: Feeding various levels of ground Sesbania macrocarpa Muhl. seed to Bobwhite quail. Poult Sci 70(3):658-660.

SESBANIA GRANDIFLORA (L.) Pers. [Fabaceae]

Common Names:

agathi; corkwood; parrot flower; scarlet wistaria tree; sesban; vegetable hummingbird

Citations:

Prasad DA, Reddy KJ, Reddy CS, et al. (1970) Agathi leaf meal (Sesbania grandiflora) in the rations of growing chickens. Indian Vet J 47(2):164-169.

SESBANIA LONGIFOLIA DC. [Fabaceae]

Synonyms:

Daubentonia longifolia DC.; ***Sesbania cavanillesii*** S. Watson

Common Names:

coffeebean

Citations:

Marsh CD (1920) A new sheep-poisoning plant of the southern states. U S Dep Agric Circ #82:4 pp.
Marsh CD, Clawson AB (1920) Daubentonia longifolia (coffee bean), a poisonous plant. J Agric Res 20(6):507-513.
Shealy AL, Thomas EF (1928) Daubentonia seed poisoning of poultry. Florida Agric Exp Sta Bull #196:337-342.

Sesbania macrocarpa Muhl., nom. nud. = Sesbania exaltata (Raf.) Rydb. ex A. W. Hill.

sesbania pea –see– *Sesbania cannabina* (Retz.) Pers.

SESBANIA PUNICEA (Cav.) Benth. [Fabaceae]

Synonyms:

Daubentonia punicea (Cav.) DC.

Common Names:

Brazil rattlebox; coffee tree; coffee weed; coffeebean; daubentonia; false poinciana; macaw tree; purple rattlebox; purple rattlebush; purple sesban; purple sesbane; purple sesbania; rattlebox; rattlebush; royal poinciana; sesbane; wild macaw plant

Citations:

Duncan WH, Piercy PL, Starling RJ (1955) Toxicological studies of southeastern plants. I. Leguminosae. Econ Bot 9(3):243-255.

Emmel MW (1943) Daubentonia punicea (Cav.) DC. poisoning in pigeons. J Am Vet Med Assoc 102(Apr):294-295.

Terblanche M, de Klerk WA, Smit JD, et al. (1966) A toxicological study of the plant Sesbania punicea Benth. J S Afr Vet Assoc 37(2):191-197.

SESBANIA SESBAN (L.) Merr. [Fabaceae]

Common Names:

sesban

Citations:

Shqueir AA, Brown DL, Klasing KC (1989) Canavanine content and toxicity of Sesbania leaf meal for growing chicks. Anim Feed Sci Technol 25(1-2):137-147.

Shqueir AA, Brown DL, Taylor SJ, et al. (1989) Effects of solvent extractions, heat treatments and added cholesterol on Sesbania sesban toxicity in growing chicks. Anim Feed Sci Technol 27(1-2):127-135.

SESBANIA VESICARIA (Jacq.) Elliott [Fabaceae]

Synonyms:

Glottidium vesicarium (Jacq.) R. M. Harper

Common Names:

bagpod; bagpod sesbane; bagpod sesbania; bladderpod; castlebean; coffeebean weed; coffeeweed; glottidium; mole bean; sesban

Citations:

Anonymous (1936) Investigations of stock-poisoning plants. U S Dep Agric Bur Anim Indus Chief Rep 1936:44-45.

Anonymous (1942) Investigations of stock-poisoning plants. U S Dep Agric Bur Anim Indus Chief Rep 1942:29.

Boughton IB, Hardy WT (1939) Hard yellow livers of sheep and cattle and feeding trials of suspected plants. Texas Agric Exp Sta Annu Rep 52:238-240.

Boughton IB, Hardy WT (1939) Toxicity of the coffee bean (Sesbania vesicaria) for sheep. J Am Vet Med Assoc 95(1030):239-241.

Duncan WH, Piercy PL, Starling RJ (1955) Toxicological studies of southeastern plants. I. Leguminosae. Econ Bot 9(3):243-255.

Emmel MW (1935) The toxicity of Glottidium vesicarium (Jacq.) Harper seeds for the fowl. J Am Vet Med Assoc 87:13-21.

Emmel MW (1944) The toxicity of Glottidium vesicarium (Jacq.) Harper for cattle. J Am Vet Med Assoc 104:222-223.

Featherly HI, Harnden EE, Dermer OC, et al. (1943) Glottidium vesicarium, a poisonous plant in the southwest. Vet Med 38(12):478-479.

McConnell WC (1945) Cattle losses caused by Glottidium vesicarium. North Am Vet. May:276-277.

Newberne JW (1953) Bladder pod poisoning in cattle. Auburn Vet 9(Spring):169-170.

Reagor JC, Jones LP (1980) Sesbania vesicaria poisoning. Southwestern Vet 33(2):112.

Specht TE, Clay BR, Ward BC (1973) Bagpod (Glottidium vesicarium) toxicity in southeastern Oklahoma cattle. Oklahoma Vet 25(2):13-15.

seso vegetal –see– *Blighia sapida* K. D. Koenig

SESSEA BRASILIENSIS Toledo [Solanaceae]

Common Names:

canela-de-veado; paroba d'água

Citations:

Andrade SO (1960) Estudos sôbre a toxicidade de Sessea brasiliensis Toledo. Arq Inst Biol (Sao Paulo) 27:191-196.

Andrade SO, Retz L, Linardi MG, et al. (1970) Estudos sôbre a atividade tóxica da Sessea brasiliensis (Toledo) (Solanaceae). Arq Inst Biol (Sao Paulo) 37(1):37.

Canella CF, Tokarnia CH, Döbereiner J (1968) Intoxicação por Sessea brasiliensis Toledo em bovinos. Pesq Agric Bras 3:333-340.

setaria –see– *Setaria sphacelata* (Schrum.) Stapf & C. E. Hubb.

SETARIA ITALICA (L.) P. Beauv. [Poaceae]

Synonyms:

Setaria viridis (L.) P. Beauv.

Common names:

green bristle grass

Citations:

Nakamura T (1989) Contact dermatitis from Setaria viridis Beauv. (green bristle grass) in Japanese children. Contact Dermatitis 20(7):156-157.

Setaria lutescens (Stuntz) F. T. Hubb. = Setaria pumila (Poir.) Roem. & Schult.

SETARIA PUMILA (Poir.) Roem. & Schult. [Poaceae]

Synonyms:

Setaria lutescens (Stuntz) F. T. Hubb.

Common Names:

bristly foxtail; foxtail millet; foxtailgrass; pigeongrass; yellow bristlegrass

Citations:

Bankowski RA, Wichmann RW, Stuart EE (1956) Stomatitis of cattle and horses due to yellow bristle grass (Setaria lutescens). J Am Vet Med Assoc 129:149-152.

SETARIA SPHACELATA (Schumach.) Stapf & C. E. Hubb. [Poaceae]

Common Names:

setaria

Citations:

Groenendyk S, Seawright AA (1974) Osteodystrophia fibrosa in horses grazing Setaria sphacelata. Aust Vet J 50(3):131-132.

Jones RJ, Seawright AA, Little DA (1970) Oxalate poisoning in animals grazing the tropical grass Setaria sphacelata. J Aust Inst Agric Sci 36:41-43.

Seawright AA, Groenendyk S, Silva KI (1970) An outbreak of oxalate poisoning in cattle grazing Setaria sphacelata. Aust Vet J 46(7):293-296.

Setaria viridis (L.) P. Beauv. = Setaria italica (L.) P. Beauv.

setterwort –see– *Helleborus foetidus* L.

seven-years'-bean –see– *Phaseolus coccineus* L.

sevenbark –see– *Hydrangea macrophylla* (Thunb.) Ser.

Seville orange –see– *Citrus aurantium* L.

shack shack –see– *Crotalaria retusa* L.

shadscale –see– *Atriplex confertifolia* (Torr. & Frem.) S. Watson

shallot –see– *Allium cepa* L. var. aggregatum G. Don

shammah –see– *Nicotiana tabacum* L.

shamouti orange –see– *Citrus sinensis* (L.) Osbeck

sharp dock –see– *Rumex acetosa* L.

Shasta daisy –see– *Leucanthemum* ×*superbum* (Bergmans ex J. W. Ingram) D. H. Kent

shavegrass –see– *Equisetum arvense* L.

she oak –see– *Casuarina equisetifolia* L.

sheeh –see– *Artemisia herba-alba* Asso

sheep bur –see– *Xanthium strumarium* L. var. canadense (Mill.) Torr. & A. Gray

sheep laurel –see– *Kalmia angustifolia* L.; *Kalmia latifolia* L.

sheep-lice –see– *Cynoglossum officinale* L.

sheep poison –see– *Kalmia angustifolia* L.

sheep poison laurel –see– *Kalmia angustifolia* L.

sheep sorrel –see– *Rumex acetosella* L.

sheepkill –see– *Kalmia angustifolia* L.

sheep's-sorrel –see– *Rumex acetosa* L.

shepherd's-calender –see– *Anagallis arvensis* L.

shepherd's-clock –see– *Anagallis arvensis* L.

shepherd's-delight –see– *Anagallis arvensis* L.

shepherd's-purse –see– *Capsella bursa-pastoris* (L.) Medik.

shepherd's-weatherglass –see– *Anagallis arvensis* L.

shin oak –see– *Quercus havardii* Rydb.; *Quercus sinuata* Walter var. breviloba (Torr.) C. H. Müll.

shinnery oak –see– *Quercus havardii* Rydb.

shipmast locust –see– *Robinia pseudoacacia* L.

shoe flower –see– *Hibiscus rosa-sinensis* L.

shoofly plant –see– *Nicandra physalodes* (L.) Gaertn.

shore apple –see– *Hippomane mancinella* L.

shore buttercup –see– *Ranunculus cymbalaria* Pursh

shore pine –see– *Pinus contorta* Douglas ex Loudon

SHOREA ASSAMICA Dyer [Dipterocarpaceae]

Citations:

Oleffe JA, Sporcq J, Hublet P (1975) Epidemiological study of the wood industry in Belgium. Contact Dermatitis 1(5):315-316.

SHOREA ROBUSTA C. F. Gaertn. [Dipterocarpaceae]

Common Names:

sal

Citations:

Garg AK, Agrawal DK, Nath K (1984) Effect of sal (Shorea robusta) seed meal tannins on serum enzymes, nutrient utilization and growth in growing calves. Agric Wastes 11(4):307-317.

Mohapatra HC, Panda NC (1978) Effect of feeding sal (Shorea robusta) seed and sal oil meal to chicks. Indian Vet J 55(Jul):559-566.

short-crown milkweed –see– *Asclepias brachystephana* Engelm.

short ragweed –see– *Ambrosia artemisiifolia* L.; *Ambrosia eliator* L.

shoti –see– *Curcuma zedoaria* (Christm.) Roscoe

shou-wu-pian –see– *Fallopia multiflora* (Thunb.) Haraldson

shouwu –see– *Fallopia multiflora* (Thunb.) Haraldson

showy crotalaria –see– *Crotalaria spectabilis* Roth

showy false hellebore –see– *Veratrum californicum* Durand

showy lady's-slipper –see– *Cypripedium reginae* Walter

showy milkweed –see– *Asclepias speciosa* Torr.

showy mocassin flower –see– *Cypripedium spectabile* Salisb.

shrub chokecherry –see– *Prunus virginiana* L. var. melanocarpa (A. Nelson) Sarg.

shrub verbena –see– *Lantana camara* L.

shrubby morning-glory –see– *Ipomoea carnea* Jacq.; *Ipomoea carnea* Jacq. subsp. fistulosa (C. Mart. ex Choisy) D. F. Austin

shu-mang-tsao –see– *Illicium anisatum* L.

Siam teak –see– *Tectona grandis* L. f.

Siberian ginseng –see– *Eleutherococcus senticosus* (Rupr. & Maxim.) Maxim.

Siberian snow rose –see– *Rhododendron aureum* Georgi

sick weed –see– *Isocoma plurifolia* (Torr. & A. Gray) Greene

sickle senna –see– *Senna tora* (L.) Roxb.

sicklepod –see– *Senna obtusifolia* (L.) H. S. Irwin & Barneby; *Senna tora* (L.) Roxb.

sicklepod milk vetch –see– *Astragalus falcatus* Lam.

sicklepod senna –see– *Senna obtusifolia* (L.) H. S. Irwin & Barneby

sicopira –see– *Bowdichia nitida* Spruce ex Benth.

SIDA ACUTA Burm. f. [Malvaceae]

Synonyms:

Sida carpinifolia L. f.; ***Sida frutescens*** Cav.; ***Sida ulmifolia*** Mill.

Citations:

Driemeier D, Colodel EM, Gimeno EJ, et al. (2000) Lysosomal storage disease caused by Sida carpinifolia poisoning in goats. Vet Pathol 37(2):153-159.

Loretti AP, Colodel EM, Gimeno EJ, et al. (2003) Lysosomal storage disease in Sida carpinifolia toxicosis: An induced mannosidosis in horses. Equine Vet J 35(5):434-438.

Seitz AL, Colodel EM, Schmitz M, et al. (2005) Use of lectin histochemistry to diagnose Sida carpinifolia (Malvaceae) poisoning in sheep. Vet Rec 156(12):386-388.

Sida carpinifolia L. f. = Sida acuta Burm. f.

Sida frutescens Cav. = Sida acuta Burm. f.

Sida ulmifolia Mill. = Sida acuta Burm. f.

siddi –see– *Cannabis sativa* L.

Sierra laurel –see– *Leucothoe davisiae* Torr. ex A. Gray

Sieva bean –see– *Phaseolus lunatus* L.

signalgrass –see– *Urochloa brizantha* (Hochst. ex A. Rich.) R. D. Webster; *Urochloa decumbens* (Stapf) R. D. Webster

sila sila –see– *Derris trifoliata* Lour.

Silberregen –see– *Robinia pseudoacacia* L.

silk tree –see– *Albizia julibrissin* Durazz.

silkweed –see– *Asclepias curassavica* L.; *Asclepias syriaca* L.; *Asclepias tuberosa* L.

silky crazyweed –see– *Oxytropis sericea* Nutt.

silky dogwood –see– *Cornus oblonga* Wall.

silky lupine –see– *Lupinus argenteus* Pursh var. holosericeus (Nutt.) Barneby; *Lupinus sericeus* Pursh

silky oak –see– *Grevillea robusta* A. Cunn. ex R. Br.

silky oak tree –see– *Grevillea banksii* R. Br.

silky sophora –see– *Sophora nuttalliana* B. L. Turner

silver apricot –see– *Ginkgo biloba* L.

silver beet –see– *Beta vulgaris* L.

silver bill –see– *Halesia tetraptera* J. Ellis

silver lupine –see– *Lupinus argenteus* Pursh

silver oak –see– *Grevillea robusta* A. Cunn. ex R. Br.

silverberry –see– *Elaeagnus angustifolia* L.

silverleaf nightshade –see– *Solanum elaeagnifolium* Cav.

silverling –see– *Baccharis glomeruliflora* Pers.; *Baccharis halimifolia* L.

silvery bean –see– *Lupinus argenteus* Pursh

silvery horsebrush –see– *Tetradymia canescens* DC.

silvery lupine –see– *Lupinus argenteus* Pursh; *Lupinus sericeus* Pursh

SILYBUM MARIANUM (L.) Gaertn. [Asteraceae]

Synonyms:

Carduus marianus L.

Common Names:

blessed milk thistle; bull thistle; cardo asnal; cardo-de-Maria; cardo lechal; hedegar; holy thistle; lady's-thistle; marian thistle; milk thistle; our lady's-thistle; spotted thistle; St. Mary's-thistle; variegated thistle

Citations:

Kendrick JW, Tucker J, Peoples SA (1955) Nitrate poisoning in cattle due to ingestion of variegated thistle, Silybum marianum. J Am Vet Med Assoc 126(934):53-56.

Reynoso Castro HW, Selfero Audicio N (1963) Intoxicacion del ganado bovino por el cardo asnal (Uso del azul de metileno en concentración elevada). Gaceta Vet 25:429-434.

sim elfar –see– *Withania somnifera* (L.) Dunal

SIMABA CEDRON Planch. [Simaroubaceae]

Common Names:

cedron

Citations:

Ciganda C, Laborde A (2001) Herbal infusions used for induced abortion. J Toxicol Clin Toxicol 39(3):318-319.

simaltarul –see– *Manihot esculenta* Crantz

SIMAROUBA AMARA Aubl. [Simaroubaceae]

Common Names:

bitter damsoe; caixeta; marupá

Citations:

Declercq J (2004) Suspected wood poisoning caused by Simarouba amara (marupá/caixeta) shavings in two dogs with erosive stomatitis and dermatitis. Vet Dermatol 15(3):188-193.

Simmondsia californica Nutt. = Simmondsia chinensis (Link) C. K. Schneid.

SIMMONDSIA CHINENSIS (Link) C. K. Schneid. [Simmondsiaceae]

Synonyms:

Simmondsia californica Nutt.

Common Names:

jojoba; goat nut; pig nut

Citations:

Arnouts S, Buyse J, Cokelaere MM, et al. (1993) Jojoba meal (Simmondsia chinensis) in the diet of broiler breeder pullets: Physiological and endocrinological effects. Poult Sci 72(9):1714-1721.

Booth AN, Elliger CA, Waiss AC Jr (1974) Isolation of a toxic factor from jojoba meal. Life Sci 15(6):1115-1120.

Di Berardino L, Di Berardino F, Castelli A, et al. (2006) A case of contact dermatitis from jojoba. Contact Dermatitis 55(1):57-58.

Verbiscar AJ, Banigan TF, Weber CW, et al. (1980) Detoxification of jojoba meal. J Agric Food Chem 28(3):571-578.

Vermaut S, Onagbesan O, Bruggeman V, et al. (1998) Unidentified factors in jojoba meal prevent oviduct development in broiler breeder females. J Agric Fd Chem 46:194-201.

simple aster –see– *Symphyotrichum lanceolatum* (Willd.) G. L. Nesom

simpler's-joy –see– *Verbena officinalis* L.

Simpson money plant –see– *Scrophularia marilandica* L.

simson –see– *Senecio vulgaris* L.

sinalo todo –see– *Kalanchoe pinnata* (Lam.) Pers.

SINAPIS ALBA L. [Brassicaceae]

Synonyms:

Brassica alba (L.) Rabenh.

Common Names:

charlock; mostarda blanca; moutarde blanche; moutarde jaune; mustard; salad mustard; white mustard

Citations:

Eaton G (1941) A series of cases of poisoning in cattle. Vet Rec 53(10):145-146.

Eaton G (1941) Suspected poisoning of bullocks by white mustard. Vet Rec 53(10):146.

Hercus CE, Purves HD (1936) Studies on endemic and experimental goitre. J Hyg (Lond) 36:182-203.

Holmes RG (1965) A case of suspected poisoning of dairy cows by white mustard seeds (Sinapis alba). Vet Rec 77(17):480-481.

Kavli G, Moseng D (1987) Contact urticaria from mustard in fish-stick production. Contact Dermatitis 17(3):153-155.

Meding B (1985) Immediate hypersensitivity to mustard and rape. Contact Dermatitis 13(2):121-122.

Panconesi E, Sertoli A, Fabbri P, et al. (1980) Anaphylactic shock from mustard after ingestion of pizza. Contact Dermatitis 6(4):294-295.

Troxler J (1981) Intoxication moretelle de 19 génisses par la Moutarde jaune (Sinapis alba L.). Schweiz Arch Tierheilkd 123(9):495-497.

SINAPIS ARVENSIS L. [Brassicaceae]

Synonyms:

Brassica arvensis (L.) Rabenh.; ***Brassica kaber*** (DC.) L. C. Wheeler; ***Brassica sinapis*** Vis.

Common Names:

Ackersenf; charlock; corn mustard; crunchweed; field mustard; hedlock; mostaza campestre; mostaza silvestre; moutarde-des-champs; mustard; ravenelle; sanve; wild mustard

Citations:

Gallie JG, Paterson JD (1945) Charlock poisoning of lambs. Vet Rec 57(17):198.

Gwatkin R, Moynihan IW (1943) Wild mustard seed poisoning in cattle. Can J Comp Med Vet Sci 7(2):76-77.

Lanson RK, Abdulla A (1963) Effects of feeding mustard seed to immature chickens and laying hens. Poult Sci 42:1283-1284.

Makarian OA (1973) [Poisoning of cattle by charlock.] Veterinariia Moscow 50(11):95-96.

Shires A, Bell JM, Keith MO, et al. (1982) Rapeseed dockage: Effects of feeding raw and processed wild mustard and stinkweed seed on growth and feed utilization of mice. Can J Anim Sci 62:275-285.

Note:

Charlock is named *Sinapis arvensis* L. subsp. *arvensis* in some publications.

Sinapis nigra L. = Brassica nigra (L.) W. D. J. Koch

Singapore nut –see– *Aleurites moluccanus* (L.) Willd.

single-seed hawthorn –see– *Crataegus monogyna* Jacq.

singletary pea –see– *Lathyrus hirsutus* L.; *Lathyrus pusillus* Elliott

sink lupine –see– *Lupinus sericeus* Pursh

sinkle Bible –see– *Aloe vera* (L.) Burm. f.

SINOMENIUM ACUTUM (Thunb.) Rehder & E. H. Wilson [Menispermaceae]

Synonyms:

Menispermum acutum Thunb.

Common Names:

boi

Citations:

Okuda T, Umezawa Y, Ichikawa M, et al. (1995) A case of drug eruption caused by the crude drug boi (Sinomenium stem/Sinomeni caulis et rhizoma). J Dermatol 22(10):795-800.

SIPHONODON PENDULUM F. M. Bailey [Celastraceae]

Citations:

Flecker H (1945) Injuries produced by plants in tropical Queensland. Med J Aust 1(Jun 23):636-637.

siris –see– *Albizia lebbeck* (L.) Benth.

sirisha –see– *Albizia lebbeck* (L.) Benth.

sisal –see– *Agave sisalana* Perrine

Sisymbrium sophia L. = Descurainia sophia (L.) Webb ex Prantl

SISYRINCHIUM PLATENSE I. M. Johnst. [Iridaceae]

Citations:

Méndez MC, Delgago PE, Santos R, et al. (1993) Intoxicação experimental por Sisyrinchium platense (Iridaceae) em bovinos. Pesq Vet Bras 13(3-4):77-81.

sitab –see– *Ruta graveolens* L.

sitaphal –see– *Annona squamosa* L.

Sium cicutifolium Schrank = Sium suave Walter

SIUM SUAVE Walter [Apiaceae]

Synonyms:

Sium cicutifolium Schrank

Common Names:

hemlock water parsnip; water parsnip

Citations:

Hansen AA (1930) Indiana plants injurious to livestock. Purdue Agric Exp Sta Circ #175:38 pp.

Woolsey JH Jr, Shaffer MH (1952) A suspected plant poisoning in cattle. California Vet 5(3):21.

Sium thunbergii DC. = Berula erecta (Huds.) Coville

six-seed podgrass –see– *Triglochin maritima* L.

ska maria pastora –see– *Salvia divinorum* Epling & Jativa

skeleton weed –see– *Chondrilla juncea* L.

skewerwood –see– *Euonymus europaeus* L.

skin mango –see– *Mangifera indica* L.

skoke –see– *Phytolacca americana* L.

skuimbos –see– *Roepera foetida* (Schrad. & J. C. Wendl.) Beier & Thulin

skullcap –see– *Scutellaria lateriflora* L.

skunk bush –see– *Phacelia crenulata* Torr. ex S. Watson

skunk cabbage –see– *Veratrum californicum* Durand; *Veratrum viride* Aiton

skunkgrass –see– *Hordeum jubatum* L.

sky-blue lupine –see– *Lupinus diffusus* Nutt.

sky flower –see– *Duranta erecta* L.

slang nut –see– *Strychnos nux-vomica* L.

slangkop –see– *Drimia altissima* (L. f.) Ker Gawl.; *Drimia physodes* (Jacq.) Jessop; *Drimia sanguinea* (Schinz) Jessop

sleeping nightshade –see– *Atropa belladonna* L.

sleepygrass –see– *Achnatherum robustum* (Vasey) Barkworth

slender bundleflower –see– *Desmanthus leptalobus* Torr. & A. Gray

slender ice plant –see– *Mesembryanthemum nodiflorum* L.

slim larkspur –see– *Delphinium nuttallianum* Pritz.

slim nettle –see– *Urtica dioica* L.

slinkweed –see– *Gutierrezia microcephala* (DC.) A. Gray; *Gutierrezia sarothrae* (Pursh) Britton & Rusby

slipper flower –see– *Pedilanthus tithymaloides* (L.) Poit.

slipper plant –see– *Pedilanthus tithymaloides* (L.) Poit.

sloe –see– *Prunus spinosa* L.

small buckeye –see– *Aesculus pavia* L.

small burdock –see– *Xanthium strumarium* L.

small cardamon –see– *Elettaria cardamomum* (L.) Maton

small dodder –see– *Cuscuta epithymum* (L.) L.

small golden tip –see– *Goodia medicaginea* F. Muell.

small-head matchbush –see– *Gutierrezia microcephala* (DC.) A. Gray

small-head sneezeweed –see– *Helenium microcephalum* DC.

small hemlock –see– *Aethusa cynapium* L.

small jack-in-the-pulpit –see– *Arisaema triphyllum* (L.) Schott

small larkspur –see– *Delphinium menziesii* DC.

small laurel –see– *Kalmia angustifolia* L.

small-leaf clematis –see– *Clematis microphylla* DC.

small-leaf laurel –see– *Kalmia microphylla* (Hook.) A. Heller

small-leaf rubber plant –see– *Ficus benjamina* L.

small physicnut –see– *Jatropha multifida* L.

small poison bride bush –see– *Pavetta harborii* S. Moore

small-flower mallow –see– *Malva parviflora* L.

small-seed nolina –see– *Nolina microcarpa* S. Watson

smartweed –see– *Persicaria hydropiper* (L.) Spach; *Persicaria punctata* (Elliott) Small

SMODINGIUM ARGUTUM E. Mey. ex Sond. [Anacardiaceae]

Common Names:

African poison ivy; rainbow leaf; tovana; um-tovane

Citations:

Findlay GH (1963) Dermatitis of "poison ivy" type from an indigenous South African plant, Smodingium argutum E. Mey. (Anacardiaceae). S Afr Med J 37(Aug 31):883-888.

Heyl T, Gorst-Allman CP, Wells MJ, et al. (1987) Smodingium dermatitis. S Afr Med J 71(7):440-441.

Hindson C, Oliver R (1975) Eczema from Smodingium argutum shrub (Anacardiaceae). Contact Dermatitis 1(6):388-389.

smokewood –see– *Clematis vitalba* L.

smooth crotalaria –see– *Crotalaria pallida* Aiton

smooth Darling pea –see– *Swainsona galegifolia* (Andrews) R. Br.

smooth horsebrush –see– *Tetradymia glabrata* Torr. & A. Gray

smooth luffa –see– *Luffa acutangula* (L.) Roxb.

smooth pigweed –see– *Amaranthus hybridus* L.

smooth senna –see– *Senna ×floribunda* (Cav.) H. S. Irwin & Barneby smooth tobacco –see– *Nicotiana glauca* Graham

smooth witchgrass –see– *Panicum dichotomiflorum* Michx.

smoothwood –see– *Cupressus arizonica* Greene

smother weed –see– *Bassia hyssopifolia* (Pall.) Kuntze

snake nut –see– *Thevetia peruviana* (Pers.) K. Schum.

snake palm –see– *Amorphophallus konjac* K. Koch

snake plant –see– *Sansevieria trifasciata* Prain

snakebean –see– *Bobgunnia madagascariensis* (Desv.) J. H. Kirkbr. & Wiersema

snakeberry –see– *Actaea rubra* (Aiton) Willd.; *Solanum dulcamara* L.

snakegrass –see– *Equisetum arvense* L.

snakeroot –see– *Ageratina altissima* (L.) R. M. King & H. Rob.; *Cicuta douglasii* (DC.) J. M. Coult. & Rose; *Chamaesyce hirta* (L.) Millsp.; *Cicuta maculata* L.; *Cicuta maculata* L. var. angustifolia Hook.; *Cicuta virosa* L.; *Gutierrezia microcephala* (DC.) A. Gray; *Gutierrezia sarothrae* (Pursh) Britton & Rusby; *Rauvolfia serpentina* (L.) Benth. ex Kurz

sneeuwles –see– *Symphoricarpos albus* (L.) S. F. Blake

sneezeweed –see– *Helenium amarum* (Raf.) H. Rock; *Helenium autumnale* L.; *Helenium microcephalum* DC.; *Hymenoxys hoopesii* (A. Gray) Bierner

sneezewort –see– *Helenium autumnale* L.

snow-on-the-mountain –see– *Chamaesyce maculata* (L.) Small; *Euphorbia marginata* Pursh

snowberry –see– *Symphoricarpos albus* (L.) S. F. Blake

snowdrop –see– *Halesia tetraptera* J. Ellis; *Ornithogalum umbellatum* L.

snowdrop anemone –see– *Anemone sylvestris* L.

soap plant –see– *Toxicoscordion nuttallii* (A. Gray) Rydb.

soapbark tree –see– *Quillaja saponaria* Molina

soapwort –see– *Quillaja saponaria* Molina

soda apple –see– *Solanum aculeatissimum* Jacq.

soda bush –see– *Neobassia proceriflora* (F. Muell.) A. J. Scott

Sodom apple –see– *Solanum anguivi* Lam.; *Calotropis procera* (Aiton) W. T. Aiton

soft khaki weed –see– *Gomphrena celosioides* Mart.

soft lumbang –see– *Reutealis trisperma* (Blanco) Airy Shaw

soft maple –see– *Acer rubrum* L.

soft roly poly –see– *Salsola tragus* L.

soja –see– *Glycine max* (L.) Merr.

Sojabohne –see– *Glycine max* (L.) Merr.

sokujikoh –see– *Gardenia jasminoides* J. Ellis

SOLANECIO GIGAS (Vatke) C. Jeffrey [Asteraceae]

Synonyms:

Senecio gigas Vatke

Citations:

Schoental R, Coady A (1968) The hepatotoxicity of some Ethiopian and East African plants, including some used in traditional medicines. East Afr Med J 45(8):577-580.

SOLANUM ACULEATISSIMUM Jacq. [Solanaceae]

Common Names:

arrebenta boi; bull nettle; cockroachberry; devil's-apple; ground cherry; love apple; soda apple; wild tomato

Citations:

Tokarnia CH, Canella CF, Döbereiner J (1973) Intoxicação experimental em bovinos pelos frutos de Solanum aculeatissimum. Pesq Agric Bras Vet 8(6):35-39.

SOLANUM ANGUIVI Lam. [Solanaceae]

Synonyms:

Solanum sodomaeum L.

Common Names:

apple-of-Sodom; black-spine nightshade; Dead Sea apple; devil's-apple; popolo; Sodom apple

Citations:

Beardmore GL (1981) Devil apple extract and skin cancer. Med J Aust 2(2):204-205.

Ross E, Furumoto HH (1970) The effect of dried fruit of Solanum sodomaeum on Japanese quail and S. C. White Leghorn cockerel chicks. Poult Sci 49(1):13-15.

Ross E, Simpson CF, Rowland LO Jr, et al. (1971) Toxicity of Solanum sodomaeum and Solanum malacoxylon to chicks. Poult Sci 50(3):870-873.

Solanum bojeri Dunal = Solanum incanum L.

SOLANUM BONARIENSE L. [Solanaceae]

Synonyms:

Solanum fastigiatum Willd.

Common Names:

joá preto; naranjillo

Citations:

Paulovich FB, Portiansky EL, Gimeno EJ, et al. (2002) Lectin histochemical study of lipopigments present in the cerebellum of Solanum fastigiatum var. fastigiatum intoxicated cattle. J Vet Med A Physiol Pathol Clin Med 49(9):473-477.

Rech RR, Rissi DR, Rodriques A, et al. (2006) Intoxicação por Solanum fastigiatum (Solanaceae) em bovinos: Epidemiologia, sinais clínicos e morfometria das lesões cerebelares. Pesq Vet Bras 26(3):183-189.

Riet-Correa F, Mendez MC, Schild AL, et al. (1983) Intoxication by Solanum fastigiatum var. fastigiatum as a cause of cerebellar degeneration in cattle. Cornell Vet 73(3):240-256.

Verdes JM, Moraña A, Gutiérrez F, et al. (2006) Cerebellar degeneration in cattle grazing Solanum bonariense ("Naranjillo") in Western Uruguay. J Vet Diagn Invest 18(3):299-303.

SOLANUM CAROLINENSE L. [Solanaceae]

Common Names:

apple-of-Sodom; ball nettle; ball nightshade; bull nettle; Carolina horse nettle; Carolina nightshade; horse nettle; love apple; nightshade; sand briar; stickerweed; tread-softly; wild tomato

Citations:

Hansen AA (1927) Stock poisoning by plants in the nightshade family. J Am Vet Med Assoc 71:221-227.

Solanum chondropetalum Dammer = Solanum kwebense N. E. Br.

SOLANUM CINEREUM R. Br. [Solanaceae]

Common Names:

Narrawa bur; wild tomato

Citations:

Bourke CA (1997) Cerebellar degeneration in goats grazing Solanum cinereum (Narrawa burr). Aust Vet J 75(5):363-365.

Dodd S (1923) Poisoning of sheep by "narrawa burr." Agric Gaz New South Wales 34:257-260.

Dodd S (1923) Poisoning of sheep by Solanum cinereum. Vet J 79:56-59.

SOLANUM DASYPHYLLUM Schumach. & Thonn. [Solanaceae]

Synonyms:

Solanum macrocarpon L. subsp. dasyphyllum

Common Names:

African eggplant

Citations:

Bizinebyera ES (2003) Acute poisoning of Friesian heifers by Solanum macrocarpon L. ssp dasyphyllum. Vet Hum Toxicol 45(4):222-223.

SOLANUM DIMIDIATUM Raf. [Solanaceae]

Synonyms:

Solanum torreyi A. Gray

Common Names:

horse nettle; potato weed; western horse nettle

Citations:

Boughton IB, Hardy WT (1939) Hard yellow livers of sheep and cattle and feeding trials of suspected plants. Texas Agric Exp Sta Annu Rep 52:238-240.

Menzies JS, Bridges CH, Bailey EM (1978) Neurological disease of cattle associated with ingestion of Solanum dimidiatum. Vet Hum Toxicol 20:195.

Menzies JS, Bridges CH, Bailey EM Jr (1979) A neurological disease of cattle associated with Solanum dimidiatum. Southwestern Vet 32(1):45-49.

SOLANUM DUBIUM Dunal [Solanaceae]

Citations:

Barri ME, Onsa TO, Elawad AA, et al. (1983) Toxicity of five Sudanese plants to young ruminants. J Comp Pathol 93(4):559-575.

SOLANUM DULCAMARA L. [Solanaceae]

Common Names:

bitter nightshade; Bittersüß; Bittersüßer Nachtschatten; bittersweet; bittersweet nightshade; blue bindweed; blue nightshade; climbing nightshade; deadly nightshade; douce amère; European bittersweet; European nightshade; felonwood; felonwort mortal; fever twig; marriage vine; morella-douce-amère; nightshade; poison flower; poison nightshade; poisonberry; scarletberry; snakeberry; staff vine; tetonwort; vigne-de-Judée; violet bloom; wild nightshade; wolf grape; woody nightshade

Citations:

Alexander RF, Forbes GB, Hawkins ES (1948) A fatal case of solanine poisoning. Br Med J 2(Sep 11):518.

Anonymous (1948) Nightshade poisoning. Lancet 2(Sep 11):438.

Baker DC, Keeler RF, Gaffield W (1989) Pathology in hamsters administered Solanum plant species that contain steroidal alkaloids. Toxicon 27(12):1331-1337.

Borys DJ, Herrick ME, Krenzelok EP, et al. (1992) The clinical effects of Solanum dulcamara (deadly nightshade) in pediatric patients. Vet Hum Toxicol 34(4):351.

Ceha LJ, Presperin C, Young E, et al. (1994) Anticholinergic crisis from nightshade berry poisoning responsive to physostigmine. Vet Hum Toxicol 36(4):359.

Ceha LJ, Presperin C, Young E, et al. (1997) Anticholinergic toxicity from nightshade berry poisoning responsive to physostigmine. J Emerg Med 15(1):65-69.

Greer FG (1947) Poisoning in the horse by woody nightshade. Vet Rec 59(45):626.

Hornfeldt CS (1989) Determination of the toxicity of nightshade berries, Solanum dulcamara. Vet Hum Toxicol 31(4):363.

Hornfeldt CS, Collins JE (1990) Toxicity of nightshade berries (Solanum dulcamara) in mice. J Toxicol Clin Toxicol 28(2):185-192.

Manoguerra AS, Ruiz E (1976) Physostigmine treatment of anticholinergic poisoning. J Am Coll Emerg Physicians 5(2):125-127.

Polster H (1953) Zwei Fälle von Nachtschattenvergiftung. Kinderaerztl Prax 21:208-211.

Rubinfeld RS, Currie JN (1987) Accidental mydriasis from blue nightshade "lipstick." J Clin Neuroophthalmol 7(1):34-37.

Swinscow D (1953) Accidental poisoning of young children. Arch Dis Child 28(137):26-29.

Yates G (1915) Poisoning by woody nightshade. Vet Rec 28(1432):269-270.

SOLANUM ELAEAGNIFOLIUM Cav. [Solanaceae]

Common Names:

bull nettle; silverleaf nightshade; tomatillo; tomato weed; trompillo; tropillo; western horse nettle; white horse nettle

Citations:

Baker DC, Keeler RF, Gaffield W (1989) Pathology in hamsters administered Solanum plant species that contain steroidal alkaloids. Toxicon 27(12):1331-1337.

Boughton IB, Hardy WT (1939) Hard yellow livers of sheep and cattle and feeding trials of suspected plants. Texas Agric Exp Sta Annu Rep 52:238-240.

Buck WB, Dollahite JW, Allen TJ (1960) Solanum elaeagnifolium, silver-leafed nightshade, poisoning in livestock. J Am Vet Med Assoc 137(6):348-351.

Dollahite JW, Allen TJ (1960) Silverleaf nightshade poisoning in livestock. Texas Agric Exp Sta Prog Rep #2146:7 pp.

Garland T, Bailey EM Jr, Reagor JC, et al. (1998) Probable interaction between Solanum elaeagnifolium and Ivermectin in horses. In: Garland T, Barr AC (eds.) Toxic plants and other natural toxicants. CABI. New York. pp. 423-427.

SOLANUM ESURIALE Lindl. [Solanaceae]

Common Names:

potato bush; potato weed; queen's-nightshade; quena; tomato bush; wild tomato

Citations:

Dunster PJ, McKenzie RA (1987) Does Solanum esuriale cause humpyback in sheep? Aust Vet J 64:119-120.

McMeniman NP (1976) Solanum esuriale, A possible cause of humpy back in sheep. Aust Vet J 52(9):432-433.

Solanum fastigiatum Willd. = Solanum bonariense L.

SOLANUM GLAUCOPHYLLUM Desf. [Solanaceae]

Synonyms:

Solanum malacoxylon Sendtn.

Common Names:

duras milk blanco; duraznillo blanco; little peach; nightshade; white persicary

Citations:

Basudde CD (1984) Polyuria and polydipsia in rats given the calcinogenic shrub Solanum malacoxylon. Kenya J Sci Tech Series B 5(1-2):89-96.

Basudde CD, Humphreys DJ (1975) The effect of the administration of Solanum malacoxylon on the chick. Res Vet Sci 18(3):330-331.

Bingley JB, Ruksan BE, Carrillo BJ (1976) Serum calcium fractions in sheep treated with Solanum malacoxylon. Res Vet Sci 21(1):121-122.

Camberos HR (1971) Arterioesclerosis calcificante en rumiantes: "Enteque seco." Gaceta Vet 33:120-128.

Camberos HR, Davis GK (1970) Accion de "Solanum malacoxylon" sobre balance mineral en ovinos. Gaceta Vet 32:466-474.

Camberos HR, Davis GK, Djafar MI, et al. (1970) Soft tissue calcification in guinea pigs fed the poisonous plant Solanum malacoxylon. Am J Vet Res 31(4):685-696.

Campero CM, Odriozola E (1990) A case of Solanum malacoxylon toxicity in pigs. Vet Hum Toxicol 32(3):238-239.

Carrillo BJ (1973) Efecto de la intoxicación de Solanum malacoxylon en el sistema óseo. Rev Invest Agric 4th Series 10(2):65-77.

Carrillo BJ (1973) Efecto de la intoxicación de Solanum malacoxylon en la morfología de las celulas parafoliculares de la triroides. Rev Invest Agric 4th Series 10(1):41-54.

Carrillo BJ, Tilley JM, Garcés NE, et al. (1971) Intoxicacion experimental de bovinos con "Solanum malacoxylon." Gaceta Vet 33:468-484.

Carrilo BJ, Ruksan B (1975) The effect of Solanum malacoxylon toxicity on bones. Proc World Vet Cong 2:1237-1241.

Collins WT, Capen CC, Döbereiner J, et al. (1977) Ultrastructural evaluation of parathyroid glands and thyroid C cells of cattle fed Solanum malacoxylon. Am J Pathol 87(3):603-614.

Dämmrich K, Döbereiner J, Done SH, et al. (1975) Skelettveränderungen nach Vergiftungen mit Solanum malacoxylon bei Rindern. Zentralbl Veterinarmed A 22(4):313-329.

De Barros S, Tabone E, Dos Santos M, et al. (1981) Histopathological and ultrastructural alterations in the aorta in experimental Solanum malacoxylon poisoning. Virchows Arch B Cell Pathol 35(2):169-175.

Döbereiner J, Dämmrich K (1974) Skelettveränderungen bei Rindern nach Vergiftungen mit Solanum malacoxylon Sendtner. Verh Dtsch Ges Pathol 58:323-326.

Döbereiner J, Dämmrich K (1975) Skelettveränderungen bei Rindern nach Vergiftungen mit Solanum malacoxylon Sendtner. Zentralbl Allg Pathol Pathol Anat 119(1-2):126.

Döbereiner J, Done SH, Beltran LE (1975) Experimental Solanum malacoxylon poisoning in calves. Br Vet J 131(2):175-185.

Döbereiner J, Tokarnia CH, Costa JB, et al. (1971) "Espichamento," intoxicação de bovinos por Solanum malacoxylon, no Pantanal de Mato Grosso. Pesq Agric Bras Vet 6(6):91-117.

Done SH, Döbereiner J, Tokarnia CH (1976) Systemic connective tissue calcification in cattle poisoned by Solanum malacoxylon: A histological study. Br Vet J 132(1):28-38.

Done SH, Tokarnia CH, Dämmrich K, et al. (1976) Solanum malacoxylon poisoning in pigs. Res Vet Sci 20(2):217-219.

Gaggino OP (1969) Desarrollo de la lesión arteriosclerótica incipiente en el enteque seco reproducido experimentalmente en ovejas. Rev Invest Agric 4th Series 6(3):31-40.

Górniak SL, Barbosa Ferreira M, Schumaher Henrique B, et al. (2004) Fetal effects of maternal ingestion of Solanum malacoxylon: Evaluation in rats. In: Acamovic T, Stewart CS, Pennycott TW (eds.) Poisonous plants and related toxins. CABI. Wallingford, UK. pp. 465-470.

Górniak SL, Dagli ML, Arruda NV, et al. (1998) Evaluation of the toxicity of Solanum malacoxylon during the perinatal period. In: Garland T, Barr AC (eds.) Toxic plants and other natural toxicants. CABI. New York. pp. 323-328.

Górniak SL, Dagli ML, Maiorka PC, et al. (1999) Evaluation in rabbits of the fetal effects of maternal ingestion of Solanum malacoxylon. Vet Res Commun 23(5):307-316.

Köhler H, Leibetseder J, Skalicky M, et al. (1977) Zur Kalzinose der Rinder in Österreich. 4. Vergleichende experimentelle Untersuchungen über den Ca- und P-Stoffwechsel des Meerschweinchens nach Hypervitaminose D und Intoxikation mit Solanum malacoxylon. Zentralbl Veterinarmed A 24(6):441-478.

Kunz W (1976) Über den Einfluss von Solanum malacoxylon (Sendtner) auf den Kalzium-, Phosphor- und Magnesium-Gehalt im Blutserum beim Rind nach parenteraler und oraler Applikation. Cong Int Maladies Betail 2(2):1009-1014.

Kunz W (1977) Über den Einfluß von Solanum malacoxylon (Sendtner) auf den Kalcium-, Phosphor- und Magnesium-Gehalt im Blutserum beim Rind nach parenteraler und oraler Application. Berl Munch Tierarztl Wochenschr 90(4):69-72.

Kunz W, Hänichen T (1981) Solanum malacoxylon: Untersuchungen über die Verträglichkeit der zur Prophylaxe der Hypokalzämischen Gebärlähmung empfohlenen Mengen. Berl Munch Tierarztl Wochenschr 94(21):421-424.

Mautalen CA (1972) Mechanism of action of Solanum malacoxylon upon calcium and phosphate metabolism in the rabbit. Endocrinology 90(2):563-567.

Morris KM, Jenkins SA, Simonite JP (1977) The effect on egg-shell thickness of the inclusion of the calcinogenic plant Solanum malacoxylon in the diet of laying hens. Vet Rec 101(25):502-504.

Norrdin RW, Barros CS, Queille ML, et al. (1979) Acute effects of Solanum malacoxylon on bone formation rates in growing rats. Calcif Tissue Int 28(3):239-243.

O'Donnell JM, Smith MW (1973) Vitamin D-like action of Solanum malacoxylon on calcium transport by rat intestine. Nature 244(415):357-358.

Ousavaplangchai L (1973) Vergleichende Untersuchungen zur Histopathologie der Intoxikation mit Vitamin D3 und Solanum malacoxylon beim Meerschweinchen. Wien Tierarztl Monatsschr 60(10):316.

Portiansky EL, Alonso CR, Costa EF, et al. (2002) Collangenous and elastic system fibres in the aorta of cattle poisoned by Solanum glaucophyllum. Vet Rec 150(Jan 12):42-45.

Ross E, Simpson CF, Rowland LO Jr, et al. (1971) Toxicity of Solanum sodomaeum and Solanum malacoxylon to chicks. Poult Sci 50(3):870-873.

Roux R, Davicco MJ, Carrillo BJ, et al. (1979) Solanum glaucophyllum in pregnant cows. Effect of colostrum mineral composition and plasma calcium and phosphorus levels in dams and newborn calves. Ann Biol Anim Biochim Biophys 19(1A):91-101.

Rucksan BE, Wells GA, Lewis G (1978) Solanum malacoxylon toxicity to pigs. Vet Rec 103(8):153-155.

Santos MN (1979) Osteopetrose induzida pelo Solanum malacoxylon (Sendtner) em coelhos em fase de crescimento. Arq Esc Vet Univ Fed Minas Gerais 31:498-500.

Santos MN, Nunes VA, Nunes IJ, et al. (1976) Solanum malacoxylon toxicity: Inhibition of bone resorption. Cornell Vet 66(4):566-589.

Tokarnia CH, Döbereiner J (1974) "Espichamento," intoxicação de bovinos por Solanum malacoxylon, no Pantanal de Mato Grosso. II. Estudos complementares. Pesq Agric Bras Vet 9(7):53-62.

Wase AW (1972) Effect of Solanum malacoxylon on serum calcium and phosphate in laboratory animals. Fed Proc 31:708.

Woodard JC, Berra G, Ruksan B, et al. (1993) Toxic effects of Solanum malacoxylon on sheep bone. Bone 14(5):787-797.

SOLANUM GUARANITICUM A. St.-Hil. [Solanaceae]

Citations:

Barros SS, Riet-Correa F, Andujar MB, et al. (1987) Solanum fastigiatum var. fastigiatum and Solanum sp. poisoning in cattle: Ultrastructural changes in the cerebellum. Pesq Vet Bras 7(1):1-5.

Zambrano M, Riet-Correa F, Schild AL, et al. (1985) Intoxicação por Solanum fastigiatum var. fastigiatum: Evolução e reversibilidade das lesões em bovinos, e susceptibilidade de ovinos, coelhos, cóbaios e ratos. Pesq Vet Bras 5(4):133-141.

SOLANUM INCANUM L. [Solanaceae]

Synonyms:

Solanum bojeri Dunal

Common Names:

afufa oyibo; bitter apple; gautan kura; usa

Citations:

Shone DK (1966) Poisonous plants of Rhodesia. Part I. Rhodesia Zambia Malawi J Agric Res 4:81-94.

Shone DK (1967) Poisonous plants of Rhodesia. Part II. Rhodesia Zambia Malawi J Agric Res 5:29-41.

SOLANUM KWEBENSE N. E. Br. [Solanaceae]

Synonyms:

Solanum chondropetalum Dammer

Common Names:

bitterappel; rooibessie

Citations:

Pienaar JG, Kellerman TS, Basson PA, et al. (1976) Maldronksiekte in cattle: A neuronopathy caused by Solanum kwebense N. E. Br. Onderstepoort J Vet Res 43(2):67-74.

Solanum macrocarpon L. subsp. dasyphyllum = Solanum dasyphyllum Schumach. & Thonn.

Solanum malacoxylon Sendtn. = Solanum glaucophyllum Desf.

SOLANUM MELONGENA L. [Solanaceae]

Common Names:

aubergine; berenghenas; brinjal; eggplant

Citations:

Baker DC, Keeler RF, Gaffield W (1989) Pathology in hamsters administered Solanum plant species that contain steroidal alkaloids. Toxicon 27(12):1331-1337.

Gil M, Hogendijk S, Hauser C (2002) Allergy to eggplant flower pollen. Allergy 57(7):652.

Kabashima K, Miyachi Y (2004) Contact dermatitis due to eggplant. Contact Dermatitis 50(2):101-102.

SOLANUM NIGRUM L. [Solanaceae]

Common Names:

black nightshade; blackberry nightshade; deadly nightshade; duscle; garden huckleberry; hierba mora; houndberry; huckleberry; makoi; morelle noire; nasgaal; nastergal; petty morel; poisonberry; Schwarzer Nachtshatten; stubbleberry; wonderberry; yerba mora

Citations:

Bonner WG (1938) Poisoning of cattle. Black (or garden) nightshade. N Z J Agr 57(2):99,101.

Carey JC (1955) Black nightshade poisoning in swine. North Am Vet 36(Jun):446.

Casselberry NH (1939) Nightshade poisoning of swine. Vet Med 34(Jul):444-445.

Davies RC (1972) Black nightshade poisoning in the dog. Vet Rec 90(2):50.

Dugan GM, Gumbmann MR (1990) Toxicological evaluation of sicklepod and black nightshade seeds in short-term feeding studies in rats. Food Chem Toxicol 28(2):101-107.

Gunning OV (1949) Poisoning in goats by black nightshade (Solanum nigrum). Br Vet J 105:473-474.

Hansen AA (1927) Stock poisoning by plants in the nightshade family. J Am Vet Med Assoc 71:221-227.

Hubbs JC (1947) Belladonna poisoning in pigs. Vet Med 42(Nov):428-429.

Northall FS, Dauncey EA, Butler JM (2003) An overview of plant and fungal poisonings in the UK, and some interesting cases. J Toxicol Clin Toxicol 41(4):518-519.

Polster H (1953) Zwei Fälle von Nachtschattenvergiftung. Kinderaerztl Prax 21:208-211.

Towers RP (1953) A case of poisoning by Solanum nigrum. Ir J Med Sci 6(326):77-80.

Solanum panduraeforme Diege ex Dun. = Solanum panduriforme E. Mey.

SOLANUM PANDURIFORME E. Mey. [Solanaceae]

Synonyms:

Solanum panduraeforme Diege ex Dun.

Common Names:

bitter apple

Citations:

Shone DK (1966) Poisonous plants of Rhodesia. Part I. Rhodesia Zambia Malawi J Agric Res 4:81-94.

Shone DK (1967) Poisonous plants of Rhodesia. Part II. Rhodesia Zambia Malawi J Agric Res 5:29-41.

SOLANUM PSEUDOCAPSICUM L. [Solanaceae]

Common Names:

apple-of-Sodom; Carolina love nettle; Christmas cherry; Christmas pepper; coral; Jerusalem berry; Jerusalem cherry; Jerusalemkirsche; Korallenstrauch; love apple; Madeira winter cherry; Natal cherry; nightshade; nipple fruit; pommier d'amour; star potato vine; Straußkirsche; winter cherry

Citations:

Anonymous (1980) Tabulations of 1977 case reports. Bull Natl Clgh Poison Control Cent 24(6):1-4.

Oehme FW (1978) The hazard of plant toxicities to the human population. In: Keeler RF, Van Kampen KR, James LF, et al. (eds.) Effects of poisonous plants on livestock. Academic Press. New York. pp. 67-80.

SOLANUM PTYCANTHUM Dunal [Solanaceae]

Common Names:

eastern black nightshade; morelle-noire-de-l'est

Citations:

Voss KA, Chamberlain WJ, Brennecke LH (1993) Subchronic toxicity study of eastern black nightshade (Solanum ptycanthum) berries in Sprague-Dawley rats. J Food Safety 13:91-97.

SOLANUM ROSTRATUM Dunal [Solanaceae]

Common Names:

beaked nightshade; buffalo bur; Colorado bur; duraznillo; Kansas thistle; mala mujer; prickly nightshade; sand bur; Texas thistle

Citations:

Simic WJ (1943) Solanine poisoning in swine. Vet Med 38(9):353-354.

SOLANUM SARRACHOIDES Sendtn. [Solanaceae]

Common Names:

hairy nightshade

Citations:

Baker DC, Keeler RF, Gaffield W (1989) Pathology in hamsters administered Solanum plant species that contain steroidal alkaloids. Toxicon 27(12):1331-1337.

Drew ML, Fowler ME (1991) Poisoning of black and white ruffed lemurs (Varecia variegata variegata) by hairy nightshade (Solanum sarrachoides). J Zoo Wildl Med 22(4):494-496.

SOLANUM SISYMBRIIFOLIUM Lam. [Solanaceae]

Common Names:

espina colorado

Citations:

Ciganda C, Laborde A (2001) Herbal infusions used for induced abortion. J Toxicol Clin Toxicol 39(3):318-319.

Ciganda C, Laborde A (2003) Herbal infusions used for induced abortion. J Toxicol Clin Toxicol 41(3):235-239.

Solanum sodomaeum L. = Solanum anguivi Lam.

SOLANUM STURTIANUM F. Muell. [Solanaceae]

Common Names:

Sturt's-nightshade; Thargomindah nightshade

Citations:

Seddon HR, Carne HR (1925) Poisoning of stock by Solanum sturtianum. Agric Gaz New South Wales 36(Mar 1):192-194.

Seddon HR, Carne HR (1925) Poisoning of stock by Solanum sturtianum. New South Wales Dep Agric Sci Bull 24:28-33.

Solanum torreyi A. Gray = Solanum dimidiatum Raf.

SOLANUM TORVUM Sw. [Solanaceae]

Common Names:

devil's-fig; susumberberry; turkeyberry; wild tomato

Citations:

Morris KM, Simonite JP, Pullen L, et al. (1979) Solanum torvum as a causative agent of enzootic calcinosis in Papua, New Guinea. Res Vet Sci 27(2):264-266.

Thompson M, Thornton M, Verjee Z (2003) Poisoning by susumber berries. J Toxicol Clin Toxicol 41(5):729.

SOLANUM TRIFLORUM Nutt. [Solanaceae]

Common Names:

cut-leaf nightshade; nightshade; spreading nightshade; three-flower nightshade; western nightshade; wild potato

Citations:

Pammel LH (1921) Three-flowered nightshade poisonous. Vet Med 16(2):46-47.

SOLANUM TUBEROSUM L. [Solanaceae]

Common Names:

camotillo; Erdapfel; Irish potato; Kartoffel; papa; potato; white potato

Citations:

Allen JR, Marlar RJ, Chesney CF, et al. (1977) Teratogenicity studies on late blighted potatoes in nonhuman primates (Macaca mulatta and Saguinus labiatus). Teratology 15(1):17-23.

Baker D, Keeler R, Gaffield W (1987) Lesions of potato sprout and extracted potato sprout alkaloid toxicity in Syrian hamsters. J Toxicol Clin Toxicol 25(3):199-208.

Baker DC, Keeler RF, Gaffield WP (1988) Mechanism of death in Syrian hamsters gavaged potato sprout material. Toxicol Pathol 16(3):333-339.

Blount WP (1928) Potato poisoning in the adult dog. Vet Rec 8(43):924-925.

Bolin FM (1962) Green potatoes can kill sheep. North Dakota Farm Res 22(7):15.

Carmichael AJ, Foulds IS, Tan CY (1989) Allergic contact dermatitis from potato flesh. Contact Dermatitis 20(11):64-65.

Gunning OV (1950) Suspected potato poisoning in a mare, with fatal termination. Br Vet J 106:32-33.

Hansen AA (1925) Two fatal cases of potato poisoning. Science 61(1578):340-341.

Hansen AA (1927) Stock poisoning by plants in the nightshade family. J Am Vet Med Assoc 71:221-227.

Hansen AA (1928) Potato poisoning. North Am Vet 9(7):31-34.

Harris FW, Cockburn T (1918) Alleged poisoning by potatoes. Am J Pharm 90(Oct):722-726.

Harris FW, Cockburn T (1918) Alleged poisoning by potatoes. Analyst 43:133-137.

Hjorth N, Roed-Petersen J (1976) Occupational protein contact dermatitis in food handlers. Contact Dermatitis 2(1):28-42.

Inman PM (1965) Dermatitis in a crisp factory. Acta Derm Venereol 45(4):295-296.

Keeler RF, Brown D, Douglas DR, et al. (1976) Teratogenicity of the solanum alkaloid solasodine and of 'Kennebec' potato sprouts in hamsters. Bull Environ Contam Toxicol 15(5):522-524.

Keeler RF, Young S, Brown D, et al. (1978) Congenital deformities produced in hamsters by potato sprouts. Teratology 17(3):327-334.

Kline BE, von Elbe H, Dahle NA, et al. (1961) Toxic effect of potato sprouts and of solanine fed to pregnant rats. Proc Soc Exp Biol Med 107(Aug-Sep):807-809.

Kotowski K (1967) Przypadki zachorowan bydła na tle nadmiernego żywienia ziemnjakami. Med Weter 23(12):736-737.

Larko O, Lindstedt G, Lundberg PA, et al. (1983) Biochemical and clinical studies in a case of contact urticaria to potato. Contact Dermatitis 9(2):108-114.

McMillan M, Thompson JC (1979) An outbreak of suspected solanine poisoning in schoolboys: Examination of criteria of solanine poisoning. Q J Med 48(190):227-243.

Meynadier J, Meynadier JM, Guilhou JJ (1982) L'urticaire de contact chez l'atopique. A propos de deux observations. Ann Dermatol Venereol 109(10):871-874.

Milligan JB (1941) Suspected potato poisoning in cattle. Vet Rec 53(35):512.

Misra SN, Nath CP, Rao SV (1979) Toxic effect of greenish potato peels in rabbits - A preliminary trial. Indian Vet J 56(May):441-442.

Nater JP, Zwartz JA (1967) Atopic allergic reactions due to raw potato. J Allergy 40(4):202-206.

Owen RA (1985) Potato poisoning in a horse. Vet Rec 117(10):246.

Pearson RS (1966) Potato sensitivity, an occupational allergy in housewives. Acta Allergol 21(1):507-514.

Peck SM, Clare HC (1945) Dermatitis from dehydration of potatoes. Arch Derm Syphilol 52:9-10.

Petersen UK (1949) Zur klinik der Solaninvergiftung. Arch Kinderheilkd 137:24-26.

Pful E (1899) Ueber eine Massenerkrankung durch Vergiftung mit stark solaninhaltigen Kartoffeln. Dtsch Med Wochenschr 25(46):753-754.

Phillips BJ, Hughes JA, Phillips JC, et al. (1996) A study of the toxic hazard that might be associated with the consumption of green potato tops. Food Chem Toxicol 34:439-448.

Poswillo DE, Sopher D, Mitchell S (1972) Experimental induction of foetal malformation with "blighted" potato: A preliminary report. Nature 239(5373):462-464.

Poswillo DE, Sopher D, Mitchell SJ, et al. (1973) Investigations into teratogenic potential of imperfect potatoes. Teratology 8(3):339-347.

Quirce S, Díez Gómez ML, Hinojosa M, et al. (1989) Housewives with raw potato-induced bronchial asthma. Allergy 44(8):532-536.
Rühl R (1951) Beitrag zur Pathologie und Toxikologie des Solanins. Arch Pharm 284:67-74.
SAC Veterinary Services (1998) Solanine poisoning causes deaths in Scottish cattle. Vet Rec 142:125.
Swinyard CA, Chaube S (1973) Are potatoes teratogenic for experimental animals? Teratology 8(3):349-358.
Terbrüggen A (1936) Tödliche Solanin-Vergiftung durch den Genuß von Kartoffelbeeren. Sammlung Vergiftungsfallen 7(A609):101-104.
Terbrüggen A (1936) Tödliche Solaninvergiftung. Beitr Pathol Anat 97:391-395.
Willimott SG (1933) An investigation of solanine poisoning. Analyst 58(689):431-439.
Wilson GS (1959) A small outbreak of solanine poisoning. Monthly Bull Ministry Health 18(Dec):207-210.

Note:

Potato is named *Solanum tuberosum* L. subsp. *tuberosum* in some publications.

SOLANUM VIARUM Dunal [Solanaceae]

Common Names:

tropical soda apple

Citations:

Porter MB, MacKay RJ, Uhl E, et al. (2003) Neurologic disease putatively associated with ingestion of Solanum viarum in goats. J Am Vet Med Assoc 223(4):501-504.

SOLANUM VIRGINIANUM L. [Solanaceae]

Synonyms:

Solanum xanthocarpum Schrad.

Common Names:

toothed nightshade

Citations:

Kulakkattolickal AT (1987) Piscicidal plants of Nepal: Preliminary toxicity screening using grass carp (Ctenopharyngodon idella) fingerlings. J Ethnopharmacol 21(1):1-9.

Solanum xanthocarpum Schrad. = Solanum virginianum L.

soldier's-cap –see– *Aconitum napellus* L.; *Dicentra cucullaria* (L.) Bernh.

SOLENOSTEMON SCUTELLARIOIDES (L.) Codd [Lamiaceae]

Synonyms:

Coleus blumei Benth.

Common Names:

coleus

Citations:

Dooms-Goossens A, Borghus A, Degreef H, et al. (1987) Airborne contact dermatitis to coleus. Contact Dermatitis 17(6):109-110.
Saihan EM, Harman RR (1978) Coleus sensitivity in a gardener. Contact Dermatitis 4(4):234-235.

SOLIDAGO SPECIOSA Nutt. [Asteraceae]

Common Names:

goldenrod

Citations:

Freer RS (1949) Toxic effects of a Solidago species on cattle. Virginia Acad Sci Proc 27:97.

SOLIDAGO SPECTABILIS (D. C. Eaton) A. Gray [Asteraceae]

Common Names:

Basin goldenrod; goldenrod; Nevada goldenrod; western goldenrod

Citations:

Fleming CE (1920) Poisonous range plants. Nevada Agric Exp Sta Annu Rep 1919:39-43.
Fleming CE, Miller MR, Vawter LR, et al. (1931) Poisonous range plants. Nevada Agric Exp Sta Annu Rep #1928:21-22.
Fleming CE, Miller MR, Vawter LR, et al. (1931) Poisonous range plants. Nevada Agric Exp Sta Annu Rep 1930:12-13.
Fleming CE, Miller MR, Vawter LR, et al. (1934) Poisonous plants. Nevada Agric Exp Sta Annu Rep 1933:10-13.
Lockett S (1917) Sheep poisoned by western goldenrod (Solidago spectabilis). J Am Vet Med Assoc 51:214-221.

Soliva pterosperma (Juss.) Less. = Soliva sessilis Ruiz & Pav.

SOLIVA SESSILIS Ruiz & Pav. [Asteraceae]

Synonyms:

Soliva pterosperma (Juss.) Less.

Common Names:

bindii; bindy-eye; jo jo; onehunga weed

Citations:

Commens C, McGeogh A, Bartlett B, et al. (1984) Bindii (Jo Jo) dermatitis (Soliva pterosperma [Compositae]). J Am Acad Dermatol 10(5 Part 1):768-773.

Solomon Island ivy –see– *Epipremnum pinnatum* (L.) Engl.

Solomon's-seal –see– *Polygonatum multiflorum* (L.) All.

solsoldong –see– *Euphorbia tirucalli* L.

Sommeradonisröschen –see– *Adonis aestivalis* L.

Sommereiche –see– *Quercus robur* L.

Sommerwicken –see– *Vicia sativa* L.

son-before-the-father –see– *Colchicum autumnale* L.

SONCHUS OLERACEUS L. [Asteraceae]

Citations:

Purchase IF, Tustin RC, van Rensburg SJ (1975) Biological testing of food grown in the Transkei. Food Cosmet Toxicol 13(6):639-647.

Sonnenblume –see– *Helianthus annuus* L.

Sonnenwolfsmilch –see– *Euphorbia helioscopia* L.

sonnette –see– *Crotalaria retusa* L.

sonwa millet –see– *Panicum miliaceum* L.

SOPHORA ALOPECUROIDES L. [Fabaceae]

Citations:

Anonymous (1940) Investigations of stock-poisoning plants. U S Dep Agric Bur Anim Indus Chief Rep 1940:62.

SOPHORA FLAVESCENS Aiton [Fabaceae]

Common Names:

light-yellow sophora; ling lang

Citations:

Deng JF, Lin TJ, Kao WF, et al. (1997) The difficulty in handling poisonings associated with Chinese traditional medicine: A poison control center experience for 1991-1993. Vet Hum Toxicol 39(2):106-114.

Tsai KJ, Lin TJ, Lin WL, et al. (1993) The potential neurotoxicity of bitter tea drinking - A case report. Vet Hum Toxicol 35(4):330.

SOPHORA MICROPHYLLA Aiton [Fabaceae]

Common Names:

kowhai; yellow kowhai

Citations:

Clinch PG, Palmer-Jones T, Forster IW (1972) Effect on honey bees of nectar from the yellow kowhai (Sophora microphylla Ait.). N Z J Agric Res 15:194-201.

SOPHORA NUTTALLIANA B. L. Turner [Fabaceae]

Common Names:

silky sophora; white locoweed; white sophora

Citations:

Burrows GE, Schwab RP, Stein LE, et al. (1998) Comparison of the reproductive effects of Baptisia australis, Iva annua and Sophora nuttalliana in rats. In: Garland T, Barr AC (eds.) Toxic plants and other natural toxicants. CABI. New York. pp. 297-302.

Sophora secundiflora (Ortega) Lag. ex DC. = Calia secundiflora (Ortega) Yakovlev

sore-eye flower –see– *Boophone disticha* (L. f.) Herb.

sorghum –see– *Sorghum bicolor* (L.) Moench

sorghum almum –see– *Sorghum* ×*almum* Parodi

SORGHUM ×ALMUM PARODI [Poaceae]

Common Names:

Columbusgrass; sorghum almum

Citations:

Knight PR (1968) Equine cystitis and ataxia associated with grazing of pastures dominated by sorghum species. Aust Vet J 44(5):257.

SORGHUM BICOLOR (L.) Moench [Poaceae]

Synonyms:

Andropogon sorghum (L.) Brot.; ***Holcus sorghum*** L.; ***Sorghum saccharatum*** (L.) Moench; ***Sorghum vulgare*** Pers.

Common Names:

African sorghum; broom millet; broomcorn sorghum; dhurra wheat; durra; fatarita; grain sorghum; great millet; Guinae corn; Indian millet; jowar; juar; kafferkoring; Kafir corn; millet; milo; sorghum; sorgo; sweet sorghum

Citations:

Adams LG, Dollahite JW, Romane WM, et al. (1969) Cystitis and ataxia associated with sorghum ingestion by horses. J Am Vet Med Assoc 155(3):518-524.

Allen JG, Creeper JH, Forshaw D, et al. (2004) Plant-associated diseases, either new or new to the state, encountered over the last decade (1991-2001) in Western Australia. In: Acamovic T, Stewart CS, Pennycott TW (eds.) Poisonous plants and related toxins. CABI. Wallingford, UK. pp. 540-547.

Armstrong WD, Featherston WR, Rogler JC (1974) Effects of bird resistant sorghum grain and various commercial tannins on chick performance. Poult Sci 53(6):2137-2142.

Awad FI (1960) A note on Sorghum vulgare (fatarita) poisoning in cattle. Sudanese J Vet Sci Anim Husb 1(1):33-34.

Belevady B, Gopalan C (1965) Production of black tongue in dogs by feeding diets containing jowar (Sorghum vulgare). Lancet 2(Dec 11):1220-1221.

Bradley GA, Metcalf HC, Reggiardo C, et al. (1995) Neuroaxonal degeneration in sheep grazing Sorghum pastures. J Vet Diagn Invest 7(2):229-236.

Chang SI, Fuller HL (1964) Effect of tannin content of grain sorghums on their feeding value for growing chicks. Poult Sci 43(1):30-36.

Connor JK, Hurwood IS, Burton HW, et al. (1969) Some nutritional aspects of feeding sorghum grain of high tannin content to growing chickens. Aust J Exp Agric Anim Husb 9(40):497-501.

Fuller HL, Potter DK, Brown AR (1966) The feeding value of grain sorghums in relation to their tannin content. Georgia Agric Exp Sta Bull 176:1-14.

Gaggino OP, Carrillo BJ (1964) Ataxia de bovinos en pastoreo de sorgo. Idia 197(May):28-30.

Hansen AA (1924) Prussic acid poisoning in livestock. A real problem. Better Crops 2(3):26-27, 38.

Hiltner RS (1900) The fatal effect of green sorghum. Nebraska Agric Exp Sta Bull #63:73-84.

Koliev MF, Fedyushin FE (1963) [Mass poisoning of swine with sorghum.] Veterinariia Moscow 40(10):45-46.

McClymont GL, Duncan DC (1952) Studies on nutrition of poultry. III. Toxicity of grain sorghum for chickens. Aust Vet J 28(Sep):229-233.

McKenzie RA, McMicking LI (1977) Ataxia and urinary incontinence in cattle grazing sorghum. Aust Vet J 53(10):496-497.

Morgan SE, Johnson B, Brewer B, et al. (1990) Sorghum cystitis ataxia syndrome in horses. Vet Hum Toxicol 32(6):582.

Pammel LH (1919) Poisoning from sorghum and Sudan grass. Am J Vet Med 14:30-31.

Peters AT, Slade HB, Avery S (1903) Poisoning of cattle by common sorghum and Kafir corn (Sorghum vulgare.). Nebraska Agric Exp Sta Bull #77(11):1-16.

SORGHUM ×DRUMMONDII (Steud.) Millsp. & Chase [Poaceae]

Synonyms:

Holcus sudanensis (Piper) L. H. Bailey; ***Sorghum sudanense*** (Piper) Stapf; ***Sorghum vulgare*** Pers. var. sudanense (Steud.) Hitchc.

Common Names:

sorghum; Sudan sorghum; Sudangrass; sudax

Citations:

Anonymous (1935) Investigation of stock-poisoning plants. U S Dep Agric Bur Anim Indus Chief Rep 1935:39-40.
Dodgen JP (1967) An acute case of sorghum cystitis in the equine. Southwestern Vet 20:156.
Howarth JA (1931) Sudan grass as a photosensitizing agent causing dermatitis in sheep. North Am Vet 12(1):29-33.
Pammel LH (1919) Poisoning from sorghum and Sudan grass. Am J Vet Med 14:30-31.
Pammel LH (1920) Will frosted Sudan grass produce poisoning? Am J Vet Med 15:27.
Prichard JT, Voss JL (1967) Fetal ankylosis in horses associated with hybrid Sudan pasture. J Am Vet Med Assoc 150(8):871-873.
Rogers CF, Boyd WL (1936) Sudan grass and other cyanophoric plants as animal intoxicants. J Am Vet Med Assoc 88:489-499.
Seaman JT, Smeal MG, Wright JC (1981) The possible association of a sorghum (Sorghum sudanese) hybrid as a cause of development defects in calves. Aust Vet J 57(7):351-352.

SORGHUM HALEPENSE (L.) Pers. [Poaceae]

Synonyms:

Andropogon arundinaceus Scop.; ***Andropogon halepensis*** (L.) Brot.; ***Holcus halepensis*** L.

Common Names:

addar; baru; Johnsongrass

Citations:

Makridis I (1971) [Poisoning of the cattle in Greece by Sorghum halepense.] Hellenike Kteniatrike 14:10-22.
Mathews FP (1932) Johnson grass (Sorghum halepense) poisoning. J Am Vet Med Assoc 81(Nov):663-666.
Richetti A, Richetti F (1977) Sulla potenziale tossicità del Sorghum exiguum nella alimentazione del bestiame. Acta Med Vet (Napoli) 23(1-2):119-122.
Richetti F (1967) Rilievi e considerazioni su due casi di avvelenemento in equini in seguito ad ingestione de Sorghum halepense. Acta Med Vet (Napoli) 13:229-238.
Richetti F (1969) Rilievi istologici sulla tossicita' spermitale da Sorghum halepense Pers. Boll Soc Ital Biol Sper 45(10):641-643.
Vasey G (1885) Johnson grass in Montana. U S Dep Agric Rep 1885:74-75.
Wiggins AM (1953) Johnsongrass poisoning. Auburn Vet 9(Winter):77-79.

Sorghum saccharatum (L.) Moench = Sorghum bicolor (L.) Moench

Sorghum sudanense (Piper) Stapf = Sorghum ×drummondii (Steud.) Millsp. & Chase

Sorghum vulgare Pers = Sorghum bicolor (L.) Moench

Sorghum vulgare Pers. var. sudanense (Piper) Hitchc. = Sorghum ×drummondii (Steud.) Millsp. & Chase

sorgo –see– *Sorghum bicolor* (L.) Moench

soro soro –see– *Euphorbia tirucalli* L.

sorrel –see– *Oxalis pes-caprae* L.; *Rumex acetosa* L.; *Rumex acetosella* L.; *Rumex crispus* L.

sorrel dock –see– *Rumex acetosa* L.

sorrelgrass –see– *Rumex acetosa* L.

sorsaka –see– *Annona muricata* L.

sorte-de-noix-d'arec –see– *Areca catechu* L.

Sosnovsky's-cow parsnip –see– *Heracleum sosnowskyi* Manden.

sosueldo –see– *Euphorbia tirucalli* L.

sotetsu –see– *Cycas revoluta* Thunb.

soup clover –see– *Melilotus indicus* (L.) All.

sour clover –see– *Melilotus indicus* (L.) All.

sour dock –see– *Rumex acetosa* L.; *Rumex acetosella* L.; *Rumex crispus* L.

sour green –see– *Rumex venosus* Pursh

sour leek –see– *Rumex acetosa* L.

sour orange –see– *Citrus aurantium* L.

sour sop –see– *Annona muricata* L.

sour sorrel –see– *Rumex acetosa* L.

sour sweet clover –see– *Melilotus indicus* (L.) All.

sour weed –see– *Rumex acetosella* L.

sourgrass –see– *Rumex acetosa* L.; *Rumex acetosella* L.

soursob –see– *Oxalis pes-caprae* L.

South African daisy –see– *Verbesina encelioides* (Cav.) Benth. & Hook. f. ex A. Gray

South African milk bush –see– *Synadenium grantii* Hook. f.

South American air plant –see– *Kalanchoe fedtschenkoi* Raym.-Hamet & H. Perrier

South American tea –see– *Ilex paraguariensis* A. St.-Hil.

South Sea rose –see– *Nerium oleander* L.

southeastern bitterweed –see– *Helenium amarum* (Raf.) H. Rock

southern baccharis –see– *Baccharis glomeruliflora* Pers.

southern canarygrass –see– *Phalaris caroliniana* Walter

southern indigo –see– *Indigofera australis* Willd.

southern pea –see– *Vigna unguiculata* (L.) Walp.

southern poison oak –see– *Toxicodendron pubescens* Mill.

southern sandbur –see– *Cenchrus incertus* M. A. Curtis

sowbread –see– *Cyclamen persicum* Mill.

soy –see– *Glycine max* (L.) Merr.

soya bean –see– *Glycine max* (L.) Merr.

soybean –see– *Glycine max* (L.) Merr.

spadic bush –see– *Erythroxylum coca* Lam.

Spanish bayonet –see– *Yucca aloifolia* L.

Spanish daisy –see– *Helenium amarum* (Raf.) H. Rock

Spanish tea –see– *Dysphania ambrosioides* (L.) Mosyakin & Clemants

Spanish vetch –see– *Lathyrus clymenum* L.

Spanish vetchling –see– *Lathyrus clymenum* L.

spantou melkbos –see– *Sarcostemma viminale* (L.) R. Br.

Spargel –see– *Asparagus officinalis* L.

SPARRMANNIA AFRICANA L. f. [Malvaceae]

Common Names:

African hemp; indoor linden; Zimmerlinde

Citations:

Kuske H, Bandi B (1958) Dermatitis und Ekzem durch Zimmerlinde. Dermatologica 116:308-313.

Schulze W (1941) Untersuchungen über die Reizwirkung der Zimmerlindenblatter auf die Haut. Arch Derm Syphilol 181:63-77.

spartium –see– *Cytisus scoparius* (L.) Link

Späte Traubenkirsche –see– *Prunus serotina* Ehrh.

spathe flower –see– *Spathiphyllum wallisii* Regel

SPATHIPHYLLUM FLORIBUNDUM (Linden & André) N. E. Br. [Araceae]

Citations:

Cahen O, Floras P, Guerineau JM, et al. (1978) Intoxication par les fruits de Redoul. Cah Anesthesiol 26(5):693-698.

Mrvos R, Krenzelok EP, Jacobsen TD (2001) Toxidromes associated with the most common plant ingestions. Vet Hum Toxicol 43(6):366-369.

SPATHIPHYLLUM WALLISII Regel [Araceae]

Common Names:

spathe flower

Citations:

Kanerva L, Estlander T, Petman L, et al. (2001) Occupational allergic contact urticaria to yucca (Yucca aloifolia), weeping fig (Ficus benjamina), and spathe flower (Spathiphyllum wallisii). Allergy 56(10):1008-1011.

spearmint –see– *Mentha spicata* L.

speckled locoweed –see– *Astragalus lentiginosus* Douglas ex Hook.

speckled milk vetch –see– *Astragalus lentiginosus* Douglas ex Hook. var. diphysus (A. Gray) M. E. Jones

Speisebohnen –see– *Phaseolus vulgaris* L.

SPHAGNETICOLA TRILOBATA (L.) Pruski [Asteraceae]

Synonyms:

Wedelia trilobata (L.) Hitchc.

Common Names:

wedelia

Citations:

Goh CL (1986) Contact sensitivity to Wedelia trilobata. Contact Dermatitis 14(2):126.

SPHENOSCIADIUM CAPITELLATUM A. Gray [Apiaceae]

Common Names:

ranger's-buttons; swamp whiteheads; whiteheads; woolly-head parsnip

Citations:

Fowler ME, Berry LJ, Bushnell R, et al. (1970) Sphenosciadium capitellatum (whiteheads) toxicosis of cattle and horses. J Am Vet Med Assoc 157(9):1187-1192.

spice pepper –see– *Capsicum annuum* L.

spider antelopehorn –see– *Asclepias asperula* (Decne.) Woodson var. decumbens (Nutt.) Shinners

spider flower –see– *Cleome spinosa* Jacq.

spider milkweed –see– *Asclepias verticillata* L.; *Asclepias viridis* Walter

spidergrass –see– *Brachyachne convergens* (F. Muell.) Stapf

Spiesia lambertii (Pursh) Kuntze = Oxytropis lambertii Pursh

SPIGELIA ANTHELMIA L. [Loganiaceae]

Common Names:

ewe-aran; West Indian pink root

Citations:

Johnson SW (1963) The toxicity of Spigelia anthelmia for small laboratory animals. Trop Agric 40:165-167.

spiked millet –see– *Pennisetum glaucum* (L.) R. Br.

spiked riceflower –see– *Pimelea trichostachya* Lindl.

Spillbaum –see– *Euonymus europaeus* L.

spinach –see– *Spinacia oleracea* L.

SPINACIA OLERACEA L. [Chenopodiaceae]

Common Names:

espinafre; spinach; spinage; Spinat

Citations:

Sanchez I, Rodriguez F, Garcia-Abujeta JL, et al. (1997) Oral allergy syndrome induced by spinach. Allergy 52(12):1245-1246.

Sander C, Jacobi H (1967) Methämoglobinvergiftung nach Spinatgenuß bei einem zweijährigen Knaben. Z Kinderheilkd 98:222-226.

Schuller A, Morisset M, Maadi F, et al. (2005) Occupational asthma due to allergy to spinach powder in a pasta factory. Allergy 60(3):408-409.
Sinios A, Wodsak W (1965) Die Spinatvergiftung des Säuglings. Dtsch Med Wochenschr 90:1856-1863.
Zohn B (1937) An unusual case of spinach hypersensitiveness. J Allergy 8:381-384.

spinage –see– *Spinacia oleracea* L.

Spinat –see– *Spinacia oleracea* L.

Spindelbaum –see– *Euonymus europaeus* L.

spindle tree –see– *Euonymus europaeus* L.

spineless horsebrush –see– *Tetradymia canescens* DC.

spiny amaranth –see– *Amaranthus spinosus* L.

spiny clotbur –see– *Xanthium spinosum* L.

spiny cocklebur –see– *Xanthium spinosum* L.

spiny pigweed –see– *Amaranthus spinosus* L.

SPIROSTACHYS VENENIFERA (Pax) Pax
[Euphorbiaceae]

Citations:

Njiro SM, Mugera GM, Maitai CK (1984) Histopathology of tissues from albino Wistar rats poisoned with Spirostachys venenifera. Kenya Vet 8(1):19-21.

Spitzklette –see– *Xanthium strumarium* L.

Splintbaum –see– *Buxus sempervirens* L.

split-leaf philodendron –see– *Monstera deliciosa* Liebm.

split pea –see– *Pisum sativum* L.

Sponia virgata Planch. = Trema cannabinum Lour.

spoon hunt –see– *Kalmia latifolia* L.

spoon hutch –see– *Kalmia latifolia* L.

spoon lily –see– *Alocasia macrorrhizos* (L.) G. Don

spoonwood –see– *Kalmia angustifolia* L.; *Kalmia latifolia* L.

spoonwood ivy –see– *Kalmia angustifolia* L.

spotted arum –see– *Arum maculatum* L.

spotted cat's-ear –see– *Hypochaeris radicata* L.

spotted cowbane –see– *Cicuta douglasii* (DC.) J. M. Coult. & Rose; *Cicuta maculata* L.; *Cicuta maculata* L. var. angustifolia Hook.; *Conium maculatum* L.

spotted dumbcane –see– *Dieffenbachia seguine* (Jacq.) Schott

spotted emubush –see– *Eremophila maculata* (Ker Gawl.) F. Muell.

spotted fuchsia –see– *Eremophila maculata* (Ker Gawl.) F. Muell.

spotted hemlock –see– *Cicuta douglasii* (DC.) J. M. Coult. & Rose; *Cicuta maculata* L.; *Conium maculatum* L.

spotted laurel –see– *Aucuba japonica* Thunb.

spotted locoweed –see– *Astragalus lentiginosus* Douglas ex Hook.; *Astragalus lentiginosus* Douglas ex Hook. var. diphysus (A. Gray) M. E. Jones; *Astragalus lentiginosus* Douglas ex Hook. var. wahweapensis S. L. Welsh

spotted parsley –see– *Cicuta douglasii* (DC.) J. M. Coult. & Rose; *Cicuta maculata* L.; *Cicuta maculata* L. var. angustifolia Hook.; *Conium maculatum* L.

spotted smartweed –see– *Persicaria maculosa* Gray

spotted spurge –see– *Chamaesyce maculata* (L.) Small

spotted St. John's-wort –see– *Hypericum punctatum* Lam.

spotted thistle –see– *Silybum marianum* (L.) Gaertn.

spotted water hemlock –see– *Cicuta maculata* L.

sprängört –see– *Cicuta virosa* L.

sprawling cymopterus –see– *Cymopterus longipes* S. Watson

spreading nightshade –see– *Solanum triflorum* Nutt.

spreading pigweed –see– *Amaranthus blitoides* S. Watson

spring buckeye –see– *Aesculus glabra* Willd.

spring cockle –see– *Vaccaria hispanica* (Mill.) Rauschert

spring larkspur –see– *Delphinium menziesii* DC.

spring parsley –see– *Cymopterus ibapensis* M. E. Jones

spring rabbitbrush –see– *Tetradymia glabrata* Torr. & A. Gray

spring vetch –see– *Vicia sativa* L.

springbokbos –see– *Hertia pallens* (DC.) Kuntze

springbokbossie –see– *Hertia pallens* (DC.) Kuntze

Springkraut –see– *Euphorbia lathyris* L.

Springwolfsmilch –see– *Euphorbia lathyris* L.

springwort –see– *Euphorbia lathyris* L.

sprinkaanbos –see– *Senecio burchellii* DC.; *Senecio ilicifolius* L.

Spritzgurke –see– *Ecballium elaterium* (L.) A. Rich.

spur lupine –see– *Lupinus arbustus* Douglas ex Lindl.; *Lupinus argenteus* Pursh var. holosericeus (Nutt.) Barneby

spur pepper –see– *Capsicum frutescens* L.

spur vetch –see– *Vicia monantha* Retz.

spurge –see– *Chamaesyce hirta* (L.) Millsp.; *Euphorbia cyparissias* L.; *Euphorbia helioscopia* L.; *Euphorbia myrsinites* L.; *Phyllanthus abnormis* Baill.; *Reverchonia arenaria* A. Gray

spurge cactus –see– *Euphorbia antiquorum* L.

spurge flax –see– *Daphne mezereum* L.

spurge laurel –see– *Daphne mezereum* L.

spurge olive –see– *Daphne mezereum* L.

spurge tree –see– *Euphorbia tirucalli* L.

squash –see– *Cucurbita pepo* L.

squash melon –see– *Cucumis myriocarpus* Naudin

squaw mint –see– *Hedeoma pulegioides* (L.) Pers.; *Mentha pulegium* L.

squaw weed –see– *Ageratina altissima* (L.) R. M. King & H. Rob.; *Senecio plattensis* Nutt.

squawroot –see– *Actaea racemosa* L.; *Caulophyllum thalictroides* (L.) Michx.

squill –see– *Drimia maritima* (L.) Stearn; *Ledebouria cooperi* (Hook. f.) Jessop

squirrel corn –see– *Dicentra cucullaria* (L.) Bernh.

squirrel food –see– *Toxicoscordion venenosum* (S. Watson) Rydb.

squirreltail –see– *Hordeum jubatum* L.

squirreltail barley –see– *Hordeum jubatum* L.

squirreltailgrass –see– *Hordeum jubatum* L.

squirting cucumber –see– *Ecballium elaterium* (L.) A. Rich.

St. Anthony's-turnip –see– *Ranunculus bulbosus* L.

St. Barnaby's-thistle –see– *Centaurea solstitialis* L.

St. Bennet's-herb –see– *Conium maculatum* L.

St. Elmos'-feather –see– *Amianthium muscitoxicum* (Walter) A. Gray

St. George's-herb –see– *Valeriana officinalis* L.

St. Ignatius's-bean –see– *Strychnos ignatii* P. J. Bergius

St. John's-wort –see– *Hypericum perforatum* L.; *Hypericum punctatum* Lam.; *Hypericum revolutum* Vahl; *Hypericum triquetrifolium* Turra

St. Lucia's-grass –see– *Urochloa brizantha* (Hochst. ex A. Rich.) R. D. Webster

St. Mary's-thistle –see– *Silybum marianum* (L.) Gaertn.

St. Peter's-wort –see– *Symphoricarpos albus* (L.) S. F. Blake

St. Thomas'-bush –see– *Calotropis procera* (Aiton) W. T. Aiton

STACHYS ARVENSIS (L.) L. [Lamiaceae]

Common Names:

corn woundwort; dead nettle; field hedge nettle; field woundwort; mintweed; staggerweed; woundwort

Citations:

Philbey AW, Hawker AM, Evers JV (2001) A neurological locomotor disorder in sheep grazing Stachys arvensis. Aust Vet J 79(6):427-430.

Seddon HR (1925) Stachys arvensis: A cause of staggers or shivers in sheep. New South Wales Dep Agric Sci Bull #24:14-28.

Seddon HR, Hindmarsh WL, Carne HR (1926) Further observations on Stachys arvensis ("stagger weed") as a cause of staggers or shivers in sheep. New South Wales Dep Agric Sci Bull #26:25-33.

staff vine –see– *Solanum dulcamara* L.

staggergrass –see– *Amianthium muscitoxicum* (Walter) A. Gray

staggers weed –see– *Matricaria nigellifolia* DC.

staggerweed –see– *Delphinium tricorne* Michx.; *Dicentra canadensis* (Goldie) Walp.; *Dicentra cucullaria* (L.) Bernh.; *Helenium autumnale* L.; *Stachys arvensis* (L.) L.

staggerwort –see– *Helenium autumnale* L.; *Jacobaea vulgaris* Gaertn.

staghorn sumach –see– *Rhus typhina* L.

Stagmaria vermiciflua Jack = *Gluta renghas* L.

standing milk vetch –see– *Astragalus adsurgens* Pall.

star anise –see– *Illicium anisatum* L.; *Illicium verum* Hook. f.

star bur –see– *Acanthospermum hispidum* DC.

star cactus –see– *Aloe vera* (L.) Burm. f.

star flower –see– *Ornithogalum umbellatum* L.

star fruit –see– *Averrhoa carambola* L.

star lily –see– *Toxicoscordion fremontii* (Torr.) Rydb.

star-of-Bethlehem –see– *Ornithogalum thyrsoides* Jacq.; *Ornithogalum umbellatum* L.

star pine –see– *Pinus pinaster* Aiton

star potato vine –see– *Solanum pseudocapsicum* L.

star thistle –see– *Centaurea solstitialis* L.

star zydadene –see– *Toxicoscordion fremontii* (Torr.) Rydb.

starchwort –see– *Arisaema triphyllum* (L.) Schott; *Arum maculatum* L.

stargrass –see– *Cynodon aethiopicus* Clayton & J. R. Harlan; *Cynodon nlemfuensis* Vanderyst; *Cynodon plectostachyus* (K. Schum.) Pilg.

Statice brasiliensis Boiss. = *Limonium brasiliense* (Bioss.) Kuntze

Stechapfel –see– *Datura stramonium* L.

Stecheiche –see– *Ilex aquifolium* L.

Stechhulsen –see– *Ilex aquifolium* L.

Stechpalme –see– *Ilex aquifolium* L.

Steinheil –see– *Drimia maritima* (L.) Stearn

Steinklee –see– *Melilotus albus* Medik.; *Melilotus officinalis* Lam.

stemless locoweed –see– *Oxytropis lambertii* Pursh

STEMODIA KINGII F. Muell. [Scrophulariaceae]

Citations:

Allen JG, Mitchell AA (1998) A newly discovered toxic plant, Stemodia kingii, in Western Australia. In: Garland T, Barr AC (eds.) Toxic plants and other natural toxicants. CABI. New York. pp. 120-124.

Raisbeck MF, Allen JG, Colegate SM, et al. (2004) Pathology of experimental Stemodia kingii intoxication in the mouse. In: Acamovic T, Stewart CS, Pennycott TW (eds.) Poisonous plants and related toxins. CABI. Wallingford, UK. pp. 280-285.

STEPHANIA SINICA Diels [Menispermaceae]

Citations:

Haller C, Dyer JE, Ko R, et al. (1999) Toxic hepatitis induced by herbal medicines. J Toxicol Clin Toxicol 37(5):620.

STEPHANIA TETRANDA S. Moore [Menispermaceae]

Citations:

Schmeiser HH, Bieler CA, Wiessler M, et al. (1996) Detection of DNA adducts formed by aristolochic acid in renal tissue from patients with Chinese herbs nephropathy. Cancer Res 56(9):2025-2028.

Stephanorossia palustris Chiov. = Oenanthe palustris (Chiov.) C. Norman

Sterculia diversifolia G. Don = Brachychiton populneus (Schott & Endl.) R. Br.

STERCULIA FOETIDA L. [Malvaceae]

Common Names:

Australian flame tree; bangar nut; bangor nut; fetid sterculia; flame tree; Java olive; wild almond

Citations:

Abou-Ashour AM, Edwards HM Jr (1970) Effect of dietary Sterculia foetida oil on pink-white discoloration and fatty acid distribution in stored eggs. J Nutr 100(7):757-766.

Abou-Ashour AM, Edwards HM Jr (1972) Hematological changes in laying hens receiving Sterculia foetida oil supplements. Poult Sci 51(1):300-305.

Austic RE, Hill FW, Wilson BW (1971) Role of yolk in chick embryo mortality induced by the feeding of Sterculia foetida oil. Poult Sci 50(5):1320-1326.

Schneider DL, Vavich MG, Kurnick AA, et al. (1961) Effect of Sterculia foetida oil on mortality of the chick embryo. Poult Sci 40:1644-1648.

Schneider DL, Kurnick AA, Vavich MG, et al. (1962) Delay of sexual maturity in chickens by Sterculia foetida oil. J Nutr 77(Aug):403-407.

sterculia gum –see– *Sterculia urens* Roxb.

STERCULIA URENS Roxb. [Malvaceae]

Common Names:

false tragacanth; gum karaya; Indian gum; Indian tragacanth; karaya gum; karayu gum; sterculia gum

Citations:

Bowen R (1939) Karaya gum as a cause of urticaria. Arch Derm Syphilol 39:506-509.

Figley KD (1940) Karaya gum (Indian gum) hypersensitivity. JAMA 114(9):747-748.

Sterculiacea altissima = Mansonia altissima (A. Chev.) A. Chev.

Sternblume –see– *Narcissus poeticus* L.

STEVIA REBAUDIANA (Bertoni) Bertoni [Asteraceae]

Citations:

Planas GM, Kuc J (1968) Contraceptive properties of Stevia rebaudiana. Science 162(857):1007.

stickerweed –see– *Amaranthus spinosus* L.; *Solanum carolinense* L.

sticktight –see– *Cynoglossum officinale* L.

sticky agrimony –see– *Ageratina adenophora* (Spreng.) R. M. King & H. Rob.

sticky elecampane –see– *Dittrichia viscosa* (L.) Greuter

sticky eupatorium –see– *Ageratina adenophora* (Spreng.) R. M. King & H. Rob.

sticky florestin –see– *Florestina tripteris* DC.

sticky palafoxia –see– *Florestina tripteris* DC.

sticky snakeweed –see– *Gutierrezia microcephala* (DC.) A. Gray

sticky thistle –see– *Chamaeleon gummifera* (L.) Cass.

Stieleiche –see– *Quercus robur* L.

Stierkraut –see– *Euphorbia peplus* L.

stiff sickness bush –see– *Crotalaria burkeana* Benth.

stiff-stem yellow flax –see– *Linum rigidum* Pursh

Stillingia sebifera (L.) Michx. = Triadica sebifera (L.) Small

STILLINGIA TRECULIANA (Müll. Arg.) I. M. Johnst. [Euphorbiaceae]

Common Names:

queen's-root; trecul queensdelight

Citations:

Boughton IB, Hardy WT (1939) Hard yellow livers of sheep and cattle and feeding trials of suspected plants. Texas Agric Exp Sta Annu Rep 52:238-240.

Boughton IB, Hardy WT (1940) Feeding trials of suspected plants. Texas Agric Exp Sta Annu Rep 53:236-237.

stinging bean –see– *Mucuna pruriens* (L.) DC.

stinging nettle –see– *Laportea canadensis* (L.) Wedd.; *Urtica chamaedryoides* Pursh; *Urtica dioica* L.

Stinkasant –see– *Ferula assa-foetida* L.

stinkblaar –see– *Datura stramonium* L.

stinkbossie –see– *Boscia foetida* Schinz

Stinkende Nieswurz –see– *Helleborus foetidus* L.

Stinkerich –see– *Mercurialis perennis* L.

stinking chamomile –see– *Anthemis cotula* L.

stinking hellebore –see– *Helleborus foetidus* L.

stinking Mayweed –see– *Anthemis cotula* L.

stinking milk vetch –see– *Astragalus praelongus* E. Sheld.

stinking nightshade –see– *Hyoscyamus niger* L.

stinking Roger –see– *Tagetes minuta* L.

stinking rose –see– *Allium sativum* L.

stinking Willy –see– *Jacobaea vulgaris* Gaertn.

stinkingweed –see– *Senna occidentalis* (L.) Link

stinkingwood –see– *Senna occidentalis* (L.) Link

stinkweed –see– *Conium maculatum* L.; *Datura stramonium* L.; *Dysphania ambrosioides* (L.) Mosyakin & Clemants;

Gutierrezia microcephala (DC.) A. Gray; *Thlaspi arvense* L.

stinkwood –see– *Zieria arborescens* Sims; *Zieria smithii* Andrews

stinkwort –see– *Datura stramonium* L.; *Dittrichia graveolens* (L.) Greuter; *Helleborus foetidus* L.

STIPA PENNATA L. [Poaceae]

Common Names:

feathergrass

Citations:

Bashkatov GA (1969) [Economic losses caused by feather-grass diseases in sheep.] Veterinariia Moscow 46(6):69-72.

Bashkatov GA (1970) [A case of feather grass disease in the transportation of sheep.] Veterinariia Moscow 47(3):93-94.

Stipa robusta (Vasey) Scribn. = Achnatherum robustum (Vasey) Barkworth

STIPA SIBIRICA (L.) Lam. [Poaceae]

Common Names:

gobu; gogu; gumai; gur ghas

Citations:

Vaid KM (1963) A note on Stipa sibirica (L.) Lam., the poisonous grass of Kashmir. Indian Forester 89:423-424.

Stipa tenacissima Trin. = Nassela tenuissima (Trin.) Barkworth

Stipa vaseyi Scribn. = Achnatherum robustum (Vasey) Barkworth

Stipa viridula Trin. var. robusta Vasey = Achnatherum robustum (Vasey) Barkworth

Stizolobium pruriens (L.) Medik. = Mucuna pruriens (L.) DC.

Stockrose –see– *Alcea rosea* L.

stolsbossie –see– *Pteronia pallens* L. f.

stoneroot –see– *Collinsonia canadensis* L.

stootsiektebossie –see– *Matricaria nigellifolia* DC.

storax bush –see– *Styrax officinalis* L.

stork's-bill –see– *Erodium cicutarium* (L.) L'Her.

straight-stem locoweed –see– *Astragalus praelongus* E. Sheld.

stramoine –see– *Datura stramonium* L.

stramonium –see– *Datura stramonium* L.

strandrøyr –see– *Phalaris arundinacea* L.

Strathmore weed –see– *Pimelea prostrata* (J. R. Forst. & G. Forst.) Willd.

Straußkirsche –see– *Solanum pseudocapsicum* L.

strawberry –see– *Fragaria ×ananassa* Duchesne ex Rozier

streaked rattlepod –see– *Crotalaria pallida* Aiton

string bean –see– *Phaseolus vulgaris* L.

strophanthus –see– *Strophanthus gratus* (Wall. & Hook.) Baill.

STROPHANTHUS GRATUS (Wall. & Hook.) Baill. [Apocynaceae]

Common Names:

nea; ouabaia; strophanthus; Uta

Citations:

Ogunlana V, Caraccio T, Mofenson H (2000) Use of digoxin immune FAB in strophanthus poisoning. J Toxicol Clin Toxicol 38(5):519.

strychnine –see– *Strychnos nux-vomica* L.

strychnine bush –see– *Myoporum acuminatum* R. Br.

STRYCHNOS IGNATII P. J. Bergius [Loganiaceae]

Synonyms:

Strychnos ovalifolia Wall. ex G. Don

Common Names:

false angustura bark; false upas tree; horsebean; Ignatius bean; ipoh akar; St. Ignatius-bean; upas radja; upas teute

Citations:

Fasal P (1945) Cutaneous diseases in the tropics. A clinical study based on observations in Malaya. Arch Derm Syphilol 51(3):163-171.

STRYCHNOS NUX-VOMICA L. [Loganiaceae]

Common Names:

Brechnuss; crow fig; ma-qian-zi; noix vomique; nux vomica; ordeal nut; poison nut; slang nut; strychnine

Citations:

Chan TY (2002) Herbal medicine causing likely strychnine poisoning. Hum Exp Toxicol 21:467-468.

Dittrich K, Bayer MJ, Wanke LA (1984) A case of fatal strychnine poisoning. J Emerg Med 1(4):327-330.

Katz J, Prescott K, Woolf AD (1996) Strychnine poisoning from a Cambodian traditional remedy. Am J Emerg Med 14(5):475-477.

Woolf A, Prescott K, Katz J (1994) Strychnine poisoning from the use of a Cambodian traditional remedy. Vet Hum Toxicol 36(4):360.

Zhang YG, Huang GZ (1988) Poisoning by toxic plants in China. Report of 19 autopsy cases. Am J Forensic Med Pathol 9(4):313-319.

Strychnos ovalifolia Wall. ex G. Don = Strychnos ignatii P. J. Bergius

STRYPHNODENDRON ADSTRINGENS (Mart.) anon. [Fabaceae]

Synonyms:

Stryphnodendron barbadetimam (Vell.) Mart.

Common Names:
barbatimão

Citations:
Pereira CA (1985) Aspectos clínicos, laboratorias e anátomo-histopatológicos na intoxicação experimental pela fava do "Barbatimão" (Stryphnodendron barbatimao Mart.) em bovinos. Arq Bras Med Vet Zootecnia 37(3):286-289.

Stryphnodendron barbadetimam (Vell.) Mart. = Stryphnodendron adstringens (Mart.) anon.

STRYPHNODENDRON CORIACEUM Benth. [Fabaceae]

Common Names:
barbatimão

Citations:
Tokarnia CH, Peixoto PV, Döbereiner J (1991) Intoxicação experimental por Stryphnodendron coriaceum (Leg. Mimosoideae) em bovinos. Pesq Vet Bras 11(1-2):25-29.

STRYPHNODENDRON OBOVATUM Benth. [Fabaceae]

Common Names:
barbatimão

Citations:
Brito MF, Tokarnia CH, Peixoto PV, et al. (2001) Intoxicação experimental pelas favas de Stryphnodendron obovatum (Leg. Mimosoideae) em bovinos. 1. Caracterização do quadro clínico. Pesq Vet Bras 21(1):9-17.
Brito MF, Tokarnia CH, Peixoto PV (2001) Intoxicação experimental pelas favas de Stryphnodendron obovatum (Leg. Mimosoideae) em bovinos. 2. Achados anátomo e histopatológicos. Pesq Vet Bras 21(2):61-71.
Camargo W (1965) Fotossensibilização em bovinos por "barbatimão" (Stryphnodendron obovatum) Benth; fam. Leguninosae. Biologico 31(1):7-11.
Tokarnia CH, Brito MF, Driemeier D, et al. (1998) Aborto em vacas na intoxicação experimental pelas favas de Stryphnodendron obovatum (Leg. Mimosoideae). Pesq Vet Bras 18(1):35-38.

stubbleberry –see– *Solanum nigrum* L.

stump tree –see– *Gymnocladus dioicus* (L.) K. Koch

sturdy rye –see– *Lolium temulentum* L.

Sturmhut –see– *Aconitum napellus* L.

Sturt's-nightshade –see– *Solanum sturtianum* F. Muell.

stylo –see– *Stylosanthes guianensis* (Aubl.) Sw.

STYLOSANTHES GUIANENSIS (Aubl.) Sw. [Fabaceae]

Common Names:
stylo

Citations:
Shenk JS (1976) The meadow vole as an experimental animal. Lab Anim Sci 26(4):664-669.

STYPANDRA GLAUCA R. Br. [Liliaceae]

Common Names:
nodding blue lily

Citations:
Whittington RJ, Searson JE, Whittaker SJ, et al. (1988) Blindness in goats following ingestion of Stypandra glauca. Aust Vet J 65(6):176-181.

STYPANDRA IMBRICATA R. Br. [Liliaceae]

Common Names:
blindgrass; cluster-leaf blindgrass

Citations:
Colegate SM, Dorling PR, Huxtable CR, et al. (1990) An investigation of possible neurotoxicity of diospyrol, the active principle of Diospyros mollis (maklua), using Stypandra imbricata (blindgrass)-induced blindness as a model. South Asian J Trop Med Pub Health 21(1):139-141.
Huxtable CR, Dorling PR, Slatter DH (1980) Myelin oedema, optic neuropathy and retinopathy in experimental Stypandra imbricata toxicosis. Neuropathol Appl Neurobiol 6(3):221-232.
Main DC, Slatter DH, Huxtable CR, et al. (1981) Stypandra imbricata ("blindgrass") toxicosis in goats and sheep - Clinical and pathologic findings in 4 field cases. Aust Vet J 57(3):132-135.

styptic weed –see– *Senna occidentalis* (L.) Link

STYRAX OFFICINALIS L. [Styracaceae]

Common Names:
storax bush

Citations:
Hepper FN (2004) Two plant fish-poisons in Lebanon. Vet Hum Toxicol 46(6):338-339.

stywesiekte bossie –see– *Crotalaria burkeana* Benth.

stywesiektebos –see– *Crotalaria burkeana* Benth.

sub clover –see– *Trifolium subterraneum* L.

subabul –see– *Leucaena leucocephala* (Lam.) de Wit

subterranean clover –see– *Trifolium subterraneum* L.

subterranean trefoil –see– *Trifolium subterraneum* L.

succory –see– *Cichorium intybus* L.

suckleya –see– *Suckleya suckleyana* (Torr.) Rydb.

SUCKLEYA SUCKLEYANA (Torr.) Rydb. [Chenopodiaceae]

Common Names:
poison suckleya; suckleya

Citations:
Berry TJ, Gonzales P (1986) Do your cattle-owning clients know about this poisonous range plant? Vet Med 81(11):1055-1056.
Thorp F Jr, Deem AW (1939) Suckleya suckleyana a poisonous plant. J Am Vet Med Assoc 94(Mar):192-197.

Thorp F Jr, Deem AW, Harrington HD, et al. (1937) Suckleya suckleyana - A poisonous plant. Colorado Agric Exp Sta Tech Bull #22:19 pp.

sucupira –see– *Bowdichia nitida* Spruce ex Benth.

Sudan sorghum –see– *Sorghum ×drummondii* (Steud.) Millsp. & Chase

Sudangrass –see– *Sorghum ×drummondii* (Steud.) Millsp. & Chase

sudax –see– *Sorghum ×drummondii* (Steud.) Millsp. & Chase

suel-da-con-suelda –see– *Euphorbia tirucalli* L.

suerda-con-suerda –see– *Euphorbia tirucalli* L.

sugar apple –see– *Annona cherimola* Mill.; *Annona squamosa* L.

sugar bean –see– *Phaseolus lunatus* L.

sugar beet –see– *Beta vulgaris* L.

sugar cane –see– *Saccharum officinarum* L.

sugar gum –see– *Eucalyptus cladocalyx* F. Muell.

sulfur lupine –see– *Lupinus sulphureus* Douglas ex Hook.

Sumach –see– *Coriaria myrtifolia* L.; *Rhus coriaria* L.; *Rhus virens* Lindh. ex A. Gray; *Toxicodendron succedaneum* (L.) Kuntze

summer adonis –see– *Adonis aestivalis* L.

summer cypress –see– *Bassia scoparia* (L.) A. J. Scott

summer dandelion –see– *Hypochaeris radicata* L.

summer lupine –see– *Lupinus formosus* Greene

summer milkweed –see– *Asclepias syriaca* L.

summer pheasant's-eye –see– *Adonis aestivalis* L.

summer skies –see– *Ipomoea tricolor* Cav.; *Ipomoea violacea* L.

summer snowflake –see– *Ornithogalum umbellatum* L.

summer stargrass –see– *Cynodon plectostachyus* (K. Schum.) Pilg.

summergrass –see– *Digitaria sanguinalis* (L.) Scop.

Sumpfdrachenwurz –see– *Calla palustris* L.

sumpfheidebeere –see– *Vaccinium uliginosum* L.

Sumpfkalla –see– *Calla palustris* L.

Sumpfkraut –see– *Calla palustris* L.

Sumpfschachtelhalm –see– *Equisetum palustre* L.

Sumpfwolfsmilch –see– *Euphorbia palustris* L.

sun spurge –see– *Euphorbia helioscopia* L.

sunchillo –see– *Pascalia glauca* Ortega

suncho –see– *Pascalia glauca* Ortega

sunflower –see– *Helianthus annuus* L.; *Hymenoxys hoopesii* (A. Gray) Bierner

sunflower daisy –see– *Wedelia asperrima* (Decne.) Benth.

suni –see– *Colocasia esculenta* (L.) Schott

sunn hemp –see– *Crotalaria juncea* L.; *Crotalaria ochroleuca* G. Don

superb lily –see– *Gloriosa superba* L.

supplejack –see– *Ventilago viminalis* Hook.

sureau noire –see– *Sambucus nigra* L.

surelle –see– *Rumex acetosa* L.

surete –see– *Rumex acetosa* L.

Surinam greenheart –see– *Tabebuia serratifolia* (Vahl) G. Nicholson

Surinamgrass –see– *Urochloa decumbens* (Stapf) R. D. Webster

Süßkartoffel –see– *Ipomoea batatas* (L.) Lam.

susumberberry –see– *Solanum torvum* Sw.

suzon –see– *Jacobaea vulgaris* Gaertn.

SWAINSONA CANESCENS (Benth.) F. Muell. [Fabaceae]

Common Names:

gray Swainsona pea

Citations:

Dorling PR, Huxtable CR, Vogel P (1978) Lysosomal storage in Swainsona spp. toxicosis: An induced mannosidosis. Neuropathol Appl Neurobiol 4(4):285-295.

Gardiner MR, Linto AC, Aplin TE (1969) Toxicity of Swainsona canescens for sheep in Western Australia. Aust J Agric Res 20:87-97.

Locke KB, McEwan DR, Hamdorf IJ (1980) Experimental poisoning of horses and cattle with Swainsona canescens var horniana. Aust Vet J 56(8):379-383.

SWAINSONA GALEGIFOLIA (Andrews) R. Br. [Fabaceae]

Common Names:

Darling pea; indigo plant; red Darling pea; smooth Darling pea

Citations:

Hartley WJ, Gibson AJ (1971) Observations of Swainsona galegifolia poisoning in cattle in northern New South Wales. Aust Vet J 47(7):300-305.

Huxtable CR (1969) Experimental reproduction and histopathology of Swainsona galegifolia poisoning in the guinea-pig. Aust J Exp Biol Med Sci 47(3):339-347.

Huxtable CR (1970) Ultrastructural changes caused by Swainsona galegifolia poisoning in the guinea-pig. Aust J Exp Biol Med Sci 48(1):71-80.

Huxtable CR (1972) The effect of ingestion of Swainsona galegifolia on the liver lysosomes of the guinea-pig. Aust J Exp Biol Med Sci 50(1):109-118.

Huxtable CR, Gibson A (1970) Vacuolation of circulating lymphocytes in guinea-pigs and cattle ingesting Swainsona galegifolia. Aust Vet J 46(9):446-448.

Laws L, Anson RB (1968) Neuronopathy in sheep fed Swainsona luteola and S. galegifolia. Aust Vet J 44(10):447-452.

Martin CJ (1897) Report of an investigation into the effects of Darling pea (Swainsona galegifolia) upon sheep. Agric Gaz New South Wales 6:363-369.

SWAINSONA LUTEOLA F. Muell. [Fabaceae]

Common Names:

black pea; Darling pea; dwarf Darling pea

Citations:

Cleland JB, McDonald AH (1917) Experimental feeding with two poison plants. Agric Gaz New South Wales 28(Oct 2):735-739.
Laws L, Anson RB (1968) Neuronopathy in sheep fed Swainsona luteola and S. galegifolia. Aust Vet J 44(10):447-452.

swallowwort –see– *Calotropis procera* (Aiton) W. T. Aiton; *Chelidonium majus* L.

swamp camas –see– *Anticlea elegans* (Pursh) Rydb.

swamp dogwood –see– *Toxicodendron vernix* (L.) Kuntze

swamp grasstree –see– *Xanthorrhoea hastilis* Sm.; *Xanthorrhoea resinifera* (Sol. ex C. Kite) E. C. Nelson & D. J. Bedford

swamp hellebore –see– *Veratrum viride* Aiton

swamp holly –see– *Ilex decidua* Walter

swamp horsetail –see– *Equisetum fluviatile* L.

swamp laurel –see– *Rhododendron ×columbianum* (Piper) Harmaja

swamp maple –see– *Acer rubrum* L.

swamp milkweed –see– *Asclepias incarnata* L.

swamp onion –see– *Allium validum* S. Watson

swamp sumach –see– *Toxicodendron vernix* (L.) Kuntze

swamp sunflower –see– *Helenium autumnale* L.

swamp whiteheads –see– *Sphenosciadium capitellatum* A. Gray

swan plant –see– *Gomphocarpus fruticosa* (L.) W. T. Aiton

swartstorm –see– *Cadaba aphylla* (Thunb.) Wild

Swartzia madagascariensis Desv. = Bobgunnia madagascariensis (Desv.) J. H. Kirkbr. & Wiersema

Swede turnip –see– *Brassica napus* L. var. napobrassica (L.) Rchb.

Swedish clover –see– *Trifolium hybridum* L.

sweet bay –see– *Laurus nobilis* L.

sweet birch –see– *Betula lenta* L.

sweet buckeye –see– *Aesculus flava* Sol.

sweet camomille –see– *Chamaemelum nobile* (L.) All.

sweet cane –see– *Acorus calamus* L.

sweet cassava –see– *Manihot esculenta* Crantz

sweet chestnut –see– *Castanea sativa* Mill.

sweet cinnamon –see– *Acorus calamus* L.

sweet clover –see– *Melilotus albus* Medik.; *Melilotus indicus* (L.) All.; *Melilotus officinalis* Lam.

sweet flag –see– *Acorus calamus* L.

sweet lupin –see– *Lupinus albus* L.

sweet noors –see– *Euphorbia caerulescens* Haw.

sweet orange –see– *Citrus sinensis* (L.) Osbeck

sweet pea –see– *Lathyrus odoratus* L.; *Lathyrus sativus* L.

sweet pepper –see– *Capsicum annuum* L.

sweet potato –see– *Ipomoea batatas* (L.) Lam. var. batatas

sweet sop –see– *Annona cherimola* Mill.; *Annona squamosa* L.

sweet sorghum –see– *Sorghum bicolor* (L.) Moench

sweet vernalgrass –see– *Anthoxanthum odoratum* L.

sweet woodruff –see– *Galium odoratum* (L.) Scop.

sweetgrass –see– *Panicum schinzii* Hack.

sweetheart plant –see– *Philodendron hederaceum* (Jacq.) Schott

sweetroot –see– *Acorus calamus* L.

SWIETENIA MACROPHYLLA King [Meliaceae]

Common Names:

Echtes Mahagoni; Honduran mahogany; mahogany; tobasco mahogany

Citations:

Estlander T, Jolanki R, Alanko K, et al. (2001) Occupational allergic contact dermatitis caused by wood dusts. Contact Dermatitis 44(4):213-217.
Fowler JF Jr (1985) Occupational dermatitis to Honduran mahogany. Contact Dermatitis 13(5):336-337.
Hjorth N (1979) Contact dermatitis from sawdust. Contact Dermatitis 5(4):339-340.
Steiner SD, Schwartz L (1944) Dermatitis from mahogany wood (Swietenia macrophylla). Indus Med 13(Mar):234-235.

SWIETENIA MAHAGONI (L.) Jacq. [Meliaceae]

Common Names:

Amerikanisches Mahagoni; caoba; Echtes Mahagoni; Mahogany

Citations:

Raghuraman V, Raveendran M (1982) Mahogony seeds - An unusual poison. J Indian Med Assoc 78(11):186-188.

SWINTONIA SPICIFERA Hook. f. [Anacardiaceae]

Citations:

Oleffe JA, Sporcq J, Hublet P (1975) Epidemiological study of the wood industry in Belgium. Contact Dermatitis 1(5):315-316.

Swiss chard –see– *Beta vulgaris* L.

Swiss-cheese plant –see– *Monstera deliciosa* Liebm.

sword bean –see– *Canavalia ensiformis* (L.) DC.

SYMPHORICARPOS ALBUS (L.) S. F. Blake [Caprifoliaceae]

Synonyms:

Symphoricarpos racemosus Michx.

Common Names:
Knallerbsenstrauch; Schneebeere; snowberry; St. Peter's-wort; waxberry

Citations:
Amyot TE (1885) Poisoning by snowberries. Br Med J 1(May 16):986.
Lamminpaa A, Kinos M (1996) Plant poisonings in children. Hum Exp Toxicol 15(3):245-249.

Symphoricarpos racemosa Michx. = Symphoricarpos albus (L.) S. F. Blake

SYMPHYOTRICHUM LANCEOLATUM (Willd.) G. L. Nesom [Asteraceae]

Synonyms:
Aster simplex Willd.

Common Names:
simple aster

Citations:
Bergeron JM, Jodoin L (1985) Fiabilité des mesures de poids et d'examens histopathologiques dans les études d'intoxication du campagnol des champs (Microtus pennsylvanicus). Can J Zool 63(4):804-810.

SYMPHYTUM OFFICINALE L. [Boraginaceae]

Common Names:
Beinwell; blackwort; boneset; comfrey; confrei; fakete nadálytö; grande consoude; knitbone; Russian comfrey

Citations:
Bach N, Thung SN, Schaffner F (1989) Comfrey herb tea-induced hepatic veno-occlusive disease. Am J Med 87(1):97-99.
Hirono I, Mori H, Haga M (1978) Carcinogenic activity of Symphytum officinale. J Natl Cancer Instit 61(3):865-869.
Ridker PM (1989) Health hazards of unusual herbal teas. Am Fam Physician 39(5):153-156.
Ridker PM, Ohkuma S, McDermott WV, et al. (1985) Hepatic venoocclusive disease associated with the consumption of pyrrolizidine-containing dietary supplements. Gastroenterology 88(4):1050-1054.
Weston CF, Cooper BT, Davies JD, et al. (1987) Veno-occlusive disease of the liver secondary to ingestion of comfrey. Br Med J 295(6591):183.
Yeong ML, Swinburn B, Kennedy M, et al. (1990) Hepatic veno-occlusive disease associated with comfrey ingestion. J Gastroenterol Hepatol 5(2):211-214.

Synadenium arborescens Boiss. = Synadenium cupulare (Boiss.) L. C. Wheeler

SYNADENIUM CUPULARE (Boiss.) L. C. Wheeler [Euphorbiaceae]

Synonyms:
Synadenium arborescens Boiss.

Common Names:
crying tree; dead-man's-tree

Citations:
Van der Walt SJ, Steyn DG (1939) Recent investigations into the toxicity of known and unknown poisonous plants in the Union of South Africa. IX. Onderstepoort J Vet Sci Anim Indus 12:335-366.

SYNADENIUM GRANTII Hook. f. [Euphorbiaceae]

Common Names:
African milk bush; South African milk bush

Citations:
Rook AJ (1965) An unrecorded irritant plant, Synadenium grantii. Br J Dermatol 77(5):284.
Spoerke DG, Montanio CD, Rumack BH (1985) Pediatric exposure to the houseplant Synadenium grantii. Vet Hum Toxicol 28(4):283-284.

SYNGONIUM PODOPHYLLUM Schott [Araceae]

Common Names:
arrowhead vine; nephthytis; tri-leaf wonder

Citations:
Der Marderosian AH, Giller FB, Roia FC Jr (1976) Phytochemical and toxicological screening of household ornamental plants potentially toxic to humans. I. J Toxicol Environ Health 1(6):939-953.
Lamminpaa A, Kinos M (1996) Plant poisonings in children. Hum Exp Toxicol 15(3):245-249.

Syntherisma sanguinalis (L.) Dulac = Digitaria sanguinalis (L.) Scop.

Syrian rue –see– *Peganum harmala* L.

syringa –see– *Melia azedarach* L.

SYZYGIUM AROMATICUM (L.) Merr. & L. M. Perry [Myrtaceae]

Synonyms:
Eugenia caryophyllata Thunb.

Common Names:
clove; cloves

Citations:
Barkin ME, Boyd JP, Cohen S (1984) Acute allergy reaction to eugenol. Oral Surg 57:441-442.
Eisen JS, Koren G, Juurlink DN, et al. (2004) N-acetylcysteine for the treatment of clove oil-induced fulminant hepatic failure. J Toxicol Clin Toxicol 42(1):89-92.
Gaul LE (1963) Dermatitis of the hands from oil of cloves. Skin (Los Angeles) 1(10):314.
Sternberg L (1937) Contact dermatitis. Cases caused by oil of cloves and by oil of camomile tea (Anthemis cotula). J Allergy 8:185-186.

SYZYGIUM CUMINI (L.) Skeels [Myrtaceae]

Common Names:
jamuna

Citations:

Kulakkattolickal AT (1987) Piscicidal plants of Nepal: Preliminary toxicity screening using grass carp (Ctenopharyngodon idella) fingerlings. J Ethnopharmacol 21(1):1-9.

szagos müge –see– *Galium odoratum* (L.) Scop.

szalej jadowity –see– *Cicuta virosa* L.

szarvas kerep –see– *Lotus corniculatus* L.

szörös disznóparéj –see– *Amaranthus retroflexus* L.

T

tabac –see– *Nicotiana tabacum* L.

tabaco –see– *Nicotiana tabacum* L.

Tabak –see– *Nicotiana tabacum* L.

tabaquillo –see– *Nicotiana obtusifolia* M. Martens & Galeotti; *Wigandia urens* (Ruiz & Pav.) Kunth var. caracasana (Kunth) D. N. Gibson

tabaquillo-de-los-rios –see– *Persicaria punctata* (Elliott) Small

Tabasco pepper –see– *Capsicum frutescens* L.

TABEBUIA IMPETIGINOSA (Mart. ex DC.) Standl. [Bignoniaceae]

Synonyms:

Tecoma impetiginosa Mart. ex DC.

Common Names:

lapacho; pao d'arco; taheebo

Citations:

Palmer M, Haller C, McKinney P, et al. (1999) Botanicals and other dietary supplements: Adverse events by age. J Toxicol Clin Toxicol 37(5):609.

TABEBUIA SERRATIFOLIA (Vahl) G. Nicholson [Bignoniaceae]

Common Names:

bethabara; ipé; Surinam greenheart

Citations:

Hausen BM, Simatupang MH, Weitbrecht U (1971) Allergie gegen Bethabara-Holz. Derm Beruf Umwelt 19(6):324-327.

Weitbrecht U (1967) Allergie gegen Bethabara-holz. Berufsdermatosen 15:183-184.

TABERNANTHE IBOGA Baill. [Apocynaceae]

Common Names:

ibogaine

Citations:

Cienki J, Mash D, Hearn W (2001) Ibogaine fatalities. J Toxicol Clin Toxicol 39(5):547.

tacks –see– *Tribulus terrestris* L.

tagasaste –see– *Chamaecytisus prolifer* (L. f.) Link

tagayasan –see– *Mesua ferrea* L.

TAGETES ERECTA L. [Asteraceae]

Common Names:

African marigold; marigold

Citations:

Sharma SC, Kaur S (1989) Airborne contact dermatitis from Compositae plants in northern India. Contact Dermatitis 21(1):1-5.

Spoerke DG, Temple AR (1978) One year's experience with potential plant poisonings reported to the Intermountain Regional Poison Control Center. Vet Hum Toxicol 20(2):85-89.

TAGETES MINUTA L. [Asteraceae]

Common Names:

Mexican marigold; stinking Roger; wild marigold

Citations:

Verhagen AR, Nyaga JM (1974) Contact dermatitis from Tagetes minuta, A new sensitizing plant of the Compositae family. Arch Dermatol 110(3):441-444.

TAGETES PATULA L. [Asteraceae]

Common Names:

French marigold; genda

Citations:

Singh R, Siddiqui MA, Baruah MC (1978) Plant dermatitis in Delhi. Indian J Med Res 68(Oct):650-655.

taheebo –see– *Tabebuia impetiginosa* (Mart. ex DC.) Standl.

Tahitian lime –see– *Citrus aurantiifolia* (Christm.) Swingle

tailcup lupine –see– *Lupinus argenteus* Pursh; *Lupinus argenteus* Pursh var. heteranthus (S. Watson) Barneby; *Lupinus argenteus* Pursh var. holosericeus (Nutt.) Barneby

tailspur lupine –see– *Lupinus argenteus* Pursh var. heteranthus (S. Watson) Barneby

tall buttercup –see– *Ranunculus acris* L.

tall cone flower –see– *Rudbeckia laciniata* L.

tall crowfoot –see– *Ranunculus acris* L.

tall fescue –see– *Festuca arundinacea* Schreb.

tall field buttercup –see– *Ranunculus acris* L.

tall gysophyl –see– *Gypsophila paniculata* L.

tall larkspur –see– *Delphinium barbeyi* (Huth) Huth; *Delphinium glaucum* S. Watson; *Delphinium nuttallianum* Pritz.; *Delphinium* ×*occidentale* (S. Watson) S. Watson; *Delphinium scopulorum* A. Gray

tall locoweed –see– *Astragalus lentiginosus* Douglas ex Hook. var. diphysus (A. Gray) M. E. Jones

tall mannagrass –see– *Glyceria grandis* S. Watson

tall meadow fescue –see– *Festuca arundinacea* Schreb.

tall mountain larkspur –see– *Delphinium glaucum* S. Watson; *Delphinium menziesii* DC.; *Delphinium scopulorum* A. Gray

tall nettle –see– *Urtica dioica* L.

tall ragweed –see– *Ambrosia trifida* L.

tall sorrel –see– *Rumex acetosa* L.

tall yellow day lily –see– *Hemerocallis lilioasphodelus* L.

tallhue –see– *Lupinus mutabilis* Sweet

tallow tree –see– *Triadica sebifera* (L.) Small

tamarind –see– *Tamarindus indica* L.

TAMARINDUS INDICA L. [Fabaceae]

Common Names:

Kamalindo; tamarind

Citations:

Murray R, Dingwall-Fordyce I, Lane RE (1957) An outbreak of weaver's cough associated with tamarind seed powder. Br J Ind Med 14(2):105-110.

Seetharam KA, Pasricha JS (1987) Condiments and contact dermatitis of the finger-tips. Indian J Dermatol Venereol Leprol 53:325-328.

tamboril –see– *Enterolobium contortisiliquum* (Vell.) Morong

tamboril-do-campo –see– *Enterolobium gummiferum* (Mart.) J. F. Macbr.

tamier –see– *Dioscorea communis* (L.) Caddick & Wilkin

Tamus communis L. = Dioscorea communis (L.) Caddick & Wilkin

TANACETUM CINERARIIFOLIUM (Trevir.) Sch. Bip. [Asteraceae]

Synonyms:

Pyrethrum cinerariifolium Trevir.

Common Names:

Dalmatian chrysanthemum; feverfew; golden feather; insect flower; Japanese daisy; pellitory; pyrethrum

Citations:

McCord CP, Kilker CH, Minster DK (1921) Pyrethrum dermatitis. A record of the occurrence of occupational dermatoses among workers in the pyrethrum industry. JAMA 77(6):448-449.

Ramirez MA (1930) Pyrethrum: An etiologic factor in vasomotor rhinitis and asthma. J Allergy 1:149-155.

Sequeira JH (1936) Pyrethrum dermatitis. Br J Dermatol Syph 48(Oct):473-476.

TANACETUM PARTHENIUM (L.) Sch. Bip. [Asteraceae]

Synonyms:

Chrysanthemum parthenium (L.) Bernh.

Common Names:

chrysanthème-matricaire; feverfew; Mutterkraut

Citations:

Burry JN (1980) Compositae dermatitis in South Australia: Contact dermatitis from Chrysanthemum parthenium. Contact Dermatitis 6(5):445.

Hausen BM (1981) Berufsbedingte Kontakallergie auf Mutterkraut (Tanacetum parthenium (L.) Schultz-Bip.; Asteraceae). Derm Beruf Umwelt 29(1):18-21.

Paulsen E, Christensen LP, Andersen KE (2002) Do monoterpenes released from feverfew (Tanacetum parthenium) plants cause airborne Compositae dermatitis? Contact Dermatitis 47(1):14-18.

TANACETUM VULGARE L. [Asteraceae]

Synonyms:

Chrysanthemum vulgare (L.) Bernh.

Common Names:

gilisztaüzö varádics; ponso; Rainfarn; tanaisie; tanse; tansy; Wurmkraut

Citations:

Edwards EG (1906) Poisoning from oil of tansy. JAMA 47:937.

Greenhouse CA, Sulzberger MB (1932-1933) The common weed tansy (Tanacetum vulgare) as a cause of eczematous dermatitis. J Allergy 4(Nov):523-526.

Hausen BM, Schulz KH (1978) Polyvalente Kontaktallergie bei einer Floristin. Derm Beruf Umwelt 26(5):175-176.

Musson EH (1906) Oil of tansy poisoning. JAMA 47(23):1917-1918.

Whitehill NM (1906) Poisoning from oil of tansy. JAMA 47(7):509.

tanaisie –see– *Tanacetum vulgare* L.

tangan tangan –see– *Leucaena leucocephala* (Lam.) de Wit; *Ricinus communis* L.

tangena nut –see– *Cerbera odollam* Gaertn.

tangerine –see– *Citrus reticulata* Blanco

Tangier pea –see– *Lathyrus tingitanus* L.

tanglefoot –see– *Karwinskia humboldtiana* (Schult.) Zucc.

tanier –see– *Xanthosoma sagittifolium* (L.) Schott

Tankawang fat tree –see– *Triadica sebifera* (L.) Small

tannergrass –see– *Urochloa arrecta* (Hack. ex T. Durand & Schinz) Morrone & Zuloaga

tanse –see– *Tanacetum vulgare* L.

tansy –see– *Jacobaea vulgaris* Gaertn.; *Tanacetum vulgare* L.

tansy mustard –see– *Descurainia pinnata* (Walter) Britton

tansy ragwort –see– *Jacobaea vulgaris* Gaertn.

tántago-de-Valencia –see– *Euphorbia characias* L.

tao-er-qi –see– *Podophyllum hexandrum* Royle

tapioca –see– *Manihot esculenta* Crantz

tapiro –see– *Sambucus mexicana* C. Presl ex DC.

tar tree –see– *Semecarpus australiensis* Engl.

tárago mayor –see– *Euphorbia characias* L.

TARAXACUM OFFICINALE F. H. Wigg. aggr. [Asteraceae]

Common Names:

Butterblume; dandelion; Löwenzahn; Ringelblume

Citations:

Burry JN, Kuchel R, Reid JG, et al. (1973) Australian bush dermatitis: Compositae dermatitis in South Australia. Med J Aust 1(3):110-116.

Chivato T, Juan F, Montoro A, et al. (1996) Anaphylaxis induced by ingestion of a pollen compound. J Invest Allergol Clin Immunol 6(3):208-209.

Davies MG, Kersey PJ (1986) Contact allergy to yarrow and dandelion. Contact Dermatitis 14(4):256-257.

Hausen BM (1979) The sensitizing capacity of compositae plants. III. Test results and cross-reactions in compositae-sensitive patients. Dermatologica 159(1):1-11.

Hausen BM, Schulz KH (1978) Allergische Kontaktdermatitis durch Löwenzahn (Taraxacum officinale Wiggers). Derm Beruf Umwelt 26(6):198.

Janke D (1950) Durch Löwenzahn (Taraxacum officinale) verursachtes Ekzem. Ein Beitrag zur Kenntnis phytogener Dermatosen. Hautarzt 1(4):177-181.

Lundh K, Hindsén M, Gruvberger B, et al. (2006) Contact allergy to herbal teas derived from Asteraceae plants. Contact Dermatitis 54(4):196-201.

tarbush –see– *Flourensia cernua* DC.

tarbuz –see– *Citrullus lanatus* (Thunb.) Matsum. & Nakai

tares –see– *Lolium temulentum* L.

taro –see– *Colocasia esculenta* (L.) Schott

taro dasheen –see– *Colocasia esculenta* (L.) Schott

taro vine –see– *Epipremnum pinnatum* (L.) Engl.

tártago –see– *Euphorbia lathyris* L.; *Jatropha curcas* L.; *Jatropha multifida* L.

tarweed –see– *Amsinckia intermedia* Fisch. & C. A. Mey.

Tasmanian blackwood –see– *Acacia melanoxylon* R. Br.

Tasmanian ngaio –see– *Myoporum insulare* R. Br.

Tasso –see– *Taxus baccata* L.

Tatarian honeysuckle –see– *Lonicera tatarica* L.

Tatarische Heckenkirsche –see– *Lonicera tatarica* L.

Tatarisches Geißblatt –see– *Lonicera tatarica* L.

tatu –see– *Dioclea erecta* Hoehne; *Dioclea latifolia* Benth.

Taubenkröpferl –see– *Lotus corniculatus* L.

Taubenkutsche –see– *Aconitum napellus* L.

Taumelhafer –see– *Lolium temulentum* L.

Taumelkorn –see– *Lolium temulentum* L.

Taumelloch –see– *Lolium temulentum* L.

tawny day lily –see– *Hemerocallis fulva* (L.) L.

tawny lily –see– *Hemerocallis fulva* (L.) L.

TAXUS BACCATA L. [Taxaceae]

Common Names:

Eibe; Eibennadeln; English yew; European yew; ground hemlock; Ibe; If; Irish yew; l'if; Roteibe; tasso; venijnboom; yew

Citations:

Anonymous (1870) Fatal poisoning by the leaves of the Taxus baccata (Yew). Lancet 2(Oct 1):471.

Barling AS (1902) Poisoning by Irish yew. Lancet 1(Apr 19):1103-1104.

Bubien MZ, Magat A, Delatour P (1970) Fréquence des intoxications des animaux en Pologne. Bull Soc Vet Med Comp Lyon 72(2):167-173.

Burke MJ, Siegel D, Davidow B (1979) Anaphylaxis; consequence of yew (Taxus) needle ingestion. N Y State J Med 79(10):1576-1577.

Czerwek H, Fischer W (1960) Tödlicher Vergiftungsfall mit Taxus baccata. Versuche zum Nachweis der Taxusalkaloide aus Leichenorganen. Arch Toxikol 18(2):88-92.

Doyle H (1942) Yew poisoning. N Z J Agr 65(5):287-288.

Fiedler HH, Perron RM (1994) Eibenvergiftung bei australischen Emus (Dromaius novaehollandiae Latham). Berl Munch Tierarztl Wochenschr 107(2):50-52.

Frohne D, Pribilla O (1965) Tödliche Vergiftung mit Taxus baccata. Arch Toxikol 21(3):150-162.

Grandeau L (1894) Empoissonement du betail par les feuilles d'if. J Agric Pratique 58(1):229-230.

Guilhon J (1988) L'enseignement de la botanique et la mort subite au pré. Bull Acad Vet Fr 61(3):267-271.

Hurt S (1836) Poisonous effects of the berries, or seeds, of the yew (Taxus baccata). Lancet 1:394-395.

Janssen J, Peltenburg H (1985) Een klassieke wijze van zelfdoding: met Taxus baccata. Ned Tijdschr Geneeskd 129(13):603-605.

Jordan WJ (1964) Yew (Taxus baccata) poisoning in pheasants (Phasianus colchicus). Tijdschr Diergeneeskd 89(Suppl 1):187-188.

Kite GC, Lawrence TJ, Dauncey EA (2000) Detecting Taxus poisoning in horses using liquid chromatography/mass spectrometry. Vet Hum Toxicol 42(3):151-154.

Köhler H, Grünberg W (1960) Zur Pathologie der Vergiftung mit Taxus baccata unter besonderer Berücksichtigung der Vergiftung beim Känguruh. Arch Exp Veterinarmed 14:1149-1162.

Lamminpaa A, Kinos M (1996) Plant poisonings in children. Hum Exp Toxicol 15(3):245-249.

Melrose GB (1939) Cattle poisoned by garden shrubs. Care necessary in disposal of trimmings. N Z J Agr 58(Feb 20):129-130.

Musshoff F, Jacob B, Fowinkel C, et al. (1993) Suicidal yew leave ingestion - Phloroglucindimethylether (3,5-dimethoxyphenol) as a marker for poisoning from Taxus baccata. Int J Legal Med 106(1):45-50.

Neuteboom JH (1994) Arrest hoge raad der Nederlanden over aansprakelijkheid bij vergiftiging paarden door Taxus baccata. Tijdschr Diergeneeskd 119(20):612.

Panter KE, Molyneux RJ, Smart RA, et al. (1993) English yew poisoning in 43 cattle. J Am Vet Med Assoc 202(9):1476-1477.

Roberto U, Jelmoni A (1915) Un caso di avvelenamento per frutti di tasso baccato. Boll Chim Farm 54:705-709.

Schüler V (1979) Tödliche Taxusvergiftung bei Weiderindern. Dtsch Tierarztl Wochenschr 86(1):29.

Schulte T (1975) Tödliche Vergiftung mit Eibennadeln (Taxus baccata). Arch Toxikol 34(2):153-158.

Smit MP (1992) Intoxicatie van lammeren door Taxus baccata en vleeskeuring. Tijdschr Diergeneeskd 117(23):697-699.

Stanislas E, Bouissou H, Auvergne E (1964) L'intoxication expérimentale du rat par l'If commun Taxus baccata L. - Aspect anatomomopathologique. Therapie 19:1021-1026.

Stebbing J, Simmons HL, Hepple J (1995) Deliberate self-harm using yew leaves (Taxus baccata). Br J Clin Pract 49(2):101.

Szabados A (1976) Lovak tiszafa- (Taxus baccata L.) mérgezésének súlyos esetei. Magyar Allator Lapja 31(1):69.

Thompson J (1868) Poisoning by yew-berries. Lancet 2(Oct 17):530.

Van Ingen G, Visser R, Peltenburg H, et al. (1992) Sudden unexpected death due to Taxus poisoning - A report of 5 cases, with review of the literature. Forensic Sci Int 56(1):81-87.

von Dach B, Streuli RA (1988) Lidocainbehandlung einer Vergiftung mit Eibennadeln (Taxus baccata L.). Schweiz Med Wochenschr 118(30):1113-1116.

von der Werth J, Murphy JJ (1994) Cardiovascular toxicity associated with yew leaf ingestion. Br Heart J 72(1):92-93.

Willaert W, Claessens P, Vankelecom B, et al. (2002) Intoxication with Taxus baccata: Cardiac arrhythmias following yew leaves ingestion. Pacing Clin Electrophysiol 25(4 pt 1):511-512.

Yersin B, Frey JG, Schaller MD, et al. (1987) Fatal cardiac arrhythmias and shock following yew leaves ingestion. Ann Emerg Med 16(12):1396-1397.

TAXUS BREVIFOLIA Nutt. [Taxaceae]

Common Names:

Pacific yew; western yew; yew

Citations:

Cummins RO, Haulman J, Quan L, et al. (1990) Near-fatal yew berry intoxication treated with external cardiac pacing and digoxin-specific FAB antibody fragments. Ann Emerg Med 19(1):38-43.

TAXUS CUSPIDATA Siebold & Zucc. [Taxaceae]

Common Names:

Japanese yew; Japansk idegran; tree of death; yew

Citations:

Alden CL, Fosnaugh CJ, Smith JB, et al. (1977) Japanese yew poisoning of large domestic animals in the Midwest. J Am Vet Med Assoc 170(3):314-316.

Casteel SW, Cook WO (1985) Japanese yew poisoning in ruminants. Mod Vet Pract 66(11):875-877.

Cope RB, Camp C, Lohr CV (2004) Fatal yew (Taxus sp) poisoning in Willamette Valley, Oregon, horses. Vet Hum Toxicol 46(5):279-281.

Evans KL, Cook JR Jr (1991) Japanese yew poisoning in a dog. J Am Anim Hosp Assoc 27:300-302.

Helman RG, Fenton K, Edwards WC, et al. (1996) Sudden death in calves due to Taxus ingestion. Agri-Practice 17(16-18):8.

Karns PA (1983) Intoxication in horses due to ingestion of Japanese yew (Taxus cuspidata). Equine Pract 5(1):12,14-15.

Kerr LA, Edwards WC (1981) Japanese yew: A toxic ornamental shrub. Vet Med Small Anim Clin 76(9):1339-1340.

Lowe JE, Hintz HF, Schryver HF, et al. (1970) Taxus cuspidata (Japanese yew) poisoning in horses. Cornell Vet 60(1):36-39.

Rook JS (1994) Japanese yew toxicity. Vet Med 89:950-951.

Sinn LE, Porterfield JF (1991) Fatal taxine poisoning from yew leaf ingestion. J Forensic Sci 36(2):599-601.

Takebe C (1983) [Acute yew (Taxus cuspidata) poisoning of milking cows.] J Jpn Vet Med Assoc 36(9):521-524.

Thomson GW, Barker IK (1978) Case report: Japanese Yew (Taxus cuspidata) poisoning in cattle. Can Vet J 19(11):320-321.

Thompson LJ (1987) Japanese yew poisoning in cattle and other animals. Napinet Rep 1(2):3.

Veatch JK, Reid FM, Kennedy GA (1988) Differentiating yew poisoning from other toxicoses. Vet Med 83(3):298-300.

TAXUS ×MEDIA Rehder [Taxaceae]

Common Names:

Hicks yew

Citations:

Arai M, Stauber E, Shropshire CM (1992) Evaluation of selected plants for their toxic effects on canaries. J Am Vet Med Assoc 200(9):1329-1331.

Shropshire CM, Stauber E, Arai A (1992) Evaluation of selected plants for acute toxicosis in budgerigars. J Am Vet Med Assoc 200(7):936-939.

tea –see– *Camellia sinensis* (L.) Kuntze

tea oil tree –see– *Melaleuca alternifolia* (Maiden & Betche) Cheel

teaberry –see– *Gaultheria procumbens* L.

teak –see– *Tectona grandis* L. f.

teamster's-tea –see– *Ephedra viridis* Coville

teasel bur –see– *Datura stramonium* L.

Tecoma impetiginosa Mart. ex DC. = Tabebuia impetiginosa (Mart. ex DC.) Standl.

TECTONA GRANDIS L. f. [Lamiaceae]

Common Names:

Bangkok teak; Burma teak; Indian teak; Java teak; moulmain teak; Rangoon teak; Siam teak; teak; true teak

Citations:

Carrié C, Stelzer E (1955) Das Teakholz-Ekzem. Allergy 4(1):33-36.

Estlander T, Jolanki R, Alanko K, et al. (2001) Occupational allergic contact dermatitis caused by wood dusts. Contact Dermatitis 44(4):213-217.

Estlander T, Jolanki R, Kanerva L (1999) Occupational allergic contact dermatitis eczema caused by obeche and teak dusts. Contact Dermatitis 41(3):164.

Hunt E (1931) A case of teakwood dermatitis. Lancet 1:75-76.

Krogh HK (1962) Contact eczema caused by true teak (Tectona grandis) - An epidemiological investigation in a furniture factory. Br J Ind Med 19(3):42-46.

Krogh HK (1964) Contact eczema caused by true teak (Tectona grandis). A follow-up study of a previous epidemiological investigation, and a study into the sensitizing effect of various teak extracts. Br J Ind Med 21(Jan):65-68.

Schmidt H (1978) Contact urticaria to teak with systemic effects. Contact Dermatitis 4(3):176-177.

te-de-sena –see– *Senna alexandrina* Mill.

tehetu –see– *Dioclea erecta* Hoehne; *Dioclea latifolia* Benth.

Teichlilie –see– *Iris pseudoacorus* L.

tempate –see– *Jatropha curcas* L.

temple tree –see– *Plumeria rubra* L.

TEMPLETONIA EGENA (F. Muell.) Benth. [Fabaceae]

Common Names:

broom

Citations:

Cleland JB (1931) Plants, including fungi, poisonous or otherwise injurious to man in Australia. Series III. Med J Aust 2(Dec 19):775-778.

teora dahl –see– *Lathyrus sativus* L.

TEPHROSIA APOLLINA (Delile) Link [Fabaceae]

Citations:

Suliman HB, Wasfi IA, Adam SE (1982) The toxic effects of Tephrosia apollina on goats. J Comp Pathol 92(2):309-315.

TEPHROSIA PURPUREA (L.) Pers. [Fabaceae]

Common Names:

avasa; can-ja-da; pea bush

Citations:

Fianu FK, Assoku RK, Anumel P (1981) Poisonous weeds in pastures: Experimental studies in animals with Tephrosia purpurea (L) Pers. Bull Anim Health Prod Afr 29(4):341-348.

terba –see– *Chrozophora plicata* (Vahl) A. Juss. ex Spreng.

Terminalia alata B. Heyne ex Roth = Terminalia elliptica Willd.

TERMINALIA BELLIRICA (Gaertn.) Roxb. [Combretaceae]

Common Names:

bahera; barra; beleric myrobalan

Citations:

Kulakkattolickal AT (1987) Piscicidal plants of Nepal: Preliminary toxicity screening using grass carp (Ctenopharyngodon idella) fingerlings. J Ethnopharmacol 21(1):1-9.

TERMINALIA BURSARINA F. Muell. [Combretaceae]

Common Names:

yellowwood

Citations:

McIntosh KS (1934) The toxicity of yellow-wood. (Terminalia bursarina). Queensland Agric J 42(Dec 1):727-729.

TERMINALIA CHEBULA Retz. [Combretaceae]

Common Names:

aralu; black chebulic; harro; karar; myrobalan; negroés olive tree

Citations:

Arseculeratne SN, Gunatilaka AA, Panabokke RG (1985) Studies on medicinal plants of Sri Lanka. Part 14. Toxicity of some traditional medicinal herbs. J Ethnopharmacol 13:323-335.

Kulakkattolickal AT (1987) Piscicidal plants of Nepal: Preliminary toxicity screening using grass carp (Ctenopharyngodon idella) fingerlings. J Ethnopharmacol 21(1):1-9.

TERMINALIA ELLIPTICA Willd. [Combretaceae]

Synonyms:

Terminalia alata B. Heyne ex Roth

Common Names:

Indian laurel

Citations:

Kulakkattolickal AT (1987) Piscicidal plants of Nepal: Preliminary toxicity screening using grass carp (Ctenopharyngodon idella) fingerlings. J Ethnopharmacol 21(1):1-9.

TERMINALIA OBLONGATA F. Muell. [Combretaceae]

Common Names:

yellowwood

Citations:

Filippich LJ, Cao GR (1993) Experimental acute yellowwood (Terminalia oblongata) intoxication in sheep. Aust Vet J 70(6):214-218.

Hunt S, McCosker PJ (1970) Observations on serum adenosine deaminase activity in experimentally produced liver diseases of cattle and sheep: Yellow-wood, lantana, carbon tetrachloride and chronic copper poisoning. Br Vet J 126(2):74-81.

Hunt SE, McCosker PJ (1968) Bromsulphthalein fractional clearance in beef cattle suffering from yellow-wood (Terminalia oblongata) poisoning. Am J Vet Clin Pathol 2:161-170.

Legg J, Moule GR, Chester RD (1945) The toxicity of yellow-wood (Terminalia oblongata) to cattle. Queensland J Agric Sci 2(4):199-208.

testiculate buttercup –see– *Ceratocephala testiculata* (Crantz) Roth

teswa –see– *Lathyrus sativus* L.

tetonwort –see– *Solanum dulcamara* L.

TETRADENIA RIPARIA (Hochst.) Codd [Lamiaceae]

Synonyms:

Iboza riparia (Hochst.) N. E. Br.

Citations:

Bodenstein JW (1977) Toxicity of traditional herbal remedies. S Afr Med J 52(20):790.

TETRADYMIA CANESCENS DC. [Asteraceae]

Synonyms:

Tetradymia canescens DC. var. inermis (Nutt.) A. Gray; ***Tetradymia inermis*** Nutt.

Common Names:

horsebrush; silvery horsebrush; spineless horsebrush

Citations:

Anonymous (1934) Investigation of stock-poisoning plants. U S Dep Agric Bur Anim Indus Chief Rep 1934:38.

Anonymous (1935) Investigation of stock-poisoning plants. U S Dep Agric Bur Anim Indus Chief Rep 1935:39-40.

Anonymous (1936) Investigations of stock-poisoning plants. U S Dep Agric Bur Anim Indus Chief Rep 1936:44-45.

Anonymous (1936) Sheep "bighead" caused by two poisonous plants. U S Dep Agric Press Release #1498-36:3 pp.

Anonymous (1938) Investigations of stock-poisoning plants. U S Dep Agric Bur Anim Indus Chief Rep 1938:65-66.

Anonymous (1939) Investigations of stock-poisoning plants. U S Dep Agric Bur Anim Indus Chief Rep 1939:59-60.

Clawson AB, Huffman WT (1935) A report of progress in the study of bighead. Natl Wool Grower 25(1):18-20.

Clawson AB, Huffman WT (1937) Bighead in sheep caused by plant poisoning. Natl Wool Grower 27(3):13-16.

Johnson AE (1974) Predisposing influence of range plants on Tetradymia-related photosensitization in sheep: Work of Drs. A. B. Clawson and W. T. Huffman. Am J Vet Res 35(12):1583-1585.

Johnson AE (1978) Tetradymia toxicity - A new look at an old problem. In: Keeler RF, Van Kampen KR, James LF, et al. (eds.) Effects of poisonous plants on livestock. Academic Press. New York. pp. 209-216.

Tetradymia canescens DC. var. inermis (Nutt.) A. Gray = Tetradymia canescens DC.

TETRADYMIA GLABRATA Torr. & A. Gray [Asteraceae]

Common Names:

coal oil brush; horsebrush; spring rabbitbrush

Citations:

Anonymous (1935) Investigation of stock-poisoning plants. U S Dep Agric Bur Anim Indus Chief Rep 1935:39-40.

Anonymous (1936) Investigations of stock-poisoning plants. U S Dep Agric Bur Anim Indus Chief Rep 1936:44-45.

Anonymous (1936) Sheep "bighead" caused by two poisonous plants. U S Dep Agric Press Release #1498-36:3 pp.

Anonymous (1937) Investigations of stock-poisoning plants. U S Dep Agric Bur Anim Indus Chief Rep 1937:47-48.

Anonymous (1938) Investigations of stock-poisoning plants. U S Dep Agric Bur Anim Indus Chief Rep 1938:65-66.

Anonymous (1939) Investigations of stock-poisoning plants. U S Dep Agric Bur Anim Indus Chief Rep 1939:59-60.

Clawson AB, Huffman WT (1935) A report of progress in the study of bighead. Natl Wool Grower 25(1):18-20.

Clawson AB, Huffman WT (1937) Bighead in sheep caused by plant poisoning. Natl Wool Grower 27(3):13-16.

Fleming CE, Miller MR, Vawter LR (1922) The spring rabbit-brush (Tetradymia glabrata), a range plant poisonous to sheep. Nevada Agric Exp Sta Bull #104:29 pp.

Fleming CE, Miller MR, Vawter LR (1923) The spring rabbit-brush, a range plant poisonous to sheep. Am Sheep Breeder Wool Grower 43(Nov):542-544.

Fleming CE, Miller MR, Vawter LR (1924) The spring rabbit-brush, a range plant poisonous to sheep. Am Sheep Breeder Wool Grower 44:126-128, 150.

Johnson AE (1974) Experimental photosensitization and toxicity in sheep produced by Tetradymia glabrata. Can J Comp Med Vet Sci 38(4):406-410.

Johnson AE (1974) Predisposing influence of range plants on Tetradymia-related photosensitization in sheep: Work of Drs. A. B. Clawson and W. T. Huffman. Am J Vet Res 35(12):1583-1585.

Johnson AE (1978) Tetradymia toxicity - A new look at an old problem. In: Keeler RF, Van Kampen KR, James LF, et al. (eds.) Effects of poisonous plants on livestock. Academic Press. New York. pp. 209-216.

Tetradymia inermis Nutt. = Tetradymia canescens DC.

TETRAGONIA SCHENKII (Schinz) Engl. [Aizoaceae]

Common Names:

koibos

Citations:

Van der Walt SJ, Steyn DG (1946) Recent investigations into the toxicity of plants, etc. XV. Onderstepoort J Vet Sci Anim Indus 21(1):45-55.

TETRAPANAX PAPYRIFER (Hook.) K. Koch [Araliaceae]

Common Names:

Chinese-rice-paper-plant

Citations:

Giannattasio M, Pizzolongo P, Cristaudo A, et al. (1996) Contact dermatitis from Tetrapanax papyriferum trichomes. Contact Dermatitis 35(2):106-107.

TETRAPTERYS ACUTIFOLIA Cav. [Malpighiaceae]

Citations:

Tokarnia CH, Peixoto PV, Döbereiner J, et al. (1989) Tetrapterys spp. (Malpighiaceae), a causa de mortandades em bovinos caracterizadas por alterações cardíacas. Pesq Vet Bras 9(1-2):23-44.

TETRAPTERYS MULTIGLANDULOSA A. Juss. [Malpighiaceae]

Citations:

Carvalho NM, Alonso LA, Cunha TG, et al. (2006) Intoxicação de bovinos por Tetrapterys multiglandulosa (Malpighiaceae) em Mato Grasso do Sul. Pesq Vet Bras 26(3):139-146.

Melo MA, Vasconcelos AC, Datas GC, et al. (1999) Experimental intoxication with Tetrapterys multiglandulosa A. Juss in pregnant goats: Histopathological findings. Toxicol Lett 109(Suppl 1):85.

Tokarnia CH, Peixoto PV, Döbereiner J, et al. (1989) Tetrapterys spp. (Malpighiaceae), a causa de mortandades em bovinos caracterizadas por alterações cardíacas. Pesq Vet Bras 9(1-2):23-44.

TETRASTIGMA OBTECTUM (Wall. ex M. A. Lawson) Planch. ex Franch. [Vitaceae]

Citations:

Deng JF, Lin TJ, Kao WF, et al. (1997) The difficulty in handling poisonings associated with Chinese traditional medicine: A poison control center experience for 1991-1993. Vet Hum Toxicol 39(2):106-114.

tetterberry –see– *Bryonia dioica* Jacq.

tetterwort –see– *Chelidonium majus* L.

TEUCRIUM CHAMAEDRYS L. [Lamiaceae]

Common Names:

germander; germandrée petit chêne; wall germander; western herb germander; wild germander

Citations:

Castot A, Larrey D (1992) Hépatites observées au cours d'un traitement par un médicament ou une tisane contenant de la Germandrée petit-chêne. Gastroenterol Clin Biol 16(12):916-922.

Laliberte L, Villeneuve JP (1996) Hepatitis after the use of germander, a herbal remedy. Can Med Assoc J 154(11):1689-1692.

Larrey D, Vial T, Pauwels A, et al. (1992) Hepatitis after germander (Teucrium chamaedrys) administration: Another instance of herbal medicine hepatotoxicity. Ann Intern Med 117(2):129-132.

Legoux JL, Maitre F, Labarriere D, et al. (1992) Hepatite cytoytique et Germandrée petit-chêne: Un nouveau cas avec réintroduction. Gastroenterol Clin Biol 16(10):813-815.

Mattei A, Bizollon T, Charles JD, et al. (1992) Atteinte hépatique associée à la prise d'un produit de phytothérapie contenant de la Germandrée petit-chêne. Quatro cas. Gastroenterol Clin Biol 16(10):798-800.

Mostefa-Kara N, Pauwels A, Pines E, et al. (1992) Fatal hepatitis after herbal tea. Lancet 340(Sep 12):674.

TEUCRIUM POLIUM L. [Lamiaceae]

Common Names:

golden germander

Citations:

Mazokopakis E, Lazaridou S, Tzardi M, et al. (2004) Acute cholestatic hepatitis caused by Teucrium polium L. Phytomedicine 11(1):83-84.

Polymeros D, Kamberoglou D, Tzias V (2002) Acute cholestatic hepatitis caused by Teucrium polium (golden germander) with transient appearance of antimitochondrial antibody. J Clin Gastroenterol 34(1):100-101.

TEUCRIUM STOCKSIANUM Boiss. [Lamiaceae]

Citations:

Tanira MO, Wasfi IA, Homsi MA, et al. (1996) Toxicological effects of Teucrium stocksianum after acute and chronic adminstration to rats. J Pharm Pharmacol 48(10):1098-1102.

Teufelsauge –see– *Adonis annua* L.; *Hyoscyamus niger* L.

Teufelsrübe –see– *Bryonia dioica* Jacq.

Teufelszwirn –see– *Lycium barbarum* L.

Texas buckeye –see– *Aesculus glabra* Willd.

Texas desert rue –see– *Thamnosma texana* (A. Gray) Torr.

Texas locoweed –see– *Astragalus mollissimus* Torr. var. mollissimus

Texas mesquite –see– *Prosopis glandulosa* Torr.

Texas mistletoe –see– *Phoradendron leucarpum* (Raf.) Reveal & M. C. Johnst.

Texas mountain laurel –see– *Calia secundiflora* (Ortega) Yakovlev

Texas persimmon –see– *Diospyros texana* Scheele

Texas sacahuista –see– *Nolina texana* S. Watson

Texas sacahuiste –see– *Nolina texana* S. Watson

Texas sage –see– *Salvia coccinea* Buc'hoz ex Etl.

Texas skeleton plant –see– *Lygodesmia texana* (Torr. & A. Gray) Greene ex Small

Texas thistle –see– *Solanum rostratum* Dunal

Texas umbrella tree –see– *Melia azedarach* L.

THAMNOSMA TEXANA (A. Gray) Torr. [Rutaceae]

Common Names:

blister weed; Dutchman's-breeches; ruda-del-monte; Texas desert rue

Citations:

Oertli EH, Ivie GW, Bailey EM, et al. (1985) Primary photosensitization in ruminants associated with Thamnosoma texana. Am Vet Med Assoc Annu Meet 1985:115.

Oertli EH, Rowe LD, Lovering SL, et al. (1983) Phototoxic effect of Thamnosma texana (Dutchman's breeches) in sheep. Am J Vet Res 44(6):1126-1129.

Thargomindah nightshade –see– *Solanum sturtianum* F. Muell.

the-bad-seed –see– *Lophophora williamsii* (Lem. ex Salm-Dyck) J. M. Coult.

THEOBROMA CACAO L. [Malvaaceae]

Common Names:

cacahuacuahuitl; cacao; chocolate; cocoa

Citations:

Borzenkov VN, Borxendova VE, Yarmosh GU (1971) [Poisoning of cows from coca shells.] Veterinariia Moscow 48(3):91-92.

Curtis PE, Griffiths JE (1972) Suspected chocolate poisoning of calves. Vet Rec 90(11):313-314.

Day EJ, Dilworth BC (1984) Toxicity of jimson weed seed and cocoa shell meal to broilers. Poult Sci 63(3):466-468.

Drolet R, Arendt TD, Stowe CM (1984) Cacao bean shell poisoning in a dog. J Am Vet Med Assoc 185(8):902.

Hanington E (1972) Suspected chocolate poisoning of calves. Vet Rec 90(14):408-409.

Hansen S, Trammel H, Dunayer E, et al. (2003) Cocoa bean mulch as a cause of methylxanthine toxicosis in dogs. J Toxicol Clin Toxicol 41(5):720.

Hoskam EG, Haagsma J (1974) Chocoladevergiftiging bij twee Dashonden (teckels) met modelijke afloop. Tijdschr Diergeneeskd 99(10):523-525.

Hovda LR, Kingston RL (1994) Cacao bean mulch poisoning in dogs. Vet Hum Toxicol 36(4):357.

Jansson DS, Galgan V, Schubert B, et al. (2001) Theobromine intoxication in a red fox and a European badger in Sweden. J Wildl Dis 37(2):362-365.

Perfetti L, Lehrer SB, McCants M, et al. (1997) Occupational asthma caused by cacao. Allergy 52(7):778-780.

Stidworthy MF, Bleakley JS, Cheeseman MT, et al. (1997) Chocolate poisoning in dogs. Vet Rec 141(1):28.

Strachan ER, Bennett A (1994) Theobromine poisoning in dogs. Vet Rec 134:284.

Sutton RH (1981) Cocoa poisoning in a dog. Vet Rec 109(25-26):563-564.

Tarka SM Jr, Applebaum RS, Borzelleca JF (1986) Evaluation of the teratogenic potential of cocoa powder and theobromine in New Zealand white rabbits. Food Chem Toxicol 24(5):363-374.

Tarka SM Jr, Morrissey RB, Apgar JL, et al. (1991) Chronic toxicity/carcinogenicity studies of cocoa powder in rats. Food Chem Toxicol 29(1):7-19.

Yeruham I, Avidar Y (2003) Photosensitivity in feedlot calves apparently related to cocoa shells. Vet Hum Toxicol 45(5):249-250.

THERMOPSIS MONTANA Nutt. [Fabaceae]

Common Names:

false lupine; golden banner; mountain thermopsis; poison bean; yellow bean

Citations:

Baker DC, Keeler RF (1989) Thermopsis montana-induced myopathy in calves. J Am Vet Med Assoc 194(9):1269-1272.

Baker DC, Keeler RF (1992) Myopathy associated with Thermopsis montana ingestion in cattle. In: James LF, Keeler RF, Bailey JM Jr, et al. (eds.) Poisonous plants. Iowa State Univ Press. Ames, Iowa. pp. 264-270.

Chase RL, Keeler RF (1983) Mountain thermopsis toxicity in cattle. Utah Sci 44(2):28-31.

Keeler RF, Johnson AE, Chase RL (1986) Toxicity of Thermopsis montana in cattle. Cornell Vet 76(2):115-127.

THESIUM LINEATUM L. f. [Santalaceae]

Common Names:

Vaalstorm; witstorm

Citations:

Anderson LA, Joubert JP, Schultz RA, et al. (1987) Experimental evidence that the active principle of the poisonous plant Thesium lineatum L.f. (Santalaceae) is a bufadienolide. Onderstepoort J Vet Res 54(4):645-650.

Schultz RA, Kellerman TS, Van den Berg H (2005) The role of fluorescence polarization immuno-assay in the diagnosis of plant-induced cardiac glycoside poisoning livestock in South Africa. Onderstepoort J Vet Res 72(3):189-201.

THESIUM NAMAQUENSE Schltr. [Santalaceae]

Common Names:

gifbossie; poison bush

Citations:

Van der Walt SJ, Steyn DG (1940) Recent investigations into the toxicity of known and unknown poisonous plants in the Union of South Africa, X. Onderstepoort J Vet Res 15(1-2):261-277.

Thevetia neriifolia Juss. ex A. DC. = Thevetia peruviana (Pers.) K. Schum.

THEVETIA PERUVIANA (Pers.) K. Schum. [Apocynaceae]

Synonyms:

Cerbera thevetia L.; ***Thevetia neriifolia*** Juss. ex A. DC.

Common Names:

ahouai; be-still tree; campanillo amarillo; Cook tree; epile oleander; exile oleander; jorro jorro; kanir; lucky nut; milk tree; Olomiojo; pila kaner; retama; snake nut; tiger apple; tree daffodil; trumpet flower; yellow oleander; yoyote

Citations:

Ahlawat SK, Agarwal AK, Wadhwa S (1994) Rare poisoning with Cerebra thevetia (yellow oleander): A report of three cases. Trop Doct 24(1):37-38.

Ansford AJ, Morris H (1981) Fatal oleander poisoning. Med J Aust 1(7):360-361.

Bhattacharya SK, Somani PN, Srivastava PK (1976) Cardiac changes in Thevetia neriifolia poisoning. Acta Cardiol 31(2):169-174.

Bose TK, Basu RK, Biswas B, et al. (1999) Cardiovascular effects of yellow oleander ingestion. J Indian Med Assoc 97(10):407-410.

Brewster D (1986) Herbal poisoning: A case report of a fatal yellow oleander poisoning from the Solomon Islands. Ann Trop Paediatr 6(4):289-291.

de Silva HA, Fonseka MM, Pathmeswaran A, et al. (2003) Multiple-dose activated charcoal for treatment of yellow oleander poisoning: A single-blind, randomised, placebo-controlled trial. Lancet 361(9373):1935-1938.

Dev V, Wasir HS (1985) Digitalis poisoning by an indigenous plant cardiac glycoside (Thevetia neriifolia - pila kaner). Indian Heart J 37(5):321-322.

Eddleston M (2003) Management of yellow oleander poisoning. J Toxicol Clin Toxicol 41(4):426.
Eddleston M, Ariaratnam CA, Meyer WP, et al. (1999) Epidemic of self-poisoning with seeds of the yellow oleander tree (Thevetia peruviana) in northern Sri Lanka. Trop Med Int Health 4(4):266-273.
Eddleston M, Gunnell D, Karunaratne A, et al. (2005) Epidemiology of intentional self-poisoning in rural Sri Lanka. Br J Psychiatry 187(187):583-584.
Eddleston M, Rajapakse S, Rajakanthan, et al. (2000) Anti-digoxin Fab fragments in cardiotoxicity induced by ingestion of yellow oleander: A randomised controlled trial. Lancet 355(9208):967-972.
Fernando R (2002) The national poisons information centre in Sri Lanka: The first ten years. J Toxicol Clin Toxicol 40(5):551-555.
Fonseka MM, Seneviratne SL, de Silva CE, et al. (2002) Yellow oleander poisoning in Sri Lanka: Outcome in a secondary care hospital. Hum Exp Toxicol 21:293-295.
Kakrani AL, Rajput CS, Khandare SK, et al. (1981) Yellow oleander seed poisoning with cardiotoxicity. A case report. Indian Heart J 33(1):31-33.
Lucas GN (1997) Plant poisoning: A hospital-based study in Sri Lanka. Indian J Pediatr 64(4):495-502.
Mallick BK (1984) Cardiotoxicity in yellow oleander seed poisoning. J Indian Med Assoc 82(8):296-297.
Misra A (1990) Poisoning from Thevetia neriifolia (yellow oleander). Postgrad Med J 66(776):492.
Oji O, Okafor QE (2000) Toxicological studies on stem bark, leaf and seed kernel of yellow oleander (Thevetia peruviana). Phytother Res 14(2):133-135.
Pahwa R, Chatterjee VC (1990) The toxicity of yellow oleander (Thevetia neriifolia Juss) seed kernels to rats. Vet Hum Toxicol 32(6):561-564.
Pathare AV, Patil RR, Chikhalikar AA, et al. (1987) Rare poisoning with Cerebra thevetia (a case report). J Postgrad Med 33(4):216-218.
Pearn J (1987) Oleander poisoning. In: Covacevich et al. (eds.) Toxic plants and animals. A guide for Australia. pp. 37-49.
Phadke MV, Naik SG (1963) Yellow oleander (Cerebra thevetia) poisoning (Case report). Indian J Child Health 12(Mar):210-213.
Reddy DJ, Ramasubbarao M (1954) Necropsy findings in yellow oleander poisoning. J Indian Med Assoc 23(12):559-560.
Roy BM (1927) A case of toxic heart-block due to Cerbera thevetia (yellow oleander seeds). Ind Med Gaz 62(Apr):450-451.
Samal KK, Sahu HK, Kar MK, et al. (1989) Yellow oleander (Cerbera thevetia) poisoning with jaundice and renal failure. J Assoc Physicians India 37(3):232-233.
Saraswat DK, Garg PK, Saraswat M (1992) Rare poisoning with Cerebra thevetia (yellow oleander). Review of 13 cases of suicidal attempt. J Assoc Physicians India 40(9):628-629.
Saravanapavananthan N, Ganeshamoorthy J (1988) Yellow oleander poisoning - A study of 170 cases. Forensic Sci Int 36(3-4):247-250.
Sarker AK, Gosh S, Barik K (1990) A study of accidental poisoning (in children) in a rural medical college hospital of West Bengal. Indian J Public Health 34(3):159-162.
Sreeharan N, Puthrasingam S, Ranjadayalan K, et al. (1985) Yellow oleander poisoning - Clinical manifestations and prognostic criteria. Jaffna Med J 20(2):100-101.
Suresh TP (1998) A survey on yellow oleander poisoning. Nurs J India 89(2):31, 47.
Thilagar S, Thirumalaikolundusubramanian P, Gopalakrishnan S, et al. (1986) Possible yellow oleander toxicity in a neonate. Indian Pediatr 23(May):393.
Vince JD, Salamon B, Tau G, et al. (1984) Digoxin type toxicity from ingestion of Thevetia peruviana or the case of the betel nut that wasn't. Papua New Guinea Med J 27(3-4):167-169.

thick-leaf drymary –see– *Drymaria holosteoides* Benth.; *Drymaria pachyphylla* Wooton & Standl.

THILOA GLAUCOCARPA (Mart.) Eichler [Combretaceae]

Citations:

Döbereiner J, Tokarnia CH (1985) Giftpflanzenbedingte Nierenschädigungen bei Rindern in Brasilien. Dtsch Tierarztl Wochenschr 92(10):411-415.
Tokarnia CH, Peixoto PV, Döbereiner J (1988) Intoxicação experimental pelas folhas e extratos de Thiloa glaucocarpa (Combretaceae) em coelhos. Pesq Vet Bras 8(3-4):61-74.

thimble weed –see– *Rudbeckia laciniata* L.

thimbles –see– *Digitalis purpurea* L.

thistle root –see– *Argemone mexicana* L.

THLASPI ARVENSE L. [Brassicaceae]

Common Names:

fanweed; field pennycress; Frenchweed; Hellerkraut; pennycress; pennycress mustard; stinkweed

Citations:

Baksi SN, Case AA (1971) Photosensitization in guinea-pigs due to ingestion of Thlaspi arvense (penny-cress mustard) seed. Indian Vet J 48(10):1001-1006.
Hawk W (1956) Hematuria in dairy heifers probably due to a plant toxin. J Am Vet Med Assoc 128(Mar 1):261-262.
Martin T, Morgan S (1987) What caused the photosensitivity in these dairy heifers? Vet Med 82:848,850-851.
Rose SP, Bell JM, Wilkie IW, et al. (1981) Influence of weed seed oil contamination on the nutritional quality of diets containing low erucic acid rapeseed (Brassica napus, Tower cultivar) oil when fed to rats. J Nutr 111(2):355-364.
Shires A, Bell JM, Keith MO, et al. (1982) Rapeseed dockage: Effects of feeding raw and processed wild mustard and stinkweed seed on growth and feed utilization of mice. Can J Anim Sci 62:275-285.
Smith RA, Crowe SP (1987) Fanweed toxicosis in cattle: Case history, analytical method, suggested treatment, and fanweed detoxification. Vet Hum Toxicol 29(2):155-156.

Thompson's-woolly locoweed –see– *Astragalus mollissimus* Torr. var. thompsoniae (S. Watson) Barneby

thorn apple –see– *Datura ferox* L.; *Datura metel* L.; *Datura stramonium* L.

thorny amaranthus –see– *Amaranthus spinosus* L.

thorny pigweed –see– *Amaranthus spinosus* L.

thoroughwort –see– *Ageratina altissima* (L.) R. M. King & H. Rob.; *Eupatorium altissimum* L.

thousand-head kale –see– *Brassica oleracea* L. var. ramosa DC.

thousand-leaf –see– *Achillea millefolium* L.

thousand-mothers –see– *Tolmiea menziesii* Torr. & A. Gray

threadleaf broomweed –see– *Gutierrezia microcephala* (DC.) A. Gray

threadleaf groundsel –see– *Senecio flaccidus* Less. var. flaccidus; *Senecio riddellii* Torr. & A. Gray

threadleaf snakeweed –see– *Gutierrezia microcephala* (DC.) A. Gray

threadleaf sartwellia –see– *Sartwellia flaveriae* A. Gray

three-flower nightshade –see– *Solanum triflorum* Nutt.

three-leaf ivy –see– *Toxicodendron pubescens* Mill.; *Toxicodendron radicans* (L.) Kuntze

three-seed mercury –see– *Acalypha indica* L.

Threlkeldia proceriflora F. Muell. = Neobassia proceriflora (F. Muell.) A. J. Scott

throatwort –see– *Digitalis purpurea* L.

THUJA OCCIDENTALIS L. [Cupressaceae]

Common Names:

American arbor vitae; Amerikanischer Lebensbaum; arbor vitae; cédre; false white cedar; faux thuya; Lebensbaum; red cedar; thuja; thuya; white cedar; yellow cedar

Citations:

Brauch F (1932) Das klinische Bild der Thujavergiftung. Z Klin Med 119:86-91.

Millet Y, Jouglard J, Steinmetz MD, et al. (1981) Toxicity of some essential plant oils. Clinical and experimental study. Clin Toxicol 18(12):1485-1498.

Thuja orientalis L. = Platycladus orientalis (L.) Franco

THUJA PLICATA Donn ex D. Don [Cupressaceae]

Common Names:

arbor vitae; Canadian red cedar; flat cedar; giant arbor vitae; red cedar; western red cedar; yellow cedar

Citations:

Bleumink E, Mitchell JC, Nater JP (1973) Allergic contact dermatitis from cedar wood (Thuja plicata). Br J Dermatol 88(5):499-504.

Bleumink E, Nater JP (1972) Contact dermatitis from western red cedar wood (Thuja plicata). Contact Dermatol Newsl 12(Aug):340-342.

Chan-Yeung M (1982) Immunologic and nonimmunologic mechanisms in asthma due to western red cedar (Thuja plicata). J Allergy Clin Immunol 70(1):32-37.

Chan-Yeung M, Barton GM, MacLean L, et al. (1973) Occupational asthma and rhinitis due to western red cedar (Thuja plicata). Am Rev Respir Dis 108(5):1094-1102.

Chan-Yeung M, Lam S, Koener S (1982) Clinical features and natural history of occupational asthma due to western red cedar (Thuja plicata). Am J Med 72(3):411-415.

Chan-Yeung M, MacLean L, Paggiaro PL (1987) Follow-up study of 232 patients with occupational asthma caused by western red cedar (Thuja plicata). J Allergy Clin Immunol 79:792-796.

Chan-Yeung M, Vedal S, Kus J, et al. (1984) Symptoms, pulmonary functions and bronchial hyperreactivity in western red cedar workers compared with those in office workers. Am Rev Respir Dis 130(6):1038-1041.

Estlander T, Jolanki R, Alanko K, et al. (2001) Occupational allergic contact dermatitis caused by wood dusts. Contact Dermatitis 44(4):213-217.

Gandevia B, Milne J (1970) Occupational asthma and rhinitis due to western red cedar (Thuja plicata), with special reference to bronchial reactivity. Br J Ind Med 27(3):235-244.

Ishizaki T, Shida T, Miyamoto T, et al. (1973) Occupational asthma from western red cedar dust (Thuja plicata) in furniture factory workers. J Occup Med 15(7):580-585.

Milne J, Gandevia B (1969) Occupational asthma and rhinitis due to western (Canadian) red cedar (Thuja plicata) Med J Aust 2:741-744.

Mitchell C (1970) Occupational asthma due to western or Canadian red cedar (Thuja plicata). Med J Aust 57(2):233-235.

Mitchell JC, Chan-Yeung M (1974) Contact allergy from Frullania and respiratory allergy from Thuja. Can Med Assoc J 110(6):653-657.

Mue S, Ise T, Ono Y, et al. (1975) A study of western red cedar-induced asthma. Ann Allergy 34(5):296-304.

Mue S, Ise T, Ono Y, et al. (1975) A study of western red cedar sensitivity: Workers' allergy reactions and symptoms. Ann Allergy 35(3):148-152.

Ordman D (1949) Bronchial asthma caused by the inhalation of wood dust. Ann Allergy 7:492-496, 505.

thuja –see– *Thuja occidentalis* L.

thumalos –see– *Euphorbia characias* L.

thunderwood –see– *Toxicodendron vernix* (L.) Kuntze

thuya –see– *Thuja occidentalis* L.

thym –see– *Thymus vulgaris* L.

thyme –see– *Thymus vulgaris* L.

Thymian –see– *Thymus vulgaris* L.

THYMUS VULGARIS L. [Lamiaceae]

Common Names:

thym; thyme; Thymian

Citations:

Benito M, Jorro G, Morales C, et al. (1996) Labiatae allergy: Systemic reactions due to oregano and thyme. Ann Allergy Asthma Immunol 76(5):416-418.

Martínez-González MC, Goday Buján JJ, Martínex Gómez W, et al. (2007) Concomitant allergic contact dermatitis due to Rosmarinus officinalis (rosemary) and Thymus vulgaris (thyme). Contact Dermatitis 56(1):49-50.

Spiewak R, Skorska C, Dutkiewicz J (2001) Occupational airborne contact dermatitis caused by thyme dust. Contact Dermatitis 44(4):235-239.

tickberry –see– *Lantana camara* L.

tickle weed –see– *Veratrum viride* Aiton

ticklegrass –see– *Hordeum jubatum* L.

TIEGHEMELLA AFRICANA Pierre [Sapotaceae]

Common Names:

ukola; ukorro

Citations:

Dantin-Gallego J, Armayor AF, Riesco J (1952) Some new toxic woods: Some new manifestations of toxicity. Ind Med Surg 21(2):41-46.

TIEGHEMELLA HECKELII Pierre ex A. Chev. [Sapotaceae]

Common Names:

Afrikanische Birnbaum; makoré

Citations:

Oleffe JA, Sporcq J, Hublet P (1975) Epidemiological study of the wood industry in Belgium. Contact Dermatitis 1(5):315-316.

tiger apple –see– *Thevetia peruviana* (Pers.) K. Schum.

tiger lily –see– *Lilium lancifolium* Thunb.

tigerwort –see– *Scrophularia auriculata* L.

til –see– *Sesamum indicum* L.

TILIA CORDATA Mill. [Malvaceae]

Common Names:

linden

Citations:

Krakowiak A, Krêcisz B, Pas-Wyroślak A, et al. (2004) Occupational contact dermatitis with rhinoconjunctivitis due to Tilia cordata and colophonium exposure in a cosmetician. Contact Dermatitis 51(1):34.

Mur P, Freo Brito F, Lombardero M, et al. (2001) Allergy to linden pollen (Tilia cordata). Allergy 56(5):456-457.

TILIA ×EUROPAEA L. [Malvaceae]

Common Names:

lime; linden; tilleul

Citations:

Rudzki E, Rapiejko P, Rebandel P (2003) Occupational contact dermatitis, with asthma and rhinitis, from camomile in a cosmetician also with contact urticaria from both camomile and lime flowers. Contact Dermatitis 49(3):162.

TILIA ORBICULARIS hort. ex V. Engl. [Malvaceae]

Citations:

Howes FN (1979) Plants and beekeeping. Faber & Faber. London pp. 23-26.

TILIA PETIOLARIS DC. [Malvaceae]

Citations:

Howes FN (1979) Plants and beekeeping. Faber & Faber. London pp. 23-26.

tilleul –see– *Tilia ×europaea* L.

tilly –see– *Croton tiglium* L.

timbaúba –see– *Enterolobium contortisiliquum* (Vell.) Morong

timbe mirim –see– *Indigofera suffruticosa* Mill.

timber milk vetch –see– *Astragalus convallarius* Greene; *Astragalus miser* Douglas ex Hook; *Astragalus miser* Douglas ex Hook. var. hylophilus (Rydb.) Barneby; *Astragalus miser* Douglas ex Hook. var. oblongifolius (Rydb.) Cronquist; *Astragalus miser* Douglas ex Hook. var. serotinus (A. Gray ex J. G. Cooper) Barneby

timber poison vetch –see– *Astragalus convallarius* Greene

timbó –see– *Deguelia utilis* (A. C. Sm.) A. M. G. Azevedo; *Mascagnia pubiflora* (A. Juss.) Griseb.

timbó legitimo –see– *Deguelia utilis* (A. C. Sm.) A. M. G. Azevedo

timbó macaquinho –see– *Deguelia utilis* (A. C. Sm.) A. M. G. Azevedo

timborá –see– *Enterolobium timbouva* Mart.

timothy –see– *Phleum pratense* L.

timothy canarygrass see *Phalaris angusta* Nees ex Trin.

ting-kuang-teng –see– *Erycibe henryi* Prain

tinguaciba –see– *Zanthoxylum flavum* Vahl

tingui –see– *Mascagnia rigida* (A. Juss.) Griseb.

Tinnevelly senna –see– *Senna alexandrina* Mill.

tintan –see– *Cestrum diurnum* L.

Tintenbeerstrauch –see– *Ligustrum vulgare* L.

Tintenbeertraube –see– *Ligustrum vulgare* L.

tipsywood –see– *Indigofera trita* L. f.

Tipton weed –see– *Hypericum perforatum* L.

tithwan –see– *Artemisia vulgaris* L.

titimalo –see– *Euphorbia dendroides* L.

Tium michauxii (Kuntze) Rydb. = Astragalus michauxii (Kuntze) F. J. Herm.

Toano milk vetch –see– *Astragalus toanus* M. E. Jones

tobacci –see– *Nicotiana tabacum* L.

tobacco –see– *Nicotiana glauca* Graham; *Nicotiana tabacum* L.

tobacco cimarron –see– *Verbascum thapsus* L.

tobasco mahogany –see– *Swietenia macrophylla* King

tochina –see– *Lathyrus sativus* L.

toddy fishtail palm –see– *Caryota urens* L.

tolguacha –see– *Datura stramonium* L.

Tollkirsche –see– *Atropa belladonna* L.

Tollkorn –see– *Lolium temulentum* L.

Tollkraut –see– *Datura stramonium* L.; *Lolium temulentum* L.; *Scopolia carniolica* Jacq.

tollon –see– *Heteromeles salicifolia* (C. Presl) Abrams

TOLMIEA MENZIESII Torr. & A. Gray [Saxifragaceae]

Common Names:

pickaback plant; piggyback plant; thousand-mothers; youth-on-age

Citations:

Calnan CD (1969) Tolmiea menziesii. Contact Dermatol Newsl 5(Feb):98.

Hjorth N (1969) Plant dermatitis. Contact Dermatol Newsl 6(Jul):126-127.

toloache –see– *Datura stramonium* L.

tolu balsam –see– *Myroxylon balsamum* (L.) Harms

tomate –see– *Lycopersicon esculentum* Mill.

tomaten Düsche –see– *Lycopersicon esculentum* Mill.

tomates halentano –see– *Lycopersicon esculentum* Mill.

tomatillo –see– *Solanum elaeagnifolium* Cav.

tomato –see– *Lycopersicon esculentum* Mill.

tomato bush –see– *Solanum esuriale* Lindl.

tomato weed –see– *Solanum elaeagnifolium* Cav.

tonga bean –see– *Dipteryx odorata* (Aubl.) Willd.

tonka bean –see– *Dipteryx odorata* (Aubl.) Willd.

tontelbos –see– *Gomphocarpus fruticosa* (L.) W. T. Aiton

toot –see– *Coriaria arborea* Linds.

toothed nightshade –see– *Solanum virginianum* L.

toothpick ammi –see– *Ammi visnaga* (L.) Lam.

Toowoomba canarygrass –see– *Phalaris aquatica* L.

torch plant –see– *Aloe arborescens* Mill.

tordanda –see– *Euphorbia royleana* Boiss.

tornel –see– *Descurainia sophia* (L.) Webb ex Prantl

toti –see– *Dioclea erecta* Hoehne; *Dioclea latifolia* Benth.

tournesol –see– *Heliotropium europaeum* L.

toursole –see– *Heliotropium europaeum* L.

tovana –see– *Smodingium argutum* E. Mey. ex Sond.

TOXICODENDRON DIVERSILOBUM (Torr. & A. Gray) Greene [Anacardiaceae]

Synonyms:

Rhus diversiloba Torr. & A. Gray

Common Names:

California poison ivy; California poison oak; California poison sumach; hiedra; la yedra; oak-leaf poison ivy; Pacific poison oak; western poison ivy; western poison oak; yeara; yedra

Citations:

Alderson HE, Pruett HJ (1921) Poison oak dermatitis (a specific treatment). Cal State J Med 19(5):188-193.

Devich KB, Lee JC, Epstein WL, et al. (1975) Renal lesions accompanying poison oak dermatitis. Clin Nephrol 3(3):106-113.

Epstein WL, Byers VS, Frankart W (1982) Induction of antigen specific hyposensitization to poison oak in sensitized adults. Arch Dermatol 118(9):630-633.

McNair GB (1916) The pathology of dermatitis venenata from Rhus diversiloba. J Infect Dis 19:419-428.

Petrone GS (1984) Miner's itch - Report of an unusual case of poison oak dermatitis. J Am Acad Dermatol 11(1):142.

Rytand DA (1948) Fatal anuria, the nephrotic syndrome and glomerular nephritis as sequels of the dermatitis of poison oak. Am J Med 5:548-560.

Rytand DA, Burnham DK, Cox AJ Jr (1948) Periarteritis nodosa following the dermatitis of poison oak and of primrose. Stanford Med Bull 6(2):319-323.

Schwartz RS, Downham TF 2nd (1981) Erythema multiforme associated with rhus contact dermatitis. Cutis 27(1):85-86.

Spain WC, Newell JM, Meeker M (1934) Susceptibility to poison ivy and poison oak. Science 80(2065):104.

Templeton HJ, Lunsford CJ, Allington HV (1947) Poison oak dermatitis. Studies on hematologic, urinary and temperature changes. J Invest Dermatol 8:53-58.

Watts WJ (1989) Poison oak urethritis. N Engl J Med 321:194.

TOXICODENDRON PUBESCENS Mill. [Anacardiaceae]

Synonyms:

Rhus toxicodendron L.; ***Toxicodendron quercifolium*** (Michx.) Greene; ***Toxicodendron toxicarium*** (Salisb.) Gillis

Common Names:

eastern oakleaf poison ivy; eastern poison oak; Gift-sumach; oakleaf poison ivy; poison oak; southern poison oak; three-leaf ivy; yeara

Citations:

Anonymous (1848) Poisoning by infusion of poison oak. West J Med Surg 2:364.

Gochfeld M, Burger J (1983) Sexual transmission of nickel and poison oak contact dermatitis. Lancet 1(12):589.

Kishaba RG, Losek JD (1989) Toxic shock syndrome associated with poison oak dermatitis. Pediatr Emerg Care 5(1):40-42.

Laur WE, Posey RE, Waller JD (1978) Rhus dermatitis. An unusual example of exposure. Cutis 22(5):613-614.

Spain WC, Newell JM, Meeker MG (1934) The percentage of persons susceptible to poison ivy and poison oak. J Allergy 5:571-574.

Toxicodendron quercifolium (Michx.) Greene = Toxicodendron pubescens Mill.

TOXICODENDRON RADICANS (L.) Kuntze [Anacardiaceae]

Synonyms:

Rhus radicans L.

Common Names:

black mercury; climath; climbing ivy; climbing sumach; creeping sumach; Giftbaum; Giftefeu; Gifteiche; Giftsumach; markery; markweed; mercury; picry; poison ash; poison creeper; poison ivy; poison mercury; poison sumach; poison vine; poor-man's-liquid-amber; three-leaf ivy; trailing sumach; zumaque venenoso

Citations:

Anonymous (1960) Transmission of ivy poisoning. JAMA 173(Jul 23):1398.

Barefoot SW (1952) Complications of poison ivy (Rhus toxicodendron) dermatitis. N C Med J 13(Jan):40-42.

Beeaff D (1985) Poison ivy dermatitis in Arizona. Ariz Med 42(4):223-226.

Benedict RC (1933) A poison ivy experiment. Torreya 33(May-Jun):65-66.

Beurey J, Mougeolle JM, Weber M, et al. (1980) Rhus dermatitis ou «dermite au poison ivy» en Lorraine. Ann Dermatol Venereol 107(1-2):65-67.

Biberstein H (1929) Über Hautreaktionen bei Applikation von verschiedenen Rhusarten. Klin Wochenschr 8:99-102.

Block SH (1973) Rhus hyposensitization dermatitis. JAMA 224(5):627.

Brown ED (1922) Experiments on the variability in susceptibility to poison ivy. Arch Derm Syphilol 5:714-722.

Burry JN (1969) The value of patch testing: A review of 363 cases of allergic contact dermatitis. Med J Aust 1(Jun 14):1124-1231.

Cohen LM, Cohen JL (1998) Erythema multiforme associated with contact dermatitis to poison ivy: Three cases and a review of the literature. Cutis 62(3):139-142.

Cronk GA (1952) Zirconium salts in prevention and treatment of Rhus toxicodendron dermatitis. Arch Derm Syphilol 66(2):282-288.

Dvorak HF, Mihm MC Jr (1971) Basophilic leukocytes in a case of poison ivy. N Engl J Med 285(1):54-55.

Gaillard GE (1950) Poison ivy, a summary of one hundred cases treated with aqueous (Alum precipitated pyridine) extract. J Allergy 21:55-62.

Gold H, Masucci P (1941) Prophylactic oral therapy against poison ivy. J Allergy 13:157-165.

Gold H, Masucci P (1942) Oral prophylaxis against poison ivy. J Allergy 13:606-610.

Goldman L (1942) Oral prophylaxis for poison ivy dermatitis in children. Am J Dis Child 64:241-247.

Guin JD (1980) Reaction time in experimental poison ivy dermatitis. Contact Dermatitis 6(6):289-290.

Guin JD (1983) Poison ivy dermatitis in winter with an example of filial contact dermatitis. J Indiana State Med Assoc 76(3):184.

Guin JD, Reynolds R (1980) Jewelweed treatment of poison ivy dermatitis. Contact Dermatitis 6(4):287-288.

Harshberger JW (1895) When is Rhus toxicodendron most active? Garden Forest 8(381):238-239.

Hoagland RJ (1950) Prophylaxis of Rhus toxicodendron dermatitis. N Engl J Med 242(4):130-132.

Howell JB (1943) Evaluation of measures for prevention of ivy dermatitis. Arch Derm Syphilol 48:373-378.

Hurwitz RM, Rivera HP, Guin JD (1984) Black-spot poison ivy dermatitis. An acute irritant contact dermatitis superimposed upon an allergic contact dermatitis. Am J Dermatopathol 6(4):319-322.

Kaidbey KH, Kligman AM (1976) Assay of topical corticosteroids. Efficacy of suppression of experimental Rhus dermatitis in humans. Arch Dermatol 112(6):808-810.

Keil H (1958) The prophylactic treatment of poison ivy dermatitis with 3-n-pentadecyl catechol using the wheal method. N Y State J Med 58(1):57-59.

Kligman AM (1958) Poison ivy (Rhus) dermatitis: An experimental study. Arch Dermatol 77(2):149-180.

Knowles FC, Decker HB, Pratt AG, et al. (1938) Susceptibility of allergic and nonallergic persons to Rhus toxicodendron. Arch Derm Syphilol 38:773-779.

Langs RJ, Fuchs AM, Strauss MB (1959) Poison ivy dermatitis. Ind Med Surg 28(6):257-261.

Laur WE, Posey RE, Waller JD (1978) Rhus dermatitis. An unusual example of exposure. Cutis 22(5):613-614.

Lejman E, Stoudemayer T, Grove G, et al. (1984) Age differences in poison ivy dermatitis. Contact Dermatitis 11(3):163-167.

Lipton RA (1958) The use of Impatiens biflora (jewelweed) in the treatment of rhus dermatitis. Ann Allergy 16(5):526-527.

Maiden JH (1909) On some plants which cause inflammation or irritation of the skin. Agric Gaz New South Wales 20(Feb 2):111-117.

Maisel F (1932) Poison ivy. New method of immunization. J Allergy 4(1):35-37.

Mallory SB, Miller OF, Tyler WB (1982) Toxicodendron radicans dermatitis with black lacquer deposit on the skin. J Am Acad Dermatol 6(3):363-368.

Marks JG Jr, Trautlein JJ, Epstein WL, et al. (1987) Oral hyposensitization to poison ivy and poison oak. Arch Dermatol 123(4):476-478.

Meeham J (1889) Poisonous plants. Garden Forest 2(57):154.

Mitchell WF (1959) Poison ivy prophylaxis: Folklore or scientific medicine? Ohio State Med J 55(6):797-798.

Moorman JW (1866) Poisoning by eating the fruit of the Rhus toxicodendron. Am J Med Sci 51:560.

Nurse DS (1966) Poison ivy in Melbourne. Med J Aust 1(13):528-529.

Oliveira Lima A (1953) Über das antigene Verhalten der Ölharze einiger Gattungen der Familie Anacardiaceae. Int Arch Allergy 4:169-174.

Orchard S, Fellman JH, Storrs FJ (1986) Poison ivy/oak dermatitis. Use of polyamine salts of a linoleic acid dimer for topical prophylaxis. Arch Dermatol 122(7):783-789.

Passenger RE (1963) A clinical evaluation of the prophylactic treatment of poison ivy dermatitis with an alum-precipitated pyridine extract of Rhus toxicodendron. J Allergy 34(3):270-274.

Proujan BJ (1988) New treatment to prevent poison ivy and poison oak dermatitis. Res Resour Rep 12(3):5-7.

Reginella RF, Fairfield JC, Marks JG Jr (1989) Hyposensitization to poison ivy after working in a cashew nut shell oil processing factory. Contact Dermatitis 20(12):274-279.

Ross CM (1959) Poison ivy dermatitis: The first South African cases. S Afr Med J 33:657-660.

Sassaman RK (1945) Stomatitis due to chewing leaves of poison ivy. Am J Orthodon Oral Surg 31:695-696.

Silvers SH (1941) Stomatitis venenata and dermatitis of the anal orifice from chewing poison ivy leaves (Rhus toxicodendron). JAMA 116(20):2257.

Sompayrac LM (1938) Negative results of Rhus antigen treatment of experimental ivy poisoning. Am J Med Sci 195:361-362.

Spain WC, Newell JM, Meeker M (1934) Susceptibility to poison ivy and poison oak. Science 80(2065):104.

Spain WC, Newell JM, Meeker MG (1934) The percentage of persons susceptible to poison ivy and poison oak. J Allergy 5:571-574.

Spencer MC (1961) Occupational dermatitis and eczema among farmers. Ill Med J 119:136-138.

Spencer MC (1971) Poison ivy dermatitis - A simple, safe and effective treatment. Cutis 7(4):443-435.

Stokes J (1867) Peculiar cases of poisoning from the root of the Rhus toxicodendron. Med Surg Reper 17:373.

Turner T (1972) Poison ivy and poor-man's liquid amber. Med J Aust 2(3):166-167.

Vollmer H (1936) Hauterkrankung nach peroraler Zuführing von Rhus toxicodendron L. Sammlung Vergiftungsfallen 7(B68):21-22.

Whitfield A (1914) On three cases of eczematous dermatitis produced by poisonous plants. Lancet 1(Feb 28):607-608.

Whiting DA (1971) Plant dermatitis in the southern Transvaal. S Afr Med J 45(7):163-167.

Zink BJ, Otten EJ, Rosenthal M, et al. (1991) The effect of jewel weed in preventing poison ivy dermatitis. J Wilderness Med 2(3):178-182.

Zisserman L (1940) Oral immunization for poison ivy dermatitis. J Allergy 12:474-476.

Zisserman L (1940) Susceptibility to poison ivy dermatitis. J Allergy 11:600.

Zisserman L, Birch L (1939) The prophylaxis and treatment of poison ivy dermatitis with an extract of Rhus toxicodendron. J Allergy 10:596-604.

TOXICODENDRON RYDBERGII (Small ex Rydb.) Greene [Anacardiaceae]

Common Names:

Canadian poison ivy; nonclimbing poison ivy; northern poison ivy; Rydberg's-poison ivy; western poison ivy

Citations:

Spoerke DG, Temple AR (1978) One year's experience with potential plant poisonings reported to the Intermountain Regional Poison Control Center. Vet Hum Toxicol 20(2):85-89.

TOXICODENDRON STRIATUM (Ruiz & Pav.) Kuntze [Anacardiaceae]

Synonyms:

Rhus striata Ruiz & Pav.

Common Names:

amche; amte; hinchador; hinchahuevos; mala mujer; manganillo-de-cerro; manzanillo; palo-de-compadre; palo-de-sarna; palo-de-viruela; palo sarno; poison sumach; yagalache

Citations:

Hurtado I (1965) Contact dermatitis caused by the "manzanillo" (Rhus striata) tree. Report of three cases. Int Arch Allergy Appl Immunol 28(6):321-327.

Hurtado I (1967) "Manzanillo." Dermatitis venenata causada por el "Rhus striata." Med Cutanea 11(3):253-258.

Hurtado I (1968) Studies on the biological activity of Rhus striata ('manzanillo'). II. Skin response to patch test in humans. Int Arch Allergy 33(3):209-216.

TOXICODENDRON SUCCEDANEUM (L.) Kuntze [Anacardiaceae]

Synonyms:

Rhus succedanea L.

Common Names:

Indochinese lacquer; Japanese wax tree; oriental sumach; oriental wax tree; red lac; rhus; rhus tree; scarlet rhus; scarlet sumach; sumach; wax tree

Citations:

Burry JN (1969) The value of patch testing: A review of 363 cases of allergic contact dermatitis. Med J Aust 1(Jun 14):1124-1231.

Nakamura T (1985) Contact dermatitis to Rhus succedanea. Contact Dermatitis 12(5):279.

Rademaker M, Duffill MB (1995) Allergic contact dermatitis to Toxicodendron succedaneum (Rhus tree) - An autumn epidemic. N Z Med J 108:121-123.

Rademaker M, Duffill MB (1995) Toxicodendron succedaneum (Rhus tree), New Zealand's poison ivy. Contact Dermatitis 33(8):357-358.

Whiting DA (1971) Plant dermatitis in the southern Transvaal. S Afr Med J 45(7):163-167.

TOXICODENDRON SYLVESTRE (Siebold & Zucc.) Kuntze [Anacardiaceae]

Citations:

Mitchell JC, Guin JD, Maibach HI, et al. (1985) Allergenicity of Toxicodendron sylvestre (Anacardiaceae). Contact Dermatitis 12(2):113-114.

Toxicodendron toxicarium (Salisb.) Gillis = Toxicodendron pubescens Mill.

TOXICODENDRON VERNICIFLUUM (Stokes) F. A. Barkley [Anacardiaceae]

Synonyms:

Rhus vernicifera DC.; ***Rhus verniciflua*** Stokes

Common Names:

Chinese lacquer tree; Chinese shellac; Japanese lacquer tree; Japanese shellac; Japanese sumach; Japanese varnish; Japanische Lackbaum; Japanlack; lacquer; lacquer tree; ningpo varnish; oriental lacquer tree; poison lac; varnish lacquer tree; varnish tree

Citations:

Ames O (1931) Rhus verniciflua and Japanese damascence ware. J Arnold Arbor 12(1):1-3.

Barrett DK (1985) A case history of acute dermatitis caused by Rhus verniciflua Stokes and brief review of toxic sumachs in Britain. Arboricultural J 9:115-120.

Bauer H (1940) Über Sumachdermatitis. Med Klin 36(2):48-49.

Biberstein H (1929) Über Hautreaktionen bei Applikation von verschiedenen Rhusarten. Klin Wochenschr 8:99-102.

Etter RL (1951) Dermatitis caused by Japanese lacquer. U S Armed Forces Med J 2(3):505-507.
Grevenstuk A (1937) Vergiftung durch Renghas und durch Japanlack. Sammlung Vergiftungsfallen 8(A664):35-38.
Hinman F (1946) Contact dermatitis from Japanese rifles. Ann Allergy 4:384-387.
Kawai K, Nakagawa M, Kawai K, et al. (1991) Hyposensitization to urushiol among Japanese lacquer draftsmen: Results of patch tests on students learning the art of lacquerware. Contact Dermatitis 25(5):290-295.
Kobayashi Y (1935) Sensitization of guinea-pigs to Rhus venicifera and to Japanese and Chinese lacquers. Jpn J Dermatol Urol 37(Apr):479.
Levin OL (1924) Dermatitis venenata from the lacquer on the boxes of Mah Jong sets. JAMA 82(6):465.
Powell SM, Barrett DK (1986) An outbreak of contact dermatitis from Rhus verniciflua (Toxicodendron verniciflum). Contact Dermatitis 14(5):288-289.
Powell SM, Barrett DK, Venning VA (1987) Contact dermatitis from Rhus verniciflua - Further evidence concerning the hazard of domestic planting. Arboricultural J 11(2):165-168.
Zeisler EP (1924) Dermatitis venenata from the lacquer on the boxes of Mah Jong sets. JAMA 82(6):466.

TOXICODENDRON VERNIX (L.) Kuntze [Anacardiaceae]

Synonyms:

Rhus vernix L.

Common Names:

poison ash; poison dogwood; poison elderberry; poison sumach; poison swamp sumach; poison tree; poison weed; poisonwood; swamp dogwood; swamp sumach; thunderwood

Citations:

Benedict RC (1933) A poison ivy experiment. Torreya 33(May-Jun):65-66.
Jacobziner H, Raybin HW (1962) Intoxications due to tranquilizing drugs and plants. N Y State J Med 62(Oct 1):3130-3132.
Sherer JW (1916) Conjunctivitis and keratitis from poison ivy. Ophthal Record 25:191-193.

TOXICOSCORDION FREMONTII (Torr.) Rydb. [Melanthiaceae]

Common Names:

chaparral death camas; death camas; star lily; star zydadene

Citations:

Quortrup ER, McFarland RJ (1956) Animal losses involving noxious weeds in San Diego County. California Vet 9(5):14-17.

TOXICOSCORDION INTERMEDIUM (Rydb.) Rydb. [Melanthiaceae]

Synonyms:

Zigadenus intermedius Rydb.

Common Names:

death camas

Citations:

Mitchell PH, Smith G (1911) The physiological effects of alkaloids of Zygadenus intermedius. Am J Physiol 28:318-329.

TOXICOSCORDION NUTTALLII (A. Gray) Rydb. [Melanthiaceae]

Synonyms:

Zigadenus nuttallii A. Gray

Common Names:

alkaligrass; death camas; grassy death camas; Nuttall's-death camas; poison camas; poison sego; soap plant; water lily; wild onion

Citations:

Marsh CD, Clawson AB, Roe GC (1926) Nuttall's death camas (Zygadenus nuttallii) as a poisonous plant. U S Dep Agric Bull #1376:13 pp.
Morris MD (1944) Nuttall death camas poisoning in horses. Vet Med 39(12):462.

TOXICOSCORDION PANICULATUM (Nutt.) Rydb. [Melanthiaceae]

Synonyms:

Zigadenus paniculatus (Nutt.) S. Watson

Common Names:

alkaligrass; death camas; foothill death camas; panicle death camas; poison sego; wild sago

Citations:

Fleming CE (1920) Poisonous range plants. Nevada Agric Exp Sta Annu Rep 1919:39-43.
Fleming CE, Peterson NF, Miller MR, et al. (1921) Death camas (Zygadenus paniculatus and Zygadenus venenosus). Plants poisonous to sheep and cattle. Nevada Agric Exp Sta Bull #101:31 pp.
Heilpern KL (1995) Zigadenus poisoning. Ann Emerg Med 25(2):259-262.
Hilman FH (1896) Zygadenus paniculatus is mentioned as poisonous. Nevada Agric Exp Sta Annu Rep #9:25.
Marsh CD, Clawson AB (1922) The death camas species, Zygadenus paniculatus and Z. elegans, as poisonous plants. U S Dep Agric Bull #1012:25 pp.
Marsh CD, Clawson AB, Marsh H (1915) Zygadenus, or death camas. U S Dep Agric Bull #125:46 pp.
Panter KE, Ralphs MH, Smart RA, et al. (1987) Death camas poisoning in sheep: A case report. Vet Hum Toxicol 29(1):45-48.
Peterson MC, Rasmussen GJ (2003) Intoxication with foothill camas (Zigadenus paniculatus). J Toxicol Clin Toxicol 41(1):63-65.

TOXICOSCORDION VENENOSUM (S. Watson) Rydb. [Melanthiaceae]

Synonyms:

Zigadenus gramineus Rydb.; ***Zigadenus venenosus*** S. Watson

Common Names:

death camas; grassy death camas; meadow death camas; mysterygrass; poison camas; poisongrass; poison sego; poison wild onion; squirrel food

Citations:

Avery RJ, Niilo L, Kramer T, et al. (1961) Two cases of poisoning in livestock presenting difficulties in diagnosis. I. Death camas (Zygadenus gramineus, Rydb.) poisoning in sheep. Can Vet J 2(7):250-252.
Cameron K (1952) Death camas poisoning. Northwest Med 51(Aug):682-683.
Collett S, Grotelueschen D, Smith R, et al. (1996) Deaths of 23 adult cows attributed to intoxication by the alkaloids of Zygadenus venenosus (meadow death camas). Agri-Practice 17(7):5-9.
Fleming CE (1920) Poisonous range plants. Nevada Agric Exp Sta Annu Rep 1919:39-43.
Fleming CE, Peterson NF, Miller MR, et al. (1921) Death camas (Zygadenus paniculatus and Zygadenus venenosus). Plants poisonous to sheep and cattle. Nevada Agric Exp Sta Bull #101:31 pp.
Hitchcock JD (1959) Poisoning of honey bees by death camas blossoms. Am Bee J 99(Oct):418-419.
Marsh CD, Clawson AB (1922) The death camas species, Zygadenus paniculatus and Z. elegans, as poisonous plants. U S Dep Agric Bull #1012:25 pp.
Marsh CD, Clawson AB (1924) The meadow death camas (Zygadenus venenosus) as a poisonous plant. U S Dep Agric Bull #1240:13 pp.
Marsh CD, Clawson AB, Marsh H (1915) Zygadenus, or death camas. U S Dep Agric Bull #125:46 pp.
Nelson SB (1906) Feeding wild plants to sheep. Washington Agric Exp Sta Bull #73:64 pp.

toyon –see– *Heteromeles salicifolia* (C. Presl) Abrams

TRACHELIUM CAERULEUM L. [Campanulaceae]

Citations:

Banerjee P, Rycroft RJ (2002) A second florist with dermatitis from Trachelium caeruleum. Contact Dermatitis 46(4):241.

TRACHYANDRA DIVARICATA (Jacq.) Kunth [Asphodelaceae]

Common Names:

branched onion weed

Citations:

Huxtable CR, Chapman HM, Main DC, et al. (1987) Neurological disease and lipofuscinosis in horses and sheep grazing Trachyandra divaricata (branched onion weed) in south Western Australia. Aust Vet J 64(4):105-108.
Newsholme SJ, Schneider DJ, Reid C (1985) A suspected lipofuscin storage disease of sheep associated with ingestion of the plant, Trachyandra divaricata (Jacq.) Kunth. Onderstepoort J Vet Res 52(2):87-92.

TRACHYANDRA LAXA (N. E. Br.) Oberm. [Asphodelaceae]

Citations:

Grant RC, Basson PA, Kidd AB (1985) Paralysis and lipofuscin-like pigmentation of farm stock caused by the plant, Trachyandra laxa var. laxa. Onderstepoort J Vet Res 52(4):255-259.

TRACHYANDRA SALTII (Baker) Oberm. [Asphodelaceae]

Citations:

Shone DK (1966) Poisonous plants of Rhodesia. Part I. Rhodesia Zambia Malawi J Agric Res 4:81-94.
Shone DK (1967) Poisonous plants of Rhodesia. Part II. Rhodesia Zambia Malawi J Agric Res 5:29-41.

TRACHYMENE GLAUCIFOLIA (F. Muell.) Benth. [Apiaceae]

Synonyms:

Didiscus glaucifolius F. Muell.

Common Names:

carrot weed; native parsnip; wild carrot; wild parsnip

Citations:

Edgar G, Ropert J (1942) The toxicity of Didiscus glaucifolius, "wild parsnip," for sheep. Aust Vet J 18:125-126.
Philbey AW (1990) Trachymene glaucifolia associated with bentleg in lambs. Aust Vet J 67(12):468.

TRACHYMENE OCHRACEA L. A. S. Johnson [Apiaceae]

Common Names:

parsley; parsnip; wild parsnip

Citations:

Clark L (1977) Trachymene spp and infertility in sheep. Aust Vet J 53(5):249-250.

TRADESCANTIA FLUMINENSIS Vell. [Commelinaceae]

Common Names:

wandering Jew

Citations:

Kunkle GA, Gross TL (1983) Allergic contact dermatitis to Tradescantia fluminensis (wandering Jew) in a dog. Compend Cont Educ Pract Vet 5(11):925-930.
Wüthrich B, Johansson SG (1997) Allergy to the ornamental indoor green plant Tradescantia (Albifloxia). Allergy 52(5):556-559.

trailing indigo –see– *Indigofera hendecaphylla* Jacq.; *Indigofera spicata* Forssk.

trailing mahonia –see– *Berberis aquifolium* Pursh

trailing sumach –see– *Toxicodendron radicans* (L.) Kuntze

Transvaal camelthorn –see– *Acacia ×giraffae* Willd.

Transvaal kameeldoring –see– *Acacia ×giraffae* Willd.

Transvaal slangkop –see– *Drimia sanguinea* (Schinz) Jessop

Transvaal yellow tulp –see– *Moraea pallida* (Baker) Goldblatt

Traubenkirsche –see– *Prunus padus* L.

traveler's-joy –see– *Clematis vitalba* L.

tread-softly –see– *Solanum carolinense* L.

trébol acedo –see– *Oxalis acetosella* L.

trébol blanco –see– *Trifolium repens* L.

trébol dulce –see– *Melilotus albus* Medik.

trébol loco –see– *Melilotus albus* Medik.

trébol rojo –see– *Trifolium pratense* L.

trecul queen's-delight –see– *Stillingia treculiana* (Müll. Arg.) I. M. Johnst.

tree daffodil –see– *Thevetia peruviana* (Pers.) K. Schum.

tree datura –see– *Brugmansia arborea* (L.) Lagerh.

tree lucerne –see– *Chamaecytisus prolifer* (L. f.) Link

tree morning-glory –see– *Ipomoea carnea* Jacq.

tree nettle –see– *Urtica ferox* G. Forst.

tree-of-death –see– *Taxus cuspidata* Siebold & Zucc.

tree-of-heaven –see– *Ailanthus altissima* (Mill.) Swingle

tree rosewood –see– *Dalbergia nigra* (Vell.) Allemão ex Benth.

tree tobacco –see– *Nicotiana glauca* Graham

tree tutu –see– *Coriaria arborea* Linds.

tree zamia –see– *Cycas media* R. Br.

trèfle aigrelet –see– *Oxalis acetosella* L.

trèfle blanc –see– *Trifolium repens* L.

trèfle commun –see– *Trifolium pratense* L.

trèfle-des-prés –see– *Trifolium pratense* L.

trèfle-porte-fraise –see– *Trifolium repens* L.

trèfle souterrain –see– *Trifolium subterraneum* L.

trefoil –see– *Medicago polymorpha* L.; *Medicago sativa* L.

Trema aspera (Brongn.) Blume = Trema tomentosa (Roxb.) H. Hara

TREMA CANNABINUM Lour. [Cannabaceae]

Synonyms:

Sponia virgata Planch.

Common Names:

wild peach

Citations:

Hindmarsh WL (1940) Poisonous to stock. Peach leaf poison bush (Trema cannabina syn. T. aspera). Agric Gaz New South Wales 51(8):429-430.

TREMA MICRANTHA (L.) Blume [Cannabaceae]

Common Names:

Grandiúva

Citations:

Traverso SD, Corrêa AM, Pescador CA, et al. (2002) Intoxicação experimental por Trema micrantha (Ulmaceae) em caprinos. Pesq Vet Bras 22(4):141-147.

Traverso SD, Corrêa AM, Schmitz M, et al. (2004) Intoxicação experimental por Trema micrantha (Ulmaceae) em bovinos. Pesq Vet Bras 24(4):211-216.

Traverso SD, Driemeier D (2000) Experimental Trema micrantha (Ulmaceae) poisoning in rabbits. Vet Hum Toxicol 42(5):301-301.

TREMA TOMENTOSA (Roxb.) H. Hara [Cannabaceae]

Synonyms:

Trema aspera (Brongn.) Blume

Common Names:

peach-leaf poison bush; poison peach

Citations:

Hill BD, Wills LD, Dowling RM (1985) Suspected poisoning of horses by Trema aspera (poison peach). Aust Vet J 62(3):107-108.

Mulhearn CR (1942) Poison peach (Trema aspera): A plant poisonous to stock. Aust Vet J 18(Apr):68-72.

Trueman KF, Powell MW (1991) Suspected poisoning of camels by Trema tomentosa (poison peach). Aust Vet J 68(6):213-214.

trembling stagger –see– *Dicentra canadensis* (Goldie) Walp.

trevo-doce-amnelo –see– *Melilotus officinalis* Lam.

tri-leaf wonder –see– *Syngonium podophyllum* Schott

tria –see– *Erythrophleum succirubrum* Gagnep.

TRIADICA SEBIFERA (L.) Small [Euphorbiaceae]

Synonyms:

Sapium sebiferum (L.) Roxb.; ***Stillingia sebifera*** (L.) Michx.

Common Names:

arbre-à-suif; Chinese tallow tree; Japanese tallow tree; momchina; pa-teou-seou; popcorn tree; tallow tree; Tankawang fat tree; vegetable tallow tree; wu chiu

Citations:

Russell LH, Schwartz WL, Dollahite JW (1969) Toxicity of Chinese tallow tree (Sapium sebiferum) for ruminants. Am J Vet Res 30(7):1233-1238.

Soifer FK (1973) Tallow tree poisoning in a monkey. J Zoo Anim Med 4(2):13.

triangle-leaf bursage –see– *Ambrosia deltoidea* (Torr.) W. W. Payne

Trianthema monogyna L. = Trianthema portulacastrum L.

TRIANTHEMA PORTULACASTRUM L. [Aizoaceae]

Synonyms:

Trianthema monogyna L.

Common Names:
baura; black pigweed; gadon machiji; giant pigweed; horse purslane; itsit; sea purslane; verdolaga blanca

Citations:

Gupta BK, Gupta SC, Thind IS (1983) Effect of feeding 'itsit' (Trianthema monogyna L.) to goats. J Res Punjab Agric Univ 20(4):539-543.

Mathams RH, Sutherland AK (1952) The oxalate content of some Queensland pasture plants. Queensland J Agric Sci 9:317-334.

TRIBULUS MICROCOCCUS Domin [Zygophyllaceae]

Common Names:
yellow vine

Citations:

Bourke CA, MacFarlane JA (1985) A transient ataxia of sheep associated with the ingestion of Tribulus micrococcus (yellow vine). Aust Vet J 62(8):282.

TRIBULUS TERRESTRIS L. [Zygophyllaceae]

Common Names:
abrojos; bullhead; bur nut; caktrio; caltrops; cat's-head; croix-de-malte; devil's-thorn; dubbelljie; duwweljie; espigon; goat head; goat weed; gokharu; Mexican sandbur; puncture vine; puncture weed; tacks; yellow vine

Citations:

Amjadi AR, Ahourai P, Baharsefat M (1977) First report of geeldikkop in sheep in Iran. Arch Inst Razi 29(Dec):71-78.

Aslani MR, Movassaghi AR, Mohri M, et al. (2003) Experimental Tribulus terrestris poisoning in sheep: Clinical, laboratory and pathological findings. Vet Res Commun 27(1):53-62.

Bourke CA (1983) Hepatopathy in sheep associated with Tribulus terrestris. Aust Vet J 60(6):189.

Bourke CA (1984) Staggers in sheep associated with the ingestion of Tribulus terrestris. Aust Vet J 61(11):360-363.

Bourke CA (1987) A novel nigrostriatal dopaminergic disorder in sheep affected by Tribulus terrestris staggers. Res Vet Sci 43(3):347-350.

Bourke CA (2006) Abnormal turning behaviour, GABAergic inhibition and the degeneration of astrocytes in ovine Tribulus terrestris motor neuron disease. Aust Vet J 84(1-2):53-58.

Brown JM (1959) Advances in "geeldikkop" (Tribulosis ovis) research. 3. The epizootology of "geeldikkop." J S Afr Vet Assoc 30(4):403-417.

Brown JM, le Roux JM, Tustin RC (1960) Advances in "geeldikkop" (Tribulosis ovis) research. 4. The pathology of Geeldikkop - Part I. J S Afr Vet Assoc 31(2):179-193.

Coetzer AW, Kellerman TS, Sadler W, et al. (1983) Photosensitivity in South Africa. 5. A comparative study of the pathology of the ovine hepatogenous photosensitivity diseases, facial eczema and geeldikkop (tribulosis ovis), with special reference to their pathogenesis. Onderstepoort J Vet Res 50(1):59-71.

Glastonbury JR, Boal GK (1985) Geeldikkop in goats. Aust Vet J 62(2):62-63.

Glastonbury JR, Doughty FR, Whitaker SJ, et al. (1984) A syndrome of hepatogenous photosensitisation, resembling geeldikkop, in sheep grazing Tribulus terrestris. Aust Vet J 61(10):314-316.

Jacob RH, Peet RL (1987) Poisoning of sheep and goats by Tribulus terrestris (caltrop). Aust Vet J 64(9):288-289.

Kellerman TS, Van der Westhuizen GC, Coetzer JA, et al. (1980) Photosensitivity in South Africa. 2. The experimental production of the ovine hepatogenous photosensitivity disease geeldikkop (Tribulosis ovis) by the simultaneous ingestion of Tribulus terrestris plants and cultures of Pithomyces chartarum containing the mycotoxin sporidesmin. Onderstepoort J Vet Med 47(4):231-256.

McDonough SP, Woodbury AH, Galey FD, et al. (1994) Hepatogenous photosensitization of sheep in California associated with ingestion of Tribulus terrestris (puncture vine). J Vet Diagn Invest 6(3):392-395.

Pachalag SV, Patil BD (1977) Effect of gokharu (Tribulus terrestris Linn) as sole feed to weaner lambs. Indian Vet J 54(Jul):586-587.

Rimington C, Quin JI (1933) Studies on the photosensitization of animals in South Africa. II. The presence of a lethal factor in certain members of the plant genus Tribulus. Onderstepoort J Vet Sci Anim Indus 1(2):469-489.

Tapia MO, Giordano MA, Gueper HG (1994) An outbreak of hepatogenous photosensitization in sheep grazing Tribulus terrestris in Argentina. Vet Hum Toxicol 36(4):311-313.

van Tonder EM, Basson PA, van Rensburg IB (1972) Geeldikkop: Experimental induction by feeding the plant Tribulus terrestris L. (Zygophyllaceae). J S Afr Vet Assoc 43(4):363-375.

TRICHILIA HAVANNENSIS Jacq. [Meliaceae]

Citations:

Figueroa V, Sutherland TM (1972) "Muerte súbita" (sudden death) in cattle. 5. The role of toxic plants. Rev Cubana Cienc Agric Eng Ed 6(1):53-59.

TRICHODESMA INCANUM (Bunge) A. DC. [Boraginaceae]

Common Names:
camel bush

Citations:

Abdullaev NK, Azimov RK (1972) [Effect of Trichodesma incanum seeds on the development of fatty dystrophy of the liver.] Bull Eksp Biol Med 73(8):28-29.

Ibadullaev F (1967) [Pathological changes in pups experimentally poisoned with Trichodesma incanum poisoning of fowls.] Veterinariia Moscow 44(8):87-88.

Ibadullaev FI, Rasulev KK (1970) [Pathomorphological changes in experimental trichodesmotoxicosis in poultry.] Veterinariia Moscow 47(8):75-77.

Ibragimov KZ (1960) [Toxic effect of Trichodesma incanum on cattle.] Dokl Akad Nauk USSR 4:56-57.

TRICHOSANTHES KIRILOWII Maxim. [Cucurbitaceae]

Citations:

Zhang YG, Huang GZ (1988) Poisoning by toxic plants in China. Report of 19 autopsy cases. Am J Forensic Med Pathol 9(4):313-319.

trifoglio –see– *Trifolium pratense* L.

TRIFOLIUM ALEXANDRINUM L. [Fabaceae]

Common Names:

Alexandrian clover; barseem; berseem clover; Egyptian clover

Citations:

Abd El-Latif K, Awad FI (1964) Haemoglobinuria of buffaloes associated with the excessive feeding of Tripholium alexandrinum (barseem). J Vet Sci United Arab Republic 1(2):69-74.

Adler JH (1969) The effect of estrogenic activity in forage on the reproductive processes of dairy cattle. Acta Vet Brno 38:201-213.

Shehata MN, Hassan A, El-Shazly K, et al. (1978) Oestrogenic activity of fresh, wilted, dried and fermented berseem (Trifolium alexandrinum). J Agric Sci (Cambridge) 91(part 2):359-363.

TRIFOLIUM HYBRIDUM L. [Fabaceae]

Common Names:

Alsatian clover; alsike clover; bastard clover; Bastardklee; Hybridklee; Schweden Klee; Swedish clover

Citations:

Fincher MG, Fuller HK (1942) Case report. Photosensitization - Trifoliosis - light sensitization. Cornell Vet 32:95-98.

Hansen AA (1928) Trifoliosis and similar stock diseases. North Am Vet 9(8):34-36.

Morgan HA, Jacob M (1905) I. Alsike clover. II. Ill effects sometimes produced on horses and mules pastured exclusively upon alsike. Tennessee Agric Exp Sta Bull #18(3):24-30.

Pammel LH (1919) Alsike clover poisoning. Am J Vet Med 14:437-438.

Schofield FW (1933) Liver disease of horses (big liver) caused by the feeding of alsike clover. Ontario Vet Coll Circ #52:4 pp.

Traub JL, Potter KA, Bayly WM, et al. (1982) Alsike clover poisoning. Mod Vet Pract 63(4):307-309.

TRIFOLIUM INCARNATUM L. [Fabaceae]

Common Names:

crimson clover; Inkarnatklee; Italian clover

Citations:

Coville FV (1896) Crimson clover hair balls. U S Dep Agric Div Bot Circ #8:4 pp.

Trifolium officinale L. = Melilotus officinalis Lam.

TRIFOLIUM PRATENSE L. [Fabaceae]

Common Names:

cowgrass; grand trèfle rouge; peavine clover; purple clover; red clover; Rotklee; trèfle commun; trèfle-des-prés; tribol rojo; trifoglio

Citations:

Ballarini G, Rizzi E (1977) Turbe della riproduzione nella scrofa da alimentazione con trifoglio. Clin Vet (Milano) 100(11):761-769.

Barrett JF, George JM, Lamond DR (1965) Reproductive performance of Merino ewes grazing red clover (Trifolium pratense L.), improved pasture, or native pasture. Aust J Agric Res 16:189-200.

Chang TS (1961) Reproductive performance of New Zealand Romney sheep grazed on red clover (Trifolium pratense) pastures. J Agric Sci (Cambridge) 57(1):123-127.

Fox CW, Kaufmes J, Mason RW, et al. (1957) Effects of feeding red clover (Trifolium pratense) on reproduction in laboratory mice. Proc West Sec Am Soc Anim Prod 8(XLIV):1-6.

Kallela K (1964) On the oestrogenic effects of red clover fodder on sheep. Nord Vet Med 16:731-743.

Morley FH, Axelsen A, Bennett D (1964) Effects of grazing red clover (Trifolium pratense, L.) during the joining season on ewe fertility. Proc Aust Soc Anim Prod 5:58-61.

Morley FH, Axelsen A, Bennett D (1966) Recovery of normal fertility after grazing on oestrogenic red clover. Aust Vet J 42(6):204-206.

Newton JE, Betts JE (1973) The effects of red clover (Trifolium pratense var. redhead), white clover (Trifolium repens var. S100) or perennial ryegrass (Lolium perenne var. S23) on the reproductive performance of sheep. J Agric Sci 80:323-327.

Niezen JH, Barry TN, Wilson PR, et al. (1992) Red urine from deer grazed on pure red clover swards. N Z Vet J 40:164-167.

O'Dell BL, Regan WO, Beach TJ (1959) A study of the toxic principle in red clover. Missouri Agric Exp Sta Res Bull #702:12 pp.

Thomson DJ (1975) The effect of feeding red clover conserved by drying or ensiling on reproduction in the ewe. Grass Forage Sci 30(2):149-152.

TRIFOLIUM REPENS L. [Fabaceae]

Common Names:

Dutch clover; ladino clover; Ladinoklee; rastrero; trébol blanco; trèfle blanc; trèfle porte fraise; triolet; Weißklee; white clover; white trifolium

Citations:

Bergeron JM, Jodoin L (1985) Fiabilité des mesures de poids et d'examens histopathologiques dans les études d'intoxication du campagnol des champs (Microtus pennsylvanicus). Can J Zool 63(4):804-810.

Flux DS, Butler GW, Johnson JM, et al. (1956) Goitrogenic effect of white clover (Trifolium repens L.). N Z J Sci Technol A 38:88-102.

Newton JE, Betts JE (1973) The effects of red clover (Trifolium pratense var. redhead), white clover (Trifolium repens var. S100) or perennial ryegrass (Lolium perenne var. S23) on the reproductive performance of sheep. J Agric Sci 80:323-327.

Rainey JW (1961) Fixed vegetable oil as antidote to clover poisoning. Vet Rec 73(11):281-282.

Sanger VL, Bell DS (1961) Comparative effect of ladino clover and bluegrass pasture on fertilization of ova in sheep. Cornell Vet 51(Apr):204-210.

Wright PA (1960) Infertility in rabbits induced by feeding ladino clover. Proc Soc Exp Biol Med 105(Nov):428-430.

TRIFOLIUM SUBTERRANEUM L. [Fabaceae]

Common Names:

Erdklee; sub clover; subterranean clover; subterranean trefoil; trèfle souterrain; yarloop clover

Citations:

Adams NR (1977) Morphological changes in the organs of ewes grazing oestrogenic subterranean clover. Res Vet Sci 22(2):216-221.

Adams NR (1979) Masculinisation of the external genitalia in ewes with clover disease. Aust Vet J 55(1):22-24.

Chamberlain HV, Habel JD (1957) Clover infertility disease of sheep. J Dep Agric South Australia 60:238-247.

Clark NT (1964) Lactation in non-pregnant ewes grazing subterranean clover in summer. Aust Vet J 40(Dec):422-423.

Clark NT (1965) A comparison of the reproductive performance of ewes grazing Dwalganup or Bacchus Marsh subterranean clover or a mixture of lucerne and burr medic before and during mating. Aust J Exp Agric Anim Husb 5(May):106-109.

Davies HL (1976) Reduced production in sheep due to subterranean clover disease. Wool Technol Sheep Breeding 23(2):33-37.

Davies HL, Bennett D (1962) Studies on the oestrogenic potency of subterranean clover (Trifolium subterraneum L.) in South-Western Australia. Aust J Agric Res 13:1030-1040.

Davies HL, Rossiter RC, Maller R (1970) The effects of different cultivars of subterranean clover (T. subterraneum L.) on sheep reproduction in the south-west of Western Australia. Aust J Agric Res 21(2):359-369.

Donaldson LE (1983) Clover disease in two Mississippi cattle herds. J Am Vet Med Assoc 182(4):412-413.

Gardiner MR, Nairn ME, Meyer EP (1966) Urinary calculi associated with oestrogenic subterranean clover. Aust Vet J 42(9):315-320.

Kaltenbach CC, Davies HL (1970) Fertilization, sperm transport, and early embryonic loss in ewes grazed on cultivars of subterranean clover (T. subterraneum). Aust J Agric Res 21(1):107-114.

Lightfoot RJ, Wroth RH (1974) The mechanism of temporary infertility in ewes grazing oestrogenic subterranean clover prior to and during joining. Proc Aust Soc Anim Prod 10:130-134.

Meyer EP (1970) Lactation in rams grazing subterranean clover. Aust Vet J 46(7):305-307.

Thain RI (1965) Bovine infertility possibly caused by subterranean clover: A preliminary report. Aust Vet J 41(9):277-281.

Thain RI (1966) Bovine infertility possibly caused by subterranean clover. Further report and herd histories. Aust Vet J 42(6):199-203.

Wroth RH, Lightfoot RJ (1976) A study of reproductive wastage among commercial sheep flocks grazing oestrogenic pastures in south Western Australia. Proc Aust Soc Anim Prod 11:225-228.

TRIGLOCHIN MARITIMA L. [Juncaginaceae]

Common Names:

arrowgrass; goosegrass; graceful arrowgrass; low arrowgrass; marsh arrowgrass; sea arrowgrass; seaside arrowgrass; six-seed podgrass; sourgrass

Citations:

Anonymous (1935) Investigation of stock-poisoning plants. U S Dep Agric Bur Anim Indus Chief Rep 1935:39-40.

Anonymous (1936) Investigations of stock-poisoning plants. U S Dep Agric Bur Anim Indus Chief Rep 1936:44-45.

Beath OA, Draize JH, Eppson HF (1933) Arrow grass, chemical and physiological considerations. Wyoming Agric Exp Sta Bull #193:36 pp.

Clawson AB, Moran EA (1937) Toxicity of arrowgrass for sheep and remedial treatment. U S Dep Agric Tech Bull #580:16 pp.

Fleming CE (1920) Poisonous range plants. Nevada Agric Exp Sta Annu Rep 1919:39-43.

Fleming CE, Miller MR, Vawter LR, et al. (1931) Poisonous range plants. Nevada Agric Exp Sta Annu Rep 1930:12-13.

Fleming CE, Peterson NF, Miller MR, et al. (1920) Arrowgrass, a new stock-poisoning plant (Triglochin maritima). Nevada Agric Exp Sta Bull #98:22 pp.

Marsh CD, Clawson AB, Roe GC (1929) Arrow grass (Triglochin maritima) as a stock-poisoning plant. U S Dep Agric Tech Bull #113:14 pp.

trigo sarraceno –see– *Fagopyrum esculentum* Moench

TRIGONELLA FOENUM-GRAECUM L. [Fabaceae]

Common Names:

chilbe; fenugreek; foenugreek; Greek hay seed; methi; methika

Citations:

Adler JH, Egyed M (1961) Trigonella foenum-graecum poisoning in sheep. Refu Vet 18:20-21, 45.

Adler JH, Nobel TA, Egyed M, et al. (1960) [Some effects of feeding Trigonella foenum-graecum straw to cattle.] Refu Vet 17:122-125, 171.

triolet –see– *Trifolium repens* L.

Tripetaleia paniculata Siebold et Zucc. = Elliottia paniculata (Siebold & Zucc.) Benth. & Hook. f.

TRIPLOCHITON SCLEROXYLON K. Schum. [Malviaceae]

Common Names:

abachi; African maple; African whitewood; mahogany; obeche; Samba; wawa

Citations:

Estlander T, Jolanki R, Kanerva L (1999) Occupational allergic contact dermatitis eczema caused by obeche and teak dusts. Contact Dermatitis 41(3):164.

Halpin B (1961) Toxicity in fowl due to hardwood litter. Vet Rec 73(18):454-455.

Hinojosa M, Losada E, Moneo I, et al. (1986) Occupational asthma caused by African maple (obeche) and ramin: Evidence of cross reactivity between these two woods. Clin Allergy 16(2):145-153.

Hinojosa M, Moneo I, Dominguez J, et al. (1984) Asthma caused by African maple (Triplochiton scleroxylon) wood dust. J Allergy Clin Immunol 74:782-786.

Hinojosa M, Subiza J, Moneo I, et al. (1990) Contact urticaria caused by obeche wood (Triplochiton scleroxylon). Report of eight patients. Ann Allergy 64(5):476-479.

Kanerva L, Tuppurainen M, Keskinen H (1998) Contact urticaria caused by obeche wood (Triplochiton scleroxylon). Contact Dermatitis 38(3):170-171.

Kespohl S, Sander I, Merget R, et al. (2005) Identification of an obeche (Triplochiton scleroxylon) wood allergen as a class I chitonase. Allergy 60(6):808-814.

Oleffe JA, Sporcq J, Hublet P (1975) Epidemiological study of the wood industry in Belgium. Contact Dermatitis 1(5):315-316.

Venturini M, Gastaminza G, Kespohl S, et al. (2004) Cross-reactivity between obeche wood (Triplochiton scleroxylon) and natural rubber latex. Allergy 59(2):225-228.

TRIPTERYGIUM WILFORDII Hook. f. [Celastraceae]

Common Names:

lei-gong-teng

Citations:

Chan WY, Ng TB, Lu JL (1995) Effects of decoctions prepared from Aconitum carmichaeli, Aconitum kusnezoffii and Tripterygium wilfordii on serum lactate dehydrogenase activity and histology of liver, kidney, heart, and gonad. Hum Exp Toxicol 14(6):489-493.

Huang H (1982) [Clinical analysis of 31 cases of poisoning with Tripterygium wilfordii Hook. f., a Chinese traditional herb medicine.] Chin J Internal Med 21(6):363-365.

Li ZM (1983) [Treatment and prevention of Tripterygium wilfordii poisoning.] Chung Hsi I Chieh Ho Tsa Chih 3(6):368.

Takei A, Nagashima G, Suzuki R, et al. (1997) Meningoencephalocele associated with Tripterygium wilfordii treatment. Pediatr Neurosurg 27(1):45-48.

Zhang YG (1983) [An experimental pathologic study on acute Tripterginm wilfordii poisoning in rats.] Chung Hsi I Chieh Ho Tsa Chih 3(6):360-362.

Zhang YG, Huang GZ (1988) Poisoning by toxic plants in China. Report of 19 autopsy cases. Am J Forensic Med Pathol 9(4):313-319.

TRISETUM FLAVESCENS (L.) P. Beauv. [Poaceae]

Common Names:

golden oats; Goldhafer; oatgrass; yellow false oat; yellow oatgrass

Citations:

Braun U, Diener M, Camenzind D, et al. (2000) Enzootic calcinosis in goats caused by golden oat grass (Trisetum flavescens). Vet Rec 146(6):161-162.

Dirksen G, Plank P, Hänichen T, et al. (1972) Über eine enzootische Kalzinose beim Rind. V. Experimentelle Untersuchungen an Kaninchen mit selektiver Verfütterung von Knaulgras (Dactylis glomerata), Goldhafer (Trisetum flavescens) und einem Gräsergemisch. Dtsch Tierarztl Wochenschr 79(4):77-79.

Dirksen G, Plank P, Hänichen T, et al. (1973) Über eine enzootische Kalzinose beim Rind. VI. Experimentelle Kalzinose beim Kaninchen durch selektiver Verfütterung von Goldhafer (Trisetum flavescens). Dtsch Tierarztl Wochenschr 80(7):148-152.

Dirksen G, Plank P, Simon U, et al. (1974) Über eine enzootische Kalzinose beim Rind. VII. Nachweis der kalzinogenen Wirkung von Goldhafer (Trisetum flavescens [L.] P. B.) beim Wiederkäuer. Dtsch Tierarztl Wochenschr 81(1):1-5.

Dirksen G, Sachs M, Held T, et al. (1983) Untersuchungen über den Einfluß von Aluminumhydroxid und von 1-Hydroxyäthan-1,1-diphosphonat (EHDP) auf den kalzinogenen Effekt von Trisetum flavescens (L., P. B.) beim Schaf. Zentralbl Veterinarmed A 30(9):645-655.

Dirksen G, Simon U, Plank P, et al. (1975) Über eine enzootische Kalzinose beim Rind. VIII. Untersuchungen über die mögliche Bedeutung des Sonnenlichtes (UV-Strahlen) bei der Entstehung der Kalzinose sowie Nachweis der kalzinogenen Wirkung von getrocknetem Goldhafer (Trisetum flavescens (L.) P. B.). Dtsch Tierarztl Wochenschr 82(10):387-390.

Heinritzi K, Kragenings G, Hänichen T (1977) Untersuchungen über die kalzinogene Aktivität von siliertem Goldhafer (Trisetum flavescens [L.] [P. B.]). Z Tierphysiol Tierernahr Futtermittelkd 39(3):139-145.

Köhler H, Libiseller R, Schmid S, et al. (1978) Zur Kalzinose der Rinder in Österreich. VII. Untersuchungen zur Bedeutung der Aufwuchsstadien sowie der Gewinnung (Silage, Heu) von Goldhafer (Trisetum flavescens) für die Entstehung der Kalzinose. Zentralbl Veterinarmed A 25(8):617-633.

Libiseller R, Glawischnig E, Köhler H, et al. (1974) Zur Kalzinose der Rinder in Österreich. II. Experimentelle Auslösung einer Kalzinose bei Schafen und Kaninchen. Zentralbl Veterinarmed A 21:705-730.

Libiseller R, Glawischnig E, Köhler H, et al. (1976) Zur Kalzinose der Rinder in Österreich. III. Experimentelle Auslösung einer Kalzinose bei Schafen und Kaninchen durch grünen Goldhafer (Trisetum flavescens) aus dem pannonischen Klimagebiet. Zentralbl Vetinarmed A 23(1):1-30.

Libiseller R, Köhler H, Glawischnig E, et al. (1978) Zur Kalzinose der Rinder in Österreich. V. Experimentelle Untersuchungen zur Frage der Bedeutung der Aufwuchsstadien und der Menge des Goldhafers auf die Entstehung der Kalzinose. Zentralbl Veterinarmed A 25(1):1-22.

Simon U, Daniel P, Hanichen T, et al. (1978) Kalzinogene Wirkung verschiedener Sorten des Goldhafers (Trisetum flavescens (L.) P.B.) bei Schafen. Z Wirtschaftseigene Futter 24[(3-4)]:209-213.

Simon U, Daniel P, Hänichen T, et al. (1978) Über eine enzootische Kalzinose Rind. XI. Untersuchungen über den Einfluß unterschiedlich hoher Goldhaferanteile im Grünfutter auf Gewebsverkalkungen bei Schafen. Dtsch Tierarztl Wochenschr 85(9):363-366.

TRITICUM AESTIVUM L. [Poaceae]

Common Names:

bran; Weizen; wheat

Citations:

Dempster JG (1981) Contact dermatitis from bran and oats. Contact Dermatitis 7(2):122.

Dorrance GM, Ciccone EF (1937) Production of sarcoma in albino rats as a result of feeding crude wheat germ oil. Proc Soc Exp Biol Med 36:426-427.

Frigg M, Brubacher G (1976) Biotin deficiency in chicks fed a wheat-based diet. Int J Vitam Nutr Res 46:314-321.
Kanny G, Chenuel B, Moneret-Vautrin DA (2001) Chronic urticaria to wheat. Allergy 56(4):356-357.
Majamaa H, Moisio P, Holm K, et al. (1999) Wheat allergy: Diagnostic accuracy of skin prick and patch tests and specific IgE. Allergy 54(8):851-856.
McManus WR, Lee GJ, Robinson VN (1977) Microlesions of rumen papillae of sheep fed diets of wheat grain. Res Vet Sci 22:135-137.

Note:

Wheat is named *Triticum aestivum* L. subsp. *aestivum* in some publications.

troène –see– *Ligustrum vulgare* L.

trollhegg –see– *Frangula alnus* Mill.

trombeteira –see– *Brugmansia arborea* (L.) Lagerh.

trompeta blanca –see– *Brugmansia arborea* (L.) Lagerh.

trompeta-de-oro –see– *Allamanda cathartica* L.

trompetero –see– *Brugmansia ×candida* Pers.

trompillo –see– *Solanum elaeagnifolium* Cav.

tromso palm –see– *Heracleum stevenii* Manden.

tronador –see– *Hura crepitans* L.

TROPAEOLUM MAJUS L. [Tropaeolaceae]

Common Names:

Indian cress; Kapuzinerkresse; nasturtium

Citations:

Derrick E, Darley C (1997) Contact dermatitis to nasturtium. Br J Dermatol 136(2):290-291.

tropical sage –see– *Salvia coccinea* Buc'hoz ex Etl.

tropical soda apple –see– *Solanum viarum* Dunal

tropillo –see– *Solanum elaeagnifolium* Cav.

true saffron –see– *Crocus sativus* L.

true teak –see– *Tectona grandis* L. f.

trumpet flower –see– *Thevetia peruviana* (Pers.) K. Schum.

trumpet lily –see– *Brugmansia arborea* (L.) Lagerh.; *Lilium longiflorum* Thunb.; *Zantedeschia aethiopica* (L.) Spreng.

trumpet narcissus –see– *Narcissus pseudonarcissus* L.

trumpet vine –see– *Brugmansia arborea* (L.) Lagerh.

Trunkelbeere –see– *Vaccinium uliginosum* L.

trusty male fern –see– *Dryopteris affinis* (Lowe) Fraser-Jenk. subsp. borreri (Newman) Fraser-Jenk.

tschat –see– *Catha edulis* (Vahl) Forssk. ex Endl.

tu-chiao-lien –see– *Alocasia macrorrhizos* (L.) G. Don

tuá tuá –see– *Jatropha gossypiifolia* L.

tuba –see– *Croton tiglium* L.; *Jatropha curcas* L.; *Paraderris elliptica* (Wall.) Adema

tuba root –see– *Paraderris elliptica* (Wall.) Adema

tuba tuba –see– *Jatropha curcas* L.

tuber root –see– *Asclepias tuberosa* L.

tuber water hemlock –see– *Cicuta douglasii* (DC.) J. M. Coult. & Rose

tuber yam –see– *Dioscorea convolvulacea* Schltdl. & Cham.

tuca –see– *Bryonia dioica* Jacq.

tufted fishtail palm –see– *Caryota mitis* Lour.

tula-de-mate –see– *Lagenaria siceraria* (Molina) Standl.

tulip –see– *Moraea polystachya* (Thunb.) Ker Gawl.; *Tulipa gesneriana* L.

tulip onion –see– *Tulipa gesneriana* L.

TULIPA GESNERIANA L. [Liliaceae]

Common Names:

tulip; tulip onion; tulpe

Citations:

Wolf P, Blanke HJ, Wohlsein P, et al. (2003) Tulpenzwiebeln mit Blatt (Tulipa gesneriana) - Eine sehr ungewöhnliche und risikoreiche Komponente in einer Ration für Rinder. Dtsch Tierarztl Wochenschr 110(7):302-305.

Note:

There are many reports of adverse effects such as "tulip fingers" where only the genus name is stated.

tullidora –see– *Karwinskia calderonii* Standl.; *Karwinskia humboldtiana* (Schult.) Zucc.

tulp –see– *Moraea miniata* Andrews; *Moraea pallida* (Baker) Goldblatt; *Moraea setacea* Ker Gawl.; *Moraea thomsonii* Baker

tulpe –see– *Tulipa gesneriana* L.

tumbleweed –see– *Salsola tragus* L.

tumin –see– *Citrullus colocynthis* (L.) Schrad.

tung nut –see– *Vernicia fordii* (Hemsl.) Airy Shaw

tung tree –see– *Vernicia fordii* (Hemsl.) Airy Shaw

tupakiki –see– *Coriaria arborea* Linds.

turbina –see– *Turbina corymbosa* (L.) Raf.

TURBINA CORDATA (Choisy) D.F. Austin & Staples [Convolvulaceae]

Citations:

Dantas AF, Riet-Correa F, Gardner DR, et al. (2007) Swainsonine-induced lysosomal storage disease in goats caused by the ingestion of Turbina cordata in northeastern Brazil. Toxicon 49(1):111-116.

TURBINA CORYMBOSA (L.) Raf. [Convolvulaceae]

Synonyms:

Rivea corymbosa (L.) Hallier f.

Common Names:

badoh; morning-glory; ololiuqui; turbina

Citations:

Fink PJ, Goldman MJ, Lyons I (1966) Morning glory seed psychosis. Arch Gen Psychiatry 15:209-213.

Osmond H (1955) Ololiuqui: The ancient Aztec narcotic. J Ment Sci 101:526-537.

Turkestan thistle –see– *Acroptilon repens* (L.) DC.

Turkey boxwood –see– *Buxus sempervirens* L.

turkey bush –see– *Eremophila deserti* (A. Cunn. ex Benth.) Chinnock

turkey corn –see– *Dicentra canadensis* (Goldie) Walp.

turkey oak –see– *Quercus incana* W. Bartram

turkey pea –see– *Dicentra canadensis* (Goldie) Walp.

turkeyberry –see– *Solanum torvum* Sw.

Turkische Bohne –see– *Phaseolus coccineus* L.

Turk's-cap –see– *Aconitum napellus* L.; *Gloriosa superba* L.

turmeric –see– *Curcuma longa* L.

turmus –see– *Lupinus albus* L. var. albus

turnip –see– *Brassica rapa* L. subsp. rapa

turnip rape –see– *Brassica rapa* L. subsp. campestris (L.) A. R. Clapham

turpentine broomweed –see– *Gutierrezia sarothrae* (Pursh) Britton & Rusby

turpentine weed –see– *Gutierrezia microcephala* (DC.) A. Gray; *Gutierrezia sarothrae* (Pursh) Britton & Rusby

TURRAEANTHUS AFRICANUS (Welw. ex C. DC.) Pellegr. [Meliaceae]

Common Names:

avordiré; Westafrikanisches Satinholz

Citations:

Calnan CD (1970) Avordiré wood sensitivity. Contact Dermatol Newsl 8(Aug):190.

Oleffe JA, Sporcq J, Hublet P (1975) Epidemiological study of the wood industry in Belgium. Contact Dermatitis 1(5):315-316.

tussilage –see– *Tussilago farfara* L.

TUSSILAGO FARFARA L. [Asteraceae]

Common Names:

colt's-foot; farfara; Huflattich; kan-to-ka; Knospen; pas-d'Ane-commun; tussilage

Citations:

Hirono I, Mori H, Culvenor CC (1976) Carcinogenic activity of coltsfoot, Tussilago farfara L. Gann 67(Feb):125-129.

tutu –see– *Coriaria arborea* Linds.; *Coriaria ruscifolia* L.

twin-leaf –see– *Roepera ammophila* (F. Muell.) Beier & Thulin

twin-leaf senna –see– *Senna roemeriana* (Scheele) H. S. Irwin & Barneby

two-groove locoweed –see– *Astragalus bisulcatus* (Hook.) A. Gray

two-groove milk vetch –see– *Astragalus bisulcatus* (Hook.) A. Gray

two-groove poison vetch –see– *Astragalus bisulcatus* (Hook.) A. Gray

two-leaf Cape-tulip –see– *Moraea miniata* Andrews

two-leaf senna –see– *Senna roemeriana* (Scheele) H. S. Irwin & Barneby

TYLECODON GRANDIFLORUS (Burm. f.) Toelken [Crassulaceae]

Common Names:

rooisuikerblom

Citations:

Anderson LA, Joubert JP, Prozesky L, et al. (1983) The experimental production of krimpsiekte in sheep with Tylecodon grandiflorus (Burm. f.) Toelken and some of its bufadienolides. Onderstepoort J Vet Res 50(4):301-307.

TYLECODON VENTRICOSA (Burm. f.) Toelken [Crassulaceae]

Common Names:

c'nenta; plakkies

Citations:

Botha CJ, Kellerman TS, Schultz RA, et al. (1998) Krimpsiekte in a sheep following a single dose of Tylecodon ventricosus (Burm. f.) Toelken and the isolation of tylecoside D from this plant species. Onderstepoort J Vet Res 65(1):17-23.

Botha CJ, Van der Lugt JJ, Erasmus GL, et al. (1998) Krimpsiekte, a paretic condition of small stock poisoned by bufadienolide-containing plants of the Crassulaceae in South Africa. In: Garland T, Barr AC (eds.) Toxic plants and other natural toxicants. CABI. New York. pp. 407-412.

Schultz RA, Kellerman TS, Van den Berg H (2005) The role of fluorescence polarization immuno-assay in the diagnosis of plant-induced cardiac glycoside poisoning livestock in South Africa. Onderstepoort J Vet Res 72(3):189-201.

TYLECODON WALLICHII (Harv.) Toelken [Crassulaceae]

Common Names:

bandjiebos; kandlaarbos; krimpsiektebos; nenta; plakkies

Citations:

Botha CJ, Van der Lugt JJ, Erasmus GL, et al. (1998) Krimpsiekte, a paretic condition of small stock poisoned by bufadienolide-containing plants of the Crassulaceae in South Africa. In: Garland T, Barr AC (eds.) Toxic plants and other natural toxicants. CABI. New York. pp. 407-412.

Schultz RA, Kellerman TS, Van den Berg H (2005) The role of fluorescence polarization immuno-assay in the diagnosis of plant-induced cardiac glycoside poisoning livestock in South Africa. Onderstepoort J Vet Res 72(3):189-201.

U

ubi kayu –see– *Manihot esculenta* Crantz

ui –see– *Allium cepa* L.

uie –see– *Allium cepa* L.

uien –see– *Allium cepa* L.

ukola –see– *Tieghemella africana* Pierre

ukorro –see– *Tieghemella africana* Pierre

ulcardo melon –see– *Cucumis melo* L. subsp. agrestis (Naudin) Pamgalo

ului –see– *Anacardium occidentale* L.

um-tovane –see– *Smodingium argutum* E. Mey. ex Sond.

um-tovane-tovanna –see– *Smodingium argutum* E. Mey. ex Sond.

umbellate aster –see– *Doellingeria umbellata* (Mill.) Nees

umbrella arum –see– *Amorphophallus konjac* K. Koch

umbrella leaf –see– *Podophyllum peltatum* L.

umbrella plant –see– *Podophyllum peltatum* L.; *Schefflera actinophylla* (Endl.) Harms

umbrella tree –see– *Melia azedarach* L.; *Schefflera actinophylla* (Endl.) Harms

umbu –see– *Phytolacca dioica* L.

umkoko –see– *Embelia ribes* Burm. f.

umutambasha–see– *Dichapetalum michelsonii* Hauman

unicorn root –see– *Chamaelirium luteum* (L.) A. Gray

upas radja –see– *Strychnos ignatii* P. J. Bergius

upas teute –see– *Strychnos ignatii* P. J. Bergius

upas tree –see– *Antiaris toxicaria* Lesch.

upland cotton –see– *Gossypium hirsutum* L.

upland larkspur –see– *Delphinium nuttallianum* Pritz.

upland willow oak –see– *Quercus incana* W. Bartram

upright meadow crowfoot –see– *Ranunculus acris* L.

Urechites luteus (L.) Britton = Pentalinon luteum (L.) B. F. Hansen & Wunderlin

Urginea altissima (L. f.) Baker = Drimia altissima (L. f.) Ker Gawl.

Urginea burkei Baker = Drimia sanguinea (Schinz) Jessop

Urginea capitata (Hook.) Baker = Drimia depressa (Baker) Jessop

Urginea macrocentra Baker = Drimia macrocentra (Baker) Jessop

Urginea maritima (L.) Baker = Drimia maritima (L.) Stearn

Urginea physodes (Jacq.) Baker = Drimia physodes (Jacq.) Jessop

Urginea sanguinea Schinz = Drimia sanguinea (Schinz) Jessop

UROCHLOA ARRECTA (Hack. ex T. Durand & Schinz) Morrone & Zuloaga [Poaceae]

Synonyms:

Brachiaria radicans Napper

Common Names:

tannergrass

Citations:

Andrade SO, Cosentino JR, Scott WN (1972) Estudos sobre novilhas alimentadas diariamente com milho ou melaço e Brachiaria sp. (Tanner grass). Arq Inst Biol (Sao Paulo) 39(4):329-333.

Andrade SO, Figueiredo AL, Nothenberg MS (1972) Estudos sobre a Brachiaria sp. (Tanner grass). 4. Investigação sobre o efeito da gramínea, apresentando teor baixo de nitrato, em ovinos, bovinos e eqüinos. Arq Inst Biol (Sao Paulo) 39(4):257-261.

Andrade SO, Nothenberg MS, Retz L, et al. (1975) Estudos sobre cobaias alimentadas com Brachiaria sp. (Tanner grass). Arq Inst Biol (Sao Paulo) 42:243-246.

Andrade SO, Peregrino CJ, Aguiar AA (1971) Estudos sôbre Brachiaria sp. (Tanner grass). I. Efeito nocivo para bovinos. Arq Inst Biol (Sao Paulo) 38(3):135-150.

Andrade SO, Retz L, Marmo O (1971) Estudos sôbre Brachiaria sp. (Tanner grass). 3. Ocorrências de intoxicações de bovinos durante um ano (1970-1971) e níveis de nitrato em amostras da gramínea. Arq Inst Biol (Sao Paulo) 38(4):239-252.

Andrade SO, Retz L, Velloso CA (1971) Estudos sôbre Brachiaria sp. (Tanner grass). 2. Dosagem de nitrato em sôro de bovinos. Arq Inst Biol (Sao Paulo) 38(3):151-161.

Oschita M, Andrade SO, Bueno P (1972) Intoxicação de búfalos alimentados com Brachiaria sp; (Tanner grass). Arq Inst Biol (Sao Paulo) 39(3):209-211.

UROCHLOA BRIZANTHA (Hochst. ex A. Rich.) R. D. Webster [Poaceae]

Synonyms:

Brachiaria brizantha (Hochst. ex A. Rich.) Stapf

Common Names:

signalgrass; St. Lucia'sgrass

Citations:

Costa EO, Górniak SL, Benites NR, et al. (1998) Fungi microbiota and oxalate formation on Brachiaria brizantha and their possible toxic effects on calves kept at pasture without supplementation of calcium and magnesium. In: Garland T, Barr AC (eds.) Toxic plants and other natural toxicants. CABI. New York. pp. 174-178.

Gomar MS, Driemeier D, Colodel EM, et al. (2005) Lectin histochemistry of foam cells in tissues of cattle grazing Brachiaria spp. J Vet Med A Physiol Pathol Clin Med 52(1):18-21.

Torres MB, Coelho KI (2003) Foamy macrophages in the liver of cattle fed Brachiaria brizantha hay. Vet Hum Toxicol 45(3):163-164.

UROCHLOA DECUMBENS (Stapf) R. D. Webster [Poaceae]

Synonyms:

Brachiaria decumbens Stapf

Common Names:

signalgrass; Surinamgrass

Citations:

Abdullah AS, Nordin MM, Rajion MA (1988) Signal grass (Brachiaria decumbens) toxicity in sheep: Changes in motility and pH of reticulo-rumen. Vet Hum Toxicol 30(3):256-258.

Abdullah AS, Nordin MM, Rajion MA (1989) Neurological disorders in sheep during signal grass (Brachiaria decumbens) toxicity. Vet Hum Toxicol 31(2):128-129.

Abdullah AS, Rajion MA (1990) Signal grass (Brachiaria decumbens) toxicity in sheep: Changes in rumen microbial populations and volatile fatty acid concentrations. Vet Hum Toxicol 32(5):444-445.

Barrera J, Ochoa R (1977) Brachiaria decumbens y fotosensibizacion. Rev Inst Colombiano Agr 12(3):231-240.

Döbereiner J, Tokarnia CH, Monteiro MC, et al. (1976) Intoxicação de bovinos e ovinos em pastos de Brachiaria decumbens contaminados por Pithomyces chartarum. Pesq Agropec Bras Ser Vet 11:87-94.

Driemeier D, Colodel EM, Seitz AL, et al. (2002) Study of experimentally induced lesions in sheep by grazing Brachiaria decumbens. Toxicon 40(7):1027-1031.

Gomar MS, Driemeier D, Colodel EM, et al. (2005) Lectin histochemistry of foam cells in tissues of cattle grazing Brachiaria spp. J Vet Med A Physiol Pathol Clin Med 52(1):18-21.

Graydon RJ, Hamid H, Zahari P, et al. (1991) Photosensitisation and crystal-associated cholangiohepatopathy in sheep grazing Brachiaria decumbens. Aust Vet J 68(7):234-236.

Hasiah AH, Elsheikh HA, Abdullah AS, et al. (2000) Effect of phenobarbitone treatment against signal grass (Brachiaria decumbens) toxicity in sheep. Vet J 160(3):267-272.

Hasiah AH, Elsheikh HA, Khairi HM, et al. (2003) Effect of griseofulvin on the toxicity of signal grass (Brachiaria decumbens) in sheep. Vet Hum Toxicol 45(2):68-71.

Lemos RA Salvador SC, Nakazato L (1997) Photosensitization and crystal-association cholangiohepatopathy in cattle grazing Brachiaria decumbens in Brazil. Vet Hum Toxicol 39(6):376-377.

Noordin MM, Salam Abdullah A, Rajion MA (1989) Experimental Brachiaria decumbens toxicity in cattle. Vet Res Commun 13(6):491-494.

Opasina BA (1985) Photosensitization jaundice syndrome in West African dwarf sheep and goats grazed on Brachiaria decumbens. Trop Grasslands 19(3):120-123.

Othman AM, Haron S, Yusuff MK, et al. (1983) Hepatogenous photosensitivity in indigenous sheep of Malaysia grazing on pure stand of leafy regrowth of Brachiaria decumbens. Malaysian Vet J 7(4):271.

Riet-Correa G, Riet-Correa F, Schild AL, et al. (2002) Wasting and death in cattle associated with chronic grazing of Brachiaria decumbens. Vet Hum Toxicol 44(3):179-180.

Zhang SS, Noordin MM, Rahman SO, et al. (2001) The effect of zinc supplementation on antioxidant and lipid peroxidation status during Brachiaria decumbens intoxication in sheep. Vet Hum Toxicol 43(2):83-87.

UROCHLOA HUMIDICOLA (Rendle) Morrone & Zuloaga [Poaceae]

Citations:

Barbosa JD, Oliveira CM, Tokarnia CH, et al. (2006) Fotossensibilização hepatógena em eqüinos pela ingestão de Brachiaria humidicola (Gramineae) no Estado do Pará. Pesq Vet Bras 26(3):147-153.

UROCHLOA MUTICA (Forssk.) T. Q. Nguyen [Poaceae]

Synonyms:

Brachiaria purpurescens (Raddi) Henry

Common Names:

giant panicum; Paragrass

Citations:

Pereira J (1952) Transient oxytocic depression of immature guinea pigs fed on grass Brachiaria purpurescens. Endocrinology 50:124-133.

UROCHLOA RUZIZIENSIS (R. Germ. & C. M. Evrard) Crins [Poaceae]

Synonyms:

Brachiaria ruziziensis R. Germ. & C. M. Evrard

Citations:

Pierre F (1984) Dermatose de photosensibilisation sur des moutons dans le Centre de la Côte d'Ivoire. Rev Elev Med Vet Pays Trop 37(3):277-285.

urtica –see– *Urtica dioica* L.

URTICA CHAMAEDRYOIDES Pursh [Urticaceae]

Common Names:

heart-leaf nettle; nettle; ortiguilla; stinging nettle; weak nettle

Citations:

Edwards WC, Remer JC (1983) Nettle poisoning in dogs. Vet Med Small Anim Clin 78(3):347,350.

Greene JE (1958) "Bull-nettle syndrome" in hunting dogs. Mod Vet Pract 39(Dec 1):50-51.

Peterson DR (1968) Nettle poisoning in the dog. Oklahoma Vet 20(2):4-6.

URTICA DIOICA L. [Urticaceae]

Common Names:

American stinging nettle; Brennessel; British nettle; dwarf nettle; grande ortie; great nettle; Große Brennessel; hoary nettle; nettle; perennial nettle; slim nettle; stinging nettle; tall nettle

Citations:

Bathe AP (1994) An unusual manifestation of nettle rash in three horses. Vet Rec 134(1):11-12.

Cauna N (1977) Fine morphological changes in the penicillate nerve endings of human hairy skin during prolonged itching. Anat Rec 188(1):1-11.

Hughes RE, Ellery P, Harry T, et al. (1980) The dietary potential of the common nettle. J Sci Food Agric 31:1279-1286.

Krauskopf J, Fadrhoncová A (1979) Fytodermatózy způsobené rostlinami kopřivovitými (Urticaceae). Cesk Dermatol 54(5):274-279.

Oliver F, Amon EU, Breathnach A, et al. (1991) Contact urticaria due to the common stinging nettle (Urtica dioica) - Histological ultrastructural and pharmacological studies. Clin Exp Dermatol 16(1):1-7.

URTICA FEROX G. Forst. [Urticaceae]

Common Names:

giant nettle; nettle tree; New Zealand stinging nettle; ongaonga; tree nettle

Citations:

Aston BC (1909) Researches on poisonous and other plants. N Z Dep Agric Annu Rep 17:178-184.

Clark FP (1993) Tree nettle (Urtica ferox) poisoning. N Z Med J 106(957):234.

Urticastrum divaricatum (L.) Kuntze = Laportea canadensis (L.) Wedd.

uta –see– *Strophanthus gratus* (Wall. & Hook.) Baill.

Utah juniper –see– *Juniperus osteosperma* (Torr.) Little

Utah white oak –see– *Quercus gambelii* Nutt.

utupa –see– *Euphorbia tirucalli* L.

V

Vaalbos –see– *Hertia pallens* (DC.) Kuntze

Vaalstorm–see– *Thesium lineatum* L. f.

VACCARIA HISPANICA (Mill.) Rauschert [Caryophyllaceae]

Synonyms:

Gypsophila vaccaria Sibth. & Sm.

Common Names:

bladder soapwort; China cockle; cockle; corn cockle; cow cockle; cowherb; dairy pink; field soapwort; Nancy; pink cockle; spring cockle; wheat cockle

Citations:

Orozco Piñán LR (1965) Caso de intoxicación en pollos de carne por ingestión de semillas de Gypsophila vaccaria. Rev Nutr Anim 3:166-168.

VACCINIUM ULIGINOSUM L. [Ericaceae]

Common Names:

Moorbeere; Rauschbeere; Sumpfheidelbeere; Trunkelbeere

Citations:

Kreuder F (1937) Vergiftungserscheinungen nach reichlichem Genuß von Rauschbeeren (Vaccinium uliginosum, Sumpfheidelbeere, Trunkelbeere). Sammlung Vergiftungsfallen 8(A663):33-34.

Lamminpaa A, Kinos M (1996) Plant poisonings in children. Hum Exp Toxicol 15(3):245-249.

Zipf K (1943-1944) Vergiftungen durch Rauschbeeren. Sammlung Vergiftungsfallen 13:139 140.

valerian –see– *Valeriana officinalis* L.

VALERIANA OFFICINALIS L. [Valerianaceae]

Common Names:

all heal; cat's-valerian; cut finger; garden heliotrope; herb bennett; medicinal valerian; St. George's-herb; valerian

Citations:

Caldwell SH, Feeley JW, Wieboldt TF, et al. (1994) Acute hepatitis with use of over-the-counter herbal remedies. Va Med Q 121(1):31-33.

Chan TY (1998) An assessment of the delayed effects associated with valerian overdose. Int J Clin Pharmacol Ther 36(10):569.

Chan TY, Tang CH, Critchley JA (1995) Poisoning due to an over-the-counter hypnotic, Sleep-Qik (hyoscine, cyproheptadine, valerian). Postgrad Med J 71(834):227-228.

Mady WL, Cobaugh S, Wax D (1994) Valerian overdose: A case report. Vet Hum Toxicol 36(4):360.

Willey LB, Mady SP, Cobaugh DJ, et al. (1995) Valerian overdose: A case report. Vet Hum Toxicol 37(4):364-365.

valley oak –see– *Quercus lobata* Nee

Van Zyl sugar bean –see– *Phaseolus vulgaris* L.

Van Zyl suikerboontjie –see– *Phaseolus vulgaris* L.

VANGUERIA PYGMAE Schltr. [Rubiaceae]

Synonyms:

Pachystigma pygmaeaum (Schltr.) Robyns

Common Names:

gifappel; gousiektebossie

Citations:

Adelaar TF, Terblanche M, Smit JD (1966) A report on negative experiments with ferric chloride as a prophylactic agent against gousiekte. J S Afr Vet Assoc 37(2):199-201.

Fourie N, Schultz RA, Prozesky L, et al. (1989) Clinical pathological changes in gousiekte, a plant-induced cardiotoxicosis of ruminants. Onderstepoort J Vet Res 56(1):73-80.

Pretorius PJ, Terblanche M (1967) A preliminary study on the symptomatology and cardiodynamics of gousiekte in sheep and goats. J S Afr Vet Assoc 38(1):29-53.

Pretorius PJ, Terblanche M, Van der Walt JD, et al. (1973) Cardiac failure in ruminants caused by gousiekte. Recent Adv Stud Cardiac Struct Metab 2:385-397.

Pretorius PJ, Van Rooyen JM, Van Ryssen JC, et al. (1988) Experimental gallop rhythm in sheep with gousiekte: Correlation of changes in amplitude with haemodyamic parameters. Onderstepoort J Vet Res 55(4):221-225.

Prozesky L, Bastianello SS, Fourie N, et al. (2005) A study of the pathology and pathogenesis of the myocardial lesions in gousiekte, a plant-induced cardiotoxicosis of ruminants. Onderstepoort J Vet Res 72(3):219-230.

Prozesky L, Fourie N, Neser JA, et al. (1988) A field outbreak in Ile-de-France sheep of a cardiotoxicosis caused by the plant Pachystigma pygmaeum (Schltr) Robyns (Rubiaceae). Onderstepoort J Vet Res 55(4):193-196.

Schutte PJ, Els HJ, Booyens J, et al. (1984) Ultrastructure of myocardial cells in sheep with congestive heart failure induced by Pachystigma pygmaeum. S Afr J Sci 80(8):378-380.

Van der Walt JJ, Van Rooyen JM, Lotter AP, et al. (1990) A comparison of haemodynamic and vasoconstrictory responses in sheep with a toxic fraction from Pachystigma pygmaeum and with the plant material. Onderstepoort J Vet Res 57(3):157-161.

Visser JH (1964) Iron supplementation as a means of preventing poisoning in livestock by Pachystigma pygmaeum. South African J Agric Sci 7(1):173-175.

VANGUERIA THAMNUS (Robyns) Lantz [Rubiaceae]

Synonyms:

Pachystigma thamnus Robyns

Citations:

Adelaar TF, Terblanche M (1967) A note on the toxicity of the plant Pachystigma thamnus Robyns. J S Afr Vet Assoc 38(1):25-26.

vanilla –see– *Vanilla planifolia* Andrews

VANILLA PLANIFOLIA Andrews [Orchidaceae]

Common Names:

vanilla; Vanille

Citations:

Hutchinson J (1892) An eruption caused by vanilla. Arch Surg 4:49-50.

Niinimäki A (1984) Delayed-type allergy to spices. Contact Dermatitis 11(1):34-40.

Sincke GE (1934) Überempfindlichkeit gegen vanille. Dermatol Wochenschr 99(45):1480-1481.

Vanille –see– *Vanilla planifolia* Andrews

varaire –see– *Veratrum album* L.

variegated ivy –see– *Hedera helix* L. subsp. canariensis (Willd.) Cout.

variegated laurel –see– *Aucuba japonica* Thunb.

variegated philodendron –see– *Epipremnum pinnatum* (L.) Engl.

variegated spurge –see– *Euphorbia marginata* Pursh

variegated thistle –see– *Silybum marianum* (L.) Gaertn.

varnish bush –see– *Flourensia cernua* DC.

varnish lacquer tree –see– *Toxicodendron vernicifluum* (Stokes) F. A. Barkley

varnish tree –see– *Aleurites moluccanus* (L.) Willd.; *Toxicodendron vernicifluum* (Stokes) F. A. Barkley

Vasey's-stipa grass –see– *Achnatherum robustum* (Vasey) Barkworth

vegetable calomel –see– *Acorus calamus* L.; *Podophyllum peltatum* L.

vegetable hummingbird –see– *Sesbania grandiflora* (L.) Pers.

vegetable mercury –see– *Podophyllum peltatum* L.

vegetable tallow tree –see– *Triadica sebifera* (L.) Small

Veilchenwurzel –see– *Iris pseudoacorus* L.

veined dock –see– *Rumex venosus* Pursh

velvet bean –see– *Mucuna pruriens* (L.) DC.; *Mucuna pruriens* (L.) DC. var. utilis (Wall. ex Wight) Baker ex Burck

velvet brittle-stem –see– *Psathyrotes annua* (Nutt.) A. Gray

velvet lupine –see– *Lupinus leucophyllus* Douglas ex Lindl.

velvet rosettes –see– *Psathyrotes annua* (Nutt.) A. Gray

velvet sumach –see– *Rhus typhina* L.

velvet weed –see– *Abutilon theophrasti* Medik.

velvetleaf –see– *Abutilon theophrasti* Medik.

venijnboom –see– *Taxus baccata* L.

venoplant –see– *Aesculus hippocastanum* L.

VENTILAGO VIMINALIS Hook. [Rhamnaceae]

Common Names:

supplejack; vine tree

Citations:

Pryor WJ, McDonald WJ, Seawright AA (1972) Supplejack (Ventilago viminalis) feeding of sheep. Nutritional and toxicological investigations. Aust Vet J 48(6):339-344.

Venus'-chariot –see– *Aconitum napellus* L.

veratro –see– *Veratrum album* L.

veratrum –see– *Veratrum album* L.; *Veratrum californicum* Durand; *Veratrum viride* Aiton

VERATRUM ALBUM L. [Melanthiaceae]

Common Names:

Champagnerwurz; elléboro blanco; European white hellebore; false hellebore; fehér zászpa; Germer; green hellebore; hellebore; hellebore blanc; hierba ballestera; langwort; lyngwort; Niespulver; Nieswurz; varaire; veratro; veratrum; Weißer Germer; Weißer Nieswurz; white false hellebore; white hellebore

Citations:

Azzarone G, Ciammitti B, Mariani L, et al. (1984) Intossicazione da ingestione di infuso acquoso di radici di veratro. Rass Med Sper 31(12):441-447.

Berezhnitskii MN, Parashchak AP (1969) [Two cases of poisoning with veratrum album tincture.] Sov Med 32(2):141.

Carlier P, Efthymiou ML, Garnier R, et al. (1983) Poisoning with veratrum-containing sneezing powders. Hum Toxicol 2(2):321-325.

Fogh A, Kulling P, Wickstrom E (1983) Veratrum alkaloids in sneezing-powder - A potential danger. J Toxicol Clin Toxicol 20(2):175-179.

Gaillard Y, Pepin G (2001) LC-EI-MS determination of veratridine and cevadine in two fatal cases of Veratrum album poisoning. J Anal Toxicol 25(6):481-485.

Garnier R, Carlier P, Hoffelt J, et al. (1985) Intoxication aiguë alimentaire par l'ellebore blanc (Veratrum album L.) - Données cliniques et analytiques. A propos de 5 cas. Ann Med Interne (Paris) 136(2):125-128.

Garnier R, Hoffelt J, Carlier P, et al. (1982) Veratrum poisoning with home-made gentian-wine: Clinical and analytical findings. Vet Hum Toxicol 24(Suppl):138-141.

Hruby K, Lenz K, Krausler J (1981) Vergiftung mit Veratrum album (weißer Germer). Wien Klin Wochenschr 93(16):517-519.

Marinov A, Koev P, Mirchev N (1987) [Electrocardiographic studies in patients with acute hellebore (Veratrum album) intoxication.] Vutr Boles 26(6):36-39.

Philippe JM, Béraud P, Delort P (1996) Two cases of white hellebore poisoning. Vet Hum Toxicol 38(6):466.

Quatrehomme G, Bertrand F, Chauvet C, et al. (1993) Intoxication from Veratrum album. Hum Exp Toxicol 12(2):111-115.
Richter HE (1966) Schadensfälle durch den Weißen Germer, Veratrum album L. Wien Tierarztl Monatsschr 53:547-549.
Rottinghaus D, Kurta D, Krenzelok EP, et al. (2003) Pontine stroke after acute hellebore poisoning. J Toxicol Clin Toxicol 41(5):721.
Seeliger J (1956-1957) Über eine seltene Vergiftung mit weißer Nieswurz. Arch Toxikol 16(1):16-18.
Sieben P, Rørdam AM, Thomsen ES (1982) Forgiftning med nysepulver. Ugeskr Laeger 144(24):1780.

VERATRUM CALIFORNICUM Durand [Melanthiaceae]

Common Names:

California false hellebore; California hellebore; California swamp hellebore; corn lily; cow cabbage; Durand's-false hellebore; false hellebore; false white hellebore; hellebore; Indian corn; showy false hellebore; skunk cabbage; veratrum; western corn lily; western false hellebore; western hellebore; white hellebore; wild corn; wild hellebore

Citations:

Allen S (1970) A cyclopian fetus resulting from oral ingestion of Veratrum californicum. Iowa State Univ Vet 32(2):62-65.
Anonymous (1935) Investigation of stock-poisoning plants. U S Dep Agric Bur Anim Indus Chief Rep 1935:39-40.
Anonymous (1936) Investigations of stock-poisoning plants. U S Dep Agric Bur Anim Indus Chief Rep 1936:44-45.
Anonymous (1937) Investigations of stock-poisoning plants. U S Dep Agric Bur Anim Indus Chief Rep 1937:47-48.
Babbott FL, Binns W, Ingalls TH (1962) Field studies of cyclopian malformations in sheep. Arch Environ Health 5(Aug):109-113.
Binns W, James LF, Keeler RF, et al. (1968) Effects of teratogenic agents in range plants. Cancer Res 28(11):2323-2326.
Binns W, James LF, Shupe JL (1964) Congenital malformations in lambs reproduced by feeding a poisonous range plant. Southwestern Vet 17(3):221-224.
Binns W, James LF, Shupe JL (1964) Toxicosis of Veratrum californicum in ewes and its relationship to a congenital deformity in lambs. Ann N Y Acad Sci 111(2):571-576.
Binns W, James LF, Shupe JL, et al. (1962) Cyclopian-type malformation in lambs. Arch Environ Health 5(Aug):106-108.
Binns W, James LF, Shupe JL, et al. (1963) A congenital cyclopian-type malformation in lambs induced by maternal ingestion of a range plant, Veratrum californicum. Am J Vet Res 24(103):1164-1175.
Binns W, Keeler RF, Balls LD (1972) Congenital deformities in lambs, calves and goats resulting from maternal ingestion of Veratrum californicum: Hare lip, cleft palate, ataxia and hypoplasia of metacarpal and metatarsal bones. Clin Toxicol 5(2):245-261.
Binns W, Shupe JL, Keeler RF, et al. (1965) Chronologic evaluation of teratogenicity in sheep fed Veratrum californicum. J Am Vet Med Assoc 147(8):839-842.
Keeler RF (1990) Early embryonic death in lambs induced by Veratrum californicum. Cornell Vet 80(2):203-207.
Keeler RF, Stuart LD (1987) The nature of congenital limb defects in lambs by maternal ingestion of Veratrum californicum. J Toxicol Clin Toxicol 25(4):273-286.
Keeler RF, Young S, Smart R (1985) Congenital tracheal stenosis in lambs induced by maternal ingestion of Veratrum californicum. Teratology 31(1):83-88.
Keeler RF, Young S, Stuart LD (1986) Characteristics of congenital limb and tracheal defects and of embryonic death induced in lambs by Veratrum californicum. Teratology 33(3):33C-34C.
Van Kampen KR, Binns W, James LF, et al. (1969) Early embryonic death in ewes given Veratrum californicum. Am J Vet Res 30(4):517-519.
Van Kampen KR, Ellis LC (1972) Prolonged gestation in ewes ingesting Veratrum californicum: Morphological changes and steroid biosynthesis in the endocrine organs of cyclopic lambs. J Endocrinol 52(3):549-560.

VERATRUM JAPONICUM (Baker) Loes. [Melanthiaceae]

Common Names:

false hellebore

Citations:

Nelson DA (1954) Accidental poisoning by Veratrum japonicum. JAMA 156(1):33-35.

VERATRUM VIRIDE Aiton [Melanthiaceae]

Common Names:

American false hellebore; American hellebore; American white hellebore; bear corn; bugbane; crow poison; devil's-bite; duckretter; earth gall; false hellebore; green hellebore; hellebore; Indian hellebore; Indian poke; Indian uncus; itchweed; meadow poke; poison lily; puppet root; skunk cabbage; swamp hellebore; tickle weed; veratrum; white hellebore

Citations:

Courtemanche LA, Gephardt D, Jaffe AM (1989) Poisoning due to Veratrum viride (false hellebore) ingestion. Vet Hum Toxicol 31(4):360.
Crummett D, Bronstein D, Weaver Z 3rd (1985) Accidental Veratrum viride poisoning in three "ramp" foragers. N C Med J 46(9):469-471.
Jaffe AM, Gephardt D, Courtemanche L (1990) Poisoning due to ingestion of Veratrum viride (false hellebore). J Emerg Med 8(2):161-167.
Kauntze R, Trounce J (1951) The hypotensive action of veriloid (Veratrum viride), a clinical investigation. Lancet 260(Mar 10):549-555.
Petropoulos PS, Stork CM, O'Neil N, et al. (1999) Prolonged cardiotoxicity from poison lily (Veratrum viride). J Toxicol Clin Toxicol 37(5):619.
Prince LA, Stork CM (2000) Prolonged cardiotoxicity from poison lily (Veratrum viride). Vet Hum Toxicol 42(5):282-285.
Schuetz E, Adams J (1998) Multi-system toxicity from intentional hellebore ingestion. J Toxicol Clin Toxicol 36(5):454.
Underhill JE (1959) A case of hellebore poisoning. Can Field Nat 73:128-129.

verbasco –see– *Verbascum thapsus* L.

VERBASCUM THAPSUS L. [Scrophulariaceae]

Common Names:

barbasco; candelaria; flannel mullein; gordolobo; gordolobo yerba; great mullein; moth mullein; mullein; punchon; tobacco cimarron; verbasco

Citations:

Romaguera C, Grimwalt F, Vilaplana J (1985) Occupational dermatitis from gordolobo (mullein). Contact Dermatitis 12(3):176.

verbena –see– *Verbena officinalis* L.

Verbena elegans Kunth = Glandularia elegans (Kunth) Umber

Verbena ×hydrida hort. ex Groenl. & Rümpler = Glandularia ×hydrida (hort. ex Groenl. & Rümpler) G. L. Nesom & Pruski

VERBENA OFFICINALIS L. [Verbenaceae]

Common Names:

blue vervain; enchanter's-herb; European vervain; herb-of-grace; herb-of-the-cross; holy herb; Juno's-tears; pigeon grass; simpler's-joy; verbena; vervain; wild verbena

Citations:

Del Pozo MD, Gastaminza G, Navarro JA, et al. (1994) Allergic contact dermatitis from Verbena officinalis L. Contact Dermatitis 31(3):200-201.

VERBESINA ENCELIOIDES (Cav.) Benth. & Hook. f. ex A. Gray [Asteraceae]

Common Names:

American dogweed; butter daisy; cowpen daisy; crownbeard; golden crownbeard; South African daisy; wild sunflower; yellow daisy

Citations:

Keeler RF, Baker DC, Panter KE (1992) Concentration of galegine in Verbesina encelioides and Galega officinalis and the toxic and pathologic effects induced by the plants. J Environ Pathol Toxicol Oncol 11(2):75-81.

King RO (1937) "Crown beard" (Verbesina encelioides). A plant causing pneumonia in sheep. Agric Gaz New South Wales 48:364-366.

King RO (1938) Verbesina encelioides (crown beard). A plant toxic to sheep. New South Wales Dep Agric Vet Res Rep #7(1937):89-94.

Legg J (1938) Wild sunflower. A plant poisonous to stock. Queensland Agric J 49:113-114.

Lopez TA, Campero CM, Chayer R, et al. (1996) Experimental toxicity of Verbesina encelioides in sheep and isolation of galegine. Vet Hum Toxicol 38(6):417-419.

verdolaga blanca –see– *Trianthema portulacastrum* L.

vermeerbos –see– *Geigeria aspera* Harv.; *Geigeria filifolia* Mattf.

vermeerbossie –see– *Geigeria aspera* Harv.

VERNICIA FORDII (Hemsl.) Airy Shaw [Euphorbiaceae]

Synonyms:

Aleurites fordii Hemsl.

Common Names:

Chinawood oil tree; Chinese tung oil tree; Chinese-wood-oil tree; kalo nut; tung; tung nut; tung tree

Citations:

Anonymous (1933) Investigation of stock-poisoning plants. U S Dep Agric Bur Anim Indus Chief Rep 1933:34-35.

Anonymous (1965) Tung nuts don't agree with visitors. Gainesville Sun. Nov 2

Balthrop E (1952) Tung nut poisoning. South Med J 45(9):864-865.

Balthrop E, Gallagher WB, McDonald TF, et al. (1954) Tung nut poisoning. J Fla Med Assoc 40(11):813-820.

Davis GK, Mehrhof NR, McKinney RS (1946) Effect of tung meal in rations for growing chicks. Poult Sci 25:74-79.

Deng JF, Lin TJ, Kao WF, et al. (1997) The difficulty in handling poisonings associated with Chinese traditional medicine: A poison control center experience for 1991-1993. Vet Hum Toxicol 39(2):106-114.

Edwards HM Jr (1964) Observations on feeding tung oil to chickens. J Nutr 83(Aug):365-368.

Edwards HM Jr, Suso FA, Mason J (1967) Observations on feeding tung oil to laying hens with a note on identification of elaeostearic by GLC. Poult Sci 46:564-569.

Emmel MW (1947) The toxic principle of the tung tree. Florida Agric Exp Sta Bull #431:35 pp.

Emmel MW, Sanders DA, Swanson LE (1943) The toxicity of foliage of Aleurites fordi for cattle. J Am Vet Med Assoc 101:136-137.

Erickson JL, Brown JH Jr (1942) A study of the toxic properties of tung nuts. J Pharmacol Exp Ther 74:114-117.

Holmes RL, Crovetto AJ, Browning CB (1967) Effect of detoxified tung meal on ration acceptability and growth in dairy calves. J Dairy Sci 50(3):390-391.

Jacobziner H, Raybin HW (1960) Nitroglycerin, thallium, antifreeze, and tung seed poisonings. N Y State J Med 60(2):2587-2590.

Lin TJ, Hsu CI, Lee KH, et al. (1996) Two outbreaks of acute tung nut (Aleurites fordii) poisoning. J Toxicol Clin Toxicol 34(1):87-92.

Sanders DA, Emmel MW, Swanson LE (1942) Tung tree (Aleurites fordii Hemsl.) foliage poisoning of cattle. Florida Agric Exp Sta Bull #376:8 pp.

VERNICIA MONTANA Lour. [Euphorbiaceae]

Synonyms:

Aleurites montanus (Lour.) E. H. Wilson

Common Names:

candlenut tree; mu-oil tree; wood-oil tree

Citations:

Emmel MW (1947) The toxic principle of the species Aleurites. J Am Vet Med Assoc 111(Nov):386-387.

Emmel MW (1947) The toxic principle of the tung tree. Florida Agric Exp Sta Bull #431:35 pp.

Vernonia ampla O. Hoffm. = ***Vernonia myriantha*** Hook. f.

VERNONIA MOLLISSIMA D. Don ex Hook. & Arn. [Asteraceae]

Citations:

Döbereiner J, Tokarnia CH, Purisco E (1976) Vernonia mollissima, planta tóxica responsável por mortandades de bovinos no sul de Mato Grasso. Pesq Agric Bras Vet 11(9):49-58.

Gava A, Peixoto PV, Tokarnia CH (1987) Intoxicação experimental por Vernonia mollissima (Compositae) em ovinos e bovinos. Pesq Vet Bras 7(2):33-41.

Stolf L, Gava A, Tokarnia CH (1987) Intoxicação experimental por Vernonia mollissima (Compositae) em caprinos. Pesq Vet Bras 7(3):67-77.

Tokarnia CH, Döbereiner J, Amorim PR, et al. (1986) Intoxicação experimental por Vernonia mollissima (Compositae) em coelhos. Pesq Vet Bras 6(1):5-10.

VERNONIA MYRIANTHA Hook. f. [Asteraceae]

Synonyms:

Vernonia ampla O. Hoffm.

Common Names:

Zimbabwe pride

Citations:

Shone DK (1966) Poisonous plants of Rhodesia. Part I. Rhodesia Zambia Malawi J Agric Res 4:81-94.

VERNONIA NOVEBORACENSIS (L.) Michx. [Asteraceae]

Common Names:

New York ironweed

Citations:

Fortina AB, Magno FM, Cappelletti EM, et al. (2002) Contact dermatitis from Vernonia noveboracensis. Contact Dermatitis 46(6):357-358.

VERNONIA NUDIFLORA Less. [Asteraceae]

Citations:

Döbereiner J, Tokarnia CH (1984) Intoxicação experimental por Vernonia nudiflora (Compositae) em bovinos e ovinos. Pesq Vet Bras 4(1):5-10.

VERNONIA SQUARROSA (D. Don) Less. [Asteraceae]

Citations:

Tokarnia CH, Döbereiner J (1983) Intoxicação experimental por Vernonia squarrosa (Compositae) em ovinos e bovinos. Pesq Vet Bras 3(2):45-52.

vervain –see– *Verbena officinalis* L.

vervions –see– *Rosmarinus officinalis* L.

vesce cultivée –see– *Vicia sativa* L.

VESTIA FOETIDA (Ruiz & Pav.) Hoffmanns. [Solanaceae]

Common names:

huevil

Citations:

Brevis C, Quezada M, Bustamante P, et al. (2005) Huevil (Vestia foetida) poisoning of cattle in Chile. Vet Rec 156(14):452-453.

McKeough VL, Collett MG, Parton KH (2005) Suspected Vestia foetida poisoning in young goats. N Z Vet J 53(5):352-355.

VESTIA LYCIOIDES Willd. [Solanaceae]

Common names:

huevil

Citations:

Araya O, Hitschfeld M, Cubillos V (1982) Lesiones histológicas y cambios plasmáticos en ovinos intoxicados experimentalmente con huevil (Vestia lycioides). Zentralbl Veterinarmed A 29(9):694-703.

vetch –see– *Lathyrus sylvestris* L.; *Vicia benghalensis* L.; *Vicia ervilia* (L.) Willd.; *Vicia sativa* L.

vetchling –see– *Lathyrus hirsutus* L.; *Lathyrus pusillus* Elliott; *Lathyrus sativus* L.

vi vi –see– *Leucaena leucocephala* (Lam.) de Wit

VIBURNUM RHITIDOPHYLLUM Hemsl. [Adoxaceae]

Common Names:

Schneeball

Citations:

Dambra P, Nettis E, Loria MP, et al. (2000) Hypersensitivity to Viburnum rhytidophyllum. Allergy 55(5):512-513.

VICIA BENGHALENSIS L. [Fabaceae]

Common Names:

vetch

Citations:

Green JR, Kleynhans JE (1989) Suspected vetch (Vicia benghalensis L) poisoning in a Friesland cow in the Republic of South Africa. J S Afr Vet Assoc 60(2):109-110.

Harper PA, Cook RW, Gill PA, et al. (1993) Vetch toxicosis in cattle grazing Vicia villosa spp dasycarpa and V. benghalensis. Aust Vet J 70(4):140-144.

VICIA ERVILIA (L.) Willd. [Fabaceae]

Synonyms:

Ervum ervilia L.

Common Names:

bastard lentil; bitter vetch; ers; ers ervillier; lentille bâtarde; lentille ers; vetch; yeros

Citations:

Almazán Gil F (1974) Aportaciones al estudio sobre la toxicidad experimental del yero (Ervum ervilia). Rev Nutr Anim 12(3):143-162.

Almazán Gil FF (1974) Aportaciones al estudio sobre la toxicidad experimental del yero (Ervum ervilia). Rev Nutr Anim 12(2):65-74.

Palacios J, de Lemus C (1982) Algunos efectos fisiológicos de la ingestión de Vicia ervilia W. (yeros) en pollos. Rev Esp Fisiol 38(3):311-316.

VICIA FABA L. [Fabaceae]

Common Names:

broad bean; double bean; English bean; faba bean; fabismo; fava; fava bean; Favabohne; favismo; fèves; field bean; horse bean; Italian bean; Pferdebohne; Saubohne; Windsor bean; yewdow

Citations:

Abbey BW, Neale RJ, Norton G (1979) Nutritional effects of field bean (Vicia faba L) proteinase inhibitors fed to rats. Br J Nutr 41(1):31-38.

Abbey BW, Norton G, Neale RJ (1979) Effects of dietary proteinase inhibitors from field bean (Vicia faba L.) and field-bean meal on pancreatic function in the rat. Br J Nutr 41(1):39-45.

Aherne FX, Lewis AJ, Hardin RT (1977) An evaluation of faba beans (Vicia faba) as a protein supplement for swine. Can J Anim Sci 57(2):321-328.

Andersson PO, Jacobsson S, Wadenvik H (1998) Bortglömd bedräglig böna - Ett fall av favism. Bondbönan kan orsaka livshotande hemolys hos disponerade. Lakartidningen 95(23):2712-2714.

Bertoni G (1967) Il fondo oculare nel favismo. Ann Ottalmol Clin Ocul 93(11):1235-1244.

Boon WH (1972) Favism in Singapore. J Singapore Paediat Soc 14(1):17-25.

Brodribb HS (1966) Favism from pollen. Br Med J 5514:642.

Brodribb HS, Worssam AR (1961) Favism in an Englishwoman. Br Med J 1(May 13):1367-1368.

Brooks EA, James GA, Stubber LA (1958) Favism in Western Australia. Med J Aust 2(14):455-458.

Busse K (1967) Zur klinik des Favismus. Med Klin 62(40):1539-1541.

Casper J, Shulman J (1956) Bilateral cortical necrosis of the kidneys in an infant with favism. Am J Clin Pathol 26(1):42-47.

Castanon JI, Perez-Lanzac, J (1990) Substitution of fixed amounts of soybean meal for field beans (Vicia faba), sweet lupines (Lupinus albus), cull peas (Pisum sativum) and vetches (Vicia sativa) in diets for high performance laying Leghorn hens. Br Poult Sci 31(1):173-180.

Chung SF (1965) Studies on favism in Kwangtung Province. Paediatr Indones 5(3 Suppl):880-885.

Collombel C, Kissin C, Cotte J (1970) A propos de quatre cas d'anémie hémolytique après ingestion de fèves. Eur J Toxicol 3:36-41.

Discombe G, Mestitz W (1956) Favism in an English-born child. Br Med J 1(4974):1023.

Gower ND, Frommer E (1960) Favism in a Cypriot child. Lancet 1(Mar 19):628-629.

Hodge JV, Nye ER, Emerson GW (1964) Monoamine-oxidase inhibitors, broad beans, and hypertension. Lancet 1(May 16):1108.

Holt JM, Sladden RA (1965) Favism in England - Two more cases. Arch Dis Child 40(211):271-273.

Hove EL, King S, Hill GD (1978) Composition, protein quality, and toxins of seeds of the grain legumes Glycine max, Lupinus spp., Phaseolus spp., Pisum sativum, and Vicia faba. N Z J Agric Res 21:457-462.

Hutton JE (1937) Favism. An unusually observed type of hemolytic anemia. JAMA 109(20):1618-1620.

Jablonska-Skwiecinska E, Lewandowska I, Plochocka D, et al. (1999) Several mutations including two novel mutations of the glucose-6-phosphate dehydrogenase gene in Polish G6PD deficient subjects with chronic nonspherocytic hemolytic anemia, acute hemolytic anemia, and favism. Hum Mutat 14(6):477-484.

Jacobs AH (1950) Favism in two children in California. Pediatrics 6:51-54.

Josephs HW (1944) Favism. Bull Johns Hopkins Hosp 74:295-298.

Kahn A, Marie J, Desbois JC, et al. (1976) Favism in a Portuguese family due to a deficient glucose-phosphate dehydrogenase variant of 'Canton' or 'Canton-like' type. Acta Haematol 56(1):58-64.

Kanter SZ, Pinkhas J, Djaldetti M, et al. (1962) Immunologic studies in a case of favism. J Allergy 33:390-396.

Kattamis CA (1969) Some clinical and biochemical aspects of favism in childhood. Ann Soc Belg Med Trop 49(3):298-302.

Kattamis CA, Kyriazakou M, Chaidas S (1969) Favism: Clinical and biochemical data. J Med Genet 6(1):34-41.

Larkin VD (1953) Favism. Report of a case and brief review of the literature. J Pediatr 42(4):453-456.

Larralde J (1982) Estudio de algunos trastornos que se presentan en los animales por la ingestión de semillas de Vicia faba L. Rev Esp Fisiol 38(Suppl):345-347.

Lecks HI (1949) Favism. Report of a case in a child. J Pediatr 34:309-314.

Lim F, Vulliamy T, Abdalla SH (2005) An Ashkenazi Jewish woman presenting with favism. J Clin Pathol 58(3):317-319.

Mansoor S (1955) Two cases of favism in Israel. Br Med J 1(4906):149-150.

Marquardt RR, Campbell LD (1973) Raw and autoclaved faba beans in chick diets. Can J Anim Sci 53(Dec):741-746.

Marquardt RR, Campbell LD, Stothers SC, et al. (1974) Growth responses of chicks and rats fed diets containing four cultivars of raw or autoclaved faba beans. Can J Anim Sci 54(Jun):177-182.

Marquardt RR, Campbell LD, Ward T (1975) Studies with chicks on the growth depressing factor(s) in faba beans (Vicia faba L. var. minor). J Nutr 106(2):275-284.

Marquardt RR, Ward AT, Campbell LD, et al. (1977) Purification, identification and characterization of a growth inhibitor in faba beans (Vicia faba L. var minor). J Nutr 107(7):1313-1324.

Martin-Tanguy J, Guillaume J, Kossa A (1977) Condensed tannins in horse bean seeds: Chemical structure and apparent effects on poultry. J Sci Food Agric 28(8):757-765.

McCarthy OR (1955) A case of favism. Lancet 1(Apr 9):748-749.

McCrae T, Ullery JC (1933) Favism. Report of a case. JAMA 101(18):1389-1391.

McPhee WR (1956) Acquired hemolytic anemia caused by ingestion of fava beans. Report of a case and review of cases reported in American literature. Am J Clin Pathol 26(11):1287-1302.

Meloni T, Forteleoni G, Dore A, et al. (1983) Favism and hemolytic anemia in glucose-6-phosphate dehydrogenase-deficient subjects in North Sardinia. Acta Haematol 70(2):83-90.

Mentzer WC, Collier E (1975) Hydrops fetalis associated with erythrocyte G-6-PD deficiency and maternal ingestion of fava beans and ascorbic acid. J Pediatr 86(4):565-567.

Nik-Akhtar B, Khakpour M, Rashed MA (1972) Recovery from acute renal failure from favism by means of dialysis. Trans R Soc Trop Med Hyg 66(5):801-802.

Plass R, Schneider C, Lewerenz HJ, et al. (1990) Studies on estrogenic effects of faba bean (Vicia faba L. minor) products using the mouse uterine bioassay. Nahrung 34(6):571-573.

Ramadan TZ, Sexauer CL, Wenzl JE (1978) Favism in Oklahoma. J Oklahoma Med Assoc 71(3):91-94.

Rodríguez-Cuartero A, Navas A, Cerezo S (1999) Insuficiencia renal aguda complicación de fabismo. An Med Interna 16(9):491-492.

Rosen AP, Scanlan JJ (1948) Favism. N Engl J Med 239(10):367-368.

Rubio LA, Brenes A, Castaño M (1989) Histological alterations to the pancreas and the intestinal tract produced by raw faba bean (Vicia faba L. minor) diets in growing chicks. Br Poult Sci 30(1):101-114.

Russo G, Mollica F, Pavone L, et al. (1972) Hemolytic crises of favism in Sicilian females heterozygous for G-6-PD deficiency. Pediatrics 49(6):854-859.

Shibuya A, Hirono A, Ishii S, et al. (1999) Hemolytic crisis after excessive ingestion of fava beans in a male infant with G6PD Canton. Int J Hematol 70(4):233-235.

Sorcinelli R, Guiso G (1979) Vitreoretinal hemorrhages after ingestion of fava beans in a G-6-PD-deficient subject. Ophthalmologica 178(5):259-262.

Stewart AG, Koenigsloew EV, Pabst H (1966) Favism in a 4-yr-old-boy. Can Med Assoc J 94(6):292-294.

Taj-Eldin S (1971) Favism in breast-fed infants. Arch Dis Child 46(245):121-123.

Tolmas HC (1957) Favism: Case report showing heredofamilial tendency. J Pediatr 51(4):445-447.

Wharton HJ, Duiesselmann W (1947) Favism. A short review and report of a case. N Engl J Med 236(Jun 26):974-977.

Wong WY, Powars D, Williams WD (1989) 'Yewdow'-induced anemia. West J Med 151:459-460.

VICIA MONANTHA Retz. [Fabaceae]

Common Names:

algarroba; spur vetch

Citations:

Franco-Vicario R, Gamboa P, Escalante M, et al. (1997) Hypersensitivity pneumonitis induced by exposure to the legume algarroba. Allergy 52(4):478-479.

VICIA SATIVA L. [Fabaceae]

Common Names:

Ackerwicke; ervilhaca; Futterwicke; Saatwicke; Sommerwicken; spring vetch; vesce cultivée; vetch; wilde wicke

Citations:

Arscott GH, Harper JA (1963) Relationship of 2,5-diamino-4,6-diketopyrimidine, 2,4-diaminobutyric acid and a crude preparation of β-cyano-L-alanine to the toxicity of common and hairy vetch seed fed to chicks. J Nutr 80:251-254.

Arscott GH, Harper JA (1964) Evidence for a difference in toxicity between common and hairy vetch seed for chicks. Poult Sci 43(1):271-273.

Barros CS, Fighera RA, Rozza DB, et al. (2001) Doença granulomatosa sistêmica em bovinos no Rio Grande do Sul associado ao pastoreio de ervilhaca (Vicia spp). Pesq Vet Bras 21(4):162-171.

Castanon JI, Perez-Lanzac, J (1990) Substitution of fixed amounts of soybean meal for field beans (Vicia faba), sweet lupines (Lupinus albus), cull peas (Pisum sativum) and vetches (Vicia sativa) in diets for high performance laying Leghorn hens. Br Poult Sci 31(1):173-180.

Darre MJ, Minior DN, Tatake JG, et al. (1998) Nutritional evaluation of detoxified and raw common vetch seed (Vicia sativa L.) using diets of broilers. J Agric Food Chem 46:4675-4679.

Fighera RA, Barros CS (2004) Systematic granulomatous disease in Brazilian cattle grazing pasture containing vetch (Vicia spp). Vet Hum Toxicol 46(2):62-66.

Harper JA, Arscott GH (1962) Toxicity of common and hairy vetch seed for poults and chicks. Poult Sci 41:1968-1974.

Pandey JN, Pal AK (1965) A study on growth rate of albino rats fed with 'akra' (Vicia villosa and V. sativa). Indian J Physiol Allied Sci 19:65-72.

Piron SJ, Blanco Carmona JG, Garces Sotillos MD (1991) Occupational asthma caused by vetch (Vicia sativa). J Allergy Clin Immunol 88(1):135-136.

Suter RJ (2002) Suspected cyanide poisoning in cows fed vetch (Vicia sativa) hay. Aust Vet J 80(5):282.

VICIA TETRASPERMA (L.) Schreb. [Fabaceae]

Common Names:

four-seed vetch

Citations:

Wine HA, Johnson PN (1993) Vicia tetrasperma (four-seeded vetch) ingestion by a 3-year-old child. Vet Hum Toxicol 35(5):436-437.

VICIA VILLOSA Roth [Fabaceae]

Common Names:

hairy vetch; Roth vetch; Zottelwicke

Citations:

Anderson CA, Divers TJ (1983) Systemic granulomatous inflammation in a horse grazing hairy vetch. J Am Vet Med Assoc 183(5):569-570.

Arscott GH, Harper JA (1963) Relationship of 2,5-diamino-4,6-diketopyrimidine, 2,4-diaminobutyric acid and a crude preparation of β-cyano-L-alanine to the toxicity of common and hairy vetch seed fed to chicks. J Nutr 80:251-254.

Arscott GH, Harper JA (1964) Evidence for a difference in toxicity between common and hairy vetch seed for chicks. Poult Sci 43(1):271-273.

Barros CS, Fighera RA, Rozza DB, et al. (2001) Doença granulomatosa sistêmica em bovinos no Rio Grande do Sul ao pastoreio de ervilhaca (Vicia spp). Pesq Vet Bras 21(4):162-171.

Claughton WP, Claughton HD (1954) Vetch seed poisoning. Auburn Vet 10(Winter):125-126.

Fighera RA, Barros CS (2004) Systematic granulomatous disease in Brazilian cattle grazing pasture containing vetch (Vicia spp). Vet Hum Toxicol 46(2):62-66.

Harper JA, Arscott GH (1962) Toxicity of common and hairy vetch seed for poults and chicks. Poult Sci 41:1968-1974.

Johnson B, Moore J, Woods LW, et al. (1992) Systemic granulomatous disease in cattle in California associated with grazing hairy vetch (Vicia villosa). J Vet Diagn Invest 4(3):360-362.

Kerr LA, Edwards WC (1982) Hairy vetch poisoning of cattle. Vet Med Small Anim Clin 77(2):257-258.

Kienholz EW, Jensen LS, McGinnis J (1962) Evidence for chick growth inhibitors in several legume seeds. Poult Sci 41:367-371.

Odriozola E, Paloma E, Lopez T, et al. (1991) An outbreak of Vicia villosa (hairy vetch) poisoning in grazing Aberdeen Angus bulls in Argentina. Vet Hum Toxicol 33(3):278-280.

Panciera RJ (1978) Hairy vetch (Vicia villosa Roth) poisoning in cattle. In: Keeler RF, Van Kampen KR, James LF, et al. (eds.) Effects of poisonous plants on livestock. Academic Press. New York. pp. 555-563.

Panciera RJ, Johnson L, Osburn BI (1966) A disease of cattle grazing hairy vetch pasture. J Am Vet Med Assoc 148(7):804-808.

Panciera RJ, Mosier DA, Ritchey JW (1992) Hairy vetch (Vicia villosa Roth) poisoning in cattle: Update and experimental induction of disease. J Vet Diagn Invest 4(3):318-325.

Pandey JN, Pal AK (1965) A study on growth rate of albino rats fed with 'akra' (Vicia villosa and V. sativa). Indian J Physiol Allied Sci 19:65-72.

Vicia villosa Roth subsp. dasycarpa (Ten.) Cavill. = *vicia villosa* Roth subsp. varia (Host) Corb.

VICIA VILLOSA Roth subsp. varia (Host) Corb. [Fabaceae]

Synonyms:

Vicia villosa Roth subsp. dasycarpa (Ten.) Cavill.

Common Names:

hybrid vetch; woolly-pod vetch

Citations:

Burroughs GW, Neser JA, Kellerman TS, et al. (1983) Suspected hybrid vetch (Vicia villosa crossed with Vicia dasycarpa) poisoning of cattle in the Republic of South Africa. J S Afr Vet Assoc 54(2):75-79.

Harper PA, Cook RW, Gill PA, et al. (1993) Vetch toxicosis in cattle grazing Vicia villosa spp dasycarpa and V. benghalensis. Aust Vet J 70(4):140-144.

Peet RL, Gardner JJ (1986) Poisoning of cattle by hairy or woolly pod vetch, Vicia villosa subspecies dasycarpa. Aust Vet J 63(11):381-382.

vidanga –see– *Embelia ribes* Burm. f.

Vielblütige Weißwurz –see– *Polygonatum multiflorum* (L.) All.

Vigna sinensis (L.) Savi ex Hassk. = Vigna unguiculata (L.) Walp.

VIGNA UNGUICULATA (L.) Walp. [Fabaceae]

Synonyms:

Vigna sinensis (L.) Savi ex Hassk.

Common Names:

black-eye pea; cowpea; kunde; southern pea

Citations:

Grant G, Dorward PM, Buchan W (1995) Consumption of diets containing raw soya beans (Glycine max), kidney beans (Phaseolus vulgaris), cowpeas (Vigna unguiculata) or lupin seeds (Lupinus angustifolius) by rats up to 700 days: Effects on body composition and organ weights. Br J Nutr 73(1):17-29.

Grant G, Dorward PM, Pusztai A (1993) Pancreatic enlargement is evident in rats fed diets containing raw soybeans (Glycine max) or cowpeas (Vigna unguiculata) for 800 days but not in those fed diets based on kidney beans (Phaseolus vulgaris) or lupinseed (Lupinus angustifolius). J Nutr 123(12):2207-2205.

Montgomery RD (1964) Observations on the cyanide content and toxicity of tropical pulses. West Indian Med J 13(1):1-11.

vigne-de-Bakel –see– *Cissus quadrangularis* L.

vigne-de-Judée –see– *Solanum dulcamara* L.

vignette –see– *Mercurialis annua* L.

vinagrera –see– *Rumex acetosa* L.

vinca –see– *Catharanthus roseus* (L.) G. Don; *Vinca major* L.

VINCA MAJOR L. [Apocynaceae]

Common Names:

blue periwinkle; greater periwinkle; periwinkle; running-myrtle; vinca

Citations:

Van der Walt SJ, Steyn DG (1946) Recent investigations into the toxicity of plants, etc. XV. Onderstepoort J Vet Sci Anim Indus 21(1):45-55.

Vinca pusilla Murray = Catharanthus pusillus (Murray) G. Don

vine gungo pea –see– *Mucuna pruriens* (L.) DC.

vine tree –see– *Ventilago viminalis* Hook.

vinette –see– *Rumex acetosa* L.

violet bloom –see– *Solanum dulcamara* L.

violet heliotrope –see– *Heliotropium amplexicaule* Vahl

viooltje –see– *Ornithogalum thyrsoides* Jacq.

viper's-bugloss –see– *Echium plantagineum* L.; *Echium vulgare* L.

Virginia creeper –see– *Parthenocissus quinquefolia* (L.) Planch.

Virginia ivy –see– *Parthenocissus quinquefolia* (L.) Planch.

Virginia poke –see– *Phytolacca americana* L.

Virginia sumach –see– *Rhus typhina* L.

virgin's-bower –see– *Clematis vitalba* L.

VIROLA CALOPHYLLOIDEA Markgr. [Myristicaceae]

Common Names:

Epéna

Citations:

Seitz GJ (1965) Einige Bemerkungen zur Anwendung und Wirkungsweise des Epena-Schnupfpulvers der Waika-Indianer. Etnol Stud 28:117-132.

Seitz GJ (1967) Epéna, the intoxicating snuff powder of the Waika Indians and the Tucano medicine man, Agostino. Efron DH et al. (eds.) Ethnopharmacologic search for psychoactive drugs. PHS #1645:315-338.

virvi –see– *Leucaena leucocephala* (Lam.) de Wit

viscum –see– *Viscum album* L.

VISCUM ALBUM L. [Santalaceae]

Common Names:

European mistletoe; Mistel; misteltee; mistletoe; viscum

Citations:

Dixin J (1874) Case of poisoning by berries of the mistletoe: Recovery. Br Med J 1:224.

Greatorex JC (1966) Some unusual cases of plant poisoning in animals. Vet Rec 78(21):725-727.

Harvey J, Colin-Jones DG (1981) Mistletoe hepatitis. Br Med J 282(6259):186-187.

O'Farrell NM (1943) Dermatitis venenata due to mistletoe. Arch Derm Syphilol 43:416.

Ramón MF, Ballesteros S, Martínez-Arrieta R, et al. (2003) Plant and herb exposures in Spain. J Toxicol Clin Toxicol 41(4):514.

Seidemann W (1984) Allergische Rhinitis durch Misteltee (Viscum album). Allergologie 7(12):461-463.

VITEX AGNUS-CASTUS L. [Lamiaceae]

Common Names:

agnus castus; chaste tree

Citations:

Cahill DJ, Fox R, Wardle PG, et al. (1994) Multiple follicular development associated with herbal medicine. Hum Reprod 9(8):1469-1470.

VITIS LABRUSCA L. [Vitaceae]

Common Names:

grape

Citations:

Quinby GE (1979) Grape pressers dermatitis. Clin Toxicol 15(4):478.

VITIS VINIFERA L. [Vitaceae]

Common Names:

currant; grape; raisin

Citations:

Campbell A, Bates N (2003) Raisin poisoning in dogs. Vet Rec 152:376.

Eubig PA, Brady MS, Gwaltney-Brant SM, et al. (2005) Acute renal failure in dogs after the ingestion of grapes or raisins: A retrospective evaluation of 43 dogs (1992-2002). J Vet Intern Med 19(5):663-674.

Giannoccaro F, Munno G, Riva G, et al. (1998) Oral allergy syndrome to grapes. Allergy 53(4):451-452.

Gwaltney-Brant S, Holding JK, Donaldson CW, et al. (2001) Renal failure associated with ingestion of grapes or raisins in dogs. J Am Vet Med Assoc 218:1555-1556.

Mazzaferro EM, Eubig PA, Hackett TB, et al. (2004) Acute renal failure associated with raisin or grape ingestion in 4 dogs. J Vet Emerg Crit Care 14:203-212.

Penny D, Henderson SM, Brown PJ (2003) Raisin poisoning in a dog. Vet Rec 152:308.

Portway B (1957) The characteristic syndrome following excessive consumption of grapes by cows. Aust Vet J 33:210-212.

Rodríguez A, Matheu V, Trujillo MJ, et al. (2004) Grape allergy in paediatric population. Allergy 59(3):364.

Senna G, Mistrello G, Roncarolo D, et al. (2001) Exercise-induced anaphylaxis to grape. Allergy 56(12):1235-1236.

vizi harmatkása –see– *Glyceria maxima* (Hartm.) Holmb.

Vogelleimdistel –see– *Chamaeleon gummifera* (L.) Cass.

voinicel –see– *Descurainia sophia* (L.) Webb ex Prantl

vomiting bush –see– *Geigeria aspera* Harv.

voodoo bean –see– *Mucuna pruriens* (L.) DC.

vuiboom –see– *Frangula alnus* Mill.

vulgaire –see– *Juniperus communis* L.

vulgarmente ruda –see– *Ruta graveolens* L.

vuudoorn –see– *Pyracantha coccinea* M. Roem.

vuursiektebos –see– *Asaemia minuta* (L. f.) K. Bremer

vygie –see– *Drosanthemum floribundum* (Haw.) Schwantes; *Malephora smithii* (L. Bolus) H. E. K. Hartmann

W

wabaio –see– *Acokanthera oppositifolia* (Lam.) Codd; *Acokanthera schimperi* (A. DC.) Oliv.

Wacholder –see– *Juniperus communis* L.

Wahweap milk vetch –see– *Astragalus lentiginosus* Douglas ex Hook. var. wahweapensis S. L. Welsh

wake robin –see– *Arisaema triphyllum* (L.) Schott; *Arum maculatum* L.

Waldbingelkraut –see– *Mercurialis perennis* L.

Walddistelstrauch –see– *Ilex aquifolium* L.

Waldknoblauch –see– *Allium ursinum* L.

Waldmeister –see– *Galium odoratum* (L.) Scop.

Waldnachtschatten –see– *Atropa belladonna* L.

Waldschelle –see– *Digitalis purpurea* L.

wall germander –see– *Teucrium chamaedrys* L.

wallflower poison –see– *Gastrolobium grandiflorum* F. Muell.

Wallia nigra Alef. = Juglans nigra L.

Walnuß –see– *Juglans regia* L.

walnut –see– *Juglans nigra* L.

wandary curse –see– *Heliotropium europaeum* L.

wandering Jew –see– *Tradescantia fluminensis* Vell.

wannakai –see– *Rhodomyrtus macrocarpa* Benth.

warabi –see– *Pteridium aquilinum* (L.) Kuhn var. latiusculum (Desf.) Underw. ex A. Heller

warmot –see– *Artemisia absinthium* L.

wartweed –see– *Euphorbia peplus* L.

wartwort –see– *Euphorbia helioscopia* L.; *Chelidonium majus* L.

warty caltrop –see– *Kallstroemia parviflora* Norton

Warzengraß –see– *Euphorbia cyparissias* L.

Warzenkraut –see– *Chelidonium majus* L.

Wasatch milk vetch –see– *Astragalus miser* Douglas ex Hook. var. oblongifolius (Rydb.) Cronquist

Wasserchierling –see– *Cicuta maculata* L.

Wasserpfeffer–see– *Persicaria hydropiper* (L.) Spach

Wasserschierling –see– *Cicuta virosa* L.

Wasserschwertlilie –see– *Iris pseudoacorus* L.

water arum –see– *Calla palustris* L.

water betony –see– *Scrophularia auriculata* L.

water bush –see– *Myoporum acuminatum* R. Br.

water crowfoot –see– *Ranunculus sceleratus* L.

water dragon –see– *Calla palustris* L.

water dropwort –see– *Oenanthe crocata* L.

water dropwort hemlock –see– *Oenanthe crocata* L.

water fern –see– *Histiopteris incisa* (Thunb.) J. Sm.

water figwort –see– *Scrophularia auriculata* L.

water flag –see– *Iris pseudoacorus* L.

water gromwells –see– *Glyceria maxima* (Hartm.) Holmb.

water hemlock –see– *Cicuta douglasii* (DC.) J. M. Coult. & Rose; *Cicuta maculata* L.; *Cicuta maculata* L. var. angustifolia Hook.; *Cicuta virosa* L.

water horsetail –see– *Equisetum fluviatile* L.

water hyacinth –see– *Eichhornia crassipes* (Mart.) Solms

water kerwal –see– *Matricaria nigellifolia* DC.

water lily –see– *Nymphaea odorata* Aiton subsp. tuberosa (Paine) Wiersema & Hellq.

water maple –see– *Acer rubrum* L.

water nerve root –see– *Asclepias incarnata* L.

water parsnip –see– *Berula erecta* (Huds.) Coville; *Cicuta douglasii* (DC.) J. M. Coult. & Rose; *Cicuta maculata* L. var. angustifolia Hook.; *Sium suave* Walter

water pepper –see– *Persicaria hydropiper* (L.) Spach

water plant –see– *Rheum ×hybridum* Murray

water scrophularia –see– *Scrophularia auriculata* L.

water shamrock –see– *Menyanthes trifoliata* L.

water smartweed –see– *Persicaria hydropiper* (L.) Spach; *Persicaria punctata* (Elliott) Small

watermelon –see– *Citrullus lanatus* (Thunb.) Matsum. & Nakai

Waterscheerling –see– *Cicuta virosa* L.

wawa –see– *Triplochiton scleroxylon* K. Schum.

wax bean –see– *Phaseolus vulgaris* L.

wax flower –see– *Hoya australis* R. Br. ex J. Traill; *Hoya carnosa* (L. f.) R. Br.

wax-leaf privet –see– *Ligustrum vulgare* L.

wax tree –see– *Toxicodendron succedaneum* (L.) Kuntze

waxberry –see– *Symphoricarpos albus* (L.) S. F. Blake

waxplant –see– *Hoya carnosa* (L. f.) R. Br.

waxy larkspur –see– *Delphinium glaucum* S. Watson

way thorn –see– *Rhamnus cathartica* L.

weak nettle –see– *Urtica chamaedryoides* Pursh

weather plant –see– *Abrus precatorius* L.

wedding bells –see– *Ipomoea tricolor* Cav.; *Ipomoea violacea* L.

wedelia –see– *Sphagneticola trilobata* (L.) Pruski

WEDELIA ASPERRIMA (Decne.) Benth. [Asteraceae]

Common Names:

sunflower daisy; yellow daisy

Citations:

Mulhearn CR (1939) Yellow daisy (Wedelia asperrima) - A Plant toxic to sheep. Queensland Agric J 52(Oct 1):397-400.

Wedelia glauca (Ortega) O. Hoffm. ex Hicken = Pascalia glauca Ortega

Wedelia trilobata (L.) Hitchc. = Sphagneticola trilobata (L.) Pruski

wedge-leaf rattlepod –see– *Crotalaria retusa* L.

weed –see– *Cannabis sativa* L.

weedy milk vetch –see– *Astragalus miser* Douglas ex Hook.

weedy rattlebox –see– *Crotalaria sagittalis* L.

weeping ficus –see– *Ficus benjamina* L.

weeping fig –see– *Ficus benjamina* L.

Wegdorn –see– *Frangula alnus* Mill.

Wegwarte –see– *Cichorium intybus* L.

Weihnachtsstern –see– *Euphorbia pulcherrima* Willd. ex Klotzsch

Weinkraut –see– *Ruta graveolens* L.

Weinraute –see– *Ruta graveolens* L.

Weißdorn –see– *Crataegus laevigata* (Poir) DC.; *Crataegus monogyna* Jacq.

Weiße Bohne –see– *Vicia faba* L.

Weiße Nieswurz –see– *Veratrum album* L.

Weiße Robinie –see– *Robinia pseudoacacia* L.

Weißer Germer –see– *Veratrum album* L.

Weißer Maulbeerbaum –see– *Morus alba* L.

Weißer Nieswurz –see– *Veratrum album* L.

Weißer Stechapfel –see– *Datura stramonium* L.

Weißer Steinklee –see– *Melilotus albus* Medik.

Weißklee –see– *Trifolium repens* L.

Weißkraut –see– *Brassica oleracea* L. var. capitata L.

Weizen –see– *Triticum aestivum* L.

Welches Weidelgras –see– *Lolium multiflorum* Lam.

wengé –see– *Millettia laurentii* De Wild.

Wermut –see– *Artemisia absinthium* L.

wermuth –see– *Artemisia absinthium* L.

west African boxwood –see– *Sarcocephalus diderrichii* De Wild.

West Australian blue lupin –see– *Lupinus cosentinii* Guss.

West Indian lead tree –see– *Leucaena leucocephala* (Lam.) de Wit

West Indian pink root –see– *Spigelia anthelmia* L.

West Indian sandbox tree –see– *Hura crepitans* L.

West Indian satinwood –see– *Zanthoxylum flavum* Vahl

Westafrikanisches Satinholz –see– *Turraeanthus africanus* (Welw. ex C. DC.) Pellegr.

Western Australian blue lupin –see– *Lupinus angustifolius* L.; *Lupinus digitatus* Forssk.

western baneberry –see– *Actaea rubra* (Aiton) Willd.

western bitterweed –see– *Hymenoxys odorata* DC.

western buckeye –see– *Aesculus glabra* Willd.

western chokecherry –see– *Prunus virginiana* L.

western corn lily –see– *Veratrum californicum* Durand

western false hellebore –see– *Veratrum californicum* Durand

western goldenrod –see– *Solidago spectabilis* (D. C. Eaton) A. Gray

western groundsel –see– *Senecio integerrimus* Nutt.

western hellebore –see– *Veratrum californicum* Durand

western herb germander –see– *Teucrium chamaedrys* L.

western horse nettle –see– *Solanum dimidiatum* Raf.; *Solanum elaeagnifolium* Cav.

western larkspur –see– *Delphinium barbeyi* (Huth) Huth; *Delphinium glaucum* S. Watson

western laurel –see– *Kalmia microphylla* (Hook.) A. Heller

western locoweed –see– *Astragalus wootonii* E. Sheld.

western lupine –see– *Lupinus leucophyllus* Douglas ex Lindl.

western mountain laurel –see– *Calia secundiflora* (Ortega) Yakovlev

western mugwort –see– *Artemisia ludoviciana* Nutt.

western nightshade –see– *Solanum triflorum* Nutt.

western poison ivy –see– *Toxicodendron diversilobum* (Torr. & A. Gray) Greene; *Toxicodendron rydbergii* (Small ex Rydb.) Greene

western poison oak –see– *Toxicodendron diversilobum* (Torr. & A. Gray) Greene

western purple larkspur –see– *Delphinium menziesii* DC.

western ragweed –see– *Ambrosia artemisiifolia* L.; *Ambrosia psilostachya* DC.

western red cedar –see– *Thuja plicata Donn* ex D. Don

western rhododendron –see– *Rhododendron macrophyllum* D. Don ex G. Don

western rosewood –see– *Alectryon oleifolius* (Desf.) S. T. Reynolds

western serviceberry –see– *Amelanchier alnifolia* (Nutt.) Nutt. ex M. Roem.

western sneezeweed –see– *Hymenoxys hoopesii* (A. Gray) Bierner

western swamp kalmia –see– *Kalmia microphylla* (Hook.) A. Heller western swamp laurel –see– *Kalmia microphylla* (Hook.) A. Heller

western virgin's-bower –see– *Clematis ligusticifolia* Nutt.

western water hemlock –see– *Cicuta douglasii* (DC.) J. M. Coult. & Rose; *Cicuta maculata* L. var. angustifolia Hook.

western white oak –see– *Quercus garryana* Douglas ex Hook.

western whorled milkweed –see– *Asclepias subverticillata* (A. Gray) Vail

western wild cherry –see– *Prunus virginiana* L.

western yellow pine –see– *Pinus ponderosa* C. Lawson

western yew –see– *Taxus brevifolia* Nutt.

Westindische Atlasholz –see– *Chloroxylon swietenia* DC.

Westindische Satinholz –see– *Chloroxylon swietenia* DC.

Westindische Seidenholz –see– *Zanthoxylum flavum* Vahl

wet-the-bed –see– *Senecio integerrimus* Nutt.

wheat –see– *Triticum aestivum* L.

wheat cockle –see– *Vaccaria hispanica* (Mill.) Rauschert

wheel mallow –see– *Modiola caroliniana* (L.) G. Don

whip-poor-will-shoes –see– *Cypripedium reginae* Walter

whiskey cherry –see– *Prunus serotina* Ehrh.

white amaranth –see– *Amaranthus viridis* L.

white arum lily –see– *Zantedeschia aethiopica* (L.) Spreng.

white back –see– *Crotalaria berteroana* DC.

white bean –see– *Phaseolus vulgaris* L.

white broomweed –see– *Parthenium hysterophorus* L.

white bryony –see– *Bryonia dioica* Jacq.

white buckeye –see– *Aesculus glabra* Willd.

white camas –see– *Anticlea elegans* (Pursh) Rydb.

white cedar –see– *Melia azedarach* L.; *Thuja occidentalis* L.

white chameleon –see– *Chamaeleon gummifera* (L.) Cass.

white charlock –see– *Raphanus raphanistrum* L.

white clover –see– *Trifolium repens* L.

white death camas –see– *Anticlea elegans* (Pursh) Rydb.

white eye –see– *Gomphrena celosioides* Mart.

white false hellebore –see– *Veratrum album* L.

white foxglove –see– *Digitalis lanata* Ehrh.

white goosefoot –see– *Chenopodium album* L.

white hairare –see– *Deguelia utilis* (A. C. Sm.) A. M. G. Azevedo

white hearts –see– *Dicentra cucullaria* (L.) Bernh.

white hellebore –see– *Veratrum album* L.; *Veratrum californicum* Durand; *Veratrum viride* Aiton

white henbane –see– *Hyoscyamus albus* L.

white holly –see– *Ilex opaca* Aiton

white horehound –see– *Marrubium vulgare* L.

white horse nettle –see– *Solanum elaeagnifolium* Cav.

white Indian bean –see– *Phaseolus lunatus* L.

white Indian hemp –see– *Asclepias incarnata* L.

white jasmine –see– *Jasminum officinale* L.

white-leaf lupine –see– *Lupinus leucophyllus* Douglas ex Lindl.

white lily –see– *Diplarrhena moraea* Labill.

white locoweed –see– *Oxytropis campestris* (L.) DC. var. spicata Hook.; *Oxytropis lambertii* Pursh; *Oxytropis sericea* Nutt.; *Sophora nuttalliana* B. L. Turner

white locust –see– *Robinia pseudoacacia* L.

white luck –see– *Crotalaria berteroana* DC.

white lupin –see– *Lupinus albus* L.

white lupine –see– *Lupinus albus* L.

white-man's-plant –see– *Datura stramonium* L.

white-margin spurge –see– *Euphorbia marginata* Pursh

white melilot –see– *Melilotus albus* Medik.

white millet –see– *Panicum miliaceum* L.

white morning-glory –see– *Ipomoea lacunosa* L.

white moth plant –see– *Araujia sericifera* Brot.

white mustard –see– *Sinapis alba* L.

white oak –see– *Quercus alba* L.

white passion vine –see– *Passiflora alba* Link & Otto

white pepper –see– *Piper nigrum* L.

white persicary –see– *Solanum glaucophyllum* Desf.

white-point locoweed –see– *Oxytropis sericea* Nutt.

white popinac –see– *Leucaena leucocephala* (Lam.) de Wit

white poppy –see– *Papaver somniferum* L.

white potato –see– *Solanum tuberosum* L.

white radish –see– *Raphanus sativus* L.

white root –see– *Asclepias tuberosa* L.

white saccharin –see– *Hyoscyamus albus* L.

white sage –see– *Artemisia ludoviciana* Nutt.

white sandalwood –see– *Santalum album* L.

white sanicle –see– *Ageratina altissima* (L.) R. M. King & H. Rob.

white snakeroot –see– *Ageratina altissima* (L.) R. M. King & H. Rob.; *Hedeoma pulegioides* (L.) Pers.

white snakeweed –see– *Ageratina altissima* (L.) R. M. King & H. Rob.

white sophora –see– *Sophora nuttalliana* B. L. Turner

white squill –see– *Drimia maritima* (L.) Stearn

white stramonium –see– *Datura stramonium* L.

white sun lily –see– *Narcissus poeticus* L.

white sweet clover –see– *Melilotus albus* Medik.

white tepary bean –see– *Phaseolus acutifolius* A. Gray

white thorn –see– *Crataegus laevigata* (Poir) DC.

white trifolium –see– *Trifolium repens* L.

white turnip –see– *Brassica rapa* L. subsp. rapa

white vetch –see– *Lathyrus sativus* L.

white wall rocket –see– *Diplotaxis erucoides* DC.

white water lily –see– *Nymphaea odorata* Aiton subsp. tuberosa (Paine) Wiersema & Hellq.

white wild indigo –see– *Baptisia alba* (L.) Vent. var. macrophylla (Larisey) Isely

white wild vine –see– *Bryonia dioica* Jacq.

white willow –see– *Salix alba* L.

white yam –see– *Dioscorea alata* L.

whitebrush –see– *Junellia ligustrina* (Lag.) Moldenke

whitehead–see– *Parthenium hysterophorus* L.

whiteheads –see– *Sphenosciadium capitellatum* A. Gray

whitetop –see– *Ageratina adenophora* (Spreng.) R. M. King & H. Rob.; *Ageratina altissima* (L.) R. M. King & H. Rob.; *Lepidium draba* L.; *Parthenium hysterophorus* L.

whiteweed –see– *Ageratum conyzoides* L.; *Leucanthemum vulgare* Lam.

whitewood –see– *Atalaya hemiglauca* (F. Muell.) F. Muell. ex Benth.

whitsun lily –see– *Narcissus poeticus* L.

whorehouse tea –see– *Ephedra viridis* Coville

whorled milkweed –see– *Asclepias pumila* (A. Gray) Vail; *Asclepias subverticillata* (A. Gray) Vail; *Asclepias verticillata* L.

wicky –see– *Kalmia angustifolia* L.; *Kalmia latifolia* L.

Wielandbeere –see– *Daphne mezereum* L.

Wiesenbärenklau –see– *Heracleum sphondylium* L.

Wiesenkerbel –see– *Anthriscus sylvestris* (L.) Hoffm.

Wiesenraute –see– *Ruta graveolens* L.

Wiesensafran –see– *Colchicum autumnale* L.

Wiesenschlüsselblume –see– *Primula veris* L.

Wiesenwucherblume –see– *Leucanthemum vulgare* Lam.

Wigandia caracasana Kunth = Wigandia urens (Ruiz & Pav.) Kunth var. caracasana (Kunth) D. N. Gibson

WIGANDIA URENS (Ruiz & Pav.) Kunth var. caracasana (Kunth) D. N. Gibson [Boraginaceae]

Synonyms:

Wigandia caracasana Kunth

Common Names:

chocon; hoja-de-San Pablo; ortiga; palo-de-San Pablo; pringamoza; San Pablo; tabaquillo

Citations:

Anderson NP, Ayres S Jr (1931) Dermatitis venenata due to Wigandia caracasana. A hitherto unrecognized cause. Report of a case. Cal West Med 34(4):278-279.

Reynolds GW, Gafner F, Rodriguez E (1989) Contact allergens of an urban shrub Wigandia caracasana. Contact Dermatitis 21(2):65-68.

WIKSTROEMIA OVATA C. A. Mey. [Thymelaeaceae]

Common Names:

arandon; dopnit; malasampaga; salago

Citations:

Garcia F (1933) The cathartic effects in man of the leaves of Wikstroemia ovata Meyer (salago leaves). Philippine J Sci 51:485-494.

wild allamanda –see– *Pentalinon luteum* (L.) B. F. Hansen & Wunderlin

wild almond –see– *Sterculia foetida* L.

wild amaranth –see– *Amaranthus cruentus* L.

wild aprico –see– *Prunus cocomilia* Ten.

wild artichoke –see– *Cynara cardunculus* L.; *Helianthus annuus* L.

wild arum –see– *Arum maculatum* L.

wild balsam apple –see– *Momordica charantia* L.

wild barley –see– *Hordeum jubatum* L.

wild beach cherry –see– *Prunus serotina* Ehrh.

wild bean –see– *Lupinus argenteus* Pursh

wild begonia –see– *Rumex venosus* Pursh

wild bitter almond –see– *Prunus dulcis* (Mill.) D. A. Webb

wild black cherry –see– *Prunus serotina* Ehrh.

wild bleeding heart –see– *Dicentra cucullaria* (L.) Bernh.

wild bryony –see– *Bryonopsis laciniosa* (L.) Naudin; *Diplocyclos palmatus* (L.) C. Jeffrey

wild buck foot –see– *Podophyllum peltatum* L.

wild buckwheat –see– *Fagopyrum esculentum* Moench

wild calla –see– *Calla palustris* L.

wild canarygrass –see– *Phalaris minor* Retz.

wild caper–see– *Euphorbia lathyris* L.

wild carrot –see– *Cicuta douglasii* (DC.) J. M. Coult. & Rose; *Cicuta maculata* L.; *Cicuta virosa* L.; *Conium maculatum* L.; *Cymopterus ibapensis* M. E. Jones; *Trachymene glaucifolia* (F. Muell.) Benth.

wild cassava –see– *Jatropha gossypiifolia* L.

wild castor seed –see– *Jatropha curcas* L.

wild celery –see– *Apium graveolens* L.

wild cherry –see– *Karwinskia humboldtiana* (Schult.) Zucc.; *Prunus serotina* Ehrh.; *Prunus virginiana* L.

wild chervil –see– *Anthriscus sylvestris* (L.) Hoffm.

wild chicory –see– *Cichorium intybus* L.

wild chokecherry –see– *Prunus virginiana* L.

wild coffee –see– *Senna occidentalis* (L.) Link

wild coleus –see– *Perilla frutescens* (L.) Britton

wild corn –see– *Veratrum californicum* Durand

wild cotton –see– *Apocynum cannabinum* L.; *Calotropis procera* (Aiton) W. T. Aiton; *Gomphocarpus fruticosa* (L.) W. T. Aiton; *Gossypium hirsutum* L.

wild cottonbush –see– *Gomphocarpus fruticosa* (L.) W. T. Aiton

wild cucumber –see– *Citrullus colocynthis* (L.) Schrad.; *Cucumis africanus* L. f.; *Cucumis myriocarpus* Naudin; *Ecballium elaterium* (L.) A. Rich.; *Marah oregonus* (Torr. & S. Watson) J. T. Howell; *Momordica charantia* L.

wild cyclamen –see– *Cyclamen persicum* Mill.

wild date –see– *Fadogia homblei* De Wild.

wild down–see– *Calotropis procera* (Aiton) W. T. Aiton

wild feverfew –see– *Parthenium hysterophorus* L.

wild fig –see– *Ficus pumila* L.

wild flax –see– *Pimelea trichostachya* Lindl.

wild foxtail –see– *Hordeum jubatum* L.

wild garlic –see– *Allium canadense* L.; *Allium sativum* L.; *Allium ursinum* L.

wild gentian –see– *Chironia transvaalensis* Gilg

wild germander –see– *Teucrium chamaedrys* L.

wild gherkin –see– *Cucumis africanus* L. f.

wild gooseberry –see– *Nicandra physalodes* (L.) Gaertn.

wild grape –see– *Cissus cuneifolia* Eckl. & Zeyh.

wild heliotrope –see– *Heliotropium europaeum* L.

wild hellebore –see– *Veratrum californicum* Durand

wild hemlock –see– *Cicuta maculata* L.; *Conium maculatum* L.

wild hop –see– *Bryonia dioica* Jacq.

wild hyacinth –see– *Hyacinthoides non-scripta* (L.) Chouard ex Rothm.

wild hydrangea –see– *Rumex venosus* Pursh

wild indigo –see– *Baptisia alba* (L.) Vent. var. macrophylla (Larisey) Isely; *Indigofera australis* Willd.; *Indigofera hirsuta* L.

wild jalap –see– *Podophyllum peltatum* L.

wild jasmin –see– *Cestrum diurnum* L.

wild jasmine –see– *Cestrum diurnum* L.

wild jessamine –see– *Gelsemium sempervirens* (L.) J. St.-Hil.

wild larkspur –see– *Consolida regalis* Gray

wild leek –see– *Toxicoscordion venenosum* (S. Watson) Rydb.

wild lemon –see– *Podophyllum peltatum* L.

wild lettuce –see– *Lactuca sativa* L.

wild licorice –see– *Abrus precatorius* L.

wild lima bean –see– *Phaseolus lunatus* L.

wild lucerne –see– *Crotalaria dura* J. M. Wood & M. S. Evans

wild lupin –see– *Lupinus angustifolius* L.

wild macaw plant –see– *Sesbania punicea* (Cav.) Benth.

wild mandrake –see– *Podophyllum peltatum* L.

wild marigold –see– *Tagetes minuta* L.

wild mint –see– *Salvia reflexa* Hornem.

wild mustard –see– *Raphanus raphanistrum* L.; *Sinapis arvensis* L.

wild nard –see– *Asarum europaeum* L.

wild nightshade –see– *Solanum dulcamara* L.

wild onion –see– *Allium canadense* L.; *Allium cepa* L.; *Allium validum* S. Watson; *Anticlea elegans* (Pursh) Rydb.; *Ornithogalum magnum* (Baker) J. C. Manning & Goldblatt; *Toxicoscordion nuttallii* (A. Gray) Rydb.; *Toxicoscordion venenosum* (S. Watson) Rydb.

wild parsley –see– *Anthriscus sylvestris* (L.) Hoffm.

wild parsnip –see– *Cicuta douglasii* (DC.) J. M. Coult. & Rose; *Cicuta maculata* L.; *Cicuta maculata* L. var. angustifolia Hook.; *Heracleum mantegazzianum* Sommier & Levier; *Pastinaca sativa* L.; *Trachymene glaucifolia* (F. Muell.) Benth.; *Trachymene ochracea* L. A. S. Johnson

wild passion fruit –see– *Passiflora alba* Link & Otto

wild passionvine –see– *Passiflora alba* Link & Otto

wild pea –see– *Crotalaria sagittalis* L.; *Crotalaria spectabilis* Roth; *Lupinus leucophyllus* Douglas ex Lindl.

wild peach –see– *Trema cannabinum* Lour.

wild pepper –see– *Arisaema triphyllum* (L.) Schott; *Capsicum frutescens* L.; *Daphne mezereum* L.

wild persimmon –see– *Diospyros virginiana* L.

wild physicnut –see– *Jatropha gossypiifolia* L.

wild pineapple –see– *Macrozamia communis* L. A. S. Johnson; *Macrozamia lucida* L. A. S. Johnson; *Macrozamia miquelii* (F. Muell.) A. DC.; *Macrozamia spiralis* (Salisb.) Miq.

wild poinsettia –see– *Euphorbia heterophylla* L.

wild potato –see– *Cestrum laevigatum* Schltdl.; *Solanum triflorum* Nutt.

wild primrose –see– *Primula farinosa* L.

wild radish –see– *Raphanus raphanistrum* L.

wild rhubarb –see– *Heracleum mantegazzianum* Sommier & Levier

wild rue –see– *Peganum harmala* L.

wild saffron –see– *Colchicum autumnale* L.

wild sage –see– *Lantana camara* L.

wild sago –see– *Toxicoscordion paniculatum* (Nutt.) Rydb.

wild savages –see– *Agrostemma githago* L.

wild senna –see– *Senna didymobotrya* (Fresen.) H. S. Irwin & Barneby; *Senna italica* Mill.

wild sieva bean –see– *Phaseolus lunatus* L.

wild spinach –see– *Chenopodium album* L.; *Mercurialis perennis* L.

wild strawberry –see– *Fragaria ×ananassa* Duchesne ex Rozier

wild sunflower –see– *Helianthus annuus* L.; *Verbesina encelioides* (Cav.) Benth. & Hook. f. ex A. Gray

wild tamarind –see– *Leucaena leucocephala* (Lam.) de Wit

wild taro –see– *Colocasia esculenta* (L.) Schott

wild tobacco –see– *Nicotiana attenuata* Torr. ex S. Watson; *Nicotiana glauca* Graham; *Nicotiana obtusifolia* M. Martens & Galeotti

wild tomato –see– *Solanum aculeatissimum* Jacq.; *Solanum carolinense* L.; *Solanum cinereum* R. Br.; *Solanum esuriale* Lindl.; *Solanum torvum* Sw.

wild tree tobacco –see– *Nicotiana glauca* Graham

wild verbena –see– *Heliotropium amplexicaule* Vahl; *Verbena officinalis* L.

wild watermelon –see– *Citrullus lanatus* (Thunb.) Matsum. & Nakai

wild winter pea –see– *Lathyrus hirsutus* L.

wild wormwood –see– *Parthenium hysterophorus* L.

wilde Christwurz –see– *Helleborus foetidus* L.

wilde druif –see– *Cissus cuneifolia* Eckl. & Zeyh.

wilde Kastanie –see– *Aesculus hippocastanum* L.

wilde Komkommer –see– *Cucumis africanus* L. f.

wilde ui –see– *Ornithogalum magnum* (Baker) J. C. Manning & Goldblatt

wilde Wicke –see– *Vicia sativa* L.

wildekapok –see– *Gomphocarpus fruticosa* (L.) W. T. Aiton

wildemagriet –see– *Callilepis laureola* DC.

Wildetabak –see– *Nicotiana glauca* Graham

Wilder Spinat –see– *Atriplex hortensis* L.

William's-grass –see– *Festuca arundinacea* Schreb.

willow –see– *Salix alba* L.

willow-leaf jassamine –see– *Cestrum parqui* L'Her.

willow weed –see– *Persicaria maculosa* Gray

Wimmera ryegrass –see– *Lolium rigidum* Gaudin

Windenknöterich –see– *Fallopia convolvulus* (L.) Á. Löve

windowleaf –see– *Monstera deliciosa* Liebm.

windroot –see– *Asclepias tuberosa* L.

Windsor bean –see– *Vicia faba* L.

wine palm –see– *Caryota urens* L.

wine plant –see– *Rheum ×hybridum* Murray

wing bean –see– *Psophocarpus tetragonolobus* (L.) DC.

wing-leaf clitoria –see– *Clitoria ternatea* L.

wing milk vetch –see– *Astragalus pterocarpus* S. Watson

wing sorrel –see– *Rumex venosus* Pursh

wing-stalk yam –see– *Dioscorea alata* L.

wingscale –see– *Atriplex canescens* (Pursh) Nutt.

winter cherry –see– *Solanum pseudocapsicum* L.; *Withania somnifera* (L.) Dunal

winter cress –see– *Barbarea vulgaris* R. Br.

winter fern –see– *Conium maculatum* L.

winter rape –see– *Brassica napus* L. var. napus

winter ryegrass –see– *Lolium temulentum* L.

winterberry –see– *Gaultheria procumbens* L.; *Ilex decidua* Walter

wintergreen –see– *Gaultheria procumbens* L.; *Kalmia angustifolia* L.; *Kalmia latifolia* L.

Winterhaube –see– *Colchicum autumnale* L.

wireweed –see– *Polygonum aviculare* L.

wistaria –see– *Wisteria sinensis* (Sims) DC.

wisteria –see– *Wisteria floribunda* (Willd.) DC.; *Wisteria sinensis* (Sims) DC.

Wisteria chinensis Curt. = Wisteria sinensis (Sims) DC.

WISTERIA FLORIBUNDA (Willd.) DC. [Fabaceae]

Common Names:

Japanese wisteria; Reichblütige Glyzinie; wisteria

Citations:

Randeau ES (1993) Wisteria toxicity. J Toxicol Clin Toxicol 31(1):107-112.

WISTERIA SINENSIS (Sims) DC. [Fabaceae]

Synonyms:

Wisteria chinensis Curt.

Common Names:

Blauregen; Chinese kidney bean; Chinese wisteria; Chinesische Glyzinie; glicine; glycine; glycine chinensis; Glyzinie; wistaria; wisteria

Citations:

Picchioni AL (1965) Control of poisonings - Poisoning from wisteria seeds and pods. Am J Hosp Pharm 22(Nov):633.

Ventura R (1956) Su due casi di avvelenamento da glicine (Wistaria chinensis Curt.). Minerva Medicoleg 76(4):180-181.

witgatbossie –see– *Pteronia pallens* L. f.

WITHANIA SOMNIFERA (L.) Dunal [Solanaceae]

Common Names:

sim elfar; winter cherry

Citations:

Arseculeratne SN, Gunatilaka AA, Panabokke RG (1985) Studies on medicinal plants of Sri Lanka. Part 14. Toxicity of some traditional medicinal herbs. J Ethnopharmacol 13:323-335.

witloof –see– *Cichorium intybus* L.

witstorm –see– *Thesium lineatum* L. f.

witte ganzenrooet –see– *Chenopodium album* L.

wode whistle –see– *Conium maculatum* L.

wolf grape –see– *Solanum dulcamara* L.

wolfbane –see– *Arnica montana* L.

Wolfs-Eisenhut –see– *Aconitum lycoctonum* L. subsp. vulparia (Rchb.) Nyman

wolf's-milk –see– *Euphorbia lathyris* L.

wolfsbane –see– *Aconitum lycoctonum* L. subsp. vulparia (Rchb.) Nyman; *Aconitum napellus* L.

Wolfsbast –see– *Daphne mezereum* L.

Wolfskirsche –see– *Atropa belladonna* L.

Wolfsmilch –see– *Euphorbia gerardiana* Jacq.; *Euphorbia helioscopia* L.; *Euphorbia peplus* L.; *Euphorbia pilosa* L.; *Euphorbia salicifolia* Host

Wollfingerhut –see– *Digitalis lanata* Ehrh.

wonder bean –see– *Canavalia ensiformis* (L.) DC.

wonder bulb –see– *Colchicum autumnale* L.

wonder flower –see– *Ornithogalum thyrsoides* Jacq.

wonder tree –see– *Ricinus communis* L.

wonderberry –see– *Solanum nigrum* L.

Wonderbohne –see– *Ricinus communis* L.

wonderboon –see– *Ricinus communis* L.

wood laurel –see– *Kalmia latifolia* L.

wood nettle –see– *Laportea canadensis* (L.) Wedd.

wood-oil tree –see– *Vernicia montana* Lour.

wood pea –see– *Lathyrus sylvestris* L.

wood rose –see– *Ipomoea tricolor* Cav.; *Merremia tuberosa* (L.) Rendle

wood sorrel –see– *Oxalis acetosella* L.; *Oxalis pes-caprae* L.; *Rumex acetosella* L.

woodbine –see– *Gelsemium sempervirens* (L.) J. St.-Hil.; *Parthenocissus quinquefolia* (L.) Planch.

woodruff –see– *Galium odoratum* (L.) Scop.

woody aster –see– *Xylorhiza glabriuscula* Nutt.

woody nightshade –see– *Solanum dulcamara* L.

woolly-everlasting daisy –see– *Helichrysum blandowskianum* Steetz ex Sond.

woolly-groundsel –see– *Senecio flaccidus* Less. var. flaccidus; *Senecio riddellii* Torr. & A. Gray

woolly-head parsnip –see– *Sphenosciadium capitellatum* A. Gray

woolly-leaf lupine –see– *Lupinus leucophyllus* Douglas ex Lindl.

woolly locoweed –see– *Astragalus mollissimus* Torr. var. mollissimus; *Oxytropis campestris* (L.) DC. var. spicata Hook.

woolly lupine –see– *Lupinus leucophyllus* Douglas ex Lindl.

woolly morning-glory –see– *Argyreia nervosa* (Burm. f.) Bojer

woolly paper flower –see– *Psilostrophe tagetina* (Nutt.) Greene

woolly-pod milkweed –see– *Asclepias eriocarpa* Benth.

woolly-pod vetch –see– *Vicia villosa* Roth subsp. varia (Host) Corb.

woolmat –see– *Cynoglossum officinale* L.

Wooton's-milk vetch –see– *Astragalus wootonii* E. Sheld.

worm killer –see– *Aristolochia bracteolata* Lam.

wormseed goosefoot –see– *Dysphania ambrosioides* (L.) Mosyakin & Clemants

wormwood –see– *Artemisia absinthium* L.; *Artemisia vulgaris* L.

wortgrass –see– *Euphorbia helioscopia* L.

woundwort –see– *Stachys arvensis* (L.) L.

Wright's-datura –see– *Datura wrightii* Regel

Wright's-eupatorium –see– *Ageratina wrightii* (A. Gray) R. M. King & H. Rob.

wu chiu –see– *Triadica sebifera* (L.) Small

Wunderbaum –see– *Ricinus communis* L.

Wunderstrauch –see– *Codiaeum variegatum* (L.) A. Juss.

Wurmfarn –see– *Dryopteris filix-mas* (L.) Schott

Wurmkraut –see– *Tanacetum vulgare* L.

Wurmsamen –see– *Tanacetum vulgare* L.

Wurstkraut –see– *Origanum majorana* L.

wutou –see– *Aconitum carmichaelii* Debeaux

Wutschierling –see– *Conium maculatum* L.

Wyoming water hemlock –see– *Cicuta maculata* L. var. angustifolia Hook.

X

Xanthium californicum Greene = Xanthium strumarium L. var. canadense (Mill.) Torr. & A. Gray

Xanthium cavanillesii Schouw = Xanthium strumarium L. var. canadense (Mill.) Torr. & A. Gray

Xanthium chinense Mill. = Xanthium strumarium L. var. glabratum (DC.) Cronquist

Xanthium echinatum Murray = Xanthium strumarium L. var. canadense (Mill.) Torr. & A. Gray

Xanthium italicum Moretti = Xanthium strumarium L. var. canadense (Mill.) Torr. & A. Gray

Xanthium occidentale Bertol. = Xanthium strumarium L.

Xanthium pungens Wallr. = Xanthium strumarium L.

XANTHIUM SIBERICUM Patrin [Asteraceae]

Common Names:

cocklebur

Citations:

Zhang YG, Huang GZ (1988) Poisoning by toxic plants in China. Report of 19 autopsy cases. Am J Forensic Med Pathol 9(4):313-319.

Xanthium speciosum Kearney = Xanthium strumarium L. var. canadense (Mill.) Torr. & A. Gray

XANTHIUM SPINOSUM L. [Asteraceae]

Common Names:

Bathurst bur; clotbur; cocklebur; spiny clotbur; spiny cocklebur

Citations:

Burry JN, Kuchel R, Reid JG, et al. (1973) Australian bush dermatitis: Compositae dermatitis in South Australia. Med J Aust 1(3):110-116.

Rowe AH (1939) Contact allergy to cocklebur (Xanthium spinosum). Preliminary report. Arch Derm Syphilol 39:149.

Smith JP (1962) These weeds can poison your poultry. Agric Gaz New South Wales 73(Mar):136-142.

XANTHIUM STRUMARIUM L. [Asteraceae]

Synonyms:

Xanthium occidentale Bertol.; ***Xanthium pungens*** Wallr.

Common Names:

Bathurst bur; broad cocklebur; bur thistle; burweed; cat's-eggs; clotbur; cocklebur; ditchbur; heart-leaf cocklebur; Noogoora bur; small burdock; Spitzklette

Citations:

Broerman A (1926) Poisoning of swine by cocklebur. Vet Alumni Quarterly 14(2):39-41.

Forrest GP (1938) Cocklebur poisoning. J Am Vet Med Assoc 93:42-43.

Hansen AA (1924) Cocklebur sprouts are deadly. Better Crops 2(1):28-29, 43.

Kenny GC, Everist SL, Sutherland AK (1950) Noogoora burr poisoning of cattle. Queensland Agric J 70:172-177.

Martin T, Johnson BJ, Sangiah S, et al. (1992) Experimental cocklebur (Xanthium strumarium) intoxication in calves. In: James LF, Keeler RF, Bailey JM Jr, et al. (eds.) Poisonous plants. Iowa State Univ Press. Ames, Iowa. pp. 489-494.

Martin T, Stair EL, Dawson L (1986) Cocklebur poisoning in cattle. J Am Vet Med Assoc 189(5):562-563.

Oelkers S, Oehme F (1982) Cocklebur poisoning in swine. Bovine Pract 3(2):11-14.

Pammel LH (1919) Hogs poisoned by young cockleburs. Am J Vet Med 14:385-386.

Pammel LH (1922) Cockle bur. Vet Med 17(6):295.

Witte ST, Osweiler GD, Stahr HM, et al. (1990) Cocklebur toxicosis in cattle associated with the consumption of mature Xanthium strumarium. J Vet Diagn Invest 2(4):263-267.

XANTHIUM STRUMARIUM L. var. canadense (Mill.) Torr. & A. Gray [Asteraceae]

Synonyms:

Xanthium californicum Greene; ***Xanthium cavanillesii*** Schouw; ***Xanthium echinatum*** Murray; ***Xanthium italicum*** Moretti; ***Xanthium speciosum*** Kearney

Common Names:

American cocklebur; California bur; clot bur; cocklebur; duraznillo blanco; Italian cocklebur; sacha sandia; sheep bur

Citations:

Driemeier D, Irigoyen LF, Loretti AP, et al. (1999) Intoxicação espontânea pelos frutos de Xanthium cavanillesii (Asteraceae) em bovinos no Rio Grande do Sul. Pesq Vet Bras 19(1):12-18.

Hansen AA (1928) Stock poisoning plants. Cocklebur. North Am Vet 9(4):46-49.

Horváth Z, Modor V (1970) A malacok szerbtövis okozta mérgezése. Magyar Allator Lapja 25:349-354.

Kinsley AT (1909) Gastro-enteritis in hogs caused by eating young cockleburs (Xanthium canadense). Am Vet Rev 35:576-578.

Loretti AP, Bezerra PS, Ilha MR, et al. (1999) Intoxicação experimental pelos frutos de Xanthium cavanillesii (Asteraceae) em ovinos. Pesq Vet Bras 19(2):71-78.

Marsh CD, Roe GC, Clawson AB (1923) Livestock poisoning by cocklebur. U S Dep Agric Circ #283:4 pp.

Méndez MC, Santos RC, Riet-Correa F (1998) Intoxication by Xanthium cavanillesii in cattle and sheep in Southern Brazil. Vet Hum Toxicol 40(3):144-147.
Mureşan G, Nan S, Peride E (1971) O intoxicaţie la porcine cu Xanthium italicum. Rev Zootehnie Med Vet 21(8):63-65.
Tudor G, Anton E, Lupa M (1971) Cercetări epizootologice si de laborator privind intoxicatia cu Xanthium italicum la animale în Delta Dunării si zonele ei exterioare. Rev Zoot Med Vet 21(5):49-56.

XANTHIUM STRUMARIUM L. var. glabratum (DC.) Cronquist [Asteraceae]

Synonyms:
Xanthium chinense Mill.

Common Names:
Noogoora bur

Citations:
Seddon HR, King RO (1938) Noogoora burr (Xanthium chinense). Poisonous for stock in very early stage of growth. New South Wales Dep Agric Vet Res Rep 7(1937):101-108.
Smith JP (1962) These weeds can poison your poultry. Agric Gaz New South Wales 73(Mar):136-142.

XANTHORRHOEA AUSTRALIS R. Br. [Xanthorrhoeaceae]

Common name:
yacca

Citations:
Munday BL, Mason RW, Hartley WJ (1976) Encephalopathies in cattle in Tasmania. Aust Vet J 52:92-96.

XANTHORRHOEA HASTILIS Sm. [Xanthorrhoeaceae]

Common Names:
swamp grasstree

Citations:
Hall WT (1956) A note on Macrozamia and Xanthorrhoea poisoning of cattle. Aust Vet J 32:173-175.
Hall WT (1956) Xanthorrhoea hastile poisoning of cattle. Queensland J Agric Sci 13:97-106.
Hall WT (1965) Grasstree poisoning of cattle. Queensland Agr J 91:504-506.

XANTHORRHOEA MEDIA R. Br. [Xanthorrhoeaceae]

Common Names:
forest grasstree

Citations:
Hall WT (1965) Grasstree poisoning of cattle. Queensland Agric J 91:504-506.

XANTHORRHOEA MINOR R. Br. [Xanthorrhoeaceae]

Citations:
Harrison MA, Beilby CA, Friend SC, et al. (1978) Redwater in cattle associated with ingestion of Xanthorrhoea minor. Aust Vet J 54(1):40.

XANTHORRHOEA QUADRANGULATA F. Muell. [Xanthorrhoeaceae]

Common names:
yacca

Citations:
Fearn JT (1961) Yacca poisoning. Aust Vet J 37:161.

XANTHORRHOEA RESINOSA Pers. [Xanthorrhoeaceae]

Common Names:
swamp grasstree; yaca gum

Citations:
Hall WT (1965) Grasstree poisoning of cattle. Queensland Agric J 91:504-506.

XANTHORRHOEA SEMIPLANA F. Muell. [Xanthorrhoeaceae]

Common names:
yacca

Citations:
Fearn JT (1961) Yacca poisoning. Aust Vet J 37:161.

XANTHOSOMA ATROVIRENS K. Koch & C. D. Bouché [Araceae]

Common Names:
cocoyam

Citations:
Clark A (1936) Report on the effects of certain poisons contained in food-plants of West Africa upon the health of the native races. J Trop Med Hyg 39(23-24):269-295.
Saha BP, Hussain M (1983) A study of the irritating principle of aroids. Indian J Agric Sci 53: 833-836.

XANTHOSOMA SAGITTIFOLIUM (L.) Schott [Araceae]

Common Names:
cocoyam; elephant's-ear; malanga; tanier; yautia

Citations:
Osisiogu IU, Uzo JO, Ugochukwu EN (1974) The irritant effect of cocoyams. Planta Med 26(2):166-169.

xibata –see– *Arrabidaea bilabiata* (Sprague) Sandwith

XYLORHIZA GLABRIUSCULA Nutt. [Asteraceae]

Synonyms:

Xylorhiza parryi (A. Gray) Greene

Common Names:

Parry's-aster; woody aster

Citations:

Knight HG (1913) Two important western poisonous plants. Proc Soc Promot Agric Sci 33:51-58.

Xylorhiza parryi (A. Gray) Greene = Xylorhiza glabriuscula Nutt.

Y

yacca –see– *Xanthorrhoea quadrangulata* F. Muell.

yaca gum –see– *Xanthorrhoea resinifera* (Sol. ex C. Kite) E. C. Nelson & D. J. Bedford

yagalache –see– *Toxicodendron striatum* (Ruiz & Pav.) Kuntze

yage –see– *Banisteriopsis caapi* (Spruce ex Griseb.) C. V. Morton

yam bean –see– *Pachyrhizus erosus* (L.) Urb.

yangjinhua –see– *Datura metel* L.

yarloop clover –see– *Trifolium subterraneum* L.

yarrow –see– *Achillea millefolium* L.

yasmin –see– *Jasminum officinale* L.

yautia –see– *Xanthosoma sagittifolium* (L.) Schott

ye huai –see– *Sophora flavescens* Aiton

ye-huai-geng –see– *Sophora flavescens* Aiton

yeara –see– *Toxicodendron diversilobum* (Torr. & A. Gray) Greene; *Toxicodendron pubescens* Mill.

yedra –see– *Toxicodendron diversilobum* (Torr. & A. Gray) Greene

yee tho –see– *Nerium oleander* L.

yellow allamanda –see– *Allamanda cathartica* L.

yellow bean –see– *Thermopsis montana* Nutt.

yellow bird-of-paradise –see– *Caesalpinia gilliesii* (Hook.) D. Dietr.

yellow bristlegrass –see– *Setaria pumila* (Poir.) Roem. & Schult. subsp. pumila

yellow buckeye –see– *Aesculus flava* Sol.

yellow bur –see– *Centaurea solstitialis* L.

yellow burweed –see– *Amsinckia intermedia* Fisch. & C. A. Mey.

yellow cedar –see– *Thuja occidentalis* L.; *Thuja plicata Donn* ex D. Don

yellow centaurea –see– *Centaurea solstitialis* L.

yellow daisy –see– *Verbesina encelioides* (Cav.) Benth. & Hook. f. ex A. Gray; *Wedelia asperrima* (Decne.) Benth.

yellow dock –see– *Rumex crispus* L.

yellow-eye cerbera –see– *Cerbera odollam* Gaertn.

yellow false oat –see– *Trisetum flavescens* (L.) P. Beauv.

yellow flag –see– *Iris pseudoacorus* L.

yellow flax –see– *Linum rigidum* Pursh

yellow flower pea –see– *Lathyrus aphaca* L.

yellow forget-me-not –see– *Amsinckia intermedia* Fisch. & C. A. Mey.

yellow foxglove –see– *Digitalis lanata* Ehrh.

yellow foxtail –see– *Setaria pumila* (Poir.) Roem. & Schult.

yellow gowan–see– *Ranunculus acris* L.

yellow grey box –see– *Sarcocephalus diderrichii* De Wild.

yellow hat tree –see– *Duranta erecta* L.

yellow heads –see– *Gnidia kraussiana* Meisn.

yellow heart –see– *Zanthoxylum flavum* Vahl

yellow-heart prickly ash –see– *Zanthoxylum flavum* Vahl

yellow iris –see– *Iris pseudoacorus* L.

yellow jasmine –see– *Gelsemium sempervirens* (L.) J. St.-Hil.

yellow jessamine –see– *Gelsemium sempervirens* (L.) J. St.-Hil.

yellow kowhai –see– *Sophora microphylla* Aiton

yellow lady's-slipper –see– *Cypripedium reginae* Walter

yellow-leaf mother-in-law's-tongue –see– *Sansevieria trifasciata* Prain var. laurenti (De Wild.) N. E. Br.

yellow locust –see– *Robinia pseudoacacia* L.

yellow lupin –see– *Lupinus luteus* L.

yellow lupine –see– *Lupinus luteus* L.; *Lupinus sulphureus* Douglas ex Hook.

yellow maple –see– *Acer rubrum* L.

yellow milk vetch –see– *Astragalus flavus* Nutt. ex Torr. & A. Gray

yellow milkweed –see– *Asclepias tuberosa* L.

yellow nightshade –see– *Pentalinon luteum* (L.) B. F. Hansen & Wunderlin

yellow oatgrass –see– *Trisetum flavescens* (L.) P. Beauv.

yellow oleander –see– *Thevetia peruviana* (Pers.) K. Schum.

yellow oxeye –see– *Helenium autumnale* L.

yellow pine –see– *Pinus ponderosa* C. Lawson

yellow pine flax –see– *Linum neomexicanum* Greene

yellow prickly poppy –see– *Argemone mexicana* L.

yellow riceflower –see– *Pimelea flava* R. Br.

yellow rocket –see– *Barbarea vulgaris* R. Br.

yellow sage –see– *Lantana camara* L.

yellow sarson –see– *Brassica rapa* L. subsp. triocularis (Roxb.) Hamelt

yellow star –see– *Helenium autumnale* L.

yellow star thistle –see– *Centaurea solstitialis* L.

yellow sweet clover –see– *Melilotus officinalis* Lam.

yellow tarweed –see– *Amsinckia intermedia* Fisch. & C. A. Mey.

yellow thistle –see– *Argemone mexicana* L.

yellow tulip –see– *Moraea pallida* (Baker) Goldblatt

yellow tulp –see– *Moraea pallida* (Baker) Goldblatt; *Moraea spathulata* (L. f.) Klatt

yellow vetchling –see– *Lathyrus aphaca* L.

yellow vine –see– *Tribulus micrococcus* Domin; *Tribulus terrestris* L.

yellow water iris –see– *Iris pseudoacorus* L.

yellow wax bean –see– *Phaseolus vulgaris* L.

yellow weed –see– *Gutierrezia sarothrae* (Pursh) Britton & Rusby; *Hymenoxys hoopesii* (A. Gray) Bierner

Yellowstone milk vetch –see– *Astragalus miser* Douglas ex Hook. var. hylophilus (Rydb.) Barneby

yellowtop –see– *Senecio glabellus* Poir.

yellowwood –see– *Terminalia bursarina* F. Muell.; *Terminalia oblongata* F. Muell.; *Zanthoxylum flavum* Vahl

yerba buena –see– *Mentha ×piperita* L. nothosubsp. piperita

yerba-de-la-flecha –see– *Sapium biloculare* (S. Watson) Pax

yerba-de-la-perdiz –see– *Margyricarpus pinnatus* (Lam.) Kuntze

yerba-de-pasmo –see– *Baccharis pteronioides* DC.

yerba-del-diablo –see– *Persicaria punctata* (Elliott) Small

yerba-del-lobo –see– *Hymenoxys hoopesii* (A. Gray) Bierner

yerba-del-peco –see– *Actaea rubra* (Aiton) Willd.

yerba loca –see– *Astragalus bergii* Hieron.

yerba manza –see– *Baccharis pteronioides* DC.

yerba mora –see– *Solanum nigrum* L.

yerba picanta –see– *Persicaria punctata* (Elliott) Small

yerba-maté –see– *Ilex paraguariensis* A. St.-Hil.

yeros –see– *Vicia ervilia* (L.) Willd.

yesterday-and-today –see– *Brunfelsia pauciflora* (Cham. & Schltdl.) Benth.

yesterday-today-tomorrow –see– *Brunfelsia australis* Benth.; *Brunfelsia latifolia* (Pohl) Benth.; *Brunfelsia pauciflora* (Cham. & Schltdl.) Benth.

yew –see– *Taxus baccata* L.; *Taxus brevifolia* Nutt.; *Taxus cuspidata* Siebold & Zucc.; *Taxus ×media* Rehder

yewdow –see– *Vicia faba* L.

yewwood –see– *Cupressus arizonica* Greene

yo yo –see– *Pausinystalia johimbe* (K. Schum.) Pierre ex Beille

yohimbé –see– *Pausinystalia johimbe* (K. Schum.) Pierre ex Beille

York Road poison –see– *Gastrolobium grandiflorum* F. Muell.

youth-on-age –see– *Tolmiea menziesii* Torr. & A. Gray

yoyote –see– *Thevetia peruviana* (Pers.) K. Schum.

yuca –see– *Manihot esculenta* Crantz

yucca –see– *Yucca aloifolia* L.; *Yucca filamentosa* L.

YUCCA ALOIFOLIA L. [Agavaceae]

Common Names:

Spanish bayonet; yucca

Citations:

Kanerva L, Estlander T, Petman L, et al. (2001) Occupational allergic contact urticaria to yucca (Yucca aloifolia), weeping fig (Ficus benjamina), and spathe flower (Spathiphyllum wallisii). Allergy 56(10):1008-1011.

YUCCA FILAMENTOSA L. [Agavaceae]

Common Names:

yucca

Citations:

Munno G, Giannoccaro F, Riva G, et al. (2001) Allergy to yucca. Allergy 56(9):921-922.

yucuyahui –see– *Montanoa tomentosa* Cerv.

yuyo colorado –see– *Amaranthus quitensis* Kunth

yuyo-da-sapo –see– *Pascalia glauca* Ortega

Z

zabila –see– *Aloe vera* (L.) Burm. f.

Zahnkraut –see– *Hyoscyamus niger* L.

zakami –see– *Datura metel* L.

zallouh –see– *Ferula hermonis* Boiss.

zamang –see– *Samanea saman* (Jacq.) Merr.

zamia –see– *Cycas circinalis* L.; *Cycas media* R. Br.; *Macrozamia communis* L. A. S. Johnson; *Macrozamia lucida* L. A. S. Johnson; *Macrozamia miquelii* (F. Muell.) A. DC.; *Macrozamia pauli-guilielmi* W. Hill & F. Muell.; *Macrozamia riedlei* (Fisch. ex Gaudich.) C. A. Gardner; *Macrozamia spiralis* (Salisb.) Miq.

ZAMIA DEBILIS L. f. [Zamiaceae]

Common Names:

guayiga

Citations:

Mason MM, Whiting MG (1968) Caudal motor weakness and ataxia in cattle in the Caribbean area following ingestion of cycads. Cornell Vet 58(4):541-554.

Zamia floridana A. DC. = Zamia integrifolia L. f.

Zamia fraseri Van Houtte ex Regel = Macrozamia spiralis (Salisb.) Miq.

ZAMIA INTEGRIFOLIA L. f. [Zamiaceae]

Synonyms:

Zamia floridana A. DC.; ***Zamia media*** Jacq.

Common Names:

arrowroot-de-Floride; comptie; coontie; cycad; Florida arrowroot; Florida coontie; graines-d'arrow-root-de-Floride; guayiga; konti; maranguey; Seminole bread

Citations:

Albretsen JC, Safdar AK, Richardson JA (1998) Cycad palm toxicosis in dogs: 60 cases (1987-1997). J Am Vet Med Assoc 213(1):99-101.

Mason MM, Whiting MG (1968) Caudal motor weakness and ataxia in cattle in the Caribbean area following ingestion of cycads. Cornell Vet 58(4):541-554.

Roberts GA (1941) Paraplegia (wobbles) in cattle. Vet Med 36(Oct):507-509.

Senior DF, Sundlof SF, Buergelt CD, et al. (1985) Cycad intoxication in the dog. J Am Anim Hosp Assoc 21(Jan-Feb):103-109.

Zamia media Jacq. = Zamia integrifolia L. f.

zamia palm –see– *Cycas media* R. Br.; *Macrozamia miquelii* (F. Muell.) A. DC.; *Macrozamia riedlei* (Fisch. ex Gaudich.) C. A. Gardner; *Macrozamia spiralis* (Salisb.) Miq.

ZAMIA PORTORICENSIS Urb. [Zamiaceae]

Common Names:

maranguey

Citations:

Mason MM, Whiting MG (1968) Caudal motor weakness and ataxia in cattle in the Caribbean area following ingestion of cycads. Cornell Vet 58(4):541-554.

Reams RY, Janovitz EB, Robinson FR, et al. (1993) Cycad (Zamia puertoriquensis) toxicosis in a group of diary heifers in Puerto Rico. J Vet Diagn Invest 5(3):488-494.

ZANHA AFRICANA (Radlk.) Exell [Sapindaceae]

Synonyms:

Dialiopsis africana Radlk.

Citations:

Mugera GM (1970) Toxic and medicinal plants of East Africa. I. Bull Epizootic Dis Afr 18(4):377-387.

ZANTEDESCHIA AETHIOPICA (L.) Spreng. [Araceae]

Common Names:

arum lily; calla lily; lily-of-the-Nile; lirio cala; lords-and-ladies; trumpet lily; white arum lily

Citations:

Manriquez O, Varas J, Rios JC, et al. (2002) Analysis of 156 cases of plant intoxication received in the toxicologic information center at Catholic University of Chile. Vet Hum Toxicol 44(1):31-32.

Miniciullo PL, Fazio E, Patafi M, et al. (2007) Allergic contact dermatitis due to Zantedeschia aethiopica. Contact Dermatitis 56(1):46-47.

Mrvos R, Krenzelok EP, Jacobsen TD (2001) Toxidromes associated with the most common plant ingestions. Vet Hum Toxicol 43(6):366-369.

Zanthoxylum cribosum Spreng. = Zanthoxylum flavum Vahl

ZANTHOXYLUM FLAVUM Vahl [Rutaceae]

Synonyms:

Fagara flava (Vahl) Krug & Urb.; ***Zanthoxylum cribosum*** Spreng.

Common Names:

páo blanco; satinwood; tinguaciba; West Indian satinwood; Westindisches Seidenholz; yellow heart; yellowheart prickly ash; yellowwood

Citations:

Freise FW (1936) Vergiftungen durch brasilianische Werkhölzer. II. Jacareúba-Holz und Seidenholz. Sammlung Vergiftungsfallen 7(C33):61-72.

Zauke –see– *Convallaria majalis* L.

Zaunriegel –see– *Ligustrum vulgare* L.

zavila –see– *Aloe vera* (L.) Burm. f.

ZEA MAYS L. [Poaceae]

Common Names:

corn; Grünmais; Indian corn; macca; Mais; maize; mays; mealie; Sara

Citations:

Brightwell AH (1972) "Silo gas" poisoning in cattle. Can Vet J 13(9):224-225.

Davenport DF, Kerr LA (2001) Atypical rumen acidosis in a dairy herd from whiskey distillery by-products. Vet Hum Toxicol 43(3):165-167.

Gonzalo-Garijo MA, Pérez-Calderón R, Muñoz-Rodríguez A, et al. (2004) Hypersensitivity to maize pollen. Allergy 59(3):365.

Lochkarev VA (1974) [Poisoning of cattle by sugar beet and maize cobs.] Veterinariia Moscow 51(9):97-99.

Lotthammer KH, Grunert E, Elghamry MI (1970) Fruchbarkeitsstörungen in einem Rinderbestand nach Verfütterung östrogenhaltiger Mais- und Kleegrassilage. Berl Munch Tierarztl Wochenschr 83(18):353-357.

Lowry T, Schuman LM (1956) Silo-filler's disease - A syndrome caused by nitrogen dioxide. JAMA 162(3):153-160.

Seligman EJ, Key MM (1968) Corn dermatitis. Arch Dermatol 97(6):664-666.

Note:

Corn or maize is named *Zea mays* L. subsp. *mays* in some publications.

Zehrwurz –see– *Arum maculatum* L.

Zeller –see– *Apium graveolens* L.

ZELTNERA BEYRICHII (Torr. & A. Gray) G. Mans. [Gentianaceae]

Synonyms:

Centaurium beyrichii (Torr. & A. Gray) B. L. Rob.

Common Names:

mountain pink; rock centaury

Citations:

Dollahite JW, Allen TJ (1962) Poisoning of cattle, sheep, and goats with Lobezla [Lobelia] and Centarium species. Southwestern Vet 15(Winter):126-130.

ZELTNERA CALYCOSA (Buckley) G. Mans. [Gentianaceae]

Synonyms:

Centaurium calycosum (Buckley) Fernald

Common Names:

Buckley's-centaury; mountain pink; rosita

Citations:

Dollahite JW, Allen TJ (1962) Poisoning of cattle, sheep, and goats with Lobezla [Lobelia] and Centarium species. Southwestern Vet 15(Winter):126-130.

zerdali –see– *Prunus cocomilia* Ten.

zhi-cao-wu –see– *Aconitum kusnezoffii* Rchb.

zhi-chuan-wu –see– *Aconitum carmichaelii* Debeaux

Ziegelbeeren –see– *Daphne mezereum* L.

Zieria arborescens Sims = Zieria smithii Andrews

ZIERIA SMITHII Andrews [Rutaceae]

Synonyms:

Zieria arborescens Sims

Common Names:

lanoline bush; sandfly zieria; stinkwood

Citations:

Munday BL (1968) Zieria arborescens (stinkwood) intoxication in cattle. Aust Vet J 44(11):501-502.

Munday BL, Cummings R, Wilson BJ (1974) Experimental Zieria arborescens (stinkwood) poisoning in rabbits. Res Vet Sci 17(2):270-272.

Philp T (1930) Panting disease of cattle. Poisoning by Zieria smithii (stinkwood). Tasmanian J Agr 1:161-166.

Stephens WH (1967) Panting disease of cattle. Tasmanian J Agr 38:84-86.

Zigadenus coloradensis Rydb. = Anticlea elegans (Pursh) Rydb.

Zigadenus elegans Pursh = Anticlea elegans (Pursh) Rydb.

Zigadenus glaucus (Nutt.) Nutt. = Anticlea elegans (Pursh) Rydb.

Zigadenus gramineus Rydb. = Toxicoscordion venenosum (S. Watson) Rydb.

Zigadenus intermedius Rydb. = Toxicoscordion intermedium (Rydb.) Rydb.

Zigadenus muscitoxicum (Walter) Regel = Amianthium muscitoxicum (Walter) A. Gray

Zigadenus nuttallii A. Gray = Toxicoscordion nuttallii (A. Gray) Rydb.

Zigadenus paniculatus (Nutt.) S. Watson = Toxicoscordion paniculatum (Nutt.) Rydb.

Zigadenus venenosus S. Watson = Toxicoscordion venenosum (S. Watson) Rydb.

Zigeunerlauch –see– *Allium ursinum* L.

zigzag plant –see– *Pedilanthus tithymaloides* (L.) Poit.

Zimbabwe pride –see– *Vernonia myriantha* Hook. f.

Zimmerlinde –see– *Sparrmannia africana* L. f.

ZINGIBER MIOGA (Thunb.) Roscoe [Zingiberaceae]

Common Names:

mioga

Citations:

Hirono I, Mori H, Kato K, et al. (1982) Carcinogenicity examination of inflorescence of Zingiber mioga Roscoe. Cancer Lett 15(3):203-208.

ZINGIBER OFFICINALE Roscoe [Zingiberaceae]

Synonyms:

Zingiber zingiber (L.) H. Karst.

Common Names:

East Indian ginger; ginger; gingibre; luya; red ginger

Citations:

Hjorth N, Roed-Petersen J (1976) Occupational protein contact dermatitis in food handlers. Contact Dermatitis 2(1):28-42.

Niinimäki A (1984) Delayed-type allergy to spices. Contact Dermatitis 11(1):34-40.

Seetharam KA, Pasricha JS (1987) Condiments and contact dermatitis of the finger-tips. Indian J Dermatol Venereol Leprol 53:325-328.

Sinha SM, Pasricha JS, Sharma RC, et al. (1977) Vegetables responsible for contact dermatitis of the hands. Arch Dermatol 113(6):776-779.

Stäger J, Wüthrich B, Johansson SG (1991) Spice allergy in celery-sensitive patients. Allergy 46(6):475-478.

Zimt –see– *Cinnamomum verum* J. Presl

Zingiber zingiber (L.) H. Karst. = Zingiber officinale Roscoe

Zitrone –see– *Citrus limon* (L.) Burm. f.

Zitronenholz –see– *Chloroxylon swietenia* DC.

ZIZIPHUS JUJUBA Mill. [Rhamnaceae]

Common Names:

azufaifa; dazao; jujube

Citations:

Alvarado MI, Moneo I, Gonzalo MA, et al. (2002) Allergy to azufaifa fruit and latex. Allergy 57(5):460-461.

Chan TY, Chan JC, Tomlinson B, et al. (1994) Poisoning by Chinese herbal medicines in Hong Kong: A hospital-based study. Vet Hum Toxicol 36(6):546-547.

zoapatle –see– *Montanoa tomentosa* Cerv.

zobb elkeh –see– *Orobanche minor* Sm.

Zottelwicke –see– *Vicia villosa* Roth

zucchini –see– *Cucurbita pepo* L.

Zuckermelde –see– *Atriplex hortensis* L.

Zuckerrohr –see– *Saccharum officinarum* L.

Zuckerrübe –see– *Beta vulgaris* L.

zuidewindlelie –see– *Ornithogalum thyrsoides* Jacq.

zumaque venenoso –see– *Toxicodendron radicans* (L.) Kuntze

zurma –see– *Ricinus communis* L.

Zwiebel –see– *Allium cepa* L.

Zygadenus –see– *Zigadenus*

Zygophyllum ammophilum F. Muell. = Roepera ammophila (F. Muell.) Beier & Thulin

ZYGOPHYLLUM FABAGO L. [Zygophyllaceae]

Common Names:

bean caper

Citations:

Belchí-Hernández J, Moreno-Grau S, Sánchez-Gascón F, et al. (1998) Sensitization to Zygophyllum fabago pollen. A clinical and immunologic study. Allergy 53(3):241-248.

Zygophyllum foetidum Schrad. & J. C. Wendl. = Roepera foetida (Schrad. & J. C. Wendl.) Beier & Thulin

ZYGOPHYLLUM MORGSANA L. [Zygophyllaceae]

Citations:

Van der Walt SJ, Steyn DG (1939) Recent investigations into the toxicity of known and unknown poisonous plants in the Union of South Africa. IX. Onderstepoort J Vet Sci Anim Indus 12:335-366.

Zypressen-Wolfsmilch –see– *Euphorbia cyparissias* L.